The chemistry of

organic silicon compounds

Part 1

THE CHEMISTRY OF FUNCTIONAL GROUPS

A series of advanced treatises under the general editorship of
Professor Saul Patai

The chemistry of alkenes (2 volumes)
The chemistry of the carbonyl group (2 volumes)
The chemistry of the ether linkage
The chemistry of the amino group
The chemistry of the nitro and nitroso groups (2 parts)
The chemistry of carboxylic acids and esters
The chemistry of the carbon–nitrogen double bond
The chemistry of amides
The chemistry of the cyano group
The chemistry of the hydroxyl group (2 parts)
The chemistry of the azido group
The chemistry of acyl halides
The chemistry of the carbon–halogen bond (2 parts)
The chemistry of the quinonoid compounds (2 volumes, 4 parts)
The chemistry of the thiol group (2 parts)
The chemistry of the hydrazo, azo and azoxy groups (2 parts)
The chemistry of amidines and imidates
The chemistry of cyanates and their thio derivatives (2 parts)
The chemistry of diazonium and diazo groups (2 parts)
The chemistry of the carbon–carbon triple bond (2 parts)
The chemistry of ketenes, allenes and related compounds (2 parts)
The chemistry of the sulphonium group (2 parts)
Supplement A: The chemistry of double-bonded functional groups (2 parts)
Supplement B: The chemistry of acid derivatives (2 parts)
Supplement C: The chemistry of triple-bonded functional groups (2 parts)
Supplement D: The chemistry of halides, pseudo-halides and azides (2 parts)
Supplement E: The chemistry of ethers, crown ethers, hydroxyl groups and their sulphur analogues (2 parts)
Supplement F: The chemistry of amino, nitroso and nitro compounds and their derivatives (2 parts)
The chemistry of the metal–carbon bond (4 volumes)
The chemistry of peroxides
The chemistry of organic selenium and tellurium compounds (2 volumes)
The chemistry of the cyclopropyl group (2 parts)
The chemistry of sulphones and sulphoxides

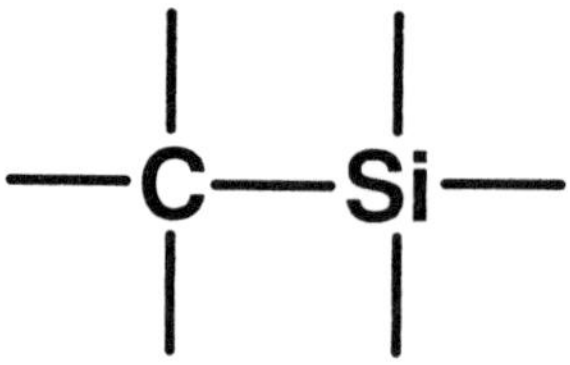

The chemistry of
organic silicon compounds

Part 1

Edited by

SAUL PATAI

and

ZVI RAPPOPORT

The Hebrew University, Jerusalem

1989

JOHN WILEY & SONS

CHICHESTER–NEW YORK–BRISBANE–TORONTO–SINGAPORE

An Interscience® Publication

Library of Congress Cataloging-in-Publication Data:

Patai, Saul
The chemistry of organic silicon compounds.
(The Chemistry of functional groups)
'An Interscience publication.'
1. Organosilicon compounds. I. Rappoport, Zvi.
II. Title. III. Series.
QD412.S6P37 1989 547'.08 87-31709

ISBN 0 471 91993 4 (set)
ISBN 0 471 91441 X (v. 1)
ISBN 0 471 91992 6 (v. 2)

British Library Cataloguing in Publication Data:

Patai, Saul
The chemistry of organic silicon compounds.
1. Organic silicon compounds.
I. Title II. Rappoport, Zvi
547'.08

ISBN 0 471 91993 4
ISBN 0 471 91441 X V. 1
ISBN 0 471 91992 6 V. 2

Printed in Great Britain at The Bath Press, Avon

Contributing authors

Y. Apeloig	Department of Chemistry, Technion-Israel Institute of Technology, 32000 Haifa, Israel
D. A. ('Fred') Armitage	Department of Chemistry, King's College London (KQC), Strand, London WC2R 2LS, UK
A. R. Bassindale	PORCG, Department of Chemistry, The Open University, Walton Hall, Milton Keynes MK7 6AA, UK
L. Birkofer	Institute of Organic Chemistry, University of Düsseldorf, D-4000 Düsseldorf 1, FRG
H. Bock	Institute of Inorganic Chemistry, Johann Wolfgang Goethe University, Niederurseler Hang, D-6000 Frankfurt (M) 50, FRG
A. G. Brook	Lash Miller Chemical Laboratories, University of Toronto, Toronto M5S 1A1, Canada
J. Y. Corey	Department of Chemistry, University of Missouri–St. Louis, St. Louis, Missouri 63121-4499, USA
R. J. P. Corriu	Institut de Chimie Fine, Heterochimie et Amino-Acides, Université des Sciences et Techniques du Languedoc, Place Eugene Bataillon, 34060 Montpelier Cedex, France
T. R. C. Crompton	Beechcroft, Wittingham Lane, Goosnargh, Preston, Lancashire, UK
C. Guerin	Institute de Chimie Fine, Heterochimie et Amino-acides, Universite des Sciences et Techniques du Languedoc, Place Eugene Bataillon, 34060 Montpelier Cedex, France
T. C. Kendrick	Dow Corning Ltd, Barry, S. Glamorgan CF6 7YL, UK
J. B. Lambert	Department of Chemistry, Northwestern University, Evanston, Illinois 60208, USA
G. L. Larson	Dynamit Nobel Chemicals, Inc., Petrarch Systems, Bartram Road, Bristol, Pennsylvania 19007, USA
H. Linoh	Institut für Anorganische und Analytische Chemie, Technische Universität Braunschweig, D-3300 Braunschweig, FRG
J. Michl	Center for Structure and Reactivity, Department of Chemistry, University of Texas at Austin, Austin, Texas 78712-1167, USA

J. J. E. Moreau	Institut de Chimie Fine, Heterochimie et Amino-acides, Universite des Sciences et Techniques du Languedoc, Place Eugene Bataillon, 34060 Montpelier Cedex, France
I. Ojima	Department of Chemistry, State University of New York at Stony Brook, Stony Brook, New York 11794-3400, USA
B. Parbhoo	Dow Corning Ltd, Barry, S. Glamorgan, CF6 7YL, UK
G. Raabe	Lehr und Forschungsgebiet Theoretische Chemie der Rheinisch Westfalischen Technischen Hochschule Aachen, D-5100 Aachen, FRG
W. J. Schulz	Department of Chemistry, Northwestern University, Evanston, Illinois 60208, USA
H. Schwarz	Institut für Organische Chemie der Technischen Universität Berlin, Strasse des 17. Juni 135, D-1000 Berlin 12, FRG
W. S. Sheldrick	Fachbereich Chemie der Universität Kaiserslautern, Erwin-Schrödinger-Strasse, D-6750 Kaiserslautern, FRG
B. Solouki	Institute of Inorganic Chemistry, Johann Wolfgang Goethe University, Niederurseler Hang, D-6000 Frankfurt (M) 50, FRG
O. Stuhl	Institute of Organic Chemistry, University of Düsseldorf, D-4000 Düsseldorf 1, FRG
R. Tacke	Institut für Anorganische und Analytische Chemie, Technische Universität Braunschweig, D-3300 Braunschweig, FRG
P. G. Taylor	Dêpartment of Chemistry, The Open University, Walton Hall, Milton Keynes MK7 6AA, UK
T. D. Tilley	Department of Chemistry, D-006, University of California at San Diego, La Jolla, California 92093, USA
R. Walsh	Department of Chemistry, University of Reading, Whiteknights, P.O. Box 224, Reading, Berkshire RG6 2AD, UK
R. West	Department of Chemistry, University of Wisconsin, Madison, Wisconsin 53706, USA
J. W. White	Dow Corning Ltd, Barry, S. Glamorgan, CF6 7YL, UK
E. A. Williams	General Electric Company, Corporate Research and Development, P.O. Box 8, Schenectady, New York 12301, USA
J. C. Young	Edward Davies Chemical Laboratories, The University College of Wales, Aberystwyth SY23 1NE, UK

Foreword

This volume deals with organic compounds containing carbon–silicon bonds, including also derivatives with other heteroatoms.

The chemistry of organic silicon compounds is one of the fastest growing fields in organic chemistry. In this subject, theory often precedes and sometimes even directs experimental work and this is reflected in a comprehensive and detailed theoretical chapter. Several exciting new developments of organosilicon chemistry of the last decade have been reviewed, such as multiple bonds to silicon.

Only two chapters among those planned did not materialize, namely that on 'Silyl radicals, silylenes and rearrangements' and that on 'Silicon–nitrogen bonds' but at least part of these topics is covered in various sections of other chapters.

The literature coverage in this volume is mostly up to mid-1987, with some chapters containing references up to the beginning of 1988.

We will be very grateful to readers who will call our attention to mistakes or omissions and who would send to us proposals relating to this volume or to other volumes in 'The Chemistry of Functional Groups' series.

Jerusalem
April 1988

SAUL PATAI
ZVI RAPPOPORT

The Chemistry of Functional Groups
Preface to the Series

The series 'The Chemistry of Functional Groups' is planned to cover in each volume all aspects of the chemistry of one of the important functional groups in organic chemistry. The emphasis is laid on the functional group treated and on the effects which it exerts on the chemical and physical properties, primarily in the immediate vicinity of the group in question and secondarily on the behaviour of the whole molecule. For instance, the volume *The Chemistry of the Ether Linkage* deals with reactions in which the C–O–C group is involved, as well as with the effects of the C–O–C group on the reactions of alkyl or aryl groups connected to the ether oxygen. It is the purpose of the volume to give a complete coverage of all properties and reactions of ethers in as far as these depend on the presence of the ether group but the primary subject matter is not the whole molecule, but the C–O–C functional group.

A further restriction in the treatment of the various functional groups in these volumes is that material included in easily and generally available secondary or tertiary sources, such as Chemical Reviews, Quarterly Reviews, Organic Reactions, various 'Advances' and 'Progress' series as well as textbooks (i.e. in books which are usually found in the chemical libraries of universities and research institutes) should not, as a rule, be repeated in detail, unless it is necessary for the balanced treatment of the subject. Therefore each of the authors is asked not to give an encyclopaedic coverage of his subject, but to concentrate on the most important recent developments and mainly on material that has not been adequately covered by reviews or other secondary sources by the time of writing of the chapter, and to address himself to a reader who is assumed to be at a fairly advanced postgraduate level.

With these restrictions, it is realized that no plan can be devised for a volume that would give a complete coverage of the subject with no overlap between chapters, while at the same time preserving the readability of the text. The Editor set himself the goal of attaining reasonable coverage with moderate overlap, with a minimum of cross-references between the chapters of each volume. In this manner, sufficient freedom is given to each author to produce readable quasi-monographic chapters.

The general plan of each volume includes the following main sections:

(a) An introductory chapter dealing with the general and theoretical aspects of the group.

(b) One or more chapters dealing with the formation of the functional group in question, either from groups present in the molecule, or by introducing the new group directly or indirectly.

(c) Chapters describing the characterization and characteristics of the functional groups, i.e. a chapter dealing with qualitative and quantitative methods of determination including chemical and physical methods, ultraviolet, infrared, nuclear magnetic resonance and mass spectra: a chapter dealing with activating and directive effects exerted by the group and/or a chapter on the basicity, acidity or complex-forming ability of the group (if applicable).

(d) Chapters on the reactions, transformations and rearrangements which the functional group can undergo, either alone or in conjunction with other reagents.

(e) Special topics which do not fit any of the above sections, such as photochemistry, radiation chemistry, biochemical formations and reactions. Depending on the nature of each functional group treated, these special topics may include short monographs on related functional groups on which no separate volume is planned (e.g. a chapter on 'Thioketones' is included in the volume *The Chemistry of the Carbonyl Group*. In other cases certain compounds, though containing only the functional group of the title, may have special features so as to be best treated in a separate chapter, as e.g. 'Polyethers' in *The Chemistry of the Ether Linkage*, or 'Tetraaminoethylenes' in *The Chemistry of the Amino Group*.

This plan entails that the breadth, depth and thought-provoking nature of each chapter will differ with the views and inclinations of the author and the presentation will necessarily be somewhat uneven. Moreover, a serious problem is caused by authors who deliver their manuscript late or not at all. In order to overcome this problem at least to some extent, it was decided to publish certain volumes in several parts, without giving consideration to the originally planned logical order of the chapters. If after the appearance of the originally planned parts of a volume it is found that either owing to non-delivery of chapters, or to new developments in the subject, sufficient material has accumulated for publication of a supplementary volume, containing material on related functional groups, this will be done as soon as possible.

The overall plan of the volumes in the series 'The Chemistry of Functional Groups' includes the Titles listed below:

The Chemistry of Alkenes (two volumes)
The Chemistry of the Carbonyl Group (two volumes)
The Chemistry of the Ether Linkage
The Chemistry of the Amino Group
The Chemistry of the Nitro and Nitroso Groups (two parts)
The Chemistry of Carboxylic Acids and Esters
The Chemistry of the Carbon–Nitrogen Double Bond
The Chemistry of the Cyano Group
The Chemistry of Amides
The Chemistry of the Hydroxyl Group (two parts)
The Chemistry of the Azido Group
The Chemistry of Acyl Halides
The Chemistry of the Carbon–Halogen Bond (two parts)
The Chemistry of the Quinonoid Compounds (two volumes, four parts)
The Chemistry of the Thiol Group (two parts)
The Chemistry of Amidines and Imidates
The Chemistry of the Hydrazo, Azo and Azoxy Groups (two parts)
The Chemistry of Cyanates and their Thio Derivatives (two parts)
The Chemistry of Diazonium and Diazo Groups (two parts)
The Chemistry of the Carbon–Carbon Triple Bond (two parts)
Supplement A: The Chemistry of Double-bonded Functional Groups (two parts)
The Chemistry of Ketenes, Allenes and Related Compounds (two parts)
Supplement B: The Chemistry of Acid Derivatives (two parts)
Supplement C: The Chemistry of Triple-bonded Functional Groups (two parts)

Supplement D: The Chemistry of Halides, Pseudo-halides and Azides (two parts)
Supplement E: The Chemistry of Ethers, Hydroxyl Groups and their Sulphur Analogues (two parts)
The Chemistry of the Sulphonium Group (two parts)
Supplement F: The Chemistry of Amino, Nitroso and Nitro Groups and their Derivatives (two parts)
The Chemistry of the Metal–Carbon Bond (four volumes)
The Chemistry of Peroxides
The Chemistry of Organic Selenium and Tellurium Compounds (two volumes)
The Chemistry of the Cyclopropyl Group (two parts)
The Chemistry of Sulphones and Sulphoxides

Titles in press:

The Chemistry of Enones
Supplement A2: The Chemistry of Double-bonded Functional Groups

Titles in preparation:

The Chemistry of Enols
The Chemistry of Sulphinic Acids, Esters and Derivatives
The Chemistry of Sulphenic Acids, Esters and Derivatives

Advice or criticism regarding the plan and execution of this series will be welcomed by the Editor.

The publication of this series would never have started, let alone continued, without the support of many persons. First and foremost among these was the late Dr Arnold Weissberger, whose reassurance and trust encouraged me to tackle this task. The efficient and patient cooperation of several staff-members of the Publishers also rendered me invaluable aid (but unfortunately their code of ethics does not allow me to thank them by name). Many of my friends and colleagues in Israel and overseas helped me in the solution of various major and minor matters, and my thanks are due to all of them, especially to Professor Z. Rappoport. Carrying out such a long-range project would be quite impossible without the non-professional but none the less essential participation and partnership of my wife.

The Hebrew University
Jerusalem, ISRAEL

SAUL PATAI

Contents

List of abbreviations used

Ac	acetyl (MeCO)
acac	acetylacetone
Ad	adamantyl
All	allyl
An	anisyl
Ar	aryl
Bz	benzoyl (C_6H_5CO)
Bu	butyl (also *t*-Bu or Bu^t)
CD	circular dichroism
CI	chemical ionization
CIDNP	chemically induced dynamic nuclear polarization
CNDO	complete neglect of differential overlap
Cp	η^5-cyclopentadienyl
DBU	1,8-diazabicyclo[5.4.0]undec-7-ene
DME	1,2-dimethoxyethane
DMF	*N,N*-dimethylformamide
DMSO	dimethyl sulphoxide
ee	enantiomeric excess
EI	electron impact
ESCA	electron spectroscopy for chemical analysis
ESR	electron spin resonance
Et	ethyl
eV	electron volt
Fc	ferrocene
FD	field desorption
FI	field ionization
FT	Fourier transform
Fu	furyl(OC_4H_5)
Hex	hexyl(C_6H_{11})
c-Hex	cyclohexyl(C_6H_{11})
HMPA	hexamethylphosphortriamide
HOMO	highest occupied molecular orbital

i-	iso
Ip	ionization potential
IR	infrared
ICR	ion cyclotron resonance
LCAO	linear combination of atomic orbitals
LDA	lithium diisopropylamide
LUMO	lowest unoccupied molecular orbital
M	metal
M	parent molecule
MCPBA	*m*-chloroperbenzoic acid
Me	methyl
MNDO	modified neglect of diatomic overlap
MS	mass spectrum
n	normal
Naph	naphthyl
NBS	*N*-bromosuccinimide
NMR	nuclear magnetic resonance
Pen	pentyl(C_5H_{11})
Pip	piperidyl($C_5H_{10}N$)
Ph	phenyl
ppm	parts per million
Pr	propyl (also *i*-Pr or Pr^i)
PTC	phase transfer catalysis
Pyr	pyridyl (C_5H_4N)
R	any radical
RT	room temperature
s-	secondary
SET	single electron transfer
SOMO	singly occupied molecular orbital
t-	tertiary
TCNE	tetracyanoethylene
THF	tetrahydrofuran
Thi	thienyl(SC_4H_3)
TMEDA	tetramethylethylene diamine
Tol	tolyl(MeC_6H_4)
Tos	tosyl (*p*-toluenesulphonyl)
Trityl	triphenylmethyl(Ph_3C)
Xyl	xylyl($Me_2C_6H_3$)

In addition, entries in the 'List of Radical Names' in *IUPAC Nomenclature of Organic Chemistry*, 1979 Edition, Pergamon Press, Oxford, 1979, pp. 305–322, will also be used in their unabbreviated forms, both in the text and in structures.

We are sorry for any inconvenience to our readers. However, the rapidly rising costs of production make it absolutely necessary to use every means to reduce expenses—otherwise the whole existence of our Series would be in jeopardy.

The Chemistry of Organic Silicon Compounds
Edited by S. Patai and Z. Rappoport

CHAPTER 1

Historical overview and comparison of silicon with carbon

JOYCE Y. COREY
University of Missouri-St. Louis
St. Louis
MO 63121-4499, USA

I. PROPERTIES AND CHARACTERISTICS OF SILICON AND CARBON

Just how closely do the properties of carbon and silicon and their respective compounds resemble one another? The differences can be small but they can also be wonderfully dramatic. It is the purpose of this chapter to develop an introduction to silicon chemistry and to demonstrate the analogies to carbon where they exist, but also to demonstrate the uniqueness of the chemistry, structure and reactivity of silicon. Today the inorganic and organic chemistry of silicon has a broad impact on technology but the chemistry of this element was a laboratory curiosity for more than 100 years after its discovery.

> 'Even after a very short experience, it was evident that corresponding derivatives of the two elements in question showed very considerable differences in their chemical properties; it may now be said that the principal if not the only case in which they exhibit a really close resemblance is that of the paraffins and those particular silicohydrocarbons, containing a silicon atom directly united to four alkyl radicals' (Kipping, Bakerian Lecture, 1936)[1].

At the turn of the century Frederic Stanley Kipping began his study of silicon compounds with the intention of synthesizing an optically active compound based on analogy to carbon chemistry. A brief communication was published in 1899 followed in 1901 by the first paper in a series of 51 papers[2] published in the *Journal of the Chemical Society*. These papers slowly laid the foundation of modern organosilicon chemistry as Kipping began to realize that silicon and carbon did not produce identical series of compounds. His investigations were summarized in a not particularly positive tone in the Bakerian Lecture whose conclusion reads[1]:

> 'Most if not all of the known types of organic derivatives of silicon have now been considered and it may be seen how few they are in comparison with those which are entirely organic; as moreover the few which are known are very limited in their reactions, the prospect of any immediate and important advance in this section of organic chemistry does not seem to be very hopeful'.

Kipping's investigations spanned 45 years and his last paper was published when he was 81, a remarkable achievement for an investigator who had few students. The painstaking detail of the experimental work is obvious on reading these papers. Only a few of the conclusions reached by Kipping on the new classes of compounds that he discovered have failed the test of time. With no spectroscopic tools available, Kipping used melting points and chemical analyses to determine the formation of new compounds and on occasion could be vitriolic, in print[3], about the results of less careful experimentalists who were working in the same area. Since there were expectations, at least in the early phases of the work, that a close relationship between carbon and silicon existed, the results that Kipping obtained must have appeared baffling at times. The sticky, viscous products that are slow to crystallize are commonplace and expected by today's investigators of silicon chemistry, but just as frustrating as they must have been for Kipping who had little precedent at the time from such materials in organic chemistry.

Kipping, and probably other earlier investigators of silicon chemistry, seemed to view the element as a 'failure' for not following the model set by carbon. The uniqueness that organosilicon compounds provide was not appreciated nor deemed desirable until after Kipping retired. At that time a new method for forming organosilicon compounds was discovered by Eugene Rochow which was to form the basis of a new industry and provide the impetus behind an expanded interest in the chemistry of silicon. Before continuing with a discussion of the fundamental features of silicon chemistry, some modern hindsight on the basic features of carbon and silicon will be useful.

A. Basic Properties of Silicon and Carbon

Some comparisons between silicon and carbon are illustrated in Figure 1. There are two major properties that distinguish silicon from carbon. Silicon atoms are about 50% larger than carbon and this increased size will have several ramifications, including lower barriers to silicon–element bond rotations and also less stable π-bonds. The smaller electronegativity will result in more polar silicon–element bonds compared to carbon, and it is this bond polarity that contributes to strong silicon bonds and provides the driving force in reactions. The Pauling electronegativity of silicon is actually similar to that of some of the first-row transition metals.

	Size	χ_P
C	0.77	2.5
Si	1.77	1.8

Common environments

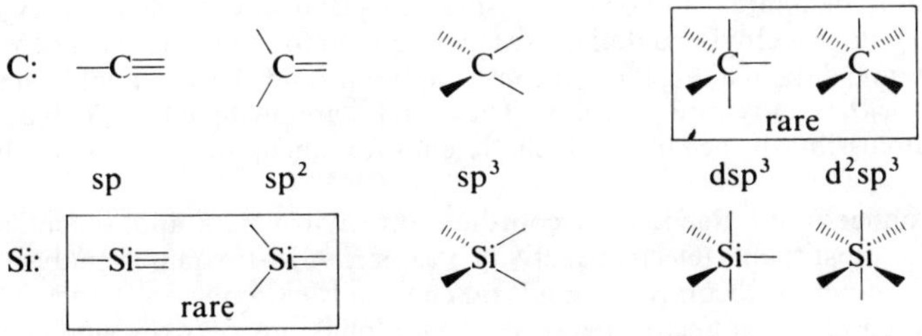

When carbon is replaced by silicon

C–C	1.54 Å
$\overset{\delta-}{\mathrm{C}}$–$\overset{\delta+}{\mathrm{Si}}$	1.90 Å
Si–Si	2.34 Å

FIGURE 1. Comparison of carbon and silicon

The common environments exhibited by Group IV elements is also shown in Figure 1. Silicon differs from carbon in forming far fewer stable multiple bonds than carbon but also in the formation of stable derivatives with more than four bonds. Carbon exhibits a maximum of four single covalent bonds, but is known to exist in some organometallic (transition metal) carbide derivatives with five or more nearest neighbors. The lack of multiple bonds seriously hampers synthetic strategies in silicon chemistry in comparison to possibilities available to the organic chemist. This is counterbalanced by the fact that the ease of formation of hypervalent species translates into relatively low activation energies in nucleophilic substitutions at silicon compared to carbon, as well as reaction pathways at tetrahedral silicon centers that are not accessible to carbon.

The only environment that silicon and carbon have in common is the tetrahedral state, supporting the conclusion of Kipping which was quoted at the beginning of the introduction to this section. Although Figure 1 uses a hybridization state to describe the expanded octet geometries of silicon, this is by no means the exclusive bonding view for hypervalent species. The participation of d orbitals in the formation of silicon compounds is a subject of continuing debate and an alternative explanation based on multicenter bonding has been proposed[4]. Multicenter bonding in hypercarbon compounds has also been proposed and the types of systems where such bonding may be important include bridging alkyls in electron-deficient Group I–III metals, carboranes, carbonyl carbides and metal clusters, and certain carbocations[5]. Continuing efforts of theoreticians and experimentalists will be required to resolve this issue. This topic is treated in more detail in chapter 2.

B. Bond Lengths and Energies in Saturated Systems

The bond lengths and bond energies of selected silicon–element and carbon–element bonds are provided in Tables 1 and 2. In this and subsequent tables the data are organized in terms of bonds of carbon and silicon to first-row (hydrogen), second-row (carbon through fluorine) and third-row (silicon through chlorine) elements. Bonds to hydrogen and nonmetals were chosen, since these form the basis of most organic derivatives. Whenever possible, the simplest compounds with hydrogen substituents were selected on the premise that these might provide standard values. However, some of these simple compounds are unknown for silicon, specifically H_3SiNH_2, H_3SiOH and H_3SiSH, therefore methylated derivatives are required for comparison. There are limited thermodynamic data available for silicon compounds and those available are for silicon derivatives with methyl substituents. These data are included in Table 2. A more complete discussion of the thermodynamic data for silicon derivatives will be found in chapter 5.

Bond distances are influenced by coordination environment and, secondarily, by the nature of the substituents (electronegativity and size). A least-squares analysis of approximately two-hundred silicon compounds taken from the Cambridge structural data base has provided a consistent set of reference values for bonds between silicon and various nonmetals and these values are shown in Table 3, as well as the values expected for single bonds to silicon corrected for electronegativity differences.

A comparison of Tables 1 and 3 shows some differences. In particular, the distances for silicon bonded to elements with electronegativity values greater than carbon tend to be less than those exhibited by the hydride derivatives shown in Table 1. In comparing carbon–element and silicon–element bonds, on progression from left to right in the Periodic Table a decrease is exhibited for both elements. For bonds to second-row elements, the decrease in length in this progression for silicon is almost double that observed for carbon. For third-row elements the shrinkage for bonds to carbon is about half that observed for the second-row elements but the trend for silicon remains the same (the decrease is now about three times that observed for carbon). The bond shortening is more than would be predicted from the use of the Schomaker–Stevenson equation, which includes an adjustment due to electronegativity differences (see Table 3 for the predicted values). Early in the history of silicon chemistry such bond shortening was rationalized on the basis of the contribution of multiply bonded structures to the ground state (d–π conjugation). These multiple bonds result from 'back-bonding' of lone pairs from substituent atoms into empty d orbitals on silicon (Figure 2).

FIGURE 2

TABLE 1. Comparison of selected carbon–element and silicon–element bond distances and barriers to rotation

Derivative	Bond distance (tetrahedral centers)		Barrier to rotation		Reference[a]
	Obs.[b] (Å)	Calc.[c]	Obs. (kcal mol^{-1})	Calc.[c]	
First Row					
D_3C–H	1.0919 (IR)	1.083			7
D_3Si–H	1.481 (IR)	1.475			8
Second Row					
H_3C–CH_3	1.534 (IR)	1.542	2.74	2.9	9, 10
H_3Si–CH_3	1.867 (M)	1.883	1.7; 1.51[d]	1.5	11, 12
H_3C–NH_2	1.474 (M)	1.472	1.96	2.3	13
[H_3Si–NH][e]	*f*	1.710		0	
H_3C–OH_2	1.414 (M)[g]	1.440	1.06	1.0	17
[H_3Si–OH][e]	*h*	1.633		0.4	
H_3C–F	1.383 (M)[i]	1.404			20
H_3Si–F	1.593 (M)[j]	1.593			21
Third Row					
H_3Si–SiH_3	2.327	2.342	1.22	0.9	22, 23
H_3C–PH_2	1.863 (M)	1.856	1.96	1.8	24
H_3Si–PH_2	2.249 (ED)	2.256	1.51	0.1 (1.48)	25, 26
H_3C–SH	1.819 (M)[k]	1.822	1.27	1.1	28
[H_3Si–SH][e]	*l*	2.139		0.9	
H_3C–Cl	1.781 (M)[m]	1.806			31
H_3Si–Cl	2.048 (M)	2.056			21, 32

[a] Second reference refers to measurement of barrier to rotation.
[b] Key to method of determination: IR = infrared; M = microwave; ED = electron diffraction.
[c] Calculated values from 3-21G and 3-21G* basis sets as reported in Ref. 6.
[d] Measured by molecular-beam electric-resonance method.
[e] Unknown compound.
[f] SiN in $(H_3Si)_3N$ is 1.738 (ED) (Ref. 14) and in H_3SiNMe_2 is 1.713 (ED) (Ref. 15).
[g] C–O = 1.428 (ED) (Ref. 16).
[h] Si–O in $(H_3Si)_2O$ is 1.634 (ED) (Ref. 18).
[i] Si–F is 1.594 (IR) (Ref. 19).
[j] Measured for CH_2DF.
[k] C–S is 1.8177 (M) (Ref. 27).
[l] Si–S in $(H_3Si)_2S$ is 2.136 A (ED) (Ref. 29).
[m] C–Cl is 1.783 (ED) (Ref. 30).

If (d–p)π bonding occurs then it is necessary to identify a model system where such "back-bonding" does not occur and thus to evaluate a single bond distance. In the silylamines described in Tables 1 and 3 the Si–N distance is 1.71 to 1.73 Å (calculated and observed) whereas the predicted single bond distance is 1.90 Å. Recently, the Si–N

TABLE 2. Comparison of selected carbon–element and silicon–element bond energies

Me_3El–X [Ref.]	kcal mol^{-1a}	H_3El–X [Ref.]	Exptl.	Theory[b]
First Row				
Me_3C–H [33, 34]	92[c]	H_3C–H [34]	105	
Me_3Si–H [35, 36]	90.3[b]	H_3Si–H [37]	90.3	90.6
Second Row				
Me_3C–CH_3 [34b]	82	H_3C–CH_3 [35]	90.4	
Me_3Si–CH_3 [37]	89.4	H_3Si–CH_3 [35]	82.1	85.5
Me_3C–NHMe [34b]	80	H_3C–NH_2 [35]	84.9	
Me_3Si–NHMe	100	H_3Si–NH_2		98.7
Me_3C–OH [34b]	91	H_3C–OH [35]	92.3	
Me_3Si–OH [37]	128	H_3Si–OH [38]	108	118.9
Me_3C–F [39]	104	H_3C–F [35]	109.9	
Me_3Si–F	—	H_3Si–F [38]	135	147.8
F_3C–F [34b]	130			
F_3Si–F [37]	160			
Third Row				
Me_3C–$SiMe_3$				
Me_3Si–$SiMe_3$ [40]	80.5[c]	H_3Si–SiH_3 [37]	74	70.6
Me_3C–PH_2	—			
Me_3Si–PH_2	—			
H_3C–PMe_2 [41]	63			
H_3Si–PH_2 [42]	88.3[d]			
Me_3C–SBu [37]	71	H_3C–SH [35]	74	
Me_3Si–SBu [37]	99	H_3Si–SH [38]	70	83.1
Me_3C–Cl [34b]	80[c]	H_3C–Cl [35]	84.6	
Me_3Si–Cl [37]	113[c]	H_3Si–Cl [38]	91	104.9

[a] Values are derived unless otherwise specified.
[b] Theoretical values from Ref. 7.
[c] Experimental values.
[d] With parameter used by Walsh for heat of formation of SiH_3, value would be 84.3

distance in $Me_3SiN(SO_2CH_3)_2$ was found to be 1.851 Å with an essentially planar nitrogen center[47a]. This bond distance is reasonably close to the single bond distance predicted by the Stevenson-Schomaker equation and suggests the absence of any π-bonding effects. However, evaluation of these effects is not always straightforward. Recently, the gas phase structures of the series of methylfluorosilanes, $(Me)_{4-n}SiF_n$ ($n = 1$ to 3), have been determined (or redetermined) by electron diffraction[47b]. The results show a linear decrease in Si–F bond distance from Me_3SiF to SiF_4 and at the same time Si–C bond distance shortens with increasing fluorination. Although $(d\text{–}p)\pi$ bonding may explain the shortening of the Si–F bonds it does not account for the shortening of the Si–C bonds. The authors favoured an explanation of increasing polar contributions and

TABLE 3. Comparison of selected silicon–element bond lengths at different coordination numbers (CN)[a]

Second Row CN	C–Si	N–Si	O–Si	F–Si
4	1.872(1)[b]	1.721(5)	1.610(4)	1.500(6)
	1.872(1)[c]	1.718(4)	1.613(4)	1.520(5)
	[1.94][d]	[1.90][d]	[1.91][d]	[1.88][d]
5	1.973(1)	1.823(5)	1.711(4)	1.602(6)
	1.979(1)	1.817(4)	1.706(4)	1.607(5)
6	2.056(1)	1.906(5)	1.794(4)	1.685(6)
	2.071(1)	1.901(4)	1.785(4)	1.682(5)
Third Row CN	**Si–Si**			**Cl–Si**
4	2.391(11)			2.026(7)
	2.391(15)			2.022(8)
	[2.34][d]			[2.16][d]
5	2.492(11)			2.128(7)
	2.528(15)			2.137(8)
6	2.576(11)			2.211(7)
	2.645(15)			2.236(8)

[a] Derived by the least-squares analysis of experimental data, employing the bond strength/bond length relationships given in *b* and *c*. Values are taken from Ref. 43.
[b] $n_i/n_0 = \exp[-(d_{iN} - d_{i(0)})/c]$ (Ref. 44).
[c] $n_i/n_0 = (d_i/d_{i(0)})$ (Ref. 45).
[d] Values from sum of covalent radii adjusted for electronegativity differences using values developed by Peter in Ref. 46.

contraction of the silicon valence shell through the series. The shorter bonds to silicon with more electronegative elements may be explained[6] by greater contributions from ionic structures of the type $R_3Si^+X^-$. Calculations do not appear to support the participation of d orbitals in bonding to silicon[6,47]. The problem of "back-bonding" is another of the factors that distinguishes silicon chemistry from carbon chemistry.

Although bonds to silicon are longer than bonds to carbon, the bonds are stronger than expected as illustrated for methyl and hydrido derivatives in Table 2. Thermodynamic data for silicon compounds are sparse, but the general conclusions from the most recent data analyses suggest that Si–C/C–C as well as Si–H/C–H are close in energy whereas bonds to elements from Groups V, VI and VII (the more electronegative nonmetals) are *stronger* to silicon than to carbon[37]. The methyl effect (replacement of H– with CH_3–) which weakens C–H bonds in organic derivatives is absent in silanes; however a silyl-weakening effect has been demonstrated[37]. A recent determination of Si–H bond dissociation energies by a photoacoustic technique gives the following bond energies: R_3Si–H, 90.1 kcal mol^{-1}; $Me_3SiSi(R_2)$–H, 85.3 kcal mol^{-1}; $(Me_3Si)_3Si$–H, 79.0 kcal mol^{-1} [48]. Substitution of Si–H by halides will strengthen a remaining Si–H bond. Methyl substituents tend to strengthen silicon–silicon bonds when compared to hydride substituents. The formally single bonds to silicon are stronger than might be expected, on the basis of a premise that longer bonds are weaker bonds.

Although carbon derivatives are usually chosen as the norm, there is a suggestion that the bonds between carbon and electronegative elements may actually be weak and thus do not provide a basis of comparison for comparable bonds to silicon[37]. The problem may be due to the small size of second-row elements and not to the 'large' size of third-row elements[49].

C. Multiple Bonds

1. Bond distances in π(p–p) derivatives

The controversial d_π–p_π bonds between silicon and electronegative elements to the right of Group IV in the Periodic Table have already been mentioned. The existence of π-bonds involving overlap of p-orbitals between silicon and other elements was questioned until this decade. Certainly reactions interpreted as proceeding through double bonds between silicon and carbon had been performed in the 1960s, but no multiply-bonded systems were isolated and reported until 1981[50]. It is still true, however, that the simplest derivatives of silicon (all hydrogen substituents) are unknown and their features can only be calculated. When the calculations are performed on the silicon and carbon derivatives of the same empirical formula, some very interesting results emerge. Selected calculations with 3–21G and 3–21G* basis sets (when available) are shown in Table 4 for the two lowest energy structures for El_2H_4, $ElNH_3$, $ElPH_3$, $ElOH_2$ and $ElSH_2$ (El = C, Si). For carbon, without exception, the form with a multiple bond between the two second-row elements is preferred. The calculated bond distances, both rigorous (3–21G) and crude (Stevenson–Schomaker), agree with those observed for the simplest derivatives. The carbene, related by a 1,2-hydride shift, is always significantly higher in energy in the organic derivatives. However, in the silicon case there is frequently little energy difference between the two lowest energy forms, the double bonded species and the isomeric silylene. The energy differences are larger for silicon third-row elements. The small energy differences calculated between double bonded and silylene forms appears to be reflected in the chemistry of these species, particularly in cases where the changes from reactant to product are consistent with both intermediates. Examples of these problems will be supplied in a later section. Curiously, in the one case where the silylene is calculated to be the lowest in energy and separated by 20 kcal mol^{-1} from the silanimine, it is a derivative of the latter that has been actually isolated[54]. All Si=X multiply bonded derivatives that have been isolated and whose structures have been determined have more complicated substituents than the hydrides. Nonetheless, the bond distances that are calculated are quite similar to the observed distances. The Stevenson–Schomaker-type corrections tend to overestimate the Si=N and Si=O distances relative to the 3–21G values.

2. π-Bond energies

The >C=C< double bond is calculated to be 105 kcal mol^{-1} more stable than >Si=Si<[51]. That the >Si=Si< 3pπ bond is weaker than the >C=C< 2pπ bond is also reflected in the emission spectra of $R_2Si{=}SiR_2$ ($R{=}CH_2Bu$-*t*) where the π–π^* transition energy is approximately 103 kcal mol^{-1} less than the corresponding transition in ethylene[69]. The π-bond strength is defined as the barrier to *cis/trans* isomerization and for ethylene[70] this value is 65 kcal mol^{-1}. The rotational barrier for the parent disilene, H_2SiSiH_2, is calculated to be 22 ± 2 kcal mol^{-1} [71]. Since alkyl and aryl groups do not appear to influence Si–Si bond energies significantly, it is expected that substituted disilenes will not exhibit π-bond strengths that are significantly different from the parent. The values based on experimental data for ArRSi=SiRAr are 26 ± 5 kcal mol^{-1} [71] and

TABLE 4. Lowest energy structures for unsaturated derivatives of silicon and carbon[a]

General formula	Lowest energy form[b]	Distance between heavy atoms (Å) Calc. (Obs.)	Isomer related by 1,2-hydride shift[c]	Distance between heavy atoms (Å) Calc. (Obs.)	ΔE kcal mol^{-1}
Second Row					
C_2H_4	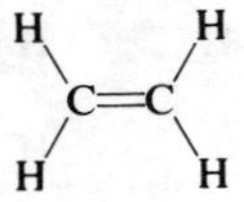	1.315(1.330)[d] [1.33][e]	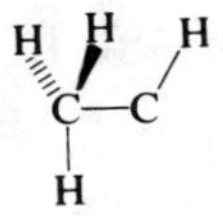	1.496	66.7
$SiCH_4$	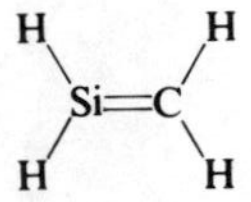	1.718[f](1.702)[g] [1.70][e]	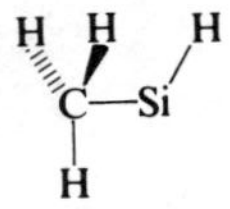	1.907	2.2
CNH_3	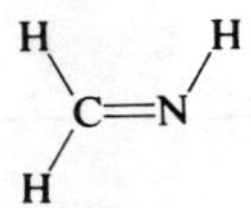	1.256, 1.283[h] (1.273)[j] [1.25][e]	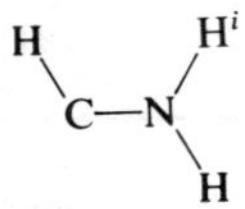	1.332	39.9
$SiNH_3$	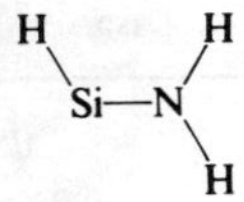	1.689	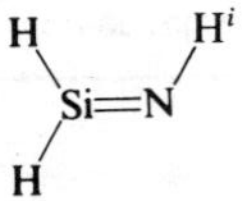	1.531, 1.585[h] (1.585)[k] [1.62][e]	20.2

TABLE 4. (*continued*)

General formula	Lowest energy form[b]	Distance between heavy atoms (Å) Calc. (Obs.)	Isomer related by 1,2-hydride shift[c]	Distance between heavy atoms (Å) Calc. (Obs.)	ΔE kcal mol^{-1}
COH_2	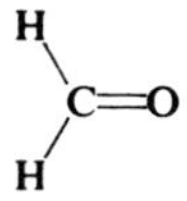	1.207(1.203)[l] [1.22][e]	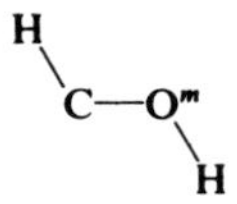	1.330	57.9
$SiOH_2$	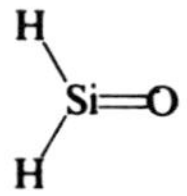	1.509 [1.60][e]	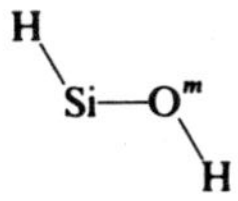	1.638(1.591)[n]	4.6
Third Row					
Si_2H_4[o]	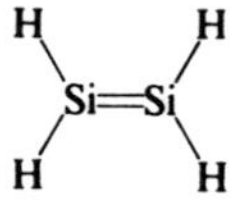	2.116(2.143)[p] [2.15][e]	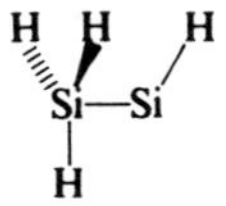	2.391	6.2
CPH_3[q]	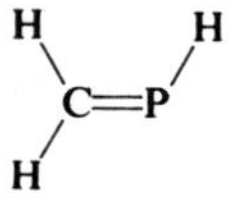	1.645, 1.672[h] (1.673)[r] [1.65][e]	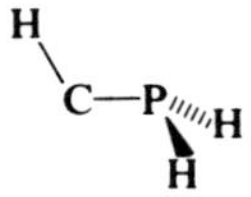	1.699	63.4
$SiPH_3$[s]	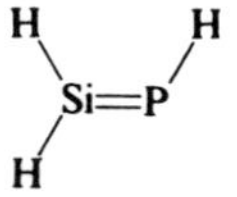	2.044[t], 2.057[h] [2.06][e]	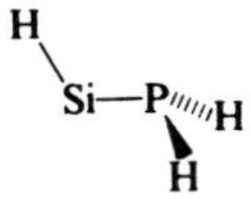	2.238	13

CSH_2 [u]	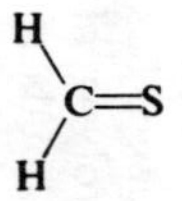	1.600(1.611) [v] [1.60] [e]	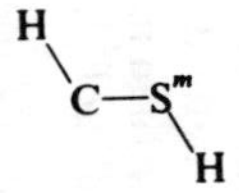	1.675	44.5
$SiSH_2$ [w]	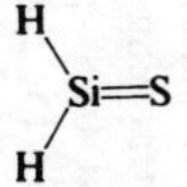	1.936 [x] [1.98] [e]	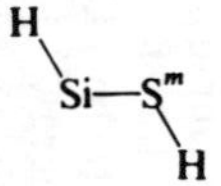	2.133	8.9

[a] Calculated structural parameters from optimized 3-21G and C-21G* basis sets are selected from Ref. 51 unless otherwise specified.
[b] All forms are singlets and are planar.
[c] Isomer lowest in energy. All structures listed are singlets.
[d] Ref. 52.
[e] Distances calculated by a modified Stevenson–Schomaker equation according to Ref. 46.
[f] Ref. 53.
[g] Value in $Me_2Si{=}C(SiMe_3)(SiMe_2Bu\text{-}t)$, Ref. 54.
[h] MRD-CI calculations according to Ref. 56.
[i] Planar form.
[j] Ref. 55.
[k] Value in $(t\text{-}Bu)_2Si{=}NSi(Bu\text{-}t)_3$, Ref. 57.
[l] Ref. 58.
[m] *Trans*, planar.
[n] Assigned species from reaction of Si atoms with H_2O, Ref. 59.
[o] Planar only when polarization functions are included.
[p] Value in Mes(*t*-Bu)Si=Si (Bu-*t*)Mes, Ref. 60.
[q] 3-21G calculations reported in Ref. 61. CEPO-PNO calculations give a C=S value of 1.613.
[r] Ref. 62.
[s] SCF 3-21G* calculations reported in Ref. 63.
[t] Some $ArP{=}SiAr'_2$ derivatives are stable in solution but have not yet been isolated, Ref. 64.
[u] 3-21G* calculations, Ref. 65.
[v] Ref. 66.
[w] HF/6-31G* calculations reported in Ref. 67.
[x] Evidence only as an intermediate. For example, see Ref. 68.

about 30 kcal mol^{-1} [72]. Because the Si=C bond will have significant polar character, the π-bond strength should be higher and appears to be about 40 kcal mol^{-1} in $Me_2Si=CH_2$ [71] and about 43 kcal mol^{-1} for 1-methyl-1-phenyl-2-neopentylsilene[73]. The π-bond strengths for $H_2X=CH_2$ (X = C, Si, Ge, Sn) have been calculated from the direct energy differences between planar and perpendicular biradical forms and from energies of disproportionation of the products of hydrogen atom addition. These values are 64–68 kcal mol^{-1} for C=C and 35–36 kcal mol^{-1} for C=Si[74]. In a different calculation of the intrinsic reaction coordinate for rotation of the planar to twisted forms, a value of about 37 kcal mol^{-1} was obtained for C=Si[75]; other calculated values range from 37–47 kcal mol^{-1} [50].

In contrast to derivatives with single bonds, C=X double bonds are stronger and shorter than the Si=X counterparts. The bond energy differences (calculated) between sets of compounds decrease (as the electronegativity of X increases) from a value of 60 kcal mol^{-1} between $\rangle$Si=C$\langle$ and $\rangle$C=C$\langle$ to 33 kcal mol^{-1} between $\rangle$Si=O and $\rangle$C=O[51]. The difference in electronegativity between Si and C also results in a larger percentage of the π-charge residing on X in $\rangle$Si=X compared to $\rangle$C=X, including silaethylene[51]. The increased bond polarity in Si=X relative to C=X should be reflected in the reactivity of the silicon derivatives as described in Section V.

Recent calculations of the π-bond energies in $\rangle$C=O, $\rangle$Si=O and $\rangle$Si=S generated values of 63 kcal mol^{-1}, 33 kcal mol^{-1} and 42 kcal mol^{-1}, respectively[67]. This suggests that the π-bond in Si=S is stronger than the π-bond in Si=O. Since silanones have been observed only in low-temperature matrices, it may be possible to employ the methods used to stabilize multiple bonds to silicon centers more effectively in the Si=S case than in the Si=O case. The Si=N bond dissociation energy is calculated to be 143 kcal mol^{-1}, which is only 45 kcal mol^{-1} greater than the Si–N single bond energy[76].

There are several consequences owing to the lack of energy gain associated with pπ interactions in silicon derivatives. The resonance energy of Si_6H_6 is calculated to be 30% less than that of benzene but another form, silaprismane, is even lower in energy by 14 kcal mol^{-1}. Differences between silicon and carbon can be driven by the fact that carbon atoms prefer sp^2 over sp^3 hybridization. This preference is absent in silicon chemistry[77].

3. *sp-Silicon*

The possibility of generating derivatives with sp–Si has been explored more in calculations than in experiment (at least in the published work)[50]. Two types of sp-Si centers can be anticipated in units such as X=Si=X or in the silicon analogs of alkynes. The only example of X=Si=X for which both spectroscopic and trapping evidence has been provided is $Me_3SiN=Si=NSiMe_3$ [78], which was generated from photolysis of $(Me_3Si)_2Si(N_3)_2$ in an argon matrix. The ground-state geometries for triple bonds to silicon differ from those calculated for carbon and the only comparison that can realistically be made is for Si≡N. The isonitrile, HNSi, was studied in matrix in 1966, and is the first compound with a SiN multiple bond[79]. Calculations suggest that HNSi does not contain a triple bond and that silicon is divalent. The first SiN triple bond was observed during the pyrolysis of phenyltriazidosilane, where PhNSi is claimed based on comparison to the PE spectrum of PhNC as well as geometry-optimized MNDO calculations[80].

The history of the study of multiple bonds in silicon chemistry is rather short if one discounts the claims in the very early studies of organosilicon chemistry. Usually the presence of reactive species is suggested by trapping experiments before the eventual isolation of a stable derivative. A formal 9,10-adduct of dimethyldisilyne with anthracene has been isolated but to date the experiment only suggests the possibility of formation of disilyne and no spectroscopic support for this intermediate has yet been provided[81]. Additional experiments will be required before disilyne chemistry is on more firm ground.

D. Bond Angles

The geometries at saturated carbon and silicon centers are essentially tetrahedral with no surprises. When a methyl group in CH_3XH_n is replaced by a silyl group, there is a definite effect on the angles about X when X is N or O. In the hydrides and alkyls of X the lone pairs are stereochemically active, but this does not appear to be the case with silylamines or disiloxanes. The classic example, trisilylamine (and other silylamines), exhibits a planar geometry at N and is very weakly basic[82,83]. A redetermination of the structure of H_3SiNMe_2 has shown that the sum of the angles about N is 354.6° indicating some pyramidalization of the nitrogen[15]. The lone pair in trisilylamine is bound more tightly than in trimethylamine as indicated by photoelectron spectroscopy data[84] and silylamines are very weak donors[85]. The SiOSi angle in disilyl ether is 144°[86]. Such observations were taken as support for (p–d)π bonding in addition to the arguments presented in the previous paragraphs. The unusual aspects of simple silicon derivatives of nitrogen and oxygen in different phases have been summarized[87] and recent work on solid-state structures of both H_3SiNMe_2 and $(H_3Si)_2O$ have revealed some fascinating features. The amine, H_3SiNMe_2, has an unusually high melting point (276 K) relative to the related $(H_3Si)_2NMe$ (below 170 K) which suggests that intermolecular interactions may be important in the solid. Indeed, dimethylsilylamine is a pentamer in the solid forming a ten-membered ring containing five-coordinate silicon with nitrogen ligands at the two axial positions[88]. The angle at O in disilyl ether is about the same in the solid and in the gas phase[89]. There appear to be secondary Si · · · O interactions at 3.05 Å (shorter than the sum of the van der Walls radii). The Si_2O and nonbonded Si atom are coplanar and the authors favor a description that implies the presence of a pure p orbital at O which may be involved in intramolecular donor π-bonding to a d orbital on silicon. The remaining lone pair at oxygen lies in the SiO_2 plane along the O · · · Si vector. Thus, the lone pairs on oxygen do not play the same role as in the carbon counterparts. The halides, SiH_3F and SiH_3I, also exhibit intermolecular interactions forming zigzag-like chains. These secondary bonding interactions may provide models for nucleophilic attack at silicon centers. There are no short intermolecular contacts involving carbon in related systems[87].

Other interesting features of olefin-like derivatives of silicon (and other group IV elements) also differ from the carbon analogs. Although disilene exhibits planar geometry, pyramidalization up to 20° in a *trans*-bent form can occur at very small energies (Figure 3). From a simple valence bond model the extent of distortions from planarity or linearity can be predicted on the basis of the singlet–triplet separation of the interacting carbenes (or fragments) that form the multiple bond[90]. The results from this model suggest that $H_2Si{=}CH_2$ will be planar but $H_2Si{=}CF_2$, $H_2Si{=}SiH_2$ and $F_2Si{=}SiF_2$ will be nonplanar. The authors of this study also included an analysis of cumulenones and predicted that $H_2Si{=}C{=}CH_2$ would be nonlinear but that $H_2C{=}Si{=}CH_2$ would be linear.

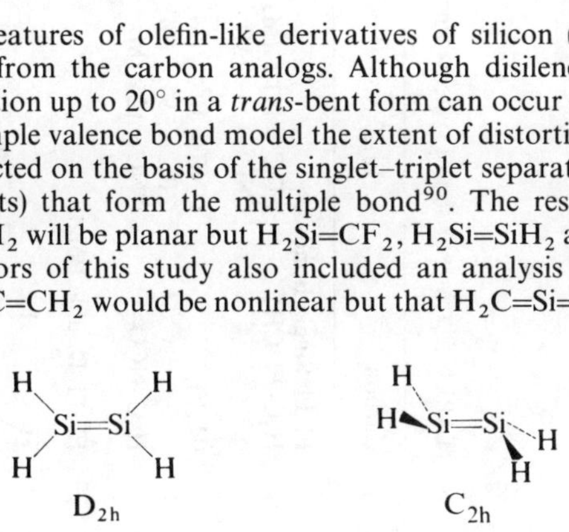

FIGURE 3

TABLE 5. Comparison of systems where bond angles at C and Si differ

Carbon	C–El–X Angle[a]	Ref.	Silicon	Si–El–X Angle[a]	Ref.
El–El′ Single Bonds					
H_3C–El–CH_3	COC = 111.5°(M)	16	H_3Si–El–SiH_3	SiOSi = 144.1°(M)	18
	CSC = 99°(M)	91		SiSSi = 97.4°(M)	92
$(H_3C)_3$El	CNC = 108.7°(M)	93	$(H_3Si)_3$El	SiNSi = 119.6°(M)	94
	CPC = 99.1°(M)	95		SiPSi = 96.4°(M)	96
H_3C–NCO	CNC = 125°(M)	97	H_3Si–NCO	SiNC = 180°(M)	98
El=El′ Double Bonds					
$H_2C{=}NH$	CNH = 117.9°(M)[b]	55	$H_2Si{=}NH$	SiNH = 127°(C)[c]	51
	= 119.3(C)	56		119.5°(C)	
			$H_2Si{=}NSiH_3$	SiNSi = 175.6(C)	99
			$R_2Si{=}NSi(Bu\text{-}t)_3$	SiNSi = 180°(M)	57
$H_2C{=}PH$	CPH = 98.9°(C)	56	$H_2Si{=}PH$	SiPH = 92°	56
	97.4°(M)	62		93.7°(C)	63

$F_2C{=}CF_2$	planar (C)	90	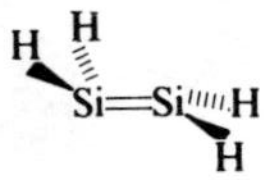	nonplanar (C)	90

=El= and El=El′					
$H_2C{=}C{=}CH_2$	CCC = 180°(C)	90	$H_2Si{=}C{=}CH_2$(C)	SiCC < 180°	90
			$H_2C{=}Si{=}CH_2$(C)	CSiC = 180°	90
$H_2C{=}C{=}O$	CCO = 180°(C)	90	$H_2Si{=}C{=}O$(C)	SiCO < 180°	90
HC≡CH	CCH = 180°(C)	90	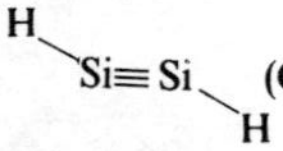 (C)	SiSiH < 180°	90

[a] M = measured; C = calculated.
[b] *trans*-NCH in CHDNH.
[c] Barrier to linearization is calculated to be 6 kcal mol^{-1}. The calculated SiNSi angle in $H_2Si{=}NSiH_3$ is 175.6° (Ref. 13a).
[d] R = *t*-Bu.

Another interesting comparison was provided by $H_2Si{=}C{=}O$ and $H_2C{=}Si{=}O$, where the former is predicted to be nonplanar and the latter planar. If $H_2Si{=}C{=}O$ is indeed strongly bent, the barrier to inversion should be large and unsymmetrically substituted derivatives would be chiral[90]. Table 5 summarizes the derivatives where CE1X and SiE1X angles differ. A more detailed discussion of the structural chemistry of silicon appears in chapter 3.

E. CLASSES OF SILICON AND CARBON COMPOUNDS

A comparison of classes of silicon and common carbon derivatives is shown in Table 6. As can be seen from this table the common single-bonded organic functional groups have

TABLE 6. Comparison of classes of silicon and carbon compounds

Carbon		Silicon	
Derivatives Based on Single Bonds			
—C—H	alkane	—Si—H	silane
—C—Cl	chloroalkane	—Si—Cl	chlorosilane
—C—OH	alcohol	—Si—OH	silanol
—C—O—C—	ether	—Si—O—C—	alkoxysilane
		—Si—O—Si—	siloxane
—C—SH	thiol	—Si—SH	silanethiol
—C—NH_2	amine	—Si—N⟨	silylamine
		—Si—N—Si—	silazane
Derivatives Based on Multiple Bonds			
⟩C=C⟨	olefin	⟩Si=C⟨	silene[a]
	alkene	⟩Si=Si⟨	disilene
—C≡C—	alkyne	[-Si≡C—][b]	silyne[c]
		[-Si≡Si—]	disilyne
—C(=O)H	aldehyde	—	
⟩C=O	ketone	[⟩Si=O][d]	silanone
⟩C=O OH	acid	[-Si(=O)O-][d]	silacarboxylate
—C(=O)—O—	ester		
⟩C=N—	imine	⟩Si=N—	silanimine
⟩C=P—	phosphimine	⟩Si=P—	silanephosphimine
⟩C=S	thione	[⟩Si=S][d]	silanethione

[a] Also silaalkene, silaolefin and silaethylene.
[b] Currently unknown.
[c] Also silaalkyne, silaacetylene.
[d] Generated as reactive intermediates.

a counterpart in silicon chemistry, although the chemistry may be quite different. Although all multiply bonded derivatives of carbon do not have an isolable silicon analog, the method for generation of such species is currently being developed and thus the listing shown in Table 6 will change in the next few years.

The nomenclature for organic derivatives can be quite complex. In 1912, Kipping suggested nomenclature rules for silicon compounds,[100] but these rules have not survived. The IUPAC rules used to describe organosilicon compounds are still in preparation and therefore current compound names are informal but quite simple. Tetravalent silicon compounds are silanes and the substituents should be given in alphabetical order. However, the common practice is to follow the order: substituted organic fragments, simple organic fragments, alkoxy, then halogen (hydride is inferred). Certain functional groups are emphasized, such as OH (silanol) and SH (silthiane) and SiOSi (siloxane), and the silicon–oxygen polymers (oligomers) have their own shorthand notation which will be described later. Some examples are given in Scheme 1. Silicon chains and rings are named according to the number of silicon atoms, disilane, trisilane, etc. When silicon groups are named as units, the general names listed in Scheme 2 are used. A more extensive discussion of nomenclature for organosilicon compounds can be found in Reference 101, but the simple examples given should suffice to understand the nomenclature in this chapter.

Compound	Common Practice
Me_4Si	tetramethylsilane
Ph_2SiH_2	diphenylsilane
Ph_2SiCl_2	diphenyldichlorosilane
$ICH_2CH_2CH_2Si(OMe)_3$	3-iodopropyltrimethoxysilane
$Ph_2Si(OH)_2$	diphenylsilanediol
$Me_3SiOSiMe_3$	hexamethyldisiloxane
$HMe_2SiNHSiMe_2H$	1,1,3,3-tetramethyldisilazane
$Ph_2MeSiOSiMe_2OSiMePh_2$	1,1,5,5-tetraphenyl-1,3,3,5-tetramethyltrisiloxane
$Me_2SiHSiH_2Me$	1,1,2-trimethyldisilane
$[Ph_2SiO]_3$	hexaphenylcyclotrisiloxane
$[Me_2Si]_6$	dodecamethylcyclohexasilane

SCHEME 1

Unit	Name
H_3Si–	silyl
Me_3Si–	trimethylsilyl
$PhMe_2Si$–	phenyldimethylsilyl
H_3SiO–	siloxanyl (siloxy)
Me_3SiO–	trimethylsiloxy
H_3SiSiH_2–	disilanyl

SCHEME 2

In the remaining sections of this chapter chemical comparisons between silicon and carbon are developed. Since all organosilicon chemistry essentially starts with the element, the discussion begins with the discovery and isolation of the element and some of the simple chemistry of silicon that was described in the eighteenth century. The basic starting materials for organosilicon chemistry were available in the past century, but their development and exploration (with the exception of a few investigators) has taken place in the past thirty to forty years. Relative to carbon, the chemistry of silicon is in its 'infancy' with more questions to be resolved than ready answers. Formation of silicon–carbon bonds is a much more important process in silicon chemistry than in organic chemistry because *all* organosilanes are manmade. Therefore the developments in this area are summarized before the simple functional group chemistries of silicon and of carbon are compared. Oligomers and polymers with a silicon backbone are discussed separately.

II. EARLY HISTORY

Silicon does not exist free in nature and the problem of extracting silicon from its natural products is only the beginning of the differences between silicon and carbon. The major source of silicon, SiO_2, was recognized as a separate substance in the 17th century and referred to as vitrifiable earth[102]. When silica is combined with metal oxides, silicates are formed and exhibit a rich variation in structural chemistry, but the majority have tetrahedral silicon centers with bonds to four oxygens. The name for the element, 'silicium' from the latin, *silex*, meaning rock, was suggested by Davy to reflect the sources of silicon[102a]. The name was later changed by Thompson to the present form, silicon, to draw attention to the relationship to the elements carbon and boron[102b]. The chemical symbol for Si was proposed by Berzelius[102c].

Although it was recognized that reduction of earths could provide elements, 'earths' were metal oxides and their reduction products were expected to be shiny. The first metallization of vitrifiable earth was reported in 1790[102d] and led to a flurry of experiments to reproduce the results, but unfortunately the material was later identified as iron. While studying the reaction of potassium with glass in the early 1800s Davy observed the formation of an amorphous powder, but since it was not shiny the identity of this new material was not recognized by the experimenter[102a,e]. The reduction of the first isolated, volatile silicon compound, SiF_4, with K gave amorphous brown powders as reported by Gay-Lussac and Thenard in 1808[102f]. Although Ampere pointed out in a letter in 1812 that this amorphous powder could only be silicon, the letter was not published until 1885[102g], thus he has not been honored for the discovery of the element. Berzelius was the first to characterize the amorphous powder formed from the reduction of potassium hexafluorosilicate with potassium in 1824 and thus receives the credit for the isolation of the element[102h]. The first crystalline samples of silicon were probably produced by Wohler in the reduction of $SiCl_4$ with H_2, but he did not publish this observation (the observations are recorded in a letter to Liebig[102i]) and the credit goes to the experimentalist, Sainte-Claire Deville, who was actually studying the chemistry of aluminum. Deville reacted $SiCl_4$ with H_2 in the presence of red hot aluminum and characterized the black crystalline product as rhombohedral silicon[102j].

Nature provides us with many sources of silicon. Silicon is second in terrestrial abundance and the earth's crust is 75% silicon and oxygen. However, the 'natural products' for silicon, silica and silicates, contain only silicon–oxygen bonds and are not particularly useful sources for synthesizing and studying new compounds of silicon. It is interesting that carbon is not among the top ten most abundant elements. In spite of the fact that carbon is not particularly abundant in the earth's crust, it is present with a rich variety of bonding types and classes of compounds. Sources of carbon compounds include coal and petroleum and all the cracking products, as well as the natural products, 'living' plant and animal materials. This range of sources for study is in part responsible for the highly developed discipline or organic chemistry today. To begin the study of organosilicon chemistry requires man-made materials, which means extraction of silicon from its natural sources at a reasonable price.

The earliest use of carbon as a reductant for silica was reported by Moissan[102k] and the earliest commercial processes involved reaction of quartz and siliceous iron ores with coke and an aluminiferous material in a blast furnace (similar to a method used by Berzelius in 1810). Such conditions generally provide ferrosilicon/silicon mixtures. For large quantities of silicon a process was developed at the turn of the century using quartz and lampblack. Extensive purification is required by procedures that do not differ significantly from that used by Berzelius[102l]. If pure quartz and pure lampblack are briquetted with cane sugar, it is possible to obtain 99.99% silicon[102m]. Presently SiO_2 is reduced with carbon in an electric arc furnace at 3000 °C. The silicon produced is about 98% pure, the major impurities are aluminum and iron. For very pure silicon, reduction of volatile silicon derivatives is required. Silicon tetrachloride is reduced with high purity zinc[102n] (Batelle Process) or by sodium[102o] (Westinghouse Process) by reactions that are essentially similar to those used to make a elemental silicon in the 18th century. The Union Carbide process involves thermal decomposition of SiH_4[102o]. A recent one-step process for production of high-purity silicon for solar cells uses the reaction that Berzelius studied, the reduction of Na_2SiF_6 with sodium. Hexaflurosilicate salts are waste products from the phosphate fertilizer industry and the reduction reaction is exothermic and self-sustaining[102p].

At the turn of the century the cost of elemental silicon was $1600/lb. Improvement of reduction conditions allowed this price to fall to $0.10/lb by 1931 and made the use of the element more practical. Elemental silicon is employed as an additive to iron to give silicon-steels, which are more acid resistant, to alloy copper for production of silicon bronzes and as an additive to aluminum or magnesium to improve the strength of these structural materials and provide resistance to corrosion. The use of silicon as a semiconductor had begun by 1940 and after discovery of the direct process reaction became the necessary starting point for the production of silicone polymers[103].

From a very simple standpoint, a major difference between silicon and carbon chemistry is the availability and range of naturally occurring materials. Silica and most silicates are insoluble in water and organic solvents and therefore are not particularly attractive candidates for conversion to other silicon derivatives. Furthermore, as described in the previous section, the silicon–oxygen bond is stronger than any other silicon–element bond except silicon–fluorine. Therefore only a limited number of synthetic methods will result in the replacement of Si–O with a bond to another element. A few other simple silicon derivatives that are suitable choices for starting materials in the study of silicon compounds were first produced in the 19th century. While studying the amorphous brown powder from Na_2SiF_6 and K, Berzelius observed that it ignited in the presence of chlorine gas[102h]. The product was $SiCl_4$ and it is surprising that this key reactive intermediate was not recognized until much later. In 1846, Ebelman reported the reaction of $SiCl_4$ with ethanol, but the liquid formed was actually identified by Mendeleev several years later[104]. By the end of the 19th century a few compounds that contained Si–C bonds were generated, (the first organosilanes) and these experiments will be described in a later section. In 1857, HCl was reacted with crystalline silicon to give several compounds, one of which was formulated as $Si_2Cl_3 \cdot 2HCl$[102q]. Purification problems deferred the correct assignment of this product as $HSiCl_3$ until ten years later. Chlorination of ferrosilicon alloys also provided silicon tetrachloride and higher members of the Si_xCl_{2x+2} series[102r]. Silicon–hydrogen derivatives were probably generated by Wohler and Buff in 1857 while studying the electrochemical behavior of Al with high Si content. Gas bubbles were produced that spontaneously ignited on contact with air and burned with a white smoke[102s]. The first identification of SiH_4 is attributed to Friedel and Ladenburg, who unexpectedly obtained this product when they treated $HSi(OEt)_3$ with sodium metal (1867)[102t]. Much of the work on these simpler systems was complicated by incorrect analyses and formulation of the products as subhalides and subhydrides. The reactive nature of silicon halides was recognized and the production of fogs from $SiCl_4$ and water was used in World War I for screening ships and trenches with smoke. Clearly the silicon–chlorine bond was considerably more reactive to water than the carbon–chlorine bond. It is probable that the more primitive experimental methods used in the last century were insufficient to produce the starting materials required for the study of organosilicon compounds that will be described in later sections. The situation was further complicated by the expectation that silicon would have classes of compounds just like carbon. The development of silicon chemistry was slow to be realized not only because brown powders were supposed to be shiny!

Silicon–carbon bonds are not found in nature, therefore, such bonds must be formed through some process in the laboratory. The basic starting materials in organosilicon chemistry are summarized in Table 7 with a brief comment concerning a difference between the silicon and carbon derivative. How can the limited number of starting materials described in the previous paragraphs be converted into the many types of organosilicon compounds presently known and how do these compounds relate to organic analogs both chemically and physically? This will form the major theme of the remainder of this chapter.

III. SILICON–CARBON BOND FORMATION[105]

Since silicon–carbon bonds do not occur in nature the key to organosilicon chemistry must be in the formation of this bond. In the previous section the major starting materials

TABLE 7. Comparison of common silicon reagents to carbon counterparts

Compound	m.p. or b.p. (in °C)	Comments
C Si	m.p. 3827 + 20[a] m.p. 1420	Silicon is more volatile than C(dia) and has lower energy of vaporization which reflects the smaller Si–Si bond energies
CO_2 SiO_2	m.p. 56.6[b] m.p. 1920	Strong single bonds between Si and O and relatively weak π (p–p) bonds favor giant lattice for silicon derivative
CH_4 SiH_4	b.p. −161.3 b.p. −11.9	Silane ignites or explodes on contact with air
CCl_4 $SiCl_4$	b.p. 76.7 b.p. 57.6	Silicon derivative hydrolyzes rapidly on contact with water
$HCCl_3$ $HSiCl_3$	b.p. 61.5 b.p. 31.8	Hydrogen is hydridic in silicon derivative
$C(OEt)_4$ $Si(OEt)_4$	b.p. 158 b.p. 165.5	Silicon derivative is source for pure silica and is also a good starting material for silicon–carbon bond formation[c]

[a] At 125 kbar.
[b] At 5.2 atm.
[c] $Si(OEt)_4$ is readily absorbed and has a hemolytic effect. $Si(OMe)_4$ vapor is said to cause severe irritation or ulceration of the cornea of the eye.

for this process were described. The materials available in large abundance as raw materials are silica and silicates but these are not particularly useful, a fact that was recognized by Kipping.

> 'One of the main factors that has retarded the development of the chemistry of silicon compounds, is undoubtedly, the insolubility in neutral solvent of all those inorganic derivatives which contain combined oxygen. This insolubility, of course, may be attributed, at any rate in part, to the great molecular complexity of all such compounds, which apparently is the result of a process of condensation brought about by the elimination of the elements of water from two or more molecules, such as those of silicic acid, containing the group $=SiOH$' (Kipping)[106].

The major problem became that of converting Si, $SiCl_4$ and $Si(OEt)_4$ to organosilicon compounds. Reactions of $SiCl_4$ and $Si(OEt)_4$ were the first to be developed. In this case it was necessary to find a reagent that could transfer an alkyl or an aryl group to the silicon center easily and with control. Actually, the most useful organic group transfer agents were discovered shortly after 1900—but the problem of control continues to be a difficulty even today. Conversion of the formally simplest reagent, elemental silicon, into haloorganosilanes did not occur until almost the middle of this century. The use of organic transfer agents provides the greatest flexibility in terms of types of derivatives that can be produced, whereas the reaction of elemental silicon provides a limited number of organosilicon compounds but in commercial quantities. It is the availabilty of relatively inexpensive starting materials that was important for the growth of organosilicon chemistry.

A. Early Organic Transfer Reagents

The growth of organosilicon chemistry parallels the history of organometallic chemistry. Two of the earliest organometallic reagents to be prepared and exploited as synthetic reagents were ethylzinc iodide and diethylzinc, which were reported by Frankland in 1849[107]. These reagents are prepared simply by reacting the element with ethyl iodide at the boiling point of the organic reactant to give EtZnI followed by heating at 120° to give Et_2Zn. To transfer an organic group from zinc to some other metal/metalloid requires somewhat rigorous conditions. The first organosilicon derivative was generated by Friedel and Crafts by heating Et_2Zn and $SiCl_4$ in a sealed tube at 140–160 °C[108]. Tetramethylsilane was made similarly from Me_2Zn at 200 °C[109]. The reaction conditions required to transfer an ethyl group from zinc to silicon are not particularly attractive and improvements would ultimately involve changing the metal that functions to transfer the organic group, and this would take another 50 years to discover. To avoid the necessity of separation or isolation of the organometallic a form of 'in situ' generation was reported by Pape, who reacted a mixture of $SiCl_4/Zn/PrI$ to give Pr_4Si[110]. The problem of using sealed tubes was overcome by reacting Et_2Zn and $Si(OEt)_4$ in the presence of sodium[111]. Unlike the case of $SiCl_4$, incomplete substitution of functional groups on silicon was observed and the products isolated were $Et_xSi(OEt)_{4-x}$. It was probably not recognized that the reactive agent in this case was most probably EtNa generated from metal exchange with Et_2Zn. Perhaps if this general organometallic reaction method[112] had been recognized at that time the organic chemistry of many main group metals/metalloids would have developed much earlier. Aryl-zinc derivatives were not useful for forming arylsilanes and the first such silicon derivative was prepared by the reaction of a close relative, Ph_2Hg, with $SiCl_4$ to give $PhSiCl_3$, but at a temperature of 300° [109,113,114].

The experiments described in the previous paragraphs actually represent the following basic reaction methods:

(1) Transfer of an organic group from an active organometallic reagent, MR (M is electropositive metal), to $SiCl_4$ or $Si(OEt)_4$.
(2) Transfer of an organic group from the combination RM/M′ (M′ is more electropositive than M) to the silicon reagent.
(3) Reaction of a mixture RX/M/[$SiCl_4$ or $Si(OEt)_4$], where the active organometallic is not generated in a separate step.

The transfer of organic groups to $SiCl_4$ is difficult to control and the products obtained are usually of the type R_4Si. Even though improvements over the use of zinc alkyls occurred with combinations such as Na/$SiCl_4$/BuCl (in the presence of small amounts of EtOAc), the product isolated was $SiBu_4$. In one of his earliest papers Kipping describes his attempt to favor formation of Ph_3SiCl by altering the proportions of $SiCl_4$/ClC_6H_5/Na, but he was not successful[115]. The problem of stepwise substitution was only partly resolved by using $Si(OEt)_4$ as starting material, as described in the reaction with $ZnEt_2$. Since $Si(OEt)_4$ is generated from $SiCl_4$, this also represents an additional reaction step which is undesirable.

In 1900, Grignard published his discovery of the reagent named after him[116] and shortly thereafter Kipping had reacted $SiCl_4$ with EtMgI to obtain a mixture of $EtSiCl_3$, Et_2SiCl_2, Et_3SiCl and Et_4Si[117]; Dilthey and Eduardoff had performed a similar reaction with PhMgBr to obtain $PhSiCl_3$, Ph_2SiCl_2 and Ph_3SiCl[118]. Kipping stated in 1907 that he made little progress on silicon compounds until the autumn of 1903, when he discovered the use of Grignard reagents. In this same paper he described a series of successive Grignard additions to obtain PhMeEtPrSi[119] (equation 1). Although successive substitution appears to be conquered, it is still an extremely difficult reaction sequence to control. During this early period Kipping usually did not record yields, but it is unlikely that single substitution products were obtained and in fact the reaction of EtMgBr with $SiCl_4$ always gave mixtures[120]. The Grignard route is still a major improvement over zinc derivatives, since the transfer of organic groups from magnesium can be performed under mild conditions in ether solvents and organic iodides are not required to successfully generate the organometallic.

$$SiCl_4 \xrightarrow{EtMgBr} EtSiCl_3 \xrightarrow{PhMgBr} EtPhSiCl_2 \xrightarrow{PrMgBr} Et(Pr)PhSiCl \xrightarrow{MeMgBr} PhMeEtPrSi \quad (1)$$

By 1908 Kipping was also able to show that the Grignard reagent did not always have to be generated separately, when $PhCH_2Si(Et)Cl_2$ was prepared by mixing $EtSiCl_3/Et_2O/Mg$ and then adding benzyl chloride[121]. Although ease of silicon–carbon bond formation had finally been achieved and would be improved still further with the use of organolithium reagents, the control of successive substitution reactions remains a problem still. There are some improvements from another reaction route which is developed much later, but before describing this new advance there are some practical aspects to be considered if organosilanes are to be available on a large scale.

Although the formation of silicon–carbon bonds was solved in the early 1900s in a modest way, the requirement for the use of an active organometallic has some distinct disadvantages. If one considers the steps that will be required to get to a simple derivative such as Me_2SiCl_2 (selected for its obvious importance in the silicones industry) by a Grignard route, the task is certainly energy intensive. Silica must be converted to the elemental form and then silicon is chlorinated to silicon tetrachloride. Methyl chloride must be prepared from chlorination of methane (or some other route) and magnesium must be obtained from magnesium chloride. One form of presenting the transformation of silica to Me_2SiCl_2 is shown in equation 2[122].

$$SiO_2 + 2C + 4Cl_2 + 2CH_4 + 2Mg \rightarrow 2CO + (CH_3)_2SiCl_2 + 2HCl + 2MgCl_2 \quad (2)$$

The mass of the products, HCl and $MgCl_2$, is double that of the desired dimethyldichlorosilane. The problem of removing these 'by-products' would be enormous, although recycling should be possible. The fact that the weight of the by-products is more than double that of the desired product still makes this a less than desirable reaction for large-scale production. The solution to this problem was provided by experiments performed in 1940 but published in 1945[123]. The discovery of the direct process reaction by Rochow was to provide a firm footing for the future of organosilicon chemistry.

B. The Direct Process Reaction[124]

A variation of a direct process reaction was described in the previous section in the reaction of ethyl iodide with elemental zinc. The real problem in silicon chemistry was to find the correct conditions for the interaction of the element with an organic halide. The trick was to find the lowest temperature that would be successful for incorporating the organic group. Above this minimum temperature the rate of the reaction increases, but the mixture of organosilanes produced becomes richer in halogen and poorer in organic groups[123]. If an appropriate catalyst could be found, the reaction could be run at a lower temperature and the side-reactions due to pyrolysis would be minimized. The catalyst that was found to be the most effective was copper, and this remains the case in present production. At the time of the publication of his successful direct process Rochow[123] wrote:

> 'The role of the catalyst is as obscure as the mechanism of the reaction and seems closely related to it · · ·

Compared to the Grignard route, the direct process reaction requires only the reduction of silica to silicon and the production of methyl chloride. A form of the overall reaction (related to equation 2) is shown in equation 3. On the basis of this equation there

$$SiO_2 + 2C + 2CH_4 + Cl_2 \rightarrow Me_2SiCl_2 + 2HCl + 2CO \qquad (3)$$

are 1.8 grams of desired silane for every 1 gram of chlorine containing by-products, a much more promising result for large-scale preparations. The basic reaction between elemental silicon and an organic halide is influenced by temperature, concentration of catalyst, form of the contact mass and rate of flow of the organic substrate. The original report covered the reactions of chloro- and bromomethane, chloro- and bromoethane and chlorobenzene, but many other examples have been studied[105,124]. The fundamental purpose of the direct process reaction is to produce useful silicon reagents in large quantities. This would be true whether or not the products contain silicon–carbon bonds. When other organic groups have been employed as reagents in the direct process reaction, useful products have been obtained but the variations are somewhat limited and silicon–carbon bonds are not always produced. A few examples of conditions and product distribution for the direct process method are shown in Table 8. Basically, the direct reaction with silicon is most successfully conducted with HCl, Cl_2 and MeCl. Higher alkyl chlorides give lesser yields of the more commercially important R_2SiCl_2 derivatives and elimination of HCl from RCl competes with reaction of RCl with silicon. Addition of HCl to either RCl or PhCl causes the formation of increased amounts of the bifunctional hydrochlorosilanes. Reactions with organic derivatives containing other functional groups have met with limited success. With ether reactants alkoxysilanes are produced if alkyl halides are present, and the only amine that has been found to react is dimethylamine[137,138].

As indicated in Table 8, the direct process reaction provides derivatives with less than four organic groups and, by adjusting the conditions, hydrosilanes (derivatives with SiH

TABLE 8. Products in the direct process reaction of silicon

Substrate	Catalyst	Temp. (°C)	Products	Ref.
HCl	Cu	380	$HSiCl_3$(75–80%), $SiCl_4$ (20–25%)	125
Cl_2	Ca		$SiCl_4$(65%), Si_2Cl_6(30%), Si_3Cl_8(4%)	126
MeCl	Cu(10%)	300	Me_2SiCl_2(42%), $MeSiCl_3$(37%), $MeSiHCl_2$(12%), $SiCl_4$(9%)	127
	a, *b*	250	Me_2SiCl_2	130
	Cu/Zn/Sn	300	Me_2SiCl_2(90%)	128
$MeCl/H_2$	Cu	332	$MeSiHCl_2$ or Me_2SiHCl(>80%)	129
MeCl/HCl(5%)	Cu(10%)[b]	300	Me_3SiCl(62%), Me_2SiCl_2(13%), $MeSiCl_3$(8%), Me_4Si(8%), $MeHSiCl_2$(5%)	130
MeBr	Cu(30%)	400	$MeSiBr_3$(71%), Me_2SiBr_2(11%), Me_3SiBr(2.2%)	131
EtCl	Cu(10%)	300–25	$SiCl_4$(37%), $EtSiCl_3$ (27%), Et_2SiCl_2(26%)	123
i-PrCl	Cu(20%)	350–60	*i*-$PrSiHCl_2$(38%), *i*-$PrSiCl_3$(21%), *i*-Pr_2SiCl_2(11%)	132
BuBr	Cu(20%)	300–10	C_4H_8(80%), $SiBr_4$(46%), $HSiBr_3$(21%)	133
PhCl	Cu(30%)	475	$PhSiCl_3$(48%), Ph_2SiCl_2(7%), PhH(23%), Ph_2(12%)	134
C_6H_6/HCl(50%)	Ag(9%)	400	$PhSiHCl_2$(40%), Ph_2SiHCl(10%)	135
MeOMe	alloy[c]		$Me_2Si(OMe)_2$(13–64%), Me_3SiOMe(3–5%), $MeSi(OMe)_3$(3–5%)	136
Me_2NH	Cu	250	$(Me_2N)_3SiH$(>90%), $(Me_2N)_2SiH_2$(2–5%)	137
MeOH	CuCl[d]	240	$(MeO)_3SiH$(80%), $(MeO)_4Si$(10%)	138

[a] Si pretreated with HCl.
[b] Ca(27%), Zn(0.2%).
[c] Alloy contains, by weight, 5.5% Cu, 3.3% Fe, 2.9% Al, 0.27% Ca, 0.05% Zn and 0.01% Sn and the presence of an alkyl halide is required.
[d] Solvent: hydrogenated terphenyls.

bonds) are also formed. The products that can be obtained from the direct process reaction become the silyl derivatives that can be incorporated into commercial silicones and inexpensive organosilicon reagents for development.

There have been thousands of papers that have addressed the direct process reaction[124]. As with many heterogeneous reactions, reproducibility is very difficult. One of the few points of agreement is that copper is necessary for a successful direct process reaction. With the development of modern surface-science techniques it has been possible to show that silicon alone does not have sites that produce Me_2SiCl_2, but that Cu_3Si has an active surface that selectively forms this desired product[139]. In a recent study of methylchlorosilane production in a laboratory-scale fluidized bed reactor using CuCl as the source of copper, further support for reaction at Cu_3Si sites[140] has been obtained. It was speculated that when silicon is exposed to air, a silicon oxide film forms which functions as a barrier to Cu–Si contact. The CuCl migrates to and then deposits in defects in the oxide film. A reaction occurs between CuCl and Si to form $SiCl_4$ and copper which diffuses into the silicon layer, possibly forming η-Cu_3Si. This reaction between copper and silicon appears to be anisotropic and alloying is slower in the $\langle 111 \rangle$ direction than either the $\langle 100 \rangle$ or $\langle 110 \rangle$ planes, since $\langle 111 \rangle$ planes are exposed as reaction with MeCl proceeds. Although this is a plausible explanation, more detailed mechanistic information is still required before the direct process reaction can be understood.

C. Laboratory Methods for Silicon–Carbon Bond Formation

One of the major differences in the synthesis of organosilicon compounds compared to organic derivatives is the rather limited range of starting materials for the former. In general, synthetic studies in organosilicon chemistry involve either the formation of silicon–carbon bonds or functional group conversion at silicon. The direct process route is not a practical laboratory method for silicon–carbon bond formation. Most commercially available silicon compounds contain rather simple organic substituents in addition to functional groups that can be converted to silicon–carbon bonds. These functional groups include Cl, OR and H. To build more complicated organosilanes these functional groups must be exploited. The number of methods are small and the most common is the use of organometallic reagents with halosilanes or alkoxysilanes and hydrosilylation with hydrosilanes. This later process will be described shortly. The organic chemist must become familiar with the transformations of a variety of functional groups and the reagents which have been developed that are specific to certain functional groups[141]. The organosilicon chemist must become familiar with the generation, handling and reaction of organomagnesium and organolithium reagents and with luck gain control over the minimum number of functional groups on silicon that react to form a silicon–carbon bond.

The basic organometallic route to silicon–carbon bond formation is summarized in equation 4:

$$-\overset{|}{\underset{|}{\mathrm{Si}}}-\mathrm{X} \;+\; \mathrm{RM} \xrightarrow{\mathrm{S}} -\overset{|}{\underset{|}{\mathrm{Si}}}-\mathrm{R} \;+\; \mathrm{MX} \qquad (4)$$

The factors that can influence the course of this reaction include the substituents on the starting halosilane (alkoxysilane), the nature of the halogen or the alkoxy group, the organic group on the metal, the nature of the metal, the solvent, the temperature and the stoichiometric ratio between the reactants. In general, the ease of replacement of Si–halogen (or alkoxide) bond decreases with increasing number of organic groups on silicon, a fact that is important in building a series of different organic groups onto a single silicon center. The approach to the construction of an organosilicon derivative will depend on how many organic groups and how many functional groups will be required at silicon in the target molecule. A major difference in developing synthetic routes in silicon chemistry compared to carbon chemistry is that the organosilicon chemist must deal with all functional groups in a single center. The organic chemist seldom deals with two or three functional groups at one carbon center, but may have to control reactions of similar functional groups at different positions in a reacting molecule.

A few examples will illustrate the problems involved in forming silicon–carbon bonds. More expanded discussion is given by Eaborn and Bott[105]. There are two extremes in making organosilanes. The first is associated with incomplete conversion of replaceable groups in $SiCl_4$, $RSiCl_3$ and R_2SiCl_2 (or their alkoxide derivatives) and the second is associated with incorporation of the last organic group (particularly a bulky group) in R_3SiX. When organometallic reagents are reacted with polychlorosilanes, product mixtures are usually obtained. A simple example is given in equation 5. A common tactic to avoid multiple substitution is to increase the ratio of the silyl chloride to organometallic as illustrated in equation 6. The solvent or solvent mixture may effect the distribution of products. The type of solvent affects the nature of the organometallic reagent, and basic solvents such as THF can convert the tetrameric or hexameric $(RLi)_x$ reagent to the more reactive dimer (or solvated monomer). This in turn provides a more active nucleophile for substitution at the silicon center. The outcome of the change in solvent is illustrated in equation 7.

$$EtMgCl + SiCl_4 \rightarrow Et_4Si + Et_3SiCl + Et_2SiCl_2 + EtSiCl_3 \qquad (5)^{142}$$

Mole ratios	Et_4Si		Et_3SiCl		Et_2SiCl_2		$EtSiCl_3$
$EtMgCl/SiCl_4 = 2.25$	0	:	1.0	:	0.4	:	1.9
3.6	0.24	:	1.0	:	0.89	:	0.19

$$MeSiCl_3 + RMgX \rightarrow RMeSiCl_2 \qquad (6)^{143}$$

6 8 molar excess; $R = Me(CH_2)_n$ (85%), $n = 5, 6, 7$

$$RSiCl_3 + RMgCl \xrightarrow{R = \text{vinyl}} Cl_3SiR + Cl_2SiR_2 + ClSiR_3 + SiR_4 \qquad (7)^{144}$$

	Cl_3SiR		Cl_2SiR_2		$ClSiR_3$		SiR_4
THF	49.9	:	19.5	:	10.3	:	20.3
THF:2 hexane(1:2)	32.3	:	42.0	:	19.2	:	6.5

Addition of a bulky substituent can be difficult, particularly in cases where the coupling involves the formation of a fourth Si–C bond. This problem is illustrated in the length of time it takes to form the dimesitylsilacyclobutane shown in equation 8. Here,

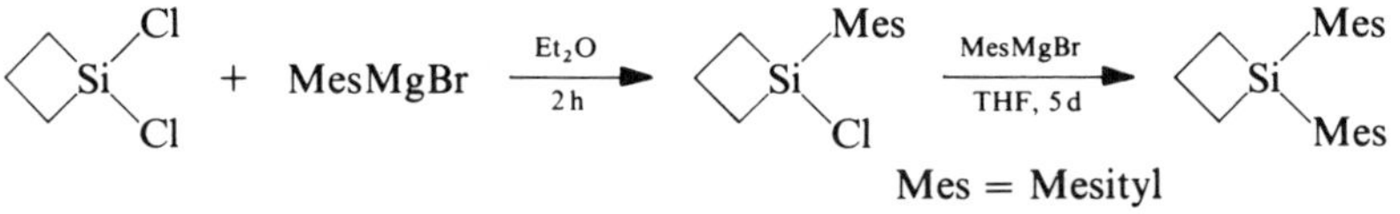

(8)[149]

Mes = Mesityl

the more nucleophilic solvent, THF, is required for successful coupling. Attachment of bulky substituents to silicon has become an important aspect in forming precursors to Si=X systems. In general, silyl fluorides are used for the introduction of large groups. This is well illustrated for the reaction of *t*-butyllithium with $SiCl_4$ and SiF_4 in equation 9. Steric parameters for substituents at silicon have only recently been reported and these values are: H(1.33); Me(0); Et(−0.149); Pr (−0.216); Bu(−0.225); *i*-Bu(−0.405); *i*-Pr (−0.556); *c*-C_6H_{11} (−1.02); *t*-Bu (−1.46)[145]. These values suggest that there is a greater effect of branching alpha to silicon than is observed in carbon. Parameters were

$$SiX_4 \xrightarrow{t\text{-BuLi}} \begin{cases} \xrightarrow{X=Cl;\ >100\,^\circ C,\ \text{overnight}} (t\text{-Bu})_2SiH_2 + (t\text{-Bu})_3SiH \\ \xrightarrow{X=F;\ 0\,^\circ C} (t\text{-Bu})_2SiF_2 \xrightarrow[39\ h,\ 63\,^\circ C]{t\text{-BuLi}} (t\text{-Bu})_3SiH \end{cases} \qquad (9)^{150,151}$$

developed only for alkyl substituents, but on the basis of the results for alcoholysis of fluorosilanes the following order of decreasing bulkiness of groups bound to silicon has been proposed: Tsi (tristrimethylsilylmethyl) $>$ 2,4,6-t-$Bu_3C_6H_2$ $>$ 2,4,6-i-$Pr_3C_6H_2$ $>$ t-Bu $>$ Mes(mesityl) $>$ i-Pr $>$ Et $>$ Me[146]. This ordering suggests that coupling of Tsi and 2,4,6-tri-t-butylphenyl groups to silicon should be the most difficult. In the reaction of $(Me_3Si)_3CLi$ with R_nCiCl_{4-n} only on Tsi group was incorporated[147]. When 2,4,6-tri-t-butylphenyllithium is added to SiF_4, one group can be added at low temperatures and a second at room temperature[148]. Even in simple systems where an organolithium reagent is quenched with R_3SiCl, the size of R influences the yields of product formed as shown in equation 10.

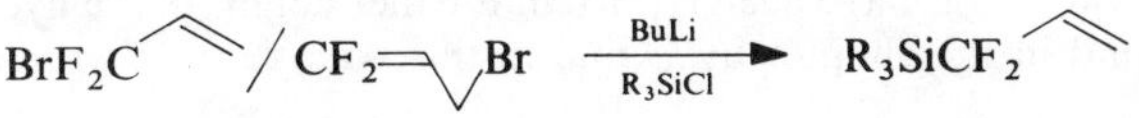

(10)[152]

$R_3 = Me_3$ (89%); $PhMe_2$ (71%); Et_3 (51%); Pr_3 (50%)

Temperature is an important factor in some coupling reactions and this was illustrated in equations 8 and 9. Lowering the temperature should reduce the rate at which coupling occurs and allow other chemical processes to occur. Such is the case for the reaction of BuLi with CO in the presence of Me_3SiCl as shown in equation 11.

$$Me_3SiCl + BuLi \xrightarrow[-110\,^\circ C]{CO,\ THF/Et_2O/\text{pentane}} \underset{(77\%)}{BuC(=O)SiMe_3} + \underset{(21\%)}{BuSiMe_3} \qquad (11)^{153}$$

These few examples illustrate some of the synthetic factors that are important in the generation of organosilicon compounds. There are basically three types of functional groups that react with organometallic reagents: Si–X (X = halogen), Si–OR and Si–H. Organolithium reagents couple with all three groups and Grignard reagents with Si–X and Si–OR. The rates of reaction of Grignard reagents with Si–X are faster than with Si–OR, for instance, MeMgCl and $(EtO)_2SiCl_2$ give $Me_2Si(OEt)_2$. In the coupling of organolithium reagents with R_3SiOR' the reactivity decreases in the order: R' = Me $>$ Et $>$ Pr $\geqslant$ Bu $>$ t-Bu[154]. Even Grignard reagents react slowly with certain RO–Si substituents as is the case with $(i\text{-PrO})_3SiCH_2MgCl$, where coupling with an organic halide (with catalysis) occurs more rapidly than does intermolecular condensation[155]. This suggests that in systems with mixed functional groups some substitution control should be possible. There are also other deficiencies of the organometallic route in that certain organic functional groups are incompatible with active organometallics. Therefore, another silicon–carbon bond formation method is desirable.

Free radical addition of hydrosilanes to olefins was discovered in the late 1940s[156]. The reaction was initiated thermally, photolytically or by free radicals and was essentially limited to $HSiCl_3$ and to olefins that did not readily polymerize. It was not until the 1950s that a suitable transition-metal catalyst was discovered for this general reaction, which is

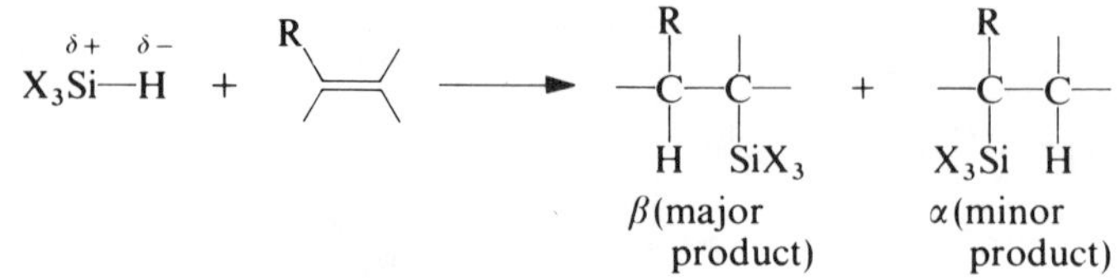

(12)

depicted in equation 12 and referred to as hydrosilylation[157]. There are few restrictions on the transition-metal-catalyzed process compared to the thermal and radical processes. If Si is considered the positive end, then hydrosilylation follows Markovnikov addition to give the β-isomer as the major product (the free radical addition is anti-Markovnikov). The principal use of this method is to introduce organic groups that are incompatible with organometallic procedures. The most common catalyst is H_2PtCl_6, called Speier's catalyst. The catalyst/SiH ratios are normally about 10^{-5} and solvents are not required. The major advantage is that the reaction is tolerant of functionality such as halogens, NO_2, CN, NH_2, RSO_3, RSO_2NH, etc. Many other transition-metal catalysts have been developed for hydrosilylation including those based on rhodium, cobalt, nickel and iron. Hydrosilylation of dienes and alkynes also occurs in the presence of transition metals, although several product variations have been reported. Further details of the hydrosilylation reaction will be found in the chapter by Ojima in this volume.

D. Future Directions in Silicon–Carbon Bond Formation

The most energy-intensive step in the preparation of organosilicon compounds involves the reduction of silica to silicon. A major improvement would be the development of a process to bypass this step and would essentially involve some way to convert silica to derivatives with silicon–carbon bonds. However, it is probable that some organometallic compound would still be required to transfer an organic group to the silicon center. At least this is true in the two cases that have been reported. When chromatographic silica or some silicate minerals are reacted with aluminum sesquihalides, tetraalkylsilanes are formed in up to 70% yield as shown in equation 13[158]. If tetrachlorosilane is added, the aluminum chloride present catalyzes a disproportionation reaction and silicon halides, R_xSiCl_{4-x}, are produced. This route would be limited to commercially available aluminum sesquihalides.

$$4Me_3Al_2Cl_3 + 3SiO_2 \rightarrow 3R_4Si + 6AlOCl + 2AlCl_3 \tag{13}$$

A second method that exploits silica has been developed by Corriu and coworkers. In this case, silica is reacted with catechol and the resultant catecholate complex is converted to an organosilane by reaction with a Grignard reagent[159]. Triorganosilanes are formed directly by reaction with an alkyl Grignard and tetraorganosilanes are produced from allyl, phenyl or alkynyl Grignards. An example is shown in equation 14.

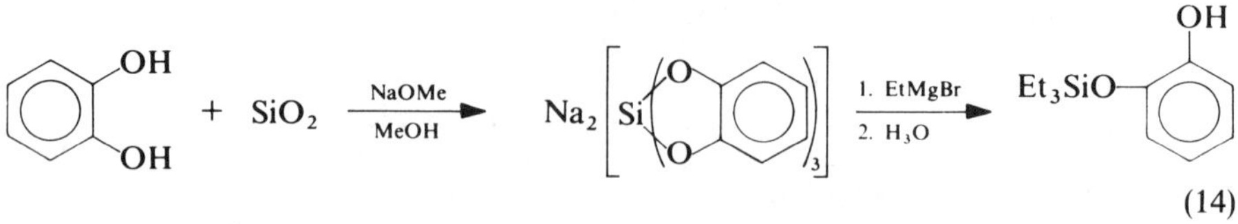

(14)

Synthetic routes to arylsilanes are quite limited. The high temperatures required for the direct process reaction from aryl chlorides produces undesirable side-reactions and the organometallic route is incompatible with certain functional groups on the aromatic ring.

A new route to arylsilanes which can be adapted to industrial scale utilizes chloromethyldisilanes, which are byproducts in the reaction of methyl chloride with silicon. This new method is the most general of those currently available and is compatible with functional groups that cannot be used with Grignard reagents. The reagent required for this route is an acid chloride and substituents that are electron withdrawing increase the rate of the reaction. The general process is shown in equation 15[160].

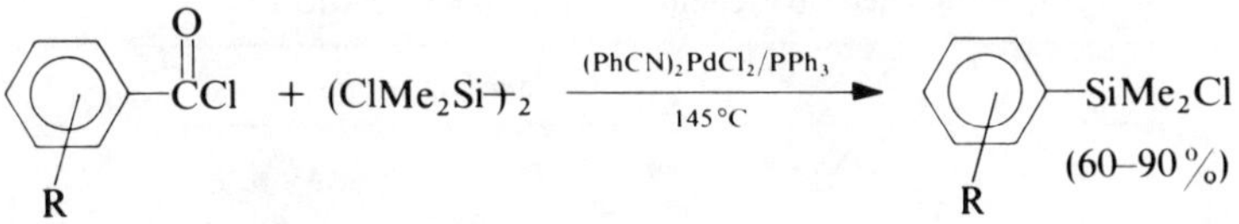

(15)

Only the basic silicon–carbon bond formation methods have been described. The discovery of the use of Grignard and later organolithium reagents provided the basis for modern organosilicon chemistry and the direct process route provided access to organochlorosilanes on an industrial scale, although with limited variations. Hydrosilylation is an excellent complement to the organometallic route and provides better substitution control. No other routes have been developed that are generally applicable and this remains a challenge for the future.

IV. THE CHEMISTRY OF FUNCTIONAL GROUPS

In the previous section one of the two major aspects of silicon syntheses was described. A major problem for the organosilicon chemist is producing a silicon–carbon bond. A second synthetic area involves functional-group transformations. Most useful reactions at silicon centers are substitution processes. This is an important difference between the chemistry of silicon compounds and that of carbon compounds. The synthetic or reaction processes in organic compounds are increased enormously by the availability of double bonds present in olefins and carbonyl derivatives. In fact, substitution at sp^3-carbon centers may represent the least interesting of organic reactions.

The functional groups at silicon that are the most useful and also those that will be found in the commercially available organosilicon derivatives are silicon–hydrogen, silicon–halogen (primarily chloride) and silicon–oxygen (primarily alkoxy) moieties. Unlike hydrocarbons, a silicon–hydrogen bond is a very useful functional group and can be the slowest to hydrolyze of the three groups listed. The reactivity of silyl chlorides to water has already been mentioned and this makes the storage and handling of such derivatives more difficult. Examples of conversion of these functional groups in silicon chemistry are shown in Table 9 with similar transformations (or lack of reactivity) in the related carbon system. Table 9 represents the more general processes and certainly is not an exhaustive list of the reactions that can take place for silicon compounds[161]. The important differences in silicon and carbon derivatives with the same functional groups are the rate at which the substitution processes take place. Nucleophilic substitutions occur rapidly at silicon, except in cases were there are two or more bulky substitutents. The transformations shown in Table 9 are primarily bimolecular processes and, in contrast to carbon, nucleophilic substitution can occur at silicon either with inversion or retention.

Many of the transformations shown in Table 9 were known early in the history of silicon chemistry. Alcoholysis of $SiCl_4$ provided one of the first volatile, covalent derivatives with all silicon–oxygen bonds[104]. The properties of this silicon–oxygen derivative were certainly quite different from silica and silicates, which also contain all silicon–oxygen bonds. Successive substitution by alcohols was one of the earliest reaction sequences reported by Kipping. Kipping, who began his study of silicon compounds trying to synthesize an optically active derivative, prepared ClSi(OMe)(OEt)(OPh) and tried to resolve it with menthol, needless to say, without success[115,162]. The conversion of alkoxysilanes to chlorosilanes by reaction with acid chlorides was demonstrated by Ladenburg in 1872[163].

TABLE 9. Functional group chemistry: comparison of common reactions

≡Si-Functional		≡C-Functional[a]	
≡Si–H +	X_2[b,c] → ≡Si–X	≡C–H +	X_2[c] —hν→ ≡C–X
	HI → ≡Si–I		HI → No reaction
	RLi → ≡Si–R		RLi → ≡C[−1 d]
	>=< → ≡Si–C–C–		>=< → No reaction
	$L_nM' \to L_{n-2}M(H)(Si\equiv)$[e]		$L_nM' \to L_{n-2}M(H)(C\equiv)$[f]
	>=O → ≡Si–O–<H[g]		>=O → No reaction
≡Si–Cl +	H_2O → ≡Si–OH[h]	≡C–Cl +	H_2O → ≡C–OH[i]
	ROH → ≡Si–OR		ROH → ≡C–OR
	NH_3[j] → ≡Si–NH_2		NH_3 → ≡C–$NH_3^+Cl^-$[k]
	HY[l] → ≡Si–Y		HY[1] → ≡C–Y
	RLi → ≡Si–R		RLi →[m]
	Mg → ≡SiMgX[n]		Mg → ≡C–MgCl
	Li or Na → ≡Si–Si–		Na[o] → ≡C–C–
	$LiAlH_4$ → ≡Si–H		$LiAlH_4$ → ≡C–H
	ZnF_2 → ≡Si–F		ZnF_2 →[p]
≡Si–OC≡ +	OH^- → ≡Si–OH	≡C–O–C≡ +	OH^- → No reaction
	HF → ≡Si–F		HF →[q]
	RCOCl → ≡Si–Cl		RCOCl → No reaction
	RLi → ≡Si–R		RLi →[r]
	RMgX → ≡Si–R		RMgX → No reaction
	$LiAlH_4$ → ≡Si–H		$LAlH_4$ → No reaction

TABLE 9. (*continued*)

$-\overset{|}{\underset{|}{Si}}-OH$	+	$NaOH \rightarrow -\overset{|}{\underset{|}{Si}}-O^-Na^+$	$-\overset{|}{\underset{|}{C}}-OH$	+	$NaOH \rightarrow$ No reaction
		$H_2SO_4 \rightarrow -\overset{|}{\underset{|}{Si}}-O-\overset{|}{\underset{|}{Si}}-$[s]			$H_2SO_4 \rightarrow -\overset{|}{\underset{|}{C}}-O-\overset{|}{\underset{|}{C}}-$[t]
		$ROCl \rightarrow -\overset{|}{\underset{|}{Si}}-Cl$			$ROCl \rightarrow -\overset{|}{\underset{|}{C}}-O\overset{O}{\overset{\|}{C}}-$

[a] In all cases the carbon is not otherwise activated.
[b] X_2 represents the halogens, Cl_2, Br_2, I_2.
[c] Other halogenating agents may be used.
[d] Occurs when hydrocarbon is more acidic than RH.
[e] There are several product variations from hydrosilanes and transition metals.
[f] Carbon–hydrogen insertion (activation) is a difficult process.
[g] Requires a suitable catalyst.
[h] More complicated reactions can occur, especially if there is more than one Si–Cl bond (see text).
[i] Reaction is very slow in most cases.
[j] Acid acceptor is usually added.
[k] More common for other halides.
[l] HP⟨or HS^-.
[m] Coupling to –C–R is not practical from a chloride; deprotonation can occur.
[n] Coupling to –Si–Si– can occur.
[o] Coupling is not usually performed with lithium.
[p] Not a viable fluorination reagent for carbon chemistry.
[q] Ether cleavage is performed with HI, not HF.
[r] Deprotonation of CH adjacent to oxygen can occur.
[s] Several product variations if there is more than one Si–OH per Si.
[t] *Inter alia*.

The real problems in early silicon chemistry resulted from a lack of understanding of the complexity of reactions of silanols produced from the hydrolysis of chlorosilanes and particularly of dichlorosilanes. The first products identified from hydrolysis of a dichlorosilane were formulated as R_2SiO, and called silicones by analogy to ketones (the name has stuck even though these compounds are quite different in structure)[164]. In 1908, Kipping reported the synthesis of benzyl ethyl silicone, which boiled considerably higher that the related benzyl ethyl ketone. The silicone did not react with reagents that were known to reduce ketones, or with hydroxylamine or phenylhydrazine and, furthermore, the molecular weight corresponded to a trimer[165]. He concluded: '· · · silicones, as a class, differ from ketones in readily forming comparatively stable molecular aggregates, but whether the latter are to be regarded as composed of loosely associated, or of chemically united, molecules we have as yet no satisfactory evidence before us'[165]. There was some reluctance to surrender this analogy to carbon, and in a study reported three years later the following premise was made: 'That a silicone would react with a magnesium alkyl or aryl halogen compound giving ultimately a tertiary silicol was, of course, a natural deduction to base on the known relationship of silicon to carbon'. The experiment was performed and the reaction proceeded in the normal manner giving the first instance in which 'silicones' showed an analogy to ketones in chemical behavior, but the analogy did not hold for long. One of the first thorough studies of hydrolysis of a dichlorosilane involved Ph_2SiCl_2[3,106]. In the first of two papers there are ten pages of experimental detail which describe the problem of isolating the hydrolysis products. The identification of the hydrolysis products of dichlorosilanes was the topic of six consecutive papers published from 1912 to 1914[3,106,166,167]. To illustrate some of the difficulties faced by Kipping, the relationships between the starting silicon dihalide and a series of

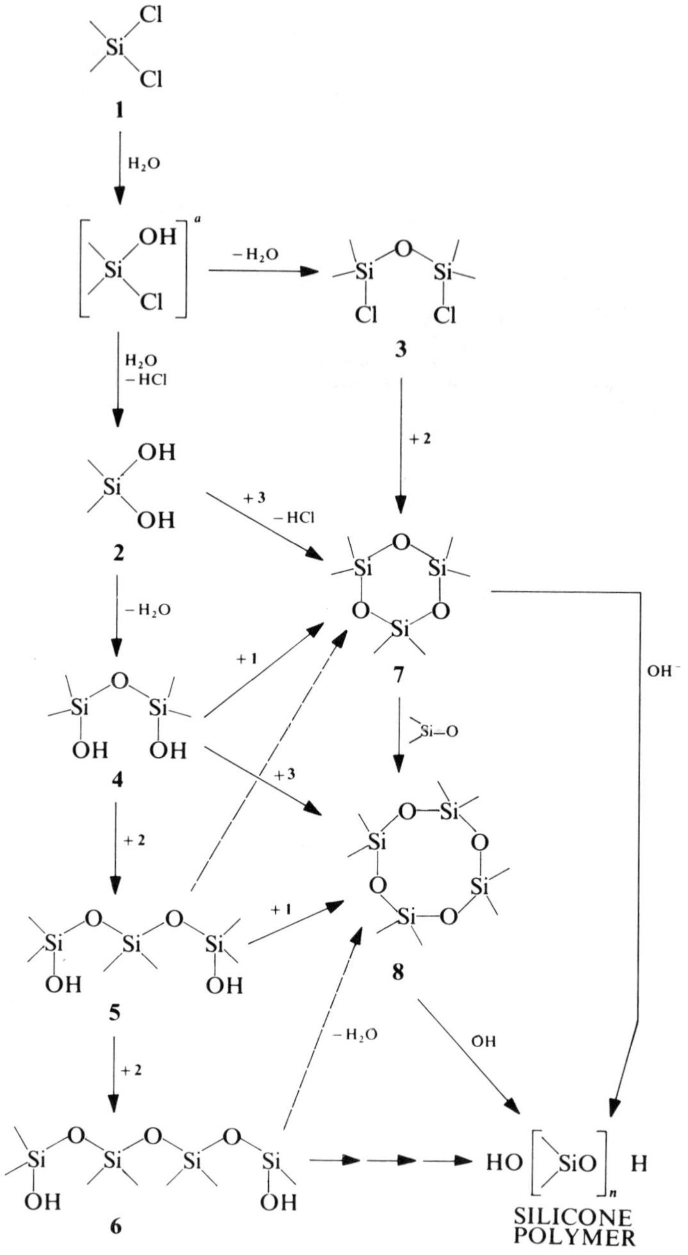

SCHEME 3

—— known; - - - speculative

[a] Hydroxyfluoro derivative has been isolated.

hydrolysis products is shown in Scheme 3. In the diphenyldichlorosilane case, the compounds that Kipping eventually isolated in pure form are related to **2**, **4**, **5**, **6**, **7** and **8**. All but two of the transformations shown in Scheme 3 have been observed and have been involved in selective synthesis of mixed cyclotrisiloxanes and cyclotetrasiloxanes. The cylcopolysiloxanes when heated with small amounts of KOH generate polymers [$\bar{M}$ about one million], which are the gum stock for elastomers. Kipping recognized that the four-membered ring, cyclodisiloxane, was missing in this series. It would take about seventy more years before this system would be identified and then from a totally different reaction process. Many of the variations of silicon-oxygen chemistry are discussed in the chapter on siloxanes in this volume.

The transformations shown in Scheme 3 are difficult to control as well as to reproduce. Anyone attempting synthetic procedures from dichlorosilanes is most likely to generate a few of these derivatives by design or by accident. It is clear why hydrolysis of unreacted chlorosilanes gives rise to gums and results in the source of some of the difficulties in trying to purify or to crystallize desired organosilicon products.

On the basis of an expected analogy to carbon, the hydrolysis of trichlorosilanes was expected to give siliconic acids and EtSiOOH was claimed by Friedel and Ladenburg in 1870[168]. Very good analyses were actually claimed for several of these derivatives. On the basis of observations concerning the condensation of diphenylsilanediol, Meade and Kipping predicted that condensation of silicon triols by elimination of water was more likely than the formation of monomeric siliconic acids and investigated the hydrolysis of $RSiCl_3$[169]. When R = Ph, they obtained soluble materials with analyses that correspon-ded to $PhSiO_2H$ and insoluble fractions whose analyses correspond more closely to that of an 'anhydride', ($PhSiO_2O$). Since the soluble fractions exhibited molecular weights corresponding to tetramers and pentamers and the more insoluble fractions to pentamers and hexamers, they reached the obvious conclusion that siliconic acids were not generated in the hydrolysis of trichlorosilanes. They assumed by analogy to work with $R_2Si(OH)_2$ that the silicon triols eliminated water intermolecularly rather than intramolecularly, but were unable to identify specific products. Evidence for a species related to a siliconic acid has only recently been found in the reaction of $ArCH_2SiMe(OH)_2$ with NaOH in Me_2SO/H_2O, where unimolecular dissociation of the dianion $ArCH_2SiMe(O^-)_2$ to $MeSiO_2^-$, a silaacetate, occurs[170].

The silanetriol, $RSi(OH)_3$, is not often isolated from the hydrolysis of $RSiCl_3$, but in one case where $R = [(Me_3Si)_3C]_3$, the structure has been determined and consists of six molecular units that form a hydrogen-bonded cage[171]. One of the next lowest molecular weight derivatives from the hydrolysis of a trichlorosilane is the sesquioxide, $[RSiO_{1.5}]_8$, with a cubane-like structure of silicon atoms with oxygens bridging all edges. Clearly, the silicon–oxygen derivatives exhibit more extensive variations than the corresponding carbon–oxygen systems.

The most important silicon–oxygen systems of course, are the linear polysiloxane polymers. These important materials are resistant to heat or oxidation, exhibit little change in properties with temperature, are water repellent and biologically inert. The history of the discovery and development of silicones has been summarized[172].

V. MULTIPLE-BONDED DERIVATIVES

Organic derivatives that contain multiple bonds either between two carbon atoms or between carbon and another element, particularly oxygen and nitrogen, are common. Not surprisingly, early investigators in silicon chemistry tried to generate similar species for silicon. The earliest claim to the simplest member of the disilene series, $H_2Si{=}SiH_2$, appears to be in 1909[173] although no such rash claims have been made since except as an intermediate. The general methods for forming olefins involve elimination reactions:

(1) water from an alcohol; (2) HX from a halocarbon; or (3) X_2 from a ⟩CX–CX⟨. Not surprisingly, at least the first two were first used in the earliest reported efforts to make ⟩Si=C⟨ bonds. Schlenk and Renning added MeMgI to Ph_2SiCl_2 to give $Ph_2MeSiCl$, which was then reacted with water and distilled to give '$Ph_2Si{=}CH_2$'[174]. The experiment was unlikely to yield a silaolefin and Kipping and Sands had some pithy remarks concerning the experimental results[175]. On the basis of the chemistry described in the previous section, it is most likely that the product obtained was $Ph_2MeSiOSiMePh_2$. Nonetheless, Kipping described his own efforts to eliminate water from a tertiary silicol (silanol), $EtPrSi(OH)CH_2Ph$, in which he obtained only the oxide (disiloxane)[176]. In the same report he mentioned his attempts to eliminate HCl from $(PhCH_2)_3SiCl$, but without success. He did not give up this approach, however, and five years later described attempting intramolecular elimination of water from Ph(*c*-$C_6H_{11})_2SiOH$, and concludes that 'this confirms the view that substances containing a Si=C bonding are probably incapable of existence'[177]. It was almost 50 more years before a ⟩Si=C⟨ derivative was in a bottle. This incredible achievement, against the odds, stemmed from an experiment reported in 1967 by Gusel'nikov and Flowers. These investigators reported the thermolysis of 1,1-dimethylsilacyclobutane, which gave 1,1,3,3-tetramethyl-1,3-disilacyclobutane through head-to-tail dimerization of $Me_2Si{=}CH_2$[178]. In 1985, the most definitive review to date in this area listed in excess of 400 references concerning ⟩Si=X chemistry[50]. The key to success in this area is to provide sufficiently bulky substituents on silicon to prevent dimerization or oligomerization of the multiple bonded derivative. Another surprising approach, based on the unique character of multiple bonds to silicon (described in Section V.B), will be illustrated in the section on silenes. Only selected aspects of the multiple-bonded chemistry of silicon will be described. Since Michl's chapter in this volume is devoted to this topic. The focus will be on ⟩Si=X species that can be isolated and the relationships to the chemical reactions observed for related ⟩C=X derivatives.

A. Disilenes

One of the first attempts to make a disilene involved the condensation of neat Ph_2SiCl_2 with sodium[175]. A number of reaction products were obtained and three crystalline materials were isolated with a total yield of less than 50%. One of these products was formulated as $Ph_2Si{=}SiPh_2$ because it reacted readily with iodine. However, when this 'disilene' reacted with iodine only the equivalent of one atom of iodine was consumed; therefore it was reformulated as Ph_8Si_4. Kipping made some novel suggestions concerning the products from this condensation reaction, some of which will be discussed in the next section, but he did not continue to consider seriously that multiply-bonded derivatives were actually formed. For ten years Kipping and coworkers also tried to prepare triple bonds between silicon atoms by condensing $PhSiCl_3$ with sodium, but this also failed[179]. Essentially, the synthetic approach attempted by Kipping is referred to as reductive coupling. The first successful example of reductive coupling to a disilene was only recently reported through reaction of R_2SiI_2 [R = $C(SiMe_3)_3$] with $C_{10}H_8Li$[180]. The success of this example may result from the use of the silicon diiodide, but it may also be related to the fact that the process is conducted under homogeneous and not heterogeneous conditions.

Removal of halogen from a *vicinal* dihalide with zinc can be used to form olefins, although it is not a common method for alkene synthesis. Similarly, dehalogenation of ⟩SiX–SiX⟨ with alkali metals appears to be successful in a limited sense. When $R_2Si(X)Si(X)R_2$ (R = *t*-Bu; X = Cl, Br) or RR′Si(Cl)Si(Cl)RR′ (R = *t*-Bu; R′ = *c*-C_6H_{11}) are reacted with lithium (sodium or potassium), the radical anion of the disilene is formed which is particularly long-lived in the latter case[181]. However, several intermediate radicals are formed from $Ar_2Si(Cl)Si(Cl)Ar_2$ (Ar = mesityl) before the stable radical anion of tetramesityldisilene is observed[175].

The dehalogenation route is not the most general route to disilenes. The first isolated disilene was formed by a method that is not used in carbon chemistry. This route involves the dimerization of singlet silylenes, R_2Si:[182]. Silylenes can be generated from photolysis of a trisilane[182], flash pyrolysis of a 1-methoxydisilane or 1,2-dimethoxysilane[183], photolysis of $(Me_3Sn)_2SiR_2$[184], photolysis of a silyldiazide[185], thermolysis of a benzosilanorbornadiene[186] or photolysis of cyclopolysilanes, such as $(R_2Si)_3$ or $(R_2Si)_6$[187–189]. The first-reported isolable solid disilene was discovered through photolysis of $Mes_2Si(SiMe_3)_2$ (Mes = mesityl) in a 3-methylpentane matrix at 77 K which gave the dark blue silylene, Mes_2Si:. Warming of the matrix resulted in the formation of a bright yellow solution from which the chrome yellow tetramesityldisilene was eventually isolated. The key to the successful isolation was the presence of bulky substituents at silicon, in this case provided by the two *o*-methyl substituents of the mesityl. Several routes to silylene precursors are required, because not all sources undergo photolysis under the usual irradiation conditions (low temperature). An alternative approach is laser flash photolysis at room temperature. Careful interpretation of results is required and the problems have recently been discussed and summarized[190]. Except for dimethylsilylene[185], comparison and/or development of precursors to the same silylene have not been evaluated. This will be an area for future growth. The silylene route provides disilenes in which the substituents at the two silicon centers are the same, and this is independent of the source of silylene. A way to bypass this dilemma is to generate silylenes from two different sources. If the codimerization competes with self-condensation, unsymmetrical disilenes can be produced. This tactic was indeed successful in the photolysis of a mixture of $Mes_2Si(SiMe_3)_2$ and $Xyl_2Si(SiMe_3)_2$, where a statistical mixture of disilenes was formed with 50% $Mes_2Si{=}SiXyl_2$[191]. There are certainly limitations of this route and entry into unsymmetrical disilenes will probably require the development of elimination reactions from disilanes a route less explored even for symmetrical disilenes.

The isolable derivatives of disilenes contain bulky substituents, but the basic reactivity patterns probably do not vary considerably except that reactions are probably faster in less stable derivatives. The number of disilenes whose reactions chemistry has been described in some detail is very small. Mechanistic information is sparse and that which is available has been obtained from *trans*-1,2-di-*tert*-butyl-1,2-dimesityldisilene. A comparison of the reactions of disilenes and olefins is shown in Table 10. The mechanisms of many reactions of disilenes have yet to be determined (or verified) in detail and the table provides preliminary conclusions. Since the stereochemistry of reactions can be probed only by studying *cis* or *trans* disilenes, the lack of variation of these derivatives represents a problem. The results in Table 10 include only one disilene of the type, $RR'Si{=}SiRR'$. Certain addition reactions of isolable disilenes occur much more rapidly than additions to olefins, but the related reaction products are similar. Disilenes are hydrogenated by lithium aluminum hydride, a reaction that does not take place with an olefin, although hydrogenations with H_2 are certainly known. The most notable differences between the reactivity of the two unsaturated derivatives is probably toward the reactants oxygen and sulfur. The reaction chemistry of disilenes with oxygen is surprisingly rich in variation. Disilenes react with oxygen at varying rates from virtually instantaneous to days[202]. The products are analogous to dioxetanes and epoxides but a major product is also cyclodisiloxane, long thought to be nonexistent since it was never isolated from hydrolysis of dihalosilanes. Disilenes have been observed in solution for sufficient lengths of time for physical measurements to be obtained. Such disilenes are normally characterized by trapping reactions such as those included in Table 10. The standard trapping agent seems to be methanol, but reactions of dienes are often used. Additional trapping agents include anthracene, which does not react with the disilenes that contain bulky substituents.

TABLE 10. Comparison of disilene and olefin reaction chemistry

Disilene					Olefin		
Reactant/ Conditions	$>Si{=}Si<$[b]	Product	Reaction type (speculated)	Ref.	Reactant/ Conditions	Product[a]	Reaction mechanism
HX	A	$-Si(X)-Si(H)-$[c]	Nucleophilic addition	193	HX/<RT	$-C(X)-C(H)-$	Electrophilic addition (regioselective)
HOMe	A	$-Si(MeO)-Si(H)-$[c]	Nucleophilic addition	193	$HOMe/H^+$	$-C(MeO)-C(H)-$	Electrophilic addition (regioselective; *inter alia*)
H_2O	A	$-Si(HO)-Si(H)-$[c]	Nucleophilic addition	193	H_2O/H^+	$-C(HO)-C(H)-$	Electrophilic addition (regioselective)
Cl_2/C_6H_6	A	$-Si(Cl)-Si(Cl)-$ (one isomer)	Concerted?	194	$Cl_2/<RT$	$-C(Cl)-C(Cl)-$	Electrophilic addition [Regioselective; stereospecific (anti)]
Cl_2/C_6H_{12}	A	$-Si(Cl)-Si(H)-$[c]	Free radical	193	HBr/O_2	$-C(H)-C(Br)-$	Free radical addition (Regioselective: anti-Markownifkov)
LAH	B	$-Si(H)-Si(H)-$	Nucleophilic addition	194	$H_2/Pd/C$	$-C(H)-C(H)-$	[e] Stereospecific (*syn*)
$O_2/-78°$	A	$-Si(O)-Si(O)-$ (O—O bonded)	Free radical	195	O_2/RT	allylic oxidation	Free radical autoxidation
$O_2/25°/C_5H_{12}$	A	$>Si-Si<$ bridged by O	Free radical	195	$ArCO_3H$	$>C-C<$ bridged by O	Electrophilic oxidation (stereospecific)

S_8	A, B	 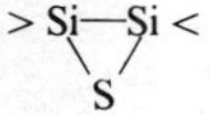	?	196	1. $ArCO_3H$ 2. SCN		1. Electrophilic oxidation 2. Nucleophilic substitution (two inversion steps)
$HgX_2/< -15°$[d]	A	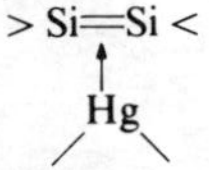	Complexation	197	HgX_2/H_2O^-	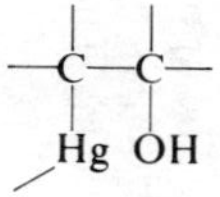	Electrophilic addition [regioselective; stereospecific (*anti*)]
HgX_2/RT[d]	A	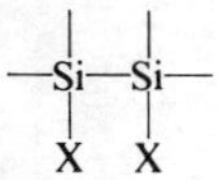		197			
Ph_2CO	B	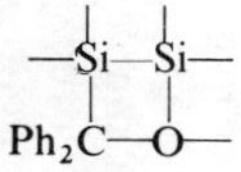	Nucleophilic cycloaddition	198	$Ph_2CO/h\nu$	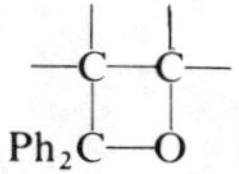	Paterno–Buchi reaction
PhC≡CH	A, B	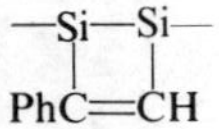	Pericyclic addition (radical process)	199	RC≡CR/$h\nu$	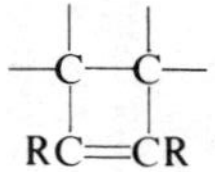	Photochemical [2+2] with activated alkenes
	A	No reaction[g]		198	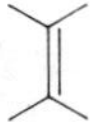	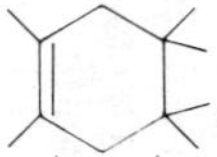	Diels–Alder reaction
$h\nu$ or Δ	A	No reaction		200	$h\nu$	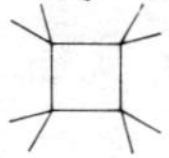	[2+2]cycloaddition (dimerization)
	C	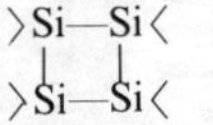	Cycloaddition	201			
$h\nu$/MeOH	D	⟩SiH(OMe)[h]		180	$h\nu$/MeOH	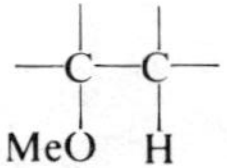	Addition of MeOH to strained photochemically produced products

[a] Examples are given in general form and are based on material taken from Ref. 192.
[b] Key to disilenes: A = 1,2-di-*tert*-butyl-1,2-dimesityldisilene; B = tetramesityldisilene; C = tetraisopropyldisilene; D = tetrakis[bis(trimethylsilyl)methyl]disilene.
[c] Two diastereomers obtained.
[d] $X = OCOCF_3$.
[e] Reaction mechanism involves an organometallic intermediate.
[f] $X = OCOCH_3$.
[g] Unhindered disilenes are good Diels–Alder partners.
[h] Through a silylene intermediate.

The chemistry of disilenes can only be considered to be in its infancy. Some of the differences between the reactions of disilenes and olefins are revealed in Table 10. However, one of the more fascinating aspects that has thus far been revealed is the 1,2-sigmatropic rearrangement shown in equation 16. Such reactions are bound to provide the impetus for future studies.

(16)

B. Silenes

At almost the same time that isolation of tetramesityldisilene was reported, the first crystalline silene, stable in the absence of air, was also isolated[203]. The approach to the first isolated silene involved a type of chemistry that could not be anticipated on the basis of the simple elimination chemistry that has been used to generate olefins. The basic route is shown in equation 17, which for several R groups actually involves an equilibrium

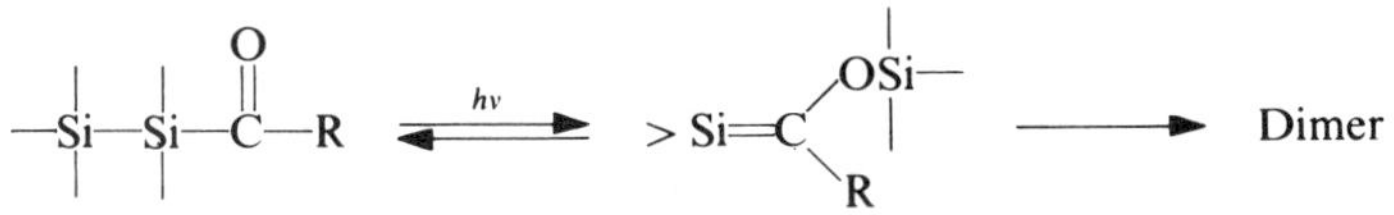
(17)

between the acyl silane and the isomeric silene. The key was finding the right set of substituents that would shift the equilibrium toward the silene and also prevent the formation of the silene dimer. The successful experiment involved the choice of adamantyl as the organic substituent on the carbonyl carbon atom[203].

Many precursors to unstable silenes have been developed since the first experiments of Gusel'nikov and Flowers were reported. The more general reactions involve thermolyses, such as retro Diels–Alder and 'ene' reactions as well as 1,2-elimination processes. These have been described in some detail[50,204]. The chemistry is rich in its variations and has been studied by many talented investigators. The routes to stable silenes, however, have been slow in developing. At this writing only one other method has proved successful and it is based on a 1,2-elimination of XY from the unit, $\rangle$SiXCY$\langle$. The successful precursor is based on α-lithiated halosilanes, but since this is also an equilibrium process the right combinations of substituents and conditions also had to be developed. The experiments performed on $Me_2Si(X)C(Li)(SiMe_3)_2$ provided the foundation for the approach. Elimination of LiX provided the silene which rapidly dimerized. Other trapping reactions were also consistent with the formation of the silene. A simple modification of the substituents on silicon in an approach similar to that used by Brook and coworkers[203] provided the bulk required to stabilize the silene. The removal of the LiF actually required addition of Me_3SiCl as shown in equation 18. In the absence of the silyl chloride the elimination reaction required weeks and the thermal conditions required to induce the elimination result in secondary products from decomposition of the silene. The effective method which is to treat the adduct of $(t\text{-Bu})_2Si(F)C(Li)\,(SiMe_3)_2\cdot 4THF$ with Me_3SiCl/Et_2O results in formation of the THF adduct of the silene shown in equation 18[207]. The THF can be removed by azeotropic distillation with benzene and the resultant silene is kinetically stable at room temperature. It is clear from equation 18 that a 1,3-rearrangement of a methyl group occurs in this system prior to the formation of the silene. Even in the silene, fluxional behavior is exhibited both by $Me_2Si{=}C(SiMe_3)_2$[208] and $Me_2Si{=}C(SiMe_3)[SiMe(Bu\text{-}t)_2$[207] and this property is a unique feature of silenes.

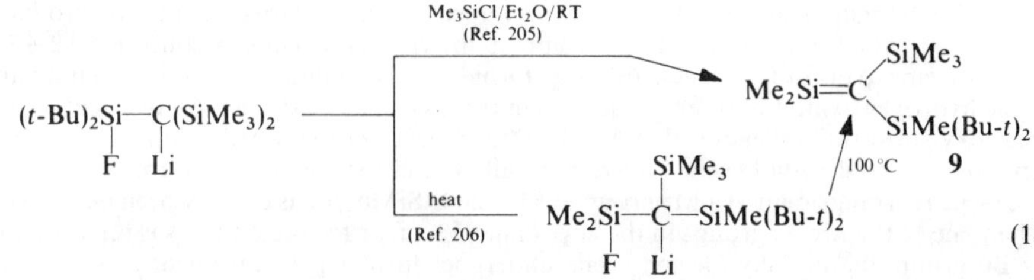

(18)

The formation of adducts from silenes also represents a reactivity not observed in olefins. This chemistry is a consequence of the fact that silicon bears a positive charge in silene (such a charge distribution is obviously not observed in disilenes and olefins). The structure of both the silene, **9**, and the adduct with THF have been reported[54, 209]. The consequence of coordination to the Lewis base is a slight elongation of the Si=C bond by 0.045 Å and pyramidalization at silicon. The bond distance in the adduct is still consistent with considerable double-bond character and poses an interesting problem for the bonding description. The favored description involves zwitterionic and no-bond resonance structures as shown in Figure 4. The formation of adducts suggests possible

FIGURE 4

method for stabilizing more reactive silenes. Indeed, $Me_2Si{=}C(SiMe_3)_2$ which is unstable at $-100\,^\circ C$ forms an adduct with NMe_3, which is stable at room temperature[210]. The order of Lewis basicity toward silenes has been determined: $F^- > NMe_3 > NEt_3 > Br^- > THF$. The resistance to decomposition of the adducts, $\rangle Si{=}C\langle \cdot D$, to silene dimer and free donor D decreases through this sequence[210]. The fluoride adduct of **9** has been isolated with the $[Li(12\text{-crown-}4)_2]^+$ counterion. In this carbanion the Si–C bond distance increases still further (1.777 and 1.822 Å in the two independent anions) but the carbon center remains strictly trigonal planar. The fluoride adduct does not decompose up to 100 °C[211]. These observations suggest that the more reactive silenes could be stabilized and isolated through adduct formation.

The reactions of silenes exhibit some of the same patterns described for disilenes. Addition of water, alcohols and hydrogen halides occurs either by stepwise or concerted processes, but involves nucleophilic addition as the rate-determining step[50]. Halogens also add readily, but so do $Me_3SiY(Y{=}Cl, OMe, NMe_2)$, a reactivity which is not exhibited by disilenes. Jones and coworkers have demonstrated that the addition of alkoxysilanes to silenes occurs by stereospecific *syn*-addition[212]. The nucelophilic additions of carbonyls to disilenes described in the previous section also occur with silenes, although the siloxetanes that are formed are less stable. Silenes, like olefins, also exhibit Diels–Alder reactions. Most of the reported cases have involved the trapping of transient silenes. However, Brook and coworkers have recently shown that the stable silenes, $(Me_3Si)_2Si{=}C(OSiMe_3)R$, exhibit [2+2] and [2+4] reactions with dienes and alkenes[213]. The [2+2] reaction of a silene with an alkene is formally forbidden. A stepwise process through the intermediacy of a diradical was eliminated by comparison of the results of the reaction of silene/butadiene in the presence and absence of the radical trap,

Bu_3SnH. Changing solvent polarity did not appear to affect either the rate or product distribution which mitigates against a dipolar intermediate. Thus, a concerted [2+2] process seems to occur for silenes, although forbidden for olefins. Alkenes that contain an allylic hydrogen exhibit only 'ene' reactions in the alkoxysilene system. A novel feature of the siloxy-substituted silenes is their head-to-head dimerization, which contrasts with the formation of head-to-tail cyclic dimers normally found for the more unstable silenes[214]. The facile rearrangement of a Me group in $Me_2Si{=}C(SiMe_3)_2$ has already been described. When one of the Me_3Si groups in the acylsilane precursor to stable silenes is replaced by a *t*-Bu group, the initially formed silene undergoes further rearrangement as shown in equation 19[215].

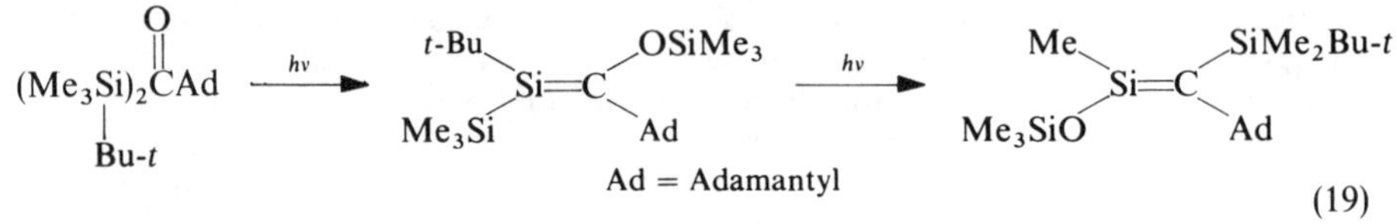

(19)

Silene adducts probably react through the free silaethene, although the reaction with bromine may be an exception. Addition of bromine to **9** gives the expected dibromide, but the adduct, **9**·THF, gives a mixture of the dibromide and the product $\rangle Si[O(CH_2)_4Br]C(Br)\langle$. The formation of the latter suggests electrophilic addition to the Si=C bond before the dissociation of THF[211].

C. Silicon–Heteroatom Multiple Bonds

The generation of stable, isolable $\rangle$Si=X bonds, where X is a Group V or VI element, is a more difficult task. To isolate disilenes and silenes, bulky substituents are required and if moving one group to the right in the Periodic Table one of the four substituents is 'removed', which suggests that even more bulky substituents are required for stabilization and isolation. A second move to the right removes both substituents from the heteroatom and free $R_2Si{=}X$ probably cannot be isolated. This does not imply, however, that such systems cannot be studied under matrix conditions, only that it is unlikely that stable isolable systems can be obtained.

Silene derivatives may be viewed as a case of silicon–heteroatom multiple bonding, since carbon is more electronegative than silicon. Such a view provides a starting point for developing the synthetic methods that would be useful to generate other Si=X bonds. Isolable silenes are formed from rearrangements of acylsilanes and from 1,2-elimination reactions of α-lithiohalosilanes. The task in silanimine chemistry is to develop a suitable precursor for the rearrangement process and this precursor is the azide. When Et_3SiN_3 is photolyzed in the presence of an alkoxysilane $RSi(OMe)_3$, the product obtained is $Et_2Si(OMe)$–NEt–$SiR(OMe)_2$ which may be explained as the trapping product of the intermediate silanimine, $Et_2Si{=}NEt$[216]. The most obvious attempt based on disilene chemistry would be to use mesityl substituents in an effort to stabilize the $\rangle$Si=N–bond. Indeed, irradiation of Mes_3SiN_3 in 3-methylpentane at $-140\,^\circ C$ provided $Mes_2Si{=}NMes$ but the silanimine decomposed at temperatures above $-120\,^\circ C$[217]. A similar photolysis of $Mes_2(Me_3Si)SiN_3$ also provided a silanimine isolated in matrix but not stable on warming[218]. It is again a problem of choosing the right substituent that is the key in the successful experiment. The ordering of bulky substituents at silicon (Section III.C) indicates that alkyl groups larger than methyl must be located in positions *ortho* to silicon for aryl substituents or that a *t*-Bu substituent might be sufficient. Indeed, when $NaSi(Bu\text{-}t)_3$ is added to $t\text{-}Bu_2Si(N_3)Cl$ in Bu_2O at $-78\,^\circ C$, a quantitative conversion to the silanimine occurs[57]. The silanimine was characterized not only by the expected trapping reactions with water and methanol, but also by the solid-state structure determination. The silanimine forms a complex with THF in which the ether is more strongly bound than in the silene case. This would be expected on the basis of the more polar Si=N linkage and in this case THF cannot be removed by azeotropic distillation as was observed for the related silene complex.

The second basic route, 1,2-elimination of an α-lithiohalosilane, proved to be a considerable challenge. The elimination of LiF from $(i\text{-Bu})_2Si(F)N(Li)Ar$ [$Ar = 2,4,6\text{-}(t\text{-Bu})_3C_6H_2$] gave the LiF adduct of the silanimine[219]. This result is not surprising in view of the fact that F^- is the best Lewis base studied thus far toward a silene, combined with the observation that THF coordinates so strongly to $t\text{-Bu}_2Si{=}NSi(Bu\text{-}t)_3$ that it could not be removed. The Si–N bond distance in the fluoride complex is intermediate between that observed for the noncomplexed silanimine and an Si–N single bond, a trend also observed in silene complexes. Attempts to remove LiF resulted in the formation of a dimer, again a reactivity that might be expected based on analogy to silene chemistry. The α-lithiofluoride also functions as a delivery agent for silanimine when trapping agents are present.

The solution to the dilemma of elimination of LiF was solved by employing the same method Wiberg used to generate silenes, that is, the addition of Me_3SiCl[220]. These transformations are summarized in Scheme 4.

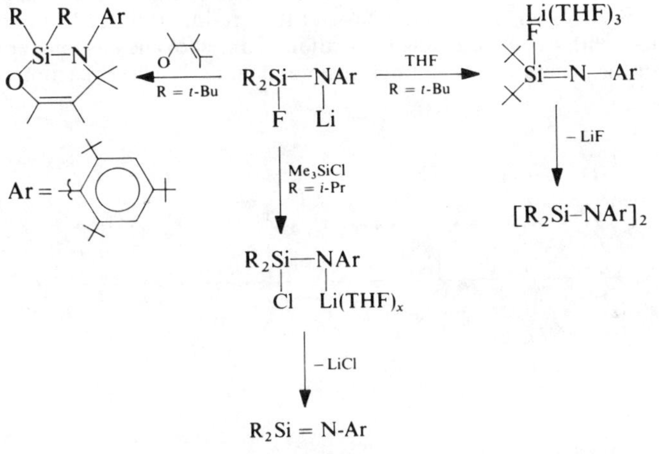

SCHEME 4

A novel approach to silanimine involves a form of double bond 'exchange' using a silene[57,211]. It is probable that a cycloaddition product of the silene is first formed, and subsequently the silanimine is eliminated (equation 20). Although the approach relies on the formation of silene, it still may provide a viable route to additional silanimines.

Silanephosphimines have thus far eluded isolation although stable species in solution have been generated[64]. The basic method is similar to that just described but has two variations. Deprotonation of $Ar'P(H)Si(Cl)Mes_2$ with BuLi in THF at $-60°$ provided a mixture of $Ar'P{=}SiMes_2$ and $Ar'P(H)Si(Bu)Mes_2$. The silanephosphimine was characterized by addition of methanol. The same silanephosphimine was generated by condensation of $Ar'P(SiMe_3)Li$ with Cl_2SiMes_2[221].

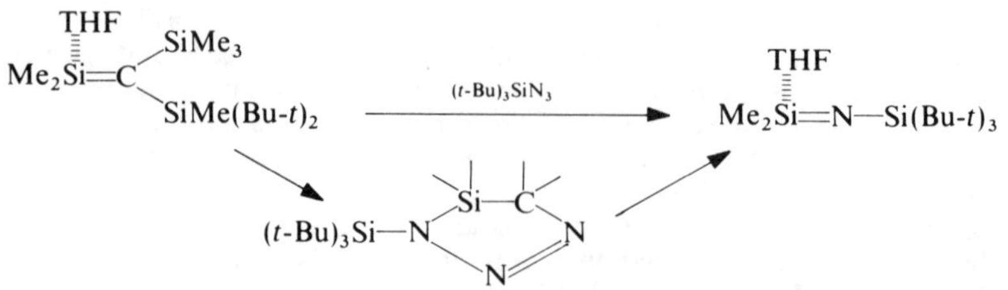

(20)[211]

The challenge of isolating a silanone or a silanethione still remains. Calculations have suggested that the Si=S π–bond energy may be greater than that of Si=O, which may mean that it is more likely that a silathione will be stable at room temperature than a silanone. Since an effective way to stabilize silenes and silanimines appears to be through complexation, it may be that this tactic will lead to the successful isolation of multiply-bonded derivatives between silicon and Group VI elements. As is often the case, discoveries can be made by accident. In his elegant studies of hypervalent silicon species, Corriu and coworkers have developed the chemistry of intramolecularly coordinated ligands. In one case the reaction of a coordinated dihydrosilane with sulfur has led to the isolation of a solid derivative whose analyses and properties are consistent with a complexed silathione[222]. These preliminary results were only recently reported but hold out a promise for this type of approach. Although less definitive, a coordinated silanone intermediate is also suggested on the basis of the product isolated from reaction of the same substrate with carbon dioxide. Formation of a trisiloxane has long been used as an argument for the generation of an intermediate silanone[50]. These reactions are shown in equation 21[222].

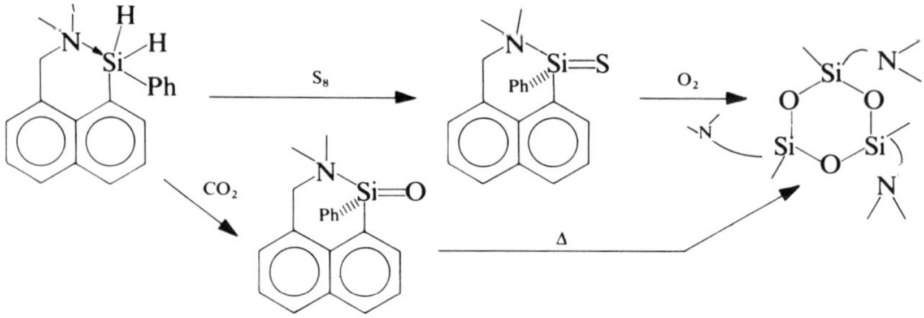

(21)

Multiple bonds between silicon and other elements comprise a new area with many surprises. Considerable development in methods still remains. Only recently have Si=M (M = transition metal derivatives) been generated. The approach used in the first reported case was based on simple condensation chemistry, similar to that used for the formation of silicon–carbon bonds, and is shown in equation 22. Not surprisingly, the

$$(OC)_4FeNa + Cl_2Si(OR)_2 \xrightarrow{HMPA} (OC)_4Fe{=}\overset{OP(NMe_2)_3}{\overset{\vdots}{Si}}(OR)_2 \qquad (22)$$

product was isolated as a complex with HMPA[223]. A second 'silylene' complex was generated by a different route, but also isolated as the solvent complex as shown in equation 23[224]. These results suggest a whole new area in silicon–multiple bond chemistry.

$Cp^*L_2RuSiPh_2Cl + Me_3SiOTf = Cp^*L_2RuSiPh_2(OTf)$

$Cp^* = C_5Me_5;\ L = PMe_3$

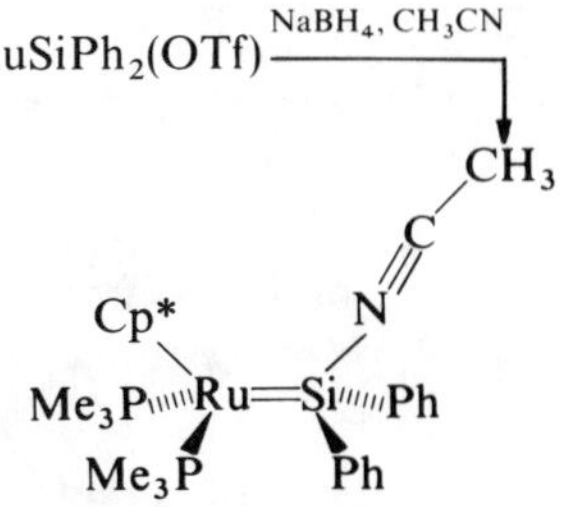

(23)

VI. CATENATED DERIVATIVES

Catenated derivatives of carbon occur naturally. The corresponding compounds of silicon must be constructed, and such derivatives were considered limited and of no interest until recently. The first silicon–silicon bonded compound was prepared as early as 1869[225] but the chemistry of organopolysilanes was largely ignored for nearly a century. Some of the early chemistry of silicon–silicon bonded compounds provides a clue as to why this area of silicon chemistry was so slow to develop. Hexachlorodisilane is formed as one of the products from the reaction of silicon with chlorine and thus was one of the earlier precursors in the study of derivatives with silicon–silicon bonds. Although hexachlorodisilane could be alkylated with zinc alkyls[226] and Grignard reagents[227], some cleavage of the silicon–silicon bond occurred. Similarly, when Si_3Cl_8 was reacted with phenyl magnesium bromide a mixture of Ph_4Si and Ph_6Si_2 was obtained[228]. These observations led to the conclusion that silicon–silicon bonds were 'unstable'. Kipping most likely generated polymers with Ph_2Si repeat units, but he did not characterize them[229] and the next polymer was not claimed until 1949 and was based on Me_2Si[230]. A major discovery was reported in 1975 that the 'useless' polymer based on Me_2Si was a good precursor for silicon carbide fibers[231]. Since then the study of catenated derivatives of silicon has nearly exploded, leading to new uses for long chains containing silicon in the backbone. In addition to functioning as precursors to silicon carbide, other uses include: (a) photoresists for microelectronics; (b) photoinitiators for radical reactions; (c) photoconducting films for electrophotography; and (d) nonlinear optical materials.

Derivatives with silicon–silicon bonds include cycles, chains and polymers. The relationships between derivatives that contain two or more contiguous silicon centers is shown in Scheme 5. In most cases the origin of the catenated derivatives may be traced back to a dihalosilane.

The only well-developed method for the formation of silicon–silicon bonds involves alkali or alkaline earth metal condensation of a halosilane, and in most cases this is a chlorosilane. The reaction is analogous to Wurtz coupling in organic chemistry. In the simplest case a monochlorosilane is condensed to a disilane, whereas a dichlorosilane provides either polymers or cyclopolysilanes whose ring size will depend on the substituents on silicon and the specific conditions employed. Condensation of trichlorosilanes is even more compex and products are essentially not characterized. The chemistry of short chains, cyclopolysilanes and linear polysilanes is interrelated, as shown in Scheme 5. The Wurtz coupling route is essentially ineffective for producing short chains with functional groups which terminate the chains. A functional group at the chain end is required if additional chemistry is to be performed. A Wurtz coupling reaction is a brute-force, nondiscriminating synthetic method, which is a problem with many heterogeneous reactions. There are no general alternatives, although dehydrogenative coupling of a hydrosilane may have some promise in this direction. The chemistry of polysilanes will be covered in more detail in West's chapter in this volume.

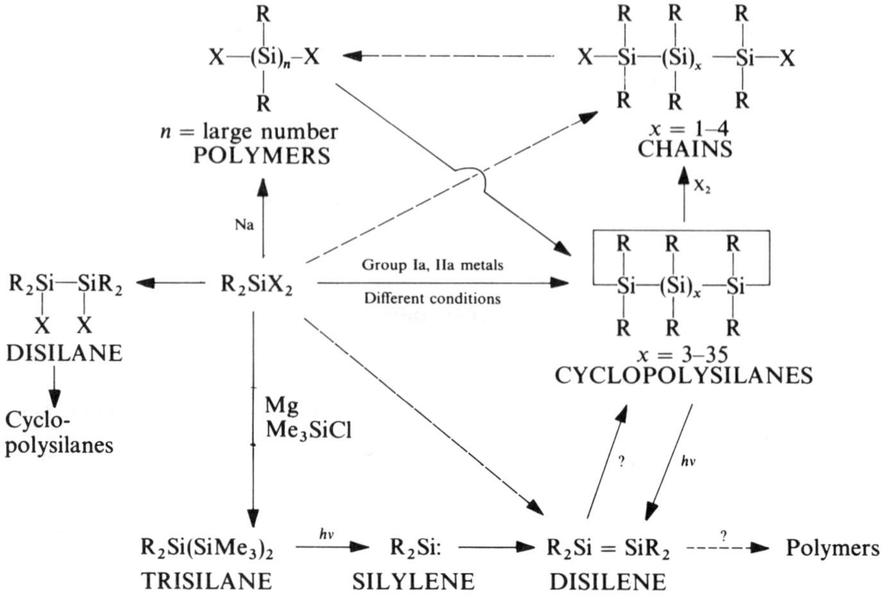

SCHEME 5

A. Cyclopolysilanes[232,233]

The simplest cycloalkanes are based on methylene units, $(CH_2)_n$. However, the related cyclic silanes are surprisingly difficult to prepare and most cyclopolysilanes are based on $(R_2Si)_n$ where R is an alkyl or aryl substituent. Cyclopolysilanes are formed primarily from condensation of dichlorosilanes with an alkali metal. The principal ring size produced appears to be related to the substituents on silicon, the reductant, solvent, temperature and halide. It is not possible to evaluate each factor, since it is probable that the products are a result of a combination of effects. Furthermore, most reactions are run under kinetic conditions where a series of ring sizes is often produced. Table 11 illustrates some of the typical conditions and substituents that have been studied. Watanabe and coworkers[243] have shown that in the condensation of dialkyldichlorosilanes, the ring size is related to the steric parameters developed by Cartledge (Section I)[145]. Most reactions are run with a slight excess of alkali metal (10–20%) to produce the highest yield of combined cyclic products. It is probable that with two equivalents of alkali metal nonequilibrium distribution of cyclic products is obtained. The use of potassium may produce the equilibrium mixtures at a faster rate than does lithium. In the case of condensation of Et_2SiCl_2 with potassium the major product isolated was $(Et_2Si)_5$ (96%), although the kinetic product is the tetramer. The equilibrium distribution for $(Et_2Si)_n$ was established from $(Et_2Si)_4$ and found to be $n=5$ (95%), $n=4$ (2%) and $n=6$ (2%)[244]. The sequence of steps in the reductive coupling of halosilanes is essentially unknown. It is possible that polymeric product is produced first and that depolymerization is then required to provide the cyclics. Alternatively, condensation of one unit at a time occurs followed by cyclization in which case it should be possible to intercept the short chain. Only in one case does an intermediate disilane appear to be reported and that is for *t*-Bu(Ph)$SiCl_2$, where condensation with a deficiency of lithium was shown to produce significant quantities of *t*-Bu(Ph)Si(Cl)Si(Cl)(Ph)Bu-*t*[245]. The disilane

TABLE 11. Formation of Cyclopolysilanes from Dihalosilanes

R in R_2SiX_2	X	General conditions[a]	n in $(R_2Si)_n$ (%)	Ref.
$(Me_3Si)_2CH$	I	Li/naphthalene/$-78°$	2	180
$2,6-Me_2C_6H_3$	Cl	Li/naphthalene	3(10)	187
$t-BuCH_2$	Cl	Li/0°C	3(14)	189
t-Bu	I	Li^+ $Naph^-$	3(72)	234
Et_3Si	Br	Na	3	235
Me_3Si	Br	Na	4	235
i-Pr	Cl	Li	4(74)	236
$EtMe_2Si$	Br	Na/toluene	4	237
Ph	Cl	Li or Na/toluene	4(11–95)	238
		Li/THF	5(39–75)	
Et	Cl	K	5(91)	239
Pr	Cl	Li	5(49)	236
$-(CH_2)_4-$	Cl	$Li/Ph_3SiSiMe_3$	6(80)	240
Me	Cl	Li/anthracene[b]	6(70)	238b
Me	Cl	Na/K	5–36[c]	241
Me	Cl	Na/xylene	Large (90)[d]	242

[a] Unless specified otherwise reactions are conducted in THF.
[b] Ultrasound conditions.
[c] Major product was $n=6$.
[d] Depolymerization of the polymer occurs in the presence of Na/naphthalene/THF to give cyclohexasilane (75%).

$Ph_3SiSiMe_3$ appears to be useful as an equilibrating agent in catalytic amounts[246]. Another interesting, but essentially unexplored effect of $Ph_3SiSiMe_3$ is in the condensation of $RR'SiCl_2$. Of the four isomeric cyclic tetramers produced when t-Bu(Me)$SiCl_2$ and Li are condensed in the presence of $Ph_3SiSiMe_3$, a single isomer (*trans-trans-trans*) is obtained in 68% yield. In the absence of the disilane four isomers are obtained[246, 247].

The effect of changing the alkali metal and the stoichiometry has been studied only in rare instances and the best two cases are the condensation of Et_2SiCl_2[244] and MeEt-$SiCl_2$[248]. Another area that has not been explored is the effect of changing the halogen in R_2SiX_2. This is most likely due to the fact that the preparation of silyl bromides and iodides requires additional synthetic steps from a dichloride or dihydride precursor. Other routes to cyclopolysilanes have been developed but are less general and require additional synthetic steps to obtain the precursor[249].

Cyclopolysilanes carry substituents on every ring atom and are thus quite different from the usual cycloalkanes. The silicon analogs of cyclopentane and cyclohexane can be prepared by indirect routes. It is useful in polysilane chemistry to view a phenyl substituent as a functional group, since aryl groups can be cleaved by electrophilic reagents, a process that is not usually observed for alkyl groups. There is a balance in conditions for these aryl cleavage reactions, since silicon–silicon bonds may also react. A good compromise is the use of $HCl/AlCl_3$. This approach was successfully employed by Hengge and Bauer to prepare $(SiH_2)_n$ as shown in equation 24. It has also been demonstrated that p-tolyl groups are cleaved faster than phenyl groups[250] although this fact has not been exploited yet in polysilane chemistry. Homogeneous, amorphous Si:H films may be obtained at low temperatures by heating cyclic $(SiH_2)_n$[251].

Three models could be used to fit the electron-diffraction structure of cyclohexasilane, but the best overall fit corresponded to a mixture of chair (62%), twist (25%) and

$$Si_nPh_{2n} \xrightarrow{HCl/AlCl_3} Si_nCl_{2n} \xrightarrow{LiAlH_4} Si_nH_{2n} \qquad (24)$$

$n = 4$[252], 5[253], 6[253] $\qquad n = 5$[254], 6[255]

boat (13%) forms[256]. This mixture corresponds to an energy difference of 1.6 and 2.1 kcal mol^{-1} for the twist and boat forms, respectively, relative to the chair form. These values correspond closely to values calculated by molecular mechanics[257]. The differences in energies between the various conformers is smaller than that for cyclohexane, which exists almost entirely in the chair form at room temperature. The structure of cyclopentasilane (by electron diffraction) is nonplanar, but it was not possible to distinguish between the envelope and chain-like conformations[258] and there appears to be little energy difference between the two forms. Such an observation is consistent with the fact that Si–Si bonds are longer than C–C bonds. The cyclotetrasilanes exhibit folded or planar structures depending on substituents. Bond distances are longer and Si–Si stretching force constants are lower in the cyclotetrasilanes compared to the cyclopenta- and cyclohexasilanes[232b]. Bond distances in cyclotrisilanes are similar to those for cyclotetrasilanes but in one remarkable case, that of $[(t\text{-Bu})_2Si]_3$, the Si–Si bond distance is 2.51 Å, the longest Si–Si bond distance reported in a cycle. This distance corresponds to a bond order of about one-half[259]. The ring strain energies for peralkylcyclopolysilanes seem to be in good agreement with those of the corresponding cycloalkanes, $(CH_2)_n$ (except $n = 2$)[260]. Although the structures of small and medium-sized cyclopolysilanes are similar to the carbon analogs, larger rings such as $(Me_2Si)_{13}$ and $(Me_2Si)_{16}$ exhibit conformations that are different from the cycloalkane counterparts[261].

The chemistry of cyclopolysilanes exhibits some remarkable patterns that have no analogy in cycloalkane chemistry. Reduction to anion radicals has been reported for cyclopolysilanes, for ring sizes from three to six. The cyclopentasilane anion radicals exhibit only single values for α-^{13}C hfsc which suggests that they are either planar or rapidly pseudorotating[262]. Cyclotetrasilanes with small substituents also exhibit a single α-^{13}C hfsc, but when bulky substituents are present there are two values which support a bent structure on the ESR time scale for the latter system. The cyclopolysilanes bear some analogy to aromatic systems[263]. In aromatic anion radicals the SOMO may be represented by using only carbon $2p_z$ orbitals, but in the silicon anion radicals the SOMO is described by mixing of Si–Si σ^* orbitals constructed from 3p silicon orbitals and Si–C σ^* orbitals[264].

A second unusual feature of cyclopolysilanes is that, upon irradiation, silylenes are extruded (in some but not all cases) and in the three- and four-membered rings disilenes are also observed[232]. Similar ring contraction and elimination of carbenes is not a feature of cycloalkanes.

Although the structures of cyclopolysilanes can be similar to the carbon analogs, the methods of synthesis are not . The majority of the systems studied result from the self-condensation of R_2SiCl_2 or $RR'SiCl_2$ which provide cycles of different sizes and with identical substituents at each silicon center. Of course, it is possible to cocondense two different dichlorosilanes, however then not only are rings of different sizes produced but isomers within each ring size are generated[265] and separation can be a real problem. Clearly this area of catenated silicon derivatives would benefit from other reaction methods.

B. Short Chains

In contrast to the difficulty in obtaining the simplest derivatives of cyclopolysilanes, $(SiH_2)_n$, the short-chain hydrosilanes can be generated in simple reactions and were probably formed by Wohler in his study of the compounds formed from addition of acids to silicides[266]. However, the isolation and characterization of hydrides above SiH_4 was performed by Stock and coworkers[267] who also developed the vacuum line techniques required to study the hydrides of boron and silicon. Stock's preferred method for the generation of silicon hydrides was through the addition of magnesium silicide to an acid. Fractionation of the mixture of hydrides provided monosilane, disilane, trisilane, tetrasilane and an inseparable mixture of penta- and hexasilane. Isomerism is possible for the higher hydrides and Stock suspected that isomers of Si_5H_{12} and Si_6H_{14} were indeed formed, but this property was not demonstrated unitl the advent of gas chromatography[268]. The higher silanes decompose at room temperature over long time periods forming hydrogen and a solid that is a mixture of silicon and polymeric (SiH_2). The most careful early decomposition studies were performed not by Stock but by Stokland, who began his experimental work in 1938 on silanes and deuteriosilanes. His work was curtailed during World War II and was not published until 1948, unfortunately in a rather obscure location[269]. Stokland showed that the decomposition of SiH_4 and Si_2H_6 was first order, providing silylene and producing amorphous silicon. The decomposition of hydrosilanes is still of considerable interest because such reactions are a source of pure silicon. An alternative to the generation of short-chain hydrosilanes has been provided in the $Hg(^3P_1)$ sensitized photolysis of H_2/SiH_4 which provides Si_nH_{2n+2} (n=2,3,4)[270].

In contrast to formation of hydrosilane chains, the formation of short silicon chains with organic substituents is a more difficult problem. Condensation of R_2SiCl_2 derivatives (which are often commercially available) cannot be used in a general way to form short chains directly. As indicated in Scheme 5, the route to chains from a dichlorosilane is indirect. A two-step process of condensation to the cyclopolysilane followed by ring opening is required to form chains from a dichlorosilane. Actually, Kipping and Sands were the first to demonstrate this in the condensation of Ph_2SiCl_2 when the benzene insoluble product, Ph_8Si_4, was treated with iodine to give the ring-opening product $I(SiPh_2)_4I$[271]. There are obvious disadvantages to this route that involve not only the number of steps and the attendant isolation and purification, but also the fact that only derivatives of the type $X(SiR_2)_nX$ can be prepared. Alternative starting materials can be appropriately substituted disilanes, including $Me_3Si_2Cl_3$ which is a direct process residue. Other functional disilanes can be prepared by condensing $R_2Si(H)Cl$ with an alkali metal[272] although this sometimes results in reaction of the Si–H bond. Cleavage of phenyl groups in a manner described in the previous section is extensively used in building chains. Methods for cleaving methyl groups have also been developed, as is illustrated in the formation of $ClMe_2SiSiMe_2Cl$ from Me_6Si_2 by treatment with $Me_3SiCl/AlCl_3$[273] or $HSiCl_3/H_2PtCl_6$[274]. Photolysis of $(R_3Si)_2Hg$ derivatives also generates silicon–silicon bonds. How these methods can be combined to generate short chains is illustrated in equations 25–28. Routes leading to a chloro derivative are the most useful, since this represents a functional group that can lead to other derivatives (Section IV).

$$MeSiCl_3 + Me_3SiLi \rightarrow MeSi(SiMe_3)_3 \xrightarrow{Me_3SiCl/AlCl_3} MeSi(SiMe_2Cl)_3 \qquad (25)$$ [275]

$$2Me_3SiCl + Cl_2SiPh_2 \xrightarrow[HMPA]{Mg} (Me_3Si)_2SiPh_2 \xrightarrow[C_6H_6]{HCl/AlCl_3} (Me_3Si)_2SiCl_2 \qquad (26)$$ [276]

$$PhMe_2SiSiMe_2Cl + Na/K \rightarrow PhMe_2Si(SiMe_2)_2SiMe_2Ph \xrightarrow{HCl/AlCl_3} Cl(SiMe_2)_4Cl \qquad (27)$$ [277]

$$(Ph_2Si)_4 + Li \rightarrow Li(SiPh_2)_4Li \xrightarrow{HPh_2SiCl} H(SiPh_2)_6H \xrightarrow{HCl/AlCl_3} H(SiCl_2)_6H \qquad (28)$$ [278]

The methods for formation of short chains are rather primitive and still rely on a metal condensation somewhere in the sequence. Improvements in synthetic strategies should provide an alternative to this step. Some promise in this direction is provided by the observation that $PhSiH_3$ undergoes dehydrogenative coupling to give short chains when catalytic amounts of $Cp_2Ti(Me)_2$ are added[279]. With secondary silanes such as Ph_2SiH_2, disilanes are produced. Since metal condensation of Ph_2SiCl_2 cannot be controlled to the disilane stage, this still represents an improvement since in $Ph_2Si(H)Si(H)Ph_2$ both hydrogen and phenyl are functional groups which can be replaced by different reactions. Both Me_5Si_2H and $HMe_2SiSiMe_2H$ disproportionate in the presence of $(Et_3P)_2PtCl_2$ to give $H(SiMe_2)_nH$ where $n = 1$–6[280]. Both of these approaches require future development.

The study of short silicon chains is underdeveloped and this may in part be due to the lack of synthetic methods. The chemistry of disilanes and trisilanes has been explored in more detail, since they are precursors to silylenes and thus to disilenes. The physical properties of short chains can be quite interesting and might provide models for the study of the longer chains (Section VI.C). The wavelength of the absorption band in $Me(SiMe_2)_nMe$ moves to lower energy as the chain length increases to about 25 silicon atoms when λ_{max} reaches about 290 Å. The electronic transition that gives rise to this absorption involves electrons in the Si–Si σ-bonds. Structural information on acyclic catenated silicon derivatives is sparse, but what is available hints at the possibility of unusual species. The structure of $(t\text{-Bu})_3SiSi(Bu\text{-}t)_3$ has been determined and the silicon–silicon bond distance is a remarkable 2.697 Å, even longer than that observed for $[(t\text{-Bu})_2Si]_3$[281]. In spite of this extraordinary distance, the disilane is stable to air and moisture. The trisilane, $I[(t\text{-Bu})_2Si]_3I$, also exhibits longer than normal single bonds and a linear relation between the Si–Si bond distances and the bulk of the substituents was proposed[282]. Details, such as the conformations of short chains with four to six silicon atoms, are lacking. It will be interesting to see if these long silicon–silicon bond distances can also exist in longer chains and what properties such molecules might exhibit.

C. Long Chains

Probably Kipping and coworkers obtained the first polymers with silicon atoms in the backbone, but these noncrystalline derivatives that were products of the condensation of Ph_2SiCl_2 with sodium were not characterized[229]. The first polymer that was recognized as such was $(Me_2Si)_n$[230]. This material was extremely insoluble in organic solvents and decomposed above 250 °C, and therefore was not useful for anything. When Yajima and coworkers had shown that $[(Me_2Si)]_n$ could be converted to SiC fibers by the sequence shown in equation 29, applications for polysilanes began to hold some promise. If a soluble polysilane could be produced, the carbosilane step could be avoided. The discovery came by accident from an investigation of the condensation of a mixture of Me_2SiCl_2 and $PhMeSiCl_2$ the aim of generating mixed cyclopolysilanes (described in Section VI.A)[283]. When the ratio of the two dichlorosilanes is about one, soluble polymers are formed. The idealized structure contains one Me_2Si and one MeSiPh unit analogous to the organic polymer, polysilastyrene. Although these polymers could be spun into fibers, the yields of ceramic materials on pyrolysis were low in the initial experiments but recent work provides results near the theoretical limit. Incorporation of organic groups on silicon that cross-link more easily[284] or irradiation of a mixture of the polymer with a reagent such as $PhSi(CH{=}CH_2)_3$[285] also provide precursors to ceramics.

$$(Me_2Si)_n \xrightarrow[\text{Ar}]{400\,^\circ\text{C}} \begin{matrix}\text{Branched}\\ \text{carbosilanes}\end{matrix} \xrightarrow{\text{Melt/spin}} \text{Fibers} \xrightarrow[\text{Air}]{200\text{–}350\,^\circ\text{C}} \text{Crosslinked} \xrightarrow[N_2]{1200\,^\circ\text{C}} \begin{matrix}\text{SiC}\\ \text{fiber}\end{matrix} \quad (29)$$

There is only one basic method for the preparation of linear polysilanes and that is the condensation of a dichlorosilane with sodium in a hydrocarbon solvent ($>100\,^{\circ}C$). Condensation of R_2SiCl_2 or $RR'SiCl_2$ provides homopolymers $(RR'Si)_n$ (R = Me; R' = *n*-Pr, *n*-Bu, *n*-$C_{12}H_{25}$, Ph and others)[283a] and condensation of mixtures of dichlorosilanes provides copolymers $(R^1R^2Si)_n(R^3R^4Si)_m$. The polymeric products obtained by Kipping from Ph_2SiCl_2 were insoluble. The first soluble aromatic polysilanes were recently reported from the condensation of $(p\text{-}RC_6H_4)_2SiCl_2$ (R = Me_3Si, *t*-Bu, *i*-C_4H_9, C_6H_{13})[286]. In principle, it should be possible to alter the properties of the polymer by tailoring the substituents on silicon. The steps in the condensation process are not well known, but preliminary studies have been reported by Zeigler and coworkers. The polymer is formed preferentially or solely at the sodium surface[287]. Several intermediates have been proposed, including silylenes and silyl anions, but Zeigler's studies show that silyl radicals are the most probable species responsible for polymerization.

The polymers generated with only silicon atoms in the backbone exhibit unique properties. Preliminary light-scattering measurements are consistent with a 'semiflexible' polymer with properties intermediate between those of rigid rods and flexible coils[283a]. Semiconducting polysilane polymers are formed on doping with SbF_5. The cation-radical produced contains a 'hole' that can be delocalized through the σ-framework. Delocalization of σ-electrons is also suggested by the ultraviolet absorption spectra. The position of the UV absorption bands depends on the substituent group, with a λ_{max} of about 305 nm for small alkyl substituents to 326 nm for (*c*-C_6H_{11}), whereas aryl-substituted polysilanes absorb as high as at 400 nm. The molar absorbance of the polymers is extraordinarily high, and when exposed to sunlight the polymers degrade by extrusion of silylenes. This property has led to the study of polysilanes as photoresists and photoinitiators for polymerization of olefins. For catenated silicon compounds, it is not the similarities to carbon that are important but the new properties that are introduced by the presence of the silicon–silicon bonds. These new properties will provide directions for future applications.

VII. REFERENCES

1. F. S. Kipping, *Proc. R. Soc. London, Ser. A*, **159**, 139 (1937).
2. There are 51 numbered papers, but there are actually 53 since four papers are published under two numbers.
3. For an example see the footnote in F. S. Kipping and R. Robison, *J. Chem. Soc.*, **105**, 484 (1914) and F. S. Kipping, *J. Chem. Soc.*, **101**, 2108 (1912).
4. For a discussion of these two approaches to bonding in main group elements see, W. Kutzelnigg, *Angew. Chem., Int. Ed. Engl.*, **23**, 272 (1984). For an early summary of participation of d orbitals in bonds to silicon, see E. A. V. Ebsworth, in *The Bond to Carbon*, Vol. 1, pt. 1 (Ed. A. G. MacDiarmid), Marcel Dekker, Inc., New York, 1968 and H. Kwart and K. G. King, *d-Orbitals in the Chemistry of Silicon, Phosphorus and Sulfur*, Springer-Verlag, Berlin, 1977. Additional commentary may be found in S. N. Tandura, M. G. Voronkov and N. V. Alekseev, *Topics in Current Chemistry*, Vol. 131 (Ed. F. L. Boschke), p. 99. Springer-Verlag, Berlin, 1986.
5. a. G. A. Olah and G. K. S. Prakash, *Chem. Br.*, **10**, 917 (1983).
   b. G. A. Olah, G. K. S. Prakash, L. D. Field and K. Wade, *Hypercarbon Chemistry*, J. Wiley & Sons, New York, 1987.
6. B. T. Luke, J. A. Pople, M.-B. Krogh-Jespersen, Y. Apeloig, J. Chandrasekhar and P. v. R. Schleyer, *J. Am. Chem. Soc.*, **108**, 260 (1986).
7. L. F. Bovey, *J. Chem. Phys.*, **21**, 830 (1953).
8. D. R. J. Boyn, *J. Chem. Phys.*, **23**, 922 (1955).
9. H. C. Allen and E. K. Pyler, *J. Chem. Phys.*, **31**, 1062 (1959).
10. S. Weiss and G. F. Leroi, *J. Chem. Phys.*, **48**, 962 (1968).
11. R. W. Kilb and L. Pierce, *J. Chem. Phys.*, **27**, 108 (1957).
12. W. L. Meerts and I. Ozier, *J. Mol. Spectrosc.*, **94**, 38 (1982).

13. a. D. R. Lide, *J. Chem. Phys.*, **27**, 343 (1957).
b. T. Nishikawa, T. Itoh and K. Shimoda, *J. Chem. Phys.*, **23**, 1735 (1955).
14. K. Hedburg, *J. Am. Chem. Soc.*, **77**, 6491 (1955).
15. G. Gundersen, R. A. Mayo and D. W. H. Rankin, *Acta Chem. Scand., Ser. A*, **38**, 579 (1984).
16. K. Kimura and M. Kubo, *J. Chem. Phys.*, **30**, 151 (1959).
17. M. C. L. Gerry, R. M. Less and G. Winnewisser, *J. Mol. Spectrosc.*, **61**, 231 (1976).
18. A. Almenningen, O. Bastiansen, V. Ewing, K. Hedberg and M. Traetteberg, *Acta Chem. Scand.*, **17**, 2455 (1963).
19. C. Newman, J. K. O'Loane, S. R. Polo and M. K. Wilson, *J. Chem. Phys.*, **25**, 855 (1956).
20. N. W. Clark and F. C. DeLucia, *J. Mol. Struct.*, **32**, 29 (1976).
21. R. Kewley, P. M. McKinney and A. G. Robiette, *J. Mol. Spectrosc.*, **34**, 390 (1970).
22. J. H. Callomon, E. Hirota, K. Kuchitsu, W. J. Lafferty, A. G. Maki and C. S. Pote, *Structure Data on Polyatomic Molecules*, Landolt-Bornstein, New Series, Group II, Vol. 7 (Eds. K. H. Hellwege and A. M. Hellwege), Springer-Verlag, Berlin, 1976.
23. M. Pfeiffer and H. J. Spangenberg, *Z. Phys. Chem. (Leipzig)*, **232**, 47 (1966).
24. T. Kojima, E. L. Breig and C. C. Lin, *J. Chem. Phys.*, **35**, 2139 (1961).
25. C. Glidewell, P. M. Pinder, A. G. Robiette and G. M. Sheldrick, *J. Chem. Soc., Dalton Trans.*, 1402 (1973).
26. J. R. Durig, Y. S. Li, M. M. Chen and J. D. Odom, *J. Mol. Spectrosc.*, **59**, 74 (1976).
27. R. W. Kilb, *J. Phys. Chem.*, **23**, 1736 (1955).
28. T. Kojima and T. Nishikawa, *J. Phys. Soc. Jpn.*, **12**, 680 (1957); *Chem. Abstr.*, **51**, 16019a (1957).
29. A. Allmenningen, K. Hedberg and R. Seip, *Acta Chem. Scand.*, **17**, 2264 (1963).
30. L. S. Bartell and L. O. Brockway, *J. Chem. Phys.*, **23**, 1860 (1955).
31. S. L. Miller, L. C. Aamodt, G. Dousmanis, C. H. Townes and J. Kraitchman, *J. Chem. Phys.*, **20**, 1112 (1952).
32. B. Bak, J. Bruhn and J. Rastrup-Andersen, *Acta Chem. Scand.*, **8**, 367 (1954).
33. D. M. Golden and S. W. Benson, *Chem. Rev.*, **69**, 125 (1969).
34. a. K. W. Egger and A. T. Cocks, *Helv. Chim. Acta*, **56**, 1516, 1537 (1973).
b. S. W. Benson, *Thermochemical Kinetics*, 2nd edn., Wiley, New York, 1976.
35. D. R. McMillen and D. M. Golden, *Annu. Rev. Phys. Chem.*, **11**, 493 (1982).
36. a. A. M. Doncaster and R. Walsh, *J. Chem. Soc., Faraday Trans.*, **75**, 1126 (1979).
b. R. Walsh and J. M. Wells, *J. Chem. Soc. Faraday Trans.*, **72**, 100 (1976).
37. R. Walsh, *Acc. Chem. Res.*, **14**, 246 (1981).
38. J. H. Huheey, *Inorganic Chemistry*, 3rd edn., Harper and Row, New York, 1983, Appendix E, pp. A-29 ff.
39. V. I. Vendeneyev, L. V. Gurvich, V. N. Knodrat'yev, V. A. Medvedev and Ye. L. Frankevich, *Bond Energies, Ionization Potentials and Electron Affinities*, St. Martin's Press, New York, 1966.
40. I. M. T. Davidson and A. V. Howard, *J. Chem. Soc., Faraday Trans.*, **71**, 69 (1975).
41. T. L. Cottrell, *Strengths of Chemical Bonds*, 2nd edn., Butterworth, London, 1958.
42. F. E. Saalfeld and H. J. Svec, *Inorg. Chem.*, **3**, 1442 (1964).
43. G. Klebe, *J. Organomet. Chem.*, **293**, 147 (1985).
44. L. Pauling, *The Nature of the Chemical Bond*, Cornell Univ. Press, Ithaca, New York, 1960.
45. a. G. Donnay and R. Allmann, *Am. Mineral.*, **55**, 1003 (1970).
b. I. D. Brown and R. D. Shannon, *Acta Crystallogr., Sect. A*, **29**, 266 (1973).
46. L. Peter, *J. Chem. Ed.*, **63**, 123 (1986).
47. a. D. Schomburg, A. Blaschette and E. Wieland, *Z. Naturforsch. B: Anorg. Chem. Org. Chem.*, **41B**, 1112 (1986).
b. B. Rempfer, H. Oberhammer and N. Auner, *J. Am. Chem. Soc.*, **109**, 3893 (1987).
48. J. M. Kanabus-Kaminska, J. A. Hawari, D. Griller and C. Chatgilialoglu, *J. Am. Chem. Soc.*, **109**, 5267 (1987).
49. M. J. S. Dewar and E. Healy, *Organometallics*, **1**, 1705 (1982).
50. For the best summary of this area see G. Raabe and J. Michl, *Chem. Rev.*, **85**, 419 (1985).
51. B. T. Luke, J. A. Pople, M.-B. Krogh-Jespersen, Y. Apeloig, M. Karni, J. Chandresekhar and P. v. R. Schleyer, *J. Am. Chem. Soc.*, **108**, 270 (1986).
52. K. Kuchitsu, *J. Chem. Phys.*, **44**, 906 (1966).
53. Y. Apeloig and M. Karni, *J. Am. Chem. Soc.*, **106**, 6677 (1984).

54. N. Wiberg, G. Wagner, J. Riede and G. Muller, *Organometallics*, **6**, 32 (1987).
55. R. Pearson and F. J. Lovas, *J. Chem. Phys.*, **66**, 4149 (1977).
56. P. J. Bruna, V. Krumbach and S. Peyerimhoff, *Can. J. Chem.*, **63**, 1594 (1985).
57. N. Wiberg, K. Schurz, G. Reber and G. Muller, *J. Chem. Soc., Chem. Commun.*, 591 (1986).
58. H. W. Kroto, J. N. Murrell, A. Al-Derzi and M. F. Guest, *Astrophys. J.*, **219**, 886 (1976).
59. Z. K. Ismail, R. H. Haigue, L. Fredin, J. W. Kauffman and J. L. Margrave, *J. Chem. Phys.*, **77**, 1617 (1982).
60. M. J. Fink, M. J. Michalczyk, K. J. Haller, R. West and J. Michl, *Organometallics*, **3**, 793 (1984).
61. M. T. Nguyen, M. A. McGinn and A. F. Hegarty, *Inorg. Chem.*, **25**, 2185 (1986).
62. H. W. Kroto, J. F. Nixon and K. Ohno, *J. Mol. Spectrosc.*, **90**, 367 (1981).
63. K. J. Dykema, T. N. Truong and M. S. Gordon, *J. Am. Chem. Soc.*, **107**, 4535 (1985).
64. C. N. Smit and F. Bickelhaupt, *Organometallics*, **6**, 1156 (1987).
65. a. S. A. Pope, I. H. Hillier and M. F. Guest, *J. Chem. Soc., Chem. Commun.*, 623 (1984).
    b. R. Jaquet, W. Kutzelnigg and V. Staemmler, *Theor. Chim. Acta*, **54**, 205 (1980).
66. D. R. Johnson, F. X. Powell and W. H. Kirchhoff, *J. Mol. Spectrosc.*, **39**, 136 (1971).
67. T. Kudo and S. Nagase, *Organometallics*, **5**, 1207 (1986).
68. T. J. Barton and G. C. Paul, *J. Am. Chem. Soc.*, **109**, 5293 (1987).
69. H. Shizuka, H. Tanaka, K. Okazaki, M. Kato, H. Watanabe, Y. Nagai and M. Ishikawa, *J. Chem. Soc., Chem. Commun.*, 748 (1986).
70. J. E. Douglas, B. S. Rabinovitch and F. S. Looney, *J. Chem. Phys.*, **23**, 315 (1955).
71. G. Olbrich, P. Potzinger, B. Reimann and R. Walsh, *Organometallics*, **3**, 1267 (1984).
72. M. J. Michalczyk, R. West and J. Michl, *Organometallics*, **4**, 826 (1985).
73. P. R. Jones and M. E. Lee, *Rev. Silicon, Germanium, Tin, Lead Compd.*, **9**, 11 (1986).
74. M. W. Schmidt, M. S. Gordon and M. Dupuis, *J. Am. Chem. Soc.*, **107**, 2585 (1985).
75. K. D. Dobbs and W. J. Hehre, *Organometallics*, **5**, 2057 (1986).
76. P. v. R. Schleyer and P. D. Stout, *J. Chem. Soc., Chem. Commun.*, 1373 (1986).
77. A. Sax and R. Janoschek, *Angew. Chem., Int. Ed., Engl.*, **25**, 651 (1986).
78. S. S. Zigler, K. M. Welsh, R. West and J. Michl, *J. Am. Chem. Soc.*, **109**, 4392 (1987).
79. J. F. Ogilvie and S. Cradock, *J. Chem. Soc., Chem. Commun.*, 364 (1966).
80. H. Bock and R. Dammel, *Angew. Chem., Int. Ed. Engl.*, **24**, 111 (1985).
81. A. Sekiguchi, S. S. Zigler, R. West and J. Michl, *J. Am. Chem. Soc.*, **108**, 4241 (1986).
82. A. Stock and C. Somieski, *Chem. Ber.*, **54**, 740 (1921).
83. B. Beagley and A. R. Conrad, *Trans. Faraday Soc.*, **66**, 740 (1970).
84. S. Cradock, E. A. V. Ebsworth, W. J. Savage and R. A. Whiteford, *J. Chem. Soc., Faraday Trans. 2*, **68**, 934 (1972).
85. E. A. V. Ebsworth and H. J. Emeleus, *J. Chem. Soc.*, 2150 (1958).
86. J. Durig, M. J. Flanagan and V. F. Kalasinsky, *J. Chem. Phys.*, **66**, 2775 (1977).
87. E. A. V. Ebsworth, *Acc. Chem. Res.*, **20**, 295 (1987).
88. A. J. Blake, E. A. V. Ebsworth aand A. Welch, *Acta Crystallogr., Sect. C*, **C40**, 895 (1984).
89. M. J. Barrow, E. A. V. Ebsworth and M. M. Harding, *Acta Crystallogr., Sect. B*, **B35**, 2093 (1979).
90. G. Trinquier and J. P. Malrieu, *J. Am. Chem. Soc.*, **109**, 5303 (1987).
91. L. Pierce and M. Hayashi, *J. Chem. Phys.*, **35**, 479 (1961).
92. A. Almenningen, K. Hedberg and R. Seip, *Acta Chem. Scand.*, **17**, 2264 (1954).
93. D. R. Lide, Jr. and D. E. Mann, *J. Chem. Phys.*, **28**, 572 (1958).
94. K. Hedberg, *J. Am. Chem. Soc.*, **77**, 6491 (1955).
95. D. R. Lide, Jr. and D. E. Mann, *J. Chem. Phys.*, **29**, 914 (1958).
96. B. A. Beagley, A. G. Robiette and G. M. Sheldrick, *Chem. Commun.*, 601 (1967).
97. E. H. Eyster, R. H. Gillette and L. O. Brockway, *J. Am. Chem. Soc.*, **62**, 3236 (1940).
98. M. C. L. Gerry, J. C. Thompson and T. M. Sugden, *Nature*, **211**, 846 (1966).
99. P. v. R. Schleyer and P. D. Stout, *J. Chem. Soc., Chem. Commun.*, 1373 (1986).
100. F. S. Kipping, *J. Chem. Soc.*, **101**, 2106 (1912).
101. C. Eaborn, *Organosilicon Compounds*, University Microfilms International, 1977.
102. *Gmelin Handbook of Inorganic Chemistry*, Si Vol. Al, 8th edn., Springer-Verlag, Berlin, 1984. The early history of silicon has been reviewed and translated from the original articles in German and/or French. The observations included in this chapter are taken from this volume. The original sources are given below.

a. H. Davy, *Phil. Trans. R. Soc. London*, **98**, 333 (1808).
b. According to J. R. Partington, *A History of Chemistry*, Vol. 4, MacMillan, London, 1964, p. 150.
c. J. J. Berzelius, *Versuch uber die Theorie der Chemischen Proportionen*, published in 1819.
d. A. v. Ruprecht, *Chem. Ann. Crell.*, **11**, 13 (1790).
e. H. Davy, *Phil. Trans. R. Soc., London*, **100**, 16 (1810).
f. L. J. Gay-Lussac and L. J. Thenard, *Rech. Phys. Chim. (Paris)*, **1**, 240 and 313 (1811); **2**, 265 (1811). The original report appeared in *Nouveau Bulletin de la Societe Philomatique de Paris*, 156 (1808).
g. Joubert, *Ann. Chim. Phys.*, **4**, 5 (1885).
h. J. J. Berzelius, *Ann. Phys. Chem.*, **1**, 169 and 204 (1824).
i. A. W. Hofmann, from Justus Liebig's and F. Wohler's Briefwechesel in den Jahren 1829–73, Vol. 1, Braunschweig, 1888, pp. 230–231.
j. H. Sainte-Claire Deville, *Compt. Rend.*, **39**, 321 (1854); *Ann. Chim. Phys.*, **43**, 1 (1855).
k. H. Moissan, *Bull. Soc. Chim. Fr.*, **13**, 972 (1895).
l. H. G. Stever, in *Crystal Rectifiers* (Eds. H. C. Torrey and C. A. Whitmer), 1st edn., 3rd reprint, Published New York–London, 1948, p. 303.
m. J. Dietl, D. Helmreich and E. Sirtl, in *Silicon* (Ed. J. Grabmaier), Springer-Verlag, Berlin, 1981, pp. 57–58.
n. D. W. Lyon, C. M. Olson and E. D. Lewis, *Trans. Am. Electrochem. Soc.*, **96**, 359 (1949).
o. J. Diett, D. Helmreich and E. Sirtl, in *Silicon* (Ed. J. Grabmaier), Springer-Verlag, Berlin, 1981, pp. 49–54.
p. *Chem. Eng. News*, 8, 3 Sept., 1979.
q. H. Buff and F. Wohler, *Liebigs Ann. Chem.*, **104**, 94 (1857).
r. G. Martin, *J. Chem. Soc.*, **105**, 2835 (1914).
s. F. Wohler and H. Buff, *Liebigs Ann. Chem.*, **103**, 218 (1857).
t. C. Friedel and A. Ladenburg, *Liebigs Ann. Chem.*, **143**, 118 (1867).

103. E. Rochow, *Silicon and Silicones*, Springer-Verlag, Berlin, 1987.
104. M. Ebelmen, *Bull. Chem. Phys., Series 3*, **16**, 129 (1846), cited in E. G. Rochow, *Silicon*, '*Comprehensive Inorganic Chemistry*', Vol. 1, Pergamon Press, Oxford, 1973, p. 1424.
105. The most useful summary of silicon–carbon bond formation may be found in C. Eaborn and R. W. Bott, in *The Bond to Carbon*, Vol. 1, Pt. 1 (Ed. A. G. MacDiarmid), Marcel Dekker, Inc., New York, 1968.
106. F. S. Kipping, *J. Chem. Soc.*, **101**, 2125 (1912).
107. E. Frankland, *J. Chem. Soc.*, **2**, 263 (1848–9).
108. C. Friedel and J. M. Crafts, *Ann.*, **127**, 28 (1863); *Compt. Rend.*, **56**, 590 (1863).
109. C. Friedel and J. M. Crafts, *Ann.*, **136**, 203 (1865).
110. C. Pape, *Ann. Chem.*, **222**, 354 (1884).
111. A. Ladenburg, *Ann. Chem.*, **164**, 300 (1872); *Ber.*, **4**, 901 (1871).
112. For a discussion of basic organometallic reaction types in main group organometallic chemistry see E.-I. Negishi, *Organometallics in Organic Synthesis*, Vol. 1, Wiley, New York, 1980.
113. C. Friedel and J. M. Crafts, *Bull. Soc. Chim. Paris*, **4**, 356 (1865); *Ann. Chim. (Phys.)*, **19**, 337 (1870).
114. A. Ladenburg, *Ber.*, **6**, 379 (1873); **7**, 387 (1874); *Ann. Chem.*, **173**, 143 (1874).
115. F. S. Kipping and L. L. Lloyd, *J. Chem. Soc.*, **79**, 449 (1901). (Paper 1 in the series).
116. V. Grignard, *Compt. Rend.*, **130**, 1322 (1900).
117. F. S. Kipping, *Proc. Chem. Soc.*, **20**, 15 (1904).
118. W. Dilthey and F. Eduardoff, *Ber. Deut. Chem. Ges.*, **37**, 1139 (1904).
119. F. S. Kipping, *J. Chem. Soc.*, **91**, 209 (1907).
120. G. Martin and F. S. Kipping, *J. Chem. Soc.*, **95**, 302 (1909).
121. B. D. W. Luff and F. S. Kipping, *J. Chem. Soc.*, **93**, 2004 (1908).
122. This is a variation of the argument presented by E. Rochow in *The Chemistry of Silicones*, Wiley, New York, 1946.
123. E. G. Rochow, *J. Am. Chem. Soc.*, **67**, 963 (1945).
124. The best compilation of data on the direct process reaction can be found in R. J. H. Voorhoeve, *Organosilanes: Percursors to Silicones*, Elsevier, Amsterdam, 1967.
125. H. S. Booth and W. O. Stillwell, *J. Am. Chem. Soc.*, **56**, 1529 (1934).
126. W. C. Schumb and E. L. Gamble, *Inorg. Synth.*, **1**, 42 (1939) (Ed. H. S. Booth).

127. R. A. Turetskaya, N. T. Ivanova, R. A. Andrianov, E. A. Chernyshev and G. I. Belik, *J. Gen. Chem. USSR (Engl. Transl.)*, **46**, 1264 (1976).
128. W. J. Ward, A. Ritzer, K. M. Carroll and J. W. Flock, *J. Catal.*, **100**, 240 (1986).
129. M. G. R. T. DeCooker, J. H. N. DeBruyn and P. J. Van den Berg, *J. Organomet. Chem.*, **99**, 371 (1975).
130. B. A. Bluestein, U.S. Pat. 2,887,501; *Chem. Abstr.*, **53**, 18865 (1959).
131. K. Moedritzer and J. R. Van Wazer, *Inorg. Chem.*, **5**, 547 (1966).
132. A. D. Petrov, N. P. Smetankina and G. I. Nikishin, *Zh. Obshch. Khim.*, **25**, 2332 (1955); *Chem. Abstr.*, **50**, 9280 (1956).
133. N. S. Nametkin, A. V. Topchiev and L. I. Kartasheva, *Dokl. Akad. Nauk SSSR*, **101**, 885 (1955); *Chem. Abstr.*, **50**, 3216 (1956).
134. G. Seitz, Ger. (East) Pat. 20,712; *Chem. Abstr.*, **56**, 504 (1962).
135. S. Yamada, Japan. Pat. 3617 (1952); *Chem. Abstr.*, **48**, 3992 (1954).
136. J. R. Malek, J. L. Speier and A. P. Wright, U.S. 4,088,669; *Chem. Abstr.*, **89**, 129699b (1978).
137. B. Kanner, XV Organosilicon Symposium, Duke University, March 27, 1981.
138. Kanegafuchi Chemical Industry Co., Ltd. Jpn. Kokai Tokkyo Koho JP 82 99, 593; *Chem. Abstr.*, **97**, 127804y (1982).
139. T. C. Frank, K. B. Kester and J. L. Falconer, XIX Organosilicon Symposium, Louisiana State University, April 26, 1985.
140. W. F. Banholzer, N. Lewis and W. Ward, *J. Catal.*, **101**, 405 (1986).
141. One only has to consider the 12-volume series of L. F. Fieser and M. Fieser, *Reagents in Organic Synthesis*, Wiley, New York, to understand the enormity of the possibilities.
142. H. W. Kohlschutter, W. Scheele and G. Jaekel, cited by H. W. Kohlschutter, *Fortschr. Chem. Forsch.*, **1**, 1 (1949).
143. K. A. Andrianov, B. A. Izmailov, N. P. Minyaeva, N. V. Tsvetkova and E. A. Prozumentova, *J. Gen. Chem. USSR (Engl. Transl.)*, **46**, 2255 (1976).
144. V. D. Sheludyakov, VR. I. Zhun', G. A. Turkel'taub, M. V. Polyakova, M. G. Kuznetsova and V. F. Mironov, *J. Gen. Chem. USSR (Engl. Transl.)*, **50**, 700 (1980).
145. F. K. Cartledge, *Organometallics*, **2**, 425 (1983).
146. Y. Nakadaira, K. Oharu and H. Sakurai, *J. Organomet. Chem.*, **309**, 247 (1986).
147. F. Glockling, P. Harriott and W. K. Ng, *J. Chem. Res.*, 40 (1979).
148. M. Weidenbruch and K. Kramer, *J. Organomet. Chem.*, **291**, 159 (1985).
149. P. Jutzi and P. Langer, *J. Organomet. Chem.*, **132**, 45 (1977).
150. E. M. Dexheimer and L. Spialter, *Tetrahedron Lett.*, 1771 (1975).
151. E. M. Dexheimer, L. Spialter and L. D. Smithson, *J. Organomet. Chem.*, **102**, 21 (1975).
152. D. Seyferth, R. E. Mammarella and H. A. Klein, *J. Organomet. Chem.*, **194**, 2273 (1980).
153. D. Seyferth and R. M. Weinstein, *J. Am. Chem. Soc.*, **104**, 5534 (1982).
154. Yu. N. Baryshnikov, N. N. Kaloshina and G. I. Vesnovskaya, *J. Gen. Chem. USSR (Engl. Transl.)*, **49**, 134 (1979).
155. K. Tamao, N. Ishida and M. Kumada, *J. Org. Chem.*, **48**, 2120 (1983).
156. a. L. H. Sommer, E. W. Pietrusza and F. C. Whitmore, *J. Am. Chem. Soc.*, **69**, 188 (1947).
b. C. A. Burkhard and R. H. Krieble, *J. Am. Chem. Soc.*, **69**, 2687 (1947).
157. For a review of this reaction see J. Speier, *Adv. Organomet. Chem.*, **17**, 407 (1979).
158. R. C. Anderson, Brit. Pat. 983,435 (1965); *Chem. Abstr.*, **62**, 16298 (1965).
159. A. Boudin, G. Cerveau, C. Chuit, R. J. P. Corriu and C. Reye, *Angew. Chem., Int. Ed. Engl.*, **25**, 474 (1986).
160. J. Rich, VIIIth International Symposium on Organosilicon Chemistry, Washington University, St. Louis, MO 63121, June 1987, Abstr. A41.
161. The chemistry of silafunctional derivatives is reviewed each year in the 'Annual Surveys in Silicon Chemistry' which is published in the *Journal of Organometallic Chemistry Library Series* or *J. Organomet. Chem.*, both published by Elsevier, Amsterdam.
162. Resolution of functional optically active silicon compounds was not actually reported until 1959. L. H. Sommer, "Stereochemistry, Mechanism and Silicon", McGraw-Hill, New York, 1965.
163. A. Ladenburg, *Ann. Chem.*, **164**, 309 (1872).
164. a. C. Friedel, *Ann. Chim. Phys.*, 95 (1866).
b. W. Dilthey, *Ber.*, **38**, 4132 (1905).
165. R. Robison and F. S. Kipping, *J. Chem. Soc.*, **93**, 439 (1908).
166. R. Robison and F. S. Kipping, *J. Chem. Soc.*, **101**, 2142 and 2156 (1912).

167. R. Robison and F. S. Kipping, *J. Chem. Soc.*, **105**, 40, 484 (1914).
168. C Friedel and A. Ladenburg, *Ber.*, **3**, 15 (1870).
169. J. A. Meade and F. S. Kipping, *J. Chem. Soc.*, **105**, 679 (1914); **107**, 459 (1915).
170. J. Chmielecka, J. Chojnowski, C. Eaborn and W. A. Stancyzk, *J. Chem. Soc., Chem. Commun.*, 1337 (1987).
171. N. H. Buttrus, R. I. Damja, C. Eaborn, P. B. Hitchcock and P. D. Lickiss, *J. Chem. Soc., Chem. Commun.*, 1385 (1985).
172. H. A. Liebhafsky, *Silicones on the Monogram*, Wiley, New York, 1978.
173. P. Lebeau, *Compt. Rend.*, **148**, 43 (1909).
174. W. Schlenk and J. Renning, *Annalen*, **394**, 221 (1912).
175. F. S. Kipping and J. E. Sands, *J. Chem. Soc.*, **119**, 830 (1921).
176. F. S. Kipping, *J. Chem. Soc.*, 104 (1927).
177. N. W. Cusa and F. S. Kipping, *J. Chem. Soc.*, 2205 (1932).
178. L. E. Gusel'nikov and M. Flowers, *J. Chem. Soc., Chem. Commun.*, 864 (1967).
179. F. S. Kipping, A. G. Murraay and J. G. Maltby, *J. Chem. Soc.*, 1180 (1929).
180. S. Masamune, Y. Eriyama and T. Kawase, *Angew. Chem., Int. Ed. Engl.*, **26**, 584 (1987).
181. a. M. Weidenbruch, K. Dramer, A. Schafer and J. K. Blum, *Chem. Ber.*, **118**, 107 (1985).
b. M. Weidenbruch and K.-L. Thom, *J. Organomet. Chem.*, **308**, 177 (1986).
182. R. West, M. J. Fink and J. Michl, *Science*, **214**, 1343 (1981).
183. a. G. Maier, G. Mihm, H. P. Reisenauer and D. Littmann, *Chem. Ber.*, **117**, 2369 (1984).
b. P. P. Gaspar, Y.-S. Chen, A. P. Helfer, S. Konieczny, E. C. L. Ma and S. H. Mo, *J. Am. Chem. Soc.*, **103**, 7344 (1981).
184. T. J. Drahnak, J. Michl and R. West, *J. Am. Chem. Soc.*, **103**, 1845 (1981).
185. H. Vancik, G. Raabe, M. J. Michalczyk, R. West and J. Michl, *J. Am. Chem. Soc.*, **107**, 4097 (1985).
186. H. Sakurai, H. Sakaba and Y. Nakadaira, *J. Am. Chem. Soc.*, **104**, 6156 (1982).
187. A. S. Masamune, Y. Hanzawa, S. Mureakami, T. Bally and J. F. Blount, *J. Am. Chem. Soc.*, **104**, 1150 (1982).
188. S. Masamune, S. Murakami, J. T. Snow, H. Tobita and D. J. Williams, *Organometallics*, **3**, 333 (1984).
189. H. Watanabe, T. Okawa, M. Kato and Y. Nagai, *J. Chem. Soc., Chem. Commun.*, 781 (1983).
190. P. P. Gaspar, D. Holten, S. Konieczny and J. Y. Corey, *Acc. Chem. Res.*, **20**, 337 (1987).
191. a. H. B. Yokelson, J. Maxka, D. A. Siegel and R. West, *J. Am. Chem. Soc.*, **108**, 4239 (1986).
b. R. West, G. R. Gillette and H. B. Yokelson, *Inorg. Syn.*, in press.
192. K. P. C. Vollhardt, *Organic Chemistry*, W. H. Freeman, New York, 1987.
193. D. J. DeYoung, M. J. Fink, R. West and J. Michl, *Main Group Metal Chem.*, **10**, 15 (1987).
194. a. R. West, M. J. Fink, M. J. Michalczyk, D. J. DeYoung and J. Michl in *Organosilicon and Bioorganosilicon Chemistry* (Ed. H. Sakurai), Pt. I, Chapter 1, Ellis Horwood, Sussex, UK, 1985, p. 3ff.
b. R. West, *Science*, **225**, 1109 (1984).
195. M. Michalczyk, R. West and J. Michl, *J. Chem. Soc., Chem. Commun.*, 1525 (1984); M. J. Michalczyk, M. J. Fink, K. J. Haller, R. West and J. Michl, *Organometallics* **5**, 531 (1986).
196. R. West, D. J. DeYoung and K. J. Haller, *J. Am. Chem. Soc.*, **107**, 4942 (1985).
197. C. Zybill and R. West, *J. Chem. Soc., Chem. Commun.*, 857 (1986).
198. M. J. Fink, D. J. DeYoung, R. West and J. Michl, *J. Am. Chem. Soc.*, **105**, 1070 (1983).
199. D. J. DeYoung and R. West, *Chem. Lett.*, 883 (1986).
200. M. J. Michalczyk, R. West and J. Michl, *Organometallics*, **4**, 724 (1985).
201. H. Matsumoto, T. Arai, H. Watanabe and Y. Nagai, *J. Chem. Soc., Chem. Commun.*, 724 (1984).
202. R. West, H. B. Yokelson, G. R. Gillette and A. J. Millevolte, in *Silicon Chemistry* (Eds., J. Y. Corey, E. R. Corey and P. Gaspar), Pt. V, Chapter 26, Ellis Horwood, Sussex, UK, 1988, p. 269.
203. A. G. Brook, F. Abdesaken, B. Gutekunst, G. Gutekunst and R. K. Kallury, *J. Chem. Soc., Chem. Commun.*, 191 (1981).
204. A. Brook and K. M. Baines, *Adv. Organomet. Chem.*, **25**, 1 (1986).
205. N. Wiberg and G. Wagner, *Angew. Chem., Int. Ed. Engl.*, **22**, 1005 (1983).
206. N. Wiberg and G. Wagner, *Chem. Ber.*, **119**, 1455 (1986).
207. N. Wiberg and G. Wagner, *Chem. Ber.*, **119**, 1467 (1986).
208. N. Wiberg and H. Kopf, *Chem. Ber.*, **120**, 653 (1987).
209. N. Wiberg, G. Wagner, G. Muller and J. Riede, *J. Organomet. Chem.*, **271**, 381 (1984).

210. N. Wiberg and H. Kopf, *J. Organomet. Chem.*, **315**, 9 (1986).
211. N. Wiberg, G. Wagner, G. Reber, J. Riede and G. Muller, *Organometallics*, **6**, 35 (1987).
212. a. A. H. B. Cheng, P. R. Jones, M. E. Lee and P. Roussi, *Organometallics*, **4**, 581 (1985).
b. P. R. Jones, T. F. Bates, A. F. Cowley and A. M. Arif, *J. Am. Chem. Soc.*, **108**, 3122 (1986).
213. A. G. Brook, K. Vorspohl, R. R. Ford, M. Hesse and W. J. Chatterton, *Organometallics*, **6**, 2128 (1987).
214. A. G. Brook, in *Organosilicon and Bioorganosilicon Chemistry* (Ed. H. Sakurai), Pt. I, Chapter 2, Horwood, Sussex, UK, 1985, p. 15.
215. A. G. Brook, K. D. Safa, P. D. Likiss and K. M. Baines, *J. Am. Chem. Soc.*, **107**, 4338 (1985).
216. L. E. Gusel'nikov and N. S. Nametkin, *Chem. Rev.*, **79**, 529 (1979).
217. S. S. Zigler, R. West and J. Michl, *Chem. Lett.*, 1025 (1986).
218. A. Sekiguchi, W. Ando and K. Honda, *Chem. Lett.*, 1029 (1986).
219. U. Klingebiel, VIIIth International Organosilicon Symposium, St. Louis, MO, June 1987.
220. M. Hesse and U. Klingebiel, *Angew. Chem., Int. Ed. Engl.*, **25**, 638 (1986).
221. V. D. Romanenko, A. V. Ruban, A. B. Drapailo and L. N. Markovskii, *Zh. Obschch. Khim.*, **55**, 2793 (1985).
222. R. J. P. Corriu, in *Silicon Chemistry* (Eds., J. Y. Corey, E. R. Corey and P. Gaspar), Pt. IV, Chapter 22, Ellis Horwood, Sussex, UK, 1988, p. 225.
223. C. Zybil and G. Muller, *Angew. Chem. Int. Ed. Engl.*, **26**, 669 (1987).
224. D. A. Strauss, T. D. Tilley, A. L. Rheingold and S. J. Grib, *J. Am. Chem. Soc.*, **109**, 5872 (1987).
225. C. Friedel and A. Ladenburg, *Compt. Rend.*, **68**, 920 (1869).
226. C. Friedel and A. Ladenburg, *Compt. Rend.*, **68**, 923 (1869); *Ann. Chim. Phys.*, **19**, 401 (1880); *Ann.*, **203**, 251 (1880).
227. B. Martin, *Ber.*, **46**, 2442, 3294 (1913).
228. W. C. Schumb and C. M. Saffer, Jr., *J. Am. Chem. Soc.*, **61**, 363 (1939).
229. a. F. S. Kipping, *J. Chem. Soc.*, **125**, 2291 (1924).
b. A. R. Steele and F. S. Kipping, *J. Chem. Soc.*, **131**, 1431 (1928).
c. A. R. Steele and F. S. Kipping, *J. Chem. Soc.*, **132**, 2545 (1929).
230. C. A. Burkhard, *J. Am. Chem. Soc.*, **71**, 963 (1949).
231. a. S. Yajima, J. Hayashi and M. Omori, *Chem. Lett.*, 931 (1975).
b. S. Yajima, K. Okamura and J. Hayashi, *Chem. Lett.*, 1209 (1975).
232. For general reviews of cyclopolysilanes see R. West, *Comprehensive Organometallic Chemistry* (Eds. G. Wilkinson, F. G. A. Stone and E. W. Abel), Pergamon, Oxford, 1982.
233. E. Hengge, *Phosphorous Sulfur*, **28**, 43 (1986).
234. A. Schafer, M. Weidenbruch, K. Peters and H. G. von Schnering, *Angew. Chem., Int. Ed. Engl.*, **23**, 302 (1984).
235. a. H. Matsumoto, N. Yokoyama, A. Sakamoto, Y. Sramaki, R. Endo and Y. Nagai, *Chem. Lett.*, 1643 (1986).
b. H. Matsumoto, A. Sakamoto and Y. Nagai, *J. Chem. Soc., Chem. Commun.*, 1768 (1986).
236. H. Watanabe, T. Muraoka, M. Kageyama, K. Yoshizumi and Y. Nagai, *Organometallics*, **3**, 141 (1984).
237. Y. Nagai, VIIIth International Organosilicon Meeting, St. Louis, MO, June 1987; Abstr. B27.
238. a. A. W. P. Jarvie, H. J. S. Winkler, D. J. Peterson and H. Gilman, *J. Am. Chem. Soc.*, **83**, 1921 (1961).
b. P. Boudjouk and B. H. Han, *Tetrahedron Lett.*, **22**, 3813 (1981).
c. M. F. Lemanshi and E. P. Schram, *Inorg. Chem.*, **15**, 2515 (1976).
239. C. W. Carlson and R. West, *Organometallics*, **2**, 1792 (1983).
240. C. W. Carlson, K. J. Haller, X. H. Zhang and R. West, *J. Am. Chem. Soc.*, **106**, 5521 (1984).
241. L. F. Brough and R. West, *J. Am. Chem. Soc.*, **103**, 3049 (1981).
242. M. Kubo, Jpn. Kokai Tokkyo Koho 79,130,541; *Chem. Abstr.*, **96**, 2155385 (1980).
243. H. Watanabe, T. Muraoka, M. Kageyama, K. Yoshizumi and Y. Nagai, *Organometallics*, **3**, 141 (1984).
244. C. W. Carlson and R. West, *Organometallics*, **2**, 1792 (1983).
245. H. Matsumoto, A. Sakamoto, M. Minemura, K. Sugaya and Y. Nagai, *Bull. Chem. Soc. Jpn.*, **59**, 3314 (1986).
246. S.-M. Chen, L. D. David, K. J. Haller, C. L. Wadsworth and R. West, *Organometallics*, **2**, 409 (1983).
247. M. Biernbaum and R. West, *J. Organomet. Chem.*, **131**, 179 (1977).

248. A. Katti, C. W. Carlson and R. West, *J. Organomet. Chem.*, **271**, 353 (1984).
249. For an example of the condensation of 1,2-dichlorodisilanes see H. Watanabe, J. Inose, K. Fukushima, Y. Kougo and Y. Nagai, *Chem. Lett.*, 1711 (1983).
250. E. Hengge and G. Bauer, *Monatsh. Chem.*, **106**, 503 (1975).
251. M. Hirooka, Ger. Offen. DE 3,426,822; *Chem. Abstr.*, **102**, 159072w (1984).
252. E. Hengge and D. Kovar, *Z. Anorg. Allg. Chem.*, **458**, 163 (1979).
253. E. Hengge and D. Kovar, *J. Organometal. Chem.*, **125**, C29 (1977).
254. E. Hengge and G. Bauer, *Monatsh. Chem.*, **106**, 503 (1975).
255. E. Hengge and D. Kovar, *Z. Anorg. Allg. Chem.*, **459**, 123 (1979).
256. Z. Smith, A. A. Menningen, E. Hengge and D. Kovar, *J. Am. Chem. Soc.*, **104**, 4362 (1982).
257. J. P. Hummel, J. Stackhouse and K. Mislow, *Tetrahedron*, **33**, 1925 (1977).
258. Z. Smith, H. M. Seip, E. Hengge aand G. Bauer, *Acta Chem. Scand., Ser. A*, **30**, 697 (1976).
259. M. Weidenbruch, *Comments Inorg. Chem.*, **5**, 247 (1986).
260. Y. Nagai, H. Watanabe and H. Matsumoto, in *Silicon Chemistry* (Eds., J. Y. Corey, E. R. Corey and P. Gaspar), Pt. V, Chapter 24, Ellis Horwood, Sussex, UK, 1988, p. 247.
261. F. Shafiee, K. J. Haller and R. West, *J. Am. Chem. Soc.*, **108**, 5478 (1986).
262. C. L. Wadsworth, R. West, Y. Nagai, H. Watanabe and T. Muraoka, *Organometallics*, **4**, 1659 (1985).
263. R. West, *Pure Appl. Chem.*, **54**, 1041 (1982).
264. C. L. Wadsworth and R. West, *Organometallics*, **4**, 1664 (1985).
265. B. J. Helmer and R. West, *J. Organomet. Chem.*, **236**, 21 (1982).
266. F. Wohler, *Ann. Chem.*, **107**, 112 (1858).
267. A. Stock, *Hydrides of Boron and Silicon*, Cornell University Press, 1933.
268. K. Borer and C. S. G. Phillips, *Proc. Chem. Soc.*, 189 (1959).
269. K. Stokland, *D. K. n. v. s. Skrifter*, **3**, 4 (1948). I thank P. Gaspar for pointing out this work.
270. G. Olbrich, P. Potzinger, B. Reimann and R. Walsh, *Organometallics*, **3**, 1267 (1984).
271. F. S. Kipping and J. E. Sands, *J. Chem. Soc.*, **119**, 830 (1921).
272. a. K. Triplett and M. D. Curtis, *J. Organometal. Chem.*, **107**, 23 (1976).
b. P. Boudjouk and B. H. Han, *Tetrahedron. Lett.*, **22**, 3813 (1981).
273. E. Hengge and S. Waldhor, *Monatsch. Chem.*, **105**, 671 (1974).
274. E. Hengge and W. Peter, *J. Organomet. Chem.*, **148**, C22 (1978).
275. K. Hassler, *Monatsh. Chem.*, **117**, 613 (1986).
276. a. M. Ishikawa and M. Kumada, *J. Organomet. Chem.*, **42**, 325 (1972).
b. M. Ishikawa, K. Nakagawa, M. Ishiguro, F. Ohi and M. Kumada, *J. Organomet. Chem.*, **152**, 155 (1978).
277. M. Kumada, M. Ishikawa and S. Maeda, *J. Organomet. Chem.*, **2**, 478 (1964).
278. E. Hengge and G. Miklau, *Z. Anorg. Allg. Chem.*, **508**, 33 (1984).
279. C. Aitken, J. F. Harrod and E. Samuel, *J. Organomet. Chem.*, **279**, C11 (1985).
280. K. Yamamoto, H. Okinoshima and M. Kumada, *J. Organomet. Chem.*, **23**, C7 (1970).
281. N. Wiberg, H. Schuster, A. Simon and K. Peters, *Angew. Chem., Int. Ed. Engl.*, **25**, 79 (1986).
282. M. Weidenbruch, B. Blintjer, K. Peter and H. G. von Schnering, *Angew. Chem., Int. Ed. Engl.* **25**, 1129 (1986).
283. a. R. West, *J. Organomet. Chem.*, **300**, 327 (1986).
b. L. D. David, *Chem. Br.*, **23**, 553 (1987).
284. R. West *et al.*, In *Science of Ceramic Chemical Processing* (Eds. L. L. Hench and D. Ulrich), Wiley–Interscience, New York, 1986.
285. R. West, J. Maxka, R. Sinclair and P. M. Cotts, *Polym. Prepr.*, **28**, 387 (1987).
286. R. D. Miller, J. R. Rabolt, R. Sooriyakumaran, G. N. Fickes, B. L. Farmer and H. Kuzmany, *Polym. Prepr.*, **28**, 387 (1987).
287. a. J. M. Zeigler, *Polym. Prepr., Am. Chem. Soc., Div. Polym. Chem.*, **28**, 109 (1987).
b. J. M. Zeigler, L. A. Harrah and A. W. Johnson, *Polym. Prepr., Am. Chem. Soc., Polym. Chem.*, **28**, 424 (1987).

The Chemistry of Organic Silicon Compounds
Edited by S. Patai and Z. Rappoport

CHAPTER **2**

Theoretical aspects of organosilicon compounds

YITZHAK APELOIG
Department of Chemistry, Technion—Israel Institute of Technology, Haifa 32000, Israel

I. INTRODUCTION

The application of theory, particularly quantum-chemical methods, to silicon chemistry is a relatively recent endeavor. Being a member of the third row of the periodic table, silicon was considered to be a difficult problem to handle in the early applications of quantum-chemical methods. This is true both for *ab initio* methods, for which the relatively large number of electrons posed severe limitations on the calculations, and for semiempirical methods (and even purely empirical methods such as force-field calculations) where the scarcity of experimental data prevented the development of reliable parametrization. Consequently, the theoretical study of silicon compounds lagged by a 'phase' of 10–15 years behind the analogous theoretical developments for the chemistry of carbon compounds. Recently, the situation has changed dramatically due to the technological breakthroughs in computer hardware and the development of new theoretical methods and related computer programs and software. In the last decade and particularly in the last five years it has become possible to carry out *ab initio* calculations (using appropriate levels of theory) for silicon compounds of sizable complexity, which can serve as models for compounds which are of interest to chemists working in this field. These calculations played an extremely important (and in some cases even a crucial) role in the development of silicon chemistry, especially in laying the foundations for the understanding of compounds that have multiple bonding to silicon. The fact that stable compounds with multiple bonds to silicon were unknown experimentally until ten years ago placed theory in a special, possibly unique, position in the development of this field. Thus, in many cases the theoretical predictions preceded the experimental studies and in other cases theory identified and pointed out erroneous interpretations of experimental data.

In general, the experimental and theoretical aspects of the field have grown together in an interactive manner. In this respect the history of organosilicon and of organic chemistry are quite different. Organic chemistry was a mature and well-developed field when reliable computational methods for tackling small organic compounds became available. On the other hand, many areas in the field of organosilicon chemistry, such as molecules with multiply bonded silicon, were experimentally in their infancy at the point when computational methods in chemistry came of age and could be applied reliably to a variety of problems. Furthermore, given the scarcity of quantitative experimental fundamental data, such as thermochemistry, geometry and so on, for many organosilicon compounds, quantum-mechanical calculations have been almost the sole source of primary information employed to derive patterns of understanding. These circumstances placed theory at a special position in the development of organosilicon chemistry.

This chapter reviews various theoretical aspects of silicon chemistry. The emphasis will be on 'organosilicon' compounds but formal 'inorganic' compounds of silicon such as $H_2Si{=}SiH_2$, H_2Si:, H_3SiNH_2, polysilanes, etc., which are closely related to the theme of this book and which we consider to be of interest to the potential reader, will also be discussed. However, we have omitted from the discussion (or mention only briefly) 'purely inorganic' silicon compounds such as silicates.

The intensive research activity in organosilicon chemistry in the last decade has led to a flourish in the number of reviews and books, both experimental and theoretical, that were devoted to this field. We mention here only the most general reviews. More specific reviews will be cited at the appropriate places in the text. Excellent general comprehensive reviews of organosilicon chemistry appeared in the series of *Comprehensive Organic Chemistry*[1] and *Comprehensive Organometallic Chemistry*[2]. These excellent reviews include references to books and literature published before 1980. Organosilicon chemistry is surveyed regularly (mostly annually) in various series of monographs[3]. Two excellent recent books by Weber[4a] and Colvin[4b] are devoted mainly to the use of organosilicon reagents in organic synthesis. Two other recent books[5,6] discuss a variety of aspects of organosilicon chemistry and include also theoretical chapters. The molecular and electronic structure of penta- and hexa-coordinate silicon compounds were reviewed in an excellent paper by Tandura, Alekseev and Voronkov[7]. This review is an excellent source for both experimental and theoretical work in this field. Theoretical[8] and experimental[9] aspects of the stereochemistry of reactions at silicon, in particular nucleophilic attacks, were discussed in two other reviews[8,9]. Several recent reviews dealing with multiple bonding to silicon have appeared since the publication of the now classical 1979 review by Gusel'nikov and Nametkin[10]. These include two reviews that appeared in 1981, one by Bertrand, Trinquier and Mazerolles[11], and the other by Coleman and Jones[12]. These reviews are no longer of current interest because first-landmark reports on the isolation of stable compounds having carbon–silicon[13] and silicon–silicon[14] double bonds were published in 1981, after the above-mentioned reviews were submitted for publication. More recent reviews on this topic are those by West and Michl (1984)[15], Wiberg (1984)[16], Cowley (1984 and 1986)[17] and Brook (1986)[18]. Although the aforementioned reviews are concerned mainly with the experimental aspects of the research on multiple bonding to silicon, they usually include also short discussions of the available theoretical calculations. An excellent comprehensive review by Raabe and Michl which covers all aspects, experimental as well as theoretical, of multiply bonded organosilicon compounds was published in 1985 and covers the literature through the end of 1984[19].

Several reviews which emphasize the theoretical aspects of organosilicon chemistry are also available. Most of them are concerned mainly with multiply bonded silicon compounds. The first theoretically inclined review in this field was published in 1982 by Schaefer[20]. Schaefer's paper was particularly important, because it drew the attention of many theoreticians, including this author, to some of the interesting open questions in silicon chemistry and pointed out the important role that theory can play in this rapidly developing field of chemistry. Two excellent theoretical reviews were published by Gordon in the last two years[21,22]. The first discusses 'multiple bonding to silicon'[21]. The second review, entitled 'Theoretical studies of silicon chemistry', is more general and includes an extensive discussion of theoretical studies of silylenes, cyclic silicon compounds and several reaction surfaces[22]. The important review by Raabe and Michl[19], which covers also the theoretical work on multiply bonded silicon, was mentioned above.

The chapter reviews calculations for many types of organosilicon compounds and some overlap (which might even be helpful to the reader) with the theoretical discussion in other chapters of this book is unavoidable. The field of organosilicon chemistry is vast and so is the related theoretical literature. We do not intend to review all the theoretical work that was published in this field. The selection of topics and the emphasis of particular papers reflects the author's own interests and preferences. In selecting the material I have also tried to avoid extensive overlap with recent theoretical reviews[19, 21, 22], and topics which in previous reviews received little attention were emphasized. As topics which have been previously reviewed extensively (e.g., multiple bonding to silicon[19, 21]), we will briefly scan the older literature and the main discussion will concentrate on recent work, published after these reviews had appeared. In general, this review will concentrate on theoretical studies published in the last decade and will cover the literature through the middle of 1987. Due to the help and courtesy of colleagues (see Acknowledgements), to whom I am deeply indebted, it was possible to include results that were still in press when this manuscript was submitted.

On scanning the literature one finds considerable repetition of calculations, especially for the smaller molecules. This repetition is a natural outcome of the way in which computational chemistry develops. Thus, calculations for small molecules are often used to test the operation of new theoretical methods and techniques. As better theoretical methods were developed (see Section II) they were applied again to the same fundamental set of simple small molecules. Since calculations which are performed with the more sophisticated theoretical methods are generally more reliable, we will discuss mostly the more recent theoretical calculations. When high-level calculations for a certain molecule are available, we will include calculations that use inferior theoretical methods (e.g. smaller basis sets) only if they serve to demonstrate the operation of the different theoretical methods, or if they include discussion of aspects that were not studied in the more elaborate calculations. We apologize to the authors of contributions that will be omitted from this review as a result of this policy, which is essential in light of the space limitations. These earlier papers are discussed in previous reviews as well as in the primary literature.

II. THEORETICAL METHODOLOGY

The emphasis in this review is on *ab initio* methodologies, although semiempirical and force-field calculations will be mentioned occasionally. We will concentrate on recent studies which generally use elaborate *ab initio* methods. Such methods are better suited to tackle, with a reasonable level of confidence, problems where experimental data are scarce and cannot be used to guide and to calibrate the calculations.

The short section on theoretical methodology that follows is included for the convenience of readers who are not familiar with quantum-mechanics theory, but it does not intend to review the available theoretical methods. The interested reader is referred to the monographs and books listed below. The purpose of this short discussion is to familiarize these readers with some of the common names, abbreviations and jargon used by quantum chemists in the majority of the cited literature and also in this chapter.

A vast literature which describes and reviews the various quantum-mechanical methods in common use is available, and the reader is referred to these sources for details. Of the rich quantum-mechanics literature we single out a few recent monographs: (a) The excellent series on 'Methods of Electronic Structure Theory' edited by Schaefer[23]. (b) The two-volume monograph *Ab Initio Methods in Quantum Chemistry* edited by Lawley[24]. (c) The excellent book by Hehre, Radom, Schleyer and Pople[25] that describes in detail

'Pople's methodology', which is by far the most popular among computational chemists. The book by Clark[26] is a clear practical guide that describes how to use the most popular computer programs and, in particular, the popular Gaussian series of programs[27] developed by Pople's group.

A. *Ab Initio* Methods

In general there are two major points which should be defined regarding any calculation, namely the type of *method* and the *basis set* which are used. An extensive account of these topics is given in References 23–26 and in the primary theoretical literature. Here we will review these issues shortly. We will discuss first *ab initio* methods and then semiempirical and force-field methods.

Ab initio calculations include methods based on either molecular orbitals or valence bonds. Although several papers mentioned in this review use the Generalized Valence Bond (GVB) method developed mainly by Goddard[28], most investigators use molecular-orbital methods. These methods are generally divided into two categories: (1) single determinant Hartree–Fock (HF) self-consistent field (SCF) calculations[23–25], and (2) methods which apply some sort of correction for electron correlation to the SCF energy[23–25].

Let us say a little more about these methods. The HF methods are divided into spin-restricted HF (RHF) and spin-unrestricted HF (UHF) methods. Closed-shell systems are almost always calculated using RHF. In this procedure, one set of molecular orbitals is calculated and pairs of electrons are entered to the lowest-energy orbitals. If the molecule has an odd number of electrons, one orbital will be singly occupied and the species is a radical (spin = 0.5, expectation value $<S^2>$ of the spin-squared operator = 0.75). In most cases, however, radicals are calculated using the UHF formalism. UHF calculations determine two sets of molecular orbitals, one for each type of spin named alpha and beta. These MO sets are similar but not identical. A radical, for instance, has one more α than β electron. The UHF procedure is more flexible than RHF because the paired α and β orbitals, which correspond to doubly occupied MOs in the RHF formalism, need not be identical. So UHF allows for spin polarization but, on the other hand, spin-contamination occurs (i.e., states of higher spin are mixed into the wave function).

1. Basis sets

The first step in performing a calculation with any of these methods is to choose a basis set. The basis set refers to the set of functions, usually atomic orbitals (AO), which are used in the molecular orbital calculations. In almost all calculations the AOs are represented by linear combinations of Gaussian functions.

By far the most widely used basis sets are those developed by Pople's group[25]. This is mainly because these basis sets have been developed carefully to give a balanced inner-consistent electronic picture for various atoms and molecular environments[25]. Furthermore, those basis sets were tested extensively in the last decade and their advantages as well as deficiencies are well documented[25]. The experience which has accumulated allows one, in many cases, to predict which is the smallest basis set that can be used to describe reasonably a particular molecule or molecular property and to estimate which changes are to be expected when the basis set is enlarged. A detailed evaluation of Pople's basis sets and their applications is presented in References 25 and 26. These basis sets exist as internal information within the Gaussian series of programs[27]. The types of basis sets which are most commonly used are described below. Within each group of basis sets we also mention the specific basis set which is the most popular.

a. Minimal basis sets (mbs). This group includes the smallest number of atomic orbitals required to properly describe the atoms involved. For example, a 1s atomic orbital (AO) on hydrogen, 1s, 2s and three 2p AOs on carbon; 1s, 2s, three 2p, 3s and three 3p AOs on Si, and so on. Note that d-orbitals are not included in the minimal basis set for Si and other elements of the third row of the periodic table. The most commonly used basis set in this group is STO-3G[29], where each AO (which in this case is an STO, i.e., an approximate Slater-type orbital) is expanded as a linear combination of three Gaussians. Minimal basis sets and STO-3G in particular were very popular at the early stages of computational chemistry, due to their small size and thus the relatively little computer time that they required. These basis sets are rarely used nowadays, particularly with third-row elements, except for very large molecules where other calculations are impractical.

b. Double-zeta (dz) basis sets. This group includes exactly twice the number of AOs in the minimal basis set. The two AOs describing the same AO have different sizes, one more contracted and closer to the nuclei and one more diffuse, thus allowing the electron to have higher flexibility in space. The effect of splitting the AO is shown schematically in **1**. For example, for H there are two 1s orbitals, for carbon there are two sets of 1s, 2s and $3 \times 2p$ orbitals, for Si two sets of 1s, 2s, $3 \times 2p$, 3s and $3 \times 3p$ orbitals, and so on. The most popular basis set of this type were reviewed recently by Dunning and Hay[30a], Ahlrichs and Taylor[30b], Huzinga and coworkers[30c] and Wilson[30d].

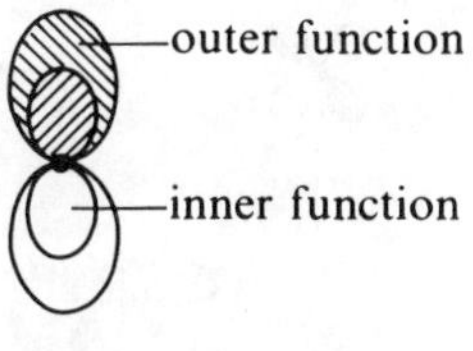

(1)

c. Split valence (sv) basis sets. In these basis sets a compromise is made between the minimal and dz basis sets. The inner shells are kept minimal, but the valence shells are split (i.e., they are of dz quality). This allows increased flexibility in the valence region which is believed to be the more important in forming chemical bonds. These basis sets are smaller and more compact than the dz basis sets and the calculations are therefore less time-consuming. Experience shows that the price for economizing, paid in terms of the reliability of the results, is small. Thus, a sv basis set for H is the same as in a minimal basis set, but for Si it includes 1s, 2s, $3 \times 2p$, and two sets of the valence shell ($3s + 3 \times 3p$). The most popular basis sets in this category are Pople's N-M1G[25–27], in particular 4-31G[31a,b], 6-31G[31c,d] and 3-21G[32]. Of these, the smallest basis set (3-21G) is probably the most widely used, especially for geometry optimizations. In these basis sets each of the inner-shell AOs is expanded as a linear combination of N Gaussians (e.g., four Gaussians in 4-31G for the 1s orbital of C), and the valence shell is split into inner and outer shells. The AOs in the inner part of the valence shell are expanded as a linear combination of M Gaussians, and those in the outer part consist only of a single Gaussian. Thus, for Si and other third-row elements, the 3-21G basis set (actually more fully designated as 33-21G) includes: 1s, 2s, $3 \times 2p$ AOs (the inner shell) each expanded as a linear combination of three Gaussians, and $3s + 3 \times 3p$ orbitals (the valence shell) that are split each into inner and outer shells expanded as linear combinations of two and one Gaussians, respectively. In some cases it may be useful to split the valence shell further, such as the 6-311G basis set[33], where the valence shell is composed of three regions: the inner, middle and outer, which are expanded as linear combinations of 3, 1 and 1 Gaussians, respectively.

d. Polarized (p) basis sets (pbs). One feature common to the basis sets discussed above is that they comprise functions constrained to be centered at the nuclear positions. This is not the case in reality, particularly with highly polar molecules and small-ring systems. The polarized basis sets are designed to correct this deficiency by augmenting the dz or sv basis sets with functions that are not required to accommodate the electrons in the ground state of the free atom. Usually these functions are represented by AOs with a quantum number l higher by one than the highest ones occupied in the ground state of the atoms. Thus, for H one adds a set of 3 × 2p AOs and for C or Si one adds a set of 5 (or 6) d-AOs. These functions are termed *polarization functions*, because they allow the electrons to acquire the needed directionality for forming bonds, particularly in small rings. The effect on the bond directionality of adding polarization functions is shown schematically in **2** for the case of a d-type function. Thus, with a pure p-orbital, the electron density enclosed within the two hemispheres defined by the dashed line is the same. Addition of a single d-orbital (the d-orbital contributes less than the p-orbital to the linear combination) allows more charge to concentrate in one of these hemispheres as exemplified in **2**. Similarly, the effect of the addition of a p-orbital to a s-orbital is shown schematically in **3**. Thus, in the polarized basis sets the centers of positive and negative charges do not coincide.

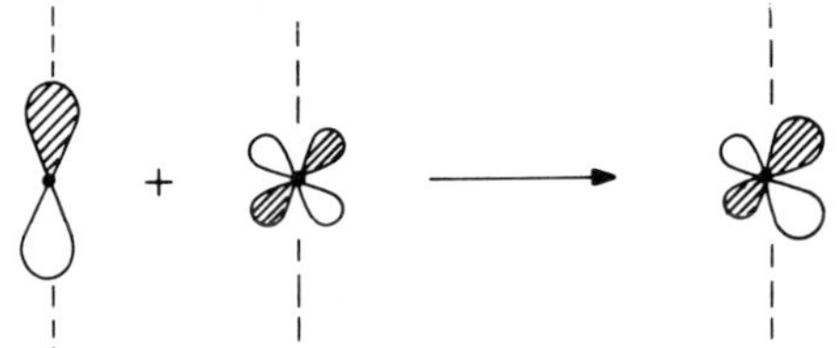

(2)

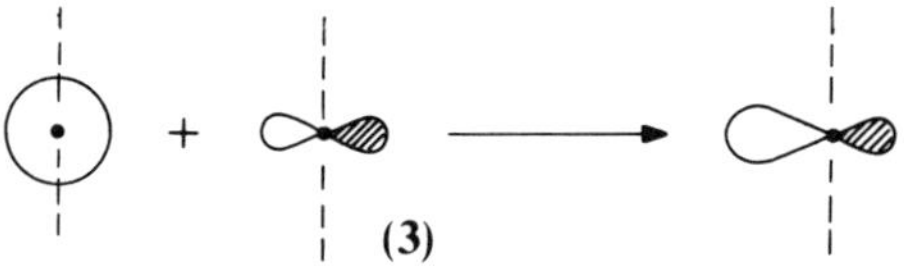

(3)

According to Pople's nomenclature the polarized basis sets are denoted by adding a superscript asterisk to the basis set, e.g. the popular 6-31G* [34] [other authors prefer the notation 6-31G(d)]. When p-orbitals are added to H as well, a second asterisk is added (e.g. 6-31G**)[34]. The first basis set developed in this series for third-row elements, STO-3G* [35], was only *partially* polarized, i.e., it included polarization functions (5 pure d-type Gaussians) only on third-row elements, e.g. Si, but not on second-row elements. This basis set was later replaced by the 3-21G$^{(*)}$ basis set (parentheses around the asterisk indicate that only third-row elements are augmented by d-functions)[36]. Both STO-3G* and 3-21G$^{(*)}$ should not be viewed as full polarized basis sets. The philosophy in using these 'half-polarized' basis sets is similar to that used in going from a dz to a sv basis set. Thus, polarization functions are added only on the atoms for which it is believed that they play a larger role. This tactic may be helpful in cases where calculations with a fully polarized basis set are not feasible.

e. Extended basis sets. This group includes various basis sets which are larger than the polarized basis sets, up to the Hartree–Fock limit. Examples are basis sets where some of the functions are represented by AOs with quantum numbers l higher by two than the highest ones occupied in the ground state of the atoms (e.g., f-functions on C or Si)[37].

Another possibility of extending the basis set is to add diffuse functions on some of the atoms. This is usually obtained by adding to a standard basis set such as 6-31G* a single set of very diffuse s- and p-orbitals on the heavy atoms. In the Pople series such basis sets are designated by the addition of plus signs, e.g., the 6-31+G* basis set includes diffuse p-orbitals on all heavy atoms, and 6-31G++G* includes also diffuse s-functions on hydrogen[38]. The similar 3-21+G and 3-21+G$^{(*)}$ basis sets are also available. Addition of diffuse functions improves the description of the electron at large distances from the nucleus. These basis sets are mostly intended for use in calculations on anions and molecules that require very good descriptions of nonbonding electrons. A recent paper evaluates the performance of diffuse function-augmented basis sets for third-row elements and concludes that their effect is smaller than for second-row elements[39].

f. Pseudopotentials[40,41]. This represents a different approach to the computational problem. Here the inner shells, which are believed to be less important than the valence shells in determining the bonding properties of the atoms, are not included explicitly in the calculations. Instead, the inner shells are replaced by a model potential (the pseudopotential) that represents them. The valence shells can be represented by any of the basis sets described above.

2. Post Hartree–Fock calculations

Even when using the largest basis set and reaching the so-called Hartree–Fock (HF) limit[23–25] the method accounts for only 95–99% of the total energy of the electrons. What is missing in the HF description is electron correlation. Although electron correlation is a relatively small fraction of the total energy, it may be extremely important chemically. In the HF method each electron 'feels' the average field of all other electrons, but this level of theory takes no account of correlation between the motions of electrons with opposite spins, in a way that will minimize their mutual repulsions as much as possible. For example, the calculations do not include information which will force two electrons of opposite spin occupying the same molecular orbital to occupy at a certain instance different regions in space, remote as possible within this orbital.

Mathematically, the problem of electron correlation is tackled either by perturbation theory or by using the variational principle. The most effective perturbation solution for this problem was developed by Pople and his coworkers[42] following a procedure originally suggested by Moller and Plesset (MP)[43]. In the MP method, electron correlation is estimated by adding successive corrections to the HF wave function. The perturbational corrections can be terminated at the second (MP2), third (MP3) or fourth order (MP4). The order of substitution is sometimes given also. Thus MP4SDTQ means an MP4 calculation including all single (S), double (D), triple (T) and quadruple (Q) substitutions. The major advantage of the MP methods is their speed and economy in comparison with other classical methods based on the variational principle. In general, the MP methods work reasonably well in cases where the wave functions are dominated by a single electronic configuration.

The other approach to the electron-correlation problem is based on the variational principle and includes two major procedures: configuration interaction (CI) and multi-configurational (MCSCF) methods[44]. In both methods the wave function is described by more than one determinant (configuration). Usually, excited configurations are mixed into the ground-state configuration. In a CI calculation the MO coefficients are obtained from a previous HF calculation (e.g. 6-31G*) and the CI coefficients are optimized to give the lowest energy. In a MCSCF procedure one optimizes both the MO and the CI coefficients. Sometimes, the two approaches are combined in a MCSCF–CI procedure, where one starts with a small MCSCF calculation to obtain a good set of orbitals, and this is followed by a CI calculation using the MCSCF orbitals.

B. Geometry Optimization

Geometry optimization is a crucial step in any calculation. In the early days of computational chemistry, geometry optimizations were carried out by finite-difference methods, i.e. by changing stepwise and independently each geometrical parameter (i.e. bond lengths, bond angles and torsion angles) until the lowest energy was reached. The nature of the geometry obtained was not analyzed analytically. In the early eighties analytical gradient techniques for carrying out geometry optimization and for characterizing the nature of various points on the potential-energy surface became widely available also for *ab initio* calculations[45] (these methods were available earlier in some of the semiempirical methods, e.g. MNDO). These analytical methods use the first and second derivatives of the energy with respect to all degrees of freedom (these form a matrix called the Hessian), to locate and characterize minima, maxima or saddle points (i.e. transition structures≡transition states) on the potential-energy surface.

In classical mechanics, the first derivative of the potential energy for a particle is equal to minus the force acting on this particle, and the second derivative (for a quadratic potential) is the force constant. For all stationary points (e.g. minima or transition states) on the potential-energy surface all atomic forces (i.e. the first derivatives of the energy) are zero. The nature of the stationary point can be determined from the second derivatives. In one dimension, the second derivative is positive for a minimum, negative for a maximum and zero for a point of inflection. In more than one dimension, the eigenvalues of the second-derivative matrix, rather than individual elements of the matrix, determine the characteristic of the stationary points. A minimum on the potential-energy surface, which represents an equilibrium structure, is characterized by a Hessian matrix which has only positive eigenvalues. A local maximum has a Hessian matrix with only negative eigenvalues. A saddle point or a transition state has one (and only one) negative eigenvalue (i.e. imaginary frequency), and all the rest are positive. Thus, a transition state is a maximum in one direction and a minimum in all perpendicular directions. The eigenvector corresponding to the single negative eigenvalue of a first-order saddle point is termed the transition vector, and its direction corresponds to the reaction path between reactants and products. It is important to note that structures with more than one negative value in the force-constant matrix are neither minima or maxima nor transition states. They represent 'regular' non-stationary points on the surface, such as arbitrary points along the reaction paths from reactants to products.

The development of these analytical gradient techniques and their implementation into the *ab initio* programs has revolutionized the way in which the quantum chemist probes a potential surface, and it has played a very important role in the calculations of silicon compounds. Many structures which were believed to be minima (usually due to the use of assumed symmetry) were found by these methods to be false minima, by showing that the corresponding Hessian matrix has more than one negative eigenvalue. Reoptimization of such structures using analytical techniques and elevation of the previous imposed symmetry constraints lead to new structures which are the real minima (at that specific level of theory) on the potential-energy surface. A typical example is HSi≡SiH. Early calculations, which were carried out before the gradient techniques were available, assumed the traditional linear structure analogous to acetylene[46]. More recent calculations, which used gradient techniques, determined that the linear geometry does not correspond to a minimum and that the optimum geometry is a doubly bridged C_{2v} structure[47] (more details are given in Section V.B.2).

C. Frequency and Zero-point Energy (ZPE) Calculations[48]

The calculations of the second derivatives of the energy with respect to the mass-weighted cartesian displacements, evaluated at the equilibrium nuclear configuration, provide additional invaluable information. These second derivatives give, as mentioned above, the quadratic force constants and thus the normal-mode vibrational frequencies. These frequencies are particularly valuable in a field such as silicon chemistry, where many of the synthesized novel compounds are short-lived molecules which in many cases are identified by spectroscopy of matrices. Comparison of the calculated and experimental infrared lines may be most valuable as a means of identification. In some cases it may even be the only indication that a new molecule has actually been formed. A beautiful recent example, although not from silicon chemistry, that demonstrates this point is the identification of $H_2C{=}C{=}C$: by a comparison of experimental and calculated IR frequencies and intensities of this species[49]. The calculated normal-mode vibration frequencies also provide thermodynamic properties of stable molecules or of transition structures. Thus reaction entropies, entropies of activation and isotope effects can be calculated. Finally, theoretical vibrational frequencies may also be used to evaluate zero-point vibrational energies (ZPE) and to correct experimental thermochemical data to 0 K.

D. Semiempirical Methods[50]

Semiempirical methods use experimental data and various approximations for evaluating some of the integrals that are required in the MO calculations[50]. Some of these methods have been quite successful in organic chemistry, but their use in silicon chemistry is more problematic due to the scarcity of reliable quantitative experimental data which can be used for 'calibration'. Owing to these limitations most of the theoretical studies in silicon chemistry used *ab initio* methods; semiempirical methods were used much less frequently. The major advantage of semiempirical methods is their high efficiency, which allows one to calculate much larger molecules than possible with even the simplest *ab initio* methods. Unfortunately, the reliability of semiempirical calculations in silicon chemistry has not yet been evaluated systematically for a wide variety of molecules containing silicon. Such a study is highly desirable, as it will allow one to establish to which type of problems semiempirical methods can be applied with some confidence. Such a critical evaluation might be possible now, when a significant number of high-level *ab initio* calculations and some experimental data have become available. Note that gradient-optimization techniques, frequency calculations and corrections for electron correlations are available also in the modern semiempirical programs.

The semiempirical methods used most frequently in organosilicon chemistry are CNDO/2[51], MINDO/3[52] and MNDO[53], which are available from the Quantum Chemistry Program Exchange (QCPE). It is important to note that the original published MINDO/3 parameters for silicon[54] (which are also included in the QCPE computer program) are erroneous, as they do not reproduce the published results[54] for a variety of silicon compounds. It was later recommended by Dewar and coworkers that instead of MINDO/3, the MNDO method (with the parameters given in Reference 55, 3d-orbitals are not included) should be used for silicon-containing molecules[55]. However, Verwoerd[56] compared the operation of MNDO and MINDO/3 for a limited set of silicon-containing molecules and concluded that even with the published parameters (officially erroneous[55]) the MINDO/3 method gives more reliable results than MNDO. For unsaturated molecules particularly large discrepancies in the geometries and stabilities are found between the two methods, and in comparison with *ab initio* calculations[56]. More recently, Dewar and coworkers[57] have reoptimized the MNDO parameters for silicon. The results are now in much better agreement with experiment and with high-level *ab initio* calculations than was the case with the previous parameters. However, in my opinion the errors and disagreements with experiment are still far too large and the set of molecules examined is too limited to allow a reliable routine application of MNDO to silicon-containing molecules. Additional work, which will test more fully the operation of these methods and establish for which types of silicon-containing molecules MINDO/3 or MNDO can be used with some confidence, is highly desirable. Some revision in the MNDO parameters or the inclusion of d-orbitals might help to achieve a satisfactory representation of silicon compounds at the semiempirical level.

E. Force-field Methods[58]

Force-field or molecular-mechanics methods are entirely empirical calculational methods, designed to give accurate *a priori* structures and energies for molecules. The method is based on the basic idea that bonds have 'natural' lengths and angles, so that molecules will adjust their geometries in order to adopt these values in simple cases. In addition, steric interactions are included using van der Waals potential functions. Molecular-mechanics calculations employ an empirically derived set of equations, called the force-field, which contain adjustable parameters that are optimized to obtain the best fit of calculated and experimental parameters of the molecules, such as geometries, conformational energies, heats of formation or other properties[58]. Molecular mechanics was applied very successfully to hydrocarbons, but its application to silicon-containing molecules is limited by the scarcity of experimental data which can be used to generate a reliable silicon force-field. Two parametrizations for a sp^3-hybridized silicon have been reported (no parameters are available for sp^2-hybridized silicon atoms)[59]. A short description of the application of these force fields to silanes and polysilanes is given in References 58 and 59. Other applications can be found in Reference 60.

F. Practical Considerations and Nomenclature

A short comment that describes the normal practice usually followed when carrying out MO calculations might be of help to the experimentalist reader. One starts by assuming a geometry for the molecule of interest and then optimizes its structure (see Section I.B) using an appropriate basis set (usually the largest practical) for the problem at hand. In most cases this would be 3-21G for relatively large molecules, and a partially polarized (e.g. 3-21G$^{(*)}$) or a fully polarized (e.g. 6-31G*) basis set for smaller molecules. The optimized geometry obtained is used for performing single-point calculations with better basis sets or including a correction for electron correlation, in order to obtain a better description of the energy of the molecule. In many cases, the results obtained in this manner are good approximations to those that would have been obtained through a full geometry optimization at the higher level. Naturally, in some cases this procedure may be misleading, e.g. if geometry optimization is carried out at 3-21G for a molecule where the inclusion of d orbitals is crucial. To describe the computational procedure employed in a particular calculation, we will use the convenient labelling introduced by Pople[25]. Thus, a single-point UHF calculation using the 6-31G* basis set and the 3-21G$^{(*)}$ optimized geometry is designated UHF/6-31G*//3-21G$^{(*)}$. An analogous MP3 spin-unrestricted calculation is labelled UMP3/6-31G*//3-21G$^{(*)}$. Similarly, 6-31G*//6-31G* denotes a RHF/6-31G* energy calculated at the 6-31G* optimized geometry (the method is assumed to be RHF unless stated otherwise).

To complete the introduction to the theoretical methodology it might be of interest to compare the required computer times and the results obtained from different types of calculations. As an illustrative example we have chosen H_3SiF. The calculations were carried out on the Technion's IBM 3081D computer, using the Gaussian 82 series of programs[27]. The calculations, which include geometry optimizations, used throughout the same starting geometry, namely r(S–F) = 1.615Å, roughly midway between the 3-21G and 3-21G$^{(*)}$ values: r(Si–H) = 1.47Å, θ(H–Si–X) = 109.47°. The results of the calculations are given in Table 1.

TABLE 1. Computer times and geometries of H_3SiF calculated using various theoretical methods

Method[a]	CPU time[b]	r(Si–F)[c]	r(Si–H)[c]	θ HSiF[d]	E (hartrees)
STO-3G, FO	1.10	1.624	1.423	109.33	−385.39137
3-21G, FO	1.12	1.635	1.478	109.19	−388.07406
3-21G$^{(*)}$, FO	3.21	1.593	1.469	109.06	−388.19334
6-31G*, FO	14.91	1.594	1.470	108.76	−390.14840
6-31G*, SP	1.51	*e*	*e*	*e*	−390.14840
MP2/3-21G$^{(*)}$, SP	1.03	*f*	*f*	*f*	−388.38912
MP3/3-21G$^{(*)}$, SP	2.23	*f*	*f*	*f*	−388.39777
MP2/6-31G*, SP	2.78	*e*	*e*	*e*	−390.39970
MP3/6-31G*, SP	5.12	*e*	*e*	*e*	−390.41053
Experimental[g]	—	1.594	1.485	108.4	

[a] FO = geometry fully optimized assuming C_{3v} symmetry; SP = single-point calculation at a specified geometry.
[b] In minutes.
[c] In Å.
[d] In degrees.
[e] At the 6-31G* optimized geometry.
[f] At the 3-21G$^{(*)}$ optimized geometry.
[g] From R. Kewley, P. M. McKinney and A. G. Robiette, *J. Mol. Spectrosc.*, **34**, 390 (1970).

Finally, it may be appropriate to add a note of caution for the nonexpert. Even when the most sophisticated techniques are used, the assumptions and simplifications that have to be made for a molecule of any complextiy are still quite severe. Thus, the absolute errors in total energies are huge, by far larger than any chemical phenomena that we may wish to describe. However, the vast experience that has accumulated in the last two decades has shown that in spite of the many uncertainties involved in the results of quantum-mechanical computations, a careful use of the methods can produce extremely useful results. In fact, in some cases discussed below the calculations will serve to point out erroneous experimental interpretations. This review will emphasize the usefulness, strength and credibility of the computational methods. However, as in carrying out experiments, also when performing calculations one must keep in mind the assumptions and pitfalls involved in the specific method used.

An additional important point that one should bear in mind is that the calculations describe explicitly only the *isolated molecule in the gas phase* (in some cases, simulation of solvation is attempted but such calculations are still relatively rare). It is now well established that in many cases solvation affects drastically a wide variety of molecular and chemical properties and could even reverse experimental trends, relative stabilities, etc[61]. This point should be kept is mind whenever comparing calculations with experimental data, particularly when charged or highly polar molecules are involved. For energies we use hartrees or kcal mol^{-1} throughout. 1 hartree = 2625.54 kJ mol^{-1} = 627.52 kcal mol^{-1}.

III. COMPARISON OF THE CARBON AND SILICON ATOMS

Silicon is a contiguous group 4 (group 14 in the more recent notation adopted by IUPAC and ACS nomenclature committees) element. Yet, it is now well established and widely recognized that there are dramatic differences between analogous silicon and carbon compounds. Several obvious examples are the existence of stable hypervalent compounds for silicon but not for carbon, the difficulties in forming multiple bonds to silicon, and the different geometries and other physical properties of analogous silicon and carbon compounds. Despite these striking differences it is the common practice to compare silicon compounds with the analogous carbon compounds and we will also make such comparisons in this review. An appropriate place to start is therefore to compare some of the fundamental properties of the carbon and silicon atoms. In Table 2 are listed some fundamental properties of C and Si atoms. The interested reader is also referred to an excellent detailed discussion by Kutzelnigg of differences between isolated atoms of homologous main-group elements[62].

TABLE 2. Some fundamental properties of C and Si atoms

Property	C	Si
Atomic number	6	14
Atomic weight[a]	12.011	28.0855
Atomic radius (Bragg–Slater), pm[b]	70	110
Covalent radius, pm[c]	77.2	116.9
Ionic radius, pm[d]	260(−4)	41(+4)
Radii of the principal maxima of outer orbital, pm[e]	62.0 (2s); 59.6 (2p)	90.4 (3s); 106.8 (3p)
van der Waals radius, pm		
Pauling[f]	170	217
Glidewell[g]	125	155
Ionization energies, eV[h]	260(−4)	41(+4)
E_1	11.26	8.15
E_2	24.38	16.34
E_3	47.89	34.49
Electron affinity, eV[i]	1.12	1.39
Electronegativity[j]		
Pauling, 1932[f]	2.5	1.8
Mulliken, 1934[k]	2.63	2.44
Allred-Rochow, 1958[l]	2.60	1.90
Voronkov-Kovalev, 1965[m]	2.46	1.89
Mande-Deshmukh, 1977[n]	2.73	1.87
Sanderson, 1983[c]	2.746	2.138
Dipole polarizabilities[o]	11.8	36.3

[a] Commission on Atomic Weights and Isotopic Abundances, *Pure Appl. Chem.*, **55**, 1101 (1983); **56**, 653 (1984).
[b] J. C. Stater, *J. Chem. Phys.*, **41**, 3199 (1964).
[c] R. T. Sanderson, *J. Am. Chem. Soc.*, **105**, 2259 (1983).
[d] R. D. Shannon, *Acta Crystallogr., Sect.* **A 32**, 751 (1976).
[e] J. T. Waber and D. T. Cromer, *J. Chem. Phys.*, **42**, 4116 (1965).
[f] L. Pauling, *The Nature of the Chemical Bond*, 3rd edn., Cornell University Press, Ithaca, New York, 1960.
[g] C. Glidewell, *Inorg. Chim. Acta*, **36**, 135 (1979).
[h] C. E. Moore, Ionization Potentials and Ionization Limits Derived from the Analysis of Optical Spectra, NSRDS—NBS34, National Bureau of Standards, Washington DC, 1970.
[i] M. C. Day, Jr. and J. Selbin, *Theoretical Inorganic Chemistry*, Reinhold, New York, 1969.
[j] In the sp^3 valence state.
[k] R. S. Mulliken, *J. Chem. Phys.*, **2**, 782 (1934).
[l] A. L. Allred and E. G. Rochow, *Inorg. Nucl. Chem.*, **5**, 269 (1958).
[m] M. G. Voronkov and I. F. Kovalev, *Izv. Akad. Nauk Latv. SSR, Ser. Khim.*, 158 (1965).
[n] C. Mande, P. Deshmukh and P. Deshmukh, *J. Phys.* **B10**, 2293 (1977).
[o] In atomic units (1 au = 0.1482 $Å^3$), from Ref. 54.

The following differences between Si and C are apparent (for references see Table 2): (1) The covalent and van der Waals radii of silicon are significantly larger than those of carbon. (2) On all scales the electronegativity of silicon is lower than that of carbon. (3) The ionization energies of Si are lower (by 3.1 eV for the first ionization energy) and the electron affinity is higher than that of C. (4) The silicon atom is much more polarizable than the carbon atom. (5) The valence atomic orbitals of Si have a larger size and are more diffuse than those of C. (6) Silicon, in contrast to carbon, has relatively low-lying vacant d orbitals which *may* participate in bonding.

The role of d orbitals in bonding to silicon is the subject of lively debate and has been reviewed[63]. In general, it is very difficult to separate experimentally the effect of 3d orbitals from other effects such as ionic contributions. Furthermore, even if 3d orbitals contribute to bonding, it is not clear how one can establish to what extent this effect is of structural, spectroscopic or energetic significance[63]. The application of theory to resolve this problem is not straightforward either[63-67]. It is now well established that experimental quantities, such as bond lengths or dipole moments, are reproduced better if d-functions are included in the Si basis set (see discussion below). However, the same is true also for the analogous carbon compounds[25]. It is therefore not obvious whether such functions just serve as polarization functions and compensate for an inadequate sp basis set, or whether they imply actual participation of Si 3d orbitals. Pitt was the first to note that most (if not all) of the features attributed to d orbital participation can be explained alternatively in terms of the hyperconjugative effect of a silicon-centered σ^*-orbital of proper symmetry[64]. The controversy over the importance of d orbitals in bonding to silicon continues, and some of the more recent papers dealing with this question are given in References 65–67. The recent paper by Janes and Oldfield[66b] contains an extensive set of references to the relevant literature up to 1986. Janes and Oldfield reach the conclusion, based on the analysis of the oxygen-17 quadrupole coupling constants, that in silicates (d–p) π-bonding is important[66b]. On the other hand, Schleyer and coworkers have argued that (d–p) conjugation is a minor contributor even in carbanions, e.g. $H_3SiCH_2^-$, where a high-lying p-lone-pair is available and conjugation with low-lying d-orbitals on silicon may be expected to be particularly efficient[67]. Further evidence that supports the conclusion that (d–p) π-bonding is of minor importance in organosilicon chemistry is discussed in Sections IV.A.2 and IV.B.1.

IV. SINGLY BONDED SILICON COMPOUNDS

A. Monosubstituted Singly Bonded Silicon Compounds

In this section we will discuss calculations for compounds of the general formula H_3SiX. The simpler silicon hydrides such as H_2Si, H_3Si and the corresponding charged species will be discussed elsewhere with the corresponding substituted derivatives (e.g., H_2Si will be discussed in the paragraph on silylenes, H_3Si^+ in the paragraph on silicenium ions, etc.). The most extensive and systematic theoretical study on the H_3SiX molecules was published in 1986 by Luke, Pople, Krogh-Jespersen, Apeloig, Chandrasekhar and Schleyer (LPKACS)[68] and this section is based mostly on this paper, which also includes extensive references to previous calculations, mostly of lower quality[69].

LPKACS studied at various levels of theory the H_3SiX series, where X spans the entire standard series of second- and third-row substituents (X = H, Li, BeH, BH_2, CH_3, NH_2, OH, F, Na, Mg, AlH_2, SiH_3, PH_2, SH, Cl), and compared them with the analogous CH_3X series. The H_3SiX series is so fundamental that, in contrast to the general practice in this review, we bring in Table 3 the total energies of these molecules at 3-21G//3-21G, 3-21G$^{(*)}$//3-21G and MP4SDTQ/6-31G*//3-21G. The calculated ZPE are also given. This basic set of energies is required in many comparisons with other molecules and we therefore think that their inclusion here will prove useful to the reader.

1. *Structural trends*

Table 4 lists the Si–X and the Si–H bond lengths and the HSiX angles in the H_3SiX molecules as calculated at 3-21G$^{(*)}$. The last column gives the change in the Si–X

TABLE 3. HF/3-21G energy, zero-point energy (ZPE), HF/3-21G$^{(*)}$ and MP4/6-31G* total energies and relative energy (RE) for silicon compounds[a]

Molecule	Sym	State	HF/3-21G	ZPE[b]	HF/3-21G$^{(*)}$	MP4SDTQ/6-31G*	RE[c]
SiH	$C_{\infty v}$	$^2\Pi$	−287.907 21	2.9 [0]	−287.971 42	−289.482 24	0.0
SiH	$C_{\infty v}$	$^4\Sigma^-$	−287.878 56	3.2 [0]	−287.938 14	−289.428 23	34.1
SiH_2	C_{2v}	$^1A_1$	−288.484 29	7.5 [0]	−288.560 28	−290.089 96	0.0
SiH_2	C_{2v}	$^3B_1$	−288.480 90	7.9 [0]	−288.553 63	−290.064 34	16.4
SiH_2	$D_{\infty v}$	$^3\Sigma_u^-$	−288.445 74	6.7 [2]	−288.508 32	−290.023 95	40.5
SiH_2	$D_{\infty v}$	$^1\Delta_u$	−288.374 36	8.1 [1]	−288.436 06	−289.970 92	75.1
SiH_3	C_{3v}	$^2A_1$	−289.080 54	13.8 [0]	−289.166 16	−290.695 72	0.0
SiH_3	D_{3h}	$^2A_2''$	−289.072 49	12.8 [1]	−289.153 99	−290.687 06	4.4
SiH_4	T_d	$^1A_1$	−289.686 98	20.2 [0]	−289.784 26	−291.331 26	
H_3SiLi	C_{3v}	$^1A_1$	−296.509 92	15.1 [0]	−296.590 96	−298.196 80	
H_3SiBeH	C_{3v}	$^1A_1$	−304.231 69	20.7 [0]	−304.318 44	−305.991 75	
H_3SiBH_2	C_s, **7**[d]	$^1A'$	−314.783 31	28.5 [0]	−314.876 68	−316.651 68	0.0
H_3SiBH_2	C_s, **7a**[e]	$^1A'$	−314.783 29	28.4 [1]	−314.876 65	−316.651 69	−0.1
H_3SiCH_3	C_{3v}, **5**[d]	$^1A_1$	−328.526 12	40.5 [0]	−328.622 74	−330.524 62	0.0
H_3SiCH_3	C_{3v}, **6**[d]	$^1A_1$	−328.524 32	40.5 [1]	−328.620 47	−330.522 22	1.5
H_3SiNH_2	C_s, **7a**[e]	$^1A'$	−344.450 39	33.3 [0]	−344.555 48	−346.559 84	0.0
H_3SiNH_2	C_s, **7**[d]	$^1A'$	−344.450 35	33.2 [1]	−344.555 48	−346.559 93	−0.2

H_3SiOH	C_s, **8**[d]	$^1A'$	−364.180 60	25.2 [0]	−364.293 37	−366.414 42	0.0
H_3SiOH	C_s, **8a**[f]	$^1A'$	−364.179 80	24.8 [1]	−364.292 64	−366.413 08	0.4
H_3SiF	C_{3v}	$^1A_1$	−388.074 06	18.1 [0]	−388.193 34	−390.422 77	
H_3SiNa	C_{3v}	$^1A_1$	−449.968 32	14.6 [0]	−450.051 28	−452.587 07	
H_3SiMgH	C_{3v}	$^1A_1$	−488.152 40	18.7 [0]	−488.259 54	−490.947 98	
H_3SiAlH_2	C_s, **7a**[e]	$^1A'$	−530.845 48	24.0 [0]	−530.990 88	−533.866 63	0.0
H_3SiAlH_2	C_s, **7**[d]	$^1A'$	−530.845 48	24.0 [1]	−530.990 88	−533.866 63	0.0
H_3SiSiH_3	D_{3d}	$^1A_{1g}$	−578.241 21	31.9 [0]	−578.427 52	−581.508 28	0.0
H_3SiSiH_3	D_{3h}	$^1A_1'$	−578.239 93	31.8 [1]	−578.426 09	−581.506 63	0.9
H_3SiPH_2	C_s, **7**[d]	$^1A'$	−629.265 70	27.5 [0]	−629.465 45	−632.760 15	0.0
H_3SiPH_2	C_s, **7b**[g]	$^1A'$	−629.264 02	27.4 [1]	−629.463 21	−632.759 85	0.1
H_3SiPH_2	C_s, **7a**[e]	$^1A'$	−629.230 71	26.6 [1]	−629.424 52	−632.720 41	24.0
H_3SiSH	C_s, **8**[d]	$^1A'$	−685.277 10	22.0 [0]	−685.487 90	−689.007 45	0.0
H_3SiSH	C_s, **8a**[f]	$^1A'$	−685.275 85	21.8 [1]	−685.486 17	−689.005 65	0.9
H_3SiCl	C_{3v}	$^1A_1$	−746.457 79	17.0 [0]	−746.669 85	−750.423 00	

[a] From Ref. 68. The Hartree–Fock and MP4 total energies are in hartrees while the ZPE and RE are in kcal mol^{-1}.
[b] The number in brackets is the number of imaginary frequencies.
[c] The relative energy (RE) is determined by differencing the MP4 total energies and correcting for changes in the ZPE.
[d] The numbers refer to the formulas in the text.
[e] Structure **7** with HXSiH angle = 0°. For X = B and AP the $SiXH_2$ unit is planar.
[f] Structure **8** with HXSi angle = 0°.
[g] Rotated from conformation **7** by 180°.

TABLE 4. HF/3-21G(*) optimized Si–H and Si–X bond lengths and H–Si–X bond angles for H_3SiXH_n molecules[a,b]

XH_n	r(Si–H)[c]	θ(H–Si–X)[d]	r(Si–X)[c]	Δr(Si–X)[e]
H	1.475	109.5	1.475	0.012
Li	1.495	115.1	2.542	−0.011
BeH	1.483	111.8	2.204	−0.009
BH_2	1.479	110.7	2.025	0.013
CH_3	1.478	110.6	1.883	0.034
NH_2	1.478	111.2	1.710	0.032
OH	1.468	110.5	1.633	0.042
F	1.469	109.1	1.593	0.042
Na	1.496	115.2	2.837	−0.002
MgH	1.486	112.8	2.607	0.005
AlH_2	1.481	111.5	2.467	0.014
SiH_3	1.477	110.3	2.342	0.041
PH_2	1.475	109.8	2.256	0.091
SH	1.471	109.2	2.139	0.116
Cl	1.467	108.5	2.056	0.135

[a] From Reference 68.
[b] When more than one distinct H–Si–X angle is present, the average value is reported.
[c] Distances are in angstroms.
[d] Angles are in degrees.
[e] This difference is the HF/3-21G optimized bond length minus the HF/3-21G(*) separation (caused by adding d-functions to the third-row atoms), in angstroms.

distances caused by adding d-functions to the third-row atoms (i.e., on changing the basis set from 3-21G to 3-21G(*)).

The d-functions generally decrease the bond lengths (exceptions are Si–Li, Si–Be and Si–Na). The bond-length decrease is much larger between two third-row atoms, than between second- and third-row atoms. Unfortunately few experimental structures are available for comparison, but the few available are in quite good agreement with the 3-21G(*) results. For example, the 3-21G(*) calculated Si–P distance in H_3SiPH_2 is 2.256 Å, to be compared with the experimental value of 2.249 Å[70]. In H_3SiF and H_3SiCl, the calculated 3-21G(*) Si–X bond lengths (1.593 Å and 2.056 Å respectively) are almost identical with the experimental values (1.594 Å[71] and 2.048 Å[72], respectively). The Si–H distances and the HSiX bond angles tend to decrease on moving from left to right across either the second or the third row of the periodic table. Similar trends are found for the analogous methyl species, CH_3X[68].

2. H_3Si–X bond strengths

The knowledge of bond strengths is central to chemistry. Unfortunately, thermodynamic data for silicon molecules are limited and only a few Si–X bond energies are known, most of them for substituted compounds where the analysis is complicated by the additional substitution. In such a situation theory is extremely valuable, being able to provide with reasonable reliability data not otherwise available. Even if the accuracy of the calculated bond energies is not very good, the trends which are observed are most probably real and useful.

At the level of theory used in the LPKACS study (MP4SDTQ/6-31G*)[68] directly calculated bond energies are not accurate and the authors have therefore used equation

1. This equation compares the Si–X and C–X bond strengths in saturated silyl and methyl species, and the calculated energies are given in Table 5. These energies can be combined with the known experimental dissociation energies of CH_3–X to give the theoretical H_3Si–X bond dissociation energies (Table 5). Where available, the experimental Si–X bond energies[73] [such values are not available for H_3SiX and the cited values are taken from polysubstituted systems, such as $(CH_3)_3$ SiX] are also given in Table 5. Unfortunately, a meaningful comparison between theory and experiment cannot be undertaken due to the large uncertainties in the experimental data (e.g., dissociation energies differing by 20 kcal mol^{-1} were reported for the Si–Si and Si–O bonds[73]) and the fact that the experimental values are average bond strengths. Note that missing values for $D(H_3Si–X)$ in Table 5 are due to the lack of experimental data for the corresponding $D(CH_3–X)$ bonds.

$$H_3SiX + \dot{C}H_3 \rightarrow CH_3X + \dot{S}iH_3 \qquad (1)$$

The theoretical Si–X bond strengths in Table 5 provide very interesting and useful information. Two conclusions are immediately apparent: (a) The Si–X bond dissociation energies increase on going from left to right in the periodic table (e.g., compare Si–C with

TABLE 5. Calculated heats (ΔH_{rxn}) (equation 1), experimental H_3C–X bond energies, and theoretical and experimental H_3Si–X bond strengths[a]

X	ΔH_{rxn}[b]	Exptl $D^{\circ}_{298.15}(H_3C–X)$[c]	Theory[d] (Exptl) $D(H_3Si–X)$
H	−14.5	105.1 ± 0.2	90.6 (90.3)[e]
Li	6.2	39.2	45.4
BeH	−14.8		
BH_2	−19.6	*f*	
CH_3	−4.9	90.4 ± 0.3	85.5 (82.1)[g]
NH_2	13.8	84.9 ± 1.1	98.7 (100)[h]
OH	26.6	92.3 ± 1	118.9 (128)[i] (108)[j]
F	37.9	109.9 ± 1	147.8 (135)[j]
Na	8.3		
MgH	−1.6		
AlH_2	−10.7	*f*	
SiH_3	−11.5	82.1 ± 3	70.6 (74)[e] (53)[j]
PH_2	−2.5	*f*	
SH	9.1	74.0 ± 3	83.1 (70?)[j]
Cl	20.3	84.6 ± 0.2	104.9 (113)[i] (91)[j]

[a] From Ref. 68. All energies are in kcal mol^{-1}.
[b] Calculated from MP4/6-31G* energies, corrected for zero-point energy differences.
[c] Except where noted, data are from D. F. McMillen and D. M. Golden, *Annu. Rev. Phys. Chem.*, **11**, 493 (1982).
[d] Calculated by adding the energies of columns 2 and 3.
[e] From Ref. 73b.
[f] Average bond strengths for trimethyl compounds are available in Ref. 73a.
[g] From the reference in footnote *c*.
[h] From $(CH_3)_3SiNHCH_3$, Ref. 73b.
[i] From $(CH_3)_3SiX$, Ref. 73b.
[j] From Ref. 73c.

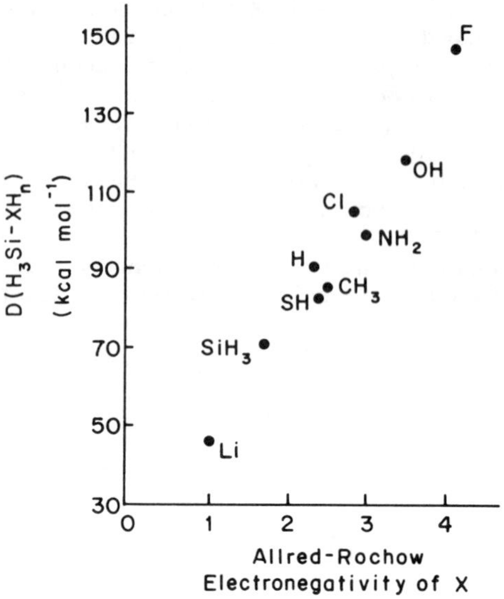

FIGURE 1. Plots of the calculated $H_3Si{-}XH_n$ bond strengths vs. the Allred–Rochow electronegativity of atom X[68]. Reprinted with permission from *J. Am. Chem. Soc.*, **108**, 260 (1986). Copyright (1986) American Chemical Society[68].

Si–F). (b) Si–X bonds with second-row elements are stronger than those with corresponding third-row elements (e.g., compare Si–O and Si–S). The plot of the Si–X bond energies against the Allred–Rochow electronegativities shown in Figure 1 is particularly instructive. The bond energies correlate very nicely with the electronegativity of X, suggesting that it is the major factor which determines the strengths of single bonds to silicon. As the electronegativity of X increases, so does the Si–X bond strength. Stronger bonds are formed with more electronegative groups because of greater contributions from ionic structures of the type $H_3Si^+X^-$. The bonds to the most electronegative elements are consequently extremely strong, as is well known experimentally. The calculated $H_3Si{-}F$ and $H_3Si{-}OH$ bond dissociation energies are 148 and 119 kcal mol^{-1}, respectively.

The correlation in Figure 1 explains in a straightforward manner why, for example, the energy of the Si–S bond is similar to that of the Si–C bond (S and C have similar electronegativities), but much lower than the isoelectronic Si–O bond energy. The correlation in Figure 1 argues strongly against an important contribution of d-π conjugation[63–67] to the strength of certain Si–X bonds, even with good π-donors such as NH_2 or OH possessing lone-pair electrons (see **4**)[74]. The fact that bonds between silicon and substituents of similar electronegativity but different chemical character have similar dissociation energies also argues against the importance of d-π bonding[63–67]. Thus,

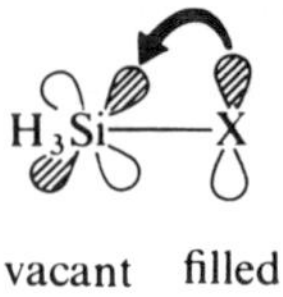

(4)

$D(H_3Si–SH)$ approximately equals $D(H_3Si–CH_3)$, although sulfur has lone pairs that can participate in d-π conjugation as shown in **4** (or represented by the resonance forms $H_3Si–\ddot{X} \leftrightarrow H_3Si^-{=}X^+$ and the methyl group does not.

Another point of interest is the comparison of the Si–X and C–X bond dissociation energies. These are given directly by the calculated energies of equation 1, ΔH_{rxn}, and are listed in the second column of Table 5, and also plotted against the Allred–Rochow electronegativities in Figure 2.

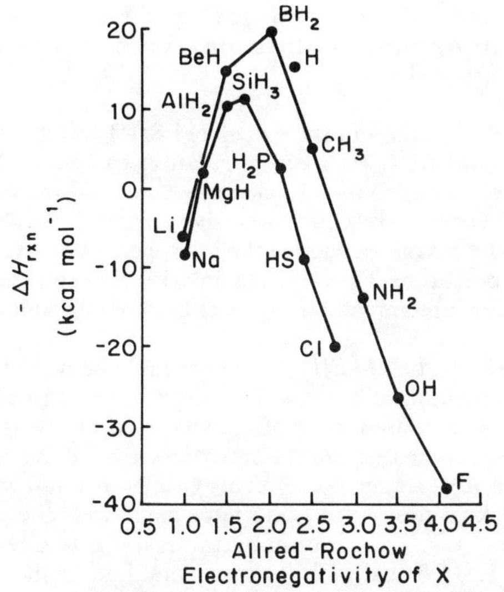

FIGURE 2. Plot of the heats of reaction for exchange between $H_3Si–XH_n$ and $H_3C–XH_n$ (equation 1) vs. the Allred–Rochow electronegativity of atom X. The energies are the negative of the values in Table 5[68]. Reprinted with permission from *J. Am. Chem. Soc.*, **108**, 260 (1986). Copyright (1986) American Chemical Society[68].

Figure 2 provides further insight into the relations between Si–X and C–X bonds. It reveals two inverted V-shaped curves, one for second-row and the other for third-row substituents, but these merge on the electropositive side. Thus all the Si–X bonds which appear below the zero-energy line in Figure 2 are stronger and those above the line are weaker than the corresponding C–X bonds. About half of the Si–X bonds are stronger than the corresponding C–X bonds, and vice versa. The Si–H bond in silane is by 14.5 kcal mol^{-1} weaker than the C–H bond in methane. The inverted V-shaped curves can be understood in terms of the ambielectronic nature of H_3Si, which supports both a negative and a positive charge better than CH_3. Thus, the ionization potential is smaller and the electron affinity larger for the silyl radical than for the methyl radical[68]. Several examples will illustrate the point. The Si–Li and Si–Na bonds are stronger than their carbon counterparts because contribution of ionic structures to the bonding are better for the $H_3Si^-X^+$ ion pairs than for $H_3C^-X^+$. The same is true for electronegative elements such as OH and Cl, where $H_3Si^+–X^-$ character is favored over $H_3C^+–X^-$ character. Particularly remarkable are the much stronger Si–X bonds compared with C–X bonds, for R = F, Cl and OH, i.e., by 38, 20 and 27 kcal mol^{-1}, respectively. These huge differences in bond energies are expected to play a major role in comparisons of substitution reactions at silicon and at carbon, when these groups are used as the leaving groups or as nucleophile. The substituents with electronegativities closer to silicon have stronger bonds to carbon than to silicon due to higher ionic character in the C–X bonds. For example, the C–Si bond in H_3CSiH_3 is by 11.5 kcal mol^{-1} stronger than the Si–Si bond in H_3SiSiH_3. The preference of CH_3CH_3 over CH_3SiH_3 is smaller, 4.9 kcal mol^{-1}.

3. *Individual H_3SiX molecules*

Several individual H_3SiX molecules require additional discussion as they are the prototypes of large classes of compounds; e.g., H_3SiNH_2 is the prototype for other silylamines, and so forth. Again, calculations are taken mainly from Reference 68, unless otherwise stated or referenced.

a. SiH_4. Silane is tetrahedral and the calculated Si–H bond length of 1.475 Å agrees well with experiment (1.481 Å)[75]. Mulliken population analysis (3-21G$^{(*)}$) shows that the hydrogens are negatively charged (−0.13 electron each) and the silicon atom is positive (∼ +0.52), as would be expected by electronegativity arguments. This polarity is opposite to that in CH_4, where the charge on each of the hydrogens is ∼ +0.2, resulting in a highly negatively charged carbon atom. The opposite polarity of Si and C in these molecules and other silanes is one of the fundamental differences between organic and silicon chemistry.

b. CH_3CH_3, CH_3SiH_3 and H_3SiSiH_3. The agreement between theory and experiment for the geometries of these molecules is good. As expected, all three compounds prefer the staggered geometry **5**. As a consequence of the increased heavy-atom separations along this series, there is a decrease in the rotational barriers around the central bond. In ethane the calculated eclipsed (**6**)—staggered (**5**) energy difference (all values at MP4SDTQ/6-31G*//3-21G) is 2.9 kcal mol,$^{-1}$ in good agreement with the experimental rotation barrier of 2.74 kcal mol^{-1} [76]. The rotation barrier is lower in CH_3SiH_3 (1.5 kcal mol^{-1} calculated, 1.7[77a] and 1.51[77b] kcal mol^{-1} experimental and still lower in H_3SiSiH_3 (0.9 kcal mol^{-1} calculated, 1.2 kcal mol^{-1}, experimental[78]).

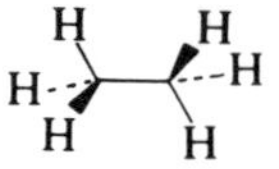

(5)

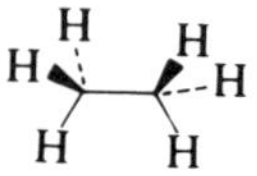

(6)

There is also an interesting change in the valence-orbital energies ordering along this series. In ethane, the a_{1g} orbital is 10 kcal mol^{-1} lower in energy than the degenerate e_g orbitals. In methylsilane this order is reversed, with the a_1 orbital 8 kcal mol^{-1} higher in energy. This trend continues in disilane, where the e_g orbitals are considerably lower (43 kcal mol^{-1}) in energy than the a_{1g} orbital. This trend can be understood by examining the nodal properties of the orbitals pictured in Scheme 1. The e_g orbital has antibonding

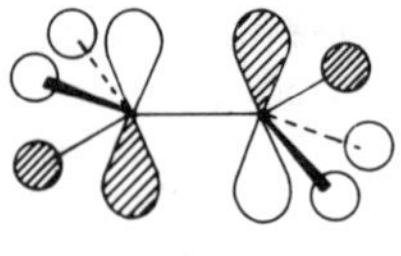

e_g (e in CH_3SiH_3)

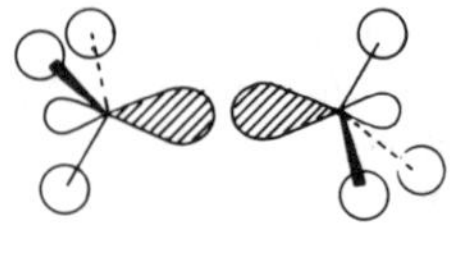

a_{1g} (a_1 in CH_3SiH_3)

SCHEME 1

character between the two heavy atoms and its energy is therefore lowered when the central distance increases. The a_1 orbital involves σ-bonding and it is destabilized due to loss in orbital overlap as the central bond lengthens. Further discussion of ethane and CH_3SiH_3 and a complete list of references through 1980 can be found in the book by Mulliken and Ermler[79].

c. H_3SiNH_2 and H_3SiPH_2. Silylamine is not known experimentally, but both $(CH_3)_3SiNH_2$[80a] and $H_3SiN(CH_3)_2$[80b] are known. Both 3-21G and 3-21G$^{(*)}$ calculations predict that the nitrogen in H_3SiNH_2 has essentially a planar geometry, the angle between the NH_2 bisector and the Si–N axis being 176.7° at 3-21G$^{(*)}$. Planar arrangements at nitrogen have indeed been found in amines with several silyl substituents, e.g. $(H_3Si)_2NH$[81] and $(H_3Si)_3N$[82]. The calculations also predict essentially free rotation about the Si–N bond.

Silylphosphine is known experimentally and the good agreement for the geometry between theory and experiment[70] has been noted above. The molecule adopts a bent staggered conformation **7**. The bond angles at phosphorus are small; the calculated Si–P–H and H–P–H values of 96.8° and 95.1°, respectively, are only slightly larger than reported experimentally (92.8° and 93.9°, respectively)[70]. The calculated rotation barrier in H_3SiPH_2 of 0.1 kcal mol^{-1} is smaller than the experimental value of 1.51 kcal mol^{-1}[83]. The calculated inversion barrier in H_3SiPH_2 is 24.0 kcal mol^{-1}, much smaller than that in H_3CPH_2 (35.5 kcal mol^{-1}). The ability of the H_3Si group to lower inversion barriers, demonstrated here for H_3SiNH_2 and H_3SiPH_2, has been noted in other cases[67], and is probably due to the lower electronegativity of silicon.

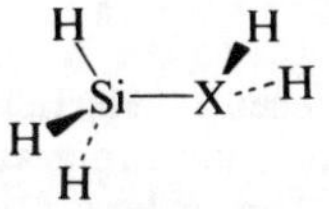

(7)

d. H_3SiOH and H_3SiSH. Both molecules are calculated to prefer the staggered equilibrium structures, **8**, in analogy to their carbon analogues.

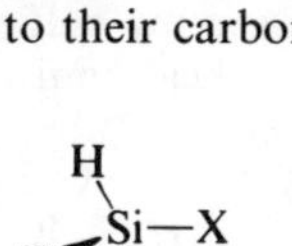

(8)

H_3SiOH is unstable towards disproportionation and has never been isolated experimentally. Substituted silanols, on the other hand, are well known[1,2]. The 3-21G$^{(*)}$ calculated bond length of 1.633 Å is in excellent agreement with the experimental distance of 1.63 Å in the diol and triol[84]. The calculated Si–O–H bond angle is exceptionally large, 128.8°. Even larger angles at oxygen are observed experimentally and theoretically in disiloxanes, e.g. $H_3SiOSiH_3$ (see Section IV.B.2). The long Si–O bond length and the large Si–O–H bond angle help to reduce the rotational barrier to 0.4 kcal mol^{-1} compared to 1.0 kcal mol^{-1} in methanol.

Silyl mercaptan has also not been characterized experimentally. The 3-21G$^{(*)}$ calculations predict a Si–S bond length of 2.139 Å, a Si–S–H bond angle of 98.2° and a rotational barrier of 0.9 kcal mol^{-1} (1.27 kcal mol^{-1} in CH_3SH).

e. H_3SiLi. This molecule, the prototype of organosilyllithium reagents, exists as a monomer in solution[85], and theory predicts that it has a remarkable structure[86]. Following previous work on inverted tetracoordinate carbon atoms, Schleyer and Clark showed in a beautiful study that silyllithum is likely to be the simplest molecule that prefers an inverted C_{3v} geometry **9** over the conventional van't Hoff C_{3v} structure **10**. At 6-31G*, **9** and **10** are both indicated, on the basis of the Hessian matrix, to be minima. The transition structure **11** connecting **9** and **10** has C_1 symmetry and was characterized analytically. The final energies for all these H_3SiLi species were calculated at MP4SDTQ/6-31G**//6-31G* and included a correction for zero-point energy. The inverted H_3SiLi structure **9** is found to be by 2.4 kcal mol^{-1} more stable than the conventional tetrahedral structure **10**. Furthermore, the two H_3SiLi isomers are separated by a significant barrier. The best interconversion pathway involves transition structure **11**, which lies 13.5 kcal mol^{-1} higher in energy than **9**. It was thus predicted that it should be possible to observe both isomers **9** and **10**, in matrix isolation. However, the authors also speculate that this stability order should be reversed in solution due to a better solvation of **10**.

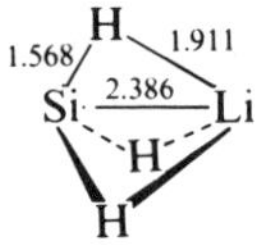

(**9**) C_{3v}

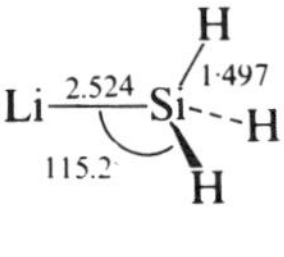

(**10**) C_{3v}

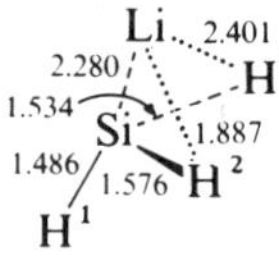

$< H^1SiH^2 = 98.7°$
$< H^1SiH^3 = 101.4°$
$< H^2SiH^3 = 90.2°$

(**11**) C_1

f. H_3SiCN and H_3SiNC. Very recently Hertler and coworkers[87] studied the silyl cyanide/isocyanide isomerization and found that at the SCF level H_3SiCN is less stable than H_3SiNC by 1.5 kcal mol^{-1}, but this order is reversed at the correlated level and with MP2 H_3SiCN is 7.8 kcal mol^{-1} more stable than H_3SiNC, in agreement with experiment[87]. It is interesting to compare this energy difference to ΔE for CH_3CN/CH_3NC which is 22.7 kcal mol^{-1} at the correlated level of theory. Substitution of the three hydrogens by hydroxy groups reduces somewhat the cyanide/isocyanide energy difference: $(HO)_3SiCN$ is 3.7 kcal mol^{-1} more stable than $(HO)_3SiNC$, in agreement with experiment[87].

The calculated geometry for H_3SiCN has r(C–Si) = 1.872 Å, r(C–N) = 1.141 Å, r(Si–H) = 1.467 Å and $\angle$(HSiH) = 111.5°. For H_3SiNC r(Si–N) = 1.745 Å, r(N–C) = 1.165 Å, r(Si–H) = 1.466 Å and $\angle$(HSiH) = 111.0°.

g. H_3SiCH_2OH and $H_2C{=}C(SiH_3)OH$. These molecules were studied by Apeloig, Stanger and Arad primarily as models for compounds which were studied solvolytically; they are discussed in Section VII.B.2.a.

B. Larger Non-cyclic Singly Bonded Silicon Compounds

1. Multiply substituted silicon systems of the type H_nSiX_{4-n}

A few molecules of this class have been studied, but in general the data are rather limited and this class of compounds deserves more theoretical attention.

a. H_2SiLi_2 and H_2SiNa_2. These molecules have been studied very recently by Rajca and Streitwieser[88] and the results will be discussed in Section VIII.

b. Fluorosilanes. Roelandt and coworkers, using a basis set of approximately double-zeta quality, have carried out SCF calculations for H_2SiF_2 (also for H_3SiF), and they give tables of orbital energies and discuss the nature of the MOs, using a population analysis[65b]. STO-3G, 3-21G, 6-31G and 6-31G* calculations for SiF_4 were reported by Hess, McMillan and O'Keeffe[89]. The Si–F bond length is 1.584 Å at 3-21G, 1.625 Å at 6-31G, 1.557 Å at 6-31G* and 1.552 Å experimentally[90]. These results clearly demonstrate the crucial role of d-orbitals on silicon in producing the correct Si–F bond length. Harmonic force constants and frequencies have been calculated and the results were shown to be in good agreement with experiment[89, 90].

c. $H_2Si(OH)_2$ and $H_2Si(SH)_2$. These systems are of special interest for several reasons: (a) they may serve as simple models for silicone polymers and for silicates; (b) they are analogous to carbon compounds, such as $CH_2(OH)_2$, that exhibit the anomeric effect[91].

The anomeric effect at carbon was investigated extensively, both experimentally and theoretically[91]. Surprisingly, the anomeric phenomena at silicon received relatively little attention. The first to address this question, although with little detail and using relatively simple theoretical models, were Hargittai and Seip[92] who studied the conformational preferences of $H_2Si(OH)_2$ and $HSi(OH)_3$ [92]. These authors used a partially polarized basis set (d-orbitals on Si), but the geometries used were assumed and no geometry optimizations were carried out[92].

A more complete study of $H_2Si(OH)_2$ was carried out recently by Apeloig and Stanger[93, 94], and the following discussion is based mostly on this paper. The geometries of $H_2Si(OH)_2$ and $H_2Si(SH)_2$ were fully optimized at 3-21G, 3-21G$^{(*)}$ and 6-31G* and more reliable energies were obtained at these geometries, using MP3/6-31G*. Both $H_2Si(OH)_2$ and $H_2Si(SH)_2$ prefer the *gauche, gauche* conformation **12** (i.e. $\angle$HXSiX $\sim$ 120°), and the optimized structures are given in Figure 3. Hargittai and Seip also found **12** to be the most stable conformation of $H_2Si(OH)_2$[92]. A *gauche, gauche* conformation **12** was found to be preferred also for $H_2Si(XCH_3)_2$, X = O, S. This is also the preferred conformation in the carbon analogues. The Si–O and Si–S bonds in **12** are somewhat shorter than in H_3SiOH (1.647 Å, 6-31G*).

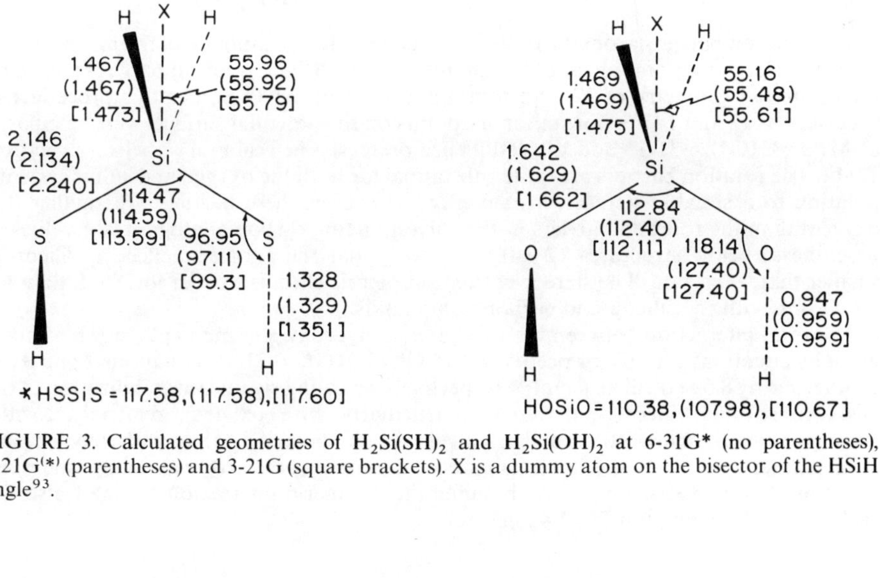

FIGURE 3. Calculated geometries of $H_2Si(SH)_2$ and $H_2Si(OH)_2$ at 6-31G* (no parentheses), 3-21G$^{(*)}$ (parentheses) and 3-21G (square brackets). X is a dummy atom on the bisector of the HSiH angle[93].

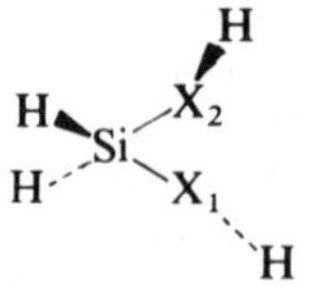

(12) g, g

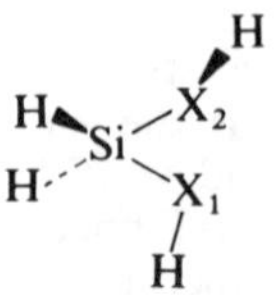

(13) a, g

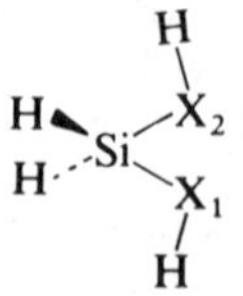

(14) a, a

	(12) g, g	(13) a, g	(14) a, a
X=O	0.0 (0.0)	2.5 (4.8)	6.3 (10.8)
X=S	0.0 (0.0)	2.1 (2.3)	5.3 (6.0)

In simple molecular-orbital terms the anomeric effect results from interaction of the oxygen lone pairs with the *gem*-Si–Oσ^* orbital, as shown schematically in **15**[91]. According to this model conformation **12**, in which two such interactions are present, is expected to be the most stable because in this stereochemical arrangement the σ^*(Si–O) orbital is properly aligned to interact with the 2p-lone pair of oxygen, while in **13** ($\measuredangle HX_1SiX_2 = 180°$) it interacts with the much higher-lying σ^* (Si–H) orbital, as shown in **16**. In **14** ($\measuredangle HX_1SiX_2 = \measuredangle HX_2SiX_1 = 180°$) the 2p-lone pairs of both oxygens are out of conjugation with the adjacent σ^*(Si–O) orbitals. The fact that $H_2Si(OH)_2$ and $H_2Si(SH)_2$ adopt the *gauche* conformation is therefore a strong indication of the operation of an anomeric effect.

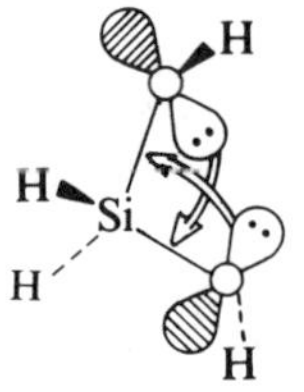

(15)

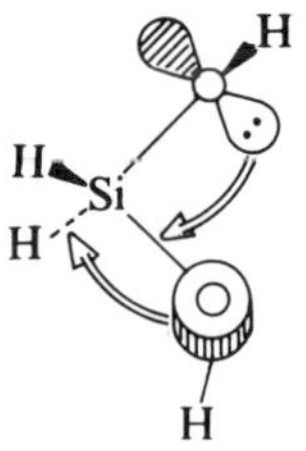

(16)

The rotation barriers around the Si–O bonds provide one estimate of the magnitude of the stereoelectronic component of the anomeric effect. 'Rigid' rotation barriers (i.e. using the optimized geometry of **12**, and rotating around the Si–O bonds; this procedure is necessary as **13** and **14** are not stationary points on the potential surface) were calculated at MP3/6-31G*//6-31G* and the results are presented in kcal mol^{-1} below structures **12**–**14**. The rotation barriers are quite substantial for both the oxygen and sulfur systems, pointing to a sizeable stereoelectronic effect. However, these barriers are smaller (for oxygen) than the rotation barriers in the corresponding carbon analogues (see values in parentheses below structures **12**–**14**), suggesting that the anomeric effect at silicon is smaller than at carbon. The stereoelectronic anomeric effect is smaller for X = S than for X = O, for both the silicon and carbon compounds.

The total interaction between the two geminal hydroxy and mercapto substituents is given by equations 2 and 3, respectively. At MP3/6-31G*//6-31G*, equations 2 and 3 are exothermic by 8.6 and 2.2 kcal mol^{-1} respectively revealing significant stabilization of the *gem*-disubstituted silicon compounds (particularly with oxygen), relatively to the corresponding monosubstituted derivatives. Again, one finds that with oxygen this *geminal* interaction is smaller than for carbon; the energy of the corresponding isodesmic equation 4 is 15.8 kcal mol^{-1}. With sulfur the anomeric interaction is small also for carbon, i.e., ΔE (equation 5) = 1.4 kcal mol^{-1}.

$$H_2Si(OH)_2 + SiH_4 \rightarrow 2H_3SiOH \quad (2)$$

$$H_2Si(SH)_2 + SiH_4 \rightarrow 2H_3SiSH \quad (3)$$

$$H_2C(OH)_2 + CH_4 \rightarrow 2H_3COH \quad (4)$$

$$H_2C(SH)_2 + CH_4 \rightarrow 2H_3CSH \quad (5)$$

In conclusion, the anomeric effect at silicon, although smaller than for carbon, is significant especially for oxygen and it probably plays an important role in determining the preferred conformations of relevant silicon compounds such as disilaoxiranes.

d. $HSi(OH)_3$ and $HSi(SH)_3$. Trihydroxysilane and trimercaptosilane were also studied by Apeloig and Stanger[93,94]. At 3-21G both $HSi(OH)_3$ and $HSi(SH)_3$ prefer conformation **17**, which has a nearly *gauche* conformation around all three Si–O and Si–S bonds. This is expected if the anomeric effect dictates the torsional arrangement in this compound. Hargittai and Seip[92], on the other hand, did not find any indication of an anomeric effect in $HSi(OH)_3$. However, as mentioned above the geometries which these authors used were assumed[92]. The isodesmic equations 6 reveal again the dramatic difference between the effects of oxygen and sulfur. Thus, at 6-31G*//3-21G the energy of equation 6a is 10.1 kcal mol^{-1} and that of equation 6b only 0.4 kcal mol^{-1}. Thus a third hydroxy substituent stabilizes significantly the organosilane while a third SH substituent has nearly no stabilizing effect[94]. The significance of these results for analyzing the dissociation energies of these molecules to the corresponding cations are discussed in Section VII.B.1.

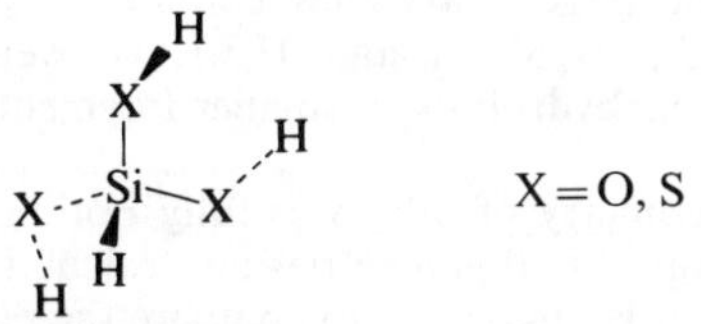

(17)

$$HSi(OH)_3 + SiH_4 \rightarrow H_2Si(OH)_2 + H_3SiOH \quad (6a)$$

$$HSi(SH)_3 + SiH_4 \rightarrow H_2Si(SH)_2 + H_3SiSH \quad (6b)$$

Calculations for $(HO)_3SiCN$ and $(HO)_3SiNC$ have been reported recently, but the anomeric effect was not discussed[87].

e. $Si(OH)_4$. Several recent papers by O'Keeffe and coworkers[89,95] and by Sauer[96] report calculations for tetrahydroxysilane, mainly as a model for silicates. These calculations provide a nice illustrative example for the power of modern analytical optimization techniques (see Section II.B). In their first studies O'Keeffe and coworkers optimized the structure of $Si(OH)_4$ at 6-31G* and found that the minimum-energy geometry has D_{2d} symmetry[95a]. However, gradient techniques were not used to analytically characterize the optimized structure. In a following paper harmonic force constants were calculated and a negative eigenvalue (i.e., an imaginary frequency) was present in the force constant matrix indicating that the D_{2d} geometry is a saddle point rather than a global minimum[89]. Hess, McMillan and O'Keeffe in a recent paper determined that the global minimum is a structure with S_4 symmetry (hydrogen staggered), which is 13.8 kcal mol^{-1} (at 6-31G**) lower in energy than the D_{2d} structure[95b]. Simultaneously and independently, Sauer reached the same conclusions[96]. Sauer also analyzed in detail the effect of the different basis sets on the geometry of the molecule and concluded that the inclusion of d-orbitals on oxygen is a prerequisite for a correct description of the geometry, especially of the HOSi bond angle (see below)[96a]. Thus, the 6-31G* basis set should be used, in preference to the 6-31G$^{(*)}$ basis set which was used in an earlier study by O'Keeffe and coworkers[95a] [these authors have simultaneously published a 6-31G** optimized geometry of $Si(OH)_4$[95b]]. The major geometrical parameters of $Si(OH)_4$ as calculated by the different basis sets are shown in Table 6.

Table 6 summarizes an extensive collection of calculations and provides a good illustration of the changes in the geometrical parameters as a function of the basis set. One point of specific interest is the influence of silicon d-orbitals on the SiO bond length. Sauer[96a] concluded that the quality of the basis set to which the silicon d-functions are added is a critical factor. Thus, shortening of the SiO bond is as much as 0.05 Å for STO-3G, 0.041 Å for 3-21G and 0.042 Å for 6-31G, but it drops to only 0.023 Å if d-functions are already present on the oxygen atoms. An important point is that all these shortenings are much smaller than the 0.16 Å difference ascribed by Pauling[97] to (p–d) π-interactions. These results therefore do not support the (p–d) π-bonding picture. On the other hand, opposite conclusions supporting the (p–d) π-bonding model were reached by analysis of the oxygen-17 nuclear quadrupole coupling constants in silicates[66]. Further discussion on the role of (p–d) π-bonding is given in Section III. Sauer[96a] also notes that omission of the d-functions on silicon has little effect on the geometry, except for shortening of the SiO bond. In contrast, omission of d-orbitals on the oxygen atoms has a major effect on the bond angles at the oxygen atoms and on the conformation[96a]. O'Keeffe and coworkers concluded that Hartree–Fock calculations for $Si(OH)_4$ can provide realistic force constants for silicate molecules and clusters[95b].

Calculations for larger molecules containing silicon and oxygen atoms, such as $H_4Si_2O_6$, $H_2Si_2O_5$, $H_6Si_3O_3$ and $H_6Si_2O_7$, were also reported by O'Keeffe and coworkers and their hydrolysis to smaller fragments was discussed[95a,c].

f. SiF_4. The geometry of SiF_4 was fully optimized at STO-3G, 3-21G, 6-31G and 6-31G*, and the optimized geometries and calculated harmonic vibrational frequencies were reported[95b]. It is interesting to compare the experimentally determined Si–F bond length of 1.552 Å with the calculated values (Å): 1.557 (6-31G*), 1.625 (6-31G), 1.584 (3-21G), 1.585 (STO-3G)[95b]. The crucial role that the d-orbitals on silicon have in producing the correct Si–F bond length is clearly demonstrated.

g. $H_2Si(SiH_3)_2$. This molecule was studied recently by Schaefer and is discussed in Section IV.C.1.a.

2. *Disiloxane ($H_3SiOSiH_3$) and methyl silyl ether (H_3COSiH_3)*

Disiloxane, the simplest model of the siloxane bond, has attracted considerable interest, both experimental and theoretical. The theoretical interest stems mainly from two unusual geometrical features of disiloxane, namely the unusually short Si–O bond and the large Si–O–Si bond angle. Thus, in $H_3SiOSiH_3$, the Si–O bond length is 1.634 Å[98], much shorter than the value of 1.77 Å which is expected from the Schomaker–Stevenson Rule (i.e., the sum of the covalent radii of Si and O, corrected for differences in the electronegativities)[97]. The experimentally determined Si–O–Si bond angle in disiloxane is 144.1°[98], much larger than in dimethyl ether (111.5°)[99]. A variety of calculations were reported for $H_3SiOSiH_3$, aiming at testing which level of theory is required to reproduce its peculiar geometry. The calculations included semiempirical calculations (CNDO/2[100], MINOO/3[101] and MNDO[101]) and *ab initio* calculations using a variety of basis sets[101–104]. The calculated Si–O bond lengths and the Si–O–Si bond angles are given in

TABLE 6. Calculated molecular geometries of orthosilicic acid $Si(OH)_4$ at various theoretical levels[a]

Parameter	STO-3G[b]	3-21G[b]	6-31G[b]	6-31G$^{(*)}$[c]	6-31G$^{(+)}$[e]	6-31G*[g]	6-31G*[h,i]	6-31G**[b]
r(SiO)	1.655	1.642	1.662	1.622	1.652	1.635	1.629	1.629
r(OH)	0.983	0.957	0.941	0.995[d]	0.948	0.947	0.947	0.947
∢SiOH	109.1	128.4	134.6	129.5	116.4	116.5	117.1	117.2
∢ OSiO (α)	112.4	111.5	111.9	114.3	116.6	(109.5)[g]	115.8	112.6
∢ OSiO (β)	103.7	105.5	104.7	107.1	*f*	*f*	106.4[j]	103.3
∢ HOSiO	60.0	60.0	60.0	60.0	−31.8	0.0	−33.3	60.0
Point group	D_{2d}	D_{2d}	D_{2d}	D_{2d}	S_4	S_4	S_4	D_{2d}

[a] Distances are in Å and angles in degrees.
[b] From Reference 89b.
[c] d-orbitals on silicon only, from Reference 95a.
[d] This is probably a misprint in Reference 95a.
[e] d-orbitals on oxygen only, from Reference 96a.
[f] Not reported.
[g] From Reference 89b; in these calculations the SiO_4 skeleton was assumed to be a regular tetrahedron, i.e., with all OSiO angles fixed at 109.47°.
[h] Fully optimized, Reference 96a.
[i] Essentially the same results were simultaneously reported in Reference 95b using the 6-31G** basis set.
[j] From Reference 95b.

TABLE 7. Optimized Si–O Bond lengths (Å) and Si–O–Si Bond Angles (degrees) in $H_3SiOSiH_3$

Method	Si–O	Si–O–Si	Reference
CNDO/2[a]	(1.63)[b]	150	100
MINDO/3	c	180	101
NMDO	c	180	101
3-21G	1.6516	180.0	101
Split valence[d]	1.617	143.3	102
Split valence[e]	(1.634)[b]	180.0	103
Split valence[f] + d	(1.634)[b]	154.0	103
3-21G + d[g]	1.6516	149.48	101
MP2/6-31G*	c	161.0	104
Experimental	1.634	144.1	98

[a] d-functions are not included in the basis set.
[b] Not optimized.
[c] Not reported.
[d] Using 21, 4-21 and 3-3-21 basis sets for H, C and Si, respectively.
[e] Using an effective potential for the inner shell and a 31G basis set for the valence shell.
[f] Using an effective potential for the inner shell and a 61G basis set for the valence shell.
[g] The d-shell exponents are 0.3 for Si and 0.8 for O.

Table 7 and compared with the experimental values. The experimental geometrical parameters are reproduced with reasonable accuracy only when d-functions are added to the basis set. However, the contribution of the d-functions to the molecular wave function is minor and this was taken as evidence against the concept that (p–d) π back-bonding is responsible for the unusual geometry of $H_3SiOSiH_3$. Most theoretical studies prefer to describe the unusual geometry of $H_3SiOSiH_3$ as a consequence of its ionic character, which is considerably higher than that estimated from electronegativity differences. The effect on various molecular properties of $H_3SiOSiH_3$ of different uncontracted polarization functions was extensively tabulated and presented graphically[101]. It was concluded that variation of the polarization functions on oxygen has only a minor influence on the molecular properties of disiloxane, but the presence of polarization functions is required to obtain a bent geometry for the disiloxane bond. When the polarization functions on silicon were varied, the best agreement with experiment was obtained with exponent values of 0.3 and 0.9[101]. When the d-functions are removed from the basis set, $H_3SiOSiH_3$ adopts a linear geometry.

The relatively large changes in the Si–O–Si bond angle as a function of the basis set is not surprising in light of the very flat potential surface around the minimum. The experimental barrier to linearization at oxygen is only 0.32 kcal mol^{-1} [105]; calculations produce similar values[101]. A nearly linear Si–O–Si bond angle of 178.1° was calculated at 3-21G also in $H_2C{=}SiH(OSiH_3)$[106].

H_3COSiH_3 exhibits at all levels of calculation[100–102,104], and also experimentally, a behavior that is intermediate between H_3COCH_3 and $H_3SiOSiH_3$. For example, at 3-21G$^{(*)}$ (experimental values are given in parentheses) r(Si–O) = 1.647 Å (1.640 Å), r(C–O) = 1.410 Å (1.418 Å), ∢ CSiO = 122.6° (120.6°)[101]. In general, acceptable agreement between experiment and theory is reached at lower levels of theory for H_3COSiH_3 than for $H_3SiOSiH_3$.

C. Singly Bonded Cyclic Compounds

The general interest of chemists in the nature of bonding in strained-ring systems has drawn attention to such cyclic compounds which contain one or more silicon atoms. Here too the analogy between the organic and organosilicon compounds proved to be only partial and the investigators in this field, both experimentalists and theoreticians, faced many surprises and unexpected behavior. Also, in this field, theory played (and still plays) an important role as will be demonstrated in the following discussion. We will first examine cyclopolysilanes, i.e., rings that contain only silicon atoms, and then consider the hetero-substituted cyclosilanes. Ring-containing multiply bonded silicon atoms or aromatic rings are discussed in Section V.

1. Saturated cyclopolysilanes

In addition to the general interest in the strain and structure of these molecules, special interest in this class of compounds arises from their unusual electronic properties, which are believed to arise from electron delocalization in the σ-framework[107]. A variety of cyclopolysilanes, up to $(Me_2Si)_{35}$, have been synthesized in the last two decades mainly by West's group in Wisconsin[107b]. Quite recently, derivatives of the small-ring silicon compounds, cyclotrisilane (trisilacyclopropane) **18**[108a] and cyclotetrasilane (tetrasilacyclobutane) **19**[108b], have been synthesized and their X-ray determined. Even the highly strained bicyclo[1.1.0]tetrasilane derivative could be prepared[108c,d]. However, no experimental thermochemical data are available for these compounds and here theory has made an important contribution.

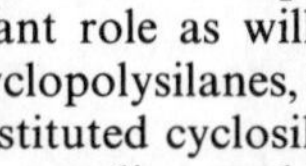

(18)

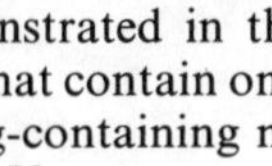

(19)

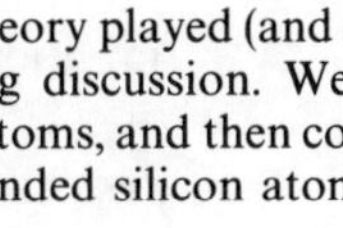

(20)

a. Monocyclic compounds

(i) Geometries. A systematic study of the geometries of cyclotrisilane (**18**), cyclotetrasilane (**19**) and cyclopentasilane (**20**) was published very recently by Grev and Schaefer[109]. Previous calculations for **18** and **19** include pseudopotential calculations using a dz + p valence basis set[110], and *ab initio* dz calculations[111, 112]. Gordon and Bartol[113] have also reported recently the optimized geometry of **18** at the $3\text{-}21G^{(*)}$ level. The geometries in these studies are generally in agreement with those calculated by Grev and Schaefer and which are shown in Figure 4[109].

Figure 4 demonstrates that polarization functions must be included in the basis set to describe correctly the structures of compounds **18**–**20**. Using a nonpolarized dz basis set the calculated Si–Si bond distances are too long. We have noted this phenomenon above, in the discussion of the bond length in Si_2H_6. A direct comparison of the calculated structures of **18**–**20** with experiment cannot be carried out because the structures of only heavily substituted systems were determined experimentally[108].

Let us now examine the structures of each ring individually. In cyclotrisilane, **18**, the Si–Si bond length is only slightly shorter than in disilane. The experimentally determined Si–Si bond lengths in the available, heavily substituted systems are significantly longer[108]. Although electronic effects may be responsible for this observation, steric crowding undoubtedly plays a major role. Cyclotetrasilane, **19**, is found to have a puckered

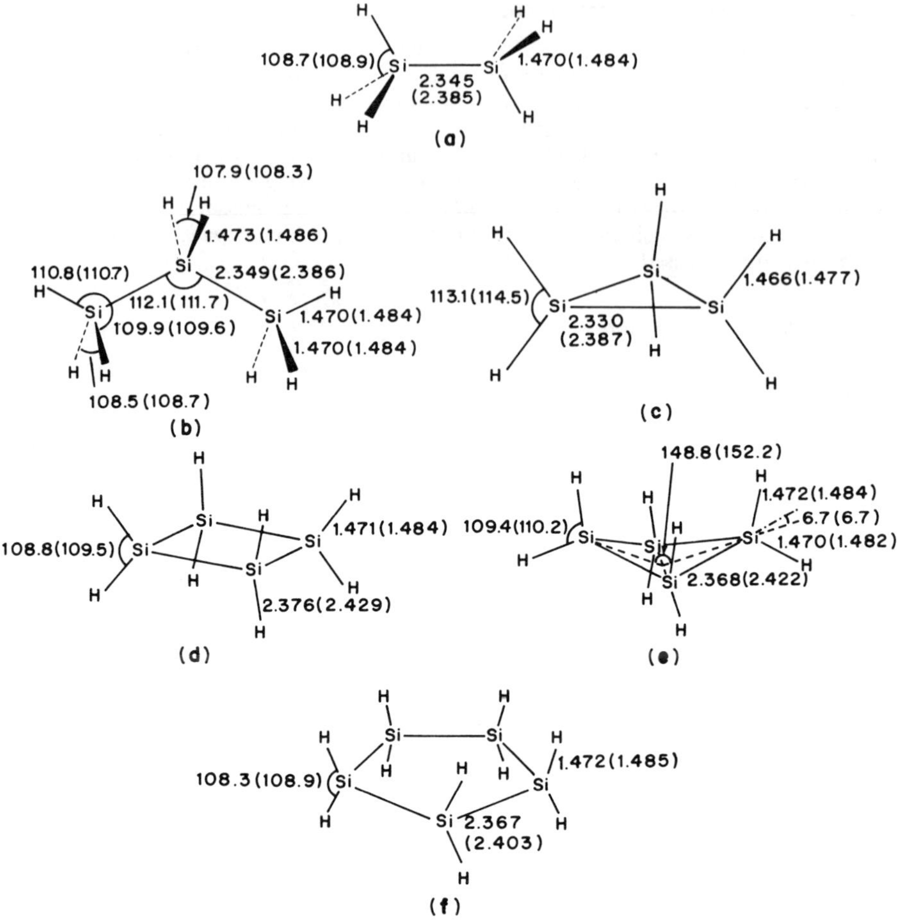

FIGURE 4. Optimized geometries of disilane (**a**, D_{3d}), trisilane (**b**, C_{2v}), cyclotrisilane (**c**, D_{3h}), cyclotetrasilane (**d**, D_{4h} and **e**, D_{2d}) and cyclopentasilane (**f**, D_{5h}) using dz (in parentheses) and dz +d basis sets. Reprinted with permission from *J. Am. Chem. Soc.*, **109**, 6569 (1987). Copyright (1987) American Chemical Society[109].

structure but the barrier to planarity is very low, only 1.5–1.75 kcal mol^{-1} using various levels of correlated wave functions[109,112a]. The folding (or puckering) angle is around 30°[109,112a]. Cyclopentasilane, **20**, is known experimentally[114] to be puckered, but Grev and Schaefer optimized only the planar structure (Figure 4). The Si–Si bond lengths are intermediate between those in **18** and **19**.

The theoretical conclusion from the comparison of these structures is that the Si–Si bond distances increase in the order **19** > **20** > **18**, while the experimental order (of substituted derivatives) is **18** > **19** > **20**. The discrepancy has been attributed[109] to the presence of bulky substituents in the experimentally determined smaller rings[108a,b]. Grev and Schaefer have also calculated the vibrational frequencies and the absolute infrared intensities of the compounds[109].

(ii) Strain. The ring strain in cyclopolysilanes has been determined via the use of the homodesmic reactions 7. Grev and Schaefer[109] and Schleyer[115] obtained similar results and those from the latter paper are given in Table 8. This table also includes the strain energies in the corresponding carbon ring systems and a breakdown of these strain energies into various energetic contributions to the total strain. These strain energies agree with those obtained previously by Sax using a local pseudopotential SCF procedure[112] but contradict those reported by Schoeller and Dabisch[111]. However, the latter results are probably not valid[115] as the authors used isodesmic equations instead of the more appropriate homodesmic equations 7. Nagase and coworkers have recently reached similar conclusions[116].

$$(SiH_2)_n + nH_3SiSiH_3 \rightarrow nH_3SiSiH_2SiH_3 \quad (7)$$

It is interesting that the strain energies do not vary much as a function of the basis set, despite the large changes in the Si–Si bond distances[109,115]. That d-orbitals have little effect on the strain energy of **18** suggests that 'bent bonding' is less important for silicon than for carbon[115].

The most striking feature of Table 8 is that the strain energies of the silicon rings differ very much from their carbon counterparts. Thus, the strain follows the order: 3-membered > 4-membered > 5-membered in the cyclosilanes, and 3-membered ~4-membered ≫ 5-membered in the cycloalkanes. This change results from the fact that **18** is 10 kcal mol^{-1} less stable (or more strained) than cyclopropane, while **19** is 9 kcal mol^{-1} more stable than cyclobutane. The strain in **20** of 6 kcal mol^{-1} is similar to that in cyclopentane. An interpretation of these facts was offered by both Grev and Schaefer[109] and by Schleyer[115]. Both papers agree that the strain changes within the cyclosilane series can be understood according to the traditional model of angle deformation and torsional strain. Both papers also reject a significant contribution from σ-aromaticity proposed recently by Dewar[117]. However, angle and torsional effects alone can account for only part of the differences between the carbon and the silicon series, and Schleyer has devised a beautiful explanation to explain the dilemma[115]. He suggested that, in silicon rings, angle deformation effects dominate, but that in cyclopropane the weakness of the C–C bonds due to angle strain is compensated somewhat by rehybridization, which strengthens the C–H bonds. In cyclotrisilane, on the other hand, silicon rehybridization on bonding is much less important, and Si–H bond strengthening does not compensate for Si–Si bond weakening. These arguments are supported by the crude decomposition of the total strain to various energetic contributions given in Table 8 and by the uniquely high C–H bond dissociation energies in cyclopropane.

Sax studied four Si_3H_6 isomers including **18** and found that **18** is the lowest in energy, lying 12 kcal mol^{-1} below $H_3SiSiH{=}SiH_2$, in contrast to the analogous carbon compounds where the stability order is reversed[112b].

TABLE 8. Roughly estimated contributions to small-ring strains (kcal mol^{-1})[a]

Contribution	c-$(CH_2)_3$	c-$(CH_2)_4$	c-$(SiH_2)_3$ (**18**)	c-$(SiH_2)_4$ (**19**)
XH bond eclipsing	5	7	1	2
1,3-Repulsions	0	10	0	5
Angle strain	33	10	38	8
Rehybridization	−10	0	−1	0
Total strain	28	27	38	15

[a] From Reference 115.

Grev and Schaefer proposed a theoretical model for the electronic structures of the radical anions of **18–20** and other similar cyclopolysilanes[109]. It was suggested that the singly occupied molecular orbital of cyclopolysilane anion radicals is π-bonding between the ring atoms but silicon-substituent (i.e. hydrogen) antibonding. This π^* model explains available evidence from ESR and UV spectroscopy, but cannot easily explain results from recent ^{29}Si ENDOR studies. The authors point to the fact that the alternative σ^* model (involving the σ^* Si–Si orbitals[118]) which does account for the ^{29}Si ENDOR results is, on the other hand, not consistent with the ESR and UV intensity data[109]. A decision as to which of these alternative models is correct could not be made.

Ab initio calculations for rings larger than five are not available. Force-field calculations are available for the six-, seven- and eight-membered cyclopolysilanes. The permethylated cyclohexasilane adopts the familiar chair structure[119a], $(Me_2Si)_7$ adopts the twist–chair structure[119b] (as does cycloheptane) and $(Me_2Si)_8$ adopts a twist–chair–chair structure[119c] as the lowest-energy conformation (cyclooctane, on the other hand, adopts a boat–chair conformation). References to previous molecular-mechanics calculations of cyclopolysilanes can be found in Reference 119c.

b. Polyhedral silanes. Several polyhedral silanes have been studied very recently (in 1986 and 1987) by *ab initio* methods revealing intriguing properties. Here theory is leading experiment, and many of the theoretical predictions are awaiting experimental testing. Much still remains to be learned about these interesting molecules, theoretically as well as experimentally.

Bicyclo[1.1.0]tetrasilane. This molecule, **21**, the sila-analogue of bicyclo[1.1.0]butane, has been studied in several papers[120–123]. Similar geometries were obtained in all studies which used a polarized basis set and the geometries obtained using 3-21G$^{(*)}$ are shown in Figure 5, structure **21a**. The calculated structures are in reasonable agreement with the X-ray structure of **22**[108d], taking into account the heavy substitution in **22**.

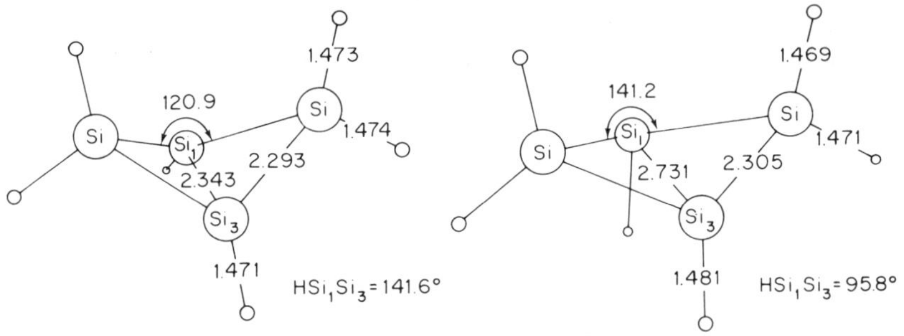

(21a) (21b)

FIGURE 5. 3-21G$^{(*)}$ structures of bicyclo[1.1.0]tetrasilanes. Reproduced by permission of VCH Verlagsgesellschaft from Ref. 121.

A careful study of **21** by Schleyer and coworkers has revealed the existence of an unusual 'bond-stretched' isomer **21b**[121]. **21b** is different from **21a** in three main features: (a) the central Si–Si bond is by about 0.4 Å longer in **21b**; (b) the interflap angles between the three-membered ring moieties is larger in **21b** (140° vs. 120°); and (c) the Si–Si–H angles at the bridgehead positions are much smaller in **21b** (90° vs. 140°). Furthermore, at the MC-SCF level the unusual 'bond-stretched' isomer **21b** is 17 kcal mol^{-1} more stable than **21a** with a barrier of only 0.12 kcal mol^{-1} separating the two isomers! **21b** can be regarded as a singlet diradical[120a, 121] (Dabisch and Schoeller[120a] located only **21b** but did not find **21a**; Collins and coworkers[120b] on the other hand located **21a** but not **21b**). It was further shown that methyl substitution at the bridgehead positions reverses the relative stabilities of the dimethylated **21a** and **21b**, the former being now the more stable[121]. Schleyer and coworkers have therefore concluded that **22** adopts a structure of type **21a** only because of the special requirements of the bulky bridgehead substituents[121]. Collins and coworkers studied the inversion barrier in **21a** and found it to be extremely low, 18 kcal mol^{-1} (MP3/6-31G*//6-31G* + ZPE). The bond connecting the bridgehead atoms is not broken during the inversion process, but changes from a bent σ-bond at equilibrium geometry to a π-bond in the transition structure[120b]. Vibrational frequencies for **21a** were also reported[120b].

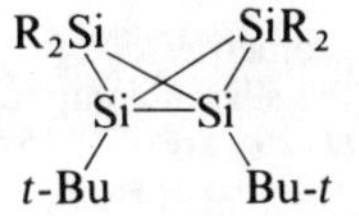

$R = 2,6\text{-}Et_2C_6H_3$

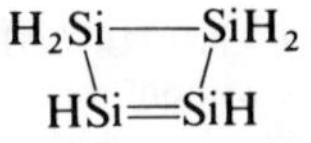

(21) **(22)** **(23)**

The strain energy of **21a** (using appropriate homodesmic reactions) is estimated at MP2/6-31G*//3-21G$^{(*)}$ + ZPE as 60.3 kcal mol^{-1}, by 6.6 kcal mol^{-1} lower than in bicyclo[1.1.0]butane[122] (at 3-21G$^{(*)}$ this conclusion is reversed[121]). Boatz and Gordon also find that **21a** is by 3.4 kcal mol^{-1} more stable than the isomeric **23**[122].

Boatz and Gordon's paper includes also a comprehensive study of the structures, relative energies, bent bonds, and strain energies of silabicyclo[1.1.0]butanes and silacyclobutanes of the general formula $C_nSi_{4-n}H_6$ ($n = 0$–4)[122]. Some of these molecules are discussed in Section IV.C.2.

c. Silicon analogues of tetrahedrane, prismane and cubane. An interesting systematic comparison (at the 6-31G* level) of tetrahedrane, prismane and cubane and their silicon analogues **24**–**26** has been carried out recently by Nagase and coworkers[116]. The 6-31G* calculated geometries of both the carbon and silicon compounds, **24**–**26**, are given in Figure 6.

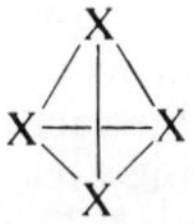

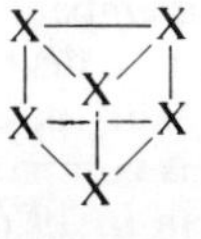

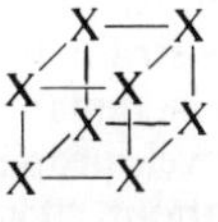

(24) **(25)** **(26)**

(a) X = SiH **(b)** X = CH

The Si–Si bond lengths increase in the order **26a** (2.396 Å) > **25a** (2.359 Å and 2.375 Å) > **24a** (2.314 Å). The Si–Si bond lengths in the three-membered rings are shorter in **24a** than in cyclotrisilane **18** (2.341 Å, 6-31G*), while those in the four-membered rings are longer in **26a** than in cyclotetrasilane **19** (2.373 Å, 6-31G*). Clabo and Schaefer[123] reported a similar geometry for tetrasilatetrahedrane using a dz + p basis set[123].

Table 9, which summarizes the strain energies of **24**–**26** as calculated by the homodesmic reactions 8, reveals an interesting contrast between the carbon (X = C) and silicon (X = Si) compounds. The good agreement between the calculated and experimental strain energies for X = C suggests that these strain energies are reliable also for X = Si where

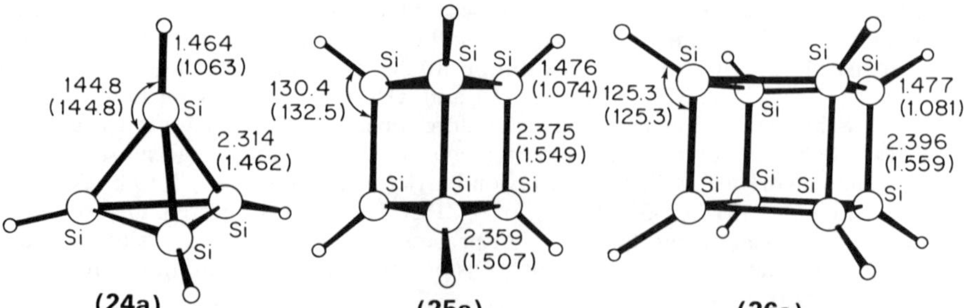

(24a) **(25a)** **(26a)**

FIGURE 6. The HF/6-31G* optimized geometries of **24** (T_d), **25** (D_{3h}) and **26** (O_h) in Å and deg. The values without parentheses refer to the silicon compounds and in parentheses to the carbon compounds. The total energies of **24**–**26** are −1157.82123 (−153.59789), −1736.88739 (−230.50328) and −2315.94264 (−307.39391) hartrees, respectively. Reproduced by permission of the Royal Society of Chemistry from Ref. 116.

TABLE 9. Strain energies (kcal mol^{-1}) calculated from homodesmic reactions 8 at the HF/6-31G* level[a]

Compound	Homodesmic	Experimental[b]
24 (T_d)	140.9 (141.4)	(140.0)
25 (D_{3h})	113.8 (145.3)	
26 (O_h)	93.5 (158.6)	(154.7)
18 (D_{3h})	38.9 (28.7)	(27.5)
19 (D_{2d})	16.7 (26.7)	(26.5)

[a] From Reference 116. Values in parentheses refer to the analogous carbon compounds.
[b] Taken from K. B. Wiberg, *Angew. Chem., Int. Ed. Engl.*, **25**, 312 (1986).

experimental data are unavailable. The strain energies of the sila- and carbon tetrahedranes are very similar, around 140 kcal mol^{-1} (a similar strain energy for **24a** was reported in Ref. 115). However, as three-membered rings are replaced by four-membered rings, the strain in the silicon compounds decreases while the strain in the carbon compounds remains roughly constant. This is expected in light of the fact that the strain in cyclopropane and cyclobutane is comparable, while the strain in cyclotrisilane **18** is by about 20 kcal mol^{-1} higher than in **19** (see discussion above)[115]. Thus **25a** is 32 kcal mol^{-1} less strained than **25b** and silacubane **26a** is 65 kcal mol^{-1} more stable (less strained) than cubane **26b**. This suggests that **26a** is a reasonable target for synthesis and isolation[116].

$$(XH)_n + (3n/2)H_3XXH_3 \rightarrow nHX(XH_3)_3 \qquad (8)$$

The nature of tetrasilatetrahedrane **24a** has been studied in more detail by both Sax and Kalcher[124] and Clabo and Schaefer[123]. At the HF level the two groups find **24a** to be a minimum on the Si_4H_4 potential-energy surface, although it lies higher in energy than either tetrasilacyclobutadiene (by about 6 kcal mol^{-1} at the CI level[123, 124]) or the dissociation products (2 molecules of Si_2H_2). Yet Clabo and Schaefer point to the fact that both conversions are symmetry-forbidden (4π electron) processes and are therefore probably associated with a larger barrier. Thus, these authors conclude that **24a** may be observed experimentally.

2. *Hetero-substituted saturated cyclopolysilanes*

Cyclopolysilanes, in which one or more SiR_2 units are replaced by isovalent units such as O, S, CH_2, etc., exhibit unusual structural features. For example, the X-ray structure of 1,1,2,2-tetramesityldisilaoxirane shows an extremely short Si–Si bond of 2.227 Å, which is closer to that of a 'normal' Si–Si double-bond length (ca. 2.16 Å) than to a 'normal' Si–Si single bond (ca. 2.38 Å), and a nearly planar arrangement around the silicon atoms[125]. Similar unique structural 'puzzles' are exhibited by other hetero-substituted three- and four-membered rings (see below). The structural-bonding questions raised by these compounds attracted considerable interest and have stimulated considerable theoretical activity, which is discussed below. In general, it can be stated that theory has stood up to the challenge posed by experiment and provided a satisfactory bonding model for these intriguing molecules. The two most comprehensive studies of these molecules, also the most recent ones, are those by Grev and Schaefer[126] and by Boatz and Gordon[127]. We will discuss first three-membered rings, then monocyclic four-membered rings and then compounds with more than one ring.

a. Three-membered ring compounds. Early studies of three-membered rings containing silicon concentrated around silacyclopropanes and their substituted derivatives used semiempirical[128] and simple *ab initio* methods[129, 130]. Spiro-substituted silacyclopropanes were studied by both extended-Hückel[110] and CNDO/2 calculations[128b]. The discussion below is based however on two recent extensive *ab initio* studies of the entire series of monosubstituted cyclopolysilanes **27**[126, 127] (**27**, X=SiH_2 has been discussed above). These studies also compare the sila-**27** with the corresponding carbon analogues **28**, and also report calculated vibrational frequencies for these compounds.

(27)

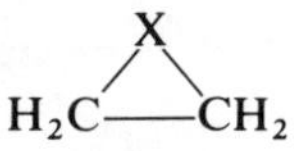

(28)

X = CH_2, NH, O, SiH, PH_2, S

In both studies the structures of compounds **27** were optimized using gradient techniques and a polarized basis set[126, 127]. The optimized geometries of compounds **27** obtained by Grev and Schaefer[126] are shown in Table 10 (those obtained by Boatz and Gordon[127] are similar). The parameters used to describe the SiH_2 group orientations are shown in Figure 7.

The most striking feature of compounds **27** is the unusually short Si–Si distances found in all the minimum-energy structures. The values range from 2.195 Å in $(H_2Si)_2O$ to 2.270 Å in C_s symmetry $(H_2Si)_2PH$, and are all significantly shorter than the 'normal' Si–Si bonds, e.g. 2.345 Å in disilane (at the same level of calculations). The oxygen-substituted ring is unique in that the Si–Si distance in it is much closer to a typical double-bond value (e.g. 2.165 Å found in *trans*-disilene) than it is in disilane. In the other compounds the Si–Si distance is intermediate between the above values, but always shorter than in Si_3H_6. The Si–Si bond lengths correlate well with the electronegativity of the X substituent, so that it decreases with increasing electronegativity of X. In addition, the bonding around the silicon atoms is nearly planar. The available crystal structure data for X=CH_2[131a], S[131b] and O[125] are for molecules that possess large substituents. Nonetheless, the experimental structures agree quite well with the above calculations.

How can one understand the peculiar geometries of the three-membered rings in **27**? Two, somewhat different, viewpoints have been proposed by the two groups which

TABLE 10. Optimized geometries of monosubstituted three-membered rings $(H_2Si)_2X$[a]

	PH						
Parameter	C_s	C_{2v} (planar)	NH	S	O	SiH_2[j]	CH_2
r(Si–Si)	2.270	2.344	2.232	2.249	2.195	2.330	2.251
r(Si–H)[b]	1.4624	1.466	1.467	1.461	1.462	1.466	1.464
r(Si–H)[c]	1.4618						
r(Si–X)[d]	2.281	2.155	1.724	2.156	1.674		1.901
r(X–H)	1.406	1.389	0.999				1.084
⟨HSiH, deg	113.3	108.0	108.8	111.6	110.2	113.1	112.6
θ,[e] deg	22.8	27.2	12.7	19.1	6.7	30	17.5
⟨SiXSi, deg	59.7	65.9	80.7	62.9	81.9	60	72.6
ε,[f] deg	7.3	1.3	12.1	10.2	17.8	0	9.4
θ_r,[g] deg	0.8						
ϕ_t,[h] deg	3.5						
β,[i] deg	83.6						55.3

[a] From Reference 126. Bond distances in Å. For definition of angles see Figure 7.
[b] SiH is on the same side of the ring plane as PH.
[c] SiH is on the opposite side of the plane to PH.
[d] X = PH, NH, S, O, SiH_2, CH_2.
[e] θ = angle between H_2Si plane and Si–Si bond.
[f] Wagging angle: $\varepsilon = 1/2\ (\measuredangle XSiSi) - \theta$.
[g] Rocking angle: $\phi_r = (\alpha_{41} + \alpha_{43}) - (\alpha_{51} + \alpha_{53})$. For definition of α_{ij} see Figure 7.
[h] Twisting angle: $\phi_t = (\alpha_{41} - \alpha_{43}) - (\alpha_{51} - \alpha_{53})$.
[i] β = angle between ring plane and X–H bond.
[j] Reference 109. D_{3h} symmetry.

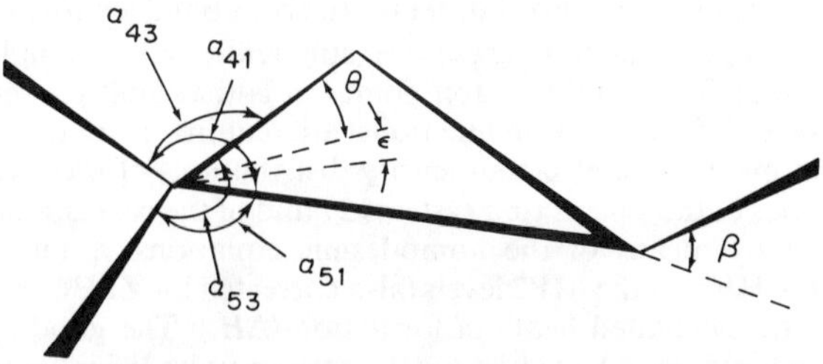

FIGURE 7. Definition of angles used to describe the SiH_2 group orientations in monosubstituted three-membered rings[126]. Reprinted with permission from *J. Am. Chem. Soc.*, **109**, 6577 (1987). Copyright (1987) American Chemical Society[126].

have studied the problem[126, 127]. Grev and Schaefer adopted the well-known Dewar–Chatt–Duncanson bonding model[132] for metal–olefin complexes shown schematically in **29**.

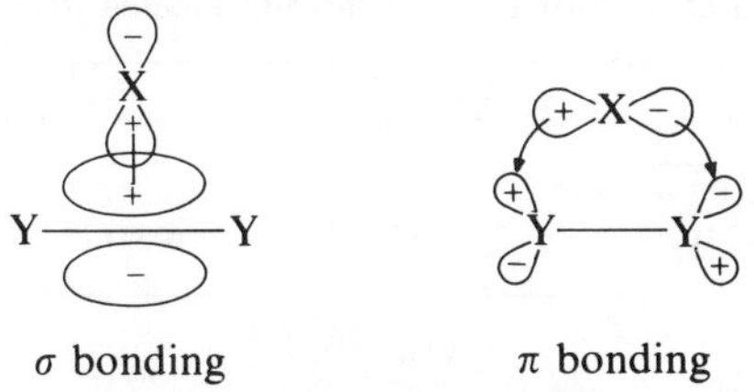

(29)

According to this model the bonding is described in terms of two components. First, a σ-bond is created between the metal and the disilene by donation of π-electron density from the disilene into an empty orbital of appropriate symmetry on the X group (a metal in the original Dewar model). Second, the π* orbital of the disilene is the recipient of electron density from a filled orbital on X of appropriate symmetry. In terms of this model the unusual geometries of **27** and the related, experimentally studied systems are a direct result of decreased back-donation from the substituent into the π*-orbital of the Si_2H_4 fragment. The relative back-donating capacity of the substituent determines whether the basal bond length is shorter or longer than in the unsubstituted ring. More electronegative substituents will lead to shorter distances, as a result of decreased repulsions in the basal fragment, and electropositive substituents will have the opposite effect[126].

Boatz and Gordon[127] suggested a different explanation for the unusual geometries of **27**. They have calculated 'bent' bond lengths according to the definition of Bader and coworkers[133]. In this formalism a bond is defined as the path of maximum electron density (MED) connecting two atoms. Normally, the MED path coincides with the internuclear axis, but in highly strained systems it may curve away from the internuclear axis[133]. It was suggested that in such cases the length of the MED path (λ) may be regarded as the proper measurement of the bond length, rather than the internuclear distance (L)[133]. The λ values for **27** and for the analogous carbon compounds were calculated. It was found that the Si–Si bonds in the rings **27** are more bent than the C–C bonds in the analogous carbon rings. The displacement of the MED path away from the Si–Si internuclear axis can be as large as 0.3 Å and the differences between λ and L as large as 0.15 Å It was concluded[127] that the short Si–Si internuclear distances in **27** are a consequence of severe bond bending, rather than the result of significant π character in the

Si–Si bond as suggested in Reference 126. Thus, the bent bond lengths λ all fall within the range 2.31–2.38 Å, suggesting that these are still really single bonds. An important operational point that Boatz and Gordon point out is that multiple sets of polarization functions (i.e., two sets of polarization functions) are required in all heavy-atom basis sets in order to obtain meaningful electron-density distributions of these compounds[127].

The strain energies of the silicon ring systems **27** and of the corresponding carbon rings **28** were evaluated by means of the homodesmic equations 9. The calculated strain energies, both at the HF and the MP2 levels (also corrected for ZPE), are given in Table 11 which also gives the calculated heats of formation (ΔH_f). The good agreement for the calculated and experimental ΔH_f values for the carbon rings **28** lends support also to the calculated values for the silicon rings **27**, where experimental data are not available. Also, the differences between the HF and MP2 strain energies are small.

$$\mathbf{27}\ (\text{or}\ \mathbf{28}) + 3\ H_3SiSiH_3\ (\text{or}\ H_3CCH_3) \rightarrow H_3SiXSiH_3\ (\text{or}\ H_3CXCH_3) + 2\ H_3SiSiH_2SiH_3\ (\text{or}\ H_3CCH_2CH_3) \tag{9}$$

TABLE 11. Strain energies and heats of formation for three-membered rings[a, b]

Molecule		Strain energy, ΔE^c			Heat of formation, ΔH_f		
		6-31G*//6-31G*	MP2/6-31G*//6-31G*	Obs.[d]	6-31G*//6-31G*	MP2/6-31G*//6-31G*	Obs.[d]
	X=						
27	CH_2	44.3	42.6		49.1	47.7	
	NH	47.2	44.5				
	O	56.4	50.8				
	SiH_2	37.5	35.6		70.2	68.3	
	PH	30.8	28.4				
	S	28.9	26.7				
	X=						
28	CH_2	25.9	27.3	26.0	16.6	18.0	16.6
	NH	22.9	23.2	23.1	33.9	34.2	34.1
	O	20.9	19.9	19.8	−8.6	−9.6	−9.7
	SiH_2	43.1	40.9		38.6	36.4	
	PH	22.4	20.0				
	S	16.3	15.5	17.1	21.6	20.8	22.4

[a] From Reference 122.
[b] All values are in kcal mol^{-1}.
[c] As calculated from ΔE of the homodesmic reactions 9, including the corrections of zero-point vibrational energies.
[d] References to the experimental values can be found in Reference 122.

The three-membered silicon rings **27** generally have higher strain energies than their carbon counterparts **28**. The strain difference between **27** and **28** for X=O is particularly large, 29 kcal mol^{-1}. Generally, the strain energy decreases as the electronegativity of X increases (for X within a given period), except for **27**, X=CH_2, NH and O. The highest strain energy in both series, 56 kcal mol^{-1}, is calculated for **27**, X=O. An interesting observation is that, for all silicon rings, the HOMO is a Si–Si σ-bond, while in the carbon rings the HOMO is a lone pair on X (for X=NH, O, PH and SH).

b. Four-membered ring compounds. Four-membered ring compounds containing one heteroatom of the general formula **30**, X = CH_2, NH, O, SiH_2, PH and S, were studied by Grev and Schaefer[126]. They report optimized geometries and vibrational frequencies using a polarized basis set. In contrast to the analogous three-membered ring compounds **27**, the four-membered ring compounds **30** have fairly normal structures and do not appear to be exceptional in any way[126].

In contrast to **30**, the series **31** containing two heteroatoms in the 1,3-positions exhibit remarkable structures. The interest in these systems started when West solved the X-ray structure of compound **32** and noted that the Si · · · Si distance is extremely short, 2.31 Å[134a], the normal Si–Si single bond length of 2.34 Å being only slightly longer than this formally nonbonded Si–Si distance. West proposed two possible models to explain the short Si–Si distance. According to the first model, severe lone-pair repulsions between the oxygens force a contracted Si–O–Si angle and thus 'push' the two Si atoms closer together. The second bonding model suggests the existence of a localized two-electron Si–Si bond with four-center–six-electron delocalized bonding about the periphery (**33b**). Combination of these two possibilities, as represented by the resonance between the classical structure **33a** and **33b**, can represent the structure of the molecule.

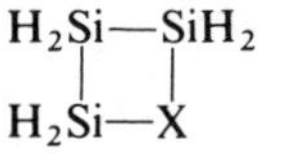

(30)

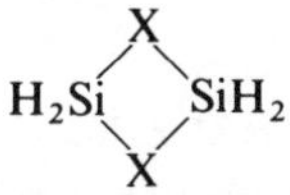

(31)

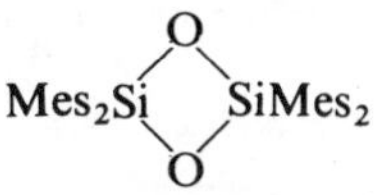

(32) Mes = Mesityl

(33a) ⟷

(33b)

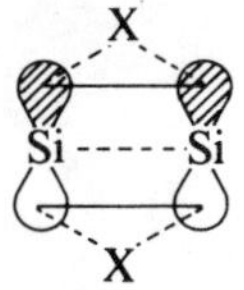

(33c)

The structural puzzle posed by the 1,3-hetero-substituted cyclotetrasilanes is further complicated by the fact that, in other derivatives (**31**, X = P–R[135a], N–R[135b], S[135c] and CR_2[135d]), the nonbonded Si · · · Si distances vary widely, but all of them are significantly longer than normal single-bond ranges (with the possible exception of X = NR). Table 12 lists the experimentally determined Si · · · Si distances and Si–X–Si bond angles in a variety of rings of type **31**.

The first to calculate a derivative (i.e. tetrahydoxy) of **33**, i.e. tetrahydroxy, and to reproduce the peculiar structure of the ring, were O'Keeffe and Gibbs in 1984[95c]. This paper preceded the experimental report by West and coworkers[134] and serves to demonstrate the predictive power of *ab initio* calculations. O'Keeffe and Gibbs did not analyze the bonding in this compound. In 1985 Bachrach and Streitwieser[136] and Kudo and Nagase[137] addressed the bonding question almost simultaneously. Calculations which included, in addition to O and S, also other X substituents followed later[138]. In 1987, Grev and Schaefer[126] also analyzed the bonding in **31**. In addition to structures, the above papers[136, 137] also report the calculated vibrational frequencies.

The first important conclusion from all the theoretical studies is that the calculations can reproduce fairly well the experimental structures, provided that polarized basis sets are used for geometry optimizations. The large effect that the addition of polarization

TABLE 12. Experimentally determined nonbonded Si . . . Si distances (in Å) and angles (in degrees) in substituted planar 1,3-disubstituted four-membered rings $(R_2SiX)_2$ (**31**) and calculated Si–Si distances in the parent monosubstituted three-membered rings $(H_2Si)_2X$ (**27**)[a].

X	$(R_2SiX)_2$ [b]			$(H_2Si)_2X$[a]
	r(Si · · · Si)	∡ (Si–X–Si)	Ref.	r(Si–Si)
O	2.31–2.40	86.2–91.1	134a,b	2.195
NH	2.48–2.56	91.7–94.3	135b[c]	2.232
CH_2	2.62	85.9	135d	2.251
S	2.81–2.84	81.1–82.5	135c[c]	2.249
PH	3.2	89.2–92.9	135a	2.270
SiH_2	3.36	90.0	109	2.330

[a] From Reference 126.
[b] A range of values is given in cases for which the structure of more than one derivative is available.
[c] See also references cited therein.

TABLE 13. Geometrical parameters and energies of cyclodisiloxane (**33**)[a]

Parameter	3-21G	3-21G$^{(*)}$	3-21G*	6-31G*[b]	(expt.)[c]
		Bond distance (Å)			
Si–O	1.717	1.666	1.653	1.671	1.66, 1.72
Si–Si	2.539	2.420	2.359	2.394	2.31
O–O	2.312	2.290	2.313	2.332	2.47
		Bond angles (deg)			
Si–O–Si	95.4	93.2	91.1	91.5	86
O–Si–O	84.6	86.8	88.8	88.5	94
Energy[d]	−726.131706	−726.298447	−726.262679		

[a] From Reference 136.
[b] See Reference 137.
[c] See Reference 134.
[d] In hartrees.

functions has on the geometry, particularly on the Si · · · Si distance, can be seen clearly in Table 13, which presents the results of calculations for **33**.

Having established the relevance of the calculations, let us analyze the bonding in these compounds. Bachrach and Streitwieser[136] attributed the peculiar structure of 1,3-cyclodisiloxane **33** to the strong repulsions between the oxygens. These repulsions result from electrostatic interactions (each oxygen carrying more than a full negative charge) coupled with extreme lone-pair–lone-pair repulsions[136]. Furthermore, there is very little silicon–silicon bonding or attraction. This can be demonstrated by examining the projected electron-density maps of the highest six occupied MOs which are shown in Figure 8[136]. None of these MOs is a Si–Si bonding orbital; they are either Si–Si nonbonding or Si–Si antibonding. MOs 22 and 23 show large oxygen–oxygen lone-pair repulsions, one in the molecular plane (i.e. σ-interactions) and one perpendicular to this plane (i.e. π-interactions). Thus overall, Bachrach and Streitwieser's calculations prefer the bonding picture represented in the resonance structure **33a**[136]. O'Keeffe and Gibbs also

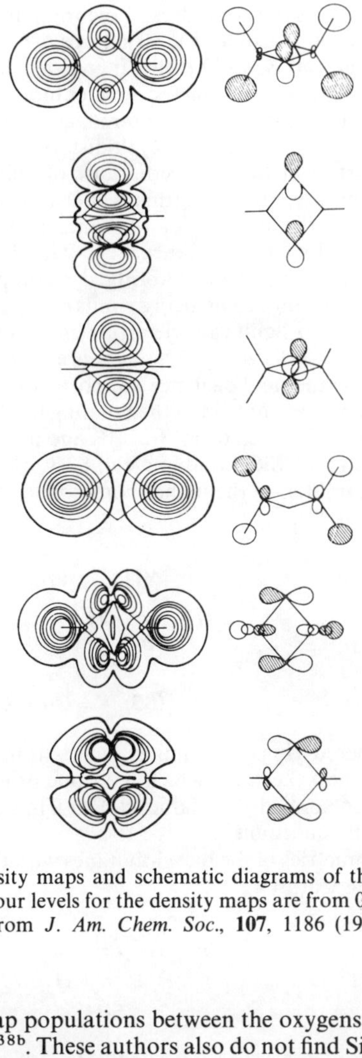

FIGURE 8. Projected electron-density maps and schematic diagrams of the highest six MOs of cyclodisiloxane **33** at 3-21G$^{(*)}$. Contour levels for the density maps are from 0.001 to 0.101 by 0.01 e au^{-2}. Reprinted with permission from *J. Am. Chem. Soc.*, **107**, 1186 (1985). Copyright (1985) American Chemical Society[136].

report strong antibonding overlap populations between the oxygens in **33**, and conclude that there is no Si to Si bonding[138b]. These authors also do not find Si to Si bonding in **31**, X=S.

Kudo and Nagase reached slightly different conclusions, although they also find that there is very little silicon–silicon bonding in **31**, X=O or in **31**, X=S[137]. Kudo and Nagase reject the lone-pair–lone-pair repulsions model, because in their view such interactions are expected to lead to unusual C–C distances in $(H_2CO)_2$, which is not the case. They suggested that the short Si $\cdots$ Si distance in **31**, X=O is the result of the very strong attraction between the silicon and oxygen atoms and the high ionic character of the Si–O bonds[137]. The fact that in **31**, X=S (where the Si–X bonding is much weaker), the Si–Si distance is 0.46 Å longer than that in **31**, X=O was cited as supporting this model[137].

The most recent paper that analyzes the bonding in **31** reached yet another conclusion[126]. Thus, Grev and Schaefer concluded that the Dewar–Chatt–Duncanson model[132] that rationalizes the structures of the three-membered ring systems (see above) can be extended to account also for the structures of compounds **31**[126]. According to the authors, the advantage of this model is that it gives a coherent bonding picture for both **27** and **31**. The model is, in essence, a one-parameter model; only a knowledge of the relative electronegativities of the ring atoms is required, although the sizes of the atoms may also be important. According to this model, cyclodisiloxane **33** is an extreme case in series **31** and exhibits the shortest Si–Si distance because of the large electronegativity difference between the ring atoms. According to this model an unusual amount of electron density should concentrate in the interior of the ring, and this receives support from the deformation density plots in Reference 136. The Dewar model also suggests that compounds **31** are best described as 'dibridged π-complexes', as shown in **33c**, with the bonding between the silicon atoms being similar to that of a π-bond unsupported by a σ-bond, so that there is no significant electron density directly between the silicon atoms. Recent experiments by West's group, who measured the ^{29}Si NMR coupling constants for several 1,3-cyclodisiloxanes and disilenes, support either the 'dibridged π-complex' **33c** or the **33a** bonding models but contradict the bonding model suggested by structure **33b**[134c].

1,2-silaoxetane **34** was studied using 3-21G*, and it was found to be planar with a Si–O distance of 1.643 Å and a CSiO bond angle of 82.3°[136]. Kudo and Nagase also reported the structures and stabilities of the higher homologues of **33**, namely the trimer $(H_2SiO)_3$ and tetramer $(H_2SiO)_4$[137].

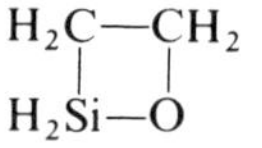

(**34**)

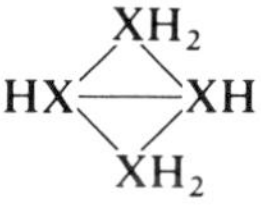

(**35**) X=C or Si

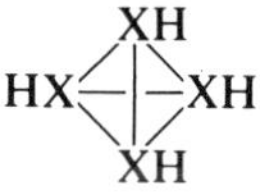

(**36**) X=C or Si

Other four-membered rings containing heteroatoms have been studied. These are mostly polycyclic carbon skeletons which have one or more silicon atoms. For example, Boatz and Gordon[122] studied the silabicyclo[1.1.0]butanes **35** and the silacyclobutenes **36**, containing 1–4 silicon atoms.

The calculated geometries of the bicyclobutanes with two heteroatoms (**37a–c**) exhibit a number of interesting features.

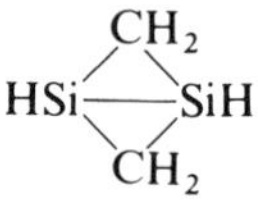

(**37a**)

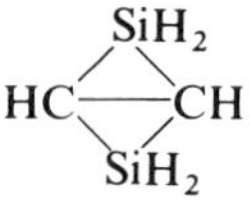

(**37b**)

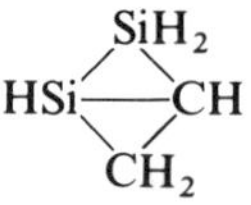

(**37c**)

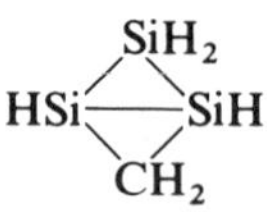

(**37d**)

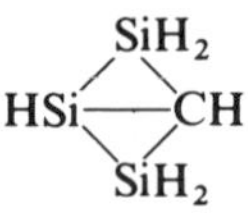

(**37e**)

In **37a** the calculated (3-21G*) Si–Si distance is 2.114 Å, less than the Si–Si double-bond length of 2.12 Å in disilene!! This contrasts with the predicted C–C bridge bond length of **37b** which is over 0.2 Å longer than the C–C bond in the ethane. Compound **37c** also has an unusual geometry. The Si–C bridge bond of 2.225 Å is stretched by about 0.3 Å compared to the Si–C bond in methylsilane. The two peripherial Si–C bonds in this molecule are quite different, 1.828 Å and 2.019 Å. The calculated strain energies of these compounds are (MP2/6-31G*//3-21G*, kcal mol^{-1}): **37a** (89.0) > **37b** (74.1) > **37c** (65.3)[122].

The two possible trisilabicyclo[1.1.0]butanes **37d** and **37e** were also studied[122]. In **37d**, the Si–Si bridge bond distance is 2.22 Å, intermediate between the representative lengths of 2.12 Å and 2.34 Å of Si–Si double and single bonds, respectively. The calculations show that all bonds in **37d**, in particular the bridge bond, are highly bent. In **37e** only the peripheral Si–Si bond (2.247 Å) is highly bent. **37d** and **37e** have very similar stabilities and their calculated strain energies of 72.9 and 71.1 kcal mol^{-1}, respectively, are similar to that of **37c**[122]. The higher homologue tetrasilabicyclo[1.1.0]butane, **21**, is discussed is Section IV.C.1.b.

Mono- (**38a**) and di-silatetrahedrane (**38b**)[139b] were also studied. The mono-silatetrahedrane is predicted to be higher in energy than silacyclobutadiene by over 30 kcal mol^{-1} at 6-31G*//3-21G. However, the dissociation products, acetylene and silaacetylene, are calculated to be 23 kcal mol^{-1} less stable than silatetrahedrane[139a]. The disilatetrahdrane (**38b**) was found to be a local minimum on the $Si_2C_2H_4$ surface, but it lies 87 kcal mol^{-1} (MP3/6-31G*//3-21G) higher than the global minimum[139b]. Tetrasilatetrahedrane (**24**, X = SiH) is discussed in Section IV.C.1.b.

(38a)

(38b)

c. Miscellaneous silicon ring compounds. Aromatic and antiaromatic compounds containing silicon, and rings containing multiply bonded silicon atoms are discussed in Sections VI and V.A.1.c, respectively. Here we mention briefly other miscellaneous ring compounds.

Molecular mechanics was used to calculate several rings containing silicon, emphasis being placed on the prediction of conformation and geometry[58–60]. Unfortunately, due to scarcity of reliable experimental data it is difficult to determine the reliability of the calculated heats of formation[58, 59a]. In general, the predicted structures agree well with experiment, e.g. the geometries of silacyclohexane and silacyclopentane[58, 59]. 1-methylsilacyclohexane was predicted to prefer the *axial* over the *equatorial* conformation[59a]. Quellette and coworkers[59b] found that even a *t*-butyl group will not ensure conformational stability to a silacyclohexane, in contrast to the case with cyclohexane derivatives. On the other hand, the geometries of highly strained systems such as silacyclobutane are not accurately predicted using molecular mechanics[59c].

Several strained organosilicon compounds were studied by semiempirical and *ab initio* methods. Silacyclobutane and its 1,1-dimethyl derivative were studied by MINDO/3[54]. Substituent effects on the ring stability of silacyclopropane were studied with CNDO/2[128b]. It was concluded that a spiro substituent stabilizes silacyclopropane (and this was established experimentally[140a]) by increasing the s-character in the sp-hybrids forming the C–C bond of the silirane ring[128b]. On the other hand, Mollere and Hoffmann[140b] using EH suggested that a σ–σ hyperconjugative effect is responsible for the enhanced stability[140b]. It may be interesting to reinvestigate these systems using more reliable *ab initio* techniques. Several *ab initio* calculations are available for silacyclopropane and silacyclopropene[129, 141]. These studies usually investigated (in addition to the geometry) the stability of these rings relative to other C_2H_4Si or C_2H_6Si isomers. Silacyclopropane, for example, is at 6-31G*//3-21G more stable than $CH_2{=}CHSiH_3$ and $CH_3\ddot{S}iCH_3$ by 16 and 2.5 kcal mol^{-1}, respectively[141a]. Silacyclopropene was studied by both pseudopotential[141b] and *ab initio* methods[129, 141c]. All studies find silacyclopropene to be substantially (ca. 15–17 kcal mol^{-1}) higher in energy than $H_3SiC{\equiv}CH$. An interesting finding pointed out by Gordon is that the introduction of unsaturation in the silicon ring increases the strain by only 3.7 kcal mol^{-1}(i.e., silacyclopropene vs. silacyclopropane), while in the case of cyclopropene the strain was increased by nearly 30 kcal mol^{-1} relative to cyclopropane[129]. CNDO/2 calculations for silacyclopropene, 1,1-difluorosilacyclopropene and various methyl derivatives are also available[128a].

Ab initio calculations at the 6-31G*//3-21G level of the geometry of silaspiropentanes, with either carbon or silicon atoms at the spiro-position, all other ring atoms being silicon, have also been performed[142a]. The interesting conclusion is that the spiro carbon atom is highly nucleophilic and this may account for the difficulty in its preparation. When the spiro atom is silicon, the charge separation is much smaller and the compound is stable. It was also shown that these silaspiropentanes are quite stable to dissociation to disilene and cyclic methylene or silylene[142a]. Other ring systems studied computationally include silacyclobutadiene[139] and various cyclic $C_2H_4Si_2$ isomers[142b]. Some of these molecules will be discussed in Section VI.D in relation to their antiaromatic character. Sila-substituted Dewar benzenes are discussed in Section VI.A.

Mixed clusters of silicon and carbon atoms and pure silicon clusters have recently attracted some interest. Grev and Schaefer[143a] have found that the linear and cyclic forms of SiC_2 are very similar in energy, the cyclic structure being slightly favored with the inclusion of correlation effects. Si_2C was predicted to have a C_{2v} structure with a very long Si–Si 'bond' of 2.926 Å[143a, b]. For the Si_3 cluster, the cyclic C_{2v} and D_{3h} (triplet) structures are predicted to be nearly isoenergetic[143b–d]. Ragavachari[143e] studied systematically several larger silicon clusters, Si_n, where $n = 2$–7, 10, and found all to be ground-state singlets.

Polysilanes, although not ring compounds, should also be mentioned here[107]. Due to their large size, only semiempirical methods can be applied to these intriguing compounds. The main feature of theoretical interest in polysilanes is the delocalization of σ-electrons as apparent from their ultraviolet absorbtion spectra[107]. A qualitative discussion of the electronic excited states in polysilanes, which are not well understood, has recently been presented by Wadsworth and West[118]. Bigelow[144a] performed a CNDO/2 (S + DESCI) calculation on *trans* $H(SiMe_2)_5H$ as a model for longer polysilane chains. It was shown that the computations yield energies and transition intensities which parallel the observed optical spectrum. Si to Si bond stabilization due to methyl substitution is attributed in part to ionic bonding enhancements[144a]. Halevi and coworkers[144b] have emphasized the likely importance of d-orbital contributions to higher-lying excitations of hydrogenated and methylated silane chains. Gleiter and coworkers combined photoelectron spectroscopy and MINDO/3 and MNDO calculations of several cyclic ring systems containing Si–Si single bonds and C≡C bonds, and demonstrated that the Si–Si σ-bond can exhibit strong σ/π interactions[144c]. In a very recent paper Michl and coworkers[144d] explored the nature of the electronic excitation in poly(di-*n*-hexylsilane) by INDO/S calculations for $[(CH_3)_2Si]_n$ and $(H_2Si)_m$ ($n \leq 20$, $m \leq 40$), using molecular-mechanics calculations. They found that the first transition corresponds predominantly to σ(HOMO)$\rightarrow\sigma^*$(LUMO) excitation; both orbitals have Si–Si and virtually no Si–H or Si–C character. The computed excitation energy is lower for the all-*trans* than for the all-*gauche* backbones of equal *m* and this is explained in terms of the stronger 1–4 interactions in the *trans* orientation[144d]. It is also found that the polysilane chain is effectively separated into a series of chromophores communicating by rapid energy transfer. The chromophores are all-*trans* chain segments and they are separated by *gauche* links[144d]. In another recent paper, semiempirical band calculations were reported for $-(SiXY)_n-$ polysilane model compounds in which X and Y are H, Me, Et, Pr and Ph substituents, and the electronic structure of these poly(organosilanes) including the conduction band was discussed[144e]. Further theoretical studies of this interesting and practically important group of compounds will be most welcomed.

V. MULTIPLE BONDS TO SILICON

For many years the synthesis of stable compounds containing multiple bonds to silicon was one of the major challenges of organosilicon chemistry. The many failures to synthesize such compounds attracted the attention of theoreticians who tried to understand the reasons for the dramatically different behavior of carbon and silicon in forming multiple bonds. In the seventies, considerable evidence pointing to the existence of compounds with multiple bonds of silicon as reactive intermediates was accumulated[10–12]. The breakthroughs occurred in 1981 when the syntheses of stable compounds containing C=Si and Si=Si bonds were reported by Brook and coworkers[13] and by West and coworkers[14], respectively. The synthesis of stable compounds containing Si=N bonds followed in 1986[145]. At the same time there was a surge in the theoretical activity in this field and these calculations played an extremely important role in the shaping and development of this field. The subject of multiply bonded silicon compounds has been reviewed extensively, most recently in 1986 by Gordon (theory)[21] and in 1985[19] as well as in the present volume by Raabe and Michl (theory and experiment). As these reviews are very recent this allows us to discuss mainly the newer calculations which were published between 1985 and mid-1987 and to emphasize points which, in our opinion, are important but have received little coverage in the previous reviews. In general, the reader should consult the above-mentioned reviews[19, 21] for a comprehensive discussion of the older literature. We will review first doubly bonded molecules, then triply bonded species and then the aromatic silicon compounds. Most recently the chemistry of Si=Si bonds was reviewed by West[376].

A. Doubly Bonded Silicon Compounds

In this section we first review some of the common properties and the general trends that are found for Si=X double bonds, where X = C, Si, N, P, O, S, and then discuss individually, in more detail, each of these double bonds. There are two recent extensive systematic studies of such bonds: one by Pople, Apeloig, Schleyer (PAS) and their coworkers[146] and the second by Gordon's group[147]. PAS studied the series H_2SiX (X = Be, BH, CH_2, SiH_2, NH, O) and Schmidt, Troung and Gordon studied the compounds H_2SiX (X = CH_2, SiH_2, NH, PH, O, S). Both studies concluded that d-functions on silicon are essential for describing correctly the geometry of these molecules and in some cases (e.g. $H_2Si{=}NH$) a polarized basis set on X is also required. It was found that, in general, addition of polarization functions on silicon considerably shortens the Si=X double-bond distances.

The most interesting general conclusions reached from these studies are the following:

(1) All the Si=X bond lengths are considerably shorter than the corresponding Si–X single-bond distances, indicating significant double-bond character in these molecules. Representative Si=X and Si–X bond lengths are given in Table 14.

(2) While the structural differences in saturated silyl and methyl compounds are not large[68], for unsaturated compounds there are often large (even dramatic) structural differences between unsaturated carbon and silicon molecules, both in the ground state and in the excited state. See, for example, the discussion on HSi≡SiH in Section V.B below.

TABLE 14. Bond lengths and distance shortening on going from single to double bonds[a,b]

Bond	Single bond	Double bond	Shortening
C–C	1.534	1.315	0.219
Si–C	1.889	1.695	0.194
Si–Si	2.342	2.116	0.226
C–N	1.460	1.248	0.212
Si–N	1.721	1.572	0.149
C–P	1.858	1.650	0.208
Si–P	2.256	2.050	0.206
C–O	1.404	1.181	0.223
Si–O	1.636	1.494	0.142
C–S	1.815	1.593	0.222
Si–S	2.139	1.927	0.212

[a] From Reference 147.
[b] At 3-21G*, all values in Å. Similar values are obtained at 6-31G*, see Reference 175b.

(3) There are dramatic differences between unsaturated carbon and silicon compounds in energy separations between various states and possible isomers. For the carbon species, forms with π-bonds are favored over those possessing nonbonding electrons (e.g. acetylene is favored over vinylidene carbene). On the other hand, the ground states of most of the silicon species prefer to have one more nonbonding electron on silicon rather than a double bond (e.g. H_2Si_2 prefers the silylene structure $H_2Si{=}Si$: to the conventional linear $HSi{\equiv}SiH$ structure)[146].

The small energy differences between double-bonded and divalent species is a general important feature of silicon compounds. It indicates that silicon is reluctant to form doubly bonded compounds and this is responsible for many of the differences between carbon and silicon chemistry. It can be generalized that for $n=0$–2, *$H_2C{=}XH_n$ is much more stable than $HCXH_{n+1}$, while $H_2Si{=}XH_n$ and $HSiXH_{n+1}$ have similar stabilities.*

(4) The energetic separation between various stable isomeric species is smaller for most of the silicon molecules than for the carbon analogues. An example is $H_2Si{=}O$ for which there are two other stable species within 5.1 kcal mol^{-1} of the ground state[146]. Such small energy separations make the assignment of the ground-state structure uncertain, and without doubt contributes to the high reactivity of the multiply bonded silicon molecules.

(5) Multiple bond energies can be evaluated in several ways, but all reasonable evaluations reach the conclusion that Si=X bonds are considerably weaker than the corresponding C=X bonds[62, 73, 146, 147]. Probably the two best measures of multiple bond energies are the barriers to rotation around the π-bond (not available, of course, for X = O or S) and hydrogenation thermochemical cycles[147]. These values together with those for the corresponding C=X bonds are presented in Table 15[147]. Previous estimates made by Kutzelnigg[62], who used mainly diatomic data, are also given in Table 15. In light of the assumptions made in constructing Table 15, the quoted bond energies can be expected to be of only moderate accuracy. However, the general trends and conclusions reached from the table are expected to remain unchanged also when more reliable and accurate bond energies become available.

Table 15 shows clearly that in a doubly bonded compound the strength of the π-bond decreases when a single second-period atom is substituted for a third-period atom, and it falls even further when a second third-period atom is involved. Thus, the π-bond in $H_2Si{=}SiH_2$ is much weaker than in $H_2C{=}SiH_2$. Exceptions exist, thus the π-bond

TABLE 15. $H_2Si{=}X$ and $H_2C{=}X$ bond strengths[a, b]

Bond	K[c]	Rotation	Hydrogenation	Experiment	Recommended
	Estimated from				
C=C	70	65.4	62	59, 65	65
C=Si	33	35.6	37	38, 39	38
Si=Si	28	22.7	23	26	25
C=N	65	63.3	65	*d*	63
Si=N	*d*	36.4	36	*d*	36
C=P	48	43.1	45	*d*	43
Si=P	*d*	28.7	35	*d*	29
C=O	90		77	*d*	77
Si=O	40		50	*d*	50
C=S	63		53	52	52
Si=S	*d*		50	*d*	50

[a] Calculated data from Reference 147 unless otherwise stated.
[b] All values in kcal mol^{-1}.
[c] Kutzelnigg values from Reference 62.
[d] Not available.

strengths in Si=S and Si=O are about the same. Of the three third-period elements studied (S, P, Si), *Si forms the weakest π-bonds*. It is interesting to note that several π-bonds involving Si which are very poorly characterized experimentally[19], such as Si=O, Si=P and Si=S, are actually stronger bonds than the better characterized Si=Si bond.

(6) In general, it is found that a Si=X double bond is much weaker than two corresponding Si–X single bonds. While this is true also for the analogous carbon compounds, the double–single bond strength differences are much smaller in the carbon than in the silicon compounds. This is certainly one factor that is responsible for the high reactivity of doubly (or triply) bonded silicon compounds. A typical example is the Si=O double bond. The dimerization reaction of $H_2Si{=}O$ (converting two Si=O to two Si–O bonds) is exothermic by about 110 kcal mol^{-1} while the dimerization of $H_2C{=}O$ is nearly thermoneutral (see Section V.A.5).

(7) The high polarity of most Si=X double bonds is another important factor that contributes to their high reactivity.

(8) Most of the studies which were published since 1985 include calculations of the vibrational frequencies and these are now available for all the prototype $H_2Si{=}X$ double bonds. Calculated stretching frequencies (3-21G*) for a variety of C–X and Si–X single and multiple bonds are given in Table 16. In general the agreement with experiment, when available, is encouraging, considering the relatively crude level of theory that was used.

1. Silenes

Quantum-mechanical calculations played a particularly important role in the development of our understanding of silenes, i.e. carbon–silicon double bonded compounds (**39**). Schaefer's excellent review published in 1982 outlined clearly points of disagreement between theory and experiment[20]. This important review played a central role in the development of theory in organosilicon chemistry as it attracted many theoreticans, including this author, to this area of chemistry. We now know that in all the three points of disagreement between theory and experiment which were pointed out by Schaefer it was theory which gave the more reliable answers (see below).

Table 16. Stretching frequencies for C–X and Si–X single and double bonds (in cm^{-1})[a]

Bond	Exp.[b]	Calc.[c]	Bond	Exp.[b]	Calc.[c]
$H_3C–CH_3$	995	1046	$H_2C=CH_2$	1623	1853
$H_3Si–CH_3$	700	736	$H_2Si=CH_2$	985	1080
$H_3Si–SiH_3$	432	466	$H_2Si=SiH_2$	630[d]	653
$H_3C–NH_2$	1044	1133	$H_2C=NH$	1638	1898
$H_3Si–NH_2$	671[d]	907	$H_2Si=NH$	*e*	1260
$H_3C–PH_2$	676	747	$H_2C=PH$	*e*	1109
$H_3Si–PH_2$	454	485	$H_2Si=PH$	*e*	673
$H_3C–OH$	1033	1186	$H_2C=O$	1746	2033
$H_3Si–OH$	859	951	$H_2Si=O$	1202	1394
$H_3C–SH$	710	778	$H_2C=S$	1059	1196
$H_3Si–SH$	501[d]	545	$H_2Si=S$	*e*	787

[a] From Reference 147.
[b] References can be found in the footnotes to Table 4 in Reference 147.
[c] At 3-21G*.
[d] Taken from a substituted compound.
[e] Not available.

This very impressive success of theory stimulated additional theoretical studies and certainly played a major role in bringing the power of molecular-orbital calculations to the attention of many of the experimentalists working in this field. It also played an important role in shaping the way in which research in this field is conducted.

$R^1R^2C=SiR^3R^4$ **(39)** $H_2C=SiH_2$ **(40)**

a. The parent $H_2C=SiH_2$. Numerous calculations have been reported for the parent silene **40**[106, 146–165]. Some of the results are presented in Table 17 and it is (in my view) interesting to follow the historical development of the understanding of this basic system.

The first *ab initio* calculations of $H_2C=SiH_2$ which were published in 1976 used an STO-4G basis set and found the triplet state to be 1.4 kcal mol^{-1} more stable than the singlet state[148]. In later calculations the same authors estimated that the inclusion of correlation energy would place the singlet about 16 kcal mol^{-1} lower than the triplet[149]. More elaborate calculations that followed established unequivocally that the ground state of **40** is a singlet state, with the triplet state lying much higher in energy, i.e. by approximately 36 kcal mol^{-1} (see Table 17). Thus the ground-state properties of **40** should be discussed in terms of the singlet state.

Schaefer[20] pointed out three major disparities between theory and experiment regarding the C=Si bond. The first concerns the C=Si bond length, the second concerns the isomerization barrier for its rearrangement to the corresponding silylene and the third is the question of the relative stabilities of **40** and of the isomeric silylene $H\ddot{S}iCH_3$. Let us discuss first the problem of the geometry.

(i) Geometry. A short survey of the interplay of experiment and theory regarding this problem is instructive. As can be seen in Table 17, even relatively simple calculations which use split-valence basis sets predict that the singlet silaethylene is planar with a C–Si bond length ranging around 1.69–1.72 Å. Schaefer pointed out that these results are in conflict with an electron–diffraction report of Mohaffy, Gutowski and Montgomery[166]

TABLE 17. Calculated geometries[a] and the lowest singlet–triplet energy differences [E(T–S)] in the parent silene $H_2C{=}SiH_2$, **40**

State	E(T–S)[b]	r(Si=C)	∡HSiC	∡HCSi	∡HSiH	∡HCH	Twist angle	Pyramidalization	METHOD	Ref.
S		1.637	122.7	123.1			0	0	STO-4G	148
T	−1.4	1.846	109.0	123.0			90.0	30.0	UHF/STO-4G	148, 154
S		1.69	122.5	124.0	115.0	112.0	0	0	CI/dz + d(Si,C) + p(H)	150
T	227.9	1.85			115.0	112.0	90.0	45.0	dz + d(Si,C) + p(H)	150
S			Geometry as in Ref. 148 (see above)						tz for valence, dz for inner shells	149
T	9.6		Geometry as in Ref. 148 (see above)							149
S			Geometry as in Ref. 148 (see above)						CI/STO-3G + d(Si)	151
T	17.0		Geometry as in Ref. 148 (see above)						CI/STO-3G + d(Si)	151
S		1.715	122.5	122.5			0	0	CI/dz + d(Si,C)	152
T	34.7	1.880		122.7	109.1	114.6	90.0	49.7	CI/dz + d(Si,C)	153
S			Geometry as in Ref. 150						CI/dz	155
T	34.7		Geometry as in Ref. 150						CI/dz	155
S		1.705	122.7	122.5			0	0	CI/dz + d(Si,C)	156
S		1.698	123.1	122.5			0	0	psuedopotential + CI/dz + d(Si,C)	157
S		1.692	122.9	122.5			0	0	CI/dz + d(Si,C) + p(H)	159
S		1.718	122.2	122.6			0	0	MP2/6-31G*//3-21G	160
S		1.686	123.2	122.4			0	0	CEPA/dz + d(Si,C) + p(H)	161
T	36.5	1.860	111.4		107.8	114.5	90.0	51.7	CEPA/dz + d(Si,C) + p(H)	161
S		1.718	122.2	122.6			0	0	3-21G	106
S		1.769	122.3	121.7			0	0	MCSCF/3-21G	164
T	36.0	1.916			109.0	116.2	90.0	49.2	MCSCF/3-21G	164
S		1.692			114.3	114.8	0	0	MCSCF[g]	163
T	33.7[c,d]	1.859					90.0		MP4/SDTQ/6-31G*//3-21G(*)	146
T	34.2[c,e,f]	1.885	119.0	122.5		115.0	90.0	50.1	MCSCF/3-21G*//3-21G*	147
S		1.703	122.8	122.4	114.5	115.1	0	0	CI/dz + d	165
S		1.691	122.9	122.5	114.2	115.0	0	0	dz + d	165

[a] Bond lengths in Å, bond angles in deg.
[b] In kcal mol^{-1}.
[c] Including contributions from ZPE.
[d] E(T–S) is 6.5 and 8.2 kcal mol^{-1} at 3-21G and 3-21G(*), respectively (including contributions from ZPG).
[e] E(T–S) at MCSCF/6-31G*//6-31G* is 34.5 kcal mol^{-1} and at SOCI/6-31G*//3-21G(*) it is 34.2 kcal mol^{-1} [147].
[f] r(Si=C) of the singlet at 3-21G* and MCSCF/3-21G* are 1.695Å and 1.743Å, respectively[147].
[g] The basis set used included Dunning–Hay's contracted Gaussian functions augmented with polarization functions on Si and C.

who concluded that the C=Si bond length in $(CH_3)_2Si{=}CH_2$ is 1.83 ± 0.04 Å. In an attempt to resolve the discrepancy Schaefer repeated the calculations using a polarized basis set, but the change in *r*(C=Si) was small[20]. The use of a highly correlated wave function (including all single and double excitations) did not lead to a dramatic effect either. The theoretical *r*(C=Si) lengthened slightly to 1.705 Å[20]. Schaefer also showed, using a polarized basis set, that *r*(C=Si) is practically the same in $H_2C{=}SiH_2$ and $(CH_3)_2C{=}SiH_2$, so that the methyl substituents cannot be responsible for the discrepancy. On the basis of these calculations Schaefer concluded that in $H_2Si{=}CH_2$, *r*(C=Si) = 1.705 ± 0.03 Å, and that the experimental–theoretical discrepancy is so large that the unavoidable conclusion is that the interpretation of the electron-diffraction data[166] was erroneous[20]. Another related difficulty was presented by the X-ray structure of Brook's stable silene **41**[13] which showed *r*(C=Si) = 1.764 Å[167]. However, as **41** is heavily substituted and is also twisted by 16° about the Si=C bond, it is difficult to extrapolate from its structure to the structure of **40**. Therefore, Schaefer did not consider the relatively long C=Si bond in **41** to be in conflict with the theoretical results. In 1984, Apeloig and Karni showed that if the effects of substituents are taken into consideration, the relatively long C=Si bond in **41** is fully consistent with theory (see Section V.A.1.b below)[106,162].

The theoretical–experimental dispute was finally resolved in favor of theory a year later in 1985 when the crystal structure of **42**, which was reported by Wiberg, Wagner and Muller[168], showed *r*(C=Si) = 1.702 ± 0.005 Å. This bond length fits very well the calculated C=Si bond length in $H_2C{=}SiH_2$ and the C=Si bond length predicted for **42** by Apeloig and Karni[106,162] (see below). So, it can be concluded with confidence that high-level molecular-orbital calculations are an extremely reliable tool for obtaining structural information pertaining to organosilicon compounds. This conclusion is very important, because the available structural information in this field is still very limited, and many compounds are unstable and their structures cannot be determined experimentally. This course of events should convince even the most skeptical experimentalists that calculations are a highly reliable and useful tool for obtaining structural information.

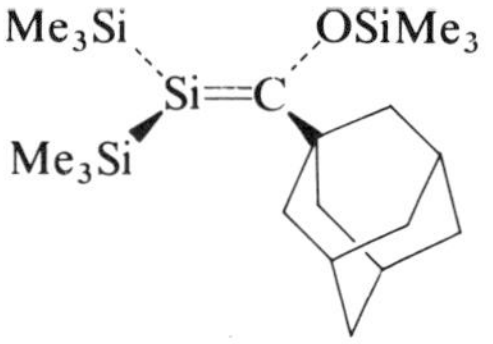

(41)

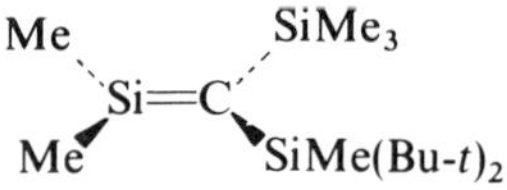

(42)

The highest level of theory applied so far to calculate the geometry of $H_2C{=}SiH_2$ used a CI method and a dz + d basis set. The calculated structure is shown in Figure 9, together with the SCF/dz + d structure[165]. The molecule is planar and the calculated *r*(C=Si) bond length is 1.703 Å (1.691 Å at the SCF level). The HSiC and the HCSi bond angles are very similar, 122.8° and 122.4°, respectively.

Colvin and coworkers[165] have also calculated the infrared spectrum of $H_2C{=}SiH_2$ and the results are presented in Table 18 together with the results of previous lower-level calculations and the experimental results of Maier and coworkers[169]. From this comparison it was concluded that the correlation between theory (harmonic) and experiment (anharmonic) is quite good. All seven predicted harmonic vibrational frequencies lie above their observed anharmonic counterparts, the average percent error being 7.4%[165]. This average error is twice as high as the average error for a series of molecules composed only of H, C, N and O atoms. Thus, it seems that the CI/dz + d level

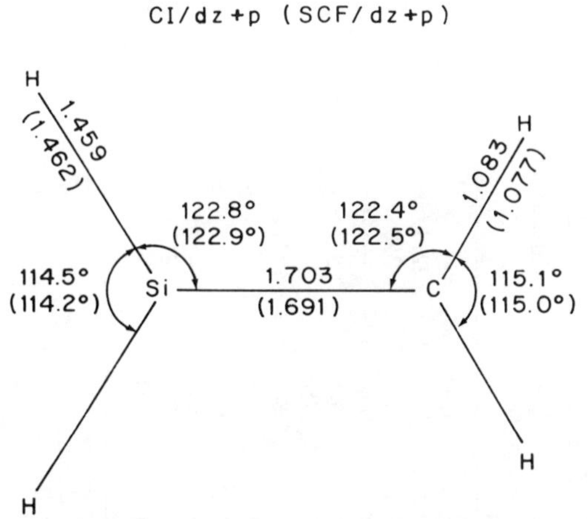

FIGURE 9. Predicted equilibrium geometries for silaethylene. Bond distances are in Å, bond angles in deg. Values without parentheses are from a CI calculation (including all single and double excitations) using a dz + d basis set. Values in parentheses are at the SCF/dz + d level. Reproduced by permission of the American Institute of Physics from Ref. 165.

of theory is not as reliable for H_2C=SiH_2 as for the systems comprised of second-row atoms. Nevertheless, it is clear that such calculated harmonic frequencies can be very helpful in supporting suggested assignments for spectra of molecules whose structures have not been fully established[165].

(ii) Higher energy states. The lowest triplet state of silaethylene has been the subject of several theoretical studies[146–151, 153–155, 161, 164] and the calculated triplet (T)–singlet (S) (S and T always refer to the lowest singlet and triplet states S_0 and T_1, respectively) energy differences are given in Table 17. As mentioned above, the earlier investigations[148] predicted very small T–S separations, but this was due to the relatively low-level theory which was used[149]. More flexible basis sets, polarization functions and correlation energy all increase the T–S energy difference (Table 17). The more recent calculations place the singlet 35–37 kcal mol^{-1} lower than the triplet[147, 153, 155, 161, 164].

As expected, at all levels of calculation the triplet state has a perpendicular geometry and the C–Si bond is significantly elongated relative to the singlet geometry (Table 17)[147, 150, 153, 154, 161, 164]. The best available geometry of the triplet state is probably the one calculated at the MCSCF/3-21G* level and it is shown in Figure 10, which also gives the MCSCF/3-21G geometry[147, 164]. Thus, at MCSCF/3-21G* the C–Si bond length in the triplet is 1.885 Å[147] and, as expected, at MCSCF/3-21G this distance is much longer, 1.916 Å[164]. The carbon end of the triplet is planar and the silicon end pyramidal (the pyramidalization angle being 50.1°). This geometry is consistent with the diradical character of the triplet and the fact that methyl radical is planar, while the silyl radical is pyramidal (see Section VII.C).

The geometry of the perpendicular singlet lowest excited ($^1A''$) state, obtained by a 90° torsion around the C=Si bond and also shown in Figure 10, is similar to that of the triplet having a long C–Si bond and a pyramidal H_2Si group. The perpendicular singlet state is calculated to lie 1 kcal mol^{-1} above the triplet[147, 164].

(iii) The strength of the π-bond in H_2C=SiH_2. The strength of the π-bond in H_2C=SiH_2 can be estimated by calculating (or measuring, this experiment has not been done) the barrier to rotation around the C=Si bond. The $^1A''$ state is the highest point

TABLE 18. Theoretical and experimental vibrational frequencies (in cm^{-1}) for silaethylene, $H_2Si=CH_2$

Symmetry	Normal-mode description	Split valence[a]	Köhler –Lischka[b]	dz + p[c,d]	dz + p[c] CI	Experiment[e]
a_1	sym CH stretch	3096	3325	3320 (2.8)	3254	—
	sym SiH stretch	2479	2304	2411 (45.3)	2388	2219 (m)
	CH_2 bend	1590	1558	1534 (14.0)	1476	1350 (w)
	Si=C stretch	1151	1069	1078 (39.1)	1041	985 (w)
	SiH_2 bend	1061	996	1023 (26.9)	989	927 (w)
a_2	torsion	—	789	783 (0.0)	750	—
b_1	CH_2 wag	—	909	896 (56.8)	801	741 (s)
	SiH_2 wag	—	564	565 (50.0)	438	—
b_2	asym CH stretch	3160	3408	3411 (1.1)	3352	—
	asym SiH stretch	2483	2308	2419 (117.8)	2404	2239 (m)
	CH_2 rock	969	882	915 (102.4)	876	817 (s)
	SiH_2 rock	556	498	532 (9.8)	501	—

[a] H. B. Schlegel, S. Wolfe and K. Mislow, *J. Chem. Soc., Chem. Commun.*, 246 (1975).
[b] From Reference 161.
[c] From Reference 165.
[d] The values in parentheses are the IR intensities in km mol^{-1}.
[e] From References 158 and 169.

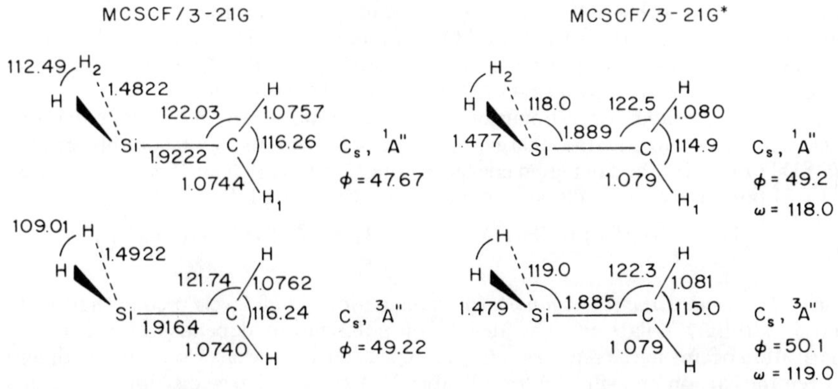

FIGURE 10. MCSCF/3-21G[147] and MCSCF/3-21G* [164] geometries of the twisted singlet ($^1A''$) and triplet ($^3A''$) silaethylenes. Bond distances are in Å, bond angles in deg, ω is the dihedral angle HCSiH and ϕ is the flap angle between the C–Si bond and the H_2Si plane[147, 164].

along the rotation coordinate as shown schematically in Figure 11. The first estimations of these barriers which were based on semiempirical calculations (EH, CNDO) were very low (even less than 10 kcal mol^{-1}), indicating very little stability of the C=Si bond[170, 171] When using *ab initio* methods it was clear that the use of closed-shell HF wave functions for the rotated singlet would lead to erroneously large rotation barriers, which is indeed the case. The first *ab initio* calculations, which used methods appropriate for this problem, were carried out by Ahlrichs and Heinzmann[150]. These authors used spin-restricted singlet open-shell HF to treat the rotated species and two-configuration CI for the planar singlet and obtained a rotational barrier of 45 kcal mol^{-1}. A detailed MCSCF study of the minimum energy pathway for the rotation of the silicon–carbon double bond was reported recently by Schmidt, Gordon and Dupuis[164]. The rotational barrier is found to be around 35 kcal mol^{-1} (35.6[147] and 35.4 kcal mol^{-1} [164] at MCSCF/3-21G* and

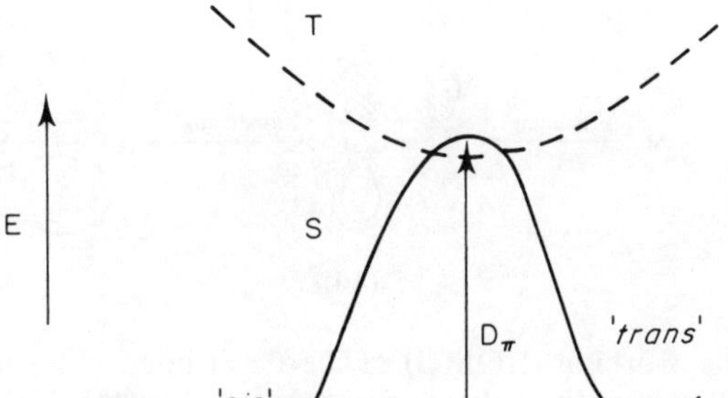

FIGURE 11. A qualitative scheme of the rotational potential-energy surfaces of the lowest singlet and triplet states of $H_2Si{=}CH_2$

MCSCF/3-21G, respectively, including vibration effects), in good agreement with the experimental estimates of 39 ± 5[73b] or 38[172] kcal mol^{-1} for the strength of the π-bond in $(CH_3)_2Si{=}CH_2$. This bond strength is intermediate between that found for the C=C bond (65 kcal mol^{-1}) and the Si=Si bond (22 kcal mol^{-1})[73b] strengths.

A different method for estimating the C=Si bond strength is via equation 10, which compares directly the energies of the C=Si and the C=C double bonds. At MP4SDTQ/6-31G* the calculated energy of equation 10 is 59.6 kcal mol^{-1} showing that the C=Si bond is dramatically less stable than a C=C bond[146].

$$H_2C{=}CH_2(^1A') + SiH_2(^1A') \longrightarrow H_2Si{=}CH_2(^1A') + CH_2(^1A') \qquad (10)$$

(iv) Electronic structure and polarity. Calculations at all levels predict that the C=Si bond is strongly polarized. The degree of polarization depends on the basis set. Polarization occurs in the sense $Si^+{-}C^-$, as expected from the electronegativity difference between the carbon and silicon atoms (Table 2). A typical charge distribution calculated at 6-31G*//3-21G is given in **43**.

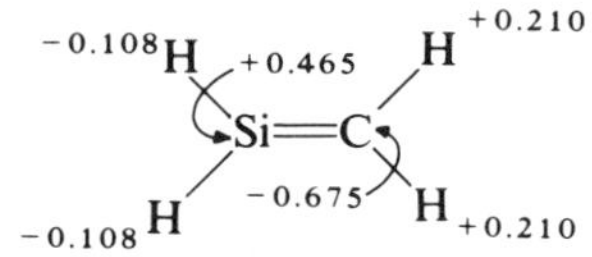

(43)

Ahlrichs and Heinzmann[150] pointed out that this polarity leads to an increase in the π-overlap and the strength of the π-bond (relative to a hypothetical less polarized molecule), since the increased charge expands the valence orbitals at carbon while the positive charge contracts the orbitals of silicon. This reduces the 'mismatching' of the π AOs on C and Si which has been thought to be the main reason for the weakness of the Si–C π-bond. These arguments are presented schematically in **44**, where the dotted and solid lines denote the orbitals sizes before and after the formation of the C–Si bond, respectively. An actual contour plot of the C–Si π-bond is shown in Figure 12.

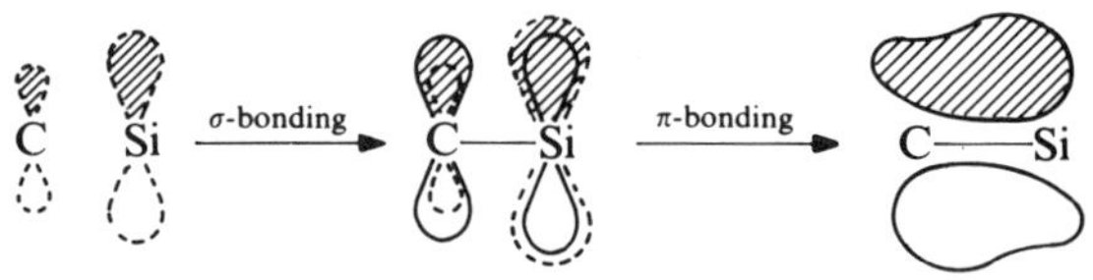

(44)

The energy of the π-orbital (HOMO) of the C=Si bond is higher, and that of the π^*-orbital (LUMO) lower, than those in ethylene. Thus the first measured vertical ionization potential of $H_2C{=}SiH_2$ is at 8.85 eV, in agreement with calculations which predict 8.95 eV, about 2.5 eV below that of ethylene[19]. The band in the electronic spectrum near 245–260 nm assigned to the $\pi \rightarrow \pi^*$ (HOMO–LUMO) transition is at significantly lower energies than the analogous transition in simple alkenes, reflecting the weaker π-interaction in $CH_2{=}SiH_2$[19]. This lowering of the HOMO–LUMO gap is probably related to the higher reactivity of silaethylene, as discussed below in Section V.A.1.a.vi.

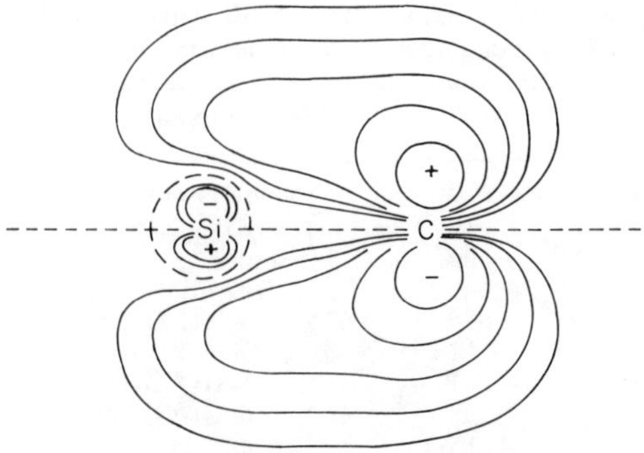

FIGURE 12. Contour plot of the bonding carbon–silicon π-orbital in $H_2Si{=}CH_2$. Lines plotted correspond to the following values: $\psi = 0.0$ (broken), $\pm$ 0.05, ± 0.07, ± 0.1, ± 0.14, ± 0.2 (in au). Reprinted with permission from *J. Am. Chem. Soc.*, **99**, 7452 (1985). Copyright (1985) American Chemical Society[150].

(v) Relative stability with respect to other isomers. Silaethylene can rearrange by a 1,2 hydrogen shift to either methylsilylene **45** or silylmethylene **46** (equation 11). The questions regarding the thermodynamics and kinetics of these intramolecular rearrangements form the second and third points of major disparity between theory and experiment and were pointed out and discussed in detail in Schaefer's review[20]. These questions were also clearly resolved in favor of theory (see below).

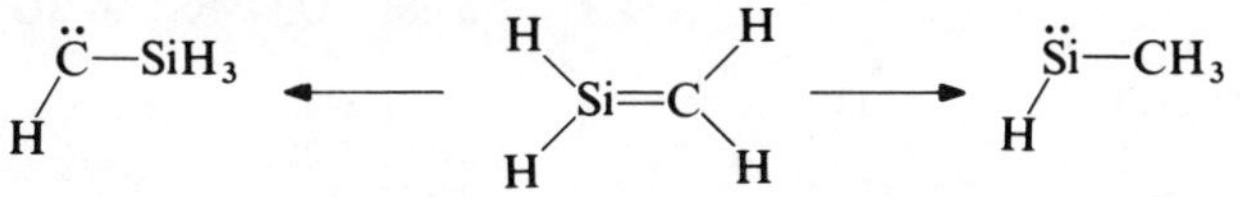

(11)

(46) **(40)** **(45)**

Let us examine first the thermochemistry of these reactions. A variety of calculations at various levels of theory, summarized in Table 19, show that $H_3Si\ddot{C}H$, **46**, is much higher in energy than silaethylene. On the other hand, at all levels of calculation (Table 19) the rearrangement of silaethylene to methylsilylene, **45**, is found to be nearly *thermoneutral*. Gordon was the first to point out this astonishing result[151], in light of our experience with carbon chemistry, namely that a *divalent* silicon compound is as stable as the isomeric compound containing a tetravalent multiply bonded silicon atom. This unusual feature of silicon (which is even more pronounced with germanium) is responsible for many of the differences between multiply bonded organosilicon and organic compounds, as we have already pointed above. The value of the energy difference between **40** and **45** is dependent on the computational method. At MP4SDTQ/6-31G* silaethylene is more stable than $H_3C\ddot{Si}H$ by 2.2 kcal mol^{-1}, of which 1.4 kcal mol^{-1} comes from the difference in zero-point energy[146].

At the time that Schaefer's review[20] was published, the computational results disagreed with the experimental findings of Pau, Pietro and Hehre[173], who claimed that 1-methylsilaethylene is 28 kcal mol^{-1} more stable than dimethylsilylene. Additional calculations have shown that replacement of a hydrogen by a methyl group has little effect on the energies of these compounds, so that substitution could not explain the discrepancy. Now we know that it is highly unlikely that the above experimental result is valid. New experimental evidence, including a recent pyrolytic work, is in line with theory,

TABLE 19. Calculated energies (in kcal mol^{-1}) for the isomerization of $H_2Si{=}CH_2$ to $CH_3\ddot{S}iH$ (**45**) and $H\ddot{C}SiH_3$ (**46**) and the corresponding isomerization barrier heights.

Reaction	ΔE^a	$\Delta E^{\neq b}$	Method	Reference
$H_2Si{=}CH_2 \rightarrow CH_3SiH$	−23.2		STO-4G	151
	−9.2		STO-3G*	151
	−11.6		dz	153
	−4.9		dz+d	153
	−14.9	42.9	3-21G//3-21G	175a
	−4.1		3-21G$^{(*)}$//3-21G$^{(*)}$	146
	−14.1	43.4	6-31G//6-31G	175a
	−5.8	43.4	6-31G*//6-31G*	175a
	−12.5		CI/STO-4G	151
	−6.8		CI/STO-3G*	151
	−2.3		CI/dz	153
	−0.4	41.0	CI/dz+d	153
		40.6	CI/dz+d(Si,C)+p(H) + Davidson correction	159
	3.5		Pseudopotential + CI/ dz+d(Si,C)//dz+d(Si)	157
	0.6		CEPA/extended	161
	−3.8	41.6	CISD/6-31G*//6-31G*	175a
	−3.4	39.3	CISDQ/6-31G*//6-31G*	175a
	2.1		MP2/6-31G*//4-31G	130
	2.1		MP2/6-31G*//3-21G	160
	−0.8	42.2	MP3/6-31G*//6-31G*	175a
	0.8^c		MP4SDTQ/6-31G*//3-21G$^{(*)}$	146
$H_2Si{=}CH_2 \rightarrow H_3Si\ddot{C}H$	62.7		3-21G1/3-21G	146
	56.4		3-21$^{(*)}$1/3-21G$^{(*)}$	186
	54.4		6-31G*//3-21G	186
	47.9		CI/STO-3G$^{(*)}$	151
	49.4		MP4SDTQ/6-31G*//3-21G$^{(*)}$	146

[a] Heats of isomerization.
[b] Barriers for isomerization.
[c] 2.2 kcal mol^{-1} after corrections for ZPE.

indicating that the stabilities of both isomers are indeed comparable[19]. However, see Section X Addendum and Reference 377.

The third theoretical–experimental dilemma concerned the estimation of the height of the barrier that separates $H_2Si{=}CH_2$ and $H_3C\ddot{S}iH$, and which therefore determines the lifetime of the silene in the absence of collisions. The first theoretical study of this barrier by Schaefer and coworkers used a dz basis set. The transition state which was analytically characterized was predicted to lie 45.0 kcal mol^{-1} above silaethylene[153]. Schaefer[20] estimated that even if all possible deficiencies of the calculations are taken into account, the barrier still would exceed 25 kcal mol^{-1}. Following the theoretical predictions two experimental communications[174] were published, concluding that such isomerizations occur quite easily, i.e. that these reactions face a much lower barrier (estimated to be only about 5 kcal mol^{-1}) than those predicted theoretically. Consequently, Schaefer used higher-level calculations to compute the barrier to rearrangement, but the theoretical–experimental disagreement remained[20]. The possibility that the methyl substituent present in the experimentally studied molecule (i.e. $CH_3SiH{=}CH_2$) is responsible for the dilemma was also discarded by Schaefer. Subsequent calculations by other

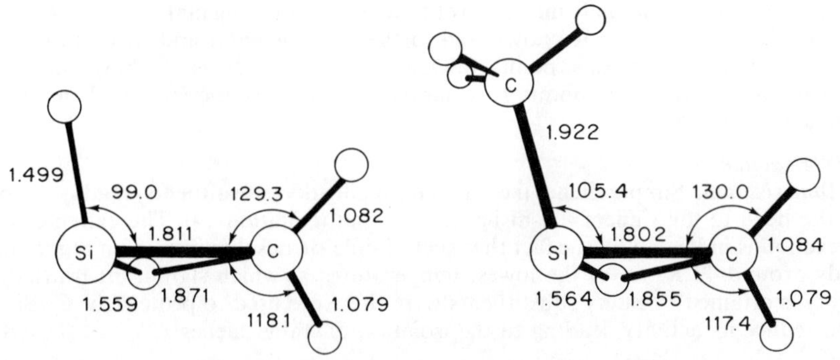

FIGURE 13. 6-31G* transition structures (bond lengths in Å, bond angles in deg) for the isomerization of $H_2Si{=}CH_2$ to $CH_3\ddot{S}iH$ (equation 12) and of $CH_3SiH{=}CH_2$ to $CH_3\ddot{S}iCH_3$ (equation 13). Reproduced by permission of Kluwer Academic Publishers from Ref. 175b.

groups using a variety of computational techniques did not change the theoretical conclusion[161, 175]. For example, Nagase, Kudo and Ito used the 6-31G* basis set to calculate the transition structures for the rearrangement of both silaethylene (equation 12) and methylsilaethylene (equation 13) and the calculated geometries of these transition states are shown in Figure 13[175].

$$H_2Si{=}CH_2 \rightleftharpoons H\ddot{S}iCH_3 \quad (12)$$

$$CH_3SiH{=}CH_2 \rightleftharpoons CH_3\ddot{S}iCH_3 \quad (13)$$

The calculated energies associated with equations 12 and 13 are given in Table 20. The best estimates of the barrier heights included corrections for correlation energy using either MP3 or a CI calculation including all single and double excitations and the Davidson correction. As can be seen in Table 20 the energy barrier of both processes is very high, around 40 kcal mol^{-1} [175b]. The two processes are nearly thermoneutral as already discussed above.

This particular experimental–theoretical controversy was also recently resolved in favor of theory. Arrington, West and Michl[176] have re-examined their report[174a] of a fast low-energy isomerization of $CH_3SiH{=}CH_2$ to $CH_3\ddot{S}iCH_3$ and concluded that the

TABLE 20. Barrier heights and heats of reaction (in kcal mol^{-1}) for reactions (12) and (13) calculated at several levels of theory[a, b]

Level of theory[a]	Barrier height (12)	Barrier height (13)	Heat of reaction (12)	Heat of reaction (13)
HF//3-21G	42.9	45.5	−14.9	−15.7
HF//6-31G	43.4	46.0	−14.1	−14.8
HF//6-31G*	43.5	47.4	−5.8	−5.1
MP3/6-31G*	42.2	43.5	−0.8	−1.9
CI(S+D)/6-31G*	41.6	44.9	−3.8	−2.7
CI(S+D+QC)/6-31G*	39.3	41.4	−3.4	−2.0

[a] Correlation calculations were carried out at the 6-31G* HF optimized geometries. In the MP and CI calculations, all single (S) and double (D) excitations were included, with core-like orbitals frozen. The energies obtained by the CI calculations were further improved with the Davidson formula to allow for unlinked cluster quadruple correction (QC).
[b] From Reference 175b.

hydrogen migration is actually induced by the act of the trapping matrix[176]. However, the nature of this reaction is not known and further experimental and theoretical studies are necessary to clarify this point. Barton, Burns and Burns[177] have also found experimental evidence that points to a larger barrier to isomerization than previous experiments have shown.

(vi) Reactions

α. Dimerization. Simple silenes (i.e. carrying small alkyl substituents) readily couple to yield the head-to-tail dimers, 1,3-disilacyclobutanes (equation 14). The extreme ease of this reaction is indicated by the fact that such simple silenes dimerize in an argon matrix already around 40 K, i.e. at the lowest temperatures at which significant translational mobility is attained[19]. Bulky substituents were therefore used to protect the C=Si bond and to reduce its activity, leading to the isolation of stable silenes[13].

(14)

The available calculations for the dimerization of $H_2C{=}SiH_2$ produce a similar picture. Ahlrichs and Heinzmann[150] were the first to perform *ab initio* calculations for the dimerization of $H_2Si{=}CH_2$. Using a split-valence basis set they found that the head-to-tail dimerization is exothermic by 71 kcal mol^{-1}; the addition of d-functions on Si and C increased the exothermicity to 76 kcal mol^{-1} (the geometry of the product was assumed). Ahlrichs and Heinzmann also estimated (making various assumptions regarding the reaction path and the transition structure) that the barrier for the dimerization is lower than 14 kcal mol^{-1}. The authors note that the dimerization reaction is formally Woodward–Hoffmann forbidden and attributed the low barrier mainly to the high polarity of the C=Si bond, which also relieves the symmetry restriction and the 'forbiddenness' of the dimerization reaction. The low-lying π^*-orbital also contributes[150]. These conclusions are similar to the qualitative results obtained in previous semiempirical investigations[170, 178]. More recent calculations by Morokuma[179] used gradient techniques and a 3-21G basis set to actually calculate and characterize the transition state for the dimerization of $CH_2{=}SiH_2$. These calculations confirm the previous conclusions,

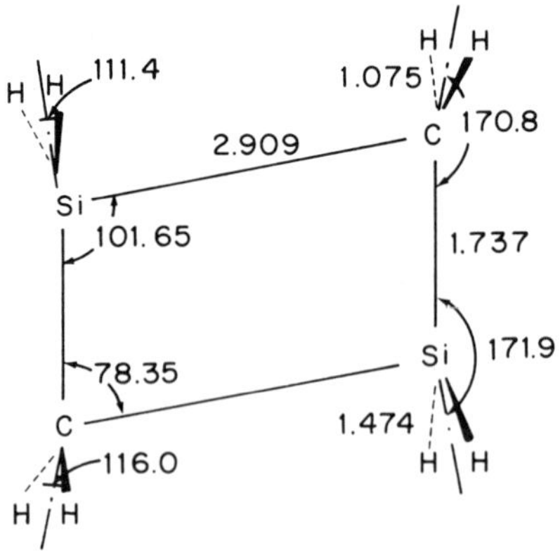

FIGURE 14. 3-21G geometry of the transition state for the dimerization of $H_2Si{=}CH_2$. Bond distances are in Å, bond angles in deg[179]. Reproduced with permission from *Organosilicon and Bioorganosilicon Chemistry* by Sakurai, published by Ellis Horwood Ltd, Chichester, England, 1985.

predicting an energy barrier of only 8.5 kcal mol^{-1} [179]. The calculated transition structure is shown in Figure 14 and can be characterized as 'early', as expected for a very exothermic reaction. Morokuma has also calculated the barriers for dimerization of various substituted silenes and those calculations are discussed below in Section V.A.1.b.

β. Addition reactions. In addition to dimerization, silenes undergo facile addition of polar reagents such as HCl across the double bond. Nagase and coworkers[175b, 180] have studied theoretically the addition of HCl and of H_2O to $H_2C{=}SiH_2$. They considered the addition of HCl and of H_2O to be a good model for electrophilic and nucleophilic reactions, respectively. Both reactions are highly exothermic and follow the general reaction path shown in Figure 15, where the formation of a complex precedes the addition step.

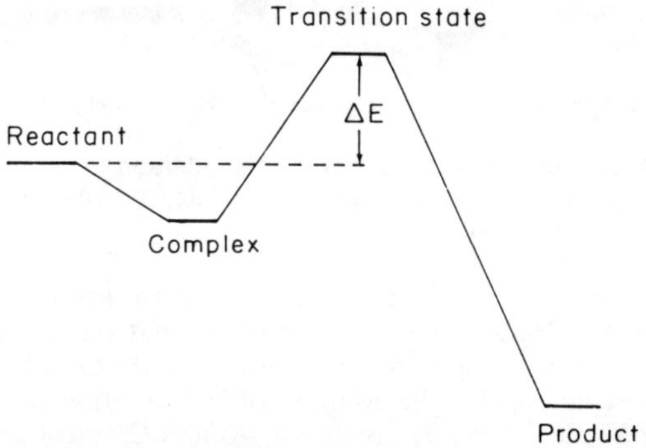

FIGURE 15. Schematic description of the energy profiles for addition reactions to doubly bonded silicon compounds. Reproduced by permission of Kluwer Academic Publishers from Ref. 175b.

The transition structures for these reactions were calculated and characterized at 6-31G* and they are presented in Figure 16 for H_2O addition and in Figure 17 for HCl addition[175b]. The analogous transition structures for addition of these reagents to $H_2C{=}CH_2$ are also included in Figures 16 and 17[175b]. The transition structures for the

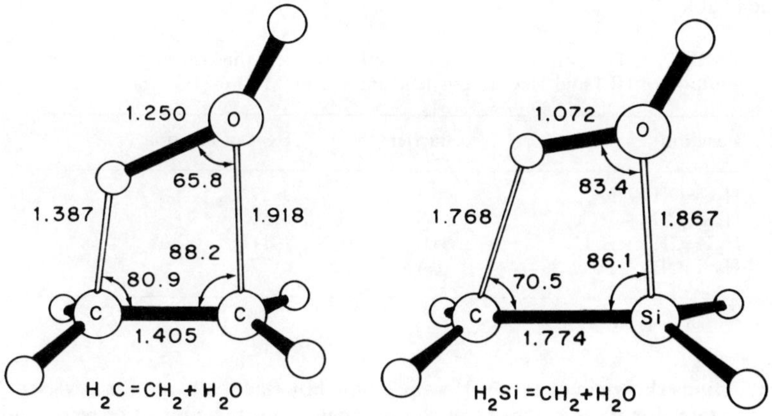

FIGURE 16. 6-31G* transition state structures for the addition of H_2O to silaethylene and to ethylene. Bond distances are in Å, bond angles in deg. Reproduced by permission of Kluwer Academic Publishers from Ref. 175b.

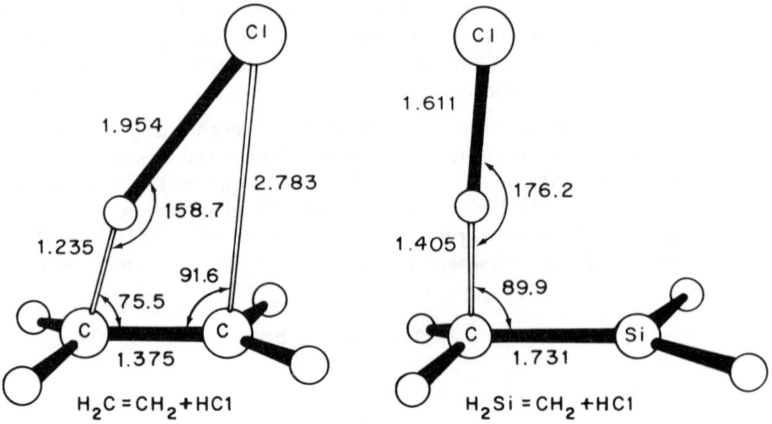

FIGURE 17. 6-31G* transition state structures for the addition of HCl to silaethylene and to ethylene. Bond distances are in Å, bond angles in deg. Reproduced by permission of Kluwer Academic Publishers from Ref. 175b.

addition of water to ethylene and silene are very similar to each other (Figure 16). In the transition structures, the H_2O moiety is oriented so that one of the lone pairs has a favorable nucleophilic interaction with the π^*-orbital of the double bond. On the other hand, the transition structures for the addition of HCl to ethylene and silene are quite different (Figure 17). While the former involves a 4-center C_s transition state, the addition of HCl to silene proceeds via a 2-center-like transition structure. Thus, the addition of HCl to silene proceeds via an electrophilic attack of the positively charged hydrogen at the negative end (i.e. carbon) of the C=Si bond. Apparently, in this transition state, the interaction between Si and Cl is relatively unimportant[175b].

The energetics of these reactions have been evaluated at MP3/6-31G*//6-31G* and the results are shown in Table 21. The major conclusion from these calculations is similar to that reached above for the dimerization reactions, i.e. the reactions of silenes are much more exothermic and the barriers are much lower than for the analogous reactions of ethylene (Table 21).

TABLE 21. Heats of reaction and reaction barriers (in kcal mol^{-1}) for the addition of HCl and H_2O to ethylene and silene (MP3/6-31G*//6-31G*)[a]

Reaction	Barrier	Heat of reaction
$H_2C=CH_2 + HCl$	46.5	−27.8
$H_2Si=CH_2 + HCl$	13.8	−80.1
$H_2C=CH_2 + H_2O$	64.0	−11.8
$H_2Si=CH_2 + H_2O$	8.4	−77.0

[a] From Reference 175b.

The reactions shown in equation 15 were studied by Bachrach and Streitwieser[136]. The transition states for these reactions were not located but the energetics were studied by using the 3-21G, 3-21G$^{(*)}$ and 3-21G* basis sets. The overall reaction of silene with $H_2C=O$ to produce ethylene and $H_2Si=O$ is exothermic by approximately 30 kcal mol^{-1}.

However, the proposed intermediate, 1,2-silaoxetane **34**, is more stable than both the reactants and the products by at least 50 kcal mol^{-1}.

$$H_2C{=}O + H_2Si{=}CH_2 \rightarrow \begin{matrix} H_2C{-}CH_2 \\ | \quad\;\; | \\ H_2Si{-}O \end{matrix} \rightarrow CH_2{=}CH_2 + H_2Si{=}O \qquad (15)$$

(34)

The addition of H_2 to silene was also studied[181]. At MP4SDTQ/6-31G**//MP2/6-31G** the energy released in the hydrogenation reaction is predicted to be 58.5 kcal mol^{-1}, nearly twice the exothermicity of the analogous reaction for ethylene. The transition state for the addition of H_2 to silene shown in Figure 18 is unsymmetrical, with both attacking hydrogens closer to the less electronegative silicon, probably in order to minimize repulsions. It is interesting that, in contrast to the polar reactions discussed above, the barrier for the addition of H_2 to silene is calculated to be higher (47.6 kcal mol^{-1}) than the barrier for the addition of H_2 to ethylene (36.1 kcal mol^{-1}).

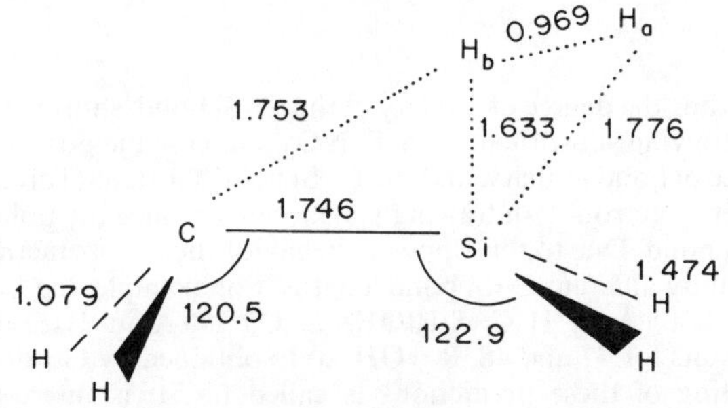

FIGURE 18. Calculated transition structure (MP2/6-31G**) for the addition of hydrogen to $H_2C{=}SiH_2$. Dihedral angles are: $(HCSiH_a) = 98.6°$, $(HSiCH_b) = 97.4°$. Reproduced by permission of North-Holland Physics Publishing from Ref. 181.

b. Substituted silenes. Most of the theoretical studies of substituted silenes evaluated the effect of methyl substitution in relation to the above-mentioned theoretical–experimental discrepancies (see Section V.A.1.a). Relatively few papers have investigated the effect of substituents other than methyl. Gordon has investigated fluorine substitution[160], Gordon and George studied hydroxy substitution[182] and Nagase's group studied fluorine[175], hydroxy[175] and silyl[175, 183] substitution. The only systematic studies which included a wide range of substituents have been carried out by Apeloig and Karni[106, 162] and this section is based mainly on their results.

(i) Geometries. Apeloig and Karni optimized the geometries of the substituted silenes **47** and **48** at 3-21G. In Table 22 we give only the calculated C=Si bond lengths. The full details of the geometries can be found in Reference 106.

$$H_2C{=}SiHR \qquad RCH{=}SiH_2$$

(47) **(48)**

$$R = H,\ CH_3,\ SiH_3,\ OH,\ OSiH_3,\ F,\ CN,\ NO_2$$

In agreement with other calculations the effect of methyl substitution is found to be small. However, some substituents such as hydroxy induce quite large geometrical effects. The effect of the substituents on the C=Si bond length is best understood in terms of the inherent polarity of the C=Si bond shown in **43**. Substituents that increase the

TABLE 22. 3-21G optimized Si=C bond lengths (in Å) in the substituted silenes $H_2C{=}SiHR$ (**47**) and $RCH{=}SiH_2$ (**48**)[a]

R	**47**	**48**
H	1.718	1.718
CH_3	1.716	1.725
SiH_3	1.725	1.721
OH	1.705	1.746
$OSiH_3$	1.705	1.749
F	1.698	1.730
CN	1.711	1.727
NO_2	1.702	1.731

[a] From Reference 106.

polarization, and thus the degree of ionicity of the C=Si bond, shorten this bond. Thus, in **47**, electron-withdrawing substituents (e.g. F, NO_2) increase the positive charge at silicon and thus the degree of bond ionicity, and the C=Si bond shortens (Table 22). The opposite is true for **48**, where electron-withdrawing substituents reduce the polarization and thus lengthen the C=Si bond. Due to this opposite behavior the two isomeric silenes **47** and **48** may have significantly different C=Si bond lengths. For example, the C=Si bond length is calculated to be 1.705 Å in $H_2C{=}SiH(OH)$ and 1.746 Å in $H_2Si{=}CH(OH)$. Similar computational results for **47** and **48**, R = OH, were obtained by Gordon and George[182]. Experimental testing of these predictions is called for. It is interesting to note that π-conjugation between the substituent and the double bond is of little importance in determining the C=Si bond length, as demonstrated by the fact that the C=Si bond distance is longer in **48**, R = OH that in $H_2Si{=}CH_2$, while in **47**, R = OH it is shorter than in $H_2Si{=}CH_2$[106]. If π-conjugation effects contribute significantly to the structure, then the C=Si distance in both **47** and **48**, R = OH, is expected to be longer than in $H_2C{=}SiH_2$, due to contributions from resonance structures with C–Si single bonds such as $H\overset{+}{O}{=}SiH\bar{C}H_2$ or $H\overset{+}{O}{=}CH{-}\bar{Si}H_2$.

These calculations had an important bearing on the experimental–theoretical discrepancy regarding the carbon–silicon double bond length (Section V.A.1.a.i). Apeloig and Karni also found that the effect of substituents on the C=Si distance is additive[106, 162]. This finding is important because it allows the prediction of the double bond lengths of highly substituted silenes which are normally inaccessible theoretically, owing to the prohibitive cost of a reliable calculation for such molecules. On the basis of this additivity and the substituent effects in Table 22 it was possible to show that the long C=Si bond length of 1.764 Å, measured for Brook's silene **41**[167], is fully consistent with an unperturbed r(C=Si) of 1.70 Å[106, 162], in contrast to the measured electron-diffraction value of 1.832 Å[166] (see above). Thus, the estimated C=Si bond length (in Å) in $(H_3Si)_2Si{=}CCH_3(OSiH_3)$, which serves as a close electronic model of **41**, is given by: $r(Si{=}Si) = \{1.718$, [r(C=Si) in $H_2C{=}SiH_2$] + $[3 \times (1.725 - 1.718) + (1.749 - 1.718)$, i.e. changes in r(C=Si) due to the effect of two $H_3Si + CH_3 + OSiH_3$ substituents, respectively]$\} = 1.770$ Å[106, 162]. Corrections for the addition of polarization functions and correlation effects are expected to shorten this bond by approximately 0.013 Å[20] to about 1.757 Å. This calculated C=Si bond length is very close to Brook's experimental value of 1.764 Å[167]. The excellent experimental–theoretical agreement also suggests that elongation of the C=Si bond by steric repulsions between the bulky substituents or as a result of the observed 16° twisting about the C=Si bond are apparently of small importance.

The additivity model coupled with the substituent effects in Table 22 also reproduce nicely the C=Si bond length in **42**, which was reported after the publication of the calculations[168]. Thus, the predicted r(C=Si) in **42** is: {1.718, [r(C=Si) in $H_2C{=}SiH_2$] + $2 \times (1.721 - 1.718)$, [change due to two H_3Si substituents] + $2 \times (1.716 - 1.718)$, [change due to two CH_3 substituents] = 1.720 Å. Corrections for the deficiencies of the 3-21G basis set (see above) bring the predicted C=Si distance to 1.707 Å. The experimental C=Si bond length in **42** is almost the same, 1.702 Å[168].

(ii) Thermodynamic stabilities. The calculated relative stabilities of the isomeric **47** and **48** are given in Table 23[106]. Note that the results at 3-21G//3-21G and at 6-31G*//3-21G are not very different. The relative thermodynamic stabilities of **47** and **48** are determined primarily by the differences in the bond energies of the Si–R vs. the C–R bonds and of the Si–H vs. the C–H bonds (see Table 5). Consequently, substitution at silicon is generally more stabilizing than substitution at carbon (except for $R{=}SiH_3$), especially for electronegative substituents which form particularly strong Si–R bonds. For example, $H_2C{=}SiH(OSiH_3)$ is calculated to be (at 6-31G*//3-21G) 48.6 kcal mol^{-1} more stable than $H_2Si{=}CH(OSiH_3)$. Note the exceptional behavior of the silyl substituent, where **48**, $R = SiH_3$ is more stable than **47**, $R = SiH_3$ by 3.6 kcal mol^{-1}.

TABLE 23. Relative energies of $H_2C{=}SiHR$ (**47**) and $RCH{=}SiH_2$ (**48**) at 3-21G//3-21G and 6-31G*//3-21G[a]

Substituent R	Relative energies[b] (kcal mol^{-1}) 3-21G	6-31G*
H	0.0	0.0
CH_3	13.4	8.5
SiH_3	−2.5	−3.6
OH	45.0	41.2
$OSiH_3$	55.3	48.6
F	54.0	53.9
CN	7.2	3.0
NO_2	13.5	2.7

[a] From Reference 106.
[b] Between isomers, i.e. $E(H_2Si{=}CHR) - E(H_2C{=}SiHR)$. Positive values indicate that **47** is more stable than **48.**

The effects of the substituents on the thermodynamic stability of each of the isomeric silenes were evaluated by Apeloig and Karni by means of the isodesmic equations 16 and 17, in which the substituent R is transferred from a C=Si double bond to a C–Si (or C–C) single bond. The results are given in Table 24[106]. The substituent effects are relatively small, usually only a few kcal mol^{-1}, especially when R is bonded to silicon. The largest effects are exerted by R = F (destabilizing by 7.8 kcal mol^{-1}) and SiH_3 (stabilizing by 6.0 kcal mol^{-1}) when attached to carbon.

$$H_2C{=}SiHR + SiH_4 \rightarrow H_2C{=}SiH_2 + H_3SiR \qquad (16)$$

$$H_2Si{=}CHR + CH_4 \rightarrow H_2C{=}SiH_2 + H_3CR \qquad (17)$$

TABLE 24. Substituent effects on the thermodynamic stabilities of $H_2C{=}SiHR$ (**47**) and $RCH{=}SiH_2$ (**48**) as evaluated by the isodesmic equations 16 and 17[a, b]

Substituent R	Equation 16 ($RSiH{=}CH_2$)[c] 3-21G	6-31G*	Equation 17 ($RCH{=}SiH_2$)[c] 3-21G	6-31G*
H	0.0	0.0	0.0	0.0
CH_3	0.5	0.1	3.7	0.3
SiH_3	2.0	2.9	7.6	6.0
OH	0.8	0.3	1.3	0.4
$OSiH_3$	0.3	−0.2	2.2	−2.8
F	−2.3	−2.0	−3.9	−7.8
CN	−0.8	−1.4	6.4	3.8
NO_2	−1.5	−3.4	7.5	2.4

[a] From Reference 106.
[b] In kcal mol^{-1}.
[c] Positive values indicate stabilization of the substituted silene.

The substituent effects on the charge distribution and on the frontier orbitals is of interest in relation to the reactivity of these silenes. It was found that substituents have a strong effect on the charge distribution within the C=Si bond and that polarization of the σ-electrons is generally larger than that of the π-electrons[106]. In silenes with strong π-donors attached to carbon, the 'natural' $C^{\delta-}{=}Si^{\delta+}$ π-polarity vanishes and the charge distribution becomes more balanced. In more extreme cases (e.g. OH), the polarity is even reversed to $C^{\delta+}{=}Si^{\delta-}$ [106]. The effect of substituents on the energies of the π- and π^*-orbitals of silenes and olefins are qualitatively similar. The π–π^* energy difference is however much smaller in silenes. An interesting contrast between olefins and silenes is that, in olefins, substituent effects on the energies and on the coefficients of the frontier orbitals are generally reinforcing[184] while in silenes they often operate in opposite directions[106]. The full details of the charge distributions, HOMO and LUMO energies and coefficients can be found in Table 5 of Reference 106.

(iii) Reactions. The major obstacle in isolating silenes is believed to be their high reactivity and, in particular, their fast dimerization to disilacyclobutanes. The dimerization reaction of the parent silene (equation 14) was discussed in Section V.A.1.a.vi. The dimerization reactions of substituted silenes such as **47** or **48** can proceed either 'head-to-head' to give substituted 1,2-disilacyclobutanes, or 'head-to-tail' to give substituted 1,3-disilacyclobutanes. The latter products are usually observed. Apeloig and Karni[106] discussed these reactions using qualitative Frontier Molecular Orbital (FMO) theory. They concluded that in most cases polarity plays a more important role than the energies of the frontier orbitals in controlling the reactivity of silenes and their dimerization products.

According to this model, in cases where the substituent enforces a 'reversed polarity' of the C=Si bond, a reduced reactivity and higher yields of 'head-to-head' dimers are expected. Brook's silene **41** is an example of such an electronically stabilized silene, and it indeed yields 'head-to-head' products[13, 18, 19, 167]. Based on these ideas it was also predicted that silenes, such as $(NC)_2Si{=}CR(OSiMe_3)$, $(Me_3Si)_2Si{=}C(OSiMe_3)SiMe_3$, $(Me_3Si)_2Si{=}C(OBu\text{-}t)R$, etc., should be electronically stabilized and probably also kinetically stable[106]. These qualitative conclusions were recently confirmed by quantitative calculations by Morokuma[179] and by Nagase and coworkers[175b]. Another interesting conclusion concerns the thermodynamics of the dimerization reactions. It was found that the relative stability of various substituted silenes runs parallel to the stability of the corresponding dimers, so that the effect of substituents on the overall exothermicity of the dimerization reaction is small[106].

More recently Morokuma calculated at 3-21G the barriers for dimerization of a variety of substituted silenes and the results are given in Table 25[179]. Although the calculations are very approximate, as the structures of the transition states were not optimized and various assumptions and simplifications were introduced, the trends in Table 25 are expected to hold also when more complete calculations become possible.

TABLE 25. 3-21G barriers (ΔE, kcal mol^{-1}) for the dimerization of $H_2C{=}SiHR$ (**47**) and $RCH{=}SiH_2$ (**48**)[a]

R	ΔE (**47**)	ΔE (**48**)
H	8.5	8.5
CH_3	4.7	13.0
SiH_3	12.2	4.4
NH_2	−3.2	31.4
OH	−3.1[b]	22.1
F	−4.8[b]	15.5
SH	5.5	8.6
CN	7.6	0.5
NO_2	5.3	−3.4[b]
CHO	8.2	1.2

[a] Data from Table I of Reference 179. However, we have assumed that in Referrence 179 the roles of X (substituent at C) and Y (substituent at Si) have been erroneously interchanged, as there is a conflict between the text and the values in Table I.
[b] The energy of the transition state is lower than that of the reactants. This artifact results from the fact that geometry optimizations were carried out for the silene but not for the transition structure.

In general, Morokuma's calculations[179] agree with the earlier qualitative predictions of Apeloig and Karni[106]. Morokuma finds[179] that the substituent effects on the dimerization barriers are quite substantial, being 8.5 kcal mol^{-1} for $H_2C{=}SiH_2$ and 31.4 kcal mol^{-1} for $H_2Si{=}CHNH_2$. The polarity of the C=Si bond indeed seems to play a major role in determining these energy barriers, as shown by the dimerization barriers for $H_2C{=}SiH_2$, $H_2C{=}SiH(OH)$ and $H_2Si{=}CH(OH)$ (Table 25). As the charge in the silene becomes more balanced as in $H_2Si{=}CH(OH)$, the barrier to dimerization increases. Morokuma finds a good correlation between the calculated barrier heights and the σ_R value of the substituent on the carbon atom (i.e. in **48**); a strongly π-donating substituent increases the barrier height. There is also a correlation, although less profound, between the σ_I value of the substituent attached to silicon (i.e. in **47**) and the barrier height; a σ-donating inductive substituent on the silicon atom increases the barrier height. The effect of double substitution on the barrier height is approximately additive[179]. Although the studies of Apeloig and Karni[106] and of Morokuma[179] add important insights to the understanding of the dimerization process, further theoretical studies are needed to clarify and quantify the factors that determine the barrier height and the regiochemistry of this important reaction.

The additions of HCl and H_2O to $H_2C{=}SiH_2$, $Me_2Si{=}CH_2$, $F_2Si{=}CH_2$ and $H_2Si{=}CH(OH)$ were studied by Nagase and coworkers and the calculated reaction barriers are given in Table 26[175b]. Nagase's studies also support the earlier qualitative conclusions[106] that charge distribution is more important than the energies of the frontier orbitals in controlling the reactivity of silenes. Thus, both $Me_2Si{=}CH_2$ and $F_2Si{=}CH_2$ react with HCl with lower barriers than $H_2Si{=}CH_2$, although the energies of the HOMO follow the order $Me_2Si{=}CH_2 > H_2Si{=}CH_2 > F_2Si{=}CH_2$ (see Table 26). The barrier for the addition of H_2O to $H_2Si{=}CH(OH)$ is by 5.6 kcal mol^{-1} higher than for $H_2Si{=}CH_2$ although their LUMOs have similar energies, probably because of the 'inverse polarity' of the hydroxysilene[175b].

TABLE 26. Calculated barriers (6-31G*//3-21G) for the addition of HCl and H_2O to substituted silenes, and the calculated π and π^* energies (eV) and charge densities of the substituted silenes[a]

	Reaction barrier (kcal mol^{-1})		Orbital energies (eV)		Charge density	
Silene	HCl	H_2O	π	π^*	at Si	at C
$H_2Si{=}CH_2$	19.2	12.0	−8.4	2.4	0.46	−0.65
$(CH_3)_2Si{=}CH_2$	5.2	*b*	−7.8	3.6	0.74	−0.68
$F_2Si{=}CH_2$	10.1	no barrier	−9.1	3.5	1.28	−0.76
$H_2Si{=}CH(OH)$	*b*	17.6	−7.6	2.7	0.15	−0.05

[a] From Reference 175b.
[b] Not given.

(iv) Silene–silylene isomerism. Substituted silenes, depending on the substitution site, can rearrange by a 1,2-hydrogen shift to either an α- or a β-substituted silylene (equations 18 and 19, respectively). Migration of the substitutent R is also possible (equation 20).

$$H_2C{=}SiHR \rightarrow H_3C\ddot{S}iR \tag{18}$$

$$H_2Si{=}CHR \rightarrow RCH_2\ddot{S}iH \tag{19}$$

$$H_2C{=}SiHR \rightarrow RCH_2\ddot{S}iH \tag{20}$$

The parent $H_2Si{=}CH_2$ and $H_3C\ddot{S}iH$ have nearly the same energy (see Section V.A.1.a.v). We have seen in Section V.A.1.b. ii that the effect of substituents on the stability of silenes is usually small[106]. On the other hand, the effect of α-substituents on the energy of silylenes can be substantial (see Section VII.A and, for example, Reference 146). Due to these very different substituent effects the silene–silylene energy difference can vary substantially as a function of the substituent, with the α-substituted silylene being usually more stable than the isomeric silaolefin. For example, an hydroxy substituent stabilizes the singlet silylene by 15 kcal mol^{-1} [146] while the silene is stabilized by only 0.3 kcal mol^{-1} [106]. $H_3C\ddot{S}iOH$ is therefore expected to be more stable than $H_2C{=}SiH(OH)$ by about 12 kcal mol^{-1}. The effect of β-substituents on the thermodynamic stability of silylenes has not been studied systematically, but it is reasonable to assume that in general it is much smaller than the α-substituent effect. Thus, the effect of substituents on the thermodynamics of equations 19 and 20 is expected to be smaller than that for equation 18.

The effect of substituents on the barrier for the 1,2 hydrogen shift that interconverts silenes and silylenes has been studied only for methyl and silyl substitution. As noted in Section V.A.1.a.v the effect of methyl on the barrier for 1,2-hydrogen shift (equation 18,

$R = CH_3$) is small, and so also is the effect of silyl[183]. The calculated barriers are 42.2 kcal mol^{-1} in $H_2C{=}SiH_2$, 43.5 kcal mol^{-1} in $H_2C{=}SiH(CH_3)$ and 42.8 kcal mol^{-1} in $H_2C{=}SiH(SiH_3)$, all at the MP3/6-31G*//6-31G* level of theory[183]. The barriers may be smaller with substituents such as OH, which stabilize the silylene strongly, thus making equation 18 more exothermic. However, the isomerization barriers are still expected to be substantial so that at low temperatures substituted silenes are predicted to be kinetically stable towards 1,2-hydrogen shifts. Systematic studies on the effect of substituents on these isomerization barriers are needed for more quantitative estimations.

Nagase and Kudo[183] have studied also the barrier heights and the transition structures for the migration of methyl and silyl groups (equation 20, $R = CH_3$, SiH_3). The calculated geometries of the reactants, the products and the transition structures connecting them are shown in Figure 19 for methyl and in Figure 20 for silyl[177, 183].

At MP3/6-31G*//6-31G* the following barrier heights (relative to the silenes) were calculated for equations 20 (kcal mol^{-1}): 42.2 ($R = H$), 54.7 ($R = CH_3$), 26.2 ($R = SiH_3$)[183]. Thus, the migration of methyl is more difficult while silyl migration is significantly more facile than that of hydrogen. Still, even for silyl the barrier is relatively large so that isomerization according to equation 20 is predicted[183] to occur only at relatively high temperatures in agreement with experiment[185].

c. Other systems containing C=Si bonds. Dewar and coworkers[54] were the first to study (using MINDO/3) silicon analogues of butadienes such as **49**, **50** and $H_2Si{=}SiHSiH{=}SiH_2$ and extended conjugated systems such as $H_2Si{=}SiHSiH{=}SiHSiH{=}SiH_2$. They found that delocalization in butadiene and in the silicon analogues of butadiene is comparable.

$H_2Si{=}CH{-}CH{=}CH_2$ **(49)** $H_2C{=}SiH{-}CH{=}CH_2$ **(50)**

Trinquier and Malrieu[157] used a pseudopotential method combined with a polarized basis set for the valence electrons to study **49** and **50** at the *ab initio* level. They find that **50** is more stable than **49** by 5.8 kcal mol^{-1}. Their analysis also suggests that conjugation in these molecules is comparable with that occurring in butadiene, but π-conjugation is not responsible for the higher stability of **50** compared with **49**. It was also estimated that **50** is by approximately 5 kcal mol^{-1} more stable than the isomeric silylene, $CH_3\ddot{S}iCH{=}CH_2$[157].

The isomeric monosilallenes **51** and **52** were studied by several groups[141b,c, 186–188].

$H_2Si{=}C{=}CH_2$ **(51)** $CH_2{=}Si{=}CH_2$ **(52)**

The highest level calculations reported by Krogh-Jespersen included geometry optimizations at 3-21G and energy comparisons at the MP3/6-31G*//3-21G level[188]. The C=Si bond lengths in both **51** and **52** are 'normal', 1.702 Å and 1.701 Å, respectively[188]. **51** is by 15.5 kcal mol^{-1} more stable than **52**[188], a lower energy difference than found in the previous studies that did not include corrections for correlation energy[141, 186, 187]. Note, however, that the various isomeric silylenes are generally more stable than **51**[141, 186, 187]. The barriers to internal rotation in **51** and **52** are 35 and 20 kcal mol^{-1}, respectively, much smaller than in $H_2C{=}C{=}CH_2$ where, at the same level of theory (MP3/6-31G*//3-21G), it is 53 kcal mol^{-1} [188]. The transition states for internal rotation are predicted to possess bent geometries in all three molecules with open-shell singlet-three-π-electron configurations in allene and **51**, but a closed-shell singlet-two-π-electron configuration in **52**. It is concluded that the thermodynamic stabilities of the silaallenes, particularly **51**, are probably sufficient to allow its experimental isolation[188]. Silaallenes were suggested as potential reactive intermediates[19].

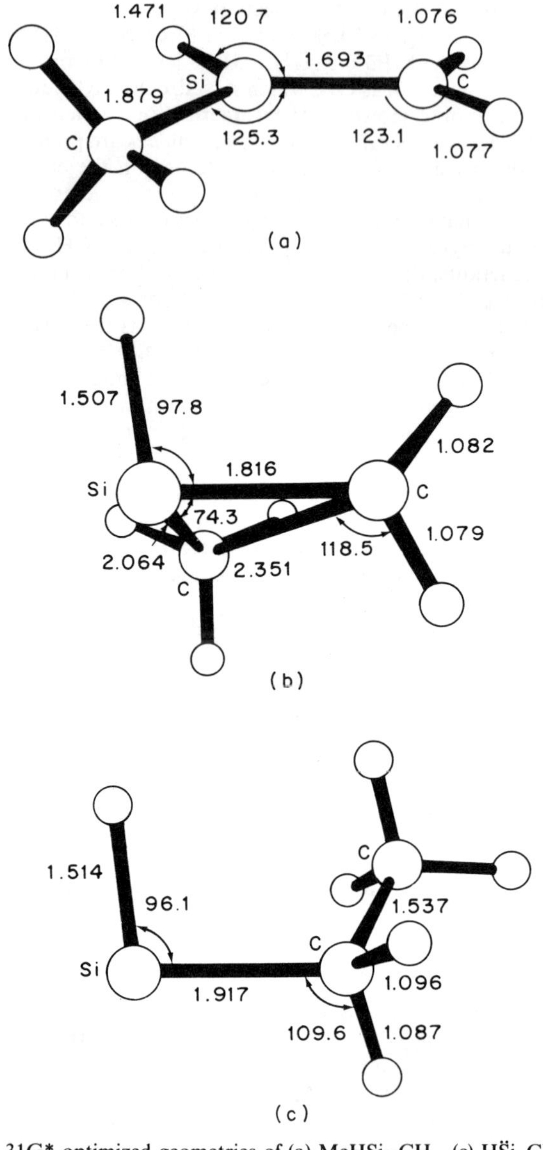

FIGURE 19. The 6-31G* optimized geometries of (a) $MeHSi{=}CH_2$, (c) $H\ddot{S}i{-}CH_2Me$, and (b) the transition state connecting them. Bond lengths are in Å, bond angles in deg. Reproduced by permission of the Royal Society of Chemistry from Ref. 183.

Barthelat and coworkers[141b], Gordon and Koob[186] and Lien and Hopkinson[141c] studied also other C_2SiH_4 isomers. At 6-31G*//3-21G the most stable C_2SiH_4 isomer is $HC{\equiv}CSiH_3$, which lies 29 kcal mol^{-1} below **51**[186]. 1-silacyclopropene **53a** is also more stable than **51**, but it lies 15 kcal mol^{-1} above $HC{\equiv}CSiH_3$[186]. As expected, the isomeric **53b** that contains a C=Si bond is 50 kcal mol^{-1} less stable than **53a**. The contrast between

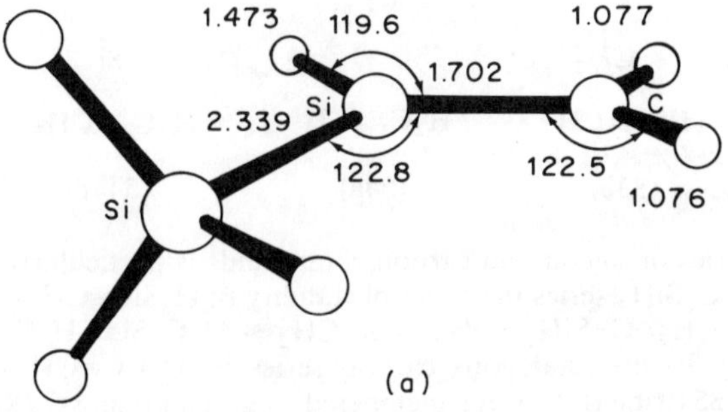

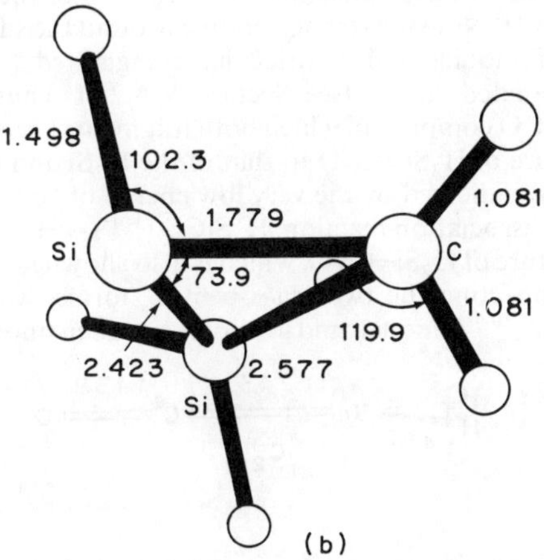

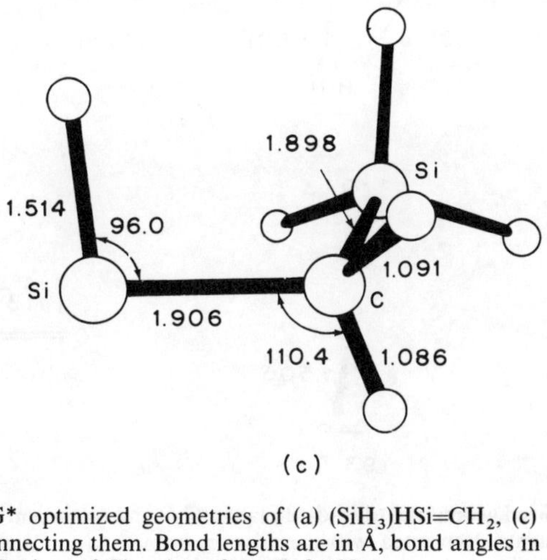

FIGURE 20. The 6-31G* optimized geometries of (a) $(SiH_3)HSi{=}CH_2$, (c) $H\ddot{S}i{-}CH_2SiH_3$, and (b) the transition state connecting them. Bond lengths are in Å, bond angles in deg. Reproduced by permission of the Royal Society of Chemistry from Ref. 183.

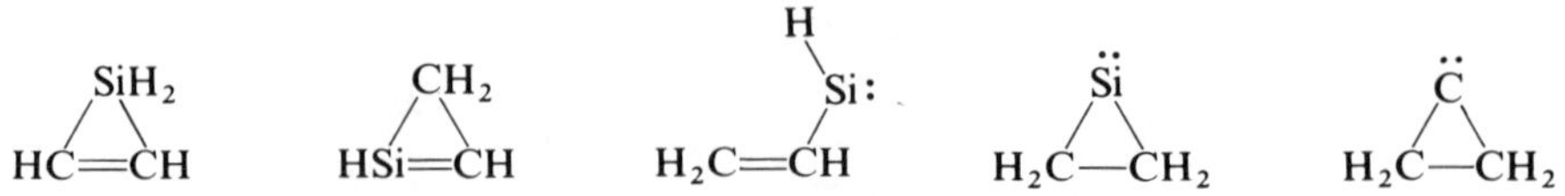

(**53a**) (**53b**) (**54a**) (**54b**) (**55**)

the analogous series of silicon and carbon compounds is particularly interesting in this case. Thus in the C_2SiH_4 series the order of stability is: $H_3SiC{\equiv}CH >$ **54a** (singlet) ~ **54b** (singlet) ~ **53a** > $CH_2{=}C{=}SiH_2 > H_2C{=}Si{=}CH_2 \gg H_3C{-}Si{\equiv}CH$[186]. In contrast, the order of stability in the analogous carbon series is: $H_3CC{\equiv}CH \sim H_2C{=}C{=}CH_2 >$ cyclopropene ≫ **55** (triplet)[186]. Four-membered rings containing C=Si bonds such as **36** were studied by Boatz and Gordon[122].

Trinquier and Malrieu[189] have calculated the structure of the silaketene $H_2Si{=}C{=}O$ which is given in Figure 21. Note the strong bending around the silicon which is consistent with the model that Trinquier and Malrieu have suggested to predict the degree of bending at multiply bonded silicon (see Section V.A.2.a). Thus, $H_2Si{=}C{=}O$ is better described as a silylene–CO complex, in which both fragments keep their identity (compare in Figure 21 the structure of $H_2Si{=}C{=}O$ to that of the H_2Si and CO fragments), than as 1-silaketene. This is also indicated by the very low energy of 10 kcal mol^{-1} calculated (at the dz + p level) for the dissociation reaction: $H_2SiCO\ (^1A') \rightarrow H_2\ddot{S}i\ (^1A') + CO\ (^1\Sigma^+)$. The classic planar C_{2v} structure of $H_2Si{=}C{=}O$, which is also shown in Figure 21, is found to be just a saddle point connecting the two *trans*-bent C_s forms, which are 23 kcal mol^{-1} (dz + p) lower in energy[189]. Trinquer and Malrieu use their model to predict qualitat-

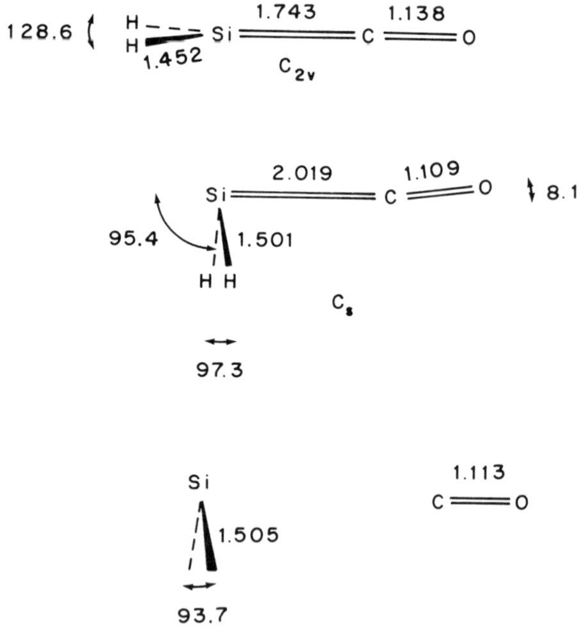

FIGURE 21. Optimized geometries (dz + d) of planar (C_{2v}) and *trans*-bent (C_s) silaketenes and of the SiH_2 and CO fragments. Reprinted with permission from *J. Am. Chem. Soc.*, **109**, 5303 (1987). Copyright (1987) American Chemical Society[189].

ively (calculations are not provided) that $H_2Si{=}C{=}S$ has also a bent geometry at silicon, although the bending angle is probably smaller and the surface flatter than for $H_2Si{=}C{=}O$. They have also predicted that $H_2C{=}Si{=}O$ and $H_2C{=}Si{=}S$ are planar, whereas their difluoro-derivatives are not[189].

Compounds with C=Si double bonds that formally are either silaaromatic (e.g. silabenzene) or silaantiaromatic (e.g. silacyclobutadiene) are discussed in Section VI.

2. *Disilenes*

Disilenes **56**, and in particular the parent **57**, have attracted considerable interest from the theoretical arena[146, 147, 189–206]. The work of Kohler and Lischka was particularly important, because it was the first to study the Si_2H_4 system at a level of theory (flexible basis set and including corrections for electron correlation) which gives a realistic picture of the potential energy surface[161, 192]. The theoretical interest in Si=Si double bonds has naturally intensified after West[14] and later Masamune[207] and Boudjouk[208] and their coworkers reported the synthesis of stable disilenes (i.e. kinetically stabilized by bulky substituents, e.g. **56**, $R^1 = R^2 =$ 2,4,6-trimethylphenyl).

$R^1R^2Si{=}SiR^1R^2$ **(56)** $H_2Si{=}SiH_2$ **(57)**

The theoretical work in this field has been reviewed recently[19, 21] and we bring here only the major conclusions from the earlier work up to 1984 and discuss in some more detail the more recent papers.

a. The singlet state. The calculations have yielded agreement that the ground state of disilene is a singlet state (see below). Until very recently the calculations seemed to have converged to the singlet geometry shown in Figure 22a. The specific geometry shown in Figure 22a was obtained at the MP2/6-31G* level and other CI-type calculations give similar results. The calculated Si=Si bond length is 2.169 Å[199], to be compared with the values of 2.16 Å and 2.14 Å observed experimentally in heavily substituted derivatives[19]. Previous calculations which did not include corrections for correlation energy have given,

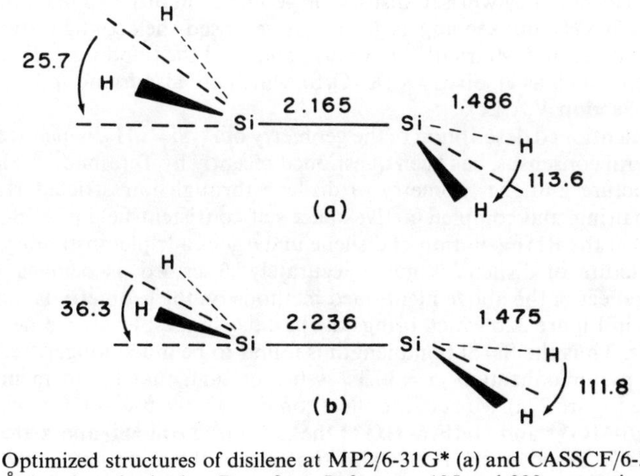

FIGURE 22. Optimized structures of disilene at MP2/6-31G* (a) and CASSCF/6-31G* (b). Bond lengths are in Å, bond angles in deg. Data from References 195 and 202, respectively.

as expected, somewhat shorter (by about 0.03–0.05 Å) Si=Si bond lengths[146, 190–194]. The Si=Si distance in $H_2Si{=}SiH_2$ is much shorter than the Si–Si single bond length of 2.342 Å calculated for disilane at the same level of theory, indicating considerable π-bonding and double-bond character.

A much more delicate and computationally demanding question is the degree of pyramidalization at silicon. Various calculations have predicted for **57** a peculiar C_{2h} *trans*-bent geometry, **58**[189, 195–202] with variable degrees of pyramidalization at silicon, up to α of 38.5°[198]. It is clear, however, from all calculations that the planar **59** and *trans*-bent **58** structures are very close in energy, within 1–2 kcal mol^{-1}. Thus the bending potential in **58** is very shallow and other factors (e.g. substituent effects, neighboring molecules in the matrix or crystal) could have a decisive effect on the observed geometry. In agreement with this picture, the available X-ray structures of disilenes[14, 207, 208] show exact or approximate planarity around the Si=Si bond, with a variable degree of deviation in the twisting and pyramidalization angles. The flatness of the bending and twisting potential in disilene is probably responsible for the thermochromic behavior of substituted disilenes[19].

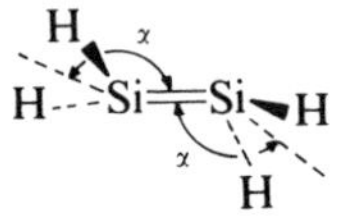

(58)

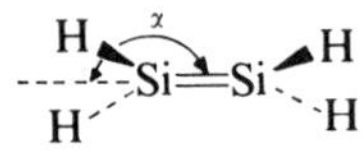

(59)

The peculiar C_{2h} *trans*-bent structure **58** of disilene is apparently related to its intermediate position between the planar ethylene and the strongly bent *trans*-doubly pyramidalized structures of digermene and distannene[209]. In a recent paper which discusses nonclassical distortions at multiple bonds for a wide variety of compounds, Trinquier and Malrieu[189] concluded that the occurrence and extent of these distortions are related to the singlet–triplet separation of the interacting fragments forming the multiple bond. According to this model, as the singlet–triplet energy difference in the fragments (in this case XH_2 units) increases, as is the case along the series X = C, Si, Ge, Sn, the molecule becomes more strongly pyramidalized. Thus, when singlet configurations are strongly preferred, they work to distort the geometry in order to maximize their own interactions, each XH_2 unit keeping its ground-state closed-shell configurations[189]. Using this model Trinquier and Malrieu[189] have also analyzed the bonding in HSi≡SiH and in extended systems such as $H_2Si{=}C{=}X$ (X=O, S), which are also found to be strongly bent at silicon (see Section V.A.1.c).

The above-mentioned description of the geometry of $H_2Si{=}SiH_2$, which seemed to have reached a general consensus, has been questioned recently by Teramae[202]. He studied the electronic structure and the geometry of disilene through unrestricted Hartree–Fock, GVB perfect pairing, and complete-active-space self-consistent-field (CASSCF) methods. It was found that the RHF solution of disilene undergoes a triplet instability and that the electronic structure of disilene is more accurately described as being a weak singlet diradical. The effect of the above mentioned methods on the geometry is quite dramatic, as can be seen in Figure 22b which brings the best (i.e. highest level) geometry calculated by Teramae[202]. Thus the Si=Si bond length is found to be much longer, i.e. 2.236 Å, and the degree of pyramidalization ($\alpha = 36.3°$) is higher than that found in most previous studies (e.g. see Figure 22a). Most of the effect comes from the use of UHF wave functions. Thus at RHF/6-31G** and UHF/6-31G** the optimized r(Si=Si) and α are 2.128 Å, 5.8° and 2.277 Å, 43.7°, respectively. It was also shown that the basis-set dependencies of the optimized geometries of disilenes noted in previous work is an artifact of the RHF treatment which does not include contributions from the diradical character. The diradical character increases from 2.4% for the RHF/6-31G** structure to 13% for the CASSCF/6-31G** structure. The *trans*-bent angle α increases so as to stabilize the diradical electronic structure. Teramae concluded that the electronic structure of disilene involves important contributions from the diradical structures **60**. The actual contour plots of the GVB pair orbitals representing this 'diradical' character are shown in Figure 23[202]. We wish to point, however, to the fact that the structure calculated by Teramae[202] is in poor agreement with the experimental structures that show Si=Si bond lengths of 2.14–2.16 Å and α of 18° in tetramesityldisilene and *trans*-1,2-di(tert-butyl)dimesityldisilene (**42**). Although the above disilenes are heavily substituted and extrapolation to the smaller calculated molecules should be done with caution the disagreement between the structures of **41** and **42** and Teramae's calculation is, in our view, quite alarming.

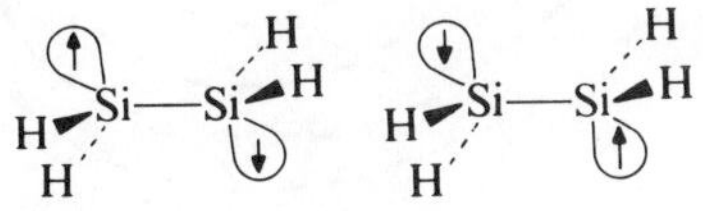

(60)

b. The triplet state. All recent *ab initio* calculations agree that the ground state of disilenes is the closed-shell singlet, but a reliable estimate of the singlet–triplet splitting can be obtained only if corrections for the correlation energy are included. While at the HF level these two states are nearly degenerate[192a], the current estimates are that the triplet lies 20–25 kcal mol^{-1} higher than the ground state singlet[146, 147, 161, 191]. For example, at the MCSCF/6-31G* level the singlet–triplet splitting is 21.5 kcal mol^{-1} [147].

As expected, the triplet state is characterized by a large twist angle and a high degree of pyramidalization on both silicon atoms, and a Si–Si distance which is considerably longer than that in the singlet ground state. The most recent structure was reported at the MCSCF/3-21G* level and it is shown in Figure 24 together with the geometry of the corresponding transition state for π-bond rotation[147]. Note that the Si–Si π-bond is essentially broken in both the singlet and the triplet states and it is even slightly longer than the Si–Si single bond length in disilane.

c. π-Bond strength. On the basis of the calculated rotation barriers and hydrogenation energies, Gordon and coworkers[147] recommended a value of 25 kcal mol^{-1} for the π-bond strength in $H_2Si{=}SiH_2$ (see also Section V.A). Previous calculations reached similar conclusions[194]. This estimate is considerably smaller than the 36–38 kcal mol^{-1} π-bond energy in $H_2C{=}SiH_2$. Another way to demonstrate the weak nature of the double bond in disilene is to compare the dissociation energies of disilene (equation 21) and of disilane (equation 22)[147, 199].

$$H_2Si{=}SiH_2 \rightarrow 2H_2Si: \qquad (21)$$

$$H_3Si{-}SiH_3 \rightarrow 2H_3Si\cdot \qquad (22)$$

At MP4/6-31G**//MP2/6-31G** the calculated energies of reactions 21 and 22 are 61.8 and 74.3 kcal mol^{-1}, respectively[21, 147, 199]. Thus it is easier to dissociate the Si–Si double bond in disilane than the single bond in disilane. Although the inherent stability of silylene contributes to this surprising result, the inherent weakness of the Si–Si π-bond is certainly also an important factor.

d. Relation to isomers. As in the case of silene (Section V.A.1.a), the parent disilene **57** and the isomeric silylsilylene, $H_3Si\ddot{S}iH$ (**61**), are found to be quite close in

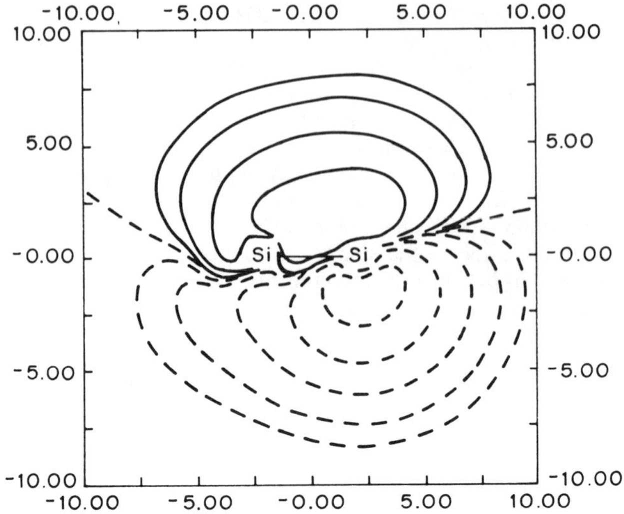

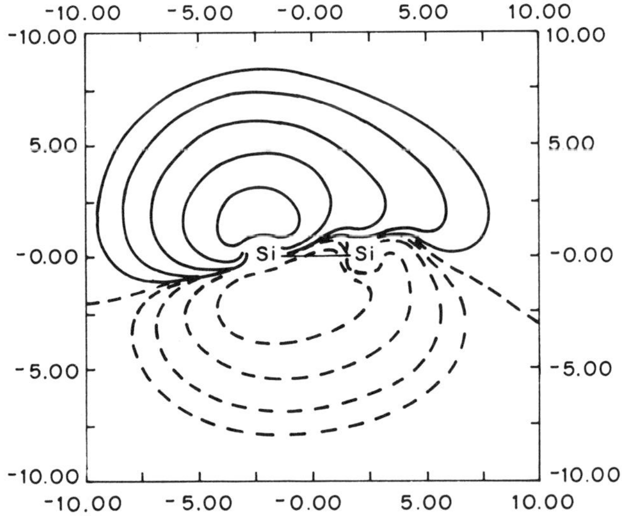

FIGURE 23. Contour plots of the GVB pair orbitals of $H_2Si{=}SiH_2$ representing 'diradical' orbitals. Calculation carried out at GVB(4)/6-31G**//GVB(4)/6-31G**. Reprinted with permission from *J. Am. Chem. Soc.*, **109**, 4140 (1987). Copyright (1987) American Chemical Society[202].

energy[146, 147, 161, 191, 195–197], in agreement with experiment[19]. As expected, the calculated **57**–**61** energy difference is sensitive to the level of theory used. At the HF level, the silylene is found to be lower in energy than the disilene by 2–9 kcal mol^{-1}, depending on the size of the basis set[190c, 192]. Addition of polarization functions and correlation energy reduces this energy difference. Poirier and Goddard[191], using the 3-21G basis set augmented by singles and doubles CI, predicted disilene to be lower in energy by 8–10 kcal mol^{-1}. More accurate calculations at the MP3/6-31G**//MP2/6-31G*,

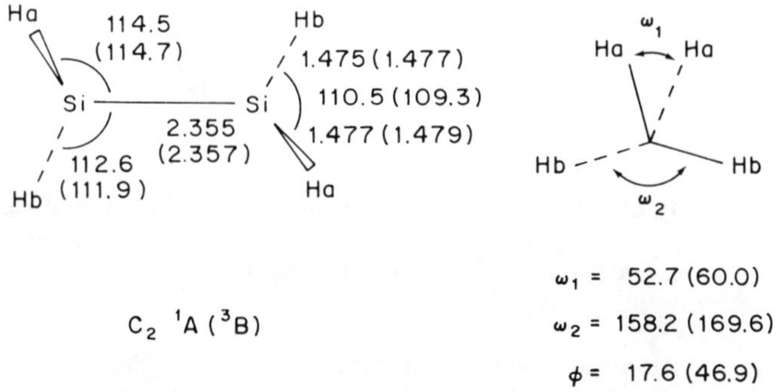

FIGURE 24. Transition state structure for π-bond rotation and equilibrium triplet structure (in parentheses) of disilene at MCSCF/3-21G*. Bond lengths are in Å, bond angles in deg. ϕ is the flap angle between the Si–Si bond and the SiH_2 plane[147].

MP4/6-31G*//3-21G(*) and MCSCF levels place the silylene **61** 5[195,197], 6.2[146] and 3.9[147] kcal mol^{-1}, respectively, higher in energy than *trans*-bent singlet $H_2Si{=}SiH_2$.

An important difference between silene and disilene is the much lower barrier that separates the doubly bonded species from the corresponding silylenes in the latter case. Krogh-Jespersen[197] and Gordon and coworkers[199] located the transition structure for hydrogen migration at 6-31G** and MP2/6-31G** (Figure 25), respectively. At MP3/6-31G**//6-31G** the transition structure is 17.3 kcal mol^{-1} above **57** and 12.3 kcal mol^{-1} above **61**[197], while at MP4/6-311G* the barrier is somewhat lower, 9.9 kcal mol^{-1} above **61**[199]. Thus, the barrier separating **57** and **61** is only about a third of the barrier that separates $H_2Si{=}CH_2$ and $H\ddot{S}iCH_3$ (about 40 kcal mol^{-1}). Yet, the isomerization barrier in disilene is substantial and a low isomerization rate to silylsilylene is expected at low temperatures.

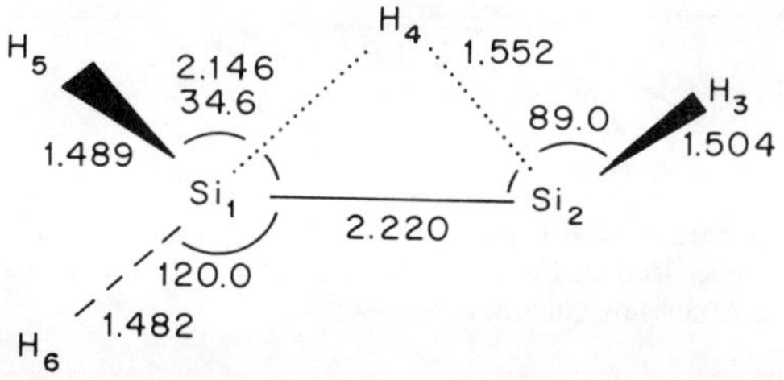

FIGURE 25. Transition structure (MP2/6-31G**) for the isomerization of $H_2Si{=}SiH_2$ to $H_3Si\ddot{S}iH$. Bond lengths are in Å, bond angles in deg. The dihedral angles are: $H_4Si_1Si_2H_3 = 99.2°$, $H_5Si_1Si_2H_3 = -21.5°$, $H_6Si_1Si_2H_3 = 156.0°$. Reprinted with permission from *J. Am. Chem. Soc.*, **108**, 1421 (1986). Copyright (1986) American Chemical Society[199].

In addition to **57** and **61**, other local minima were located on the lowest singlet Si_2H_4 surface. The closest in energy to disilene are the hydrogen bridged structures **62** and **63**[203]. Using a small polarized basis set, **62** and **63** are found to be local minima on the surface lying 23 kcal mol^{-1} higher in energy than **57**[203]. Unfortunately these structures were not reinvestigated with more reliable theoretical methods, which are expected to further lower their energies relative to disilene. A structure analogous to **62**, with the hydrogens replaced by fluorines, was found to be the lowest energy structure of Si_2F_4[189]. The classical disilene structure $F_2Si{=}SiF_2$ is not a minimum on the potential surface[175b,189].

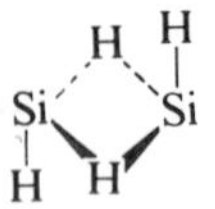

(62)

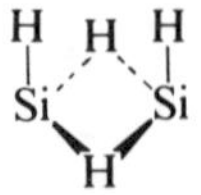

(63)

The proton affinity of disilene was calculated to be 207[192b] kcal mol^{-1} and 223 kcal mol^{-1} [193] according to two different estimations. Both values are substantially higher than that of ethylene. The hydrogen bridged and the classical structures of the protonated species are very close in energy[192b, 193].

e. Reactions. Gordon and coworkers have recently studied in detail the hydrogenation reaction of $H_2Si{=}SiH_2$. At the MP4/6-31G**//MP2/6-31G** level the hydrogenation energy of disilene is 52.2 kcal mol^{-1} [199], only slightly less than that found for $H_2Si{=}CH_2$. The calculated activation energy for the reaction is 36.5 kcal mol^{-1}. Because the least motion approach is symmetry forbidden, the transition state structure is quite interesting, with both approaching hydrogens closer to one of the silicon atoms. The calculated MP2/6-31G** transition structure is shown in Figure 26 and can be compared with Figure 18, that shows the transition structure for the addition of H_2 to $H_2Si{=}CH_2$. The progression of the reaction in Figure 27 shows that the initial approach of H_2 appears to form a silane–silylene complex. Only after the transition state is transversed does one of the hydrogens move towards the second silicon atom forming the saturated product. This suggests that the Si=Si π-bond is first transformed into a lone pair and only later into a Si–H bond[199].

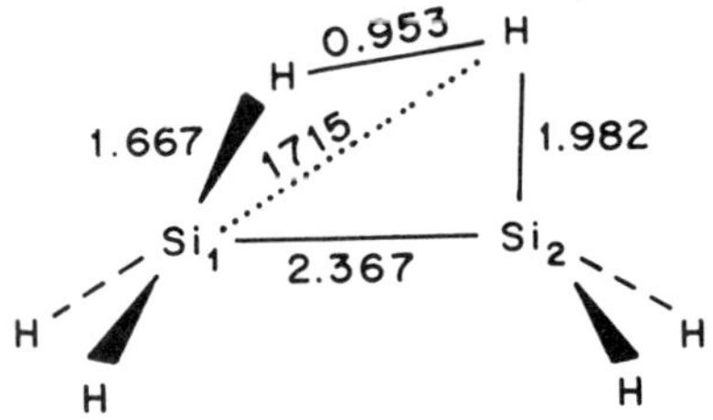

FIGURE 26. Transition state structure (MP2/6-31G**) for the addition of H_2 to $H_2Si{=}SiH_2$. Bond lengths are in Å, bond angles in deg. Reprinted with permission from *J. Am. Chem. Soc.*, **108**, 1421 (1986). Copyright (1986) American Chemical Society[199].

The addition of HCl and H_2O to $H_2Si{=}SiH_2$ was studied by Nagase and coworkers[175b]. Both reactions are highly exothermic with heats of reactions of -69.8 kcal mol^{-1} and -65.9 kcal mol^{-1}, respectively. The calculated 6-31G* transition structures are shown in Figure 28. Both reactions involve a four-center-like transition state and proceed in a concerted manner. Note the contrast in the transition structures for the addition of HCl to $H_2Si{=}SiH_2$ (four centered) and the addition of HCl to $H_2Si{=}CH_2$ (two-centered, Figure 17). At MP3/6-31G*//6-31G* the calculated barriers for the additions of HCl and H_2O to disilene are 16.2 kcal mol^{-1} and 0.2 kcal mol^{-1}, respectively. These barriers, particularly for the addition of H_2O, are much smaller than for the same reactions with $H_2C{=}SiH_2$ (8.4 and 13.8 kcal mol^{-1} for H_2O and HCl, respectively) and with $H_2C{=}CH_2$ (64.0 and 46.5 kcal mol^{-1} for H_2O and HCl, respectively, Table 21)[175b].

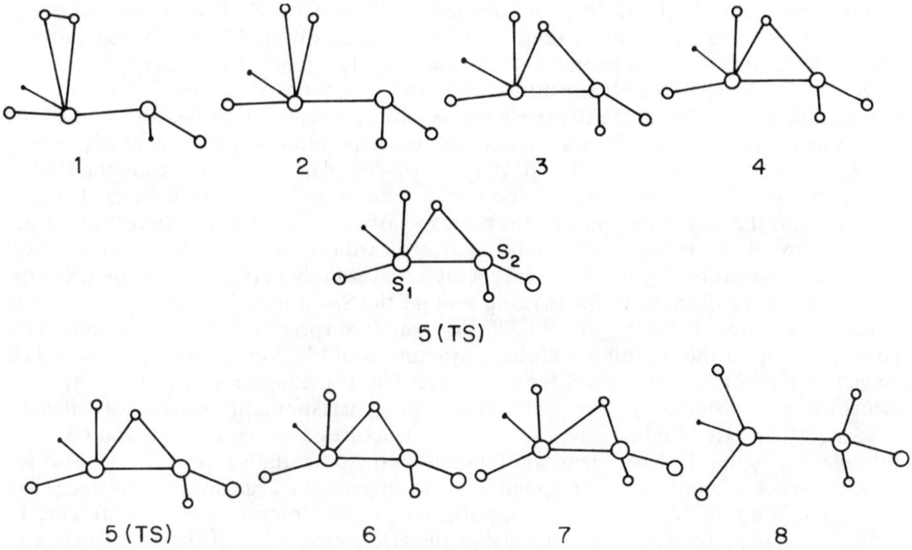

FIGURE 27. Schematic representation of structural changes along the $H_2 + SiH_2{=}SiH_2$ (1) ⟶ H_3SiSiH_3(8) intrinsic reaction coordinate (IRC) at 3-21G*. Reprinted with permission from *J. Am. Chem. Soc.*, **108**, 1421 (1986). Copyright (1986) American Chemical Society[199].

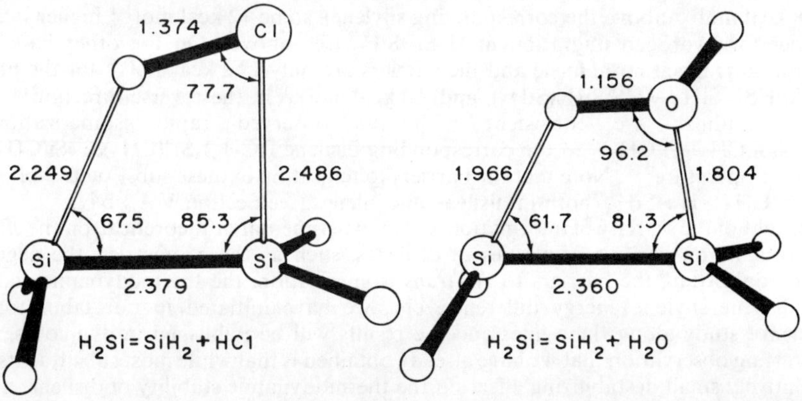

FIGURE 28. Transition state structures (6-31G*) for the addition of HCl and H_2O to $H_2Si{=}SiH_2$. Bond lengths are in Å, bond angles in deg. Reproduced by permission of Kluwer Academic Publishers from Ref. 175b.

The high reactivity of disilene was attributed to its high-lying π-orbital (-7.5 eV) and the low-lying π^*-orbital (1.2 eV) relatively to silene and ethylene ($\pi = -8.5$ eV and -10.2 eV, $\pi^* = 2.5$ eV and 5.0 eV, respectively).

Finally, the dissociation of disilene to two silylenes has been studied and this reaction is discussed (as the dimerization of silylene in Section VII.A.3.a.

f. Substituted disilenes. Only a limited number of studies has been devoted to substituted disilenes. Early studies used mostly semiempirical methods and included combinations of methyl and chlorine substituents[204]. It was found (with one exception) that the disilenes are more stable than the corresponding silylene[204], but the reliability of these calculations is questionable. *Ab initio* studies included only Li[196], CH_3[175b, 196, 205], H_3Si[175b, 205], NH_2[206] and F substitution[196], and concentrated on the disilene–silylene energy differences. The most extensive study is that by Krogh-Jespersen, who studied the tetra-substituted disilenes X_2SiSiX_2 and the corresponding silylenes $X\ddot{S}iSiX_3$, where X = Li, CH_3 and F[196]. $(CH_3)_2Si{=}Si(CH_3)_2$ reveals 'regular' behavior resembling that of the analogous Si_2H_4 species with respect to structural and energetic features. Thus, in both systems the singlet disilene and the corresponding silylsilylene are almost isoenergetic, in the two disilenes the dissociation energies toward two silylenes are comparable and both disilenes feature very flat potential energy surfaces for bending of the geminal groups in a mutual *trans*-fashion or for twisting around the Si=Si bond[196]. In contrast to this 'regular' behavior, F_2SiSiF_2 and $Li_2SiSiLi_2$ exhibit surprising features. No minimum corresponding to the traditional disilene structure could be located for the closed-shell singlet of F_2SiSiF_2. A minimum for a diradical-like triplet species was found, but this structure is considerably less stable than the corresponding singlet silylsilylene, $F\ddot{S}iSiF_3$[196]. In later studies it was found that structures analogous to **62** and **63** (with fluorines replacing the hydrogens) are minima on the potential surface[175b,189]. Also for Si_2Li_4, there is no minimum corresponding to a conventional disilene or silylsilylene. The global minimum for singlet Si_2Li_4 is a distorted, planar structure with two bridging Li atoms[196]. Krogh-Jespersen concluded that the electronegativity of the substituent can have a dramatic effect on the stabilities and structures of disilenes.

Nagase and Kudo have studied the isomerization of $(H_3Si)HSi{=}SiH_2$ to $H\ddot{S}iSiH_2(SiH_3)$ and of $(CH_3)HSi{=}SiH_2$ to $H\ddot{S}iSiH_2(CH_3)$[175b,205]. They found that the barrier for methyl migration is as large as 34.7 kcal mol^{-1} above $(CH_3)HSi{=}SiH_2$ (27.8 kcal mol^{-1} above the corresponding silylene), some 12 kcal mol^{-1} higher than the barrier for hydrogen migration in $H_2Si{=}SiH_2$ (see above). On the other hand, silyl migration is much more facile and the barriers are only 17.2 kcal mol^{-1} for the process $(H_3Si)HSi{=}SiH_2 \rightarrow H\ddot{S}iSiH_2(SiH_3)$, and 8.4 kcal mol^{-1} in the reverse direction[175b, 205]. These findings are consistent with the observed rapid isomerization of $CH_3\ddot{S}iSi(CH_3)_2Si(CH_3)_3$ to the corresponding disilene $[(CH_3)_3Si](CH_3)Si{=}Si(CH_3)_2$, at room temperature[210]. Note that the barriers to migration of these substituents follow the order: $CH_3 > H > SiH_3$, both in disilene and silenes (see Section V.A.1.b.iv).

In light of the scarcity of information, either experimental or theoretical, on the effect of substituents on various properties of disilenes, such as the geometries, the degree of pyramidalization, the barriers to *cis–trans* isomerization, the thermodynamic stability, the disilene–silylene energy difference, etc., we have initiated in our laboratory an extensive study along these lines and the results will be published in due course. One interesting observation that we have already obtained is that while most substituents have a relatively small destabilizing effect on the thermodynamic stability of disilene, a SiH_3 substituent is strongly stabilizing (by about 5 kcal mol^{-1} at MP3/6-31G*//3-21G). The use of four such substituents can lead to a substantial electronic stabilizing effect, which might prove to be of practical importance in preparing novel stable disilenes.

3. *Silanimines*

Silanimines **64** containing double bonds between nitrogen and silicon have been studied much less extensively than C=Si or Si=Si double bonds. The recent successful synthesis[145] of sterically protected substituted silanimines, e.g. **64**, $R^1 = R^2 = t$-Bu, $R^3 = (t\text{-}Bu_3)Si$, has stimulated the theoretical interest in this group of compounds.

$R^1R^2Si{=}NR^3$	$H_2Si{=}NH$	$H_2Si{=}NSiH_3$	$H\ddot{S}iNH_2$
(64)	**(65)**	**(66)**	**(67)**

Several calculations have been reported for the parent silanimine **65**[146,147,211,212]. The best available structure of **65** has been calculated with the 6-31G** basis set and it is shown in Figure 29. The Si=N bond length is 1.576 Å[211] (1.531 Å at 3-21G$^{(*)}$[146]), much shorter than the Si–N distance of 1.721 Å in $H_3Si–NH_2$. At the MCSCF level the Si=N bond distance is longer, 1.624 Å[147]. These values may be compared with that of 1.568 Å measured for **64**, $R^1 = R^2 = t$-Bu, $R^3 = (t\text{-Bu})_3Si$[145a]. Thus, the MCSCF procedure gives bond lengths which are too long, probably due to the mixing of some antibonding character into the bond description[147]. The HNSi angle varies with the basis set and it is 132.2°, 180°, 126.6° and 125.2° at 3-21G, 3-21G(*), 6-31G* and 6-31G**, respectively[146,211,212]. The conclusion is that a balanced basis set with polarization functions on both Si and N is required to describe correctly the geometry of this molecule (see below).

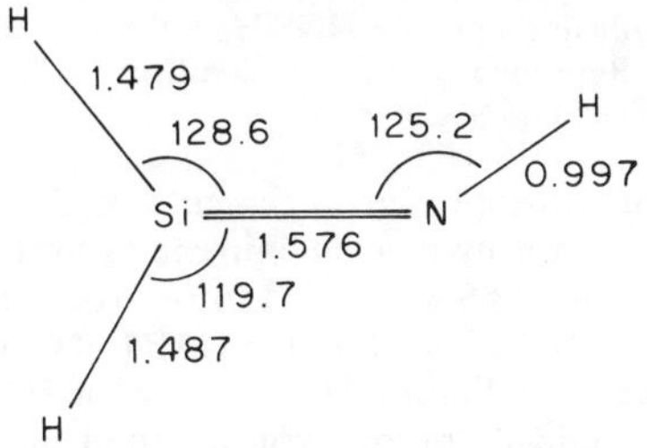

FIGURE 29. Optimized geometry of $H_2Si=NH$ (6-31G**). Bond lengths are in Å, bond angles in deg. Reprinted with permission from *J. Am. Chem. Soc.*, **108**, 1775 (1986). Copyright (1986) American Chemical Society[211].

The barrier for the inversion process at nitrogen in $H_2Si=NH$ is only 5.6 kcal mol^{-1} (at MP4/MC-311G**//6-31G**), much lower than the barrier to internal rotation around the Si=N bond (37.9 kcal mol^{-1}, see below)[211]. This implies a great ease of *syn–anti* isomerization relative to the carbon analogue $H_2C=NH$, where the barrier is ca. 25 kcal mol^{-1}. As the inversion barrier in $H_2Si=NH$ is low, it is conceivable that substitution at the nitrogen may favor a linear structure at nitrogen. Schleyer and Stout indeed found that attachment of SiH_3 to nitrogen as in **66** results in a nearly linear structure (SiNSi angle of 175.6°)[212a]. It was therefore concluded that electronic rather than steric effects are responsible for the nearly linear SiNSi arrangement found in the X-ray structure of **64**, $R^1 = R^2 = t$-Bu, $R^3 = (t\text{-Bu})_3Si$[145a].

The strength of the π-bond in $H_2Si=NH$ was estimated to be 36 kcal mol^{-1}, from the rotation barriers around the Si=N bond and from hydrogenation reactions[147]. This value can be compared with 63 kcal mol^{-1}, the estimated strength of the C=N π-bond[147]. The Si=N π-bond strength is similar to that in silaethylene (37 kcal mol^{-1}), but this bond is significantly stronger than the π-bond in disilene (22 kcal mol^{-1})[147]. However, the Si=N double bond is 54.1 kcal mol^{-1} weaker than two Si–N single bonds whereas the corresponding difference for C=N vs. two C–N bonds is only 1.6 kcal mol^{-1}[212a]. This large difference in bond strengths between Si=N and C=N bonds must have a dramatic effect on the reactivities of these analogous molecules.

The global minimum on the H_3SiN potential surface is not the silanimine **65**, but the aminosilylene **67** which lies 18 kcal mol^{-1} lower in energy[146,211]. However, the activation energy for the 1,2-hydrogen shift converting **65** to **67** is predicted to be high (60 kcal mol^{-1}), so that **65** and substituted silanimines are expected to be kinetically stable towards isomerization. The calculated transition structure for the **65**→**67** rearrangement is shown in Figure 30[211].

The model silylnitrene–silanimine isomerization ($H_3SiN \longrightarrow H_2Si=NH$) was studied at MP4SDQ/6-31G++G*//3-21G$^{(*)}$ in four possible electronic states[212b]. The

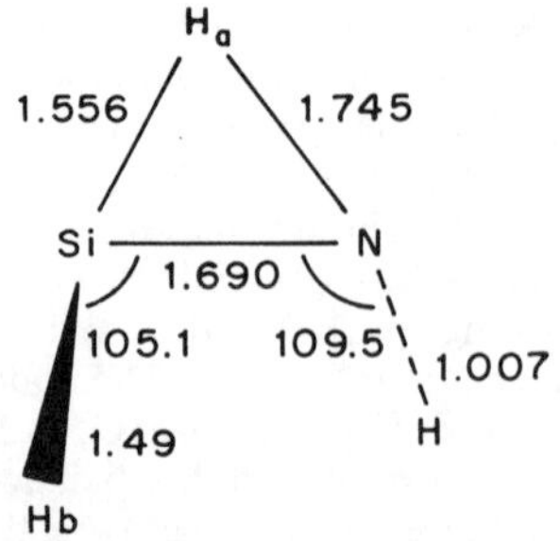

FIGURE 30. Optimized geometry of the transition state structure (6-31G**) for the rearrangement of $H_2Si{=}NH$ to $H\ddot{S}i{-}NH_2$. Dihedral angles: ∢$HNSiH_b = 188.0°$, ∢$HSiNH_a = 107.1°$. Bond lengths are in Å, bond angles in deg. Reprinted with permission from *J. Am. Chem. Soc.*, **108**, 1775 (1986). Copyright (1986) American Chemical Society[211].

conclusion was that the Curtius-type rearrangement occurring during the thermolysis or photolysis of silyl azide does *not* involve silylnitrene as an intermediate[212b].

The hydrogenation reaction of **65** was found to be exothermic by 38.7 kcal mol^{-1} and to have an activation energy of 89.7 kcal mol^{-1} (MP4/MC-311G**//6-31G**)[211]. The transition structure is shown in Figure 31. Addition of H_2 to the silylene **67** is both thermodynamically and kinetically more favorable than the addition of H_2 to **65**[211].

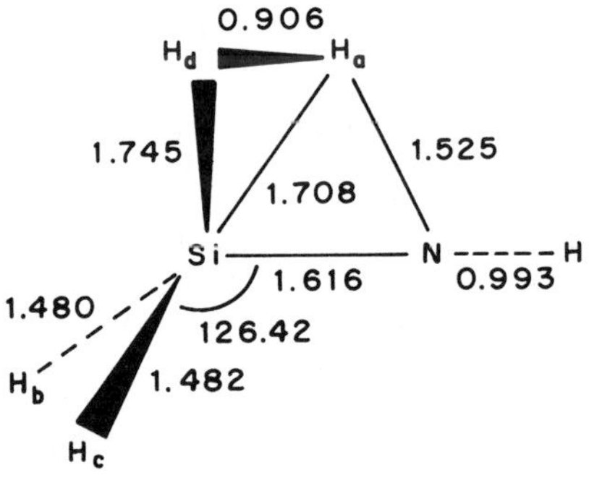

FIGURE 31. Transition state structure for the addition of H_2 to $H_2Si{=}NH$. ∢ $H_eNSi = 135.4°$, ∢$H_bSiN = 125.4°$, the $HNSiH_a$ dihedral angle = 203.1°, the $SiNH_aH_d$ dihedral angle = 7.4°, the H_bSiNH_a dihedral angle = −101.7° and the H_cSiNH_a dihedral angle = 94.9°. Bond lengths are in Å, bond angle in deg. Reprinted with permission from *J. Am. Chem. Soc.*, **108**, 1775 (1986). Copyright (1986) American Chemical Society[211].

4. Silanephosphimines

Unlike silanimines solid stable silanephosphimines **68** have not been yet isolated, but a sterically hindered silanephosphimine which is stable in solution was recently reported[19].

$$R^1R^2Si{=}PR^3 \qquad H_2Si{=}PH \qquad H\ddot{S}iPH_2$$

(68) **(69)** **(70)**

Three theoretical studies of $H_2Si{=}PH$ (**69**) have been reported[147, 213]. The calculated geometry of $H_2Si{=}PH$ is given in Figure 32. The calculated Si=P bond length is 2.044 Å at 3-21G* and 2.106 Å at the MCSCF/3-21G* level, both distances being significantly shorter than the Si–P single bond in H_3SiPH_2 of 2.256 Å at 3-21G*[147]. The calculated

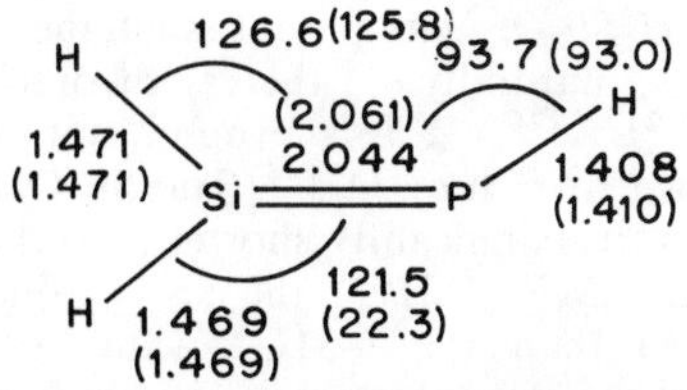

FIGURE 32. Optimized structure of $H_2Si{=}PH$. Bond lengths are in Å, bond angles in deg. 6-31G** values in parentheses[213b], 3-21G(*) values without parentheses.

stretching frequencies of the Si=P and the Si–P bonds are 673 and 485 cm^{-1}, respectively[147].

The strength of the Si=P double bond, as measured by the rotation barrier, is 29 kcal mol^{-1} compared with 43 kcal mol^{-1} for the C=P bond[147]. As the inversion barrier at phosphorous is higher than 30 kcal mol^{-1}, the preferred mechanism for *syn–anti* isomerization is via rotation[147]. This contrasts with the behavior of $H_2Si{=}NH$, where the inversion barrier is lower in energy (see Section V.A.3). **69** is found to be about 13 kcal mol^{-1} lower in energy than the isomeric phosphinosilylene **70** and the barrier to the 1,2-hydrogen shift is 40 kcal mol^{-1} [213a]. Note that silanimine **65** exhibits the opposite behavior, being less stable than the silylene **67**. The opposite relative stability in the isomeric pairs **69–70** and **65–67** results mainly from the fact that the silylene is stabilized more effectively by NH_2 than by PH_2. The transition structure for the **69–70** isomerization is shown in Figure 33. The potential energy surface for the triplet approximately parallels the singlet surface, but it is higher in energy.

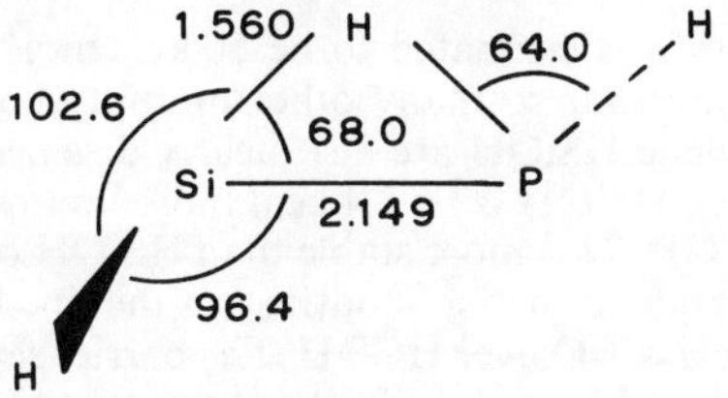

FIGURE 33. Optimized geometry of the transition state structure for the rearrangement of $H_2Si{=}PH$ to $H\ddot{S}i{-}PH_2$ (3-21G(*)). Bond lengths are in Å, bond angles in deg. Reprinted with permission from *J. Am. Chem. Soc.*, **107**, 4535 (1985). Copyright (1985) American Chemical Society[213].

5. *Silanones*

No silanones **71**, i.e. compounds with silicon–oxygen double bonds, have been isolated yet either neat or even in dilute solution, but matrix-isolation techniques have recently allowed their direct observation. Consequently, most of our experimental knowledge of silanone properties still originates in studies on transients[19]. Theory, being a primary source of reliable fundamental information, is therefore extremely valuable to the study of these species.

The simplest molecules with silicon–oxygen double bonds are SiO and SiO_2[146, 214]. The calculated bond lengths are 1.494 Å in SiO[214a] (experimental 1.51 Å[214b]) and 1.484 Å in SiO_2 which is calculated to be linear[214a]. Similar results were reported for these compounds in Reference 146.

$R^1R^2Si{=}O$ $H_2Si{=}O$

(71)   **(72)**

a. The parent silanone, $H_2Si{=}O$. The parent silanone **72**, the silicon analogue of formaldehyde, is naturally the theoretically most studied silanone[137,146,147,175b,182,215–220]. **72** is a ground state singlet and its optimized geometry at 6-31G* is shown in Figure 34[217a]. The Si=O bond length at 6-31G* is 1.498 Å (1.509 Å at 3-21G$^{(*)}$[146]), significantly shorter (by 0.149 Å) than the Si–O single bond length of silanol[217a]. At MP2/6-31G* the Si=O bond distance is found to be significantly longer (1.545 Å) than the 6-31G* value or the CEPA-PNO value (1.507 Å)[215a]. This may reflect the general overcorrection of multiple bond lengths at the MP2 level[25].

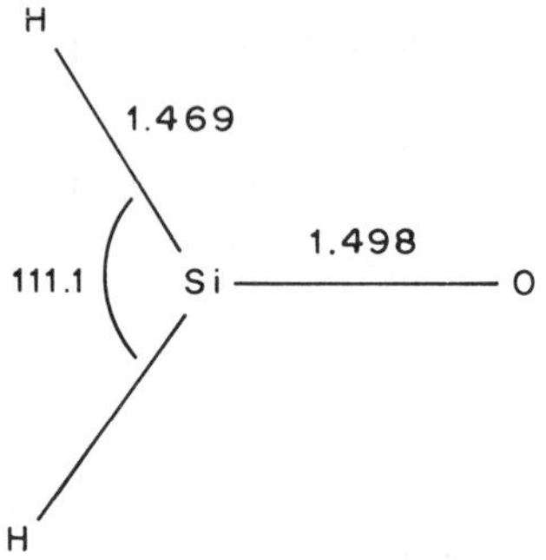

FIGURE 34. Optimized structure (6-31G*) of $H_2Si{=}O$[217a]. Bond lengths are in Å, bond angles in deg.

The Si=O bond strength was estimated to be 50 kcal mol^{-1} compared with 77 kcal mol^{-1} for C=O[147]. As found for so many other multiply bonded silicon compounds, $H_2Si{=}O$ and hydroxy silylene HS̈iOH are also nearly degenerate. At the CI(SDQC)/6-31G**//6-31G + ZPE level, HS̈iOH is by 2.4 kcal mol^{-1} more stable than $H_2Si{=}O$[217a], but at MP4/6-31G*//3-21G$^{(*)}$ **72** is more stable than HS̈iOH by 4.6 kcal mol^{-1}[146]. This small energy difference stands in distinct contrast to the much larger energy difference (53.9 kcal mol^{-1}) favoring $H_2C{=}O$ over HC̈OH. The barrier separating **72** and HS̈iOH is large, 60.8 kcal mol^{-1} at the CI level[217a]. This suggests that silanone is kinetically stable towards a 1,2-hydrogen shift. As will be demonstrated below, **72** is also kinetically stable towards dissociation to $H_2 + SiO$, either via a one-step or a stepwise mechanism[217a].

The triplet ($^3A''$) state of H_2SiO is calculated (MP4/6-31G*) to be 69.3 kcal mol^{-1} above the singlet, only slightly less than the 74.2 kcal mol^{-1} singlet–triplet separation found in formaldehyde. The Si–O bond distance in the triplet is much longer (1.692 Å) than in the singlet state, and this increase in the bond distance is comparable to that found in comparing singlet and triplet formaldehyde[146].

The low-lying valence and Rydberg excited states and the singlet vertical excitation energies of $H_2Si{=}O$ have been calculated using the multiple-reference CI methods[218]. Both the excitation order of the absorptions and their magnitudes are considerably different in $H_2Si{=}O$ and in the analogous $H_2C{=}O$[218]. Glinski and coworkers[216] have also calculated several of the excitation energies from the ground state of H_2SiO and used the theoretical data to conclude that the dominant chemiluminescent features in the range 4700–6500 Å, observed in gas-phase oxidation processes in the silane–ozone system, are due to H_2SiO[216].

b. Substituted silanones. Several substituted silanones **71** have been studied, in particular, those with R^1 and (or) $R^2 = CH_3$[182], OH[95c], F[217b] and SiH_3[221]. In these studies attention was concentrated on the effect of the substituent(s) on the relative energies of the various possible isomers. For example, Gordon and George[182] studied, in addition to methyl silanone **71**, $R^1 = H$, $R^2 = CH_3$ five other CH_4SiO isomers. At MP3/6-31G*, assuming additivity of polarization and correlation effects, they find that methyl silanone is the most stable structure, $CH_3\ddot{Si}OH$ and $H_3SiCH{=}O$ being 7.4 and 19 kcal mol^{-1}, respectively, higher in energy. $H_2C{=}SiH(OH)$, the enol form of $CH_3SiH{=}O$, lies 21.2 kcal mol^{-1} higher in energy. The isomeric enol $H_2Si{=}CH(OH)$ is as expected (Si–O vs. C–O bonding), even higher in energy, lying 35.1 kcal mol^{-1} above $H_2C{=}SiH(OH)$ and 37.3 kcal mol^{-1} above the corresponding ketone $H_3SiCH{=}O$[182].

Kudo and Nagase found that fluorine substitution reverses the order of stability of silanone and of the corresponding silylene, so that at MP3/6-31G*//3-21G, $H\ddot{Si}OF$ is by 126.5 kcal mol^{-1} less stable than $FHSi{=}O$[217b]. This result could have been anticipated considering the dramatic differences in the strengths of the Si–F and O–F bonds.

Kudo and Nagase also found that silyl substitution has a significant effect on the energetics of the 1,2-isomerization process[221]. Thus, the 1,2-silyl shift in **73**, shown in equation 23, is calculated [MP4SDTQ/6-31 + G(2d,p)// 6-31G* + ZPE] to be exothermic by 18 kcal mol^{-1} and the isomerization barrier is 22.7 kcal mol^{-1}. This barrier is considerably lower than the 61 kcal mol^{-1} calculated barrier for the 1,2-hydrogen shift in $H_2Si{=}O$[217a]. Thus, while **72** is kinetically stable towards isomerization, the theoretical prediction is that **73** may rearrange to silylene under relatively mild conditions, in agreement with recent experimental studies[222a].

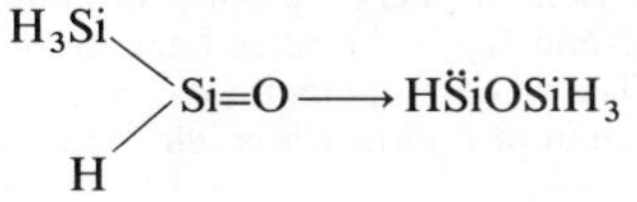

(23)

(73)

c. Reactions. Considerable attention has been devoted to the unimolecular decomposition of silanone to either $H_2 + SiO$ (equation 24) or $H + HSiO$[217a]. These reactions were found to have a barrier which is higher than the barrier for the 1,2-hydrogen shift that converts silanone to the hydroxysilylene (see above).

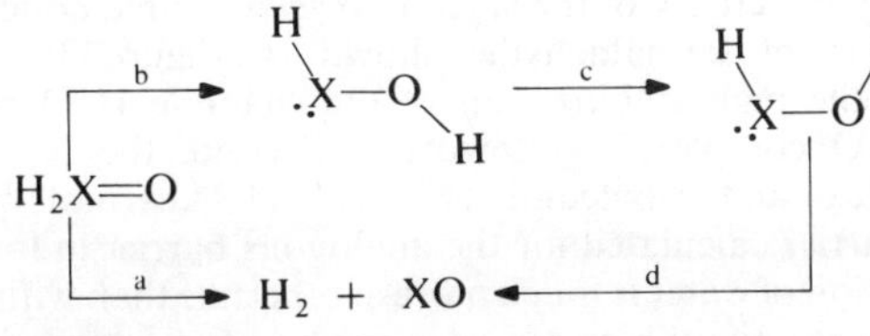

(24)

$X = C, Si$

Comparison of the reaction steps in equation 24 for silanone ($X = Si$) and for formaldehyde ($X = C$) shows that the two molecules decompose to hydrogen + XO via different mechanisms (Figure 35)[217a]. The single-step elimination of H_2 (process *a* in equation 24) is highly endothermic for both $H_2Si{=}O$ and $H_2C{=}O$. However, the two systems behave differently in the multistep reaction (steps *b–d* in equation 24). With $H_2Si{=}O$ the highest barrier along the path $b \rightarrow c \rightarrow d$, of 60.8 kcal mol^{-1}, is associated with the isomerization step *b*. With $H_2C{=}O$, on the other hand, the highest barrier in the multistep path is for step *d*. This barrier, which is in excess of 105 kcal mol^{-1}, is much higher than calculated for reaction *d* for $H_2Si{=}O$ or for reaction *a* of $H_2C{=}O$ (Figure 35). Thus, $H_2Si{=}O$ decomposes to $H_2 + SiO$ via a multistep reaction (path $b \rightarrow c \rightarrow d$) with a

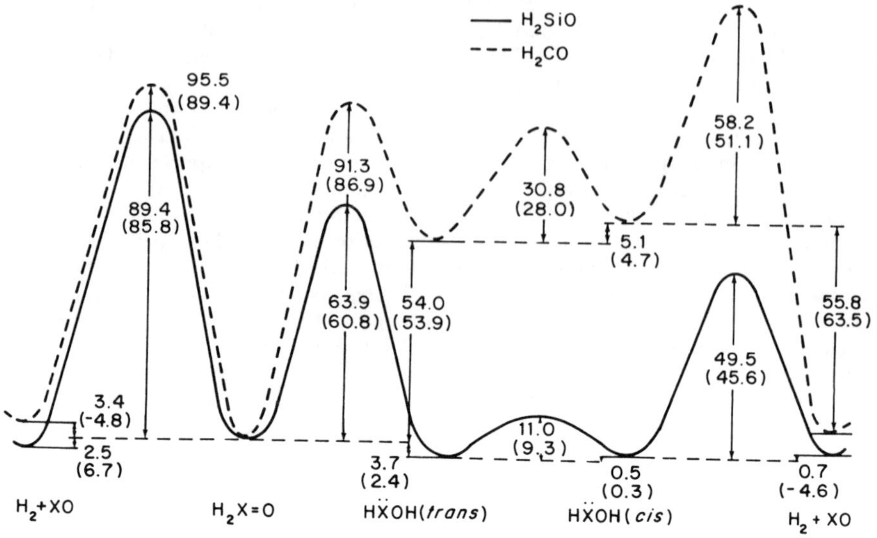

FIGURE 35. Schematic comparison of energy profiles for unimolecular rearrangements and dissociations of $H_2Si{=}O$ (full line) and $H_2C{=}O$ (dotted line). Relative energies in kcal mol^{-1} at CI (S + D + QC)/6-31G**//6-31G*. The values in parentheses are corrected for zero-point vibrational energies. Reprinted with permission from *J. Phys. Chem.*, **88**, 2833 (1984). Copyright (1984) American Chemical Society[217a].

barrier of 60.8 kcal mol^{-1}, while formaldehyde dissociates to $H_2 + CO$ in a single-step reaction (i.e. path *a*) with a higher barrier of 89.4 kcal mol^{-1} [217a]. Dynamic aspects of the unimolecular dissociation reactions of silanone were also analyzed recently by Tachibana, Fueno and Yamabe[219].

The addition of water to silanone was also studied in detail by Kudo and Nagase[217a]. The reaction proceeds through an intermediate complex, but with no overall barrier, indicating the very high reactivity of the silicon–oxygen double bond. Comparison with the analogous reaction of formaldehyde, shown in Figure 36, reveals interesting differences[217a]. First, the high stabilization of the silanone–H_2O complex relative to the formaldehyde–H_2O complex is noteworthy. Second, the barrier separating the silanone–H_2O complex and silanediol is small (4.8 kcal mol^{-1}), in contrast to the 39.1 kcal mol^{-1} barrier calculated for the analogous barrier in formaldehyde. Third, with silanone the addition of water is much more exothermic than with formaldehyde[217a]. The large difference in reactivity between silanone and formaldehyde was explained in terms of the higher polarity and the lower HOMO–LUMO gap in silanone. Thus, the high reactivity of doubly bonded silicon compounds is both 'frontier orbital controlled' as well as 'charge-controlled'[217a].

The potential energy surface for the 'head-to-tail' dimerization of silanone (equation 25) was also studied[137]. The fascinating structure of the product **33** was discussed in Section IV.C.2.b. The dimerization is calculated to be highly exothermic (109.4 kcal mol^{-1} at

(72) **(33)** (25)

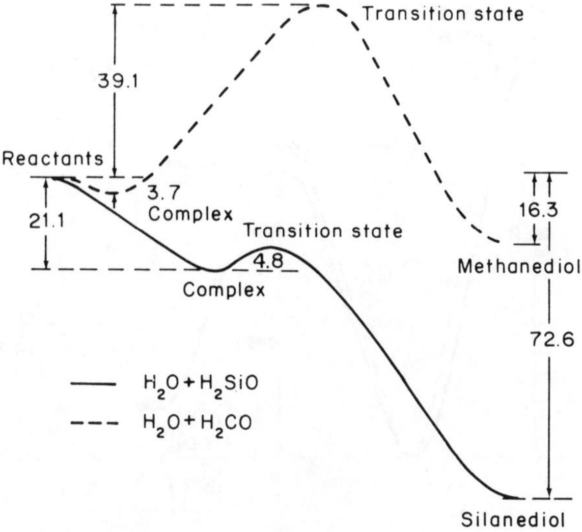

FIGURE 36. Energy profiles (in kcal mol^{-1}) for the addition of H_2O to $H_2Si{=}O$ (full line) and $H_2C{=}O$ (dotted line), at MP3/6-31G*//6-31G*. Reprinted with permission from *J. Phys. Chem.*, **88**, 2833 (1984). Copyright (1984) American Chemical Society[217a].

MP2/6-31G*//6-31G*) and to proceed with no barrier, via a stepwise mechanism. In contrast, the 'head-to-tail' dimerization of formaldehyde involves a high energy barrier. The 'head-to-head' dimerization of **72**, in contrast to the 'head-to-tail' process, does involve a considerable barrier[137].

The possible metathesis reaction 26 was also studied and at MP3/6-31G* is exothermic by 25 kcal mol^{-1} [182]. This value was used to suggest that the C=O bond is about 10 kcal mol^{-1} stronger than the Si=O bond[182]. This reaction was also studied by Bachrach and Streitwieser[136] and is discussed further in Section V.A.1.a.vi.

$$H_2Si{=}CH_2 + H_2C{=}O \longrightarrow H_2Si{=}O + H_2C{=}CH_2 \qquad (26)$$

6. *Silanethiones*

No silanethiones $R_2Si{=}S$ have yet been reported to be stable in pure form or in solutions of any kind and even the evidence for their transient existence is still limited[19], although recent developments are very promising[222b]. Only a few calculations have been reported, all for the parent silanethione, $H_2Si{=}S$[137,147,220].

The singlet ground state of $H_2Si{=}S$ is calculated to be planar, with C_{2v} symmetry and a Si=S bond length of 1.936 Å (6-31G*)[220], which is 0.216 Å shorter than the Si–S single bond length, e.g. 2.152 Å in $H_3Si{-}SH$[220]. This indicates the existence of a strong π-bond between the Si and S atoms. The strength of the π-bond is estimated to be 50 kcal mol^{-1} [147], as strong as the C=S and Si=O π-bonds[147]. According to a different estimate the Si=S π-bond strength is 42 kcal mol^{-1}, i.e. the bond is significantly stronger than the Si=O bond (33 kcal mol^{-1}) but weaker than the C=O π-bond (63 kcal mol^{-1})[220]. The open-shell triplet states of $H_2Si{=}S$, both $^3A''$ (n–π^*) and $^3A'$(π–π^*), were also studied[220].

Various unimolecular reactions of $H_2Si{=}S$, such as the 1,2-hydrogen shift to produce $H\ddot{S}iSH$ and the dissociation reactions to $H_2 + SiS$ or to $H + HSiS$, have been studied[220]. The calculated potential energy surfaces of analogous reactions of $H_2Si{=}S$ and of $H_2Si{=}O$

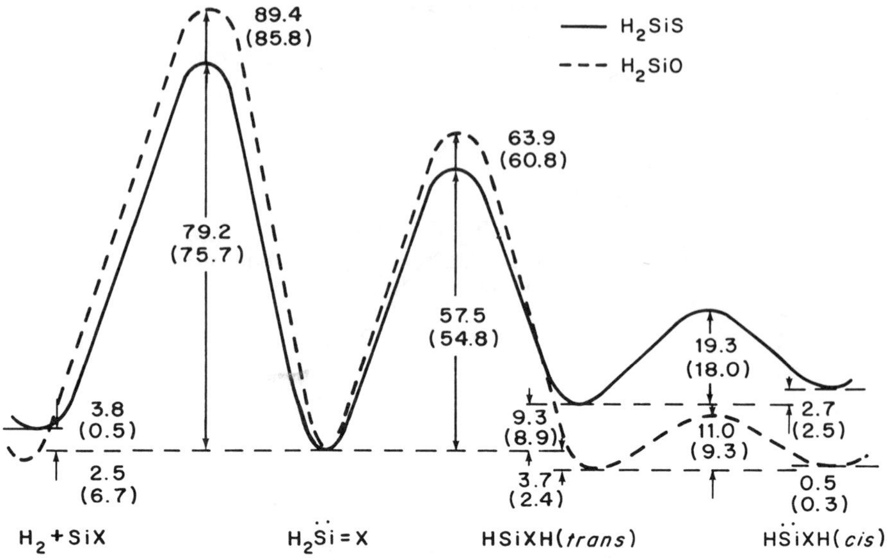

FIGURE 37. Schematic comparison of the energy profiles for unimolecular rearrangements and dissociations of $H_2Si{=}S$ (full line) and $H_2Si{=}O$ (dotted line). Relative energies in kcal mol^{-1} at CI (S + D + QC)/6-31G**//6-31G*. The values in parentheses are corrected for zero-point vibrational energies. Reprinted with permission from *Organometallics*, **5**, 1207 (1986). Copyright (1986) American Chemical Society[220].

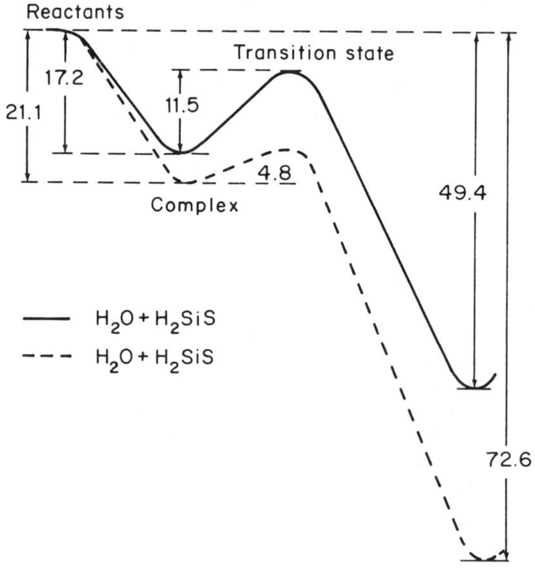

FIGURE 38. Energy profiles (in kcal mol^{-1}) for the addition of H_2O to $H_2Si{=}S$ (full line) and $H_2Si{=}O$ (dotted line), at MP3/6-31G*//6-31G*. Reprinted with permission from *Organometallics*, **5**, 1207 (1986). Copyright (1986) American Chemical Society[220].

are very similar, as shown in Figure 37. The addition of H_2O to $H_2Si{=}S$ (Figure 38) and to $H_2Si{=}O$ (Figure 36) also have similar energy profiles, although the former is some 23 kcal mol^{-1} less exothermic, and the barrier for the decomposition of the intermediate complex is somewhat higher for $H_2Si{=}S$[220].

More dramatic differences in reactivity between $H_2Si{=}S$ and $H_2Si{=}O$ are found in their dimerization reactions[137]. The calculated 'head-to-tail' dimerization energies are (MP2/6-31G*//6-31G*+ZPE) −104.0, −67.7 and 6.1 kcal mol^{-1} for $H_2Si{=}O$, $H_2Si{=}S$ and $H_2C{=}O$, respectively[137]. The calculated thermodynamic quantities, such as enthalpies, entropies and free energies, have been also reported and ΔG° for the above reactions are −92.1, −55.8 and 17.1 kcal mol^{-1} [137]. Kudo and Nagase have also studied the protonation reactions of $H_2Si{=}S$ and $H_2Si{=}O$ leading to the corresponding silicenium ions (see Section VA.B.1)[220].

Kudo and Nagase concluded that the π-bond in silanethione is significantly stronger than that in silanone and it is also more stable kinetically. They suggested that the synthesis and isolation of silanethiones might be achieved by replacing the hydrogen in $H_2Si{=}S$ by substituents that reduce the polarity of the silicon–sulfur double bond[220].

B. Triply Bonded Silicon Compounds

Both experimentally and theoretically triply bonded silicon compounds proved to be a 'harder nut to crack' than doubly bonded silicon compounds. Strong experimental evidence for the existence of triply bonded silicon compounds such as RSi≡CR or RSi≡SiR is not yet available (however, HN≡Si has been identified, see below), although some indications that they may exist as transient species have been presented[19]. In this situation, theory remains the only reliable source of information for these molecules that allow one to study their fundamental properties and reactions. Furthermore, the calculations can be used to develop and test ideas, which may eventually lead to the synthesis of stabilized molecules of this family that may have a better chance to be observed experimentally. As will be demonstrated in the discussion below, theory has faced many difficulties in studying these molecules and it seems that only in the last two years the calculations are converging towards a consensus about the structures and relative stabilities of these molecules.

1. Silynes

Silynes (or silaethynes or silaacetylenes) are the silicon analogues of acetylenes and are thus expected to possess a carbon–silicon triple bond in the classical structure **74**. There is as yet no strong experimental evidence for the existence of such molecules[19]. The molecule which was studied theoretically most extensively is the parent silaacetylene H_2CSi, but even for this simple molecule much still remains to be done, as most of the studies were carried out more than five years ago when the state-of-the-art methods were much less accurate and reliable than those available now.

Several possible structures **75**–**78** were considered for H_2CSi in an attempt to establish which structure is the global minimum on the potential energy surface[146,160,186,223–226]. However, only in two of these studies were the structures properly characterized using gradient techniques[146,226].

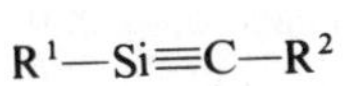

(74)

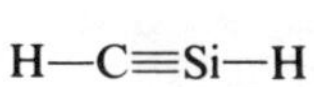

(75)

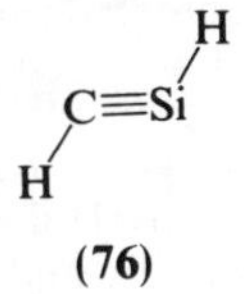

(76)

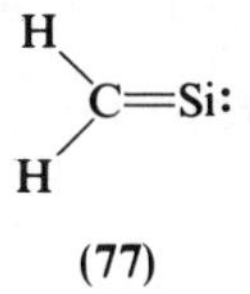

(77)

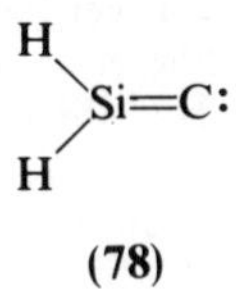

(78)

At the HF level of theory the classical linear structure **75** is not a minimum on the potential energy surface. At this level of theory, the only minimum on the surface is the silylidene **77**[225,226]. The first to predict this surprising result, considering the analogy with carbon chemistry, were Murrell, Kroto and Guest[223]. Gordon and Pople[225] suggested that if electron correlation is included, an equilibrium *trans* structure **76** can be found in a shallow minimum (however, the structure of **76** has not been characterized analyically). This suggestion was verified later by Hoffmann, Yoshioka and Schaefer[226], who characterized **76** analytically and showed that it is a minimum at the CI/dz+p level of theory.

The best available geometries of **76** and **77**, the analytically characterized minima on the H_2CSi surface, are shown in Figure 39. The calculated C≡Si triple bond length in **76** of 1.635 Å is significantly shorter than the C=Si double bond in $H_2C{=}SiH_2$ (~1.70 Å) indicating shortening of the C=Si bond upon the introduction of the second π-bond. The C≡Si bond is even shorter in the linear structure **75** (Figure 39). However, as indicated above, **75** is not a minimum on the potential energy surface and it lies some 9 kcal mol^{-1} above the *trans*-bent structure **76**[226]. The *cis*-bent silaacetylene is also not a minimum on the potential energy surface.

Although the triply bonded structure **76** was found to be a local minimum on the potential surface, it is not very stable, either thermodynamically or kinetically. **76** lies considerably higher in energy than the silylidene isomer **77**, i.e. by approximately 50 kcal mol^{-1} at either MP3/6-31G*[225] or CI/dz+p[226]. Furthermore, the barrier separating **76** and **77** is small, only 5–8 kcal mol^{-1} [225,226] and the calculated transition structure is also shown in Figure 39. The silavinylidene **78** which lies in a very shallow minimum is by about 90 kcal mol^{-1} less stable than **77**[226]. The large **77**–**78** energy difference reflects to a large degree the much higher stability of a lone-pair on silicon in comparison to a lone-pair on carbon. Gordon and Koob[186] also pointed out the extremely low stability of silaacetylenes, as expressed in the dramatic increase in hydrogenation energy when a second π-bond to silicon is added. This increase was considerably larger than would have been expected by comparing C≡C with C=C.

In conclusion, the available calculations suggest that the singlet silyne resides in a very shallow minimum on the H_2CSi potential energy surface. However, on the basis of previous experience it is reasonable to predict that stage-of-the-art calculations will further reduce the isomerization barrier, and that **76** might not even be a minimum on the energy surface. Consequently, our conclusion is that the prospects to observe the parent singlet silyne in a matrix or even as a transient intermediate are small.

The triplet surface of the H_2CSi species was studied recently by Luke and coworkers[146]. The lowest-energy triplet state ($^3A_2$) adopts the silylidene structure **77** and it results from the promotion of an electron from the π-orbital to the nonbonding silicon p-orbital in the plane of the molecule. The triplet **77** lies 35.6 kcal mol^{-1} above the singlet at MP4SDTQ/6-31G*//3-21G$^{(*)}$[146] (see also Reference 224b). The lowest triplet ($^3A''$) which has the *trans*-bent structure **76** lies 61.1 kcal mol^{-1} above singlet **77**, and still higher in energy is a $^3A'$ triplet with a *cis*-bent structure[146].

The effect of methyl, fluorine and lithium substitution at both the carbon and the silicon ends of silyne was studied by Gordon[160,186]. Hopkinson and coworkers studied methyl substitution[224]. The most interesting conclusion from Gordon's studies[160,186] is that both fluorine and lithium substitution at silicon (i.e. **74**, $R^2 = H$, $R^1 = F$ or Li) stabilizes the silaethyne relative to the isomeric silylidene, whereas the effect of methyl substitution is relatively small. On the other hand, substitution at carbon has little effect on this energy difference. Thus, at 6-31G*//3-21G linear HC≡SiF is less stable by 21.5 kcal mol^{-1} than HFC=Si: compared to the **76**–**77** energy difference of 56.3 kcal mol^{-1} at the same level of theory[160]. This is not surprising in view of the typical high strength of the Si–F bond which stabilizes HC≡SiF relative to HFC=Si:. Distortion of the linear structure of HC≡SiF lowers its energy substantially and a minimum is found (∡ FSiC = 104° and

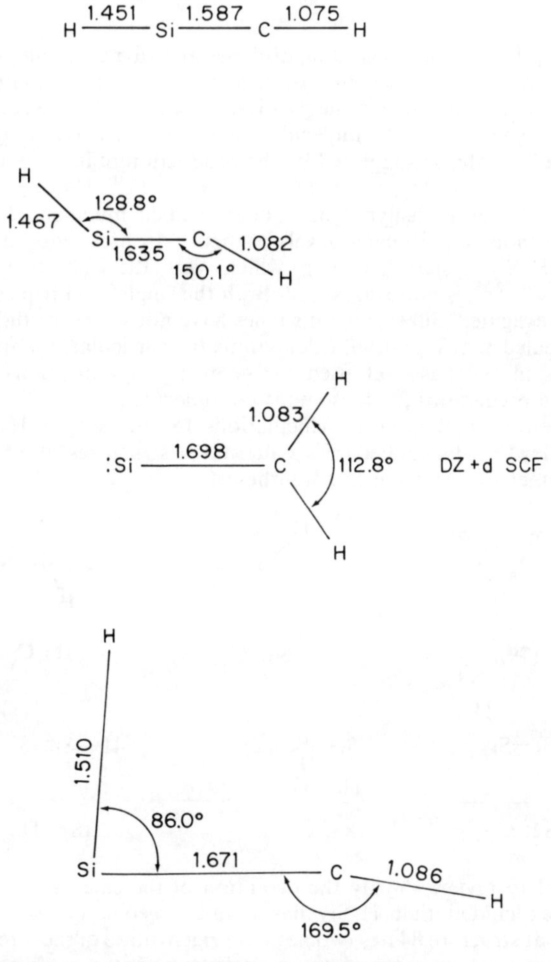

FIGURE 39. Optimized structures of H_2CSi in the linear (**75**) and *trans*-bent (**76**) conformation (CI/dz + d), and in $H_2C{=}Si$: (**77**) (dz + d) and the transition state structure for their interconversion (dz + d.) Bond lengths are in Å. Reprinted with permission from *J. Am. Chem. Soc.*, **105**, 1084 (1983). Copyright (1983) American Chemical Society[226].

$\angle$ HCSi = 164°) which is 1.5 kcal mol^{-1} lower in energy than HFC=Si:[160]. Furthermore, fluorine substitution not only stabilizes the formal triple-bond structure relative to the silylidene, but it also raises the barrier to the silyne–silylidene isomerization to 16.8 kcal mol^{-1}[160], thus increasing the kinetic stability of the silaethyne.

The above calculations of substituent effects show interesting trends, which may provide important clues for designing molecules with more stable C≡Si triple bonds. However, in light of the present experience, it is clear that these calculations (e.g. Reference 160) used levels of theory which are too crude for reaching reliable conclusions (e.g. minima and transition structures were not characterized analytically, relatively small basis sets and correlation techniques were used, etc.). We therefore think that it is highly desirable to study these molecules with the presently available more accurate theoretical methods.

2. Disilynes

Experimentally, the situation regarding disilynes and silynes is quite similar. No direct experimental evidence for the existence of disilynes is available[19], although Si_2H_2 has been proposed as an intermediate in the pyrolysis of silane[227, 228]. Recently, West, Michl and coworkers designed a beautiful molecule which, upon thermolysis, gives the synthetic equivalent of MeSi≡SiMe, as suggested by the production of its trapping products with anthracene[229].

Theoretically , however, disilynes have been studied much more extensively than silynes and using more sophisticated levels of theory. Consequently our understanding of the parent H_2Si_2 system (singlet), which was the subject of all theoretical studies[46, 47, 146, 230 – 235], is more advanced. Both the singlet and triplet potential energy surfaces were investigated. Substituted disilynes have not yet been studied. The study of H_2Si_2 again revealed that high-level calculations (in particular, the inclusion of polarization functions in the basis set used for geometry optimizations) are essential in obtaining reliable predictions for these unusual molecules.

The most extensive and accurate calculations for the singlet H_2Si_2 surface were reported by Binkley[47b], who studied the six alternative structures **79**–**84** shown below (the point-group symmetries are given in parentheses).

(79) C_{2v}

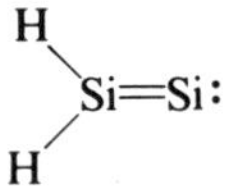

(80) C_{2v}

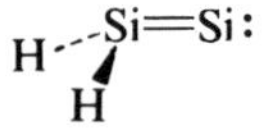

(81) C_s

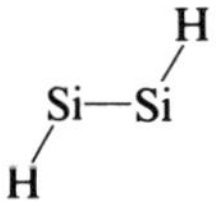

(82) C_{2h}

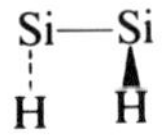

(83) C_2

H—Si≡Si—H

(84) $D_{\infty h}$

It is of interest to review shortly the evolution of the calculations for H_2Si_2. Early calculations[46b] concluded that H_2Si_2 has a linear geometry (**84**), but it was later determined[230] that structure **84** has two negative eigenvalues of the force constant matrix and therefore does not correspond to a minimum on the surface. Snyder and coworkers[231] determined the equilibrium geometries of five prototypical conformations using HF theory and the 4-31G basis set. Calculations which included electron correlation (PP-GVB), but used the 4-31G geometries, concluded that the most stable Si_2H_2 is **80**[231]. The 'classical' linear disilyne **84** was found to be the least stable conformer[231]. Kawai and coworkers also reported (at the same time) that the linear geometry **84** is not a minimum and suggested that the equilibrium geometry of Si_2H_2 is twisted (**83**)[232]. Lischka and Kohler[47a] studied Si_2H_2 with analytical force-constant techniques and optimized the geometries of the various structures with inclusion of electron correlation (CEPA) and a dz basis set. In contrast to the previous studies they found that the global minimum is the bridged structure **79**, with **80** and **82** being higher in energy by 11.8 and 14.3 kcal mol^{-1}, respectively[47a]. These conclusions did not change when more extensive and accurate calculations were applied to this problem by Kalcher and coworkers[233], by Binkley[47b] and by Luke and coworkers[146]. These studies used polarized basis sets for the geometry optimizations and relative energies were evaluated by subsequent single point calculations which included correlation energy. The inclusion of polarization functions in the basis set was found to be essential in describing these molecules[47b, 146]. It was further shown that the energy ordering predicted for the isomeric Si_2H_2 singlets is unaltered by inclusion of electron correlation, provided that the basis set contains polarization functions[47b].

Let us now examine in more detail the different H_2Si_2 species **79**–**84**, using mainly Binkley's[47b] MP4/6-31++G**//6-31G** results (geometry optimizations at MP2/6-31G* had little effect[47b]). At this level of theory two minima **79** and **80** were located on the singlet surface. The *trans*-bent structure **82** has one imaginary frequency and it is therefore a transition structure. The calculated relative energies of the various H_2Si_2 isomers are shown in Table 27.

TABLE 27. Calculated (MP4/6-31++G**//6-31G**) relative energies (ΔE) of various H_2Si_2 isomers[a]

Isomer	Electronic state	ΔE^b
Bridged disilyne (**79**)	$^1A_1$	0.0
Planar silasilylene (**80**)	$^1A_1$	11.3
trans-Bent disilyne (**82**)	^1A_g	14.2
Bent silasilylene (**81**)	$^3A''$	22.0
Twisted disilyne (**83**)	^3B	32.1
Linear disilyne (**84**)	$^1\Sigma_g^+$	40.0

[a] From Reference 47b.
[b] Relative energies in kcal mol^{-1}, including 6-31G** zero-point corrections.

Probably the most striking result in Table 27 is that the linear disilyne **84**, the structure analogous to acetylene, is the *least stable* isomer of Si_2H_2. Furthermore, **84** has two imaginary frequencies and it is therefore not a minimum on the singlet potential surface. The most stable Si_2H_2 isomer is the bridged form **79**, which is more stable than the silasilene **80** by 11.3 kcal mol^{-1} [47b] (Luke and coworkers calculated 9.1 kcal mol^{-1} at MP4SDTQ/6-31G*//3-21G$^{(*)}$ [146]). The *trans*-bent form **82** (not a stable structure) lies 14.2 kcal mol^{-1} above **79**. A comparison with the analogous carbon compounds is instructive. Linear acetylene is, of course, the global minimum on the C_2H_2 surface. Vinylidene, $H_2C{=}C$:, is 39.6 kcal mol^{-1} higher in energy and structures analogous to **79** and **82** (triplet) are higher in energy (by 72.1 and 93.2 kcal mol^{-1}, respectively)[47b].

The calculated geometries of **79** and **80** at 6-31 G, 6-31 G** and MP2/6-31 G* are given in Table 28. The Si–Si distances in **79** and **80** are not very different, being 2.181 Å and

TABLE 28. Optimized geometrical parameters at 6-31G, 6-31G** and MP2/6-31G* of the bridged disilyne (**79**) and silasilylene (**80**)[a, b]

Geometrical parameter	6-31G		6-31G**		MP2/6-31G*	
	79	**80**	**79**	**80**	**79**	**80**
r(Si–Si)	2.299	2.314	2.181	2.187	2.202	*d*
r(X–H)[c]	1.319	*e*	1.276	*e*	1.253	*e*
∢ (HXH)[c]	101.1	*e*	102.2	*e*	104.5	*e*
r(Si–H)	1.750	1.490	1.679	1.475	1.668	*d*
∢(HSiSi)	48.9	123.4	49.5	123.4	48.7	*d*

[a] From Reference 47b.
[b] Bond distances in Å, bond angles in deg.
[c] X denotes the midpoint of the Si Si bond.
[d] Not given.
[e] Not relevant in **80**.

2.187 Å, respectively (at 6-31G**). Both distances are similar to the double bond length in $H_2Si{=}SiH_2$ (2.128 Å at 6-31G**[202]). Binkley also gives harmonic frequencies for the Si_2H_2 isomers at these levels of theory[47b].

Binkley interpreted the instability of linear disilyne and the peculiar preference of the bridged structure in terms of reduced s–p mixing in silicon relative to carbon. This leads to a Si–Si σ-bond which is dominated by the s-orbitals, which in turn causes a shortening of this bond and induces excessive repulsions due to crowding in the π-space. These repulsions are relieved by bending of the bond and twisting it into the bridged geometry **79**[47b]. Baird[234] has similarly attributed the preference of the nonclassical bridged structure to the weakness of the π-bonding between the silicon atoms and to the high energetic cost of s- to p-electron promotion. A different interpretation was recently proposed by Trinquier and Malrieu[189], who suggested that the occurrence and extent of such distortions is related to the singlet–triplet separations in the fragments that form the multiple bond, in this case two HSi units. According to this analysis the energy difference between the $^2\pi_r$ and $^4\Sigma^-$ states increases along the series: HC≡CH < HSi≡CH < HSI≡SiH, and with it the structure changes from linear (HC≡CH), to borderline (HC≡SiH), to nonlinear (HSi≡SiH)[189].

The triplet surface of H_2Si_2 was studied by several groups[47,146,233,235]. The most recent and accurate study is that by Koseki and Gordon[235]. The global minimum on the triplet potential surface is probably the planar silasilene (**80**,$^3A_2$), but the bent silasilene (**81**,$^3A''$) is only 2 kcal mol^{-1} higher in energy (at MP4SDTQ/6-31G++G//6-31G**)[235]. Previous studies found the $^3A''$ state to be somewhat lower in energy than the $^3A_2$ state[47b,146]. Koseki and Gordon also found that the most stable triplet disilyne is the *trans*-bent HSiSiH (**82**, ^3A_u), which is more stable than the twisted structure **83**, reported earlier[47b] as being the lowest in energy. Furthermore, the *trans*-bent **82**, ^3A_u state lies only 3–4 kcal mol^{-1} above the global minimum on the triplet surface (**80**, $^3A_2$). The *cis*-bent HSiSiH structure (C_s, $^3A''$) was also found to be a minimum on the triplet surface and it was suggested that it might be observable[235]. Due to the small energy differences between the various structures we believe that the triplet calculations have not yet converged to a definitive answer and further calculations are needed to probe more carefully this shallow energy surface. It should also be noted that the triplet and singlet H_2Si_2 surfaces are quite close in energy. At MP4SDTQ/6-31G*//3-21G$^{(*)}$ the $^1A_1$ and $^3A''$ states of **80** and **81**, respectively, are separated by only 10 kcal mol^{-1} [146]. At higher levels of theory (and in reality), the two surfaces might be even closer in energy.

We are unaware of any study of the effect of substituents on the shape of the H_2Si_2 potential energy surface. Such studies, which are most desirable, can be expected to reveal many interesting features and will hopefully assist the experimentalists in designing a more stable disilyne.

3. *Silanitriles, silaisonitriles and their phosphorous analogues*

The isomeric parent silanitriles (HSi≡N) and silaisonitriles (HN≡Si), which formally possess Si–N triple bonds, were the subject of several theoretical studies[146,236–238]. In the analogous carbon molecules HCN is substantially more stable than HNC (i.e. by 10–25 kcal mol^{-1}, for references to experimental and theoretical work see Reference 146). When carbon is replaced by silicon, the order of relative stability is reversed. The lowest energy structure is linear HNSi ($^1\Sigma^+$), with linear HSiN lying much higher in energy (e.g. 55 kcal mol^{-1} at MP4SDTQ/6-31G*//3-21G$^{(*)}$ [146], 75[236] and 68[237] kcal mol^{-1} according to other CI calculations). The barrier separating HSiN from HNSi was not evaluated. The calculations show that HNSi probably does not contain a true triple bond, as there is no promotion of the 3s pair, i.e. the silicon atom is practically divalent[236]. Additional support for this suggestion comes from the calculated Si–N bond distance in HNSi

(1.528 Å at 3-21G$^{(*)}$ [146]) which is almost identical to the Si–N distance in singlet H_2Si $=NH$[146]. The triplet states of both HNSi and HSiN lie 80–90 kcal mol^{-1} above the corresponding singlets[146].

The isonitrile HNSi has been observed in matrix isolation and its reported structure[239] is in good agreement with the calculated CI geometries[237, 238]. On the other hand, HSiN or CH_3SiN are not known experimentally[19]. The phenyl-substituted systems C_6H_5NSi and C_6H_5SiN were studied by MNDO in conjuction with related photoelectron studies[240].

Calculations for the phosphorus analogues of silanitrile and silaisonitrile, HSi≡P and HP≡Si, have been performed with geometry optimizations at the 3-21G$^{(*)}$ level[213a]. HSiP was calculated to be linear while HPSi was calculated to be highly bent (∢ HPSi = 74°) and lower in energy than HSiP at all levels of calculation (e.g. by 9.4 kcal mol^{-1} at MP3/6-31G*//3-21G$^{(*)}$). At the same level of theory the barrier separating the two isomers is 16.8 kcal mol^{-1}, relative to HSiP[213a].

Both the Si–N and the Si–P triply bonded molecules deserve in our view more theoretical attention (such as a study of substituent effects), in particular as some of these molecules have already been shown to have sufficiently long lifetimes to be observed experimentally[19, 239].

VI. AROMATIC AND ANTIAROMATIC COMPOUNDS

Aromatic compounds containing silicon have attracted considerable interest, both experimental and theoretical, and efforts in this area have intensified since the first synthesis of stable silenes and disilenes[19]. Despite considerable effort no stable silaaromatic compounds have yet been isolated, pure or even in a solution. All direct observations to date have been in the gas phase or in matrix isolation. Experimental knowledge about these compounds is therefore quite limited[19] (see also below), leaving theory as one of the main sources of reliable information.

A. Silabenzenes

The first to generate and trap a silaaromatic compound were Barton and Banasink[241a], who reported in 1977 on silatoluene, **85**[241]. The discovery was soon followed by a matrix isolation and IR, UV and PE spectral characterization of both **85**[241b, c] and silabenzene **86**[242].

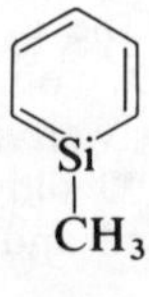

(85)

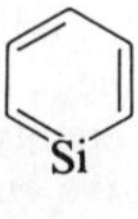

(86)

1. Silabenzene

Theory preceded experiment in predicting that compounds such as silabenzene should be stable with respect to other isomers and to possible decomposition products. The first calculations for **86**, and the first predictions that it is a molecule capable of existence, were made by Dewar and coworkers[54] on the basis of MINDO/3 calculations. The first *ab initio* calculations for silabenzene were reported by Schlegel and coworkers[243], who used the STO-3G basis set. Later calculations for **86** included floating spherical Gaussian orbital (FSGO) calculations[244], MNDO[245], STO-2G[246a, 247], STO-3G[243, 245], 3-21G[246b] and 3-21G$^{(*)}$[245, 246b] calculations.

The major problem addressed in these studies naturally concerned the question of the degree of aromaticity of these compounds. According to gradient STO-3G and 3-21G$^{(*)}$ geometry optimizations[243, 245, 246] **86** is a planar molecule with a Si–C bond length of 1.722 Å[245] (1.760 Å at 3-21G$^{(*)}$ [246b]). This distance is intermediate between the values for a single and a double C–Si bond, displaying almost the same relative shortening found in benzene. Thus, the calculated structure exhibits the expected partial double-bond character and suggests that silabenzene possesses some degree of aromaticity. Further support for the aromatic character of **86** can be found in the molecular orbital energies and coefficients which show the typical benzenoid pattern[243]. The electrons in the σ-framework are polarized away from the electropositive silicon and toward the neighboring carbons[243, 244].

A better criterion for aromaticity is probably energetics. A simple method for evaluating aromaticity was suggested by Pople and coworkers[248]. According to this method the energy of the aromatic compound is compared to that of its component isolated single and double bonds by a reaction (e.g. reaction 27 for benzene and reaction 28 for silabenzene) in which the number of each formal type of bond is conserved[248]. Such reactions are expected to be relatively insensitive to changes in the basis set or to correlation errors.

$$C_6H_6 + 6CH_4 \longrightarrow 3C_2H_6 + 3C_2H_4 \qquad (27)$$

$$\mathbf{86} + 5CH_4 + SiH_4 \longrightarrow 2C_2H_6 + 2C_2H_4 + CH_3SiH_3 + CH_2{=}SiH_2 \qquad (28)$$

Using the 3-21G basis set, reactions 27 and 28 are found to be endothermic by 59 kcal mol^{-1} and 47 kcal mol^{-1}, respectively[246b]. Similar energies are obtained at 3-21G$^{(*)}$[246b]. Thus, the aromaticity (resonance energy) of silabenzene is about 75% that of benzene. Similar estimates (slightly lower) were obtained (STO-3G) earlier by Schlegel and coworkers[243] by comparing the isomerization energies of reaction 29.

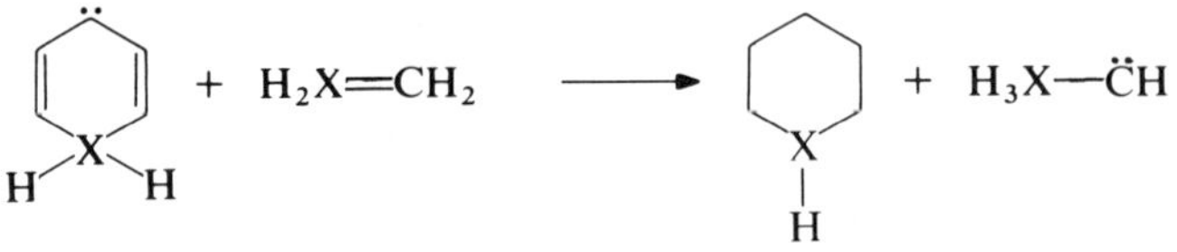

(29)

$$\Delta E(X{=}Si) = -22 \text{ kcal mol}^{-1}$$
$$\Delta E(X{=}C) = -32 \text{ kcal mol}^{-1}$$

Schleyer and coworkers[245] studied in detail the energetic relationships between silabenzene and four of its isomers, **87–90**. Geometry optimizations were carried out at STO-3G (MNDO for **89**) and energy comparisons were made at 3-21G$^{(*)}$. It should be noted that the calculated relative energies of **86–90** should be regarded as being only approximate, due to the basic level of theory used and the diverged structural types considered.

(87)

(88)

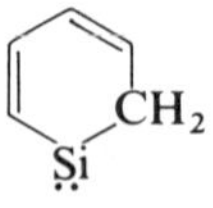

(89)

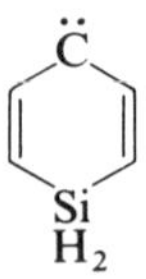

(90)

These calculations revealed that the silylene isomers **88** and **89** are approximately 20–25 kcal mol^{-1} higher in energy than **86**. **90** is a ground-state triplet and lies about 60 kcal mol^{-1} above **86**. The **86–88** (or **89**) energy difference is much smaller than between the analogous isomers of benzene, reflecting the preference of silicon to adopt divalent structures instead of forming multiple bonds (see Section V.A). Yet, this energy difference is unusually large compared to other multiply bonded silicon systems (e.g. $H_3C\ddot{S}iH$ vs. $H_2C{=}SiH_2$). The 20–25 kcal mol^{-1} energy difference between **86** and **88** was therefore used by Schleyer and coworkers as a measure of the aromaticity in silabenzene. This estimate is in close agreement with the 22 kcal mol^{-1} estimate of aromaticity (equation 29) suggested previously by Schlegel and coworkers[243].

Dewar silabenzene **87** is 38 kcal mol^{-1} less stable than **86**, but this energy difference is expected to be reduced at higher levels of theory[245] (see also discussion below). This value is still much smaller than the experimental energy difference of 60 kcal mol^{-1} between benzene and Dewar benzene. The lower energy difference in the silicon system results from two major factors: (1) the weakness of the π-bond involving silicon relative to the corresponding σ-bond (see Section V.A.1), and (2) the reduced strain in Dewar silabenzene compared to that in Dewar benzene[245] (see Section IV.C), As shown below, the energy difference between the planar aromatic ring and its Dewar valence isomer is further reduced when additional carbon atoms in **86** are substituted for silicon.

Considering the relatively basic level of theory used in most of the above studies it is probably appropriate, as recent calculations for the analogous Si_6H_6 demonstrate (see below), to repeat some of these calculations using more reliable levels of theory.

2. Substituted silabenzenes

As mentioned above, silatoluene (**85**) is the only substituted silabenzene that has been characterized experimentally[241]. Schleyer and coworkers[245] pointed out that the relative stabilities of structures **86**–**90** are influenced also by the weakness of Si–H bonds relative to C–H bonds. By this argument, substitution at silicon may therefore alter the relative energies of the various silabenzene isomers. Hanamura and coworkers[147] have indeed found that methyl substitution at silicon favors multiply bonded silicon structures including **86**, by nearly 20 kcal mol^{-1} relative to the silylene isomers, in comparison to the energy difference in silabenzene itself.

An extensive systematic study of substituted silabenzenes was carried out recently by Baldridge and Gordon[247]. In their study the 12 substituents F, Cl, OH, SH, OCH_3, NH_2, PH_2, CH_3, SiH_3, CN, NO_2 and COOH were attached to the four unique positions on the ring (i.e., overall, 48 compounds were studied). The analogous substituted benzenes were also considered for comparison. The problems studied included the relative isomer energies and stabilities, the degree of aromaticity, the electron density distributions and the dipole moments. In general, geometries were partially optimized at STO-2G and single-point calculations were carried out using 3-21G, and in some cases 3-21G$^{(*)}$ [247]. A summary of the major findings is given below.

Baldridge and Gordon determined the π-electron directing effects by these substituents from density difference plots. Typical density difference plots are shown in Figure 40 for 1-hydroxysilabenzene and phenol, in both the planar and perpendicular conformations. These plots were constructed by substracting the electron density of silabenzene (or benzene) from that of substituted silabenzene (or benzene) in a plane parallel to and 0.7 bohr above the plane of the aromatic ring. Positive contours (solid lines) correspond to a buildup in π-electron density, while negative contours (dashed lines) correspond to a depletion of π-electron denstiy relative to that in silabenzene. In both cases, the buildup of electron density in the *ortho* and *para* positions is clearly seen and, as expected, is more pronounced when the OH group is in the planar conformation. Using such density difference plots Baldridge and Gordon conclude that the substituents have the same directing tendencies in silabenzene and in benzene (see Table 29). A note of caution is, however, that charge distributions are not always a reliable guide for predicting the

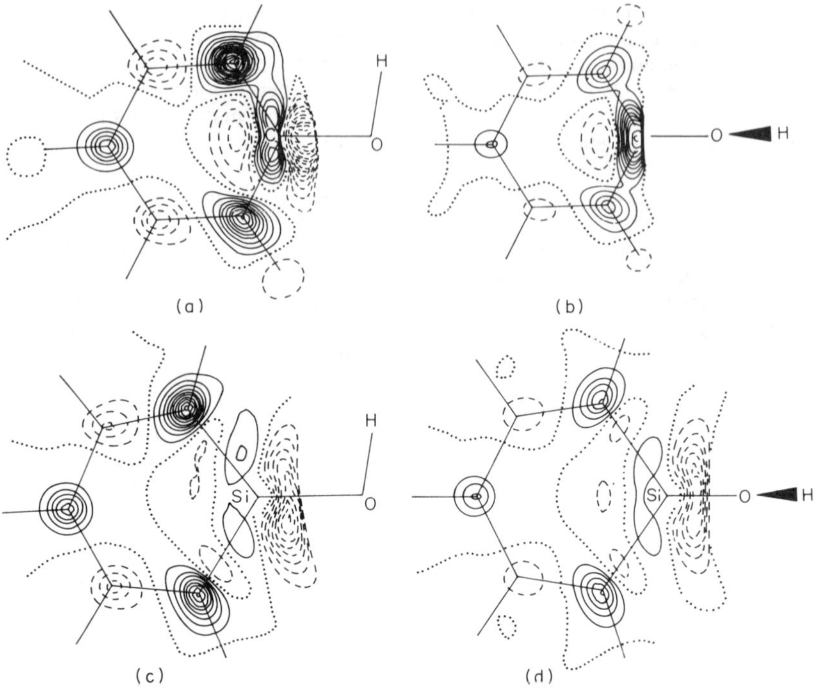

FIGURE 40. Density difference plots (substituted aromatic–parent aromatic), taken in a plane 0.7 bohr above the plane of the ring of phenol (*a, b*) and of 1-hydroxysilabenzene (*c, d*) in planar (*a, c*) and perpendicular (*b, d*) conformations (3-21G//STO-2G). Reprinted with permission from *Organometallics*, **7**, 144 (1988). Copyright (1988) American Chemical Society[247].

substitution site in electrophilic reactions. In principle, the relative energies of the appropriate intermediates should be used for making such predictions.

A general feature of all the density difference plots is the lack of any negative charge buildup on silicon, whether or not the particular substituent is *ortho*, *para* or *meta* directing. For all molecules considered, silicon is the least electronegative atom and therefore prefers to remain positive. This feature appears to determine also the relative stabilities of the *ortho*-, *meta*- and *para*-substituted isomers (see below).

The second important conclusion is that, of the four possible isomers, substitution at silicon gives the most stable isomer (except for COOH), as shown in Table 29. This is to be expected since all substituents (except silyl) are more electronegative than silicon, so that substitution at the 1-position will produce a stabilizing positive charge at the silicon. Similar conclusions were reached by Apeloig and Karni in their study of substituted silenes[106] (see Section V.A.1.b.ii).

Among the other ring positions, the relative stability of the *ortho*, *meta* and *para* isomers appears to be controlled by the effect of the substituent on the charge density at silicon, favoring positions in which the substituent directs electron density away from the silicon. Thus, among all *ortho*, *para* directors the *meta*-substituted species is the next lowest in energy after the *ipso*-substituted isomer. Among all *meta*-directors (except NO_2), the *ortho*-substituted species is the second most stable (Table 29). Using appropriate bond-separation reactions (e.g. similar to equations 27 and 28) it was found that substitution affects the degree of aromaticity of the silaaromatic molecule to a larger degree than in the corresponding benzenes, but the effects are not very large. Most substituted silabenzenes

TABLE 29. Relative energies (3-21G, in kcal mol^{-1}) of the four isomers of substituted silabenzenes[a]

Substituent	Directing tendency[b]	Position of substitution 1	2	3	4
F	*o, p*	0.0	45.2	35.2	38.1
Cl	*o, p*	0.0	33.8	31.9	32.6
OH	*o, p*	0.0	38.8	29.1	33.6
SH	*m*	0.0	24.0	22.3	24.3
NH_2	*o, p*	0.0	31.7	27.3	28.6
PH_2	*m*	0.0	11.7	13.3	12.9
CN	*m*	0.0	8.6	12.9	10.2
NO_2	*m*	0.0	6.9	10.7	6.4
CH_3	*o, p*	0.0	14.0	13.0	14.2
SiH_3	*m*	0.0	2.3	6.2	5.6
OCH_3	*o, p*	0.0	35.2	29.4	32.7
COOH	*m*	0.0	−7.0	−2.3	−6.4

[a] From Reference 247.
[b] As predicted by density difference plots (see text). *o, p* = *ortho, para* directing substituent; *m* = *meta* directing substituent.

have 70–80% of the aromatic character of silabenzene. Several substituted silabenzenes (e.g. 3-fluorosilabenzene) have an even higher degree of aromaticity (by 17%) than silabenzene, becaube in this position fluorine directs electrons away from the silicon. According to the bond-separation reaction that Baldridge and Gordon[247] have used, the aromaticity of benzene and silabenzene is estimated to be 60 and 47 kcal mol^{-1}, respectively.

The effect of substituents X in the *ipso*-position on the stability of the ring has been evaluated by means of equation 30, which compares the effect of X in the aromatic system with that in H_3SiX (a comparison with $H_2C{=}SiHX$ would have been a better choice). The results are given in Table 30[247].

$$ipso\text{-}XSiC_5H_5 + SiH_4 \rightarrow SiC_5H_6(\mathbf{86}) + H_3SiX \qquad (30)$$

F, Cl, SH, NO_2 and CN destabilize the ring (or alternatively stabilize H_3SiX more than the ring) and the remaining *ipso* substituents stabilize the ring only slightly. In general, the energies in Table 30 are similar to those calculated for substitution at silicon in $H_2C{=}SiH_2$ (see Table 24). Note that in contrast to benzene, all substituents for which lone-pair conjugation to the ring is possible have a destabilizing effect and, where possible, they prefer the conformation that reduces the π-interaction with the ring (e.g. the preference of orthogonal conformations for OH, SH, OCH_3, etc.; NH_2 is exceptional, preferring the planar π-conjugated conformation)[247]. The analogous substituent effects for benezene, which are also given in Table 30, are always stabilizing and the effect is generally much larger in magnitude than for silabenzene. This reflects the smaller conjugating ability of silabenzene (and of a multiply bonded silicon atom in general) relative to benzene.

3. *Disilabenzenes*

There are three possible isomeric disilabenzenes, **91**–**93**, neither of which is known experimentally.

TABLE 30. Calculated substituent effects on the thermodynamic stabilities of substituted benzenes and of *ipso*-substituted silabenzenes (3-21G; ΔE in kcal mol^{-1})[a]

Substituent X	Conformation	$\Delta E^{b,c}$ Silabenzenes	$\Delta E^{c,d}$ Benzenes
F		− 1.2	12.9
Cl		− 2.3	3.5
OH	planar[e]	− 1.5	12.3
	orthogonal[f]	1.9	10.1
OCH_3	planar[e]	− 0.8	12.5
	orthogonal[f]	1.4	11.0
SH	planar[e]	− 1.0	2.7
	orthogonal[f]	− 0.1	4.4
NH_2	pyramidal (1)[g]	3.0	5.8
	pyramidal (2)[h]	1.3	9.2
	planar[e]	10.4	*j*
PH_2	pyramidal (1)[g]	1.9	3.3
	pyramidal (2)[h]	3.5	4.0
	planar[e]	−18.5	*j*
CN		− 1.2	1.9
NO_2	planar	− 4.6	7.4
CH_3		1.1	2.1
SiH_3		3.4	1.4
COOH	planar	13.1	11.3

[a] From Reference 247.
[b] Calculated by equation 30.
[c] Positive values indicate stabilization.
[d] Calculated by the equation: $XC_6H_5 + CH_4 \rightarrow C_6H_6 + H_3CX$, which is analogous to equation 30.
[e] ∢HYSiC (or ∢HYCC in benzene)[i] = 0°.
[f] ∢HYSiC (or ∢HYCC in benzene)[i] = 90°.
[g] ∢HYSiC (or ∢HYCC in benzene)[i] = 60°.
[h] ∢HYSiC (or ∢HYCC in benzene)[i] = 150°.
[i] Y is the atom in group X directly bonded to the ring.
[j] Not given.

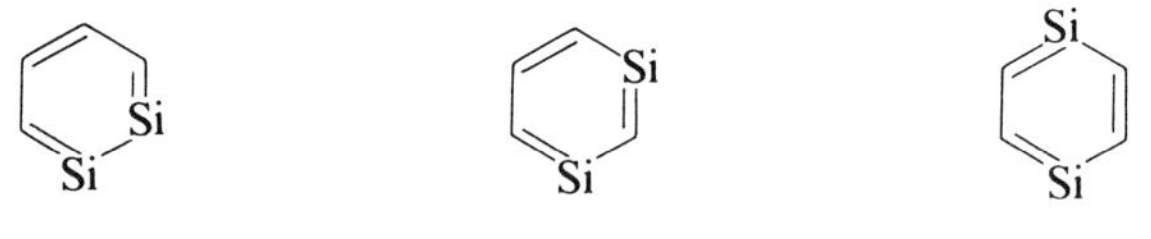

(91) **(92)** **(93)**

Calculations show that all three isomers are stable structures in the planar geometry with positive force constant matrices at the STO-2G levels of theory[249]. The fact that *ortho*-disilabenzene is planar whereas disilene is slightly nonplanar is indicative of electron delocalization in the ring which stabilizes the planar Si=Si moiety. The geometries of **91**–**93** reveal, as expected, that the bond lengths, including the bonds to silicon, are intermediate between single and double bond lengths, reflecting electron delocalization and aromaticity.

The three disilabenzenes are within 10 kcal mol^{-1} of each other (at 3-21G), the order of stability being: **92** (most stable) > **91** > **93**[249]. Using the appropriate bond-separation reactions, aromatic stabilization for the three isomers is found to be 22.5, 47.2 and 36.4 kcal mol^{-1} for **91**, **92** and **93**, respectively, compared to 47 kcal mol^{-1} for silabenzene[249]. Baldridge and Gordon suggested that the relative stabilities of **91**–**93** may be rationalized by considering the contributing ionic Kekulé structures for the three isomers. Only for the *meta* isomer **92** do all contributing structures (e.g. **92a**) have a formal positive charge on silicon. In **91** and **93** there are structures which place a negative charge on silicon (e.g. **91a** and **93a**), which is undesirable[249]. We believe that other factors, such as the relative energies of C=Si vs. Si=Si bonds, might be even more important than these charge effects. Furthermore, the above order of stability of the isomeric disilabenzene was obtained at 3-21G and it is questionable whether it will remain the same at higher levels of theory. Additional calculations are required for a more definitive analysis of this problem.

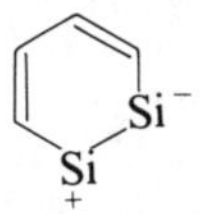

(**91a**)

(**92a**)

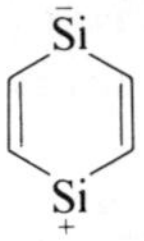

(**93a**)

Following an earlier qualitative prediction that *para*-disilabenzene **93** and the corresponding Dewar structure **94** should be close in energy[245], Chandrasekhar and Schleyer[250] studied this system using the 3-21G$^{(*)}$//STO-3G level of theory. In addition to **94** the isomeric silylenes **95** and **96** were also studied.

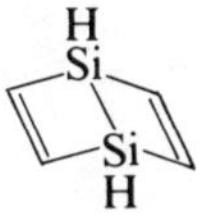

(**94**)

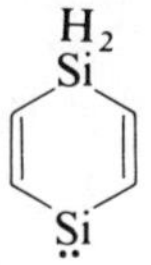

(**95**)

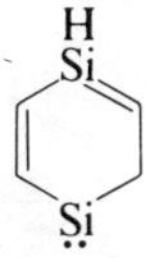

(**96**)

The results confirm the trends which were discussed above for the isomers of silabenzene. At 3-21G$^{(*)}$//STO-3G the Dewar structure **94** is only 6 kcal mol^{-1} higher in energy than **93**. Furthermore, the silylene structure **95** is found to be more stable than **93** by 10 kcal mol^{-1} [250]. **96** is higher in energy than the other isomers. It is expected that higher levels of theory would stabilize **94** relative to **95**. The **93**–**94** energy difference is considerably smaller than the difference between Dewar silabenzene **87** and silabenzene (see above) and it is much smaller than the Dewar benzene–benzene energy difference. The explanation for these surprising energetic trends is along the lines discussed above for silabenzene, but in disilabenzene the effect is larger. Two factors are mainly responsible for the observed energy differences: (1) The ring strain is expected to decrease in the order: Dewar benzene > Dewar silabenzene > Dewar disilabenzene, because silicon is more capable of accommodating the smaller valence angles (see Section IV.C.1). (2) Successive replacement of unsaturated carbon centers by silicon destabilizes the planar 'aromatic' structure due to the lower stability of C=Si bonds relative to C=C bonds. Similar trends (and rationalizations) were reported also for Si_6H_6 (see Section VI.A.4 below). The remarkable stability of the silylene isomer **95** relative to **93** can be understood as resulting mainly from the fact that C=C bonds are by at least 25 kcal mol^{-1} stronger than Si=C bonds, thus compensating for the loss of aromaticity which is estimated to stabilize **93** by only about 23 kcal mol^{-1} [250]. Also, in this case more reliable up-to-date calculations are recommended.

4. Hexasilabenzene

Hexasilabenzene **97**, being the complete silicon analogue of benzene, has attracted considerable theoretical interest and, as a result, our knowledge about this fascinating molecule is extensive although it had not yet been observed experimentally. It is interesting to follow the procession of the calculations for this molecule.

Dewar and coworkers[54] were the first to study hexasilabenzene. They used MINDO/3 and predicted that benzene, silabenzene and hexasilabenzene have comparable aromaticities. However, as will become evident in the following discussion, far more sophisticated theoretical methods than MINDO/3 are required to make reliable predictions about this molecule[251–255]. Nagase, Kudo and Aoki[251] were the first to carry out *ab initio* calculations for **97**. They found that the 3-21G equilibrium gometry of **97** (characterized by only positive eigenvalues of the force constant matrix) is planar with D_{6h} symmetry, as in the case of benzene. The Si–Si bond length of 2.220 Å (which is shortened by about 0.015 Å upon the addition of polarization functions[254, 255]) is 0.077 Å longer than the Si=Si double bond in $H_2Si{=}SiH_2$, but is 0.162 Å shorter than the Si–Si single bond in $H_3Si{-}SiH_3$, reflecting the aromatic character of **97**. The calculations which followed[252–255] revealed that this conclusion regarding the equilibrium geometry of hexasilabenzene may have been premature.

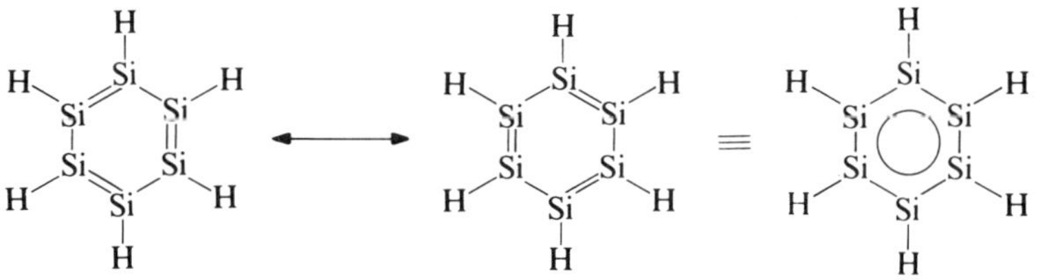

(**97**)

Almost simultaneously with Nagase's group[251], Sax and Janoschek[2, 253] also studied Si_6H_6 using pseudopotential methods and a nonpolarized basis set for the valence electrons. They also found that **97** is a planar D_{6h} molecule, but they pointed out that the b_{2g} out-of-plane ring deformation mode has an extremely low wave number of only 60 cm^{-1}. The b_{2g} mode allows the silicon and hydrogen atoms to move up and down the molecular plane as shown in Figure 41a. This implied that a nonplanar molecule would have almost the same energy. Clabo and Schaefer[254] studied the structure of **97** with a larger dz basis set augmented with polarization functions. These calculations confirmed that the equilibrium structure of **97** is planar with D_{6h} symmetry, but at the dz level of theory the b_{2g} deformation mode was characterized by a wave number of only 10 cm^{-1} (compared with 796 cm^{-1} in benzene!). This very low vibration frequency pointed to the possibility that **97** may actually prefer a nonplanar geometry. Very recently, Nagase and coworkers[255] examined this possibility by carrying out more accurate and elaborate calculations. They found that when d-functions are added to the basis set, the b_{2g} vibrational frequency becomes imaginary (74 cm^{-1}) (due to size limitations this point could not be checked by Clabo and Schaefer[254]), i.e. the D_{6h} planar structure becomes a *maximum* on the Si_6H_6 potential energy surface.

Geometry optimization at 6-31G* without the constraints of D_{6h} symmetry indeed leads to a minimum which is a chair-like puckered structure of D_{3d} symmetry, with equal

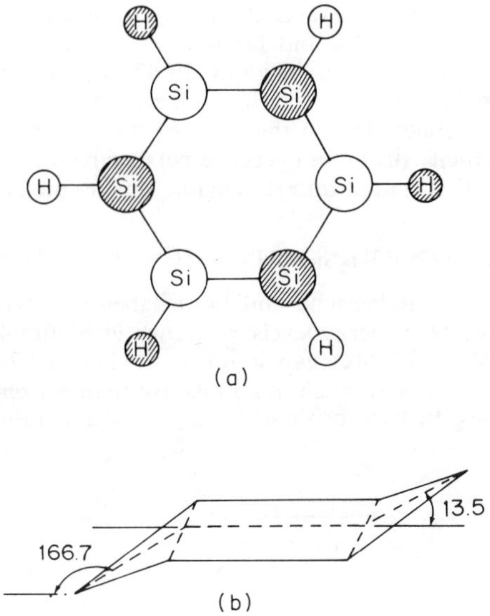

FIGURE 41. The b_{2g} vibrational mode (a) for the deformation of D_{6h} hexasilabenzene to the D_{3d} nonplanar structure (b, 6-31G**). Silicon and hydrogen atoms move alternately up (○) and down (◍). Reproduced by permission of the American Institute of Physics from Ref. 255.

Si–Si and Si–H bond lengths[255]. The deviation from planarity is however relatively small, as is the gain in energy (0.12 kcal mol^{-1}). Similar results are obtained also with the 6-31G and 6-31G** basis sets[255]. However, the use of more extended basis sets (including double sets of d-functions [6-31G+ +G(2d, p)] or a set of f-functions [6-31G(df, p)] on Si) reverses the relative stability of the two structures and the D_{3d} minimum disappears. On the other hand, the addition of electron correlation makes the Si_6H_6 potential energy surface less flat, and leads again to a greater preference of the D_{3d} puckered structure. Nagase and coworkers[255] conclude that *hexasilabenzene adopts a chair-like puckered structure of D_{3d} symmetry* (Figure 41b). The preference of the D_{3d} structure over the D_{6h} structure is explained in terms of localization of electrons onto each Si, generating at each silicon a 'radical' character.

We believe that due to the large sensitivity of the results to the computational method, further studies should be carried out before a definitive answer regarding the equilibrium geometry of **96** can be reached. It is clear, however, that the potential energy surface around the minimum is very flat and that hexasilabenzene is a very flexible molecule, in sharp contrast to benzene. The above discussion therefore demonstrates the great caution which should be taken when making predictions regarding the structures of silicon compounds. Due to their high flexibility and the many structures which are generally close in energy, reliable predictions can be made in many cases only when very large basis sets are used and when correlation energy is included in the calculations (see also the discussion on Si_2H_2, Section V.B). In analogous carbon compounds, calculations at the SCF level with relatively modest basis sets are in most cases sufficient to reach reliable conclusions.

In addition to the question of geometry a second point of interest is the degree of aromatic stabilization in hexasilabenzene. Using appropriate bond-separation reactions hexasilabenzene is found to possess approximately half the aromatic stabilization energy of benzene[251]. Thus, at 6-31G*//3-21G the energy of equation 31 is 14.2 kcal mol^{-1}, about half that in benzene[251]. Sax and Janoschek[252], using the same bond-separation equation, estimated the resonance stabilization in **97** as being 70% that of benzene, but their basis set was smaller than that of Nagase and coworkers[251]. On the other hand, Clabo and Schaefer[254] concluded on the basis of bond energies that hexasilabenzene has little or no aromaticity (from an energetic perspective). They also found that **97** is stable with respect to dissociation into reasonable smaller molecules (e.g. $3Si_2H_2$, 6SiH, $3Si_2 + 3H_2$)[254].

$$Si_6H_6(\mathbf{97}) + 3H_2Si{=}SiH_2 \rightarrow 3H_2Si{=}SiH{-}SiH{=}SiH_2 \quad (31)$$

Large differences between benzene and hexasilabenzene can also be found in the pattern of their valence MO energy levels, as shown in Figure 42. The HOMO is 2 eV higher while the HOMO–LUMO gap is much smaller in **97** than in benzene. This suggests that hexasilabenzene is much more reactive than benzene toward both electrophiles and nucleophiles. In view of these findings Sax and Janoschek[252, 253] correctly

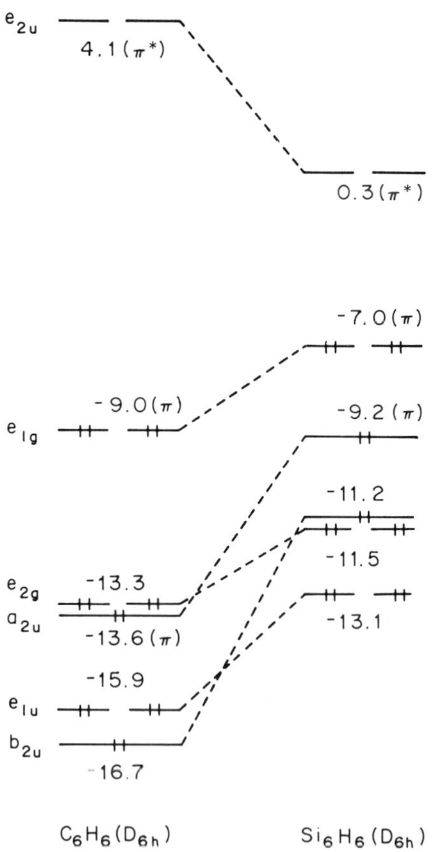

FIGURE 42. Energy levels (eV) of hexasilabenzene and benzene (6-31G*). Reproduced by permission of the American Institute of Physics from Ref. 255.

point to the fact that although **97** exhibits a relatively high delocalization energy, it is expected to behave chemically very differently from benzene.

A third point of interest is the energy of hexasilabenzene relative to other isomeric Si_6H_6 structures such as **98, 99** and **25a**[116] (see also Section IV.C.1.b.ii). The calculated relative energies of **97**–**99** and **25a** (at 6-31G*//3-21G) are given in Table 31.

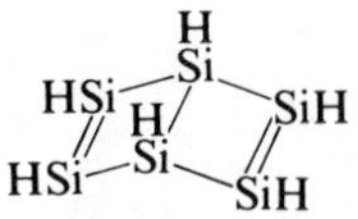

(98) C_{2v}

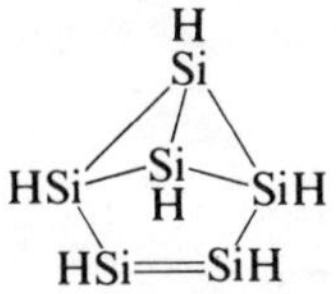

(99) C_{2v}

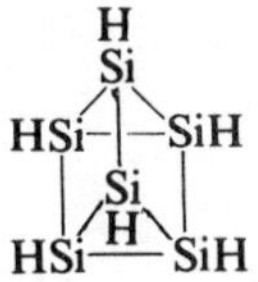

(25a) D_{3h}

TABLE 31. Relative energies[a] (6-31G*//3-21G) of Si_6H_6 and C_6H_6 isomers

Si_6H_6 isomer	Si_6H_6	C_6H_6[b]
97	0.0	0.0
98	3.7	88.1
99	0.5	84.5
25a	−9.5	127.6

[a] From Reference 252; in kcal mol^{-1}.
[b] Benzene structures analogous to structures **97, 98, 99** and **25a**.

Nagase and coworkers[251] and Sax and Janoschek[252] reached independently and almost simultaneously the fascinating conclusion that hexasilaprismane **25a** is actually more stable than hexasilabenzene **97**! At 6-31G*//3-21G, **25a** is more stable than **97** by 9.5 kcal mol^{-1} [251]. Sax and Janoschek calculated a slightly smaller energy difference of 4 kcal mol^{-1} [253]. To put this surprising conclusion in perspective the reader is reminded that benzene is more stable than prismane by 128 kcal mol^{-1} (Table 31). Note that **98** and **99** are also very close in energy to **97** while their carbon analogues lie some 85–88 kcal mol^{-1} above benzene.

The reasons for the dramatic differences between the Si_6H_6 and C_6H_6 systems are already familiar to the reader (e.g. see the discussion regarding silabenzene and disilabenzene). The main factor which controls the relative energies of the isomeric Si_6H_6 and C_6H_6 structures is the energy difference between sp^2- and sp^3-hybridized C and Si atoms. Thus, the marked energetic preference of carbon for sp^2-hybridization compared to sp^3-hybridization, which makes benzene stabler than prismane, is absent in the case of silicon. The difference in ring strain between prismane and hexasilaprismane also contributes, but to a much smaller degree (i.e. **25a** is less strained than prismane **25b** by 32 kcal mol^{-1}, see Table 9). The order of stability: **25a** (most stable) > **99** > **98** (Table 31) also reflects the fact that silicon is reluctant to form double bonds; stability increases as the number of the Si–Si double bonds decreases. In contrast, in the analogous carbon series prismane **25b** is by far the most strained (least stable) compound (Table 31).

Finally, we note that all the Si–Si bond lengths in **25a** are identical at 2.370 Å, similar to the bond distances in cyclotetrasilane, **19** (2.371 Å), but longer than in cyclotrisilane, **18** (2.336 Å)[253]. Sax and Janoschek also reported the harmonic vibrational frequencies and ZPE of **25a** and **97**[253].

B. Silicocene

The synthesis of decamethylsilicocene, **100**, $R=CH_3$, the first stable silanediyl, was recently reported by Jutzi and coworkers[256]. The single crystal structure analysis indicates two structures **100a**, $R=CH_3$ and **100b**, $R=CH_3$ in the ratio $a:b=1:2$. In **100a**, $R=CH_3$ the rings are parallel while in **100b**, $R=CH_3$ they form an interplanar angle of 25.3°[256].

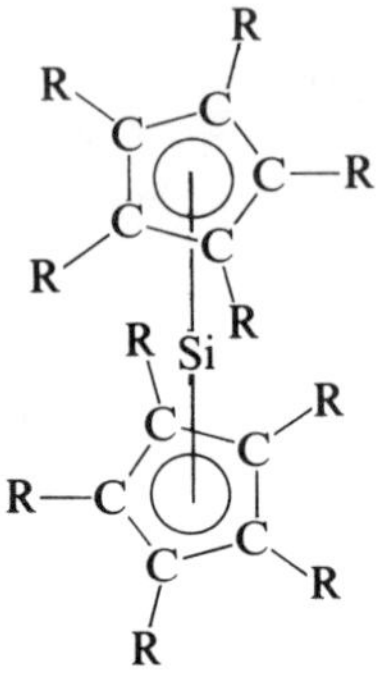

(100a)

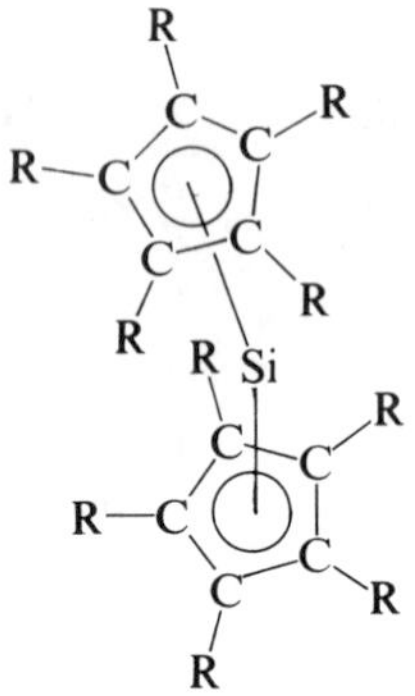

(100b)

Following this report Baldridge and Gordon[257] used the 3-21G$^{(*)}$ basis set to study the parent silicocene **100a**, R=H in D_{5d} symmetry. The calculated geometrics arc in reasonable agreement with experiment, the calculated bond distances being about 4% longer. The calculated force field for the D_{5d} structure reveals a degenerate pair of imaginary frequencies leading to structure **100b**, R=H. Baldridge and Gordon have therefore concluded that the preferred gas-phase conformation of **100**, R=H has a small interplanar angle between the two cyclopentadienyl rings. This parallels the experimental findings in the solid phase where structure **100b**, $R=CH_3$ dominates over structure **100a**, $R=CH_3$[256]. The calculations also show that the binding in silicocene results primarily from a weak π-interaction between the silicon atom and the cyclopentadienyl rings, and that due to the long silicon-ring distances silicocene is an extremely floppy molecule[257].

C. Charged Systems

A natural extension of the above studies is to charged aromatic molecules containing silicon. The potential for aromaticity in the silacyclopentadienyl anion **101** and the silacyclopropenium cation **102**, both formally possessing an aromatic $(4n+2)$ π-electron network, was studied by Gordon and coworkers[246a]. **101** is isoelectronic with the cyclopentadienyl anion and **102** with the cyclopropenium cation, both well established aromatic systems with high delocalization energies.

(101a) **(101b)**

(102a) **(102b)**

Calculations at the 3-21G//STO-2G level reveal that the delocalization (aromatic) stabilization of both **101** and **102** is very low. Using appropriate bond-separation reactions it is found that the stabilization energy in **101** is only 25% of that in the cyclopentadienyl anion, and in **102** it is practically nonexistent[246a]. **102** is actually slightly destabilized by the presence of the 2π-electrons. These conclusions are only slightly modified when diffuse functions are added to the basis set. The stabilization energies of **101** and **102** are much smaller than in silabenzene, which is two-thirds as aromatic as benzene (Section VI.A.1). Apparently, the presence of a charge substantially reduces the effectiveness of electron delocalization. Thus both **101** and **102** are better represented by the localized structures **101a** and **102a**, than by the aromatic delocalized structures **101b** and **102b**.

The different behavior of the analogous, potentially aromatic, silicon and carbon anions (**101** and cyclopentadienyl, respectively) appears to result from the inability of the silicon ring in **101** to redistribute the excess electron throughout the molecule. The increase in negative charge on the formal ionic center is much greater for silicon than for carbon (at STO-2G the total charges are −0.071 electrons at each carbon atom of cyclopentadienyl and −0.470 electrons at Si in **101**)[246a]. The alternative, to distribute the negative charge in **101** onto the carbons, would increase the double-bond character of the C–Si bond and decrease that of the C–C bond, an energetically undesirable process.

The low stability of **102**, despite the fact that it possesses 2π-electrons, has been attributed to ring strain (see geometry in Figure 43)[246a]. The calculated low stability of

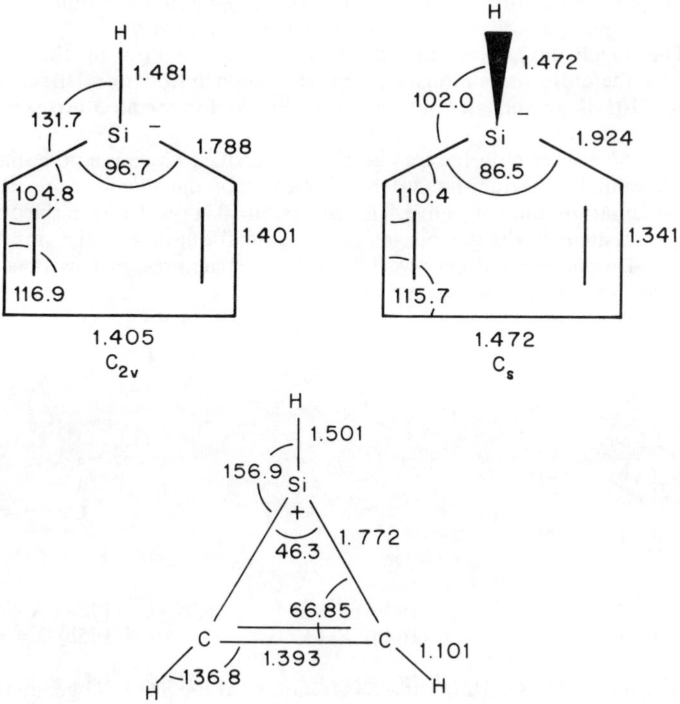

FIGURE 43. Optimized structures (6-31G*) of the C_{2v} and C_s silacyclopentadienyl anion (**101**)[259] and of the silacyclopropenium cation (**102**)[246a]. Bond lengths are in Å, bond angles in deg. Reprinted with permission from *J. Org. Chem.*, **51**, 5028 (1986). Copyright (1986) American Chemical Society[259].

102 suggests that silicenium ions cannot be stabilized to a substantial degree by π-conjugation. This is consistent with the many failures to generate in solution silicenium ions which are stabilized by conjugation (see Section VII.B.1 below)[258].

The small aromatic stabilization of **101** was further emphasized in a recent study by Damewood[259] who studied, using 6-31G*, the planar (C_{2v}) and pyramidal (C_s) geometries of **101**. The calculated geometries of both forms are given in Figure 43.

Only the C_s structure is a minimum on the **101** potential surface. The planar C_{2v} structure is the transition state for pyramidal inversion between equivalent C_s structures, the inversion barrier being 16.2 kcal mol^{-1} [259]. Note the large differences in the geometries of the C_{2v} and C_s structures, in particular the much longer C–Si bond and the butadiene-like geometry of the carbon skeleton in the C_s structure. The fact that **101** is not planar indicates that electron delocalization is weak and relatively unimportant. A quantitative measure of the extent of electron delocalization in the C_s **101** is given by isodesmic equation 32.

(32)

The energy of equation 32 for C_s **101** is only 2.2 kcal mol^{-1} (6-31G*), to be compared with 73.4 kcal mol^{-1} for the cyclopentadienyl anion[259]. Thus, in the ground state, **101** has only approximately 3% of the resonance stabilization exhibited by the cyclopentadienyl anion[259]. The much higher calculated aromatic stabilization of the planar C_{2v} structure[246a] is therefore misleading. One practical conclusion from these theoretical studies is that **101** is a much less attractive η^5-ligand for metal complexes than the cyclopentadienyl anion[259].

Theoretical studies were reported also for the cation **103a**, which is a potential 'pseudo' or 'three-dimensional' 6π-aromatic system[260]. Note that the silicon atom in **103a** has formally a coordination number of five. The interest in **103a** results from the observation of a $C_5H_5Si^+$ fragment in the gas phase[261]. The crucial question is the stability of the pyramidal ion **103a** relative to other possible $C_5H_5Si^+$ structures, such as **104a** and **105a**, and a comparison with the analogous $C_6H_5^+$ isomers.

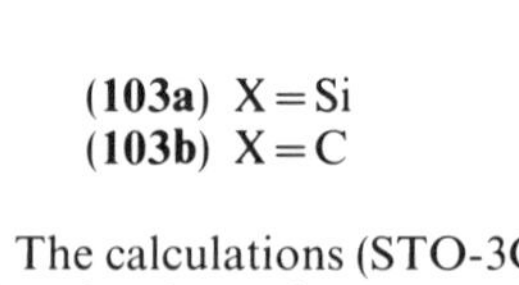

(**103a**) X=Si
(**103b**) X=C

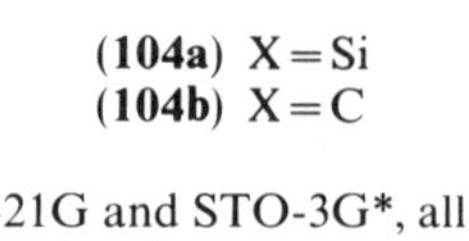

(**104a**) X=Si
(**104b**) X=C

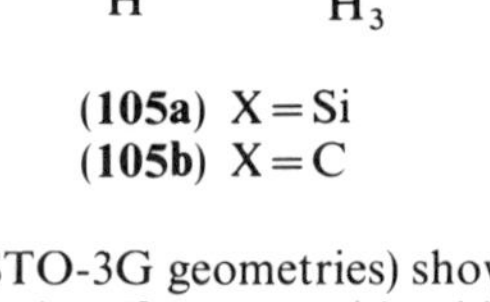

(**105a**) X=Si
(**105b**) X=C

The calculations (STO-3G, 3-21G and STO-3G*, all at the STO-3G geometries) show that, for the carbon compounds, the planar structures **104b** and **105b** are considerably lower in energy than the pyramidal structure **103b**. In the silicon series, on the other hand, the pyramidal structure **103a** is favored over the two planar structures (however, the authors caution that higher-level calculations are required to establish this fact). The reluctance of silicon to participate in multiple bonding destabilizes the planar structures containing silicon relative to the all-carbon compounds. The ring dimensions in **103a** and **103b** are very similar, but the apical–basal distance is, as expected from simple arguments on atomic sizes, considerably larger in **103a** (1.745 Å) than in **103b** (1.357 Å)[260].

D. Potential Antiaromatic Systems

Several sila-analogues of cyclobutadiene, the prototype of Hückel 4*n* antiaromatic systems, were studied computationally. Silacyclobutadiene **106** was first studied by Gordon[139a] using the 6-31G*//3-21G level of theory. More extensive calculations for the lowest singlet and triplet states of **106** were later carried out by Colvin and Schaefer[142b] and the optimized geometries using a dz+d basis set are given in Figure 44 (large differences were noted between the dz and dz+d geometries). Vibrational frequencies were also reported[142b].

HC=SiH
| |
HC=CH

(106)

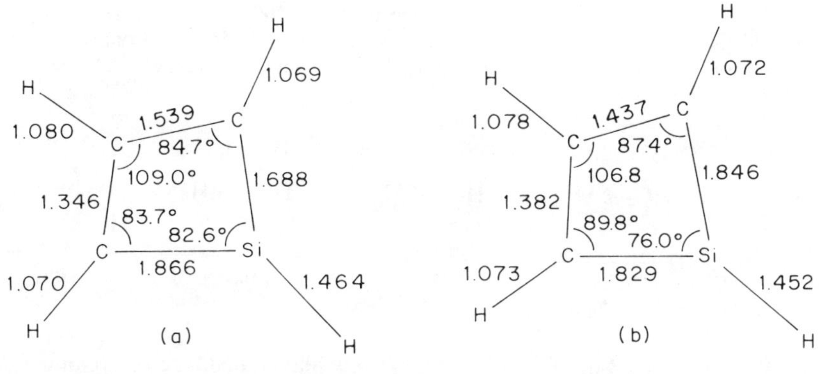

FIGURE 44. Optimized geometries (dz+d) of silacyclobutadiene (**106**) in the lowest $^1A'$ singlet (a) and $^3A'$ triplet (b) states. Bond lengths are in Å[142b].

The singlet ground state is predicted to possess a planar geometry with alternating single and double bonds. The C–Si bonds are calculated to be 1.866 Å and 1.688 Å, close to those in methylsilane and silaethylene, respectively. This indicates little π-conjugation, as suggested also by the vibrational frequencies. The structure of the $^3A'$ triplet is very different. The shortest C–C and C–Si bonds are not on opposite sides of the ring as they are in the singlet. Furthermore, the Si–C bonds have lengths comparable to a Si–C single bond, while the C–C bond lengths are closer to that of a C=C double bond. Although the silicon substitution rules out a truly square structure, the triplet structure is much more 'square-like' in geometry than is the singlet silacyclobutadiene. The singlet–triplet energy difference is, as expected, strongly dependent on the level of the calculations. At the MCSCF+CI level the triplet is found to lie 8 kcal mol^{-1} above the singlet[142b]. This singlet–triplet splitting is about 14 kcal mol^{-1} smaller than in the case of cyclobutadiene[142b].

The singlet **106** is, at 6-31G*//3-21G, by 33.3 kcal mol^{-1} more stable than monosilatetrahedrane **38a**, and by 56.3 kcal mol^{-1} more stable than its decomposition products, acetylene and silaacetylene[139a]. However, more recent findings lead to a dramatic downward revision of this estimate. Taking into account the fact that the lowest minimum on the H_2SiC surface is 1-silavinylidene **77** (an even lower minimum was located later, see Section V.B.1) and not silaacetylene, Colvin and Schaefer[142b] concluded that **106** is actually less stable (by about 5 kcal mol^{-1}) than its decomposition products. However, since the initial concerted reaction of silacyclobutadiene to form acetylene and silaacetylene is symmetry forbidden, the barrier to unimolecular dissociation should be significant.

The antiaromatic destabilization of **106** was estimated using equation 33 to be 53.5 kcal mol^{-1} (6-31*//3-21G; 49.1 kcal mol^{-1} using a correlated wave function[142b]), by 14.6 kcal mol^{-1} less than calculated for cyclobutadiene at the same level of theory[139a]. The conclusion that **106** is less antiaromatic than cyclobutadiene is consistent with the smaller aromatic character of silabenzene relative to benzene (Section VI.A.1). Colvin and Schaefer note, however, that equation 33 cannot distinguish destabilizing conjugation effects from angle strain or 1,3-interactions[142b].

$$\mathbf{106} + 3CH_4 + SiH_4 \rightarrow CH_3SiH_3 + CH_3CH_3 + H_2C{=}CH_2 + H_2C{=}SiH_2 \qquad (33)$$

Gordon and Holmes[139b] studied the disilacyclobutadienes **107** and **108** along with other $C_2Si_2H_4$ isomers at the MP3/6-31G*//3-21G level. **107a** and **107b** are not minima on the energy hypersurface and, upon optimization, decay to a disilylene structure **107c** with a very long Si–Si bond of 2.860 Å. **108** is a minimum on the RHF surface and it is estimated to be 20 kcal mol^{-1} higher in energy than **109**—the most stable $C_2Si_2H_4$ isomer. **108** possesses in analogy to cyclobutadiene a high diradical character and it therefore cannot be adequately described at the RHF level of theory[139b].

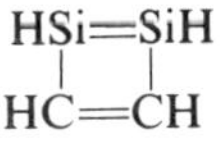

(**107a**)

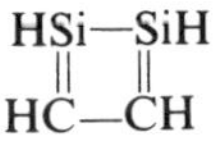

(**107b**)

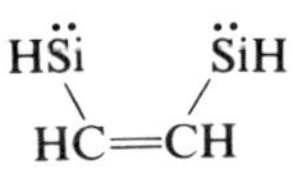

(**107c**)

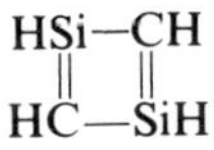

(**108**)

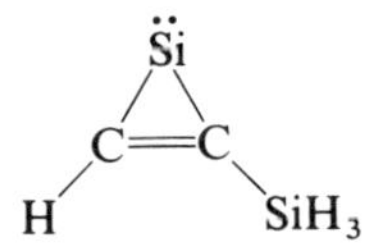

(**109**)

Tetrasilacyclobuta-1,3-diene, **110**[123,124], the full silicon analogue of cyclobutadiene, does not correspond to a local minimum on the Si_4H_4 energy hypersurface[123]. Upon geometry optimization the planar D_{2h} structure collapses to a C_{2h} structure, **111**, which lies 20 kcal mol^{-1} below the D_{2h} structure. **111** is however by 12.1 kcal mol^{-1} less stable than the singlet disilylene **112**, which is the lowest energy isomer of Si_4H_4[123]. The low stability of **110** is also reflected in the fact that tetrasilatetrahedrane **24a** is only about 6–10 kcal mol^{-1} less stable than planar **110**[123,124] (Section IV.C.1.b.ii).

(**110**)

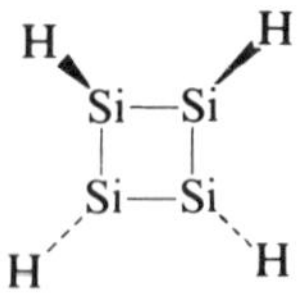

(**111**)

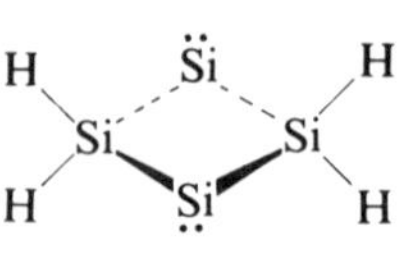

(**112**)

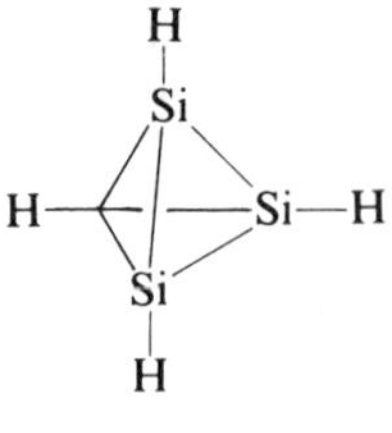

(**24a**)

VII. REACTIVE INTERMEDIATES

In this section we will discuss various reactive intermediates that contain silicon. In particular, the silicon analogues of the classic organic reactive intermediates such as carbenes (silylenes), carbenium (silicenium) ions and carbanions (silyl anions) have attracted the interest of experimentalists and theoreticians. In light of the elusive nature of these species it is not surprising that much of what we know about their properties comes from theory, in particular *ab initio* molecular orbital calculations.

A. Silylenes

Silylenes **113** are the silicon analogues of carbenes and formally contain a divalent silicon atom. We will use here the term silylene which is commonly used in the literature in preference to the systematic IUPAC term, silanediyl. Although silylenes of type **113** are the major subject of this section, other types of silylenes, such as the unsaturated **114** (1-silavinylidene) or **115** (2-silavinylidene) or cyclic silylenes, are possible reactive intermediates and we have discussed some of them in previous sections. In Section VII.A.4 silylenes other than **113** are mentioned.

R^1R^2Si:	$R^1R^2C{=}Si$:	$R^1R^2Si{=}C$:
(113)	**(114)**	**(115)**

Silylenes have attracted strong interest of both experimentalists and theoreticians and their chemistry has been reviewed[1–3, 262]. The theoretical studies of silylenes have been reviewed most recently by Gordon and coworkers[22].

Theoretical interest in silylenes has centered around four major problems: (1) The singlet–triplet energy difference in the parent H_2Si: and in substituted silylenes. (2) The UV-visible spectra of silylenes. Interest in this aspect was prompted by the recent exciting developments due mainly to West, Michl and their coworkers[176, 263], who developed methods for isolating silylenes in matrices and for measuring their spectra[176, 263, 264]. (3) Insertion and addition reactions of silylenes. (4) The energetic relationship between silylenes and the isomeric multiply bonded species (e.g. $CH_3\ddot{Si}H$ vs. $H_2C{=}SiH_2$, $H_3Si\ddot{Si}H$ vs. $H_2Si{=}SiH_2$). This question has already been discussed in detail in Section V and will not be repeated here. However, the reader should be reminded that silylenes are in general very close in energy to their multiple-bond isomers, in contrast with carbenes which are substantially higher in energy than the isomeric multiply bonded compounds.

1. Structures and singlet–triplet energy differences

a. The parent H_2Si:. The parent H_2Si: has been studied by a variety of theoretical methods and we will concentrate on the most recent results, which are also the most reliable[68, 146, 265c, d, g, 266, 267]. Experimentally it has been established that the ground state of H_2Si: is the closed-shell singlet ($^1A_1$), with the $^3B_1$ triplet state lying higher in energy[268]. This is in contrast with the triplet ground state of methylene[269]. The different multiplicity of the ground state of H_2C: and H_2Si: can be understood qualitatively in terms of the energies of the highest occupied (HOMO) and lowest unoccupied (LUMO) molecular orbitals in these species. In both species the singlet ground state has two electrons in an orbital of σ-symmetry as shown schematically for H_2Si: in **116**. In the triplet state this electron pair is unpaired and one electron resides in an orbital of π-symmetry (**117**). In methylene, the energy gain resulting from the unpairing of the electrons is larger than the HOMO–LUMO gap and the triplet is the ground state. In H_2Si: the HOMO–LUMO gap is much larger than in H_2C: (52 and 27 kcal mol^{-1}, respectively[68]) and the singlet configuration is the ground state.

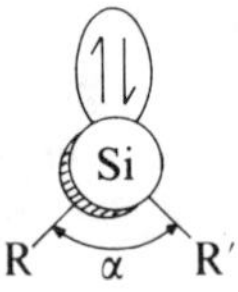

(116)

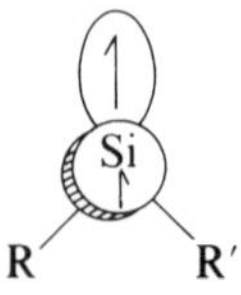

(117)

A significant, although not dramatic, disparity still existed until very recently between the theoretical predictions and the measured $^1A_1$–$^3B_1$ energy difference. Except for the early calculations which used small basis sets and only modest correlation correction[191, 265a], most theoretical estimates of this S–T gap lie in the range of 17–20 kcal mol^{-1} [265b–g, 266]. The most accurate calculations for this problem were carried out by Balasubramanian and McLean[266]. The calculations used the same level of accuracy as the best calculations done for methylene and included a very large extended basis set (triple zeta + d+f) and extensive MCSCF corrections for electron correlation. These calculations predict a singlet–triplet splitting of 21.0 ± 1 kcal mol^{-1}, significantly higher than the experimental estimates which are lower than 14 kcal mol^{-1} [268a]. Considering the very high level of theory used by Balasubramanian and McLean and the precise singlet–triplet splitting that such methods give for methylene, it is probably safe to predict that in the case of silylene the calculated singlet–triplet splitting is also correct. A very recent study brings theory and experiment closer together. Berkowitz and coworkers[268b] found that the singlet–triplet splitting in H_2Si: is either 21.0 ± 0.7 or 18.0 ± 0.7 kcal mol^{-1}. Additional experimental determinations of the $^1A_1$–$^3B_1$ energy difference are required in order to resolve unambigously the remaining disparity[268a] with the calculations.

The $^1A_1$ and $^3B_1$ states both have the expected bent geometries. At the best available level of theory the $^1A_1$ ground state is calculated to have a Si-H bond length of 1.519 Å and a HSiH angle of 92.5°[266]. This is in excellent agreement with the experimental values of 1.516 Å and 92.8°, respectively[270]. For the $^3B_1$ triplet state the calculations predict a Si–H bond length of 1.480 ± 0.005 Å and HSiH angle of 118.5 ± 1°[266], but experimental data for comparison are not available. Comparison of the equilibrium structures of H_2C: and H_2Si: shows that the bond angles are ~10° smaller in H_2Si:. In the $^1A_1$ state the HCH and HSiH bond angles are 102.4° and 92.8°, respectively, while in the $^3B_1$ state these angles are 130° and 118°, respectively. This is consistent with the general observation that silicon prefers to have nonbonding electrons in atomic orbitals with a high percentage of s-character, so that the Si–H bonds have higher p character than the C–H bonds[146]. Schleyer, Apeloig, Pople and coworkers[146] calculated that in the singlet $^1A_1$ state the lone pair is in a hybrid orbital with 88.8% s-character in H_2Si: compared with 59.6% s-character in H_2C:. In the $^1B_1$ triplet states the s-character used in the in-plane singly occupied molecular orbital is 36.2% in H_2Si: compared to only 20.8% in H_2C:[146].

Higher excited states of H_2Si: were also studied[265a–d, 266]. The lowest excited singlet, the $^1B_1$ state, is of special interest in relation to the UV-visible spectra of silylenes (see below). The most accurate calculations place the $^1B_1$ state 44.2 kcal mol^{-1} above the $^1A_1$ state[266], in excellent agreement with the experimental value of 44.4 kcal mol^{-1} [270]. The other theoretical estimates place the $^1B_1$ state a few kcal mol^{-1} higher in energy[265a–d]. The calculated geometry of the $^1B_1$ state with r(Si–H) = 1.493 Å (1.487 Å) and ∢ HSiH = 122.8° (123°) are also in excellent agreement[266] with experiment (values in parentheses)[270]. The next two higher states, $^1A_2$ and $^3A_2$, have strongly bent structures with HSiH angles of less than 60°[265a, c], and they are found to lie 59.3 and 42.1 kcal mol^{-1} above the ground state[265c].

The heat of formation of H_2Si: is another point where, until very recently, considerable disparity between theory and experiment existed. The theoretical values ranged from 63.4

to 68.1 kcal mol^{-1}, with the most recent predictions favoring the higher values[68, 146, 267]. On the other hand, the early experimental estimates were in the range of 58–59 kcal mol^{-1} [73b, 271]. This discrepancy has recently been resolved in favor of theory; recent experiments now favor values between 65–69 kcal mol^{-1} [268b, 272].

b. Substituted silylenes

(i) Substituent effects. The most extensive study of substituted silylenes was carried out by Pople, Apeloig, Schleyer and coworkers[146], who examined the series $R\ddot{S}iH$ where R was systematically varied along the second-row substituents: Li, BeH, BH_2, CH_3, NH_2, OH and F. The calculations used the MP4SDTQ/6-31G//3-21G$^{(*)}$ level of theory and systematic comparisons were made with the corresponding carbenes. The individual silylenes are discussed in detail in the original paper and here we discuss mainly the general features and trends. A similar study of third-row substituents (which is not yet available) is highly desirable.

The effect of the substituents R on the stabilities of the singlet and triplet silylenes can be evaluated by the isodesmic bond separation equations 34 and 35, respectively. The MP4SDTQ/6-31G*//3-21G results are presented in Table 32 together with the data for the analogous carbenes, derived from equations 36 and 37.

$$R\ddot{S}iH\ (\text{singlet}) + SiH_4 \longrightarrow H_3SiR + H_2Si\text{: (singlet)} \quad (34)$$

$$R\ddot{S}iH\ (\text{triplet}) + SiH_4 \longrightarrow H_3SiR + H_2Si\text{: (triplet)} \quad (35)$$

$$R\ddot{C}H\ (\text{singlet}) + CH_4 \longrightarrow H_3CR + H_2C\text{: (singlet)} \quad (36)$$

$$R\ddot{C}H\ (\text{triplet}) + CH_4 \longrightarrow H_3CR + H_2C\text{: (triplet)} \quad (37)$$

TABLE 32. Stabilization energies[a] (kcal mol^{-1}) for singlet and triplet substituted silylenes $R\ddot{S}iH$ and substituted carbenes $R\ddot{C}H$ (MP4SDTQ/6-31G*//3-21G)[b] as calculated by equations 34–37

Substitutent R	Silylenes		Carbenes	
	Singlet[c]	Triplet[d]	Singlet[e]	Triplet[f]
H	0.0	0.0	0.0	0.0
Li	− 1.2	25.2	9.2	27.0
BeH	− 1.8	23.7	− 1.6	18.5
BH_2	7.8	16.1	36.8	15.6
CH_3	0.7	− 2.7	12.0	5.4
NH_2	22.3	− 1.6	61.5	13.0
OH	15.0	− 6.7	51.2	11.0
F	9.3	−12.0	35.0	5.5

[a] positive values denote stabilization.
[b] From Reference 146.
[c] For equation 34.
[d] For equation 35.
[e] For equation 36.
[f] For equation 37.

Two features are immediately obvious from the data in Table 32. The singlet and triplet states respond differently to substitution and the effects are mostly larger (especially in the singlet state) in the carbenes than in the silylenes. Let us examine first the substituent effects in the singlet state. Most substituents stabilize the singlet silylene, but the most effective stabilization is achieved with π-donors and it decreases with the decrease in the

π-donating ability of R, i.e. NH_2 (22.3 kcal mol^{-1}) > OH (15.0 kcal mol^{-1}) > F (9.3 kcal mol^{-1}). This suggests that resonance form **118b** is important for these substituents, as also suggested by the significant charge that is transferred into the 3p-orbital of the silylene and by the shortening of the Si–R bond relative to the corresponding saturated silyl derivative (e.g. the Si–N bond length in $H\ddot{S}iNH_2$ is 1.531 Å compared with 1.710 Å in H_3SiNH_2 at 3-21G$^{(*)}$ (see also Table 33)[146].

(**118a**) (**118b**)

M = C, Si

The singlet carbenes behave similarly, except that the stabilization energies are usually much greater than for the corresponding silylenes, especially for π-donors (Table 32). The smaller substituent effects in silylenes result mainly from the lower tendency of silicon to form π-bonds (Section V.A). The similar qualitative response of silylenes and carbenes to substituent effects makes it possible to use the extensive knowledge on substituted carbenes to predict qualitatively the effect of other substituents on singlet silylenes.

The substituent effects in the triplet silylenes are very different from those in the singlets. The triplet stabilization energies increase roughly as the electronegativity of the substituent decreases. Thus Li, BeH and BH_2 are highly stabilizing while OH and F substitution destabilizes the triplet state significantly (Table 32). This behavior can be understood by the tendency of silicon to have nonbonding electrons in atomic orbitals with high s-character. Electropositive substituents transfer charge from the Si–R bonding orbital with much p-character to the singly occupied s type orbital on silicon (see **117**), thus stabilizing the silylene. In the carbene, the C–R bonding and the carbon lone-pair orbitals both have sp^2 character, and such electron transfer is not as important. The substituent effects in the triplet carbenes are therefore very different from those in the triplet silylenes, both the σ- and π-donors being effective stabilizing substituents in the carbene[146].

(ii) Singlet–triplet energy differences. In the parent H_2Si: the singlet is the ground state and the $^3B_1$ triplet state is 21.0 ± 1 kcal mol^{-1} higher in energy, according to the best theoretical estimates[266] (Section VII.A.1.a). How is the singlet–triplet (S–T) energy difference affected by substitution? Table 33 lists the calculated S–T gaps between the lowest-energy singlet and triplet in both substituted silylenes and in the corresponding carbenes, along with the Si–R bond lengths and H–Si–R bond angles. Note that the S–T energy differences in Table 33 are determined by the S–T gaps in the parent divalent molecules and by the substituent effects on these states, discussed above (Table 32). The data in Table 33 are based mostly on the systematic MP4SDTQ/6-31G* study of Pople, Apeloig, Schleyer and coworkers[146] and other values for the same substituents are not listed even when additional calculations (even at better levels of theory) are available. In this way most values can be compared at a uniform level of theory. For substituents not considered in Reference 146, the best available theoretical values from other studies are also included in Table 33.

Examination of the systematic second-row series R = Li to F reveals that in the silylene series there is a continual increase in the stabilization of the singlet relative to the triplet (i.e. an increase in the S–T gap) as the electronegativity of the atom bound to silicon increases. This observation is in agreement with the qualitative predictions made previously by Harrison and coworkers[273]. Thus, in the monosubstituted silylenes the S–T

TABLE 33. Relative energies (MP4SDTQ/6-31G*//3-21G) of the lowest singlet (S) and triplet (T) states of RS̈iR′ and RC̈R′ and Si–R bond lengths and RS̈iR′ bond angles [a]

				RS̈iR′	S–T gap (kcal mol^{-1})	
R	R′	State	r(Si–R) (Å)	(deg)	silylene	carbene
H	H	$^1A_1$	1.506	93.4	0.0	0.0
		$^3B_1$	1.471	118.0	16.4	−16.8
Li	H	$^1A'$	2.635	94.4	0.0	0.0
		$^3A''$	2.406	140.5	− 9.7	−34.7
Li	Li[b]	$^1A_1'$	2.533	92.0	0.0	—
		$^3\Sigma_g^-$	2.429	180.0	−10.3	—
BeH	H	$^1A'$	2.257	94.2	0.0	0.0
		$^3A''$	2.128	128.6	− 9.4	−37.3
BH_2	H	$^1A'$	1.994	91.5	0.0	0.0
		$^3A''$	1.961	123.1	8.8	4.3
CH_3	H	$^1A'$	1.907	95.8	0.0	0.0
		$^3A''$	1.899	118.6	19.8	−10.7
CH_3	CH_3[b]	$^1A_1$	1.926	98.8	0.0	—
		$^3B_1$	1.914	117.7	22.1	—
t-Bu	t-Bu[c]	$^1A_1$	1.945	117.7	0.0	—
		$^3B_1$	1.926	124.7	3.7	—
SiH_3	H	$^1A'$	d	d	0.0	0.0
		$^3A''$	d	d	10.3	−25.8
SiH_3	SiH_3[e]	$^1A_1$	2.392	93.7	0.0	—
		$^3B_1$	2.327	125.2	5.8	—
NH_2	H	$^1A'$	1.698	95.9	0.0	0.0
		$^3A''$	1.710	120.7	40.3	31.7
OH	H	$^1A'$	1.638	97.1	0.0	0.0
		$^3A''$	1.645	118.5	38.0	23.3
F	H	$^1A'$	1.596	98.3	0.0	0.0
		$^3A''$	1.602	116.0	37.8	12.9
F	F[b]	$^1A_1$	1.598	99.6	0.0	—
		$^3B_1$	1.598	113.6	73.8	—

[a] From Reference 146 unless stated otherwise.
[b] At CISD/6-31G*//6-31G*, from Reference 196.
[c] At 3-21G*//3-21G*. The S–T gap was estimated to be larger than 10 kcal mol^{-1} at a correlated level. From Reference 276.
[d] Not given.
[e] At MP4/6-31G*/3-21G*, from Reference 113.

separation is the largest in HS̈iNH$_2$, HS̈iOH and HS̈iF[146]. Two such substituents further increase the S–T splitting. For example, in F_2Si: the calculated S–T gap is 73.5[265b] (or 74.0[196]) kcal mol^{-1} compared with the experimental value of 75.2 kcal mol^{-1} [274]. On the other hand, electropositive substituents stabilize the triplet silylene more effectively than the singlet and thus reduce the S–T gap. A similar trend is found for the carbenes, where the triplet is the ground state for HC̈Li, HC̈BeH and $HC̈CH_3$, while $HC̈NH_2$, HC̈OH and HC̈F prefer singlet ground states[146].

The possibility that substitution with electropositive elements may lead to a ground-state triplet silylene is interesting, because triplet silylenes are expected to exhibit entirely different chemistry from that of the experimentally known singlet silylenes[262]. It is therefore not surprising that other groups in addition to Pople and coworkers[146] searched for triplet ground-state silylenes. Colvin and coworkers[275] used polarized basis sets and configuration interaction methods to study the different low-lying electronic states of HS̈iLi. In addition to geometries and relative energies, vibrational frequencies were predicted for all stationary points. The triplet is indeed calculated to be the ground state with the lowest singlet lying about 7 kcal mol^{-1} higher in energy[275]. A surprising feature of the HS̈iLi potential-energy surfaces, for both the singlet and triplet, is the existence of high-lying states with very small HS̈iLi bond angles of approximately 50° [275]. The structure was interpreted as a weakly bound complex between silicon and LiH. Krogh-Jespersen[196], and Gordon and Schmidt[276] studied the dilithiosilylene, Li_2Si:, and found it to be a triplet ground state with a linear geometry. The bent singlet excited state is 10.3[196] kcal mol^{-1} (CISD/6-31G*//6-31G*) or 11.6[276] kcal mol^{-1} (MP4/MC-311 +G(3d)) higher in energy.

Lithium, however, is not a very attractive substituent for practically generating a triplet silylene. Another electropositive substituent that has the potential to reduce, or even to reverse, the S–T gap is silyl. In $H_3SiS̈iH$ the S–T gap is calculated to be 10.3 kcal mol^{-1} at MP4/6-31G*, significantly lower than in H_2Si: and $CH_3S̈iH$ (16.4 and 19.8 kcal mol^{-1} respectively at the same level of theory)[146]. Kohler and Lischka[161] predicted a S–T gap of 15.8 kcal mol^{-1} using a correlated wave function while Poirier and Goddard[191] predicted an even lower splittings of 5.8 kcal mol^{-1} using CISD/3-21G[191]. Gordon and Bartol[113] have gone one step further and studied the effect of disilyl substitution on the S–T energy difference. At the SCF level the $^3B_1$ state of $(H_3Si)_2Si$: is predicted to be about 8 kcal mol^{-1} lower in energy than the $^1A_1$ state. However, as expected, the addition of correlation corrections preferentially stabilizes the singlet state, but even at the highest level of theory available (MP4/6-31G*//3-21G$^{(*)}$) the closed-shell $^1A_1$ state is still about 6 kcal mol^{-1} below the lowest triplet. So even $(H_3Si)_2Si$: is a ground-state singlet. A second triplet state (^3A) with a very small Si–Si–Si angle is found to be more than 30 kcal mol^{-1} above the ground singlet state. The global minimum on the Si_3H_6 ground-state surface is predicted to be trisilacyclopropane (see Section IV.C.1.a) $H_3SiSiH{=}SiH_2$ is also calculated to be by 8.6 kcal mol^{-1} more stable than $(H_3Si)_2Si$:[113].

Steric effects can also influence the S–T splitting. This stems from the fact that the potential curves for the singlet and triplet states have different curvature as a function of the angle around silicon and the two curves therefore cross[277]. Figure 45 shows that for the parent H_2Si: the singlet–triplet crossing occurs at an H–Si–H angle of about 130°. Thus, if the angle at silicon can be forced to be larger than 130°, the triplet may become the ground state. Gordon and Schmidt[276] tested this idea and carried out detailed calculations for $[(CH_3)_3C]_2Si$:, including geometry optimization at RHF/3-21G$^{(*)}$ and UHF/3-21G$^{(*)}$ for the singlet and triplet states, respectively. This is an extremely large computational task requiring about 5 CPU hours on a CRAY 1 computer. In the singlet state, di-*t*-butyl substitution causes an increase of 18.3° in the angle around silicon bringing it to 111.7°. In the triplet the SiSiSi angle is 124.7°, by 6° larger than in $(CH_3)_2Si$:. Thus, the optimal SiSiSi angle in the singlet is far below the SiH_2 singlet–triplet crossing angle of approximately 130°. Indeed, Gordon and Schmidt find a singlet ground state also for $[(CH_3)_3C]_2Si$:, lying at least 10 kcal mol^{-1} below the lowest triplet state[276]. As far as we know the goal of generating a ground-state triplet has not been yet achieved, and additional calculations might direct the experimentalists to the desired molecule. One possibility is to combine electronic effects (electropositive substituents) and steric effects

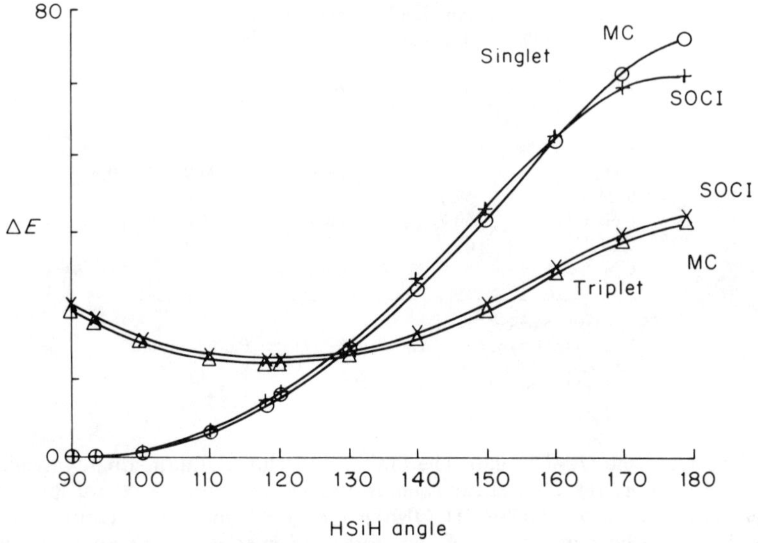

FIGURE 45. MCSCF and SOCI potential-energy curves for singlet and triplet silylene. Relative energies are in kcal mol^{-1}, angles in deg. The geometries were optimized at 6-31G**. Reproduced by permission of North-Holland Physcis Publishing from Ref. 277.

(bulky substituents that will enforce opening of the angle at silicon) as, for example, in $(R_3Si)_2Si$:, where R is very bulky, e.g. *t*-Bu.

2. The UV-visible spectra

A major development in the study of silylene has been the ability to isolate these species in inert matrices and to measure their spectra[176, 263, 264]. The UV-visible absorbtion maxima are used as a spectroscopic fingerprint to identify the presence of a silylene. A major problem in the field has been the limited knowledge on the effect of substituents on the spectra of silylenes, so that in many cases it was difficult to decide if the observed spectra are indeed of the desired silylene. One example is the recent controversy concerning the UV-visible spectrum of $(CH_3)_2Si$: (see below). To assist the interpretation of silylene spectra Apeloig and Karni[278] carried out a systematic study of the effect of substituents R on the lowest vertical electronic transition in silylenes. It was assumed, in analogy with the parent H_2Si:[270, 279], that the lowest electronic transition involves the promotion of one of the n(Si) electrons to the empty 3p(Si) orbital [i.e. **116**→**117** (singlet), namely a $^1A'$–$^1A''$ transition]. The calculated n(Si)→3p(Si) transition energies (UHF/6-31G*//3-21G) (using a correction for the contribution of the triplet to the 'mixted state' obtained for the excited singlet with the UHF procedure) are given in Table 34[278].

The calculated vertical transitions (ΔE) are consistent with expectations based on resonance and PMO arguments. For example, π-donors such as NH_2, OH, SH, etc. increase ΔE substantially and are thus predicted to induce a 'blue shift' in the absorption maximum. This is because the resonance stabilization of the ground-state singlet (empty 3p-orbital on Si) as in **118b** is larger than in the excited singlet (with a half-filled 3p-orbital on Si). Electronegative substituents increase ΔE (compare R = F and Cl in Table 34), while

TABLE 34. Calculated n(Si) → 3p(Si) transition energies (ΔE, 6-31G*//6-31G, kcal mol^{-1}) for $R\ddot{S}iH^{a}$

R	ΔE^{b}	R	ΔE^{b}
H	53.6	NH_2	87.8
CH_3	56.1	OH	74.9
$(CH_3)_2^{c}$	57.9	F	72.6
$CH{=}CH_2$	60.7	H_3Si	36.6
C≡CH	61.8	SH	70.4
CN	62.4	Cl	66.5

[a] From Reference 278.
[b] Corrected for the contributions of the triplet to the 'mixed state' (see text).
[c] For $(CH_3)_2Si$:.

σ-donors decrease the $^1A'$–$^1A''$ gap. The limited experimental data already available[263] qualitatively support the theoretical conclusions. For example, a 40 nm 'blue shift' is found in the spectra of $H_3C\ddot{S}iC_6H_4OMe$-*p* (i.e. a π-donor substituent) relative to $H_3C\ddot{S}iC_6H_5$[263c]. The dependence of the transition energy on the angle around the divalent silicon (α) was also studied[278]; α is larger in the excited singlet (120° in H_2Si:) than in the ground-state singlet (93.2° in H_2Si:). Therefore, as α increases ΔE decreases and a 'red shift' is calculated[278]. This prediction is also supported experimentally, such as in the series: *t*-$Bu\ddot{S}iCH_3$ (470 nm) > $CH_3CH_2\ddot{S}iCH_3$ (459 nm) > $(CH_3)_2Si$: (453 nm)[263c].

Dimethylsilylene is of special interest in this context. West, Michl and coworkers determined that the absorption maximum is at 450 nm[176, 263]. This assignment was later questioned by Griller and coworkers[264] who found that flash photolysis of known precursors of $(CH_3)_2Si$: produces species absorbing at 350 nm in hydrocarbon solution and at 300 nm in tetrahydrofuran and which yield insertion products expected of $(CH_3)_2Si$: when scavengers were added to the solution. No absorbtion at 450 nm was observed. Griller and coworkers concluded that either the UV-visible spectrum has been incorrectly assigned or that there is a large (~100 nm) shift in the absorption maximum of $(CH_3)_2Si$: in going from matrices to solution[264]. Apeloig and Karni were the first to evaluate this discrepancy theoretically[278]. Although the level of theory which they used could not reproduce the absolute excitation energies, it was possible to use the calculated effect of CH_3 on ΔE (Table 34) to conclude that the calculations are more consistent with an absorption at 450 nm, as reported by West and Michl's groups[176, 263]. Shortly afterwards the issue was resolved both experimentally and theoretically in favor of the 450 nm value. West, Michl and coworkers carried out additional experiments which confirmed their original assignment[280a]. Griller and coworkers have repeated their experiments and concluded that the absorption at 350 nm was probably due to photolysis of trace impurities[280b] and they have therefore withdrawn their original assignment[264]. Simultaneously Grev and Schaefer[281] carried out high-level calculations for $(CH_3)_2Si$: using a dz + p basis set in conjunction with two-configuration SCF methods for geometry optimizations and a (triple-zeta + two sets of d) basis set in conjunction with two-configuration SCF methods for energy comparisons. At the best available level of theory they calculate ΔE to be 62.7 kcal mol^{-1}, which corresponds to an electronic transition of 456 nm[281], in excellent agreement with the measurements by West, Michl and coworkers[176, 263, 280a], confirming that the observed species is indeed a ground-state $(CH_3)_2Si$:. The infrared intensities of ground-state dimethylsilylene were also calculated and used to assign some features of the experimental IR spectrum[281]. The possibility that the species observed by Griller and coworkers[264] is the $^3B_1$ dimethylsilylene was also examined but the results of the calculations in this respect were inconclusive[281]. Finally, the Franck–Condon overlaps and the transition moments between the vibrational states of the $^1A_1$ and $^1B_1$ states of H_2Si: were recently calculated at the MCSCF/6-31G* level[265d].

3. Reactions

In this section we will review the computational studies on reactions of silylenes, mainly addition and insertion reactions. Some reactions, in particular the following isomerizations of silylenes to the corresponding multiply bonded species, were discussed above: (a) to silaethylene in Section V.A.1.a.v, (b) to substituted silenes in Section V.A.1.b.iv, (c) to disilenes in Sections V.A.2.d and f, (d) to silanimines in Section V.A.3, (e) to silanephosphimines in Section V.A.4, (f) to silanones in Section V.A.5, (g) to silanethiones in Section V.A.6, (h) to silynes in Section V.B.1, (i) to disilynes in Section V.B.2, (j) to aromatic compounds in Section VI.A and to antiaromatic compounds in Section VI.D.

a. Dimerizations. The dimerization of two silylenes to produce disilene (equation 38, X=Y=Si) is observed experimentally when a matrix containing silylenes is warmed up[262]. This is also the reaction used for synthesizing stable disilene[14, 15]. The mechanism of this reaction is therefore of considerable interest.

$$H_2X: + H_2Y: \longrightarrow H_2X{=}YH_2 \qquad (38)$$

Both H_2C: and H_2Si: have the same frontier orbitals, denoted by σ and π. Two possible reaction paths can be considered for the dimerization reaction: a 'least-motion' path where both molecules approach each other in a plane and a 'non-least-motion' path where at large distances the fragments approach each other in perpendicular planes (Figure 46).

The dimerization of two singlet methylenes to form ground-state ethylene (equation 38, X=Y=C) is symmetry-forbidden along the 'least-motion' path[282a] and therefore proceeds along a 'non-least motion' path[282b]. Two ground-state triplet methylenes, on the

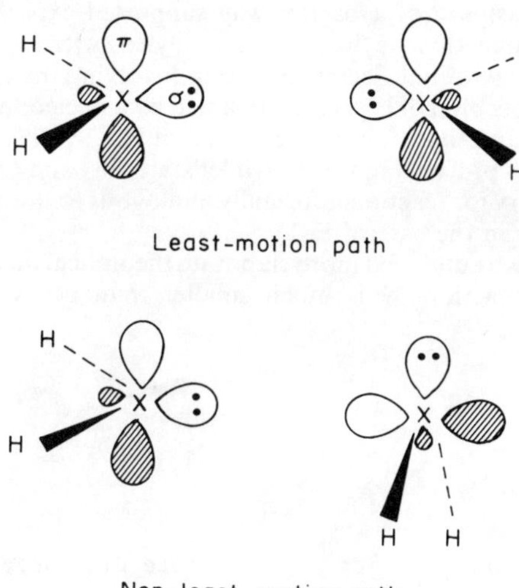

FIGURE 46. 'Least' and 'non-least-motion' approaches for the dimerization of H_2X: (X=C, Si)

other hand, can dimerize via the path of 'least motion', without a barrier to give ground-state ethylene[282]. Do the same arguments apply to the coupling reactions of H_2Si: ?

Davidson, Morokuma and coworkers studied the coupling reactions of H_2C: and H_2Si: fragments to ethylene, disilene and silaethylene (equation 38, X or Y=C or Si)[163, 283]. Potential curves for these reactions were obtained at the MCSCF level using a dz+p basis set, but the calculations used either fixed or assumed geometries. Both 'least' and 'non-least-motion' paths were calculated. Note that a major difference between H_2C: and H_2Si: is that methylene is a ground-state triplet while silylene is a ground-state singlet (see Section VII.A.1.a). Two ground-state $^1A_1$ silylenes give a ground-state disilene with a barrier in the 'least-motion' path, and without a barrier in the 'non-least-motion' path. As mentioned above, the excited state $^1A_1$ methylene behaves similarly. A ground-state $^3B_1$ methylene and an excited state $^1A_1$ methylene behaves similarly. A ground-state $^3B_1$ methylene and an excited-state $^3B_1$ silylene give a ground-state silene without a barrier in the 'least-motion' path. The same is found also for the dimerization of an excited-state $^1A_1$ methylene and a ground-state $^1A_1$ silylene[163, 283]. Thus, the dimerization of either H_2C: or H_2Si: and the reaction of H_2C:+H_2Si: proceed without a barrier (either via a 'least-motion' or via a 'non-least-motion' path), in both the ground and excited states[163, 283]. Similar studies which led to similar conclusions were also carried out by Bernardi and Robb[284].

b. Insertion reactions. The major emphasis of this section is on insertion reactions, but it is important to realize that the reverse of these reactions describes dissociation processes. For example, the reverse of reaction 39, which describes the insertion of silylene into hydrogen, is the thermal decomposition of silane to silylene and molecular hydrogen.

$$H_2Si{:} + H_2 \longrightarrow SiH_4 \qquad (39)$$

Reaction 39 was studied by using a variety of theoretical methods[285–291]. Qualitatively all these calculations show the same features, i.e. a 'non-least-motion' reaction path, similar to that found for the insertion of H_2C: into H_2. However, while insertion of singlet methylene into hydrogen has no energy barrier, substantial barriers, as high as 10.4 kcal mol^{-1} were found for reaction 39 using the earlier calculations[285, 286, 290]. The theoretical prediction of the existence of a barrier was supported experimentally[271a, 292a]. In analogy to the reaction of methylene the reaction of silylene with H_2 is envisioned to occur in two steps (Figure 47): an initial electrophilic step involving the empty p-orbital of the silylene and the bonding orbital of H_2 is accompanied by a nucleophilic step involving the lone pair of the silylene and the antibonding orbital of H_2[287, 290]. Although the reaction need not be thought of as proceeding via the triplet state, Sax and Olbrich[290] argued that the valence state necessary for reaction is formally analogous to the triplet, which required some promotion energy in the case of H_2Si:.

A recent study[291], which employed more elaborate theoretical methods, concluded that the actual barrier for reaction 39 is much smaller than believed earlier[285–290]. At

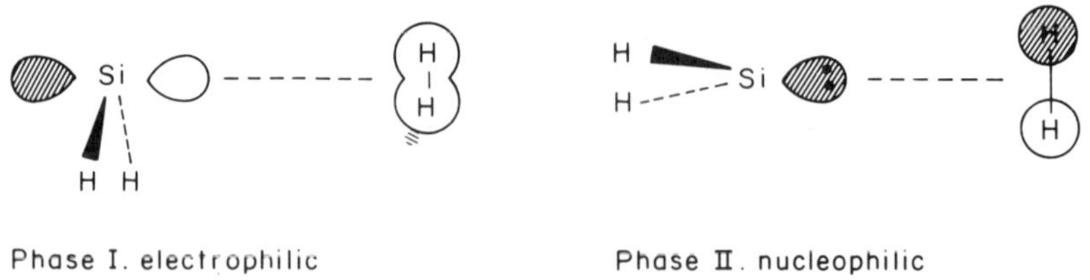

FIGURE 47. Schematic representation of the electrophilic and nucleophilic steps in the insertion of H_2Si: into the H–H bond.

MP4SDTQ/MC-311 + +G(3df, 3pd)//MCSCF/6-31G* the transition state for the insertion of H_2Si: into H_2 is only 1.7 kcal mol^{-1} above the separated fragments, and most of this barrier actually results from the addition of zero-point vibrational energies[291]. This calculated barrier is much smaller than the earlier experimental estimates[271a, 292a] but is in agreement with the more recent work of Jasinski[292b]. The disagreement with earlier experiments may result from the similar disparity between the experimental and theoretical estimates of the heat of formation of silylene (see Section VII.A.1.a).

An interesting observation regarding the mechanism of reaction 39 is the initial formation of a silylene–hydrogen complex which lies in a long-range shallow potential well[289, 291]. The depth of the well is only about 1 kcal mol^{-1} at the SCF level of theory, but it increases when correlation is introduced, and with the best available calculations it is 3.1 kcal mol^{-1} below H_2Si: + H_2 and 4.8 kcal mol^{-1} below the transition state of reaction 39 which is shown in Figure 48[291]. The reverse of reaction 39, the molecular elimination of H_2 from silane, is endothermic by 52.2 kcal mol^{-1} and the transition state lies 56.9 kcal mol^{-1} above silane[291].

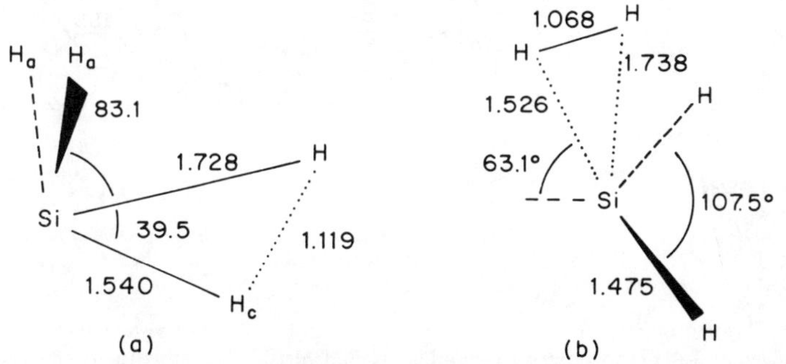

FIGURE 48. Calculated transition structures for the reaction $H_2Si: + H_2 \rightarrow SiH_4$: (a) at MCSCF/6-31G*. Reprinted with permission from *J. Am. Chem. Soc.*, **108**, 2191 (1986). Copyright (1986) American Chemical Society[291], (b) at TCSCF/dz + d. Reproduced by permission of the Royal Society of Chemistry from Ref. 288. Bond lengths are in Å, bond angles in deg.

Sosa and Schlegel[289] studied the insertion into H_2 of the silylene series H_2Si:, HFSi: and F_2Si: and compared it with a similar series of substituted methylenes. The calculated transition structures for the insertion reaction of both the silylenes and the carbenes are shown in Figure 49. It was found that fluorine substitution has a large effect on the insertion barriers which, at MP4SDQ/6-31G*//6-31G*, are (kcal mol^{-1}): 12.2, 31.1 and 65.2 for H_2Si:, HFSi: and F_2Si:, respectively; and 1.9, 15.3 and 47.1 for H_2C:, FHC: and F_2C:, respectively[289]. Although the calculated barriers at this level of theory are probably too high (see discussion above for H_2Si:[291]), the differences resulting from fluorine substitution probably give a realistic idea about the relative reactivities of these silylenes. The increase in the insertion barriers upon fluorine substitution was explained with a PMO model. Several effects of fluorine substitution on orbital energies act in the same direction, the most important being the destabilizing interaction between the silylene lone pair and the σ-orbital of the hydrogen molecule[289]. All the above-mentioned reactions are highly exothermic, but the agreement between theory and experiment is good only for H_2C: + H_2 and for H_2Si: + H_2. For substituted carbenes and silylenes, the errors in the calculated reaction exothermicities are approximately 3.6 kcal mol^{-1} per C–F bond and approximately 9.6 kcal mol^{-1} per Si–F bond[289].

The insertion of $H(CH_3)Si$: into H_2 was found to have a barrier of 10.3 kcal mol^{-1} at MP4SDTQ/6-311G**//MP2/6-31G** + ZPE[181]. It was predicted that further improve-

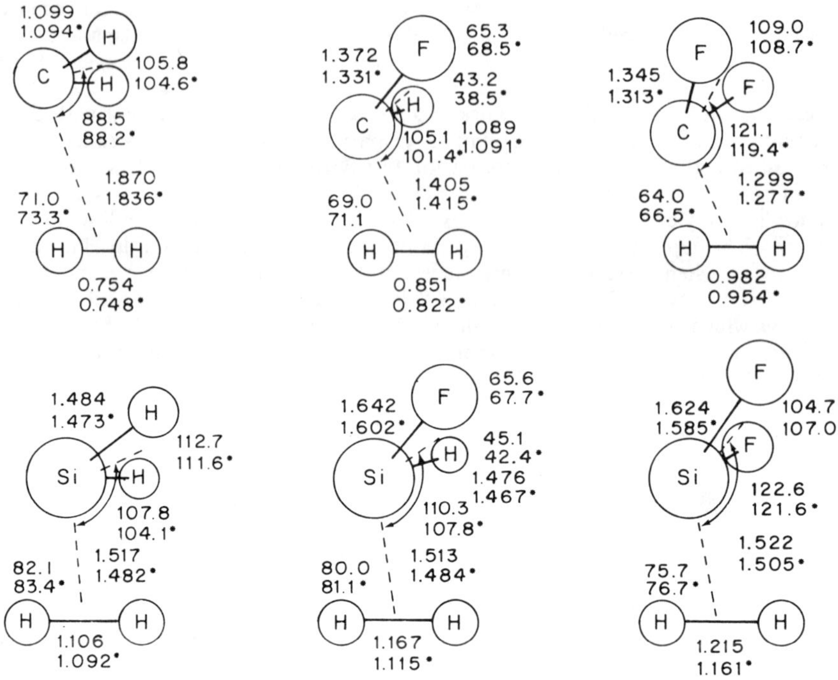

FIGURE 49. Transition structures for the insertion reactions of carbene and silylene into the H–H bond, optimized at 3-21G (no superscript) and 6-31G* (asterisk). The dihedral angles defined by the FCX (FSiX) and the C-hydrogen (Si-hydrogen) planes are 98.9° and 99.4°* for HFC:, and 101.7° and 102.4°* for HFSi:. X is a dummy atom in the CHF (SiHF) and the C-hydrogen (Si-hydrogen) planes. Bond lengths are in Å, bond angles in deg. Reprinted with permission from *J. Am. Chem. Soc.*, **106**, 5847 (1984). Copyright (1984) American Chemical Society[289].

ment in the calculations will reduce the barrier height by as much as 5 kcal mol^{-1} [291]. The calculated transition structure is presented in Figure 50a and it shows a considerable lengthening of the H–H bond[293].

The transition structure for the insertion of $H(H_3Si)Si$: into H_2 is shown in Figure 50b. The calculated barrier is very low, 1.9 kcal mol^{-1} at MP4SDTQ/MC-311G**//MP2/6-31G**, and the reaction is exothermic by 53.4 kcal mol^{-1} [199]. The reverse reaction, i.e. the elimination of H_2 from disilane, is the lowest energy process for the decomposition of H_3SiSiH_3[199].

Insertions into a variety of other σ-bonds were also studied. Gordon and Gano[287] studied insertion of H_2Si: and H_2C: into methane and silane. Transition structures were determined at 3-21G and energies at MP3/6-31G*//3-21G. While all four reactions have nonzero SCF barriers, only the insertion of silylene into methane retains a nonzero barrier of 27.5 kcal mol^{-1} at MP3/6-31G* [287]. More recent calculations for this reaction, at the MP4SDTQ/6-311G**//MP2/6-31G** + ZPE level, predict a barrier of 22.1[181] kcal mol^{-1}, somewhat higher than the experimental activation energy of 16.6 kcal mol^{-1} [293]. The calculated transition structure is shown in Figure 51. Recent calculations[199] at the MP4SDTQ/6-31G** level find that the insertion of H_2Si: into SiH_4 is exothermic by 50 kcal mol^{-1} and confirm that this reaction proceeds without a barrier. Looking at the reverse of the above reactions it was concluded that the lowest energy

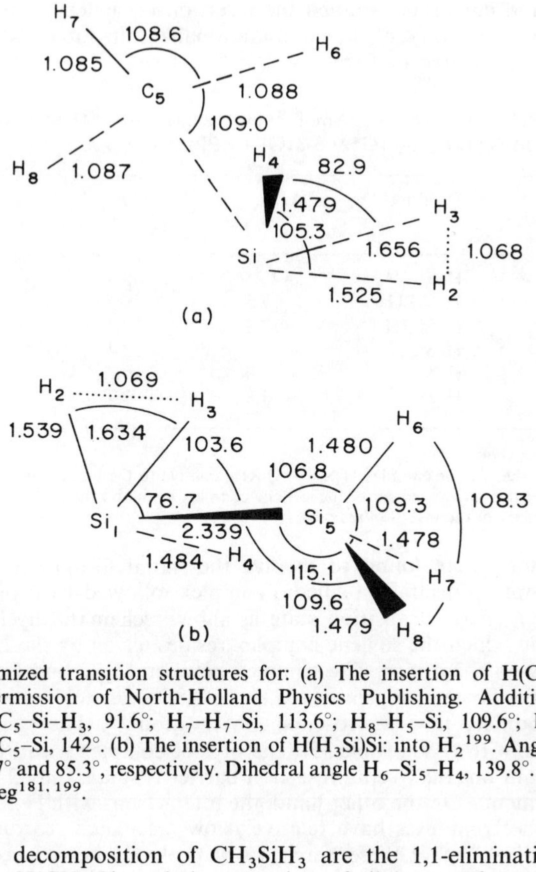

FIGURE 50. Optimized transition structures for: (a) The insertion of $H(CH_3)Si$: into H_2[181]. Reproduced by permission of North-Holland Physics Publishing. Additional bond angles: C_5–S:–H_2, 110.8°; C_5–Si–H_3, 91.6°; H_7–H_7–Si, 113.6°; H_8–H_5–Si, 109.6°; H_8–H_5–H_6, 107.2°. Dihedral angle H_6–C_5–Si, 142°. (b) The insertion of $H(H_3Si)Si$: into H_2[199]. Angles H_4–Si_1–H_2 and H_4–Si_1–H_3 are 101.7° and 85.3°, respectively. Dihedral angle H_6–Si_5–H_4, 139.8°. Bond lengths are in Å, bond angles in deg[181, 199].

processes for the decomposition of CH_3SiH_3 are the 1,1-elimination of molecular hydrogen to form $H(CH_3)Si$: and the extrusion of silylene to form methane (via the transition structure shown in Figure 51). The 1,2-hydrogen elimination to produce silene (see Section V.A.1) is predicted to be higher in energy.

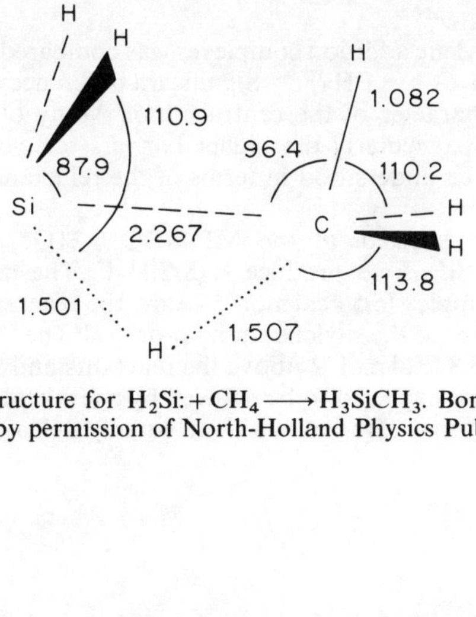

FIGURE 51. Transition structure for $H_2Si: + CH_4 \longrightarrow H_3SiCH_3$. Bond lengths are in Å, bond angles in deg. Reproduced by permission of North-Holland Physics Publishing from Ref. 181.

Raghavachari and coworkers studied the insertion of silylene into NH_3, H_2O, HF, PH_3, H_2S and HCl and some of the computational results (at MP4SDTQ/6-31G**//6-31G* + ZPE) are presented in Table 35 [104, 294].

TABLE 35. Relative energies (kcal mol^{-1}) for the insertion of H_2Si: into X–H bonds calculated at MP4SDTQ/6-31G**//6-31G* + ZPE[a]

Substrate XH_n	Product	$\Delta E_{complex}$ [b]	ΔE_{rear} [c]	ΔH_r [d]
NH_3	H_3SiNH_2	−25.1	38.3	−60.0
PH_3	H_3SiPH_2	−17.5	19.7	−53.1
H_2O	H_3SiOH	−13.3	22.0	−70.2
H_2S	H_3SiSH	− 8.6	13.4	−60.2
HF	H_3SiF	− 7.0	9.7	−84.3
HCl	H_3SiCl	− 1.8	7.7	−68.8

[a] From Reference 294.
[b] The energy of the silylene complex $H_2Si \cdots\cdots XH_n$ relative to the separated $H_2Si: + XH_n$.
[c] The barrier for the rearrangement of the silylene complex to the product.
[d] The exothermicity of the insertion reaction.

All these reactions were found to involve the initial formation of a donor (lone pair)–acceptor (empty p-orbital on silicon) complex followed by a proton shift via an unsymmetrical high-energy transition state as shown schematically in Figure 52. The depth of the well in which the silylene complex resides, given by the $\Delta E_{complex}$ values in Table 35, depends on the donor. The complex between H_2Si: and NH_3 is particularly stable, with a barrier for rearrangement (ΔE_{rear}) of 38.3 kcal mol^{-1} and it is predicted to be a suitable candidate for spectroscopic observation[294]. The same applies to the interaction of H_2Si: with H_2O (Table 35). The silylene complex with PH_3 also involves a fairly deep minimum but the overall insertion barrier is small, only 2 kcal mol^{-1} above the separated fragments. On the other hand, the interactions with H_2S, HF and HCl are fairly weak and the complexes have relatively low calculated rearrangement barriers. Experience shows[181, 291] that these barriers will probably be reduced further or even disappear at higher levels of theory.

In an earlier study Raghavachari and coworkers[104] studied the insertion of H_2Si: into the O–H bond of both H_2O and H_3SiOH. They found that the latter reaction proceeds in a manner qualitatively analogous to the first reaction, with the initial formation of a complex. The overall activation energy is, however, lower in the reaction of H_3SiOH than for the reaction with water[104].

The bonding in the silylene addition complexes was compared with that in their carbon analogues (i.e. such as $H_2\bar{C}: \cdots \overset{+}{P}H_3$)[294]. Significant differences were found, particularly in the multiple bond character of the central atom. Many of the differences between silylenes and carbenes, particularly the higher barriers to hydrogen migrations in the silylene complexes, can be understood in terms of the reluctance of silicon to rehybridize[294].

Schlegel and Sosa[295] studied, at the MP4SDQ/6-31G*//6-31G + ZPE level, the insertion of H_2Si: into SiH_3F to produce H_3SiSiH_2F. The insertion proceeds by the formation of a stable complex (5.5 kcal mol^{-1} below the reactants) in which the fluorine lone pairs donate electrons to the silylene empty p-orbital. The transition structure shown in Figure 53 lies about 3.8 kcal mol^{-1} above the reactants and 9.3 kcal mol^{-1} above the complex, but the barrier is expected to be reduced by a few kcal mol^{-1} at higher levels of theory[295]. The transition structure can be viewed equally well as a H_2Si: insertion into the

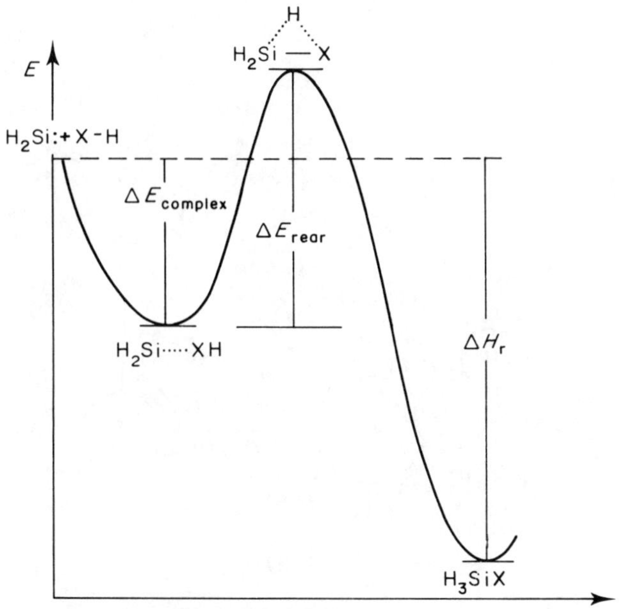

FIGURE 52. A schematic drawing of an energy profile for the insertion reaction of H_2Si: into an X–H bond

Si–F bond or as a [1,2]-fluorine shift in H_3SiSiH_2F (see equation 40). The authors note that the insertions of H_2Si: into the Si–F bond of H_3SiF[295] and into HF[289] are quite similar.

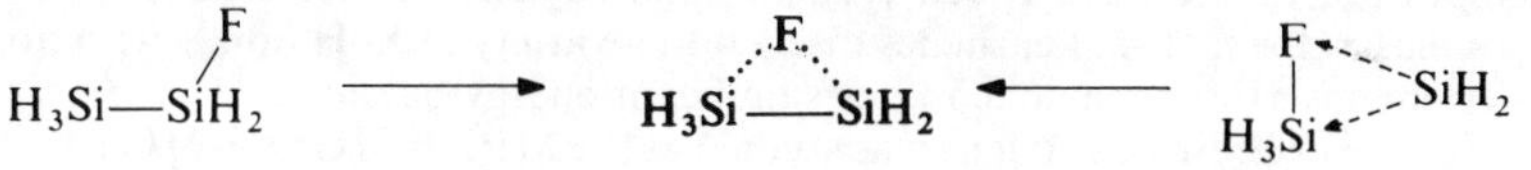

(40)

The energy surface for the interaction of silylene with BH_3 and B_2H_6 was explored at 6-31G* in connection with experimental studies on the incorporation of boron into amorphous silicon[296]. Both reactions have similar exothermicities at the MP4SDQ/6-31G*//6-31G* level of theory (48.5 kcal mol^{-1} for BH_3 and 49.1 kcal mol^{-1} for B_2H_6)[296]. No barrier is found for the reaction of H_2Si: with BH_3, but a small barrier of 4.2 kcal mol^{-1} is calculated at MP4SDTQ/6-31G*//6-31G* for the insertion of H_2Si: into a B–H nonbridge bond of B_2H_6. (The addition of correlation energy reduces the barrier dramatically: it is 25.3 kcal mol^{-1} at 6-31G*//6-31G*.)[296]

Hydrogen abstractions by triplet silylene (excited state) and methylene (ground state) from CH_4 and SiH_4 were studied by Gordon[297]. UMP3/6-31G**//3-21G calculations give the energetics and the transition-state structures shown in Figure 54. The general conclusions are as follows: (1) The reactions of triplet silylene and methylene have much higher reaction barriers than the insertion reactions of the corresponding singlet species. (2) The reactions with silylene are more endothermic (or less exothermic) than the same reactions of methylene, reflecting the weaker Si–H vs. C–H bond. (3) Triplet silylene is less effective than methylene as a hydrogen abstractor and SiH_4 is a better hydrogen donor than CH_4 (see Figure 54).

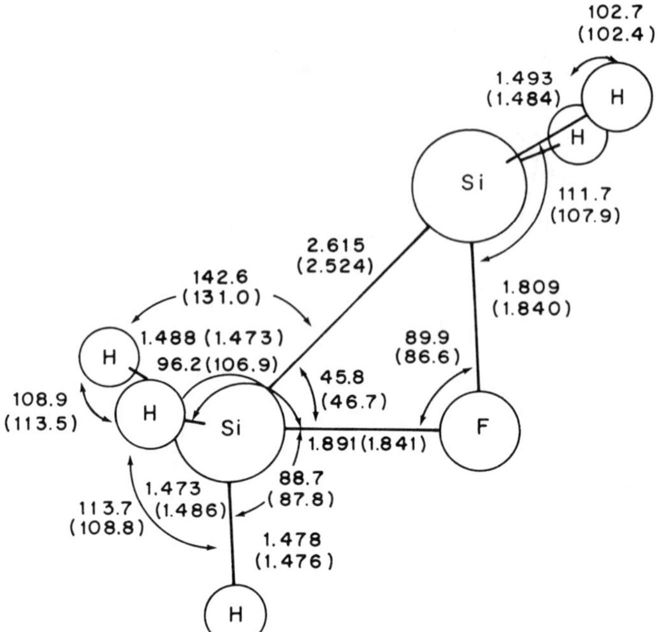

FIGURE 53. The transition structure for the insertion of H_2Si: into H_3Si–F, optimized at the 3-21G (no parentheses) and 6-31G* (in parentheses) levels. Bond lengths are in Å, bond angles in deg. Reprinted with permission from *J. Phys. Chem.*, **89**, 537 (1985). Copyright (1985) American Chemical Society[295].

An important difference between carbenes and silylenes, which was revealed by the above studies, is the fact that carbenes insert into a variety of X–H bonds with no barrier while silylene insertions may encounter significant energy barriers.

The addition of H_2Si: to ethylene was studied at the MP2/6-31G*//3-21G level[298]. The C_{2v} path is symmetry forbidden[282a] and the reaction proceeds with no barrier via a 'non-least-motion' C_s path. The overall exothermicity of the reaction is 47.9 kcal mol^{-1} at the above level of theory[298].

The dissociation of H_2Si: according to equation 41, where M is an inert molecule in the reaction mixture, requires a barrier of 55 kcal mol^{-1} (at 3-21G//3-21G) along the C_{2v} ('least-motion') path[299a]. Gordon and coworkers[21] studied this process at MCSCF/6-31G** and found that the $^1A_1$ and $^3B_1$ states dissociate to Si(^{1}D) + $H_2(^1\Sigma g^+)$ and to Si(^{3}p) + $H_2(^1\Sigma g^+)$, respectively[21].

$$H_2Si\colon + M \rightarrow Si + H_2 + M \qquad (41)$$

The dissociation of $H\ddot{S}iOH$ to silicon atom and water was studied by Tachibana and coworkers at the MP4SDTQ/6-31G**//MP2/6-31G* level[299b]. Looking at the reverse reaction (i.e. the insertion of silicon atom into water) the following reaction mechanism was found: First, a triplet silicon atom forms a triplet silicon–water adduct. Then a hydrogen migrates to silicon by proton tunnelling to form the triplet $H\ddot{S}iOH$, which may be converted to the singlet state by intersystem crossing. The overall barrier is 31.1 kcal mol^{-1} and the reaction is exothermic by 52.1 kcal mol^{-1} relative to the silicon–water adduct[299b]. This picture is in close agreement with matrix isolation experiments[299c].

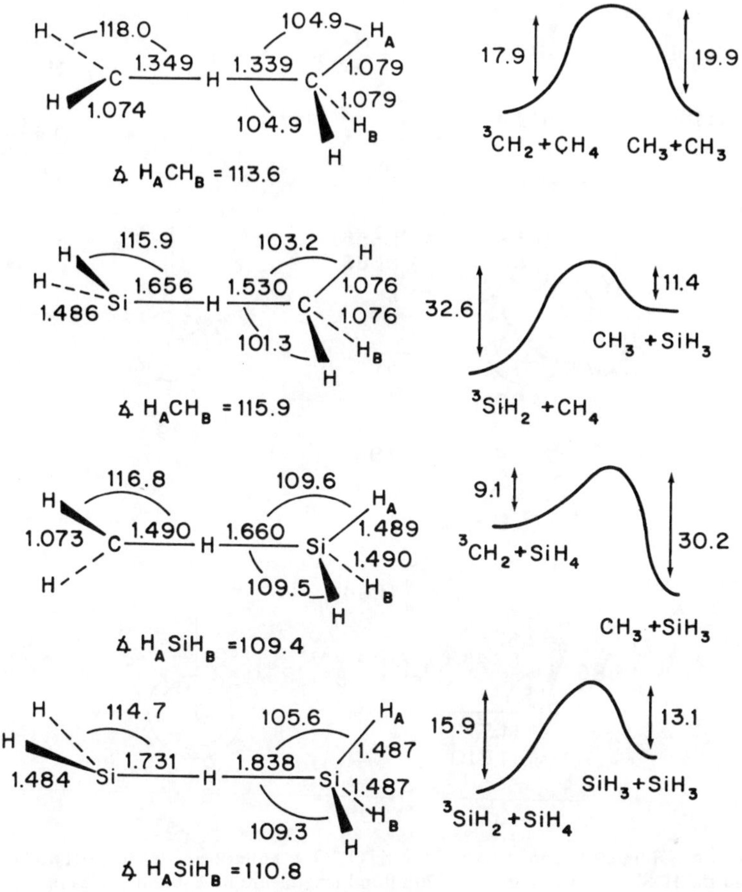

FIGURE 54. Transition structures and reaction profiles for hydrogen abstraction from methane and silane by triplet silylene and methylene. Bond lengths are in Å, bond angles in deg, energy in kcal mol^{-1}. Reprinted with permission from *J. Am. Chem. Soc.*, **106**, 4054 (1984). Copyright (1984) American Chemical Society[297].

H_2SiLiF has been studied[300] as a model for 'silylenoids' of the type R_2SiMX (M = alkali metal, X = halogen), analogous to carbenoids which exist in very unusual structural forms[301] and play an important role in carbene chemistry. Alternatively, SiH_2LiF can be viewed as resulting from the insertion of silylene into lithium fluoride. At MP2/6-21G//3-21G, the two most stable forms of SiH_2LiF are the $SiH_2Li^+F^-$ ion pair **119** and the H_2Si: –FLi complex **120**, analogous to the similar structures previously found for carbenoids[301]. The energies of **119** and **120** are within 1.3 kcal mol^{-1} of each other and more sophisticated calculations are required to decide which is the most stable. The optimized structures of **119** and **120** are shown in Figure 55. Two additional species, a H_2Si: –LiF complex (**121**) and the 'classical' tetrahedral H_2SiFLi (**122**), are also local minima on the potential surface, but they are significantly higher in energy, i.e. by 35.9 and 18.8 kcal mol^{-1} respectively (higher-level calculations are necessary for more reliable conclusions).

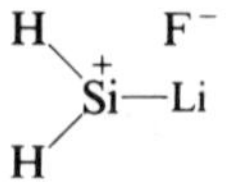

(**119**)

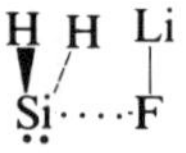

(**120**)

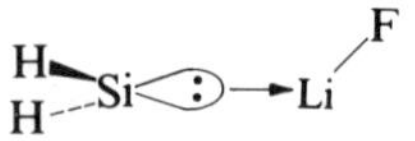

(**121**)

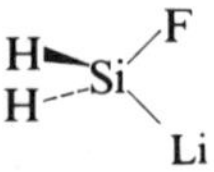

(**122**)

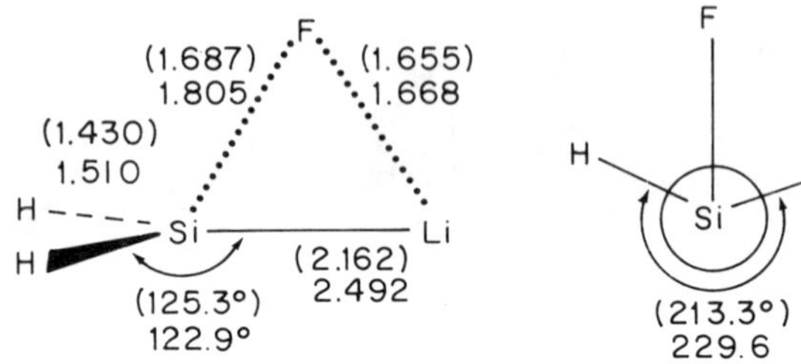

(119)

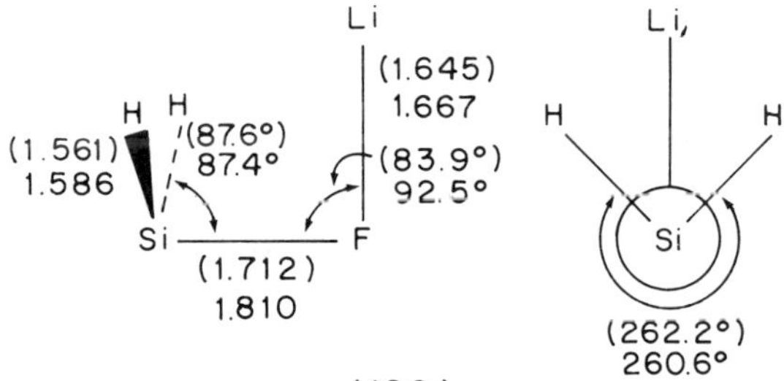

(120)

FIGURE 55. Optimized geometries (3-21G, STO-3G in parentheses) of the $SiH_2Li^+F^-$ ion-pair (**119**) and the H_2Si: · · · · FLi complex (**120**). Bond lengths are in Å, bond angles in deg[300].

4. Miscellaneous silylenes

Most of the 'simple' silylenes of the type RR′Si: were discussed in this section and at the beginning of Section VII.A.3, where citations were given to isomerization of such silylenes to multiply bonded species. In Table 36 and Figure 56 we list all the silylenes not mentioned in Section VII.A, and refer the reader to other sections in which these silylenes are discussed. For silylenes which were not mentioned in earlier sections, references to the original literature are given.

B. Cationic Species

1. Silicenium ions

Molecules with trivalent positive silicon (R_3Si^+) are termed either silicenium ions (in analogy to carbenium ions) or silylenium ions (derived from silylene as carbenium is derived from carbene). Silicenium ions have been known and studied in the gas phase for over 20 years[302]. In solution, silicenium ions have been sought for more than 40 years[258]

TABLE 36. Citations to miscellaneous silylenes

Silylene	Section or Ref.	Silylene	Section or Ref.
Noncyclic		*Cyclic*	
R^1R^2Si:[a]	VII.A	**54b**	V.A.1.c
54a	V.A.1.c, VII.A	**109**	VI.D
$H_2C{=}CHSiCH_3$	V.A.1.c	**128**	142a
$H_2C{=}CHCH_2\ddot{S}iH$	157	**124**	139b
$H_3SiSiH_2\ddot{S}iH$	V.A.2.f	**125**	139b
$HO\ddot{S}iCH_3$	V.A.5.b VII.A	**88**	VI.A.1
$H_3SiO\ddot{S}iH$	V.A.5.b	**89**	VI.A.1
$H_3SiO\ddot{S}iCH_3$	182	**95**	VI.A.3
$FO\ddot{S}iH$	V.A.5.b	**96**	VI.A.3
$R^1_3Si\ddot{S}iR^2$[b]	V.A.2.f		
123	139b	*Multisilylenes*	
77	V.B	**107c**	VI.D
$R^1HC{=}Si$:[c]	V.B.1	**112**	VI.D
$F_2C{=}Si$:	V.B.1	**126**	139b
$LiFC{=}Si$:	V.B.1	**127**	253b
80	V.B.2		
$HN{=}Si$:	V.B.3		

[a] $R^1 = H$, $R^2 = Li$, BeH, BH_2, CH_3, SiH_3, NH_2, OH, F, C≡CH, CN, PH_2; $R^1 = R^2 = CH_3$; $R^1 = R^2 = Li$; $R^1 = R^2 = F$; $R^1 = R^2 = t$-Bu; $R^1 = R^2 = SiH_3$.
[b] R^1 or $R^2 = CH_3$, Cl, F.
[c] $R^1 = Me$, F, Li.

and in an important review published in 1974 Corriu and Henner concluded that silicenium ions do not exist in solution[258b]. However, recent studies are beginning to change this consensus although the data are still limited[258a]. Lambert's group has brought evidence for the observation under stable ion conditions of two specific silicenium ions[258a, 303], although their conclusions were questioned recently by Olah and coworkers[304]. Eaborn's group has provided extensive evidence for the solvolytic generation of a bridged silicenium cation **129**[305]. Apeloig and Stanger demonstrated that silicenium ions can be generated solvolytically via a 1,2-methyl shift of a properly designed carbenium ion (equation 42)[306]. Some additional details of these experiments are discussed below together with the theoretical studies.

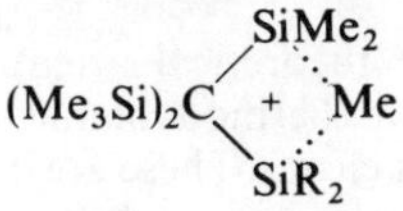

(129)

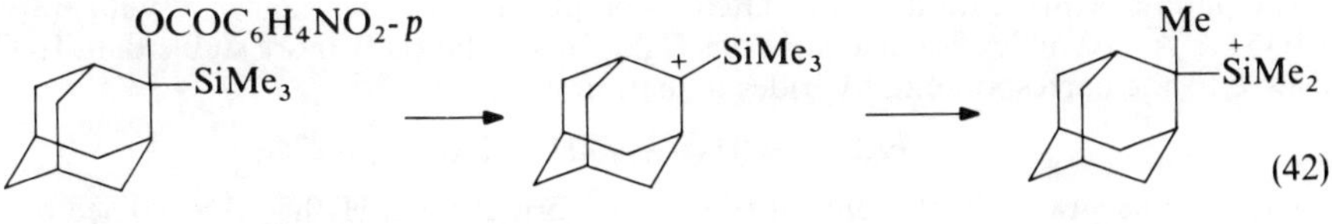

(42)

(130a) **(131)** **(132)**

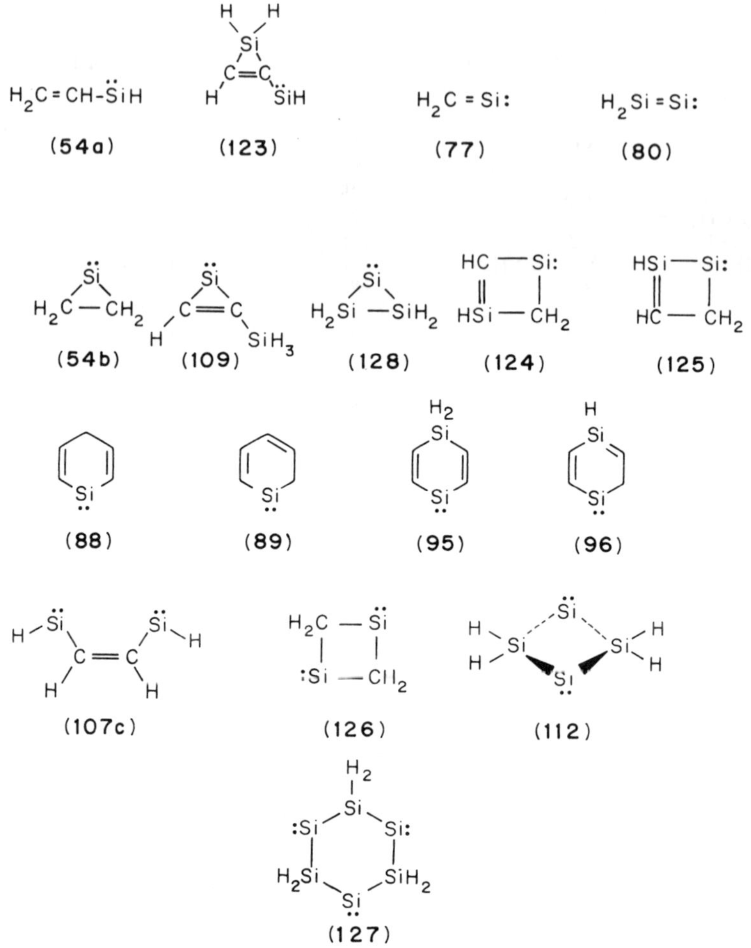

FIGURE 56. Structures of silylenes cited in Table 36

Silicenium ions received some theoretical attention, but to a smaller extent than silylenes. Interest centered on their thermodynamic stabilities relative to the analogous carbenium ions and on substituent effects. These are important questions, because the low thermodynamic stability of silicenium ions was believed by many workers to be one of the major factors frustrating their experimental preparation[258].

The parent H_3Si^+ cation is calculated to be planar (D_{3h} symmetry)[307] with r(Si–H) = 1.454 Å[307c]. At all levels of calculations H_3Si^+ is substantially more stable than H_3C^+, relative to the corresponding hydrides (equation 43).

$$H_3Si^+ + CH_4 \longrightarrow SiH_4 + H_3C^+ \qquad (43)$$

The most accurate calculations for the heat of formation of H_3Si^+ were carried out at the MP4SDTQ/6-31G + G(2d, f)//6-31G* level giving $\Delta H^\circ_{f\,298} = 234.1$ kcal mol^{-1} [307b, c], in excellent agreement with the recent experimental value of 235.7± 1.6 kcal mol^{-1} [272b, 308]. Using this value, and the calculated heats of formations at 298 K of H_3C^+ (261.7 kcal mol^{-1})[307b], CH_4 (−19.8 kcal)[267a] and SiH_4 (6.3 kcal mol^{-1})[267a], we calculate that at 298 K equation 43 is endothermic by 53.7 kcal mol^{-1} (53.4 kcal mol^{-1} at MP4SDTQ/6-31G**//6-31G*, only 49.2 kcal mol^{-1} at 6-31G*//3-21G[94]). The above theoretical calculations and the recent experiments[272b, 308] are in conflict with previous experiments[309] by Lampe and coworkers, who reported for equation 43 an energy of 47 kcal mol^{-1} [309a]; we therefore believe that the older value is erroneous. Thus, according to both theory and experiment, the parent silicenium ion is substantially more stable than the parent carbenium ion. This is reasonable in view of the higher polarizability and lower electronegativity of silicon relative to carbon (see Section III).

What are the relative stabilities in the gas phase of substituted silicenium ions and the analogous substituted carbenium ions? To answer this question Apeloig and coworkers carried out systematic studies on substituted silicenium ions $RSiH_2^+$, where R was varied along the standard second-row (i.e. Li to F)[310a] and third-row (i.e. Na to Cl)[310b] substituents. The effect of second-row substituents was also studied by Hopkinson and Lien[311]. Although these calculations used relatively simple levels of theory (STO-3G and 3-21G), semiquantitative conclusions could be reached with confidence. Using isodesmic comparisons (equation 44 for silicenium ions and equation 45 for carbenium ions) it was found that the substituents examined are all less effective in stabilizing silicenium ions than in stabilizing carbenium ions[310]. This is most pronounced for substituents which are good π-donors, such as NH_2 and OH. For example, an amino substituent stabilizes the methyl cation by 93.8 kcal mol^{-1} and the silyl cation by only 33.3 kcal mol^{-1} [310a]. This is probably a result of poorer orbital overlap for silicon than for carbon and the reluctance of silicon to participate in multiple bonding.

$$H_2SiR^+ + SiH_4 \longrightarrow SiH_3^+ + SiH_3R \qquad (44)$$

$$H_2CR^+ + CH_4 \longrightarrow CH_3^+ + CH_3R \qquad (45)$$

The weak conjugation between the silicenium center and the lone pair on the substituent is nicely demonstrated by the fact that $H_2PSiH_2^+$ prefers a pyramidalized structure at phosphorous, although conjugation with the empty cationic orbital is much more effective in the all-planar structure[310b]. Godleski and coworkers have studied in more detail (by MNDO) the phosphino- and amino-substituted silicenium ions[312]. They found that NH_2 stabilizes silicenium ions more effectively than PH_2 and have attributed this behavior[312] to the higher energy required to planarize P relative to N[310]. If the planarization energy of phosphorous is reduced by substitution, the stabilizing power of the PR_2 substituent increases substantially. For example, while PH_2 stabilizes SiH_3^+ (equation 44) by only 2.5 kcal mol^{-1}, $HP(SiH_3)$ stabilizes it by 26.1 kcal mol^{-1} [312]. Multiple substitution provides higher stabilizations[312].

Other conjugating substituents such as vinyl and phenyl, which strongly stabilize carbenium ions, were also found to be only 30–40% as effective in stabilizing silicenium ions as in stabilizing their carbenium ion analogs[313]. A demonstrative example is the potentially aromatic silacyclopropenium ion **102**[246a], which is slightly destabilized by conjugation (see Section VI.C). Electron-withdrawing substituents (CHO, CN and CF_3) destabilize SiH_3^+ while C≡CH is weakly stabilizing (dz calculations)[314].

These studies show that the traditional approach for stabilizing silicenium ions by conjugating substituents (such as aryl groups[258b]) is not very effective. On the other hand, stabilization by electropositive substituents such as Li and BeH is very effective[310]. This is nicely exemplified by the fact that H_2SiLiF is more stable as a silicenium ion-pair $H_2SiLi^+ F^-$, where the positive center is stabilized by lithium, than in the covalent tetrahedral structure[300].

An important conclusion from all the calculations is that the same types of substituents stabilize both silicenium ions and carbenium ions. However, as substituents stabilize carbenium ions more effectively than silicenium ions, the much higher inherent stability of H_3Si^+ relative to CH_3^+ is reduced significantly in the substituted cations (equation 46). For example, SiH_3^+ is more stable than CH_3^+ by 73.5 kcal mol^{-1}, but $H_2SiNH_2^+$ is more stable than $H_2CNH_2^+$ by only 18.0 kcal mol^{-1} (STO-3G)[310a].

$$RSiH_2^+ + RCH_3 \longrightarrow RCH_2^+ + RSiH_3 \qquad (46)$$

The effect of substitution by methyl, a substituent which is commonly encountered experimentally, was studied by several groups[310a, 311, 314, 315]. Particularly extensive studies have been carried out recently by our group[94]. At MP4/6-31G*//3-21G a methyl stabilizes SiH_3^+ by 15.2 kcal mol^{-1} (equation 44, $R=CH_3$), while the stabilization is 40.3 kcal mol^{-1} for CH_3^+ (equation 45, $R=CH_3$)[94]. Stabilization per methyl is slightly reduced upon multiple substitution: at 6-31G* the values are 13.8, 11.9 and 10.0 kcal mol^{-1} for the first, second and third methyl substitution, respectively[94]. These results are in reasonable agreement with the experimental results of Lampe and coworkers[309]. At MP2/6-31G* $(CH_3)_3Si^+$ is more stable than $(CH_3)_3C^+$ by 22.4 kcal mol^{-1} [94], in agreement with the more recent gas-phase experiments which predict 10–20 kcal mol^{-1} [302, 316], but the value is much higher than the value of 7 kcal mol^{-1} reported earlier by Lampe and coworkers[309b]. Considering the fact that Lampe and coworkers also found that H_3Si^+ is more stable than H_3C^+ by only 47 kcal mol^{-1} [309a] while highly reliable calculations[307b, c] and new experimental measurements[272b, 308] yield a value of about 55 kcal mol^{-1}, we suggest that Lampe's measurements have a constant error of approximately 10 kcal mol^{-1} [94].

The fact that alkyl-substituted tertiary silicenium ions are more stable than the isomeric tertiary carbenium ions was utilized predictively by Apeloig and Stanger to design an experiment, which lead to the first demonstration of a solvolytic generation of a silicenium ion[306]. Thus, the calculations showed that $Me_3CSiMe_2^+$ is more stable than $Me_3SiCMe_2^+$ by 9.7 kcal mol^{-1} (3-21G//3-21G), and that the barrier for methyl migration is only a few kcal mol^{-1}. If this is so, then $Me_3SiCMe_2^+$ should rearrange easily to $Me_3CSiMe_2^+$. The analogous adamantyl derivative **130a** was solvolyzed and product studies indeed suggested that the initially generated carbenium ion **131** rearranged to the silicenium ion **132** (equation 42)[306].

Sulfur substituents are of particular interest in relation to the recent report by Lambert and coworkers on the observation of a stable $(i\text{-PrS})_3Si^+$ cation[303a, c]. Similar experiments did not produce the analogous oxy-silicenium ions, $(i\text{-PrO})_3Si^+$, but it was not clear if this failure is due to thermodynamic or kinetic factors[317]. Lambert has attributed the successful generation of $(i\text{-PrS})_3Si^+$ to its high thermodynamic stability, resulting from effective conjugation between the empty 3p-orbital on Si and the 3p-lone pair of sulfur and to the high polarizability of sulfur[303a, c]. Is sulfur really unique? Initial calculations did not give a clear answer, since at STO-3G they indicated that SH and OH stabilize silicenium ions to a similar extent while OH was substantially more stabilizing than SH at 3-21G. In a more detailed study it was first established that (at MP2/6-31G*//6-31G*) OH and SH stabilize SiH_3^+ practically to the same degree, i.e. by 19.7 and 20.1 kcal mol^{-1} according to equation 44, R=OH and SH, respectively[94]. This conclusion remains practically unchanged with R=MeS vs. MeO and with R=*i*-PrS vs. *i*-PrO. In order to model more closely Lambert's experimental system, the corresponding di- and tri-substituted silicenium ions were also studied. The results, which are based on MP2/6-31G*//3-21G calculations of appropriate isodesmic equations (e.g. equations 47 and 48 for the di-substituted cations and neutrals, respectively), are shown schematically in Figure 57.

$$H_2SiR_2 + H_2SiR^+ \longrightarrow HSiR_2^+ + H_3SiR \qquad (47)$$

$$H_2SiR_2 + SiH_4 \longrightarrow 2H_3SiR \qquad (48)$$

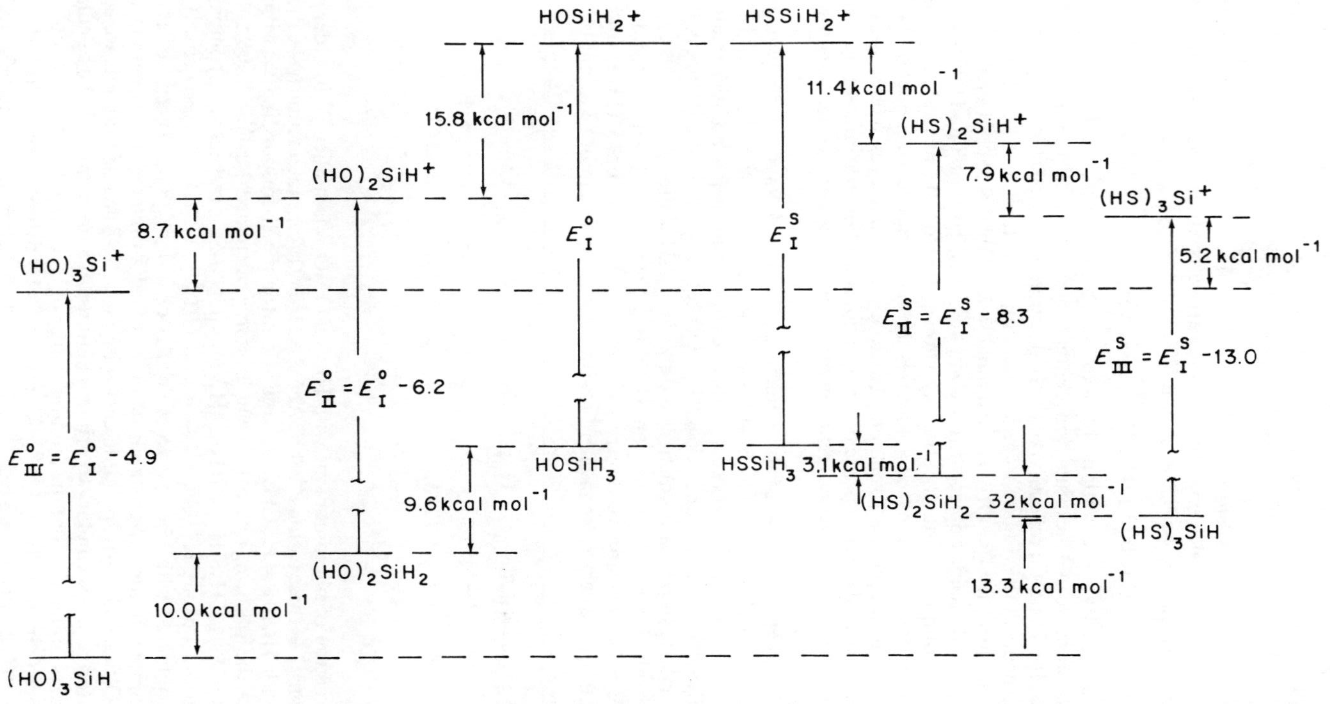

FIGURE 57. Schematic description of the stabilization of a silicenium center and of a saturated silicon center by mono-, di- and tri-OH and SH substitutions, based on MP2/6-31 G*//3-21 G calculations of appropriate isodesmic equations (e.g. equations 47 and 48 for the di-substituted systems). E^{o} and E^{s} are the dissociation energies of the neutral oxygen and sulfur substituted neutrals, respectively, to the corresponding cations and hydride ion. Subscripts I, II and III denote mono-, di- and tri-substitution, respectively. H_3SiOH and H_3SiSH are placed artificially at the same energy level.

To assist the comparison between the systems, H_3SiSH and H_3SiOH were placed artifically in Figure 57 at the same energy level. The relative dissociation energies of H_3SiOH and H_3SiSH to the corresponding cations and hydride ion (E_I^o and E_I^s, respectively) are nearly equal (reaction 44). However, hydride transfer from the sulfur-substituted silane to the oxygen-substituted silicenium ion becomes successively more exothermic on going from mono- to di- to the tri-substituted system (Figure 57). Thus, reaction 49 is exothermic by 9.2 kcal mol^{-1}.

$$(HS)_3SiH + (HO)_3Si^+ \longrightarrow (HS)_3Si^+ + HSi(OH)_3 \quad (49)$$

The exothermicity of equation 49 *does not* result from the fact that three SH substituents stabilize a silicenium ion center much more effectively than three OH substituents. It is mainly a reflection of the stronger anomeric effect of oxygen than that of sulfur in the neutrals (see Figure 57 and discussion in Section IV.B.1.c). Thus, reaction 48 is more endothermic for R = OH (9.6 kcal mol^{-1}) than for R = SH (3.1 kcal mol^{-1}) by 6.5 kcal mol^{-1}, and the same applies for the comparison of $(HO)_3SiH$ and $(HS)_3SiH$. Part of the anomeric ground-state stabilization is lost on going from the neutral precursor, which adopts a *gauche* conformation that maximizes the anomeric effect (see Section IV.B.1.c), to the cation which adopts an all-planar structure that maximizes interactions with the empty cationic orbital at the expense of losing favorable anomeric interactions. The loss of such anomeric interactions is energetically more important for oxygen substitution than for sulfur substitution.

H_2SiSH^+ and H_2SiOH^+ were also studied by Kudo and Nagase mainly in relation to the protonation of $H_2Si{=}S$ and $H_2Si{=}O$ (see Sections V.A.5 and V.A.6)[220]. They studied the rotation and inversion processes in these cations and their stabilities relative to the isomeric silylenes $H\ddot{S}iSH_2^+$ and $H\ddot{S}iOH_2^+$, respectively. It was concluded that both H_2SiSH^+ and H_2SiOH^+ are kinetically stable to 1,2-H shifts[220].

The silyl-substituted silicenium ion $H_3SiSiH_2^+$ was studied with a polarized basis set and CEPA-2 inclusion of electron correlation[192b]. The open (**133a**) and bridged (**133b**) structures are very close in energy, with **133a** being less than 1 kcal mol^{-1} lower in energy[192b].

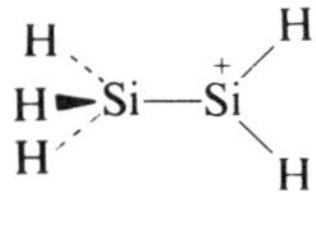

(**133a**)

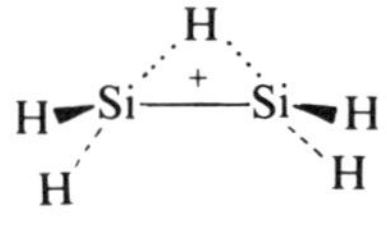

(**133b**)

The interaction of silicenium ions with HCl and H_2O was studied[318] in order to model the solvation[318] of these cations[94, 306b]. At 6-31G*//3-21G the interaction (i.e. solvation[318]) of R_3Si^+ with HCl (R = H, CH_3) is larger than for the corresponding carbenium ions R_3C^+. Thus, the following interaction energies were calculated (kcal mol^{-1}): 9.5, 14.9, 10.9 and 8.5 for H_3Si^+, $CH_3SiH_2^+$, $(CH_3)_2SiH^+$ and $(CH_3)_3Si^+$, respectively, compared with 25.3, 9.6, 3.9 and 4.6 for CH_3^+, $CH_3CH_2^+$, $(CH_3)_2CH^+$ and $(CH_3)_3C^+$, respectively. Similar results are obtained with one or two molecules of water[94]. Thus except for H_3Si^+, the other silicenium ions are better stabilized or 'solvated' by interaction with donors such as HCl or H_2O than the corresponding carbenium ions[94]. The interaction is weakened on going from tertiary to secondary to primary cations. In agreement with the calculations, it was found experimentally that the binding energy of H_2O to $(CH_3)_3Si^+$ is about 20.0 kcal mol^{-1} higher than for $(CH_3)_3C^+$ [302]. Methyl and silyl cation affinities for other donors, NH_3, H_2O, PH_3 and H_2S were calculated (dz + d + ZPE, kcal mol^{-1}) to be as follows: for SiH_3^+, 68.7, 49.2, 50.5 and 37.6, respectively, and for CH_3^+, 86.3, 48.7, 83.4 and 55.9, respectively[319]. However, the generally higher affinities of CH_3^+, relative to SiH_3^+ might be expected to be reversed with substituted cations, as found for the interaction with HCl and H_2O[94].

The entire series of small H_nSi^+ cations ($n = 1$–5) was studied by Pople and Curtiss using MP4 theory and a series of extended basis sets[307b, c]. Ionization potentials and heats of formations were calculated (Table 37) and evidence was presented that the theoretical results are accurate to about ± 2 kcal mol^{-1}.

TABLE 37. Theoretical (MP4/6-31G*+ZPE)[a] and experimental[b] heats of formation (kcal mol^{-1}) and ionization potentials (eV) of SiH_n^+ cations

Cation	$\Delta H^\circ_{f\ 298}$	IP[c]
Si^+	—	8.13 (8.15)
SiH^+	271.5 (274.3; 273.3)	7.86 (<7.98)
SiH_2^+	276.6 (277.7)	9.18 (9.48)[d]
SiH_3^+	234.1 (234.2)	8.03 (8.14)
SiH_4^+	264.9	11.12 (11.6)
SiH_5^+[e]	220.5 (220.5)	

[a] From Reference 307b.
[b] Values in parentheses, cited in the tables of Reference 307b.
[c] Adiabatic ionization potentials of the corresponding neutrals.
[d] $SiH_2(^1A_1) \rightarrow SiH_2^+$.
[e] The calculated proton affinity of SiH_4 is 152.9 kcal mol^{-1} (155 kcal mol^{-1}, experimental).

Protonated silane, SiH_5^+, is of special interest as a model for other pentacoordinated siliconium ions which are believed to play a role in solvolytic reactions[258b]. Like CH_5^+, SiH_5^+ adopts a C_s structure (6-31G*)[307c, 320]. However, the differences between SiH_5^+ and CH_5^+ are more significant than the similarities. The D_{3h}, C_{3v} and C_{4v} forms of SiH_5^+ are much higher in energy than the C_s form and SiH_5^+ is predicted not to undergo hydrogen scrambling (pseudorotation) readily, in contrast to CH_5^+. The proton affinity of SiH_4 exceeds that of CH_4 by about 25 kcal mol^{-1}, but the heat of hydrogenation of SiH_3^+ is smaller than that of CH_3^+ by nearly the same amount[307c, 320]. Thermodynamically, SiH_5^+ is more stable than CH_5^+ and proton transfer from CH_5^+ to SiH_4 is predicted to occur[320].

The four isomeric $CSiH_3^+$ cations (**134**–**137**) were studied with a dz basis set[224a]. As the charge becomes more localized on silicon the structure becomes more stable regardless of the valence of silicon, so that the stability order is: **134** (most stable) $\gg$ **135** > **136** > **137**[224a]. **136** collapses without a barrier to **135**, while the rearrangement of **135** to **134** and of **137** to **136** involve a barrier[224a].

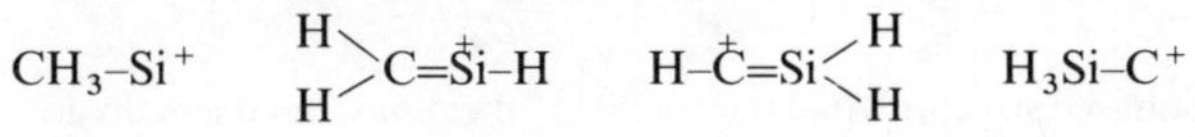

(**134**) (**135**) (**136**) (**137**)

High-level calculations (MP4/6-31G**//6-31G*) were used in conjunction with charge-stripping experiments to characterize SiH_n^{2+} dications ($n = 1$–5)[321a–c]. SiH_n^{2+} ($n = 1$–3) are observable species[321a–c]. The 6-31G* structures of these cations are given in Figure 58. The minimum-energy structures of the SiH_n^{2+} dications ($n = 4,5$) are best described as complexes between a doubly-ionized SiH_{n-2}^{2+} dication and H_2. Comparison of SiH_n^{2+} with the analogous CH_n^{2+} cations shows that deprotonation energies according to the

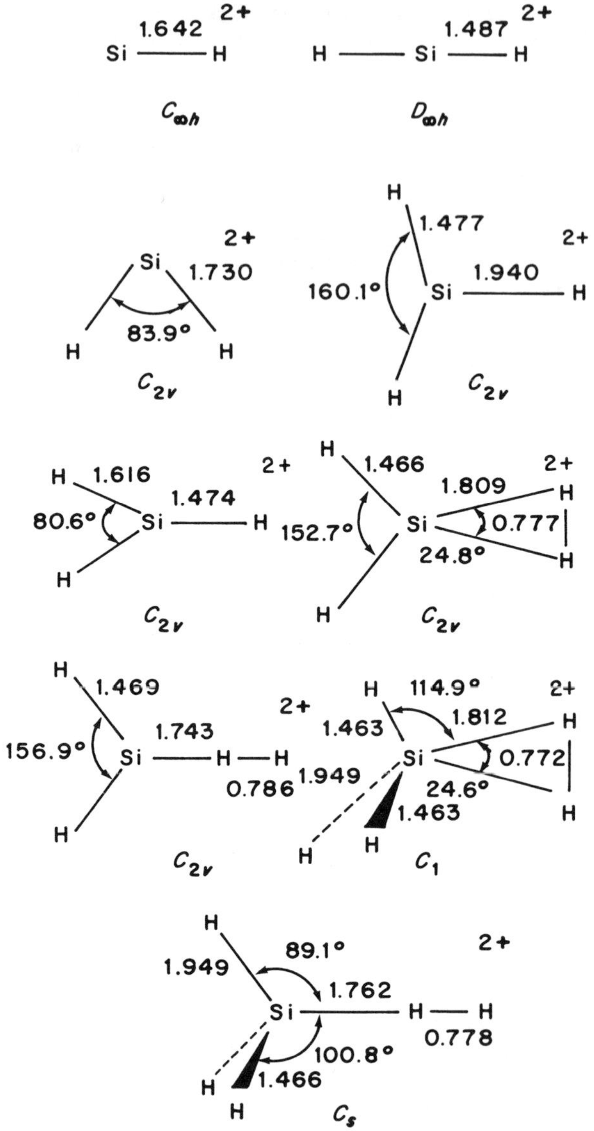

FIGURE 58. Optimized structures (6-31G*) of SiH_n^{2+} dications. Bond lengths are in Å, bond angles in deg[320].

reactions $XH_n^{2+} \rightarrow XH_{n-1}^{+} + H^{+}$ are significantly more exothermic for X=C than for X=Si. This is a direct consequence of the better capability of silicon to accomodate the positive charge.

Radom and coworkers obtained remarkable results for the $CSiH_4^{2+}$ dications[321d]. Optimization of geometry at MP2/6-31G* shows that the C_{3v} structure **138** is not a minimum on the potential-energy surface and it collapses without a barrier to **139**, which is the lowest $CSiH_4^{2+}$ structure, lying 106 kcal mol^{-1} (MP4/6-31G**) below **138**. Another

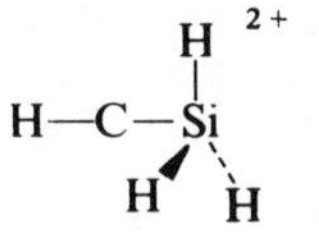

(138)

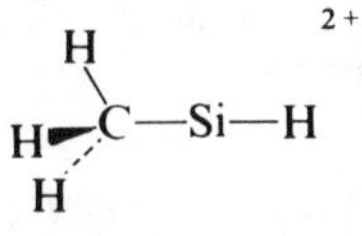

(139)

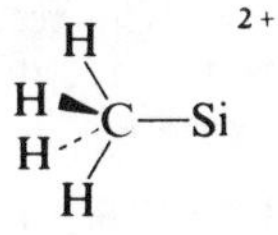

(140)

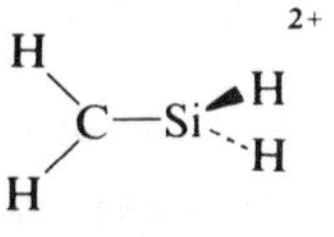

(141)

low-lying isomer is **140**, which is only 3.1 kcal mol^{-1} less stable than **139**. Dication **141** is by 23 kcal mol^{-1} higher in energy than **139**[321d].

2. Silyl-substituted carbenium ions

Silyl-substituted carbenium ions have attracted considerable experimental interest because they are believed to be intermediates in electrophilic additions to vinyl, ethynyl and aryl silanes, in solvolytic reactions and in cationic cyclization reactions[1–4, 322]. Despite this wide interest, knowledge concerning the effect of silyl substitution on the stabilities of carbenium ions was rather qualitative and only recently more quantitative data became available. Theoretical studies centered mainly around α- and β-silyl-substituted carbenium ions, but γ-silyl effects have been also studied.

a. α-Substitution. The simplest α-silyl-substituted carbenium ion $H_3SiCH_2^+$ is not a minimum on the potential-energy surface and it rearranges without a barrier to the more stable (by 49.1 kcal mol^{-1} at MP4/6-31G*//3-21G) silicenium ion $H_3CSiH_2^+$ [315]. However, calculations using a fixed geometry which prevents the rearrangement allow one to compare the α-effects of silyl and methyl substitution (equation 50).

$$H_3SiCH_2^+ + CH_3CH_3 \longrightarrow H_3SiCH_3 + CH_3CH_2^+ \qquad (50)$$

According to equation 50, $CH_3CH_2^+$ is more stable than $H_3SiCH_2^+$ by 13.2 kcal mol^{-1} [323a], yet an α-H_3Si substituent is more stabilizing than hydrogen by 16.1 kcal mol^{-1} (6-31G*//3-21G)[323]. At MP3/6-31G*//6-31G* the energy difference between ethyl cation and $H_3SiCH_2^+$ increases slightly to 16.7 kcal mol^{-1} [194]. The silyl substituent is less effective than methyl also in stabilizing secondary and tertiary carbenium ions. Thus, in a hydride transfer reaction $CH_3(SiH_3)CH^+$ is less stable by 14.3 kcal mol^{-1} than $(CH_3)_2CH^+$, and $H_3SiC(CH_3)_2^+$ is less stable than $(CH_3)_3C^+$ by 8.6 kcal mol^{-1} (6-31G*//3-21G). These relative stabilities change only slightly for the more frequently used $(CH_3)_3Si$ group. Thus, $(CH_3)_3CC(CH_3)_2^+$ is estimated to be more stable than $(CH_3)_3SiC(CH_3)_2^+$ by 6–8 kcal mol^{-1} [194]. In order to test these theoretical predictions, Apeloig and Stanger[324] measured the solvolysis rates of **130a** and **130b** and found them to be practically the same. This apparent contradiction with the above calculations was resolved by additional calculations.

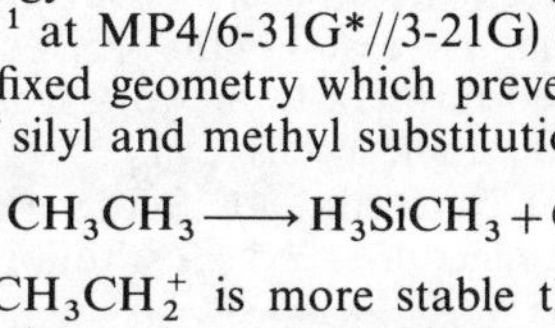

(130a) R = $SiMe_3$
(130b) R = CH_3

The differences in the ground-state energies of the silyl and alkyl systems as modeled by the corresponding alcohols were evaluated by reactions 51, which reveal that H_3SiCH_2OH and Me_3SiCH_2OH are significantly destabilized relative to CH_3CH_2OH

and Me_3CCH_2OH, respectively. This results from the fact that alkyl and hydroxy have a *stabilizing* anomeric effect, while the interaction of the more electropositive silyl group with hydroxy leads to *destabilization*[94].

$$R_3SiCH_2OH + R_3CCH_3 \longrightarrow R_3SiCH_3 + R_3CCH_2OH \quad (51)$$

$$R{=}H \qquad \Delta E = -7.3 \text{ kcal mol}^{-1} \quad (MP2/6\text{-}31G^*//3\text{-}21G)$$

$$R{=}CH_3 \qquad \Delta E = -6.3 \text{ kcal mol}^{-1} \quad (6\text{-}31G^*//3\text{-}21G)$$

The ground-state effects, as reflected by the data for equations 51, place the energy of the silyl derivative above that for the corresponding alkyl derivative. This compensation for the differences in ion stabilities results in solvolysis of **130a** and **130b** at the same rate[324]. It was concluded that both in solution and in the gas phase an α-silyl substituent at a carbenium ion center is less stabilizing than a methyl substituent but more stabilizing than hydrogen[324]. The destabilizing effect of α-silyl substitution was used to affect the rearrangement of **131** to the more stable silicenium ion **132**[306]. Note that the barriers for hydrogen or methyl migration in an α-silycarbenium ion to a silicenium ion are so small that in the gas phase α-silylcarbenium ions are predicted either to exist in a very shallow energy minimum, or not to be stationary points on the energy surface.

The destabilizing α-effect of silyl relative to methyl is unexpected in light of the lower electronegativity of silicon compared to carbon. The calculations show that this effect results from a weaker hyperconjugation of the silyl group and from electrostatic repulsions in $H_3SiCH_2^+$ between the adjacent positively charged cationic carbon and silicon[323a]. Figure 59 shows that as a result of the weaker hyperconjugation with the silyl group less charge is transferred in the π-framework from the substituent to the empty cationic 2p-orbital in $H_3SiCH_2^+$ than in $CH_3CH_2^+$ [94].

In conjunction with solvolytic studies of Stang and coworkers[325], Apeloig and Stanger[326] calculated also the effect of α-silyl substitution on the stability of two families of vinylic cations, **142** and **143**.

$$H_2C{=}\overset{+}{C}{-}R \qquad\qquad H_2C{=}\overset{+}{C}{-}C{\equiv}CR$$

(142) **(143)**

α-silyl and α-methyl substituents stabilize $H_2C{=}CH^+$ by 25.0 and 23.9 kcal mol^{-1}, respectively (6-31G*//3-21G). Thus, in contrast to the alkyl cations where α-silyl is less stabilizing than α-methyl, in **142** the two groups are stabilizing to comparable extents. In analogy with the saturated systems discussed above, $H_2C{=}C(OH)SiH_3$ is destabilized by 3.8 kcal mol^{-1} relative to $H_2C{=}(OH)CH_3$ due to geminal unfavorable interactions (see above)[326]. Thus, according to the calculations α-silylvinyl triflates are predicted[326] to solvolyze faster than α-alkylvinyl triflates, as indeed is the case[325]. A triple bond was found to transmit the electronic effects of alkyl substituents more effectively than those of silyl substituents due to the poorer hyperconjugative ability of silyl. Thus **143**, $R{=}CH_3$ and **143**, $R{=}SiH_3$ are more stable than the parent cation **143**, R = H by 13.7 and 8.3 kcal mol^{-1}, respectively (6-31G*//3-21G)[326]. Consequently **143**, $R{=}CH_3$ is more stable than **143**, $R{=}SiH_3$[326], in agreement with the solvolytic results[325]. Jorgensen and coworkers[327] have also studied (at MP3/6-31G*//3-21G$^{(*)}$) the α-silyl effect in both $H_3SiCH_2^+$ and $H_2C{=}\overset{+}{C}SiH_3$ and their results are very similar to those discussed above.

Recently Rappoport, Apeloig and coworkers reported an interesting application of the unusual destabilizing effect of the silyl group[328]. The analogy between a carbenium ion and the positively charged carbon of a carbonyl group suggested that α-silyl ketones should be destabilized relatively to the analogous alkyl ketones. Indeed, at MP2/6-31G**//6-31G* equation 52, which denotes the effect of replacing one of the methyl groups of acetone by a silyl substituent, is endothermic by 12.0 kcal mol^{-1}. On the other

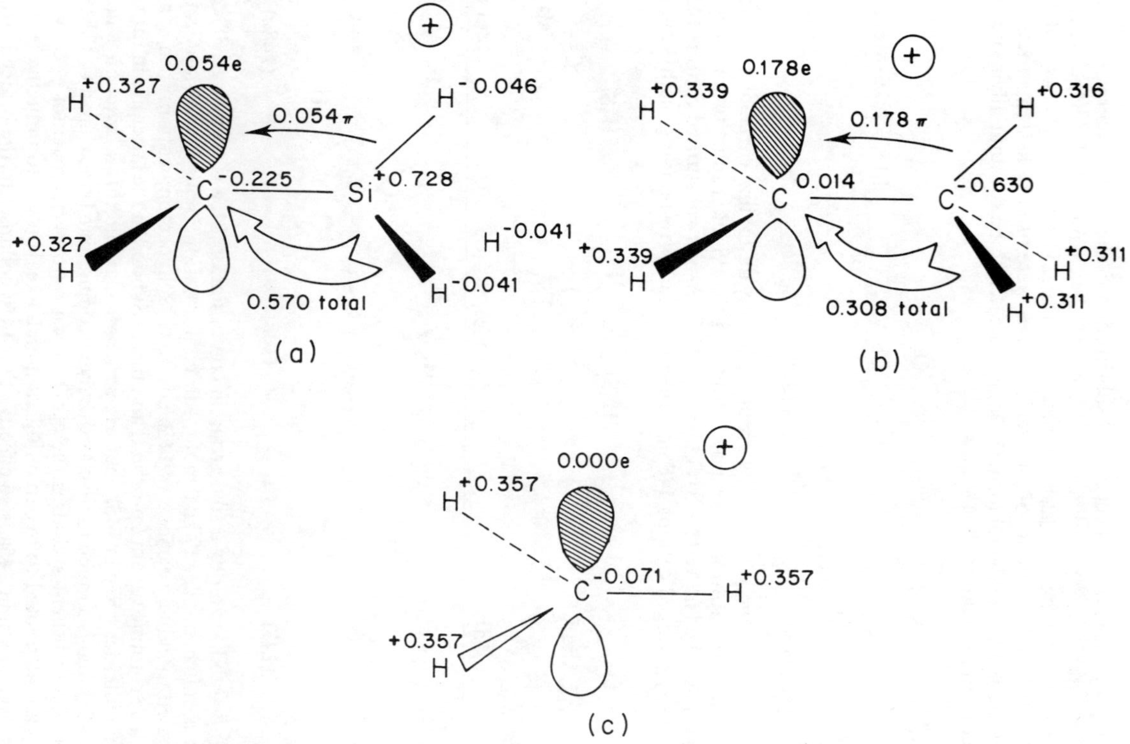

FIGURE 59. Orbital populations and atomic charges in $H_3SiCH_2^+$ (a), $CH_3CH_2^+$ (b) and CH_3^+ (c) (6-31G*//6-31G*).

hand, the effect of silyl substitution on the enol form of acetone (equation 53) is small (2.3 kcal mol^{-1}).

$$CH_3COCH_3 + H_3SiCH_3 \rightarrow H_3SiCOCH_3 + CH_3CH_3 \quad (52)$$

$$H_2C{=}C(OH)CH_3 + H_2C{=}CHSiH_3 \rightarrow H_2C{=}C(OH)SiH_3 + H_2C{=}CHCH_3 \quad (53)$$

This larger destabilizing effect of silyl substitution in the ketone than in the enol reduces the keto–enol energy difference from 18.8 kcal mol^{-1} for the CH_3COCH_3/$H_2C{=}C(OH)CH_3$ pair to 12.1 kcal mol^{-1} for the CH_3COSiH_3/$H_2C{=}C(OH)SiH_3$ pair (MP2/6-31G*). The results of these calculations were used to design an unusual system **144**, where the isolated enol form (**144a**) is indicated to be substantially more stable than the keto form (**144b**)[328].

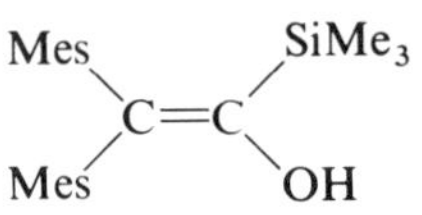
(**144a**)

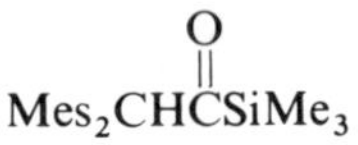
(**144b**)

Mes = 2,4,6-$Me_3C_6H_2$—

b. β-Substitution. While α-silyl groups generally destabilize carbenium ions relative to alkyl, β-silyl substitution is far more stabilizing than β-alkyl substitution. Jorgensen and coworkers studied at MP3/6-31G*//3-21G$^{(*)}$ the β-silyl-substituted cations **145**, **146** and **147**, the silyl bridged cations **148** and **148a** and the corresponding methyl analogues[327].

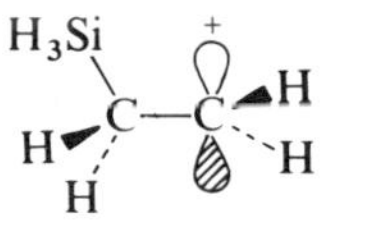
(**145a**)

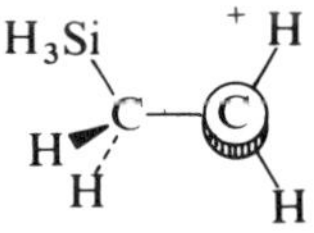
(**145b**)

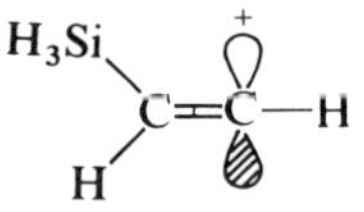
(**146**)

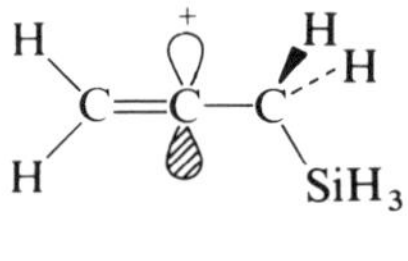
(**147**)

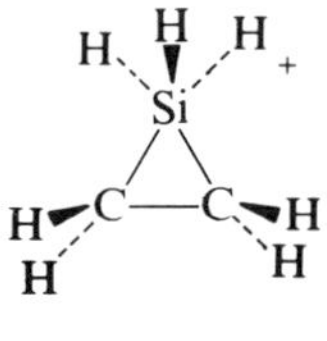
(**148**)

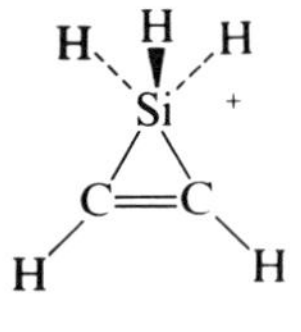
(**148a**)

The β-stabilization by a silyl group in conformation **145a** is very large, 38 kcal mol^{-1}, but it is much smaller (9 kcal mol^{-1}) in conformation **145b**. These stabilizations were attributed to inductive effects in **145b** and to C–Si hyperconjugation[329] in conformation **145a**. The cationic vacant 2p-orbital and the C–Si bond in the latter are in the same plane and can interact while in **145b** they are perpendicular and this interaction is not possible. The C–Si bond is a much more effective hyperconjugating bond than a C–C bond, since a β-methyl substituent stabilizes the ethyl cation by only 13 kcal mol^{-1} of which 9 kcal mol^{-1} are attributed to hyperconjugation and 4 kcal mol^{-1} to inductive effects. In fact, hyperconjugation in **145a** is so effective that **145a** collapses to the bridged structure **148**. Similarly, **146** and **147** are stabilized by 28.6 and 42.8 kcal mol^{-1} respectively, compared with 8 and 29.5 kcal mol^{-1} for the corresponding β-methyl-substituted cations[327]. Similar results were obtained by Arad and Apeloig who have also found that $(H_3Si)_2C{=}\overset{+}{C}H$ is more stable than $(H_3C)_2C{=}\overset{+}{C}H$ by 25.2 kcal mol^{-1} (6-31G*//3-21G) and that the bridged **148a** is more stable than the open **146** by 2.2 kcal mol^{-1} (MP2/6-31G*//3-21G$^{(*)}$[331a]. The theoretical predictions of a very large stereoelectronic (hyperconjugation) stabilizing effect by a β-silyl group have been recently supported by solvolytic experiments[330]. Thus **149**, where the β-Me_3Si group and the leaving group X are properly aligned for hyperconjugation, solvolyzes 7.2×10^7 times faster than **150**, where these groups have a 60° angle between them[330].

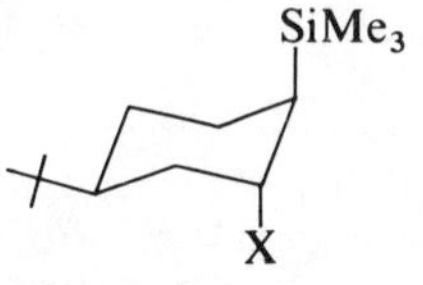

(149)

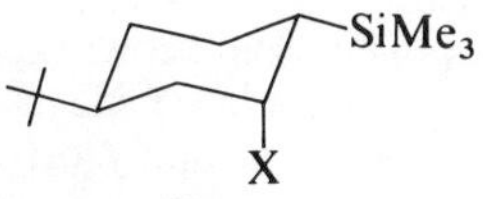

(150)

Simultaneously with Jorgensen's group[327], Apeloig and Arad[331] studied the effect of β-silyl substitution on the stabilities of vinyl and phenyl cations. They found that two β-silyl groups (**151**, R = SiH_3) stabilize the phenyl cation **151**, R = H by about 20 kcal mol^{-1} more than two methyl groups (**151**, R = CH_3). Consequently, while $H_2C{=}\overset{+}{C}CH_3$ is substantially more stable than **151**, R = H, it is calculated to be of comparable stability to **151**, R = SiH_3. As $H_2C{=}\overset{+}{C}CH_3$ can be generated solvolytically, it was predicted that in contrast to other aryl cations it should be possible to obtain **151**, R = SiH_3 by solvolysis[331b]. This prediction was confirmed recently be Sonoda and coworkers[332], who demonstrated for the first time that an aryl derivative (**152**) can be solvolyzed with C–O bond cleavage via an S_N1 mechanism to produce a phenyl cation, i.e. **151**, R = $Si(CH_3)_3$.

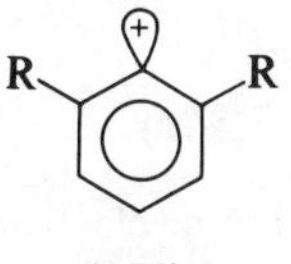

(151)

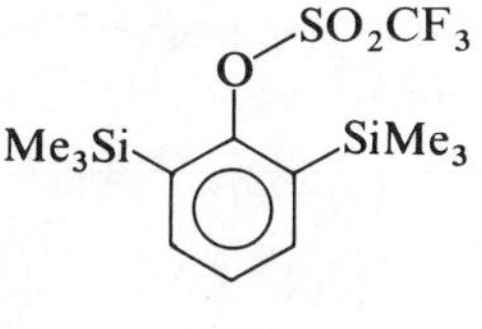

(152)

Recently Apeloig, Schwarz and coworkers studied by tandem mass spectrometry the isomeric $C_5H_{13}Si^+$ ions, including α- and β-silyl-substituted carbocations and silicenium ions and the results were complemented by calculations at the 6-31G*//3-21G level[333]. A complex scheme of rearrangements that converts the isomeric cations to one another was found necessary to explain the results. It was possible to demonstrate that upon collisional activation the bridged cation **153** dissociated to C_2H_4 and Me_3Si^+. The α-silyl cation **154** was not a minimum on the potential surface and it rearranged prior to fragmentation to either **153** or to the silicenium ion **155**. The fact that the two silicenium ions **155** and **156** dissociate to both ($C_2H_4 + Me_3Si^+$) and ($C_3H_6 + Me_2HSi^+$) requires

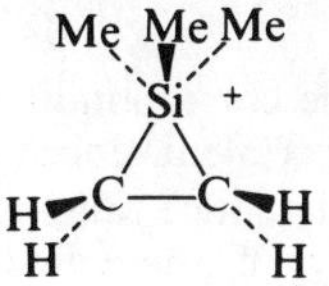

(153)

$Me\overset{+}{C}HSiMe_3$

(154)

$Me_2CH\overset{+}{Si}Me_2$

(155)

$MeCH_2CH_2\overset{+}{Si}Me_2$

(156)

that the fragmentation is preceded by isomerization steps in which 1,2-methyl and 1,2-hydrogen migrations must occur. The calculations are consistent with the experimental results and provide further insights into the energetics of the various rearrangements[333], but additional work is desirable in order to understand these processes fully.

The regio- and stereochemistry of electrophilic additions to allylsilanes were explained in terms of attack of the electrophile at the terminal olefin position and *anti* to the silyl group, allowing better stabilization by the β-silyl substituent. The model was based on 6-31G*//3-21G calculations for two conformations of $CH_2{=}CHCH(CH_3)SiH_3$[334].

c. γ-Substitution. Shiner and coworkers have shown experimentally that silyl substitution at the γ-carbon causes large effects on the stability of carbenium ion intermediates and that these effects are strongly orientation-dependent[335a, b]. Following this observation Davidson and Shiner analyzed theoretically the γ-silicon effect and calculated at 6-31G* and 6-31G levels several conformations of the cations **157**, where R = SiH_3, CH_3 and H[335c].

The γ-silyl-substituted cation strongly favors the *trans*-perpendicular structure actually shown in **157a**. The adaptation of this geometry and the γ-stabilization effect and its orientation dependence were attributed to a strong contribution from a closed cyclopropane-like structure **157b**. Silyl substitution stabilizes structure **157b** more than methyl substitution because of the higher stability of SiH_3^+ relative to CH_3^+[335c]. It was predicted that silyl substitution may exert a strong effect even for very long intervening chains, provided that the carbon adjacent to silicon can get close to the carbon attached to the leaving group[335c].

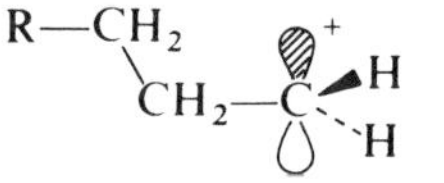

(**157a**)

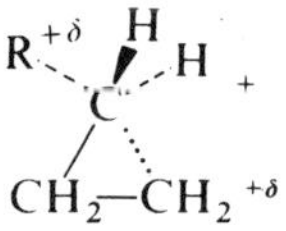

(**157b**)

Recently, evidence was found that a γ-Si–C bond can donate electrons to a metal vacant d-orbital resulting, as shown by *ab initio* calculations, in a short Ti $\cdots\cdots$ C distance and a long Si–C^{γ} bond in **158**[335d].

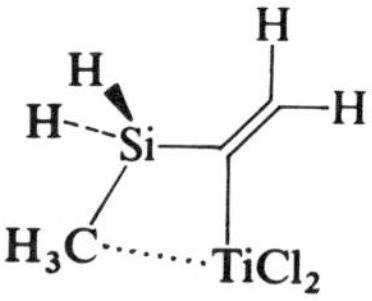

(**158**)

C. Radical Species

Silyl radicals have been studied theoretically less extensively than the corresponding cationic species[336]. Naturally, the most studied species is the parent $H_3Si^{\cdot}$. Calculations at all levels of theory find $H_3\dot{S}i$ to be pyramidal (C_{3v} symmetry) with a HSiH angle ranging from 107.2°[265g] to 115.1°[68, 281, 307, 337, 338]. An excellent summary of the calculations published before 1986 is given in a recent study by Allen and Schaefer[338]. The best calculated geometries (obtained at the CISD/dz + d level)[281, 338] are r(Si–H) = 1.47680 Å and ∡ HSiH = 111.10°, in excellent agreement with experiment[339]. Allen and Schaefer[338] also report the calculated harmonic and anharmonic vibrational frequencies and infrared intensities for $H_3\dot{S}i$ and list the results of previous studies where these properties were calculated. At the highest level of theory available the calculated classical inversion barrier for $H_3\dot{S}i$ is 5.8 kcal mol^{-1} [338]. This contrasts with $\dot{C}H_3$ which is planar[267a] (or pyramidal with a tiny inversion barrier). Pople and coworkers[267a] calculated for $H_3\dot{S}i$ $\Delta H^\circ_{f\,298}$ of 47.3 kcal mol^{-1}, in reasonable agreement with the experimental value of 46.4 kcal mol^{-1} [340]. The calculated[307b,c] ionization potential of $H_3\dot{S}i$ is 8.03 eV, in excellent agreement with a recent experimental value of 8.01 ± 0.02 eV[268b]. A previous higher experimental estimate of 8.14 eV[308b] resulted from the uncertainty associated with the observed long Franck–Condon series[268b, 308b]. The latter observation is consistent with geometry change from pyramidal in the neutral species to planar in the ion[307c]. It is interesting to note that, due to the high stability of silylene, the dissociation energy of the Si–H bond in $H_3\dot{S}i$ (66.8 kcal mol^{-1}) is much lower than in SiH_4 (91.7 kcal mol^{-1})[267a].

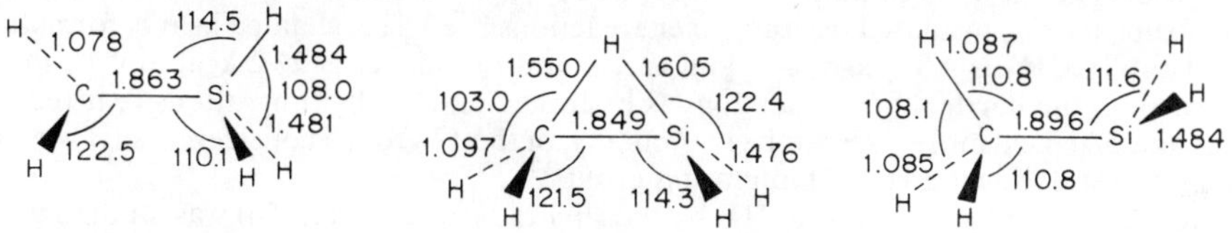

FIGURE 60. Optimized geometries (6-31G**) of the radicals $H_3Si\dot{C}H_2$ and $CH_3\dot{S}iH_2$ and the transition structure for their interconversion. Bond lengths are in Å, bond angles in deg[343].

A systematic study of substituent effects on silyl radicals is not available. Calculations for $CH_3\dot{S}iH_2$ have been reported[341, 342]. The UHF/6-31G** structure of $CH_3\dot{S}iH_2$, $H_3Si\dot{C}H_2$ and the transition structure connecting them are shown in Figure 60. The silicon centered radical is pyramidal while the carbon centered radical is planar at the radical centers. At UMP4/6-31G** + ZPE, $H_3C\dot{S}iH_2$ is more stable than $H_3Si\dot{C}H_2$ by 10.8 kcal mol^{-1} and the activation barrier is 42.6 kcal mol^{-1} above $H_3Si\dot{C}H_2$[342]. These calculated values are in agreement with experiment[342].

$Me_3\dot{S}i$ is also pyramidal[343–345] with an out-of-plane angle of 14.6° at UHF//3-21G[345], and the inversion barrier through the higher-energy planar C_{3h} structure is 13.3 kcal mol^{-1} at UHF//3-21G[344] (20.5 kcal mol^{-1} at STO-3G[344]). The planar C_{3h} structure lies still higher in energy[343]. Concerted methyl rotations require 3.3 kcal mol^{-1} [344]. Charge and spin density properties and harmonic vibrational frequencies (UHF/3-21G) have been also reported[344, 345]. A linear correlation was found between the UHF/3-21G estimates of the C–H bond lengths in $Me_3Si\cdot$ and the theoretically derived harmonic frequencies for deuterium-isolated C–H stretches[345].

Fluorine substitution increases the degree of pyramidalization and raises considerably the inversion barrier from 5 kcal mol^{-1} in $H_3\dot{S}i$ to 68 kcal mol^{-1} in $F_3\dot{S}i$[343]. Schlegel studied the series $H_3\dot{S}i$, $H_2\dot{S}iF$, $F_2\dot{S}iH$ and $F_3\dot{S}i$ at MP4/6-31G**//6-31G* and reported optimized geometries, vibrational frequencies, heats of formation and Si–H and Si–F dissociation energies[265g]. The Si–F bond lengths and the Si–F dissociation energies decrease with increasing fluorine substitution. A good correlation is found between the experimental vibrational frequencies and the theoretical harmonic frequencies, the latter being uniformly 7.4% higher than the value observed[265g]. The following heats of formation were calculated (kcal mol^{-1}): $\dot{S}iH_2F$, −42.5; $\dot{S}iHF_2$, −140.0 and $\dot{S}iF_3$, −239.0[265g].

The barrier to rotation around the C–C bond in $H_3SiCH_2\dot{C}H_2$ is 2.0 kcal mol^{-1} at 3-21G//3-21G, the orthogonal conformation **159**, R = SiH_3 being the more stable[346]. The

(159) **(160)**

same was found for other third-row β-substituents ($R=PH_2$, SH,Cl), in contrast to second-row substituents ($R=CH_3$, NH_2, OH and F) that prefer the eclipsed conformation **160** (e.g. by 0.14 kcal mol^{-1} for $R=CH_3$)[346]. PMO analysis suggests that the difference between $H_3SiCH_2\dot{C}H_2$ and $CH_3CH_2\dot{C}H_2$ results from the more efficient electron delocalization (i.e. hyperconjugation) from the bonding orbital of the C–Si bond into the half-filled 2p-orbital at the radical center[346].

The degenerate 1,2-silyl migration in $H_3SiCH_2\dot{C}H_2$ was studied by MNDO as a model for coenzyme B_{12} mediated rearrangement reactions[347]. The calculated heat of formation of $H_3SiCH_2\dot{C}H_2$ is 4.8 kcal mol^{-1} and the migration barrier is 21.2 kcal mol^{-1}, much lower than the barriers for the analogous hydrogen and methyl migrations that require 52.3 and 58.6 kcal mol^{-1}, respectively. Actually, of the 17 groups considered, silyl was the second best (following H_2O^+) migrating group[347].

Abstraction of hydrogen from HF by the silyl radical (equation 54) was studied using MNDO and *ab initio* methods, and the results were compared with those for $\dot{C}H_3$[348]. The geometry of the transition state (3-21G) has C_{3v} symmetry with a linear Si–H–F bond angle. At MP3/6-31G**(6-311G** for silicon) the abstraction reaction 54 is endothermic by 35.6 kcal mol^{-1} and the reverse reaction has practically no barrier. The analogous reaction with $\dot{C}H_3$ is about 11 kcal mol^{-1} less endothermic, reflecting the stronger C–H bond compared with the Si–H bond[348].

$$H_3\dot{S}i + HF \longrightarrow H_4Si + \dot{F} \qquad (54)$$

Elimination of a hydrogen atom from $H_2Si{=}S$ produces two distinct HSiS equilibrium structures with the same $^2A'$ symmetry. The electronic configurations of the two radicals are described by **161** and **162**, which is consistent with the fact that the Si–S bond length is shorter in **161** than in **162**[220].

(161) **(162)**

The H$\dot{S}$iS radical **161** is 6.2 kcal mol^{-1} more stable than HSi$\dot{S}$, **162** (MP3/6-31G**//6-31G*). Similarly, H$\dot{S}$iO is by 12.3 kcal mol^{-1} more stable than HSi$\dot{O}$[220], but only one minimum was located on the H$\dot{C}$O potential-energy surface[220]. Frenking and Schaefer[349] found that H$\dot{S}$iO and HSi$\dot{O}$ have nearly the same energy at the SCF/dz+d level but they predicted, in agreement with the above studies, that at a higher level of theory H$\dot{S}$iO is preferred. They have also predicted that H$\dot{S}$iO might be the only true minimum for HSiO[349]. Another minimum, 11 kcal mol^{-1} lower than for H$\dot{S}$iO, was located for $\dot{S}$iOH. The barrier separating H$\dot{S}$iO and $\dot{S}$iOH is substantial (40.9 kcal mol^{-1}, CI) and both species should be observable[349].

D. Anionic Species

The study of anionic species presents especially difficult theoretical challenges, as discussed in Section II. In general, larger basis sets (including diffuse functions) are required to describe anions than for cations or neutral molecules. Consequently, the computational results are in most cases less accurate and reliable than for neutrals and cations, especially when relatively small basis sets (e.g. split valence) are used. Yet, it is probably safe to assume that conclusions regarding comparisons between similar compounds, even when based on relatively simple methods, will remain qualitatively unchanged at higher levels of theory.

1. Silyl anions

Calculations for the parent SiH_3^-[69, 307a, 311, 314, 350–353] are not as extensive as for $H_3\dot{Si}$ radical, probably because of the difficulties in calculating very accurately various molecular properties. The most accurate calculations used the 6-31+G* basis set and many-body perturbation theory for calculating correlated total energies (calculations using effective core potentials on silicon were also reported)[350]. At 6-31+G*, H_3Si^- is pyramidal with r(Si–H) = 1.534 Å and ∡ HSiH = 96.7° (94.5°, experimental[354]) compared with r(Si–H) = 1.47 Å and ∡ HSiH = 110.9° in the $\dot{Si}H_3$ radical[350]. Similar results were obtained with dz and dz + d basis sets[350–352]. Adiabatic and vertical ionization energies were calculated (1.31 eV and 1.79 eV, respectively)[350] and the adiabatic energies were found to be in excellent agreement with photoelectron spectroscopy measurements[354]. The corresponding values for the methyl anion are much lower: 0.51 eV (vertical) and 0.19 eV (adiabatic)[350]. The proton affinity of H_3Si^- was calculated to be 394 kcal mol^{-1} compared with 446 kcal mol^{-1} for CH_3^-, using a dz + dp basis set[69b, 311]. Thus, silicon accommodates a negative charge much more effectively than carbon. The inversion barrier in SiH_3^- is 26.0 kcal mol^{-1} (dz + d)[351b] or 34.3 kcal mol^{-1} (dz)[69b], much larger than for CH_3^- (2 kcal mol^{-1}, dz + d)[351b]. The vertical electron affinity of the silyl radical to give the silyl anion is calculated to be 1.65 eV, a much higher value than for the $H_3\dot{C}$ radical (0.08 eV)[350].

The effect of second-row substituents (BH_2 to F) on SiH_3^- was studied by two groups[69] and the calculated geometries of the anions are given in Figure 61[69a]. All molecules are pyramidal, except for $H_2SiBH_2^-$ which is planar, reflecting the strong conjugation between the anionic lone pair and the empty p-orbital of boron[69]. The bonds between silicon and the substituent are up to 0.1 Å longer than in the corresponding neutrals[69] and the dz values[69b] are longer than the 3-21G* values[69a].

The acidities of substituted silanes relative to that of SiH_4 were calculated at 6-31+G** (in kcal mol^{-1}, a positive value indicates higher acidity): H_2BSiH_3 (13.4), CH_3SiH_3 (−6.8), H_2NSiH_3 (−5.9), $HOSiH_3$ (−2.1), $FSiH_3$ (5.8)[69a]. A similar trend with somewhat larger variations was obtained with a dz basis set[69b]. Thus, π-donor substituents lower the acidity while σ- (e.g. F) and π-acceptors (e.g. H_2B) increase the acidity (the effect of H_2B is, however, much smaller than in H_2BCH_3)[69].

The α-substituents also have a strong effect on the inversion barriers of Si^- which, at the dz level, are calculated to be (in kcal mol^{-1}): SiH_3^- (34.3), $CH_3SiH_2^-$ (40.4), $H_2NSiH_2^-$ (42.8), $HOSiH_2^-$ (50.7), $FSiH_2^-$ (57.3), $H_2BSiH_2^-$ (planar)[69b]. These inversion barriers are probably too high, but there is no doubt that they are much larger than in the corresponding carbanions[69b].

A variety of other silyl anions has been studied. Hopkinson and Lien, using a dz basis set, found that the potentially conjugated $NCSiH_2^-$, O=$CHSiH_2^-$ and HC≡$CSiH_2^-$ anions[314, 352] are pyramidal at silicon and have substantial barriers to inversion (35.2, 19.9 and 34.3 kcal mol^{-1}, respectively)[314]. This indicates that contributions

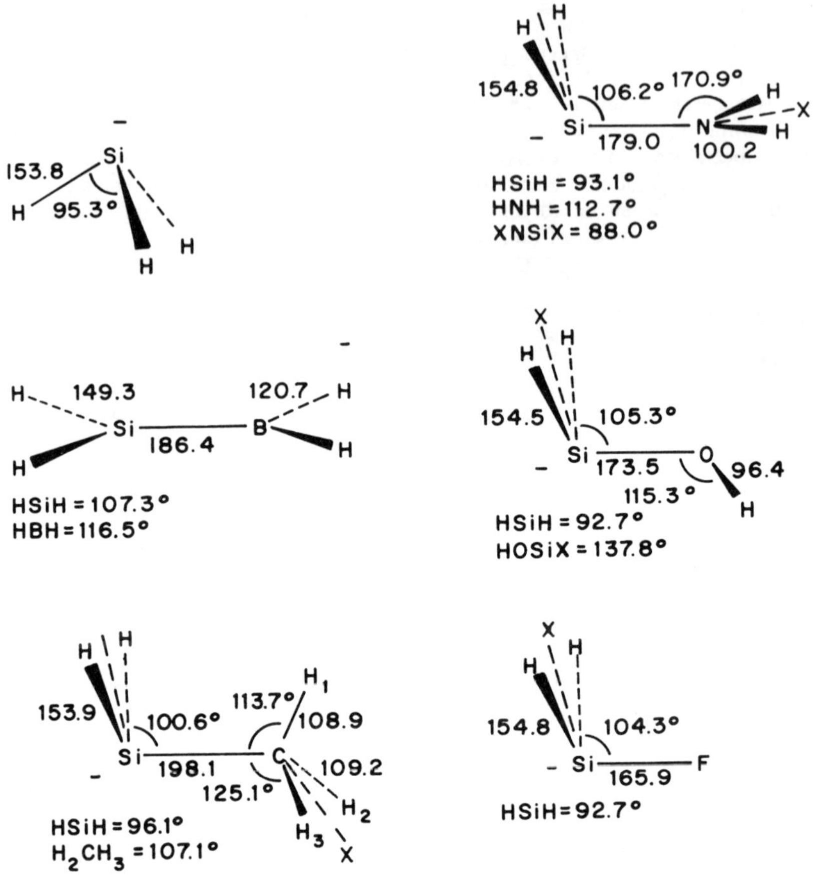

FIGURE 61. Optimized geometries (3-21G$^{(*)}$) of SiH_2X^- (X = H, BH_2, CH_3, NH_2, OH, F). Bond lengths are in pm (1 Å = 100 pm), bond angles in deg. Reproduced with permission from Magnusson, *Tetrahedron*, **41**, 2945 (1985). Copyright (1985) Pergamon Journals Ltd.

from resonance structures of type **163b** are not sufficient to overcome the barrier to planarization at silicon, in contrast with the corresponding carbanions which are planar. All three substituents increase the acidity of the conjugate acid i.e. by 20.7 kcal mol^{-1} ($O{=}CHSiH_2^-$), 28.5 kcal mol^{-1} ($NCSiH_2^-$) and 10.7 kcal mol^{-1} ($HC{\equiv}CSiH_2^-$) at the dz level[314].

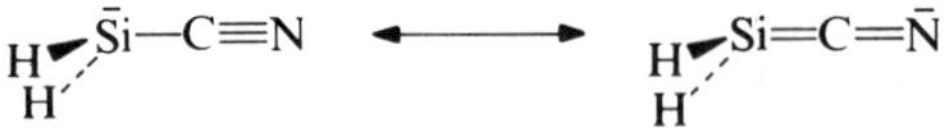

(163a) **(163b)**

Only one minimum is found on the $CSiH^-$ energy surface corresponding to $H{-}C{\equiv}Si^-$, the isomeric $H{-}Si{\equiv}C^-$ being a maximum (dz)[224a]. The $CSiH_3^-$ and $C_2SiH_3^-$ anions were studied with a dz + dp basis set[352]. The most stable $CSiH_3^-$ isomer **164** lies 25 kcal mol^{-1}

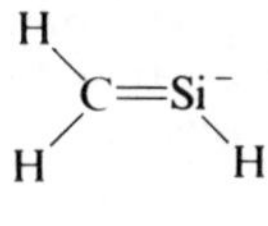
(164)

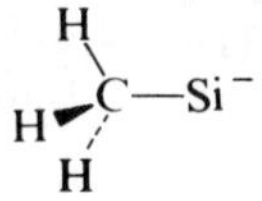
(165)

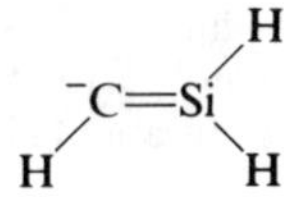
(166)

below **165** and 49.5 kcal mol^{-1} below **166**. So also at multiply bonded centers hydrogens bonded to silicon are more acidic than hydrogens bonded to carbon.

The most stable $C_2SiH_3^-$ anion is **167**, but **168** lies only 5.7 kcal mol^{-1} higher in energy. **169** and **170** are 16.0 and 17.2 kcal mol^{-1} higher in energy than **167**[352].

$H_3SiC{\equiv}C^-$ **(167)** $HC{\equiv}CSiH_2^-$ **(168)** $H_2C{=}C{=}\bar{Si}H$ **(169)** $H_3C{-}C{\equiv}Si^-$ **(170)**

The potential aromatic silacyclopentadienyl anion **101**[246a, 259] was discussed in Section VI.C. Despite the potential aromaticity of the planar C_{2v} structure, **101** adopts a pyramidal C_s structure[259]. Silacyclopentadiene is also by only 2.2 kcal mol^{-1} more acidic than SiH_4[259]. Significant stabilization is however present in the planar structure, as reflected in the lower inversion barrier of 16.2 kcal mol^{-1} in C_s **101** compared with an inversion barrier of about 26 kcal mol^{-1} in SiH_3^- at a comparable level of theory[352].

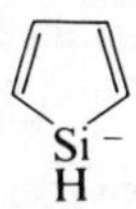
(101)

It is well established experimentally that an α-silyl substituent stabilizes an adjacent carbanionic center, but the degree of stabilization was not evaluated quantitatively [355]. According to MP2/6-31G*//3-21 + G$^{(*)}$ calculations $H_3SiCH_2^-$ is more stable than CH_3^- (relative to the corresponding hydrocarbons) by 24.9 kcal mol^{-1}, whereas $CH_3CH_2^-$ is less stable than CH_3^- by 3.1 kcal mol^{-1} [67]. The stabilizing effect of silyl was attributed mainly to effective negative hyperconjugation with the σ*(Si–H) orbitals and to the higher polarizability of silicon relative to carbon[67]. (p–d) π-conjugation is unimportant[67, 181]. This α-silyl stabilization is, however, not sufficient to overcome the inherent stability differences between silyl and methyl anions (see above) and $CH_3SiH_2^-$ is therefore more stable than $H_3SiCH_2^-$ by 15.8 kcal mol^{-1} (dz + dp)[181]. A silyl group also enhances the acidity of alkynyl protons[352]. At STO-3G, $H_3SiC{\equiv}CH$ is more acidic than $HC{\equiv}CH$ by 5.1 kcal mol^{-1} in contrast to $CH_3CC{\equiv}CH$, which is less acidic than $HC{\equiv}CH$ by 3 kcal mol^{-1}. The acidities of the acetylenic protons of $HC{\equiv}CCH_2SiH_3$ and $HC{\equiv}CCH_2CH_3$ are practically the same[352].

A silyl substituent also stabilizes adjacent N^- and P^- anions as shown by the increased calculated acidities of H_3SiNH_2, $(H_3Si)_2NH_2$, H_3SiPH_2 and $(H_3Si)_2PH$ relative to the corresponding parent and methyl substituted amines and phosphines[356].

Apeloig and Stanger found that α-silyl substitution stabilizes also the negatively charged transition state of an S_N2 reaction at an adjacent center. This suggests enhanced S_N2 reactivity of compounds of the type R_3SiCH_2X (X = leaving group), as was indeed observed experimentally[323a]. The calculated barriers for the S_N2 hydride exchange reactions 55a, 55b, and 55c are 47.4, 50.1 and 35.9 kcal mol^{-1}, respectively (at 6–31G*//

3-21G). Thus, SiH_3 stabilizes the model transition state for the S_N2 reaction (**171**) by 14.2 kcal mol^{-1} relatively to CH_3. The energy difference between the transition state models **171d** and **171e** is even higher: 20 kcal mol^{-1} in favor of the silylated system.

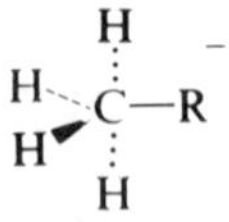

(**171**)

$$H^- + CH_3R \rightarrow [H \cdots CH_2R \cdots H]^- \rightarrow H^- + CH_3R \qquad (55)$$

(**171**)

(**a**) R = H; (**b**) R = CH_3; (**c**)R = SiH_3; (**d**) R = $C(CH_3)_3$; (**e**) R = $Si(CH_3)_3$

Bowie, DePuy and coworkers found by gas-phase flowing afterglow experiments that the ions H_3Si^- and Me_3Si^- undergo Si–O and/or Si–S bond-forming reactions with CO_2, COS, CS_2, SO_2, N_2O, MeNCO and MeNCS, forming H_3SiO^-, Me_3SiO^-, H_3SiS^- or Me_3SiS^- ions as appropriate[353]. To answer some of the questions raised by the experiments, 6-21G calculations were carried out for the reactions of H_3Si^- with CO_2, CS_2, SO_2 and N_2O (saddle points were not characterized analytically)[353]. The computational results suggest different reaction pathways in each case. The reaction of H_3Si^- with CO_2 proceeds by initial attack at carbon, and subsequent rearrangements are required to form H_3SiO^-. H_3SiS^- is formed from $CS_2 + H_3Si^-$ by a simpler path as initial attack occurs at sulfur and rearrangements are not required. These different calculated pathways account for the varying rates of the reactions of R_3Si^- anions with CO_2 and CS_2. The channels forming R_3SiO^- from SO_2 proceed through a five-coordinated intermediate in which both oxygens are bound to silicon, which subsequently decomposes. With N_2O reaction of H_3Si^- at the terminal nitrogen is calculated to be the lowest energy pathway[353].

2. *Pentacoordinated silicon*

In contrast to carbon, silicon forms many stable penta- and hexacoordinate compounds, many of them anionic, e.g. SiF_6^{2-} [7]. Two simple models have been proposed to account for the bonding in such species (as well as in the isoelectronic PF_5). Pauling suggested that the 3d-orbitals participate to form sp^3d^2 hybrids[357a]. A different approach for hypervalent bonding was suggested by Rundle[357b] and Musher[357c]. They proposed that in the trigonal bipyramid structure which is usually observed, the equatorial bonds may be regarded as normal while in the axial bonds to silicon 4-electrons are delocalized over three-centers in orbitals which are constructed from the Si 3p-orbital and the s- and/or p-orbitals of the axial ligand. The resulting five bonding orbitals are shown in Figure 62[357d]. From the four valence electrons that are available to bond the silicon and the two axial atoms, two occupy the bonding three-center orbital ($1a_2''$) and two reside in an approximately nonbonding orbital ($2a_1'$).

All recent calculations cited in this section suggest that the degree of 3d-orbital participation in bonding is small and that the Rundle–Musher model provides a better description of bonding in pentacoordinate derivatives of silicon and of other third-row elements (see also References 7 and 62). Streitwieser and coworkers[358] have recently emphasized the important role that the ionic nature of the bonds between silicon and common nonmetals, such as H, C, N, O and F, plays in allowing silicon to expand its coordination sphere, in contrast to carbon which strongly prefers covalent bonding.

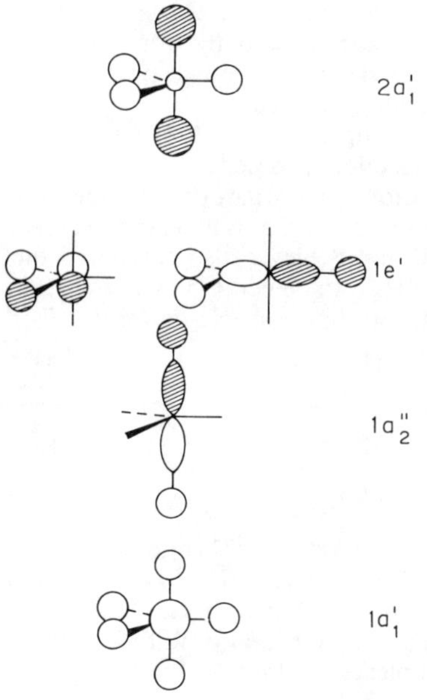

FIGURE 62. Schematic representation of the occupied molecular orbitals of SiH_5^- (D_{3h}).

Quantum-mechanical calculations have centered on anionic pentacoordinate silicon species, mainly as models for the transition states or intermediates in bimolecular S_N2 substitution reactions at silicon—a topic of current experimental interest[7–9, 359].

The prototype of a pentacoordinate silicon anion, SiH_5^-, has recently been observed experimentally in the gas phase[360]. Early studies[361] showed that SiH_5^- is stable with respect to SiH_4 and H^-, but that it is unstable (although marginally so) towards decomposition into SiH_3^- and H_2 (a large basis set + doubles CI was used)[351a]. However, recent calculations (MP4SDTQ/6-31 + +G*//6-31G + +G*)[362] by Reed and Schleyer showed that although SiH_5^- (D_{3h}) is thermodynamically unstable by 11.3 kcal mol^{-1} with respect to loss of H_2, the barrier for this reaction is 17.5 kcal mol^{-1}, in agreement with the experimental observation[360]. Hydride ion loss from SiH_5^- is endothermic by 16.0 kcal mol^{-1} at MP4/6-31 + +G*//6-31 + G* + ZPE[362]. The most stable structure of SiH_5^- is a trigonal bipyramid (D_{3h} symmetry) **172** with r(Si–H_a) = 1.622 Å and r (Si–H_e) = 1.521 Å (6-31 + +G*)[362]. Thus, the bond to the axial hydrogens is by 0.1 Å longer than the bond to the equatorial hydrogens. A second minimum on the SiH_5^- potential-energy surface is the tetragonal–pyramidal (C_{4v} symmetry) form **173**, which is 2.2 kcal mol^{-1} higher in energy[362]. The relevance of the small energy difference between **172** and **173** to nucleophilic substitution at silicon is discussed below.

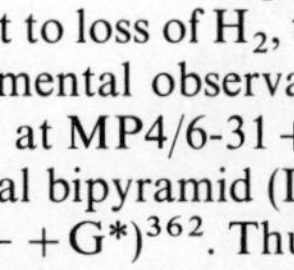

(172)

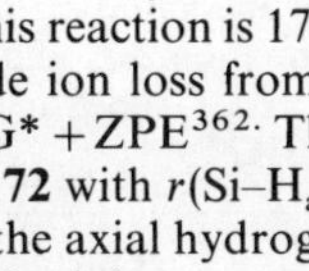

(173)

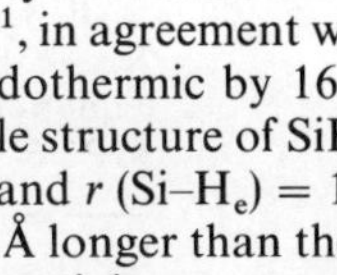

(174)

Loss of H_2 from SiH_5^- occurs through a C_{3v} transition structure **174**. In **174** only one Si–H bond is being broken; the rupture of the other Si–H bond originally present in SiH_5^- (**172**) occurred at an earlier stage[362]. The two possible modes of H^- loss from **172**, or, looking at the reverse reaction, the exothermic nucleophilic attack of H^- on SiH_4, were studied very carefully[362]. Along a C_{3v} path **175** (corresponding to elimination of an axial hydride from **172**) the reaction proceeds without a barrier. Along a C_{2v} path **176** (corresponding to elimination of an equatorial hydride from **172**) a stationary point with two imaginary frequencies is found that is at least 5 kcal mol^{-1} higher in energy than the separate species[362]. Reed and Schleyer also compared the behavior of SiH_5^- with the isoelectronic PH_5 and SH_5^+. In general, the above calculations do not support a description of these pentacoordinate species in terms of sp^3d^2 hybridization[362].

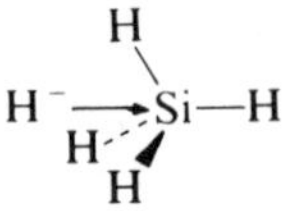

(**175**)

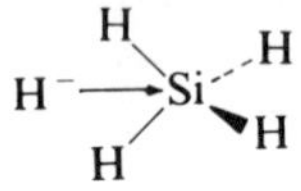

(**176**)

Recently, Gordon and coworkers[348, 363] conducted a systematic study of equation 56, where X = H, CH_3, NH_2, OH, F, SiH_3, PH_2, SH and Cl, using MP4 with extended basis sets (MNDO calculations were also performed). Stable pentacoordinated silicon anions are found for X = H, CH_3, NH_2, OH and F. With X = SiH_3, SH and Cl, only marginally stable charge–dipole complexes are found[348, 363].

$$X^- + SiH_4 \longrightarrow [XSiH_4]^- \ (D_{3h}) \tag{56}$$

For X = CH_3, OH, F and SiH_3, minima could be located for structures with either an axial or an equatorial substituent. For X=SiH_3, SH and Cl, no stable structures were found and the axial substituents are only marginally stable as charge-dipole complexes[363]. In general, the bond lengths to axial substituents are by 0.05 to 0.10 Å longer than those to equatorial groups and this trend appears to become increasingly pronounced as the electronegativity of the substituent decreases[348, 363]. The energy differences (MP2/6-31++G**/6-31G*) between the structures with axial and equatorial substituents are (in kcal mol^{-1}): 0.2 (X=CH_3), −0.7 (X=OH), 7.5 (X = F) and 3.5 (X = SiH_3), positive values indicating that the axial position is favored. The effect of

TABLE 38. Calculated energies (kcal mol^{-1}) for the reactions: $X^- + SiH_4 \rightarrow SiHX^-$ (equation 56) and $SiH_4X^- \rightarrow SiH_3^- + HX$ (equation 57), for axial substituents X, calculated at MP2/6-31++G**//6-31G*[a]

Axial substituent X	Calculated energy for equation 56	Calculated energy for equation 57
H	−16.7	− 8.1
CH_3	−30.7	−14.0
OH	−29.9	20.4
F	−29.1	42.0
SiH_3	0.2[b]	− 0.9
SH	− 4.0	30.2
Cl	− 4.5	51.6

[a] From Reference 363.
[b] SiH_4X^- is, however, bound at 6-31G*.

the substituent X on the stability of the pentacoordinated $XSiH_4^-$ anion according to equation 56 is summarized in Table 38.

Second-row substituents (X = CH_3, OH, F) are more effective in stabilizing $XSiH_4^-$ than is hydride, and much more effective than third-row substituents[363]. Similar calculations for $FSiH_4^-$ were reported recently by Streitwieser and coworkers[358].

The stabilities of the pentacoordinated $XSiH_4^-$ anions relative to elimination of HX (equation 57) were also evaluated and the results are also given in Table 38. Expulsion of HX is exothermic for X = H and CH_3 but endothermic for X = OH, F, SH and Cl and nearly thermoneutral for X = SiH_3[363].

$$SiH_4^- \ (D_{3h}) \rightarrow SiH_3^- + HX \qquad (57)$$

Much effort has been devoted to the theoretical elucidation of the mechanism of nucleophilic substitution at silicon. The calculations tried to answer two major questions: (1) Does substitution occur by a one-step reaction via transition state **177** or by a two-step process via intermediate **178** (equation 58 where Nu = nucleophile, LG = leaving group)? (2) What is the stereochemistry of the substitution and what are the factors which control it?

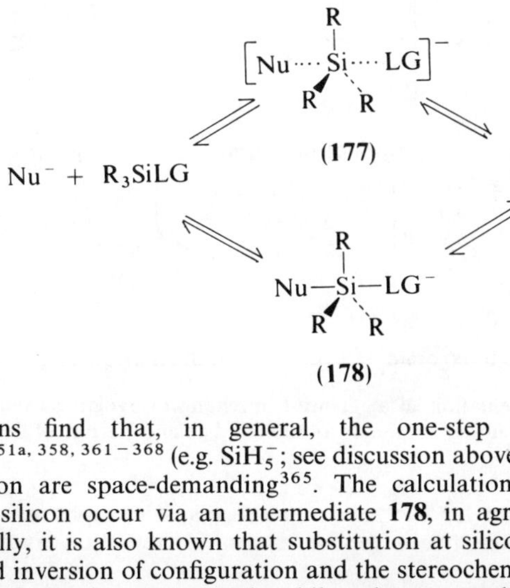

(177)

(178)

(58)

Calculations find that, in general, the one-step reaction via **177** is not favored[348, 351a, 358, 361–368] (e.g. SiH_5^-; see discussion above), even when the five ligands around silicon are space-demanding[365]. The calculations predict that nucleophilic reactions at silicon occur via an intermediate **178**, in agreement with experiment[359]. Experimentally, it is also known that substitution at silicon in solution leads to both retention and inversion of configuration and the stereochemical outcome depends upon the nature of the leaving group, nucleophile, solvent, complexing agents, and whether or not silicon is part of a ring[8, 9, 359]. Some of these factors were studied theoretically.

Anh and Minot[366] used a frontier orbital treatment to show that retention of configuration is favored by: (1) a reaction center of low electronegativity (e.g. Si vs. C); (2) a leaving group of high electronegativity; (3) a high percentage of s character in the hybrid atomic orbital connecting silicon and the leaving group; (4) a 'hard' nucleophile (i.e. having a low HOMO)[8, 366]. Calculations for the 'front-side' (leading to retention) and 'back-side' (leading to inversion) attack of H^- on SiH_3LG (LG = F, Cl) (Figure 63) supported the qualitative analysis, but the level of calculations was low (STO-3G, fixed geometries). The mechanisms in Figure 63 were applied mainly by Corriu and coworkers to interpret their experimental findings[9].

However, Anh and Minot[8, 366] have neglected the possibility that there are other mechanisms in addition to 'front-side' attack that can lead to retention. For example, a mechanism involving axial entry and axial departure accompanied by pseudorotation of the initially formed pentacoordinated intermediate, as shown in Figure 64. This is the

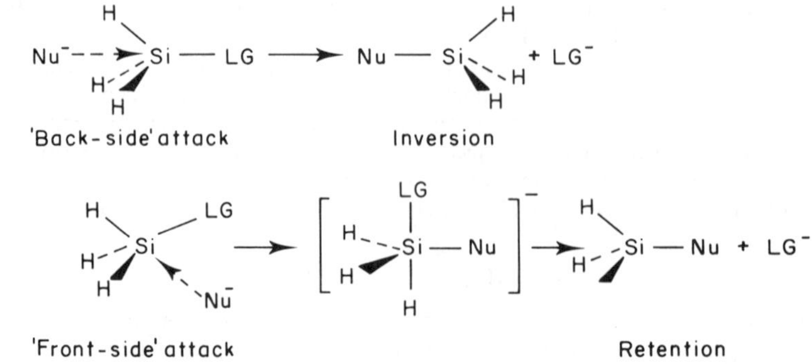

FIGURE 63. 'Back-side' (or axial) and 'front-side' (or equatorial) approaches for the reaction of nucleophiles (Nu) with H_3SiLG (LG = leaving group).

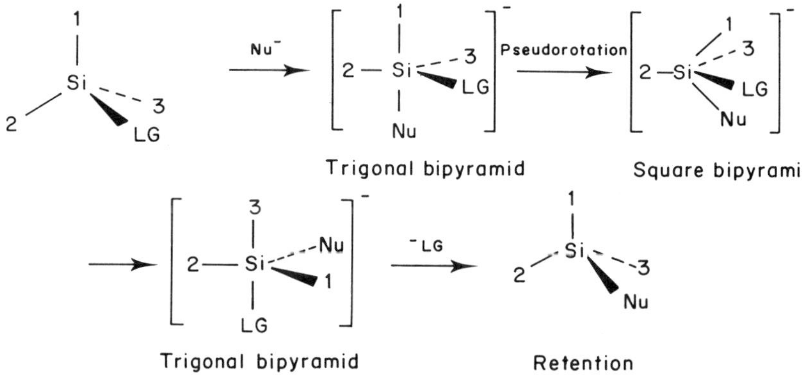

FIGURE 64. Schematic representation of a retention mechanism involving axial attack of the nucleophile (Nu) on SiH_3LG, followed by pseudorotation and axial departure of the leaving group (LG).

generally accepted mechanism for the retention process in phosphorus chemistry[359]. As will be demonstrated below, this is the mechanism for retention favored by recent theoretical studies.

Most recently, Deiters and Holmes studied computationally, in great detail, the relative merits of the alternative retention mechanisms in Figures 63 and 64[367]. They investigated the hydride attack on H_3SiF and H_3SiCl using the STO-3G, STO-6G* and 6-31+G* basis sets. In general, the results with the two larger basis sets verified the trends obtained with STO-3G. To provide more realistic pathways, a geometry associated with a minimum strain energy conformation was determined at each point of approach of H^- by a molecular mechanics approach. It is found that for both SiH_3F and SiH_3Cl, axial ('back-side') entry is a lower-energy pathway than equatorial ('front-side') entry. Furthermore, axial departure is the preferred route rather than equatorial departure, and retention via pseudorotation is of lower energy than retention via equatorial attack[367a]. These findings show that the PMO arguments of Anh and Minot[8, 366] are irrelevant to the question of retention in nucleophilic reactions at silicon.

In comparing H_3SiF and H_3SiCl, it is found that H_3SiCl favors inversion while H_3SiF is calculated to give retention. The difference in the behavior of the two molecules results from the fact that fluorine is more tightly bound than chlorine in the trigonal bipyramidal intermediate, thus enhancing the possibility of ligand rearrangement before departure[367a]. These computational conclusions are in agreement with experimental evidence[359].

In a subsequent publication[367b] Deiters and Holmes extend their study to include a series of nucleophiles other than H^- and ligands other than Cl and F. The following regular trigonal-bipyramidal and square-pyramidal conformations of $[SiH_3XY]^-$ intermediates (see Figure 64 and **172** and **173** for the prototype structures) were studied: X = H and Y = H, OH, SH, Cl, F; X = Cl, F, OH, SH and Y = Cl and F[367b]. The results of the calculations (mostly STO-3G with some points rechecked at 6-31 + G*) were correlated with experimental data on retention and inversion at silicon centers. For a given nucleophile, the experimentally established tendency towards retention in terms of the leaving group ($H^- > OH^- > SH^- \sim F^- > Cl^-$) was shown to correlate with a corresponding decrease in the calculated overlap population of the bond between silicon and the leaving group, as shown in Figure 65.

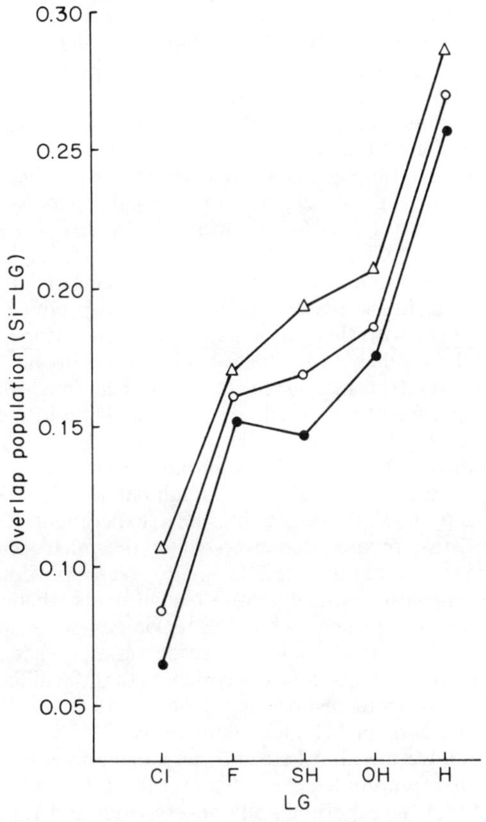

FIGURE 65. Overlap populations of the Si–LG bond (LG = leaving group), in the trigonal bipyramid $[SiH_3(Nu)(LG)^-]$ (Nu = nucleophile) as a function of the leaving group for the following nucleophiles: H^-(●), F^-(○), and Cl^- (Δ). Reprinted with permission from *J. Am. Chem. Soc.*, **109**, 1692 (1987). Copyright (1987) American Chemical Society[367b].

Calculated barriers to retention support the retention mechanism shown in Figure 64. The 'apicophilicity' of the leaving group (i.e. the energy difference between **179** and **180**) was found to influence the relative tendency of a particular system towards a retention vs. an inversion pathway. The 'apicophilicity' varies with the type of nucleophile being used (for example, for $Nu=H^-$ it follows the order: $Cl \gg SH > F > OH$). A softer nucleophile increases the 'apicophilicity' of the leaving group. We note, however, that the correlation is not very good and the changes in the 'apicophilicity' are rather small and thus tend to favor an inversion process, and vice versa, in accord with experimental data[367b]. The two excellent studies of Deiters and Holmes[367] rule out equatorial attack at silicon (Figure 63) as a feasible mechanism for retention in nucleophilic attack at silicon.

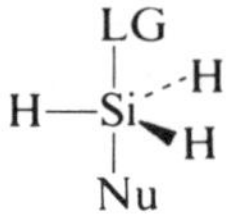

(**179**)

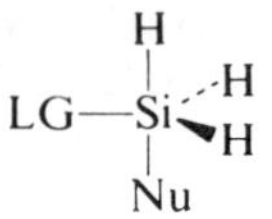

(**180**)

Bowie and coworkers[368] have studied the mechanism of the reactions of nucleophiles with organosilanes both theoretically and by gas-phase experiments. In the most recent study the reactions of HO^- and MeO^- with Me_4Si were studied experimentally and model 6-21G calculations were carried out for the $[Me_2H_3Si^-]$, a model intermediate[368a]. It was found that interconversion of apical (or axial) and equatorial methyl groups in the trigonal-bipyramidal intermediate occurs via a square-pyramidal transition state with a barrier of only 2.9 kcal mol^{-1}, in agreement with the conclusions of Deiters and Holmes[367]. Calculations for the $MeO^- + MeSiH_3$ system suggest that the observed 1, 2-eliminations of CH_4 or of C_2H_6 in the reaction of OH^- or MeO^- with Me_4Si occur by a stepwise mechanism with large internal barriers, the key being the cleavage of the Si–C bond in the reactive intermediate to produce a reacting solvated methyl anion. In agreement with this suggestion deuterium isotope effects are observed for these reactions[368a]. In another study Sheldon, Hayes and Bowie[368b] studied, at the 6-21G levels and by ion cyclotron resonance (ICR), the reactions of H^- or MeO^- with silane and the conclusion from both methods was that H_2 and methanol respectively should result and that the reactions occur without an energy barrier. The calculations also indicate that with CH_3SiH_3, OH^- and MeO^- should initially give a complex of the type $[NuH.\cdots \bar{C}H_2SiMe_3]$ which is separated by a small barrier of 5–8 kcal mol^{-1} from the more stable trigonal-bipyramidal species, but ICR experiments failed to detect this complex. Calculations also predict that even when the nucleophile and the groups attached to silicon are large, as in the system $I^- + SiF_4$, pentacoordinated silicon adducts are still stable. So, even with relatively bulky substituents, there is no resemblance between the S_N2 mechanisms at silicon which reacts via pentacoordinated intermediates (e.g. **178**) and at carbon where the reaction proceeds via a pentacoordinated transition state[368b]. Based on their results these authors conclude that the different S_N2 behavior of carbon and silicon is unlikely to result solely from steric factors[368b] as suggested by Dewar and Healy on the basis of MNDO calculations[365].

Although not anionic, we mention here in brief calculations for neutral penta- and hexacoordinated silicon compounds. The 1:1 and 1:2 adducts of SiF_4 and NH_3 were calculated as models[369] for the experimentally observed 1:1 and 1:2 adducts of SiF_4 and amines[370]. What is remarkable about the 1:1 pentacoordinated adducts is that the amine occupies an axial site[370a, b]. The calculations (both dz[369a] and STO-3G[369b]) also find that $F_4Si{\cdot}NH_3$ prefers a trigonal-bipyramidal structure with the NH_3 group axial, the structure with the equatorial NH_3 being about 11 kcal mol^{-1} higher in energy[369a].

Calculated Mulliken charge densities suggest that the axial nitrogen is more negative even than the axial F, and this was suggested as a possible explanation for NH_3 occupying the axial site in preference to fluorine[369a]. STO-3G calculations[369b] for the $SiF_4 \cdot 2NH_3$ adduct find it to be *cis* in accord with experiment[370c]. The calculations[369b] also rule out the steric hindrance between ligands as the reason for $SiF_4 \cdot 2NH_3$ being *cis*[370c], while complexes of bulkier ligands are *trans*[370d]. The Si–N bond in $SiF_4 \cdot 2NH_3$ has some degree of covalency but it is mostly electrostatic[369b].

VIII. PLANAR TETRACOORDINATED SILICON

Generally, tetracoordinate eight-valence-electron silicon compounds (as well as other main group ZH_4 molecules) prefer to be tetrahedral[357d]. The possibility to violate this rule and to prepare an organosilicon compound where the silicon atom and the four atoms directly bonded to it are all in the same plane, is intriguing. The analogous question for carbon has been studied extensively by Schleyer and his group[371] and has attracted considerable interest[372].

Calculations predict that the energy difference between the tetrahedral (**181**) and square-planar (**182**) forms of SiH_4 is large, but smaller than for CH_4[373] (93 and 160 kcal mol^{-1}, respectively, at 6-31G*//6-31G*[373c]). For carbon, it was demonstrated that appropriate substitution and geometrical constraints reduce dramatically the tetrahedral-square planar energy difference (ΔE) and molecules with ground-state planar geometries were predicted[371, 372]. Substituents can be used also to stabilize planar silicon compounds, but the requirements from the substituent are fundamentally different from that in CH_4. This results from the fact that the HOMO and LUMO in CH_4 and in SiH_4 are interchanged. In CH_4, the HOMO is **183** and **184** is the LUMO. The opposite is true for SiH_4, **183** and **184** being the LUMO and HOMO, respectively.

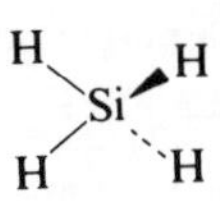

(181)

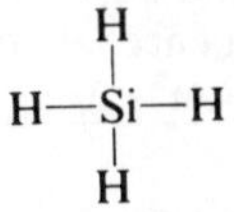

(182)

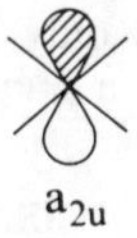

(183)

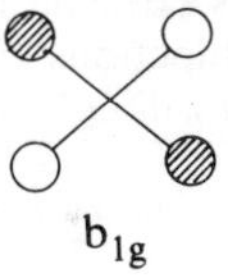

(184)

This interchange of orbital ordering has fundamental consequences for the stabilization of planar structures: π-acceptor–σ-donor substituents (e.g. BeH and BH_2) preferentially stabilize planar tetracoordinate carbon[371, 372], but silicon is predicted to require π-donor–σ-acceptor groups (e.g. F, OR, NR_2)[373]. Calculations for SiH_3OH indeed showed a reduced ΔE (84.3 kcal mol^{-1} compared with 114 kcal mol^{-1} for SiH_4 at STO-3G*)[373a]. MNDO calculations for **185** indicate a low barrier to planarization of only 33 kcal mol^{-1} [373a]. Evidence of planar tetracoordinated silicon in **185** was reported by

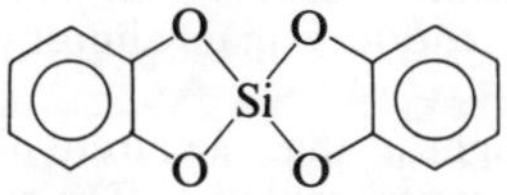

(185)

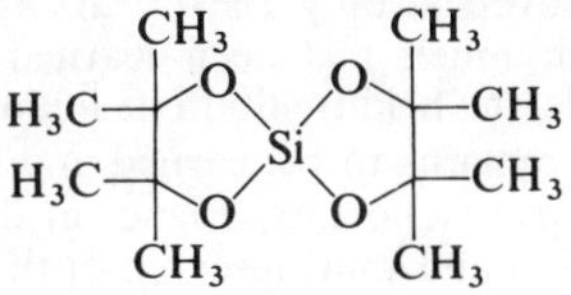

(186)

Meyer and Nagorsen[374] based on the fact that, in the solid state, the space group adopted by **185** is completely different from that of the analogous carbon compound. Additional coordination and/or intermolecular interactions in the crystal have been suggested as factors which would induce the unusual structural preference in this molecule[373a, 374]. An X-ray structure of the analogous **186** showed, however, that no significant planarization at silicon has occurred[375].

Geometrical constraints appear to be more effective in planarizing carbon than silicon. The tetrahedral–planar energy difference in the spiro derivatives **187** is 32.3 kcal mol^{-1} for M = C but 66.2 kcal mol^{-1} for M = Si (6-31G*)[142a]. The greater resistance of the silicon species in this system to undergo twisting towards planarity has been explained by the large increase in the negative charge on the central silicon which causes destabilization, in going from the tetrahedral to the planar geometry[142a].

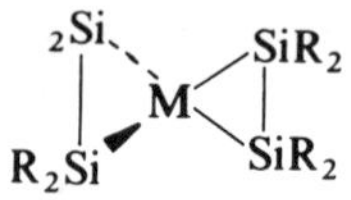

(**187**)

Dilithiosilane and disodiosilane were also studied in this context[88] and found to have the smallest calculated tetrahedral–planar energy separation yet found for tetracoordinate silicon. The values are 21 kcal mol^{-1} in SiH_2Li_2 and 24 kcal mol^{-1} in H_2SiNa_2 (3-21G), compared with 10 kcal mol^{-1} in dilithiomethane. Thus the qualitative prediction that π-acceptor–σ-donor substituents do not stabilize planar silicon[373] does not apply to lithium and sodium. The planar singlet SiH_2Li_2 has a π-symmetric HOMO (e.g. similar to **183**) analogous to that of planar CH_2Li_2. This means that lithium substitution in silane causes a drastic change in the electronic structure and the orbital ordering[88]. The addition of d-functions on silicon has only a minor influence on the relative energies of the dilithiosilane[88], in contrast to a previous suggestion[373].

IX. CONCLUSIONS

In the last decade great strides have been made in organosilicon chemistry, both experimental and theoretical. Experimentally, many new types of compounds were synthesized, characterized, isolated as stable molecules and the first steps to unravel their chemistry have been taken. Theoretical progress was also extremely exciting. Calculations have evolved from techniques using relatively simple methods to highly sophisticated computations which are capable of predicting with high reliability and accuracy many molecular properties. To a theoretician organosilicon chemistry was, and still is, a fascinating playground because many new developments could be predicted theoretically prior to their experimental confirmation. One of the most satisfying and exciting aspects of this field of research is the strong ongoing interplay of theory and experiment that has now led to close agreement between theory and experiment for many problems which were controversial only a few years ago.

Although much has been learned about the chemical behavior of organosilicon compounds, the field is still in its infancy and many territories remain almost untouched and much remains to be learned and discovered.

The relatively short experience in this field has taught us that the future lies in close collaboration and joint research of theoreticians and experimentalists. The models that can be generated by calculations have become indispensible in understanding and evaluating experimental results and interpretations, and in suggesting synthetic targets which may have useful and interesting properties. There is no doubt in my mind that this strategy would lead to many interesting new developments which will maintain organosilicon chemistry as a fruitful, active field of research for many years to come.

X. ADDENDUM

When this chapter was in press, Beauchamp and coworkers reopened the discussion on the relative stabilities of silaethylene (**40**) and methylsilylene (**45**)[377]. They found, using Fourier transform ion cyclotron resonance spectroscopy, that silaethylene (**40**) is by 10 ± 3 kcal mol^{-1} *more stable* than the isomeric methylsilyene **45**. The experiments were corroborated by *ab initio* GVB-CI calculations which found that **40** is more stable than **45** by 11.6 kcal mol^{-1}. The discrepancy between these recent results[377] and the previous high level MO calculations, which predict a **40**–**45** energy difference of nearly zero (see Table 19), is puzzling. Further calculations and experiments are required to resolve this problem.

XI. ACKNOWLEDGEMENTS

It is a special pleasure to acknowledge Professor Robert West who, during his visit to Israel eight years ago, introduced me to this fascinating field of chemistry and to thank him for the many enlightening discussions and suggestions during this period. I am especially indebted to my coworkers Drs D. Arad, M. Karni and A. Stanger who have carried out our own research in this field. Our research in this field and the collaboration with Professor West was supported in the last decade by the United States–Israel Binational Science Foundation (BSF) to whom we are indebted.

I am grateful to the following colleagues who have kindly provided me with collections of their published papers and with unpublished manuscripts and data: Professors M. S. Gordon, W. J. Hehre, R. Janoshchek, W. L. Jorgensen, K. Krogh-Jespersen, K. Morokuma, J. Michl, S. Nagase, H. F. Schaefer III, H. B. Schlegel, P. v. R. Schleyer, A. Streitwieser, A. Tachibana, G. Trinquier and R. West.

I wish to thank Dr Miriam Karni for her invaluable assistance through the various stages of the writing of this review, and Mrs Esther Avital for typing the manuscript.

XII. REFERENCES

1. I. Fleming, in *Comprehensive Organic Chemistry* (Eds. D. Barton and W. D. Ollis), Vol. 3, Chap. 13, Pergamon Press, Oxford, 1979, p. 541.
2. (a) D. A. Armitage, in *Comprehensive Organometallic Chemistry* (Eds. G. Wilkinson, F. G. A. Stone and E. W. Abel), Vol. 2, Chap. 9.1, Pergamon Press, Oxford, 1982.
   (b) T. J. Barton, in Ref. 2a, Chap. 9.2.
   (c) F. O. Stark, J. R. Falender and A. P. Wright, in Ref. 2a, Chap. 9.3.
   (d) R. West, in Ref. 2a, Chap. 9.4.
3. For example: (a) *Annual Reports on the Progress of Chemistry*, Sections A and C, Royal Society of Chemistry, London.
   (b) *Annual Survey of Organometallic Chemistry.*
   (c) *Organometallic Chemistry Reviews.*
   (d) *J. Organometallic Chemistry Library.*
   (e) *J. Organomet. Chem.*, annual surveys, the last to appear is Vol. 313 (1986).
4. (a) W. P. Weber, *Silicon Reagents for Organic Synthesis*, Springer-Verlag, Berlin, 1983.
   (b) E. W. Colvin, *Silicon in Organic Synthesis*, Butterworths, London, 1981.
5. V. Chvalovsky and J. M. Bellama, Eds., *Carbon-Functional Organosilicon compounds*, Plenum Press, New York, 1984.

6. H. Sakurai, Ed., *Organosilicon and Bioorganosilicon Chemistry: Structure, Bonding, Reactivity and Synthetic Applications*, Ellis Horwood, Chichester, 1985.
7. St. N. Tandura, N. V. Alekseev and M. G. Voronkov, *Top. Curr. Chem.*, **131**, 99 (1986).
8. N. T. Anh, *Top. Curr. Chem.*, **88**, 145 (1980).
9. R. J. P. Corriu, C. Guerin and J. J. E. Moreau, *Top. Stereochem.*, **15**, 43 (1984).
10. L. E. Gusel'nikov and N. S. Nametkin, *Chem. Rev.*, **79**, 529 (1979).
11. G. Bertrand, G. Trinquier and P. Mazerolles, *J. Organomet. Chem. Libr.*, **12**, 1 (1981).
12. B. Coleman and M. Jones, Jr., *Rev. Chem. Intermed.*, **4**, 297 (1981).
13. A. G. Brook, F. Abdesaken, B. Gutekunst, G. Gutekunst and R. K. Kallury, *J. Chem. Soc., Chem. Commun.*, 191 (1981).
14. R. West, M. J. Fink and J. Michl, *Science (Washington, D.C)*, **214**, 1343 (1981).
15. (a) R. West, *Pure Appl. Chem.*, **56**, 163 (1984).
(b) R. West, M. J. Fink, M. J. Michalczyk, D. J. De Young and J. Michl, in Ref. 6, Part I, Chap. 1, p. 3.
(c) R. West, *Science (Washington, D.C)*, **225**, 1109 (1984).
16. (a) For a review of the older literature concerning Si=N and Si=C bonds see: N. Wiberg, *J. Organomet. Chem.*, **273**, 141 (1984).
(b) For two recent preparations of stable silanimines which have appeared in the meanwhile, see Ref. 145.
17. (a) A. H. Cowley, *Polyhedron*, **3**, 389 (1984).
(b) A. H. Cowley, *Acc. Chem. Res.*, **17**, 386 (1984).
(c) A. H. Cowley and N. C. Norman, *Prog. Inorg. Chem.*, **34**, 1 (1986).
18. A. G. Brook and K. M. Baines, *Adv. Organomet. Chem.*, **25**, 1 (1986).
19. G. Raabe and J. Michl, *Chem. Rev.*, **85**, 419 (1985).
20. H. F. Schaefer III, *Acc. Chem. Res.*, **15**, 283 (1982).
21. M. S. Gordon, in *Molecular Structure and Energetics* (Eds. J. F. Liebman and A. Greenberg), Volume 1, Chap. 4, VCH Publishers, Deerfield Beach, Florida, 1986, p. 101.
22. K. K. Baldridge, J. A. Boatz, S. Koseki and M. S. Gordon, *Annu. Rev. Phys. Chem.*, **38**, 211 (1987).
23. (a) H. F. Schaefer III, Ed., *Modern Theoretical Chemistry*, Vols. 3 and 4, Plenum Press, New York, 1977.
(b) G. A. Segal, Ed., *Modern Theoretical Chemistry*, Vols. 7 and 8, Plenum Press, New York, 1977.
24. Ab Initio Methods in Quantum Chemistry—Part I (Ed. K. P. Lawley), *Adv. Chem. Phys.*, Volume LXVII, 1987; Part II is in press.
25. W. J. Hehre, L. Radom, P. v. R. Schleyer and J. A. Pople, *Ab Initio Molecular Orbital Theory*, John Wiley & Sons, New York, 1986.
26. T. Clark, *A Handbook of Computational Chemistry. A Practical Guide to Chemical Structure and Energy Calculations*, John Wiley & Sons, New York, 1985.
27. The latest version available from Carnegie-Mellon University is Gaussian 82: J. Binkley, R. A. Whiteside, K. Raghavachari, R. Seeger, D. J. DeFrees, H. B. Schlegel, M. J. Frisch, L. R. Kahn and J. A. Pople, *Gaussian* 82, Carnegie-Mellon University, Pittsburgh 1982. By the time this book is published, a more recent version (*Gaussian* 86) will be available. Previous, widely used versions included *Gaussian 80*, *Gaussian 76* and *Gaussian 70*.
28. W. A. Goddard, III, T. H. Dunning, W. J. Hunt and P. J. Hay, *Acc. Chem. Res.*, **6**, 368 (1973).
29. W. J. Hehre, R. F. Stewart and J. A. Pople, *J. Chem. Phys.*, **51**, 2657 (1969).
30. (a) T. H. Dunning, Jr. and P. J. Hay, in Ref. 23a, Vol. 3.
(b) R. Ahlrichs and P. R. Taylor, *J. Chim. Phys. Phys.-Chim. Biol.*, **78**, 315 (1981).
(c) S. Huzinaga, J. Andzelm, M. Klobukowski, E. Radzio-Andzelm, Y. Sakai and H. Tatewski, *Gaussian Basis Sets for Molecular Calculations*, Elsevier, Amsterdam, 1984.
(d) S. Wilson, Ref. 24, p. 439.
31. 4-31G:
(a) Carbon to fluorine: W. J. Hehre, R. Ditchfield and J. A. Pople, *J. Chem. Phys.*, **54**, 724 (1971).
(b) Phosphorus to chlorine: W. J. Hehre and W. A. Lathan, *J. Chem. Phys.*, **56**, 5255 (1972).
6-31G:
(c) Carbon to fluorine: W. J. Hehre, R. Ditchfield and J. A. Pople, *J. Chem. Phys.*, **56**, 2257 (1972).
(d) Third row: M. M. Francl, W. J. Pietro, W. J. Hehre, J. S. Binkley, M. S. Gordon, D. J. DeFrees and J. A. Pople, *J. Chem. Phys.*, **77**, 3654 (1982).

32. (a) Second Row: J. S. Binkley, J. A. Pople and W. J. Hehre, *J. Am. Chem. Soc.*, **102**, 939 (1980).
(b) Third row: M. S. Gordon, J. S. Binkley, J. A. Pople, W. J. Pietro and W. J. Hehre, *J. Am. Chem. Soc.*, **104**, 2797 (1982).
33. R. Krishnan, J. S. Binkley, R. Seeger and J. A. Pople, *J. Chem. Phys.*, **72**, 650 (1980).
34. (a) Second row: P. C. Hariharan and J. A. Pople, *Theor. Chim. Acta.*, **28**, 213 (1973).
(b) Third row: Ref. 31d.
35. J. B. Collins, P. v. R. Schleyer, J. S. Binkley and J. A. Pople, *J. Chem. Phys.*, **64**, 5142 (1976).
36. W. J. Pietro, M. M. Francl, W. J. Hehre, D. J. DeFrees, J. A. Pople and J. S. Binkley, *J. Am. Chem. Soc.*, **104**, 5039 (1982).
37. J. A. Pople, M. J. Frisch, B. T. Luke and J. S. Binkley, *Int. J. Quantum Chem., Symp.* **17**, 307 (1983).
38. (a) T. Clark, J. Chandrasekhar, G. W. Spitznagel and P. v. R. Schleyer, *J. Comput. Chem.*, **4**, 294 (1983).
(b) M. J. Frisch, J. A. Pople and J. S. Binkley, *J. Chem. Phys.*, **80**, 3265 (1984).
(c) Z. Latajka and S. Scheiner, *Chem. Phys. Lett.*, **105**, 435 (1984).
39. G. W. Spitznagel, T. Clark, P. v. R. Schleyer and W. J. Hehre, *J. Comput. Chem.*, **8**, 1109 (1987).
40. P. Durand and J.-P. Malrieu in Ref. 24, p. 396.
41. For the application of this method to silicon compounds see: R. Janoschek, A. Sax and E. A. Halevi, *Isr. J. Chem.*, **23**, 58 (1983), and references cited therein.
42. (a) J. S. Binkley and J. A. Pople, *Int. J. Quantum Chem.*, **9**, 229 (1975).
(b) J. A. Pople, J. S. Binkley and R. Krishnan, *Int. J. Quantum Chem., Symp.*, **10**, 1 (1976).
43. C. Moller and M. S. Plesset, *Phys. Rev.*, **46**, 618 (1934).
44. A. Szabo and N. S. Ostlund, *Modern Quantum Chemistry*, MacMillan, New York, 1982. See also Refs. 23 and 24.
45. For general reviews see:
(a) R. D. Amos, in Ref. 24, p. 99.
(b) F. Bernardi and M. A. Robb, in Ref. 24, p. 155.
(c) H. B. Schlegel, in Ref. 24, p. 249.
(d) For a short description of the methods available in the Gaussian series of programs, see Ref. 26, p. 240.
46. (a) P. H. Blustin, *J. Organomet. Chem.*, **105**, 161 (1976).
(b) B. Wirsam, *Theor. Chim. Acta*, **25**, 169 (1972).
47. (a) H. Lischka and H.-J. Kohler, *J. Am. Chem, Soc.*, **105**, 6646 (1983).
(b) J. S. Binkley, *J. Am. Chem. Soc.*, **106**, 603 (1984).
48. For an excellent discussion and evaluation of the reliability of the calculations for this purpose, see Ref. 25, pp. 226–261.
49. G. Maier, H. P. Reisenauer, W. Schwab, P. Carsky, B. A. Hess, Jr. and L. J. Schaad, *J. Am. Chem. Soc.*, **109**, 5183 (1987).
50. For a general discussion see Ref. 23b.
51. J. A. Pople, D. P. Santry and G. A. Segal, *J. Chem. Phys.*, **43**, S129 (1965); J. A. Pople and G. A. Segal, *J. Chem. Phys.*, **43**, S136 (1965); **44**, 3289 (1966); the program is available from QCPE: **11**, 91 (1966).
52. R. C. Bingham, M. J. S. Dewar and D. H. Lo, *J. Am. Chem. Soc.*, **97**, 1285 (1975); the program is available from QCPE, **11**, 279 (1975).
53. M. J. S. Dewar and W. Thiel, *J. Am. Chem. Soc.*, **99**, 4899 (1977); the program is available from QCPE, **10**, 353 (1978).
54. M. J. S. Dewar, D. H. Lo and C. A. Ramsden, *J. Am. Chem. Soc.*, **97**, 1311 (1975).
55. M. J. S. Dewar, M. L. McKee and H. S. Rzepa, *J. Am. Chem. Soc.*, **100**, 3607 (1978).
56. W. S. Verwoerd, *J. Comput. Chem.*, **3**, 445 (1982).
57. M. J. S. Dewar, J. Friedheim, G. Grady, E. F. Healy and J. J. P. Stewart, *Organomet.*, **15**, 375 (1986).
58. U. Burkert and N. L. Allinger, *Molecular Mechanics*, ACS Monograph 177, American Chemical Society, Washington, D.C., 1982.
59. (a) M. T. Tribble and N. L. Allinger, *Tetrahedron*, **28**, 2147 (1972).
(b) R. J. Quellette, D. Baron, J. Stoflo, A. Rosenblum and P. Weber, *Tetrahedron*, **28**, 2163 (1972).
60. (a) H. M. Seip, *J. Chem. Phys.*, **54**, 440 (1971).
(b) Q. Shen, R. L. Hilderbrandt and V. S. Mastryukov, *J. Mol. Struct.*, **54**, 121 (1979).

(c) F. K. Cartledge, *J. Organomet. Chem.*, **225**, 131 (1982).
61. See for example: T. H. Lowry and K. S. Richardson, *Mechanism and Theory in Organic Chemistry*, third edn., Harper & Row, New York, 1987, pp. 299–316.
62. W. Kutzelnigg, *Angew. Chem. Int. Ed. Engl.*, **213**, 272 (1984).
63. H. Kwart and K. G. King, *d-orbitals in the Chemistry of Silicon, Phosphorous and Sulfur*, Springer-Verlag, Berlin, 1977.
64. (a) C. G. Pitt, *J. Organomet. Chem.*, **61**, 49 (1973); **23**, C35 (1970).
(b) See also: T. Veszpremi and J. Nagy, *J. Organomet. Chem.*, **255**, 41 (1983).
65. For two opposing interpretations (from the seventies) of the 3d-orbital involvement in bonding to silicon see for example:
(a) 3d-functions serve merely as polarization functions: A. Breeze, G. A. D. Collins and P. W. J. Cruickshank, *J. Chem. Soc., Chem. Commun.*, 445 (1971).
(b) 3d-orbital involvement is important: F. F. Roelandt, D. F. Van de Vondel and G. P. Van der Kelen, *J. Mol. Struct.*, **50**, 147 (1978).
66. See for example:
(a) (d–p) bonding in fluorosilanes in negligible: B. Rempfer, H. Oberhammer and N. Auner, *J. Am. Chem. Soc.*, **108**, 3893 (1986).
(b) (d–p) bonding is important in silicates: N. Janes and E. Oldfield, *J. Am. Chem. Soc.*, **108**, 5743 (1986).
67. P. v. R. Schleyer, T. Clark, A. J. Kos, G. W. Spitznagel, C. Rohde, D. Arad, K. N. Houk and N. G. Rondan, *J. Am. Chem. Soc.*, **106**, 6467 (1984).
68. B. T. Luke, J. A. Pople, M.-B. Krogh-Jespersen, Y. Apeloig, J. Chandrasekhar and P. v. R. Schleyer, *J. Am. Chem. Soc.*, **108**, 260 (1986).
69. For similar systematic studies of second-row substituents which used only the HF method, see:
(a) E. Magnusson, *Tetrahedron*, **41**, 2945 (1985).
(b) A. C. Hopkinson and N. H. Lien, *Tetrahedron*, **37**, 1105 (1981).
70. C. Glidewell, P. M. Pinder, A. G. Robiette and G. M. Sheldrick, *J. Chem. Soc., Dalton Trans.*, 1402 (1972).
71. R. Kewley, P. M. McKinney and A. G. Robiette, *J. Mol. Spectrosc.*, **34**, 390 (1970).
72. (a) J. M. Mays and B. P. Dailey, *J. Chem. Phys.*, **20**, 1695 (1952).
(b) B. Bak, J. Bruhn and J. Rastrup-Andersen, *Acta Chem. Scand.*, **8**, 872 (1954).
73. The dissociation energies of various bonds to silicon have been collected in three recent reviews:
(a) A. Pilcher and H. A. Skinner, in *The Chemistry of the Carbon–Metal Bond*, Vol. 1 (Eds. F. R. Hartley and S. Patai), John Wiley & Sons, New York, 1982.
(b) R. Walsh, *Acc. Chem. Res.*, **14**, 246 (1981).
(c) J. E. Huheey, *Inorganic Chemistry*, 3rd edn., Harper and Row, New York, 1983, Appendix E.
74. For an earlier similar conclusion based on the poor correlation between the ionization potential of X and the difference between $D(Me_3Si–X)$ and $D(Me–X)$ see: B. G. Gowenlock and J. Stevenson, *J. Organomet. Chem.*, **13**, P13 (1968). This paper challenges an earlier conclusion that (p–d) bonding is important, see: S. J. Bond, I. M. T. Davidson and C. A. Lambert, *J. Organomet. Chem.*, **12**, P3 (1967). See also Refs. 63–67.
75. (a) D. R. J. Boyd, *J. Chem. Phys.*, **23**, 922 (1955).
(b) S. R. Polo and M. K. Wilson, *J. Chem. Phys.*, **22**, 1559 (1954).
76. S. Weiss and G. E. Leroi, *J. Chem. Phys.*, **48**, 962 (1968).
77. (a) R. W. Kilb and L. Pierce, *J. Chem. Phys.*, **27**, 108 (1957).
(b) J. P. Lowe, *Prog. Phys. Org. Chem.*, **6**, 1 (1968).
78. M. Pfeiffer and H. J. Spangenberg, *Z. Phys. Chem. (Leipzig)*, **232**, 47 (1966).
79. R. S. Mulliken and W. C. Ermler, *Polyatomic Molecules, Results of Ab Initio Calculations*, Academic Press, New York, 1981, pp. 182–184.
80. (a) C. H. J. Van der Brekel and P. J. Saverin, *J. Electrochem. Soc.*, **119**, 372 (1972).
(b) S. Cradock, E. A. V. Ebsworth, W. J. Savage and R. A. Whiteford, *J. Chem. Soc., Faraday Trans.* 2, 934 (1972).
81. D. W. L. Rankin, A. G. Robiette, G. M. Sheldrick, W. S. Sheldrick, B. J. Aylett, I. A. Ellis and J. J. Monaghan, *J. Chem. Soc. (A)*, 1224 (1969).
82. B. Beagley and A. R. Conrad, *Trans. Faraday Soc.*, **66**, 2740 (1970).
83. J. R. Durig, Y. S. Li, M. M. Chen and J. D. Odom, *J. Mol. Spectrosc.*, **59**, 74 (1976).
84. M. Kakudo, W. Kasai and T. Watase, *J. Chem. Phys.*, **21**, 1894 (1953). See also Ref. 92.
85. J. A. Morrison and M. A. Ring, *Inorg. Chem.*, **6**, 100 (1967).

86. P. v. R. Schleyer and T. Clark, *J. Chem. Soc., Chem. Commun.*, 1371 (1986).
87. W. R. Hertler, D. A. Dixon, E. W. Matthews, F. Davidson and F. G. Kitson, *J. Am. Chem. Soc.*, **109**, 6532 (1987).
88. A. Rajca and A. Streitwieser, Jr., submitted (1987).
89. A. C. Hess, P. F. McMillan and M. O'Keeffe *J. Phys. Chem.*, **90**, 5661 (1986).
90. R. S. McDowell, M. J. Reisfeld, C. W. Patterson, B. J. Krohn, M. C. Vaśquez and G. A. Laguna, *J. Chem. Phys.*, **77**, 4337 (1982).
91. A. J. Kirby, *The Anomeric Effect and Related Stereoelectronic Effects at Oxygen*, Springer-Verlag, Berlin, 1983.
92. I. Hargittai and H. M. Seip, *Acta Chim. Scand. Ser. A*, **30A**, 153 (1976).
93. Y. Apeloig and A. Stanger, *J. Organomet. Chem.*, submitted (1987).
94. A. Stanger, Ph.D. Thesis, Technion—Israel Institute of Technology, Haifa, Israel, 1985.
95. (a) M. O'Keeffe, B. Domenges and G. V. Gibbs, *J. Phys. Chem.*, **89**, 2304 (1985).
(b) A. C. Hess, P. F. McMillan and M. O'Keeffe, *J. Phys. Chem.*, **91**, 1395 (1987).
(c) M. O'Keeffe and G. V. Gibbs, *J. Phys. Chem.*, **81**, 876 (1984).
96. (a) J. Sauer, *J. Phys. Chem.*, **91**, 2315 (1987).
(b) J. Sauer, *Chem. Phys. Lett.*, **97**, 275 (1983).
97. L. Pauling, *The Nature of the Chemical Bond*, 3rd edn., Cornell University Press, Ithaca, New York, 1960.
98. A. Almenningen, O. Bastiansen, V. Ewing, K. Hedberg and M. Traetteberg, *Acta Chem. Scand.*, **17**, 2455 (1963).
99. K. Kimura and M. Kubo, *J. Chem. Phys.*, **30**, 151 (1959).
100. J. A. Tossell and G. V. Gibbs, *Acta Crystallogr.*, **A34**, 463 (1978).
101. S. Grigoras and T. H. Lane, *J. Comput. Chem.*, **8**, 84 (1987).
102. H. Oberhammer and J. E. Boggs, *J. Am. Chem. Soc.*, **102**, 7241 (1980).
103. C. A. Ernst, A. L. Allred, M. A. Ratner, M. D. Newton, G. V. Gibbs, J. W. Moskowitz and S. Topiol, *Chem. Phys. Lett.*, **81**, 424 (1981).
104. K. Raghavachari, J. Chandrasekhar and M. J. Frisch, *J. Am. Chem. Soc.*, **104**, 3779 (1982).
105. J. R. Durig, M. J. Flanagan and V. F. Kalasinsky, *J. Chem. Phys.*, **66**, 2775 (1977).
106. Y. Apeloig and M. Karni, *J. Am. Chem. Soc.*, **106**, 6676 (1984).
107. See for example:
(a) A. L. Rheingold, Ed., *Homoaromatic Rings, Chains and Macromolecules of Main-Group Elements*, Elsevier, Amsterdam, 1977.
(b) R. West, *Pure Appl. Chem.*, **54**, 1041 (1982).
(c) H. Watanabe and Y. Nagai, Chap. 9 in Ref. 6.
108. (a) J. C. Dewan, S. Murakami, J. T. Snow, S. Collins and S. Masamune, *J. Chem. Soc., Chem. Commun.*, 892 (1985).
(b) H. Matsumoto, K. Takasuna, M. Minemura, Y. Nagai and M. Goto, *J. Chem. Soc., Chem. Commun.*, 1366 (1985).
(c) S. Masamune, Y. Kobe, S. Collins, D. J. Williams and R. Jones, *J. Am. Chem. Soc.*, **107**, 5552 (1985).
(d) R. Jones, D. J. Williams, Y. Kobe and S. Masamune, *Angew. Chem., Int. Ed. Engl.*, **25**, 173 (1986).
109. R. S. Grev and H. F. Schaefer III, *J. Am. Chem. Soc.*, **109**, 6569 (1987).
110. J. Rubio and F. Illas, *Theochem.*, **19**, 131 (1984).
111. W. W. Schoeller and T. Dabisch, *J. Chem. Soc., Chem. Commun.*, 1706 (1985).
112. (a) A. F. Sax, *Chem. Phys. Lett.*, **127**, 163 (1986).
(b) A. F. Sax, *Chem. Phys. Lett.*, **129**, 66 (1986).
113. M. S. Gordon and D. Bartol, *J. Am. Chem. Soc.*, **109**, 5948 (1987).
114. Z. Smith, H. M. Seip, E. Hengge and G. Bauer, *Acta Chem. Scand, Ser. A*, **30**, 697 (1976).
115. P. v. R. Schleyer, in *Substituent Effects in Radical Chemistry* (Eds. H. G. Viehe, R. Janoschek and R. Merenyi), D. Reidel Publ. Co., Dordrecht, 1986, p. 69.
116. S. Nagase, M. Nakano and T. Kudo, *J. Chem. Soc., Chem. Commun.*, 60 (1987).
117. M. J. S. Dewar, *J. Am. Chem. Soc.*, **106**, 669 (1984).
118. C. L. Wadsworth and R. West, *Organometallics*, **4**, 1664 (1985).
119. (a) J. P. Hummel, J. Stackhouse and K. Mislow, *Tetrahedron*, **33**, 1925 (1977).
(b) F. Shafiee, J. R. Damewood, Jr., K. J. Haller and R. West, *J. Am. Chem. Soc.*, **107**, 6950 (1985).
(c) J. R. Damewood, Jr. and R. Gambogi, *Tetrahedron*, **42**, 6411 (1986).

120. (a) T. Dabisch and W. W. Schoeller, *J. Chem. Soc., Chem. Commun.*, 896 (1986).
(b) S. Collins, R. Dutler and R. Rauk, *J. Am. Chem. Soc.*, **109**, 2564 (1987).
121. P. v. R. Schleyer, A. F. Sax, J. Kalcher and R. Janoschek, *Angew. Chem., Int. Ed. Engl.*, **26**, 364 (1987).
122. J. A. Boatz and M. S. Gordon, *J. Phys. Chem.*, **92**, 3037 (1988).
123. D. A. Clabo, Jr. and H. F. Schaefer III, *J. Am. Chem. Soc.*, **108**, 4344 (1986).
124. A. F. Sax and J. Kalcher, *J. Chem. Soc., Chem. Commun.*, 809 (1987).
125. H. B. Yokelson, A. J. Millevolte, G. R. Gillette and R. West, *J. Am. Chem. Soc.*, **109**, 6865 (1987).
126. R. S. Grev and H. F. Schaefer III, *J. Am. Chem. Soc.*, **109**, 6577 (1987).
127. J. A. Boatz and M. S. Gordon, *J. Am. Chem. Soc.* (submitted).
128. (a) P. R. Jones and D. D. White, *J. Organomet. Chem.*, **154**, C33 (1978).
(b) P. R. Jones and D. O. White, *J. Organomet. Chem.*, **181**, 399 (1979).
(c) M. J. S. Dewar and G. P. Ford, *J. Am. Chem. Soc.*, **101**, 783 (1979).
129. M. S. Gordon, *J. Am. Chem. Soc.*, **102**, 7419 (1980).
130. M. S. Gordon, *Chem. Phys. Lett.*, **76**, 163 (1980).
131. (a) S. Masamune, S. Murakami, H. Tobita and D. J. Williams, *J. Am. Chem. Soc.*, **105**, 7776 (1983).
(b) R. West, D. J. Deyoung and K. Haller, *J. Am. Chem. Soc.*, **107**, 4942 (1985).
132. (a) M. J. S. Dewar, *Bull. Soc. Chim. Fr.*, **18**, C71 (1951).
(b) J. Chatt and L. A. Duncanson, *J. Chem. Soc.*, 2939 (1953). See also:
(c) D. Cremer and E. Kraka, *J. Am. Chem. Soc.*, **107**, 3800 (1985).
133. R. F. W. Bader, T. S. Slee, D. Cremer and E. Kraka, *J. Am. Chem. Soc.*, **105**, 5061 (1983), and references cited therein.
134. (a) M. J. Fink, K. J. Haller, R. West and J. Michl, *J. Am. Chem. Soc.*, **106**, 822 (1984).
(b) M. Michalczyk, M. J. Fink, K. J. Haller, R. West and J. Michl, *Organometallics*, **5**, 531 (1986).
(c) H. B. Yokelson, A. J. Millevolte, B. R. Adams and R. West, *J. Am. Chem. Soc.*, **109**, 4116 (1987).
135. (a) W. Clegg, M. Haase, U. Klingebiel and G. M. Sheldrick, *Chem. Ber.*, **116**, 146 (1984).
(b) W. Clegg, M. Hasse, U. Klingebiel, J. Neeman and G. M. Sheldrick, *J. Organomet. Chem.*, **251**, 281 (1983) and references cited therein.
(c) W. Wojnowski, K. Peters, D. Weber and H.-G. von Schnering, *Z. Anorg. Allg. Chem.*, **519**, 134 (1984), and references cited therein.
(d) K. Peters, E. -M. Peters and H.-G. von Schnering, *Z. Anorg. Allg. Chem.*, **502**, 61 (1983).
136. S. M. Bachrach and A. Streitwieser, Jr., *J. Am. Chem. Soc.*, **107**, 1186 (1985).
137. T. Kudo and S. Nagase, *J. Am. Chem. Soc.*, **107**, 2589 (1985).
138. (a) R. J. Berenstein and S. Scheiner, *Int. J. Quantum Chem.*, **29**, 1191 (1986).
(b) M. O'Keeffe and G. V. Gibbs, *J. Phys. Chem.*, **89**, 4574 (1985).
139. (a) M. S. Gordon, *J. Chem. Soc., Chem. Commun.*, 1131 (1980).
(b) T. A. Holmes and M. S. Gordon, *Organometallics*, **3**, 583 (1984).
140. (a) P. R. Jones and D. D. White, *J. Organomet. Chem.*, **181**, 399 (1979).
(b) P. D. Mollere and R. Hoffmann, *J. Am. Chem. Soc.*, **97**, 3680 (1975).
141. (a) M. S. Gordon, *Chem. Phys. Lett.*, **76**, 163 (1980).
(b) T. C. Barthelat, G. Trinquier and G. Bertrand, *J. Am. Chem. Soc.*, **101**, 3785 (1979).
(c) M. H. Lien and A. C. Hopkinson, *Chem. Phys. Lett.*, **80**, 114 (1981).
142. (a) M. S. Gordon and P. Boudjouk, *J. Am. Chem. Soc.*, **107**, 1439 (1985).
(b) M. E. Colvin and H. F. Schaefer III, *Faraday Symp. Chem. Soc.*, **19**, 39 (1984).
143. (a) R. S. Grev and H. F. Schaefer III, *J. Chem. Phys.*, **80**, 3552 (1984).
(b) R. S. Grev and H. F. Schaefer III, *J. Chem. Phys.*, **82**, 4126 (1985).
(c) R. S. Grev and H. F. Schaefer III, *Chem. Phys. Lett.*, **119**, 111 (1985).
(d) K. Ragavachari, *J. Chem. Phys.*, **83**, 3520 (1985).
(e) K. Ragavachari, *J. Chem. Phys.*, **84**, 5672 (1986).
144. (a) R. W. Bigelow, *Organometallics*, **5**, 1502 (1986).
(b) E. A. Halevi, G. Winkelhofer, M. Meisl and R. Janoschek, *J. Organomet. Chem.*, **294**, 151 (1985).
(c) R. Gleiter, W. Schafer and H. Sakurai, *J. Am. Chem. Soc.*, **107**, 3046 (1985).
(d) K. A. Klingensmith, J. W. Downing, R. D. Miller and J. Michl, *J. Am. Chem. Soc.*, **108**, 7438 (1986).

(e) K. Takeda, H. Teramae and N. Matsumoto, *J. Am. Chem. Soc.*, **108**, 8186 (1986).
145. (a) N. Wiberg, K. Schurz, G. Reber and G. Muller, *J. Chem. Soc., Chem. Commun.*, 591 (1986).
(b) M. Hesse and U. Klingebiel, *Angew. Chem., Int. Ed., Engl.*, **25**, 649 (1986).
146. B. T. Luke, J. A. Pople, M.-B. Krogh-Jespersen, Y. Apeloig, M. Karni, J. Chandrasekhar and P. v. R. Schleyer, *J. Am. Chem. Soc.*, **108**, 270 (1986).
147. M. W. Schmidt, P. W. Troung and M. S. Gordon, *J. Am. Chem. Soc.*, **109**, 5217 (1987).
148. O. P. Strausz, L. Gammie, G. Theodorakoupoulos, P. G. Mezey and I. G. Csizmadia, *J. Am. Chem. Soc.*, **98**, 1622 (1976).
149. O. P. Strausz, M. A. Robb, G. Theodorakoupoulos, P. G. Mezey and I. G. Csizmadia, *Chem. Phys. Lett.*, **48**, 162 (1977).
150. R. Ahlrichs and R. Heinzmann, *J. Am. Chem. Soc.*, **99**, 7452 (1977).
151. M. S. Gordon, *Chem. Phys. Lett.*, **54**, 9 (1978).
152. D. M. Hood and H. F. Schaefer III, *J. Chem. Phys.*, **68**, 2985 (1978).
153. J. D. Goddard, Y. Yoshioka and H. F. Schaefer III, *J. Am. Chem. Soc.*, **102**, 7644 (1980).
154. R. K. Gosavi, H. E. Gunning and O. P. Strausz, *Chem. Phys. Lett.*, **59**, 321 (1978).
155. K. Vasudevan and F. Grein, *Chem. Phys. Lett.*, **75**, 75 (1980).
156. Y. Yoshioka, J. D. Goddard and H. F. Schaefer III, *J. Am. Chem. Soc.*, **103**, 2452 (1981).
157. G. Trinquier and J.-P. Malrieu, *J. Am. Chem. Soc.*, **103**, 6313 (1981).
158. P. Rosmus, H. Bock, B. Solouki, G. Maier and G. Mihm, *Angew. Chem., Int. Ed. Engl.*, **20**, 598 (1981).
159. Y. Yoshioka and H. F. Schaefer III, *J. Am. Chem. Soc.*, **103**, 7366 (1981).
160. M. S. Gordon, *J. Am. Chem. Soc.*, **104**, 4352 (1982).
161. H. J. Kohler and H. Lischka, *J. Am. Chem. Soc.*, **104**, 5884 (1982).
162. Y. Apeloig and M. Karni, *J. Chem. Soc., Chem. Commun.*, 768 (1984).
163. K. Ohta, E. R. Davidson and K. Morokuma, *J. Am. Chem. Soc.*, **107**, 3466 (1985).
164. M. W. Schmidt, M. S. Gordon and M. Dupuis, *J. Am. Chem. Soc.*, **107**, 2585 (1985).
165. M. E. Colvin, J. Kobayashi, J. Bicerano and H. F. Schaefer III, *J. Chem. Phys.*, **85**, 4563 (1986).
166. P. G. Mohaffy, R. Gutowski and L. K. Montgomery, *J. Am. Chem. Soc.*, **102**, 2854 (1980).
167. A. G. Brook, S. C. Nyburg, F. Abdesaken, B. Gutenkunst, G. Gutekunst, R. K. M. R. Kallury, U. C. Poon, Y. M. Chang and W. Wong-Ng, *J. Am. Chem. Soc.*, **104**, 5667 (1982).
168. N. Wiberg, G. Wagner and G. Muller, *Angew. Chem., Int. Ed. Engl.*, **24**, 229 (1985).
169. (a) G. Maier, G. Mihm and H. P. Reisenauer, *Angew. Chem., Int. Ed. Engl.*, **20**, 597 (1981).
(b) G. Maier, G. Mihm and H. P. Reisenauer, *Chem. Ber.*, **117**, 2351 (1984).
170. M. D. Curtis, *J. Organomet. Chem.*, **60**, 63 (1973).
171. Y. A. Ustynyuk, P. I. Zakharov, A. A. Azizov, G. A. Shchembelov and I. P. Gloriozov, *J. Organomet. Chem.*, **96**, 195 (1975).
172. W. J. Pietro and W. J. Hehre, *J. Am. Chem. Soc.*, **104**, 4329 (1982).
173. C. F. Pau, W. J. Pietro and W. J. Hehre, *J. Am. Chem. Soc.*, **105**, 16 (1983).
174. (a) T. J. Drahnak, J. Michl and R. West, *J. Am. Chem. Soc.*, **103**, 1845 (1981).
(b) R. T. Conlin and D. L. Wood, *J. Am. Chem. Soc.*, **103**, 1843 (1981).
175. (a) S. Nagase and T. Kudo, *J. Chem. Soc., Chem. Commun.*, 141 (1984).
(b) S. Nagase, T. Kudo and K. Ito, in *Applied Quantum Chemistry* (Eds. V. H. Smith Jr., H. F. Schaefer III and K. Morokuma), Reidel, Dordrecht, 1986, pp. 249–267.
176. C. A. Arrington, R. West and J. Michl, *J. Am. Chem. Soc.*, **105**, 6176 (1983).
177. T. J. Barton and G. T. Burns, *Organometallics*, **1**, 210 (1982).
178. R. Damrauer and D. R. Williams, *J. Organomet. Chem.*, **66**, 241 (1974).
179. K. Morokuma, in Ref. 6, p. 40.
180. S. Nagase and T. Kudo, *J. Chem. Soc., Chem. Commun.*, 363 (1983).
181. M. S. Gordon and T. N. Truong, *Chem. Phys. Lett.*, **142**, 110 (1987).
182. M. S. Gordon and C. George, *J. Am. Chem. Soc.*, **106**, 609 (1984).
183. S. Nagase and T. Kudo, *J. Chem. Soc., Chem. Commun.*, 1392 (1984).
184. I. Fleming, *Frontier Orbitals and Organic Chemical Reactions*, Wiley, London, 1976.
185. (a) A. Sekiguchi and W. Ando, *Tetrahedron Lett.*, **24**, 2791 (1983).
(b) T. J. Barton and S. A. Jacobi, *J. Am. Chem. Soc.*, **102**, 7979 (1980).
186. M. S. Gordon and R. D. Koob, *J. Am. Chem. Soc.*, **103**, 2939 (1981).
187. M. Hanamura, S. Nagase and A. Morokuma, *Tetrahedron Lett.*, **22**, 1813 (1981).
188. K. Krogh-Jespersen, *J. Comput. Chem.*, **3**, 572 (1982).

189. G. Trinquier and J.-P. Malrieu, *J. Am. Chem. Soc.*, **109**, 5303 (1987).
190. (a) R. Daudel, R. E. Kari, R. A. Poirier, J. D. Goddard and I. G. Csizmadia, *J. Mol. Struct.*, **50**, 115 (1978).
(b) F. F. Roelandt, D. F. van de Vondel and G. P. van der Kelen, *J Organomet. Chem.*, **165**, 151 (1979).
(c) L. C. Snyder and Z. R. Wasserman, *J. Am. Chem. Soc.*, **101**, 5222 (1979).
191. R. A. Poirier and J. D. Goddard, *Chem. Phys. Lett.*, **80**, 37 (1981).
192. (a) H. Lischka and H.-J. Kohler, *Chem. Phys. Lett.*, **85**, 467 (1982).
(b) H.-J. Kohler and H. Lischka, *Chem. Phys. Lett.*, **98**, 454 (1983).
193. S. Nagase and T. Kudo, *Theochem.*, **103**, 35 (1983).
194. G. Olbrich, P. Potzinger, B. Reimann and R. Walsh, *Organometallics*, **3**, 1267 (1984).
195. K. Krogh-Jespersen, *J. Phys. Chem.*, **86**, 1492 (1982).
196. K. Krogh-Jespersen, *J. Am. Chem. Soc.*, **107**, 537 (1985).
197. K. Krogh-Jespersen, *Chem. Phys. Lett.*, **93**, 327 (1982).
198. G. Olbrich, *Chem. Phys. Lett.*, **130**, 115 (1986).
199. M. S. Gordon, N. T. Truong and E. K. Bonderson, *J. Am. Chem. Soc.*, **108**, 1421 (1986).
200. P. Ho, M. E. Coltrin, J. S. Binkley and C. F. Melius, *J. Phys. Chem.*, **90**, 3399 (1986).
201. J. A. Tossel and P. Lazzeretti, *Chem. Phys. Lett.*, **128**, 420 (1986).
202. H. Teramae, *J. Am. Chem. Soc.*, **109**, 4140 (1987).
203. H.-J. Kohler, *Z. Chem.*, **24**, 155 (1984).
204. T. N. Bell, A. F. Kieran, K. A. Perkins and P. G. Perkins, *J. Phys. Chem.*, **88**, 1334 (1984).
205. S. Nagase and T. Kudo, *Organometallics*, **3**, 1320 (1984).
206. W. W. Schoeller and V. Staemmler, *Inorg. Chem.*, **23**, 3369 (1984).
207. S. Masamune, Y. Hanazawa, S. Murakami, T. Bally and J. F. Blount, *J. Am. Chem. Soc.*, **104**, 1150 (1982).
208. P. Boudjouk, B. Han and K. R. Anderson, *J. Am. Chem. Soc.*, **104**, 4992 (1982).
209. (a) P. J. Davidson, D. H. Harris and M. F. Lappert, *J. Chem. Soc., Dalton Trans.*, 2268 (1976).
(b) J. D. Cotton, P. J. Davidson, M. F. Lappert, J. D. Donaldson and J. Silver, *J. Chem. Soc., Dalton Trans.*, 2286 (1976).
210. H. Sakurai, Y. Nakadaira and H. Sokaba, *Organometallics*, **2**, 1484 (1983).
211. T. N. Troung and M. S. Gordon, *J. Am. Chem. Soc.*, **108**, 1775 (1986).
212. (a) P. v. R. Schleyer and P. D. Stout, *J. Chem. Soc., Chem. Commun.*, 1373 (1986).
(b) M. T. Nguyen, M. Faul and N. J. Fitzpatrick, *J. Chem. Soc., Perkin Trans. 2*, 1289 (1987).
213. (a) K. J. Dykema, T. N. Troung and M. S. Gordon, *J. Am. Chem. Soc.*, **107**, 4535 (1985).
(b) J.-G. Lee, J. E. Boggs and A. H. Cowley, *J. Chem. Soc., Chem. Commun.*, 773 (1985).
214. (a) J. Pacansky and K. Hermann, *J. Chem. Phys.*, **69**, 963 (1978).
(b) G. Herzberg, *Spectra of Diatomic Molecules*, D. Van Nostrand, Princeton, N. J., 1950.
215. (a) R. Jaquet, W. Kutzelnigg and V. Staemmler, *Theor. Chim. Acta*, **54**, 205 (1980).
(b) R. Z. Ponec, *Z. Phys. Chem. (Leipzig)*, **264**, 964 (1980).
216. R. J. Glinski, J. L. Gole and D. A. Dixon, *J. Am. Chem. Soc.*, **107**, 5891 (1985).
217. (a) T. Kudo and S. Nagase, *J. Phys. Chem.*, **88**, 2833 (1984).
(b) T. Kudo and S. Nagase, *J. Organomet. Chem.*, **253**, C23 (1983).
218. T. Kudo and S. Nagase, *Chem. Phys. Lett.*, **128**, 507 (1986).
219. A. Tachibana, H. Fueno and T. Yamabe, *J. Am. Chem. Soc.*, **108**, 4346 (1986).
220. T. Kudo and S. Nagase, *Organometallics*, **5**, 1207 (1986).
221. T. Kudo and S. Nagase, *J. Am. Chem. Soc.*, submitted.
222. (a) L. Linder, A. Revis and T. J. Barton, *J. Am. Chem. Soc.*, **108**, 2742 (1986) and references cited therein.
(b) T. J. Barton and G. C. Paul, *J. Am. Chem. Soc.*, **109**, 5292 (1987).
223. J. N. Murrell, H. W. Kroto and M. F. Guest, *J. Chem. Soc., Chem. Commun.*, 619 (1977).
224. (a) A. C. Hopkinson and M. H. Lien, *J. Chem. Soc., Chem. Commun.*, 107 (1980).
(b) A. C. Hopkinson, M. H. Lien and I. G. Csizmadia, *Chem. Phys. Lett.*, **95**, 232 (1983).
225. M. S. Gordon and J. A. Pople, *J. Am. Chem. Soc.*, **103**, 2945 (1981).
226. M. R. Hoffmann, Y. Yoshioka and H. F. Schaefer III, *J. Am. Chem. Soc.*, **105**, 1084 (1983).
227. C. G. Newman, M. A. Ring and H. E. O'Neal, *J. Am. Chem. Soc.*, **100**, 5945 (1978).
228. J. Steinwandel and J. Hoeschele, *Chem. Phys. Lett.*, **116**, 25 (1985).
229. A. Sekiguchi, S. S. Zigler, R. West and J. Michl, *J. Am. Chem. Soc.*, **108**, 4241 (1986).
230. S. Obara, K. Kitaura and K. Morokuma, *Theor. Chim. Acta*, **60**, 227 (1981).

231. L. C. Snyder, Z. R. Wasserman and J. W. Moskowitz, *Int. J. Quantum Chem.*, **21**, 565 (1982).
232. K. Kawai, T. Noro, A. Murakani and K. Ohno, *Chem. Phys. Lett.*, **92**, 479 (1982).
233. J. Kalcher, A. Sax and G. Olbrich, *Int. J. Quantum. Chem.*, **25**, 543 (1984).
234. N. C. Baird, *Can. J. Chem.*, **63**, 71 (1985).
235. S. Koseki and M. S. Gordon, *J. Phys. Chem.*, in press.
236. H. W. Kroto, J. N. Murrell, A. Al-Derzi and M. F. Guest, *Astrophys. J.*, **219**, 886 (1978); see also Reference 223.
237. R. Preuss, R. J. Buenker and S. D. Peyerimhoff, *J. Mol. Struct.*, **49**, 171 (1976).
238. R. Preuss, R. J. Buenker and S. D. Peyerimhoff, *Chem. Phys. Lett.*, **62**, 21 (1979).
239. J. F. Ogilvie and S. Cradock, *J. Chem. Soc., Chem. Commun.*, 364 (1966).
240. H. Bock and R. Dammel, *Angew. Chem., Int. Ed. Engl.*, **24**, 111 (1985).
241. Silatoluene: (a) T. J. Barton and D. Banasink, *J. Am. Chem. Soc.*, **99**, 5199 (1977).
(b) C. L. Kreil, O. L. Chapman, G. T. Burns and T. J. Barton, *J. Am. Chem. Soc.*, **102**, 841 (1980).
(c) H. Bock, R. A. Bowlilng, B. Solouki, T. J. Burns and G. T. Burns, *J. Am. Chem. Soc.*, **102**, 429 (1980).
242. Silabenzene: (a) T. J. Barton and G. T. Burns, *J. Am. Chem. Soc.*, **100**, 5246 (1978).
(b) G. Maier, G. Mihm and H. P. Reisenauer, *Angew. Chem., Int. Ed. Engl.*, **19**, 52 (1980).
(c) B. Solouki, P. Rosmus, H. Bock and G. Maier, *Angew Chem., Int. Ed. Engl.*, **19**, 51 (1980).
243. H. B. Schlegel, B. Coleman and M. Jones, Jr., *J. Am. Chem. Soc.*, **100**, 6499 (1978).
244. P. H. Blustin, *J. Organomet. Chem.*, **166**, 21 (1979).
245. J. Chandrasekhar, P. v. R. Schleyer, R. O. W. Baumgartner and M. T. Reets, *J. Org. Chem.*, **48**, 3453 (1983).
246. (a) M. S. Gordon, P. Boudjouk and F. Anwari, *J. Am. Chem. Soc.*, **105**, 4972 (1983).
(b) K. K. Baldridge and M. S. Gordon, *J. Am. Chem. Soc.*, **110**, 4204 (1988).
247. K. K. Baldridge and M. S. Gordon, *Organometallics*, **7**, 144 (1988).
248. W. J. Hehre, R. Ditchfield, L. Radom and J. A. Pople, *J. Am. Chem. Soc.*, **92**, 4796 (1970).
249. K. K. Baldridge and M. S. Gordon, *J. Organomet. Chem.*, **271**, 369 (1984).
250. J. Chandrasekhar and P. v. R. Schleyer, *J. Organomet. Chem.*, **289**, 51 (1985).
251. S. Nagase, T. Kudo and M. Aoki, *J. Chem. Soc., Chem. Commun.*, 1121 (1985).
252. (a) A. F. Sax and R. Janoschek, Proceedings of the 4th International Symposium on Inorganic Ring Systems, Paris, September 2–6, 1985.
(b) A. F. Sax and R. Janoschek, *Angew. Chem., Int. Ed. Engl.*, **25**, 651 (1986).
253. A. F. Sax and R. Janoschek, *J. Am. Chem. Soc.*, submitted.
254. D. A. Clabo and H. F. Schaefer III, *J. Chem. Phys.*, **84**, 1664 (1986).
255. S. Nagase, H. Teramae and T. Kudo, *J. Chem. Phys.*, **86**, 4513 (1987).
256. P. Jutzi, D. Kanne and C. Kruger, *Angew. Chem., Int. Ed. Engl.*, **25**, 164 (1986).
257. K. K. Baldridge and M. S. Gordon, *Organometallics*, submitted.
258. (a) J. B. Lambert, Chapter 16 in this book.
(b) R. J. P. Corriu and M. Henner, *J. Organomet. Chem.*, **74**, 1 (1974).
(c) V. D. Nefedov, T. A. Kochina and E. N. Sinotova, *Russ. Chem. Rev.*, **55**, 426 (1986).
259. J. R. Damewood, Jr., *J. Org. Chem.*, **51**, 5028 (1986).
260. K. Krogh-Jespersen, J. Chandrasekhar and P. v. R. Schleyer, *J. Org. Chem.*, **45**, 1608 (1980).
261. B. J. Aylett and H. M. Colquhoun, *J. Chem. Res. (S)*, 148 (1977); *J. Chem. Res. (M)*, 1674 (1977).
262. (a) P. D. Gaspar, in *Reactive Intermediates* (Eds. M. Jones, Jr. and R. A. Moss), Vol. 3, Chapter 9, Wiley, New York, 1985. See also previous volumes of this series.
(b) Y.-N. Tang, in *Reactive Intermediates* (Ed. R. A. Abramovitch), Vol. 2, Chapter 4, Plenum Press, New York, 1982.
(c) I. M. T. Davidson, in *Annual Reports on the Progress of Chemistry*, Royal Society of Chemistry, Section C, 1985, pp. 60–70. See also previous volumes.
263. (a) T. J. Drahnak, J. Michl and R. West, *J. Am. Chem. Soc.*, **101**, 5427 (1979).
(b) C. A. Arrington, K. A. Klingensmith, R. West and J. Michl. *J. Am. Chem. Soc.*, **106**, 525 (1984).
(c) M. J. Michalczyk, M. J. Fink, D. J. De Young, C. W. Carlson, K. M. Welsh, R. West and J. Michl, *Silicon, Germanium, Tin and Lead Compounds*, **9**, 75 (1986).
264. (a) A. S. Nazran, J. A. Hawari, D. Griller, I. S. Alnaimi and W. P. Weber, *J. Am. Chem. Soc.*, **106**, 7267 (1984).

(b) J. A. Hawari and D. Griller, *Organometallics*, **3**, 1123 (1984).
265. (a) B. Wirsam, *Chem. Phys. Lett.*, **14**, 214 (1972).
(b) M. E. Colvin, R. S. Grev, H. F. Schaefer III and J. Bicerano, *Chem. Phys. Lett.*, **99**, 399 (1977).
(c) J. E. Rice an N. C. Handy, *Chem. Phys. Lett.*, **107**, 365 (1984).
(d) S. Koseki and M. S. Gordon, *J. Mol. Spect.*, **123**, 392 (1987).
(e) M. M. Heaton, *J. Chem. Phys.*, **67**, 5396 (1977).
(f) J. H. Meadows and H. F. Schaefer III, *J. Am. Chem. Soc.*, **98**, 4383 (1976).
(g) H. B. Schlegel, *J. Phys. Chem.*, **88**, 6254 (1984).
(h) Reference 79, pp. 58–65.
266. K. Balasubramanian and A. D. McLean, *J. Chem. Phys.*, **85**, 5117 (1986).
267. (a) J. A. Pople, B. T. Luke, M. J. Frisch and J. S. Binkley, *J. Phys. Chem.*, **89**, 2198 (1985).
(b) P. Ho, M. E. Coltrin, J. S. Binkley and C. F. Melius, *J. Phys. Chem.*, **89**, 4647 (1985).
(c) M. S. Gordon and J. Heitzinger, *J. Phys. Chem.*, **91**, 2353 (1987).
268. (a) A. Kasden, E. Herbst and W. C. Lineberger, *J. Chem. Phys.*, **62**, 541 (1975).
(b) J. Berkowitz, J. P. Greene, H. Cho and R. Ruscic, *J. Chem. Phys.*, **86**, 1235 (1987).
269. See, for example, A. R. W. McKellar, P. R. Bunker, T. J. Sears, K. M. Evenson, R. J. Saykally and S. R. Langhoff, *J. Chem. Phys.*, **79**, 5251 (1983).
270. I. Dubois, *Can. J. Phys.*, **46**, 2485 (1968).
271. (a) P. John and J. H. Purnell, *J. Chem. Soc., Faraday Trans. 1*, **69**, 1455 (1973).
(b) J. H. Purnell and R. Walsh, *Proc. Roy. Soc. (London)*, **A293**, 543 (1971).
(c) M. Bowrey and J. H. Purnell, *Proc. Roy. Soc. (London)*, **A321**, 341 (1971).
(d) A. J. Vanderwielen, M. A. Ring and H. E. O'Neal, *J. Am. Chem. Soc.*, **97**, 993 (1975).
272. (a) J. Jasinski, *J. Phys. Chem.*, **90**, 555 (1986).
(b) S. K. Shin and J. L. Beauchamp, *J. Phys. Chem.*, **90**, 1507 (1986).
(c) H. M. Frey, R. Walsh and I. M. Watts, *J. Phys. Chem.*, in press.
273. J. F. Harrison, P. G. Liedtke and J. F. Liebman, *J. Am. Chem. Soc.*, **101**, 7162 (1979).
274. D. R. Rao, *J. Mol. Spectrosc.*, **34**, 284 (1970).
275. M. E. Colvin, J. Breulet and H. F. Schaefer III, *Tetrahedron*, **41**, 1429 (1985).
276. M. S. Gordon and M. W. Schmidt, *Chem. Phys. Lett.*, **132**, 294 (1986).
277. M. S. Gordon, *Chem. Phys. Lett.*, **114**, 348 (1985).
278. Y. Apeloig and M. Karni, *J. Chem. Soc., Chem. Commun*, 1048 (1985).
279. I. Dubois, G. Herzberg and K. D. Verma, *J. Chem. Phys.*, **47**, 4262 (1967).
280. (a) H. Vancik, G. Raabe, M. J. Michalczyk, R. West and J. Michl, *J. Am. Chem. Soc.*, **107**, 4097 (1985).
(b) A. S. Nazran, J. A. Hawari, D. Griller, I. S. Alnaimi and W. P. Weber, *J. Am. Chem. Soc.*, **108**, 5041 (1986).
281. R. S. Grev and H. F. Schaefer III, *J. Am. Chem. Soc.*, **108**, 5804 (1986).
282. (a) R. B. Woodward and R. Hoffmann, *The Conservation of Orbital Symmetry*, Verlag Chemie, Academic Press, Weinheim, 1971.
(b) R. Hoffmann, R. Gleiter and F. B. Mallory, *J. Am. Chem. Soc.*, **92**, 1460 (1970).
283. K. Morokuma, K. Ohta, N. Koga, S. Obasra and E. R. Davidson, *Faraday Symp. Chem. Soc.*, **19**, 49 (1981).
284. (a) F. Bernardi and M. A. Robb, in Reference 24a, pp. 185–190.
(b) F. Bernardi and M. A. Robb, *Mol. Phys.*, **48**, 1345 (1983).
285. R. Viswanathan, D. L. Thompson and L. M. Raff, *J. Chem. Phys.*, **80**, 4230 (1984).
286. M. S. Gordon, *J. Chem. Soc., Chem. Commun.*, 890 (1981).
287. M. S. Gordon and D. R. Gano, *J. Am. Chem. Soc.*, **106**, 5421 (1984).
288. R. S. Grev and H. F. Schaefer III, *J. Chem. Soc., Chem. Commun.*, 785 (1983).
289. C. Sosa and H. B. Schlegel, *J. Am. Chem. Soc.*, **106**, 5847 (1984).
290. A. Sax and G. Olbrich, *J. Am. Chem. Soc.*, **107**, 4868 (1985).
291. M. S. Gordon, D. R. Gano, J. S. Binkley and M. Frisch, *J. Am. Chem. Soc.*, **108**, 2191 (1986).
292. (a) H. E. O'Neal and M. A. Ring, *Chem. Phys. Lett.*, **107**, 442 (1984).
(b) J. M. Jasinski, *J. Phys. Chem.*, in press.
293. Cited in Reference 181 as private communication by M. A. Ring and H. E. O'Neal.
294. K. Raghavachari, J. Chandrasekhar, M. S. Gordon and K. J. Dykema, *J. Am. Chem. Soc.*, **106**, 5853 (1984).
295. H. B. Schlegel and C. Sosa, *J. Phys. Chem.*, **89**, 537 (1985).

296. C. W. Bock, M. Trachtman, P. D. Maker, H. Niki and G. J. Mains, *J. Phys. Chem.*, **90**, 5669 (1986).
297. M. S. Gordon, *J. Am. Chem. Soc.*, **106**, 4054 (1984).
298. F. Anvari and M. S. Gordon, *Isr. J. Chem.*, **23**, 129 (1983).
299. (a) H. Kobayashi and T. Koga, *Theor. Chim. Acta*, **67**, 1 (1985).
(b) A. Tachibana, M. Koizumi, H. Teramae and T. Yamabe, *J. Am. Chem. Soc.*, **109**, 1383 (1987).
(c) Z. K. Ismail, R. H. Hauge, L. Fredin, J. W. Kauffman and J. L. Margrave, *J. Chem. Phys.*, **77**, 1617 (1982).
300. T. Clark and P. v. R. Schleyer, *J. Organomet. Chem.*, **191**, 347 (1980).
301. For similar calculations for the analogous CH_2LiF see T. Clark and P. v. R. Schleyer, *J. Chem. Soc., Chem. Commun.*, 883 (1979).
302. See H. Schwarz, Chapter 7 in this book.
303. (a) J. B. Lambert, W. J. Schulz, Jr., J. A. McConnell and W. Schilf, *J. Am. Chem. Soc.*, **110**, 2201 (1988).
(b) J. B. Lambert, J. A. McConnell and W. J. Schulz, Jr., *J. Am. Chem. Soc.*, **108**, 2482 (1986).
(c) J. B. Lambert and W. J. Schulz, Jr., *J. Am. Chem., Soc.*, **105**, 1671 (1983).
304. G. K. S. Prakash, S. Keyaniyan, R. Aniszfeld, L. Heiliger, G. A. Olah, R. C. Stevens, H.-K. Choi and R. Bau, *J. Am. Chem. Soc.*, **109**, 5123 (1987).
305. C. Eaborn, in Ref. 6, p. 123.
306. (a) Y. Apeloig and A. Stanger, *J. Am. Chem. Soc.*, **109**, 272 (1987).
(b) Y. Apeloig, *Studies in Organic Chemistry*, **31**, 33 (1986).
307. (a) B. Wirsam, *Chem. Phys. Lett.*, **18**, 578 (1973).
(b) J. A. Pople and L. A. Curtiss, *J. Phys. Chem.*, **91**, 3637 (1987); see references therein for previous studies.
(c) J. A. Pople and L. A. Curtis, *J. Phys. Chem.*, **91**, 155 (1987).
308. J. M. Dyke, N. Jonathan, A. Morris, A. Ridha and M. J. Winter, *Chem. Phys.*, **81**, 481 (1983).
309. (a) T. M. H. Cheng, T. Y. Yu and F. W. Lampe, *J. Phys. Chem.*, **77**, 2587 (1983); see aso
(b) G. W. Goodloe, E. A. Austin and F. W. Lampe, *J. Am. Chem. Soc.*, **101**, 3472 (1979).
(c) G. W. Goodloe and F. W. Lampe, *J. Am. Chem. Soc.*, **101**, 6028 (1979).
310. (a) Y. Apeloig and P. v. R. Schleyer, *Tetrahedron Lett.*, 4647 (1977).
(b) Y. Apeloig, S. A. Godleski, D. J. Heacock and J. M. McKelvey, *Tetrahedron Lett.*, 3297 (1981).
311. A. C. Hopkinson and M. H. Lien, *J. Org. Chem.*, **46**, 998 (1981).
312. S. A. Godleski, D. J. Heacock and J. M. McKelvey, *Tetrahedron Lett.*, 4453 (1982).
313. Y. Apeloig, unpublished results.
314. A. C. Hopkinson and M. H. Lien, *J. Mol. Struct.*, **104**, 303 (1983).
315. J. A. Pople, Y. Apeloig and P. v. R. Schleyer, *Chem. Phys. Lett.*, **85**, 489 (1982).
316. J. R. Eyler, G. Silverman and M. A. Battiste, *Organometallics*, **1**, 477 (1982).
317. J. B. Lambert, private communication.
318. W. L. Jorgensen, *J. Am. Chem. Soc.*, **100**, 1049 (1978), and references cited therein.
319. M. L. Hendewerk, R. Frey and D. A. Dixon, *J. Phys. Chem.*, **87**, 2026 (1983).
320. P. v. R. Schleyer, Y. Apeloig, D. Arad, B. T. Luke and J. A. Pople, *Chem. Phys. Lett.*, **95**, 477 (1983).
321. (a) W. Koch, G. Frenking and H. Schwarz, *J. Chem. Soc., Chem. Commun.*, 1119 (1985).
(b) W. Koch, G. Frenking, H. Schwarz, M. Maquin and D. Stahl, *J. Chem. Soc., Perkin Trans. 2*, 757 (1986).
(c) W. Koch, G. Frenking and C. C. Chang, *J. Chem. Phys.*, **84**, 2703 (1986).
(d) M. W. Wong, B. F. Yates, R. H. Nobes and L. Radom, *J. Am. Chem. Soc.*, **109**, 3181 (1987).
322. (a) For an excellent general discussion of activating and directive effects of silicon, both in carbocationic and carbanionic reactions, see: A. R. Bassindale and P. G. Taylor, Chapter 14 in this book.
(b) For a recent extensive review on cyclization reactions, see: T. A. Blumenkopf and L. E. Overman, *Chem. Rev.*, **86**, 857 (1986).
323. (a) P. J. Stang, M. Ladika, Y. Apeloig, A. Stanger, M. D. Schiavelli and M. R. Hughey, *J. Am. Chem. Soc.*, **104**, 6852 (1982).
(b) STO-3G leads to the erroneous conclusion that H_3Si is more stabilizing than CH_3. See: Y. Apeloig, P. v. R. Schleyer and J. A. Pople, *J. Am. Chem. Soc.*, **99**, 1291 (1977).
(c) Similar results to those in Ref. 323a were reported in Ref. 311.

324. Y. Apeloig and A. Stanger, *J. Am. Chem. Soc.*, **107**, 2806 (1985).
325. M. D. Schiavelli, D. M. Jung, A. K. Vaden, P. J. Stang, T. E. Fisk and D. S. Morrison, *J. Org. Chem.*, **46**, 92 (1981).
326. Y. Apeloig and A. Stanger, *J. Org. Chem.*, **47**, 1462 (1982). For additions see: *J. Org. Chem.*, **48**, 5413 (1983).
327. S. G. Wierschke, J. Chandrasekhar and W. L. Jorgensen, *J. Am. Chem. Soc.*, **107**, 1496 (1985).
328. E. B. Nadler, Z. Rappoport, D. Arad and Y. Apeloig, *J. Am. Chem. Soc.*, **109**, 7873 (1987).
329. R. Hoffmann, L. Radom, J. A. Pople, P. v. R. Schleyer, W. J. Hehre and L. Salem, *J. Am. Chem. Soc.*, **94**, 6221 (1972).
330. J. B. Lambert, G. Wang, R. B. Finzel and D. H. Teramura, *J. Am. Chem. Soc.*, **109**, 7838 (1987).
331. (a) D. Arad, Ph.D. Thesis, Technion-Israel Institute of Technology, Haifa, Israel, 1986.
(b) Y. Apeloig and D. Arad, *J. Am. Chem. Soc.*, **107**, 5285 (1985).
332. Y. Himeshida, H. Kobayashi and T. Sonoda, *J. Am. Chem. Soc.*, **107**, 5286 (1985).
333. Y. Apeloig, M. Karni, A. Stanger, H. Schwarz, T. Drewello and G. Czekay, *J. Chem. Soc., Chem. Commun.*, 989 (1987).
334. S. D. Kahn, C. F. Pau, A. R. Chamberlin and W. J. Hehre, *J. Am. Chem. Soc.*, **109**, 650 (1987).
335. (a) V. J. Shiner, M. W. Ensinger and G. S. Kriz, *J. Am. Chem. Soc.*, **108**, 842 (1986).
(b) V. J. Shiner, M. W. Ensinger and R. D. Rutkowske, *J. Am. Chem. Soc.*, **109**, 804 (1987).
(c) E. R. Davidson and V. J. Shiner, *J. Am. Chem. Soc.*, **108**, 3135 (1986).
(d) N. Koga and K. Morokuma, *J. Am. Chem. Soc.*, **110**, 108 (1988).
336. For a recent experimental review see Ref. 262c, pp. 70–73.
337. (a) M. S. Gordon, *Chem. Phys. Lett.*, **59**, 410 (1978).
(b) K. Ohta, H. Nakatsuji, I. Maeda and T. Yonezawa, *Chem. Phys.*, **67**, 49 (1982).
(c) D. S. Marynick, *J. Chem. Phys.*, **74**, 5186 (1981).
(d) P. R. Bunker and G. Olbrich, *Chem. Phys. Lett.*, **109**, 41 (1984).
338. W. D. Allen and H. F. Schaefer III, *Chem. Phys.*, **108**, 243 (1986).
339. (a) R. L. Morehouse, J. J. Christiansen and W. Gordy, *J. Chem. Phys.*, **45**, 1751 (1966).
(b) G. S. Jackel, J. J. Christiansen and W. Gordy, *J. Chem. Phys.*, **47**, 4274 (1967).
340. A. M. Doncaster and R. Walsh, *Int. J. Chem. Kinet.*, **13**, 503 (1981).
341. K. D. Dobbs and W. J. Hehre, *Organometallics*, **5**, 2057 (1986).
342. T. J. Barton, A. Revis, I. M. T. Davidson, S. Ijadi-Maghsoodi, K. J. Huges and M. S. Gordon, *J. Am. Chem. Soc.*, **108**, 4022 (1986).
343. F. K. Cartledge and R. V. Piccione, *Organometallics*, **3**, 299 (1984).
344. I. Carmichael, *J. Phys. Chem.*, **89**, 4727 (1985).
345. I. Carmichael, *J. Phys. Chem.*, **90**, 2057 (1986).
346. F. Bernardi, A. Bottoni, J. Fossey and J. Sorba, *Tetrahedron*, **42**, 5567 (1986).
347. J. J. Russell, H. S. Rzepa and D. A. Widdowson, *J. Chem. Soc., Chem. Commun.*, 625 (1983).
348. L. P. Davis, L. W. Burggraf, M. S. Gordon and K. K. Baldridge, *J. Am. Chem. Soc.*, **107**, 4415 (1985).
349. G. Frenking and H. F. Schaefer III, *J. Chem. Phys.*, **82**, 4585 (1985).
350. J. V. Ortiz, *J. Am. Chem. Soc.*, **109**, 5072 (1987).
351. (a) F. Keil and R. Ahlrichs, *Chem. Phys.*, **8**, 384 (1975).
(b) R. A. Eades and D. A. Dixon, *J. Chem. Phys.*, **72**, 3309 (1980).
352. A. C. Hopkinson and M. H. Lien, *J. Organomet. Chem.*, **206**, 287 (1981).
353. J. C. Sheldon, J. H. Bowie, C. H. DePuy and R. Damrauer, *J. Am. Chem. Soc.*, **108**, 6794 (1986).
354. (a) M. R. Nimlos and G. B. Ellison, *J. Am. Chem. Soc.*, **108**, 6522 (1986).
(b) J. Novakova, L. Kubelkova and Z. Dolejsek, *J. Catal.*, **97**, 277 (1986).
355. See for example:
(a) D. J. Ager, *J. Org. Chem.*, **49**, 168 (1984).
(b) S. Top, G. Jaouen, B. G. Sayer and M. J. McGlinchey, *J. Am. Chem. Soc.*, **105**, 6426 (1983), and references cited therein.
(c) J. C. Giordan, *J. Am. Chem. Soc.*, **105**, 6544 (1983).
356. C. Glidewell and C. Thomson, *J. Comput. Chem.*, **3**, 495 (1982).
357. (a) L. Pauling, *J. Am. Chem. Soc.*, **53**, 1367 (1931).
(b) R. E. Rundle, *J. Am. Chem. Soc.*, **85**, 112 (1963).
(c) J. I. Musher, *Angew. Chem., Int. Ed. Engl.*, **8**, 54 (1969).
(d) B. M. Gimarc, *Molecular Structure and Bonding—The Quantitative Molecular Orbital Approach*, Academic Press, New York, 1979.

358. S. Gronenrt, R. Glaser and A. Streitwieser, Jr., *J. Am. Chem. Soc.*, in press.
359. A. R. Bassindale and P. G. Taylor, Chapter 13 in this book.
360. D. J. Hajdasz and R. R. Squires, *J. Am. Chem. Soc.*, **108**, 3139 (1986).
361. (a) D. L. Wilhite and L. Spialter, *J. Am. Chem. Soc.*, **95**, 2100 (1973).
(b) P. Baybutt, *Mol. Phys.*, **29**, 389 (1975).
(c) N. M. Vitkovskaya, V. B. Mantsivoda, T. E. Moskovskaya and M. G. Voronkov, *Int. J. Quantum. Chem.*, **17**, 299 (1980).
(d) C. Minot, *Nouv. J. Chim.*, **5**, 319 (1981).
(e) W. Kutzelnigg and J. Wasilewski, *J. Am. Chem. Soc.*, **104**, 953 (1982).
(f) W. Kutzelnigg, H. Wallmeier and J. Wasilewski, *Theor. Chim. Acta*, **51**, 261 (1979).
(g) S. K. Shih, S. D. Peyerimhoff and R. J. Buenker, *J. Chem. Soc., Faraday Trans. 2*, 379 (1979).
(h) U. Brandemark and P. E. M. Siegbahn, *Theor. Chim. Acta*, **66**, 233 (1984).
(i) R. Hoffmann, J. M. Howell and E. L. Muetterties, *J. Am. Chem. Soc.*, **94**, 3047 (1972).
362. A. E. Reed and P. v. R. Schleyer, *Chem. Phys. Lett.*, **133**, 553 (1987).
363. M. S. Gordon, L. P. Davis, L. W. Burggraf and R. Damrauer, *J. Am. Chem. Soc.*, **108**, 7889 (1986).
364. F. Keil and R. Ahlrichs, *J. Am. Chem. Soc.*, **98**, 4787 (1976).
365. M. J. S. Dewar and E. Healy, *Organometallics*, **1**, 1705 (1982).
366. N. T. Anh and C. Minot, *J. Am. Chem. Soc.*, **102**, 103 (1980).
367. (a) J. A. Deiters and R. R. Holmes, *J. Am. Chem. Soc.*, **109**, 1686 (1987).
(b) J. A. Deiters and R. R. Holmes, *J. Am. Chem. Soc.*, **109**, 1692 (1987).
368. (a) J. C. Sheldon, R. N. Hayes, J. H. Bowie and C. H. DePuy, *J. Chem. Soc., Perkin Trans. 2*, 275 (1987).
(b) J. C. Sheldon, R. N. Hayes and J. H. Bowie, *J. Am. Chem. Soc.*, **106**, 7711 (1984).
369. (a) C. J. Mardsen, *Inorg. Chem.*, **22**, 3177 (1983).
(b) J. M. Chehayber, S. T. Nagy and C. S. Lin, *Can J. Chem.*, **62**, 27 (1984).
370. (a) T. J. Lorenz and B. S. Ault, *Inorg. Chem.*, **21**, 1758 (1982).
(b) B. S. Ault, *Inorg. Chem.*, **20**, 2817 (1981).
(c) R. C. Aggarwal and M. Onyszchunk, *Can. J. Chem.*, **41**, 876 (1963).
(d) E. L. Muetterties, *J. Am. Chem. Soc.*, **82**, 1082 (1960).
371. (a) J. B. Collins, J. D. Dill, E. D. Jemmis, Y. Apeloig, P. v. R. Schleyer, R. Seeger and J. A. Pople, *J. Am. Chem. Soc.*, **98**, 5419 (1976).
(b) See Ref. 25, pp. 428–431.
372. (a) For a more recent theoretical contribution see: S. M. Bachrach and A. Streitwieser, Jr., *J. Am. Chem. Soc.*, **106**, 5818 (1984).
(b) For an experimental study see: H. Kaway, J. W. Chinn, Jr. and R. Lagow, *J. Chem. Soc., Chem. Commun.*, 1664 (1984).
373. (a) E. U. Wurthwein and P. v. R. Schleyer, *Angew. Chem., Int. Ed. Engl.*, **18**, 553 (1979).
(b) M. B. Krogh-Jespersen, J. Chandrasekhar, E. U. Wurthwein, J. B. Collins and P. v. R. Schleyer *J. Am. Chem. Soc.*, **102**, 2263 (1980).
(c) See also Ref. 25, pp. 432–436.
374. N. Meyer and G. Nagorsen, *Angew. Chem., Int. Ed. Engl.*, **18**, 551 (1979).
375. D. Schomburg, *Angew. Chem., Int. Ed. Engl.*, **22**, 65 (1983).
376. R. West, *Angew. Chem., Int. Ed. Engl.*, **26**, 1201 (1987).
377. S. K. Shin, K. K. Irikura, J. L. Beauchamp and W. A. Goddard, III, *J. Am. Chem. Soc.*, **110**, 24 (1988).

The Chemistry of Organic Silicon Compounds
Edited by S. Patai and Z. Rappoport

CHAPTER **3**

Structural chemistry of organic silicon compounds

W. S. SHELDRICK
Fachbereich Chemie der Universität Kaiserslautern, Erwin-Schrödinger-Straße, D-6750 Kaiserslautern, Federal Republic of Germany

I. INTRODUCTION

The stereochemistry of silicon in the majority of its organic compounds is similar to that found in the carbon analogues. In such molecules, tetravalent silicon atoms are four-coordinate with a basically tetrahedral geometry, notwithstanding lower energetic barriers to distortion than for the lighter congener. However, in the typical lack of stability of its multiple bonds, in its more limited propensity to catenate and in its ability to expand its coordination sphere to five or even six, silicon is more reminiscent of its heavier congeners. It is therefore not surprising that many of the more intensively studied aspects of the structural chemistry of organic silicon compounds were initially stimulated by observations of stereochemical differences rather than similarities between analogous molecules containing silicon or carbon. For instance, Hedberg's finding[1] in 1955 that the central nitrogen in trisilylamine $(SiH_3)_3N$ is trigonal planar rather than pyramidal as in trimethylamine[2] led to an explosion in the number of structural investigations in the 1960s and early 1970s on compounds containing silicon bonded to the lighter elements of groups 15 and 16. In addition to the planarity of nitrogen, which was confirmed for a range of silylamines, characteristically wide Si–O–Si angles (140–180°) were established for silyl ethers. These structural parameters were in striking contrast to those observed for the corresponding methyl derivatives. An illustrative example is provided by a comparison of the bond angles at oxygen in the ethers $(CH_3)_2O$[3] and $(SiH_3)_2O$[4], which are respectively 111.5(15) and 144.1(9)°. The systematic investigation of potential electronic and steric factors controlling the oxygen geometry in silyl ethers finally culminated in the finding that the Si–O–Si angle in $(SiPh_3)_2O$ is rigorously linear in the solid state[5]. The results of earlier studies on the stereochemistry of elements of Groups 15 and 16 bonded to silicon are reviewed in two articles which appeared in 1973[6,7].

Although the fund of crystal structure determinations on compounds of tetracoordinate silicon has continued to burgeon rapidly in the past decade, it may surely be contended that the most important recent advances in the field of structural chemistry have involved derivatives containing silicon in lower or higher coordination numbers. The long experimental search for molecules with doubly bonded silicon, which are stable at room temperature, was finally rewarded in 1981 by the synthesis of the first such disilene (Si=Si bond) by West, Fink and Michl[8] and the first such silene (Si=C bond) by Brook and coworkers[9]. As we will discuss in Section III, a structural chemistry of tricoordinate silicon has begun to emerge in the past five years[10]. The available disilene and silene structures have recently been augmented by the first structural data on compounds containing Si=N bonds[11,12]. In contrast, only one stable organic derivative of divalent silicon is presently known, namely decamethylsilicocene, which was first synthesized and structurally characterised in 1986[13].

Unlike carbon, tetravalent silicon displays a pronounced acceptor ability and is capable of extending its valence shell to contain 10 or even 12 electrons. Such penta- or hexacoordinated species may either be anions or neutral coordination complexes in which the valence shell expansion is achieved by either intra- or intermolecular interaction with donor atoms belonging, as a rule, to the first (N, O and F) and second (P, S and Cl) rows. However, the Lewis acid strength of tetravalent silicon is weaker than that of the heavier members of Group 14, as dramatically illustrated by a comparison of the solid state structures of the molecules $(CH_3)_3MCN$ (M = Si, Ge or Sn), for which respective intermolecular M · · · N contacts of length 3.66[14], 3.57[15] and 2.49 Å[16] were found. In the silicon derivative, this distance corresponds to the sum of the van der Waals radii of M and N; it is 0.15 Å shorter for M = Ge. Whereas the silicon atoms are tetrahedrally coordinated in effectively discrete molecules $(CH_3)_3SiCN$, the environment of the tin atoms in the polymeric chains $[(CH_3)_3SnCN]_n$ is trigonal bipyramidal. However, two examples of intermolecular association have been reported for silicon compounds in which full pentacoordination is achieved. This is the case for the $H_3SiN(CH_3)_2$ pentamer[17] and the $F_3SiN{=}P(CH_3)_3$ dimer[18], both of which display trigonal bipyramidal coordination of the silicon atoms. In the latter example, the axial Si–N distance is 1.857(4) Å, which is only 0.06 Å longer than the Schomaker–Stevenson corrected sum of the covalent radii for these atoms[19]. The incorporation of a central silicon atom into a five-membered ring has often been used to stabilize a pentacoordinate geometry. The silatranes(I), which contain an axial nitrogen connected via three $(CH_2)_2$ bridges to the

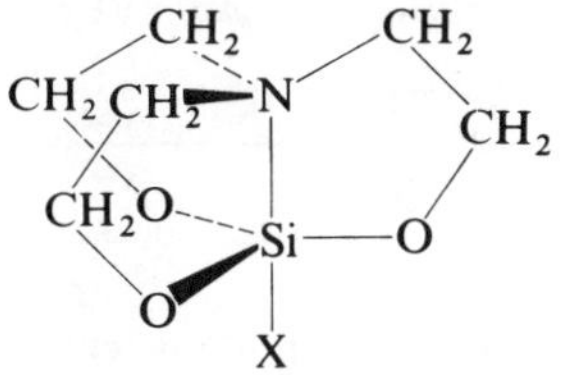

(I)

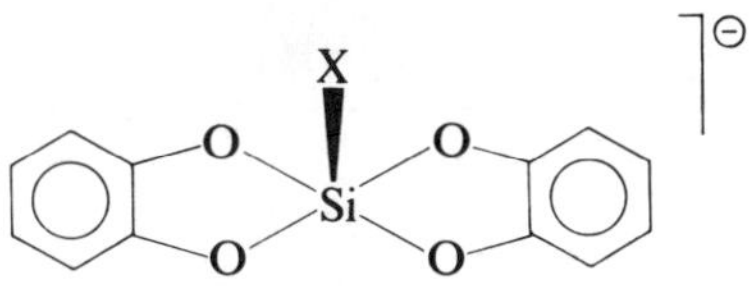

(II)

oxygen atoms of the trigonal bipyramidal equatorial plane, currently represent the class of pentacoordinate silicon derivatives for which the most structural information is available.

The principles of silatrane geometry derived from 29 molecular structures have been summarized in a recent review article[20]. Interest in this class of compounds has also been stimulated by their biological activity. The fund of experimental data on coordinative interactions to silicon in the range between 'normal' single bonds and van der Waals contacts is now so extensive that crystallographic mapping of the geometrical changes which occur on going from tetrahedral to trigonal bipyramidal coordination is possible. A juxtaposition of 46 such independent molecular fragments was interpreted as simulating the reaction pathway for S_N2 substitution reactions at tetrahedral silicon[21].

Indeed it is the variability in the environment of pentacoordinate silicon atoms which bestows this area of structural chemistry with its particular fascination. In analogy to the isoelectronic phosphoranes, rapid intramolecular ligand exchange between axial and equatorial trigonal bipyramidal sites takes place readily for many pentacoordinate silicate(IV) anions. For instance, the ^{19}F NMR spectrum of SiF_5^- contains only one fluorine resonance at room temperature. The Berry pseudorotation pathway[22], which involves a square (or rectangular) pyramid as intermediate, has been widely advanced to account for such phenomena. Theoretical calculations have indicated that the trigonal bipyramidal geometry should be energetically slightly more favourable for acyclic derivatives than the alternative square pyramid. It was, however, demonstrated in the 1970s that the latter geometry could be stabilized for solid state phosphoranes by the inclusion of phosphorus in two strained unsaturated five-membered rings containing similar atoms (generally O) in any one ring[23,24]. This principle has recently been extended to silicon and examples of rectangular pyramidal geometries have been reported for spirobicyclic silicate(IV) anions of type II or their derivatives, e.g. X = *t*-Bu[25]. A comprehensive review of X-ray structural and NMR-spectroscopic studies on penta- and hexacoordinate silicon compounds appeared in 1986[26].

TABLE 1. A comparison of the electronic properties of carbon and silicon

	C	Si
Electronic structure	[He]$2s^22p^2$	[Ne]$3s^23p^2$
Ionisation energies (eV)		
(I)	11.256	8.149
(II)	24.375	16.339
(III)	47.869	33.458
(IV)	64.474	45.129
Average distance of valence electrons from the nucleus[27]		
$\langle r \rangle_s$ (Å)	0.84	1.16
$\langle r \rangle_p$ (Å)	0.92	1.48
Electron affinity (eV)	1.268	1.385
Dipole polarizability[28](a.u.)[a]	11.8	36.3
Electronegativity (Allred-Rochow)	2.50	1.74
Covalent radius (Pauling[19]) (Å)	0.772	1.17
Hard-sphere radius (Glidewell[29]) (Å)	1.25	1.55
Van der Waals radius (Bondi[30]) (Å)	1.70	2.10

[a] 1. a.u. = 0.1482 $Å^3$

Any account of the structural chemistry of organic silicon compounds will emphasise the differences between carbon and silicon. Before we turn to individual classes of compounds, it is therefore meaningful to compare the relevant electronic properties of the two elements, as listed in Table 1. From a stereochemical point of view, the following differences are of major importance:

(1) As a result of increased shielding of the nucleus in the second row element, the electronegativity of silicon is lower than that of carbon.

(2) The silicon atom is much larger than the carbon atom as demonstrated by their respective covalent, hard-sphere and van der Waals radii.

(3) In contrast to carbon for which the $\langle r \rangle_s$ and $\langle r \rangle_p$ values are very similar, the silicon 3p-valence electrons are on average much further from the nucleus than the 3s-electrons This is a result of Fermi repulsion from the occupied 2p-orbitals[28].

(4) The 3d-orbitals are of low enough energy to be available for bonds to silicon but not to carbon.

As a result of the lower electronegativity and greater size of silicon, its valence electrons are less tightly bound than those of carbon, as evidenced by the lower ionization energies listed in Table 1. The marked increase in the $\langle r \rangle_p/\langle r \rangle_s$ ratio means that radial overlap of the s- and p-valence orbitals will be less favourable for silicon than for carbon. It follows that the degree of isovalent hybridisation for these orbitals will be lower in compounds of the heavier element. In the following sections we will see how one or more of the first three factors are responsible for what are often dramatic changes in the geometrical parameters of analogous molecules on going from carbon to silicon.

The bonding role of the silicon 3d-orbitals has fueled the fires of controversy for the past 30 years. The interaction of nitrogen 2p-orbitals normal to the NSi_3 plane in $(SiH_3)_3N$ with silicon 3d-orbitals of appropriate symmetry provided a simple and attractive explanation for both the planarity of the central atom and the shortness of the Si–N bonds [1.734(2) Å [31]] in comparison to the Schomaker–Stevenson corrected sum of the covalent radii (1.80 Å). Such (p-d)π contributions have regularly been advanced to account for the geometrical parameters in molecules containing tetravalent silicon bonded to the very electronegative elements N, O and F. The Si–X (X = N, O, F) distances in such derivatives are, as a rule, 0.05–0.15 Å shorter than predicted for such bonds on the basis of their covalent radii. However recent *ab initio* calculations provide no support for the contention that the bonds to silicon are strengthened or that bond angles at nitrogen or oxygen are widened by such p-π donation into vacant 3d-orbitals[32]. Such calculations do, however, demonstrate that the addition of d-functions to second row molecules leads to better agreement with experimentally available structures.

The earliest descriptions of bonding in penta- and hexacoordinate silicon derivatives always assumed significant involvement of the silicon 3d-orbitals. In the model, originally suggested by Pauling[33], these orbitals participate in the construction of sp^3d^2-hybrids. In total contrast to this approach was that of electron-rich three-centre bonding pioneered by Rundle for such species[34]. The equatorial Si–X bonds in an anion SiX_5^- may be regarded as normal in this model, while the delocalised three-centre four-electron orbitals are constructed for the axial Si–X bonding from the silicon $3p_z$ and the axial ligand s- and p-functions. Modern theoretical studies indicate that a reduction in steric contacts between ligands X in species MX_5^- (M = C, Si) on going from the smaller carbon to the larger silicon atom is mainly responsible for the observation of valence shell expansion in anions containing the latter atom[28]. These calculations demonstrate that the availability of 3d-orbitals is of lesser importance than was originally believed and emphasise that the Rundle model should be regarded as providing a better, when albeit simplified, description of bonding in pentacoordinate silicon species.

The aim of the present article is to provide a detailed description of all important aspects of the structural chemistry of organic silicon compounds. Experimental results from microwave spectroscopy (MW), gas phase electron diffraction (ED), single crystal X-ray diffraction (XRD) and neutron diffraction (ND) will be considered. Emphasis will be placed on the establishment of systematic structural trends and on a comparison with theoretical predictions, although the latter will not be discussed in detail. However, simple bonding models, including numerous examples with a limited theoretical basis, have always played both explanatory and predictive roles in the field of structural chemistry. Where apposite, this tradition will be continued in the present review. Although all major journals have been covered up to the end of 1986, no attempt has been made to achieve comprehensive coverage, nor indeed, in view of the fact that the fund of molecular structures of organic silicon derivatives is being currently extended by approximately 100 structures annually, would such a goal be advantageous. However, particular attention has been concentrated on those aspects which have produced the most important advances of the past decade. This means that, in addition to higher and lower coordination numbers, extensive coverage is provided for topics such as polysilanes, metal–silicon bonds and sila-drugs.

II. DIVALENT SILICON

As a result of the larger $\langle r\rangle_p/\langle r\rangle_s$ ratio, silicon prefers to place non-bonding electrons in atomic orbitals with a greater degree of s character than that observed for the analogous carbon derivatives[28,35]. The experimental value[36] of 92.5° for the H–Si–H angle in silylene SiH_2 is in accordance with that expected for divalent silicon with s^2p^2 configuration. *Ab initio* calculations using a small split valence 3-21G* basis set have yielded values of respectively 93.4° and 104.7° for the central angle in SiH_2 and methylene CH_2 in this singlet $^1A_1$ state[35], indicating that isovalent hybridization must be significant in the latter molecule, as in NH_3 or H_2O. In contrast to silylene, methylene is known to have a triplet ground state[37], for which an experimental H–C–H angle of 133.9° has been reported[38]. Whereas the energy gain on forming a triplet is large enough in CH_2 to compensate for the difference of 113 kJ mol^{-1} between the highest singly occupied orbitals a_1 and b_1 in the bent $^3B_1$ state, this is not the case for SiH_2, where this barrier is much larger (218 kJ mol^{-1})[35]. As a result the $5a_1$ orbital in SiH_2 is doubly occupied. The reader is refered to Section IV.A for a detailed discussion of the H–A–H angle in species AH_2. Experimental values of 100.76(2)° (MW), 102.8(6)° (ED) and 102.7(3)° (ED) are now available for the bond angle at silicon in SiF_2[39], $SiCl_2$[40] and $SiBr_2$[40]. When taken together with the value of 92.5° for SiH_2, these results indicate that, in addition to electronic effects, increasing ligand–ligand repulsion may also play an important role in determining the molecular geometry of such divalent silicon species.

As mentioned in the Introduction, decamethylsilicocene $(\eta^5\text{-}Me_5C_5)_2Si$ (**1**) is the only known organic derivative of divalent silicon, which is stable at room temperature. Surprisingly, the unit cell of **1** contains two independent conformers in a 1:2 ratio[13]. Both exhibit planar cyclopentadienyl (Cp) rings with a staggered conformation, but whereas the first molecule A contains a crystallographic centre of symmetry with strictly parallel Cp-rings, these are opened relative to one another in the second molecule B (Figure 1a).

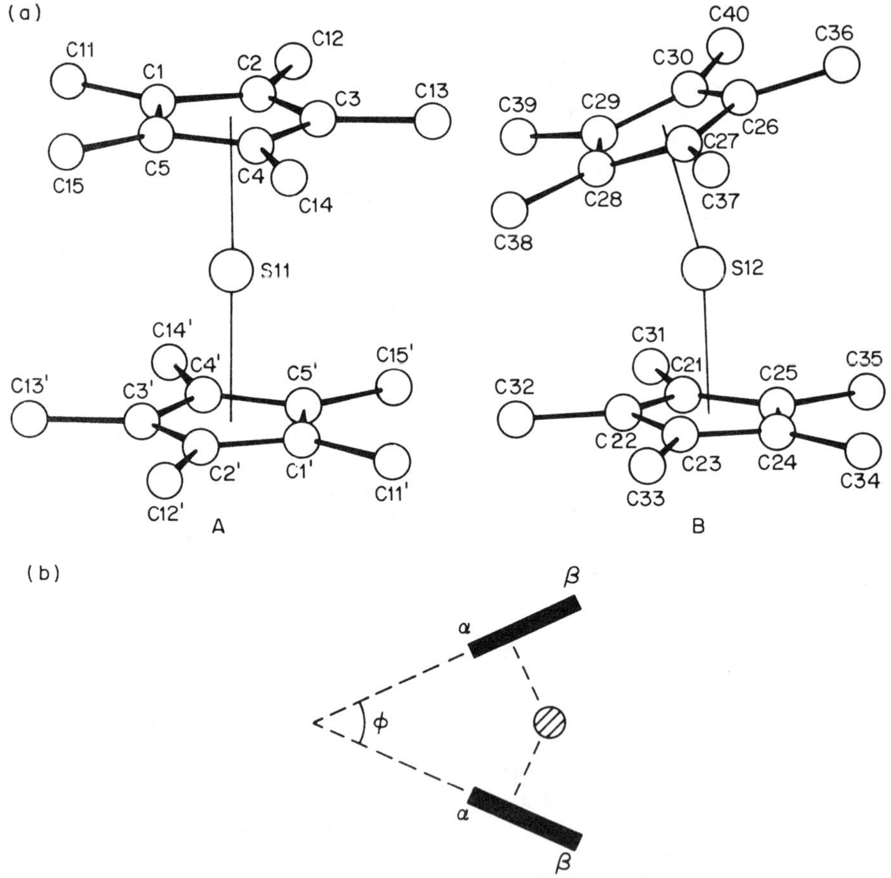

FIGURE 1. (a) Structures of the two independent molecules A and B of $(\eta^5\text{-}Me_5C_5)_2Si$ **1**. Reproduced by permission of VCH Verlagsgesellschaft, Weinheim, from Reference 13. (b) Sandwich bending and ring slippage

The interplanar angle ϕ, as defined in Figure 1(b), is 25°. A similar sandwich bending has been found in the analogous germanium and tin derivatives, which display respective ϕ angles of 22° (ED)[41] and 36° (XRD)[42]. The fact that these interplanar angles are significantly smaller for the pentamethylated species than those of 50° (XRD)[43] and 47° (XRD)[44] for germanocene and stannocene $(\eta^5\text{-}C_5H_5)_2M$ (M=Ge, Sn) has been interpreted as being in accordance with an increase in non-bonded interactions between the ligands, caused by the methyl substituents in $(\eta^5\text{-}Me_5C_5)_2M$. Extended Hückel MO calculations[42] on stannocene have demonstrated that the HOMO ($2a_{1g}$) is strongly antibonding with respect to the Sn–Cp interaction for a linear D_{5d} isomer. The experimentally observed bent sandwich structure may readily be explained by an application of the second-order Jahn–Teller effect. Upon opening of the Cp-rings, the HOMO can mix with one of the higher unoccupied antibonding $2e_{1u}$ orbitals. In terms of perturbation theory, this leads to a stabilization of the lower-lying orbital as a result of a reduction in Sn–Cp antibonding through mixing with an Sn p-orbital. In simplified terms, the resultant $3a_1$ orbital in a bent sandwich may be regarded as a non-bonding electron pair. An energy minimum is reached at $\phi = 60°$ for stannocene. It is interesting to note that, according to MNDO calculations carried out by Glidewell[45], the most stable form of $(C_5H_5)_2Si$ should be the bis-monohapto isomer. However, an underestimation of the stability of non-classical species appears to be a general failing of the MNDO method—for instance, a similar result has been reported for stannocene[46]. On the basis of these calculations, the HOMO of D_{5d} $(\eta^5\text{-}C_5H_5)_2Si$ is an e_{1g} orbital, which is non-bonding with respect to the silicon s and p orbitals. As a result of the larger energy gap between the antibonding a_{1g} and e_{1u} orbitals, sandwich bending will be expected to be less favourable in electronic terms for silicocene than for its heavier analogues. However, opening of the Cp-rings to an angle of 30° was calculated as requiring a very modest energy input (22.6 kJ mol^{-1}), indicating that the adoption of two different conformers by decamethylsilicocene must be influenced by lattice forces. As the interplanar angle ϕ will be controlled by an interplay of electronic and steric factors with such intermolecular forces, an overinterpretation of exact experimental values should be avoided. Mixing of the a_{1g} and e_{1u} orbitals as depicted in Figure 2 would be expected to lead to a shift in the position of the central atom towards the C^α atoms (Figure 1b). Ring slippage is indeed observed for molecule **B** of **1** and for the analogous tin[42] and lead[44] derivatives. Si–C bond distances in molecule **B** lie in the range 2.324(8)–2.541(7) Å; an average value of 2.42(1) Å is found for the linear molecule **A**. Further complexes of silicon will be discussed in Section VII.

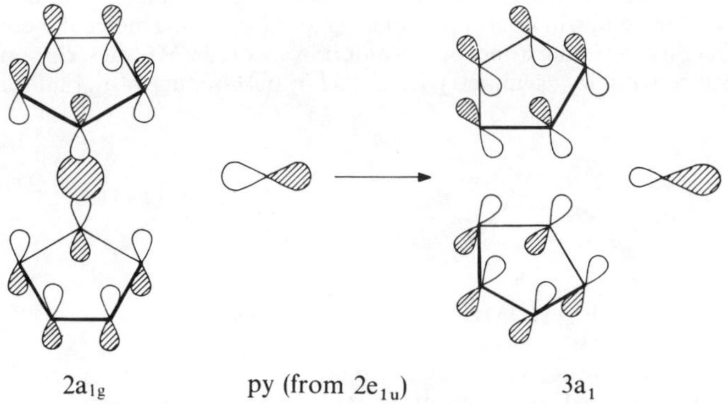

FIGURE 2. Stabilization of the antibonding a_{1g} orbital upon sandwich bending for group 14 derivatives $(\eta^5\text{-}C_5H_5)_2M$

III. DOUBLY BONDED SILICON

A. Double Bonds to Silicon

Through the use of bulky substituents a number of stable disilenes have been prepared in the recent past and X-ray diffraction studies are now available on three compounds containing Si=Si double bonds. Several recent review articles are relevant to this Section. The work of West and Michl in this field was briefly summarized in 1984[47, 48] and 1985[49]. Cowley has considered structural and theoretical aspects of disilenes, as part of two reviews of double bonds between heavier elements of groups 14 and 15[50, 51]. Finally a very extensive review of all aspects of multiple bonding to silicon appeared in 1985[52].

TABLE 2. Structural parameters for group 14 dimetallenes M_2R_4

M;R in M_2R_4	$\Sigma(M)^a$(deg)	θ^b(deg)	τ^c(deg)	$d(M{=}M)$(Å)	% shortening^d	Ref.
C; 2,4,6-$Me_3C_6H_2$	360	0	0	1.364(5)	12	53
(**2**) Si; 2,4,6-$Me_3C_6H_2$	356.3	18	6.5	2.160(1)	8	54, 55
(**3**) Si; 2,6-$Et_2C_6H_3$	360.0	0	10	2.140(3)	9	56
(**4**) Si; *t*-Bu; 2,4,6-$Me_3C_6H_2{}^e$	359.9	0	0	2.143(3)	8	55
Ge; 2,6-$Et_2C_6H_3$	358.4	15	11	2.213(3)	9	57
(**5**) Ge; $CH(SiMe_3)_2$	348.5	32	0	2.347(2)	4	58
(**6**) Sn; $CH(SiMe_3)_2$	340.6	41	0	2.768(1)	1	58

^{a} $\Sigma(M)$ = sum of angles at M.
^{b} Fold angle θ (Figure 3a).
^{c} Twist angle $\tau = \frac{1}{2}(\alpha_2 - \alpha_1)$ as defined in Figure 3b.
^{d} % shortening of the M–M distance in comparison to the Pauling single-bond distances: d(C–C) = 1.544, d(Si–Si) = 2.34, d(Ge–Ge) = 2.44, d(Sn–Sn) = 2.80 Å.
^{e} **4** is *Z*-RR'Si=SiRR' with R = *t*-Bu, R' = 2,4,6-$Me_3C_6H_2$.

Table 2 lists structural parameters for dimetallenes M_2R_4 of group 14, which have been characterised by X-ray diffraction. As these all carry bulky substituents, it is evident that this must be a factor in determining their kinetic stability with respect to oligomerization. The relative weakness of the Si=Si double bond in silicon analogues of the olefins is a result of the fact that for silicon the valence p-electrons are, on average, much further from the nucleus than the s-electrons, in contrast to carbon for which the distance is similar. Less efficient overlap of the valence p-orbitals would be expected to lead to progressive weakening of the π bonding component in olefin analogues on passing from carbon to tin. This trend will be accompanied by a change in the molecular configuration, which may readily be understood by considering a simple MO description. In planar M_2H_4 molecules with D_{2h} symmetry (e.g. C_2H_4) the bonding $\pi(b_{1u})$ and antibonding

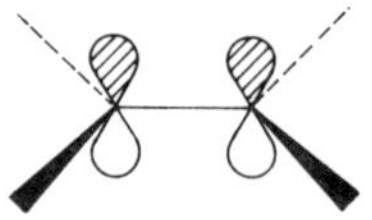

π (b_{1u}) HOMO

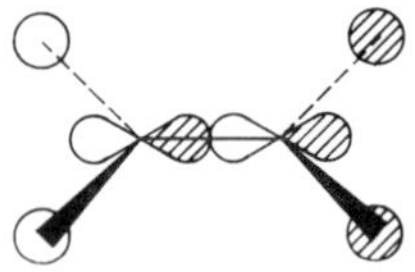

σ^* (b_{3u})

$\sigma^*(b_{3u})$ orbitals are orthogonal so that orbital interaction is precluded. Upon *trans*-folding (sometimes refered to as anti-pyramidalization) of the molecule, which yields a C_{2h} configuration (Figure 3a), these orbitals begin to mix, leading to a stabilization of the lower orbital (HOMO) with loss of conventional π-bonding[51, 58]. The π bonding orbital is

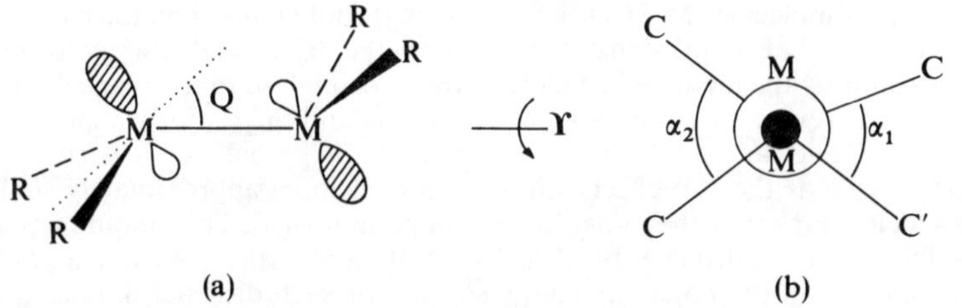

FIGURE 3. (a) *Trans*-folded structure of group 14 dimetallanes M_2R_4 (M=Si, Ge, Sn) with definition of the fold angle θ. (b) Newman projection along the M–M bond. The twist angle $\tau = \frac{1}{2}(\alpha_2 - \alpha_1)$

transformed into a delocalized sp hybrid orbital on the metal atoms. The energy gap between b_{1u} and b_{3u} orbitals decreases upon descending group 14, which means that C_{2h} distortion should become progressively more favourable, as confirmed by the experimental fold angles θ in Table 2. The second highest occupied orbital in a planar M_2H_4 molecule is the M–M σ bonding orbital with A_g symmetry. Upon *trans*-folding, the energy of this orbital will increase, mainly as a result of the loss of M–M σ bonding. The observed degree of *trans*-folding should be controlled by the relative energy change for the two highest occupied orbitals upon distortion. Figure 4 demonstrates the variation of

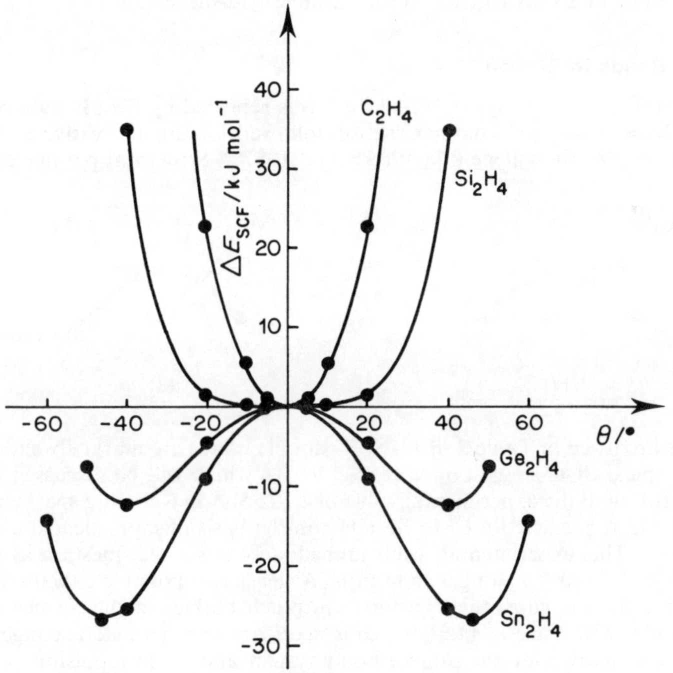

FIGURE 4. Variation of the SCF energies for group 14 dimetallenes M_2H_4 with the fold angle θ. Reproduced by permission of the Royal Society of Chemistry from Reference 58.

SCF energies for molecules M_2H_4 as a function of the fold angle θ on the basis of *ab initio* MO calculations[58]. For ethylene, the drop in the b_u orbital energy is more than compensated for by the increase in energy of the a_g orbital, so that a planar structure will be adopted. The reverse situation applies for Ge_2H_4 and Sn_2H_4 for which fold angles of respectively 40° and 46° were predicted. The total energy balance is particularly interesting for disilene as the two effects will be expected to be approximately similar when $\theta < 20°$, which means that the molecule should be non-rigid. This finding has also been confirmed for alkylated disilenes by *ab initio* calculations with a split valence 6-31G basis set[59]. The flatness of the potential energy surface for such disilenes implies that crystal packing forces may influence their molecular structures in the solid state and the limited experimental data available are in accordance with this inference. Whereas a modest degree of *trans*-folding ($\theta = 18°$) is found for disilene **2**, the geometry at the silicon atoms in **3** and **4** is planar. Indeed, this condition applies to the M_2C_4-skeleton as a whole in **4**. For disilenes **2** and **3**, the twist angle τ is non-zero. This angle will be governed by a variety of factors, including the magnitude of the barrier to torsional motion about the Si=Si double bond, substituent bulk and possibly lattice forces.

The silicon–silicon bond distances in **2**–**4** are much shorter than typical single bond lengths in disilanes, which lie in the range 2.33–2.38 Å (see Section IV.B.2 for details). The degree of shortening (8–9%) in these disilenes with respect to the single bond distance is much greater than for the germanium and tin derivatives **5** (4%) and **6** (1%). As would be expected from the MO description given above, a striking correlation is found for the group 14 dimetallenes between the percentage bond shortening and the fold angle θ. The experimental values for the Si=Si double bond are in excellent agreement with the theoretical value of 2.13 Å calculated for tetramethyldisilene[59].

B. Double Bonds to Carbon

The preparation of the first stable silene **7** was reported by Brook and coworkers in 1981[9]. An X-ray structural characterisation followed for this derivative in 1982[60] and 1985[61] and for a further silene **8** in 1985[62] and 1987[63]. Structural parameters for these

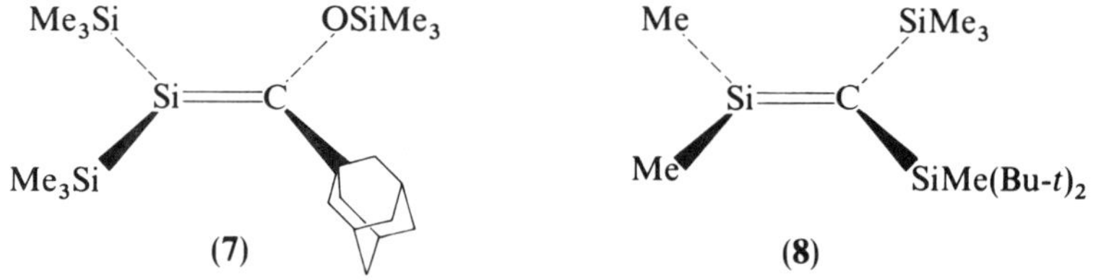

(7) **(8)**

compounds are listed in Table 3. Both Si=C double bonds are markedly shortened with respect to typical Si–C single bonds (1.86–1.91 Å), which will be discussed in detail in Section IV.B.1, with the experimental value of 1.702(5) Å in **8** coming spectacularly close to that of 1.692 Å predicted for $Me_2Si{=}CH_2$ on the basis of sophisticated *ab initio* MO calculations[64]. This experimental result immediately poses the question as to why the Si=C bond in **7** is 0.062 Å longer than in **8**? At least two possible explanations can be proposed. Firstly, the substituent groups are much bulkier in the former silene **7**, as manifest in the twist angle of 14.6°, which is necessary to relieve steric congestion. Both the loss of coplanarity for the double bond system and steric repulsions between the silicon and carbon substituents might be expected to lead to elongation of the Si=C bond. However, Apeloig and Karni have calculated that a 16° twisting about the double bond in $H_2Si{=}CH_2$ results only in a relatively modest lengthening of 0.015 Å for the central

TABLE 3. Structural parameters for silenes and their donor adducts

Compound	$\Sigma(Si)^a$(deg)	$\Sigma(C)^a$(deg)	τ^b(deg)	$d(Si=C)$(Å)	Ref.
7	360.0	359.7	14.6	1.764(3)	60, 61
8	360.0	360.0	1.6	1.702(5)	62, 63
9	348.7	358.8	36.7	1.747(5)	67
10^c	342.2	360.0	−89.2	1.782(6)	68
	341.7	360.0	−89.0	1.777(6)	

^{a} Σ(Si) and Σ(C) are the sum of angles at Si and C respectively. In calculating Σ(Si) for **9** and **10**, the donor atom is ignored.
^{b} The angle of twist $\tau = \frac{1}{2}(\alpha_1 + \alpha_2)$ as defined in Figure 6.
^{c} Two independent molecules are present in the unit cell of **10**.

bond[65,66]. Electronic effects are much more important. The Si=C double bond in silene is polarized with an atomic charge of −0.771 on carbon and +0.550 on silicon[35]. Substituents which decrease the polarization and hence the degree of bond ionicity will be expected to lengthen this bond. This is precisely the case in Brook's silene **7**, in which the negative charge on carbon will be reduced because of its attachment to the very electronegative oxygen atom in the Me_3SiO substituent. Such an electron-withdrawing atom will also produce a contraction in the size of the carbon p-orbitals, which are already much smaller than those of silicon. As a result, the degree of overlap with the more diffuse silicon p-orbitals will be further reduced. For example, d(Si=C) in $H_2Si{=}CH(OSiH_3)$ is calculated to be 0.031 Å longer than in silene itself[65,66].

The polar nature of the silicon–carbon bond is reflected in the fact that d(Si=C) in **8** is significantly shorter than the sum of the double-bond radii of the participant atoms (1.74 Å)[19]. The charge distribution in the Si=C bond leads to the ability of silene **8** to form surprisingly stable adducts with weak Lewis bases, in which the unsaturated silicon atom is coordinated. An X-ray structure is available for the adduct **9** with THF[67]. As may

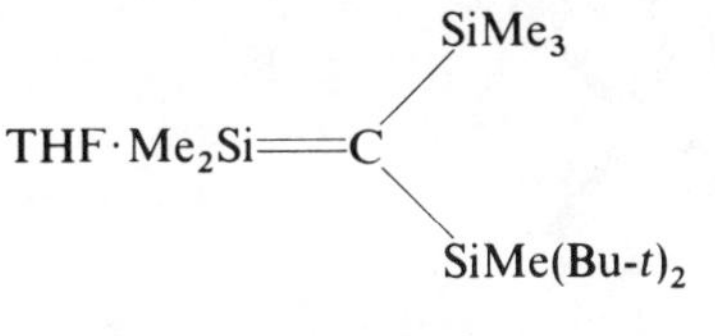

(9)

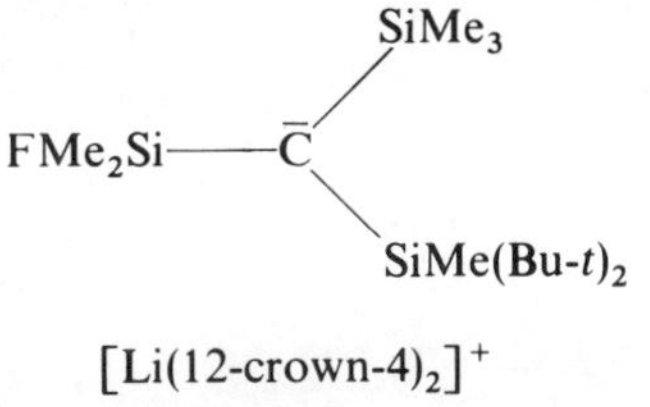

(10)

be seen from Figure 5a, the coordination geometry at silicon is no longer planar. The distortion towards a tetrahedral environment may be gauged from the value of 348.7° for Σ(Si), the sum of the C–Si–C angles (Table 3). For an ideal tetrahedral coordination geometry Σ(Si) would be 328.5°. The silicon–carbon distance of 1.747(5) Å in **9** is still drastically shorter than that for a typical single bond. In fact, it is shorter than the Si=C double bond in silene **7**. Taken together with the remarkably long Si–O bond length of 1.878(4) Å (see Section IV.D.1), this suggests that **9** may be regarded as a 'frozen' intermediate, in which the reaction coordinate of donor approach to an unsaturated silicon atom is stopped short of adoption of full tetrahedral coordination. This impression is reinforced by the structure of the carbanion **10**[68], which may formally be regarded as an adduct of silene **8** with the much stronger Lewis base F^-. Inspection of the values of

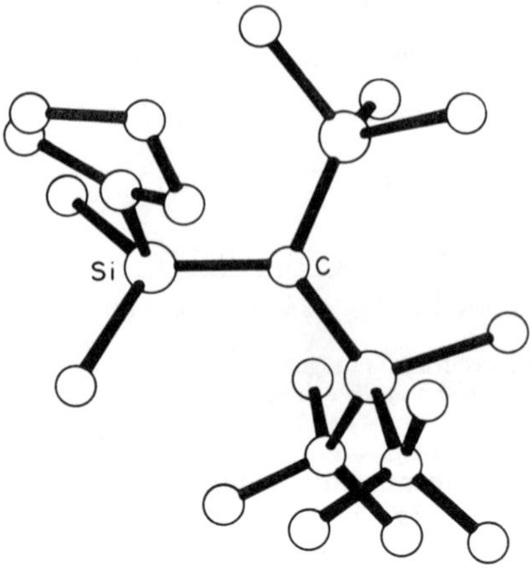

(9)

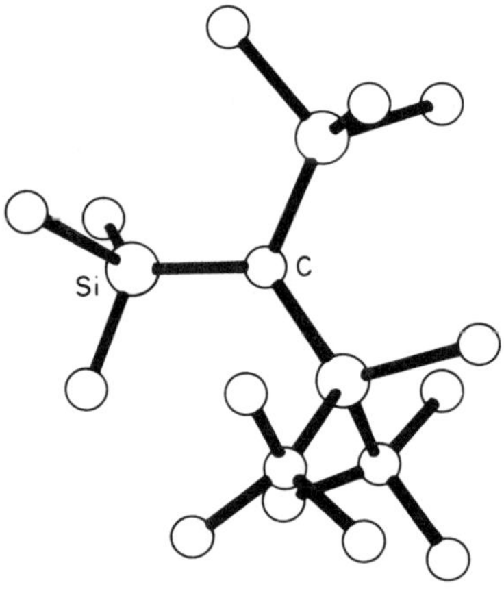

(10)

FIGURE 5. Molecular structures of silene adducts **9** and **10**. Redrawn from Reference 68

Σ(Si) and *d*(Si–C) in Table 3 indicates that donor addition is much closer to 'completion' in this anion. However, the Si–F distances of 1.640(3) and 1.647(4) Å in the two independent anions found in the unit cell are still somewhat longer than those found for typical Si–F bonds, e.g. 1.597(5) Å in CH_3SiH_2F[69]. The experimentally observed structures for **9** and **10** provide ample support for a zwitterionic/π-bonding model proposed by Wiberg and coworkers[67,68], which is essentially a resonance formulation between a no-bond structure **A** and a zwitterionic structure **B**.

As the Lewis basicity of the donor D increases (i.e. on going from THF to F^-), the relative importance of the zwitterionic form will be expected to increase. Although the former Si=C bond will be lengthened in the zwitterionic structure it should still be

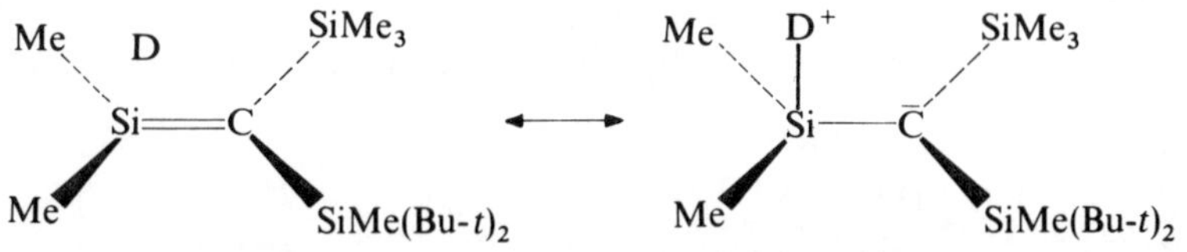

(A) **(B)**

significantly shorter than typical Si–C single bonds because of the high degree of bond ionicity. The experimental values for **8**, **9** and **10** confirm the predicted trend (Table 3). Because of the increasing negative charge on the carbon atom in this series, the C–Si bonds to the carbon substituents would be expected to shorten and this is, indeed, the case. For instance, the following values are observed for the C–Si(Me_3) bonds: 1.865(5) **8**, 1.835(5) **9** and respectively 1.812(6) and 1.796(5) Å for the two independent anions in **10**. The equivalent values for the SiMe(Bu-*t*)$_2$ substituent are 1.890(5) **8**, 1.836(5) **9** and respectively 1.806(6) and 1.822(5) Å for **10**. This means that in **10** these bonds are only 0.021–0.045 Å longer than the central silicon–carbon bond.

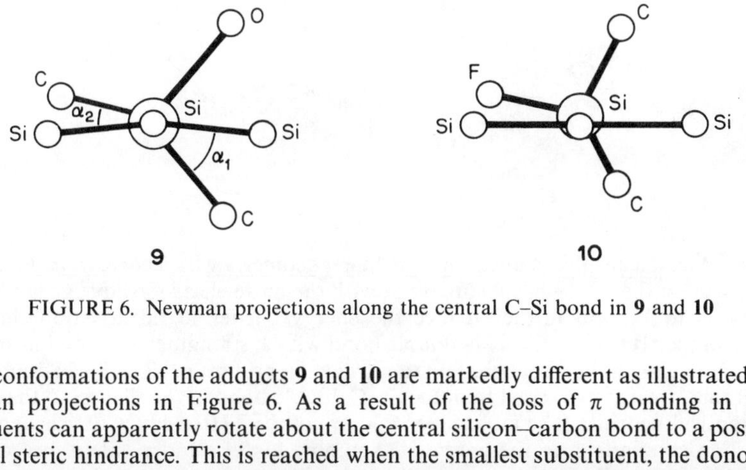

9 **10**

FIGURE 6. Newman projections along the central C–Si bond in **9** and **10**

The conformations of the adducts **9** and **10** are markedly different as illustrated by the Newman projections in Figure 6. As a result of the loss of π bonding in **10**, the substituents can apparently rotate about the central silicon–carbon bond to a position of minimal steric hindrance. This is reached when the smallest substituent, the donor atom F, adopts an essentially eclipsed position with respect to a bulky SiMe(Bu-*t*)$_2$ group. In contrast, the bulkier THF molecule in adduct **9** prefers a position far away from the carbon substituents. However, it should be noted that the energy barrier to rotation about the central silicon–carbon bond will be much larger in **9**, for which a pronounced degree of double bond character is retained.

C. Double Bonds to Nitrogen

The first stable silaketimine **11** was reported by Wiberg and coworkers in 1985[70] with an X-ray structural analysis following in 1986[11]. Contrasting distances are found for the two silicon–nitrogen bonds. Whereas the single bond distance of 1.695(3) Å is close to the typical range (1.70–1.76 Å) for such bonds, a much shorter distance of 1.568(3) Å is found for the first experimentally characterized Si–N double bond. The Si–N–Si angle is nearly linear [177.8(2)°]. It is instructive to compare these results with recent *ab initio* calculations[71]. Using a 6-31G* basis set with d-functions on both N and Si, distances of

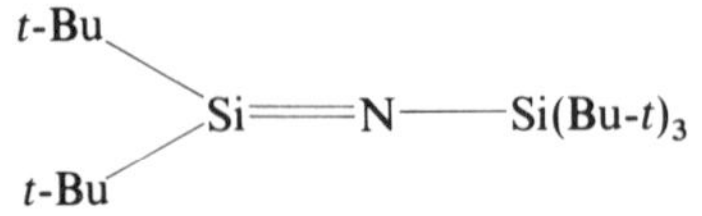

(11)

(12)

respectively 1.688 and 1.549 Å were determined for the silicon–nitrogen single and double bonds in $H_2Si{=}NSiH_3$, with an Si–N–Si angle of 175.6°. Electronic and not steric effects are clearly responsible for the wide Si–N–Si angle. In the unsubstituted silaketimine, an angle of only 126.6° was found for the same basis set. However, the potential energy surface for widening is very flat with a barrier to linearisation of only 13.8 kJ mol^{-1}, contrasting with the larger value of 136.8 kJ mol^{-1} calculated for $H_2C{=}NH$. The difference is due mainly to the larger electronegativity gap between silicon and nitrogen in comparison to carbon and nitrogen. Upon bending of a linear molecule $H_2M{=}N{-}H$, where M belongs to group 14, the HOMO, which effectively contains the nitrogen lone pair, and a σ^* orbital of higher energy can interact, leading to the stabilisation of a bent geometry provided that the energy gap between the two orbitals is small, e.g. M = C. As the electronegativity of M decreases (e.g. on going from C to Si), the energy gap will

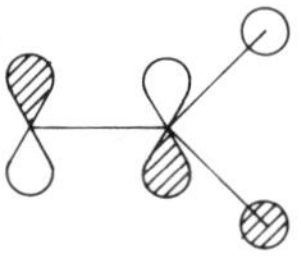

HOMO

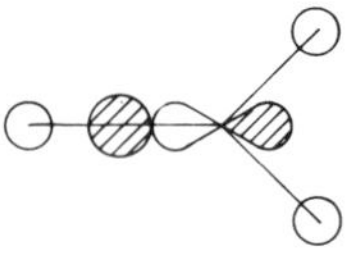

σ^*

increase, which means that a linear or near-linear geometry will become more favourable. Replacement of the nitrogen substituent H with the more electropositive group SiH_3, or $Si(Bu\text{-}t)_3$ as in **11**, will further reduce the energy barrier to linearisation through a lowering of the HOMO. The Si=N double bond will be strengthened upon linearisation, due to an increase in the s contribution to nitrogen hybridisation. The Si=N double bond is even more strongly polarised than the Si=C double bond in the silenes, so that adduct formation will be facile. The molecular structure of the THF adduct **12**[11] is depicted in Figure 7, in comparison to the parent silaketimine **11**. Two independent molecules are found in the unit cell. The changes in the immediate conformation of the unsaturated silicon atom are virtually identical to those in the analogous silene adduct **9**, so that a similar bonding model may be considered. Whereas the Si=N double bond distances increase to 1.588(9)/1.574(10) Å, the Si–N single bonds are shortened to 1.654(9)/1.667(10) Å. The Si–N–Si angle narrows somewhat to 161.5(5)/161.0(6)°. Tetrahedral distortion of the unsaturated silicon environment may be gauged by the Σ(Si) values of 349.5/349.2°, which are remarkably close to that of 348.7° in **9**.

The aminofluorosilane **A** reacts with methyllithium to yield the lithium salt **13**. An X-ray structural analysis establishes the F and not the N atom as the metal coordination site[12]. This means that, in analogy to Wiberg's THF adduct, **13** may be considered as the lithium fluoride adduct of an iminosilene, as formulated above. Support for this view comes from a comparison of the molecular geometry of **13** with that of the product **14**[12] of the reaction of **13** with Me_3SiCl. In the latter case, the lithium is once again coordinated to the most electronegative atom in the molecule, which is now the nitrogen. The Si–N–C angle in **13** is virtually linear [172.1(4)°], as is typical for an imino nitrogen

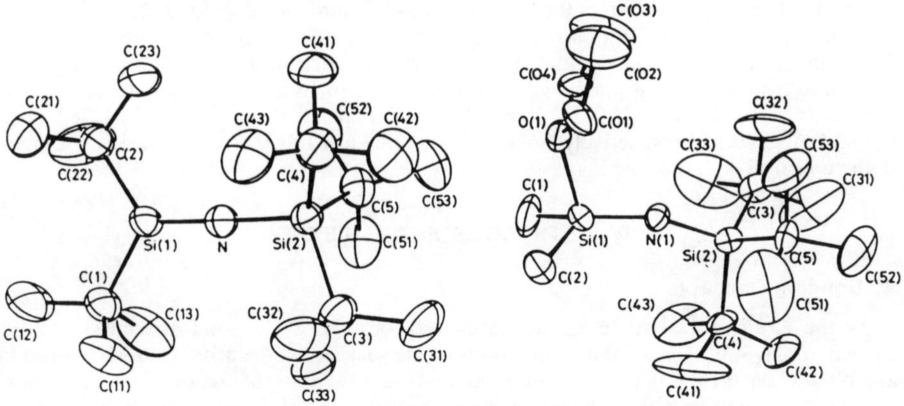

11 **12**

FIGURE 7. Molecular structures of the silaketimine **11** and its THF adduct **12**. Reproduced by permission of the Royal Society of Chemistry from Reference 11.

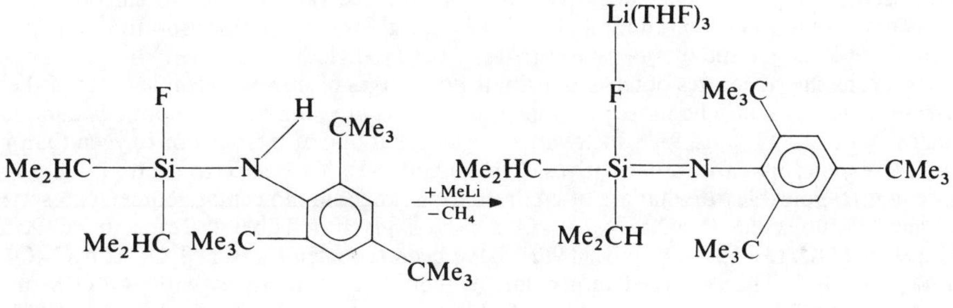

(A) **(13)**

bonded to silicon (see Section IV.C.1). For the tricoordinated N atom in **14**, this angle is only 138.7(3)°, which is similar to that previously observed for amino nitrogen atoms coordinated to alkali metals, e.g. 136(1)° in $K[N(SiMe_3)_2]\cdot 2C_4H_8O_2$[72]. However, although the silicon–nitrogen distance of 1.619(5) Å in **13** is shorter than that for the Si–N single bond in **14** [1.642(4) Å], it is significantly longer than for the Si=N double bond in

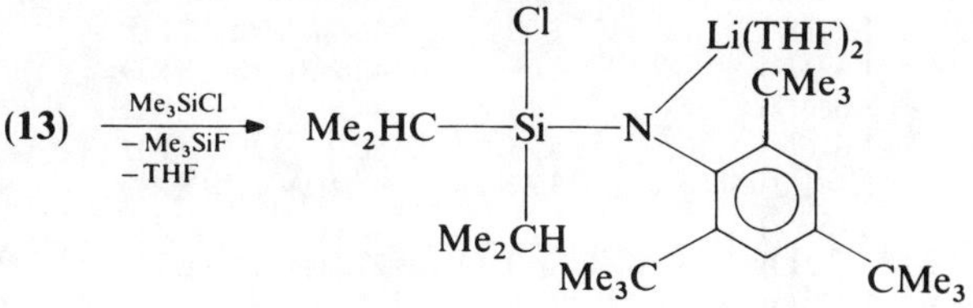

(14)

11 and **12**. This suggests that, as for the analogous lithium fluoride adduct **10** of silene **8**, a bonding model may be proposed for **13**, in which a zwitterionic structure will play a dominant role. Two further parameters are relevant to such an interpretation. Firstly, the distortion from a trigonal planar geometry at silicon in **13** is very similar to that in **10**; Σ(Si) is 342.6° in **13** and 342.2/341.7° for the two independent molecules of **10**. Secondly the Si–F distances of respectively 1.686(4) Å and 1.640(3)/1.647(4) Å are markedly longer than typical Si–F bond lengths.

IV. TETRACOORDINATE SILICON

A. Bonding Concepts

As the electronegativity of X increases, so does the Si–X bond strength for silyl derivatives H_3SiX. The fact that such bonds to the very electronegative atoms N, O and F are often 0.05–0.15 Å shorter than predicted on the basis of Schomaker–Stevenson corrected covalent radii sums has often been attributed to p-π donation from these atoms into the vacant silicon 3d-orbitals. However, *ab initio* MO calculations emphasize that the strength of these bonds is primarily a result of their polarity[32]. Stronger Si–X bonds will be formed with the more electronegative atoms because of greater contributions from ionic structures such as $H_3Si^+X^-$. The addition of silicon d-functions to the basis set for theoretical calculations often reduces the Si–X bond length and leads to better agreement with experimental results[32]. Because the electron affinity of SiH_3 is greater than CH_3, groups with very low electronegativity will also form stronger bonds to silicon in silyl derivatives than to carbon in methyl derivatives, e.g. Si–Li in comparison to C–Li. Such bonds are largely ionic involving ion pairs of the type $H_3Si^-X^+$.

As a rule, the structures of tetracoordinate derivatives of silicon resemble those of their carbon counterparts. The most prominent exceptions are found for compounds containing Si–N or Si–O bonds, for which wider angles are found at nitrogen or oxygen than in the corresponding carbon derivatives. Once again the greater electronegativity gap for silicon is responsible for what are often dramatic changes in molecular geometry. A series of selected molecules $R_3SiXSiR_3$ (X = O, S, NH) is given in Table 4. Respective C–X–C angles of 111.5(15), 127.9(1) and 98.9(2)° have been reported for $(CH_3)_2O$[3], $(Ph_3C)_2O$[81] and $(CH_3)_2S$[82]. The observed molecular geometries of derivatives with X = O, S may readily be understood in a qualitative fashion by considering the Walsh diagram for bending of a linear species XY_2 (Figure 8). For simplicity, the groups Y will be considered as contributing only a single σ-type orbital, while the central atom provides both s and p

TABLE 4. Geometry of selected molecules $R_3SiXSiR_3$

	d(Si–X) (Å)	*d*(Si ··· Si) (Å)	∠Si–X–Si (deg)	Method	Ref.
$O(SiF_3)_2$	1.58(3)	3.08	156(2)	ED	73
$O(SiCl_3)_2$	1.59(1)	3.04	146(4)	ED	74
$O(SiH_3)_2$	1.631(6)	3.09	142.2(3)	XRD	75
$O(SiMe_3)_2$	1.626(2)	3.13	148.8(2)	XRD	75
$O(SiPh_3)_2$	1.616(1)	3.23	180.0	XRD	5
$S(SiH_3)_2$	2.141(2)	3.24	98.4(1)	XRD	76
$S(SiPh_3)_2$	2.152(2)	3.57	112.0(1)	XRD	77
$HN(SiH_3)_2$	1.725(3)	3.10	127.7(1)	ED	78
$HN(SiMe_3)_2$	1.738(5)	3.16	131(1)	ED	79
$HN(SiPh_3)_2$	1.718(6)	3.21	138.1(4)	XRD	80

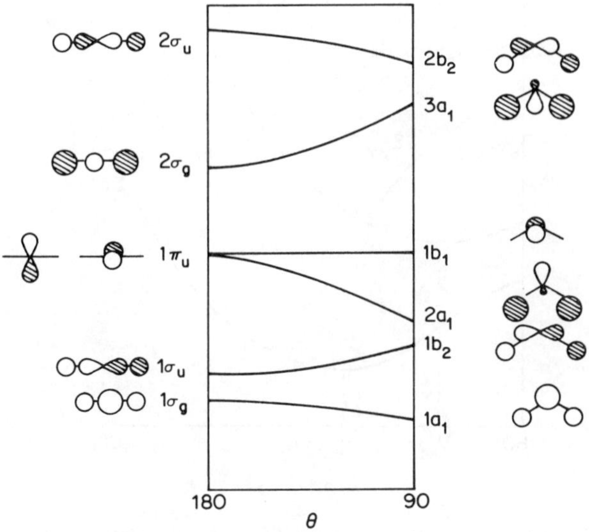

FIGURE 8. Schematic representation of energy changes on bending a linear XY_2 structure. X contributes ns and np orbitals, Y single σ-type orbitals; θ is the Y–X–Y angle

valence shell orbitals. As X bears two lone pairs, the electronic configuration will be $(1\sigma_g^+)^2(1\sigma_u^+)^2(1\pi_u)^4(2\sigma_g^+)^0(2\sigma_u^+)^0$. Upon bending, one of the non-bonding orbitals $1\pi_u$ and an antibonding orbital $2\sigma_g^+$ begin to mix leading to a stabilization of the former orbital (now $2a_1$). At the same time, a bonding orbital $1\sigma_u^+$ will become destabilized. When the $1\pi_u$ orbital is the HOMO (e.g. X=O, S), then the $1\pi_u$–$2\sigma_g^+$ gap will be of crucial importance for the establishment of the equilibrium molecular geometry[83, 84]. An increase in the size of the energy gap will lead to a reduction in the energy gain upon bending, or phrased alternatively, the barrier to linearization of a bent molecule will decrease. As the $1\pi_u$ are almost exclusively located on the central atom this result will be achieved by effecting a positive change in $\Delta\chi = \chi_x - \chi_y$, where χ_x and χ_y are the electronegativities of X and Y. For instance, on going from X=S to X=O, the $1\pi_u$ will become more tightly bound and the energy gap will increase, as reflected by the experimental Si–X–Si angles in $(SiH_3)_2S$ and $(SiH_3)_2O$. Likewise on going from Y=CH_3 to Y=SiH_3 the ligand σ-orbitals will become less tightly bound leading to a higher energy for the $2\sigma_g^+$-orbital and thereby to an increase in the crucial HOMO–LUMO gap. The lower energy barrier to linearization for Y=SiH_3 is demonstrated by the central angle of 142.2(3)° in $(SiH_3)_2O$ in comparison to that of 111.5(15)° in dimethyl ether. The change in total energy of two hypothetical molecules XY_2 (e.g. Y=C) and XZ_2 (e.g. Z=Si) is depicted in Figure 9. The increase in the Si–O–Si angle for $(R_3Si)_2O$ in the order R=H, Me, Ph mirrors the order of decreasing electronegativity $SiH_3 > SiMe_3 > SiPh_3$. For R=Ph the barrier to linearization is obviously very low and the linear structure observed in the solid state may well be influenced by lattice forces.

A limiting lower value for the central angle in molecules $(R_3Si)_2X$, X=O, NH, CH_2, may be predicted on the basis of the hard-sphere model first proposed by Bartell[85] and extended by Glidewell[86]. If the two silicon atoms are regarded as hard spheres then bending of a linear molecule will finally reach a limiting situation in which these non-bonded atoms are in contact with one another. Under these conditions $d(\text{Si}\cdots\text{Si}) = 2r$,

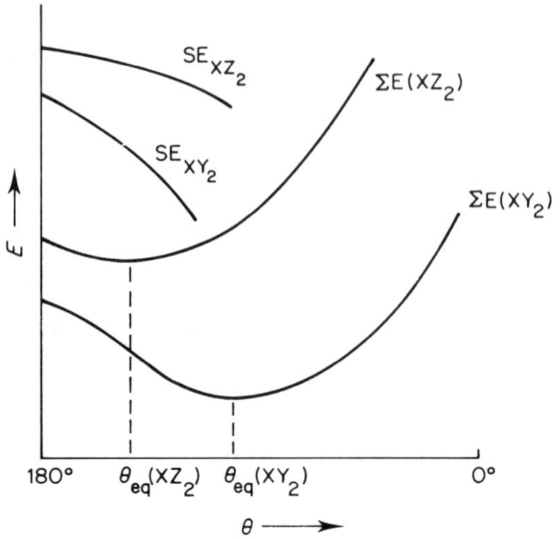

FIGURE 9. Plots of the stabilization energy (SE) and total energy (ΣE) as a function of the central angle in two hypothetical molecules XY_2 and XZ_2 (χ_x, $\chi_y \gg \chi_z$); θ is the angle at X

where r is the hard-sphere radius of silicon. By analysis of a series of simple molecules, Glidewell suggested a value for r of 1.55 Å. Further bending of a molecule $(R_3Si)_2X$ to yield values of d(Si $\cdots$ Si) lower than 3.10 Å will be expected to lead to a very rapid increase in steric strain in the molecule. If we look at the derivatives in Table 4 we notice that non-bonded Si $\cdots$ Si distances close to 3.10 Å are to be found in $(SiF_3)_2O$, $(SiCl_3)_2O$, $(SiH_3)_2O$ and $(SiH_3)_2NH$. We may assume, therefore, that the experimental value of Si–X–Si in these molecules is determined by non-bonded interaction of the two silicon atoms. We may also apply a similar approach to the molecular geometry in species XY_3, e.g. $N(SiH_3)_3$. Assuming once again that the substituents Y contribute only σ-type orbitals, a Walsh diagram as depicted in Figure 10 may be constructed. For a planar molecule XY_3 with X=N, the electronic configuration is $(1a_1')^2(1e')^4(1a_2'')^2(2a_1')^0(2e')^0$. The HOMO–LUMO gap between the $1a_2''$ and $2a_1'$ orbitals will determine whether distortion from planarity will occur. The $1a_2''$ orbital is effectively the lone pair on nitrogen. A planar molecular geometry will be expected when the electronegativity gap $\Delta_\chi = \chi_X - \chi_Y$ is large, e.g. in the case of trisilylamine. In contrast, the σ-orbitals on the carbon atoms in trimethylamine are more tightly bound, so that the $2a_1'$ orbital will be of lower energy and distortion to a pyramidal configuration, as experimentally observed[87] for $(CH_3)_3N$, should be energetically favourable. In the following Sections, individual Si–X bonds will be discussed in detail, with particular attention being placed on bonds to the second-row elements Si, P and S and those to main group and transition metals, as these were areas of pronounced expansion in the past decade. Si–H bonds will not be handled specifically, in view of the fact that their length cannot be determined reliably by X-ray diffraction. In addition the Si–H distance has often been treated as a fixed parameter in electron diffraction studies. Typical values fall in the range 1.47–1.50 Å.

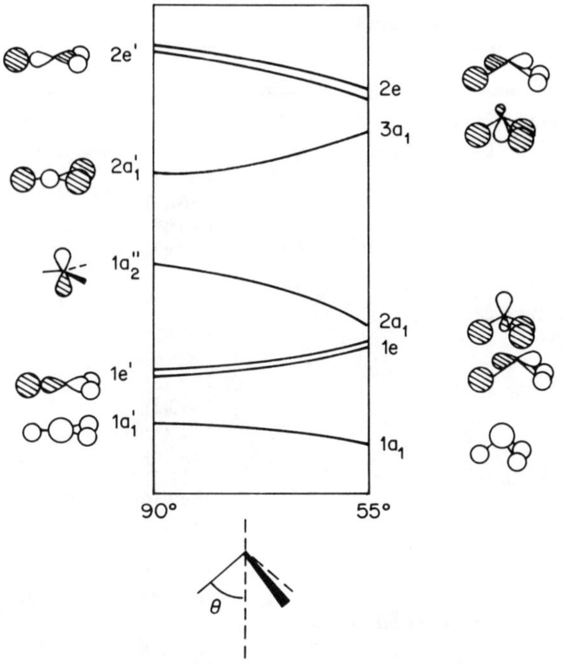

FIGURE 10. Schematic representation of energy changes on bending a planar XY_3 structure. X contributes ns and np orbitals, Y single σ-type orbitals

B. Bonds to Elements of Group 14

1. Bonds to carbon

Selected Si–C bond distances are listed in Table 5. When silicon is bonded to four sp^3 hybridized carbon atoms in molecules without a high degree of steric strain, typical Si–C bond lengths in the range 1.86–1.91 Å are found. 'Short' Si–C bonds with distances less than 1.86 Å may be observed under the following circumstances:

(1) Si bonded to an sp^2 or sp hybridized C atom, e.g. silylolefines and -acetylenes,
(2) Si participating in a three-membered ring, e.g. silacyclopropanes and -propenes,
(3) Si bonded to very electronegative elements, e.g. in methylfluoro- and chlorosilanes,
(4) Si bonded to a C atom with a degree of carbanionic character; for examples see Table 5.

Inspection of Table 5 indicates that the degree of bond shortening in silylolefines and -acetylenes is somewhat smaller than in the analogous carbon derivatives for which typical values of respectively 0.03 and 0.08 Å are found[19]. It is interesting to note that the Si–C distance of 1.872(7) Å in Ph_4Si(**47**)[119] lies very close to the value in the methylsilanes **15**–**18**. The Si–C bond is markedly polarized. Increasing the degree of bond ionicity will be expected to lead to a progressive decrease in bond length. This may be achieved by attaching silicon to very electronegative elements such as fluorine or chlorine. For instance, the Si–C distances in the methylfluorosilanes decrease progressively on going

TABLE 5. Selected Si–C bond distances

Compound	d(Si–C)(Å)	Method	Ref.
Sum of covalent radii	1.88		19
'Typical' Bonds			
15 $MeSiH_3$	1.867(1)	MW[a]	88
16 Me_2SiH_2	1.867(2)	MW	89
17 Me_3SiH	1.868(2)	MW	90
18 Me_4Si	1.875(2)	ED[a]	91
'Short' Bonds			
1. sp^2 and sp hybridized carbon			
19 $H_3SiCH{=}CH_2$	1.853(3)	MW	92
20 $H_3SiC{\equiv}CH$	1.825(6)	ED	93
21 $H_3SiC{\equiv}CCl$	1.825(8)	ED	93
22 $H_3SiC{\equiv}CMe$	1.813(5)	ED	94
23 $H_3SiC{\equiv}CSiH_3$	1.825(3)	ED	95
24 $H_3SiC{\equiv}CCF_3$	1.844(6)	ED	96
25[b] $t\text{-}Bu_2RSiC{\equiv}CSiR(Bu\text{-}t)_2$	1.851(2)	XRD	97
26 H_3SiCN	1.847(10)	MW	98
27 Me_3SiCN	1.844(22)	ED	99
2. Silacyclopropanes and -propenes			
28	1.825(3), 1.827(3)	XRD	100
29	(x)1.800(2), (y)1.839(2)	XRD	101
3. Electronegative substituents bonded to silicon			
30 $MeSiH_2F$	1.849(5)	MW	102
31 $MeSiHF_2$	1.833(2)	MW	103
32 $MeSiF_3$	1.812(14)	MW	104
33 Me_3SiCl	1.857(5)	MW	105
34 Me_2SiCl_2	1.845(5)	MW	106
	1.856(4)	ED	107
4. Carbon with carbanionic character			
10[c]	1.796(5)–1.822(5)	XRD	68
35	1.802(8)	XRD	108
36	1.790(6)–1.803(5)	XRD	109
37 $(Cl_3Si)_2C{=}PMe_3$	1.759(11), 1.795(9)	XRD	110
38[c] $(Cl_3Si)_2C{=}PMe_2Cl$	1.789(10)–1.818(12)	XRD	110
39[d]	1.793(5)–1.806(5)	XRD	111
'Long' Bonds			
5. Electronegative substituents bonded to carbon			
40 H_3SiCF_3	1.923(3)	ED	112
6. Sterically congested molecules			
41[e] $(Me_3Si)_3CPH_2$	1.941(5)	ED	113
42 $(t\text{-}Bu_2Si)_3$	1.970(5)	XRD	114
43 $I(t\text{-}Bu_2Si)_3I$	1.971(4)–2.012(4)	XRD	115
44 $(t\text{-}Bu)_3SSi(Bu\text{-}t)_3$	1.99	XRD	116
45	(x) 2.005(3)	XRD	117
46	(x) 2.028(4)	XRD	118

[a] r_s values are given for MW studies, r_a values for ED studies.
[b] $R = t\text{-}BuNH$.
[c] Two independent molecules in the unit cell.
[d] Three independent molecules in the unit cell.
[e] The value of $d(Si\text{–}CPH_2)$ is given.

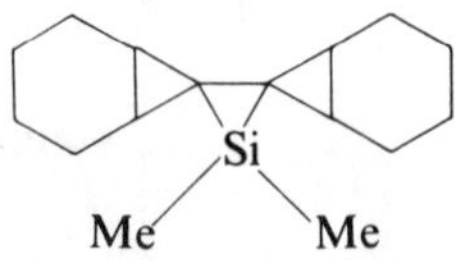

(28)

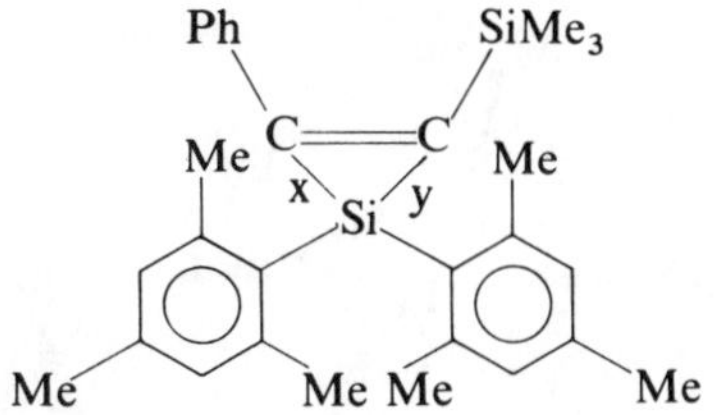

(29)

from $MeSiH_3$ to $MeSiF_3$. Trisubstitution of H by F leads to a reduction in the Si–C distance of 0.055 Å. On the basis of the X-ray diffraction data available in 1980 Hönle and von Schnering[120] proposed that successive Cl substitution of $-SiC_3$ should lead to a reduction in the effective covalent radius of silicon of respectively 0.02 ($-SiC_2Cl$), 0.05 ($-SiCCl_2$) and 0.08 Å ($-SiCl_3$). However, the microwave results for Me_3SiCl and Me_2SiCl_2 suggest that these figures may be too high. The gas-phase investigations indicate that realistic values are ca. 0.018 Å for fluorine and ca. 0.011 Å for chlorine substitution[107, 121].

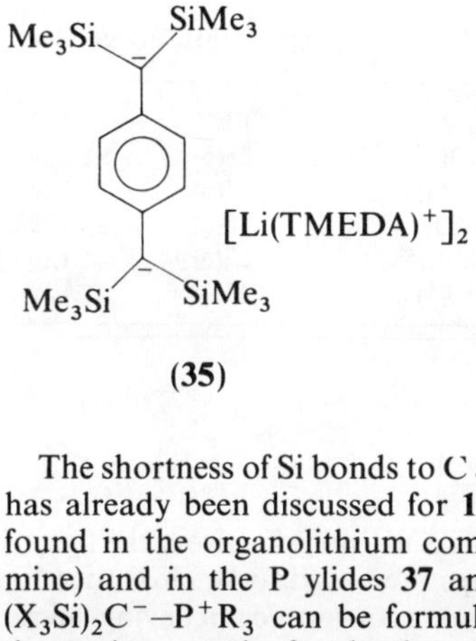

(35)

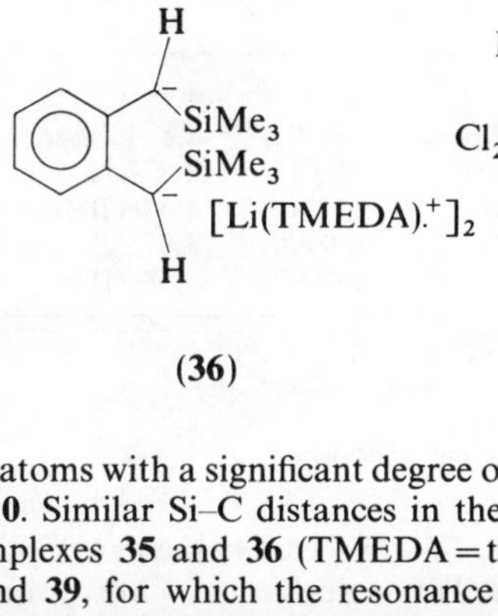

(36)

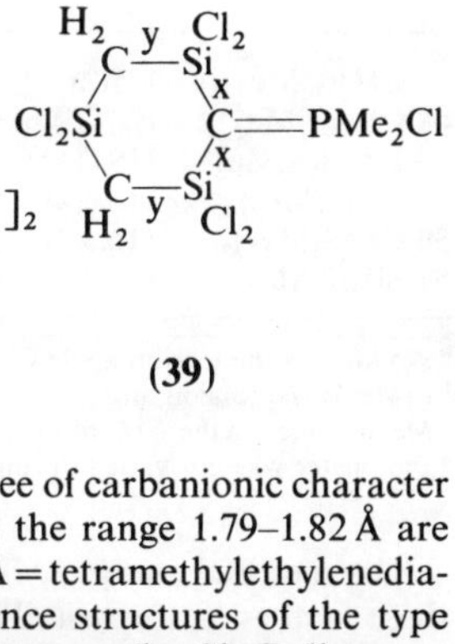

(39)

The shortness of Si bonds to C atoms with a significant degree of carbanionic character has already been discussed for **10**. Similar Si–C distances in the range 1.79–1.82 Å are found in the organolithium complexes **35** and **36** (TMEDA = tetramethylethylenediamine) and in the P ylides **37** and **39**, for which the resonance structures of the type $(X_3Si)_2C^-{-}P^+R_3$ can be formulated. It is instructive to compare the Si–C distances denoted as *x* and *y* for the three independent molecules of **39** found in the unit cell. The former bond lengths lie in the range 1.793(5)–1.806(5) Å, the latter in the range 1.834(5)–1.846(5) Å.

'Long' Si–C distances have been found under the following conditions:

(5) C bonded to very electronegative elements, e.g. H_3SiCF_3,

(6) severe steric congestion, e.g. molecules containing 2 or 3 bulky groups such as Me_3Si attached to the same C atom.

Examples are given in Table 5 of molecules in which a very pronounced degree of bond stretching is necessary to relieve steric strain. The very bulky ligand $(Me_3Si)_2CH$ has often been used to stabilize unusual low coordination numbers, e.g. 5 and 6. It is beyond the scope of this chapter to consider the numerous structural examples now available. However, the degree of steric strain in such ligands may be gauged by the comparison of geometrical parameters for molecules $XH_n(YMe_3)_{4-n}$ ($n = 2, 1$; X, Y = C, Si) presented in

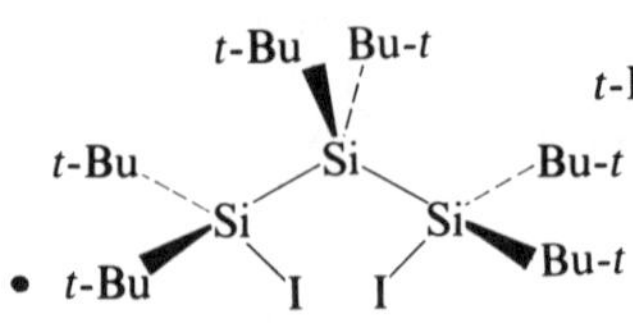

(43)

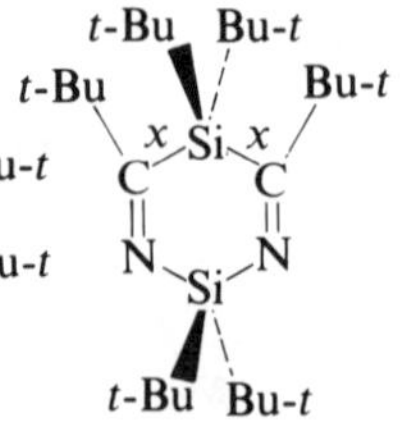

(45)

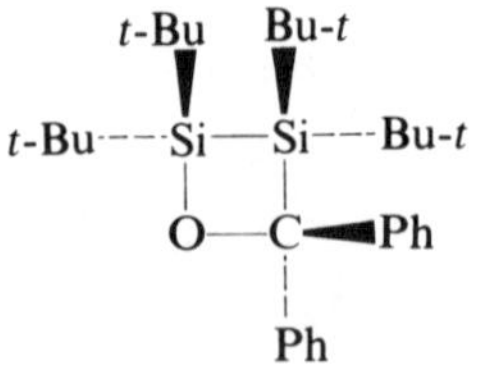

(46)

Table 6. There are several types of steric deformation available to such molecules to reduce overcrowding:

(1) torsional displacement of the YMe_3 and Me groups,
(2) tilting of the YMe_3 groups relative to the Y–X axis,
(3) widening of bond angles to values considerably greater than tetrahedral,
(4) bond stretching.

TABLE 6. Geometrical parameters for electron diffraction studies of $XH_n(YMe_3)_{4-n}$, $n = 2, 1$

Compound	∠Y–X–Y(deg)	d(X–Y)(Å)	d(Y–Me)(Å)	$\phi(YMe_3)^a$(deg)	$\phi(Me)^b$(deg)	Ref.
$CH_2(CMe_3)_2$	128(2)		1.545(2)[c]	15(2)	13(2)	122
48 $CH_2(SiMe_3)_2$	123.2(9)	1.889(4)	1.874(2)	3(2)	25(5)	123
49 $LiCH(SiMe_3)_2$	119.0(15)	1.892(2)	1.877(2)	9(3)	14(5)	124
$CH(CMe_3)_3$	116.0(4)	1.611(5)	1.548(2)	10.8(5)	18(6)	125
50 $CH(SiMe_3)_3$	116.3(4)	1.888(6)	1.873(2)	20.5(5)	27(5)	126
51 $SiH(CMe_3)_3$	113.3	1.934(2)	1.584(1)	10(1)	4.2[d]	127

[a] $\phi(YMe_3)$ is the torsion angle C–Y–X–H.
[b] $\phi(Me)$ is the torsion angle H–C–Y–X.
[c] Mean value for the C–C distances.
[d] Parameter was not varied during refinement.

The required energy will be expected to increase significantly on going from the first of these factors to the last. Tilt angles for the YMe_3 groups are found to be close to zero, indicating that this deformation mode is relatively unimportant for relieving steric congestion. The following trends can be ascertained for the series of compounds in Table 6.

(1) Steric strain increases upon introduction of more YMe_3 groups onto a central C atom.

(2) Deformation from an idealized structure with staggered substituents and tetrahedral bond angles is greater when Y=C than for Y=Si.

(3) Pronounced stretching of the X–Y bonds is apparently necessary to relieve steric strain in $XH(CMe_3)_3$; this is not the case for $CH(SiMe_3)_3$ for which torsional displacements are more important.

(4) For molecules $XH_2(YMe_3)_2$, widening of the Y–X–Y angle is the most advantageous strain relieving factor.

Electronic factors are also of significance for the central Si–C–Si angle in molecules $(R_3Si)_2CH_2$. With decreasing electronegativity of SiR_3, the Si–C–Si angle increases from

114.4(2)° ($R=SiH_3$ **52**, ED[128]) through 123.2(9)° ($R=SiMe_3$ **48**, ED[123]) to 128.8(7)° ($R=SiPh_3$ **53**, SRD[129]). Bonding aspects for such species have been considered in detail by Glidewell and Liles[129].

2. Bonds to silicon and its heavier congeners

A Si–Si bond distance of 2.340(9) Å is found in hexamethyldisilane **54**[91]. The typical range for molecules with relatively little steric strain can be taken to be 2.33–2.37 Å as demonstrated by selected examples in Table 7. Cyclopolysilanes are listed separately in Table 8. The gradual elongation of the Si–Si bond distance in the disilanes on going from **54** to **58** can be attributed to a steady increase in steric congestion upon introduction of up to two bulky groups per silicon atom. Very long Si–Si distances are found in **43** and **44**. For instance, the bond length of 2.697 Å in **44** represents a Pauling bond order of 0.26. Apparently the Si–Si bond, which is much weaker than a C–C bond, is capable of a remarkable degree of stretching in order to reduce the steric overcrowding in these molecules, caused by the extremely bulky *t*-Bu groups.

TABLE 7. Selected Si–Si bond distances

Compound	d(Si–Si) (Å)	Method	Ref.
Sum of covalent radii	2.34		19
1. Disilanes			
54 $Me_3SiSiMe_3$	2.340(9)	ED	91
55 $Me_3SiSiPh_3$	2.355(1)	XRD	130
56[a] $Mes_2HSiSiHMes_2$	2.350, 2.362	XRD	131
57[b] $RF_2SiSiFMes_2$	2.370(2)	XRD	132
58[a] $t\text{-}Bu_2HSiSiHMes_2$	2.395(2), 2.400(2)	XRD	133
44 $t\text{-}Bu_3SiSi(Bu\text{-}t)_3$	2.697	XRD	116
2. Si–Si chains			
59 $Me_3SiSiMePh(SiMe_2)_4SiMe_3$	2.335(7), 2.361(8)	XRD	134
60 $Si(SiMe_3)_4$	2.361(3)	ED	135
43 $I(t\text{-}Bu_2Si)_3I$	2.581(1), 2.644(1)	XRD	115
3. Heterocycles			
61 $(PhPSi_2Me_4)_2$	2.345(3)	XRD	136
62 $(PhP)_2(SiMe_2)_3$	2.350(1), 2.351(1)	XRD	137
63 $[PhHP(SiMe_2)_2PHPh]Mo(CO)_4$	2.378(2)	XRD	138
64	2.318(4)	XRD	139
65	2.338(2)	XRD	140
66 $[Ph_2SiC(SiMe_3)_2]_2$	2.344(9)	XRD	141
46	2.409(2)	XRD	118

[a] Two independent molecules in the unit cell; Mes = mesityl.
[b] R = 2,4,6-tri-*t*-butylphenyl.

Despite the very long Si–Si bonds, the Si–C bonds in these molecules are also significantly elongated to an average value of 1.99 Å. The sum of the bond orders for the silicon atoms in **44** is only 2.36 instead of the expected value of 4. In view of this fact, the stability of the compound is surprising. It is air stable and may be heated to its melting point of 162 °C without decomposition[116].

The cyclopolysilanes show a range of unique electronic and spectroscopic properties, which arise from electron delocalisation in the Si–Si σ framework. For instance, they

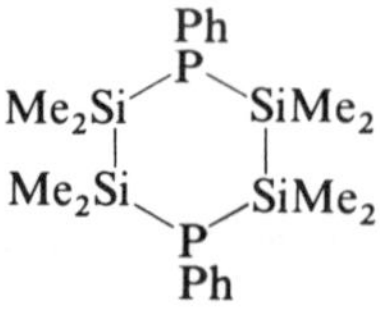

(61)

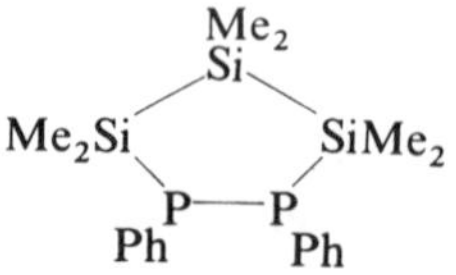

(62)

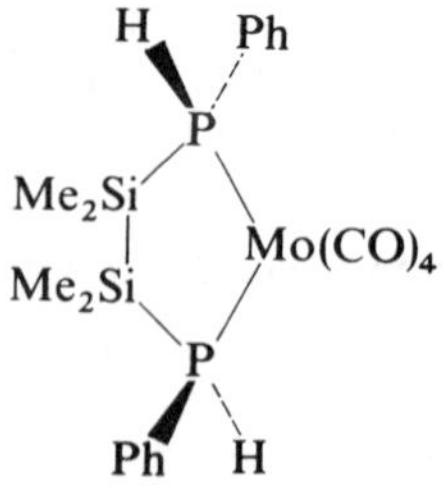

(63)

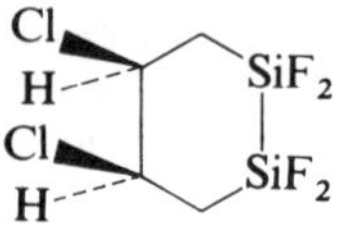

(64)

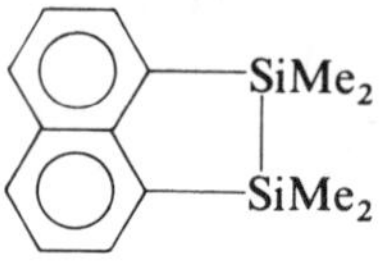

(65)

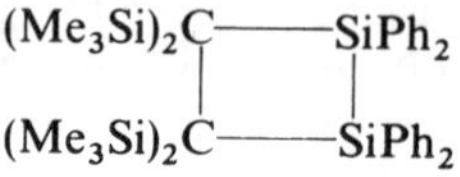

(66)

readily form charge-transfer complexes with π acceptors and may be reduced to delocalized anion and cation radicals[160]. Geometrical parameters for cyclopolysilanes are listed in Table 8. *Ab initio* calculations by Nagase and coworkers[161] for homodesmic reactions of the type

$$(H_2X)_n + nH_3X\text{–}XH_3 \rightarrow nH_3X\text{–}XH_2\text{–}XH_3, \qquad n=3,4; \qquad X=C, Si$$

have demonstrated that, whereas the strain energy decreases from 162.8 kJ mol^{-1} to 69.9 kJ mol^{-1} on going from $D_{3h}(H_2Si)_3$ to $D_{2d}(H_2Si)_4$, it remains at approximately the same level for cyclopropane and cyclobutane. The respective values for the carbon analogues are 120.1 and 110.9 kJ mol^{-1}. Geometries were optimized using a 6-31G* basis set. The calculated Si–Si bond distance in $(H_2Si)_3$ is 2.341 Å, which is effectively equal to the sum of the covalent radii. In $(H_2Si)_4$ this bond is somewhat longer (2.373 Å). The experimental average Si–Si distances in the cyclotrisilanes **67**–**70** are between 0.05 and 0.087 Å longer than the theoretical value in $(H_2Si)_3$. This elongation may be attributed once again to steric strain. The importance of this effect is dramatically illustrated by the molecular structure of $(t\text{-}Bu_2Si)_3$ **42**, in which steric repulsion between bulky substituents on neighbouring silicon atoms produce very long Si–Si distances of 2.511(3) Å.

In the first structurally characterized bicyclotetrasilane **71**, which displays an approximate C_2 symmetry (Figure 11), the outer Si–Si distances are shorter than in the cyclotrisilanes **67**–**70**. They alternate in length with Si1–Si2 and Si3–Si4 being very short [2.308(3)–2.311(3) Å] and the other two bonds close in distance to the value for hexamethyldisilane [2.334(4)–2.336(3) Å]. However, it is the relatively short transannular distance of 2.373(3) Å which is most remarkable. Recent *ab initio* calculations[162] on the parent bicyclotetrasilane Si_4H_6 with both 3-21G and 3-21G* basis sets have revealed the presence of two energy minima. Molecular dimensions for the first minimum [3-21G*, transannular d(Si–Si) = 2.343, cyclotetrasilane d(Si–Si) = 2.293 Å] are similar to those of the experimental structure. The second minimum represents a structure of lower energy with a much longer transannular Si–Si distance of 2.731 Å. It follows that the preference for the first minimum by **71** must be a result of the presence of bulky ligands on the bridgehead atoms. The transannular Si–Si bond is antibonding with a considerable degree of π-character.

TABLE 8. Geometrical parameters for cyclopolysilanes

Compound	d(Si–Si) (Å)	Average d(Si–Si) (Å)	Average Si–Si–Si(°)	Ring torsion angle τ(deg)[a]	Method	Ref.
$n=3$						
67 [(t-BuCH$_2$)$_2$Si]$_3$	2.367(8)–2.414(10)	2.391[b]	60.0	0.0	XRD	142
68 [(2,6-Me$_2$C$_6$H$_3$)$_2$Si]$_3$	2.375–2.425	2.407	60.0	0.0	XRD	143
69 *cis-trans*-[t-BuMesSi]$_3$[b]	2.395(2)–2.441(2)	2.416	60.0	0.0	XRD	144
70 *cis-cis*-[t-BuMesSi]$_3$[b]	2.425(2)–2.431(2)	2.428	60.0	0.0	XRD	144
42 [t-Bu$_2$Si]$_3$	2.511(3)	2.511	60.0	0.0	XRD	114
$n=4$						
71 [(2,6-Et$_2$C$_6$H$_3$)$_2$Si–Si(Bu-t)]$_2$	2.308(3)–2.336(3)	2.322	79.1	59	XRD	145
72 [Me$_2$Si]$_4$	2.359(1)–2.367(1)	2.363	90.0	0	XRD	146
73 [Ph$_2$Si]$_4$	2.370(1)–2.381(1)	2.377	89.6	12.8	XRD	147
74[c] [t-BuMeSi]$_4$[e]	2.377(1)	2.377	87.0	36.8	XRD	148
75[d] [Ar$_2$SiSi(R)(Cl)]$_2$	2.396–2.428	2.409	87.3	36.6	XRD	149
76 [(Me$_3$Si)$_2$Si]$_4$			90.0	0	XRD	150
$n=5$						
77 [(CH$_2$)$_4$Si]$_5$	2.343(1)–2.350(1)	2.347	104.5	27.2	XRD	151
78 [(CH$_2$)$_5$Si]$_5$	2.350(1)–2.366(1)	2.359	102.5	33.4	XRD	151
79 [Ph$_2$Si]$_5$	2.371(3)–2.413(5)	2.396	104.5	27.0	XRD	152
[H$_2$Si]$_5$	2.342(3)	2.342	104.2	26.7[e]	ED	153
$n=6$						
80 [Me$_2$Si]$_6$	2.332(1)–2.342(1)	2.338	111.9	53.5	XRD	154
81[c] [MePhSi]$_6$	2.356(1)–2.362(1)	2.359	111.1	55.2	XRD	155
82 [Ph$_2$Si]$_6$	2.393(3)–2.396(3)	2.394	113.8	47.5	XRD	156
[H$_2$Si]$_6$	2.342(5)	2.342	110.3	57.9	ED	157
$n>6$						
83 [Me$_2$Si]$_7$	2.338(1)–2.344(1)	2.340	116.2	61.5	XRD	158
84 [Me$_2$Si]$_{13}$	2.329(3)–2.377(4)	2.355	115.7		XRD	159
85 [Me$_2$Si]$_{16}$	2.348(1)–2.360(2)	2.355	111.9		XRD	159

[a] τ is the interplanar dihedral angle for $n=4$, otherwise τ is the average absolute value of the torsion angle Si–Si–Si–Si.
[b] Mes = mesityl.
[c] All-*trans* isomers.
[d] Ar = 2,6-Et$_2$C$_6$H$_3$; R = t-Bu.
[e] For C_s symmetry; an alternative C_2 model gave $\tau = 27.9°$.

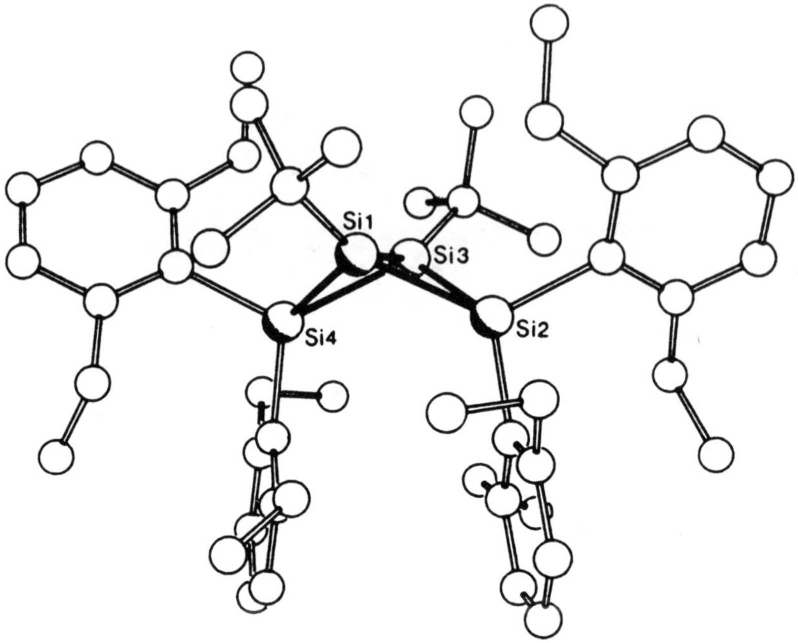

FIGURE 11. Molecular structure of the bicyclotetrasilane **71**. Reproduced by permission of VCH Verlagsgesellschaft, Weinheim, from Reference 145

Striking and not readily understandable differences are observed for the ring conformations in the cyclotetrasilanes **72–76**. A folded structure would be expected to be energetically more favourable than the planar alternative in the absence of bulky substituents. For instance, an interplanar angle of 31° was calculated for the parent compound $(H_2Si)_4$ by Nagase and coworkers[161]. Attachment of bulky substituents to the ring silicon atoms would be expected to lead to a reduction in the degree of folding in order to reduce steric strain. This argument may be applied to $(Ph_2Si)_4$ **73** and $[(Me_3Si)_2Si]_4$ **76** but surely not to $(Me_2Si)_4$ **72** for which, surprisingly, a strictly coplanar ring system is found. Although packing effects are often invoked to explain such unexpected results, it will not be possible to provide a rational explanation until gas-phase structures are available.

A conformation intermediate between C_2 and C_s symmetry is found for the Si_5 rings in **77–79** and is probably typical for cyclopentasilanes. The pseudorotation phase angle Δ[163] is respectively $-9.3°$, $-14.1°$ and $-15.9°$ for **77–79**, where Δ is zero for C_2 symmetry and $-36°$ for C_s symmetry. In the gas-phase determination of the structure of $(H_2Si)_5$ both C_2 and C_s models gave very good agreement with experimental data, this being consistent with spectral data, which indicate that cyclopentasilane undergoes pseudorotation. If the average absolute torsion angle τ is taken as a measure of the degree of planarity in these molecules, then the experimental results indicate that the Si_5 ring in **78** is less planar than in the other cyclopentasilanes. The Si_6 ring in the cyclohexasilanes **80–82** adopts a chair conformation. τ is smaller for **82** than for **80** and **81** indicating a flatter chair in the latter case, presumably as a result of steric strain. The scatter of the individual torsion angles in the homosubstituted derivatives **80** and **82** is very limited (respectively 53.2–53.7° and 47.2–47.8°), which implies highly symmetric ring systems. In contrast, these angles vary considerably between 47.9 and 61.8° in **81**. The conformation of cyclohexasilane could not be uniquely determined in the gas phase. Molecules of $(Me_2Si)_7$ exhibit an approximate C_2 symmetry and adopt a twist-chair conformation.

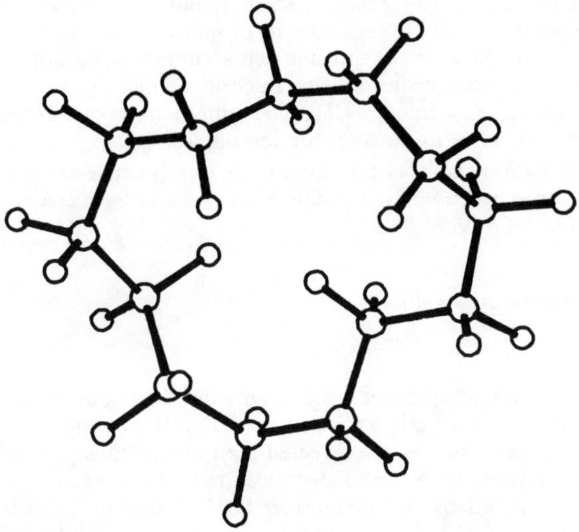

(84)

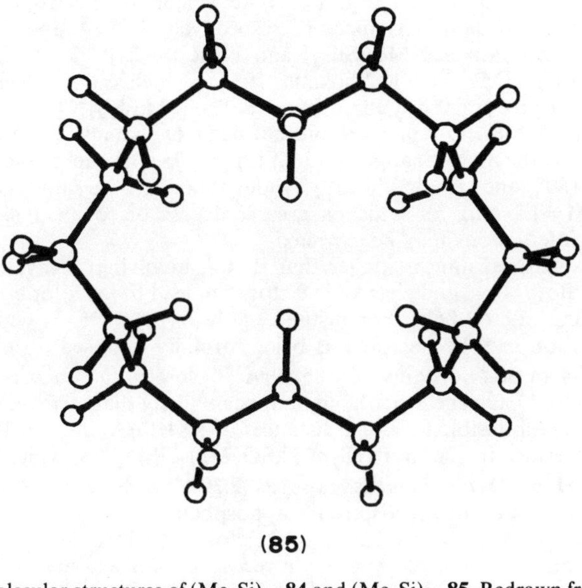

(85)

FIGURE 12. Molecular structures of $(Me_2Si)_{13}$ **84** and $(Me_2Si)_{16}$ **85**. Redrawn from Reference 159

The average Si–Si–Si angle of 116.2° is significantly larger than that reported for the cyclohexasilanes and may be a result of transannular ring strain. A similar trend is observed for medium-sized cycloalkanes. Whereas $(Me_2Si)_{13}$ also displays a similar average Si–Si–Si angle, this is not the case for $(Me_2Si)_{16}$, which exhibits a value close to that found in the cyclohexasilanes **80–82**. The barrier to rotation around an Si–Si bond is much lower than for the corresponding C–C bond. This is reflected in the abnormal torsion angles for **84** and **85**. In $(Me_2Si)_{16}$ **85** these fall into two groups 89.3–93.9° and 158.0–169.5° (absolute values) far from the ideal *gauche* (60°) and *trans* (180°) angles. There are no less than 20 C · · · C intramolecular non-bonded contacts less than 4 Å in $(Me_2Si)_{16}$ and it has been proposed that such molecules adopt a conformation (see Figure 12) which maximizes attractive CH_3–CH_3 interactions while minimizing repulsive interactions, leading thereby to unusual torsion angles.

X-ray diffraction data on compounds containing bonds between silicon and its heavier congeners are not presently available. A Ge–Si distance of 2.36(2) Å was determined for germyl silane H_3GeSiH_3 (MW[164]).

C. Bonds to Elements of Group 15

1. *Bonds to nitrogen*

As more than 700 individual Si–N bond lengths have been determined at the time of writing, a comprehensive listing is not possible within the bounds of the present work. Whereas the Schomaker–Stevenson corrected sum of the covalent radii of silicon and nitrogen is 1.80 Å, typical Si–N bond distances have been found to lie in the range 1.70–1.76 Å, as exemplified by the silylamines **86–94** listed in Table 9. This means, of course, that the polarity of the Si–N bond is strongly underestimated by the above correction. 'Short' Si–N bonds in the range 1.64–1.69 Å have been found for compounds in which the ionicity of this bond has been further enhanced either by the attachment of very electronegative substituents (e.g. F, Cl) to silicon or electropositive metals to nitrogen. For instance, bond distances of respectively 1.654 Å and 1.657(12) Å were observed for the derivatives $NMe_2(SiF_3)$ and $NMe_2(SiCl_3)$[174]. The Si–N bonds are 1.64(1) Å in length in $[(Me_3Si)_2NK.2dioxane]$ **105** for which anion and cation interact in what is certainly a predominantly ionic manner. The $[(Me_3Si)_2N]^-$ anion is, of course, isoelectronic with $(Me_3Si)_2O$, which exhibits a wide Si–O–Si angle of 148.8(2)°. As may be seen from Table 10, the Si–N–Si angle of 136.2(12)° in **105** approaches the typical Si–O–Si range of 140–180° and is significantly wider than in the analogous derivatives $(Me_3Si)_2NM$ (M = Li, Na), for which a greater degree of covalent character for the metal–nitrogen interactions may be expected.

Relatively few Si–N distances longer than 1.80 Å involving tetravalent silicon have been registered. Both steric and electronic factors can lead to the adoption of 'long' Si–N bonds in the range 1.80–1.86 Å. For instance, $[Me_3Si(py)]I$ **96**, in which the positive charge of the cation may be regarded as being formally localised on the pyridine (py) nitrogen, exhibits an Si–N distance of 1.858(9) Å[176] close to the uncorrected sum of the covalent radii (1.87 Å). On the other hand, it appears likely that steric congestion will, at least, be partially responsible for Si–N distances of 1.841(6) Å and 1.861(7) Å in the two independent molecules in the unit cell of $N(SO_2Me)_2SiMe_3$ **95**, which display several intramolecular H · · · O contacts as close as 2.30 Å. A long Si–N bond length of 1.822(9) Å was also found in the dispirodiazaphosphetidine **97**[177].

The planar geometry of the central NSi_3 skeleton in trisilylamine was established by electron diffraction more than 30 years ago. A recent flurry of activity has been concerned with the molecular architecture of mono- and dimethyl substituted silylamines. Apposite

TABLE 9. Geometrical parameters in selected silylamines

Compound	d(Si–N) (Å)	Si–N–Si (deg)	Sum of angles at N (deg)	d(Si · · · Si) Å	Method	Ref.
$N(SiH_3)_3$	1.734(2)	119.7(1)	359.1	2.997	ED	31
	1.730(5)	119.5(3)–120.4(3)	360.0	2.990–3.002	XRD	165
86 $N(SiH_2Me)_3$	1.729(13)	120	360[a]	2.995	ED	166
$NH(SiH_3)_2$	1.725(3)	127.7(1)	360	3.097	ED	78
87 $NH(SiHMe_2)_2$	1.727(3)	130.4(15)	360	3.136	ED	167
88 $NH(SiMe_3)_2$	1.738(5)	131.3(15)	360	3.17	ED	79
$NMe(SiH_3)_2$	1.726(3)	125.4(4)	360	3.068	ED	168
89 $NMe(SiH_2Me)_2$	1.718(3)	125.6(10)	360	3.057	ED	169
90 $NMe(SiHMe_2)_2$	1.727(4)	126.1(5)	360	3.079	ED	169
$NMe_2(SiH_3)$	1.713(5)		354.6		ED	170
91 $NMe_2(SiH_2Me)$	1.715(6)		355.6		ED	170
92 $NMe_2(SiHMe_2)$	1.719(5)		352.4		ED	170
93 $NMe_2(SiMe_3)$	1.710(5)		360		ED	171
	1.719(1)		357.8		XRD	171
94 $NH(SiPh_3)_2$	1.718(6)	138.1(4)	360	3.209	XRD	80
$N(SiH_3)_2BF_2$	1.737(4)		360	3.066	ED	172
$NMe_2(SiH_2Cl)$	1.689(4)		360		ED	173
$NMe_2(SiCl_3)$	1.657(12)		360		ED	174
$NMe_2(SiF_3)$	1.654(15)		360		ED	174
95[b] $N(SO_2Me)_2SiMe_3$	1.841(6)		359.8		XRD	175
	1.861(7)		360.0			

[a] Planarity was assumed for ED models with sum of angles at N equal to 360°.
[b] Two independent molecules in the unit cell.

TABLE 10. Geometrical parameters in selected $(Me_3Si)_2N$ derivatives

Compound	d(Si–N) (Å)	∠ Si–N–Si (deg)	Method	Ref.
100 $[(Me_3Si)_2NLi]_2$	1.712(7)	129.8(17)	ED	183, 184
101 $[(Me_3Si)_2NLi]_3$	1.721(4)–1.734(4)	117.7(3)–119.5(3)	XRD	185, 186
102 $[(Me_3Si)_2NLi \cdot OEt_2]_2$	1.705 (2)	121.2(2)	XRD	187, 188
103 $[(Me_3Si)_2NLi \cdot$ 12-crown-4]	1.681(2)	123.5(1)	XRD	189
104 $(Me_3Si)_2NNa$	1.687(2), 1.694(2)	125.6(1)	XRD	190
105 $[(Me_3Si)_2NK \cdot$ 2dioxane]	1.64(1)	136.2(12)	XRD	72
106 $[Me_3Si)_2N]_2Be$	1.722(7)	129.2(7)	ED	191
107 $[(Me_3Si)_2N]_2Mg$	1.703(6)	132.0(20)	ED	192
108 $[(Me_3Si)_2N]_2Zn$	1.728(7)	130.4(20)	ED	193
109 $[(Me_3Si)_2N]_2Cd$	1.727(7)	132.0(12)	ED	194
110 $[(Me_3Si)_2N]_2Hg$	1.732(9)	128.0(20)	ED	195
111 $[(Me_3Si)_2N]_3Al$	1.75(1)	118(1)	XRD	196
112 $[(Me_3Si)_2N]_3Fe$	1.731(3)	121.2	XRD	197
113 $[(Me_3Si)_2N]_3Sc$	1.715(8)	128.0(20)	ED	198
114 $[(Me_3Si)_2N]_3Ce$	1.696(15)	129(4)	ED	199
115 $[(Me_3Si)_2N]_3Pr$	1.705(15)	129(4)	ED	199

geometrical parameters are summarized in Table 9. When nitrogen is bonded to both silyl and methyl groups, the combination of two silyl and one methyl substituent seems to confer planarity to the central atom (see **89**, **90**). In the amines **91** and **92**, on the other hand, one silyl and two methyl groups appear to lead to a shallow pyramidal geometry,

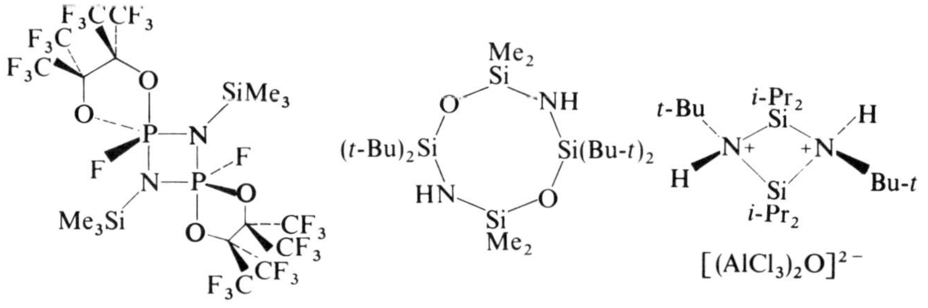

(97) **(98)** **(99)**

as evidenced by the deviation of the sum of the angles at nitrogen from 360°. The observed trend may be readily understood on the basis of the simple MO model for species XY_3 discussed in Section IV.A. Upon replacement of silyl groups by the more electronegative methyl groups distortion from a planar arrangement should become progressively more favourable. The molecular structure of $NMe_2(SiMe_3)$ **93** has been determined for both the gas and solid phases. Whereas the geometry at nitrogen is planar in the gas it is very slightly pyramidal in the crystal. The deviations from planarity in the gas-phase structures of **91** and **92** are so small that it is impossible to establish with certainty whether the average structure is truly pyramidal or whether the geometry at nitrogen is, in fact, planar with a large amplitude out-of-plane vibration. The increase in the Si–N–Si angle in the disilylamines $(R_3Si)_2NH$ for R = H, Me, Ph is in accordance with the order of decreasing electronegativity $SiH_3 > SiMe_3 > SiPh_3$. A remarkably wide Si–N–Si angle of 145.2(2)° has been recently found in the eight-membered $(SiOSiN)_2$ ring of **98**[178]. At the same time the Si–N distances [1.670(3), 1.682(3) Å] are short, as would be expected for bonds with an increased degree of s character. A number of cyclodisilazane structures have been determined (see References 179 and 180 for literature summaries). The Si_2N_2 rings are planar with N–Si–N angles between 84.4 and 89.4° and Si–N–Si angles between 90.5 and 95.6°; Si–N distances fall in the range 1.72–1.76 Å. Planarity of the Si_2N_2 ring is retained in the cyclodisilazane dication of **99**[181]. As expected for tetracoordinate protonated nitrogen atoms, the Si–N distances in the centrosymmetric cations are much longer (1.792 Å) than those in neutral cyclodisilazanes. For instance, the neutral cyclodisilazane with an identical substitution pattern to that in **99** exhibits an Si–N bond length of 1.747 Å[182]. The N–C bonds in **99** makes an angle of 16.8° with the plane of the Si_2N_2 ring.

Use of the bulky $(Me_3Si)_2N$ substituent has allowed the isolation of complexes with unusually low coordination numbers and oxidation states for many metals, for example, two- and three-coordinated transition metals. Geometrical parameters for selected $(Me_3Si)_2N$ derivatives are listed in Table 10. An early review of metal bis(trimethylsilyl)amides appeared in 1976[200]. For comprehensive coverage of individual structures the reader is refered to Reference 201. The Si–N bond length in derivative $M[N(SiMe_3)_2]_n$, $n=1$–3, has been found to fall between 1.64 Å in **105** and 1.75 Å in **111**. A general trend towards shorter Si–N distances as the ionic character of the M–N bond increases may be ascertained. This correlation is illustrated particularly well by the alkali metal complexes **100**, **101**, **104** and **105** and is also apparent for the alkaline earth complexes **106** and **107**. Four structural investigations have been carried out on the $(Me_3Si)_2NLi$ moiety. The fact that the Si–N bonds are shorter in the dimeric etherate complex **102** and in the monomeric crown ether complex **103** than in the dimer **100** or the trimer **101** indicates that the Li–N bonds possess a greater degree of ionic character when the lithium cation is stabilised by donor atoms. Although short Si–N bonds are often accompanied by wide Si–N–Si angles (e.g. **105**), the magnitude of these angles may not be regarded as a reliable guide to the degree of ionic character of M–N bonds in the metal bis(trimethylsilyl)amides, as steric effects may often influence these angles. The smaller Si–N–Si angles in the ether complexes **102** and **103** in comparison to **100** or **101** may well be attributed to non-bonded interactions between the ether and amide substituents.

All twelve compounds of the type $M[N(SiMe_3)_2]_3$, whose structures have been solved by X-ray diffraction, crystallize in the trigonal space group $P\bar{3}1c$. In all these molecules, the metal atom is coplanar with the NSi_2 plane of each of the three ligands and this plane is tilted out of the MN_3 plane by about 50°. The molecules may be divided into two classes:

(1) M is a group 13 metal (Al, Ga, In and Tl) or a 3d transition metal (Ti, V, Cr and Fe).
(2) M is Sc or a lanthanide (Nd, Eu and Yb).

Whereas molecules of class 1 exhibit a planar MN_3 skeleton, for those of class 2 the metal displays a pyramidal geometry. The following hypotheses have been proposed at various times to account for this state of affairs:

(1) dative π bonding is assumed to be of importance for class 1 but not for class 2 molecules,

(2) the M–N bonds in class 2 molecules display more ionic character than those in class 1 molecules,

(3) displacement of Sc and the other lanthanides out of the N_3 plane is due to intermolecular packing forces in the solid phase.

As subtraction of the appropriate metal atom radius from d(M–N) leads to an effectively constant value of 1.47 Å, bonding in both classes of molecule may be regarded as predominantly ionic[202]. The small differences between calculated and experimental

M–N bond lengths certainly do not warrant the drawing of any conclusions concerning the presence of dative π bonding. If the bonding in both classes of molecule is predominantly ionic, then why is the coordination of M different? An electron diffraction study[198] of $Sc[N(SiMe_3)_2]_3$ has provided revealing information on the structural changes which take place on going from the solid[203] to the gas phase. There is no significant change in the angle of twist of the NSi_2 planes in the two phases. However, whereas the ScN_3 skeleton is pyramidal in the solid state (N–Sc–N = 115.5°), it is essentially planar (N–Sc–N = 119.5°) in the gas phase. At the same time the Si–N–Si angles increase by ca. 7° and the Si–N bonds decrease by ca. 0.04 Å. These findings indicate that the energy barrier to pyramidalization of the ScN_3 skeleton must be low and that intermolecular forces in the crystal lattice will be responsible for the observed structural differences. Upon pyramidalisation the Si–N distance must increase in order to reduce interligand C · · · C contacts. Electron diffraction studies on $M[N(SiMe_3)_2]_3$ molecules with larger cations (M = Ce, Pr) have shown these to exhibit a pyramidal configuration in the gas phase. Taken together these results suggest that two factors will be of importance in establishing the metal atom geometry in such derivatives:

(1) non-bonded interligand C · · · C repulsions which favour a planar MN_3 skeleton, and

(2) dipole interactions which favour a pyramidal MN_3 skeleton.

Let us now turn our attention to Si–N bonds to two-coordinate nitrogen. Geometrical parameters for molecules containing Si–N=X fragments (X = C, Si, N, P) are listed

TABLE 11. Geometrical parameters in molecules containing Si–N=X fragments

Compound		d(Si–N)Å	∠Si–N–X(deg)	Method	Ref.
1.	X = C[a]				
	H_3SiNCO	1.703(4)	152.7(12)	ED	204
116	$Me_2HSiNCO$	1.719(5)	153.5(13)	ED	205
117	Me_3SiNCO	1.740(4)	156.9(30)	ED	206
	H_3SiNCS	1.704(6)	163.8(26)	ED	204
118	$Me_2HSiNCS$	1.723(8)	154.7(22)	ED	205
119	Me_3SiNCS	1.743(6)	158.2(10)	ED	207
120	Ph_3SiNCS	1.735(5)	170.7(2)	XRD	208
2.	X = Si				
11	$t\text{-}Bu_2Si{=}NSi(Bu\text{-}t)_3$	1.695(3)	177.8(2)	XRD	11
12[a]	$THF\cdot t\text{-}Bu_2Si{=}NSi(Bu\text{-}t)_3$	1.654(9)	161.5(5)	XRD	11
		1.667(10)	161.0(6)		
3.	X = N				
	H_3SiN_3	1.719(8)	123.8(10)	ED	209
121	Me_3SiN_3	1.734(7)		ED	210
122	Ph_3SiN_3	1.74(2)	120.5(15)	XRD	211
123	$Me_3SiN{=}NSiMe_3$	1.808(3)	120.0(4)	XRD	212
4.	X = P				
124	$Me_3SiN{=}PMe_3$	1.705(5)	144.6(11)	ED	213
125	$Me_3SiN{=}PN(SiMe_3)_2$	1.721(1)	129.9(1)	XRD	214
126	$(Me_3SiN)_2PN(SiMe_3)_2$	1.703(3)	146.8(2)	XRD	215
127		1.691(4)	172.7(3)	XRD	216
128	$Me_3CNH(Me_3SiN)PCr(CO)_5$	1.699(4)	159.4(3)	XRD	217
129		1.69(2)	173.9(19)	XRD	218
130		1.677(3)	154.8(2)	XRD	219
131		1.705(2)	125.2(2)	XRD	219

[a] Shrinkage corrections have not been applied to the ED structures.

in Table 11. We have already considered the molecular structures of linear molecules $H_2M{=}NH$ in Section III.C. Upon bending, the HOMO, which effectively contains the nitrogen lone pair, and a σ^* orbital of higher energy can interact, leading to the stabilisation of a bent geometry provided that the energy gap between the two orbitals is small (e.g. M = C). As the electronegativity of M decreases (e.g. on going from C to Si), this energy gap will increase, which means that wider, or even linear, angles at nitrogen will become more favourable. Replacement of the nitrogen substituent H with a more electropositive group will have a similar effect on the angle at nitrogen.

As the vibrational selection rules and rotational fine structures of the observed bonds were in accordance with C_{3v} symmetry, early IR studies of SiH_3NCS[220] and SiH_3NCO[221] were interpreted as suggesting that these molecules must exhibit linear or effectively linear skeletons in the gas phase. Linearity of the molecular skeleton was confirmed by MW for SiH_3NCS[222], although the observation of an extensive series of vibrational satellite spectra indicated the presence of a very low frequency bending vibration. An even more complex microwave spectrum was recorded for SiH_3NCO[223] and it could be later shown that the low frequency vibration was a two-dimensional anharmonic bend at nitrogen[224]. This means that SiH_3NCO will possess a non-rigid skeleton with many low-lying vibrational levels populated at room temperature. It exhibits a potential minimum for a Si–N=C angle of 159° but the linear structure is only 32 cm^{-1} higher in energy. As a result of the large amplitudes of the bending vibration, ED studies[204] on SiH_3NCS and SiH_3NCO yield apparently bent skeletons. When shrinkage corrections are applied good agreement with MW results (i.e. linear structures) is achieved. It seems likely that the SiNCS skeleton in the methylated pseudohalides is also pseudolinear.

Fragments Si–N=X, where X is nitrogen or phosphorus, underline the importance of the electronegativity of X for the equilibrium value of the angle Si–N=X. When X = N this angle is close to 120° (i.e. effectively sp^2 hybridization). Replacement of N by the less

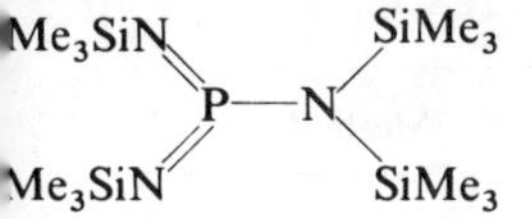

(126)

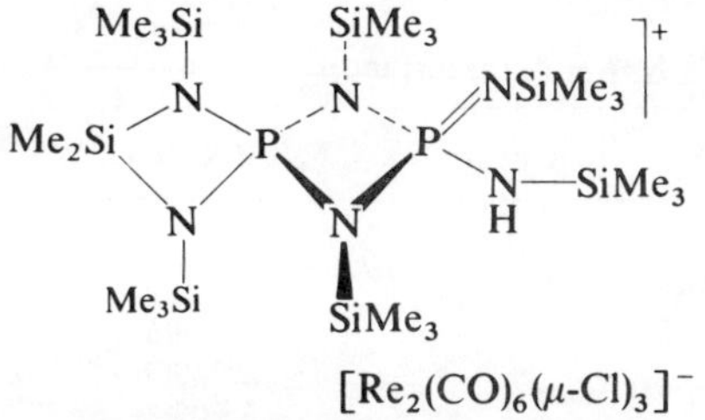

(127)

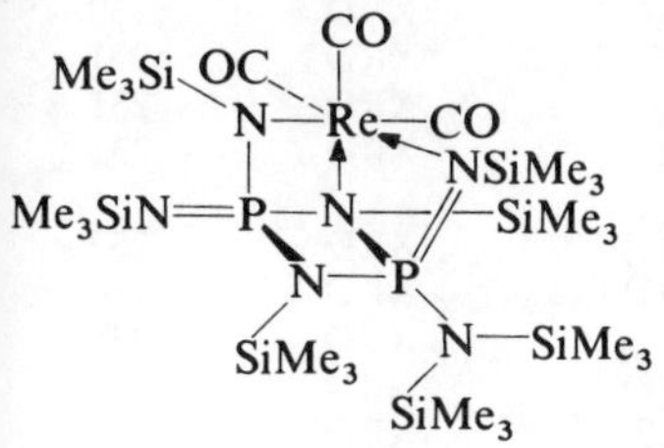

(129)

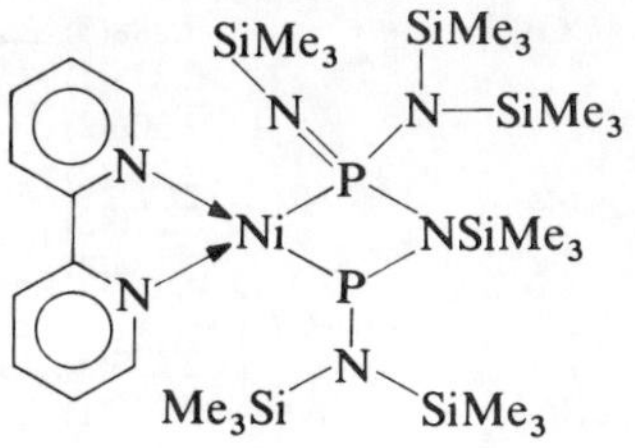

(130)

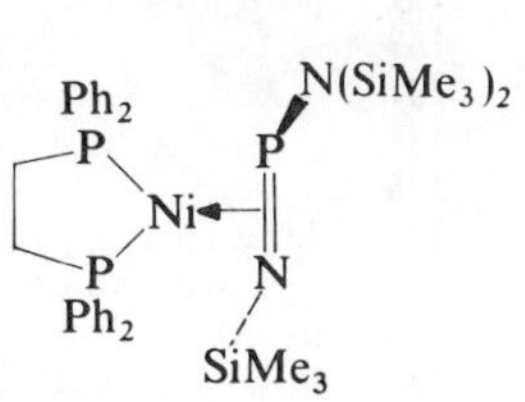

(131)

electronegative P leads to a significant lowering of the energy barrier to linearisation and Si–N–P angles in the range 130–174° have been recorded in compounds **124**–**130**. Both the N–Si and N=P bonds are shortened upon widening of the Si–N=P angle, due to an increase in the s contribution to nitrogen hybridization. For instance, a remarkably short P=N double bond length of 1.487(3) Å is found for the cation of **127**. The minimum value of the Si–N=P angle will be controlled by non-bonded interaction of the silicon and phosphorus atoms. For example, the sum of their hard-sphere radii (3.00 Å) is reached for an Si–N–P angle of 136° in **130**, using the observed bond distances. In the side-on coordinated amino(imino)phosphane ligand of complex **131**, the P=N distance increases to 1.646(2) Å and the Si–N=P angle relaxes to 125.2(2)°, indicating that a formulation as an NiPN–metallocyclus is more appropriate.

2. Bonds to phosphorus and its heavier congeners

Selected Si–X(X=P–Bi) distances are listed in Table 12. Replacement of nitrogen by the less electronegative phosphorus atom in species XY_3 should reduce the HOMO–LUMO gap, which means that the pyramidal geometry should be more favourable for silylphosphines than for the analogous nitrogen derivatives. Planar tricoordinated phosphorus has not been observed. Indeed the Si–P–Si angle of 96.5(5)° in $P(SiH_3)_3$ is similar to that of 98.6(3)° in trimethylphosphine[256]. In contrast to the silylamines, substitution of SiH_3 for H in $MePH_2$ and Me_2PH has little effect on the coordination geometry at phosphorus. The Si–P–C angles (99–101°) lie in the typical range for C–P–C angles in alkylphosphines. The experimental Si–P distance of 2.249(3) Å is close to that for the calculated 3–21G* structure[257] (2.258 Å). In comparison to the Si–N bond the ionic character of the Si–P bond is relatively limited and the typical range of 2.24–2.30 Å for Si–P bond lengths (Table 12) straddles the sum of the covalent radii of silicon and phosphorus (2.27 Å).

TABLE 12. Selected Si–X (X=P–Bi) distances

	Compound		d(Si–X)(Å)	Method	Ref.
	X=P				
	1. Silylphosphines				
	$P(SiH)_3$		2.248(3)	ED	225
	H_2PSiH_3		2.249(3)	ED	226
	H_2PSiF_3		2.207(3)	ED	227
	$MeHPSiH_3$		2.248(3)	ED	226
	Me_2PSiH_3		2.245(3)	ED	226
132	$[(Me_3Si)_2PMe_2^+][Co(CO)_4^-]$		2.288(5), 2.294(5)	XRD	228
133			2.278(1)	XRD	229
134			2.306(2)	XRD	230
135			2.264(2)	XRD	231
136			2.270(2)	XRD	231
137	$Me_3Si(t\text{-}BuP)_4SiMe_3$		2.284(2), 2.286(2)	XRD	232
	2. Rings and cages				
138	$(t\text{-}BuP)_2SiPh_2$		2.223(2), 2.226(2)	XRD	233
139	$(2,4,6\text{-}t\text{-}Bu_3C_6H_2P)_2Si(Bu\text{-}t)_2$		2.239(1), 2.243(1)	XRD	234
140[a]	$(t\text{-}BuP)_2Si(PBu\text{-}t)_2$		2.201(1)	XRD	235
141[a]	$(t\text{-}BuP)_2Si(PBu\text{-}t)_2$		2.211(3), 2.217(3)	XRD	235
142[b]	$(t\text{-}BuP)_8Si_2$	(SiP_2)	2.233(2), 2.248(2)	XRD	236
		(Si_2P_2)	2.267(2), 2.317(2)		
		(SiP_4)	2.286(2), 2.287(2)		

TABLE 12. (*Continued*)

Compound		d(Si–X)(Å)	Method	Ref.
143[c]	$[(Me_3SiNMe)_2SiPH]_2$	2.269(7), 2.275(7)	XRD	237
144[d]	$[(Me_3SiNMe)_2SiP(Bu\text{-}t)]_2$	2.241(2), 2.280(1)	XRD	238
145[d]	$[t\text{-}Bu_2SiP(Bu\text{-}t)]_2$	2.255(1)–2.263(1)	XRD	238
61	$(PhPSi_2Me_4)_2$	2.249(2), 2.256(2)	XRD	136
62	$(PhP)_2(SiMe_2)_3$	2.268(1), 2.286(1)	XRD	137
63	$[PhHP(SiMe_2)_2PHPh]Mo(CO)_4$	2.266(1)	XRD	138
146	$[Me_2Si(t\text{-}BuP)_2]_2$	2.249(3)–2.279(3)	XRD	239
147	$(SiMe_2)_5P_2(PH)_2$	2.234(3)–2.254(2)	XRD	240
148	$(SiMe_2)_6P_4$	2.249(1)	XRD	241
149	$(SiMe_2)_3P_4$	2.283(1)	XRD	242
150	$(SiMe_3)_3P_7$	2.281(3)–2.297(5)	XRD	242
151[e]	$(SiPh_3)_3P_7$	2.286(6)	XRD	243
152	$(SiMe_2)_3P_4(Bu\text{-}t)_2$	2.258(1)–2.291(1)	XRD	244
153[f]	$[(SiMe_2)_3P_4]_3[Cr(CO)_3]_2$	2.246(6)–2.284(13)	XRD	245
	X = As			
	$As(SiH_3)_3$	2.355(1)	ED	246
	Me_2AsSiF_3	2.334(9)	ED	247
154	$(SiMe_3)_4As_2$	2.362(2), 2.365(1)	XRD	248
155	$[(Me_3Si)_2AsLi \cdot DME]_2$	2.307(7)	XRD	249
156	$(SiMe_2)_6As_4$	2.351(2)	XRD	250
157[e]	$(SiMe_3)_3As_7$	2.397(3)	XRD	251
	X = Sb			
	$Sb(SiH_3)_3$	2.557(4)	ED	252
158	$(SiMe_3)_4Sb_2$	2.587(1), 2.601(1)	XRD	253
159	$[(Me_3Si)_2SbLi \cdot DME]_n$	2.532(1)	XRD	254
	X = Bi			
160	$(SiMe_3)_4Bi_2$	2.679(13), 2.687(16)	XRD	255
161	$[(Me_3Si)_2BiLi \cdot DME]_n$	2.633(14)	XRD	255

[a] Isomer **140** has point symmetry $\bar{4}$, isomer **141** point symmetry 2.
[b] Si–P distances are given for the SiP_2-, Si_2P_2- and SiP_4-rings,
[c] Isomer **143** has *cis* symmetry.
[d] Isomers **144** and **145** have *trans* symmetry.
[e] The mean value of d(Si–X) is given.
[f] **153** crystallizes in an orthorhombic and a hexagonal form (with one molecule toluene). The d(Si–P) range is given for both modifications.

Net atomic charges of -0.120(P) and $+0.430$(Si) were found for the 3–21G* structure of silylphosphine[257]. Substitution of H by F in silylphosphine leads to a reduction of the Si–P distance from 2.249(3) Å to 2.207(3) Å in H_2PSiF_3. On the other hand, the Si–P

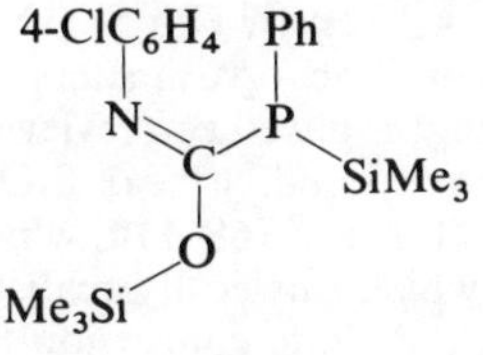

(133)

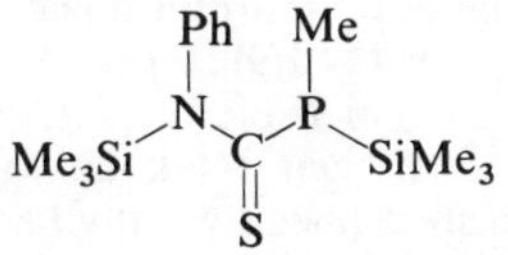

(134)

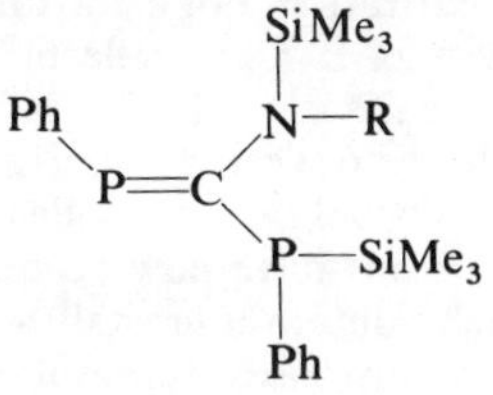

(135) (R = 4-$CF_3C_6H_4$)
(136) (R = 2-ClC_6H_4)

bond lengths in the cation $[(Me_3Si)_2PMe_2]^+$ lie close to the upper limit of the typical Si–P bond range. Short Si–P bonds in the range 2.201–2.243 Å are found in the three-membered SiP_2 rings of the diphosphasiliranes **140**–**142**. Steric strain may also lead to

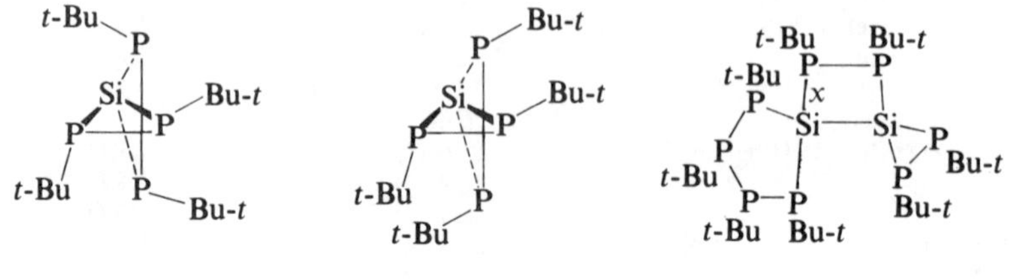

(140) **(141)** **(142)**

relatively long Si–P distances, as is well illustrated by a comparison of the Si–P bonds in **142**. The Si–P distance x in the strained four-membered Si_2P_2 ring is particularly long [2.317(2) Å], presumably as a result of interligand repulsion. Three crystal structure determinations of 1,3-diphospha-2,4-disilacyclobutanes have been reported. Whereas planar Si_2P_2 rings are found for the *trans* isomers **144** and **145**, the *cis* isomer **143** is non-planar with a fold angle of 27.2° between the two SiP_2 planes.

Fewer structural studies have been carried out on molecules containing bonds between silicon and the heavier group 15 elements. Both trisilylarsine and trisilylstibine are pyramidal with Si–X–Si angles of respectively 93.8(2)° and 88.6(2)°. The Si–X bond lengths are close to the covalent radii sums of 2.38 Å (Si–As) and 2.58 Å (Si–Sb). As for the analogous molecule $(SiMe_3)_3P_7$, steric overcrowding may be assumed to be responsible for the relatively long Si–As bonds (2.397(3) Å) to the As_7 cage of $(SiMe_3)_3As_7$. Crystal structures of $(Me_3Si)_4X_2$ and $(Me_3Si)_2XLi \cdot DME$ are available for X = As–Bi. For all three elements a pronounced shortening of the Si–X bond lengths (0.046–0.069 Å) is observed upon going from the dimetallane to the lithium derivative. The observed trend suggests that the Li–X bonds must possess a marked degree of ionic character. Whereas discrete dimers of **155** are found in the crystal lattice, the $(Me_3Si)_2X$ groups in **159** and **161** are linked via lithium atoms into helical chains.

D. Bonds to Elements of Group 16

1. *Bonds to oxygen*

We have discussed the bonding in disiloxanes in detail in Section IV.A. The HOMO–LUMO energy gap is of critical importance for the equilibrium value of the angle Y–X–Y. An increase in the magnitude of this energy gap will reduce the barrier to linearization. For disiloxanes $(R_3Si)_2O$ this may be achieved by reducing the electronegativity of R_3Si, as reflected by the adoption of a linear SiOSi skeleton by $(Ph_3Si)_2O$ in comparison to the Si–O–Si angle of 144.1(9)° for $(H_3Si)_2O$. Geometrical parameters for selected oxygen derivatives are given in Table 13. Si–O bond lengths have been reviewed by Voronkov and colleagues[278] in their book on the siloxane bond. Linear SiOSi skeletons have now been established unequivocally for the disiloxanes **168**–**170**, which each contain a crystallographic centre of symmetry and for which physically realistic Si–O distances were obtained. A very high R factor precludes a definite conclusion for $[(CF_3CH_2CH_2)_3Si]_2O$[279] **183**. The observation of apparent C_i symmetry for disiloxanes $(R_3Si)_2O$ in the solid state does not, however, necessarily allow the conclusion to be drawn that the Si–O–Si angle is strictly linear. Several examples are known in which the

TABLE 13. Geometrical parameters in selected oxygen derivatives

	Compound	d(Si–O)Å	∠ Si–O–X(deg)	Method	Ref.
	1. Disiloxanes				
	$O(SiH_3)_2$	1.634(2)	144.1(9)	ED	4
		1.630(5), 1.632(5)	142.2(3)	XRD	75
162	$O(SiH_2Me)_2$	1.642(3)	143.0(6)	ED	258
163	$O(SiHMe_2)_2$	1.635(2)	148.4(9)	ED	258
164	$O(SiMe_3)_2$	1.631(3)	148(3)	ED	259
		1.626(2)	148.8(2)	XRD	75
	$O(SiCl_3)_2$	1.59(1)	146(4)	ED	74
	$O(SiF_3)_2$	1.58(3)	156(2)	ED	73
165[a]	O[SiOBu-t)$_3$]$_2$	1.635	144.0	XRD	260
166	O[SiPh$_2$(Bu-t]$_2$	1.633(1), 1.638(1)	152.4(1)	XRD	261
167	$Ph_3SiOSiPh_2Me$	1.622(9)	158.9(6)	XRD	262
168	O(SiCH=CH$_2$)$_2$	1.614(1)	180	XRD	263
169	$O(SiPh_3)_2$	1.616(1)	180	XRD	5
170	O[Si(CH$_2$Ph)$_3$]$_2$	1.613(4)	180	XRD	264
	2. Compounds with Si–O–X fragments				
	SiH_3OCH_3	1.640(3)	120.6(10)	ED	265
	SiH_3OPh	1.648(7)	121(1)	ED	266
171	(X = P)	1.684(7)	157.8(6)	XRD	267
172	$Ph_3SiOOSiPh_3$	1.668(6)	103.6(5)	XRD	268
173	$Me_3SiOOSiMe_3$	1.681(3)	107(1)	ED	269
174	[PhCH$_2$(Me)$_2$SiO]$_2$	1.687(1)	104.5(1)	XRD	270
175[b]	$(Me_3SiO)_2TeF_4$		137.4	XRD	271
176	[Me$_3$SiOAlBr$_2$]$_2$	1.70(1)	130.4(5), 133.5(6)	XRD	272
177	$Ph_3SiOPbPh_3$	1.87(3)[c]	142(1)	XRD	273
178	$(Ph_3Si)_2CrO_4$	1.601(11), 1.605(11)	133.1(6), 162.7(7)	XRD	274
179	$Me_3SiOReO_3$	1.69(8)	164(5)	XRD	275
	3. Si–O bonds in silanolate anions				
180	$NaOSiMe_3 \cdot 3H_2O$	1.617(3)		XRD	276
181	Na$_2$[(Me$_2$SiO)$_2$O]$\cdot$4H$_2$O	1.587(9)[d]		XRD	276
182	Na$_3$[PhSi(O)O]$_3$	1.586(2)[d]		XRD	277

[a] The structure is disordered; the listed geometrical parameters were calculated on the basis of an analysis of the thermal motion of the molecule.
[b] The Si–O distance was not reported.
[c] The structure is disordered so as to give reversed occupation of ca. 16% of the Si/Pb sites; the Si–O distance cannot, therefore, be regarded as reliable.
[d] Average value.

disiloxane is disordered about a crystallographic centre of symmetry, which leads to the observation of an averaged structure by X-ray diffraction, in which the bridging O atom appears to lie at the centre of symmetry. The Si–O bond lengths will be unrealistically short in such averaged molecular structures and the O atoms will appear to exhibit unusually large thermal motions at right angles to the SiOSi axis. This is the case for **165** in which the central Si–O distance appears to be only 1.556(2) Å. However, an analysis of the anisotropic temperature factors for the atoms of the crystallographically equivalent SiO_4 tetrahedra suggests that these move as rigid bodies which exhibit pronounced librations about their respective Si atoms[260]. When the geometrical parameters are corrected to allow for this rigid body motion, all Si–O bond lengths are practically identical (1.633–1.637 Å) and the Si–O–Si angle has a value of 144.0° close to that in disilyl ether. A disordered structure has also been established for tetraphenyldisiloxane $(SiHph_2)_2O$ **184**[280].

Wide bond angles at oxygen have also been found for various SiOX linkages, where X is less electronegative than oxygen; examples are listed in Table 13. The crystal structure of $(Ph_3Si)_2CrO_4$ **178** is particularly interesting, as this molecule exhibits contrasting Si–O–Cr angles [133.1(6) and 162.7(7)°] caused presumably by lattice forces, which suggests that the energy barrier to linearization must be relatively small. A recent theoretical study[281] has shown that only ca. 4 kJ mol^{-1} are necessary for the widening of the Si–O–Si angle from 142° to 180°. In contrast the HOMO–LUMO energy gap will be small for the organosilicon peroxides (X = O) and narrow Si–O–O angles will be expected. Typical values[270] lie in the range 103–107° (**172–174** in Table 13).

On the basis of 51 organic silicon compounds available in 1976 Voronkov and coworkers[278] established that the Si–O bond length falls in the range 1.64 ± 0.03 Å for the vast majority of such derivatives. The results of the past decade have confirmed this finding. As for Si–N bonds, this value is much smaller than the Stevenson–Schomaker corrected sum of the covalent radii of silicon and oxygen (1.76 Å). A trend towards smaller Si–O distances with increasing Si–O–Si angles is apparent from Table 13. For instance, the Si–O bond length decreases from 1.634(2) Å in $(SiH_3)_2O$[144.1(9)°] to 1.616(1) Å in $(SiPh_3)_2O$ (180°). Karle and coworkers[261] obtained a least-squares line represented by $d(\text{Si–O}) = 1.6957 – 0.000455\theta$ (θ is the Si–O–Si angle in degrees) for eight disiloxanes. This correlation is in accordance with an increase in the s contribution to oxygen hybridization upon angle widening. The shortest Si–O distances (1.58–1.60 Å) are found for bonds with an increased degree of ionic character, e.g. when silicon is attached to very electronegative substituents as in $(SiF_3)_2O$[1.58(3) Å] or for terminal $Si–O^-$ bonds in silanolate anions such as **181** [1.587(9) Å]. A reduction in the degree of ionicity will be expected for the Si–O bond when the oxygen is also bonded to a very electronegative group. This is the case in the organosilicon peroxides for which Si–O distances in the

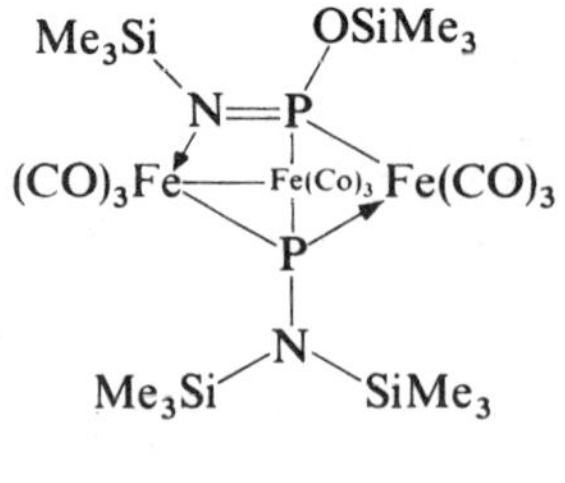

(171)

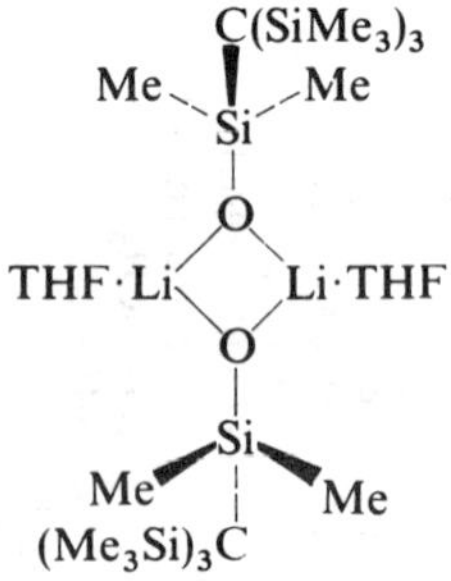

(187)

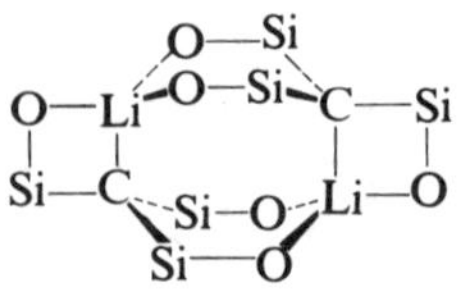

(188A)

range 1.67–1.69 Å have been reported[270]. As the Si and Pb positions in $Ph_3SiOPbPh_3$ **177** are partially disordered, the value of 1.87(3) Å for the Si–O bond length in this molecule should not be regarded as realistic. A similar disorder was observed for $Ph_3SiOGePh_3$ **185** and $Ph_3SiOSnPh_3$ **186**, for which Si–O–X angles respectively 142.5(2)° and 144.2(6)° were reported[282]. Steric strain is responsible for the long Si–O bond lengths of 1.681(3) Å in the 1,2-disilaoxetane **46**[118], and 1.663(3)–1.721(4) Å in the 1,3-cyclodisiloxanes, discussed later in this section. Several derivatives with SiOLi linkages have been characterized by X-ray diffraction. Dimeric species have been reported for silanolate $[(Me_3Si)_3CSiMe_2OLi.THF]_2$ **187**[283] and the compound $[(MeOSiMe_2)_3\,CLi]_2$ **188**[284] for which the essential skeleton (**188A**) is depicted. Each of the component molecules in the dimer **188** has one of its own OMe groups internally coordinated to its own Li atom to yield a planar four-membered ring, and its remaining OMe groups are coordinated to the Li atom of the other component molecule to give two six-membered rings in boat conformation. In contrast to **187** a cubane-type structure is found for the aminosilanolate $[t\text{-}Bu_2Si(NH_2)OLi]_4$ **189** for which the essential skeleton (**189A**) is given[285]. The central cubane-type M_4O_4 unit has also been found in the

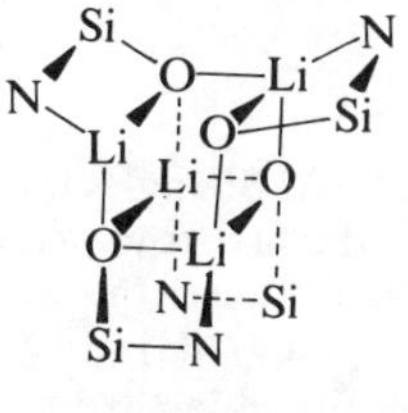

(**189A**)

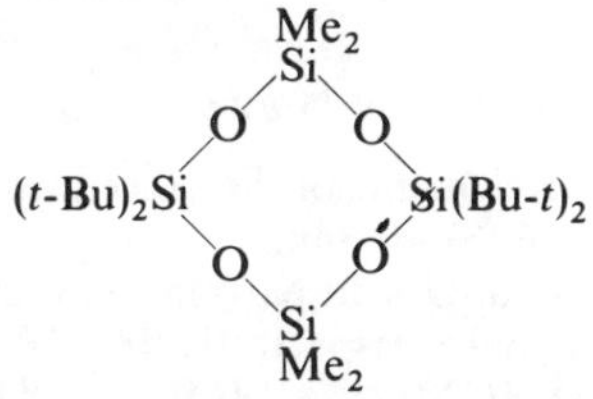

(**194**)

isotypic trimethylsilanolates $[Me_3SiOM]_4$ (M = K, Rb, Cs) **190–192**, the structures of which were determined by X-ray diffraction from powder data[286], and for $[Me_3SiOHgMe]_4$ **193**[287]. The structure of Me_3SiONa is not known but discrete tetrahedral $[Me_3SiO]^-$ anions are found for the trihydrate **180** with the sodium cations being coordinated exclusively by water oxygens.

The structural chemistry of organocyclosiloxanes have been a field of intensive study for the past 30 years. As a comprehensive review[288] of this topic appeared in 1980 only a brief account will be presented in the present article with emphasis being placed on recent developments. For cyclosiloxanes $(RR'SiO)_n$ ($n = 2$–4) the following trends may be ascertained:

(1) The Si–O bond length decreases, in general, on going from $n = 2$ to $n = 4$. The observed ranges are 1.663–1.721 Å ($n = 2$[289]), 1.64–1.65 Å ($n = 3$[288]) and 1.61–1.65 Å ($n = 4$[178,288]).

(2) The average Si–O–Si angles increase as the ring size increases. The characteristic ranges are 86.2–91.1° ($n = 2$), 132–134° ($n = 3$) and 140–149° ($n = 4$). Si–O–Si angles of 156.0(1)° and 164.2(1)° are observed in the cyclotetrasiloxane **194**[178].

(3) Both planar and slightly puckered $(SiO)_2$ rings have been found for $n = 2$. Cyclotrisiloxanes exhibit virtually planar six-membered rings with deviations from the ring plane of $\leqslant 0.10$ Å.

A variety of conformations have been found for the $(SiO)_4$ rings of the cyclotetrasiloxanes in the solid state (e.g. C_i-chair, C_2-boat, S_4-boat), indicating that energy differences are small and that conformational transitions will be possible. The eight-membered ring in **194** is effectively planar.

The recent structural characterization[289] of three cyclodisiloxanes $[Mes_2SiO]_2$ **195**, *trans*-$[t\text{-Bu(Mes)SiO}]_2$ **196** and *cis*-$\{[(Me_3Si)_2N](Mes)SiO\}_2$ **197** aroused considerable interest as these compounds exhibit transannular silicon–silicon non-bonded distances of respectively 2.306(3), 2.396(11) and 2.349(1) Å close to the typical range of 2.33–2.37 Å for Si–Si single bonds (see Section IV.B.2). Whereas the $(SiO)_2$ rings are puckered in **195** and **197** with interplanar dihedral angles of respectively 6 and 10°, the four-membered ring in **196** is strictly planar, as the molecule contains a crystallographic centre of symmetry. The short Si · · · Si distance has been attributed to strong O · · · O repulsions[289]. *Ab initio* calculations[290,291] provide no evidence for bonding interactions between the silicon atoms. At the 3–21G level[291], the electronic structures of cyclic systems $(H_2SiX)_2$ (X = O, NH, CH_2 or S) are qualitatively similar and the short non-bonded Si · · · Si distances are predominantly determined by the respective Si–X bond lengths. This accounts for the increase in the 1,3-silicon–silicon distance on traversing the series X = O, NH, CH_2, S. Representative experimental values are 2.31, 2.49, 2.70 and 2.84 Å (see Reference 291). Mulliken overlap population analysis of $(H_2SiO)_2$ indicates that the silicon–silicon interaction is slightly antibonding and that the antibonding interaction increases in magnitude for small increases of the Si · · · Si distance.

2. *Bonds to sulphur and its heavier congeners*

As established for elements of the preceding group, Si–X–Si angles for the second-row elements of group 16 (S) are much narrower than for the first-row element (O) and are close to the C–X–C angles in carbon analogues. For instance, the M–S–M angles in $(CH_3)_2S$ and $(SiH_3)_2S$ are respectively 98.9(2)° (ED[306]) and 97.4(1)° (ED[76]). As discussed in Section IV.A, the HOMO–LUMO gap will decrease dramatically for species $(R_3Si)_2X$ on going from X = O to X = S. It follows that the energy minimum will be found at much smaller Si–X–Si angles for sulphur and that the energy barrier to linearization will be much greater for this element than for the more electronegative oxygen.

Geometrical parameters for sulphur derivatives are given in Table 14. In acyclic derivatives Si–S–Si angles have been observed in the range 97–112°. Attachment of the more electronegative *t*-butoxy substituents to silicon leads to markedly wider angles (105–108°) than in SiH_3S. The Si–S bond is much less polar than the Si–O bond. From the values listed in Table 14 a typical range 2.11–2.17 Å may be established for the former bond. The sum of the covalent radii for silicon and sulphur is 2.21 Å. The degree of ionicity of a typical Si–S bond in comparison to that of a typical Si–O bond may be gauged by the shortening of these bonds relative to their respective covalent radii sums. This shortening is 0.07 Å for a typical Si–S bond (2.14 Å) but 0.19 Å for a typical Si–O bond (1.64 Å).

The cyclodisilthianes **202** and **203** exhibit crystallographic C_i symmetry so that the $(SiS)_2$ rings are strictly planar. The Si–S–Si angles [82.5(1) and 82.2(1)°] are much narrower than the S–Si–S angles. A remarkable distortion of the silicon coordination

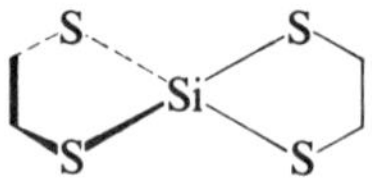

(204)

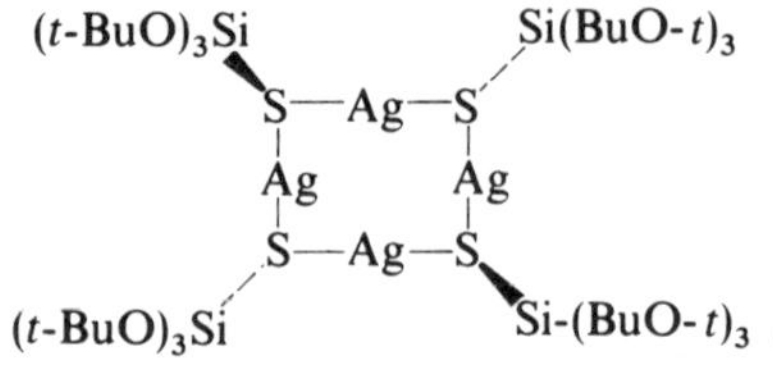

(210)

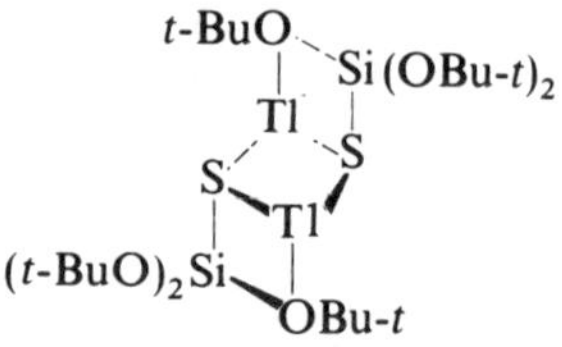

(211)

TABLE 14. Geometrical parameters in selected sulphur derivatives

	Compound	d(Si–S)(Å)	∠Si–S–X(deg)	Method	Ref.
	1. Acyclic derivatives (X = Si)				
	$S(SiH_3)_2$	2.136(2)	97.4(7)	ED	292
		2.141(2), 2.142(3)	98.4(1)	XRD	76
198[a]	$S[Si(OBu\text{-}t)_3]_2$	2.150, 2.152	108.0	XRD	260
199	$S(SiPh_3)_2$	2.150(1), 2.153(1)	112.0(1)	XRD	77
200	$(SiMe_2)_2[(t\text{-}BuO)_3SiS]_2$	2.122(1), 2.146(1)	105.3(1)	XRD	293
201	$(SiMe_2)_6[(t\text{-}BuO)_3SiS]_2$	2.134(5), 2.138(5)	107.7(2)	XRD	293
	2. Rings and cages (X = Si)				
202	$(Me_2SiS)_2$	2.152(2)	82.5(1)	XRD	294
203	$[(t\text{-}BuO)_2SiS]_2$	2.131(2), 2.142(1)	82.2(1)	XRD	295
204[b]	$(C_2H_4S_2)_2Si$	2.115(1), 2.117(1)	100.1(1), 100.5(1)	XRD	296
205[c]	$(PhMeSiS)_3$	2.143(3)	106.2(2)	XRD	297
206	$Me_{10}Si_6S$	2.169(3)	94.9	XRD	298
207	$(HSi_4)S_6$	2.133(1)–2.137(1)	102.2(1), 103.0(1)	XRD	299
	3. Acyclic derivatives (X ≠ Si)				
208	$[(Me_3SiS)_2CPMe]_2$	2.155(2)–2.166(1)	107.2(4)–109.0(3)	XRD	300
209	$(Me_3SiS)_2C{=}PPh$	2.185(1), 2.186(1)	110.4(1), 111.7(1)	XRD	301
210	$[(t\text{-}BuO)_3SiSAg]_4$	2.111(6)–2.132(4)	93.1(2)–99.5(2)	XRD	302
211	$[(t\text{-}BuO)_3SiSTl]_2$	2.078(2)	89.4(1), 90.5(1)	XRD	303
212	$[(t\text{-}BuO)_3SiS]_2Hg$	2.124(2)	95.6(1)	XRD	304
213	$\{[\{(t\text{-}BuO)_3SiS]_2Pb\}_2$	2.088(11)–2.143(10)	89.7(2), 90.5(2)	XRD	305

[a] The listed geometrical parameters were calculated on the basis of an analysis of the thermal motion of the molecule.
[b] Only the endocyclic Si–S–Si angles are given for this spirobicyclic molecule; exocyclic Si–S–Si angles are 104.6(1) and 124.7(1)° (see text).
[c] Only average values of the geometrical parameters were given in the original article.

towards a square planar geometry is found for the spirobicyclic derivative **204**. The dihedral angle between the two ring SSiS planes is 15.6° smaller than the orthogonal value (90°). MNDO/1 calculations on the species $M(XH)_4$ (M = C, Si; X = O, S) demonstrated that the energy difference between tetrahedral and planar coordination is much smaller for M = Si, as a result of the markedly greater degree of s character of the Si–X bonds[296]. The following energy differences ($kJ\ mol^{-1}$) were calculated: $Si(SH)_4$ = 204.6, $Si(OH)_4$ = 215.5, $C(SH)_4$ = 475.3, $C(OH)_4$ = 605.0. Whereas only 13.8 $kJ\ mol^{-1}$ are necessary to twist the SSiS planes in $Si(SH)_4$ by 15° relative to one another, 56.5 $kJ\ mol^{-1}$ are required for a similar distortion of $C(OH)_4$. The calculations suggest that introduction of ethylene bridges between the S atoms as in **204**, which leads to non-planar five-membered rings, will further reduce the required energy, so that only 10.0 $kJ\ mol^{-1}$ will be required to twist the SSiS planes by 15° relative to one another in this spirobicyclic derivative. Intermolecular packing forces are capable of providing an energy difference of this magnitude.

The structures of four tri-*t*-butoxysilanthiolates **210**–**213** have been reported. Whereas **210** is tetrameric with a nearly planar eight-membered ring, **211** and **212** are dimeric. In addition to the two thiolate S atoms the thallium atoms in **211** are also coordinated, albeit more weakly, by one butoxy O atom at a distance of 2.799(4) Å. A similar (2 + 1) coordination is found for the mercury atoms in the monomeric species **212** [*d*(Hg–O) = 2.878(4) Å]. Remarkably short Si–S distances are found in **213** and, in particular, **211** [2.078(2) Å], suggesting that the M–S bonds (M = Pb, T1) may exhibit a marked degree of ionicity in these compounds.

An electron diffraction study[307] on $(SiH_3)_2Se$ has yielded Si–Se distances of 2.275(4) Å with an Si–Se–Si angle of 96.6(7)°. In solid state[76] this molecule displays the following geometrical parameters: *d*(Si–Se) 2.27(2) Å, Si–Se–Si 95.7(5)°.

E. Bonds to Elements of Group 17

Silicon–halogen bond distances in selected halogenosilanes are listed in Table 15. Si–F distances are 0.10–0.13 Å shorter than the Stevenson–Schomaker corrected sum of the covalent radii for silicon and fluorine (1.69 Å), which once again demonstrates that this correction underestimates the degree of ionic character for silicon bonds to first-row elements of groups 15–17. In contrast, the Si–Cl distances are only marginally shorter (0.00–0.03 Å) than the sum of their covalent radii corrected for the electronegativity difference (2.05 Å). A progressive shortening of the Si–F distances upon substitution of H by F atoms is found for the species $SiH_{4-n}F_n$, $n = 1$–4, and $MeSiH_{3-n}F_n$, $n = 1$–3. For instance, the Si–F bond length decreases from 1.593(5) to 1.556(2) Å on going from SiH_3F to SiF_4. As expected on the basis of their relative electronegativities, the reduction of the Si–Cl distances in the analogous chlorosilanes is somewhat smaller. Whereas the difference between the Si–F bond lengths in SiH_3F and SiF_4 is 0.037 Å, this difference is only 0.029 Å for the analogous chlorosilanes. Replacement of hydrogen substituents by methyl groups leads to a small increase in the Si–X distance (X = F, Cl). A similar effect is observed for SiH_3Br and Me_3SiBr and for the analogous carbon derivatives. However, the paucity of structural data for the bromo- and iodosilanes prevents the conclusive establishment of general bond-length trends for these molecules. Steric strain leads to a lengthening of the Si–I distances in $I(t\text{-}Bu_2Si)_3I$ **43**[115] to an average value of 2.55 Å in comparison to the corrected sum of the covalent radii of silicon and iodine (2.46 Å).

F. Bonds to Other Main Group Elements

Relatively few structural studies have been carried out on compounds containing bonds between silicon and a main group metal. Table 16 contains a comprehensive listing

TABLE 15. Si–X (X = F, Cl, Br, I) bond distances in selected halogenosilanes

Compound		d(Si–X) (Å)	Method	Ref.
1.	Fluorosilanes			
	SiH_3F	1.593(5)	MW	308
	SiH_2F_2	1.577(3)	MW	309
	$SiHF_3$	1.5624(1)	MW	310
	SiF_4	1.556(2)	ED	311
30	$MeSiH_2F$	1.597(5)	MW	102
31	$MeSiHF_2$	1.583(2)	MW	103
32	$MeSiF_3$	1.574(7)	MW	104
2.	Chlorosilanes			
	SiH_3Cl	2.048(2)	MW	308
	SiH_2Cl_2	2.033(3)	MW	312
	$SiHCl_3$	2.020	MW	313
	$SiCl_4$	2.019(3)	MW	314
214	$MeSiH_2Cl$	2.052(5)	MW	315
215	$MeSiHCl_2$	2.040	MW	316
216	$MeSiCl_3$	2.026	MW	313
3.	Bromosilanes			
	SiH_3Br	2.210(2)	MW	308
	$SiHBr_3$	2.170(2)	MW	317
217	Me_3SiBr	2.235(2)	MW	318
4.	Iodosilanes			
	SiH_3I	2.437(3)	MW	308
	H_3SiSiH_2I	2.440(9)	MW	319

of Si–M bond distances to main group elements M, where M does not belong to groups 14–17. In general, the available structures are very similar to those of analogous carbon compounds. However, in the series of silylmercury derivatives **221**–**226**, two-, three- and four-coordinate mercury have been observed, whereas, for simple organomercury systems, the anionic species are not stable. Formally covalent M–Si bonds are, in general, somewhat longer than the sum of the apposite covalent radii. For instance, the tetrahedral covalent radii of Mg and Al are respectively 1.40 and 1.26 Å, leading to expected values of 2.57 and 2.43 Å for their bonds to silicon. As may be seen from Table 16, experimental values for the Mg–Si bond fall in the range 2.62–2.67 Å, those for the Al–Si bond in the range 2.44–2.49 Å. These results indicate that there is no significant bond polarity effect for such metal–silicon bond lengths.

A hexameric structure has been found for trimethylsilyllithium (Figure 13). The $(LiSiMe_3)_6$ framework can best be regarded as a six-membered lithium ring with 'folded-chair' conformation, in which each face is occupied by a μ_3-bridging trimethylsilyl group. The Li–Li bond lengths in this ring average 2.70(3) Å. In addition, there are six Li $\cdots$ Li distances between ring atoms in *meta* position to one another with an average length of 3.27(4) Å. If these contacts are regarded as bonding interactions, then the Li_6 unit should be described as a distorted octahedron. However, as the observed Li $\cdots$ Li distances are longer than twice the van der Waals radius of lithium (1.55 Å) it may be assumed that the bonding interaction is insignificant. The silicon atoms are coordinated by both three carbon and three lithium atoms. A four-centre two-electron bonding description may be proposed to describe the interaction between a silicon atom and an Li_3 face, but, in view

TABLE 16. Si–M bond distances to main group elements M (excluding groups 14–17)

	Compound	d(Si–M) (Å)	Method	Ref.
	1. M = Li			
218	$(LiSiMe_3)_6$	2.62(1)–2.78(1)	XRD	320
219[a]	$[(LiSiMe_3)_2\cdot 3TMEDA]$	2.69(1), 2.70(1)	XRD	321
220[b]	$\{[LiSi(SiMe_3)_3]_2\cdot 3DME\}$	2.630(5)	XRD	322
221	$Li_2[Hg(SiMe_3)_4]$	2.694(11), 2.869(14)	XRD	323
222	$Li_2[Hg(SiMe_2Ph)_4]$	2.90(1)–3.04(1)	XRD	323
	2. M = Mg			
223a[a]	$[Mg(SiMe_3)_2\cdot TMEDA]$	2.631(4), 2.624(4)	XRD	324
224a[c]	$[Mg(SiMe_3)_2\cdot TMDAP]$	2.651(1), 2.669(2)	XRD	325
225a[b]	$[Mg(SiMe_3)_2\cdot DME]$	2.630(2)	XRD	326
	3. M = Hg			
221	$Li_2[Hg(SiMe_3)_4]$	2.539(2), 2.548(2)	XRD	323
222	$Li_2[Hg(SiMe_2Ph)_4]$	2.493(2), 2.549(2)	XRD	323
223	$Hg(SiMe_3)_2$	2.500(5)	XRD	327
224	$Hg(SiPh_3)_2$	2.503(4)	XRD	328
225	$[(Me_2SiCH_2SiMe_2)_2Hg_2]$	2.501(4), 2.505(4)	XRD	328
226[d]	$[Mg_4(OME)_6(DME)_2][Hg(SiMe_2Ph)_3]$	2.516	XRD	329
	4. M = group 13 element			
227	1-Br-μ-$Me_3SiB_5H_7$	2.32(2)	XRD	330
228	$\{[Al(SiMe_3)_3]_2\cdot TMEDA\}$	2.470(3)–2.473(3)	XRD	326
229	$[Al(SiMe_3)_3\cdot Et_2O]$	2.464(3)–2.480(3)	XRD	331
230	$Na[Al(SiMe_3)_4]$	2.467(3)–2.490(3)	XRD	331
231	$\{Na[Al(SiMe_3)_4]\cdot 2C_6H_5CH_3\}$	2.44(2)–2.47(1)	XRD	331
232[e]	$\{[(Me_3Si)_3Si]_2Ga(\mu\text{-}Cl)_2Li\cdot 2THF\}$	2.439(5)	XRD	332
233[e]	$\{[(Me_3Si)_3Si]_2In(\mu\text{-}Cl)_2Li\cdot 2THF\}$	2.591(7)	XRD	332

[a] TMEDA = tetramethylethylene diamine.
[b] DME = 1,2-dimethoxyethane.
[c] TMDAP = 1,3-bis(dimethylamino)propane.
[d] OME = $[MeOCH_2CH_2O]^-$.
[e] Average values for d(Si–M) are given.

of recent theoretical calculations on organolithium compounds, it may reasonably be assumed that the Li–Si bonding is predominantly ionic in character. Surprisingly, the Si–Li distances to the four-coordinated lithium atoms in **219** and **220** are similar in length

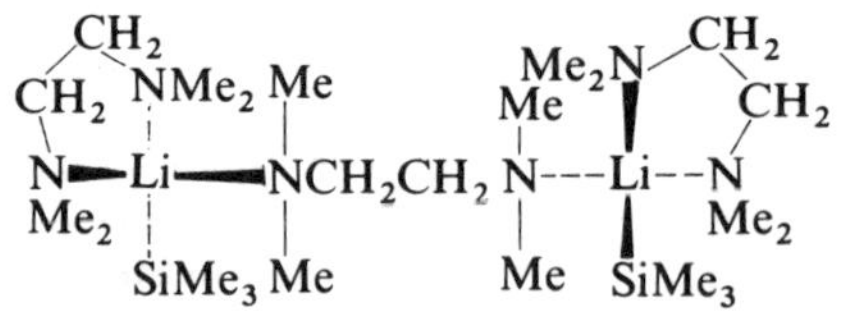

(219)

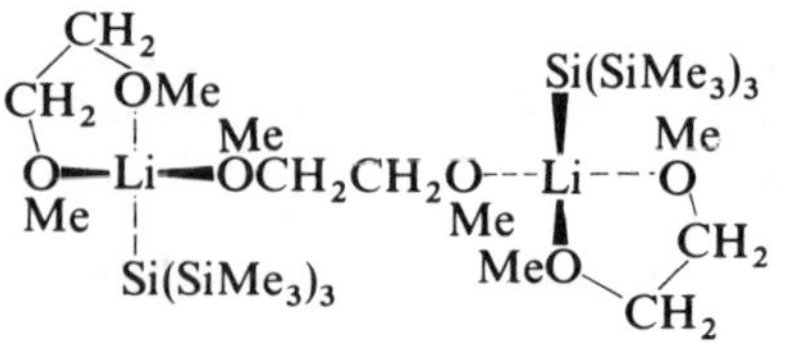

(220)

to those to the six-coordinated lithium atom in **218**. **221** exhibits discrete formula units with C_2 symmetry in which the lithium atoms are enclosed in a 'cage' of silicon and carbon atoms. Individual $[Hg(SiMe_3)_4]$ anions in **222** are linked via lithium atoms into chains. The Li–Si interactions in this compound are extremely weak [2.90(1)–3.04(1) Å].

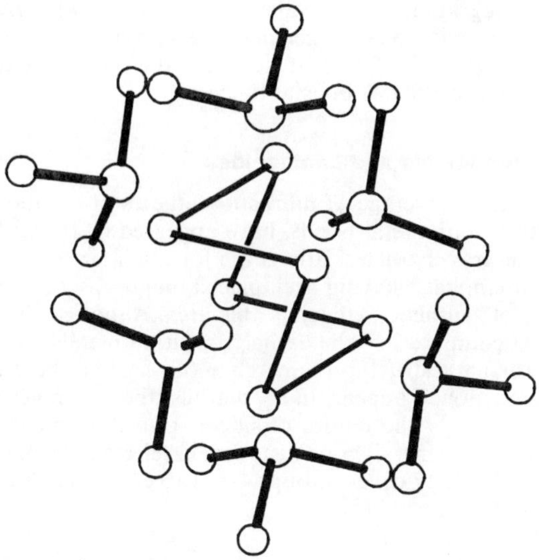

FIGURE 13. Molecular structure of the $(LiSiMe_3)_6$ hexamer **218**. Redrawn from Reference 320

Linear Si–Hg–Si skeletons are observed for the two-coordinate mercury derivatives **223–225**, e.g. Si–Hg–Si = 178.7(6)° in **225**. As **224** exhibits crystallographic C_i symmetry, the Si–Hg–Si framework is strictly linear in this molecule. The $[Hg(SiMe_2Ph)_3]^-$ anion

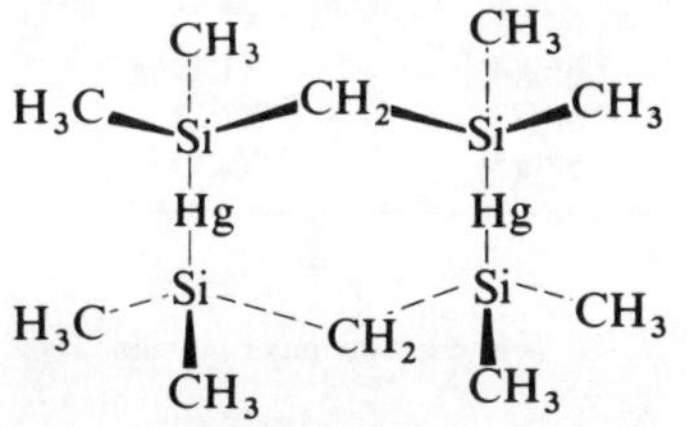

(225)

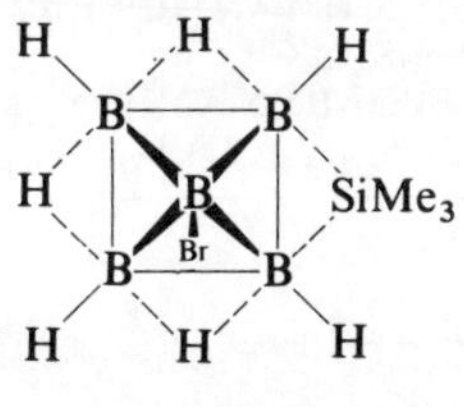

(227)

in **226** displays trigonal planar symmetry around the mercury atoms with average Si–Hg–Si angles of 119.8°. The Hg–Si bond lengths for two- (**223–225**), three- (**226**) and four-coordinated mercury (**221**, **222**) lie in the range 2.49–2.55 Å. In particular, the values for **221** and **222** are significantly less than predicted using the standard value for the tetrahedral radius of mercury (1.48 Å) and suggest that the covalent radius does not increase markedly on formation of three- or four-coordinate derivatives. **227** may be regarded as a substituted derivative of the C_{4v} pentaborane B_5H_9 in which the apicel hydrogen atom has been replaced by a bromine atom and one of the four bridging basal hydrogen atoms by a trimethylsilyl group. The resultant molecule is crystallographically restrained to display C_s symmetry. The molecular dimensions indicate that the interaction between the five-coordinate silicon atom and two basal boron atoms may be described in terms of a three-centre two-electron bond. The silicon atom is 2.32(2) Å from each of the two crystallographically equivalent boron atoms with a B–Si–B angle of only 42.6(8)°. The B–Si distance is 0.35 Å larger than the sum of the covalent radii for boron and silicon. C–Si–C angles of 110.8(6)° (twice) and 106.1(9)° suggest that the bonding silicon orbitals may be regarded as quasitetrahedral.

G. Bonds to Transition Metals and Lanthanides

Reviews, which include coverage of diffraction studies of organometallic complexes containing silicon–transition metal bonds, have appeared in 1973[380] and 1982[381]. The vast majority of structural characterizations are for the later transition metals (groups 7–10), with particular emphasis having been placed on groups 7 and 8, which reflects, of course, the intensity of synthetic activity in this area. Although there have been occasional studies of silyl complexes of the earlier transition metals (groups 4–6) it is only very recently that systematic structural research in this field has been initiated. However, as we will see in this Section, it appears likely that just these complexes will hold the key to a better understanding of the nature of silicon–transition metal bonding. For this reason Si–M (M = transition metal) bond distances in selected complexes are presented in three different Tables. Table 17 covers groups 4–6, Table 18 group 7 and Table 19 groups 8–10.

TABLE 17. Si–M bond distances where M is a group 4–6 transition metal

	Compound	*d*(Si–M) (Å)	Method	Ref.
	$[H_2SiTiCp]_2$	2.159(13)	XRD	333
234[a]	$Me_3SiTi(Cl)Cp_2$	2.67(1)	XRD	334
235	$Ph_3SiZr(Cl)Cp_2$	2.813(2)	XRD	335
236	$Me_3SiZr(S_2CNEt_2)Cp_2$	2.815(1)	XRD	336
237	$Me_2PhSiTa(H)_2Cp_2$	2.651(4)	XRD	337
238	$Me_3SiTa(PMe_3)(Cl)_2Cp$	2.642(1)	XRD	338
239[b]	$Me_3SiTa(Cl)_3Cp^*$	2.669(4)	XRD	338
240	$[F_2Si(t\text{-}Bu)C{=}C(H)SiF_2]Mo(CO)_5$	2.605(3)	XRD	339
241[c]	$(Et_2Si)_2W_2(CO)_8H_2$	2.586(5)	XRD	340

[a] $Cp = \eta^5\text{-}C_5H_5$.
[b] $Cp^* = \eta^5\text{-}C_5Me_5$.
[c] Si–W distances of 2.703(4) Å were also observed; an Si · · · H–W interaction may be assumed (see Section V.D.).

An assessment of the significance of experimental Si–M bond lengths is hampered by an underlying ambiguity in the choice of appropriate transition-metal covalent radii for comparison purposes. Either single-bond metallic radii, the effective radii of metal atoms in diatomic hydrides MH[19] or derived radii from M–X (X = C, Cl, etc.) bonds in metal complexes have, on occasion, been proposed as suitable values. For instance, suggested values[339] for the molybdenum covalent radius have ranged between 1.30 and 1.61Å. The transition metal radius in a particular complex will be influenced by the following factors:

(1) oxidation state,
(2) coordination number,
(3) stereochemistry of the complex,
(4) nature and substitution pattern of the ligands.

Furthermore, the interplay of these features is often complex, e.g. ligand electronegativity and σ/π donor and acceptor properties. This means that bonding trends may only reasonably be established when two or three of the above factors are similar or identical. This is the case for the species $R_3SiMn(CO)_5$ (M = Mn, Re) listed in Table 18. It may be seen that *d*(Si–M) decreases as R changes in the order R = Me_3Si > Me > H > F, which implies that both electronic and steric factors influence this parameter. The longest Si–M bonds are observed when bulky ligands of low electronegativity (e.g. Me_3Si) are attached to silicon. In silicon–transition metal complexes there is the possibility of (d-d)π bonding involving donation of d-electron density from filled d-levels of the transition metal into vacant silicon 3d-orbitals. If this type of interaction is of significance, then the reduction, in both size and energy, of the silicon 3d-orbitals on replacement of a silyl group by a trifluorosilyl group should lead to an increase in the degree of π bonding. Indeed, a (d-d)π bonding component has often been assigned not only to fluorosilyl but also to alkylsilyl complexes of the late or low-valent transition metals, for which experimental Si–M bond lengths have been found, which are often significantly shorter than predicted values. However, this interpretation should be approached with caution. We have emphasized the difficulties in estimating transition-metal covalent radii. An increase in the electronegativity of R in substituent R_3Si will also influence the σ donor and acceptor properties of this group. Furthermore, as we have discussed in Section IV.B.1, the covalent radius of silicon will be reduced on substitution of a hydrogen atom or a methyl group by fluorine or chlorine, e.g. by ca. 0.055 Å on going from Me_3Si to F_3Si. As a summary of these arguments, it should be emphasized that, as we have often seen for silicon, metrical data alone do not allow any conclusions to be drawn concerning the existence or significance of a π bonding component.

TABLE 18. Selected Si–M bond distances where M is a group 7 transition metal

Compound		*d*(Si–M) (Å)	Method	Ref.
	$F_3SiMn(CO)_5$	2.360(7)	ED	341
	$H_3SiMn(CO)_5$	2.407(5)	ED	342
242	$Me_3SiMn(CO)_5$	2.497(5)	XRD	343
243	$(M_3Si)_3SiMn(CO)_5$	2.564(6)	XRD	344
244	*trans*-$Me_3SiMn(CO)_4PPh_3$	2.45	XRD	345
245[a]	$(\mu\text{-}Ph_2Si)_2Mn_2(CO)_8$	2.402(2)	XRD	346
	$(\mu\text{-}H_2Si)\,[CpMn(CO)_2H]_2$	2.434(3)	XRD	347
	$H_3SiRe(CO)_5$	2.562(12)	ED	348
246	$Me_3SiRe(CO)_5$	2.600(1)	XRD	349
247	$(Me_3Si)_3SiRe(CO)_5$	2.665(9)	XRD	349
248[a]	$Ph_3SiReH(CO)_2Cp$	2.49(1)	XRD	350
249	$(\mu\text{-}Et_2Si)_2Re_2(CO)_6H_4$	2.534(3), 2.536(3)	XRD	351
250	$(\mu\text{-}Ph_2Si)_2Re_2(CO)_8$	2.542(3)	XRD	352
251	$(\mu\text{-}Ph_2Si)Re_2(CO)_8H_2$	2.544(9)	XRD	353
252	$(\mu\text{-}Et_2Si)_2Re_2(CO)_7H_2$	2.539(4)–2.549(4)	XRD	352

[a] Cp = η^5-C_5H_5.

The available experimental results indicate that M–Si bond lengths can vary within relatively wide ranges, e.g. 0.27 Å for the Fe–Si bond. The longest M–Si bonds will, of course, be expected for the larger early transition metals. Very long Zr–Si bond distances of 2.813(2) and 2.815(1) Å have been observed for **235** and **236** both of which are d^0 complexes, i.e. no (d-d)π interaction is formally possible. However, the d^0 complexes **237** and **239** exhibit Ta–Si bond lengths of 2.651(4) and 2.669(4) close to that of 2.642(1) in the

d^1 complex **238** and a predicted value of 2.61 Å[337], which indicates that $M(d^0)$–Si bonds are not inherently weak. Remarkably, the shortest known M–Si bond has also been reported for a d^0 complex, namely $[H_2SiTiCp_2]_2$ which exhibits bridging Ti–Si distances of 2.159(13) Å. In contrast, the terminal Ti–Si bond length is 2.67(1) Å in the d^0 complex **234**. When the difference in the single-bond metallic radii of titanium and zirconium (0.13 Å) is taken into account, then this distance is comparable to those in the analogous zirconium derivatives **235** and **236**. An explanation for the anomalously short Ti–Si

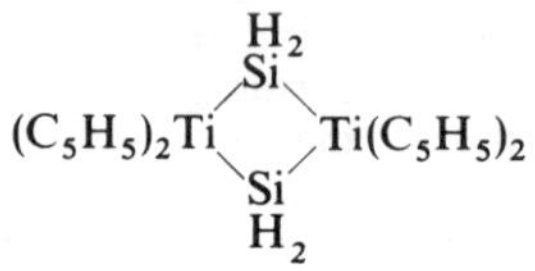

(240a)

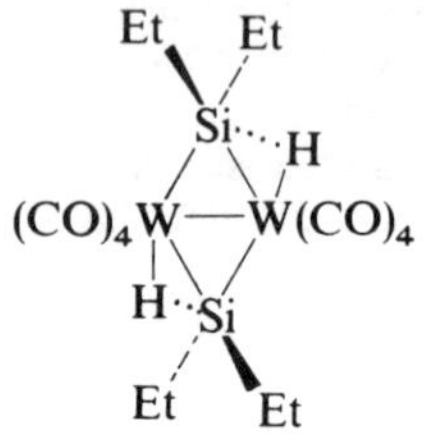

(241)

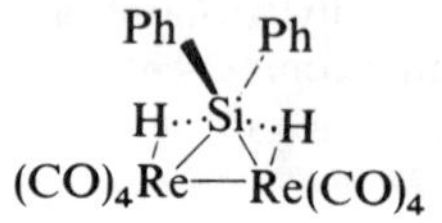

(251)

distances in **240a** is not immediately apparent. For instance, the Mn–Si bond length in $(\mu\text{-}SiH_2)[CpMn(CO)_2H]_2$[2.434(3) Å], which also contains an analogous bridging SiH_2 group, is similar to that in $H_3SiMn(CO)_5$ [2.407(5) Å]. The Ti ··· Ti distance of 3.368(10) Å precludes a significant metal–metal interaction.

The WSi_2W fragment in **241** forms a parallelogram with two distinctly different W–Si distances of 2.586(5) and 2.703(4) Å. This feature was interpreted in terms of a W–H ··· Si interaction, as steric constraints in this complex force the hydrogen atom to adopt a position close to silicon. The X-ray analysis did not allow the location of the hydrogen atom positions. This was also the case for **251**, for which IR spectral evidence lent support to a structure with some degree of H ··· Si interaction. However, the

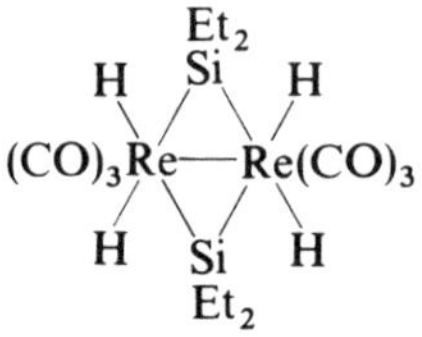

(249)

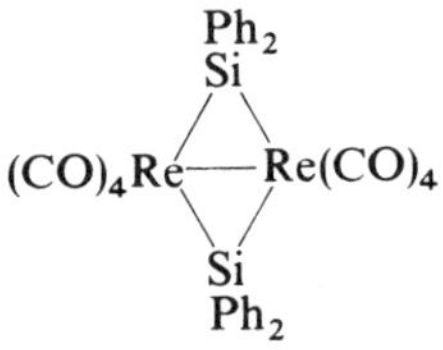

(250)

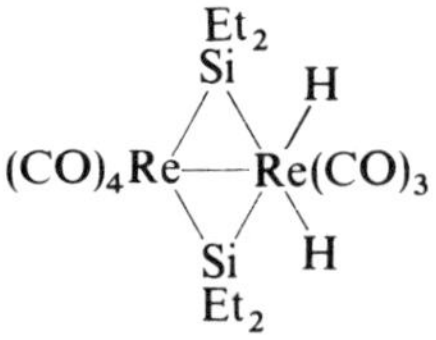

(252)

similarity of the Si–Re distances in the series **249**–**252** and, in particular, the absence of unsymmetrical bridging distances in **252** provide strong evidence for the assumption that the Re–H bonds may be regarded as normal terminal bonds. In the tungsten complex **241**, the hydrogen atoms are constrained to be close to the silicon atoms, as a result of steric overcrowding at the metal. Si–H distances between 1.76 and 1.80 Å have been confirmed by both X-ray and neutron diffraction for molecules $R_3SiMnH(CO)_2Cp$ and $R_3SiMnH(CO)_2(MeCp)$ and will be discussed in detail in Section V.D.

The shorter Fe–Si bond lengths listed in Table 19 are exhibited by halogenosilyl substitutents (e.g. Cl_3Si, MeF_2Si). Steric overcrowding may be responsible for the long Fe–Si distances in the three-coordinate anion of **256**. Unsymmetric Si–Ru bridging bonds are found in complex **257** with the Si–Ru distance in *trans* position to the carbonyl ligand being 0.10 Å shorter than that in *trans* position to the trimethylsilyl group. The terminal

TABLE 19. Selected Si–M distances where M is a group 8–10 transition metal

	Compound	d(Si–M) (Å)	Method	Ref.
1. Group 8				
	$Cl_3SiFe(CO)_2Cp$[a]	2.216(1)	XRD	354
	$[Et_4N][Cl_3SiFe(CO)_4]$	2.224(9)	XRD	355
252[a]	$(MeF_2Si)_2FeH(CO)Cp$[a]	2.249(1)	XRD	356
	$(Cl_3Si)_2FeH(CO)Cp$[a]	2.252(3)	XRD	357
253[a]	$Ph_2FSiFe(CO)_2Cp$	2.278(1)	XRD	354
	trans-$(Cl_3Si)_2Fe(CO)_4$	2.326(2)	XRD	358
254	*cis*-$Ph_3SiFeH(CO)_4$	2.415(3)	XRD	358
255	*trans*-$(Me_3Si)_2Si(CO)_4$	2.456(2)	XRD	358
256	$[Et_4N][\{(Me_3Si)_3Si\}_2FeCl]$	2.488(6), 2.491(6)	XRD	359
	$(Cl_3Si)_2Ru(CO)Cp(C_6H_4Bu\text{-}t)$	2.338(1), 2.340(1)	XRD	360
257	$(Me_3Si)_2(\mu\text{-}SiMe_2)_2Ru_2(CO)_6$	2.391(7)–2.507(8)	XRD	361
258	$[(\mu\text{-}Me_2SiCH_2PPh_2)_2Ru_2(CO)_6]$	2.426(1), 2.465(1)	XRD	362
259	$[(Ph_3P)_2N][(Et_3Si]_2Ru_3(CO)_{10}H]$	2.441(3), 2.450(2)	XRD	363
	$(Cl_3Si)_2Os_3(CO)_{12}$	2.377(3)	XRD	364
	$(Cl_3Si)_3Os_3(\mu\text{-}H_3)(CO)_9$	2.367(13)–2.394(14)	XRD	364
260[a]	$(Cl_2MeSi)_3Os_3(\mu\text{-}H)_3(CO)_9$	2.400(6), 2.420(5)	XRD	365
2. Group 9				
	$F_3SiCo(CO)_4$	2.226(5)	XRD	366
	$Cl_3SiCo(CO)_4$	2.254(3)	XRD	367
260a	$(OC)_4CoSiCo_3(CO)_9$	2.217(1)–2.222(1) 2.288(1)	XRD	368
	$H_3SiCo(CO)_4$	2.381(7)	ED	369
	$Cl_3SiRhH(Cl)(PPh_3)_2 \cdot xSiHCl_3$	2.203(4)	XRD	370
261[b]	$(Et_3Si)_2Rh(H)_2Cp^*$	2.379(2)	ND	371
262[b]	$(Et_3Si)_2Ir(H)_2Cp^*$	2.390(1)	ND	372
263	$SiMe_2OMe_2SiIrH(CO)(PPh_3)_2$	2.399(2), 2.416(2)	XRD	373
3. Group 10				
264[c]	$(\mu\text{-}Me_2Si)Pt_2[P(Hex\text{-}c)_3]_2(C{\equiv}CPh)_2$	2.255(1), 2.444(4)	XRD	374
	$N(C_2H_4O)_3SiPtCl(PMe_2Ph)_2$	2.292(1)	XRD	375
265	$[Ph_2MeSiPt(t\text{-}BuNC)(\mu\text{-}t\text{-}BuN{=}CH)]_2$	2.312(4), 2.316(4)	XRD	376
266[d]	*trans*-$Ph(Np)MeSiPtCl(PMe_2Ph)_2$	2.317(4)	XRD	377
267[c]	$[Et_3SiPt(\mu\text{-}H)P(Hex\text{-}c)_3]_2$	2.33(1), 2.34(1)	XRD	378
	trans-$H_3SiPtH[P(Hex\text{-}c)_3]_2$	2.382(3)	XRD	379

[a] $Cp = \eta^5\text{-}C_5H_5$.
[b] $Cp^* = \eta^5\text{-}C_5Me_5$.
[c] c-Hex = cyclohexyl.
[d] Np = 1-naphthyl.

Si–Ru bond exhibits a length of 2.51 Å. The three bridging Si–Co bonds in **260a** are significantly shorter [2.217(1)–2.222(1) Å] than the terminal Si–Co bond [2.288(1) Å]. Reaction of $Pt(C_2H_4)[P(c\text{-}Hex)_3]_2$ with $Me_2Si(C{\equiv}CPh)_2$ yields **264** in which the $SiMe_2$ bridging is once again markedly asymmetric. The Si–Pt bond in *trans* position to the $(c\text{-}Hex)_3P$ ligand is much longer [2.444(4) Å] than the complementary bridge bond[2.255(1) Å].

The first X-ray crystallographic characterization of an organolanthanide derivative containing a silicon–lanthanide bond has recently been reported[382]. $[Li(DME)_3]$

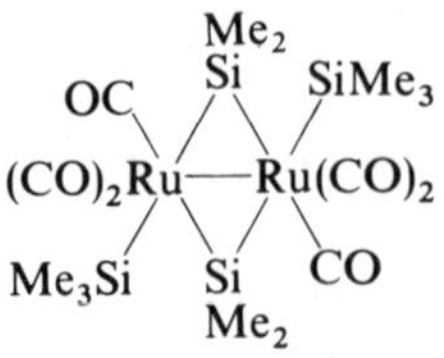

(257)

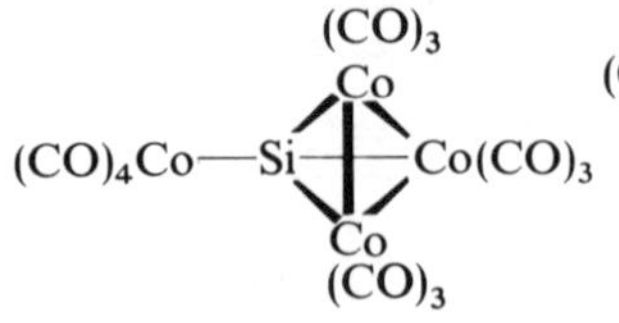

(260a)

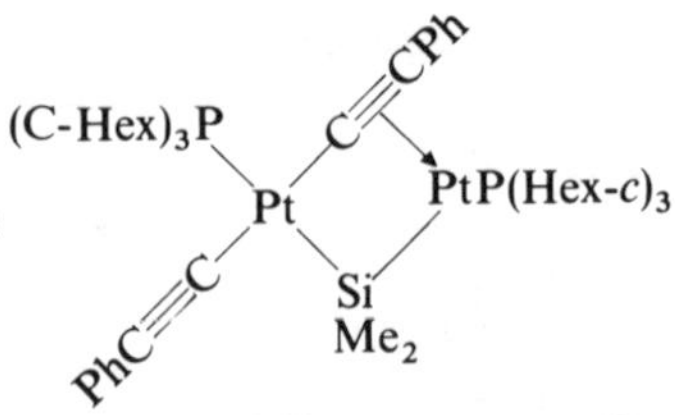

(264)

$[Cp_2Sm(SiMe_3)_2]$ exhibits an Sm–Si distance of 2.880(2) Å. Both anion and cation possess crystallographically imposed C_2 symmetry.

H. Compounds of Pharmaceutical Relevance

The discovery of the biological potential of the silatranes in the early 1960s stimulated interest in the biological activity of organic silicon compounds (see Chapter 18). One approach to the search for bioactive organosilicon derivatives is sila-substitution (C/Si-exchange) of drugs. We have discussed the most important differences between carbon and silicon in Section I. The markedly reduced electronegativity and increased size of silicon with respect to carbon will be expected to lead to changes in the chemical and physicochemical properties as well as in the molecular structures of drugs upon sila-substitution. The topography of the functional groups of a drug will determine how or whether it interacts with a particular biological receptor site. As typical Si–C bonds are ca. 0.33 Å longer than the analogous C–C bonds, replacement of carbon by silicon, in particular when the substitution is multiple, may lead to pronounced changes in the distances between functional groups in the drug, e.g. O and N atoms capable of participating in hydrogen bonds to the receptor, or hydrophobic aromatic ring systems. Furthermore, the reduced electronegativity of silicon with respect to carbon will affect the acidity of OH or the basicity of NR_2 groups directly bonded to the substituted atom. X-ray diffraction studies of both the original drug and its sila-analogue coupled with theoretical studies of conformational preference (molecular modelling) often allow the development of model concepts for the interaction of these molecules with a receptor site. We will illustrate this procedure with examples taken from the sila-analogues of three drug classes studied by Tacke and colleagues:

(1) curare-like agents related to decamethonium,
(2) antimuscarinic phenylcarbinols,
(3) nifedipine-like calcium antagonists with spasmolytic effects.

Silatranes will be considered in Section V.C.

Many muscle-relaxing agents exhibit a distance of ca. 14 Å between two quaternary nitrogen atoms, which are regarded as the binding sites for acetylcholine receptors. For instance, this distance is found in the decamethonium dication $[Me_3\overset{+}{N}(CH_2)_{10}\overset{+}{N}Me_3]$, which contains a C_{10} unit. The crystal structures of six sila-substituted derivatives of the type $[Me_3\overset{+}{N}(CH_2)_nSiMe_2(CH_2)_mSiMe_2(CH_2)_n\overset{+}{N}Me_3]2I^-$ were determined by X-ray diffraction and the N · · · N distances compared with the biological activity. Conformational data and LD_{50} (average lethal dose) values for mice are listed in Table 20. The latter values allow an estimation of the strength of the muscle-relaxing activities of **268**–**273**; the equivalent value for the standard decamethonium is 1.95 mg kg^{-1}. Replacement of two carbon atoms by silicon would be expected to lead to a increase in the

TABLE 20. Molecular geometry of sila-substituted curare-like agents $[Me_3\overset{+}{N}(CH_2)_n SiMe_2(CH_2)_m SiMe_2(CH_2)_n \overset{+}{N}Me_3]2I^-$

Compound	*n*	*m*	Conformation[a]	*d*(N · · · N) (Å)	LD_{50}(mg kg^{-1})[b]	Ref.
268	2	1	aggag̅a	9.312	6.46	383
269	2	2	agaaag̅a	11.472	4.47	383
270	2	3	aggaag̅g̅a	7.408	3.11	383
271	2	4	agaaaaag̅a	13.501	6.80	383
272	2	5	aggaaaag̅g̅a	7.747	4.60	383
273	3	2	aaggag̅g̅aa	11.81	0.42	384

[a] Sequential torsion angles are given beginning with an outer C–C bond; a = anti (180°), g = gauche$^+$ (60°), g̅ = gauche$^-$ (−60°).
[b] LD_{50} values refer to the average lethal dose for mice and allows an estimate of the strength of the muscle-relaxing properties of the derivatives.

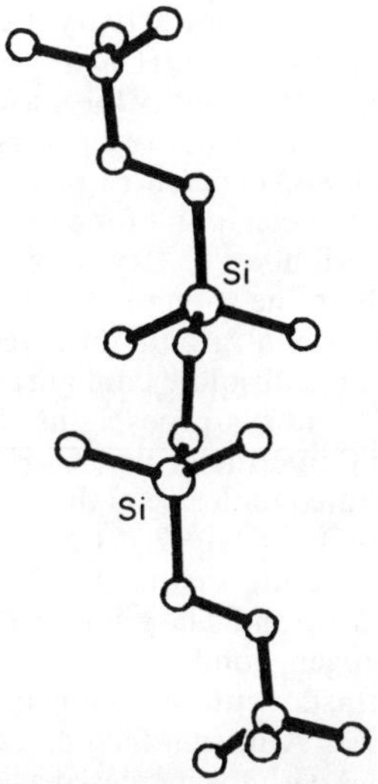

(269)

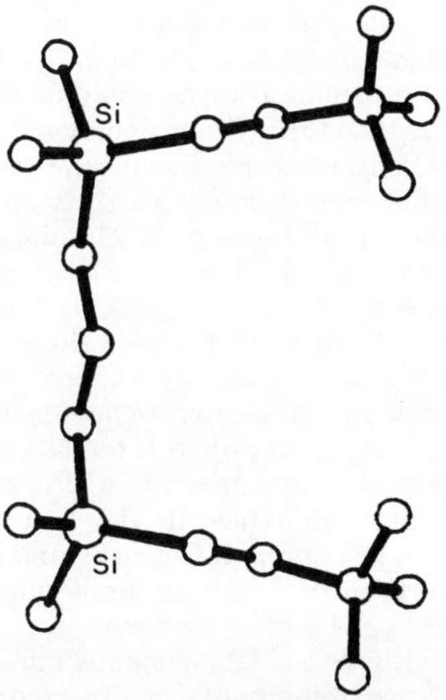

(270)

FIGURE 14. Molecular structures of the sila-substituted curare-like agents **269** and **270**. Redrawn from Reference 383

N · · · N distance if all torsion angles are ca. 180° (anti) as in decamethonium. Taking the average Si–C and C–C single bond lengths into account, maximal activity would be expected for a C_7Si_2 unit in the sila-substituted derivatives, i.e. **270**. The experimental results are surprising, as they indicate that the activities of **268**–**272** are very similar despite different numbers of chain atoms. The chain conformations in the solid state give rise to alternating short and long N · · · N distances depending on whether *m*, the number of methylene bridges between the two silicon atoms, is odd or even (Figure 14). It may be seen that at least two *gauche* C–C–Si–C torsion angles are exhibited by each chain. ^{1}H-NMR conformational analyses on solutions of **268–273** suggest that these *gauche* torsion angles will be retained in solution[383, 384]. Even allowing for the fact that energy barriers to rotation about the C–C and C–Si bonds are relatively small, so that different chain conformations will be possible in receptor complexes, these experimental results cannot be brought into agreement with the classical 14 Å model for curare activity. (2-aminoethyl)diphenylcarbinols such as pridinol (**274A**) are potent antimuscarinic agents. It is assumed that such drugs bind to the muscarinic receptors *inter alia* by a

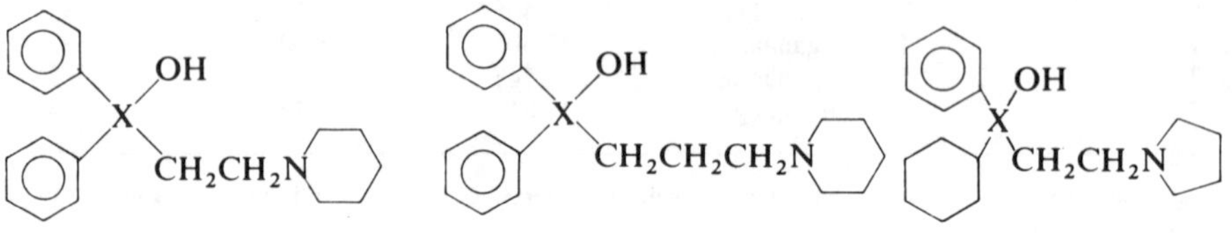

(**274**) X = Si; (**274A**) X = C (**275**) X = Si; (**275A**) X = C (**276**) X = Si; (**276A**) X = C

hydrogen bond between their OH functions and a basic centre of the receptor active site. The presence of at least one hydrophobic phenyl ring bonded to the central carbinol carbon atom and the side-chain nitrogen atom are also necessary for activity. Replacement of the carbinol-carbon atom by silicon will lead to an increase in OH acidity, so that sila-substituted derivatives would be expected to exhibit a greater affinity for muscarinic receptors than their carbon analogues. Comparative pharmacological investigators using the isolated guinea pig ileum have confirmed this hypothesis. Differences of up to an order of magnitude in potency between C/Si analogues have been found for compounds pairs such as **274** (pridinol), **275** (difenidol) or **276** (procyclidine)[385]. Crystal structure analyses on the sila-derivatives have provided information on the strength of OH · · · N hydrogen bonds and the possible topography of the functional groups in a receptor complex. In addition they have allowed the assignment of the absolute configuration of the more potent enantiomers of both procyclidine **276A**[386] and sila-procyclidine **276**[387]. These are in both cases (*R*)-configurated. Both exhibit intermolecular O–H · · · N hydrogen bonds, but whereas one (*R*)- and one (*S*)-configurated molecule of the racemate are linked in this manner into hydrogen bonded dimers [*d*(O · · · N)=2.791 Å], the (*R*)-configurated molecules in the latter compound build infinite chains [*d*(O · · · N) =2.792 Å]. A comparative structural study of pridinol **274A** and sila-pridinol **274** has been performed[388]. An intramolecular O–H · · · N hydrogen bond of length 2.739 Å (∠O–H · · · N=154°) is found for **274A**, which contrasts with an almost linear (∠O–H · · · N=173°) analogous interaction of length 2.685 Å in sila-difenidol **275**[389]. The relative strengths of these interactions are in accordance with the increased acidity of the silanol OH. Sila-pridinol exhibits hydrogen bonded dimers in the solid state [*d*(O · · · N)=2.758 Å], as found for racemic procyclidine.

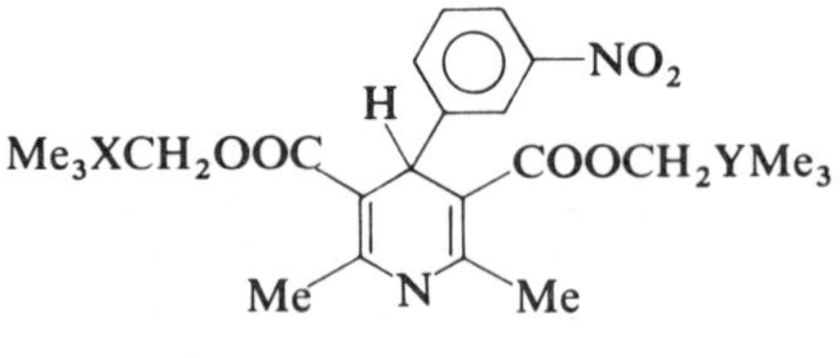

(**277A**) X = Y = C
(**277B**) X = C, Y = Si
(**277C**) X = Y = Si

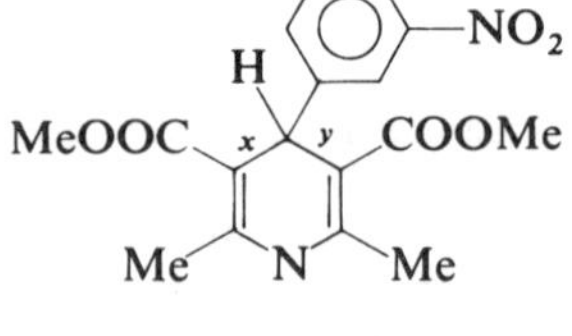

(**277D**)

Compounds **277A–277C** display similar spasmolytic activities (*in vitro*, guinea pig ileum) to that of nifedipine **277D**. Their crystal structures are isostructural[390], so that the observed similarity of their *in vitro* activities is not surprising. These correlate well with a structure-activity model proposed by Triggle and coworkers[391], who suggested that the spasmolytic activity should decrease with increasing distortion of the 1,4-dihydropyridine ring from planarity, as measured by the average value α of the absolute ring torsion angles for the bonds x and y. α lies between 21.5 and 22.4° for **277A–277C** and is similar to the value for **277D** (20°), which displays a comparable activity[391].

V. PENTACOORDINATE SILICON

A. Bonding Concepts

In contrast to carbon, tetravalent silicon displays a pronounced acceptor ability and is capable of extending its valence shell to contain 10 or even 12 electrons. A comprehensive review of penta- and hexacoordinate silicon compounds appeared in 1986[26] so that this and the following Section will be limited to a description of general structural trends. Potential pentacoordinate species are anions $[SiX_5]^-$ (**I**), neutral coordination complexes $X_4Si \cdot D$ (D = Lewis base) (**II**) or Lewis base stabilised cations $[X_3Si \cdot DD]^+$ (DD = chelate ligand) (**III**). Siliconium cations of type **III**, though proposed[392], for example, in

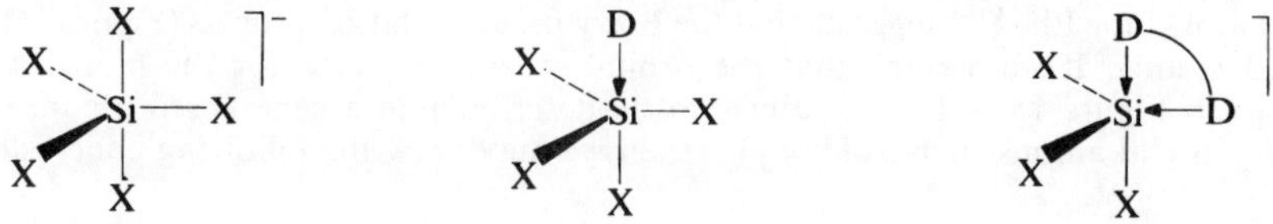

(**I**) (**II**) (**III**)

species such as $[Ph_3Si \cdot bipy]^+ I^-$, have yet to be definitively characterized by X-ray structural analysis. Furthermore, although simple adducts of type **II** are well characterized for electronegatively substituted silanes with simple amines (e.g. $F_4Si \cdot NMe_3$), no structural information is available at the time of writing. However, strong intermolecular N→Si coordinate interactions are exhibited by the $F_3SiN{=}PMe_3$ dimer[18] and the H_3SiNMe_2 pentamer[19], which display trigonal bipyramidal geometry at silicon with espective axial Si–N distances of 1.857(4) Å and 1.94–2.09(2) Å. In the vast majority of neutral derivatives of pentacoordinate silicon the coordinate interaction is intramolecular. The stability of such complexes increases with the number of bridges between the donor atom and silicon, i.e. in the order acyclic (**A**) < monocyclic(**B**) < bicyclic(**C**) < tri-

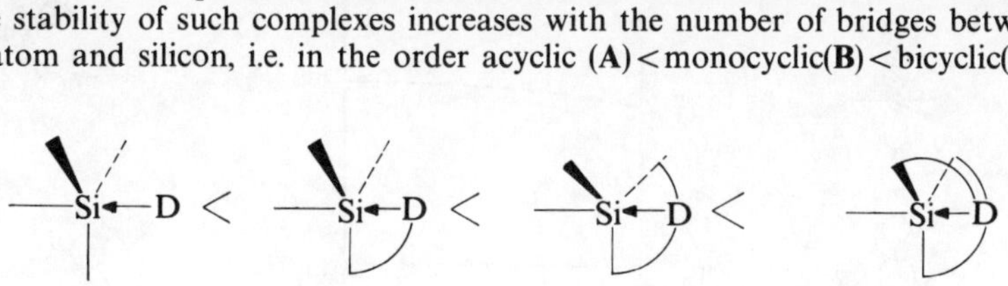

(**A**) (**B**) (**C**) (**D**)

cyclic(**D**). The latter framework **D** is found in the silatranes and analogous derivatives, for which more than 30 molecular structures are now available[20].

In comparison to phosphorus, relatively few theoretical studies have dealt with bonding in pentacoordinate silicon derivatives. Two simple models have been proposed to account for bonding in species such as PF_5 or $[SiF_5]^-$. Pauling suggested that the 3d-orbitals participate in the construction of sp^3d^2 hybrids[33]. In total contrast to this approach was that of electron-rich three-centre bonding pioneered by Rundle[34] and expressed in a more generalised form by Musher[393] for hypervalent bonding. The equatorial Si–X bonds in an anion $[SiX_5]^-$ may be regarded as normal in this model, while the delocalized three-centre four-electron orbitals are constructed for the axial Si–X bonds from the silicon $3p_z$ and the axial ligand s- and p-functions. Recent calculations suggest that the degree of 3d-orbital participation in bonding is less than previously supposed and that the Rundle model provides a better description of bonding in pentacoordinate derivatives of second-row elements[28]. If we accept this conclusion, then we may construct[23] in a qualitative fashion the molecular orbitals depicted in Figure 15. The doubly noded HOMO $2a'_1$ is separated by a large gap from the other bonding orbitals. Although *ab initio* calculations may lead to an inversion of the levels of the $1a''_2$ and $1e'$ orbitals this will be of no consequence to a qualitative understanding of bonding. The HOMO is derived primarily from the a'_1 axial hydrogen combination mixing itself into the planar SiH_3 σ and σ^* a' orbitals leading to cancellation at silicon. Bonds to the equatorial hydrogen atoms are stronger than those to the axial hydrogen atoms, which carry a greater net negative charge. Both MNDO[394] and *ab initio*[395] calculations on pentacoordinate silicon anions indicate that the trigonal bipyramidal geometry is energetically somewhat more favourable than the alternative square pyramid. However, the magnitude of the energy difference (e.g. 12.3 kJ mol^{-1} for $[SiF_5]^-$ in comparison to 16.0 kJ mol^{-1} for PF_5[394]) suggests that the Berry pseudorotation process (Figure 16) will proceed readily. If we assume that the orbital scheme depicted for the model anion $[SiH_5]^-$ in Figure 15 will be equally valid in describing, in a generalised manner, the bonding in real anions such as $[SiF_5]^-$, then we may draw the following conclusions.

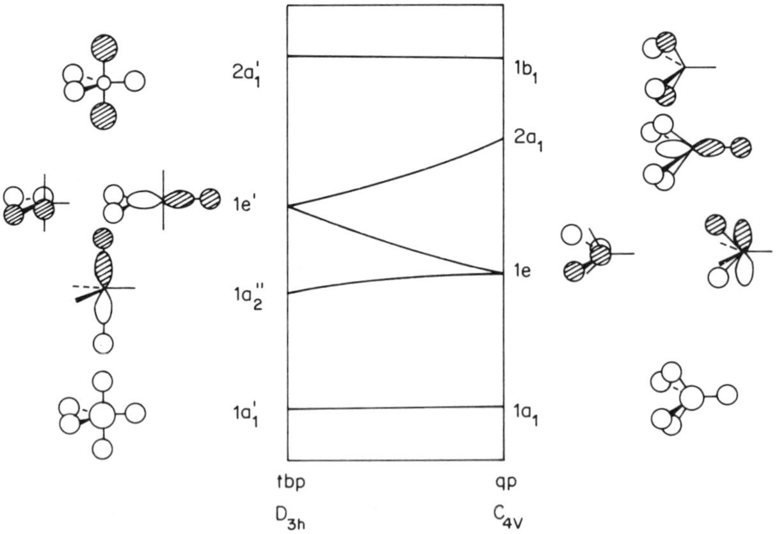

FIGURE 15. Occupied molecular orbitals (schematic) for D_{3h} and $C_{4v}[SiH_5]^-$ with the correlation diagram for the Berry pseudorotation transformation

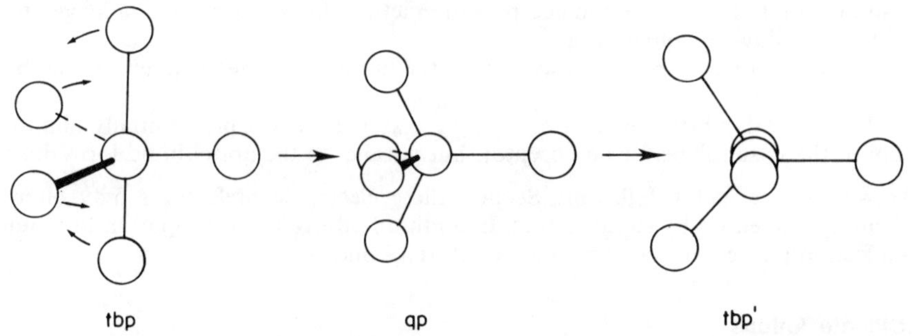

FIGURE 16. The Berry pseudorotation transformation

(1) For 10 electron systems in which the $3a_1'$ orbital (LUMO) is lower in energy than the antibonding $2e'$ and $2a_2''$ orbitals, the trigonal bipyramidal geometry (D_{3h}) is more stable than the alternative square pyramid (C_{4v}).

(2) The $1a_2''$ and one of the $1e'$ orbitals will be destabilized on distortion from a trigonal bipyramid to a square pyramid. At the same time the other $1e'$ and the $3a_1'$ orbital will be stabilized. As a result the HOMO–LUMO energy gap is much smaller for the C_{4v} geometry.

(3) More electronegative substituents will prefer to occupy those positions of highest electron density—the axial positions of a trigonal bipyramid, or the basal positions of a square pyramid. It follows that for anions $[SiX_4Y]^-$, where X is more electronegative than Y, a C_{2v} distortion from D_{3h} towards C_{4v} skeletal geometry could be favoured.

(4) If 3d-orbitals are not of importance, the only energetically significant π interactions of a substituent will be with the occupied skeletal orbitals—for acceptors stabilizing, for donors inherently destabilizing. The axial and equatorial interactions which are allowed by symmetry are shown in Scheme 1 for the model anion $[SiH_5]^-$. Interactions e_x and e_y are identical by symmetry and calculations on model donors and acceptors for PH_5 have shown the b_1 interaction to be of comparable strength[396]. The b_2 interaction is weaker than its axial counterpart e_y where the hydrogen ligands are also involved. The following conclusions may be drawn:

(i) Interactions between donor orbitals and framework σ-orbitals are stronger for axial than for equatorial substitution, so that π-acceptors will prefer axial, π-donors equatorial sites.

(ii) An equatorial single π-system will prefer to have its acceptor orbital perpendicular to the equatorial plane, or its donor orbital in that plane.

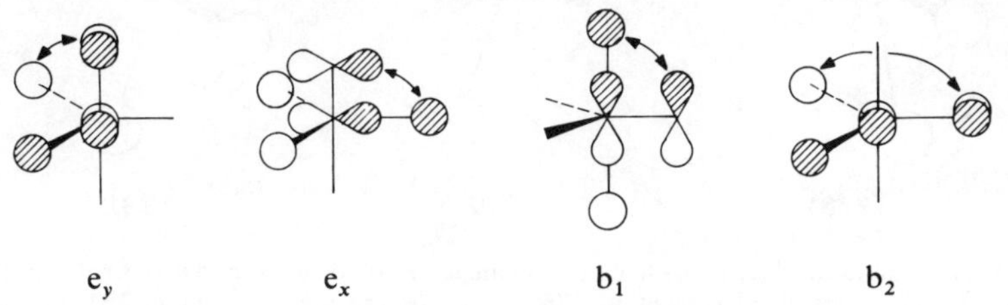

SCHEME 1

A similar analysis of donor and acceptor interactions for a square pyramidal geometry leads to the following conclusions:

(i) In a C_{4v} geometry π-donors will prefer the apicel positions, π-acceptors the basal sites.

(ii) If the basal substituent carries a single π-system, it will preferentially align that system in the basal plane for an acceptor, but parallel to the apicel bond for a donor.

As we will see in the following Section these electronic preferences may often be profoundly influenced by steric factors. In addition 3d-orbital participation in bonding will affect, in particular, the axial and equatorial bond orders.

B. Silicate Anions

X-ray diffraction studies on acyclic anions of pentacoordinate silicon are limited to the fluorosilicate anions listed in Table 21. In all cases the axial sites of the trigonal bipyramid are preferentially occupied by the more electronegative fluorine atoms; the substituted fluorosilicate anions all display small distortions from the ideal trigonal bipyramidal geometry. For the anions $[PhSiF_4]^-$, $[t\text{-}BuPhSiF_3]^-$ and $[1\text{-}NpPh_2SiF_2]^-$ the axial fluorine atoms tilt away from a bulky pivotal equatorial ligand (repectively Ph, *t*-Bu and Ph) along the Berry exchange coordinate leading to F_{ax}–Si–F_{ax} angles, which are smaller than 180°. At the same time, the non-pivot equatorial angle widens to respectively 121.4(2)° (**278**), 124.3° (**280**) and 125.8(3)° (**283**). The tilt direction for the anions $[PhMeSiF_3]^-$, $[Ph_2SiF_3]^-$ and $[Xyl_2SiF_3]^-$ is towards the pivotal equatorial fluorine atom (Figure 17), which means that the F_{ax}–Si–F_{ax} angles are larger than 180° with respect to the Berry exchange coordinate (Table 21). The non-pivotal C_{eq}–Si–C_{eq} angle narrows to 118.9(1)° in **281** but widens as a result of steric interactions with neighbouring equatorial substituents to respectively 122.8(3)° in **279** and 127.5° in **282**. Whereas the former distortion is in accordance with an anti-Berry displacement, the latter changes must be regarded as non-Berry displacements away from the energetically favourable reaction pathway, thereby emphasizing the degree of steric overcrowding in such anions.

Both axial and equatorial bond distances increase steadily upon introduction of less electronegative ligands at silicon. As may be seen from the least-squares lines in Figure 18, the average rate of increase for the equatorial bond lengths appears to be somewhat

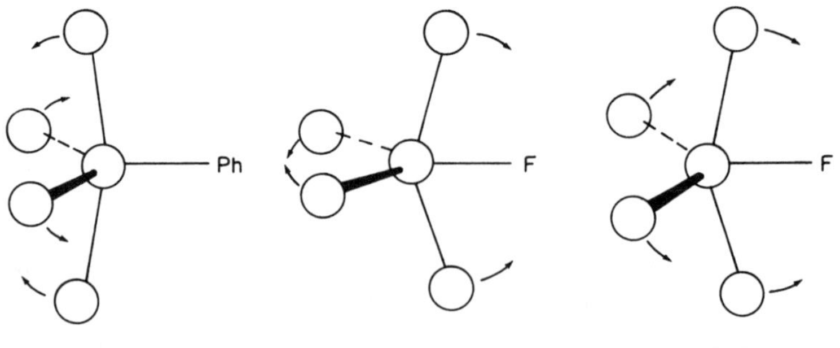

(**278**) (**281**) (**282**)

FIGURE 17. Ligand displacements from the idealised trigonal-bipyramidal geometry in the series $[R_nSiF_{5-n}]^-$: (a) Berry displacement in **278**; (b) anti-Berry displacement in **281**; (c) non-Berry displacement in **282**

TABLE 21. Molecular dimensions for trigonal-bipyramidal fluorosilicates

Anion		Cation	$d(Si–F_{ax})$ (Å)	$d(Si–F_{eq})$ (Å)	$d(Si–C_{eq})$ (Å)	$F_{ax}–Si–F_{ax}$ (°)[b]	Ref.
	$[SiF_5]^-$	$PhCH_2NMe_3^+$	1.646(1)	1.579(2) 1.602(2)		179.4(1)	397
278	$[PhSiF_4]^-$	n-Pr_4N^+	1.668(3) 1.670(3)	1.597(3) 1.606(2)	1.871(4)	174.6(1)	398
279	$[PhMeSiF_3]^-$	n-Pr_4N^+	1.689(4) 1.701(4)	1.621(4)	1.845(7)[c] 1.882(7)	184.4(2)	399
280	$[t$-$BuPhSiF_3]^-$	K^{+a}	1.682 1.718	1.639	1.897[d] 1.902	171.9	399
281	$[Ph_2SiF_3]^-$	Me_4N^+	1.688(1)	1.648(2)	1.893(2)	187.5	397
282[e]	$[Xyl_2SiF_3]^-$	K^{+a}	1.703 1.725	1.652	1.919 1.925	184.3	399
283[f]	$[1$-$NpPh_2SiF_2]^-$	$(Me_2N)_3S^+$	1.720(4)		1.915(7)[g] 1.925(7)	176.4(2)	399

[a] (18-crown-6) K^+.
[b] With respect to the Berry exchange coordinate.
[c] d(Si–Me).
[d] d(Si–Bu-t).
[e] Xyl = xylyl.
[f] 1-Np = 1-naphthyl.
[g] d(Si–Ph).

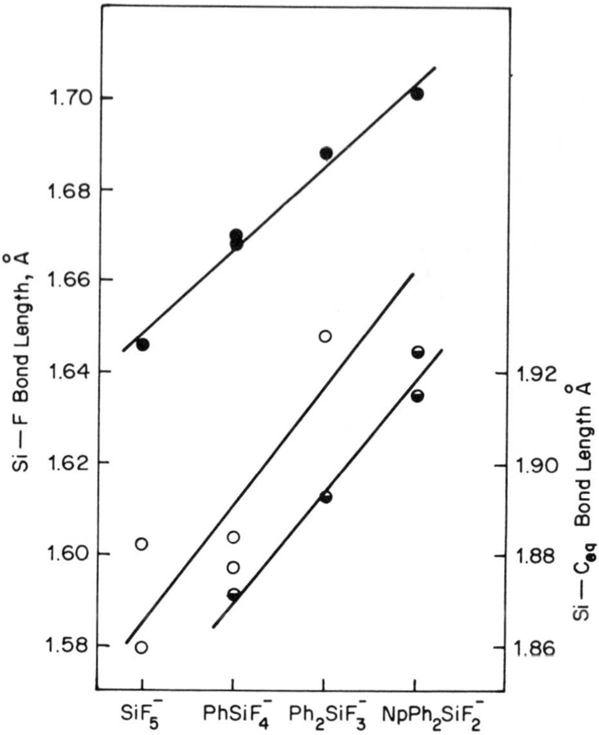

FIGURE 18. Variations in silicon–ligand bond lengths in the series $[R_nSiF_{5-n}]^-$, $n=0$–3; $d(Si–F_{ax})$ ●, $d(Si–F_{eq})$ ○, $d(Si–C_{eq})$ ◒. Reprinted with permission from *Inorg. Chem.*, **26**, 760 (1987). Copyright (1987) American Chemical Society

greater than for $d(Si–F_{ax})$ in the series $[R_nSiF_{5-n}]^-$ ($n=0$–3, R = Ph or 1-Naphthyl). This behaviour would be in contrast to that exhibited by the isoelectronic methylfluorophosphoranes Me_nPF_{5-n} ($n=0$–3)[23]. For instance, $d(P–F_{ax})$ increases by 0.066 Å on going from PF_5 to Me_2PF_3, $d(P–F_{eq})$ by only 0.019 Å for the same molecules. The equatorial P–C bond length increases by 0.033 Å between $MePF_4$ and Me_3PF_2, whereas $d(P–F_{ax})$ changes by 0.073 Å for the same pair. However, analysis of Table 21 suggests that steric crowding between equatorial aryl and alkyl groups also leads to bond elongation and it may reasonably be concluded[399] that this is an important factor in accounting for the differences in bond-length variations between the phenylfluorosilicates and the methylfluorophosphoranes. This steric effect is particularly marked for the equatorial Si–C bonds, e.g. $d(Si–C_{eq})$ increases from 1.882(7) Å in $[PhMeSiF_3]^-$ to 1.915(7) Å in $[1\text{-}NpPh_2SiF_2]^-$.

The axial Si–F bonds in the fluorosilicate anions listed in Table 21 are between 0.040 and 0.080 Å longer than the respective equatorial Si–F bonds. An increase in the silicon coordination number is associated with an elongation in the Si–X bond distances, as exemplified by a comparison of SiF_4[311] and $[SiF_5]^-$. The equatorial Si–F bonds in the pentacoordinated anion are respectively 0.023 and 0.046 Å longer than the Si–F bonds in the tetrahedral molecule SiF_4.

The orientation of the π-donor orbital of an equatorial aromatic substituent such as a phenyl group appears to be dictated by a compromise between steric interactions and electronic stabilization, so that it is not normally in the equatorial plane, i.e. the aromatic ring system is not normally perpendicular to the equatorial plane. For instance, respective dihedral angles of 54.3, 78.2 and 75° are observed between the phenyl rings of **278**, **279** and **281** and the equatorial plane. The bulkiness of the 1-naphthyl ligand in **283** is reflected in its analogous dihedral angle of 34.8°, which contrasts with values of 58.4 and 87.2° for the phenyl substituents in this anion. The latter angle is exhibited by the pivot ligand.

It has been found that the general structural features which stabilize a square pyramidal geometry (or rectangular pyramidal geometry for spirobicyclic derivatives) for phosphoranes[23] may be extended to the isoelectronic pentacoordinate silicon anions. These requirements may be summarized as follows:

(1) Four electronegative substituents (e.g. oxygen) which may occupy the basal positions of the square/rectangular pyramid must be present. In a trigonal bipyramid, two of these substituents would be forced to adopt an electronically less favourable equatorial placement.

(2) Two unsaturated five-membered rings containing similar electronegative atoms in any one ring directly bonded to the central atom must be present (i.e. a spirobicyclic system). Stabilization due to delocalized π-bonding is of greater magnitude than any relaxation of ring-strain effects which might be achieved by ring puckering. For the resultant planar five-membered ring a greater reduction of ring strain may be achieved by the assumption of a rectangular pyramidal geometry (rp), in which the Y–X linkages (Y = P, Si) are essentially equivalent.

(3) The apicel ligand should be of low electronegativity. Bulky apicel ligands should also favour the rectangular pyramid as this geometry allows a minimization of steric interactions with this substituent.

The geometrical parameters for spirobicyclic silicate anions studied by X-ray diffraction are listed in Table 22. The degree of displacement along the Berry exchange coordinate towards a rectangular pyramid may be calculated by the dihedral angle method[408]. A simple way of gauging the extent of this displacement is to calculate the dihedral angle δ_{24} (see Table 22 for definition) which varies between 53.1° (tbp) and 0° (rp). The variation of this dihedral angle is correlated with those of the bond angles between the axial ligands (θ_{15}) and the non-pivot equatorial angles (θ_{24}). In contrast to the phosphoranes, large displacements along the Berry exchange coordinate are also observed for compounds **290** (53.3%), **291** (53.6%, 70.8%) and **295** (72.1%) which contain saturated five-membered rings. The closest approach to a rectangular pyramid is found for **297** (89.8%) and **298** (97.6%). In both anions electron-pair repulsions between basal oxygen atoms will be reduced by the presence of either electron withdrawing substituents (**297**) or an extended π-system (**298**). The adoption of geometries with significantly differing degrees of rectangular pyramidal character by the independent molecules of **289** (52.8%, 68.7%) and **291** (53.6%, 70.8%) demonstrates that the energy barrier to displacement along the Berry exchange coordinate is relatively small and that lattice forces will influence the observed geometry. In contrast to the phosphoranes, hydrogen bonding to the silicate oxygens may also affect the molecular geometry. This is the case for the pyridinium salts **288**, **292** and **298**, for which Si–O bonds to hydrogen-bonded oxygen atoms suffer an increase, averaging 0.02–0.03 Å[403]. From the least-squares equations obtained for axial and equatorial Si–O distances in pentacoordinate silicate anions, Si–O bond lengths of 1.78 (axial) and 1.68 Å (equatorial) have been calculated for an ideal tbp geometry, 1.73 Å (basal) for an ideal rp geometry[403]. As for the acyclic fluorosilicates, the

TABLE 22. Geometrical parameters for spirobicyclic pentacoordinated silicate anions

Compound	δ_{24}(deg)[a]	θ_{15}(deg)[a]	θ_{24}(deg)[a]	d(Si–O$_{ax}$)(Å)	d(Si–O$_{eq}$)(Å)	%(tbp→rp)[b]	Ref.
284a[c]	41.2	176.6(2)	132.2(2)	1.782(4)		25.4	400
				1.806(4)			
284b	39.6	176.2(2)	133.7(2)	1.787(3)		28.7	400
				1.792(4)			
285	37.3	167.9(2)	127.8(3)	1.743(5)	1.723(5)	29.0	401
				1.821(5)	1.705(5)		
286	37.0	167.7(4)	127.9(4)	1.794(5)	1.700(9)	29.5	402
287	36.5	168.3(3)	129.1(3)	1.763(6)	1.700(6)	30.8	25
				1.789(5)	1.708(7)		
288	35.5	168.4(3)	130.5(3)	1.751(5)	1.698(5)	33.2	403
				1.820(5)	1.710(5)		
289a[c]	26.6	168.4(3)	140.7(3)	1.738(5)	1.699(5)	52.8	404
				1.741(5)	1.701(5)		
290	24.7	161.7(1)	136.0(1)	1.749(2)	1.711(2)	53.3	405
				1.769(2)	1.735(2)		
291a[c]	26.2	168.4(5)	141.2(5)	1.689(8)	1.683(10)	53.6	406
				1.734(8)	1.684(11)		
292	21.9	160.4(2)	137.8(2)	1.734(4)	1.722(4)	58.7	403
				1.772(4)	1.734(4)		
293	21.7	161.1(1)	138.6(1)	1.748(2)	1.720(2)	59.4	401
				1.785(2)	1.727(2)		

294	19.8	159.4(8)	139.0(8)	1.78(1)	1.729(9)	63.8	25
289b	18.0	164.0(2)	145.5(3)	1.736(5)	1.704(5)	68.7	404
				1.742(5)	1.706(5)		
291b	17.0	163.7(5)	146.4(5)	1.707(12)	1.688(9)	70.8	406
				1.732(11)	1.692(7)		
295	15.3	159.2(1)	143.6(1)	1.738(2)	1.685(2)	72.1	407
				1.757(2)	1.717(2)		
296	10.1	153.3(2)	143.0(2)	1.769(5)	1.739(5)	80.3	401
				1.769(5)	1.742(5)		
297	5.5	152.9(2)	147.3(2)	1.753(4)	1.746(4)	89.8	407
				1.759(4)	1.757(4)		
298	1.2	152.1(1)	150.9(1)	1.733(2)	1.734(2)	97.6	403
				1.750(2)	1.759(2)		

[a] δ_{24} is the dihedral angle between the triangular faces 124 and 245; it varies between 53.1° (tbp) and 0°(rp). θ_{15} and θ_{24} are the bond angles 1–Si–5 and 2–Si–4.
[b] Calculated by the dihedral angle method[408]:

$$\% \text{ (tbp} \rightarrow \text{rp)} = (\Sigma_i|\delta_i(\text{C}) - \delta_i(\text{TBP})|/\Sigma_i|\delta_i(\text{RP}) - \delta_i(\text{TBP})|) \times 100.$$

[c] Two independent molecules a and b in the unit cell.

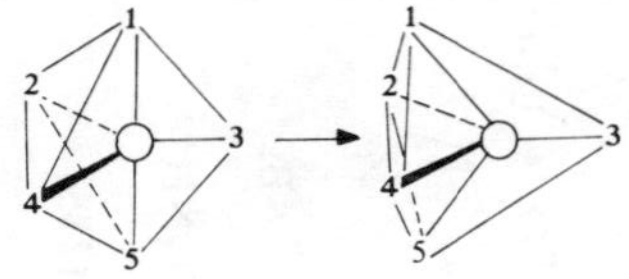

tbp rp

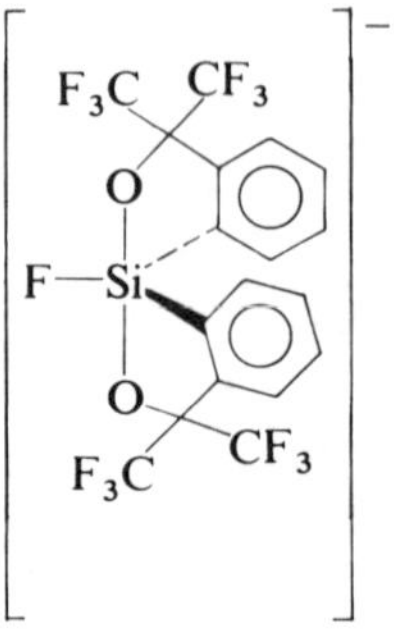

$[(Me_2N)_3Si]^+$

(284)

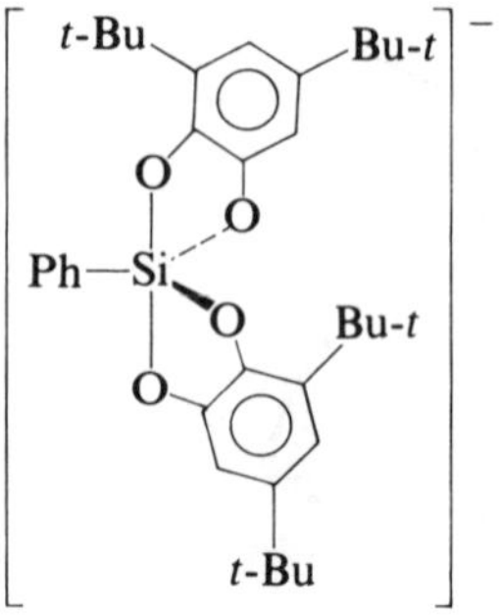

$[Et_3NH]^+$

(285)

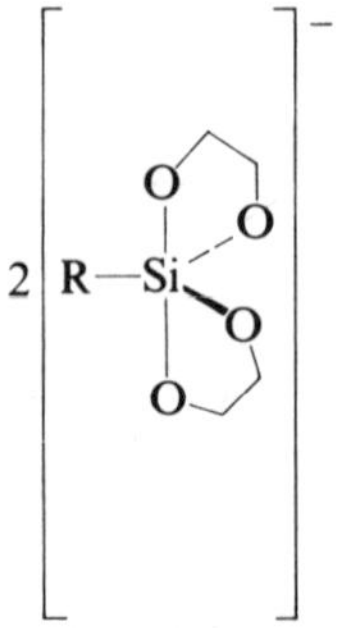

$[H_3N(CH_2)_6NH_3]^{2+}$

(290) R = Me
(295) R = Ph

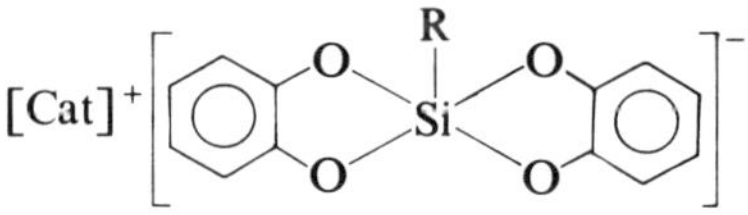

(286) [Cat] = [Me_4N]; R = Ph
(287) [Cat] = [Et_4N]; R = 1-naphthyl
(288) [Cat] = [C_5H_6N]; R = 2, 5-$Cl_2C_6H_3$
(289) [Cat] = [Et_4N]; R = F
(292) [Cat] = [C_5H_6N]; R = 1-naphthyl
(293) [Cat] = [Et_3HN]; R = Ph
(294) [Cat] = [Et_4N]; R = n-Bu
(296) [Cat] = [Et_4N]; R = t-Bu

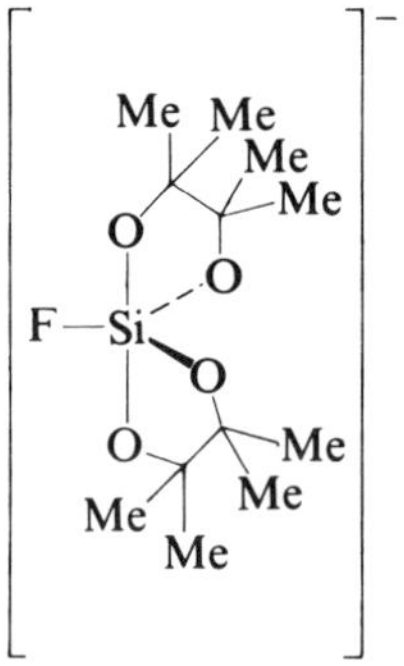

$[Me_4N]^+$

(291)

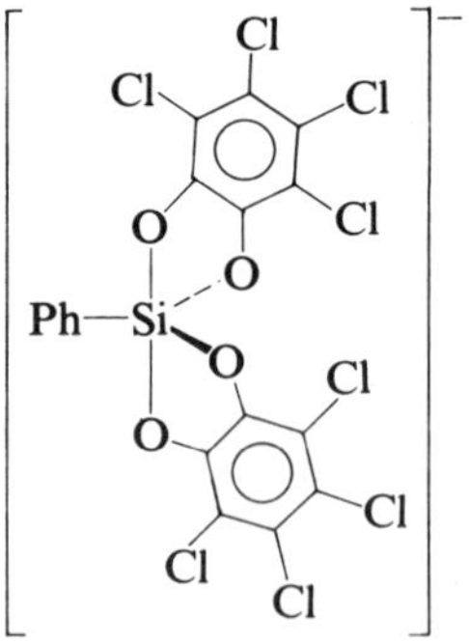

$[Et_4N]^+$

(297)

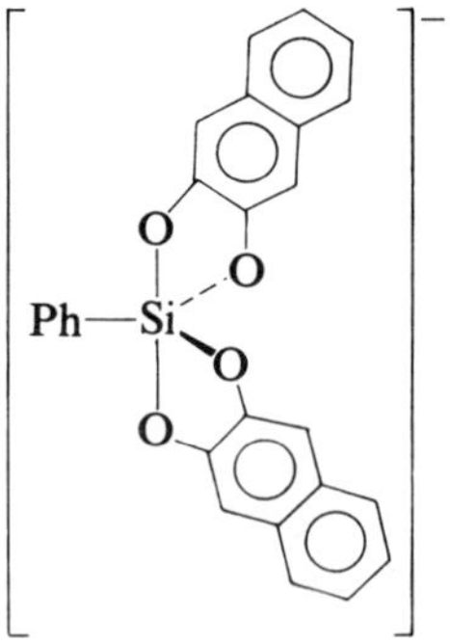

$[C_5H_6N]^+$

(298)

equatorial bond length is significantly longer than the typical value for tetrahedral derivatives [d(Si–O) = 1.64 ± 0.03 Å].

C. Neutral Complexes

As we have discussed previously, many structures of molecules which contain intramolecular dative bonds to silicon are known. The reader is referred to the review article of Tandura and colleagues for a comprehensive listing[26]. The extent to which a trigonal bipyramidal geometry is adopted by the central silicon atom will depend on the strength of the coordinate bond. Examples of relatively strong O→Si and N→Si interactions are provided by **299**[409] and **300**[410] for which respective dative bond distances of 1.918(6) and

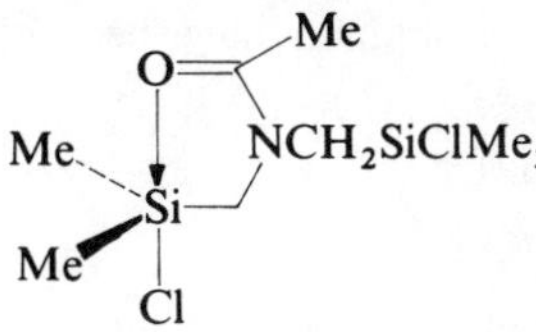

(299)

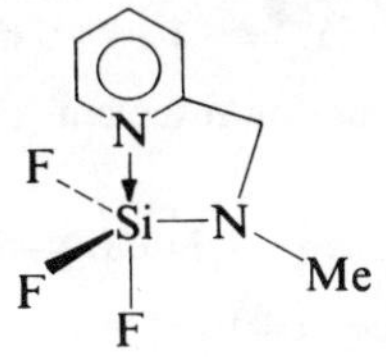

(300)

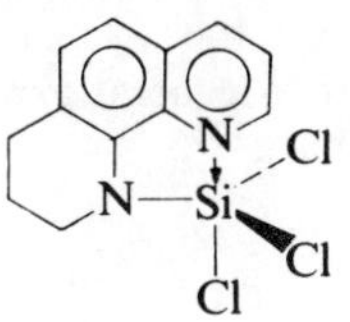

(301)

1.974(6) Å were reported (XRD). An analysis of 46 molecular fragments containing such interactions has allowed the geometrical changes which occur on going from tetrahedral to trigonal bipyramidal coordination to be mapped[21]. The electron deformational density in **301** has been studied by a combination of X-ray and neutron diffraction[411]. The X–N maps reveal an electron density maximum of 0.4e Å^{-3} between silicon and nitrogen in the equatorial bond [1.746(2) Å], and maxima of the nitrogen 'lone pair' density which are significantly displaced towards silicon. The authors suggested a polarization of the 'lone pair' density at nitrogen by silicon as an explanation. The centre of electron density for the dative N→Si bond [1.979(2) Å] is about 0.7 Å from the donor nitrogen atom. This density distribution is strongly deformed in the direction of silicon and shows a slower decrease along the Si–N axis towards silicon than towards nitrogen. Although such findings must be approached with caution, in view of inherent problems with systematic errors associated with this method, further experimental results would be desirable.

The principles of silatrane geometry have been discussed in a recent review article[20] which considered 40 molecular structures of silatranes and silatrane-like derivatives. As these features also allow more general conclusions to be drawn for neutral pentacoordinate silicon complexes as a whole they will be summarized here. Derivatives $RSi(OCR'_2CH_2)_n(XCH_2CH_2)_{3-n}N$ (X = O, OCH_2, CH_2) and $R_2Si(OCH_2CH_2)_2NR'$ were considered; the parent silatrane skeleton is given (**I**). The strength of the N→Si bond will

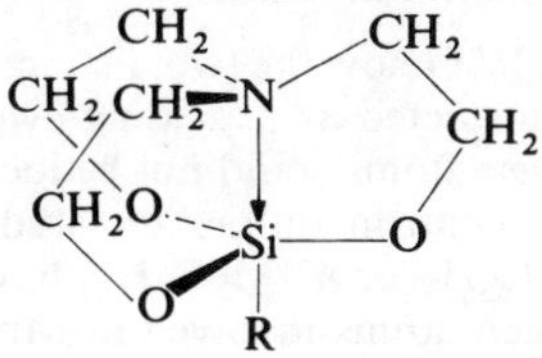

(I)

be primarily affected by the electronegativity of the substituents R and X. Electron-withdrawing groups will reduce the electron density at the silicon atom and increase its acceptor ability leading to shorter Si–N distances; e.g.: R = Cl, X = O[412], d(Si–N) = 2.023 Å, ∠NSi–O = 86.7°; R = Me, X = O = **302**[413], d(Si–N) = 2.175(4) Å, ∠N–Si–O = 82.7°; R = Me, X = CH_2, n = 2 = **303**[414], d(Si–N) = 2.336(4) Å, ∠N–Si–O = 80.1°. Other factors which may influence the length/strength of the transannular bond are as follows: (1) the basicity of nitrogen, (2) the substituents of the silatrane moiety, (3) the conformations of the five-membered rings, and (4) steric interactions.

The displacement of silicon (Δ_{Si}) from the equatorial plane of the three oxygens in silatranes is correlated with the length of the dative N→Si bond. Voronkov and coworkers[412] provided the following equation:

$$\Delta_{Si} = -1.14 + 0.62\, d(\text{N–Si}).$$

Alternative general expressions for nitrogen and oxygen adducts have been given by Schomburg[394]:

$$\cos\alpha = 1.249/d(\text{N–Si}) - 0.6915,$$

$$\cos\alpha = 1.281/d(\text{O–Si}) - 0.7288,$$

where α is the average angle between axial and equatorial substituents (Figure 19). These equations provide a mathematical description of the change in silicon geometry on going from tetrahedral to trigonal bipyramidal coordination. The approach of a donor atom along a vertical axis with retention of local C_{3v} symmetry for identical equatorial substituents, which may be assumed for an S_N2 substitution reaction at silicon, is simulated by a juxtaposition of molecular fragments provided by X-ray structural analyses.

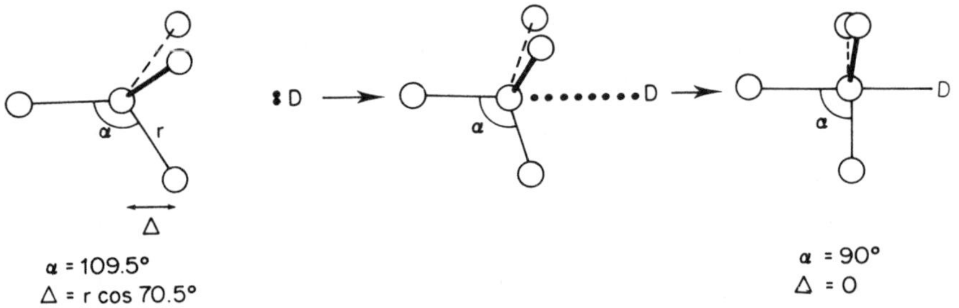

FIGURE 19. Transformation from tetrahedral to trigonal-bipyramidal silicon geometry upon adduct formation

D. Hydrogen-bridged Transition Metal–Silicon Bonds

We have discussed in Section IV.G how the two distinct W–Si distances of 2.586(5) and 2.703(4) Å in **241** have been interpreted as providing evidence for W–H ··· Si bridged bonds[340]. However, the hydrogen atoms could not be located in the course of the X-ray structural analysis. X-ray and neutron diffraction studies on molecules of the type $Cp'(CO)_2Mn(H)SiR_3$ ($Cp' = \eta^5\text{-}C_5H_5$ or $\eta^5\text{-}MeC_5H_4$) have allowed in several cases the unequivocal location of hydrogen atoms involved in Mn–H ··· Si bridge bonds. Geometrical parameters for such complexes are listed in Table 23. The neutron diffraction structure of **305** provides reliable molecular dimensions for an Mn–H ··· Si interaction.

TABLE 23. Geometrical parameters in complexes $Cp'(CO)_2Mn(H)SiR_3$ with Mn–H ··· Si interactions

Compound	Cp′	SiR_3	d(Mn–Si)(Å)	d(Mn–H)(Å)	d(Si–H)(Å)	Method	Ref.
	MeCp[a]	$SiCl_3$	2.254(1)	1.47(3)	1.79(4)	XRD	415
304	Cp	$SiCl_2Ph$	2.310(2)	1.49(6)	1.79(6)	XRD	416
305	MeCp	$SiFPh_2$	2.352(4)	1.569(4)	1.802(5)	ND	417
306[b]	MeCp	$SiHPh_2$	2.364(2)			XRD	418
307	Cp	$SiPh_3$	2.424(2)	1.55(4)	1.76(4)	XRD	419
308[b]	MeCp	SiMePh(1-Np)[c]	2.461(7)			XRD	420

[a] $Cp = \eta^5\text{-}C_5H_5$, $MeCp = \eta^5\text{-}MeC_5H_4$.
[b] Hydrogen atom positions were not located.
[c] 1-Np = 1-naphthyl.

Only one further value of an Mn–H distance characterized by neutron diffraction is available, namely 1.601(16) Å in $(CO)_5MnH$. Taken together with the less reliable X-ray diffraction values for the Mn–H bond length [1.47(3)–1.55(4) Å] listed in Table 23, the Mn–H bond length of 1.569(4) Å in **305** suggests that this parameter is relatively little affected by its participation in a short H ··· Si interaction. As a result of the Mn–H ··· Si interaction, the silicon atoms in the series $Cp'(CO)_2Mn(H)SiR_3$ achieve a formal pentacoordination. For instance, the H ··· Si–F angle in **305** is 148.8(2)°, i.e. the hydrogen and fluorine atoms adopt the axial positions of a very distorted trigonal bipyramid. In view of the acute Mn–H ··· Si angle of 88.2(2)° in **305**, a closed three-centre two-electron bonding scheme may be proposed for the interaction. The $SiFPh_2$ group is tilted away from the Mn–Si bond towards the centre of the MnHSi triangle. Similar geometrical parameters have also been established[417] for the complexes $MeCp(CO)_2Mn(H)SiFPh_2$ **309** and $C_5Me_5(CO)_2Mn(H)SiHPh_2$ **310** (XRD).

The compounds in Table 23 are listed in order of decreasing thermal stability, i.e. increasing tendency to eliminate R_3SiH. The geometrical parameters for the $Mn(H)SiR_3$ fragments allow the following trends to be established:

(1) The Mn–Si bond length decreases dramatically on passing from $SiR_3 = SiMePh$ (1-Np) **308** or $SiR_3 = SiPh_3$ **307** to $SiR_3 = SiCl_3$. Only ca. 0.03–0.04 Å of this decrease can be accounted for by the reduction in size of the covalent radius of silicon when attached to three chlorine atoms (Section IV.B.1).

(2) The angle of tilt of the SiR_3 group away from the Mn–Si axis decreases, for example, from 6.9(3)° for $SiR_3 = SiPh_3$ to 3.4(1)° for $SiR_3 = SiCl_3$.

In view of the unreliability of the hydrogen atom positions from X-ray structural analyses, the following trends must be regarded as tentative:

(3) The Mn–H distance decreases with increasing thermal stability of the complex.

(4) The Si–H distances exhibit very little variation for the listed complexes.

These findings suggest that the molecular fragments provide geometrical parameters for part of the reaction pathway for the oxidative addition of the silane R_3SiH at a

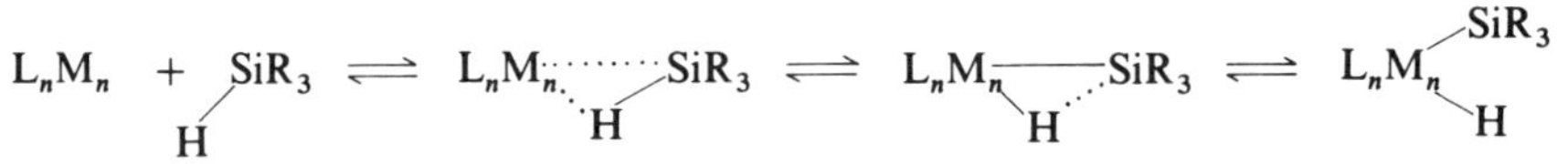

(A) (B) (C) (D)

manganese complex L_nM as depicted here. The first stage (**A→B**) leads to the building of an Mn–H interaction together with a weak Mn–Si interaction. **B** can be considered as exhibiting an open three-centre two-electron bond. As the reaction proceeds the Mn–Si bond is strengthened with concomitant weakening of the Si–H bond (**C** represents a closed three-centre two-electron bond). In the final stage (**C→D**) the Si ··· H interaction disappears and full oxidative addition is achieved.

VI. HEXACOORDINATE SILICON

We may extend the bonding principles discussed in Section V to hexacoordinate silicon derivatives. The anion $[SiF_6]^{2-}$ has been characterized in a series of salts[26]; the typical Si–F distances fall in the range 1.67–1.69 Å. The progressive elongation of the Si–F bond lengths with increasing coordination number may be gauged from the following series: SiF_4[311], d(Si–F) = 1.556(2) Å; $[SiF_5]^-$[397], $d(Si–F_{eq})$ = 1.579(2), 1.602(2) Å, $d(Si–F_{ax})$ = 1.646(1) Å; $[SiF_6]^{2-}$, typical range d(Si–F) = 1.67–1.69 Å. X-ray structural characterisations of hexacoordinate organic silicon derivatives are limited to octahedral neutral

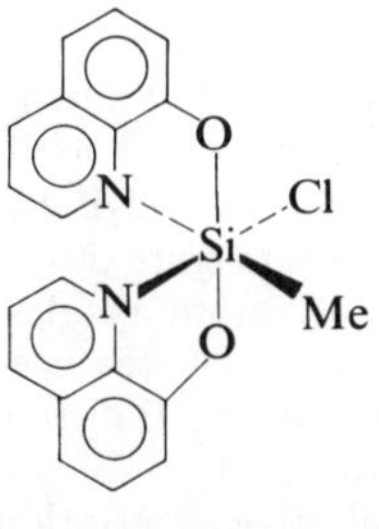

(311)

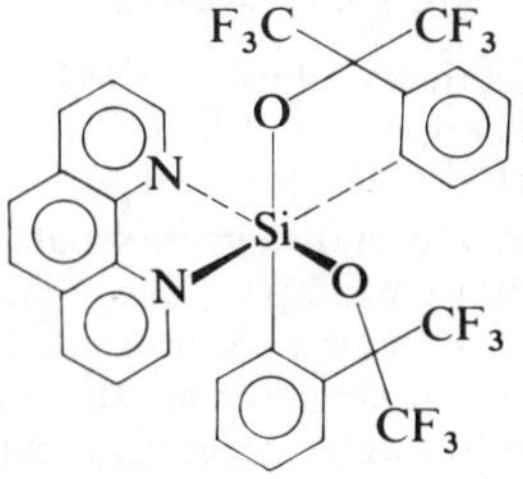

(312)

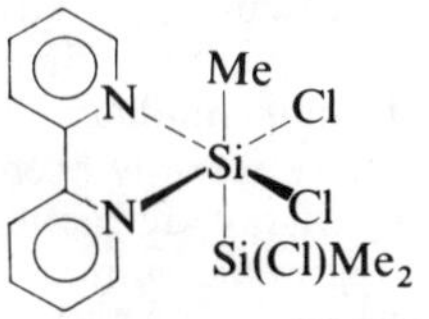

(313)

complexes. The Si–O bond lengths in **311**[421] [1.763(2), 1.766(2) Å] and **312**[422] [1.772(7), 1.772(7) Å] are ca. 0.12–0.13 Å longer than typical tetrahedral Si–O distances. The elongation of the octahedral Si–Cl bonds in **313** is even more marked. These are respectively 2.274(8) and 2.367(5) Å and may be compared with the tetrahedral Si–Cl distances of 2.075(6) and 2.087(7) Å in this complex. In contrast, the Si–C bond lengths in **311**–**313** (1.89–1.94 Å) are only slightly longer than the sum of the tetrahedral covalent radii for these atoms (1.88 Å). The Si–Si distance of 2.367(5) Å in **313**[423] falls in the typical range for tetrahedral Si–Si bond lengths (2.33–2.37 Å).

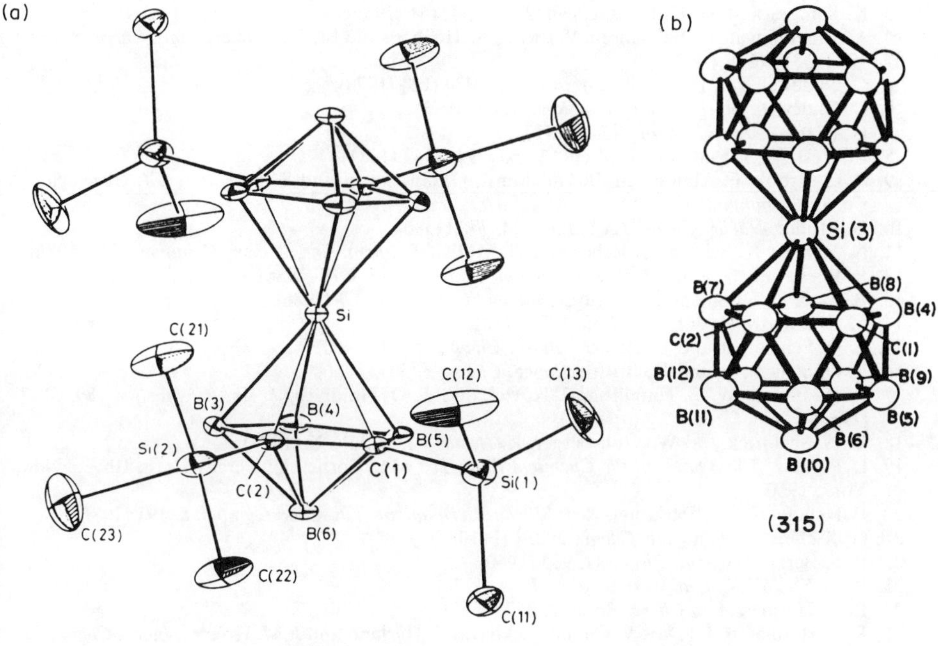

(314) (315)

FIGURE 20. (a) Molecular structure of $[(Me_3Si)_2C_2B_4H_4]_2Si$ **314** (without B and methyl H-atoms). Reproduced by permission of the Royal Society of Chemistry from Reference 424. (b) Molecular structure $[C_2B_9H_{11}]_2Si$ **315** (without H-atoms). Reprinted with permission from *J. Am. Chem. Soc.*, **108**, 5369 (1986). Copyright (1986) American Chemical Society

VII. π-COMPLEXES

We have discussed the structure of the π-complex $(\eta^5\text{-}Me_5C_5)_2Si$ **1**, the only known stable organic derivative of divalent silicon, in detail in Section II. The carbaborane analogues of bis(cyclopentadienyl)silicon(IV), $[(Me_3Si)_2C_2B_4H_4]_2Si$ **314**[424] and $[C_2B_9H_{11}]_2Si$ **315**[425] have recently been synthesized and their molecular structures (Figure 20) established by X-ray diffraction. In both examples, the central silicon atom resides at a crystallographic centre of symmetry and is η^5-coordinated by two C_2B_3 cluster faces. The Si–C distances of 2.23(1) and 2.25(1) Å are substantially longer than the Si–B distances of 2.10(1), 2.03(1) and 2.10(1) Å. Although theoretical calculations of the silicon–cage interaction are not available at the time of writing, a simple qualitative description may be advanced by analogy to bis(cyclopentadienyl) derivatives. A pair of carbaborane cages provides a total of 12 electrons for bonding in the filled ligand orbital combinations a_g, a_u, e_{1g} and e_{1u}. The a_g, a_u and e_{1u} orbitals are of correct symmetry to combine with the s, p_z and p_x/p_y atomic orbitals of silicon. Similarly the silicon d_{z^2} and d_{xz}/d_{yz} orbitals could interact significantly with the a_g and e_{1g} sets, if the energy disparity is not too large. For a large energy gap between silicon d and ligand orbitals, two essentially non-bonding ligand electron pairs must reside in the e_{1g} orbital set.

VIII. REFERENCES

1. K. Hedberg, *J. Am. Chem. Soc.*, **77**, 6491 (1955).
2. B. Beagley and T. G. Hewitt, *Trans. Faraday Soc.*, **64**, 2561 (1968).
3. K. Kimura and M. Kubo, *J. Chem. Phys.*, **30**, 151 (1959).
4. A. Almenningen, O. Bastiansen, V. Ewing, K. Hedberg and M. Traetteberg, *Acta Chem. Scand.*, **17**, 2455 (1963).
5. C. Glidewell and D. C. Liles, *Acta Cryst.*, **B34**, 124 (1978).
6. C. Glidewell, *Inorg. Chim. Acta Rev.*, **7**, 69 (1973).
7. H. Bürger, *Angew. Chem.*, **85**, 519 (1973).
8. R. West, M. J. Fink and J. Michl, *Science* **14**, 1343 (1981).
9. A. G. Brook, F. Abdesaken, B. Gutekunst, G. Gutekunst and R. K. Kallury, *J. Chem. Soc., Chem. Commun.*, 191 (1981).
10. G. Müller, *Nachr. Chem. Tech. Lab.*, **34**, 778 (1986).
11. N. Wiberg, K. Schurz, G. Reber and G. Müller, *J. Chem. Soc., Chem. Commun.*, 591 (1986).
12. R. Boese and U. Klingebiel, *J. Organomet. Chem.*, **315**, C17 (1986).
13. P. Jützi, D. Kanne and C. Krüger, *Angew. Chem.*, **98**, 163 (1986).
14. M. J. Barrow, *Acta Cryst.*, **B38**, 150 (1982).
15. E. O. Schlemper and D. Britton, *Inorg. Chem.*, **5**, 511 (1966).
16. E. O. Schlemper and D. Britton, *Inorg. Chem.*, **5**, 507 (1966).
17. R. Rudmann, W. C. Hamilton, S. Novick and T. D. Goldfarb, *J. Am. Chem. Soc.*, **89**, 5157 (1967).
18. W. S. Sheldrick and W. Wolfsberger, *Z. Naturforsch.*, **32b**, 22 (1977).
19. L. Pauling, *The Nature of the Chemical Bond*, 3rd edn., Cornell University Press, Ithaca, New York, 1960.
20. P. Hencsei and L. Parkanyi, *Rev. Silicon, Germanium, Tin, Lead Compd.*, **8**, 191 (1985).
21. G. Klebe, *J. Organomet. Chem.*, **293**, 147 (1985).
22. R. S. Berry, *J. Chem. Phys.*, **32**, 933 (1960).
23. W. S. Sheldrick, *Top. Curr. Chem.*, **73**, 1 (1978).
24. R. R. Holmes, *Acc. Chem. Res.*, **12**, 257 (1979).
25. R. R. Holmes, R. O. Day, V. Chandrasekhar, J. J. Harland and J. M. Holmes, *Inorg. Chem.*, **24**, 2016 (1985).
26. S. N. Tandura, M. G. Voronkov and N. V. Alekseev, *Top. Curr. Chem.*, **131**, 99 (1986).
27. J. P. Desclaux, *At. Data Nucl. Data Tables*, **12**, 311 (1973).
28. W. Kutzelnigg, *Angew. Chem.*, **96**, 262 (1984).
29. C. Glidewell, *Inorg. Chim. Acta*, **36**, 135 (1979).

30. A. Bondi, *J. Phys. Chem.*, **68**, 441 (1964).
31. B. Beagley and A. R. Conrad, *Trans. Faraday Soc.*, **66**, 2740 (1970).
32. B. T. Luke, J. A. Pople, M.-B. Krogh-Jespersen, Y. Apeloig, J. Chandrasekhar and P. v. R. Schleyer, *J. Am. Chem. Soc.*, **108**, 260 (1986).
33. L. Pauling, *J. Am. Chem. Soc.*, **53**, 1367 (1931).
34. R. E. Rundle, *J. Am. Chem. Soc.*, **85**, 112 (1963).
35. B. T. Luke, J. A. Pople, M.-B. Krogh-Jespersen, Y. Apeloig, M. Karni, J. Chandrasekhar and P. v. R. Schleyer, *J. Am. Chem. Soc.*, **108**, 270 (1986).
36. I. Dubois, *Can. J. Phys.*, **46**, 2455 (1968).
37. E. Wasserman, V. J. Kuck, R. S. Hutton and Y. W. Yager, *J. Am. Chem. Soc.*, **92**, 7491 (1970).
38. P. Jensen, P. R. Bunker and A. R. Hoy, *J. Chem. Phys.*, **77**, 5370 (1982).
39. H. Shoji, T. Tanaka and E. Hirota, *J. Mol. Spectrosc.*, **47**, 491 (1973).
40. I. Hargittai, G. Schultz, J. Tremmel, N. D. Kagramov, A. K. Maltsev and O. M. Nefedov, *J. Am. Chem. Soc.*, **105**, 2895 (1983).
41. L. Fernholt, A. Haaland, P. Jützi, F. X. Kohl and R. Seip, *Acta Chem. Scand.*, **A38**, 211 (1984).
42. P. Jützi, F. Kohl, P. Hofmann, C. Krüger and Y.-H. Tsay, *Chem. Ber.*, **113**, 757 (1980).
43. M. Grenz, E. Hahn, W.-W. du Mont and J. Pickardt, *Angew. Chem.*, **96**, 69 (1984).
44. J. L. Atwood, W. E. Hunter, A. H. Cowley, R. A. Jones and C. A. Stewart, *J. Chem. Soc., Chem. Commun.*, 925 (1981).
45. C. Glidewell, *J. Organomet. Chem.*, **286**, 289 (1985).
46. M. J. S. Dewar, G. L. Grady, D. R. Kuhn and K. M. Merz, Jr., *J. Am. Chem. Soc.*, **106**, 6773 (1984).
47. R. West, *Pure Appl. Chem.*, **56**, 163 (1984).
48. R. West, *Science*, **225**, 1109 (1984).
49. R. West, M. J. Fink, M. J. Michalczyk and D. J. De Young, in *Organosilicon and Bioorganosilicon Chemistry* (Ed. H. Sakurai), Ellis Horwood, Chichester, UK, 1985, p. 3.
50. A. H. Cowley, *Acc. Chem. Res.*, **17**, 386 (1984).
51. A. H. Cowley and N. C. Norman, *Progr. Inorg. Chem.*, **34**, 1 (1986).
52. G. Raabe and J. Michl, *Chem. Rev.*, **85**, 419 (1985).
53. J. F. Blount, K. Mislow and J. Jacobus, *Acta Cryst.*, **A28**, S12 (1972).
54. M. J. Fink, M. J. Michalczyk, K. J. Haller, R. West and J. Michl, *J. Chem. Soc., Chem. Commun.*, 1010 (1983).
55. M. J. Fink, M. J. Michalczyk, K. J. Haller, R. West and J. Michl, *Organometallics*, **3**, 793 (1984).
56. S. Masamune, S. Murakami, J. T. Snow, H. Tobita and D. J. Williams, *Organometallics*, **3**, 333 (1984).
57. J. T. Snow, S. Murakami, S. Masamune and D. J. Williams, *Tetrahedron Lett.*, **25**, 4191 (1984).
58. D. E. Goldberg, P. B. Hitchcock, M. F. Lappert, K. M. Thomas, A. J. Thorne, T. Fjeldberg, A. Haaland and B. E. R. Schilling, *J. Chem. Soc., Dalton Trans.*, 2387 (1986).
59. K. Krogh-Jespersen, *J. Am. Chem. Soc.*, **107**, 537 (1985).
60. A. G. Brook, S. C. Nyburg, F. Abdesaken, B. Gutekunst, G. Gutekunst, R. Krishna, M. R. Kallury, Y. C. Poon, Y.-M. Chang and W. Wong-Ng, *J. Am. Chem. Soc.*, **104**, 5667 (1982).
61. S. C. Nyburg, A. G. Brook, F. Abdesaken, G. Gutekunst and W. Wong-Ng, *Acta Cryst.*, **C41**, 1632 (1985).
62. N. Wiberg, G. Wagner and G. Müller, *Angew. Chem.*, **97**, 220 (1985).
63. N. Wiberg, G. Wagner, J. Riede and G. Müller, *Organometallics*, **6**, 32 (1987).
64. H. F. Schaefer, *Acc. Chem. Res.*, **15**, 283 (1982).
65. Y. Apeloig and M. Karni, *J. Chem. Soc., Chem. Commun.*, 768 (1984).
66. Y. Apeloig and M. Karni, *J. Am. Chem. Soc.*, **106**, 6676 (1984).
67. N. Wiberg, G. Wagner, G. Müller and J. Riede, *J. Organomet. Chem.*, **271**, 381 (1984).
68. N. Wiberg, G. Wagner, G. Reber, J. Riede and G. Müller, *Organometallics*, **6**, 35 (1987).
69. L. C. Krisher and L. Pierce, *J. Chem. Phys.*, **32**, 1619 (1960).
70. N. Wiberg, K. Schurz and G. Fischer, *Angew. Chem.*, **97**, 1058 (1985).
71. P. v. R. Schleyer and P. D. Stout, *J. Chem. Soc., Chem. Commun.*, 1373 (1986).
72. A. M. Domingos and G. M. Sheldrick, *Acta Cryst.*, **B30**, 2512 (1974).
73. W. Airey, C. Glidewell, D. W. H. Rankin, A. G. Robiette, G. M. Sheldrick and D. W. J. Cruickshank, *Trans. Faraday Soc.*, **66**, 551 (1970).
74. W. Airey, C. Glidewell, A. G. Robiette and G. M. Sheldrick, *J. Mol. Struct.*, **8**, 413 (1971).

75. M. J. Barrow, E. A. V. Ebsworth and M. M. Harding, *Acta Cryst.*, **B35**, 2093 (1979).
76. M. J. Barrow and E. A. V. Ebsworth, *J. Chem. Soc., Dalton Trans.*, 211 (1982).
77. W. Wojnowski, K. Peters, E.-M. Peters and H. G. von Schnering, *Z. Anorg. Allg. Chem.*, **525**, 121 (1985).
78. D. W. H. Rankin, A. G. Robiette, G. M. Sheldrick, W. S. Sheldrick, B. J. Aylett, I. A. Ellis and J. J. Monaghan, *J. Chem. Soc. (A)*, 1224 (1969).
79. T. Fjeldberg, *J. Mol. Struct.*, **112**, 159 (1984).
80. C. Glidewell and H. D. Holden, *Acta Cryst.*, **B37**, 754 (1981).
81. C. Glidewell and D. C. Liles, *Acta Cryst.*, **B34**, 696 (1978).
82. L. Pierce and M. Hayashi, *J. Chem. Phys.*, **35**, 479 (1961).
83. C. Glidewell, *Inorg. Chim. Acta*, **29**, L283 (1978).
84. C. Glidewell, *J. Organomet. Chem.*, **159**, 23 (1978).
85. L. S. Bartell, *J. Chem. Educ.*, **45**, 754 (1968).
86. C. Glidewell, *Inorg. Chim. Acta*, **12**, 219 (1975).
87. B. Beagley and T. G. Hewitt, *Trans. Faraday Soc.*, **64**, 2561 (1968).
88. R. W. Kilb and L. Pierce, *J. Chem. Phys.*, **27**, 108 (1957).
89. L. Pierce, *J. Chem. Phys.*, **34**, 498 (1961).
90. L. Pierce and D. H. Petersen, *J. Chem. Phys.*, **33**, 907 (1960).
91. B. Beagley, J. J. Monaghan and T. G. Hewitt, *J. Mol. Struct.*, **8**, 401 (1971).
92. J. M. O'Reilly and L. Pierce, *J. Chem. Phys.*, **34**, 1176 (1961).
93. W. Zeil, J. Haase and M. Dakkouri, *Discuss. Faraday Soc.*, **47**, 149 (1969).
94. S. Cradock, J. Koprowski and D. W. H. Rankin, *J. Mol. Struct.*, **77**, 113 (1981).
95. A. R. Green, *Ph.D. Thesis*, Edinburgh, 1976; cited in Ref. 97.
96. D. W. W. Anderson, S. Cradock, E. A. V. Ebsworth, A. R. Green, D. W. H. Rankin and A. G. Robiette, *J. Organomet. Chem.*, **271**, 235 (1984).
97. W. Clegg, U. Klingebiel, J. Neemann, G. M. Sheldrick and N. Vater, *Acta Cryst.*, **B37**, 987 (1981).
98. N. Muller and R. C. Bracken, *J. Chem. Phys.*, **32**, 1577 (1960).
99. M. Dakkouri and H. Oberhammer, *Z. Naturforsch.*, **29a**, 513 (1974).
100. G. L. Delker, Y. Wang, G. D. Stucky, R. L. Lambert, Jr., C. K. Haas and D. Seyferth, *J. Am. Chem. Soc.*, **98**, 1779 (1976).
101. K. Hirotsu, T. Higuchi, M. Ishikawa, H. Sugisawa and M. Kumada, *J. Chem. Soc., Chem. Commun.*, 726 (1982).
102. L. C. Krisher and L. Pierce, *J. Chem. Phys.*, **32**, 1619 (1960).
103. J. D. Swalen and B. P. Stoicheff, *J. Chem. Phys.*, **28**, 671 (1958).
104. J. R. Durig, Y. S. Li and C. C. Tong, *J. Mol. Struct.*, **14**, 255 (1972).
105. J. R. Durig, R. O. Carter and Y. S. Li, *J. Mol. Spectr.*, **44**, 18 (1972).
106. M. Nakata, H. Takeo and C. Matsumara, *J. Mol. Spectr.*, **82**, 117 (1980).
107. V. S. Mastryukov, A. V. Golubinskii and L. V. Vilkov, *Zh. Strukt. Khim.*, **21**, 48 (1980).
108. W.-P. Leung, C. L. Raston, B. W. Skelton and A. H. White, *J. Chem. Soc., Dalton, Trans.*, 1801 (1984).
109. L. M. Engelhardt, W.-P. Leung, C. L. Raston and A. H. White, *J. Chem. Soc., Dalton Trans.*, 337 (1985).
110. G. Fritz, U. Braun, W. Schick, W. Hönle and H. G. v. Schnering, *Z. Anorg. Allg. Chem.*, **472**, 45 (1981).
111. G. Fritz, W. Schick, W. Hönle and H. G. v. Schnering, *Z. Anorg. Allg. Chem.*, **511**, 95 (1984).
112. H. Beckers, H. Bürger, R. Eujen, B. Rempfer and H. Oberhammer, *J. Mol. Struct.*, **140**, 281 (1986).
113. A. H. Cowley, J. E. Kilduff, E. A. V. Ebsworth, D. W. H. Rankin, H. E. Robertson and R. Seip, *J. Chem. Soc., Dalton Trans.*, 689 (1984).
114. A. Schäfer, M. Weidenbruch, K. Peters and H. G. v. Schnering, *Angew. Chem.*, **96**, 311 (1984).
115. M. Weidenbruch, B. Flintjer, H. G. v. Schnering and K. Peters, *Angew. Chem.*, **98**, 1090 (1986).
116. N. Wiberg, H. Schuster, A. Simon and K. Peters, *Angew. Chem.*, **98**, 100 (1986).
117. M. Weidenbruch, A. Schäfer, K. Peters and H. G. v. Schnering, *J. Organomet. Chem.*, **314**, 25 (1986).
118. A. Schäfer, M. Weidenbruch and S. Pohl, *J. Organomet. Chem.*, **282**, 305 (1985).
119. C. Glidewell and G. M. Sheldrick, *J. Chem. Soc. (A)*, 3127 (1971).

120. W. Hönle and H. G. v. Schnering, *Z. Anorg. Allg. Chem.*, **464**, 139 (1980).
121. K. Endo, H. Takeo and C. Matsumara, *Bull. Chem. Soc. Jpn.*, **50**, 626 (1977).
122. L. S. Bartell and W. F. Bradford, *J. Mol. Struct.*, **37**, 113 (1977).
123. T. Fjeldberg, R. Seip, M. F. Lappert and A. J. Thorne, *J. Mol. Struct.*, **99**, 295 (1983).
124. T. Fjeldberg, M. F. Lappert and A. J. Thorne, *J. Mol. Struct.*, **127**, 95 (1985).
125. H. B. Burgi and L. S. Bartell, *J. Am. Chem. Soc.*, **94**, 5236 (1972).
126. B. Beagley and R. G. Pritchard, *J. Mol. Struct.*, **84**, 129 (1982).
127. S. K. Doun and L. S. Bartell, *J. Mol. Struct.*, **63**, 249 (1980).
128. A. Almenningen, H. M. Seip and R. Seip, *Acta Chem. Scand.*, **24**, 1697 (1970).
129. C. Glidewell and D. C. Liles, *J. Organomet. Chem.*, **234**, 15 (1982).
130. L. Párkányi and E. Hengge, *J. Organomet. Chem.*, **235**, 273 (1982).
131. S. G. Baxter, K. Mislow and J. F. Blount, *Tetrahedron*, **36**, 605 (1980).
132. M. Weidenbruch, K. Kramer, S. Pohl and W. Saak, *J. Organomet. Chem.*, **316**, C 13 (1986).
133. M. Weidenbruch, K. Kramer, K. Peters and H. G. von Schnering, *Z. Naturforsch.*, **40b**, 601 (1985).
134. T. A. Beineke and L. L. Martin, *J. Organomet. Chem.*, **20**, 65 (1969).
135. L. S. Bartell, F. B. Clippard, Jr., and T. L. Boates, *Inorg. Chem.*, **9**, 2436 (1970).
136. A. W. Cordes, P. F. Schubert and R. T. Oakley, *Can. J. Chem.*, **57**, 174 (1979).
137. T. H. Newman, J. C. Calabrese, R. T. Oakley, D. A. Stanislawski and R. West, *J. Organomet. Chem.*, **225**, 211 (1982).
138. W. S. Sheldrick and A. Borkenstein, *Acta Cryst.*, **B33**, 2916 (1977).
139. T.-H. Hseu, T.-L. Hwang and C.-S. Liu, *Inorg. Chem.*, **20**, 2182 (1981).
140. O. A. D'Yachenko, S. V. Soboleva and L. O. Atovmyan, *Zh. Strukt. Khim.*, **17**, 350 (1976).
141. M. Ishikawa, T. Fuchikama, M. Kumada, T. Higuchi and S. Miyamoto, *J. Am. Chem. Soc.*, **101**, 1348 (1979).
142. H. Watanabe, M. Kato, T. Okawa, Y. Nagai and M. Goto, *J. Organomet. Chem.*, **271**, 225 (1984).
143. S. Masamune, Y. Hanzawa, S. Murakami, T. Bally and J. F. Blount, *J. Am. Chem. Soc.*, **104**, 1150 (1982).
144. J. C. Dewan, S. Murakami, J. T. Snow, S. Collins and S. Masamune, *J. Chem. Soc., Chem. Commun.*, 892 (1985).
145. R. Jones, D. J. Williams, Y. Kabe and S. Masamune, *Angew. Chem.*, **98**, 176 (1986).
146. C. Kratky, H. G. Schuster and E. Hengge, *J. Organomet. Chem.*, **247**, 253 (1983).
147. L. Párkányi, K. Sasvári and I. Barta, *Acta Cryst.*, **B34**, 883 (1978).
148. C. J. Hurt, J. C. Calabrese and R. West, *J. Organomet. Chem.*, **91**, 273 (1975).
149. S. Masamune, Y. Kabe, S. Collins, D. J. Williams and R. Jones, *J. Am. Chem. Soc.*, **107**, 5552 (1985).
150. Y.-S. Chen and P. P. Gaspar, *Organometallics*, **1**, 1410 (1982).
151. C. W. Carlson, K. J. Haller, X.-H. Zhang and R. West, *J. Am. Chem. Soc.*, **106**, 5521 (1984).
152. L. Párkányi, K. Sasvári, J. P. Declerc and G. Germain, *Acta Cryst.*, **B34**, 3678 (1978).
153. Z. Smith, H. M. Seip, E. Hengge and G. Bauer, *Acta Chem. Scand.*, **A30**, 697 (1976).
154. H. L. Carrell and J. Donohue, *Acta Cryst.*, **B28**, 1566 (1972).
155. S.-M. Chen, L. D. David, K. J. Haller, C. L. Wadsworth and R. West, *Organometallics*, **2**, 409 (1983).
156. M. Dräger and K. G. Walter, *Z. Anorg. Allg. Chem.*, **479**, 65 (1981).
157. Z. Smith, A. Almenningen, E. Hengge and D. Kovar, *J. Am. Chem. Soc.*, **104**, 4362 (1982).
158. F. Shafiee, J. R. Damewood, Jr., K. J. Haller and R. West, *J. Am. Chem. Soc.*, **107**, 6950 (1985).
159. F. Shafiee, K. J. Haller and R. West, *J. Am. Chem. Soc.*, **108**, 5478 (1986).
160. H. Watanabe and Y. Nagai, in *Organosilicon and Bioorganosilicon Chemistry* (Ed. H. Sakurai), Ellis Horwood, Chichester, UK, 1985, p. 105.
161. S. Nagase, M. Nakano and T. Kudo, *J. Chem. Soc., Chem. Commun.*, 60 (1987).
162. P. v. R. Schleyer, A. F. Sax, J. Kalcher and R. Janoschek, *Angew. Chem.*, **99**, 374 (1987).
163. C. Altona, H. J. Geise and C. Romers, *Tetrahedron*, **24**, 13 (1968).
164. A. P. Cox and R. Varma, *J. Chem. Phys.*, **46**, 2007 (1967).
165. M. J. Barrow and E. A. V. Ebsworth, *J. Chem. Soc., Dalton Trans.*, 563 (1984).
166. E. A. V. Ebsworth, E. K. Murray, D. W. H. Rankin and H. E. Robertson, *J. Chem. Soc., Dalton Trans.*, 1501 (1981).

167. G. Gundersen and D. W. H. Rankin, *Acta Chem. Scand.*, **A38**, 647 (1984).
168. C. Glidewell, D. W. H. Rankin, A. G. Robiette and G. M. Sheldrick, *J. Mol. Struct.*, **4**, 215 (1969).
169. G. Gundersen, D. W. H. Rankin and H. E. Robertson, *J. Chem. Soc., Dalton Trans.*, 191 (1985).
170. G. Gundersen, R. A. Mayo and D. W. H. Rankin, *Acta Chem. Scand.*, **A38**, 579 (1984).
171. A. J. Blake, E. A. V. Ebsworth, D. W. H. Rankin, H. E. Robertson, D. E. Smith and A. J. Welch, *J. Chem. Soc., Dalton Trans.*, 91 (1986).
172. A. G. Robiette, G. M. Sheldrick and W. S. Sheldrick, *J. Mol. Struct.*, **4**, 423 (1970).
173. D. G. Anderson, A. J. Blake, S. Craddock, E. A. V. Ebsworth, D. W. H. Rankin and A. J. Welch, *Angew. Chem.*, **98**, 97 (1986).
174. W. Airey, C. Glidewell, A. G. Robiette, G. M. Sheldrick and J. M. Freeman, *J. Mol. Struct.*, **8**, 423 (1971).
175. D. Schomburg, A. Blaschette and E. Wieland, *Z. Naturforsch.*, **41b**, 1112 (1986).
176. K. Hensen, T. Zengerly, P. Pickel and G. Klebe, *Angew. Chem.*, **95**, 739 (1983).
177. J. A. Gibson, G.-V. Röschenthaler, D. Schomburg and W. S. Sheldrick, *Chem. Ber.*, **110**, 1887 (1977).
178. E. Egert, M. Haase, U. Klingebiel, C. Lensch, D. Schmidt and G. M. Sheldrick, *J. Organomet. Chem.*, **315**, 19 (1986).
179. L. Párkányi, A. Szöllösy, L. Bihátsi, P. Hencsei and J. Nagy, *J. Organomet. Chem.*, **256**, 235 (1983).
180. P. Clare, D. B. Sowerby and I. Haiduc, *J. Organomet. Chem.*, **310**, 161 (1986).
181. U. Kliebisch, U. Klingebiel, D. Stalke and G. M. Sheldrick, *Angew. Chem.*, **98**, 921 (1986).
182. W. Clegg, M. Haase, G. M. Sheldrick and N. Vater, *Acta Cryst.*, **C40**, 871 (1984).
183. T. Fjeldberg, P. B. Hitchcock, M. F. Lappert and A. J. Thorne, *J. Chem. Soc., Chem. Commun.*, 822 (1984).
184. T. Fjeldberg, M. F. Lappert and A. J. Thorne, *J. Mol. Struct.*, **125**, 265 (1984).
185. D. Mootz, A. Zinnius and B. Böttcher, *Angew. Chem.*, **81**, 398 (1969).
186. R. D. Rogers, J. L. Atwood and R. Grüning, *J. Organomet. Chem.*, **157**, 229 (1978).
187. M. F. Lappert, M. J. Slade, A. Singh, J. L. Atwood, R. D. Rogers and R. Shakir, *J. Am. Chem. Soc.*, **105**, 302 (1983).
188. L. M. Engelhardt, A. S. May, C. L. Raston and A. H. White, *J. Chem. Soc., Dalton Trans.*, 1671 (1983).
189. P. Power and X. Xiaojie, *J. Chem. Soc., Chem. Commun.*, 358 (1984).
190. R. Grüning and J. L. Atwood, *J. Organomet. Chem.*, **137**, 101 (1977).
191. A. H. Clark and A. Haaland, *Acta Chem. Scand.*, **24**, 3024 (1970).
192. T. Fjeldberg and R. A. Andersen, *J. Mol. Struct.*, **125**, 287 (1984).
193. A. Haaland, K. Hedberg and P. P. Power, *Inorg. Chem.*, **23**, 1972 (1984).
194. E. C. Alyea, K. J. Fisher and T. Fjeldberg, *J. Mol. Struct.*, **127**, 325 (1985).
195. E. C. Alyea, K. J. Fisher and T. Fjeldberg, *J. Mol. Struct.*, **130**, 263 (1985).
196. G. M. Sheldrick and W. S. Sheldrick, *J. Chem. Soc. (A)*, 2279 (1969).
197. D. C. Bradley, M. B. Hursthouse and P. F. Rodesiler, *J. Chem. Soc., Chem. Commun.*, 14 (1969).
198. T. Fjeldberg and R. A. Andersen, *J. Mol. Struct.*, **128**, 49 (1985).
199. T. Fjeldberg and R. A. Andersen, *J. Mol. Struct.*, **129**, 93 (1985).
200. D. C. Bradley and M. H. Chisholm, *Acc. Chem. Res.*, **9**, 273 (1976).
201. M. F. Lappert, P. P. Power, A. R. Sanger and R. C. Srivastava, *Metal and Metalloid Amides*, Ellis Horwood, Chichester, UK, 1980.
202. K. N. Raymond, and C. W. Eigenbrot, Jr., *Acc. Chem. Res.*, **13**, 276 (1980).
203. J. S. Ghotra, M. B. Hursthouse and A. J. Welch, *J. Chem. Soc., Chem. Commun.*, 669 (1973).
204. C. Glidewell, A. G. Robiette and G. M. Sheldrick, *Chem. Phys. Lett.*, **16**, 526 (1972).
205. S. Cradock, C. M. Huntley, D. W. H. Rankin and H. E. Robertson, *J. Chem. Soc., Dalton Trans.*, 859 (1986).
206. S. Cradock, C. M. Huntley and J. R. Durig, *J. Mol. Struct.*, **127**, 319 (1985).
207. C. M. Huntley, D. W. H. Rankin and H. E. Robertson, *J. Mol. Struct.*, **156**, 103 (1987).
208. G. M. Sheldrick and R. Taylor, *J. Organomet. Chem.*, **87**, 145 (1985).
209. C. Glidewell and A. G. Robiette, *Chem. Phys. Lett.*, **28**, 290 (1974).
210. M. Dakkouri and H. Oberhammer, *Z. Naturforsch.*, **27a**, 1331 (1982).

211. E. R. Corey, V. Codey, M. D. Glick and L. J. Radonovich, *J. Inorg. Nucl. Chem.*, **35**, 1714 (1973).
212. M. Veith and H. Bärnighausen, *Acta Cryst.*, **B30**, 1806 (1974).
213. E. E. Astrup, A. M. Bouzga and K. A. Ostoja Starzewski, *J. Mol. Struct.*, **51**, 51 (1979).
214. S. Pohl, *Chem. Ber.*, **112**, 3159 (1979).
215. S. Pohl and B. Krebs, *Chem. Ber.*, **110**, 3183 (1977).
216. O. J. Scherer, P. Quintus and W. S. Sheldrick, *Chem. Ber.*, **120**, 1183 (1987).
217. S. Pohl, *J. Organomet. Chem.*, **142**, 195 (1977).
218. O. J. Scherer, J. Kerth and M. L. Ziegler, *Angew. Chem.*, **95**, 510 (1983).
219. O. J. Scherer, R. Walter and W. S. Sheldrick, *Angew. Chem.*, **97**, 525 (1985).
220. A. G. MacDiarmid and A. G. Maddox, *J. Inorg. Nucl. Chem.*, **1**, 411 (1955).
221. E. A. V. Ebsworth and M. J. Mays, *J. Chem. Soc.*, 4844 (1962).
222. K. F. Dössel and A. G. Robiette, *Z. Naturforsch.*, **A32**, 462 (1977).
223. M. C. L. Gerry, T. M. Sugden and J. C. Thompson, *Nature (London)*, **211**, 846 (1966).
224. J. A. Duckett, A. G. Robiette and I. M. Mills, *J. Mol. Spectrosc.*, **62**, 34 (1976).
225. B. Beagley, A. G. Robiette and G. M. Sheldrick, *J. Chem. Soc. (A)*, 3002 (1968).
226. C. Glidewell, P. M. Pinder, A. G. Robiette and G. M. Sheldrick, *J. Chem. Soc., Dalton Trans.*, 1402 (1972).
227. R. Demuth and H. Oberhammer, *Z. Naturforsch.*, **28a**, 1862 (1973).
228. H. Schäfer and A. G. Macdiarmid, *Inorg. Chem.*, **15**, 848 (1976).
229. G. Becker and O. Mundt, *Z. Anorg. Allg. Chem.*, **459**, 87 (1979).
230. G. Becker, W. Massa, R. E. Schmidt and G. Uhl, *Z. Anorg. Allg. Chem.*, **520**, 139 (1985).
231. R. Appel, F. Knoch, B. Laubach and R. Sievers, *Chem. Ber.*, **116**, 1873 (1983).
232. K.-F. Tebbe and R. Fröhlich, *Z. Naturforsch.*, **37b**, 534 (1982).
233. K.-F. Tebbe, *Z. Anorg. Allg. Chem.*, **468**, 202 (1980).
234. M. Weidenbruch, M. Herrndorf, A. Schäfer, K. Peters and H. G. von Schnering, *J. Organomet. Chem.*, **295**, 7 (1985).
235. K.-F. Tebbe and T. Heinlein, *Z. Anorg. Allg. Chem.*, **515**, 7 (1984).
236. M. Baudler, T. Pontzen, U. Schings, K.-F. Tebbe and M. Fehér, *Angew. Chem.*, **95**, 803 (1983).
237. U. Klingebiel, N. Vater, W. Clegg, M. Haase and G. M. Sheldrick, *Z. Naturforsch.*, **38b**, 1557 (1983).
238. W. Clegg, M. Haase, U. Klingebiel and G. M. Sheldrick, *Chem. Ber.*, **116**, 146 (1983).
239. R. Fröhlich and K.-F. Tebbe, *Acta Cryst.*, **B38**, 115 (1982).
240. G. Fritz, R. Biastoch, W. Hönle and H. G. von Schnering, *Z. Anorg. Allg. Chem.*, **535**, 86 (1986).
241. W. Hönle, and H. G. von Schnering, *Z. Anorg. Allg. Chem.*, **442**, 91 (1978).
242. W. Hönle and H. G. von Schnering, *Z. Anorg. Allg. Chem.*, **440**, 171 (1978).
243. C. Mujica, D. Weber and H. G. von Schnering, *Z. Naturforsch.*, **41b**, 991 (1986).
244. W. Hönle and H. G. von Schnering, *Z. Anorg. Allg. Chem.*, **442**, 107 (1978).
245. G. Fritz, R. Uhlmann, K. D. Hoppe, W. Hönle and H. G. von Schnering, *Z. Anorg. Allg. Chem.*, **491**, 83 (1982).
246. B. Beagley, A. G. Robiette and G. M. Sheldrick, *J. Chem. Soc. (A)*, 3006 (1968).
247. H. Oberhammer and R. Demuth, *J. Chem. Soc., Dalton Trans.*, 1121 (1976).
248. G. Becker, G. Gutekunst and C. Witthauer, *Z. Anorg. Allg. Chem.*, **486**, 90 (1982).
249. G. Becker and C. Witthauer, *Z. Anorg. Allg. Chem.*, **492**, 28 (1982).
250. W. Hönle and H. G. von Schnering, *Z. Naturforsch.*, **35b**, 789 (1980).
251. G. Becker, A. Münch and C. Witthauer, *Z. Anorg. Allg. Chem.*, **492**, 15 (1982).
252. D. W. H. Rankin, A. G. Robiette, G. M. Sheldrick, B. Beagley and T. G. Hewitt, *J. Inorg. Nucl. Chem.*, **31**, 2351 (1969).
253. G. Becker, H. Freudenblum and C. Witthauer, *Z. Anorg. Allg. Chem.*, **492**, 37 (1982).
254. G. Becker, A. Münch and C. Witthauer, *Z. Anorg. Allg. Chem.*, **492**, 15 (1982).
255. O. Mundt, G. Becker, M. Rössler and C. Witthauer, *Z. Anorg. Allg. Chem.*, **506**, 42 (1983).
256. L. S. Bartell and L. O. Brockway, *J. Chem. Phys.*, **32**, 512 (1960).
257. K. J. Dykema, T. N. Truong and M. S. Gordon, *J. Am. Chem. Soc.*, **107**, 4535 (1985).
258. D. W. H. Rankin and H. E. Robertson, *J. Chem. Soc., Dalton Trans.*, 265 (1983).
259. B. Csákvári, Z. Wagner, P. Gömöry, F. C. Mijlhoff, B. Rozsondai and I. Hargittai, *J. Organomet. Chem.*, **107**, 287 (1976).

260. W. Wojnowski, W. Bocheńska, K. Peters, E.-M. Peters and H. G. von Schnering, *Z. Anorg. Allg. Chem.*, **533**, 165 (1986).
261. I. L. Karle, J. M. Karle and C. J. Nielsen, *Acta Cryst.*, **C42**, 64 (1986).
262. M. G. Voronkov, D. Yu. Nesterov, A. I. Gusev, N. V. Alexeev and M. B. Lotarev, *Dokl. Akad. Nauk SSSR*, **257**, 1377 (1981).
263. A. I. Gusev, M. Yu. Antipin, D. S. Yufit, Yu. T. Struchkov, V. D. Sheludyakov, V. I. Zhuń and S. D. Vlasenko, *Zh. Strukt. Khim.*, **24**, 178 (1983).
264. C. Glidewell and D. C. Liles, *J. Organomet. Chem.*, **212**, 291 (1981).
265. C. Glidewell, D. W. H. Rankin, A. G. Robiette, G. M. Sheldrick, B. Beagley and J. M. Freeman, *J. Mol. Struct.*, **5**, 417 (1970).
266. C. Glidewell, D. W. H. Rankin, A. G. Robiette, G. M. Sheldrick, B. Beagley and J. M. Freeman, *Trans. Faraday Soc.*, **65**, 2621 (1969).
267. O. J. Scherer, K. Göbel and J. Kaub, *Angew. Chem.*, **99**, 59 (1987).
268. V. A. Lebedev, Yu. N. Drozdov, E. A. Kuz'min, A. V. Ganyushkin, V. A. Yablokov and N. V. Belov, *Sov. Phys. Crystallogr.*, **26**, 106 (1981).
269. D. Käss, H. Oberhammer, D. Brandes and A. Blaschette, *J. Mol. Struct.*, **40**, 65 (1977).
270. V. E. Shklover, T. V. Timofeeva, Yu. T. Struchkov, A. V. Ganyushkin and V. A. Yabloklov, *Zh. Strukt. Khim.*, **22**, 68 (1981).
271. B. Pötter, D. Lentz, H. Pritzkow and K. Seppelt, *Angew. Chem.*, **93**, 1095 (1981).
272. M. Bonanico and G. Dessy, *J. Chem. Soc. (A)*, 1786 (1967).
273. P. G. Harrison, T. J. King, T. A. Richards and R. C. Phillips, *J. Organomet. Chem.*, **116**, 307 (1976).
274. B. Stensland and P. Kierkegaard, *Acta Chem. Scand.*, **24**, 211 (1970).
275. G. M. Sheldrick and W. S. Sheldrick, *J. Chem. Soc. (A)*, 2160 (1969).
276. I. L. Dubchak, V. E. Zhklover, M. Yu. Antipin, Yu. T. Struchkov, V. M. Kopylov, A. M. Muzafarov, P. L. Prikhod'ko and A. A. Zhdanov, *Zh. Strukt. Khim.*, **23**, 63 (1981).
277. I. L. Dubchak, V. E. Shklover, M. M. Levitskii, A. A. Zhdanov and Yu. T. Struchkov, *Zh. Strukt. Khim.*, **21**, 103 (1980).
278. M. G. Voronkov, V. P. Mileshkevich and Yu. A. Yuzhelevskii, *The Siloxane Bond*, Consultants Bureau, New York, London, 1978.
279. A. I. Gusev, Yu. D. Nesterov, N. V. Alexeev, N. E. Rodzevich, V. V. Zverev, N. V. Ivanova and M. V. Sobolevskii, *Zh. Strukt. Khim.*, **17**, 944 (1976).
280. W. Wojnowski, K. Peters, E.-M. Peters, T. Meyer and H. G. von Schnering, *Z. Anorg. Allg. Chem.*, **537**, 31 (1986).
281. G. V. Gibbs, *Am. Mineral.*, **67**, 421 (1982).
282. B. Morosin and L. A. Harrah, *Acta Cryst.*, **B37**, 579 (1981).
283. P. B. Hitchcock, N. H. Buttrus and A. C. Sullivan, *J. Organomet. Chem.*, **303**, 321 (1986).
284. N. H. Buttrus, C. Eaborn, S. H. Gupta, P. B. Hitchcock, J. D. Smith and A. C. Sullivan, *J. Chem. Soc., Chem. Commun.*, 1043 (1986).
285. O. Graalmann, U. Klingebiel, W. Clegg, M. Haase and G. M. Sheldrick, *Angew. Chem.*, **96**, 904 (1984).
286. E. Weiss, K. Hoffmann and H.-F. Grützmacher, *Chem. Ber.* **103**, 1190 (1970).
287. G. Dittmar and E. Hellner, *Angew. Chem.*, **81**, 707 (1969).
288. V. E. Shklover and Yu. T. Struchkov, *Usp. Khim.*, **49**, 518 (1980).
289. M. J. Michalczyk, M. J. Fink, K. J. Haller, R. West and J. Michl, *Organometallics*, **5**, 531 (1986).
290. T. Kudo and S. Nagase, *J. Am. Chem. Soc.*, **107**, 2589 (1985).
291. E. D. Temmis, P. N. V. Pavan Kumar and N. R. Sreenivas Kumar, *J. Chem. Soc., Dalton Trans.*, 271 (1987).
292. A. Almenningen, K. Hedberg and R. Seip, *Acta Chem. Scand.*, **17**, 2264 (1963).
293. W. Wojnowski, B. Dreczewski, K. Peters, E.-M. Peters and H. G. von Schnering, *Z. Anorg. Allg. Chem.*, **540/541**, 271 (1986).
294. W. E. Schklover, Yu. T. Strutschkow, L. E. Guselnikow, W. W. Wolkowa and W. G. Awakyan, *Z. Anorg. Allg. Chem.*, **501**, 153 (1983).
295. W. Wojnowski, K. Peters, D. Weber and H. G. von Schnering, *Z. Anorg. Allg. Chem.*, **519**, 134 (1984).
296. W. Wojnowski, K. Peters, M. C. Böhm and H. G. von Schnering, *Z. Anorg. Allg. Chem.*, **523**, 169 (1985).

297. L. Pazdernik, F. Brisse and R. Rivest, *Acta Cryst.*, **B33**, 1780 (1977).
298. W. Wojnowski, B. Dreczewski, A. Herman, K. Peters, E.-M. Peters and H. G. von Schnering, *Angew. Chem.*, **97**, 978 (1985).
299. A. Haas, R. Hitze, C. Krüger and K. Angermund, *Z. Naturforsch.*, **39b**, 890 (1984).
300. G. Becker, W. Massa, R. E. Schmidt and G. Uhl, *Z. Anorg. Allg. Chem.*, **517**, 75 (1984).
301. G. Becker, O. Mundt and G. Uhl, *Z. Anorg. Allg. Chem.*, **517**, 89 (1984).
302. W. Wojnowski, M. Wojnowski, K. Peters, E.-M. Peters and H. G. von Schnering, *Z. Anorg. Allg. Chem.*, **530**, 79 (1985).
303. W. Wojnowski, K. Peters, E.-M. Peters and H. G. von Schnering, *Z. Anorg. Allg. Chem.*, **531**, 147 (1985).
304. W. Wojnowski, M. Wojnowski, H. G. von Schnering and M. Noltemeyer, *Z. Anorg. Allg. Chem.*, **531**, 531 (1985).
305. W. Wojnowski, M. Wojnowski, K. Peters, E.-M. Peters and H. G. von Schnering, *Z. Anorg. Allg. Chem.*, **535**, 56 (1986).
306. L. Pierce and M. Hayashi, *J. Chem. Phys.*, **35**, 479 (1961).
307. A. Almenningen, L. Fernholdt and H. M. Seip, *Acta Chem. Scand.*, **22**, 51 (1968).
308. R. Kewley, P. M. McKinney and A. G. Robiette, *J. Mol. Spectr.*, **34**, 309 (1970).
309. V. W. Laurie, *J. Chem. Phys.*, **26**, 1359 (1957).
310. A. R. Hoy, M. Bertram, I. M. Mills, *J. Mol. Spectr.*, **46**, 429 (1973).
311. B. Beagley, D. P. Brown and J. M. Freeman, *J. Mol. Struct.*, **18**, 337 (1973).
312. R. W. Davies and M. C. L. Gerry, *J. Mol. Spectr.*, **60**, 117 (1976).
313. H. Takeo and C. Matsumara. *Bull. Chem. Soc. Jpn.*, **50**, 1633 (1977).
314. Y. Morino and Y. Murata, *Bull. Chem. Soc. Jpn.*, **38**, 104 (1965).
315. W. Zeil, R. Gegenheimer, S. Pferrer and M. Dakkouri, *Z. Naturforsch.*, **27a**, 1150 (1972).
316. K. Endo, H. Takeo and C. Matsumara, *Bull. Chem. Soc. Jpn.*, **50**, 626 (1977).
317. M. Mitzlaff, R. Holm and H. Hartmann, *Z. Naturforsch.*, **23a**, 65 (1968).
318. M. D. Harmony and M. R. Strand, *J. Mol. Spectr.*, **81**, 308 (1980).
319. J. R. Durig, J. S. Church and Y. S. Li, *Inorg. Chem.*, **21**, 35 (1982).
320. W. H. Ilsley, T. F. Schaaf, M. D. Glick and J. P. Oliver, *J. Am. Chem. Soc.*, **102**, 3769 (1980).
321. B. Teclé, W. H. Ilsley and J. P. Oliver, *Organometallics*, **1**, 875 (1982).
322. G. Becker, H.-M. Hartmann, A. Münch and H. Riffel, *Z. Anorg. Allg. Chem.*, **530**, 29 (1985).
323. W. H. Ilsley, M. J. Albright, T. J. Anderson, M. D. Glick and J. P. Oliver, *Inorg. Chem.*, **19**, 3577 (1980).
324. D. W. Goebel, Jr., J. L. Hencher and J. P. Oliver, *Organometallics*, **2**, 746 (1983).
325. L. Rösch, J. Pickardt, S. Imme and U. Börner, *Z. Naturforsch.*, **41b**, 1523 (1986).
326. A. R. Claggett, W. H. Ilsley, T. J. Anderson, M. D. Glick and J. P. Oliver, *J. Am. Chem. Soc.*, **99**, 1797 (1977).
327. P. Bleckmann, M. Solimann, K. Reuter and W. P. Neumann, *J. Organomet. Chem.*, **108**, C18 (1976).
328. W. H. Ilsley, E. A. Sadurski, T. F. Schaaf, M. J. Albright, T. J. Anderson, M. D. Glick and J. P. Oliver, *J. Organomet. Chem.*, **190**, 257 (1980).
329. E. A. Sadurski, W. H. Ilsley, R. D. Thomas, M. D. Glick and J. P. Oliver, *J. Am. Chem. Soc.*, **100**, 7761 (1978).
330. J. C. Calabrese and L. F. Dahl, *J. Am. Chem. Soc.*, **93**, 6042 (1971).
331. L. Rösch, G. Altnau, C. Krüger, and Y.-H. Tsay, *Z. Naturforsch.*, **38b**, 34 (1983).
332. A. M. Arif, A. H. Cowley, T. M. Elkins and R. A. Jones, *J. Chem. Soc., Chem. Commun.*, 1776 (1986).
333. G. Hencken and E. Weiss, *Chem. Ber.*, **106**, 1747 (1973).
334. L. Rösch, G. Altnau, W. Erb, J. Pickardt and N. Bruncks, *J. Organomet. Chem.*, **197**, 51 (1980).
335. K. W. Muir, *J. Chem. Soc. (A)*, 2663 (1971).
336. T. D. Tilley, *Organometallics*, **4**, 1452 (1985).
337. M. D. Curtis, L. G. Bell and W. M. Butler, *Organometallics*, **4**, 701 (1985).
338. J. Arnold, D. N. Shina, T. D. Tilley and A. M. Arif, *Organometallics*, **5**, 2037 (1986).
339. T. H. Hseu, Y. Chi and C.-S. Liu, *Inorg. Chem.*, **20**, 199 (1981).
340. M. J. Bennett and K. A. Simpson, *J. Am. Chem. Soc.*, **93**, 7156 (1971).
341. D. W. H. Rankin, A. Robertson and R. Seip, *J. Organomet. Chem.*, **88**, 191 (1975).
342. D. W. H. Rankin and A. Robertson, *J. Organomet. Chem.*, **85**, 225 (1975).

343. R. S. Hamilton and E. R. Corey, *Abstr. 156th Meet. Amer. Chem. Soc.* p. INOR 025 (1968).
344. B. K. Nicholson, J. Simpson and W. T. Robinson, *J. Organomet. Chem.*, **47**, 403 (1973).
345. M. C. Couldwell and J. Simpson, *J. Chem. Soc., Dalton Trans.*, 714 (1976).
346. G. L. Simon and L. F. Dahl, *J. Am. Chem. Soc.*, **95**, 783 (1973).
347. W. A. Herrmann, E. Voss, E. Guggolz and M. L. Ziegler, *J. Organomet. Chem.*, **284**, 45 (1985).
348. D. W. H. Rankin and A. Robertson, *J. Organomet. Chem.*, **105**, 331 (1976).
349. M. C. Couldwell, J. Simpson and W. T. Robinson, *J. Organomet. Chem.*, **107**, 323 (1976).
350. R. A. Smith and M. J. Bennett, *Acta Cryst.*, **B33**, 1113 (1977).
351. M. Cowie and M. J. Bennett, *Inorg. Chem.*, **16**, 2321 (1977).
352. M. Cowie and M. J. Bennett, *Inorg. Chem.*, **16**, 2325 (1977).
353. M. Elder, *Inorg. Chem.*, **9**, 762 (1970).
354. U. Schubert, G. Kraft and E. Walther, *Z. Anorg. Allg. Chem.*, **519**, 86 (1984).
355. P. R. Jansen, A. Oskam and K. Olie, *Cryst. Struct. Commun.*, **4**, 667 (1975).
356. R. A. Smith and M. J. Bennett, *Acta Cryst.*, **B33**, 1118 (1977).
357. L. Manojlovič-Muir, K. W. Muir and J. A. Ibers, *Inorg. Chem.*, **9**, 447 (1970).
358. L. Vancea, M. J. Bennett, C. E. Jones, R. A. Smith and W. A. G. Graham, *Inorg. Chem.*, **16**, 897 (1977).
359. D. M. Roddick, T. D. Tilley, A. L. Rheingold and S. J. Geib, *J. Am. Chem. Soc.*, **109**, 945 (1987).
360. F. W. B. Einstein and T. Jones, *Inorg. Chem.*, **21**, 987 (1982).
361. M. M. Crozat and S. F. Watkins, *J. Chem. Soc., Dalton Trans.*, 2512 (1973).
362. M. J. Auburn, R. D. Holmes-Smith, S. R. Stobart, M. J. Zaworotko, T. S. Cameron and A. Kumari, *J. Chem. Soc., Chem. Commun.*, 1523 (1983).
363. H.-P. Klein, U. Thewalt, G. Herrmann, G. Süss-Fink and C. Moinet, *J. Organomet. Chem.*, **286**, 225 (1985).
364. A. C. Willis, G. N. van Buuren, R. K. Pomeroy and F. W. B. Einstein, *Inorg. Chem.*, **22**, 1162 (1983).
365. G. N. van Buuren, A. C. Willis, F. W. B. Einstein, L. K. Peterson, R. K. Pomeroy and D. Sutton, *Inorg. Chem.*, **20**, 4361 (1981).
366. K. Emerson, P. R. Ireland and W. T. Robinson, *Inorg. Chem.*, **9**, 436 (1970).
367. W. T. Robinson and J. A. Ibers, *Inorg. Chem.*, **6**, 1208 (1967).
368. G. Schmid, V. Bätzel and G. Etzrodt, *J. Organomet. Chem.*, **112**, 345 (1976).
369. A. G. Robiette, G. M. Sheldrick, R. N. F. Simpson, B. J. Aylett and J. A. Campbell, *J. Organomet. Chem.*, **14**, 279 (1968).
370. K. W. Muir and J. A. Ibers, *Inorg. Chem.*, **9**, 440 (1970).
371. M.-J. Fernandez, P. M. Bailey, P. O. Bentz, J. S. Ricci, T. F. Koetzle and P. M. Maitlis, *J. Am. Chem. Soc.*, **106**, 5458 (1984).
372. J. S. Ricci, T. F. Koetzle, M.-J. Fernandez, P. M. Maitlis and J. C. Green, *J. Organomet. Chem.*, **299**, 383 (1986).
373. M. D. Curtis, J. Greene and W. N. Butler, *J. Organomet. Chem.*, **164**, 371 (1979).
374. M. Ciriano, J. A. K. Howard, J. L. Spencer, F. G. A. Stone and H. Wadepohl, *J. Chem. Soc., Dalton Trans.*, 1749 (1979).
375. C. Eaborn, K. J. Odell, A. Pidcock and G. R. Scollary, *J. Chem Soc., Chem. Commun.*, 317 (1976).
376. M. Ciriano, M. Green, D. Gregson, J. A. K. Howard, J. L. Spencer, F. G. A. Stone and P. Woodward, *J. Chem. Soc., Dalton Trans.*, 1294 (1979).
377. C. Eaborn, P. B. Hitchcock, D. J. Tune and D. R. M. Walton, *J. Organomet. Chem.*, **54**, C1 (1973).
378. M. Green, J. A. K. Howard, J. Proud, J. L. Spencer, F. G. A. Stone and C. A. Tsipis, *J. Chem. Soc., Chem. Commun.*, 671 (1976).
379. E. A. V. Ebsworth, V. M. Marganian, F. J. S. Reed and R. O. Gould, *J. Chem. Soc., Dalton Trans.*, 1167 (1978).
380. C. S. Cundy, B. M. Kingston and M. F. Lappert, *Adv. Organomet. Chem.*, **11**, 253 (1973).
381. B. J. Aylett, *Adv. Inorg. Radiochem.*, **25**, 1 (1982).
382. H. Schumann, S. Nickel, E. Hahn and M. J. Heeg, *Organometallics*, **4**, 800 (1985).
383. R. Tacke, H. Linoh, M. T. Attar-Bashi, W. S. Sheldrick, L. Ernst, R. Niedner and J. Frohnecke, *Z. Naturforsch.*, **37b**, 1461 (1982).
384. R. Tacke, R. Niedner, J. Frohnecke, L. Ernst and W. S. Sheldrick, *Liebigs Ann. Chem.*, 1859 (1980).

385. R. Tacke and H. Zilch, *Endeavour*, New Series, **10**, 191 (1986).
386. R. Tacke, H. Linoh, D. Schomburg, L. Ernst, U. Moser, E. Mutschler and G. Lambrecht, *Liebigs Ann. Chem.*, 240 (1986).
387. W. S. Sheldrick, H. Linoh, R. Tacke, G. Lambrecht, U. Moser and E. Mutschler, *J. Chem. Soc., Dalton Trans.*, 1743 (1985).
388. R. Tacke, M. Strecker, W. S. Sheldrick, L. Ernst, E. Heeg, B. Berndt, C.-M. Knapstein and R. Niedner, *Chem. Ber.*, **113**, 1962 (1980).
389. R. Tacke, M. Strecker, W. S. Sheldrick, E. Heeg, B. Berndt and K. M. Knapstein, *Z. Naturforsch.*, **34b**, 1279 (1979).
390. R. Tacke, A. Bentlage, W. S. Sheldrick, L. Ernst, R. Towart and K. Stoepel, *Z. Naturforsch.*, **37b**, 443 (1982).
391. A. M. Triggle, E. Shefter and D. J. Triggle, *J. Med. Chem.*, **23**, 1442 (1980).
392. J. Y. Corey and R. West, *J. Am. Chem. Soc.*, **85**, 4035 (1963).
393. J. I. Musher, *Angew. Chem.*, **81**, 68 (1969).
394. D. Schomburg, *Habilitationsschrift*, Braunschweig, 1985.
395. F. Keil and R. Ahlrichs, *Chem. Phys.*, **8**, 384 (1975).
396. R. Hoffmann, J. M. Howell and E. L. Muetterties, *J. Am. Chem. Soc.*, **94**, 3047 (1972).
397. D. Schomburg and R. Krebs, *Inorg. Chem.*, **23**, 1378 (1984).
398. D. Schomburg, *J. Organomet. Chem.*, **221**, 137 (1981).
399. J. J. Harland, J. S. Payne, R. O. Day and R. R. Holmes, *Inorg. Chem.*, **26**, 760 (1987).
400. W. G. Farnham and R. L. Harlow, *J. Am. Chem. Soc.*, **103**, 4608 (1981).
401. R. R. Holmes, R. O. Day, V. Chandrasekhar and J. M. Holmes, *Inorg. Chem.*, **24**, 2009 (1985).
402. F. P. Boer, J. J. Flynn and J. W. Turley, *J. Am. Chem. Soc.*, **90**, 6973 (1968).
403. R. R. Holmes, R. O. Day, J. J. Harland and J. M. Holmes, *Organometallics*, **3**, 347 (1984).
404. J. J. Harland, R. O. Day, J. F. Vollano, A. C. Sau and R. R. Holmes, *J. Am. Chem. Soc.*, **103**, 5269 (1981).
405. D. Schomburg, *Z. Naturforsch.*, **37b**, 195 (1982).
406. D. Schomburg, *Z. Naturforsch.*, **38b**, 938 (1983).
407. R. R. Holmes, R. O. Day, J. J. Harland, A. C. Sau and J. M. Holmes, *Organometallics*, **3**, 341 (1984).
408. R. R. Holmes and J. A. Deiters, *J. Am. Chem. Soc.*, **99**, 3318 (1977).
409. R. W. Hillyard, Jr., C. M. Ryan and C. H. Yoder, *J. Chem. Soc., Chem. Commun.*, 209 (1978).
410. G. Klebe, M. Nix and K. Hensen, *Chem. Ber.*, **117**, 797 (1984).
411. G. Klebe, J. W. Bats and H. Fuess, *J. Am. Chem. Soc.*, **106**, 5202 (1984).
412. A. A. Klemme, Ya. Ya. Bleidelis, V. A. Pestunovich, V. P. Baryshok and M. G. Voronkov, *Dokl. Akad. Nauk SSSR*, **243**, 688 (1978).
413. L. Párkányi, L. Bihátsi and P. Hencsei, *Cryst. Struct. Commun.*, **7**, 435 (1978).
414. F. P. Boer and J. W. Turley, *J. Am. Chem. Soc.*, **91**, 4134 (1969).
415. U. Schubert, K. Ackermann, G. Kraft and B. Wörle, *Z. Naturforsch.*, **38b**, 1488 (1983).
416. K. A. Simpson, *Ph.D. Thesis*, Univ. of Alberta, Edmonton, 1973.
417. U. Schubert, G. Scholz, J. Müller, K. Ackermann, B. Wörle and R. F. D. Stonsfield, *J. Organomet. Chem.*, **306**, 303 (1986).
418. K. Ackermann, *Dissertation*, Techn. Univ. München, 1982.
419. W. L. Hutcheon, *Ph.D. Thesis*, Univ. of Alberta, Edmonton, 1971.
420. F. Carré, E. Colomer, R. J. P. Corriu and A. Vioux, *Organometallics*, **3**, 1272 (1984).
421. G. Klebe and D. T. Qui, *Acta Cryst.*, **C40**, 476 (1984).
422. W. B. Farnham and J. F. Whitney, *J. Am. Chem. Soc.*, **106**, 3992 (1984).
423. G. Sawitzki and H. G. von Schnering, *Chem. Ber.*, **109**, 3728 (1976).
424. N. S. Hosmane, P. de Meester, U. Siriwardane, M. S. Islam and S. S. C. Chu, *J. Chem. Soc., Chem. Commun.*, 1421 (1986).
425. W. S. Rees, Jr., D. M. Schubert, C. B. Knobler and M. F. Hawthorne, *J. Am. Chem. Soc.*, **108**, 5369 (1986).

The Chemistry of Organic Silicon Compounds
Edited by S. Patai and Z. Rappoport

CHAPTER **4**

Dynamic stereochemistry at silicon

ROBERT J. P. CORRIU, CHRISTIAN GUERIN and JOEL J. E. MOREAU
Institut de Chimie Fine, Hétérochimie et Amino-Acides, Unité Associée CNRS U.A. n° 1097, Université des Sciences et Techniques du Languedoc, 34060 Montpellier Cedex, France

I. INTRODUCTION

The elucidation of reaction mechanisms in organic chemistry has involved both kinetic and stereochemical studies. Similarly, stereochemistry[1,2] has played a determining role in understanding the reaction mechanism in organosilicon compounds.

Unlike carbon compounds, optically active organosilanes are not natural species. Therefore, it is not surprising that studies of stereochemistry at silicon compounds were first reported by Kipping and coworkers[3] and later by Eaborn and Pitt[4] who achieved the resolution of chiral tetrasubstituted silanes. An important step was then made when Sommer and Frye[5] reported the first resolution of a functional trisubstituted silane allowing one to perform stereochemical studies at a silicon centre. Important contributions of other groups in this field[1,2] have been made since the original publication of Sommer.

Thus, this chapter aims to review stereochemistry at silicon and its mechanistic implications. The following two aspects are discussed:

1. The general features of chirality at silicon and the routes to optically active compounds, including asymmetric synthesis.
2. The mechanism and stereochemistry of nucleophilic substitution at silicon, including displacement of silyl ligands in silicon–transition metal complexes.

A few results obtained with related germanium and tin compounds are mentioned, although some aspects of their stereochemistry have been recently reviewed[6].

II. SYNTHESIS AND CHEMICAL BEHAVIOUR OF CHIRAL SILICON SPECIES

Numerous optically active organosilicon compounds have now been obtained and various resolution procedures employed. These include resolution through diastereomer formation and separation, as well as kinetic resolution and asymmetric synthesis. Moreover, the stereospecificity of substitution reaction at silicon makes possible the synthesis of various optically active compounds starting from resolved organosilicon compounds. Therefore, a large variety of chiral compounds are accessible, including polyfunctional organic and inorganic species.

A. Organic Silicon Compounds

Chiral tetraalkylsilanes R_4Si and trialkylsilanes R_3SiH exhibit high configurational stability. Functional organosilicon compounds R_3SiX are also optically stable. Racemization occurs only in the presence of nucleophilic agents. The racemization of halogenosilanes involves two molecules of nucleophile in the rate-determining step[7]. The process is controlled by the entropy of activation and takes place without substitution of X in R_3SiX. Silicon compounds, however, are stable enough in nucleophilic solvents to allow stereospecific substitution reactions.

1. Optically active silanes by resolution methods

a. From trisubstituted (−)menthoxy derivatives. The resolution through diastereomer formation and separation, widely applied in organic chemistry[8], was the first one considered for the resolution of silicon compounds. The procedure was initially used in the early work of Kipping[3] and Eaborn[4] but the work of Sommer and Frye first introduced the use of (−)menthoxysilane derivatives in the resolution of organosilicon compounds[5]. Since then, (−)menthol (MenOH) proved to be a convenient resolving agent for organosilicon compounds. Acyclic and cyclic compounds have been resolved using this method.

Sommer's original resolution involving separation of (±)-1-naphthylphenylmethyl-(−)-menthoxysilane[9] is represented in Scheme 1. The diastereomeric menthoxysilanes were obtained from the racemic methoxysilane by an exchange reaction under basic

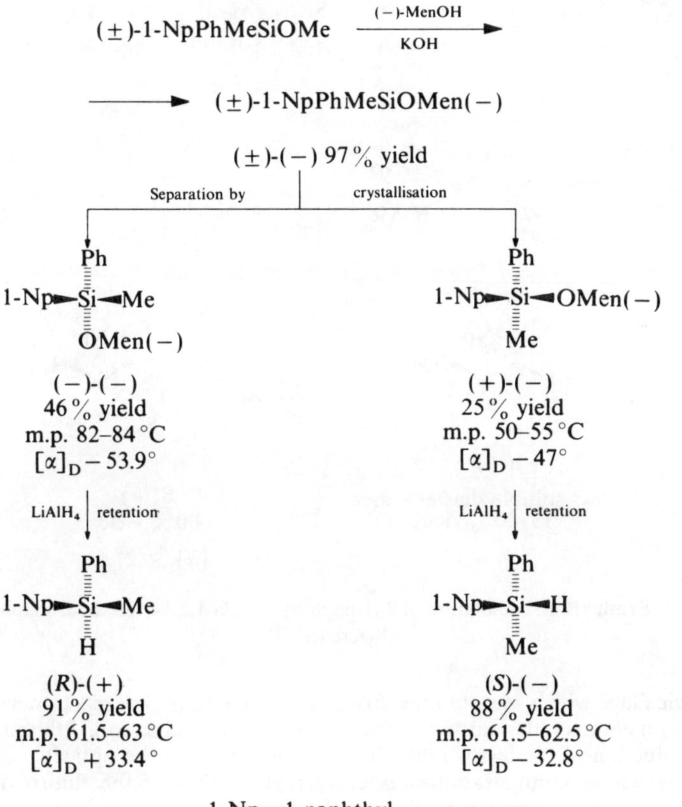

1-Np = 1-naphthyl

SCHEME 1. Resolution of 1-naphthylphenylmethylsilane (after Reference 9)

conditions. Fractional crystallisation from pentane gave the less soluble diastereomer ($[\alpha]_D - 53.9°$) in a 46% yield. The (+)-1-naphthylphenylmethylsilane ($[\alpha]_D + 33.4°$) was obtained upon lithium aluminium hydride reduction of this diastereomer. Its absolute configuration was shown to be *R* by X-ray diffraction studies[10]. Similar treatment of the more soluble diastereomer afforded the *S*-(−)-1-naphthylphenylmethylsilane ($[\alpha]_D - 32.8°$).

The method developed by Sommer proved to be general and similar reaction sequences were used to resolve, for example, 1-naphthylphenylvinyl- and ethylsilanes[11–13]. It was also extended to the resolution of compounds having Si–Si[14] or Si–Ge[15] linkages. The resolved compounds are presented in Figure 1. The relative configurations of these compounds were established by employing physical methods[16] and correlated with the known absolute configuration of Sommer's compound. In some cases, the relative configurations were determined on the basis of chemical correlations[15].

In addition to acyclic silicon compounds, this resolution method was successfully applied in the case of the cyclosilanes represented in Figure 1[17–19]. As an example, the synthesis and resolution of 2-(1-naphthyl)-2-sila-1,2,3,4-tetrahydronaphthalene is depicted in Scheme 2[17].

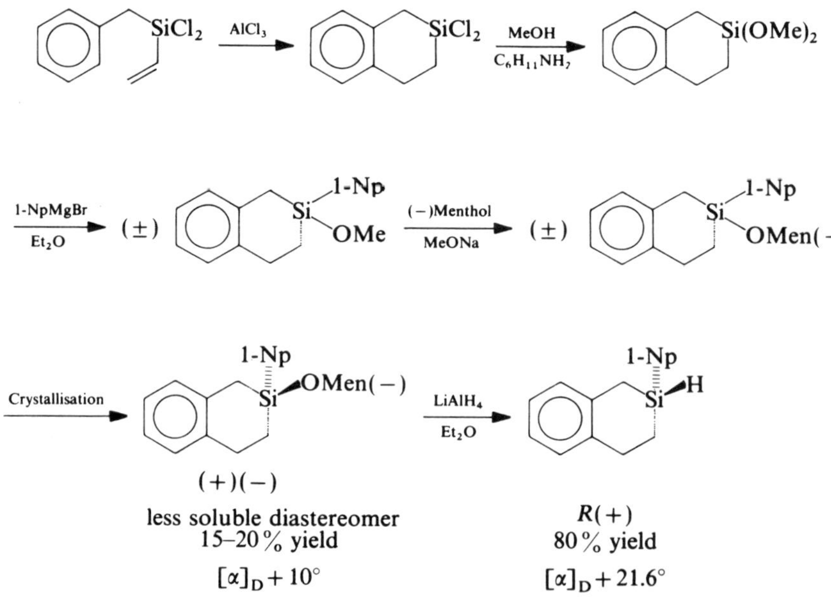

SCHEME 2. Synthesis and resolution of 2-(1-naphthyl)-2-sila-1,2,3,4-tetrahydronaphthalene (after Reference 17)

The cyclosilane which was obtained from a Friedel–Crafts-type reaction was converted to the (−)-menthoxy derivatives. After fractional crystallization, lithium aluminium hydride reduction of the less soluble diastereomer afforded the optically active *R*-(+)-hydrosilane whose configuration was correlated to that of the fluoro derivative of *S* configuration as determined by X-ray diffraction[20].

From the resolved organosilanes, other optically active silicon compounds have been prepared by stereospecific substitution reactions. As a result of stereospecific cleavage of

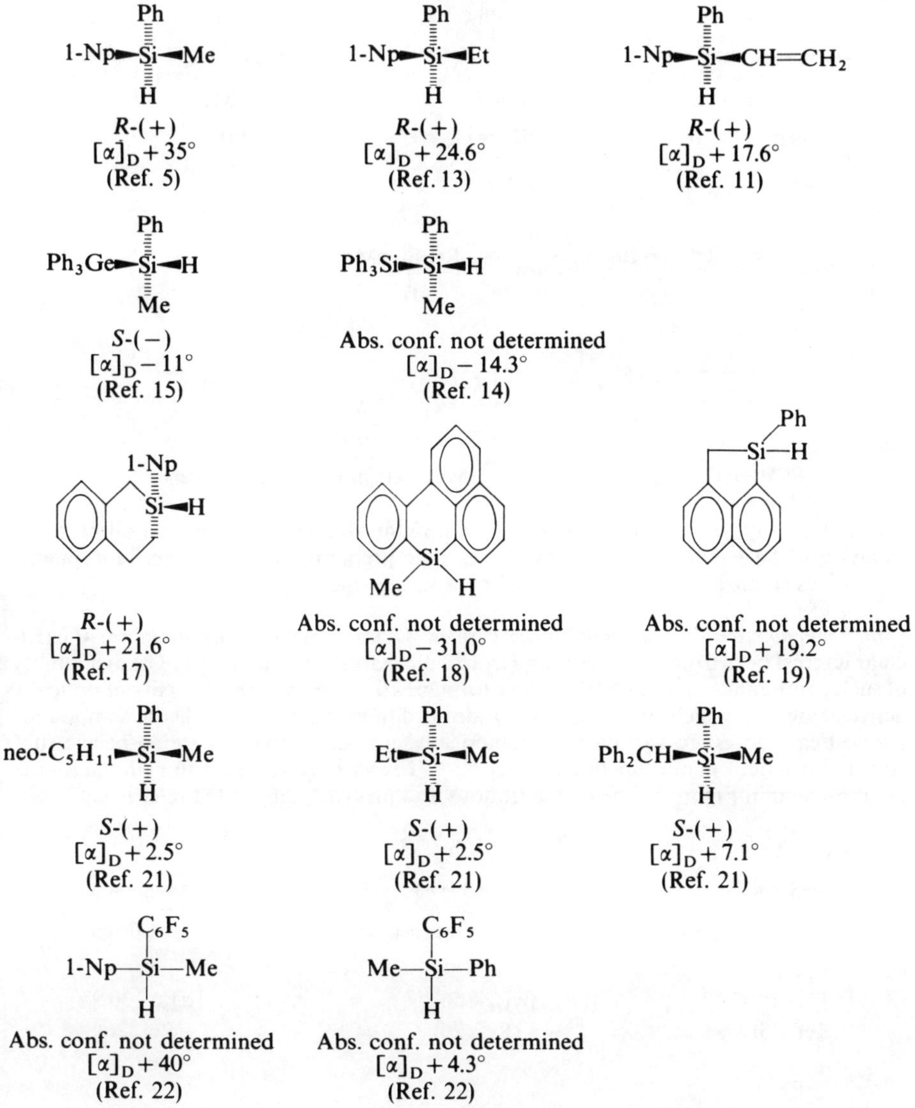

1-Np = 1-naphthyl

FIGURE 1. Optically active organosilicon compounds resolved via the separation of (−)menthoxy diastereomers

the silicon–naphthyl bond, optically active acyclic organosilanes were obtained as shown in Scheme 3[21].

Successive stereoselective chlorination and condensation with alkyllithium reagents led to tetrasubstituted silanes. The achievement of a stereospecific cleavage of the

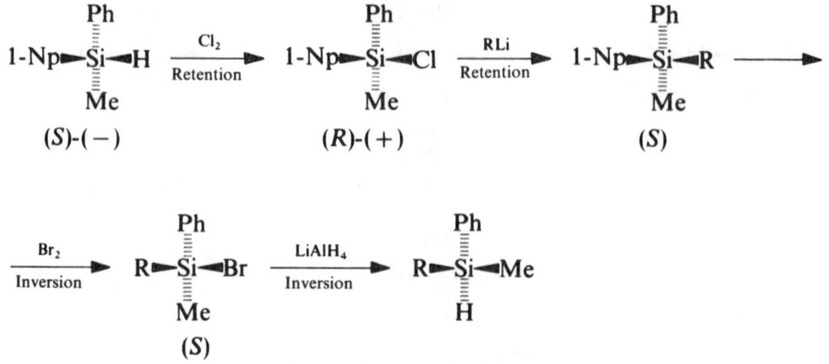

R = neo-C_5H_{11} (S)-(+)$[\alpha]_D + 2.5°$
R = Et (S)-(+)$[\alpha]_D + 2.5°$
R = Ph_2CH (S)-(+)$[\alpha]_D + 7.1°$

SCHEME 3. Optically active alkylphenylmethylsilanes (after Reference 21)

silicon–naphthyl bond with bromine, after lithium aluminium reduction, allowed the isolation of *S*-(+)-alkylphenylmethylsilanes. The preparation of some perfluorophenyl derivatives (Figure 1) was also reported by a similar method.

b. From functional (−)menthoxy derivatives. Difunctional organosilanes are of particular interest both from a synthetic and from a mechanistic point of view. The availability of such compounds in an optically active form greatly enhanced the diversity of optically active materials which are accessible. Indeed, difunctional organosilicon compounds allowed easy access to various enantiomerically pure derivatives by stereospecific substitution reactions of one functional group[23, 24]. The various resolved difunctional silanes are presented in Figure 2. The configurations and properties indicated refer in each case

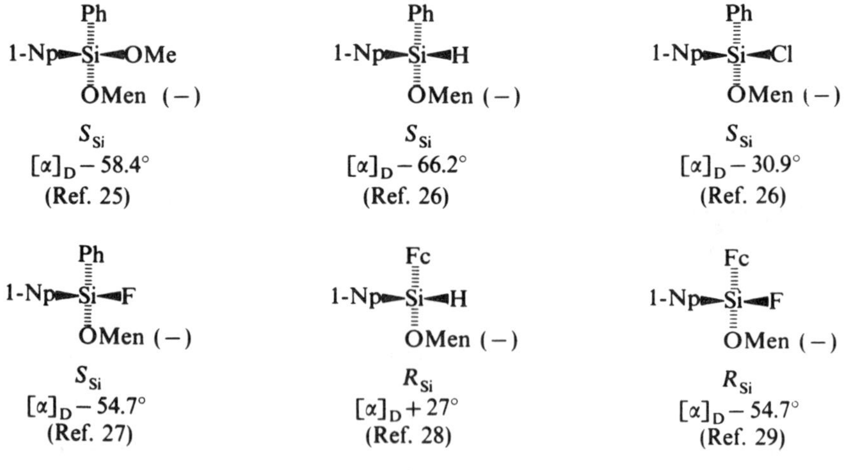

1-Np = 1-naphthyl; Fc = Ferrocenyl

FIGURE 2. Difunctional (−)-menthoxysilanes

to the less soluble diastereomer isolated by fractional crystallization of the diastereomeric (−)menthoxy silanes mixture. Chemical correlations were used to establish the relative configurations of these compounds and to correlate them to the X-ray-determined absolute configuration of (+)-(−)-1-NpPhSi(OMe)(OMen) ($[\alpha]_D - 58.4°$)[30]. Relative configurations were also correlated by use of physical correlation methods. Interestingly, an extension of the Fredga method[31] was also proposed for the determination of the relative configurations around the silicon atom in these diastereomeric compounds[32].

From a synthetic point of view, obtaining difunctional organosilanes allowed the preparation of new functional cyclosilanes. Optically active oxasilacyclopentane and thiasilacyclopentane were, for example, obtained from (−)-(−)1-NpPh(MenO)SiH by use of the stereoselective reaction sequence depicted in Scheme 4[33, 34].

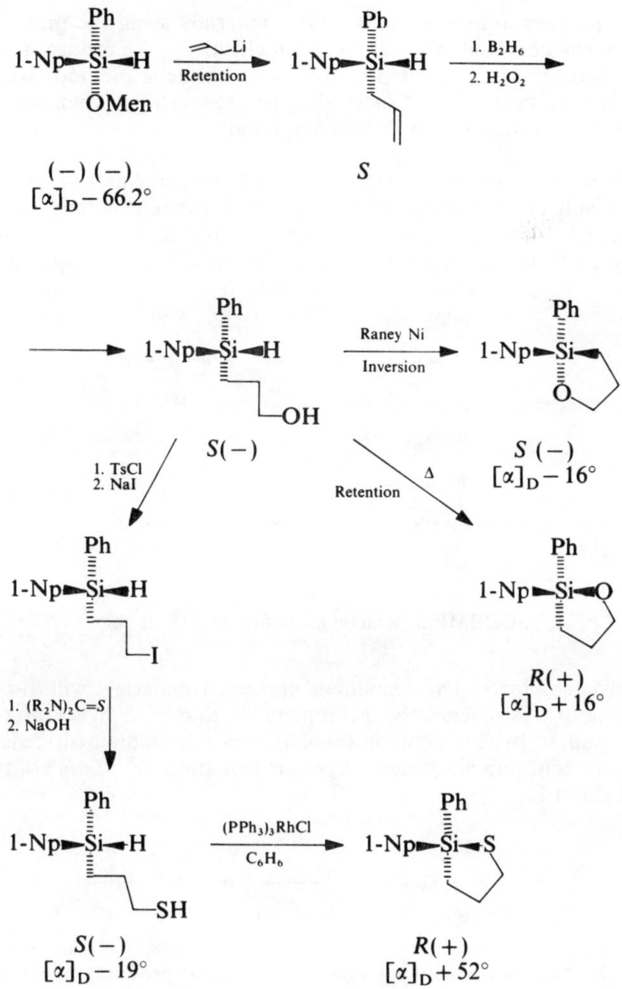

SCHEME 4. Optically active oxa- and thia-silacyclopentanes (after References 33 and 34)

The reaction of allyllithium yielded the allylhydrosilane and retained the silicon atom configuration. Hydroboration allowed the conversion to a γ-hydroxysilane from which cyclization to the (*S*)-oxasilacycloalkane was obtained by Raney Ni-catalysed intramolecular dehydroalcoholysis with inversion of configuration at the silicon centre. On the other hand, the thermal dehydrogenative cyclization with retention at silicon led to the (*R*) isomer[33]. The method was also applied to the synthesis of an approximately 50% optically pure homologous oxasilacyclohexane[35]. Similarly, the (*R*)-thiasilacyclopentane was isolated[34]. The previous γ-hydroxysilane was converted to a thiol derivative and cyclized with retention of configuration in the presence of an homogeneous rhodium catalyst.

2. *Optically active silanes by use of asymmetric synthesis methods*

Whereas the preparation of optically active materials using the principles of asymmetric synthesis has been widely used in carbon chemistry, the examples of asymmetric induction at heteroatoms are limited[36]. However, synthetic methods based on asymmetric synthesis have been applied successfuly to organosilicon chemistry and provide interesting routes to chiral organosilicon compounds.

a. Kinetic resolution. Kinetic resolution is of general interest in organosilicon chemistry, since substitution reactions at a silicon atom are generally stereospecific. As shown in Scheme 5, a partial stereospecific substitution reaction of a racemic mixture of $R^1R^2R^3SiX$ by a chiral reagent Y* will lead to the formation of unequal amounts of the

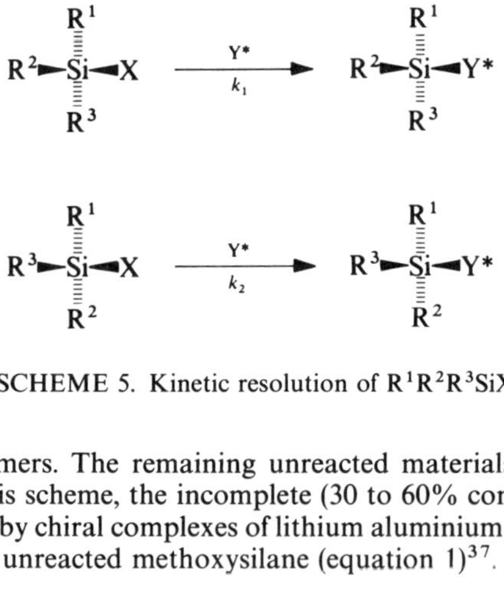

SCHEME 5. Kinetic resolution of $R^1R^2R^3SiX$

two possible diastereomers. The remaining unreacted materials will also be optically active. According to this scheme, the incomplete (30 to 60% conversion) reduction of a racemic methoxysilane by chiral complexes of lithium aluminium hydride led to optically active hydrosilane and unreacted methoxysilane (equation 1)[37]. Some of the results are presented in Table 1.

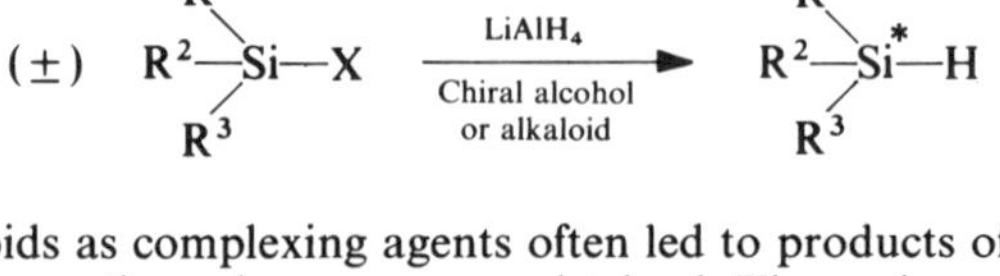

(1)

The use of alkaloids as complexing agents often led to products of low optical purity even if up to 41% enantiomeric excess was obtained. The main synthetic limit of the method is the necessity of stopping the reaction short of completion. Products are

TABLE 1. Partial reduction of racemic methoxysilane by alkaloids/$LiAlH_4$ (after Reference 37)

			Unreacted $R^1R^2R^3SiOMe$		$R^1R^2R^3SiH$	
$R^1R^2R^3SiOMe$	Alkaloid	% Reduction	$[\alpha]_D$ (deg)	Optical purity (%)	$[\alpha]_D$ (deg)	Optical purity (%)
1-NpPhMeSiOMe	(−)quinine	35	−0.24	1.4	+1.52	4.3
	(+)quinidine	33	−0.93	5.5	+1.96	5.5
	(+)cinchonine	43	−0.40	2.4	+1.24	3.5
	(+)φ-ephedrine	27	−0.86	5.1	+1.53	12.8
PhEtMeSiOMe	(−)cinchonidine	51	+0.59		−0.18	7.1
	(+)quinidine	54	−1.35		+0.44	14.4
	(+)cinchonine	56	−3.89		+0.5	41.2

therefore obtained with moderate chemical yields. This drawback was eliminated in one interesting example of kinetic resoluton with total chemical transformation of organosilane starting material. As illustrated in Scheme 6, the menthanolysis of 1-naphthylphenylchlorosilane allowed quantitative recovery of the menthoxysilane diastereomers in a 1 to 2 ratio. The key point is that the starting (±)chlorosilane undergoes a rapid racemization, faster than the alcoholysis reaction. According to this kinetic scheme, the menthoxysilane diastereomers are formed in a ratio close to that of the rate constants k_1:k_2. Fractional crystallization of the reaction mixture allowed isolation of the less soluble diastereomer in almost 70% yield.

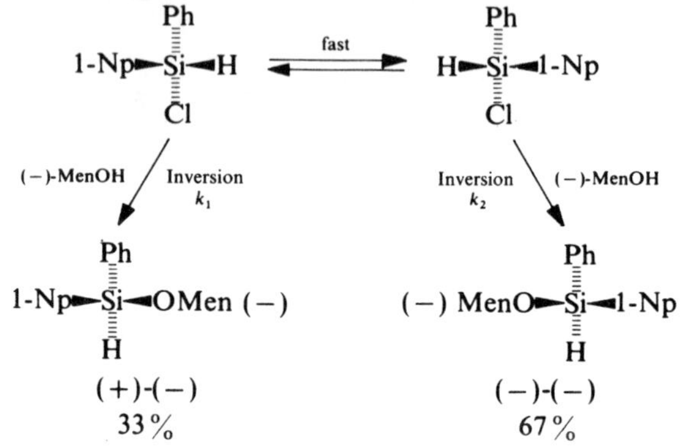

SCHEME 6. Menthanolysis of 1-naphthylphenylchlorosilane (after Reference 26)

b. Asymmetric synthesis. The development of asymmetric synthesis in carbon chemistry is largely due to the use of trigonal carbon as a prochiral moiety. Although a solid Si=C[38] and Si=Si[39] derivatives have been isolated, silicon atoms exhibit a general inability to form stable derivatives with multiple bonds. Transient silenes are, however, configurationally stable[40] and could in principle be used as prochiral substrates. An interesting example, although of limited synthetic utility, was shown in the reaction of (−)isoborneol with prochiral silene generated by photolysis of substituted silacyclobutenes (equation 2)[41]. Despite the fact that only the bulky (−)-isoborneol gave appreciable enantiomeric excess, it clearly demonstrates the prochiral nature of a transient trigonal silicon atom.

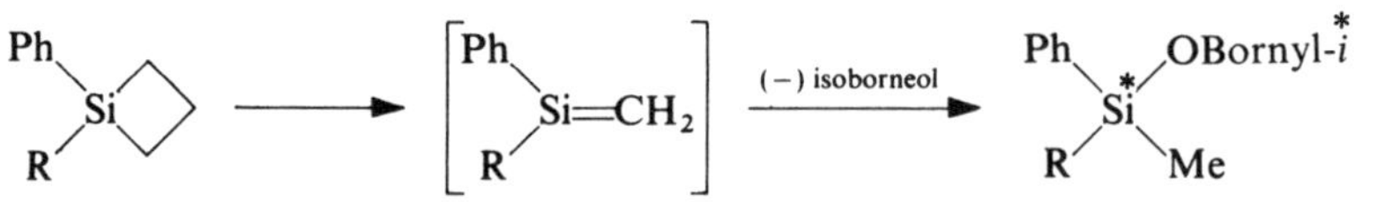

(2)

R	e.e.
Et	24%
t-Bu	30%

An asymmetric synthesis at silicon was developed starting from tetragonal silicon compounds. Prochiral compounds of type $R^1R^2SiX_2$ having two enantiotopic functional groups X are of particular interest. Owing to the many stereospecific substitution reactions that can be performed at a functional silicon atom, the preparation of optically active silicon compounds can be achieved by enantioselective substitution reaction at silicon according to Scheme 7. The stereospecific monosubstitution of X by a chiral reagent Y* will produce unequal amounts of two diastereomers of opposite configuration around the silicon atom.

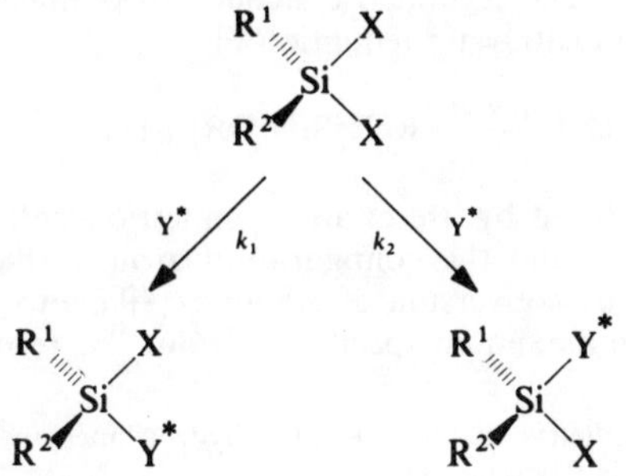

SCHEME 7. Substitution reactions at a prochiral organosilicon compound

The first example of an asymmetric induction at tetragonal silicon was reported by Klebe and Finkbeiner[42] and is shown in equation 3. The reaction of a prochiral bis-(acetamido)silane with optically active amino acids led to two diastereomeric 2-silaoxazolidones in unequal amounts. These diastereomers were shown to undergo a second-order asymmetric transformation: crystallization was accompanied by a rearrangement of the less abundant into the more abundant diastereomer. From the silaoxazolidones, alcoholysis reactions yielded an optically active dialkoxysilane.

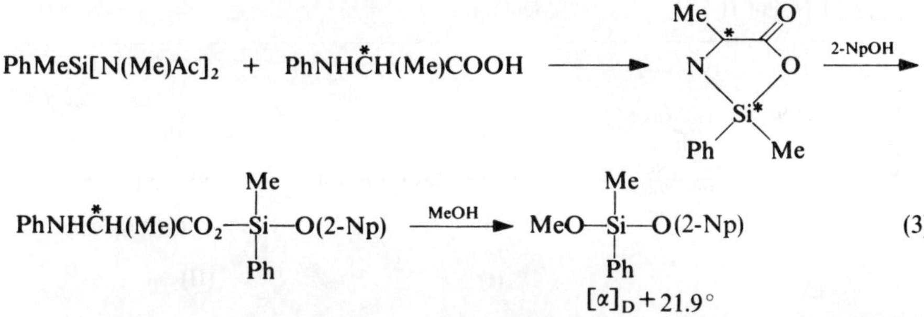

(3)

(i) Rhodium-catalysed reactions of prochiral dihydrosilanes. The preparation of optically active organosilanes by use of asymmetric reactions of prochiral dihydrosilanes appears to be of general applicability. Dihydrosilanes were shown to react selectively

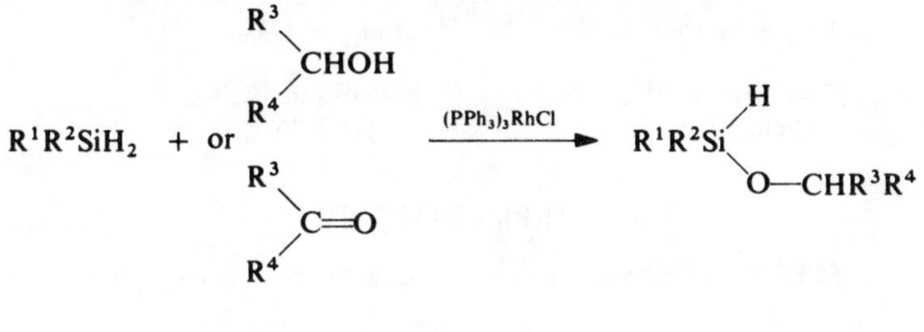

(4)

with alcohols and ketones in the presence of rhodium complexes to yield quantitatively chiral hydroalkoxysilanes[43, 44] (equation 4). According to these reactions, optically active organosilanes have been obtained by use of a chiral alcohol or ketone and also by the use of an asymmetric rhodium catalyst.

Asymmetric alcoholysis. The asymmetric alcoholysis of dihydrosilanes was studied in the presence of a rhodium catalyst[45] (equation 5).

$$R^1R^2SiH_2 + ROH \xrightarrow{\text{"}L_2RhCl\text{"}} R^1R^2SiH(OR) + H_2 \xrightarrow[\text{Retention}]{R^3MgX} R^1R^2R^3SiH \tag{5}$$

Chirality can be introduced by use of an asymmetric alcohol and/or an asymmetric catalyst. The optical yield and the configuration around silicon of the predominant isomer were determined by conversion of the alkoxysilane to a trisubstituted silane of known configuration and maximum specific rotation. The results are given in Table 2.

TABLE 2. Asymmetric alcoholysis of dihydrosilanes in the presence of $(Ph_3P)_3RhCl$ (after Reference 45)

Dihydrosilane $R^1R^2SiH_2$	Alcohol	e.e. (%)	Configuration of $R^1R^2R^3SiH$
1-Np$PhSiH_2$	(−)-Menthol	48	*R*
	(−)-Cholesterol	3	*R*
	(+)-Cinchonine	13	*S*
	(−)-Borneol	6	*S*
	(−)-Ephedrine	54	*R*
	(+)-Ephedrine	54	*S*
1-Np$EtSiH_2$	(−)-Menthol	21	*S*
1-Np$PhCH_2SiH_2$	(−)-Menthol	20	*S*

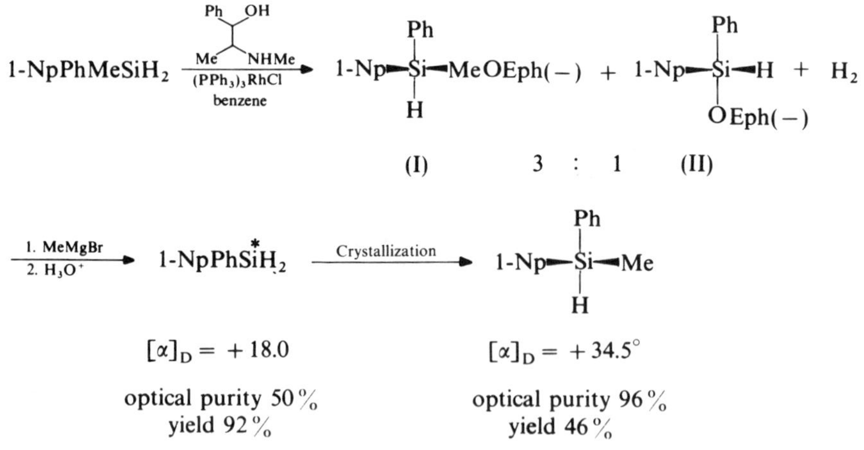

Eph = CH(Ph)CH(Me)NHMe

SCHEME 8. Preparation of *R*(+)-α-naphthylphenylmethylsilane

The nature of the organosilane and alcohol are very important. 1-NpPhSiH_2 gave the highest enantiomeric excess. It provided the basis for a one-pot synthesis of Sommer's compound *R*-(+)- or *S*(−)-1-NpPhMeSiH as described in Scheme 8[46]. The reaction of (+) or (−) ephedrine catalysed by $(PPh_3)_3$RhCl led to a 3:1 mixture of diastereomers. This was subsequently converted to *R* or *S*-1-NpPhMeSiH in 96% optical purity with a 46% chemical yield starting from 1-NpPhSiH_2.

The use of an optically active catalyst in the alcoholysis reaction was also considered[45]. The results are given in Table 3.

TABLE 3. Asymmetric alcoholysis of 1-naphthylphenylsilane (after Reference 45)

Alchohol	Ligand in catalyst L_2RhCl(S)[a]	e.e.[b] (%)	Configuration of $R^1R^2R^3$SiH
MeOH	(+)-DIOP	3	*R*
i-PrOH	(+)-DIOP	8	*R*
t-BuOH	(−)-DIOP	7	*S*
cyclo-C_6H_{11}OH	(+)-DIOP	17	*R*
(−)-Menthol	(−)-DIOP	31	*R*
(−)-Menthol	(+)-DIOP	49	*R*
(−)-Menthol	NMDPP	22	*R*
(−)-Menthol	MDPP	44	*R*
(−)-Ephedrine	(−)-DIOP	42	*R*
(−)-Ephedrine	(+)-DIOP	46	*R*

[a] S = solvent. Ligands: DIOP = 2,3-*O*-isopropylidene-2,3-dihydroxy-1,4-bis(diphenylphosphino)butane[47]; NMDPP = neomenthyldiphenylphosphine; MDPP = menthyldiphenylphosphine[48].

[b] Enantiomeric excess, determined after conversion of the alkoxysilane to a trisubstituted silane.

Optical yields of up to 54% have been observed with various dihydrosilanes. 1-NpPhSiH_2 gave the highest enantiomeric excess. It is noteworthy that in the reactions with a chiral alcohol, regardless of the asymmetric catalyst used, the optical yields are similar to those obtained with $(PPh_3)_3$RhCl. This indicates only a weak interaction

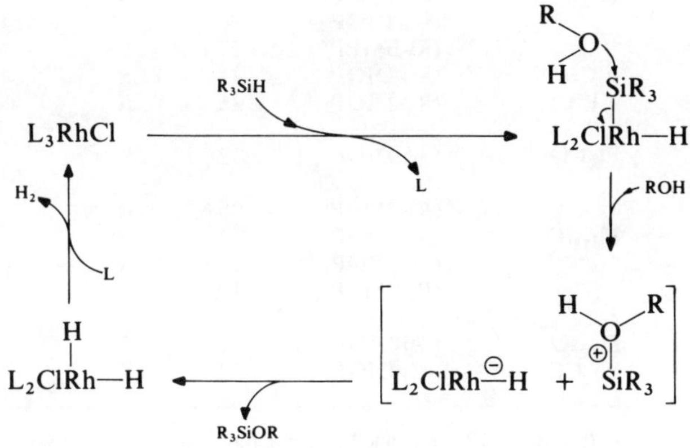

SCHEME 9. Mechanism of the alcoholysis of hydrosilanes

between alcohol and catalyst during the reaction. To account for this observation, a mechanism involving nucleophilic attack of the alcohol at the coordinated silicon has been suggested[49] (Scheme 9). The nucleophilic cleavage of the Si–Rh bond without prior coordination of the alcohol is suppored by the role of the solvent in the stereochemical course of the reaction and by the low influence of the catalyst ligands in the asymmetric reactions[49].

Asymmetric hydrosilylation of ketones. Chiral alkoxysilanes have also been obtained in the rhodium-catalysed hydrosilylation of carbonyl compounds[50–52] (equation 6).

$$R^1R^2SiH_2 + R^4R^5C{=}O \xrightarrow{\text{“}L_2RhCl\text{”}} R^1R^2(H)\overset{*}{Si}O\overset{*}{C}HR^4R^5 \xrightarrow{R^3MgX} R^1R^2R^3\overset{*}{Si}H + R^4R^5\overset{*}{C}HOH \quad (6)$$

When prochiral silane and ketone are used, hydrosilylation, in the presence of a chiral catalyst, results in asymmetric induction at both silicon and carbon centres. Treatment of the diastereomeric alkoxysilanes with a Grignard reagent leads to recovery of an organosilane and of an alcohol of different optical purity (Tables 4 and 5).

Optically active silanes were obtained from symmetrical ketones RCOR, in up to 46% enantiomeric excess, the catalyst being the only source of chirality. Asymmetric induction was also observed at the prochiral carbon of constitutionally unsymmetrical ketones:

TABLE 4. Asymmetric hydrosilylation of symmetric ketones and aldehydes by prochiral dihydrosilanes (after References 50, 52 and 53)

$R^1R^2SiH_2$	Carbonyl compound	Ligand in catalyst $L_2RhCl(S)$[a]	e.e.[b] (%)	Configuration $R^1R^2R^3SiH$	Ref.
1-NpPhSiH_2	Me_2CO	(+)-DIOP	30	*R*	50
	Et_2CO	(+)-DIOP	46	*R*	50
	n-Pr_2CO	(+)-DIOP	39	*R*	50
	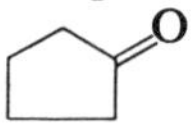	(+)-DIOP	35	*R*	50
	Ph_2CO	(+)-DIOP	31	*R*	50
		(−)-DIOP	46	*S*	53
		(*R*)-BMPP[c]	27.7	*S*	52
	EtCHO	(+)-DIOP	14	*R*	50
	n-PrCHO	(+)-DIOP	16	*R*	50
1-NpMeSiH_2	Et_2CO	(+)-DIOP	12	*S*	50
		(*R*)-BMPP[c]	8.6	*R*	52
	Ph_2CO	(+)-DIOP	7	*S*	50
		(−)-DIOP	18.8	*R*	53
		(*R*)-BMPP[c]	19.7	*R*	52
1-NpEtSiH_2	Et_2CO	(+)-DIOP	21	*S*	50
	Ph_2CO	(+)-DIOP	33	*S*	50

[a] See footnote a in Table 3; (*R*)-BMPP = (*R*)-Benzylmethylphenylphosphine.
[b] Enantiomeric excess, determined after conversion of the alkoxysilane to a trisubstituted silane.
[c] Catalyst: $[L_2RhH_2(S_2)]^+ClO_4^-$.

TABLE 5. Asymmetric hydrosilylation of prochiral ketones by a prochiral dihydrosilane (after Reference 50)

			$R^1R^2R^3SiH$		R^4R^5CHOH	
$R^1R^2SiH_2$	Ketone R^3R^4CO	Ligand in catalyst[a]	e.e.[b] (%)	Configuration	e.e.[b] (%)	Configuration
1-NpPhSiH_2	PhMeCO	(−)-DIOP	32	*S*	55	*R*
	EtMeCO	(−)-DIOP	40	*S*	42	*R*
	PhEtCO	(+)-DIOP	30	*R*	56	*S*

[a] See footnote a in Table 3.
[b] Enantiomeric excess.

RCOR′. The optical yield at the carbon atom is different from that at silicon. The hydrosilylation of optically active ketones was also studied in the presence of a chiral or an achiral rhodium catalyst. For example, the hydrosilylation of (−)-menthone gave two menthoxy and two neomenthoxysilanes (Table 6), the relative proportions of which reflect the stereoselectivity of attack on the carbonyl group. The predominant isomer formed corresponds to the attachment of the hydrogen atom to the less hindered face of the carbonyl compound. Optical purities at silicon of up to 82% were obtained. The nature of the catalyst ligands exerts a determining influence on the optical yield and configuration, even with an achiral catalyst. The combination of chiral ketone with a chiral catalyst may rise or decrease the optical yield. As previously, enantiomeric excess at carbon is different from that at silicon and also varies greatly according to reactants and catalysts.

The strong dependence of the stereochemical course of the reaction in substrates and catalyst ligands points to the importance of an intermediate rhodium complex in which both ketone and silane are coordinated to the rhodium. Moreover, enantiomeric excess at silicon has been shown to be largely influenced by which face of the chiral ketone is attacked[51]. This strongly supports π-coordination of the carbonyl group to the rhodium. The mechanism is represented in Scheme 10. The predominant configurations obtained have been tentatively explained on the basis of steric interactions in a hexacoordinated rhodium complex[54].

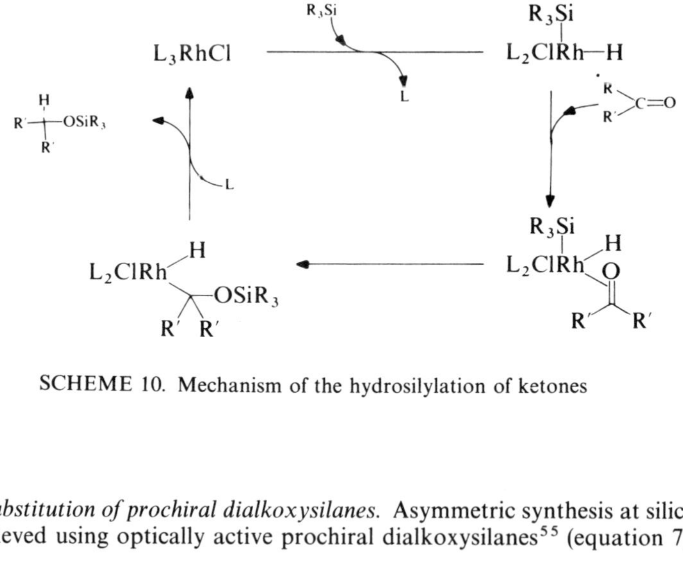

SCHEME 10. Mechanism of the hydrosilylation of ketones

(ii) Substitution of prochiral dialkoxysilanes. Asymmetric synthesis at silicon has also been achieved using optically active prochiral dialkoxysilanes[55] (equation 7).

$$R^1R^2Si(OR)^*_2 \xrightarrow{EtLi} R^1R^2Et\overset{*}{Si}OR^* \qquad (7)$$

Reaction with ethyllithium at low temperature led to unequal amounts of diastereomeric monoalkoxysilanes. Their ratio, which reflects the enantiomeric excess at silicon, was measured by NMR. Results are given in Table 7.

An interesting feature of these reactions is that they permit the synthesis of simple chiral silicon compounds in up to 69% optical purity using readily available reagents. Moreover, the method appears to possess wide applicability.

TABLE 6. Asymmetric hydrosilylation of (−)menthone by prochiral dihydrosilanes (after Reference 51)

$R^1R^2SiH_2$	Ligand in catalyst[a]	Alkoxysilanes			$R^1R^2R^3SiH$[a]	
		Neomenthoxy (%)	Menthoxy (%)	e.e. (%)[b] at Carbon	e.e.[b] (%)	Configuration
1-$NpPhSiH_2$	PPh_3	73	27	46	67	*R*
	C_8H_{14}	66	34	32	23	*S*
	(+)-DIOP	86	14	72	82	*R*
	(−)-DIOP	68	32	36	46	*R*
	MDPP	58	42	16	4	*S*
	NMDPP	69	31	38	16	*S*

[a] See footnotes to Table 3.
[b] Enantiometric excess.

TABLE 7. Reactions of chiral prochiral dialkoxysilanes with ethyllithium (after Reference 55)

Dialkoxysilne $R^1R^2Si(OR)_2$	Reaction temperature (°C)	e.e.[a] (%)
PhMeSi[OMen(−)]$_2$	−70	49
VinylMeSi[OMen(−)]$_2$	−70	69
AllylMeSi[OMen(−)]$_2$	−70	30
PhMeSi[OBornyl(−)]$_2$	−10	50
PhMeSi[ONeomenthyl(+)]$_2$	−70	36

[a] Enantiomeric excess.

B. Inorganic Silicon Compounds

Numerous silyl–transition metal complexes have been obtained and studied[56, 57], owing to the many stereospecific substitution reactions which can be performed at the silicon atom.

The cyclopentadienyliron dicarbonyl anion was reacted with an optically active chlorosilane giving rise to an optically active iron complex[58] (equation 8) which was isolated in 98% optical purity. On the basis of the known stereochemistry of nucleophilic displacement of chlorine nucleofuge, the reaction was assumed to occur with inversion of configuration at silicon.

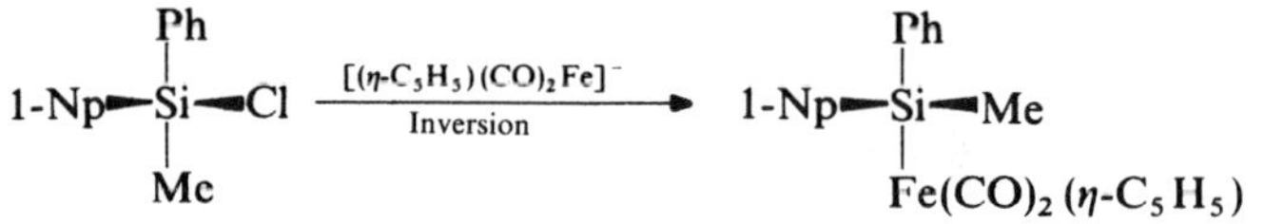

(8)

(*R*)-(+) (*S*)-(−)

$[\alpha]_D + 6.3°$ $[\alpha]_D - 41.5°$

optical purity: 98%

Interesting, optically acive complexes were obtained upon reaction of organosilicon hydrides. The reactions of R_3SiH with transition metal complexes represent a key step in catalytic hydrosilylation reactions. Catalytic activation of silicon hydrides has been proposed to arise from an oxidative addition process to a transition metal centre. Such a process has been shown to be reversible via a reductive elimination step (equation 9). The stereochemistry of addition to an iridium complex was shown to occur in a *cis* fashion[59] (equation 10).

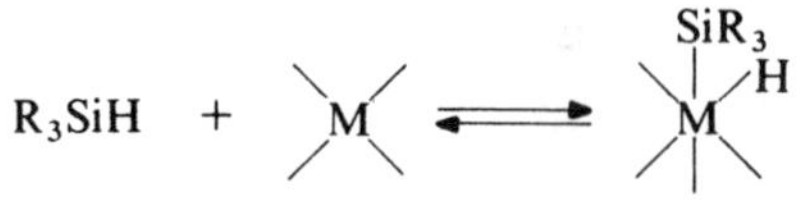

(9)

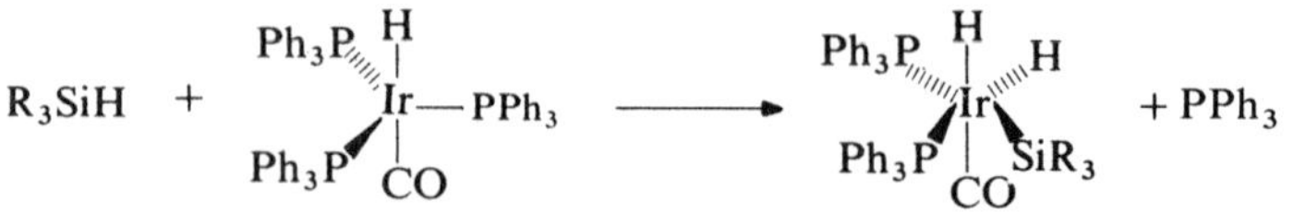

(10)

A similar *cis* addition was established in the case of a rhodium complex[60]. One of the main characteristics of oxidative addition as well as reductive elimination is that these processes occur with high stereoselectivity with an optically active organosilicon hydride. Both processes involve retention of configuration at the silicon atom.

Several optically active silylplatinum complexes have been obtained by Eaborn and coworkers[61–63]. Their preparation always involves an initial oxidative addition of a chiral hydrosilane, followed by ligand displacement. Complexes obtained according to equations 11–13 have a square-planar geometry around the platinum atom. All these reactions were proposed to proceed with retention at the silicon atom. This was confirmed in one case by X-ray structure determination.[64]

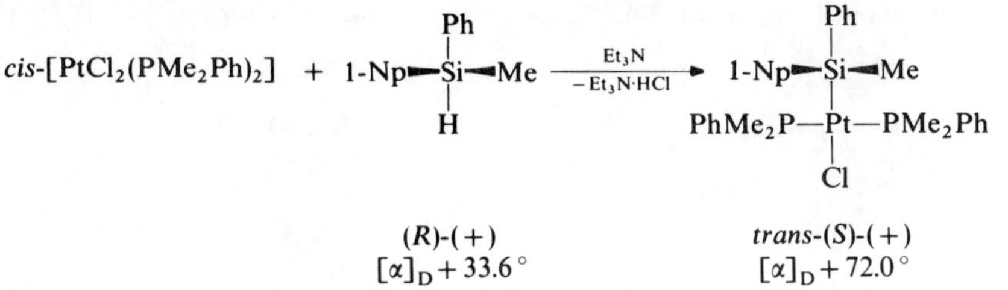

(11)

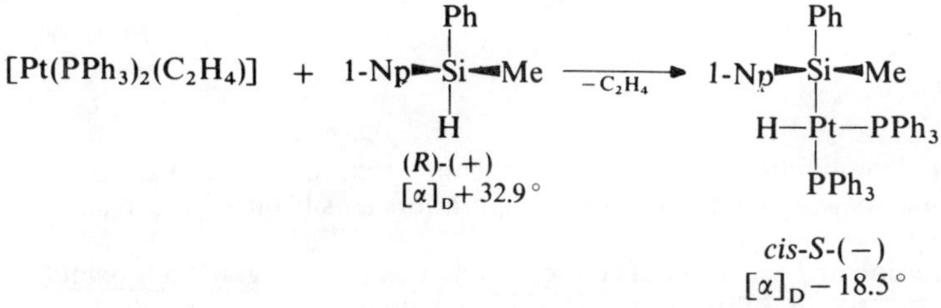

(12)

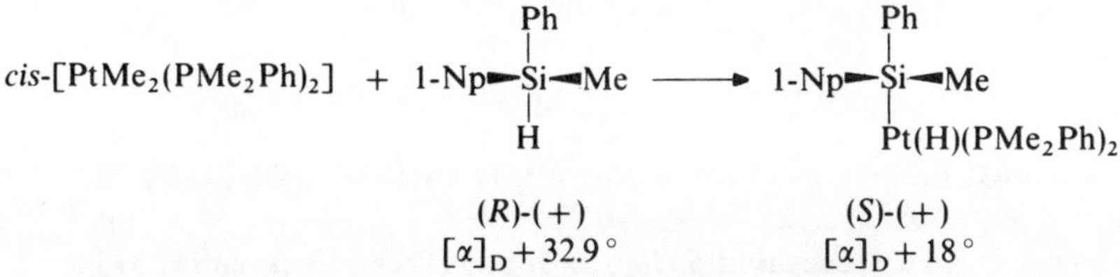

(13)

Chiral silyl–manganese complexes have also been obtained from optically active silicon hydrides. The oxidative addition of silicon hydride to methyl cyclopentadienyl

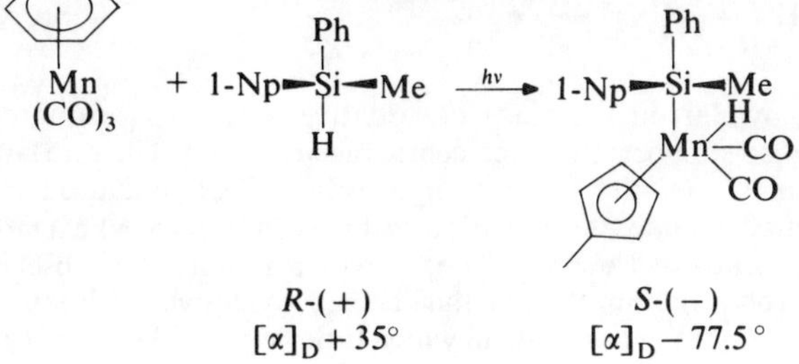

(14)

manganese tricarbonyl also provides a convenient route to chiral complexes[65,66] (equation 14). Retention at the chiral silicon was proposed.

The reaction of metal carbonyl dimers with silicon hydrides also probably involves an initial oxidative addition step. Chiral silyl–cobalt and silyl–manganese carbonyl complexes have been obtained through the reaction of optically active organosilicon hydrides with metal carbonyls[65–68] (equation 15 and 16). Phosphine-substituted cobalt complexes were similarly obtained by reaction of a chiral hydrosilane with $Co_2(CO)_6L_2$ [L = PPh_3, $P(OPh)_3$, $P(c\text{-}C_6H_{11})_3$][69].

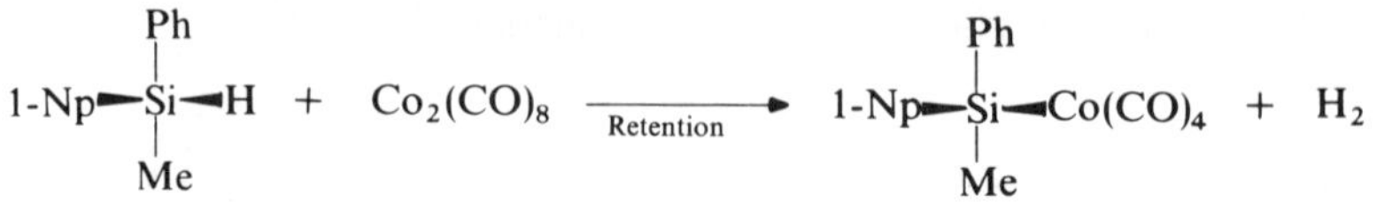

(15)

(S)-(−) $[\alpha]_D - 35°$ (R)-(−) $[\alpha]_D - 2°$

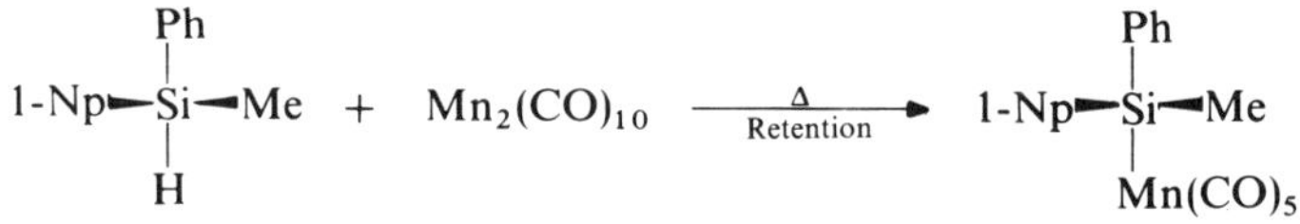

(16)

(R)-(+) (S)-(−)

X-Ray determination of the absolute configuration of the silyl–cobalt complex[70] allows one to assign retention of configuration at silicon in the reaction shown in equation 15.

Cis addition and retention of configuration at silicon suggests that oxidative addition may occur through a three-centre transition state as in Figure 3.

FIGURE 3. Three-centre transition state for oxidative addition of R_3SiH

An analogy with carbene insertion into Si–H bonds has been proposed by Eaborn and coworkers[61]. Carbene insertions are supposed to involve a three-centre mechanism and proceed with retention of configuration at silicon[71] (equation 17).

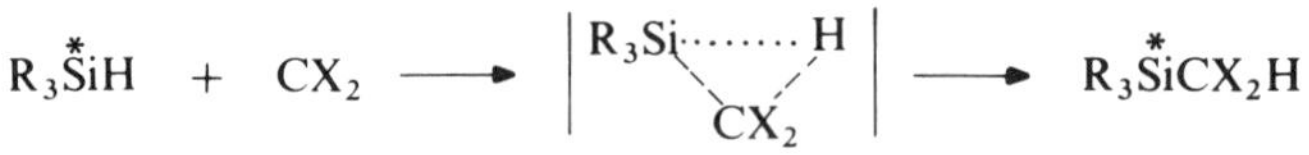

(17)

A kinetic study by Harrod and Smith of oxidative addition to a square-planar cationic iridium complex also supports the three-centre mechanism[72]. The rate law is first order in the iridium complex and first order in hydrosilane. Determination of the activation parameters indicated a moderate activation enthalpy ($\Delta H^{\neq} = 5\text{–}6$ kcal mol^{-1}) and a large negative activation entropy ($\Delta S^{\neq} = -47$ e.u.). No variations were observed on changing the solvent. Harrod and Smith concluded that oxidative addition proceeds via a concerted three-centre transition state in which little bond-making or bond-breaking had occurred. The activation enthalpy was attributed to a deformation of the square-planar complex on its approach to the transition state.

Interestingly, Harrod and coworkers[73] also compared the kinetics of triphenylphosphine and hydrosilane addition to a neutral iridium complex (equation 18). The activation enthaplies for the phosphine and silane reaction were almost identical:

$$\Delta H_{p}^{\ddagger}-\Delta H_{Si}^{\neq}=0 \text{ kcal mol}^{-1} \quad \text{and} \quad \Delta S_{p}^{\ddagger}-\Delta S_{Si}^{\neq}=13 \text{ e.u.}$$

This striking similarity is not necessarily a proof that phosphine and silane addition are mechanistically analogous processes. However, this kinetic behaviour is consistent with the above description of oxidative addition.

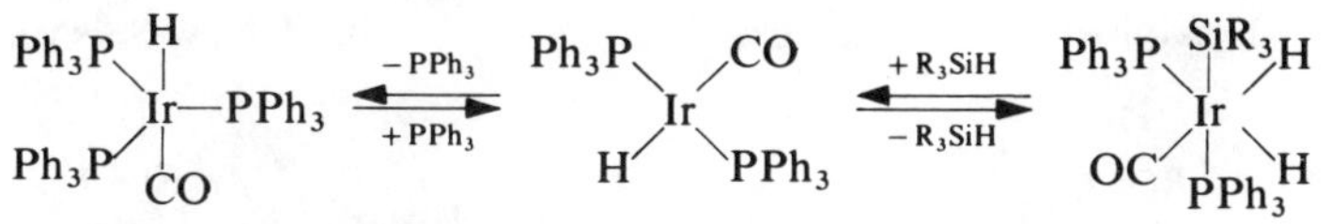

(18)

As regards the similarity with simple coordination of two-electron ligands, it is of interest to discuss the oxidative addition of silicon hybrid to methyl cyclopentadienylmanganese carbonyl and the nature of the bonding in the adduct. Hydrido silyl–manganese complexes were first prepared by Jetz and Graham according to

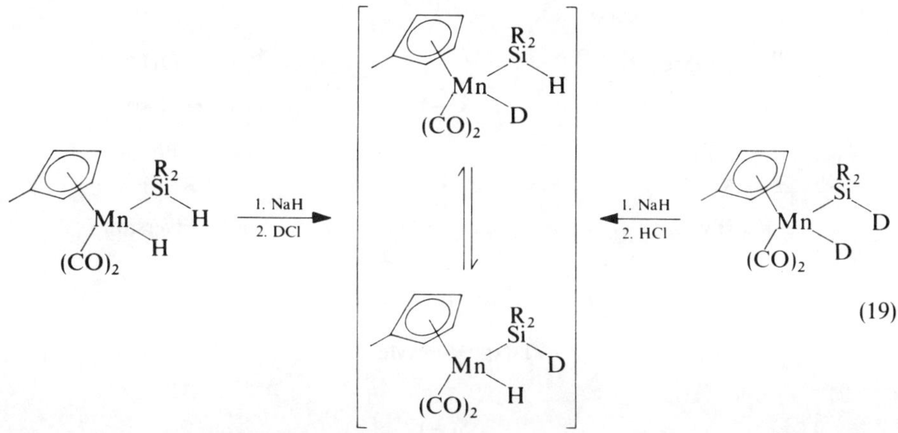

(19)

equation 14[74]. These complexes have been shown to readily undergo reductive elimination of R_3SiH upon reaction with various electrophilic or nucleophilic reagents[65, 75]. As shown in Scheme 11, the chiral hydrido silyl–manganese complex exhibits interesting reactivity. Different types of reagent were shown to react in two ways[65, 66]:

(i) Reagents giving substitution at silicon with cleavage of the Mn–Si bond; nucleophiles such as MeOH, H_2O or $LiAlH_4$ fall into this category and cause inversion of configuration at silicon.

(ii) Reagents both nucleophiles (MeLi, PPh_3) and electrophiles (Cl_2), leading to a reductive elimination of the organo-silicon hydride. As expected for such a process, retention at silicon was observed.

The easy reductive elimination process does not seem to arise from a particular lability of the hydrido silyl–manganese complex. It is stable in solution. Moreover, functionalized

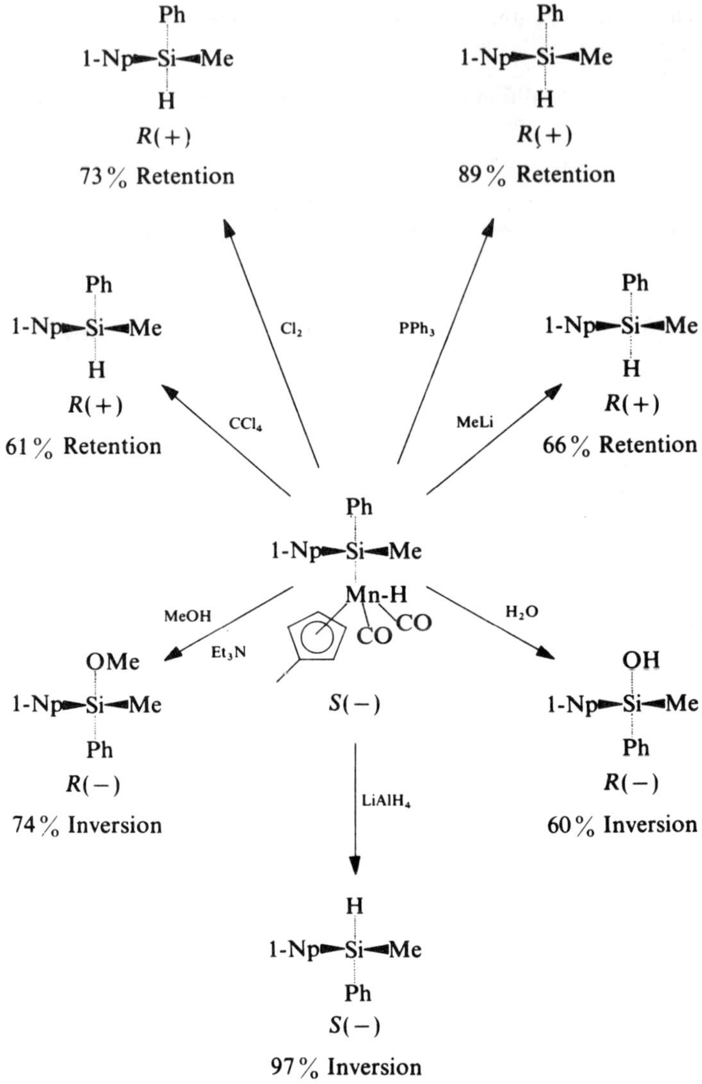

SCHEME 11. Reactions of optically active hydrido silyl–manganese complex

silyl–manganese complexes were shown to undergo substitution at silicon without Si–Mn bond cleavage[76].

The particular chemical behaviour of the hydrido silyl–manganese complex, which contrasts sharply with that of hydrido silyl–iron complexes[56, 77], has been interpreted in terms of a bonding interaction between hydrido and silyl ligands.

Of interest also is the intramolecular hydrogen–deuterium exchange invoked to explain the reactions depicted in equation 19[78]. The reactions of hydrido and deuterio complexes led to the same equimolar mixture of two complexes, indicating that these two compounds are in equilibrium. An intermolecular exchange has been rejected on the basis of the findings shown in equation 20. The above exchange thus occurs within the coordination sphere of manganese and does not involve formation of free hydrosilane by a reductive elimination process.

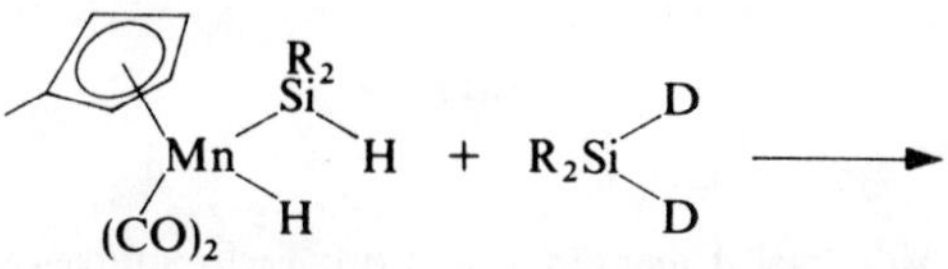
no exchange (20)

The particular chemical behaviour or hydrido silyl–manganese complexes has been related to their molecular structure determined by X-ray or neutron diffraction analysis. The hydrido and silyl ligands are *cis* oriented and the principal distances are given in Table 8. According to Jetz and Graham the hydrido ligand, while being located at a 'normal' distance from the manganese atom, was close enough to silicon to imply some degree of bonding. Support for this idea comes from a ^{29}Si NMR study[78]. J(Si–H) coupling constants of 65 and 69 Hz were measured, representing higher values than expected for a ^{2}J(Si–Mn–H) coupling constant. To explain their results, Hart-Davies and Graham had written manganese complexes in terms of resonance hybrids as shown in Figure 4[75]. However, the possibility that the close Si–H distance is due to steric constraints has been raised by Bennett and Simpson[80]. Steric effects favour a *cis* orientation of silyl and hydrido ligands and thus facilitate reactions involving reductive elimination pathways.

TABLE 8. Principal distances (Å) in 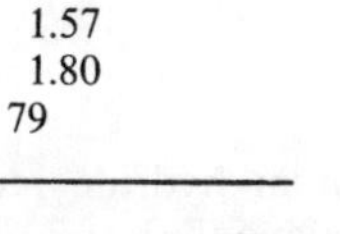

	$R_3 = Ph_3$	Cl_2Ph	FPh_2
Mn–Si	2.42	2.31	2.35
Mn–H	1.55	1.49	1.57
Si–H	1.76	1.79	1.80
Ref.	75	75	79

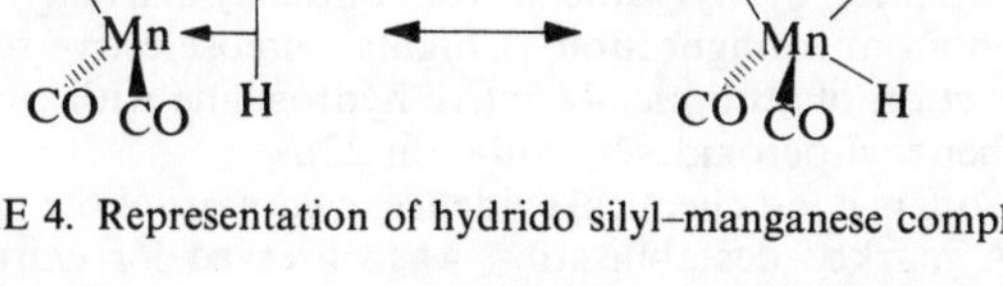

FIGURE 4. Representation of hydrido silyl–manganese complexes

The situation has also been discussed by Andrews and coworkers[81] and Harrod and coworkers[73] in terms of 'incomplete' oxidative addition of R_3SiH to the manganese atom. Owing to steric crowding, the compound arising from an oxidative addition process is proposed to represent a case where the descent to the energy minimum is arrested at some point along the reaction coordinate, before complete cleavage of the Si–H bond.

In a more recent proposal, a description of the bonding as a two-electron three-centre Mn(μH)Si has been suggested[65, 76, 78, 79] (Figure 5). Such a representation would account for the above-mentioned chemical and physical properties of these complexes. A similar suggestion has been put forward in the case of silyl–diplatinum complexes, where a Si–H distance of 1.72 Å was found[82].

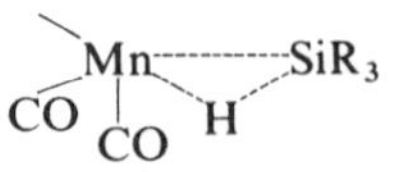

FIGURE 5. Representation of hydrido silyl–manganese as a two-electron three-centre Mn(μH)Si bond

C. Organosilicon Radicals

Reactions of optically active organosilanes that involve intermediate radicals were shown to proceed with predominant retention of configuration at silicon.

The photolysis of an optically active acetyl silane in carbon tetrachloride afforded a chlorosilane with retention of configuration[83]. As shown in equation 21, the photolysis reaction implies the formation of an asymmetric radical which was trapped by carbon tetrachloride.

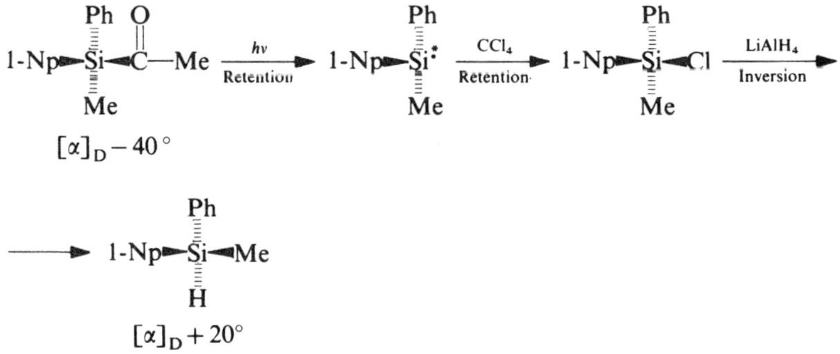

(21)

Lithium aluminium hydride reduction, with inversion of the chlorosilane formed yielded the hydrosilane ($[\alpha]_D + 20°$). Assuming a neat inversion in the reduction step, it appeared that the formation and trapping of the silyl radical occurred with 65% retention.

Similarly, the generation of silyl radicals from optically active hydrosilanes was shown to retain the silicon atom configuration. A highly stereoselective reaction sequence was observed in the reaction of an optically active hydrosilane with carbon tetrachloride in the presence of dibenzoyl peroxides[84] (equation 22).

This experiment offers good chemical evidence in support of the pyramidal structure of the silyl radical. A marked destabilisation was observed for chiral disilanyl radical[85] (equation 23).

The low stereoselectivity here is consistent with ESR data, which indicated a more planar structure for the pentamethyldisilanyl radical $Me_3Si–SiMe_2^{\bullet}$[86]. Nevertheless, the pyramidal silyl radical exhibits significant configurational stability. The inversion of the pyramidal species, is generally slow enough to allow reactions in which the configuration at silicon is retained.

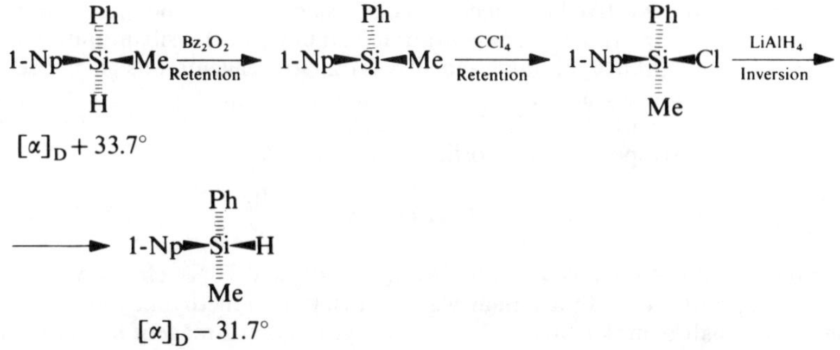

(22)

$$(Ph_3Si)PhMe\overset{*}{Si}\text{–}H \xrightarrow{Bz_2O_2} (Ph_3Si)PhMe\overset{*}{Si}\cdot \xrightarrow{CCl_4} \xrightarrow{LiAlH_4} (Ph_3Si)PhMe\overset{*}{Si}\text{–}H$$

$[\alpha]_D - 8.9°$ $\qquad$ $[\alpha]_D + 0.9°$ (23)

Evidence was also given for the pyramidal structure of the analogous germyl radical. Optically active germyl radical was generated from an optically active hydrogermane and trapped with carbon tetrachloride involving substantial retention of configuration[87]. Interestingly, a predominant retention of configuration was found in the thermal radical hydrogermylation reaction[88, 89] (equation 24).

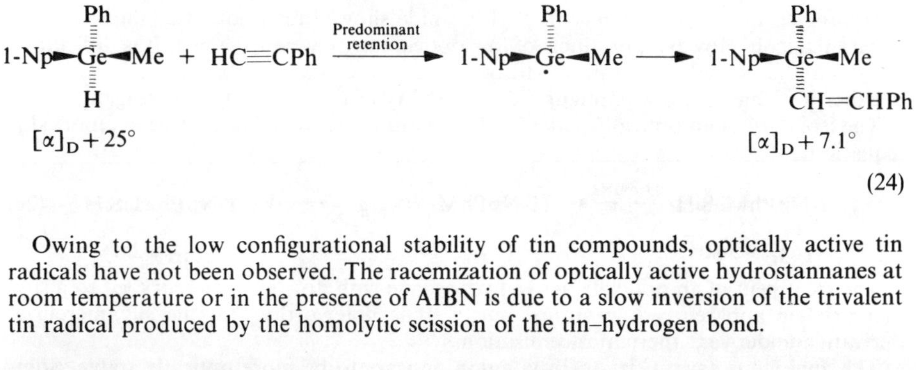

(24)

Owing to the low configurational stability of tin compounds, optically active tin radicals have not been observed. The racemization of optically active hydrostannanes at room temperature or in the presence of AIBN is due to a slow inversion of the trivalent tin radical produced by the homolytic scission of the tin–hydrogen bond.

D. Organosilicon Anions

Evidence for the configurational stability of chiral silyl anions was reported. Silylmetallic compounds[90] have a considerably enhanced optical stability relative to acyclic alkyllithium compounds. An NMR study of di-isopropylphenylsilyllithium showed that the isopropyl methyl groups are anisochronous up to 185 °C, indicating that the barrier to inversion at silicon in the silyl anion is greater than 24 kcal mol^{-1}[91].

The formation of an asymmetric silyllithium reagent from an optically active disilane has been reported[92] (equation 25).

$$(-)\text{neo-}C_5H_{11}PhMe\overset{*}{Si}\text{—}SiMePh_2 \xrightarrow[THF]{Li} \text{neo-}C_5H_{11}PhMe\overset{*}{Si}Li \xrightarrow[H_2O]{HCl} (-)\text{neo-}C_5H_{11}PhMe\overset{*}{Si}\text{–}H \qquad (25)$$

$[\alpha]_D - 5.0°$ $\qquad$ $[\alpha]_D - 2.2°$

The observed stereoselective lithium cleavage of a silicon–silicon bond and stereoselective protonolysis demonstrate the configurational stability of the silyllithium reagent.

The formation of chiral silyllithium and silyl-Grignard reagents was also observed in the reactions of an optically active silyl–cobalt carbonyl complex[93,94]. Treatment of the chiral cobalt complex with methyllithium produced optically active silyllithium which hydrolysed to the corresponding hydrosilane (equation 26).

$$\text{1-NpPhMe}\overset{*}{\text{Si}}\text{–Co(CO)}_4 \xrightarrow{\text{MeLi}} \text{1-NpPhMe}\overset{*}{\text{Si}}\text{–Li} \xrightarrow{H_2O} \text{1-NpPhMe}\overset{*}{\text{Si}}\text{–H} \quad (26)$$

The optical purity of the recovered hydrosilane indicated 70% retention of configuration in the overall process. In a similar way the reaction of methylmagnesium bromide with the silyl–cobalt complex afforded a chiral silyl-Grignard reagent (equation 27).

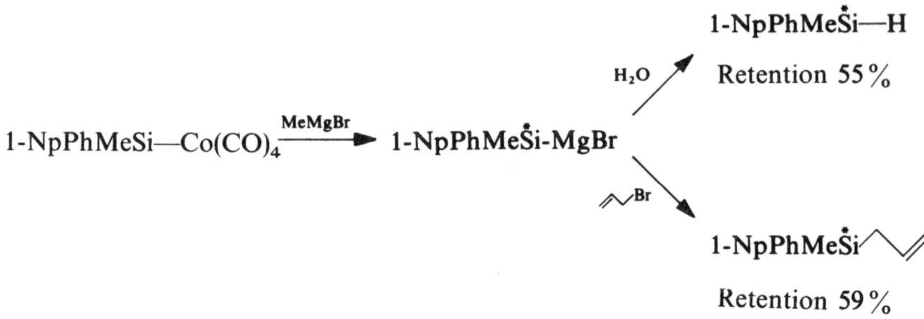

(27)

Hydrolysis or condensation with allyl bromide showed retention of configuration to occur at silicon. However, products of low optical purity were obtained. It is difficult to know whether some racemisation had occurred in the silicon–cobalt bond cleavage or whether it is due to a low configurational stability of the silyl Grignard reagent.

The isolation of an optically active silyl-mercurial derivative has also been reported[95] (equation 28).

$$\underset{[\alpha]_D + 34.0°}{\text{1-NpPhMe}\overset{*}{\text{Si}}\text{H}} \xrightarrow{t\text{-Bu}_2\text{Hg}} \underset{[\alpha]_D - 19.2°}{(\text{1-NpPhMe}\overset{*}{\text{Si}})_2\text{Hg}} \xrightarrow{\text{LiAlH}_4} \underset{[\alpha]_D + 2.8°}{\text{1-NpPhMe}\overset{*}{\text{Si}}\text{H}} \quad (28)$$

The treatment of an optically active hydrosilane with di-*t*-butyl mercury led to a silyl mercurial in apparently low optical purity. It is noteworthy that the silyl-mercurial derivative underwent thermal racemisation.

The analogous asymmetric germyl anion appears to be more optically stable. Solutions of optically active germyllithium reagent are easily prepared by treatment of a chiral hydrogermane with butyllithium[96–98] (equation 29).

$$R^1R^2R^3\overset{*}{\text{Ge}}\text{H} \xrightarrow[\text{Retention}]{\text{BuLi}} R^1R^2R^3\overset{*}{\text{Ge}}\text{Li} \xrightarrow[\text{Retention}]{H_2O} R^1R^2R^3\overset{*}{\text{Ge}}\text{H} \quad (29)$$

The germyllithium reagents have been found to undergo a variety of stereospecific reactions[99–102].

III. STEREOCHEMISTRY OF NUCLEOPHILIC DISPLACEMENT AT SILICON

A. Introduction and General Considerations

Stereochemical studies of substitution at silicon by nucleophiles were the topics of a number of reviews. The earlier work up to 1964 was the subject of an important monograph by Sommer[1a]; more recent studies have also been summarised by him[1b] and others (see Prince[103a] or Fleming[103b]). A more detailed analysis may be found in our previous comprehensive reviews.[3]

The stereochemistry of displacement of silyl ligands in silicon–transition metal complexes was also reviewed[2,56].

Unlike carbon chemistry where a similar type of reaction invariably takes place by an inversion process for chiral derivatives, both inversion and retention processes compete with one another for chiral derivatives of silicon. The factors influencing the dominance of one process over the other generally are known from studies[2] conducted under a variety of experimental conditions.

The major conclusions resulting from these studies will be discussed in this section. Before dealing with this problem, we examine briefly the following three questions:

(1) Is the R_3Si^+ cation acting as an intermediate in displacement reactions at silicon?

(2) Are the nucleophilic substitutions at silicon consistent with the occurrence of a rate-determining formation of a five-coordinate silicon intermediate?

(3) Is substitution at silicon a nucleophilic or a single electron transfer process?

1. Role of R_3Si^+ as a reaction intermediate

The role of carbenium ions R_3C^+ as reaction intermediates in S_N1 processes is well-documented[104]. In such unimolecular reactions, the carbon atom attached to the leaving group undergoes racemisation. In solution, silicenium ions have been proposed as reaction intermediates in S_N1 type reactions and rearrangements.[105] Recently, Apeloig and coworkers presented evidence that a reactive intermediate R_3Si^+ can be generated via a 1,2-methyl migration of a solvolytically produced α-silyl carbonium ion (equation 30).[106]

$$R_3Si\overset{+}{C}R^1_2X \xrightarrow{\text{Solvolysis}} R_3Si\overset{+}{C}R^1_2 \xrightarrow{\text{NR}} R_2\overset{+}{Si}CR^1_2R \rightarrow \text{products} \qquad (30)$$

Moreover, Lambert and coworkers claimed the preparation and the characterization in solution of the triphenylsilyl cation on the basis of ^1H, ^{13}C spectra and conductance measurements.

Under conditions developed for nucleophilic displacement reactions at silicon, no evidence for intermediacy of R_3Si^+ silicenium cations was provided[1,2]. The device, well-known in carbon chemistry, of placing unsaturated groups next to the developing positive charge, does not change the mechanism in the silicon series[108]. Monofunctional triarylsilanes do not solvolyse by way of a Ar_3Si^+ cationic intermediate, and a *p*-methoxy substituent retards substitution relative to unsubstituted phenyl[109]. Moreover, nucleophilic substitutions at silicon are found to be very stereoselective, occurring with a high degree of either retention or inversion of configuration. Even vinyl- or ferrocenylsilanes, the structures of which might favour sp^2 cation formation due to stabilization by the substituents, do not lead to racemic solvolysis products (Table 9)[110].

Finally, it was clearly established that, even in hydroxylic solvents, there is no S_N1-type reaction at silicon[7]. Data have been discussed in great detail in comprehensive reviews[2,7]. However Eaborn and coworkers have pointed out that the lack of evidence for rate-determining ionizations of organosilicon halides and similar species might be attributable to their normally very high reactivity in S_N2 processes.[111] One of the obvious ways to suppress the bimolecular reaction is by steric hindrance. For instance, they provided evidence for the intermediacy of bridged delocalized silicon-containing cations of the type **1** in the solvolysis of highly sterically congested precursors, such as $TSiSiR_2I$ [TSi $=(Me_3Si)_3C$].[111]

TABLE 9. Stereochemistry of solvolysis of vinyl- and ferrocenylsilanes (Reference 110)

Silane[a]	Reagent	Product	Predominant stereochemistry[b]
1-NpPhViSi–Cl $[\alpha]_D = -5.7°$ (CCl_4)	CH_3OH/pentane	$R^1R^2R^3Si–OCH_3$ $[\alpha]_D = -9.3°$	100% RN
	H_2O/Et_2O	$R^1R^2R^3Si–OH$ $[\alpha]_D = -9.2°$	89% IN
	KOH/xylene	$R^1R^2R^3Si–OK$ $[\alpha]_D = +24.9°$	98.5% IN
1-NpPhFcSi–Cl $[\alpha]_D = +33.8°$ (CCl_4)	CH_3OH/pentane	$R^1R^2R^3Si–OMe$ $[\alpha]_D = -14.8°$	IN

[a] Vi = Vinyl; Fc = ferrocenyl; 1-Np = 1-naphthyl.
[b] RN denotes retention; IN denotes inversion

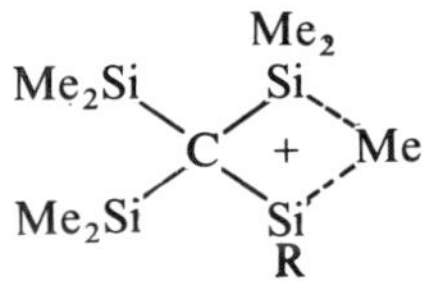

(1)

2. Role of a five-coordinate intermediate in nucleophilic displacement

a. Kinetic probes. In order to determine the energy profile of the reaction (Scheme 12), the rates of reactions occurring with either retention or inversion at silicon were studied.[112]

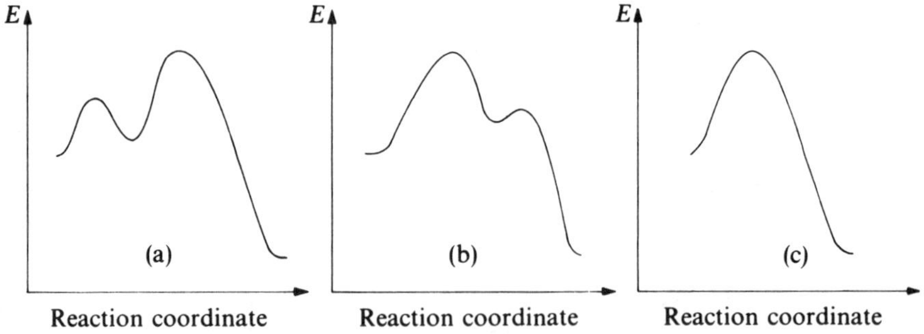

(a) Rapid formation of an intermediate complex followed by slow cleavage of the Si–X bond to give the products.
(b) Formation of an intermediate in the rate-determining step, followed by a fast breakdown to the products.
(c) Synchronous bond-forming and bond-breaking, involving a single transition state.

SCHEME 12. Energy diagrams

Some results for reactions leading to retention are summarised in Table 10, and results for reactions giving inversion are given in Table 11.

TABLE 10. Kinetic data for reactions of organosilanes with organometals (RM) in diethyl ether occurring with retention

Organosilane	X= α	or β	RM	k_α/k_β
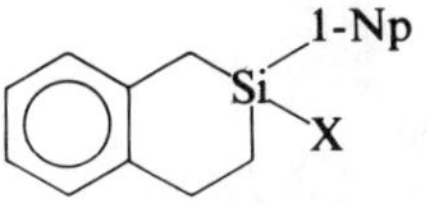	F	OMe	EtMgBr	1.5–2.5
	F	OMe	PhMgBr	2–3
	F	OMe	 Li	4–5
1-NpPhEtSi–X	OMe	H	BuLi	5–6
	H	D	BuLi	1
	F	OMe	BuLi	50

TABLE 11. Kinetic data for reactions of an organosilane with organometals (RM) in diethyl ether occurring with inversion

Organosilane	X= α	or β	RM	k_α/k_β
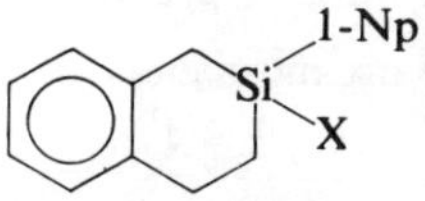	F	Cl	CH_3MgBr	1
	F	Cl	$PhCH_2MgBr$	1.6
	F	OMe	allylMgBr	$>10^5$
	F	OMe	$PhCH_2MgBr$	2×10^3–5×10^3

Two types of behaviour are observed: in one, rate constants for pairs of leaving groups are changed by a factor of 1–50, while in the other they change by a factor $>10^3$. The former reactions are the majority and apply to both retention and inversion modes, whereas the latter are only for the pair F and OMe in reactions giving inversion. Relative ratios near unity suggest that the rate-determining step does not involve Si–X bond cleavage. Thus the energy profile (Scheme 12, diagram C) involving a single transition state can be excluded. The small kinetic influence of the leaving group agrees with a rate-determining formation of a pentacoordinate intermediate; the very small (negligible) deuterium isotope effect confirms this proposal[112].

b. Chemical probes. Pentavalent silicon anions, including silicon anions with four carbon substituents, are postulated as intermediates in the reactions of organosilicon compounds with nucleophiles, but under normal circumstances have lifetimes too short for study[2]. Of particular interest are the reactions studied by Sullivan, De Puy and Damrauer[113] in the gas phase. Pentavalent silicon anions, including silicon anions with five carbon substituents, have been generated by reaction of anions with substituted silanes. For instance, direct addition of F^- or an allyl anion occurs, leading to anions formulated as pentacoordinate species (Scheme 13). These findings suggest that the formation of the pentavalent silicon anions may take place even in the case of a silicon atom bearing carbon ligands.

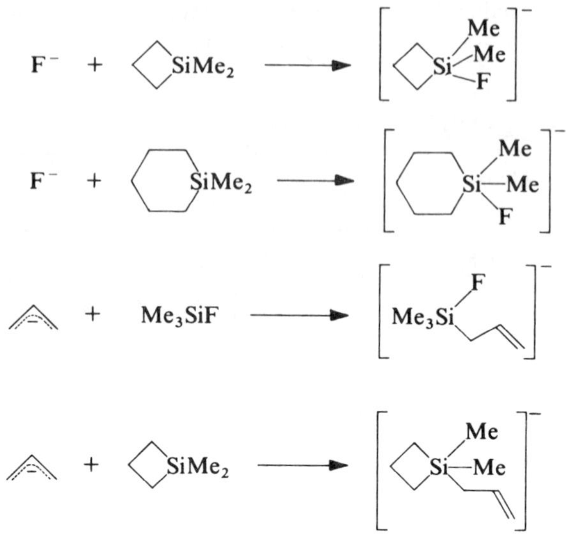

SCHEME 13. Pentavalent silicon anions in the gas phase

3. Substitution at silicon: nucleophilic or a single electron transfer process?

Coupling reactions of a nucleophilic reagent, Nu^-, with an organosilane, $R_3Si–X$, are usually represented by a one-stage nucleophilic mechanism (equation 31)[1,2].

$$Nu^- + R_3Si–X \rightarrow R_3Si–Nu + X^- \quad (31)$$

A two-stage reaction pathway might, however, be invoked as an alternative, the first step being a single electron transfer (SET) from the electron-rich nucleophile, Nu^-, to the electrophilic silicon atom (equation 32).

$$Nu^- + R_3Si–X \xrightarrow{SET} \overline{Nu\cdot + R_3Si–X^{\dot{-}}} \rightarrow R_3Si–Nu + X^- \quad (32)$$

The single electron transfer process can be eliminated on the basis of the following chemical and electrochemical data:

(1) The $CpFe(CO)_2^-$ anion reacts concurrently by S_N2 and free radical processes with alkyl halides[114] (Scheme 14, equation 33); the same ion is easily able to displace Cl from silicon, leading to the formation of a Fe–Si bond (Scheme 14, equation 34)[115]. At the opposite extreme, Cp(DPPE)FeMgBr shows very low nucleophilicity and reacts with alkyl halides only by a radical process (Scheme 14, equation 35)[115]. The same reagent was found completely unreactive towards halosilanes, which are known to be very reactive substrates in nucleophilic reactions (Scheme 14, equation 36)[115].

(2) The low ability of a Si–X bond to react by a single electron transfer process is also clearly indicated by the electrode potential data derived from the electrochemical reduction of silicon halides (Table 12)[116].

Although the electrochemical reduction of halosilanes exhibits a single irreversible wave (a fast dimerisation of the presumed intermediate radical occurs at the electrode), the half-wave reduction potentials are much more negative than the $E_{1/2}$ values measured

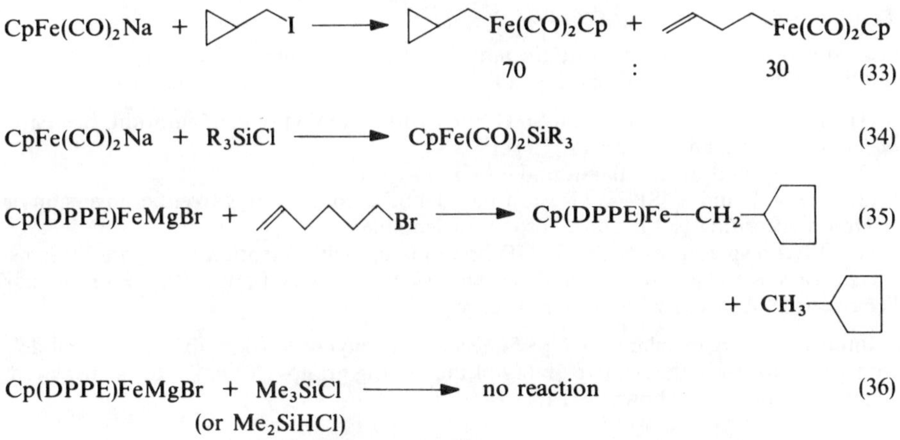

DPPE = bis(diphenylphosphino)-1,2-ethane

SCHEME 14. Coupling reactions with $Cp(CO)_2Na$ and Cp(DPPE)FeMgBR

TABLE 12. Half-wave reduction potential of some organic halides (after Reference 117) and halosilanes (after Reference 116) using a reference saturated calomel electrode

R–X or R_3Si–X	Solvent[a]	E (volts)
CH_3–Cl	DO	2.23
CH_3–Br	DO	2.01
CH_3–I	DO	1.63
Ph_3Si–F	THF	2.60
Ph_3Si–Cl	THF	2.40
Ph_3Si–Br	THF	2.37

[a] THF = tetrahydrofuran; DO = dioxane

for organic halides. This indicates a more difficult addition of an extra electron to a Si–X as compared to a C–X bond.

Thus we may conclude that most of the coupling reactions between organosilanes and nucleophiles occur by a one-stage nucleophilic process.

B. Factors Controlling the Stereochemistry

Displacement reactions proceed with either retention or inversion of configuration at silicon[1,2]. The change in stereochemistry is mainly a function of the nature of the leaving group and the nucleophile. In borderline cases other factors, such as the geometry of the organosilane, may also influence the stereochemical pathway.

1. Influence of the leaving group

a. Stereochemistry governed by the nature of the leaving group. Table 13 summarizes results for diverse nucleophiles. The following comments may be made:

(1) Hydride is displaced from a Si–H bond with retention of configuration. Inversion occurs only in reactions with Ph_2CHLi[118].

(2) Chloro- and bromosilanes undergo mainly inversion.

(3) The Si–F and Si–SR bonds show parallel behaviour, leading to either inversion or retention according to the nature of the nucleophile.

(4) OR is displaced from the Si–OR bond mainly with retention by organolithiums, except for some charge-delocalized reagents such as $PhCH_2Li$ and Ph_2CHLi. Similar behaviour is observed with Grignard reagents.

Bifunctional organosilanes, 1-NpRSi(Y)X (R = phenyl or ferrocenyl), are good models with which to study the relative ability of the leaving groups (X and Y) to be displaced. Typical reactions are shown in Table 14.

TABLE 13. Stereochemical behaviour of monofunctional organosilanes in the substitution 1-NpPh(R)Si–X + NuM → 1-NpPh(R)Si–Nu + MX

Substrate	Nucleophile (solvent)[a]	Predominant stereochemistry[c]	References
1-NpPh(R)Si–H (R = Me, Et, *i*-Pr, Vi)	R^1Li (R^1 = Me, Et, Pr, Bu, *p*-$CH_3OC_6H_4CH_2$, allyl, $PhCH_2$)	RN	118, 119
	Ph_2CHLi	IN	118
	NH_2Li or NH_2Na	RN	118
	KOH (xylene)	RN	110, 126
	t-BuOK (*t*-BuOH)	RN	1
1-NpPh(R)Si–OMe (R = Me, Et, *i*-Pr, Vi) and PhMe(neo-C_5H_{11})Si–OMe	R^1Li (R^1 = Me, Et, Bu, *p*-$CH_3OC_6H_4CH_2$, allyl[b])	RN	118, 119
	R^2Li (R^2 = $PhCH_2$, Ph_2CH)	IN	118, 119
	R^1MgX (R^1 = Me, Et)	RN	120, 123
	$PhCH_2MgX$	IN	118, 123
	$LiAlH_4$	RN	1, 110, 123
	KOH (xylene)	RN	1, 110, 127
1-NpPh(R)Si–F (R = Me, Et, *i*-Pr, Vi)	R^1Li (R^1 = Me, Et, Pr, Bu, i-Bu, *p*-MeC_6H_4, PhC≡C)	RN	119, 120
	R^2Li (R^2 = allyl, benzyl, *p*-$CH_3OC_6H_4CH_2$)	IN	119, 120
	RMgX (R = Me, Bu, allyl, benzyl)	IN	123
	$LiAlH_4$	IN	1, 124
	R_3SiOK (xylene)	RN	120

TABLE 13. (*continued*)

Substrate	Nucleophile (solvent)[a]	Predominant stereochemistry[c]	References
1-NpPh(Me)Si–X	R^1Li	RN	121, 122
X = SMe, SPh	(R^1 = Et, Bu, neo-C_5H_{11})		
	R^2Li	IN	121, 122
	(allyl, benzyl)		
	RMgX	IN	122
	(R = Et, Bu, allyl, benzyl)		
	$LiAlH_4$	IN	121, 122
	H_2O, CH_3OH	IN	121
1-NpPh(R)Si–Cl	RLi	IN	119, 120
(R = Me, Et, *i*-Pr, Vi)			
and	RMgX	IN	123
	$LiAlH_4$	IN	1, 124
PhMe(neo-C_5H_{11})SiCl	$LiNH_2$	IN	1
	PhSLi	IN	122
	$RCOO^-$	IN	1, 110, 125
	c-$C_6H_{11}NH_3^+F^-$($CHCl_3$)	IN	1, 125
	H_2O, CH_3OH	IN	1, 110, 125
	t-BuOK (*t*-BuOH)	IN	110, 125
	KOH (xylene)	IN	110, 125
	R_3SiOK (xylene)	IN	125
1-NpPh(R)Si–Br	RLi	IN	122
(R = Me, neo-C_5H_{11}) and	RMgK	IN	122
	$LiAlH_4$	IN	1, 122, 125
PhMe(R)Si–Br	H_2O, MeOH	IN	1, 125
	KOH (xylene)	IN	125
(R = neo-$C_5H_{11}Ph_2CH$, Et)	*t*-BuOK (*t*-BuOH)	IN	125
	R_3SiOK (xylene)	IN	125
	c-$C_6H_{11}NH_3^+X^-$($CHCl_3$)	IN	1, 125
	(X^- = F^-, Cl^-)		

[a] When not indicated, the solvent is anhydrous ether.
[b] Sommer and coworkers reported previously retention of configuration as the stereochemical outcome[118]. In our hands, the same reaction led, several times, to predominant retention (85% RN).
[c] RN denotes retention; IN denotes inversion.

The (H) substituent appears to be the least effective leaving group with 1-NpPhSiH(OR) and 1-NpFcSi(H)F derivatives[129]. Only displacement of F and OR occurs; so both of these are better leaving groups than (H). At the opposite extreme, the chlorine substituent is the best leaving group. With 1-NpPhSi(Cl)OMen and 1-NpFcSi(F)Cl, displacement of Cl only occurs with inversion, whatever the nucleophile. Comparison of F and OR substituents is more complex[129]. However, when the reaction takes place with inversion, only highly stereoselective displacement of the fluorine atom occurs.

Finally, we note that acyclic fluorosilanes show a general shift of the stereochemistry from retention to inversion of configuration, as indicated below.

$$R^1R^2Si(H)F > R^1R^2R^3SiF > R^1R^2Si(F)OR^3$$
$$RN \longrightarrow IN$$

TABLE 14. Influence of the leaving group: the behaviour of bifunctional organosilanes

Substrate	Nucleophile ($S = Et_2O$)	Products	Predominant stereochemistry[a]	References
1-NpPh(Cl)Si–OMen	EtLi ($R = Me$, Bu, p-$CH_3C_6H_4$, $CH_2=CHCH_2$)	1-NpPh(R)Si–OMen	IN	2, 23
	RMgX ($R = Me$, Et, p-$MeOC_6H_4$, $CH_2=CHCH_2$, $PhCH_2$)	1-NpPh(R)Si–OMen	IN	2, 23
1-NpPhSi(F)–OMen	RLi ($R = p$-MeC_6H_4, $CH_2=CHCH_2$, $PhC\equiv C$)	1-NpPh(R)Si–OMen	IN	29
	$CH_2=CHCH_2MgBr$	1-NpPh(allyl)Si–OMen	IN	29
1-NpPh(H)Si–OMen	RLi ($R = Me$, Et. Bu. p-$CH_3C_6H_4$, $CH_2=CHCH_2$)	1-NpPh(H)Si–R	RN	2, 23
	R^1MgX ($R^1 = Me$, Et, Pr, Bu, p-$MeOC_6H_4$)	1-NpPh(H)Si–R^1	RN	23
	R^2MgX ($CH_2=CHCH_2$, $PhCH_2$)	1-NpPh(H)Si–Rx	IN	2

1-NpFc(F)Si–Cl	RLi or RMgX ($R = Et$, Ph, $CH_2{=}CHCH_2$)	1-NpFc(F)Si–R	IN	24
1-NpFcSi(F)OEt	RLi ($R = Ph$, $CH_2{=}CHCH_2$, Ph–C≡C)	1-NpFc(R)Si–OEt	IN	29, 129
	$CH_2{=}CHCH_2MgX$	1-NpFc($CH_2{=}CHCH_2$)Si–OEt	IN	29, 128
1-NpFc(H)Si–F	RLi ($R = Me$, Et, Ph, $CH_2{=}CHCH_2$, Ph–C≡C)	1-NpFc(H)Si–R	RN	24
	R^1MgX ($R = Me$, Et, $CH_2{=}CHCH_2$)	1-NpFc(H)Si–R^1	IN	24
	R^2MgX ($R = Ph$, Ph–C≡C)	1-NpFc(H)Si–R^2	RN	24

[a] IN denotes inversion; RN denotes retention.

This stereochemical shift is illustrated by the data in Table 15.

TABLE 15. A comparison of the stereochemical behaviour[a] of acyclic fluorosilanes in the reaction of RM with $R^1R^2R^3SiF$ (after References 29, 119, 120 and 130)

RM	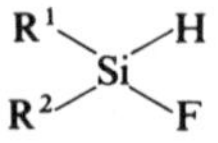	$R^1R^2R^3Si–F$	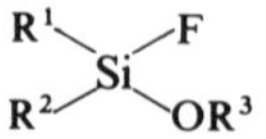
AllylLi, PhMgBr, Ph–C≡C–MgBr	RN	IN	IN
ArylLi, Ph–C≡C–Li	RN	RN	IN

[a] RN denotes retention; IN denotes inversion.

The mutual influence of one of the two leaving groups on the other changes its lability and thereby the stereochemistry of its displacement by nucleophiles.

b. A rationalization of the dominant influence of the leaving group. The experimental facts above show, in the main, parallel chemical and stereochemical behaviour for mono- and bifunctional silanes. The relative ease of displacement is: Br, Cl > F, SR > OR > H.

This empirical order parallels a general change of the stereochemistry of displacement from inversion to retention:

$$\begin{array}{ll}\text{ability to be cleaved: OTs, OCOR}^{131}\text{, Br, Cl} > \text{F, SR} > \text{OR} > \text{H} \\ \text{stereochemistry} \quad \text{IN} \longrightarrow \text{RN}\end{array}$$

Attempted rationalization of the stereochemical changes in terms of the pK_a of the conjugate acid of the leaving group[1] or in terms of its electron polarizability according to Brewster (I > Br > Cl; SH ~ Br > Cl; OH ~ OR > H > F)[132] does not give a satisfactory picture of the stereochemical changes observed:

(1) The Si–SR and Si–F bonds, for example, show very similar stereochemical behaviour leading to either retention or inversion (Table 13). Yet, the SR leaving group (the corresponding conjugate acid, RSH, has a $pK_a > 10$) should be a poor leaving group and undergo displacement mainly with retention of configuration, whereas the fluoro substituent (HF, $pK_a \simeq 3.5$–4) should be a good leaving group and undergo inversion.

(2) The highly polarizable Si–Cl and Si–SR bonds are always displaced with inversion. If Si–F is compared to Si–OR and Si–H bonds, Si–F is far less polarisable; nevertheless the Si–F bond is more easily cleaved by nucleophiles with inversion of configuration (Table 13), instead of retention as would be expected. In fact, Si–F resembles Si–SR in stereochemical behaviour, as already mentioned.

However, if polarisability is defined as the ability of the Si–X bond to be stretched under the influence of an attacking nucleophile, the relative tendency of Si–X bonds to undergo inversion and their order of polarisability run parallel. This dependence is well illustrated by the following data:

(1) The X-ray crystal structure of **2** shows the chlorine atom in apical position *trans* to the oxygen[133]; the Si–Cl bond is considerably longer (2.348 Å instead of 2.01 Å[134] in

tetravalent organosilanes). A similar pronounced lengthening of the Si–Cl_{ax} bonds was observed in the case of other pentacoordinated complexes of silicon (**3**: 2.269 Å; **4**: 2.207 Å; **5**: 2.150 Å) vs Si–Cl_{eq} bonds (**4**: 2.049 Å; **5**: 2.094 Å)[135]. In contrast, for the fluorine-containing complexes **6**[136] and **7**[135] no significant differences are observed in the bond lengths between Si–F_{ax} and Si–F_{eq}. Thus if Si–F is compared with Si–Cl and Si–Br, these observations suggest that the Si–F bond here is less able to be stretched under the influence of an incoming nucleophile; as a consequence, the stereochemistry is displaced towards retention (Table 13).

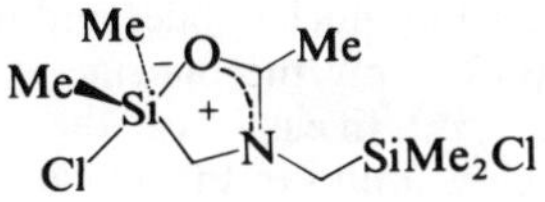

(2)

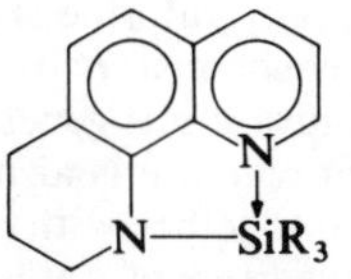

(3) $SiR_3 = SiMe_2Cl$
(4) $SiR_3 = SiMeCl_2$
(5) $SiR_3 = SiCl_3$

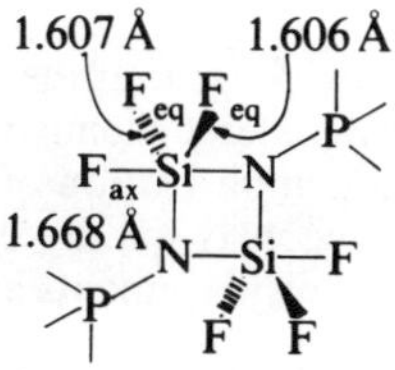

(6)

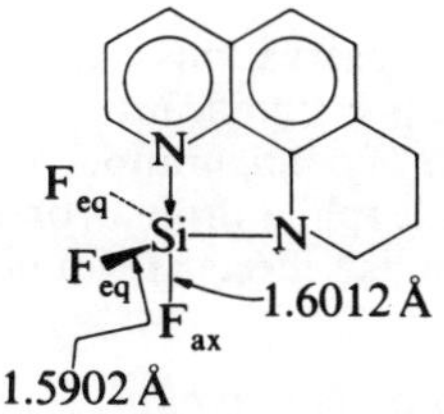

(7)

(2) The ^1H NMR spectra of *o*-$(Me_2NCH_2)C_6H_4SiX_2R$ (**8**) and *o*-$(Me_2NCH_2)C_6H_4SiXR^1R^2$ (**9**) show the expansion of coordination at silicon by an intramolecular Si←N bond.

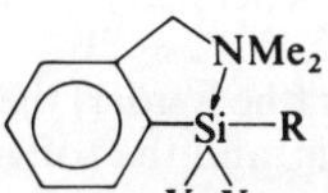

(8)

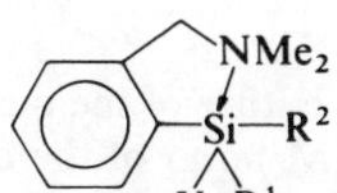

(9)

The ease of pentacoordination may be ranked as follows[137]:

$$X = H, OR < F, SR < OAc, Cl, Br$$

In the case of the *o*-$(Me_2NCH_2)C_6H_4Si(1\text{-}Np)H_2$, although the diastereotopism of the NMe_2 group was not observed, the ^{29}Si chemical shift is displaced 11.6 ppm upfield (at 30 °C) indicating the formation of a Si←N coordinative bond[138]. This behaviour can be explained if the two hydrogen atoms occupy equatorial positions. An X-ray investigation of two pentacoordinated silicon hydrides shows that the Si–H bonds occupy equatorial positions in the trigonal–bipyramidal structures[139]. These results show the great ability of the Si–H bond to be in an equatorial position: this observation has to be connected with the retention of configuration in nucleophilic displacements of the Si–H bond (Table 13).

The low-temperature ^{19}F NMR spectra of bifunctional organosilanes o-(Me_2NCH_2)C_6H_4SiXFR lead to the following relative orders of apicophilicity for groups X bonded to a pentacoordinate silicon atom versus the apicophilicity of F[140]:

$$Cl > F, \quad F > OR \quad \text{and} \quad F > H$$

It is noteworthy to find chlorine more apicophilic than fluorine, although the latter is more electronegative. This may explain the stereochemical behaviour of 1-NpFcSi(Cl)F[24]; only chlorine is displaced by carbon nucleophiles and inversion is the predominant stereochemistry. In 1-Np(Fc)SiF(OR), only the fluorine is displaced and inversion of configuration is generally observed[29, 128]. In each case, the group displaced in the substitution reaction is also the one which is found in the apical position.

The above data, together with previous data[2, 7], suggest a close relationship between the rate of racemization of halosilanes, the substitution of R_3Si–X with inversion of configuration and the ease of pentacoordination at a silicon atom:

Pentacoordination ability H, OR < F, SR < OAc, Cl, Br

Predominant stereochemistry RN ⟶ IN

Racemization ability Si–Br > Si–Cl > Si–F

Thus the above observations can be taken as good evidence that the ease of stretching of the Si–X bond is a significant factor in directing the stereochemistry of substitution towards inversion of configuration. They are a good demonstration of the fact that the attack of the nucleophile (frontal or back-side) at silicon is mainly controlled by the leaving group. Similar effects are found in nucleophilic displacements at phosphorus[141].

2. *Influence of the nucleophile*

The stereochemistry at silicon is extremely sensitive to the electronic character of the nucleophile (Table 13). The percentage of retention or inversion is quite sensitive to small changes in the structure of the anion, the metal, or the solvent (Table 16).

(1) The harder the organic anionic group (alkyl or p-methoxyphenoxide anions), the closer the stereochemistry is to pure retention; the softer the nucleophile (allyl, benzyl or p-nitrophenoxide anions), the more the stereochemistry tends towards inversion (Tables 13 and 16).

(2) Concerning the nature of the metal, the softer (the harder) the cation M^+, the more covalent (ionic) the RM bond of the organometallic, and the softer (harder) the organic anionic group R[145], the greater the extent of inversion (retention). Thus, the organolithiums, which are harder than the analogous Grignard reagents, react mainly with retention whereas the Grignard reagents react with inversion (Tables 13 and 16).

(3) An increase in the solvating power of the solvent (Et_2O < THF < DME) modifies the electronic character of the nucleophile, and hence the stereochemical course of reaction. Thus a Grignard reagent is harder in THF or DME than in ether (Scheme 15);

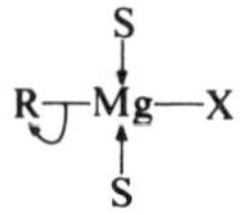

SCHEME 15. Effect of a coordinating solvent (S) in reaction with Grignard reagents

TABLE 16. Influence of the nucleophile on stereochemistry

Organosilane ($R^1R^2R^3Si$–X)	X	Nu	Solvent	Product (stereochemistry)[a]	Ref.
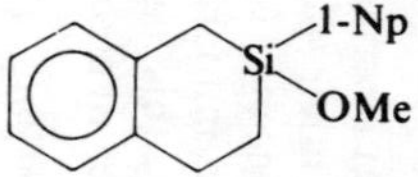	OMe	PrMgBr	Et_2O	$R^1R^2R^3Si$–Pr etc. (RN)	142
			Et_2O	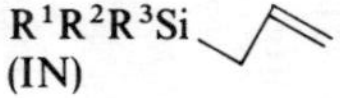	142
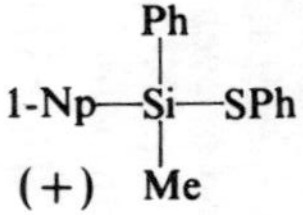	SPh	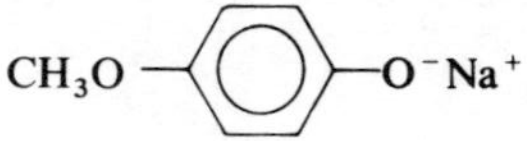	THF	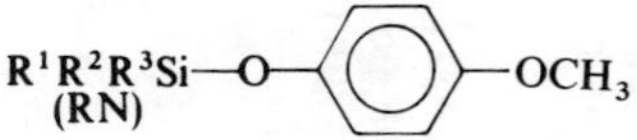	143
		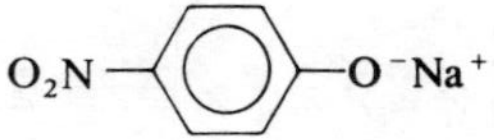	THF	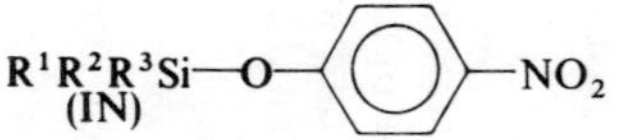	143
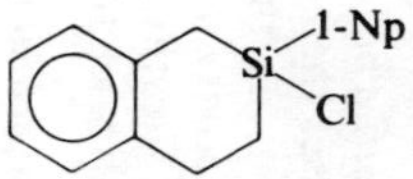	D	EtLi	Et_2O	$R^1R^2R^3Si$–Et (RN)	144
		EtMgBr	Et_2O	$R^1R^2R^3Si$–Et (IN)	142
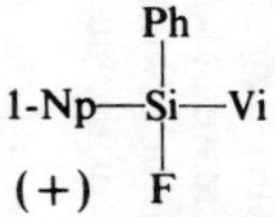	F	EtMgBr	Et_2O	$R^1R^2R^3Si$–Et (IN)	123
		EtMgBr	DME	$R^1R^2R^3Si$–Et (RN)	123

[a] RN denotes retention; IN denotes inversion.

the more coordinating solvents polarise the carbon–metal bond and increase the hardness of the nucleophile, orienting the stereochemistry towards retention of configuration[122, 146].

We will now give a more detailed analysis of the available data, using the following four cases: 1. Coupling reactions with phenoxides: influence of the electronic character and the effect of ion-pair dissociation. 2. Coupling reactions with alkoxides: influence of solvation. 3. Nucleophilic substitution at silicon by carbon nucleophiles. 4. Reduction reactions by alanes, AlH_nY_{3-n}.

a. Coupling reactions with phenoxides: influence of the electronic character. The electronic character of the negative charge on the phenoxide oxygen atom can be changed by varying the substituent Y in the *para* position. With the same nucleophilic (oxygen) atom the stereochemistry depends greatly on the *para* substituent Y and on the extent of ion-pair dissociation (Table 17).[143]

TABLE 17. Stereochemistry of reactions of R_3SiX with phenoxides in THF

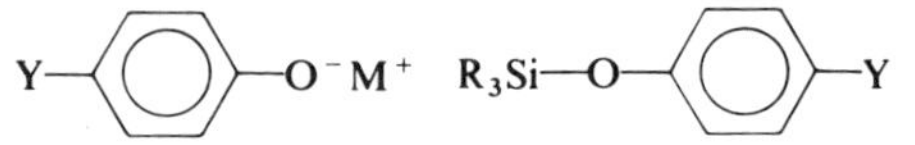

R_3Si–X	Y	M^+	$[\alpha]_D$ (deg)	Stereochemistry[b]
Ph	MeO	Na^+	+6	RN
|	MeO	Na^{+a}	+9	RN
1-Np–Si–SPh	MeO	NR_4^+	+8	RN
|	H	Na^+	+2	IN
Me	H	Na^{+n}	+5	IN
(+)	NO_2	Na^+	+3	IN
	NO_2	Na^{+a}	+8	IN
	NO_2	NR_4^+	+5	IN
Ph	MeO	Na^+	−4.5	RN
|	MeO	Na^{+a}	−6	RN
1-Np–Si–Me	H	Na^+	+3	RN
|	H	Na^{+a}	−2.5	IN
F	NO_2	Na^+	+0.5	RN
(+)	NO_2	Na^{+a}	−4	IN

[a] Reactions carried out with a cryptand specific for Na^+.
[b] RN denotes retention; IN denotes inversion.

(i) Influence of the substituent on the phenyl group. An electron-releasing substituent (such as MeO) that increases the negative charge on the nucleophilic centre and thus its hardness favours retention as the stereochemical outcome. On the other hand, a significant decrease of selectivity in retention or even predominant inversion occurs with an electron-withdrawing substituent such as NO_2. In this case, there is a greater delocalization of the negative charge, resulting in a softer nucleophile.

(ii) Influence of the dissociation of the ArO^-M^+ ion pair. The effect of a variable delocalization of the negative charge is increased when the reaction is carried out in a basic solvent[143] or with a cryptand specific for the Na^+ cation (Table 17). With *p*-methoxyphenoxide, dissociation of ion pair leads to increase of the negative charge on

the oxygen atom and the proportion of the retention increases. On the other hand, with a softer anion, such as *p*-nitrophenoxide, complexing the Na^+ cation favours the delocalization of the negative charge and thus favours inversion. The same observation is made when Na^+ is substituted by the bulky NR_4^+ cation: dissociation of the ion pair is increased and once again we see the above stereochemical shift towards greater retention with the OMe group and towards greater inversion with the NO_2 group.

b. Coupling reactions with alkoxides: influence of the solvation. Concerning the influence of solvation, pertinent data were reported by Sommer and Fujimoto[147] for nucleophilic substitution by alkoxides at Si–OMe, Si–F and Si–SMe bonds (Table 18).

TABLE 18. Coupling reactions with alkoxides

$$R_3Si{-}X + BuOH/BuOLi \xrightarrow{\text{benzene}} R_3SiOBu$$

Organosilane[a]	ROH/ROM (mol/mol)	ROH/R_3SiX (mol/mol)	Predominant stereochemistry[b]
R_3Si–OMe	10	2.6	100% RN
	72	47	65% IN
	210	79	>81% IN
R_3Si–F	2	1.31	92% RN
	14.3	1.11	62% RN
	32	1.37	88% IN
R_3Si–SMe	0.7	1.15	70% RN
	3	1.13	61% IN
	39	1.22	85% IN

[a] R_3Si = 1-NpPhMeSi.
[b] RN denotes retention; IN denotes inversion.

The most important fact is the stereochemical crossover from retention to inversion upon increasing the alcohol content of the medium. Whatever the nature of the leaving group (OMe, F, SMe), inversion becomes predominant. The same shift of the stereochemistry towards inversion is observed with BuONa[147]. The most striking data are observed with R_3Si–F and R_3Si–SMe: for nearly constant ROH/R_3Si–X proportions of 1:1 to 1:4, a change in the ROH/ROM proportion (from 1 to 40) is enough to invert the stereochemistry.

These observations suggest an explanation in terms of electronic factors. The RO^-M^+ species in benzene have a localized negative charge on the oxygen atom, that is, they act as hard anions and react with retention. On increasing the alcohol ratio, the charge is dispersed by strong hydrogen bonding interactions with the solvent. Such dispersal of the negative charge produces a softer species that reacts with predominant inversion.

c. Substitution at silicon by carbon nucleophiles. In the case of carbon organometallics, it is readily possible to modify the electronic character of the nucleophile while preserving the same reacting atomic centre.

(i) Influence of the nature of R^-. Pertinent data reported in the literature are summarized in Table 19.

TABLE 19. Influence of the nature of the organic anion

Substrate	Nucleophile	Predominant[a] stereochemistry	Ref.
R	Ph–C≡C–Li	100% RN	16
|	MeLi	93% RN	16
1-NpPhSi–F	PrLi	97% RN	16
(R = Et or Vi)	BuLi	93% RN	16
		98% IN	16
	EtMgBr	65% IN	123
	$PhCH_2Li$	82% IN	16
	MgBr ether	93% IN	123
Vi	$PhCH_2MgCl$	91% IN	123
|	PhC≡CMgBr (THF)	76% RN	123
1-NpPhSi–F	BuMgBr (THF)	59% RN	123
	MgBr THF	62% IN	123

[a] RN denotes retention; IN denotes inversion.

Aryl-, phenylethynyl- and alkylorganolithiums, namely species with a well-localized negative charge, react mainly with retention. Inversion is observed only with the best leaving groups, Cl or Br[16, 118–122].

In contrast, allyl- or benzyllithiums, in which the negative charge is delocalized over an sp^2 carbon system, always react with inversion, except with hydrogen, the poorest leaving group (Table 13). The Ph_2CH^- anion, in which the negative charge is particularly well-delocalized, leads only to inversion, even with hydrogen as leaving group[118].

Quite significant also is the case of the *p*-$MeOC_6H_4CH_2^-$ anion. *p*-$MeOC_6H_4CH_2Li$ cleaves the Si–OMe bond with retention in contrast to the inversion observed for $C_6H_5CH_2Li$[118]. The *p*-methoxybenzyl anion, in which the negative charge is more localized on the reactive carbon centre because of the electron-donor methoxy group, is a harder reagent than the benzyl anion. Consequently, the stereochemistry is shifted towards retention.

(ii) Influence of the metal. Concerning the nature of the metal, the softer (harder) the cation M^+, the more covalent (ionic) is the R–M bond of the organometallic and the softer (harder) is the nucleophile. Thus the organolithiums, RLi, will be harder reagents than the analogous Grignard reagents, RMgX. Organolithiums react mainly with retention whereas Grignard reagents react with inversion (Table 20).

As might be expected, organolithium and organosodium compounds behave similarly to each other[119].

(iii) Influence of the solvent and complexing agents (TMDA, cryptands). The stereochemistry (and mechanism) for the same leaving group is extremely sensitive to the solvent or to a complexing agent such as TMDA (tetramethylethylenediamine) or cryptands. This is specially so in the case of borderline leaving groups, such as fluoride, which can be displaced either with retention or inversion. In the case of Grignard reagents, nucleophilic solvents such as DME or THF promote retention of configuration (or decrease the percentage of inversion) (Table 21).

TABLE 20. Comparison of the stereochemical behaviour of R'Li and R'MgX

Substrate	Nucleophile	Predominant stereochemistry[a]	Ref.
1-NpPh(R)Si–F (R = Me, Et, *i*-Pr, Vi)	R'Li (R' = Me, Et, Pr, Bu)	RN	16, 86
	R'MgX (R' = Me, Bu)	IN	123
1-NpFc(H)Si–F	RLi (R = Me, Et, CH_2=CH–CH_2)	RN	24
	RMgX (R = Me, Et, CH_2=CH–CH_2)	IN	24
1-NpPhMeSi–X (X = SMe or SPh)	RLi (R = Et, Bu, neo-C_5H_{11})	RN	121, 122
	RMgX (R = Et, Bu)	IN	121, 122
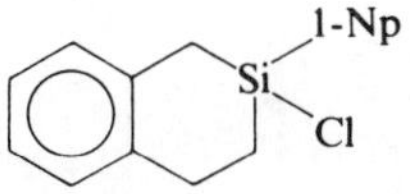	RLi (R = Me, Et, Pr)	RN	144
	RMgX (R = Me, Et, Pr, Bu)	IN	142

[a] RN denotes retention; IN denotes inversion.

TABLE 21. Influence of the solvent on the stereochemistry of substitution of 1-NpPhViSiF by Grignard reagents (after Reference 123)

Nucleophile	Solvent	Predominant stereochemistry[a]
EtMgBr	Et_2O	⩾65% IN
	THF	66% IN
	DME	59% RN
BuMgBr	Et_2O	⩾67% IN
	THF	⩾59% RN
	DME	⩾71% RN
Bu_2Mg	Et_2O	⩾69% IN
	THF	⩾75% RN
	DME	⩾74% RN

[a] RN denotes retention; IN denotes inversion.

Moreover, for all Grignard reagents used, whether there is retention or inversion or a transition from one to the other, a net rate increase is observed when the basicity of the solvent is increased in the order: $Et_2O < THF < DME$ (Table 22).

The influence of complexing agents, such as TMDA or a cryptand, is well illustrated in coupling reactions between organosilanes **12**–**14** and organolithiums[149] (Table 23).

TABLE 22. Stereochemistry and rates of reactions of **10** and **11** with Grignard reagents in various solvents (after Reference 148)

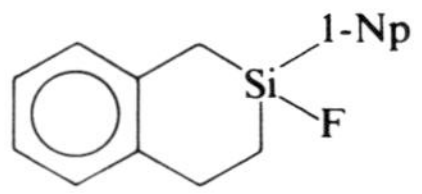

(10) and 1-NpPhViSiF **(11)**

Substrate	Organometallic (solvent)	Stereochemistry[a]	Absolute rate constant (min^{-1})	k_{THF}/k_{Et_2O}	k_{DME}/k_{THF}	k_{DME}/k_{Et_2O}
10	$PrMgBr(Et_2O)$	RN	6×10^{-4}	75	24	1800
	PrMgBr(THF)	RN	4.5×10^{-2}			
	PrMgBr(DME)	RN	1.1			
10	$BuMgBr(Et_2O)$	RN	6×10^{-4}			1300
	BuMgBr(DME)	RN	0.8			
11	$PrMgBr(Et_2O)$	IN	1.7×10^{-4}	36	2.8	100
	PrMgBr(THF)	RN	6.2×10^{-3}			
	PrMgBr(DME)	RN	1.7×10^{-2}			
11	$BuMgBr(Et_2O)$	IN	5×10^{-5}			200–300
	BuMgBr(DME)	RN	1×10^{-2}			
11	$PhCH_2MgCl(Et_2O)$	IN	5.4×10^{-4}	215	>50	$>10^{-4}$
	$PhCH_2MgCl(THF)$	IN	0.11			
	$PhCH_2MgCl(DME)$	IN	>5			

[a] RN denotes retention; IN denotes inversion.

TABLE 23. Percentage of retention in the substitution of organosilanes by organolithiums in the presence of complexing agent

	Organosilane			
Reagent	**10**	**12**	**13**	**14**
$EtLi/K^{+}_{Li}Et_2O^{a}$	100%	98%	95%	
$EtLi/TMDA/Et_2O^{b}$	99%	96%	91%	
$EtLi/Et_2O$	96%	90%	90%	
EtLi/benzene		76%	70%	
$BuLi/K^{+}_{Li}/Et_2O^{a}$		100%	98%	86%
$BuLi/TMDA/Et_2O$		90%	82%	65%
$BuLi/Et_2O$		86%	80%	65%
BuLi/heptane		68%		

[a] K^{+}_{Li} = Kryptofix 211, (tetraoxa-4,7,13,18-diaza-1,10-bicyclo[8,5,5]eicosane), specific for Li^+ cation.
[b] TMDA = tetramethylethylenediamine.
[c] Predominant sterechemistry; the $[\alpha]_D$ of optically pure R_3SiEt and R_3SiBu are known.

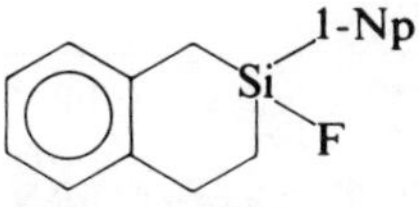
$[\alpha]_D = -34°$
(10)

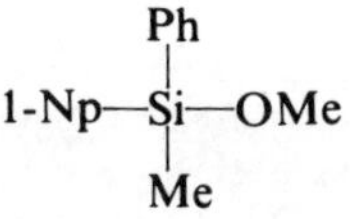
$[\alpha]_D = -16.5°$
(12)

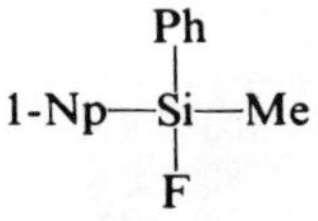
$[\alpha]_D = +47°$
(13)

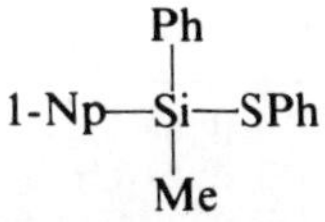
$[\alpha]_D = +27°$
(14)

Non-polar solvents, such as benzene or heptane, in which alkylithiums are more aggregated, increase the proportion of inversion. At the opposite extreme, complexing agents for the Li^+cation, such as TMDA or, better, K^{+}_{Li} which generates free anions[150], promote retention of configuration. Moreover, the addition of the lithium cryptand results in a 10^3–10^4-fold rate acceleration[149].

Clearly, these data are in contrast to a quasicyclic mechanism (S_Ni–Si, Scheme 16)[1].

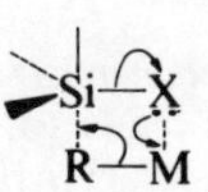

SCHEME 16. S_Ni–Si mechanism

In such a process, the electrophilic assistance of the counterion M^+ determines the apical attack of the nucleophile with the leaving group in an equatorial position, and thus governs the stereochemistry. The use of a more basic solvent or of a cryptand would lead to a shift of the stereochemistry to inversion and to a rate decrease.

In contrast, it is obvious that an increase in the solvating power of the solvent (Et_2O < THF < DME) modifies the electronic character of the nucleophile. A solvent which coordinates more strongly at the metal polarises the carbon–metal bond and increases the hardness of the nucleophile, orienting the stereochemistry towards retention of configuration (Table 21).

Similarly, a cryptand or TMDA increases the dissociation of the R^-Li^+ ion pair and consequently increases the negative charge on the reactive carbon centre. Thus when R is an alkyl group, its hardness is increased and the stereochemistry is directed to retention (Table 23).

d. Reduction. Table 24 gives stereochemical data for reductions of optically active silanes **10, 12, 13** and **15** by alanes, AlH_nY_{3-n}[151]. The substituent and solvent effects on the reactivity of alanes are discussed below.

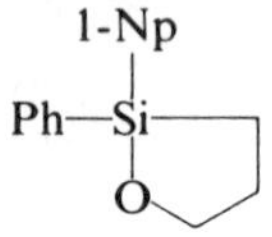

(15)

TABLE 24. Reduction reactions of optically active organosilanes by alanes

Reagent[a] (solvent)	Predominant stereochemistry[b] 10	12	13	15
Dibal (hexane)	100% RN	100% RN[152]	98% RN[152]	†α]$_D$ = +19° RN
Dibal (Et_2O)	80% RN	99% RN[152]	90% IN[152]	†α]$_D$ = +16° RN
Dibal (TMDA-hexane)	70% RN	95% RN[152]	90% IN[152]	†α]$_D$ = +15° RN
Dibal (THF)				$[\alpha]_D$ = +10° RN
AlH_3 (Et_2O)	100% RN	92% RN	96% IN	$[\alpha]_D$ = +16° RN
AlH_3(THF)	70% IN	90% RN	100% IN	$[\alpha]_D$ = + 8° RN
$(t\text{-BuO})_2AlH(Et_2O)$	75% IN	90% RN	100% IN	
$(t\text{-BuO})_2AlH$(THF)	80% IN	90% RN	100% IN	$[\alpha]_D$ = −14° IN
$(EtS)_2AlH$(THF)	85% IN	85% RN	100% IN	$[\alpha]_D$ = −16° IN
H_2AlI(THF)	80% IN	55% RN		$[\alpha]_D$ = −18° IN

[a] Dibal denotes di-isobutylaluminium hydride.
[b] RN denotes retention; IN denotes inversion.

(i) Effect of substituents on the stereochemistry of reduction. Substitution of hydrogen atoms in AlH_3 by OR or SR groups leads to increased inversion when the resulting reagent attacks at silicon. In the case of H_2AlI, there is very high selectivity towards inversion of configuration. A change in the Y group leads to a change in the electronic character of the Al–H bond and consequently to a change in the stereochemistry.

(ii) Solvent effects. The nature of solvent effects is also very clear. For AlH_3, $(t\text{-BuO})_2AlH$ and $(EtS)_2AlH$, an increase in the basicity of the solvent leads to increased inversion at silicon (Table 24). $(i\text{-Bu})_2AlH$ is especially interesting in this respect. In hexane, it substitutes all Si–X bonds (X=Cl, SR, F and OR) with retention[152]. In contrast, the percentage of inversion at silicon increases with the basicity of the solvent (hexane < Et_2O < THF) or by the use of a complexing agent such as TMDA. In the presence of a [2.1.1]cryptand, $LiAlH_4$ in Et_2O (naked AlH_4^- anions) gives inversion at silicon whereas in the absence of the macrocyclic ether $LiAlH_4$ leads to predominant retention (Table 25)[149, 153].

TABLE 25. Change in the stereochemistry of reduction of organosilanes by $LiAlH_4$ by added cryptand

Reagent (solvent)	Predominant stereochemistry[a]	
	12	**10**
$LiAlH_4$ (Et_2O)	95% RN	Rac
$LiAlH_4/K^+_{Li}$ (Et_2O)	75% RN	65% IN

[a] RN denotes retention; IN denotes inversion.

Thus, the more localized on the hydrogen atom the negative charge, the easier is the inversion at silicon. This fact is also well illustrated by the behaviour of $(i\text{-Bu})_2AlH$ or AlH_3 in hexane as solvent. The Al–H bond is slightly polarized and these species behave as hard reagents. On the other hand, when used in a basic solvent, the coordination of the solvent at the aluminium atom delocalizes the negative charge on hydrogen (Scheme 17).

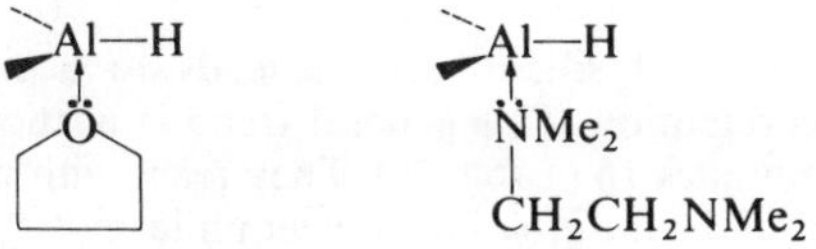

SCHEME 17. Effect of the solvent or a complexant in reactions with alanes

Consequently, under these conditions, $(i\text{-Bu})_2AlH$ and AlH_3 behave as soft species and the inversion-to-retention ratio at silicon increases.

3. *Stereochemical changes with angle strain at silicon*

The stereochemical data reported for optically active 1-naphtylphenylmethylsilanes can be generally extended to other acyclic R_3Si–X systems. The examination of the stereochemical data for derivatives in which the substituents on silicon encompass significant variation in their steric and polar effects leads to the following results:

(1) It seems clear that the introduction of R groups of moderate to large steric requirements leaves the stereochemistry unchanged[1,2] (cf. Table 13).

(2) Concerning possible polar effects on the stereochemistry, comparison of the 1-NpPhMeSiX and 1-NpC_6F_5MeSiX systems is interesting. Although the pentafluorophenyl group is a powerful electron-withdrawing group relative to phenyl, the stereochemistry of the two systems do not differ to any significant extent[1,22]. A similar comparison can be made between the 1-NpPhEtSi–X and 1-NpPhViSi–X derivatives.

On the other hand, when the silicon is included in a ring, particularly in a strained ring[154], the stereochemical behaviour is strongly influenced by the geometry of the substrate. The stereochemistry of nucleophilic displacements of both exocyclic leaving groups (attached to a silicon-containing ring) and endocyclic leaving groups (forming part of the sila-ring) has been reported[2,155].

a. Stereochemical behaviour of systems carrying exocyclic leaving groups. Important changes in stereochemistry compared to acyclic silanes were reported with the cyclic systems **16**–**20** (Table 26). These stereochemical data lead to the following conclusions:

(**16**)

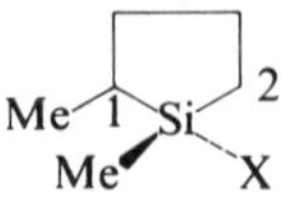

(**17**)

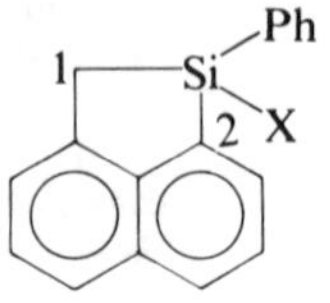

(**18**)

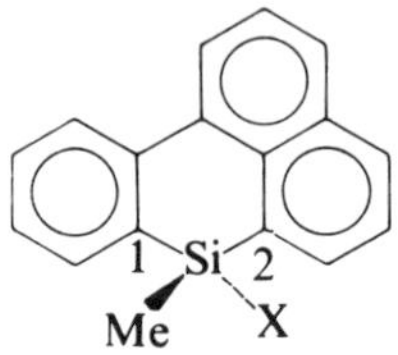

(**19**)

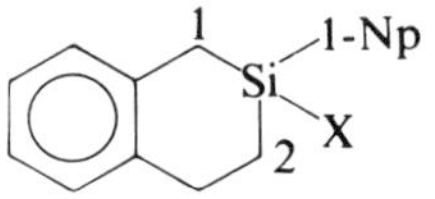

(**20**)

(1) Increased angle strain at silicon always leads to a significant change of the stereochemistry towards retention. This general trend is particularly evident with most angle-strained silacyclobutanes **16** (Table 26). They react with predominant retention of configuration, regardless of the nature of the nucleophile[155, 156]. Even coupling reactions between **16** (Si–Cl) and $LiAlH_4$ or Grignard reagents occur with complete retention of configuration, whereas the same reactions carried out with 1-NpPhMeSi–Cl lead to complete inversion (Reference 1 and Table 13).

(2) A small degree of angle strain is enough to bring about such a stereochemical change 2-(1-Napthyl)-2-sila-1,2,3,4-tetrahydronaphthalene, **20**, shows significant differences compared with 1-NpPhMeSi–X (**21**) (Table 27).

Compound **20** (X = Cl) reacts with alkyllithiums with complete retention, in contrast to the inversion observed with R_3Si–Cl; allyllithium or benzyllithium and alkyl Grignard reagents also cleave the Si–F bond of **20** (X = F) with retention rather than inversion as observed with 1-NpPhMeSi–F.

(3) Ring strain does not change the main factors which govern the stereochemistry, that is, the effects of the nature of the leaving group and the electronic character of the nucleophile.

b. Stereochemical behaviour of systems carrying endocyclic leaving groups[33, 35]. Stereochemical data for compounds **22** and **23** are summarized in Tables 28 and 29 and compared with those observed with **14**. Two points are noteworthy: (i) Cyclic compounds behave similarly to their acyclic analogues. Again we find a dominant influence of the leaving group on the stereochemistry. The SR groups in **22** and **23** (Table 29)

1-NpPhMeSi–SMe

(**22**)

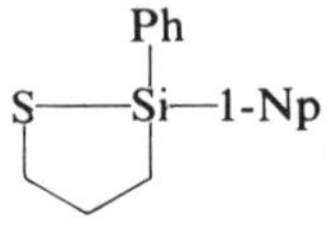

(**23**)

TABLE 26. Stereochemical behaviour of exocyclic leaving groups[a]

Substrate	Angle (C_1–Si–C_2)	$LiAlH_4$	RLi R = alkyl	RLi R = aryl	RMgX R = alkyl	RMgX R = aryl	References
16 (Si–Cl)	<90° [156]	RN	—	—	RN	RN	157, 158
16 (Si–F)	<90°	RN	RN	—	RN	—	157, 158
16 (Si–OR)	<90°	RN	RN	—	RN	—	157, 158
18 (Si–Cl)	93.4°	RN	—	—	—	—	19
19 (Si–Cl)	≃90°	IN	—	—	—	—	18
19 (Si–F)	≃90°	RN	—	—	—	—	18
17 (Si–Cl)	92°–96° [160]	IN	—	IN	IN	—	159
17 (Si–F)	92°–96°	RAC	—	—	—	—	159
17 (Si–H)	92°–96°	—	—	RN	—	—	159
20 (Si–Cl)	≃105° [161]	IN	RN	RN	IN	RN	143, 144
20 (Si–F)	≃105°	RAC	RN	RN	RN	RN	143, 144
20 (Si–OR)	≃105°	RAC	RN	RN	RN	RN	143, 144
20 (Si–H)	≃105°	—	RN	RN	—	—	143, 144
R_3Si–Cl	≃109° [161]	IN	IN	IN	IN	IN	1, 2
R_3Si–F	≃109°	IN	RN	—	IN	IN	1, 2
R_3Si–OMe	≃109°	RN	RN	RN	—	—	1, 2
R_3Si–H	≃109°	—	RN	RN	—	—	1, 2

[a] RN denotes retention; IN denotes inversion; RAC denotes racemisation.

TABLE 27. Comparison of the stereochemical behaviour of **20** and **21** with organolithium reagents RM[a]

	X = Cl		X = F		X = OMe		X = H	
RM	**20**	**21**	**20**	**21**	**20**	**21**	**20**	**21**
EtLi	63% RN	100% IN	96% RN	90% RN	94% RN	90% RN	100% RN	—
BuLi	82% RN	59% IN	98% RN	80% RN	96% RN	86% RN	100% RN	100% RN
CH_2=$CHCH_2Li$	86% RN	100% IN	95% RN	71% IN	97% RN	85% RN	85% RN	89% RN
$PhCH_2Li$	99% RN	100% IN	70% RN	85% IN	75% RN	79% IN	96% RN	90% RN

[a] RN denotes retention; IN denotes inversion.

TABLE 28. Stereochemical behaviour of the endocyclic OR leaving group. Comparison of **15** with **12** in nucleophilic substitution (after References 33 and 35)

Nucleophile	Product	Predominant stereochemistry[a]	
		12	**15**
EtLi	$R^1R^2R^3SiEt$	95% RN	—
MeLi	$R^1R^2R^3SiMe$	—	$[\alpha]_D = -\ 8°$ RN
MeMgBr	$R^1R^2R^3SiMe$	—	$[\alpha]_D = -\ 8°$ RN
$LiAlH_4$	$R^1R^2R^3SiH$	90% RN	$[\alpha]_D = +\ 13°$ RN
$LiAlH_4$/4CuI/THF	$R^1R^2R^3SiH$	55% RN	$[\alpha]_D = -\ 18°$ IN
$CH_2{=}CH{-}CH_2Li$	$R^1R^2R^3SiCH_2{-}CH{=}CH_2$	85% RN	$[\alpha]_D = -\ 5°$ IN
p-$CH_3OC_6H_4CH_2Li$	$R^1R^2R^3Si{-}CH_2C_6H_4OCH_3$-*p*	90% RN	$[\alpha]_D = -\ 4°$ IN
$C_6H_5CH_2Li$	$R^1R^2R^3SiCH_2Ph$	79% IN	$[\alpha]_D = -\ 3°$ IN
$C_6H_5CH_2MgBr$	$R^1R^2R^3SiCH_2Ph$	75% IN	$[\alpha]_D = -\ 3.7°$ IN

[a] RN denotes retention; IN denotes inversion.

TABLE 29. Stereochemical behaviour of the endocyclic SR leaving group. Comparison of reactions of **14**, **22** and **23** with nucleophiles (after reference 34)

Nucleophile	Product	Predominant stereochemistry[a]		
		22	**14**	**23**[a]
Dibal/hexane	$R^1R^2R^3Si{-}H$	RN	RN	RN
$LiAlH_4$	$R^1R^2R^3Si{-}H$	IN	IN	IN
R′Li (R′ = alkyl)	$R^1R^2R^3Si{-}R'$	RN	RN	RN
R′MgX (R′ = Me, Et)	$R^1R^2R^3Si{-}R'$		IN	IN

[a] RN denotes retention; IN denotes inversion.

favour inversion of configuration, compared to the analogous OR groups (Table 28). Charge-localized nucleophiles (alkyllithiums and alkyl Grignard reagents) and $LiAlH_4$ lead to the expected retention of configuration in the case of alkoxy leaving groups. In contrast, **23** leads to inversion with MeMgBr and $LiAlH_4$ (Table 25). Charge-delocalized nucleophiles react with inversion (Table 28). The same observations can be made with SR leaving groups (Table 29). (ii) In borderline cases, some differences are apparent. Nucleophiles, such as allyllithium, *p*-methoxybenzyllithium and the $LiAlH_4$–4CuI reagent lead to inversion in the case of the five-membered ring oxasilacyclopentane (**15**) (Table 28), whereas retention is observed with the acyclic analogue **12**.

Thus with either exocyclic or endocyclic leaving groups, ring strain does not change the dominant influences of the leaving group and of the electronic character of the nucleophile. It is only an additional factor, which leads to an overall change of the stereochemistry, only in borderline cases. A dramatic change is observed only in the case of the more highly strained four-membered rings.

C. Chiral Silyl–Transition Metal Complexes: Nucleophilic Displacement of Silyl Ligands

Stereochemical studies have brought new insights into catalytic reactions of organosilicon compounds. An approach to the mechanism of transition metal reactions has been made in the study of the stereochemistry of silyl–transition metal complexes. The chemistry of silicon, germanium and tin transition metal compounds has been the subject of several reviews[56, 57]. Optically active silyl ligands have been introduced into a transition metal complex by reaction of chiral functional organosilanes. However, the reactions are limited to a few metal centres, e.g. iron, cobalt, platinum and manganese (cf. Section II.B). The factors determining the stereochemistry of nucleophilic substitution at silicon have been discussed in Section III.B and nucleophilic displacement of silyl ligands in transition metal complexes may be regarded as a similar process with the transition metal being the leaving group (equation 37). We will discuss here experiments which seem to fit this interpretation.

$$L_nX_xM{-}SiR_3 \xrightarrow{Nu^-} L_nX_xM^- + Nu{-}SiR_3 \qquad (37)$$

The stereochemistry of the $LiAlH_4$ cleavage of Co–Si and Co–Ge bonds in $(CO)_3LCo{-}MR^1R^2R^3$ complexes has been shown to depend on the electronic environment of the transition metal centre which is induced by the nature of the ligand (Table 30)[69, 162].

The ligand L *trans* to the MR_3 group modifies the electronic character of the cobalt–silicon (germanium) bond. $(CO)_4Co^-$ is a weak nucleophile and thus behaves as a good leaving group. Nucleophilic cleavage occurs with predominant inversion at silicon and with high stereoselectivity. Substitution of one carbonyl ligand by donor ligands (which are also less effective acceptors) results in increased electron density at the cobalt atom; $(OC)_3LCo^-$ thus becomes a poorer leaving group. This results in competing substitution with retention and thus a decrease in stereoselectivity, or even in an inversion of the predominant stereochemistry, i.e. 58% retention with $L = PBu_3$ (Table 30).

In agreement with a nucleophilic process at silicon, characteristic solvent effects have been observed. $LiAlH_4$ cleavage in dimethoxyethane always took place with inversion at silicon with high stereoselectivity[69] (Table 31).

TABLE 30. Stereochemistry of the cleavage reaction

$$(OC)_3LCo{-}\overset{*}{M}R^1R^2R^3 \xrightarrow[Et_2O]{LiAlH_4} H{-}\overset{*}{M}R^1R^2R^3$$

L	Predominant stereochemistry[a] M = Si	M = Ge
CO	91% IN	87% IN
$P(OPh)_3$	91% IN	87% IN
$P(OEt)_3$		65% IN
PPh_3	68% IN	60% IN
$P(Bu\text{-}t)_3$		63% IN
$P(c\text{-}C_6H_{11})_3$	58% RN	RAC
PBu_3		58% RN
C(OEt)Bu		55% RN

[a] RN denotes retention; IN denotes inversion; RAC denotes racemisation.

TABLE 31. Percentage of inversion in the reaction

$$(CO)_3LCo\text{-}\overset{*}{M}R^1R^2R^3 \xrightarrow[\text{DME}]{\text{LiAlH}_4} H\text{-}\overset{*}{M}R^1R^2R^3$$

(after Reference 69)

L	% Inversion	
	M=Si	M=Ge
CO	94	
$P(OPh)_3$		89
$P(c\text{-}C_6H_{11})_3$	80	81
PBu_3	82	

Such behaviour would be expected for a nucleophilic substitution at silicon in which increasing solvation of the nucleophile would cause the stereochemistry to be directed towards inversion of configuration.

Cleavage with other nucleophiles (Scheme 18) has also been observed to occur with predominant inversion[67, 163].

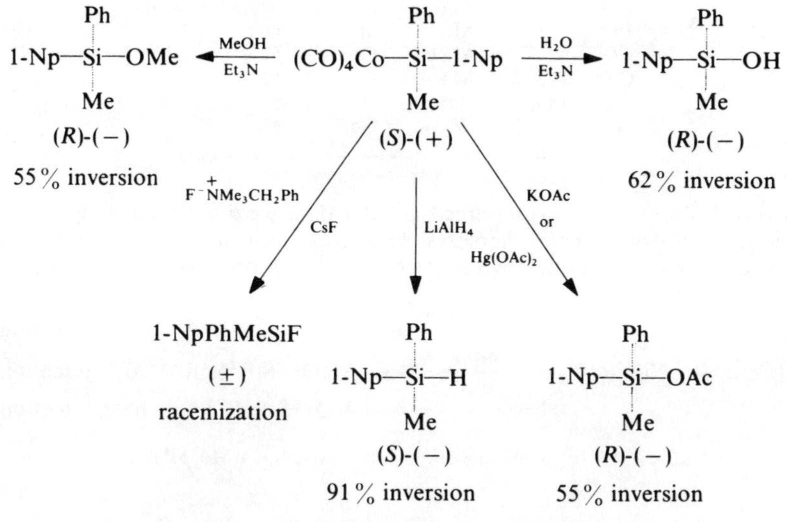

SCHEME 18. Reactions of optically active 1-NpPhMeSi*–$Co(CO)_4$

For a given transition metal complex, nucleophilic displacements of silyl ligands occur with a stereochemistry which is consistent with the results observed for nucleophilic substitution at silicon, in agreement with current mechanistic proposals. A comparison of one transition metal with another seems more difficult. Predominant retention at silicon in the cleavage of the Si–Fe bond in complex **24** is consistent with the above results with cobalt complexes, since $(\eta\text{-}C_5H_5)(CO)_2Fe^-$, a good nucleophile, is expected to be a poor leaving group[58]. However, the cleavage of the Si–Mn bond in a related manganese

1-NpPhMeSi*—Fe(CO)$_2$(η-C$_5$H$_5$)

(24)

1-NpPhMeSi*—Mn(CO)$_2$(η-C$_5$H$_5$)
|
H

(25)

complex **25** was observed to occur with inversion of configuration at silicon[65,66] (cf. Scheme 11, Section II.B). This particular behaviour has been attributed to a bonding interaction between the Si and H atoms in this complex[65,76,78].

Interestingly, a new factor determining the predominant stereochemistry at silicon has recently been pointed out[77]. The $LiAlH_4$ cleavage of hexacoordinated silyl–transition metal complexes was shown to proceed with predominant retention at silicon, regardless of the nature of the ligands around the metal atom (Table 32). The stereoselectivity was

TABLE 32. Percentage of retention in the cleavage reaction

$$(CO)_4LM_T{-}M^*R_3 \xrightarrow[Et_2O]{LiAlH_4} H\text{-}M^*R^1R^2R^3$$

(after Reference 77)

L	M_T	M	% Retention
CO	Mn	Si	68
PPh_3	Mn	Si	66
CO	Mn	Ge	57
C(OEt)Me	Mn	Ge	53
C(OEt)Me	Re	Ge	57
NO	W	Ge	67

low and no solvent effect was observed. Similar data were obtained using DME. The octahedral geometry in these complexes seems to govern the stereochemical pathway. Other types of the nucleophilic cleavage were shown to proceed similarly[164] (Scheme 19).

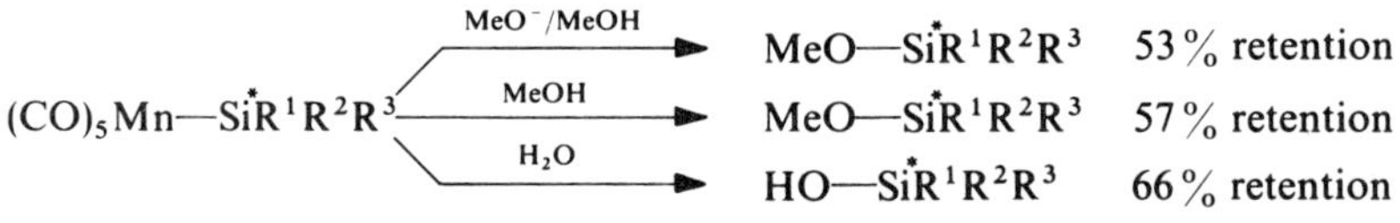

SCHEME 19. Reactions of optically active (OC)$_5$Mn–Si*R^{1}R^{2}R^3

A detailed interpretation of these results has not been given. However, nucleophilic substitution at silicon can be accepted as a frontier orbital controlled process between the HOMO of the nucleophile and the σ*Si–X antibonding orbital as LUMO[165]. There are important differences in energy levels between octahedral and pentagonal complexes, and molecular orbital calculations might suggest a rationalization of these differences.

D. Geometry of Attack of the Nucleophile and Conclusion

In the above sections, we discussed in detail the main factors which control the stereochemistry of the nucleophilic displacement at silicon. Either retention or inversion of configuration can occur and change in stereochemistry is clearly a function of the nature of the leaving group and the electronic character of the nucleophile. Two preliminary remarks can be made at this point:

(1) Stereochemical results with optically active monofunctional organosilanes, $R^1R^2R^3Si^*$–X, in which the substituents R attached to silicon encompass significant variation in their steric effects, are quite clear (Table 13, Section III.B.1). The introduction of R groups of moderate to large steric requirements leaves the stereochemistry unchanged. For a constant leaving group, both inversion and retention can be observed, depending on the nature of the nucleophile: the harder the nucleophile, the closer the stereochemistry is to a pure retention as summarized for carbon organometallics in Scheme 20.

RN ——————————————————→ IN

RLi PhC≡CLi	>	RMgX PhC≡CMgX	>	CH_2=$CHCH_2Li$ $PhCH_2Li$	>	CH_3=$CHCH_2MgBr$ $PhCH_2MgX$

(R = alkyl, aryl)

SCHEME 20. Stereochemical behaviour of organolithium and Grignard reagents.

For borderline leaving groups, such as fluorine, small changes in the nature of the solvent or added salts lead to significant shifts of stereochemistry (Table 33). Hence the stereochemistry is never controlled by steric hindrance at silicon, but rather by the nature of the leaving group and the electronic character of the nucleophile.

(2) There is a parallel between the chemical and stereochemical data for bifunctional organosilanes (Table 14, Section III.B.1). Inversion takes place selectively with the best leaving group which is the only one displaced, regardless of the nature of the other leaving group attached at silicon. The stereochemistry changes from inversion to retention when going from Si–Cl to Si–H. However, the mutual influence of one of the two leaving groups on the other can change its lability and thereby the stereochemistry of its displacement by nucleophiles (cf. Table 15, Section III.B.1). Again, charge-delocalized nucleophiles favour inversion, whereas highly charge-localized reagents lead to retention. When the two potential leaving groups have close nucleofugalities, e.g. in R_2Si(OMe)OR′ (R = Ph or Fc; R′ = Men or Et), reactions with organolithiums and with Grignard reagents are different (Table 34). (i) Soft organolithiums and Grignard reagents which lead to inversion cleave only the best leaving group (F vs. OR). (ii) Hard organolithiums show a lack of selectivity and substitute simultaneously the two leaving groups with predominant cleavage of the better one either due to electronegativity (F vs. OR group) or for steric reasons (OMen vs. OMe). (iii) Saturated Grignard reagents, which react with predominant retention, lead to the selective cleavage of the alkoxy group (F vs. OR). Morever, with dialkoxysilanes, only the less bulky group is cleaved (OMe vs. OMen). (iv) In contrast, coupling reactions of soft Grignard reagents with 1-NpPhSi(OMe)OMen lead only to cleavage of the bulkier group (OMen vs. OMe) with predominant inversion. The results are summarised in Table 34.

These comments show clearly that in borderline cases, i.e. for leaving groups having close nucleofugalities, steric reasons (OMen vs. OMe with organolithiums) and the electrophilic assistance of the cation M^+ can direct the substitution selectivity. The Li^+ cation has no great affinity for either OR or F groups, and this explains the lack of selectivity with hard organolithiums which give retention where the best leaving group is predominantly cleaved. In contrast, with Grignard reagents which react with retention

TABLE 33. Effect of the solvent or added salts on the stereochemical behavior of fluorosilanes

Organosilane	Nucleophile (solvent)	Added salt	Product[a]	Predominant stereochemistry[b]	Ref.
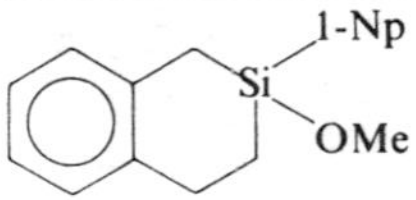	$CH_3CH{=}CHCH_2MgBr$ (Et_2O)	—	$R_3SiCH_2CH{=}CHCH_3$	92% RN	146
	$CH_3CH{=}CHCH_2MgBr$ (Et_2O)	$MgBr_2$	$R_3SiCH_2CH{=}CHCH_3$	57% RN	146
1-NpPh(MeO)SiOMen(−)	$CH_2{=}CHCH_2MgBr$ (Et_2O)	—	$R_2(OMe)SiCH_2CH{=}CH_2^*$	69% IN	146
	$CH_2{=}CHCH_2MgBr$ (Et_2O)	$MgBr_2$	$R_2(OMe)SiCH_2CH{=}CH_2^*$	88% IN	146

[a] $R_2Si = $ 1-NpPhSi.
[b] RN denotes retention; IN denotes inversion.

TABLE 34. Chemical behaviour of bifunctional organosilanes with organometallics: influence of an electrophilic assistance

Organosilane	Nucleophile	Products (yield)	Stereochemistry[a]	References
1-NpPh(F)Si–OMen(−)	MeLi	1-NpPh(Me)Si–OMen (100%)	RN	166
	EtLi	1-NpPh(Et)Si–OMen (60%)	RN	166
		1-NpPh(F)Si–Et (40%)	(not isolated)	
	PrLi	1-NpPh(Pr)Si–OMen (56%)	RN	166
		1-NpPh(F)Si–Pr (44%)	(not isolated)	
	BuLi	1-NpPh(Bu)Si–OMen (75%)	RN	166
		1-NpPh(F)Si–Bu (25%)	(not isolated)	
	neo-C_5H_{11}Li	1-NpPh(neo-C_5H_{11})Si–OMen	(not isolated)	166
		1-NpPh(F)Si-(neo-C_5H_{11}) (90%)	RN	
	RMgX (R = Me, Et, Pr, *p*-$CH_3C_6H_4$)	1-NpPh(F)Si–R	RN	166
1-NpFc(F)Si–OEt	MeLi	1-NpFc(Me)Si–OEt	RN	166
	EtLi	1-NpFc(Et)Si–OEt (52%)	IN	166
		1-NpFc(F)Si–Et (19%)	RN	

TABLE 34 (*continued* overleaf)

TABLE 34. (*continued*)

Organosilane	Nucleophile	Products (yield)	Stereochemistry[a]	References
	PrLi	1-NpFc(Pr)Si–OEt −(48%)	IN	166
		1-NpFc(F)Si–Pr (17%)	RN	
	RMgBr (R = Me, Et, Pr, Ph)	1-NpFc(F)Si–R	RN	166
1-NpPh(OMe)Si–OMen	MeLi	1-NpPh(Me)Si–OMen +(73%)	RN	25, 167
		1-NpPh(OMe)Si–Me (27%)	RN	
	EtLi	1-NpPh(Et)Si–OMen +(85%)	RN	25, 167
		1-NpPh(OMe)Si–Et (15%)	RN	
	PrLi	1-NpPh(Pr)Si–OMen +(>80%)	RN	25, 167
		1-NpPh(OMe)Si–Pr (<20%)	RN	25, 167
	BuLi	1-NpPh(Bu)Si–OMen +(87%)	RN	25, 167
		1-NpPh(OMe)Si–Bu (13%)	RN	
	R^1MgX (R^1 = Me, Et, Pr, *i*-Pr)	1-NpPh(R^1)Si–OMen	RN	25
	R^2MgX (CH_2=$CHCH_2$, $PhCH_2$)	1-NpPh(OMe)Si–R^2	IN	168

[a] RV denotes retention; IN denotes inversion.

the OR group is complexed by the magnesium and becomes a better leaving group than fluorine: it is cleaved selectively. With $R_2Si(OMe)OR'$, the complexation occurs only with the oxygen atom of the less bulky alkoxy group (OMe vs. OMen).

Nevertheless, these data show clearly that steric factors or electrophilic assistance are never the controlling factors of the stereochemistry. Softer anions (allyl, benzyl) react with inversion and the better leaving group is cleaved. For hard reagents (alkyl anions), it is the electronic character of the anion which at first determines the geometry of attack; for instance, electrophilic assistance may influence the selectivity in changing the respective cleavage ability of the two potential leaving groups. Small changes of the nucleophile, the metal (Li vs. Mg), the solvent and the salts (MgX_2) can lead to dramatic variations in the stereochemistry of substitution (Table 34). In any case, 1-naphthyl or ferrocenyl groups have no substantial influence on the stereochemistry. A steric control of the stereochemistry of the nucleophilic displacements at silicon as proposed recently[165] by comparison with phosphonium salt cleavage by OH^- [170] can be ruled out on the basis of the above observations. Influence of steric hindrance either with mono- and bifunctional derivatives or even with uncrowded cyclic systems (Table 26) was never dominant.

Our purpose is now to propose a reasonable explanation of these results and, in particular, to discuss the geometry of attack of the nucleophile at silicon which leads in the first step to the formation of a pentacoordinate intermediate.

Concerning the reaction occurring with inversion, the stereochemistry requires a back-side attack of the nucleophile from the centre of a tetrahedral face opposite to the leaving group, giving rise to a pentacoordinate intermediate with entering and leaving groups in axial positions (intermediate **26**).

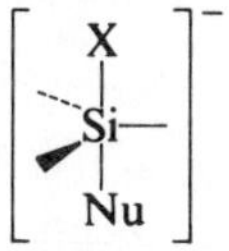

(26)

Concerning the reaction occurring with retention, it requires a nucleophilic attack at 90° with respect to the leaving group. Two different mechanisms can be advanced:

(1) A mechanism similar to that invoked for retention in nucleophilic displacement reactions at phosphorus[171–175]. It invokes axial attack in the centre of a tetrahedral face containing the leaving group and results in a five-coordinated trigonal bipyramid like **27**, with the leaving group in equatorial position. Pseudo-rotation through a square pyramid with the leaving group and the nucleophile in *cis*-basal positions brings the leaving group to an axial position and the nucleophile to an equatorial position of the resultant trigonal pyramid **28**. Departure of the leaving group takes place from the axial position (Scheme 21).

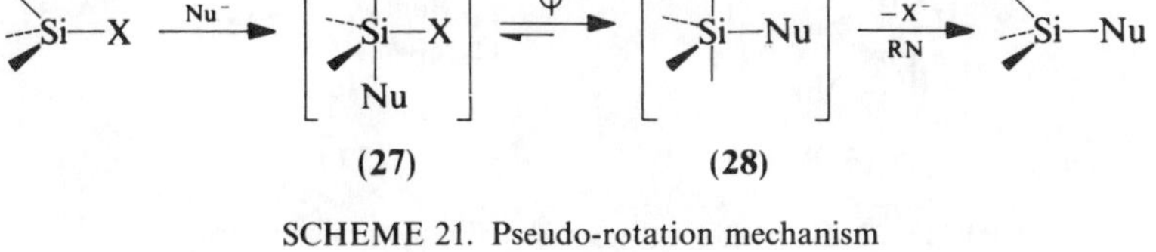

(27) **(28)**

SCHEME 21. Pseudo-rotation mechanism

(2) An alternative mechanism does not require a pseudo-rotation step. Equatorial attack by nucleophile is assumed in a tetrahedral edge between a ligand and the leaving group. It results in a five-coordinated trigonal bipyramid like **29**, with the leaving group in axial position and the nucleophile in equatorial position. Axial departure of the leaving group is invoked[2].

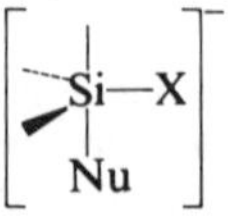

(29)

Alkyllithiums RLi (R = Me, Et, Pr, Bu) lead to competitive cleavage of the two electronegative groups attached at silicon in both fluoroalkoxysilanes, $R_2Si(F)OR'$ and 1-NpPhSi (OMen)OMe, with retention (stereoselectivity up to 97%) (Table 34). It implies a nucleophilic attack at 90° with respect to the two leaving groups. Thus we can only explain the stereochemical data either by an equatorial attack of the nucleophile leading to **30** and **30′** or by an apical attack leading to **31** and **31′** (Scheme 22). **30** and **30′** seem energetically favoured compared to **31** and **31′**, because the two electronegative groups

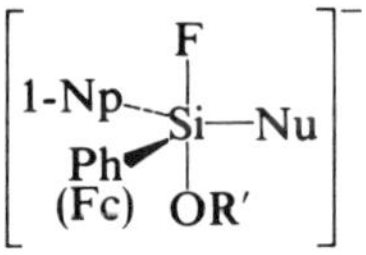

(30)

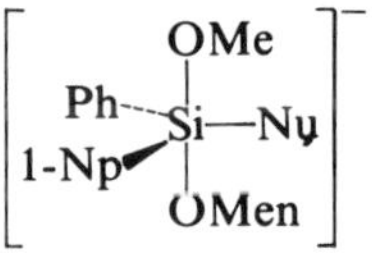

(30′)

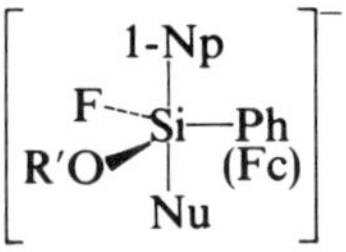

(31)

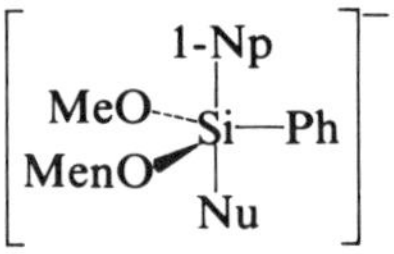

(31′)

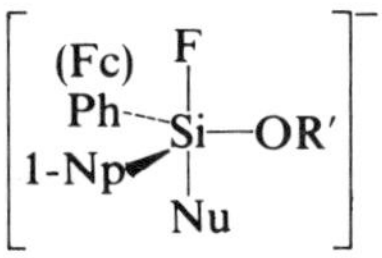

(32)

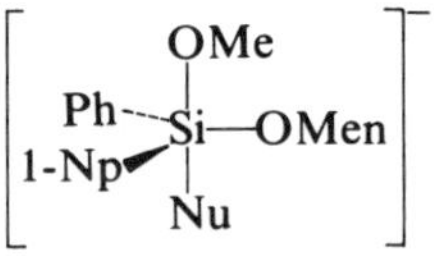

(32′)

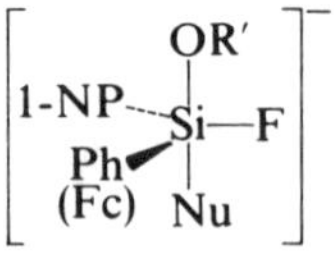

(33)

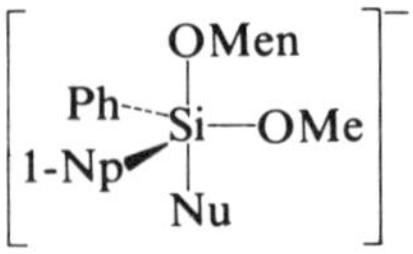

(33′)

SCHEME 22. Geometrics of the trigonal bipyramidal intermediates

(F and OR) are in apical positions. Intermediates such as **32** (and **32'**) or **33** (and **33'**) can rationalize either the displacement of the fluorine atom (and of the methoxy group) with inversion and of the OR' group (and the OMen group) with retention, or the retention for displacement of the fluorine atom (and the methoxy group) and inversion for the OR' group (and the OMen group) instead of the observed retention. From these two respects, an equatorial attack of the nucleophile would provide a reasonable description of the observed stereochemical data.

However, recent molecular mechanics calculations have indicated that a retention pathway via pseudo-rotation of an intermediate such as **31** and **31'** would be of lower energy than retention occurring via equatorial attack and axial departure[176]. The apicophicity of the leaving group is controlled by the electronic character of the nucleophile. This factor perhaps determines the relative stability of **31** and **31'** vs. **30** and **30'**. Moreover, NMR and X-ray studies of pentacoordinate organosilicon species formed by an intramolecular nitrogen coordination always show an apical entry of the incoming nitrogen atom[139, 177].

Among these species, pentacoordinated functional organosilicon compounds with two alkoxy groups attached at silicon are quite illustrative; the two oxygen atoms are always equatorial even when an equatorial entry would be more favourable owing to the geometry of the ring, as reported recently by Corey and coworkers in the case of spirocyclic pseudo-silatranes[177]. Thus, more experimental and theoretical studies seem necessary to understand in more detail the nucleophilic displacements at silicon which take place with retention.

IV. ACKNOWLEDGEMENTS

Discussions with Professor Nguyen Trong Anh and Dr C. Minot have been very fruitful. The stimulating interest of Drs H. Felkin and J. Seyden-Penne is appreciated. The help of Dr J. Jacques is gratefully acknowledged. In addition, Professor Corriu wishes to thank those coworkers whose names appear in the references.

Finally, we acknowledge support of the research by the CNRS.

V. REFERENCES

1. (a) L. H. Sommer, *Stereochemistry, Mechanism and Silicon*, McGraw-Hill, New York, 1965; (b) L. H. Sommer, *Intra-Sci. Chem. Rep.* **7**, 1 (1973).
2. R. J. P. Corriu and C. Guérin, *J. Organomet. Chem.*, **198**, 231 (1980); R. J. P. Corriu and C. Guérin, *Adv. Organomet. Chem.*, **20**, 265 (1982); R. J. P. Corriu, C. Guerin and J. J. E. Moreau, *Top. Stereochem.*, **15**, 43 (1984).
3. F. S. Kipping, *J. Chem. Soc.*, 209 (1907); F. S. Kipping, *J. Chem. Soc.*, 457 (1908); B. D. W. Duff and F. S. Kipping, *J. Chem. Soc.*, 2090 (1908) ; F. Challenger and F. S. Kipping, *J. Chem. Soc.*, 755 (1910).
4. C. Eaborn and C. Pitt, *Chem. Ind.*, 830 (1958).
5. L. H. Sommer and C. L. Frye, *J. Am. Chem. Soc.*, **81**, 1013 (1959).
6. M. Gielen, *Top. Stereochem.*, **12**, 217 (1981).
7. R. J. P. Corriu and M. Henner, *J. Organomet. Chem.*, **74**, 1 (1974).
8. S. H. Wilen, *Top. Stereochem.*, **6**, 107 (1971).
9. L. H. Sommer, C. L. Frye, G. A. Parker and K. W. Michael, *J. Am. Chem. Soc.*, **86**, 3271 (1964).
10. Y. Okaya and T. Ashida, *Acta Crystallogr.*, **20**, 461 (1966).
11. S. L. Davydova, Yu. A. Purisson, B. D. Lavrukhin and N. A. Plate, *Izv. Akad. Nauk SSSR (Engl transl.)*, 387 (1965).
12. R. J. P. Corriu and G. Royo, *J. Organomet. Chem.*, **14**, 291 (1968).
13. R. J. P. Corriu, G. Lanneau and G. Royo, *Bull. Soc. Chim. Fr.*, 458 (1968).
14. L. H. Sommer and K. T. Rosborough, *J. Am. Chem. Soc.*, **91**, 7067 (1969).

15. R. J. P. Corriu, G. F. Lanneau and S. Ould-Kada, *J. Organomet. Chem.*, **248**, 23 (1983).
16. R. J. P. Corriu and G. Royo, *Tetrahedron*, **27**, 4289 (1971).
17. R. J. P. Corriu and J. Masse, *Bull. Soc. Chim. Fr.*, 3491 (1969).
18. J. D. Citron, *J. Organomet. Chem.*, **86**, 359 (1975).
19. D. N. Roark and L. H. Sommer, *J. Am. Chem. Soc.*, **95**, 969 (1973).
20. J. P. Vidal and J. Falgueirettes, *Acta Crystallogr., Sect. B*, **29**, 2833 (1973).
21. L. H. Sommer, K. W. Michael and K. W. Korte, *J. Am. Chem. Soc.*, **89**, 868 (1967).
22. L. H. Sommer and M. A. Silverman, *J. Org. Chem.*, **38**, 636 (1973).
23. R. J. P. Corriu, G. F. Lanneau and M. Leard, *J. Organomet. Chem.*, **64**, 79 (1974).
24. C. Breliere, R. J. P. Corriu and G. Royo, *J. Organomet. Chem.*, **107**, 148 (1978).
25. R. J. P. Corriu, G. F. Lanneau and G. Royo, *J. Organomet. Chem.*, **35**, 35 (1972).
26. R. J. P. Corriu and G. F. Lanneau, *J. Organomet. Chem.*, **64**, 63 (1974).
27. R. J. P. Corriu, F. Larcher and G. Royo, *J. Organomet. Chem.*, **129**, 299 (1977).
28. R. J. P. Corriu, F. Larcher and G. Royo, *J. Organomet. Chem.*, **104**, 161 (1976).
29. C. Breliere, R. J. P. Corriu, A. De Saxe, F. Larcher and G. Royo, *J. Organomet. Chem.*, **164**, 19 (1979).
30. J. A. Kanters and A. M. Van Veen, *Cryst. Struct. Commun.*, **2**, 261 (1973).
31. A. Fredga, *Tetrahedron*, **8**, 126 (1960).
32. G. F. Lanneau, Ph.D. Thesis, University of Montpellier, 1962.
33. R. J. P. Corriu, C. Guérin and J. Massé, *J. Chem. Res. (S)*, 160 (1977); R. J. P. Corriu, C. Guérin and J. Massé, *J. Chem. Res. (M)*, 1877 (1977).
34. R. J. P. Corriu, J. M. Fernandez, C. Guérin and A. Kopton, *Bull. Soc. Chim. Belg.*, **89**, 783 (1980).
35. R. J. P. Corriu and C. Guérin, *J. Organomet. Chem.*, **195**, 261 (1980).
36. J. Mathieu and J. Weil-Raynal, *Bull. Soc. Chim. Fr.*, 1211 (1968); J. D. Morrison and H. S. Mosher, *Asymmetric Organic Reactions*, Prentice Hall, Englewood Cliffs, N.J., 1971.
37. A. Holt, A. W. P. Jarvie and G. J. Jervis, *Tetrahedron Lett.*, 4087 (1968); A. Holt, A. W. P. Jarvie and G. J. Jervis, *J. Chem. Soc., Perkin Trans. 2*, 114 (1973).
38. A. G. Brook, F. Abdesaken, G. Gutekunst and R. K. Kallury, *J. Chem. Soc., Chem. Commun.*, 191 (1981).
39. R. West, M. J. Fink and J. Michl, *Science*, **214**, 1343 (1981).
40. H. Sakurai, Y. Nakadaira and T. Kobayashi, *J. Am. Chem. Soc.*, **101**, 487 (1979); A. G. Brook, F. Abdesaken, G. Gutekunst and N. Plavac, *Organometallics*, **1**, 994 (1982); P. R. Jones and M. E. Lee, *J. Organomet. Chem.*, **232**, 33 (1982); P. R. Jones, M. E. Lee and L. T. Lin, *Organometallics*, **2**, 1039 (1982); P. R. Jones and M. E. Lee, *J. Am. Chem. Soc.*, **105**, 6725 (1983); P. R. Jones and M. E. Lee, *J. Organomet. Chem.*, **271**, 299 (1984); A. H. B. Cheng, P. R. Jones, M. E. Lee and P. Roussi, *Organometallics*, **3**, 881 (1985); P. R. Jones, T. F. Bates, A. F. Cowley and A. M. Arif, *J. Am. Chem. Soc.*, **108**, 3122 (1986).
41. G. Bertrand, J. Dubac, P. Mazerolles and J. Ancelle, *J. Chem. Soc.*, 382 (1980).
42. J. F. Klebe and H. Finkbeiner, *J. Am. Chem. Soc.*, **88**, 4470 (1966); J. F. Klebe and H. Finkbeiner, *J. Am. Chem. Soc.*, **90**, 7255 (1968).
43. I. Ojima, M. Nihonyanagi and Y. Nagai, *Chem. Commun.*, 938 (1972).
44. R. J. P. Corriu and J. J. E. Moreau, *J. Chem. Soc., Chem. Commun.*, 38 (1973).
45. R. J. P. Corriu and J. J. E. Moreau, *Tetrahedron Lett.*, 4469 (1973); R. J. P. Corriu and J. J. E. Moreau, *J. Organomet. Chem.*, **120**, 337 (1976).
46. R. J. P. Corriu and J. J. E. Moreau, *Bull. Soc. Chim. Fr.*, 901 (1975).
47. H. B. Kagan and T. P. Dang, *J. Am. Chem. Soc.*, **94**, 6429 (1972).
48. J. D. Morrison and W. F. Masler, *J. Org. Chem.*, **39**, 270 (1974).
49. R. J. P. Corriu and J. J. E. Moreau, *J. Organomet. Chem.*, **114**, 135 (1976).
50. R. J. P. Corriu and J. J. E. Moreau, *J. Organomet. Chem.*, **64**, C51 (1974); R. J. P. Corriu and J. J. E. Moreau, *J. Organomet. Chem.*, **85**, 19 (1975).
51. R. J. P. Corriu and J. J. E. Moreau, *J. Organomet. Chem.*, **91**, C27 (1975); R. J. P. Corriu and J. J. E. Moreau, *Nouv. J. Chim.*, **1**, 71 (1977).
52. T. Hayashi, K. Yamamoto and M. Kumada, *Tetrahedron Lett.*, 331 (1974).
53. I. Ojima, K. Yamamoto and M. Kumada, in *Aspects of Homogeneous Catalysis* (Ed. R. Ugo), Vol. 3, D. Reidel, Dordrecht, 1977, p. 185.
54. R. Glaser, *Tetrahedron Lett.*, 2127 (1975).
55. W. J. Richter, *J. Organomet. Chem.*, **169**, 9 (1979).

56. E. Colomer and R. J. P. Corriu, *Top. Curr. Chem.*, **96**, 79 (1981).
57. E. H. Brooks and R. J. Cross, *J. Organomet. Chem. Rev.*, *A*, **6**, 227 (1971); H. G. Ang. and P. T. Lau, *J. Organomet. Chem. Rev.*, *A*, **8**, 235 (1972); F. Hofler, *Top. Curr. Chem.*, **50**, 129 (1974).
58. R. J. P. Corriu and W. E. Douglas, *J. Organomet. Chem.*, **51**, C3 (1973).
59. J. F. Harrod, D. F. R. Gibson and R. Charles, *Can. J. Chem.*, **47**, 2205 (1969).
60. K. W. Muir and J. A. Ibers, *Inorg. Chem.*, **9**, 440 (1970).
61. C. Eaborn, D. J. Tune and D. R. M. Walton, *J. Chem. Soc., Dalton Trans.*, 2255 (1973).
62. C. Eaborn, P. N. Kapoor, D. J. Tune, C. L. Turpin and D. R. M. Walton, *J. Organomet. Chem.*, **34**, 153 (1972).
63. C. Eaborn, D. J. Tune and D. R. M. Walton, *J. Chem. Soc., Chem. Commun.*, 1233 (1972).
64. C. Eaborn, P. B. Hitchcok, D. J. Tune and D. R. M. Walton, *J. Organomet. Chem.*, **54**, C1 (1973).
65. E. Colomer, R. J. P. Corriu and A. Vioux, *J. Chem. Res. (S)*, 168 (1977); *J. Chem. Res. (M)*, 1939 (1977).
66. E. Colomer, R. J. P. Corriu and A. Vioux, *J. Chem. Soc., Chem. Commun.*, 175 (1976).
67. L. H. Sommer, J. E. Lyons and H. Fujimoto, *J. Am. Chem. Soc.*, **91**, 7051 (1969).
68. A. P. Hagen and A. G. MacDiarmid, *Inorg. Chem.*, **6**, 686 (1966); B. J. Aylett and J. M. Campbell, *Chem. Commun.*, 159 (1967).
69. G. Cerveau, E. Colomer and R. J. P. Corriu, *J. Organomet. Chem.*, **236**, 33 (1982).
70. H. Manassero, personal communication.
71. L. H. Sommer, L. A. Ulland and G. A. Parker, *J. Am. Chem. Soc.*, **94**, 3469 (1972).
72. J. F. Harrod and C. A. Smith, *J. Am. Chem. Soc.*, **92**, 2699 (1970).
73. J. F. Harrod, C. A. Smith and K. A. Than, *J. Am. Chem. Soc.*, **94**, 8321 (1972).
74. W. Jetz and W. A. G. Graham, *Inorg. Chem.*, **10**, 4 (1971).
75. A. J. Hart-Davis and W. A. G. Graham, *J. Am. Chem. Soc.*, **93**, 4388 (1971) and references cited therein.
76. E. Colomer, R. J. P. Corriu and A. Vioux, *Inorg. Chem.*, **18**, 695 (1979).
77. G. Bellachioma, G. Cardacl, E. Colomer, R. J. P. Corriu and A. Vioux, *Inorg. Chem.* (in press).
78. E. Colomer, R. J. P. Corriu, C. Marzin and A. Vioux, *Inorg. Chem.*, **21**, 368 (1982).
79. U. Schubert, K. Ackermann and B. Worle, *J. Am. Chem. Soc.*, **104**, 7378 (1982); U. Schubert, K. Akermann, G. Kraft and B. Worle, *Z. Naturforsch. B*, **38**, 1488 (1983); U. Schubert, G. Kraft and E. Walther, *Z. Anorg. Allg. Chem.*, **519**, 96 (1984); U. Schubert, G. Kraft and C. Kalbas, *Transition Met. Chem.*, **9**, 161 (1984); G. Kraft, C. Kalbas and U. Schubert, *J. Organomet. Chem.*, **289**, 247 (1985).
80. M. J. Bennet and K. A. Simpson, *J. Am. Chem. Soc.*, **93**, 7156 (1971).
81. M. A. Andrews, E. W. Kirtley and M. D. Kaesz, *Adv. Chem. Ser.*, **176**, 215 (1978).
82. M. Auburn, M. Ciriano, J. A. K. Howard, M. Murray, N. J. Pugh, J. L. Spencer, F. G. A. Stone and P. Woodward, *J. Chem. Soc., Dalton Trans.*, 659 (1980).
83. A. G. Brook and J. M. Duff, *J. Am. Chem. Soc.*, **91**, 2118 (1969).
84. H. Sakurai, M. Murakami and M. Kumada, *J. Am. Chem. Soc.*, **91**, 519 (1969).
85. H. Sakurai and M. Murakami, *Chem. Lett.*, 7 (1972).
86. S. W. Bennet, C. Eaborn, A. Hudson, R. A. Jackson and K. D. Root, *J. Chem. Soc. (A)*, 348 (1969) and references cited therein.
87. H. Sakurai and K. Mochida, *Chem. Commun.*, 1581 (1971).
88. R. J. P. Corriu and J. J. Moreau, *Chem. Commun.*, 812 (1971).
89. R. J. P. Corriu and J. J. Moreau, *J. Organomet. Chem.*, **40**, 73 (1972).
90. H. Gilman and H. J. S. Winkler, in *Organometallic Chemistry* (Ed. H. Zeiss), Reinhold, New York, 1960, p. 270.
91. J. B. Lambert, M. Urdaneta-Perez and Hs-N. Sun, *J. Chem. Soc., Chem. Commun.*, 806 (1976).
92. L. H. Sommer and R. Mason, *J. Am. Chem. Soc.*, **87**, 1619 (1965).
93. E. Colomer and R. J. P. Corriu, *J. Chem. Soc., Chem. Commun.*, 177 (1976).
94. E. Colomer and R. J. P. Corriu, *J. Organomet. Chem.*, **133**, 159 (1977).
95. C. Eaborn, R. A. Jackson, D. J. Tune and D. R. M. Walton, *J. Organomet. Chem.*, **63**, 85 (1973).
96. A. G. Brook and G. J. D. Peddle, *J. Am. Chem. Soc.*, **85**, 2338 (1963).
97. C. Eaborn, R. E. E. Hill and P. Simpson, *J. Organomet. Chem.*, **37**, 267 (1972).
98. F. Carre and R. J. P. Corriu, *J. Organomet. Chem.*, **65**, 349 (1974).
99. C. Eaborn, R. E. E. Hill and P. Simpson, *J. Organomet. Chem.*, **37**, 275 (1972).
100. E. Colomer and R. J. P. Corriu, *J. Chem. Soc., Chem. Commun.*, 435 (1978).

101. F. Carre, G. Cerveau, E. Colomer and R. J. P. Corrlu, *J. Organomet. Chem.*, **229**, 257 (1982).
102. G. Cerveau, E. Colomer, R. J. P. Corriu and A. Vioux, *Organometallics*, **1**, 866 (1982).
103. (a) R. H. Princs, in *M.T.P. Int. Rev. Sci., Inorg. Chem. Ser. One*, **9**, 353 (1972).
(b) I. Fleming, in *Comprehensive Organic Chemistry*, Vol. 3 (Ed. D. Neville Jones), 542 (1979).
104. For instance: D. Bethell and V. Gold, *Carbonium Ions: An Introduction*, Academic Press, London and New York, 1967 and references cited therein.
105. For instance: M. S. Flowers and L. E. Gusel'nikov, *J. Chem. Soc. B*, 419 (1968); R. West, *Int. Symp. Organosilicon Chem., Sci. Commun.* 1 (1965); L. H. Sommer and F. Evans, *J. Am. Chem. Soc.*, **76**, p. 801 (1954).
106. Y. Apeloig and A. Stanger, *J. Am. Chem. Soc.*, **109**, 272 (1987).
107. J. B. Lambert, J. A. McConnell and W. J. Schultz, Jr. *J. Am. Chem. Soc.*, **108**, 2482 (1986).
108. A. D. Allen, J. C. Charlton, C. Eaborn and G. Modena, *J. Chem. Soc.*, 3668 (1957); A. D. Allen and G. Modena, *J. Chem. Soc.*, 3671 (1957).
109. G. Schott and V. Bondybey, *Chem. Ber.*, **100**, 1773 (1967); K. D. Kaufmann and K. Ruhlmann, *Z. Chem.*, **7**, 391 (1967); R. L. Schowen and K. S. Latham, *J. Am. Chem. Soc.*, **89**, 4677 (1967).
110. R. J. P. Corriu and G. Royo, *J. Organomet. Chem.*, **14**, 291 (1968); R. J. P. Corriu and G. Royo, *Bull. Soc. Chim. Fr.*, **4**, 1490 (1972); R. J. P. Corriu, F. Larcher and G. Royo, *J. Organomet. Chem.*, **104**, 161 (1976).
111. C. Eaborn and F. M. S. Mahmoud, *J. Chem. Soc. Perkin Trans. 2*, 1309 (1981); C. Eaborn, P. D. Lickiss, S. T. Najim and M. N. J. Romanelli, *J. Chem. Soc., Chem. Commun.*, 1754 (1985) and earlier references cited therein.
112. R. J. P. Corriu and B. Henner, *J. Chem. Soc., Chem. Commun.*, 116 (1973); G. Chauvière, R. J. P. Corriu and B. Henner, *J. Organomet. Chem.*, **86**, Cl (1975); R. J. P. Corriu and B. Henner, *J. Organomet. Chem.*, **102**, 407 (1975).
113. S. A. Sullivan, C. H. De Puy and R. Damrauer, *J. Am. Chem. Soc.*, **103**, 408 (1981).
114. P. J. Krusic, P. J. Fagan and J. San Filippo, *J. Am. Chem. Soc.*, **99**, 250 (1977); J. San Filippo, J. Sibermann and P. J. Fagan, *J. Am. Chem. Soc.*, **100**, 4834 (1978).
115. H. Felkin, P. J. Knowles and B. Meunier, *J. Organomet. Chem.*, **146**, 151 (1978).
116. R. J. P. Corriu, G. Dabosi and M. Martineau, *J. Chem. Soc., Chem. Commun.*, 457 (1979).
117. C. K. Mann and K. K. Barnes, *Electrochemical Reactions in Nonaqueous Systems*, M. Dekker, New York, 1970; M. M. Baizer, *Organic Electrochemistry*, M. Dekker, New York, 1973; M. R. Rufi and F. H. Covitz, *Introduction to Organic Electrochemistry*, M. Dekker, New York, 1974.
118. L. H. Sommer and W. D. Korte, *J. Am. Chem. Soc.*, **89**, 5802 (1967).
119. R. J. P. Corriu and G. Royo, *Tetrahedron*, **27**, 4289 (1971); R. J. P. Corriu and G. Royo, *Bull. Soc. Chim. Fr.*, 1497 (1972).
120. L. H. Sommer, W. D. Korte and P. G. Rodewald, *J. Am. Chem. Soc.*, **89**, 862 (1967).
121. L. H. Sommer and J. McLick, *J. Am. Chem. Soc.*, **88**, 5359 (1966); L. H. Sommer and J. McLick, *J. Am. Chem. Soc.*, **89**, 5806 (1967).
122. R. J. P. Corriu, J. M. Fernandez and C. Guerin, *J. Organomet. Chem.*, **152**, 25 (1978).
123. R. J. P. Corriu and G. Royo, *J. Organomet. Chem.*, **40**, 229 (1972).
124. L. H. Sommer, C. L. Frye, G. A. Parker and K. W. Michael, *J. Am. Chem. Soc.*, **86**, 3271 (1964).
125. L. H. Sommer, G. A. Parker, N. C. Lloyd, C. L. Frye and K. W. Michael, *J. Am. Chem. Soc.*, **89**, 857 (1967).
126. L. H. Sommer, C. L. Frye, M. C. Musolf, G. A. Parker, P. G. Roderwald, K. W. Michael, Y. Okaya and R. Popinsli, *J. Am. Chem. Soc.*, **83**, 2210 (1961).
127. L. H. Sommer, C. L. Frye and G. A. Parker, *J. Am. Chem. Soc.*, **86**, 3276 (1964).
128. C. Breliere, R. J. P. Corriu and G. Royo, *J. Chem. Soc., Chem. Commun.*, 906 (1976).
129. C. Breliere, R. J. P. Corriu, A. De Saxcé and G. Royo, *J. Organomet. Chem.*, **166**, 153 (1979); R. J. P. Corriu and G. Lanneau, *Bull. Soc. Chim. Fr.*, 3102 (1973); R. J. P. Corriu, G. Lanneau and M. Léard, *J. Organomet. Chem.*, **64**, 79 (1974).
130. R. J. P. Corriu, F. Larcher and G. Royo, *J. Organomet. Chem.*, **102**, C25 (1975); C. Breliere, R. J. P. Corriu and G. Royo, *J. Organomet. Chem.*, **148**, 107 (1978).
131. L. H. Sommer, G. A. Parker and C. L. Frye, *J. Am. Chem. Soc.*, **86**, 3280 (1964).
132. J. H. Brewster, *J. Am. Chem. Soc.*, **81**, 5475 (1959).
133. K. D. Onan, A. T. McPhail, C. H. Yoder and R. W. Hillyard, *J. Chem. Soc., Chem. Commun.*, 209 (1978).
134. E. A. V. Ebsworth, *Volatile Silicon Compounds*, Pergamon, Oxford, 1963, p. 54.
135. G. Klebe, J. W. Bats and K. Hensen, *J. Chem. Soc., Dalton Trans.*, 1 (1985).

136. W. S. Sheldrick and W. Wolfsberger, *Z. Naturforsch B*, **32B**, 22 (1977).
137. R. J. P. Corriu, G. Royo and A. De Saxcé, *J. Chem. Soc., Chem. Commun.*, 892 (1980).
138. B. J. Helmer, R. West, R. J. P. Corriu, M. Poirier, G. Royo and A. De Saxcé, *J. Organomet. Chem.*, **251**, 295 (1983).
139. C. Breliére, F. Carré, R. J. P. Corriu, M. Poirier and G. Royo, *Organometallics*, **5**, 388 (1986).
140. C. Breliére, F. Carré, R. J. P. Corriu, A. De Saxcé, M. Poirier and G. Royo, *J. Organomet. Chem.*, **206**, C1 (1981).
141. R. J. P. Corriu, J. P. Dutheil, G. Lanneau and S. Ould-Kada, *Tetrahedron*, **35**, 1889 (1979).
142. R. J. P. Corriu and J. Masse, *J. Chem. Soc., Chem. Commun.*, 1373 (1968); R. J. P. Corriu and J. Masse, *J. Organomet. Chem.*, **35**, 51 (1972).
143. R. J. P. Corriu and C. Guérin, *J. Organomet. Chem.*, **225**, 141 (1982).
144. R. J. P. Corriu and J. Masse, *Tetrahedron Lett.*, 5197 (1968); R. J. P. Corriu and J. Masse, *J. Organomet. Chem.*, **34**, 221 (1972).
145. O. Eisenstein, J. M. Lefour, C. Minot, Nguyen Trong Anh and G. Soussan, *C.R. Hebd. Seances Acad. Sci., Ser. C*, **274**, 1310 (1972).
146. R. J. P. Corriu, J. Masse and G. Royo, *J. Chem. Soc., Chem. Commun.*, 252 (1971).
147. L. H. Sommer and H. Fujimoto, *J. Am. Chem. Soc.*, **90**, 982 (1968).
148. R. J. P. Corriu and B. Henner, *J. Organomet. Chem.*, **71**, 393 (1974).
149. R. J. P. Corriu, J. M. Fernandez and C. Guérin, *J. Organomet. Chem.*, **152**, 21 (1978); R. J. P. Corriu, J. M. Fernandez and C. Guérin, *Tetrahedron*, **37**, 2467 (1981).
150. J. M. Lehn, *Acc. Chem. Res.*, **11**, 49 (1978).
151. R. J. P. Corriu and C. Guérin, *J. Chem. Soc., Chem. Commun.*, 74 (1977); R. J. P. Corriu and C. Guérin, *J. Organomet. Chem.*, **144**, 165 (1978).
152. L. H. Sommer, J. McLick and C. Golino, *J. Am. Chem. Soc.*, **94**, 669 (1972).
153. R. J. P. Corriu, J. M. Fernandez and C. Guérin, *Tetrahedron Lett.*, 3391 (1978).
154. J. Dubac and P. Mazerolles, *J. Organomet. Chem. Lib.*, **12**, 149 (1981).
155. S. E. Cremer and C. Blankenship, *Organometallics*, **5**, 1329 (1986).
156. L. S. Vilkov, B. S. Mastrivukov, J. V. Baowrover, V. M. Vdovin and L. Trinberg, *Dokl. Akad. Nauk SSSR*, **117**, 1084 (1967).
157. J. Dubac, P. Mazerolles and B. Seves, *Tetrahedron Lett.*, 529 (1972); J. Dubac, P. Mazerolles and B. Seves, *Tetrahedron Lett.*, 3495 (1972); J. Dubac, P. Mazerolles and B. Seves, *Tetrahedron*, **30**, 749 (1974); J. Dubac, P. Mazerolles and B. Seves, *Tetrahedron*, **30**, 759 (1974).
158. B. G. McKinnie, N. S. Bhacca, F. K. Cartledge and J. Fayssous, *J. Am. Chem. Soc.*, **96**, 2637 (1974); B. G. McKinnie, N. S. Bhacca, F. K. Cartledge and J. Fayssous, *J. Org. Chem.*, **41**, 1534 (1976).
159. F. K. Cartledge, J. M. Wolcott, J. Dubac, P. Mazerolles and P. Fagoaga, *Tetrahedron Lett.*, 3593 (1975); J. M. Wolcott and F. K. Cartledge, *J. Organomet. Chem.*, **111**, C35 (1976); F. K. Cartledge, J. M. Wolcott, J. Dubac, P. Mazerolles and M. Joly, *J. Organomet. Chem.*, **154**, 187 (1978); J. Dubac, P. Mazerolles, M. Joly, F. K. Cartledge and J. M. Wolcott, *J. Organomet. Chem.*, **154**, 203 (1978).
160. K. G. Dzhaparidze, *Soobshch. Akad. Nauk Gruz. SSR*, **29**, 401 (1962); *Chem. Abstr.*, **58**, 9703b (1963); J. R. Durig and J. N. Willes Jr., *J. Mol. Spectrosc.*, **32**, 320 (1969); J. R. Durig, W. J. Lafferty and W. F. Kalinski, *J. Phys. Chem.*, **80**, 1199 (1976).
161. J. P. Vidal, J. Lapasset and J. Falgueirettes, *Acta Crystallogr., Sect. B*, **28**, 3137 (1972).
162. G. Cerveau, E. Colomer and R. J. P. Corriu, *Angew. Chem., Int. Ed. Engl.*, **20**, 478 (1981).
163. R. J. P. Corriu and E. Colomer, *J. Organomet. Chem.*, **133**, 159 (1977).
164. G. Cerveau, E. Colomer and R. J. P. Corriu, *Organometallics*, **1**, 867 (1982).
165. Nguyen Trong Anh and C. Minot, *J. Am. Chem. Soc.*, **102**, 103 (1980).
166. C. Breliére, R. J. P. Corriu, A. De Saxcé and G. Royo, *J. Organomet. Chem.*, **166**, 153 (1979).
167. R. J. P. Corriu and G. Lanneau, *J. Organomet. Chem.*, **67**, 243 (1974).
168. R. J. P. Corriu and G. Lanneau, *Bull. Soc. Chim. Fr.*, 3102 (1973).
169. J. A. Deiters and R. R. Holmes, *J. Am. Chem. Soc.*, **109**, 1686 (1987).
170. R. Luckenbach, *Phosphorus*, **1**, 223 (1972).
171. R. Luckenbach, *Dynamic Stereochemistry of Pentacoordinated Phosphorus and Related Elements*, Georg Thieme, Stuttgart, 1973.
172. W. E. McEwen and K. D. Berlin (Eds.), *Organophosphorus Stereochemistry*, Parts I and II, Bench Mark Papers (Vol. 3), Dowden, Hutchinson and Ross, Stroudsburg, PA, 1975.
173. F. H. Westheimer, *Acc. Chem. Res.*, **1**, 70 (1968).

174. K. Mislow, *Acc. Chem. Res.*, **3**, 321 (1970).
175. P. Gillespie, P. Hoffman, H. Keusacek, D. Marquarding, S. Pfohl, F. Ramirez, E. A. Tsolis and I. Ugi, *Angew. Chem., Int. Ed. Engl.*, **10**, 687 (1971).
176. J. A. Deiters and R. R. Holmes, *J. Am. Chem. Soc.*, **109**, 1692 (1987).
177. J. Corey, private communication.

The Chemistry of Organic Silicon Compounds
Edited by S. Patai and Z. Rappoport

CHAPTER 5

Thermochemistry

R. WALSH
Department of Chemistry, University of Reading, Whiteknights, PO Box 224, Reading RG6 2AD, UK

I. INTRODUCTION

A knowledge of molecular heats of formation and chemical bond dissociation energies has always been regarded as fundamental to the understanding of chemical bonding and reactivity. This chapter deals with the extent and reliability of our knowledge of these quantities for silicon-containing compounds. The information and material presented represent a consolidation and extension of earlier reviews by the author on bond dissociation energies[1] and thermochemistry[2,3]. In those reviews an attempt was made to assess reliability by comparison of information from as wide a range of sources as possible. The same approach will be adopted here. As well as data obtained by the traditional methods of *calorimetry* and the *study of equilibria*, results from *kinetic studies* will also be considered. We have discussed the latter method in its application to silicon-containing species in recent articles[4,5]. A major recent development has been the application of theoretical chemistry (valency theory) to the calculations of heats of formation of silicon-containing molecules. Calculations range from *ab initio* to *semi-empirical*, and while some of the results are not well matched to experimental figures, the quality of calculations is improving all the time. This approach is likely to make increasing impact in future.

This chapter concentrates on results rather than the details of experimental techniques. These can be found either in the reviews[1–5] already mentioned or in original articles. Oxygen-bomb calorimetry, the source of so much data on hydrocarbons, has proven particularly difficult for organosilicon compounds because of incomplete combustion and much of the older (pre-1970) data on heats of formation[6] are unreliable. This is well documented and has led to its exclusion from data compilations, such as those of Cox and Pilcher[7] and Pedley and Rylance[8]. A substantial amount of these older data as well as some newer results are to be found in a useful review by Tel'noi and Rabinovitch[9], who argue that it may nevertheless have some limited value. The preferred modern calorimetry method is one in which the organosilane is combusted simultaneously with an auxiliary fluorine-containing substance to ensure complete conversion of the silane to $H_2SiF_6(aq)$[10]. One of the most important results of this method is the value for $\Delta H_f^\circ(Me_4Si, g)$ obtained by Steele[11], although other studies[8,12] have produced results of an inconsistent nature (*vide infra*). A very recent calorimetric study by Voronkov and coworkers[13] exploits a technique developed by the group of Gadzhiev[14], in which the organosilicon sample is vapourised at a high temperature by prior ignition of a small quantity of a standard substance, such as benzoic acid. The organosilicon vapour/oxygen mixture then ignites and burns completely. This supposedly avoids the problems of incomplete combustion. Initial results look promising, although there may still be some problems (*vide infra*). The results of these studies and some earlier data will be considered in detail, but for the most part we omit discussion of oxygen, static-bomb calorimetric results. For some classes of compounds (e.g. polysiloxanes and polysilanes) there is a crying need for even a single item of data.

Although this chapter concentrates on organosilicon compounds, a few purely inorganic compounds are also included (the hydrides and halides), since these provide inevitable reference points. It is worth mentioning at this point that the chapter title 'Thermochemistry' is used with its traditional meaning here to signify *heats* (or *enthalpies*) of *reaction*, rather than in the looser sense of 'chemistry under the action of heat' as employed by some workers in the field of organosilicon chemistry. Unless otherwise stated, all standard heats of reaction and heats of formation refer to the gas phase at 298.2 K.

II. COMPOUNDS OF TETRAVALENT SILICON

A. General Considerations and Additivity Rules

We have argued previously[2] that, as with organic compounds, heats of formation should be assessed and judged according to how well they fit additivity rules[15]. The difficulty in the past has been that, with too few reliable data, the additivity increments (or contributions) have themselves not been sufficiently reliable. We are, however, now reaching the stage where this is becoming possible.

The additivity approach may be illustrated by consideration of some observed redistribution reactions, of the general type

$$2L_2SiXY \leftrightarrows L_2SiX_2 + L_2SiY_2$$

where L represents a fixed ligand and X and Y exchangeable ones. The law of bond additivity[16] states that for such reactions any thermodynamic change will be zero. The extent to which non-zero changes are observed represents a departure from this law, and indicates the existence of bond-interaction terms which may be incorporated into modified additivity schemes, such as that of Allen[17]. If a reaction obeys the law of bond additivity, then the equilibrium constant will be given solely by symmetry numbers (so-called statistical redistribution): in the case of the above, for example,

$$K\text{ (bond additivity)} = \frac{[L_2SiX_2][L_2SiY_2]}{[L_2SiXY]^2} = \tfrac{1}{4}$$

A number of equilibrium studies (with equilibrium brought about by the use of suitable catalysts) have been reported by Moedritzer[18]. We have also carried out such studies for the methylsilanes[3]. The results of a number of these studies are shown in Table 1. They demonstrate that departures from statistical equilibrium are not large. The intrinsic[21] Gibbs free-energy changes are small, and since entropies (again intrinsic) obey bond additivity if anything better than enthalpies of formation, this suggests that $\Delta G°$(intrinsic) are a close approximation to $\Delta H°$ for these reactions. Thus the conclusion from Table 1 is that, for the cases listed (involving Me-for-H, Cl-for-H and Me-for-Cl replacements), redistribution enthalpy changes are small and bond additivity is quite a good approximation. There is even circumstantial evidence that enthalpy changes involving Et-for-F redistribution are small (*ca* 4 kJ mol^{-1})[2]. This is important because the Si–F bond, being the most polar bond formed by silicon, should give rise to the greatest electrostatic bond–bond interactions and therefore potentially the greatest departures from additivity. It is certainly true that the fluoromethanes exhibit much greater departures from bond additivity than do the chloromethanes[22].

In terms of a criterion of assessment of consistency, therefore, bond additivity can be very useful. Small percentage errors in the determination of equilibrium constants lead to only small errors in $\Delta H°$ (usually *ca* 1–2 kJ mol^{-1}) whereas small percentage errors in

TABLE 1. Equilibrium constants and free-energy changes for some catalysed redistribution reactions

Redistribution reaction	T/°C	K(obs)	Ref.	K (random)[a]	ΔG° (intrinsic)[b] /kJ mol^{-1}
$2Me_3SiH \rightleftarrows Me_4Si + Me_2SiH_2$	124	0.43	3	0.375	−0.5
$2Me_2SiH_2 \rightleftarrows Me_3SiH + MeSiH_3$	124.5	0.66	3	0.444	−1.3
$2MeSiH_2Cl \rightleftarrows MeSiHCl_2 + MeSiH_3$	100	0.476	19	0.333	−1.1
$2MeSiHCl_2 \rightleftarrows MeSiCl_3 + MeSiH_2Cl$	100	0.041	19	0.333	+6.5
$2Me_2SiHCl \rightleftarrows Me_2SiCl_2 + Me_2SiH_2$	100	0.098	19	0.25	+2.9
$2Me_3SiCl \rightleftarrows Me_4Si + Me_2SiCl_2$	350	0.0135	20	0.375	+17
$2Me_2SiCl_2 \rightleftarrows Me_3SiCl + MeSiCl_3$	350	0.0251	20	0.444	+15

[a] K (random) may be identified with K (bond additivity), see text.
[b] For definition, see text.

enthalpies of combustion, because these are large, can lead to errors in ΔH_f° values which may be *ca* 10–20 kJ mol^{-1}. Thus one may justify the use of this kind of consistency argument in assessment of combustion-derived (and other calorimetric) data. Moreover, it suggests that a better strategy for the establishment of organosilicon heats of formation than perhaps hitherto adopted would be to limit combustion studies with their attendant difficulties to a minimum, say the symmetrical (homoleptic) SiR_4 compounds (where R are the key ligands), and to determine other ΔH_f° values via catalysed equilibrium studies.

Next in the hierarchy of additivity laws comes that of group additivity[15,16]. For the purposes of this law a 'group' consists of an atom together with all of its bonded atoms. It potentially offers a better test of fit to data, since it effectively incorporates both 'bond' and 'bond-interaction' items. The difficulty in applying it is the need for an even greater data base of compounds to establish the group contributions. There is, however, now sufficient data on organosilicon compounds to attempt a limited application. O'Neal and Ring[23] have tried this for methylsilanes and disilanes, but the data base used contained some values now known to be unreliable. Other additivity schemes[24] are also hampered by this difficulty.

B. Experimental Data and Preferred ΔH_f° Values

1. Methylsilanes

We have reviewed the data up to 1986[3] and do not repeat the discussion here. The preferred values for ΔH_f° are shown in Table 2. It should be noted that the values for $MeSiH_3$, Me_2SiH_2 and Me_3SiH have been determined via equilibrium studies and so they depend on the calorimetric values of $\Delta H_f^\circ(SiH_4)$ and $\Delta H_f^\circ(Me_4Si)$. Any change in the latter values will affect the former. $\Delta H_f^\circ(SiH_4)$ appears to be well established[25] and we have argued strongly[3] for the reliability of Steele's value[11] for $\Delta H_f^\circ(Me_4Si)$. This has been recently confirmed in the new, high-temperature combustion studies of Voronkov and coworkers[13] who obtained the value of -230.1 ± 5.7 kJ mol^{-1} in good agreement with Steele[11]. This re-enforces the view that, for reasons which are not entirely clear, the studies of Pedley and coworkers[8,12,26], although using the modern calorimetric method (use of auxiliary fluorine-containing substance), have given erroneous values for a number of substances. For $\Delta H_f^\circ(Me_4Si)$ they obtained -245.4 ± 9.9 kJ mol^{-1}. Dewar's parametized SCF MO calculations appear to be improving in this area. The most recent version (AM1)[27] gives closer agreement for the methyl silanes than an earlier one (MNDO)[28], but differences with the data of Table 2 are still up to 16 kJ mol^{-1} and not all systematically in the same sense!

TABLE 2. Standard enthalpies of formation of the methylsilanes

Silane	ΔH_f°/kJ mol^{-1}	Ref.
SiH_4	34.3 ± 2.0	25
$MeSiH_3$	-29.1 ± 4.0	3
Me_2SiH_2	-94.7 ± 4.0	3
Me_3SiH	-163.4 ± 4.0	3
Me_4Si	-233.2 ± 3.2	11

2. *Higher alkyl silanes*

For the reasons discussed in the introduction, early work[6] on these compounds was unsatisfactory. The very recent new data of Voronkov and coworkers[13] looks much more promising. The data for a variety of mixed methyl-, ethyl- and propylsilanes is shown in Table 3. For comparison purposes, group additivity estimates are also shown. These are based on group increments previously worked out by ourselves[3] and shown in the appendix. It should be noted that no non-bonded (*gauche*-type) interactions are incorporated into the additivity scheme. The agreement between the data and the estimates is extremely good. Apart from this, various pointers are encouraging: first, the agreement with Steele's value for $\Delta H_f^\circ(Me_4Si)$; secondly, the magnitude of the increment between methyl and ethyl substituents (*ca* -17 kJ mol^{-1}). This quantity tends to be a function of electronegativity or polarity of the element to which the groups are bonded. This increment ($-\Delta$) tends to be larger (>21 kJ mol^{-1}) for electronegative (or negatively charged) atoms or groups (–F, –OH, $-NH_2$) or smaller (<21 kJ mol^{-1}) for positively charged atoms as an examination of existing data[7,8] shows. This kind of correlation of enthalpy of formation increments with polarity was first carried out by Benson[22] for hydrogen- and methyl-substituted species. The same correlation for methyl and ethyl compounds enabled us to predict[3] the difference of -17 kJ mol^{-1} for methyl and ethyl silanes, exactly as now observed.

TABLE 3. Standard enthalpies of formation of alkylsilanes

Compound	ΔH_f°/kJ mol^{-1}	
	Voronkov and coworkers[13]	Group additivity[a]
Me_4Si	-230 ± 6	-233
Et_2SiMe_2	-264 ± 6	-267
Et_3SiMe	-281 ± 6	-284
Et_4Si	-297 ± 6	-301
Pr_2SiMe_2	-305 ± 7	-309
Pr_3SiMe	-350 ± 7	-346
Pr_2SiEt_2	-342 ± 6	-343
Pr_4Si	-379 ± 6	-384

[a] See appendix for group contributions.

A close examination of Table 3 shows a small systematic (average) difference of *ca* 2.5 kJ mol^{-1} between the additivity estimates and Voronkov's data. This suggests that a small revision in the group values could improve agreement further, but since it is already well within experimental uncertainties no change is made here.

Pedley and Rylance[8] include ΔH_f° data for Et_2SiH_2 (-182.5 kJ mol^{-1}), Et_3SiH (-200.9 kJ mol^{-1}) and Et_4Si (-265.4 kJ mol^{-1}). Apart from the difference for Et_4Si with Table 3, the other values fit no pattern and differ substantially from group additivity estimates, and must therefore be considered unreliable.

3. *Other Si/C/H compounds*

There are two recent studies of vinyl-substituted silanes but they are neither self-consistent nor convincing. For example, Voronkov and coworkers[13] have obtained -81.6 ± 6.8 kJ mol^{-1} for $\Delta H_f^\circ((C_2H_3)_4Si)$ whilst Gadzhiev and Gubareva[29] report

$+61.9 \pm 2.9$ kJ mol^{-1} for the same quantity. This is disappointing, since the calorimetric methods used were those claiming to have overcome the problems of incomplete combustion by use of high temperatures or explosions. That incomplete combustion is probably still a problem in these cases can be inferred by examining the implied heat of hydrogenation of a selected case, viz

$$Me_3SiC_2H_3 + H_2 \rightarrow Me_3SiC_2H_5$$

$\Delta H_f^\circ(Me_3SiC_2H_3)$ is reported[13] as -190 ± 5 kJ mol^{-1} and $\Delta H_f^\circ(Me_3SiC_2H_5)$ may be estimated (by group additivity) to be -250 kJ mol^{-1}. Thus the heat of hydrogenation is apparently *ca* -60 kJ mol^{-1}. This is very far from the values of *ca* -125 kJ mol^{-1} which may be estimated for olefins[8]. It is, of course, of great interest to know whether there is any special interaction, and if so of what magnitude, between a bonded vinyl group and silicon. There is no chemical evidence for an effect of the magnitude of *ca* 65 kJ mol^{-1} implied by the difference in hydrogenation heats. A comparison with tetraphenylsilane suggests that an interaction beyond the range 0–10 kJ mol^{-1} is unlikely, and so this casts further doubt on the published figures. A zero vinyl to silicon interaction (apart from the σ bond) would imply $\Delta H_f^\circ((C_2H_3)_4Si)$ of *ca* $+205$ kJ mol^{-1}.

The only phenyl compound which has been studied is tetraphenylsilane, for which $\Delta H_f^\circ((C_6H_5)_4Si) = 334 \pm 6$ kJ mol^{-1} has been obtained[9,30]. There appears to be no disagreement over this value. Furthermore, it corresponds to a reasonable average bond energy for the Si–C_6H_5 linkage. Based on current values for C_6H_5 and CH_3 radical heats of formation (see appendix), $\bar{D}(Si–C_6H_5) = 358$ kJ mol^{-1} and $\bar{D}(Si–CH_3) = 317$ kJ mol^{-1}. The difference of 41 kJ mol^{-1} coincides with an average bond-energy difference of 41 ± 6 kJ mol^{-1} for all the group 4 M elements as shown in Table 4. This is good evidence for the reliability of the data and the lack of any special interaction between silicon (or indeed any group 4 M element) and a bonded phenyl group. If there is no interaction with phenyl, it is hard to see why there should be any with vinyl.

There are a number of values for alkyl polysilanes[8,9]. Unfortunately they come from data sources about which doubts have already been expressed, and are therefore not considered in detail. One compound, however, hexamethyldisilane, is worth discussing. Pedley and Rylance[8] quote $\Delta H_f^\circ = -363.1 \pm 8.6$ kJ mol^{-1} obtained from Pedley's earlier work[12,26]. This quantity may also be estimated via consideration of the reaction

$$Me_3SiSiMe_3 \rightarrow 2Me_3Si$$

TABLE 4. Average bond energies for group 4M[a] tetraphenyl and tetramethyl compounds

	$\Delta H_f^\circ(R_4M)$[b]		$\bar{D}$[c]	
Element (M)	$R = C_6H_5$	$R = CH_3$	M–C_6H_5	M–CH_3
C	398	−167	408	366
Si	334[d]	−233[e]	358	317
Ge	446	−103[f]	312	265
Sn	573	−19	261	226
Pb	710	136	200	160

[a] In this context 4M = Main Group 4.
[b] All data from Reference 8, unless specified otherwise.
[c] For ΔH_f° of radicals and atoms, see appendix.
[d] Reference 30.
[e] Reference 11.
[f] Reference 31.

Rate data now exist for both forward[32] and reverse[33] rate constants for this reaction and $\Delta H_f^\circ(Me_3Si)$ is known (*vide infra*). We have analysed these data[34] to obtain $\Delta H_f^\circ(Me_3SiSiMe_3) = -361 \pm 10$ kJ mol^{-1}. This is satisfactory agreement, but in view of the uncertainties of Pedley's work and other uncertainties involved in kinetic studies, $\Delta H_f^\circ(Me_6Si_2)$ cannot yet be taken as reliably established. Dewar's calculations[27] give the somewhat different value of -277 kJ mol^{-1} and O'Neal and Ring, who have previously considered this thermochemistry[23], have queried[35] these numbers. $\Delta H_f^\circ(Me_6Si_2)$ is important, since it will effectively permit the estimation of ΔH_f° values for the partially methylated disilanes, and higher polysilanes (via the additivity rules). These compounds are of great interest to organosilicon chemists as sources of silylenes, and through their heats of formation provide the means to establish silylene thermochemistry[4,36]. Some ΔH_f° values for methylated disilanes exist in the literature based on semi-empirical (MOB1) calculations[37] or mass-spectral appearance potential measurements[38]. The latter are known to be inconsistent with data on methylated monosilanes[3] and are not considered further. The MOB1 calculations of Bell, Perkins and Perkins[39] have also not provided reliable ΔH_f° values for chloro or methylsilanes[3]. Whether their values for methyl (and chloro) disilanes[37] prove to be reasonable cannot yet be judged owing to a dearth of data. However, interpolation, using bond additivity, between $\Delta H_f^\circ(Si_2H_6)$ and $\Delta H_f^\circ(Me_6Si_2)$ is likely to give better values for ΔH_f° for the methyl disilanes, just as it does for the methylsilanes, since these are not particularly polar compounds. It would be valuable, therefore, to have a further independent determination of $\Delta H_f^\circ(Me_6Si_2)$.

A limited amount of data exists for silicon-containing ring compounds. This is shown in Table 5. All the studies involved used combustion with an auxiliary fluorine compound, and so it is disappointing that for the four-membered ring compounds (siletane, disiletane) the agreement is so poor. Steele's values are taken to be the most reliable, both

TABLE 5. Standard enthalpies of formation of some silicon-containing ring compounds

	ΔH_f°/kJ mol^{-1}		
Compound	Steele[a]	Pedley and Rylance[b]	Genchel and coworkers[c]
	-108.7 ± 6.0	-138 ± 11	-82.8 ± 5.9
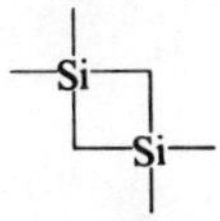	-239 ± 10	-301 ± 14	-226 ± 10
		-182 ± 12	
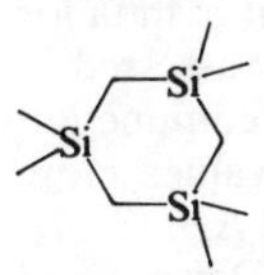			-449 ± 18

[a] Reference 40.
[b] Reference 8.
[c] Reference 41.

on the grounds of reliability of his other studies[11,30] and also because the implied ring strain in 1,1-dimethyl-1-siletane (94 ± 8 kJ mol^{-1} calculated relative to the group additivity value) appears the most reasonable. This figure fits best a thermochemical analysis[4,34] of the kinetics of decomposition of the dimethylsiletane[42] (via a biradical mechanism).

Ring-strain energies in silacycles are currently of great interest to organosilicon chemists and Gordon and his group[43] have calculated (via *ab-initio* methods) strain energies in 3-, 4-, 5- and 6-membered rings (containing one silicon atom). These are shown in Table 6 together with implied heats of formation for the compounds concerned, using the group additivity increments in the appendix. In these calculations it is assumed that ring strain is unaffected by methyl substitution at silicon. Although these numbers can only be considered as tentative, it is encouraging that the figure for 1,1-dimethyl-1-siletane is not too far from the preferred experimental value. It may be noted that the variation of ring strain is somewhat more marked in the silacycles than in their all-carbon analogues. An interesting rationalization of this (based on other theoretical estimates) has been offered by Schleyer[44]. It would be very useful to have experimental confirmation of some of these numbers.

TABLE 6. Strain energies and estimated standard enthalpies of formation of some prototype silacycles

Compound	Strain energy[a] /kJ mol^{-1}(R = H)	ΔH_f°/kJ mol^{-1} [b]	
		R = H	R = Me
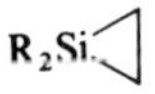	173	+130	−9
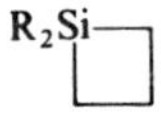	103	+39	−99
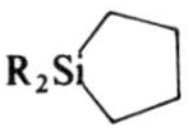	19	−66	−204
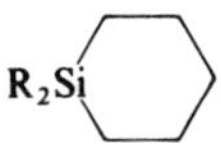	17	−88	−227

[a] Reference 43.
[b] Group additivity estimates (see appendix).

4. *Chlorine-containing compounds (Si/C/H/Cl)*

The chlorosilanes have been reviewed[2] and this is not repeated here. For completeness some preferred ΔH_f° values are given in the appendix. A considerable amount of data for the methyl chlorosilanes exists and that is considered here. It is most easily assessed in terms of the Cl-for-Me substitution effects and so these data are presented in conjunction with those for Me_4Si and $SiCl_4$ in Table 7. There is clearly some variation in values, more so amongst some compounds than others. Since $\Delta H_f^\circ(Me_4Si)$ (*vide supra*) and $\Delta H_f^\circ(SiCl_4)$ (see appendix) are well established, the semi-theoretical papers of Dewar and coworkers[27,28] and Bell, Perkins and Perkins[39] which do not fit them are not considered further. $\Delta H_f^\circ(Me_3SiCl) = -354 \pm 3$ kJ mol^{-1} appears to be well-established from solution[46,47] and combustion[14] calorimetric studies. For $\Delta H_f^\circ(Me_2SiCl_2)$ and

TABLE 7. Standard enthalpies of formation of methylchlorosilanes (comparison) $\Delta H_f^\circ/\mathrm{kJ\,mol^{-1}}$

Compound	Cox and Pilcher[a]	Pedley and Rylance[b]	Mosin and Shaulov[c]	Gadzhiev and Agarunov[d]	Dewar and coworkers[e] MNDO	Dewar and coworkers[f] (AM1)	Bell and coworkers[g]	Bond additivity
Me_4Si		−246	−220		−247	−218	−228	−233
Me_3SiCl	−354	−354	−354	−354	−366	−355	−353	−341
Me_2SiCl_2	−461	−448	−473	−464	−465	−482	−460	−448
$MeSiCl_3$		−572	−577	−585	−549	−602	−549	−555
$SiCl_4$			−663		−618	−715	−610	−663

[a] Reference 7.
[b] Reference 8.
[c] Reference 45.
[d] Reference 14.
[e] Reference 28.
[f] Reference 27.
[g] Reference 39.

$\Delta H_f^\circ(MeSiCl_3)$ the data are less satisfactory. There are combustion values for both[14] and a solution value[48] for $\Delta H_f^\circ(Me_2SiCl_2)$. The compilers[7,8] have made different selections. However, in addition to these data, there are disproportionation studies (see Table 1) by Zemany and Price[20] who obtained $+15\ kJ\ mol^{-1}$ for disproportionation enthalpies of Me_3SiCl and Me_2SiCl_2. These data were used by Mosin and Shaulov[45], who assumed a similar value for the disproportionation enthalpy of $MeSiCl_3$, to solve for individual ΔH_f° values for Me_4Si, Me_2SiCl_2 and $MeSiCl_3$ assuming the values of Me_3SiCl and $SiCl_4$. This approach, similar to that adopted by us for the methylsilanes[3], works when the system is not over-determined. Since $\Delta H_f^\circ(Me_4Si)$ is now reliably known (but was not at that time), the approach of Mosin and Shaulov needs modification. Zero disproportionare shown in Table 7. Since this does not reproduce $\Delta H_f^\circ(Me_3SiCl)$ it clearly needs modification. Assuming the disproportionation enthalpies of Zemany and Price [and $\Delta H_f^\circ(Me_4Si)$ and $\Delta H_f^\circ(SiCl_4)$] gives $\Delta H_f^\circ(Me_3SiCl) = 363\ kJ\ mol^{-1}$, which is clearly too high. Interpolation of the values between these two assumptions to fit the measured $\Delta H_f^\circ(Me_3SiCl)$ produces the recommended values shown in Table 8. The disproportionation enthalpies (*ca* $+9\ kJ\ mol^{-1}$) are a little less than those of Zemany and Price, but the latter study is nearly 40 years old and could bear re-investigation. Table 8 also includes recommended values for ΔH_f° for the three remaining methylchlorosilanes, two of which have been subjected to combustion[14], and compares these estimates with the original values obtained by Gadhziev and Agarunov[14]. Most of the authors of the studies cited in Table 7 also give values for these three additional compounds. The figures show a scatter similar in magnitude to those of Table 7 and are not reproduced here. In reaching recommended values we were guided again by the need to produce sensibly small disproportionation enthalpies. The recommended values fit precisely the ΔG°(intrinsic), assumed equal to ΔH°, of the disproportionation reactions studied by Moedritzer[19] (Table 1). Also, when combined in other sequences involving both methylsilanes and chlorosilanes, they exhibit suitably small disproportionation enthalpies, viz Me-for-H replacement -1 to $+5\ kJ\ mol^{-1}$, Cl-for-Me replacement $+5$ to $+11\ kJ\ mol^{-1}$. It was noted that marginally better fits were obtained with the lower value for $\Delta H_f^\circ(Me_4Si)$ (-230 rather than $-233\ kJ\ mol^{-1}$).

The results show considerable similarity to those of Mosin and Shaulov[45] but should represent a marginal improvement. There is, however, still room for more (and better) experimentation on these compounds.

TABLE 8. Recommended standard enthalpies of formation of methylchlorosilanes and methyl(hydrido)chlorosilanes

	$\Delta H_f^\circ/kJ\ mol^{-1}$	
Compound	Recommended[a]	Gadzhiev and Agarunov[b]
Me_3SiCl	-354 ± 3	-354 ± 3
Me_2SiCl_2	-466 ± 10	-464 ± 2
$MeSiCl_3$	-569 ± 11	-585 ± 4
$MeSiH_2Cl$	-210 ± 7	—
$MeSiHCl_2$	-393 ± 9	-414 ± 3
Me_2SiHCl	-282 ± 7	-305 ± 3

[a] See text.
[b] Reference 14.

5. Other halogen-containing compounds

The only other data worth mentioning here are $\Delta H_f^\circ(Me_3SiBr) = -293 \pm 4$ kJ mol^{-1} [8] and $\Delta H_f^\circ(Me_3SiI) = -217 \pm 4$ kJ mol^{-1} [49]. The former comes from a solution calorimetric study[47] which should be reliable and the latter from an equilibrium study by ourselves[49]. It should be mentioned that $\Delta H_f^\circ(Me_3SiI)$ is dependent on $\Delta H_f^\circ(Me_3SiBr)$. It has been also pointed out[4] that from gas-phase studies of ion breakdown there appears to be a small discrepancy (*ca* 9 kJ mol^{-1}) between $\Delta H_f^\circ(Me_3SiBr)$ and $\Delta H_f^\circ(Me_4Si)$ which could be reduced if the former were more negative (or the latter more positive). However, *in general* ionic process thermochemistry is not of the same precision as that of neutral reactions (although in organosilicon chemistry this is sometimes a debatable proposition) and so we do not include these results in this review.

6. Oxygen-containing compounds (Si/C/H/O)

We focus here on siloxanes and alkyl silyl ethers (and polyethers). There are a substantial number of data available[9] for these and other oxygen-containing compounds. However, many were determined by the old static bomb calorimetric method and are therefore not considered. Data obtained in the main by the new high-temperature calorimetry by Voronkov, Klyuchnikov and coworkers[13,50] are shown in Table 9. The enthalpy of formation of hexamethyldisiloxane has been investigated several times by the modern calorimetric method[13,26,41,51] and the experiments are discussed in detail in previous reviews[7,9]. It must be considered one of the most reliably known heats of formation. The other data may be assessed once again by their consistency with the additivity rules. The group increments shown in the appendix have to be derived from these data themselves, but there are still sufficient data to indicate a good fit. This

TABLE 9. Standard enthalpies of formation of silanol, siloxanes and silyl ethers

	ΔH_f°/kJ mol^{-1}	
Compound	Experimental[a]	Group additivity[d]
Me_3SiOH	-500 ± 3[b]	-500[e]
$(Me_3Si)_2O$	-777 ± 6[c]	-777[e]
Me_3SiOEt	-498 ± 4	-498[e]
Me_3SiOPr	-515 ± 4	-518
Me_3SiOPr-i	-524 ± 4	-537
$Me_2Si(OEt)_2$	-777 ± 5	-777[e]
$Pr_2Si(OEt)_2$	-851 ± 4	-852
$MeSi(OMe)_3$	-941 ± 4	-943
$MeSi(OEt)_3$	-1045 ± 5	-1045[e]
$Si(OMe)_4$	-1179 ± 4	-1179
$Si(OEt)_4$	-1315 ± 4	-1315[e]
$Si(OPr)_4$	-1397 ± 5	-1397

[a] References 13 and 50, unless otherwise mentioned.
[b] References 7 and 8.
[c] See text.
[d] See appendix for group contributions.
[e] Data used to fix group values.

TABLE 10. Redistribution enthalpies[a] for selected silyl polyethers and siloxane

Redistribution reaction	ΔH_f°/kJ mol^{-1}
$2Me_3SiOEt \rightleftarrows Me_4Si + Me_2Si(OEt)_2$	−12
$2Me_2Si(OEt)_2 \rightleftarrows Me_3SiOEt + MeSi(OEt)_3$	+10
$2MeSi(OEt)_3 \rightleftarrows Me_2Si(OEt)_2 + Si(OEt)_4$	−1
$2Me_3SiOH \rightleftarrows (Me_3Si)_2O + H_2O$	−20
$2Me_3SiOMe \rightleftarrows (Me_3Si)_2O + Me_2O$	−34
$2MeOH \rightleftarrows Me_2O + H_2O$	−23

[a] Data taken from Table 9 and Reference 13 (silicon compounds) and Reference 8 (others).

provides further encouragement that Voronkov and Klyuchnikov's method is reasonably reliable. Disproportionation reactions (bond additivity), however, are also worth examining here since this provides an additional criterion. A number of these are shown in Table 10. In the EtO-for-Me replacement on silicon processes, ΔH° values are appropriately small, consistent with those for other group redistribution reactions (Table 10). These redistribution equilibria have not been studied directly but are not out of line with others which have[18]. Redistribution processes centred on oxygen are more problematical. Disproportionations of Me_3SiOH and Me_3SiOMe have fairly negative ΔH° values. However, this is also true for CH_3OH, for which the data are unquestionably known. It must be concluded that bond interaction terms are fairly substantial here, favouring the symmetric compounds (with their alternating bond polarities).

Despite caveats, some of the earlier data on silyl ethers[52, 53] are in reasonable accord with that discussed here. For reasons of space, results for phenyl and vinyl silyl ethers[50] are not considered.

7. Nitrogen-containing compounds (Si/C/H/N)

Only silylamines are considered here. Their thermochemistry has been established via solution calorimetry in a study by Pedley's group[47]. The data are shown in Table 11. There are insufficient data to compare the results with group additivity (each compound requires an independent group).

Disproportionation enthalpies for group redistribution on nitrogen may be evaluated (Table 12). Unfortunately, it is hard to discern a pattern. The methylamines have negative H-for-Me redistribution enthalpies (as does MeOH). The trimethylsilylamines have one

TABLE 11. Standard enthalpies of formation of silylamines[a]

Compound	ΔH_f°/kJ mol^{-1}
$Me_3SiNHMe$	-227 ± 4
Me_3SiNMe_2	-248 ± 4
$(Me_3Si)_2NH$	-477 ± 5
$(Me_3Si)_2NMe$	-449 ± 8
$(Me_3Si)_3N$	-671 ± 12

[a] Data from Reference 47. See also References 7 and 8.

TABLE 12. Redistribution enthalpies[a] for silylamines and methylamines

Redistribution reaction	ΔH_f°/kJ mol^{-1}
$2Me_3SiNMe_2 \rightleftarrows (Me_3Si)_2NMe + Me_3N$	+23
$2(Me_3Si)_2NMe \rightleftarrows (Me_3Si)_3N + Me_3SiNMe_2$	−21
$2Me_3SiNHMe \rightleftarrows (Me_3Si)_2NH + Me_2NH$	−41
$2Me_2NH \rightleftarrows MeNH_2 + Me_3N$	−10
$2MeNH_2 \rightleftarrows NH_3 + Me_2NH$	−18

[a] Data taken from Table 11 and Reference 7 (silylamines) and Reference 8 (others).

positive and two negative values for Me_3Si-for-Me redistribution enthalpies. As with the trimethylsilyl methyl ether, maybe these all ought to be negative. In the absence of further evidence this hints at possible error. An old value for $\Delta H_f^\circ(Me_3SiNEt_2) = -335 \pm 8$ kJ mol^{-1} [54] suggests too high a value for the Et-for-Me substitution increment. More work is clearly needed on this class of compounds.

Tel'noi and Rabinovitch[9] list ΔH_f° values for a number of other nitrogen-containing compounds of uncertain quality. Voronkov and colleagues report ΔH_f° (values) for a series of silatranes[13]. Assuming the data are correct and based on an unstrained reference scheme, these [3.3.3] tricyclic systems possess apparent strain energies[13] of *ca* 40 kJ mol^{-1}, assuming no extra Si←N interaction. Unfortunately, it is impossible to separate strain and interaction effects to decide on the energetic magnitude of the Si←N interaction.

8. Sulphur-containing compounds (Si/C/H/S)

To our knowledge only one compound has been investigated. Pedley's group[47] obtained $\Delta H_f^\circ(Me_3SiSBu) = -381 \pm 3$ kJ mol^{-1} by the same hydrolysis technique as applied to amines and halides.

III. FREE RADICALS AND BOND DISSOCIATION ENERGIES

A. General Comments

Some years ago we set out on a program of measurements of Si–H bond dissociation energies using the gas-phase kinetic iodination technique pioneered by Benson[55]. The object was to establish reliable values and investigate specific effects of various substituents. The substantial part of our work was incorporated in a review in 1981[1]. In a sense it has been gratifying that in subsequent years no substantial alterations in values or perspective have occurred. This means that there is no substantial discussion required. Perhaps the largest development has been the improvement in reliability of the ΔH_f° values of saturated (tetravalent) silicon compounds reviewed in Section II and elsewhere[2]. This means that, from the original measurements[1], an increased number of radical heats of formation may be derived via the relationship

$$\Delta H_f^\circ(R\cdot) = \Delta H_f^\circ(RH) - \Delta H_f^\circ(H) + D(R\text{–}H)$$

and these, in turn, may be used more profitably to derive other bond dissociation energies via

$$D(R\text{–}X) = \Delta H_f^\circ(R\cdot) + \Delta H_f^\circ(X\cdot) - \Delta H_f^\circ(RX)$$

where X may be one of a variety of groups.

The uncertainties in bond dissociation energies are still, in general, of the order of ± 6 kJ mol^{-1}, and uncertainties in the derived $\Delta H_f^\circ(R\cdot)$ values slightly greater. There has been some concern expressed recently by Tsang[56], that the kinetic iodination technique has given rise to bond dissociation energies which are too low for *hydrocarbons* (by *ca* 8–12 kJ mol^{-1}). There is considerable evidence that slightly higher values are appropriate, e.g. $D(C_2H_5\text{–}H) = 421 \pm 2$ kJ mol^{-1} rather than 411 ± 4 kJ mol^{-1} [57]. It appears that some of the earlier kinetic studies of iodination reactions produced erroneously low activation energy (E_a) values. For instance, we have reinvestigated the kinetics of reaction of iodine with i-C_4H_{10}[58,59] and found $E_a = 99.6$ kJ mol^{-1} rather than 89.1 kJ mol^{-1} measured earlier[60] for the abstraction

$$I + i\text{-}C_4H_{10} \rightarrow i\text{-}C_4H_9 + HI$$

This is consistent with $D(i\text{-}C_4H_9\text{–}H) = 392$ kJ mol^{-1}, an increase of 11 kJ mol^{-1} over that earlier claimed[60]. The importance of this result is that it shows that there is nothing intrinsically wrong with the technique itself, but rather some of the earlier studies had systematic errors. There is thus no reason to doubt the general validity of the results derived for silane dissociation energies (care must always, however, be taken in every set of experimental measurements!).

B. Experimental Data: Measured Dissociation Energies and Radical Heats of Formation

Table 13 shows the values of our directly measured Si–H (and C–H) dissociation energies with concomitant derived radical heats of formation. There is no indication of any serious error in any of these values. Recent experimental[62–64] and theoretical studies[65–67] confirm $\Delta H_f^\circ(SiH_3)$ within experimental error. Gas-phase kinetic bromination studies[68,69] are consistent with $D(H_3Si\text{–}H)$ and $D(Me_3Si\text{–}H)$. Gas-phase studies[70] of the abstraction reaction of $Me_3Si + D_2$ are also consistent with $D(Me_3Si\text{–}H)$. Photoacoustic studies[71] have yielded $D(Et_3Si\text{–}H) = 377 \pm 4$ kJ mol^{-1}, $D(Me_3SiSiMe_2\text{–}H) = 357 \pm 4$ kJ mol^{-1} and $D((Me_3Si)_3Si\text{–}H) = 331 \pm 4$ kJ mol^{-1}. These are consistent with the values of Table 13, but show also that successive silyl substitution produces a substantial Si–H bond weakening effect, as we originally suspected[1].

TABLE 13. Measured bond dissociation energies for silanes and derived standard enthalpies of formation of associated radicals

Bond	D/kJ mol$^{-1\,a}$	$\Delta H_f^\circ(R\cdot)$/kJ mol$^{-1\,c}$
$H_3Si\text{–}H$	378 ± 5	194 ± 6
$MeSiH_2\text{–}H$	375 ± 8	128 ± 9
$Me_2SiH\text{–}H$	374 ± 8	61 ± 9
$Me_3Si\text{–}H$	378 ± 5	-3 ± 7
$H_3SiSiH_2\text{–}H$	361 ± 8	223 ± 8
$C_6H_5SiH_2\text{–}H$	369 ± 5	*ca* 260[d]
$Me_3SiCH_2\text{–}H$	415 ± 5	-36 ± 6
$Me_3SiCMe_2CH_2\text{–}H$	405 ± 5[b]	*ca* −125[d]
$Cl_3Si\text{–}H$	382 ± 5	-335 ± 8
$F_3Si\text{–}H$	419 ± 5	-1000 ± 20

[a] All data from Reference 1, unless indicated otherwise.
[b] Reference 61 (see also Reference 5).
[c] Calculated using $\Delta H_f^\circ(RH)$ from this review or Reference 2.
[d] Based on bond additivity estimate for $\Delta H_f^\circ(RH)$.

It is worth re-emphasising that there is no experimental evidence for a bond-weakening substituent effect of methyl groups[1]. This is apparently at odds with a model based on correlating isolated vibration frequencies and anharmonicities amongst the methylsilanes (and other molecules)[72].

Theoretical studies[73] slightly overestimate $\Delta H_f^\circ(Si_2H_5)$. $\Delta H_f^\circ(SiCl_3)$ and $\Delta H_f^\circ(SiF_3)$ are discussed elsewhere[2].

C. Derived Bond Dissociation Energies

A set of bond dissociation energies are derived from the molecular and radical heats of formation of previous sections. These are shown in Tables 14, 15 and 16, which are

TABLE 14. Derived silicon–carbon and silicon–silicon bond dissociation energies

Bond	D/kJ mol^{-1}	Bond	D/kJ mol^{-1}
$H_3Si–CH_3$	369 ± 7	$H_3Si–SiH_3$	308 ± 8
$MeSiH_2–CH_3$	368 ± 10	$H_3Si–Si_2H_5$	296 ± 10
$Me_2SiH–CH_3$	370 ± 10	$Si_2H_5–Si_2H_5$	283 ± 12
$Me_3Si–CH_3$	376 ± 8	$Me_3Si–SiMe_3$	354 ± 8[a]

[a] Based on the directly measured data of Reference 74 corrected to room temperature[34]

TABLE 15. Derived silicon–halogen bond dissociation energies

	D/kJ mol^{-1} [a]		
Halogen(X)	$D(H_3Si–X)$	$D(Me_3Si–X)$	$D(X_3Si–X)$
F	(628 ± 10)[b]	(665 ± 20)[c]	694 ± 20
Cl	451 ± 12	472 ± 8	449 ± 10
Br	370 ± 11	402 ± 8	368 ± 25
I	303 ± 10	321 ± 8	276 ± 30

[a] Based on data in this review and also Reference 2, unless otherwise mentioned.
[b] Based on $\Delta H_f^\circ(SiH_3F) = -355$ kJ mol^{-1} (Reference 75).
[c] A pure estimate.

TABLE 16. Derived silicon–oxygen and silicon–nitrogen bond dissociation energies[a]

Bond	D/kJ mol^{-1}	Bond	D/kJ mol^{-1}
$Me_3Si–OH$	536 ± 8	$Me_3SiO–H$	(493)
$Me_3Si–OMe$[b]	477 ± 9	$Me_3SiO–Me$	(384)
$Me_3Si–OEt$	484 ± 9	$Me_3SiO–Et$	(390)
$Me_3Si–OSiMe_3$	(549)		
$Me_3Si–NHMe$	401 ± 11	$(Me_3Si)_2N–H$	(461)
$Me_3Si–NMe_2$	390 ± 11	$(Me_3Si)_2N–CH_3$	(361)
$Me_3Si–N(SiMe_3)_2$	(434)		

[a] Figures in parentheses have been estimated (see text).
[b] Uses group additivity estimate for $\Delta H_f^\circ(Me_3SiOMe) = -464$ kJ mol^{-1}.

similar to, and an extension of, those given by Walsh[1]. Atomic and non-silicon-containing free radical heats of formation are shown in the appendix. Many of the trends and effects shown in these tables have been discussed previously[1], especially in relation to analogous bonds in organic molecules, and are not repeated here.

Supportive experimental evidence for the magnitudes of Si–C bond dissociation energies comes from studies of organosilane pyrolysis. Davidson and coworkers[76,77] have carried out pyrolyses of a number of molecules where decomposition is initiated by bond fission as the rate-determining step. Under these conditions, and provided no chain reactions are occurring, the activation energy (E_a) provides a reasonable approximation to the bond dissociation energy. For Me_3SiCl, loss of Me via Si–C bond breaking occurs with $E_a = 367 \pm 7$ kJ mol^{-1} [76]. Pyrolysis of allyltrimethylsilane[77] yields $E_a = 305 \pm 8$ kJ mol^{-1} for the step

$$Me_3SiCH_2CH{=}CH_2 \rightarrow Me_3Si + CH_2CH{=}CH_2$$

Since the allyl radical is stabilised by *ca* 75 kJ mol^{-1} (relative to methyl)[79], this is consistent with an Si–C bond strength of *ca* 380 kJ mol^{-1}.

The values for $D(Me_3SiCH_2–H)$ and $D(Me_3SiCMe_2CH_2–H)$ in Table 13 correspond to bond-weakening effects relative to $D(Me_3CCH_2–H) = 417$ kJ mol^{-1} of 2 and 12 kJ mol^{-1}, respectively, and represent α- and β-stabilization respectively of the derived alkyl radicals by silicon[5]. Measurements of C–C bond fission activation energies by Barton, Davidson and coworkers[78], in pyrolyses of appropriate organosilanes (relative to a standard), yield equal α- and β-stabilization energies of 11 ± 4 kJ mol^{-1}. Thus there is apparently agreement for β- but not for α-stabilization. However, the disagreement may be more apparent than real, since a difference can arise from a different choice of standards. The ground states of organosilanes containing the group C–(Si)(C)$(H)_2$ differ from their organic analogues [containing the group C–$(C)_2(H)_2$] by *ca* 4 kJ mol^{-1} in energy (see appendix and earlier discussion). Thus bond breaking adjacent to such a group, as occurs in the Barton and Davidson studies, 'releases' an extra 4 kJ mol^{-1} which may or may not be attributed to α-stabilization in the resulting radical (according to preference of definition). With a possible 2 kJ mol^{-1} attributable to release of steric effects, the discrepancy between these estimates of α-stabilization is reduced to only 3 kJ mol^{-1}, well within experimental error margins.

It is apparent from Table 15 that substituent methyl groups can act as bond strengtheners for silicon–halogen bonds: $D(Me_3Si–X) > D(H_3Si–X)$ for all halogens X. Bond strengthening by methyl groups has been discussed earlier[1]. By contrast, substituent halogens show two different effects: fluorine strengthens, whereas Cl, Br and I weaken (although with Cl it is only marginal). There is undoubtedly a combination of electronic effects operating here.

Si–OR and Si–NR_2 bonds, generally very strong by comparison with C–O and C–N bonds, show different effects expected according to the nature of R (H or alkyl). The figures in parentheses in Table 16 were obtained indirectly by use of estimates of the radical heats of formation, $\Delta H_f^\circ(Me_3SiO) = -225$ kJ mol^{-1} and $\Delta H_f^\circ((Me_3Si)_2N) = -233$ kJ mol^{-1}. These were obtained by assuming that sequential Si–O and Si–N dissociation energies in $(Me_3Si)_2O$ and $(Me_3Si)_3N$ follow the same proportionate relationships as O–H and N–H dissociation energies in H_2O and NH_3. This assumption attempts, empirically, to take account of the differential effect on sequential dissociation energies of changes in effective charge and electron density at the sites of the bonded atoms. Until experimental measurements are made, or a reliable *ab-initio* calculation is carried out, these represent the best available values for these dissociation energies.

Apart from the figures in these tables, the only other bond type for which experimental data permit a calculation is that for the Si–S bond. We calculate $D(Me_3Si–SC_4H_9) = 414 \pm 11$ kJ mol^{-1} as before[1].

High-level *ab-initio* calculations[66] on the exchange reaction

$$H_3SiX + CH_3 \rightarrow SiH_3 + H_3CX$$

yield ΔH° values which, when combined with known $D(CH_3–X)$ data[79], yield figures for $D(SiH_3–X)$ which are within 20 kJ mol^{-1} of the experimental values of Tables 14–16 for Si–H, Si–C, Si–N, Si–O, Si–F and Si–Cl bonds in the mixed hydrides. Only $D(H_3Si–SH) = 348$ kJ mol^{-1} is significantly out of line with the Si–S bond dissociation energy quoted above (but alkyl substitution will account for part of the difference). Theory and experiment appear to be converging in this area. However, other published bond dissociation energies derived from unreliable molecular heats of formation calculated by the MOB1 method[39] (discussed earlier) are themselves unreliable (if not actually wrong).

IV. BIVALENT SILICON, π-BONDED SPECIES AND SEQUENTIAL BOND DISSOCIATION ENERGIES

This is a very active current research area[80] and, if included here in detail, would extend this article beyond reasonable confines of length. We plan to review the data elsewhere[81]. In recent articles we have attempted to establish best values for $\Delta H_f^\circ(SiH_2)$[5,82], $\Delta H_f^\circ(SiMe_2)$[4,36,83], $\Delta H_f^\circ(Me_2Si{=}CH_2)$[4], $\Delta H_f^\circ(H_3SiSiH)$[84] and $\Delta H_f^\circ(H_2Si{=}SiH_2)$[84]. For the sake of completeness best values for these and similar species are simply listed here, with estimates of their uncertainties. They are shown in Tables 17 and 18. These tables also include values for the practically useful indices of reactivity, the π-bond energy, $D\pi$ (for the unsaturated species), and the divalent state stabilization energy, DSSE (for

TABLE 17. Standard enthalpies of formation of unsaturated silicon-containing molecules[a]

Silylene	ΔH_f°/kJ mol^{-1}	π-bond energy (D_π)[b]
$H_2Si{=}CH_2$	155 ± 20	172 ± 20
$MeHSi{=}CH_2$	88 ± 20	172 ± 20
$Me_2Si{=}CH_2$	21 ± 20	172 ± 20
$H_2Si{=}SiH_2$	$\leq 261 \pm 8$	$\geq 100 \pm 12$
$H_2Si{=}O$	-90 ± 30	250 ± 40

[a] See text for reference sources.
[b] See text for definition.

TABLE 18. Standard enthalpies of formation of silylenes[a]

Molecule	ΔH_f°/kJ mol^{-1}	DSSE[b]
SiH_2	273 ± 6	81 ± 8
MeSiH	184 ± 12	101 ± 15
Me_2Si	109 ± 8	118 ± 11
H_3SiSiH	312 ± 8	54 ± 11

[a] See text for reference sources.
[b] See text for definition.

silylenes). These are operationally defined as follows:

(i) $D_\pi(\text{Si=X}) = D_\sigma(\text{Si–H}) + D_\sigma(\text{X–H}) - D(\text{H–H}) + \Delta H_{\text{hyd}}$, where ΔH_{hyd} is the heat of hydrogenation of the species $H_2Si{=}X$ (X being the π-bonded atom or group).

(ii) $\text{DSSE}(SiR_2) = D(R_3Si\text{–}R) - D(R_2Si\text{–}R)$, where R is the substituent in the silylene.

Uncertainties are still quite high for these quantities. For useful reference the sequential bond dissociation energies in SiH_4 and $SiMe_4$ are shown in Table 19. The figures in this table supersede our earlier published values[1,3]. Data for inorganic silylenes and sequential bond dissociation energies in inorganic silanes are published elsewhere[2,5].

TABLE 19. Successive bond dissociation energies in silane and tetramethylsilane[a]

Bond	D/kJ mol^{-1}	Bond	D/kJ mol^{-1}
H_3Si–H	378 ± 5	Me_3Si–Me	376 ± 8
H_2Si–H	297 ± 8	Me_2Si–Me	258 ± 11
HSi–H	*ca* 321[b,d]	MeSi–Me	*ca* 329[c,d]
Si–H	*ca* 292[b,d]	Si–Me	*ca* 303[c,d]

[a] Based on previous tables, except where indicated.
[b] Based on $\Delta H_f^\circ(\text{SiH}) = 377$ kJ mol^{-1} (Reference 25).
[c] Based on $\Delta H_f^\circ(\text{SiMe}) = 293$ kJ mol^{-1} (estimated).
[d] These values may be uncertain by as much as 30 kJ mol^{-1}.

TABLE 20. Group increment contributions to standard heats of formation, for silicon compounds

Group	ΔH_f°/kJ mol^{-1}
Si–(C)(H)$_3$	13.7
Si–(C)$_2$(H)$_2$	−9.1
Si–(C)$_3$(H)	−35.0
Si–(C)$_4$	−62.0
C–(Si)(H)$_3$	−42.8[a]
C–(Si)(C)(H)$_2$	−17.0
O–(Si)(H)	−309.2
O–(Si)$_2$	−396.6
O–(Si)(C)	−230.5
Si–(C)$_3$(O)	−62.0[b]
Si–(C)$_2$(O)$_2$	−76.6
Si–(C)(O)$_3$	−80.8
Si–(O)$_4$	−85.8
C–(C)(H)$_3$	−42.8
C–(C)$_2$(H)$_2$	−20.7
C–(O)(C)(H)$_2$	−33.9
C–(O)(C)$_2$(H)	−30.1

[a] Arbitrary value. C–(Si)(H)$_3$ set equal to C–(C)(H)$_3$.
[b] Arbitrary value. Si–(C)$_3$(O) set equal to Si–(C)$_4$.

TABLE 21. Standard enthalpies of formation for various atoms and free radicals

Species	ΔH_f°/kJ mol^{-1}	Ref.	Species	ΔH_f°/kJ mol^{-1}	Ref.
H	218.0	86	C	715.0	86
CH_3	145.6	86	Si	450.6	86
C_2H_5	118.8	57	Ge	376.6	87
C_6H_5	329	79	Sn	302.1	87
			Pb	195.6	86
OH	39.5	79			
OMe	17.6	79	F	78.9	86
OEt	−17.2	79	Cl	121.0	86
			Br	111.9	86
			I	106.9	86
NH_2	185	79			
NHMe	177	79			
NMe_2	145	79			

TABLE 22. Standard enthalpies of formation of silicon hydrides and halides

Molecule	ΔH_f°/kJ mol^{-1} [a]	Molecule	ΔH_f°/kJ mol^{-1} [b]	Molecule	ΔH_f°/kJ mol^{-1} [b]
SiH_4	34.3 ± 1.2	SiH_3F	−355 ± 2 [c]	SiF_4	−1615 ± 1
Si_2H_6	80 ± 1.5	SiH_3Cl	−136 ± 10	$SiCl_4$	−663 ± 5
Si_3H_8	121 ± 4.4	SiH_3Br	−64 ± 9	$SiBr_4$	−415 ± 8
Si_4H_{10}	(162) [d]	SiH_3I	−2 ± 8	SiI_4	−110 ± 16

[a] Data from Reference 12, unless otherwise indicated.
[b] Data from Reference 2, unless otherwise indicated.
[c] Reference 75. This value is probably more reliable than that of Reference 2.
[d] Additivity estimate.

V. APPENDIX

Group increments for silicon compounds, derived and used in accordance with the group additivity scheme[15,85], are shown in Table 20. Radical and atomic heats of formation for non-silicon-containing species used in this chapter to derive bond dissociation energies are shown in Table 21. Miscellaneous inorganic silane heats of formation used in this chapter are included in Table 22.

VI. REFERENCES

1. R. Walsh, *Acc. Chem. Res.*, **14**, 246 (1981).
2. R. Walsh, *J. Chem. Soc., Faraday Trans. 1*, **79**, 2233 (1983).
3. A. M. Doncaster and R. Walsh, *J. Chem. Soc., Faraday Trans. 2*, **82**, 707 (1986).
4. R. Walsh, *J. Phys. Chem.*, **90**, 389 (1986).
5. R. Walsh, *Pure Appl. Chem.*, **59**, 69 (1987).
6. S. Tannenbaum, *J. Am. Chem. Soc.*, **76**, 1027 (1954).
7. J. D. Cox and G. Pilcher, *Thermochemistry of Organic and Organometallic Compounds*, Academic Press, London, 1970.
8. J. B. Pedley and J. Rylance, *Sussex-N.P.L. Computer Analysed Thermochemical Data: Organic and Organometallic Compounds*, University of Sussex, 1977.
9. V. I. Tel'noi and I. B. Rabinovitch, *Russ. Chem. Rev.*, **49**, 603 (1980).

10. Reference 7, pp. 67–68.
11. W. V. Steele, *J. Chem. Thermodyn.*, **15**, 595 (1983).
12. J. B. Pedley and B. S. Iseard, *CATCH Tables for Silicon Compounds*, University of Sussex, 1972.
13. M. G. Voronkov, V. A. Klyuchnikov, T. F. Danilova, A. N. Korchagina, V. P. Baryshok and L. M. Landa, *Bull. Acad. Sci. USSR, Div. Chem. Sci.*, 1790, 1795 (1986).
14. S. N. Gadzhiev and M. J. Agarunov, *J. Organomet. Chem.*, **11**, 415 (1968); **22**, 305 (1970).
15. S. W. Benson, F. R. Cruickshank, D. M. Golden, G. R. Haugen, H. E. O'Neal, A. S. Rodgers, R. Shaw and R. Walsh, *Chem. Rev.*, **69**, 279 (1969).
16. S. W. Benson and J. H. Buss, *J. Chem. Phys.*, **29**, 546 (1958).
17. T. L. Allen, *J. Chem. Phys.*, **31**, 1039 (1959).
18. K. Moedritzer, *Adv. Organomet. Chem.*, **6**, 171 (1968).
19. K. Moedritzer, unpublished results reported in Reference 18.
20. P. D. Zemany and F. P. Price, *J. Am. Chem. Soc.*, **70**, 4222 (1948).
21. The term *intrinsic* is here used to mean free from (or corrected for) symmetry factors.
22. S. W. Benson, *Angew. Chem., Int. Ed. Engl.*, **17**, 812 (1978).
23. H. E. O'Neal and M. A. Ring, *J. Organomet. Chem.*, **213**, 419 (1981).
24. M. M. Kanovitch, Yu. G. Papulov, V. M. Smolyakov, V. N. Poterin and V. A. Klyuchnikov, *Russ. J. Phys. Chem.*, **56**, 1070 (1982).
25. *JANAF Thermochemical Tables*, 1978 Supplement, *J. Phys. Chem. Ref. Data*, **7**, 793 (1978).
26. B. S. Iseard, J. B. Pedley and J. A. Treverton, *J. Chem. Soc. (A)*, 3095 (1971).
27. M. J. S. Dewar and C. Jie, *Organometallics*, **6**, 1486 (1987).
28. M. J. S. Dewar, J. Friedheim, G. Grady, E. F. Healy and J. J. P Stewart, *Organometallics*, **5**, 375 (1986).
29. S. N. Gadzhiev and A. I. Gubareva, *Russ. J. Phys. Chem.*, **60**, 10 (1986).
30. W. V. Steele, *J. Chem. Thermodyn.*, **10**, 445 (1978).
31. L. H. Long and C. I. Pulford, *J. Chem. Soc., Faraday Trans. 2*, **82**, 465 (1986).
32. I. M. T. Davidson and A. V. Howard, *J. Chem. Soc., Faraday Trans. 1*, **71**, 69 (1975).
33. N. Shimo, N. Nakashima and K. Yoshihara, *Chem. Phys. Lett.*, **125**, 303 (1986).
34. R. Walsh, manuscript in preparation.
35. H. E. O'Neal, private communication.
36. R. Walsh, *Organometallics*, **7**, 75 (1988).
37. T. N. Bell, K. A. Perkins and P. G. Perkins, *J. Phys. Chem.*, **86**, 3922 (1982).
38. P. Potzinger, A. Ritter and J. Krause, *Z. Naturforsch., A*, **30**, 347 (1975).
39. T. N. Bell, K. A. Perkins and P. G. Perkins, *J. Chem. Soc., Faraday Trans. 2*, **77**, 1779 (1981).
40. W. V. Steele, unpublished results (private communication).
41. V. G. Genchel', N. V. Demidova. N. S. Nametkin, L. E. Gusel'nikov, E. A. Volnina, E. N. Burdasov and V. N. Vdovin, *Izv. Akad. Nauk SSSR, Ser. Khim.*, **10**, 2337 (1976).
42. M. C. Flowers and L. E. Gusel'nikov, *J. Chem. Soc. (B)*, 419, 1396 (1968).
43. M. S. Gordon, paper presented at the 8th International Symposium on Organosilicon Chemistry, St. Louis, June 1987.
44. P. v. R. Schleyer, *Substituent Effects in Free Radical Chemistry, NATO ASI Ser. C189* (Ed. H. G. Viehe), D. Reidel Publ. Co., Dordrecht, 1986, pp. 69–81.
45. A. M. Mosin and Yu. Kh. Shaulov, *Russ. J. Phys. Chem.*, **46**, 1054 (1972).
46. A. E. Beezer and C. T. Mortimer, *J. Chem. Soc. (A)*, 514 (1966).
47. J. C. Baldwin, M. F. Lappert, J. B. Pedley and J. A. Treverton, *J. Chem. Soc. (A)*, 1980 (1967).
48. M. A. Ring, H. E. O'Neal, A. H. Kadhim and F. Jappe, *J. Organomet. Chem.*, **5**, 124 (1966).
49. A. M. Doncaster and R. Walsh, *J. Phys. Chem.*, **83**, 3037 (1979).
50. V. A. Klyuchnikov, T. F. Danilova, M. G. Voronkov and V. M. D'yakov, *Dokl. Akad. Nauk. SSSR*, **281**, 1141 (1985).
51. W. D. Good, J. L. Lacina, B. L. De Prater and J. P. McCullough, *J. Phys. Chem.*, **68**, 579 (1964).
52. J. C. Baldwin, M. F. Lappert, J. B. Pedley and J. S. Poland, *J. Chem. Soc., Dalton Trans.*, 1943 (1972).
53. P. Sellers, 4th International Conference on Chemical Thermodynamics, Montpellier (France); cited in Reference 9.
54. A. P. Claydon and C. T. Mortimer, *J. Chem. Soc.*, 3212 (1962).
55. D. M. Golden and S. W. Benson, *Chem. Rev.*, **69**, 125 (1969).
56. W. Tsang, *J. Am. Chem. Soc.*, **107**, 2872 (1985).

57. M. Brouard, P. D. Lightfoot and M. J. Pilling, *J. Phys. Chem.*, **90**, 445 (1986).
58. J. Bracey and R. Walsh, unpublished results cited in Reference 59.
59. M. J. Rossi and D. M. Golden, *Int. J. Chem. Kinet.*, **15**, 1283 (1983).
60. H. Teranishi and S. W. Benson, *J. Am. Chem. Soc.*, **85**, 2887 (1963).
61. N. Auner, R. Walsh and J. Westrup, *J. Chem. Soc., Chem. Commun.*, 207 (1986).
62. M. R. Nimlos and G. B. Ellison, *J. Am. Chem. Soc.*, **108**, 6522 (1986).
63. B. H. Boo and P. B. Armentrout, *J. Am. Chem. Soc.*, **109**, 3549 (1987).
64. J. Berkowitz, J. P. Greene, H. Cho and B. Ruscic, *J. Chem. Phys.*, **86**, 1235 (1987).
65. J. A. Pople, B. T. Luke, M. J. Frisch and J. S. Binkley, *J. Phys. Chem.*, **89**, 2198 (1985).
66. B. T. Luke, J. A. Pople, M.-B. Krogh-Jespersen, Y. Apeloig, J. Chandrasekhar and P. v. R. Schleyer, *J. Am. Chem. Soc.*, **108**, 260 (1986).
67. P. Ho, M. E. Coltrin, J. S. Binkley and C. F. Melius, *J. Phys. Chem.*, **89**, 4647 (1985).
68. N. K. Serdyuk, V. V. Gutorov and V. N. Panfilov, *React. Kinet. Catal. Lett.*, **16**, 393 (1981).
69. K. Y. Choo and M. H. Choe, *Chem. Phys. Lett.*, **89**, 115 (1982); *Bull. Korean Chem. Soc.*, **6**, 196 (1985).
70. R. Ellul, P. Potzinger, B. Reimann and P. Camilleri, *Ber. Bunsenges. Phys. Chem.*, **85**, 407 (1981).
71. J. M. Kanabus-Kaminska, J. A. Hawari, D. Griller and C. Chatgilialoglu, *J. Am. Chem. Soc.*, **109**, 5267 (1987).
72. D. C. McKean, *J. Mol. Struct.*, **113**, 251 (1984) and references cited therein.
73. P. Ho, M. E. Coltrin, J. S. Binkley and C. F. Melius, *J. Phys. Chem.*, **90**, 3399 (1986).
74. I. M. T. Davidson and A. V. Howard, *J. Chem. Soc., Faraday Trans. 1*, **71**, 69 (1975).
75. H. B. Schlegel, *J. Phys. Chem.*, **88**, 6254 (1984).
76. I. M. T. Davidson and C. E. Dean, *Organometallics*, **6**, 966 (1987).
77. T. J. Barton, S. A. Burns, I. M. T. Davidson, S. Ijadi-Maghsoodi and I. T. Wood, *J. Am. Chem. Soc.*, **106**, 6367 (1984).
78. I. M. T. Davidson, T. J. Barton, K. J. Hughes, S. Ijadi-Maghsoodi, A. Revis and G. C. Paul, *Organometallics*, **6**, 644 (1987).
79. D. M. Golden and D. F. McMillen, *Ann. Rev. Phys. Chem.*, **11**, 493 (1982).
80. See, for example, I. M. T. Davidson, *Royal Chem. Soc. Annu. Rep. C*, 47 (1985).
81. R. Walsh, article in preparation.
82. H. M. Frey, R. Walsh and I. M. Watts, *J. Chem. Soc., Chem. Commun.*, 1189 (1986).
83 J. E. Baggott, M. A. Blitz, H. M. Frey, P. D. Lightfoot and R. Walsh, *Chem. Phys. Lett.*, **135**, 39 (1987).
84. R. Becerra and R. Walsh, *J. Phys. Chem.*, **91**, 5765 (1987).
85. S. W. Benson, *Thermochemical Kinetics*, 2nd edn., Wiley–Interscience, New York, 1976.
86. *J.A.N.A.F. Thermochemical Tables*, 2nd edn., NSRDS–NBS 37 National Bureau of Standards, Washington D.C., 1971.
87. *NBS Tables of Thermochemical Properties, J. Phys. Chem. Ref. Data*, **11**, Supplement 2, 1982.

The Chemistry of Organic Silicon Compounds
Edited by S. Patai and Z. Rappoport

CHAPTER **6**

Analysis of organosilicon compounds

T. R. CROMPTON
'Beechcroft', Wittingham Lane, Goosnargh, Prestón, Lancashire, UK

I. ELEMENTS AND FUNCTIONAL GROUPS

A. Elements

1. Determination of silicon

The determination of silicon in organosilicon compounds has been reviewed[1,2].

Silicon can be determined by decomposing the organosilicon compound and then determining the silica or silicate produced. The technique generally used to bring about decomposition is wet combustion or fusion with sodium peroxide[3-11]. If a wet-combustion method is used, the resulting silica can be determined by dehydration of silicic acid by classical methods or by converting to silicofluoride ions in the presence of an excess of hydrogen ions. Silicofluoride is formed in accordance with the following equation:

$$Si(OH)_4 + 6F^- + 4H^+ = SiF_6^{2-} + 4H_2O$$

Fusion with sodium peroxide leads to the formation of sodium silicate, which can be determined colorimetrically, e.g. as molybdosilicic acid, volumetrically, or gravimetrically as the oxine or quinoline salt of molybdosilicic acid. A major problem associated with the analysis of organosilicon compounds is the successful decomposition of the sample. Some compounds give rise to silicon carbide when digested with acids. Generally, fusion with sodium peroxide in a Parr-type bomb is the most satisfactory way of dealing with a wide variety of compounds. Alternate methods of decomposition are available, but are usually more time-consuming. Thus when siloxanes are decomposed by heating with a mixture of sodium and potassium hydroxides in a sealed nickel crucible, quantitative decomposition may require more than three hours[12-15]. The most commonly used procedure is fusion with sodium peroxide and determination of the silicate so formed by precipitation and titration of the quinoline salt of molybdosilicic acid.

a. Bomb fusion with sodium peroxide. Following decomposition of the organosilicon compound by sodium peroxide, the resulting silicate can be determined spectrophotometrically as molybdosilicic acid[3], volumetrically and gravimetrically as the oxine salt of molybdosilicic acid[4] or gravimetrically as the quinoline salt of molybdosilicic acid[5]. Colorimetric determination of silicon as molybdosilicic acid is subject to error because of the formation of the metastable beta form of the acid, and a determination based on the formation of the alpha form has been described[7]. The colorimetric method based on the reduction of molybdosilicic acid to molybdenum blue involves use of sodium sulphite in acid solution as the reducing agent[8,9]. It has been suggested that reduction by stannous chloride enhances the colour produced and therefore possibly gives greater sensitivity in the determination of silicon by this method.

A rapid procedure has been described[10] based on sodium peroxide bomb fusion, for the determination of silicon and halogen in fluorine-containing organosilicon compounds and resins. The silicon is separated from the decomposition product as zinc silicate and estimated gravimetrically as silica. The filtrate is concentrated, acidified and, when necessary, reduced with sulphur dioxide. Chloride, bromide or iodide is then determined by the usual methods. Fluoride can be determined in neutral solution either gravimetrically as calcium fluoride, or volumetrically with zirconium tetrachloride or thorium nitrate, or directly in the decomposition solution by titration with zirconium tetrachloride.

Procedures have been described[11] for the gravimetric determination of silicon in organosilicon compounds on the milligram scale. The sample is fused with sodium peroxide, and silicon is determined in the extract by precipitation as quinoline silicomolybdate. This procedure is not applicable if phosphorus is present in the sample but the presence of fluorine within prescribed limits is tolerated. The same workers compared the precisions obtained in the precipitation of the blue and yellow quinoline silicomolybdate complex from standard silicate solutions and concluded that there was no statistically significant difference in the precisions obtained and that the theoretical gravimetric factor (0.0120) could be applied to the yellow complex.

b. Open and sealed fusions with alkalis. A method has been described[12] employing alkali fusion in an open nickel crucible for the determination of silicon in siloxane polymers and silicone-containing samples, utilising the Technicon Auto Analyser system. In this method[13] Oxine precipitation and reduced heteropoly blue complex methods were used to estimate silicon in the alkali digests. Generally, with linear and cyclic polydimethylsiloxanes and various other types of silicones, a higher silicon recovery was obtained by direct fusion with potassium hydroxide than with sodium hydroxide. However, recoveries of low-molecular-weight cyclic polydimethylsiloxanes were usually less than 50% using either fusion reagent. The use of an alcoholic alkali pre-treatment, using either alkali, minimised losses of such volatile silicon compounds and led to general improvement in silicon recovery, since under these mild conditions the siloxane bonds are converted to non-volatile alkali metal silanolates.

$$\equiv\text{Si–O–Si}\equiv + \text{KOH} \xrightarrow{\text{ROH}} \equiv\text{Si–O–K} + \text{HOSi}\equiv$$

$$\equiv\text{SiOH} + \text{KOH} \xrightarrow{\text{ROH}} \equiv\text{Si–O–K} + \text{H}_2\text{O}$$

After alcohol removal, fusion of the sample cleaves all organic groups from silicon, converting silanolate to water-soluble silicates.

$$\equiv\text{Si–O–K} \xrightarrow{\text{ROH}} \text{K}_4\text{SiO}_4 + \text{hydrocarbons}$$

Both the direct and the alcoholic alkali fusion methods gave recoveries of 90% or better with phenyl-containing silicones, the results obtained by the alcoholic alkali modification being generally higher.

Other workers[14,15] studied the decomposition of siloxanes with mixtures of sodium hydroxide and potassium hydroxide in a sealed[12] nickel crucible followed by gravimetric or colorimetric determination of the silicate produced. When fused at 210°, dimethylsiloxane oil, octaphenylcyclotetrasiloxane, diphenylsilanediol, siloxanes of the varnish type and siloxanes modified with alkyd resins are quantitatively decomposed within 1, 2, 4, 4 and 4 h, respectively. Another worker[15] decomposed organosilicon compounds by fusion with potassium hydroxide in a nickel microbomb or by wet combustion with fuming sulphuric acid. Silicon is subsequently determined as molybdenum blue at 812 nm. It is claimed that results are accurate to within $\pm 0.2\%$ for silanes, siloxanes and their derivatives and also for complex organosilicon compounds also containing phosphorus, arsenic, antimony, germanium, titanium, aluminium and tin. In an alternate method, the organosilicon compound is fused with potassium hydroxide in a nickel bomb followed by conversion of the silicate to the quinoline salt of molybdosilicic acid, which is then estimated by acidimetric titration[4].

c. Mineralization and digestion. Various acids and other reagents have been used for the mineralization of organosilicons preparatory to the determination of silicon. These include concentrated or fuming nitric acid and sulphuric acid[16,18], mixtures of nitric acid, sulphuric acid and hydrofluoric acid[19], ammonium persulphate[20], fuming sulphuric acid–ammonium sulphate[21], chloric–sulphuric acid[22], hydrochloric acid[23] and ethanolic hydrofluoric acid[24,25].

(*i*) *Fuming nitric–sulphuric acid*[16]. In a procedure claimed to be suitable for the determination of silicon in esters of orthosilicic acid, arylsilanes, arylalkylsilanes, polysiloxanes, alkylsilanes and alkyl- and aryl-halogenosilanes, the determination of silicon is carried out acidimetrically. A mixture of oleum and fuming nitric acid is used for the oxidation, which is followed by ignition. Different techniques are used according to the volatility and ease of oxidation of the compound. Non-volatile compounds are weighed directly into a platinum crucible to which the oxidants are added. Alkylsilanes are weighed into a crucible (cooled in solid carbon dioxide) containing the oxidants. Arylhalogenosilanes are oxidised in nitric acid–sulphuric acid containing a little water; alkylhalogenosilanes are weighed into a platinum crucible (containing pyridine cooled in solid carbon dioxide) and the oxidants are added to this. The crucible is heated gently on a sand bath, then more strongly until the contents are ignited. The residue is dissolved in the minimum amount of water containing sodium (or potassium) carbonate, transferred to a flask, acidified by dilute hydrochloric acid and boiled to remove carbon dioxide, and the analysis is carried out by acid-base titration.

In another procedure[17] the decomposition of organosilicon compounds is effected in a quartz flask using a mixture of 25% oleum and fuming nitric acid, containing 2% oxides of nitrogen. The heated mixture is treated with further amounts of fuming nitric acid until decomposition is complete, the acids are evaporated off and the flask containing the residue of silica is weighed.

(*ii*) *Nitric–sulphuric–hydrofluoric acid.* Mixtures of nitric acid, sulphuric acid and hydrofluoric acids have been used[19] for the determination of silicon in alkyl- or aryl-substituted organosilicon compounds such as trichlorophenylsilane, dichloromethylphenylsilane, diethoxydiphenylsilane and triphenylsilanol with a relative error not exceeding $\pm 1.5\%$. The sample in a platinum crucible is treated with fuming sulphuric acid and fuming nitric acid and heated slowly until sulphur trioxide evolution ceases. Then, 20 ml 30% hydrofluoric acid–anhydrous acetic acid–methanol (1 : 1 : 10) is added and the mixture cooled in solid carbon dioxide–acetone. When the residue is completely dissolved the resulting hydrofluorosilicic acid solution is transferred to a polyethylene beaker and 30 ml ethanol added. The solution is titrated conductometrically with 0.5% benzidine in ethanol and the end-point deduced graphically.

d. Ammonium persulphate. Ammonium persulphate has been used[20] as a digestion reagent for the determination of silicon in tetraalkoxysilanes and polymeric organometallic siloxanes containing aluminium or titanium. The sample containing 4–5 mg silicon is heated to fumes with sulphuric acid and then with ammonium persulphate until it is colourless. The cooled residue is treated with hydrogen peroxide and with excess solid potassium hydroxide and neutralised with sulphuric acid. The solution is then boiled with an acidic ammonium molybdate solution and the molybdosilicic acid extracted into butan-2-ol. An aliquot of the extract is mixed with acetone and titrated conductometrically with tetraethylammonium hydroxide which has been previously standardised against pure silicate solutions of known composition.

(*i*) *Fuming sulphuric acid–ammonium sulphate.* A procedure has been described[21] for the determination of silicon in polyaluminoorganosiloxane resins. The resin is decomposed with fuming sulphuric acid in the presence of ammonium sulphate, the solution is diluted and the aluminium complexed with tartaric acid. The finely divided resin and ammonium sulphate are heated in a quartz crucible with dropwise addition of fuming sulphuric acid. Silicic acid is precipitated and transferred back to the quartz crucible, the filter paper ashed at 800°, then ignited to constant weight at 1000–1050° and finally weighed as silica.

(*ii*) *Chromic–sulphuric acids.* In a rapid procedure[22] for the determination of silicon in organic silicon compounds, based on their oxidation by a mixture of chromic and sulphuric acids at 150°, the resulting silica is filtered off, dissolved in sodium hydroxide and converted into fluorosilicate ion by ammonium fluoride in the presence of an acid; the excess of acid is back-titrated with alkali.

(*iii*) *Ethanolic Hydrofluoric Acid.* Various workers[24, 26] have discussed this technique in which the hydrofluoric acid produced during the digestion is estimated by various methods such as iodometry[24], conductometric titration with benzidine, toluidine, or 2-naphthylamine[25] or gravimetrically as the benzidine–H_2SiF_6 complex followed by titration with standard potassium hydroxide solution[27]. Other workers have reported procedures for the determination of organically-bound silicon, based on formation of the fluorosilicate ion. In one procedure[28, 29] the fluorosilicate finish is accomplished by alkaline titration. A gravimetric finish[28] has been described involving precipitation of the benzidine–fluorosilicate complex, followed by ethanol dissolution of the latter and titration with standard sodium hydroxide; another method[26] describes a finish based on titration with methanol or acetone solutions of benzidine or 2-naphthylamine which, it is claimed, is applicable to the determination of silicon in alkoxysilanes such as tetramethoxysilane, tetraethoxysilane, tetrapropoxysilane, tetrabutoxysilane and in polymers and resins with organosilicon compounds.

(*iv*) *Miscellaneous methods.* Hydrochloric acid digestion has been used[23] for the determination of silicon in easily hydrolysible materials such as ethyl silicate. A spectrophotometric evaluation of the silicomolybdate complex is used in this procedure. Another method[30] describes the spectrophotometric determination of silicon in ethyl silicate and polyethylsilicate in which the sample is hydrolysed with acid in the presence of ammonium molybdovanadate to form the yellow molybdovanadosilicate. In another method[31] to analyse hydrolysible organosilicon compounds the sample is treated with potassium hydroxide solution, then in turn with ammonium molybdate, sodium sulphite and sodium sulphate and finally with a reagent consisting of ammonium oxalate, sodium carbonate and glycerol. The molybdenum blue colour formed was evaluated at a wavelength of 650 to 700 nm.

Volatile organosilicon compounds have been analysed for silicon[32]. Oxygen is bubbled through the weighed sample, the vapours are passed into a weighed silicon tube and ignited on a plug of prepared asbestos. The silica tube is removed and ignited to a constant weight at 800°.

In the oxygen flask combustion technique[33] the sample is burnt in an oxygen-filled nickel flask. The silica produced is absorbed in a sodium hydroxide solution and the silicate concentration is determined spectrophotometrically at a wavelength of 400 nm[34]. An accuracy of within $\pm 2\%$ is claimed for this method.

Another method[34] involves fusion of the organosilicon with magnesium to produce magnesium silicide. This is then decomposed to produce gaseous silicon hydrides by addition of dil. sulphuric acid. The hydrides are absorbed in bromine water and thus hydrolysed to silicic acid. The latter is converted into molybdosilicic acid and determined colorimetrically as molybdenum blue. The error is ± 0.73 to 0.54% over the range 10.53 to 37.8% of silicon. After removal of silicon hydrides, the elementary carbon which separates is deactivated with ferric or aluminium salts and filtered off, when halogens can be determined in the filtrate by Volhard's method.

In emission spectrography[35] the sample is dissolved in a suitable solvent, together with a known weight of zinc chloride. The sample is placed in a U-shaped fulgurator, one limb of which consists of a hollow graphite electrode. The test solution slowly passes to the top of this electrode and forms a film during a spark discharge between the end and a copper counter electrode. The lines of silicon (634.701 nm) and zinc (636.235 nm) are measured.

Spectrographic analysis of organosilicon compounds has been carried out[36] by the use of the emission lines at 634.7 and 637.1 nm. The sample is decomposed in acid solution by a spark discharge between platinum electrodes.

Silicon has been determined in organosilicon compounds such as arylsilanes by X-ray fluorescence with a vacuum spectrograph[37].

A method has been described[38] for the determination of trace amounts of siloxanes in water sediments, sewage sludges and biological materials. The method involves extraction into petroleum ether, evaporation to dryness, dissolution of the residue in methyl isobutyl ketone and determination by inductively coupled plasma emission spectrometry. The detection limit is 0.01 μg per ml for organosilicones. In a further method[39] the volatiles produced by the thermal or catalytic cracking (platinum) followed by oxidation of organosilicon compounds are passed over a wire electrode fed by a 12,000–15,000 V high-frequency generator on which the finely divided silica is electrodeposited and subsequently weighed. An absolute error not exceeding $\pm 0.3\%$ is claimed for this procedure.

2. *Phosphorus, aluminium, titanium and tin in organosilicon compounds*

Occasionally elements other than silicon occur in organosilicon compounds. Available methods for the determination of these elements are reviewed in Table 1.

Phosphorus has been determined[40] in phosphosilico organic compounds (POSi bonds) by dissolving the sample in water, leaving for 4–5 h to hydrolyse (or for 30–40 min at 80°) and titrating the alkylphosphinic acid so produced with 0.1 M-alkali to the thymolphthalein end-point. A method has been described[41] for the simultaneous determination of free phosphorus, combined phosphorus and silicon in the reaction products of tetraalkoxysilanes with phosphorus halides.

3. *Determination of halogens*

Halogen in organosilicon compounds may be either directly linked to silicon or may be part of a substituent group. Total halogen is usually determined by alkali fusion methods. Halogen attached directly to silicon is easily hydrolysable by aqueous alcohols and the halogen acid so produced can be estimated by direct alkali titration or by indirect procedures.

a. Total halogen. Bomb fusion methods have been described[64] for the simultaneous determination of fluorine, chlorine and silicon in organosilicon compounds. The sample mixed with 4–5 times its weight of solid potassium hydroxide is placed in a nickel bomb. If the sample contains one fluorinated phenyl radical it is fused for 40–45 min at 900–950°; if it contains two fluorinated phenyl radicals it is fused for 60 min at 1000°. The melt is then allowed to cool and is dissolved in 200 ml water. Fluoride and chloride are determined respectively in two aliquots by titration with thorium nitrate and ammonium thiocyanate using standard procedures. Silicon is determined in a further aliquot of the test solution by an acidimetric procedure with a mixed methyl red bromocresol green indicator.

A method has been described[65] for the estimation of fluorine, chlorine and silicon, based on fusion with potassium metal in a bomb. The organosilicon sample is weighed in a gelatin capsule or polyethylene ampoule. Ignition and analysis of the melt are carried out as described above.

Bomb fusion with alkali has been applied[66] to the determination of fluorine in organofluorosilicon compounds. It is recommended that the large amount of alkali present after the decomposition of the organofluoro compound is removed by the use of a cationite. The test solution followed by water is passed through a column of SDV-3 ion-exchange resin and the combined solution is titrated with standard thorium nitrate solution, using Alizarin red S indicator.

TABLE 1. Methods for determination of aluminium, silicon, phosphorus, titanium and tin in organosilicon compounds

Elements	Sample Digestion	Finish	Comments	References
P, Si	Concentrated nitric–sulphuric acid	Gravimetric phosphate using nitratopentammino cobaltinitrate reagent. Gravimetric silicon as SiO_2 in platinum crucible.		42–45
P, Si	Sodium peroxide bomb fusion	Gravimetric phosphate as quinoline phosphomolybdate.	Accuracy within $\pm 0.3\%$.	44
P, Si			Satisfactory in presence of fluorine.	43, 46–48
Si		a. Gravimetrically using lead chloride.	Satisfactory in presence of fluorine.	49, 50
		b. Gravimetrically using oxine.	Unsatisfactory in presence of phosphate.	
			Interference by fluorine in determination of silicon can be overcome by addition of boric acid.	51
P, Si	Sodium peroxide fusion in nickel bomb	Silicon, spectrophotometrically as silicomolybdate.	Satisfactory in presence of fluorine, mean recovery 101%.	52, 53
		Phosphorus, spectrophotometrically as vanadomolybdate.		
Al	Digestion with acetone–hydrochloric acid.	Titrimetric EDTA finish.		54
	Wet digestion with nitric acid–sulphuric acid–ammonium acetate.	Titrimetric EDTA finish.		55

TABLE 1. (*continued*)

Elements	Sample Digestion	Finish	Comments	References
	Digestion with hydrofluoric acid–sulphuric acid followed by fusion of residue in sodium carbonate.	Polarography	Applied to alumino–organosiloxanes.	57
Al, Si	Digestion with acetone–hydrochloric acid	Aluminium, titrimetric EDTA finish. Silicon, gravimetric.		56
Al, Si		Cumene solution of sample analysed by emission spectrography.		56
Al, Si	Digestion with concentrated sulphuric acid–chromic acid in current of oxygen.	Silicon determined volumetrically as Na_2SiF_6. Aluminium determined by titrimetric EDTA method.	Applied to alumino–organosiloxanes.	58
Al, Si, P	Digestion with diethyl ether–concentrated sulphuric acid–potassium persulphate.	Silicon determined gravimetrically. Aluminium determined by titrimetric EDTA method. Phosphorus determined by spectrophotometric molybdenum blue method.	Applied to complex polysiloxanes.	59, 60
Ti, P, Si	Digestion with concentrated sulphuric acid–potassium persulphate.	Silicon determined gravimetrically. Phosphorus and titanium separated by ion-exchange chromatography. Phosphorus determined spectrophotometrically as molybdenum blue. Titanium determined spectrophotometrically.	Absolute errors in the 10–33% element range less than 0.5%.	60, 61

Ti, P, Si	Fusion with sodium fluoride–potassium pyrosulphate–concentrated sulphuric acid.	Analysed by classical procedures.	62
Ti, Al	Digestion with fuming sulphuric acid–ammonium sulphate–concentrated nitric acid.	Aluminium, titrimetric EDTA. Titanium, EDTA titration of titanium–hydrogen peroxide complex.	55
Sn, Si, Ti	Digestion with concentrated sulphuric acid in platinum followed by addition of ammonium nitrate.	Analysis by classical methods.	63

b. Alcoholysable halogen. Non-aqueous titrimetry has been applied[67] to the determination of chloro-, bromo- and iodo-silanes and siloxanes. This procedure, which can also be applied on the micro-scale[68], involves alcoholysing the sample in isopropanol and titrating it with standard alcoholic potassium hydroxide using a tetrabromophenolphthalein ethyl ester (bromophthalein magenta E) end-point. Analyses of iodosilanes by the above methods were not quantitative. Iodosilanes can be determined, following reaction with excess ethanolic base, by back-titration with standard acid. This method precludes the formation of ethyl iodide. Standard deviations range from 0.7 parts per thousand (ppt) for dichloro (2-carbomethoxyethyl) methylsilane to 2.9 ppt for the volatile dichlorodimethylsilane, and to a better-than-expected 0.6 ppt for the iodotrimethylsilane by the back-titration technique.

In variants based on the determination of alcoholyzable halogen, chlorine and bromine have been determined[69] by solvolysing the sample in a sealed tube with sodium butoxide. The resulting chloride or bromide was determined by argentimetric titration, with visual, or preferably potentiometric, indication of the end-point.

The effectiveness has been studied[70] of various media for hydrolysing alkyl- and arylchlorosilanes. The best medium was shown to be a mixture (1:1) of ethanol and water. After hydrolysis, chloride ion was determined by neutralisation or by mercurimetry. To determine total chlorine a sodium in liquid ammonia method was used[71]. Halogen on β-carbon atoms attached to silicon can be hydrolysed with sufficient ease to be titrated with dilute solutions of bases[72]. Volumetric methods have been described[73] for the determination of chlorine in methyl chlorosilanes and in phenylchlorosilanes. In this method ethyltrichlorosilane, dimethyldichlorosilane, phenyltrichlorosilane or diphenyldichlorosilane is hydrolysed with 50% aqueous ethanol in a sealed vessel, and chloride is titrated with silver nitrate in an aqueous suspension. Dichlorofluorescein is the best indicator. The presence of ethanol and organopolysiloxanes does not interfere in the method. Chronoconductiometry has been used[74] for the analysis of mixtures of trichlorosilane and silicon tetrachloride. The method is based on the determination of hydrochloric acid formed during the hydrolysis of the substrates and hydrolysis of the reaction products of trichlorosilane with mercurous chloride. One portion of the sample mixture is hydrolysed with 50% ethanol, whilst a second portion is treated with ethanolic mercurous chloride and then hydrolysed. The hydrochloric acid formed is determined by chronoconductimetric titration with sodium acetate and the concentration of each substance in the sample is calculated from the total hydrochloric acid formed during the two treatments. The accuracy of the method is within $\pm 3\%$ for each substance. Decomposition of chlorosilanes with ammonium fluoride in glacial acetic acid has been used[75] for the determination of chlorine in organosilanes.

$$R_nSiCl_{4-n} + 6NH_4F + nCH_3COOH \rightarrow (4-n)NH_4Cl + (NH_4)_2SiF_6 + nRH + nCH_3COONH_4$$

Chloride is determined by mercurimetric titration to the sodium nitroprusside end-point. Carbonates, acetates and borates do not interfere in the titration but large amounts of ammonium salts do interfere. The method has been applied to methyltrichlorosilane, dimethyldichlorosilane, diethyldichlorosilane, phenylmethyldichlorosilane and phenyltrichlorosilane. Methods have been described[76,77] based on reaction with amines for the determination of chlorine directly linked to silicon in alkyl aryl chlorosilanes. One method[76] is based on the formation of aniline hydrochloride according to the equation:

$$R_{n'}SiCl_{4-n} + 2(4-n)C_6H_5NH_2 \rightarrow R_nSi(NHC_6H_5)_{4n} + (4-n)C_6H_5NH_2.HCl$$

The sample is placed in a glass ampoule, in a dried and weighed beaker with a porous bottom containing 7–10 ml of a mixture of anhydrous aniline with benzene or ether (1:4). The ampole is broken and after 5 min the liquid is sucked off and the precipitate washed with benzene or ether. The beaker containing the precipitate of aniline hydrochloride is dried to constant weight at 105°. The method is not suitable for the analysis of alkoxychlorosilanes. A method based on similar principles has been described[78] for the determination of chlorine in organochlorosilanes of the general formula R_nSiCl_{4-n} (R = Me, Et or Ph). These substances react with aniline in diethyl ether with 97% yield at room temperature. The sample solution in ether contains a few drops of ammonium thiocyanate and ferric chloride. It is titrated with aniline in toluene until a red colour appears. The amount of the organosilane can then be calculated by an empirical formula. The reaction of alkyl chlorosilanes and aromatic amines to form complex amine hydrochlorides, sometimes of low solubility, has been used for conductometric titration of chlorine in organosilicon compounds. During titration in acetone with benzidine, aniline or other amines, with platinum electrodes, the conductivity decreases up to the end-point, then remains steady. A further method[79] for the determination of chloride directly linked to silicon in certain alkyl- and arylchlorosilanes is based on the iodometric determination of the hydrochloric acid formed on hydrolysis of the chlorosilanes. The sample is transferred to a flask containing a potassium iodide–potassium iodate solution and the liberated iodine is titrated with sodium thiosulphate. This method gives low results when applied to the determination of chlorine in alkoxychlorosilanes. Another method[20] is based on the observation that ammonium chloride is precipitated by double decomposition when ammonium thiocyanate is heated with dimethyldichlorosilane:

$$2NH_4CNS + (CH_3)_2SiCl_2 = 2NH_4Cl + (CH_3)_2Si(CNS)_2$$

To the chlorosilane solution in ether are added a few drops of 0.3 M ammonium thiocyanate in acetone from a micro-burette. The turbid mixture is shaken and one drop of a 1% solution of ferric chloride in ether is added. The titration is continued with vigorous shaking and cooling of the vessel to below 5° in ice water until the red colour of ferric thiocyanate persists. The concentration (C) of the chlorosilane is given by $C = AV \times 354{\cdot}57/B\%$ (w/v), where A is the molar concentration of the ammonium thiocyanate solution, V is the titre in ml and B the chlorine content in % (w/w) of the chlorosilane. The method has the advantage, over methods performed in aqueous media, that no precautions against loss of hydrochloric acid are needed. Organochlorosilanes of general formula R_nSiCl_{4-n} ($R = CH_3$ or C_6H_5) react with potassium thiocyanate or ammonium thiocyanate in anhydrous ether to yield potassium chloride or ammonium chloride quantitatively[81]. An ethereal solution of the silane (containing less than 100 mg chlorine) is mixed with dry ether and ferric chloride in ether and titrated with 0.1–0.5 M potassium thiocyanate or ammonium thiocyanate in dry acetone until the solution becomes red. The error is less than 1% (relative) for methylsilanes and less than 2% for phenylsilanes and the method is applicable to the determination of total chlorine in mixed organochlorosilanes. Various workers have described methods for the determination of fluorine in organosilicon compounds[49] or in organosilicon phosphorus compounds[82,83].

4. Determination of carbon, hydrogen, oxygen and silicon

Various combustion procedures have been described for the determination of carbon[84–97], hydrogen[84–97], oxygen[98] and silicon[90,91] in organosilicon compounds.

a. Carbon and hydrogen. Rapid combustion methods using chromic oxide mixed with fibrous asbestos as a catalyst (to prevent the formation of silicon carbide) have been used[90,91,99] to determine carbon, hydrogen and silicon. The sample is placed at the closed end of a quartz tube and the tube is three-quarters filled with a chromium oxide–asbestos catalyst, prepared by ignition of ammonium chromate or ammonium dichromate mixed with asbestos and heated in oxygen at 900–950°. The tube is weighed and gently heated with a burner, which is moved gradually from the open to the closed end while oxygen is passed through the combustion tube and through the normal absorbers for water and carbon dioxide. The concentration of silicon in the sample is calculated from the weight of silica. To determine carbon, hydrogen, silicon and halogen attached to silicon[90,91] the weighed sample in a quartz vessel and previously ignited asbestos are placed in a combustion tube connected to three absorbers for halogens, water and carbon dioxide, respectively. If the substance contains nitrogen, a further absorber is placed between the second and third absorber. After combustion, the beaker and the three absorbers are weighed. The method is also suitable for many other compounds containing halogen attached to silicon, but some of them give silicon carbide instead of silica. In such cases a catalyst (chromium dioxide) has to be used. Standard micro dry combustion techniques have been used[92,94] for determining carbon and hydrogen in organosilicon compounds. It is necessary in such procedures to ensure that silicon is completely converted to silica, and that silicon carbide formation is avoided, since this will lead to low carbon analyses. It has been suggested[95,96] that a two-stage combustion, in which the sample is vaporised at as low a temperature as possible and then oxidized on a platinum gauze at 850°, will overcome this difficulty, and indeed this method gave good results for a wide variety of organosiloxane polymers. Further modifications of this method have been described[97,99] in which the use of asbestos (which is hygroscopic) is avoided, and a 100-fold amount of powdered chromium trioxide is used. The packing of the catalyst is critical: loose packing allows escape of the silica formed, whereas too-dense packing can lead to explosion.

The formation of stable silanes can be avoided by using magnesium oxide as the catalyst at 900°. The initial volatilization of the sample must be carried out very carefully[84].

In a further semi micro combustion method[85] for the determination of carbon and hydrogen in low boiling compounds the decomposition product of $AgMnO_4$ was used as combustion catalyst.

Cobaltic oxide on pumice has also been used[86] as a catalyst to avoid the formation of silicon carbide. This catalyst is heated at 650–700° and silver wool at 450–500° (to absorb halogens) can be used for the analysis of compounds containing carbon, hydrogen, oxygen, silicon, chlorine, bromine and sulphur. Explosions, likely to occur during combustion, can be avoided by conducting the combustion in a stream of nitrogen or in the presence of a large excess of oxygen[87]. Other workers[88,99] describe a combustion method for the determination of carbon and hydrogen in organo-oxyhalogenosilanes based on oxidation with mixtures of chromic, sulphuric and phosphoric acids at 150–180°. Carbon dioxide and chlorine (or bromine) are determined by a standard method. The presence of sulphur does not cause interference.

b. Carbon, hydrogen, oxygen and silicon. A method has been described[98] for the determination of hydrogen, oxygen and silicon in volatile or gaseous organosilicon compounds. After combustion, silicon remains as silica in the combustion tube, whilst hydrogen and oxygen are determined by standard methods. Halogen, if present, is removed quantitatively by a boat containing powdered silver in an additional furnace. With 30–40 mg samples, the method gives an accuracy of $\pm 0.4\%$ for silicon, $\pm 0.3\%$ for carbon, $\pm 0.2\%$ for hydrogen and $\pm 0.2\%$ for chlorine.

c. Carbon, chlorine, bromine and silicon. In this combustion method[58,59,100] chromium trioxide–sulphuric acid is used to oxidize the sample, and the products are further oxidised over chromium trioxide on pumice. Carbon is determined as carbon dioxide, chlorine or bromine are determined argentimetrically. Silicic acid is precipitated, separated, converted into fluorosilicate and determined volumetrically as previously described[59]. Polyorganoaluminosiloxanes are incompletely oxidised under the conditions cited and must first be decomposed by heating with conc. sulphuric acid, when the resulting siloxanes are readily oxidised. Carbon is determined gravimetrically, silicon volumetrically, and aluminium in the filtrate after removal of silicon.

B. Functional Groups

a. Phenyl and alkyl radicals. Spectroscopy has been used[101] for the determination of phenyl radicals in methyl phenyl-siloxane resins. The sample is dissolved in chloroform and the extinction measured at 260, 266 and 272 nm in 1 mm quartz cells. The results are referred to a calibration graph prepared using pure phenyltriethoxysilane and standard mixtures containing methyl groups. Good agreement was obtained with results obtained by IR spectroscopy.

Phenyl radicals have been estimated[102] by reacting the sample with anhydrous aluminium chloride followed by the addition of ethyl bromide, and reflux for 2 h. The hexaethylbenzene formed is extracted with diethyl ether, washed, and then ether and ethyl bromide are distilled off. The residue is dried over phosphorus pentoxide and weighed. The phenyl radical content can then be calculated using an empirical coefficient of ethylation of 0.91.

A gasometric method has been described[103] for the determination of alkyl radicals in polyalkylsiloxanes and other organosilicon compounds. Methyl and ethyl radicals are determined by heating the sample with powdered potassium hydroxide for 2 h at 250–270 °C and measuring the amount of gaseous hydrocarbon evolved. An accuracy of 1–2% is claimed. Various workers[104, 107] have investigated the application of gas chromatography to the determination of alkyl and phenyl groups in organosilicon compounds. Ethyl and phenyl groups have been determined[105] in siloxane polymers. These groups were split off with phosphorus pentoxide and water and the liberated ethane and benzene determined by gas–solid chromatography with argon as carrier gas and flame ionisation detection. The error in this procedure is claimed to be less than 10%. The technique of alkali fusion reaction–gas chromatography has been applied[106, 107] to the determination of alkyl and aryl groups in polysiloxanes. Reactions are driven to completion with no apparent decomposition, in less than 10 min, by fusing the sample with potassium hydroxide in an inert atmosphere, and the volatile products are separated and determined by gas chromatography. Fluids, gum rubbers and resins are handled with equal ease. A gas chromatogram obtained from a polymethylphenyl-siloxane shows that methane and benzene as the only volatile reaction products, originating from the silicon-bound methyl and phenyl groups.

b. Vinyl groups. Reaction-gas chromatography has been used to determine vinyl groups in orgaosilicon compounds, particularly in siloxane polymers. Generally, these methods involve reaction of the sample with hot phosphorus pentoxide[109, 110], syrupy phosphoric acid[110], 90% sulphuric acid[111] or potassium hydroxide pellets[111, 112] or by pyrolysis followed by estimation[113] of the ethylene produced by flame ionisation gas chromatography.

A method has been reported[110] for determining vinyl groups involving cleavage with phosphorus pentoxide to liberate vinyl groups as ethylene which was estimated by gas chromatography.

A similar method[108, 109] is based on the chromatographic determination of the ethylene formed during interaction between the polymer, phosphorus pentoxide and water. An alumina column (170 cm × 6 mm) is used and detection is by flame ionisation. The relative errors in determining 0.05 and 20% of vinyl groups are less than 10 and 4%, respectively.

A thorough investigation has been carried out[111] of methods for determining vinyl groups in siloxane polymers. In the range 0.1 to 2% of vinyl groups generally found in such polymers, these methods all give precisions no better than ± 5 to 10%. Reaction conditions were critical and temperatures above 400 °C were required for the complete cleavage of the sample. Sulphuric acid rather than phosphoric acid was used for the cleavage of alkyl groups in siloxanes, since the former is reactive at lower temperatures. Poropak S packing and a hot-wire thermal conductivity detector was used at 190 °C with He as the carrier gas. The conditions are not critical as long as they can be reproduced accurately from run to run.

c. Silicon–hydrogen bonds and ≡Si–H groups. The microdetermination of silicon–hydrogen bonds in compounds such as dichlorodimethylsilane has been carried out[114] by destruction of the Si–H bond in the presence of potassium hydroxide to produce an equivalent volume of hydrogen, which is determined by gas chromatography. The sample is injected onto a porous mass (e.g. Rysorb P) saturated with aqueous potassium hydroxide. The products are led into a column of activated charcoal with nitrogen as carrier gas and thermal-conductivity detection. The error is $\pm 6.5\%$ for a content of 0.1% of dichloroethylsilane, and the sensitivity is 0.005%.

Bromination and reaction with mercuric chloride are the basis of the two principal methods for estimating silidyne groups. SiH and ≡Si–H groups react quantively with bromine solution in acetic acid at room temperature, and at the boiling point of the acid[115], respectively. Addition of excess bromine, followed by addition of excess potassium iodide and back-titration of excess iodine with 0.1 n sodium thiosulphate permit quantitative determination of either group, separately or in the same molecule. The accuracy of determination of the SiH group by this method (15–120 mg samples) is within $\pm 5\%$, and for the ≡Si–H group (30–100 mg samples) is within $\pm 0.5\%$.

A bromination method has been described[116] based on amperometric titration at 0.3 V of the ≡Si–H groups in the presence of double bonds in 4 m hydrochloric acid medium with 0.1 M-methanolic bromine saturated with sodium bromide with a rotating platinum micro-electrode, then analogous titration of the double bonds after treatment with methanolic potassium hydroxide to destroy the silidyne groups. The content of silidyne groups is then found by difference.

Other workers[117] treat the sample containing down to 0.0025% H as silidyne groups with bromine in acetic acid, and the unconsumed bromine is determined iodimetrically by adding potassium iodide solution and titrating the liberated iodine with sodium thiosulphate. The infrared determination of ≡SiH in the range 0.001 to 0.03% has been described[118]. Various workers[119, 120] determine small amounts of organosilicon compounds (containing SiH groups in alkyl and aryl chlorosilanes) mercurimetrically by treating a solution of the sample in acetone, ethyl methyl ketone, or mixtures of these with benzene. The SiH-containing impurities are determined from the turbidity produced by the mercurous chloride formed in accordance with the equation:

$$\equiv SiH + 2HgCl_2 \rightarrow \equiv SiCl + Hg_2Cl_2 + HCl$$

The extinction is directly proportional to the concentration of the compounds containing ≡SiH bonds. As little as $5 \times 10^{-5}\%$ of impurity (as H) can be determined by this method. The same workers[119, 120] also determined SiH groups by amperometric titration. The organosilicon compound is dissolved in benzene; benzene:methanol (1:1) containing 0.3 M lithium chloride is added and, using a mercury pool and dropping mercury cathode, the height of the wave at -0.8 V is measured. The solution is then titrated with 3% mercuric chloride solution in benzene:methanol (1:1) until the wave at -0.8 V disappears.

Bork and Shvyrkova[121] outline various methods for the determination of ≡SiH in organosilicon compounds. The phototurbidimetric method is based on measurement of the extinction of the suspension formed by the reaction of ≡SiH with mercuric chloride in a non-aqueous solvent. The gravimetric method is based on weighing the precipitate of mercurous chloride produced under the conditions described above. The photometric method is based on the oxidation of the Si–H bond with potassium permanganate and photometric measurement of the reduced potassium permanganate solution. In a polarographic determination of ≡SiH groups in the presence of silicon–silicon bonds in polyorganosiloxanes[122], the sample is dissolved in ethyl methyl ketone, and mercuric chloride solution is added to oxidise the Si–H bonds. The decrease in the concentration of mercuric ion is determined polarographically as the mercury–iodine complex[123]. The coefficient of variation is $\pm 0.3\%$ for 1.5–7.5 mg of tetramethylcyclotetrasiloxane.

Hydrogen attached directly to silicon reacts with alkali to yield a condensed silanol and hydrogen gas is evolved. The hydrogen can be collected and its volume measured to give an estimate of the SiH hydrogen content of the sample. When aqueous alkali does not bring about quantitative evolution of hydrogen, an alcoholic solution of alkali can be used to dissolve the sample. Polymeric organosilanes have been treated with moist piperidine[124]. Non-quantitative evolution of hydrogen usually indicates that the sample has not been completely dissolved or that a stronger base than piperidine is required as catalyst. With poly(methyl hydrogen) siloxane the reaction is:

$$(CH_3HSiO)_n + \tfrac{1}{2}nH_2O \xrightarrow{\text{alkali catalyst}} (CH_3SiO_{1.5})_n + nH_2$$

The reaction is quantitative even at extremely low concentrations; it also occurs with Si–Si bonds, one such bond yielding 1 mol of hydrogen. This method cannot, therefore, be used to determine Si–H in the presence of Si–Si bonds.

d. Unsaturation. To determine unsaturation in organosilicon compounds the sample is dissolved in carbon tetrachloride. Then anhydrous acetic acid and conc. hydrochloric acid is added and the mixture stirred for 1 min. Electrodes are inserted and a current of −2.5 mA established (rotating platinum anode and standard calomel cathode) and the mixture titrated with 0.5 N potassium bromide–bromate solution. A blank determination is carried out in parallel[125].

Amperometric titration with standard bromine solution has been used[126] to estimate unsaturation in organosilicon compounds. For substances not containing phenyl radicals, the sample (4–30 mg) is mixed with hydrochloric acid and the solution diluted with methanol (or acetic acid) saturated with sodium bromide. The solution is titrated at +0.2–0.3 V (rotating platinum anode and S.C.E. as cathode) with 0.1 N bromine in methanol (or acetic acid) saturated with sodium bromide added in 0.1–0.2 ml portions. For substances containing phenyl radicals the procedure has to be modified slightly. The same workers[127] also described a procedure based on titration with iodine monochloride in a non-aqueous medium. To determine double bonds in trisubstituted silanes, they first determined ≡SiH bonds gravimetrically, turbidimetrically or volumetrically as described in the previous section. They then determined ≡SiH plus C=C bonds by amperometric titration and calculated double bonds by difference.

Double bonds can also be determined by IR spectroscopy[128], using the integral absorption coefficient calculated from the Lorenz function. This coefficient is less sensitive to intermolecular interaction than is the absorption coefficient at the absorption maximum. Equations are given for calculating the integral coefficient and the concentration of the substance being determined. A procedure is described for determining tetraethoxydivinyldisiloxane and triethoxyvinylsilane in carbon disulphide solution.

e. Determination of alkoxy and aryloxy groups. The alkoxy content of alkoxysilanes is often determined, following hydrolysis, by employing the conventional analytical methods for determining alcohols. However, the slow steps in such analyses are the hydrolysis and the separation of the alcohol from the other products. A modified Zeisel technique has been used[129, 130] for the analysis of methoxy-, ethoxy- and propoxy-silanes, but it is time-consuming and there is a molecular-weight limitation on the alkyl iodide that can be distilled off as iodide. Accurate assays of aryloxysilanes capable of stoichiometric bromination have been reported[131]. Infrared spectrophotometry[132] has been used to determine the methoxy content of siloxane polymers.

An acetylation method has been described[133, 134] for the rapid and accurate determination of alkoxy groups in alkoxysilanes. Both monomers and soluble high-molecular-weight siloxanes containing a wide variety of alkoxy and aryloxy groups can be determined quantitatively by this technique. In the method alkoxy groups bonded to silicon are acetylated by acetic anhydride with perchloric acid as catalyst in ethyl acetate solvent, in less than 2 min at room temperature. Following hydrolysis of excess acetic anhydride, the alkoxy content is calculated from the difference between the blank and sample titrations obtained with standard alcoholic potassium hydroxide.

The limitation of the methods appears to be the solubility of the material to be analysed, and possible steric hindrance effects. Substituted siloxysilyl, olefinic, nitrile, silicon hydride, γ-chloropropyl or β-chloroethyl groups do not interfere. Perchloric acid does not appear to attack the methyl ether groups in methoxyethoxysilanes. Water and silanol, within limits, do not interfere. Compounds containing acid groups, however, must first be analysed for titratable acid and appropriate corrections made in the alkoxy content calculations.

A very similar method has been described[135], in which the acetylated silicon compound is hydrolysed with water and the liberated acetic acid determined by titration. The alcohols present in the reaction mixture are identified by paper chromatography.

Hydrolysis with aqueous hydrochloric acid followed by spectrophotometric estimation of the alcohol produced (using ceric ammonium nitrate) has been used[136] to estimate C_2-C_4 alkoxy groups in various compounds including siloxanes. A spectrophotometric method[137] for the determination of phenoxy groups is based on the formation of the indophenol blue.

For the determination of phenoxy groups in phenoxyphenylsilanes, bomb fusion with potassium hydroxide is carried out and an aliquot of an aqueous solution of the melt is analysed for phenoxide by titration with potassium bromide–potassium bromate solution. Phenylphenoxy groups in phenyl(phenylphenoxy)silanes are similarly determined, but the melt is dissolved in anhydrous acetic acid[138].

A procedure based on reaction with hydriodic acid has been described[139] for the estimation of lower alkoxy groups in organosilicon compounds. This procedure has been modified[140] so that small amounts of ethoxy and butoxy groups in polysiloxanes could be determined with an error of $\pm 10\%$ of the content. For ethoxy groups, the sample is reacted with hydriodic acid and the product carried by a stream of carbon dioxide through a washing vessel into a receiver containing sodium acetate in glacial acetic acid and 5–6 drops bromine. The contents of the receiver are treated with formic acid to destroy free bromine, and then with dil. sulphuric acid and potassium iodide solution. The liberated iodine is titrated with sodium thiosulphate. For butoxy groups, longer reaction times and higher temperatures are used. In another procedure[141] the sample is decomposed with potassium iodide–phosphoric acid to form the alkyl iodide, which is then distilled off and oxidized with bromine to iodic acid which is determined iodometrically. Other methods based on digestion with hydriodic acid for the determination of alkoxy groups have also been discussed[142–145].

A procedure has been described[112] for the determination of alkoxy and vinyl substituents in amounts between 0.001 and 1% in siloxane materials using a combination of alkali fusion and gas chromatography. The alcohol and ethylene produced by cleavage of the vinyl and alkoxy groups are determined in the measured volume of gas evolved. A relative standard deviation of less that 20% is claimed in the alkoxy analysis and less than 10% in the vinyl analysis.

In general, the alkoxy method is most applicable to sample containing 0.001 to 1% ≡SiOR groups and to materials with high boiling points above 300 °C. Lower boiling materials usually will vaporise before the alkali melt reaction becomes effective at about 350 °C. The C_1 to C_4 alcohols were readily resolved by gas chromatography. Alkoxy bound to carbon (≡COR) will interfere with ≡SiOR analysis because of significant OR cleavage from carbon to give alcohols. However, the latter reaction was not quantitative, so total alkoxy content could not be obtained if both ≡COR and ≡SiOR were present. The method is capable of detecting 0.001% vinyl in siloxane fluids and gums. The upper use range generally limited to 1.0%. The relative standard deviation was usually within 10%.

f. Determination of acetoxy groups. Two methods described for the determination of acetoxy groups bound to silicon, e.g. $R_nSi(OCOCH_3)_{4-n}$[146–148], are based on titration of the organisilicon compound dissolved in ethyl methyl ketone with methanolic sodium methoxide:

$$R_nSi(OCOCH_3)_{4-n} + (4-n)CH_3ONa \rightarrow R_nSi(OCH_3)_{4-n} + (4-n)CH_3COONa$$

The end-point of the titration is found visually with thymol blue as indicator, or potentiometrically with glass and saturated calomel electrodes. The error of the determination does not exceed 1.4%. Alkoxy groups do not interfere in the procedure.

g. Silicon-bound hydroxy groups (silanols). Methods proposed for the determination of silanol groups in organosilicon compounds include manometric lithium aluminium hydride and Zerevitinoff procedures[149, 150] (discussed later), reaction with phenyl isocyanate[151], and silanol condensation procedures[152–155] using alkali, acid or iodine as catalyst. The water produced in these condensation methods is removed by azeotropic distillation with benzene or toluene and is measured either volumetrically or by Karl Fischer titration[156–158]. The condensation procedures are usually slow, incomplete and non-reproducible, particularly when applied to silicone resins.

A catalyst system[159] consisting of a mixture of boron trifluoride, acetic acid and pyridine gives rapid and reproducible total hydroxyl analysis (silanol plus water) in a variety of silicone materials. This method avoids many of the interferences, empirical calibration and miscellaneous problems such as poor solubility, incomplete reaction and interfering siloxane cleavage associated with many earlier methods. Results obtained by this procedure compare well with results obtained by the lithium aluminium hydride procedure[160].

An extensive examination has been carried out[162–164] of the solvent used in the catalysed acetylation procedure with the conclusion that, for the determination of silanols, 1:2 dichloroethane is in many ways a superior solvent to ethyl acetate.

A gasometric procedure[160] using lithium aluminium hydride gives good results for the determination of silanol groups in the presence of siloxanes which, upon cleavage, produce volatile silicon hydrides. Diglyme is recommended as a solvent instead of the usual butyl ether.

Lithium aluminium dibutylamide has been used[165] as a titrant for the determination of silanols. Silanol hydroxy groups were titrated in monomeric and polymeric silanols, and in silicone resins, using *N*-phenyl-4-phenylazoaniline as indicator.

Silanols have also been estimated[166, 167] gasometrically using a Grignard reagent. A solution of the sample in dibutyl ether is added to a standard solution of methyl magnesium iodide (in the same solvent) in the reaction flask, through which a slow stream of CO_2 is passed. The volume of methane produced is measured in a sodium hydroxide filled burette after 6–10 min. The percentage of hydroxyl can then be calculated. The method is applicable to the hydrolysis products from alkyl and aryl trialkoxysilanes, and to commercial silicone resins. For a sample of silicone resin, most of the solvent should be removed under vacuum at room temperature before dilution with dibutyl ether.

Similar methods have been described[151, 168, 169] based on the use of methyl magnesium iodide; they work well for both large and small numbers of hydroxyl groups and also for commercial silicone resins[166]. These procedures are subject to occasional error because of the limited solubility in the reagent of some of the partly condensed products of the reaction. None of them differentiates between Si–OH and C–OH; the latter group can be determined by reaction with acetyl chloride, and the Si–OH can then be determined by difference. Any $\equiv$Si–$OCOCH_3$ formed by the reaction of $\equiv$SiOH groups with acetyl chloride is readily hydrolysed with water.

To determine hydroxyl groups in silicone resins the sample[170], dissolved in chlorobenzene or dioxan, was reacted for 2 h with isocyanatobenzene, the excess of which is then reacted with isobutylamine, and the residual amine is titrated with standard hydrochloric acid. This method will determine hydroxy groups present in water as well as silanol groups. To distinguish between the two types, a Karl Fischer determination is carried out; water reacts immediately but silanol groups only slowly.

h. Alkylthiosilanes. The catalysed acetylation procedure[161] can be used for the direct determination of Si–S–C linkages. This acetylation method depends upon reaction of the alkylthiosilane with excess acetic anhydride to give a thiolacetate and an acetoxysilane. The acetoxysilane and remaining anhydride are hydrolysed and titrated with standard base.

It has been pointed out[171] that the ease of hydrolysis of alkylthiosilanes allows an indirect assay of the alkylthio functional group by reaction with excess mercuric acetate in a methanol–toluene solvent. The excess mercuric acetate is titrated with standard hydrochloric acid in butanol to a thymol blue end-point[172]. Residual moisture in the solvents and hydrochloric acid–butanol titrant suffice for the hydrolysis of the alkylthiosilane to the corresponding thiol. Steric hindrance appears to preclude a quantitative determination of the Si–S–C linkage in the *tert*-butylthiosilane in reasonable lengths of time by 0.04 M perchloric-acid-catalysed acetylation.

II. NON-AQUEOUS TITRIMETRY

Various workers[173–176] have made an extensive study of non-aqueous titrimetric methods of analysis of alkyl and aryl alkyl chlorosilanes, and other types of organosilicon compounds containing nitrogen, carboxyl and thiocyanato functional groups[177–179].

To titrate alkylchlorosilanes such as trimethylsilicon chloride, dimethylsilicon dichloride, methylsilicon trichloride and silicon tetrachloride, titration in methyl cyanide with 0.05 n phenazone or nitron in methyl cyanide, or with a solution of amidopyrine in benzene by visual and potentiometric methods is suggested[173]. The indicators used are crystal violet, dimethylaminoazobenzene, bromocresol purple, methyl orange, bromophenol blue and gallo sea blue (C.I.-Mordant Blue 14); the addition of benzene, toluene, chlorobenzene or carbon tetrachloride does not effect the results.

a. Non-aqueous conductometric titration. This technique has been used[174] for the determination of trimethylsilicon chloride, dimethylsilicon dichloride and methylsilicon trichloride. The method is based on the quantitative conversion of the chlorosilanes into the thiocyanato derivatives by the action of ammonium thiocyanate and conductometric titration in methyl cyanide–ethyl ether (2:3) solvent with 0.1 *m*-amidopyrine in benzene. The non-substituted compound is first titrated, then the di-substituted and finally the tri-substituted. The error in determining the individual chlorosilanes is $\pm 2\%$.

b. Non-aqueous electrometric methods. This technique has been used[175] for the analysis of monomeric and polymeric silicon compounds in non-aqueous media. The methods are applicable to alkyl and aryl chlorosilanes, aminosilanes, weak acids containing silicon and other organic silicon compounds. In non-aqueous media, several organic silicon compounds have strongly basic or acid properties. Solutions of tetraethylammonium hydroxide and amidopyrine in benzene or mixtures of benzene with other solvents can be used as titrants for these types of compound.

c. Non-aqueous amperometric titration. This technique has been used[176] for the determination of dimethylsilicon dichloride, methylphenyldichlorosilane and trimethylchlorosilane. Two methods are described. In *methanol–benzene medium*, the alkylchlorosilane is titrated amperometrically (mercury anode; dropping mercury cathode, potential -1.2 V) with lead acetate dissolved in methanol or in methanol-benzene. The error is less than $\pm 1.5\%$ and sensitivity is 0.01% of chlorine in the sample. In *anhydrous acetic acid medium*, the sample is titrated amperometrically (-1.4 V) with 0.15 N cadmium nitrate dissolved in anhydrous acetic acid.

To determine nitrogen-containing organosilicon compounds[177] the sample is dissolved in methyl cyanide or nitromethane, an equal volume of benzene or dioxan is added, and the solution is titrated potentiometrically with 0.1 N perchloric acid in glacial acetic acid. Alternatively, the end-point may be found visually by crystal violet or bromocresol indicators. With carboxyl-containing compounds, the solution in the mixture of organic solvents is titrated with a solution of tetraethylammonium hydroxide in benzene-methanol. A similar method[178] uses perchloric acid in methyl cyanide, methyl nitrite, or their mixtures with benzene or dioxan. The indicators used were crystal violet, bromocresol purple, bromophenol blue, cresol red, dimethylaminoazobenzene, thymol blue, methyl red and methyl orange. Compounds containing nitrogen attached directly to silicon and compounds with nitrogen in an organic radical were titrated The potential jumps for titration in methyl nitrite are 10 mV larger than those in methyl cyanide.

The non-aqueous titration of alkyl thiocyanatosilanes[179] is carried out with 0.1 n sodium methoxide in methyl cyanide or methanol, either visually with a hydroxyanthraquinone indicator, or potentiometrically.

Alkylsilyl sulphates were determined[180] by potentiometric and conductiometric titration in anhydrous alcohols, ketones, methyl cyanide or nitromethane. Titration curves were the same for all solvents. In potentiometry, titration with sodium, potassium or lithium methoxide gave curves with one step; titration with quaternary ammonium compounds gave curves with two steps. In conductiometry, titration with quaternary ammonium compounds gave curves with no inflection, whereas titration with alkali metal methoxides gave curves with a sharp break at the end-point.

Non-aqueous conductiometric titration has been used[181] to determine trialkylsilyl phosphates and bis(trialkylsilyl)sulphates. Methanol, ethanol, butanol, acetone, ethyl methyl ketone, nitromethane, nitrobenzene, methyl cyanide, dimethylformamide, tetrahydrofuran, butyl acetate and ethyl formate were used as solvents, and methanolic sodium, potassium and lithium methoxides and tetraethylammonium hydroxide as titrants. For the phosphates, characteristic curves were obtained only in nitromethane and methyl cyanide. Titrating in nitromethane with 0.15 N-methanolic potassium methoxide gave the best results. Alcohols, ketones, nitromethane and methyl cyanide were suitable solvents for the sulphates, but use of tetraethylammonium hydroxide as titrant did not give good results.

A method has been described[182] for the titration of organosilylmetallic compounds, including those of lithium. These compounds react with butyl bromide to yield an equivalent amount of metallic bromide, which is titrated argentimetrically by the Volhard procedure.

In a non-aqueous titration procedure for the analysis of alkylsilyl phosphates[183] trimethylsilyl dihydrogen phosphate [$(CH_3)_3SiOPO(OH)_2$] and its analogues are titrated in non-aqueous media with solutions of lithium, sodium, or potassium methoxides and tetraethylammonium hydroxide. The end-point is determined potentiometrically with glass and saturated mercurous chloride electrodes, or visually in the presence of quinizarin, bromophenol blue, brilliant green, alkaline blue or methyl red. Methyl cyanide, acetone, methyl ethyl ketone, methanol, ethanol, isopropyl alcohol, butanol and benzyl alcohol can be used as titration media. In the titration of trimethylsilyl dihydrogen phosphate with lithium methoxide solution in methanolic medium, the reaction proceeds stepwise, with the use of first one and then two equivalents of the titrant. In the other alcohols (except benzyl alcohol), the ketones and methyl cyanide, the reaction proceeds until three equivalents of lithium methoxide have been used per mole of trimethylsilyl dihydrogen phosphate. The titration of trimethylsilyl dihydrogen phosphate with potassium methoxide or tetraethylammonium hydroxide in all the media mentioned above gave a single potential jump, corresponding to the reaction:

$$(CH_3)_3SiOPO(OH)_2 + CH_3OK \rightarrow (CH_3)_3SiOCH_3 + KH_2PO_4$$

The titration of trimethylsilyl dihydrogen phosphate with sodium methoxide in a medium of isopropyl alcohol, butanol or benzyl alcohol proceeds similarly. In methanol and ethanol, two jumps were obtained, corresponding to two equivalents of titrant.

III. SPECTROPHOTOMETRIC PROCEDURES

A spectrophotometric procedure has been described[184] based on the formation of a coloured complex with antimony pentachloride in carbon tetrachloride, for the determination of 0.01–1.0% biphenyl in phenyltrichlorosilane. These workers also described a spectrophotometric method for the determination of 0.01–10% biphenyl in phenyltrichlorosilane based on measurement of the difference in extinction of solutions of the sample and a standard in chloroform-ethanol at 251 nm. The determination of biphenyl in phenyltrichlorosilane and its hydrolysis products have also been discussed[185]: The extinctions of pure phenyl trichlorosilane and the sample are measured at 251 nm in chloroform:ethanol 1:1; the difference is related to the content of biphenyl, which absorbs 60 times more strongly than phenyltrichlorosilane at that wavelength; 0.05–10% of biphenyl may be determined.

Spectrophotometric methods have been described for the determination of copper, iron, cobalt and manganese in trichlorosilanes and in silicon tetrachloride[186]; boron in silicon tetrachloride and in hexachlorosilane[187] and iron and aluminium in silicone polymers[189].

Trichlorosilane[190] and alkylchlorosilanes have both been estimated in air, using methods based on the formation of molybdenum blue when organosilicon compounds react with ammonium molybdate. In the method for determining alkylchlorosilanes[191] the solution is evaporated, and the residue fused with a potassium carbonate–sodium carbonate mixture prior to dissolution of the melt in dilute sulphuric acid and production of molybdenum blue by the addition of ammonium molybdate, tartaric acid and ascorbic acid. In one method[192] for the estimation of nitrogen-containing organosilicon compounds in air, the compound is absorbed in dilute sulphuric acid and, after wet oxidation, the silicon is determined colorimetrically as ammonia using Nesslers reagent, after sulphuric acid hydrolysis of the organosilicon compound. In a further method, the organosilicon compound is oxidised with potassium bromate and the liberated bromine determined using the eosin method. The applicability of ultraviolet spectroscopy to organosilicon compounds has been reviewed[193–194]. Alkyl-substituted silanes and siloxanes are not ultraviolet absorbers. Several ultraviolet absorption bands have been found[195] for methyl siloxanes; these compounds do not, however, absorb above 220 nm and the bands observed probably arise from aromatic impurities. This conclusion has been confirmed by Eaborn[196]. No absorption was found for trimethylsilanol, hexamethyldisiloxane or trimethylmethoxysilane. Phenyl substitution on silicon, however, gives rise to a series of absorption peaks located at 248, 254, 260, 264, 266 and 271 nm. These bands are common to triphenylsilane, triphenylethoxysilane and triphenylsilanol[197] and trimethylphenylsilane. *Para*-substituted phenyl groups produce an intense absorption at about 270 nm[198]. It is obvious that ultraviolet spectroscopy is not a very good tool for identifying silicones. It is, however, very useful for detecting trace amounts of aromatic contaminants in dimethylsiloxane fluids or determining traces of aryl-substituted silicones in the presence of non-absorbing materials[199].

IV. INFRARED SPECTROSCOPY

Infrared spectroscopy has certain limitations when applied to organosilicon compounds. First, non-silicon substituents present in small amounts may be overlooked. Secondly, infrared is not very sensitive to differences in molecular weight or molecular weight distribution.

The classification of silicone polymers into resins, fluids, etc., has been discussed[200]. The following remarks apply principally to pure silicone polymers. Some polymers are soluble in carbon disulfide and carbon tetrachloride, and solution in these materials is a convenient method for obtaining the IR spectra. However, high-viscosity dimethylpolysiloxanes have a limited solubility in carbon disulfide and it may be necessary to use carbon tetrachloride throughout the entire spectral region or to use some solvent such as 2,2,4-trimethylpentane in the 11- to 16-μ region. Frequently, sufficiently good spectra may be obtained by simply casting a film of the polymer on a polished salt plate from a low boiling solvent such as chloroform. Hard, brittle materials such as cured resins can be run by the mineral oil mull[201] or potassium bromide pressed plate[202, 203] techniques. In cases where a filler or a reinforcing agent is present, the pyrolysis technique[204] is very useful.

Interpretation of infrared spectra of organosilicon compounds has been discussed in some detail. Two observations are noteworthy: the characteristic vibrations of substituents vary only slightly in wavelength, regardless of the compound in which they are found; and the intensity of the absorption bands lying beyond 7μ is 5–10 times greater than is normal with most organic compounds. These characteristics make silicones relatively easy to identify by their IR spectra. Silicone polymers are distinguished by a strong absorption in the 9- to 10-μ region arising from an Si–O stretching vibration in the siloxane backbone of the polymer. Linear and cyclic polymers containing eight or fewer units can be characterized by the shape and position of the siloxane band[205].

In commercial polymers the silicon atoms are usually substituted by some combination of methyl, ethyl, phenyl, hydrogen, alkoxy or hydroxyl groups and these are discussed in more detail later on. Information on the correlation of structure and infrared spectra has been tabulated by various workers.

One group of workers[206] gave diagrams of the IR absorption spectra of eleven silicon-organic compounds between 2 and 10 μ. Wavelengths (microns) representative of various groupings are assigned: CH_3, 3.45 and 6.8–7.1; C_6H_5, 3.25, 6.25, 6.7 and 9.7; ≡Si–C, 7.95–8.0; ≡S–O, 9.5–9.65; –O–R, 8.6–8.7; and $CH{=}CH_2$, 6.15.

In one detailed review[206] it was pointed out that the concept of group frequencies is particularly useful in the interpretation of the solution IR spectrum of organosilicon compounds. This is due to the insulating effect of the silicon atom which causes the spectra of these compounds to be approximately the same as the sum of independent vibrations of the substituent groups. Small shifts from these characteristic frequencies are to some extent predictable, so the assignment of molecular structure is in many cases comparatively simple. Such perturbations of group frequencies as do occur usually arise from:

(1) The inductive effect of neighbouring groups or atoms.
(2) Coupling or interaction of vibrations.
(3) Steric effects such as bond strain.

The vibrational insulation provided by the silicon atom can probably be attributed, at least in part, to the fact that its size and mass are greater than those of carbon.

Also, characteristic of the spectra of silicones are the large absorption intensities of bands falling at wavelengths greater than 7.5 μ. This may result at least in part from the electronegativity difference between carbon and silicon which renders a greater ionic character to the Si–C bonds. The C–H stretching and bending bands in the 2–7.5 μ region are weak compared with those in conventional organic compounds, probably owing to the smaller C–H dipole resulting from withdrawal of electrons from the C–H bond by the strong Si^+C^- dipole.

Further details on the assignment of frequencies to particular structural features of organosilicon compounds are given below. These are useful for the general characterization of organosilicon compounds. This is followed by some examples of the application of infrared spectroscopy to the quantitative analysis of organosilicon compounds.

A. Quantitative Infrared Analysis

a. Alkyl and aryl groups. Several workers[207–219] have studied the applicability of infrared spectroscopy to the determination of silicon bound alkyl and aryl groups. The determination of methyl and phenyl groups in siloxane resins as 5% w/v solutions in carbon disulphide has been carried out[207] using the band at 8 μ to determine methyl groups, and that at 14.36 μ for phenyl groups. The monomers used in preparing the resins were employed for preparing the calibration graphs. Since the extinction coefficient for the phenyl groups occurring in compounds of the types $(C_6H_5)_2SiO$ and polymeric (C_6H_5SiO) varied, two calibration graphs were used when determining phenyl groups, and results were calculated as mean values from both graphs. For the determination of methyl groups, the coefficient of variation was 2.8%; for phenyl groups it was 2.0%. In a continuation of this work[208] methyl groups attached to the benzene ring and those attached to the siloxane chain were differentiated. The peak of the band at 7.90 μ shifts from 7.95 μ for diethoxydimethylsilane to 7.94 μ for diethoxymethylphenylsilane, and to 7.90 μ for triethoxymethylsilane; hence the compensation procedure gives the content of the methyl groups in the two positions. The phenyl groups are determined by measuring the extinction at 6.99 and 6.27 μ. The absorption of 6.99 μ is due to the C_6H_5-Si linkages, but compensation is required for distortion caused by the nearby ethoxy-group absorption band. Although well-shaped the 6.27 μ band is weak and the extinction coefficient is seriously affected by the direct neighbours of the phenyl group. The identification of the $Si(C_6H_5)$ and $Si(C_6H_5)_2$ groups from the absorption spectrum between 25 and 167 μ was also studied.

Reasonable agreement was obtained[209,210,212] between theoretical and determined values of methyl-to-phenyl ratio in silicone polymers by measuring the intensity of the methylsilicon and phenylsilicon bands at 7.92 and 6.97 μ, respectively.

Infrared and ultraviolet spectroscopy have been applied[213] to the determination of methyl and phenyl groups in silicones. While siloxanes containing only methyl groups have no absorption in the region above 240 nm, those containing phenyl groups adjacent to silicon have marked absorption at 270, 264, 259 and 253 nm. The molar extinction coefficient corresponding to one phenyl group is practically constant (335) at 264 nm for high-molecular-weight siloxanes. This is useful for the determination of the phenyl group. Compounds in which two phenyl groups are attached to silicon have much greater extinction above 240 nm than those in which one phenyl group is attached to silicon. The following relationship holds[215,216] in silicones: $-\log[\text{Me}]/[\text{Ph}] = 1.507 \log(E_{\text{Me}}/E_{\text{Ph}}) + 0.202$, where Me/Ph is the ratio of methyl to phenyl bound to silicon; E_{Me} is the extinction at 3.4 μ and E_{Ph} is that at 3.3 μ.

An infrared method has been described[217] for the determination of the ratio of phenyl and *p*-tolyl groups in tetraaryl silanes and hexaaryldisilanes. The absorption maxima for the *p*-tolyl group bands fall within the range 12.47 to 12.50 μ and the phenyl group bands within the ranges 13.40 to 13.45 μ and 14.30 to 14.33 μ. The molar absorptivities of the functional groups are not constant, but accurate results are possible. Certain substituent groups on the silicon atom can interfere; aliphatic groups absorb strongly in the 12–15 μ region, silanols absorb at 12.4 μ in solution and the Si–H group gives an absorption band at about 12.5 μ.

b. Methyl, phenyl and hydroxy groups. Infrared methods have been described[219] for the estimation of free hydroxyl (2.8 μ), hydrogen-bonded hydroxyl (2.9 μ), and also phenyl (3.3 μ) and methyl (3.4 μ) groups in dimethyldiphenyl silicone resins.

c. Methoxy groups. Precise conditions have been reported[221–223] for the quantitative determination of methoxy groups in siloxane polymers where measurements were carried out at 8.4 μ attributed to $SiO{-}CH_3$ groups.

d. Siloxane groupings. For the identification and determination of methylsiloxanes such as $CH_3SiO_{1.5}$, $(CH_3)_2$ SiO and $CH_3SiO_{0.5}$, characteristic absorptions at 11.7, 12.5 and 13.0 μ have been used[214,220]. Since almost all commercial silicones are binary condensation products of $CH_3SiO_{1.5}$ and $(CH_3)_2SiO$, their quantitative analysis can be effected by an empirical method.

The sample is analysed as a 2% solution in carbon disulphide using a 0.1 mm cell.

e. Silicon–hydrogen bonds. Exact conditions are described for the determination at 4.67μ of silanic hydrogen in silicone polymers[221–223]. The method gives $\pm 0.001\%$ accuracy for $\equiv$S–H contents between 0.001 and 0.03%. The reference compound was 99+ % pure 1,1,1,3,5,5,5-heptamethyltrisiloxane.

f. Halosilicon compounds. Chlorosilanes are most conveniently run as dilute solutions in carbon tetrachloride (2–7.5 μ). Chloro-, bromo- and iodosilanes hydrolyse very easily and must be handled in a dry glove box. Silanes with a high vapour pressure can be run in the gas or vapour state. Absorptions characteristic of the Si–Cl bonds appear in the potassium bromide prism region at 16–25 μ. If more than one chlorine atom is present, two bands appear, corresponding to the symmetrical and asymmetrical frequencies. In general, an asymmetrical stretch appears at a shorter wavelength and is the more intense of the two frequencies. The Si–Cl bands are the basis of an analytical method for the determination of other chlorosilanes appearing as impurities in trimethylchlorosilane[224].

Frequency correlations as applied to silicon-containing polymers and silane monomers also apply to the chlorosilanes. However, electronegative substituents such as chlorine may shift some of these vibrational bands to shorter wavelengths. The siloxane band in hexamethyldisiloxane, for example, is located at 9.50 μ. As the methyl groups are replaced by chlorine, the wavelength of the absorption gradually shifts until it is found at 8.99μ for hexachlorosiloxane. The shift of the Si–H stretching frequency in substituted silanes can be predicted precisely from the structure of the molecule[225].

Infrared studies of methylchlorosilanes[226–232] show that the wavelength of the SiCl bands varies in a regular manner and these absorptions have been used for quantitative analysis[232, 233].

SiBr and SiI fundamentals may fall beyond the range of the potassium bromide prism, but can usually be detected with caesium bromide optics[222, 234, 235, 240]. The tendency of the halosilanes to react with traces of moisture to form siloxanes must not be overlooked when working with these materials[241].

Infrared spectroscopy has also been applied to the determination of phenyltrichlorosilane in various mixtures of chlorosilanes[242], in mixtures of diphenyldichlorosilane, phenyltrichlorosilane and chlorobenzene[243], in mixtures containing diphenyldichlorosilane[243], chlorosilanes and trimethylchlorosilane[233]. IR was also used for detecting impurities in silicon tetrafluoride[244] and for trichloro-3-cyanopropylsilane in dichloro-(3-cyanopropyl)methylsilane[245].

B. Infrared–Structure Correlations

a. Methyl group. In addition to the weakness of their absorptions, the CH stretching vibrations of methyl groups on silicon may be displaced from their normal position in aliphatic compounds. Both the symmetric and antisymmetric vibrations are influenced by the presence of adjacent electron-withdrawing groups such as chlorine or oxygen. The effect of electronegative elements on frequencies in organosilicon compounds has been discussed[246]. There is also a surprisingly large difference between the intensities of the two modes, with the antisymmetric vibration absorption being the more intense. The reverse is true with the bending modes, where the symmetric deformation at 7.9 μ is much more intense than the antisymmetric one at 7.1 μ.

The band most characteristic of methyl on silicon is the symmetric deformation at 7.8–8.0 μ[207]. This band is intense, sharp and easily identified. Its exact position depends on the nature of the other substituents on the silicon[247, 248]. If three methyl groups are present on the same silicon, vibrational interactions usually split this band into two components of unequal intensity. The 8 μ is always accompanied by one or more equally intense bands in the 11.8–13.1 μ region, which arise from methyl rocking and Si–C stretch vibrations[249, 250]. This region also gives useful clues about the structure of the molecule. A single methyl group on silicon absorbs at 13.1 μ, two methyls at 11.7 and 12.5 μ and three methyls at 11.9 and 13.1μ. Tetramethylsilane has an intense band at 11.6 μ arising from a degenerate methyl rocking vibration[250, 251]. If an alkyl group as well as n methyl groups are attached to a silicon, the spectral pattern in the 11–13 μ range usually resembles that obtained from $(n+1)$ methyl groups.

Studies of the infrared and Raman spectra of organosilanes ($SiH_n(CH_3)_{(4-n)}$) suggest that coupling between bending motions of SiH and SiH_2 groups and other motions prevents the assignment of SiH bending modes. Otherwise complete assignments were made[252, 253].

b. Ethyl groups. In ethyl-substituted silicon compounds, both the CH_3 and CH_2 modes are influenced by the silicon. An exact interpretation of the band envelope is difficult because a calcium fluoride prism does not resolve it fully, and overtone and combination bands may confuse the pattern. Assignments are therefore tentative.

An ethyl group on silicon is characterized by a band of medium intensity at 8–8.2 μ, arising from a symmetrical CH_2 deformation. A pair of bands found at 9.8–9.9 and 10.3–10.6 μ further serve to identify the ethyl-silicon grouping. A strong absorption arising from the Si–C stretching mode occurs in the 12–15 μ region.

c. Higher alkyl groups. It may be expected that with longer-chain alkyl substituents, the silicon would have progressively less influence on the hydrogen motions, and this is indeed the case. It is somewhat surprising, however, that the CH stretch bands, even in butyl trichlorosilane, are considerably displaced from their normal positions in organics. As one proceeds through propyl, butyl, amyl and higher alkyl substituents, the 3.4, 6.8 and 7.2 μ bands increase in intensity[254–256]. The lower members of the series show characteristic absorption patterns, which are remarkably constant in going from one molecule to another and which are useful for identification. The $SiCH_2$-deformation absorption becomes progressively weaker, appearing at 8.2–8.38 μ in propyl–Si, 8.35–8.4 μ in butyl–Si, and at about 8.4–8.5 μ in longer-chain alkyl substituents[254, 255, 268].

Straight-chain alkyl compounds also show absorption in the 13–15 μ region, due at least in part to CH_2 rocking in the methylene chains[257]. Shorter linear- and branched-chain alkyl silicon compounds have a number of characteristic absorptions in the 7.5–13 μ range, but none of them are very intense.

The spectra of dialkyl cyclopentamethylene silanes show characteristic bands at 10.93–11.00 μ[258].

d. Vinyl and allyl groups. Alkyl groups on silicon, which absorb at 6.10 μ and give several bands in the 8–11 μ range[259–262], are easily distinguished from β-methyl vinyl substituents, which have a band at 10.3 μ as well as absorption in the CH stretching and bending regions characteristic of a methyl group[261]. Allyl groups also absorb in the 12.4–13 μ region[259]. Unsaturation farther removed from the silicon gives a conventional pattern corresponding to its substitution and symmetry[263].

Vinyl-substituted silanes always show a strong CH in-plane bending absorption at 7.1–7.2 μ, as well as a distinctive doublet at 9.8–10 and 10.2–10.5 μ[264, 259, 265, 260]. These latter bands arise from out-of-plane CH bending vibrations[266]. A weak but distinctive overtone at 5.2 μ is also characteristic.

e. $Si(CH_2)_nSi$ groups. Vibrations involving the disilyl-methylene linkage give rise to a strong absorption in the 9–10 μ range, but this band is rather narrow and thus can usually be distinguished from siloxane absorption. This band has been attributed to the CH_2 twisting frequency[267] but assignment to the CH_2 wag may be more reasonable. The wavelength of this band is affected by mass and inductive effects of the other silicon substituents, but to a lesser degree than in the siloxane absorption. A weak but distinctive band appears at 7.4 μ, which is characteristic of $SiCH_2Si$. The asymmetric SiCSi stretch falls at 13.0 μ[267]. Disilylethylene groups conveniently show two bands in the 8–10 μ region which are quite characteristic if not hidden by other absorptions. Disilylpropylene and longer-chain bridges give absorption patterns in the 8–10 μ interval which are somewhat reminiscent of propylsilicon, butylsilicon, etc., groups. This observation is in accord with expectations, since at least some of the characteristic absorptions in this region arise from motions of hydrogens on the carbon adjacent to the silicon.

f. Phenyl groups. The phenyl group when attached to silicon usually shows at least three sharp bands in the aromatic CH stretching region. Their frequencies remain quite constant from one compound to another, except where they are influenced by strongly electronegative groups attached to the silicon.

Bands characteristic of phenyl on silicon include a sharp, narrow band at 6.98 μ[209, 216], a broader band at 8.9 μ, weak absorptions at 9.7 and 10.0 μ, and two or three bands at 13.5–14.4 μ[249, 208, 257]. All absorptions except the latter are essentially invariant in wavelength, regardless of the other substituents on the silicon. The 8.9 μ band, however, often splits into two components when two phenyl groups are present on the same silicon[249] but reverts to a single band for the Ph_3Si group. Taken all together, these bands identify the phenyl-silicon linkage. Only the 8.9 μ band is truly characteristic, however, since compounds containing the groups Sn–Ph, Pb–Ph and Ge–Ph show a pattern essentially identical to that of Si–Ph except for a shift in this one band[269].

Spectra of a number of phenyl-containing silanes and siloxanes have been studied[270] and their IR and Raman bands assigned to skeletal frequencies. A partial frequency assignment of the inner vibrations of the phenyl ring has been made[271]. By analogy with normal aromatic compounds, the 13.5 and 14.3 μ bands are attributed to the out-of-plane hydrogen deformation modes. 6.25, 6.73, 7.66, 7.97, 8.95 and 10.2 μ absorptions are described as being characteristic of CH and C=C vibrations and ring deformations. Assignment of the ring vibrations is not easy, because not only are the frequencies different from those of analogous organic molecules, but the intensity distribution of the bands is also different. It has been shown that the 8.9 μ band is affected by the electronegativity of the substituent[272] and this band may arise from an 'X-sensitive' planar ring vibration which has some Si–C stretch character[273]. The three bands found at 6.27, 6.73 and 6.99 μ must arise from C–C ring stretching vibrations[273]. The usual intensity relationship of the 6.27 and 6.73 μ bands is reversed in silicon-substituted phenyl compounds, with the former band being the stronger.

In spite of the great similarity of the spectra of compounds containing the phenyl-silicon group, specific compounds can often be distinguished by slight differences in the 13–15 μ absorption bands[274].

g. Other aromatics. A number of substituted aryl trimethyl silanes have been characterized[275]. The 5–6 μ range is useful in characterizing many substituted benzenes[276] and is applicable to most aryl silane derivatives. In many such compounds the 12–13.5 μ region, often used to distinguish isomers[277], is confused by the presence of both the out-of-plane bending modes of aromatic hydrogen and the methyl rocking and SiC stretch absorptions. In some cases isomers of a given substituent, such as chlorine, will show characteristic absorption patterns. The *p*-biphenylyl group absorbs characteristically at 6.48, 7.40, 9.91 and 12.00 μ[271]. Spectra have also been published for disilyl xylenes[278]. The spectra of several naphthyl silanes have also been discussed[279, 280].

h. Siloxanes (SiOSi). Almost all siloxane compounds show at least one strong band between 9 and 10 μ arising from an asymmetric SiOSi stretch vibration[281, 282] and corresponding to the COC stretching absorption in aliphatic ethers. In simple disiloxanes, the position of this band varies from 8.9 to 9.8 μ depending on the mass and inductive effect of the substituents on the silicon. With progressively longer-chain compounds (trisiloxanes, tetrasiloxanes, etc.), the absorption splits into two or more overlapping components until, with an essentially infinite siloxane chain, the absorption occupies almost the entire interval between 9 and 10 μ with points of maximum absorption at 9.2 and 9.8 μ[281, 283].

Cyclic polysiloxanes also absorb in this region, with a characteristic cyclotrisiloxane band at 9.8 μ[281]. Cyclic tetramers and pentamers have a single band at about 9.2 μ which gradually widens and splits with larger rings[284]. The positions of these bands are essentially independent of the other silicon substituents. The shift of the siloxane band to longer wavelengths in the case of the trimer is due to ring strain, and the siloxane absorption may move to as far as 11 μ in highly strained ring systems. Such extreme shifts are rare, however, and compounds showing them are unstable. The cyclotrisiloxane ring is planar[285] but the cyclotetrasiloxane ring has a non-planar (and probably unstrained) structure[286].

The siloxane band is occasionally shifted to shorter wavelengths by the inductive effect of other substituents. Thus, in $Cl_3SiOSiCl_3$, the siloxane absorption occurs at 8.9 μ. Silsesquioxane ($RSiO_{3/2}$) structures usually exhibit their siloxane bands at about 9.0 μ[287]. The symmetric SiOSi stretching vibration lies at much longer wavelengths. In the case of $Me_3SiOSiMe_3$, it falls at about 19.23 μ[287] and for disiloxane at about 16.67 μ[288]. Symmetrical vibrations such as this usually show a large intensity in the Raman spectrum and may be missing or only weakly active in the infrared.

i. Dimethyl substituted polysiloxanes. These have strong bands at 4.6 and 11.2 μ which indicate the presence of Si–H, and the Si–Me group has a characteristic band at 7.95 μ. An all methyl silicone resin with a high ($MeSiO_{3/2}$) content shows $SiO_{3/2}$ units which can be recognised by a strong band at 9.0 μ. An ethyl-substituted resin is essentially a silsesquioxane. The Et-Si group is characterized by absorptions at 8.0, 9.9 and 10.3 μ. Silicones used to modify organic resins usually contain either the dimethylsiloxane (Me_2SiO) group which absorbs at 7.9 and 12.5 μ, or the phenylsilicon group which shows a needle-sharp band at 7.0 μ, or both. A broad absorption in the 9- to 10-μ region indicates the presence of siloxane.

Infrared spectroscopy has been used to identify silicone polymers[289], and resins[290–292], silicone wire enamels[293], silicone emulsions and elastomers[294–296]. Infrared spectroscopy in conjunction with gas chromatography has been used for the examination of organosilicon isomers[297].

j. Hydroxyl groups. Silanols exhibit the usual OH stretching bands between 2.7 and 3 μ as found in organic alcohols, corresponding to monomers and hydrogen-bonded polymers. Although these bands are not unique the broad, asymmetric SiO stretching absorption at 11 and 12 μ which is also present in the spectra of silanols serves to characterize the SiOH group.

Several investigators have studied hydrogen bonding in silanols[298, 299] and concluded that, in crystalline silanols, two hydroxyl groups interact in such a way as to form a four-membered ring. Studies of association in solutions and in silicone resins[300–303] have been carried out and assignment of the stretching and bending vibrations discussed[246, 302, 304, 305]. There have also been a number of studies concerning the interaction of SiOH groups on silica with water vapour[306, 307] and other molecules[308, 309] and their behaviour on thermal dehydration[310, 311].

k. Alkoxy groups. All components containing the group ROSi, where R is a saturated aliphatic radical, have at least one strong band between 9 and 10 μ, which arises from the C–O stretching vibration[312–314]. Although this band is often obscured by the intense SiOSi absorption which follow in the same range, other bands appear which usually permit identification of the alkoxy group.

Methoxy compounds have a sharp, distinctive band at 3.55 μ which varies only slightly in wavelength with a change in the other substituents on the silicon. Also characteristic is an absorption at 8.4 μ attributed to $SiOCH_3$[315] rocking, which has been used for quantitative estimation of CH_3OSi[316]. The C–O stretch falls at 9.1 μ and an SiO mode falls between 11.8 and 12.5 μ[315, 317, 318]. This vibrational spectra of methoxy-substituted silanes have been the subject of many investigations[317–320]. Ethoxy substituents show a doubling of the SiOC band at 9–9.3 μ. In addition, analytically useful bands occur at

8.5–8.6 and 10.35–10.65 μ. Vibrational assignments for ethoxy-substituted silanes are not at all firm in spite of the large amount of work which has been done on molecules containing this group[290, 314, 318, 321, 322].

Longer-chain alkoxy silanes show analogous absorptions at 8.5–8.8, 9.1–9.3 and 10.1–10.6 μ. Alkoxy groups with branching on the α-carbon, such as *i*-PrO and *t*-BuO, do not absorb between 10 and 11 μ[317, 321–324]. Infrared spectra have been compared[325] of the lower members of methoxy-methylmethoxy and methoxy-end-blocked dimethylpolysiloxanes with those of dimethyl ether, dimethylpolysiloxanes, etc. The characteristic absorption band of the $SiOCH_3$ group appears near 8.40 μ. Except in the case of the first members of these alkoxysiloxanes, a spectral sequence corresponding to the structural unit sequence of SiO_4, SiO_3C and SiO_2C_2 is observed in the region of 10.00–9.09 μ.

Ethoxy- and methylethoxy-polysiloxanes have characteristic bands due to the $(Si)OC_2H$ group which are located near 10.41 and 8.62 μ, and near 11.90 μ in ethoxy end-blocked dimethylpolysiloxanes, and near 12.66 μ in ethoxypolysiloxanes[326].

l. Acetoxy groups. Compounds containing the grouping SiOOCR show, besides the expected C=O absorption at 5.7–5.9 μ, bands analogous to those of the alkoxy silanes. Where R is CH_3, characteristic bands appear at 8.2 and 10.6 μ. The SiOC stretch, which absorbs strongly at 9.2 μ in aliphatic silicon ethers, is probably responsible for the 8.1 μ absorption in acetoxy compounds, by analogy with aliphatic acetates[327]. Other acetoxy compounds absorb in the region 8.4–8.8 and 10.3–10.8 μ[328].

m. Phenylsiloxy units. Phenylsiloxy monomers such as $(C_6H_5)_2SiO$ and $C_6H_5(CH_3)SiO$ can be identified by IR absorption at 20.6 μ[329]. The ratio of the units $C_6H_5SiO_{1.5}$ and $(C_6H_5)_2SiO$ is determined by measuring the extinction (*D*) at 19.5 and 20.6 μ with the aid of the empirical formula:

$$\text{molar fraction of } (C_6H_5)_2SiO = \left(\frac{1.19D_{19.5}}{D_{19.5}+D_{20.6}}\right) - 0.19$$

n. Silane hydrogen. The SiH group is usually easy to identify, since its stretching absorption is quite intense and falls in the 4.4–4.8 μ range, where there is very little interference from other bands. Its position depends on the inductive power of the other substituents on the silicon, and in many cases can be predicted exactly[330]. The bending modes of a single hydrogen on silicon fall between 10.9 and 13.1 μ[331]. The SiH_2 scissors mode has been assigned to the range 10.2–10.8 μ, the SiH_2 wagging between 10.5 and 11.8 μ, the SiH_2 twist between 13.5 and 16.0 μ, and the SiH_2 rock between 19 and 24 μ[222], although in most cases these limits are somewhat narrower[234, 235]. The SiH_3 group absorbs strongly at about 10.8 μ. Both the symmetric and antisymmetric SiH_3 deformation vibrations absorb in the 10.5–11 μ region, whereas the SiH_3 rocking mode absorbs between 15 and 17 μ[234, 332, 236, 237]. Again, the wavelengths of these absorption bands depend on the nature of the other silicon substituents[221, 223]. Owing to interferences, the longer-wavelength SiH bands are not as useful for characterization as the stretching frequency, although the 1–11 μ region is often useful. The only interferences from organosilicon compounds in this region are from SiOH, SiF, SiNSi and SiOM, where M is a metal.

o. Silicon halogen bands. As expected the Si–halogen stretching frequencies fall in the same order as the atomic masses of the halogens. Because of inductive effects, however, the shifts in frequency upon going from one member to another are larger than would be predicted by a simple Hook's law calculation. The stretching vibrations of the SiF group fall in the 10–12 μ region[222, 237, 285, 333–337]. With SiF_2 and SiF_3, as is usually the case when two or more identical single atoms are attached to silicon, the asymmetric stretching mode falls at a higher frequency and is more intense than the symmetric mode.

SiCl bonds absorb strongly between 16 and 24 μ[338, 339]. Earlier work[340] misidentified an SiH bending mode at 12.5 μ in $HSiCl_3$ as being due to SiCl. The fact is that the wavelength extremes of the SiCl stretching vibrations are represented with bands at 16.3 μ and 23.6 μ; the latter band is infrared inactive. Highly electronegative groups or atoms such as fluorine on the same silicon atom could, however, displace the Si–Cl stretch to shorter wavelengths.

p. Silicon–nitrogen compounds. Primary silyl amines show two hydrogen stretching frequencies between 2.8 and 2.95 μ, as well as a strong NH_2 deformation falling at 6.5 μ. The SiNHSi linkage is characterized by a group of bands including the NH stretch at 2.95 μ, an NH bending at 8.5 μ and the asymmetric SiNSi stretch at 10.7 μ[231, 232]. Cyclic silazanes show behaviour analogous to the cyclic siloxanes, with the SiNSi vibrations absorbing at 10.85 μ in hexamethyl cyclotrisilazane[285] and 10.65 μ in the octamethyl cyclotetrasilazane. There is also a shift of the NH deformation from 8.7 μ in the trimer to 8.5 μ in the tetramer.

Trisilylamine has a planar structure with the asymmetric SiN stretch falling at about 10 μ[341–343]. In bis(trimethylsilyl)methylamine, however, the SiN stretch is found at 11 μ[267].

The infrared spectra of silicon isocyanide compounds have been reported[222, 344, 345]. Isotopic studies[346, 347] as well as chemical evidence indicate that only the normal cyanide is formed in syntheses of Si–CN compounds. The CN stretching vibration falls between 4.5 and 4.6 μ, which is at slightly longer wavelengths than in the corresponding organic compounds. Cyanosilanes hydrolyse readily in moist air to form hydrogen cyanide.

The infrared spectra have been reported[347] of gaseous SiH_3CN and SiD_3CN at 28°, also the vibrational assignments and structural interpretations. For SiH_3CN the chief absorption bands are at about 16.67, 10.87, 7.35 and 4.54 μ, whilst for SiD_3CN they are at about 16.67, 14.28, 9.26, 6.25 and 4.52 μ. They showed that there is very strong evidence for a normal cyanide rather than isocyanide structure for SiD_3CN.

q. Silicon–sulphur compounds. The SiSH stretch in Me_3SiSH falls at 3.9 μ[326], but there are apparently no characteristic SiS bands in the NaCl prism range. In $Me_3SiSSiMe_3$, the SiS stretch vibrations are found at 20 μ[231], in $H_3SiSSiH_3$ at 19.5 μ[348, 350] and in Me_3SiSH at 22 μ[351]. Disilyl selenide[349, 356] and silyl isothiocyanates[352, 353] have also been studied.

r. Silicon–phosphorus compounds. These compounds show a broad SiOP absorption at 9.6 μ[354] and the Si–P band at 22 μ[355].

s. Silicon–Metal Compounds. SiONa, SiOK and SiOSn in metal trimethylsilanol, triethylsilanol and diphenylsilanediol show absorptions, respectively, between 10 and 11 μ. SiOHg shows a strong band at 11 μ and the SiO–Ti vibration is at 10.9 μ[356–358].

V. GAS CHROMATOGRAPHY

Gas chromatography has been used[359] to separate trisilane and the mixture resulting from the passage of an electrical discharge through trisilane. For mixed silicon–germanium hydrides, it was found that compounds with similar retention volumes gave an approximately molar response so that a gas density balance could be used to measure molecular weights. Table 2 shows observed retention volumes for a range of silicon hydrides[360].

TABLE 2. Logarithm of retention times for silicon hydrides—comparison of calculated and observed values[a]

Mobile Phases	Silicone 702				Tritolyl Phosphate			
	20 °C		64.5 °C		20 °C		64.5 °C	
	Obs.	Calc.	Obs.	Calc.	Obs.	Calc.	Obs.	Calc.
Si_3H_8	0.66	0.65	0.98	0.98	0.79	0.78	1.08	1.06
iso-Si_4H_{10}	1.25	1.24	1.41	1.42	1.32	1.31	1.47	1.47
n-Si_4H_{10}	1.41	1.40	1.54	1.55	1.48	1.48	1.59	1.60
iso-Si_5H_{12}	2.00	2.00	2.00	2.00	2.00	2.01	2.00	2.01
n-Si_5H_{12}	2.17	2.16	2.13	2.13	2.18	2.18	2.13	2.14
iso-Si_6H_{14}			2.58	2.57	2.68	2.70	2.54	2.55
n-Si_6H_{16}			2.72	2.70	2.87	2.87	2.69	2.68
iso-Si_7H_{16}			3.14	3.13			3.07	3.10
n-Si_7H_{16}			3.28	3.26			3.22	3.23

[a] Retention volumes calculated as described in References 360 and 361; iso-silanes refer to branched silanes.

Silanes up to Si_8H_{18} have been separated[361] on a column of 15% squalane on kieselguhr at 110 °C using helium as carrier gas.

a. Alkyl and aryl silanes. It has been shown[359] that for alkyl silicon compounds, log (retention time) versus the number of silicon atoms gave straight lines in a manner similar to that reported for the hydrocarbons (Figure 1). In this figure the column packing was Silicone 702 operating at 100 °C. The solid circles and line represent the straight-chain compounds. The open circles denote the branched isomers. The latter lie below the line and are analogous in behaviour to the hydrocarbons. In an alternative approach which gives approximately straight-line plots, the log of corrected retention volume is plotted versus the total number of carbon atoms in the alkyl groups. This relationship provides a method for identifying additional compounds. An extensive tabulation of retention data has been carried out[362, 363] on E 301 Silicone oil. Other column packings have been discussed[360, 364]. Retention volumes for a range of alkylsilanes have been listed[360]. An ionisation gauge detector has been used for the detection of organosilicon compounds emerging from a column comprising E 301 Silicone oil on Celite operated at various temperatures between 80 and 140 °C, with nitrogen as carrier gas. Table 3 presents the specific retention volume data obtained by Pollard and coworkers[363] for a range of organosilicon compounds. It is seen that the specific retention volume increased with the molecular weight of the alkylsilane.

The thermal decomposition of some aryltrimethylsilanes in the presence of hydrogen in a continuous-flow reactor has been studied[365] using gas chromatography for the analysis of the reaction products.

Gas-chromatographic retention data on a range of alkylsilanes have been published[366]; these show patterns which enable retention times to be predicted. A comparison was also carried out between alkylsilanes and analogous hydrocarbon systems. The alkylsilanes were analysed on columns of squalane using a flame ionisation detector and hydrogen as carrier gas. The solid support was Embacel, which had been made inactive by treatment with hexamethyldisilazane[367]. Retention data are expressed as the logarithm of the retention time relative to that of mesitylene ($\log t_R = 2$). The use of

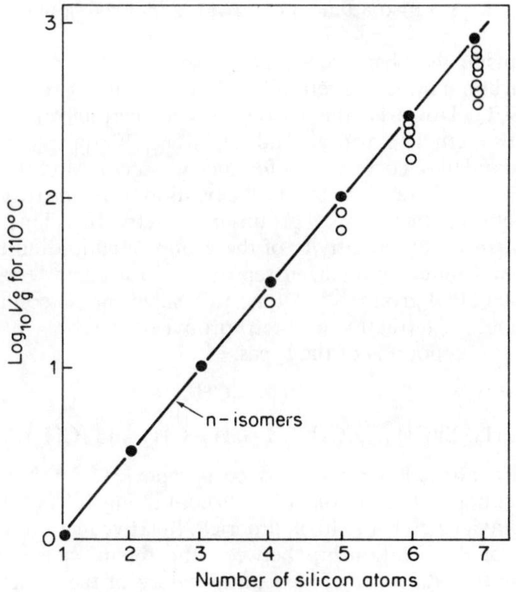

FIGURE 1

TABLE 3. Specific retention volumes [Vg] (ml) of organosilicon compounds on 15% E 301 Silicone oil

Compound	80 °C	100 °C	110.5 °C	140 °C
$SiMe_4$	11.82±0.15	7.65±0.07	6.31±0.07	3.09±0.03
Me_3SiEt	31.4±0.2	18.45±0.06	13.77±0.2	7.15±0.05
Me_2SiEt_2	79.2±0.5	43.2±0.2	32.0±0.7	16.4±0.2
Me_2SiEt_2	190.9±0.9	98.7±0.7	68.3±1.8	34.1±0.1
$SiEt_4$	446.8±2.7	216.8±3.0	147.7±2.9	69.3±0.4
Me_3SiPr	61.6±0.6	34.3±0.4	24.9±0.1	13.33±0.2
Me_3SiBu	129.3±2.0	68.0±0.2	47.7±0.2	24.3±0.1
Et_3SiH	10.8±0.4	64.6±1.1	44.8±0.6	23.2±0.1
$Me_3SiOSiMe_3$	74.4±0.5	40.1±0.1	28.4±0.4	14.7±0.05

retention parameters for the calculation of retention times works well for silanes and tetraalkylsilanes[366, 368, 369].

Equations have been developed[369, 370] for the calculation of retention indices for mixed tetraalkylsilanes using the same methods as applied for the tetra-alkoxysilanes[367, 369–372]. In investigations of tetraalkoxysilanes[370, 371], their change in retention index with temperature was found to be always negative and to increase with molecular size. For the tetraalkylsilanes, the temperature increments are considerably lower and alternate between positive and negative values in an apparently erratic manner. However, there is also a tendency towards increasing negative values for larger molecules.

A linear relationship has been reported[362] between the logarithm of the retention volume and the carbon number for tetraalkylsilanes belonging to the series R_4Si; R_3SiR'; $R_2SiR'_2$; $RSiR'_3$; SiR'_4. However, this is true only for certain series. Thus for the seven series Me_nSiR_{4-n}(R=ethyl to pentyl) and Et_nSiR_{4-n}(R=propyl to pentyl) the above statement is approximately correct only for the two series Me_nSiEt_{4-n} and Et_nSiR_{4-n}. For the other five series, there is an increased deviation from linearity with increase in the difference in size between the two alkyl groups in the tetraalkylsilane, as a consequence of the increased departure from additivity of the group retention indices.

Specific retention volumes have been reported[373] for some tetraalkylsilanes mainly containing branched alkyl groups[374]. A quantity called the 'silicon index' (ISi) has been reported for tetramethyl, tetraethyl and tetrapropyl silanes.

The behaviour of compounds of the types:

$$(CH_3)_3Si(CH_2)_nSi(CH_3)_3,\ (CH_3)_2\overset{\ulcorner\!\!-\!\!-\!\!-\ CH_2\ -\!\!-\!\!\urcorner}{Si{-}CH_2{-}CH_2}\ \text{and}\ (CH_3)_3Si(CH_2)_nCH_3$$

in comparison with normal paraffins has been reported[375, 376] during separation at 50–150 °C using helium carrier gas on columns containing 20% of polymethylphenylsiloxane or polyoxyethylene glycol 1540 on firebrick. Relative retention volumes (nonane as standard), graphs of the relationship between the retention times of the substances investigated on the two stationary phases, and values of the structural increments for calculating the relative retention times are presented.

The retention volume has been reported[377] for triethyl and tetraethyl silanes on columns packed with silanized Chromosorb W supporting 20% of Apiezon L or 15% of Carbowax 20 M and operated at 120 and 90 °C, respectively, using helium as carrier gas and thermal conductivity or flame ionisation detection. Measurements were also carried out by gas–solid chromatography on columns of Carbochrom-1 (graphitized thermal carbon black with 0.1% of Apiezon L) or Silochrom C-80.

Trichlorosilane and silicon tetrachloride mixtures in carbon tetrachloride have been separated[378] using transformer oil activated by glycerol as the liquid phase.

b. Chlorosilanes. The separation has been studied[379] of mixtures of di- and trichlorosilane, silicon tetrachloride, methyl chloride and also various methylchlorosilanes on columns containing benzyl benzoate, dibutyl phthalate or diethyl phthalate supported on Celite or on diatomite brick previously treated with methyl silicon dichloride vapour. Helium was used as a carrier gas with a katharometer detector.

Dibutyl phthalate, also tritolyl phosphate or Silicone oils KF96 and DC703 (Dow Corning) as stationary phases with helium as the carrier gas have been used to separate silicon tetrachloride and phosphorus trichloride in trichlorosilane[380, 381]. The esters vitiated results due to the production of hydrogen chloride by hydrolysis of the sample.

It has been shown[382] that chlorosilanes could not be separated by gas absorption chromatography on any of the usual solid absorbents. By gas–liquid partition chromatography, quantitative separation of the lower-boiling silicon compounds was achieved on a column comprising Silicone oil DC200/350 and dibutyl phthalate on kieselguhr at 30 °C with hydrogen as carrier gas, and of the higher-boiling compounds at 150 °C.

Mixtures of mono-, di- and trichlorosilane, silicon tetrachloride and silane (SiH_4) have been separated[383] using two separation columns, both operated at room temperature. This mixture was fully resolved on a stationary phase of silicon oil with the exception that any nitrogen in the sample elutes with silane. Nitrogen was separated from silane on a column of silica gel[384].

The elution characteristics have been reported of various silicon chlorides of the type Si_nCl_{2n+2} obtained by the pyrolysis of $(SiCl_2)_n$[385]. Separation was achieved in benzene solution on a column comprising Silicone oil 550 on Celite using nitrogen as the carrier gas and a thermal conductivity detector.

Mixtures of silicon tetrachloride, chlorine, phosgene, argon, nitrogen, carbon monoxide and carbon dioxide obtained in the reductive chlorination of metal ores are separated[386, 387] on a three-column, three-detector gas chromatograph. The three gas chromatographs were equipped with the following columns in series: column 1 was filled with Kel-F 10 on Chromosorb T and operated at 60 °C, column 2 was filled with silica gel and operated at 139 °C and column 3 was filled with molecular sieve 5A and operated at 90 °C. Thermal-conductivity detectors were used, and the carrier gas was helium. Silicon tetrachloride, chlorine and phosgene were determined on column 1, carbon dioxide on column 2 and carbon monoxide, argon and nitrogen on column 3. It was possible to detect hydrogen chloride on column 1 by operation at a lower column temperature.

c. Alkyl, aryl and vinyl chlorosilanes. Six silanes including trimethylchlorosilane and silicon tetrachloride have been separated[388] using nitrobenzene on infusorial earth as the stationary phase. Ethylchlorosilanes were separated[389] on a column packed with methylsilicone oil MSO 150 on ground unglazed tile at 90 °C, with nitrogen as carrier gas.

In the separation of mono-, di- and tri-methylchlorosilane, CH_3SiHCl_2 and silicon tetrachloride, using two four-metre columns containing liquid paraffin and H-132, respectively, the five compounds are satisfactorily separated at 70 °C[390].

The effect has been studied[391] of the nature of the stationary phase, and the solid support, the duration of stationary phase, the rate of flow and moisture content of the carrier gas, and the column temperature on the gas–liquid chromotographic separation of the components of a mixture obtained in the preparation of methylchlorosilanes including hydrogen chloride, methyl chloride, silicon tetrachloride, trimethylsilicon chloride, dimethyl silicon dichloride and methyl silicon trichloride. A column containing nitrobenzene on firebrick was used.

An apparatus has been developed[392] for the analysis of a mixture containing trichlorosilane, methyldichlorosilane, silicon tetrachloride, trimethylchlorosilane, dimethyldichlorosilane and methyltrichlorosilane by gas–liquid chromatography on a column of nitrobenzene supported on Celite 545. The column is eluted with nitrogen and the emergent gas is absorbed in flowing 0.01 *N* potassium chloride. Hydrolysis of the silanes yields hydrochloric acid which alters the electrical resistance of the potassium chloride solution and this permits quantitative analysis of the silane mixture.

Methylchlorosilanes were separated[393] on two columns (in series) containing, respectively, tritolyl phosphate and dioctyl phthalate on kieselguhr.

The separation has been carried out[394] of methylchlorosilanes on different supports and with different liquids. Celite and Sterchamol were both satisfactory supports but Termalite (firebrick) had a strong selective absorption effect, resulting in asymmetric elution curves. Of the stationary phases tested (liquid paraffin, silicone oil, 1-chloronaphthalane, dibutyl phthalate and nitrobenzene), those with the highest dipole moments were the most effective. Celite impregnated with 23% of nitrobenzene gave complete separation of a six-component mixture. Nitrile silicone oil and *m*-nitrotoluene stationary phases have been used[395] with a katharometer detector to separate a mixture of various methylchlorosilanes and chlorosilanes obtained in reactions between ferrosilicon and methyl chloride.

Mixtures of methylchlorosilanes produced in the reaction of methyl chloride with silicon copper alloys separate well on a column of FS-16 (Dow DC methyl *p*-chlorophenyl siloxane polymer) on firebrick at 70 °C using nitrogen carrier gas[396].

Methylchlorosilane mixtures have also been separated[397] on a column consisting of Rysorb impregnated with a polymethylphenylsiloxane oil.

The effect of 31 stationary phases on the relative volatility of silicon tetrachloride and methylchlorosilanes has been examined[398] on firebrick containing the various stationary phases. 2-Chloroethyl ether, ethyl chloroacetate and 1,1,3-trichloropropane were found suitable for separating multi-component mixtures.

Relative retention volumes and retention indices have been reported[399] for silicon tetrachloride, methyltrichloro- and dimethyldichloro-silane, phosphorus trichloride and phosphorus oxychloride at 50 °C on stationary phases containing vaseline oil, polysiloxane liquids VKZh-94, PFMS-2 and DS-701, fluorosilicone oil 169, dinonyl phthalate and dibutyl phthalate. The retention index for phosphorus oxychloride increased with increase in the polarity of the stationary phase; that for silicon tetrachloride remained practically constant.

Low concentrations of methyl chlorosilanes have been determined[400] using a flame ionisation detector and silicon FS-16 on silanized INZ-600 brickdust, with nitrogen as carrier gas. Contamination of the detector by silica could be avoided by venting most of the column effluent, before it reached the detector.

The separation and determination of the chlorination products of alkylchlorosilanes, such as trichloroethylsilane or dichlorodimethylsilane, are effected on a column packed with PFMS-4 poly(methylphenylsiloxane) on Celite 545, with helium as carrier gas and katharometer detection[401]. The chlorination products of trichloromethylsilane are analysed on a column packed with SKTFT-50-siloxane rubber on TND-TS-M.

With the exception of one work[400] all studies on the gas chromatography of alkylchlorosilanes discussed so far in this section utilise a katharometer detector. When using a flame ionisation detector[400] this had to be protected from silica produced by thermal decomposition of the organosilicon compound. Other workers[402] report that the flame ionisation detector is unsuitable due to adverse dissociation processes occurring in the ionisation space in the detector. Mixtures of trimethylchlorosilane, dimethyldichlorosilane, triethyltrichlorosilane, $SiHCH_3$, $MeSiHCl_2$ and carbon tetrachloride have been analysed[402] on a Pye Argon chromatograph using argon as carrier gas and a glass column packed with dioctylphthalate on Celite, as stationary phase. The same chromatograph was also used for the analysis of other organosilicon compounds[402], such as linear and cyclic polymers of low molecular weight and methylethoxysilanes. Retention indices have been tabulated[403] for alkylchlorosilanes and propylchlorosilanes on a polymethylsiloxane liquid PMS-20,000 and on a fluorosiloxane liquid FS-303.

The gas chromatographic separation of chlorosilanes, methylchlorosilanes and their associated siloxanes have been reported[404] on 11 different stationary phases. Temperature programming is necessary for the elution of the siloxanes but many of the stationary phases reported in the literature cannot be used under these conditions because of their high volatility at elevated temperatures. Among the liquids or gums which were used as stationary phases, the most effective were dimethyl, diethyl and dibutyl phthalates.

A mixture of phenyltrichlorosilane, diphenyldichlorosilane, benzene, chlorobenzene and biphenyl was separated[405] at 240 °C on a column containing a silicone elastomer on Chromosorb. The carrier gas is nitrogen, and after the separation the gases are passed into an ignition tube containing cupric oxide wire, pumice coated with reduced silver, and finally with iron filings. The organic substances are ignited at 480–600 °C; the chlorine formed is absorbed, and the water formed is decomposed with the liberation of hydrogen, which is determined.

Technical mixtures of benzene, chlorobenzene, phenyltrichlorosilane, diphenyl and diphenyldichlorosilane produced during the manufacture of silicones were separated[406] at 240 °C on a column of Chromosorb impregnated with silicone-elastomer E 301

(Griffin and George). The issuing gases are burnt to carbon dioxide and water and determined by standard methods.

A gas chromatographic method using a gas density balance detector has been described[407] for the determination of chlorinated phenylchlorosilanes using a column packed with Celite 545 supporting 10% of Lukopren G 1000 (a silicone elastomer), with nitrogen as carrier gas.

Retention data have been listed for[374] 44 alkyl-chloro and chlorophenylsilanes, 34 linear and branched dimethyl and methyl-vinylsiloxanes, 14 cyclic dimethyl and methylvinylsiloxanes and 6 linear methylphenyl- and ethoxyvinyl-siloxanes.

The separation and determination has been carried out[408, 409] of dichloromethylvinylsilanes and the chlorovinylsilane and impurities therein. Compounds in trichlorovinylsilane were separated on a column of silicone elastomer supported on Celite. For impurities in trichloromethylvinylsilane, best results were obtained with a column of nitrobenzene supported on kieselguhr.

In one method for the preparation of dichloromethylvinylsilane from dichloromethylsilane and vinyl chloride, the product is contaminated with trichlorovinylsilane, which cannot be separated by distillation. Hence a gas-chromatographic method was developed[410], using dibutylphthalate on Celite with nitrogen as carrier gas. The chromatogram shows eight characteristic peaks, including those of trichlorovinylsilane and dichloromethylvinylsilane which can be determined quantitatively from the peak areas.

A gas chromatographic procedure has been described[376] for the determination of fluoro and chlorofluoro organosilicon compounds of the type $(CF_3CH_2CH_2)_3SiH$, $(CF_3CH_2CH_2)_3SiCl$ and $(CF_3CH_2CH_2)_4Si$.

d. Alkoxy silanes. The components formed during the Grignard synthesis of the methylphenyldiethoxysilanes can be separated by gas chromatography[381]. The components of the mixtures produced in the reaction between vinyltrichlorosilanes and ethanol, 2-ethoxyethanol or 2-methoxyethanol can be separated by gas chromatography on a column of silicone E-301 supported on Chromosorb. This method was applied to the determination of vinylethoxysilanes[381]. A mixture of deuterated ethoxymethylsilanes, containing ethoxytrimethylsilane, diethoxydimethylsilane, triethoxymethylsilane, unchanged tetraethoxysilane and ethyl ether, was separated[411] on a column of dinonyl phthalate on firebrick, operated at 95 °C with nitrogen as carrier gas. The separated components are detected by thermal conductivity. Products formed during the preparation of ethoxyphenylsilanes by the Grignard reaction have been separated[412] on an aluminium spiral column packed with SE-30 on Chromosorb W, with hydrogen as carrier gas and a thermal conductivity detector.

Compounds having the general formula $Si(OC_2H_5)_n(OC_3H_7)_{4-n}$ have been separated on a column of Triton X-100 on Chromosorb P with helium as carrier gas and flame ionisation detection[413].

Methoxy- and ethoxy-chlorosilanes were separated on columns packed with Rysorb or Chemosorb coated with Se-30, FS-60, Carbowax 6000, Apiezon L or DC-550 with nitrogen as carrier gas[414].

The thermal decomposition products of tetraethoxysilane (ethylsilicate) separate on columns of polypropylene glycol on silanized Chromosorb P at 40 °C using a thermal conductivity detector. The main impurities in the sample were ethyl alcohol, acetaldehyde, ethylene and ethane[415].

The retentions of tetraethoxysilane relative to toluene on three different stationary phases have been measured[416]. The specific retention volume of tetraethoxysilane on a column containing silicone oil as the stationary phase has been reported[417]. The gas chromatographic separation has been studied[418] of a number of symmetrical and mixed tetraalkoxysilanes present in a heat transfer liquid, including retention indices for tetramethoxy and tetraethoxysilane.

Determination and calculation of Kòváts retention indices have been reported for tetraalkoxysilanes with non-branched[419, 371] and branched[420] alkyl groups. A procedure has been devised[421] involving the use of refractive index in conjunction with Kòváts retention indices for the identification of tetraalkoxysilanes with either straight or branched alkoxy groups. This work was performed using a flame ionisation detector and columns packed with either Apiezon M or cyanosilicone GE EX-60, respectively, on Chromosorb G.

The usefulness of retention index versus temperature plots for the evaluation of the optimum column temperature has been confirmed[422] when analysing mixtures of compounds with similar boiling points. For example, plots for the tetraalkoxysilanes $(MeO)_nSi(OEt)_{4-n}$ give straight lines that converge towards higher temperatures. Hence, the best separation will be obtained at the lowest possible column temperature.

A linear relationship exists between the logarithm of the retention volume and the carbon number for tetraalkylsilanes belonging to the series R_4Si, $R_3SiR'_3$ and SiR'_4[362]. For the corresponding series of tetraalkoxysilanes, an approximately linear relationship between retention index and carbon number will exist only for the series $(PrO)_nSi(OBu)_{4-n}$, but not for the series $(MeO)_nSi(OBu)_{4-n}$.

A modified equation for the calculation of retention indices of mixed tetraalkoxysilanes from the retention indices of the symmetrical counterparts has been compared with earlier equations[371]. These methods have been extended to compounds which also contained branched alkoxy groups (isopropoxy, isobutoxy and *sec*-butoxy)[371,420]. The best agreement between experimental and calculated values for retention indices was obtained with Apiezon M stationary phase[420]. With cyanosilicone EX-60 stationary phases, good agreement in retention indices was obtained only for tetraalkoxysilanes that contain isobutoxy groups.

e. Siloxanes. The *cis*- and *trans*-isomers of 2,4,6-trimethyl-2,4,6-triphenylcyclotrisiloxane have been separated[423] on a column of silicone rubber or Apiezon L grease on Chromosorb at less than 290 °C with helium as carrier gas. Partial separation of the tetramer isomers was also achieved.

Linear and cyclic polydimethylsiloxanes were determined on a column of silicone elastomer (Lukopren M) on Rysorb BLK at various temperatures according to the components to be determined. The relative error of measurement of a single component was less than $\pm 7.5\%$[424,425]. The determination of methylphenylpolysiloxanes of boiling point up to 500 °C has been reported[426] on silanised crushed firebrick containing various amounts of polymethylsiloxane liquid. A katharometer detector was used. With a mixture of ten methylphenylpolysiloxanes (b.p. 100–500 °C) satisfactory separation without decomposition was obtained.

An all-glass gas chromatographic apparatus has been used[427] for the analysis of mixtures of siloxanes and of other mixtures. The gas chromatography of methylhydrocyclosiloxanes has been studied[428].

The gas chromatographic analysis has been reported[429] of cyclic and linear products up to $[(CH_3)_2SiO]_{40}$ with molecular weight up to 2530 and 2960 formed upon heating $(CH_3)_3SiO[(CH_3)_2SiO]_8OSi(CH_3)_3$ with acid clay catalyst at 25 °C. A linear programmed temperature (50–400 °C) gas chromatograph with a thermal conductivity detector was used. The column used was packed with diphenylsiloxane-dimethylsiloxane copolymer gum (Dow Corning proprietary gum of undisclosed composition) on Chromosorb W.

Reaction of methylpolysiloxanes with boron trifluoride diethyl etherate produces methylfluorosilanes which are amenable to gas chromatography. This method determines the relative amounts of $Me_3Si{-}$, $Me_2Si{=}$ and $MeSi{\equiv}$ groups in methylsiloxane polymers. Polymers which cannot be analysed in their original condition are depolymerized to methylfluorosilane monomers with boron trifluoride etherate in an enclosed system:

$$2MeSiO_{3/2} + 2BF_3 \cdot Et_2O \rightarrow 2\,MeSiF_3 + B_2O_3 + 2\,Et_2O$$

$$3Me_2SiO + 2BF_3 \cdot Et_2O \rightarrow 3Me_2SiF_2 + B_2O_3 + 2Et_2O$$

$$6Me_3SiO_{1/2} + 2BF_3 \cdot Et_2O \rightarrow 6Me_3SiF + B_2O_3 + 2\,Et_2O$$

The method is also applicable to siloxanes containing silanic hydrogen. The gas chromatography of siloxanes associated with chlorosilanes and methyl chlorosilanes has been discussed[404].

Eight components of polymethylphenylsiloxanes were identified[430] by the use of columns of Aerosilogel A-380 or Silochrom C-80 supporting 3,3′-oxydipropionitrile and detected by ultraviolet spectroscopy.

Structural identifications have been made[431; 432] of polycyclic methylpolysiloxanes of the general formula $[(CH_3SiO_{1.5})]_x \cdot [(CH_3)_2SiO]_y$, x and y being between 2 and 5 based on retention index additivity. Retention data for a series of siloxanes were determined on squalane, Apiezon L, QF-1 or Polysorbate 60. The value of the retention index is related to the structural characteristics of the molecule, which are regarded as additive and to a stationary phase factor.

Branched low-molecular-weight methyl and methylhydrosiloxanes have been separated and identified by gas chromatography–mass spectrometry[433,434]. Separations were carried out on columns packed with Apiezon L or Silicone oil AK 30,000 on Aw-DMCS Chromosorb G. Helium was used as carrier gas and the columns were temperature programmed.

Technical hexamethylsiloxane has been analysed[435] on a column packed with Chromosorb W.Aw impregnated with dibutyl phthalate. A combined gas chromatographic–mass spectrometric procedure is also described in which a column of Chromosorb G-AW DMCS impregnated with silicone oil is temperature programmed, with helium as the carrier gas.

High boiling organopolycyclosiloxanes have been analysed[436] on a composite column of two stationary phases, Apiezon L and PMS-100 using a thermal conductivity detector.

Gas chromatography has been applied[437] to the determination of the low-molecular-weight portion in poly(dimethylsiloxane). Polymers of molecular weight up to 700 in poly(dimethylsiloxane) can be determined on a column packed with OV-1 on Chromosorb W AW-HP with helium as carrier gas and a flame ionisation detector.

The application of time sharing to data reduction in gas chromatography has been applied to the analysis of polydimethylsiloxane fluids[438].

f. Silyl ethers and silyl thioethers. Retention data have been obtained[439]on a homologous series of trimethylsilyl ethers and trimethylsilyl thioethers. The stationary phases used were on Silicel C-22 firebrick; the liquids used were polyethylene glycol 400, dinonyl phthalate, tritolyl phosphate, squalane and Apiezon L grease. Great care was taken to ensure that no moisture was present on the liquid phase or support. The two stationary phases for systematic measurements of the oxygen ethers were Apiezon L and tritolyl phosphate, neither of which caused any appreciable amount of hydrolysis of the samples injected. The only phase found suitable for gas chromatography of the silyl thioethers was Apiezon L. Complete breakdown of the thioethers was noted on tritolyl phosphate, even at 80 °C.

g. Silyl sulphides. The gas chromatography has been discussed[440] of bis(triethylsilyl)sulphide and its separation from its germanium, tellurium and tin analogues on a column packed with Chromosorb W supporting Apiezon L.

h. Silicon-nitrogen compounds. Bis(trimethylsilyl)acetamide has been analysed[441] on a column of Chromosorb W coated with 5% potassium hydroxide. The support was heated at 500 °C overnight, then coated with liquid paraffin.

i Silicon-containing phenols. Silicon-containing sterically hindered phenols and their oxidation products, including 4-*t*-butyl-2,6-bis(trimethylsilyl)phenol and 2,4-di-*t*-butyl-6-trimethylsilylphenol, were separated on columns of Apiezon L or OV-17 on Chromosorb W, with helium as carrier gas and detection by katharometer or flame ionisation[442].

j. Silicon-containing alcohols. A method has been described[443] for the qualitative and quantitative characterisation of water-borne organosilicon compounds. Silanols or functional materials are converted to trimethylsilylated derivatives which can be determined by gas–liquid chromatography. The method gives good accuracy and precision in the ppm range and with suitable precautions can be extended to the low ppb range.

k. Silicon heteroelement compounds. Retention volumes have been measured[444] for compounds of the type $(C_2H_5)_3M–M'(C_2H_5)_3$, where M and M′ are silicon, germanium or tin on columns of silanised Chromosorb W supporting Apiezon L or Carbowax 20 M, with helium as carrier gas and thermal conductivity or flame ionisation detection.

The separation of the *cis*- and *trans*-isomers of 1,2-bis(trimethylsilyl)-, 1-trimethylgermanyl-2-trimethylsilyl- and 1,2-bis(trimethylgermanyl)ethylene is possible by gas–solid chromatography on graphitised carbon black and by gas–liquid chromatography on Chromosorb W supporting Apiezon L, with helium as carrier gas and a thermal conductivity or flame ionisation detector[445].

The separation is possible[446] of compounds containing bonds between two or three unlike heteroatoms (silicon, germanium, tin or sulphur), also heterocyclic compounds with silicon atoms in the ring by gas–liquid chromatography on columns of silanised Chromosorb W coated with Apiezon L, polyoxyethylene glycol 20 M or Reoplex 400, and also by gas–solid chromatography on graphitised carbon black, with helium as carrier gas and flame ionization or thermal conductivity detection. Gas–solid chromatography provided greater selectivity than gas–liquid chromatography, especially for the *o*- and *p*-isomers of sterically hindered silicon containing phenol derivatives.

l. Silyl arenes. High boiling point bis(dimethylsilyl)arenes have been separated[447]. Mixtures containing 1,4-bis(dimethylsilyl)benzene were separated on a column of INZ-600 firebrick supporting poly(4-dimethylsiloxyphenyl-4-dimethylsilylphenyl ether) with helium as carrier gas and katharometer detection. Mixtures containing bis-(4-dimethylsilylphenyl) ether were separated on a column of TND-TSM supporting polymetacarboranesiloxane with helium as carrier gas and katharometer detection.

m. Cyclosilazanes. Gas chromatography has been used to study the interchanges involving different cyclosilazanes occurring during redistribution reactions of tris (*N*-alkylhexamethyl)cyclosilazanes[448].

n. Impurities in organosilicon compounds. Gas chromatography has been used[449] for the determination in organosilicon monomers of traces (down to 0.001%) of benzene, chlorobenzene, chlorodiphenylsilane, chlorophenylsilane, biphenyl, chlorobiphenyl, dichlorophenylsilane and *o*- and *p*-terphenyls. The column was Rysorb BLK supporting polymethylphenylsiloxane with nitrogen carrier gas.

For the determination of ethanol in triethoxysilane[450] the column comprised silicone PFMS-3 supported on firebrick with hydrogen or helium as carrier gas and a katharometer detector, or, alternately, petroleum jelly on firebrick with a flame ionisation detector and hydrogen as carrier gas. Benzene is determined[451] in trichlorosilane by separation on a column containing petroleum oil supported on brick, with a katharometer or flame ionisation detector. The sensitivity of the method is 0.05%.

Down to 0.0001%, benzene in silicon tetrachloride could be determined in a column containing petroleum jelly on firebrick with flame ionisation detector and hydrogen as the carrier gas. The minimum determinable amount of benzene is 10^{-5} mg, with a relative error of 7.2%. The method was also applied to the determination of benzene in trichlorosilane.

o. Gas chromatographic detectors for organosilicon compounds. Detailed studies have been performed on various types detectors in the gas chromatography of organosilicon compounds. These include argon ionisation[452], flame ionisation[364,453–455], alkali flame glow discharge tubes[456], microwave emission[456–463], hydrogen atmosphere flame ionisation[461,462,464,465] (HAFID) atomic emission/absorption and atmospheric-pressure microwave-induced helium-plasma[463,464] detectors.

An argon ionisation detector to the gas–liquid chromatography of organosilicon compounds[452] was unsuitable for the analysis of methylchlorosilanes. However, a pyrolysis chamber permitted the analysis of both linear and cross-linked methylsiloxane polymers of high molecular weight.

The anomalous response of the flame ionisation detector in the analysis of organosilicon compounds has been investigated[454,455]. The behaviour of the compounds was studied on columns of silicone elastomer on Chromosorb or Celite. Above a critical sample size, the top of the peak became inverted, the maximum height remaining the same, but the depth of the inverted portion increased with increasing sample size.

The inversion of peaks of organosilicon compounds observed by some workers[454] bears no relation to the slight deposition of silica which occurs on the electrodes of the flame ionisation detector.

Retention indices (ΔI) have been determined for 68 organosilicon compounds on columns containing Apiezon L, SE-30, QF-1, XF-112 and *o*-nitrotoluene as stationary phases, with flame ionisation or thermal conductivity detectors. Some generalizations and rules are reported of character relationships based on ΔI values and retention-index increments ($\Delta\delta I$), for homologous compounds[364].

A detailed study has been made[456] of the application of the alkali flame detector to organosilicon, organotin and organolead compounds.

The use of a glow discharge tube as a detector for the gas chromatography of organosilicon compounds has been discussed[466].

The microwave emission detector has been demonstrated to be useful for the detection of organosilicon compounds[457]. The applicability of helium microwave glow discharge detectors to the detection of tetravinylsilane has been studied[457]. The use of helium (or argon) plasma led to enhanced sensitivity and high resolution and selectivity. Because it is a constituent of the discharge tube, silicon exhibits unique behaviour with respect to flow rate. Not only does its response continually increase with increasing flow rate up to the value at which the plasma becomes unstable (ca. 475 ml/min), but the selectivity also increases in a similar fashion.

When silane is added to an atmosphere of hydrogen in which an oxygen-fed flame burns, organometallic compounds ionise efficiently in the flame, providing a sensitive and selective chromatographic separation[461]. The effect on organosilane response of doping the hydrogen atmosphere with organometallics has been studied[462]. As expected, it was found that the metal-doped flame system enhanced silicon ionisation, but it did not produce a linear calibration curve. However, a non-doped hydrogen atmosphere flame system was found to exhibit a linear response curve while still providing good selectivity for silicon-containing compounds. The severe peak tailing reported for silicon-containing compounds was attributed to chromatographic conditions rather than detector design. Commercial flame ionisation detector would be rendered specific for silicon-containing compounds by simply interchanging the oxygen and hydrogen inlets so that the flame burned in a hydrogen atmosphere[461]. The optimal configuration of various detector geometries was found to be a narrow cylindrical tube with the collector electrode positioned more than 10 cm from the flame. This configuration eliminated peak tailing which is prevalent in other designs. Helium was the carrier gas of choice since nitrogen reduced the sensitivity of the detector to silicon-containing compounds, but not to hydrocarbon compounds. The calibration curve for tetraethylsilane proved to be linear for about three orders of magnitude with a minimum detectable amount of 4 ng and a selectivity over decane of 2600. This corresponds to a silicon-to-carbon selectivity ratio better than 1×10^4.

This technique was also applied to the analysis of a mixture of *n*-hydrocarbons and *n*-alcohols. The alcohols were converted to their trimethylsilyl derivatives and the mixture was chromatographed on two different instruments, one equipped with a standard flame ionisation detector and the other with the preferred design of the silicon-sensitive hydrogen atmosphere flame ionisation detector.

p. Micropyrolysis–gas chromatography of organosilicon compounds. Reaction-gas chromatography has been used[467,468] to detect alkyl, chloroalkyl, phenyl, chlorophenyl, alkoxy and vinyl groups, also ≡Si–Si≡ and ≡Si–H, in organosilicon compounds. Treatment of organosilicon compounds with boron tribromide at 130 °C converts silicon-bound alkoxy and phenyl groups to the corresponding alcohols and benzene, respectively[467]. Acetaldehyde and propionaldehyde are produced from ethylene oxide and propylene oxide chains in organosilicon compounds on treatment with potassium bisulphate or persulphate at 260 °C.

The micropyrolytic gas chromatography technique was applied to the identification of siloxane polymers and as an aid to the control of the reproducibility of cross-linked siloxane resins[452]. In the case of linear polymers the column packing was Apiezon M and a devolatilized linear siloxane polymer (the Hungarian product Hu-Au-120, mol. wt. 59000) supported on Celite. With a detector voltage of 1000 V, column temperature of 100 °C and argon flow rate 20 ml/min, the sample was applied in a solvent-free state to the heated wire-coil. Cross-linked siloxane polymers were analysed as solutions in butyl acetate on a solid support containing 10% linear siloxane polymer as stationary phase under similar conditions.

Some functional groups and bond types may be identified by degradation of the sample in a microreactor, and gas chromatographic separation of the volatile reaction products[467]. Results are presented for the degradation of alkyl-aryl- and chloralkylsilanes and siloxanes (with concentrated sulphuric acid saturated with vanadium pentoxide), of alkoxysilanes (with hydriodic acid at 75 °C) and of vinylsilanes (by saturation with chlorine on a water bath followed by decomposition with concentrated sulphuric acid at 175 °C). Alkyl, chloroalkyl, phenyl, chlorophenyl, alkoxyl and vinyl groups, and ≡Si–Si≡ and ≡Si–H bonds, were detected by these procedures. A scheme has been described[469] for the qualitative analysis of silicones after cleavage of the hydrocarbon groups by sulphuric acid. A 70% yield of methane is reported from a polydimethylsiloxane fluid and lower yields from polymers containing higher alkyl groups. Hydrogen, methyl, ethyl, amyl and phenyl groups attached to silicon were successfully identified.

The proportions and types of organic groups present in polymeric siloxanes have been found by reaction with hydrofluoric acid and then separation of the resulting organofluorosilanes[470–472].

The analysis has been reported[473] of phenylmethyldichlorosilane by reaction chromatography with sodium fluorosilicate involving conversion in a pre-column to the corresponding fluorosilanes, which are then separated on the chromatographic column. Due to the similarity of their boiling points, phenylmethylsilicon dichloride and phenylsilicon trichloride cannot be separated by normal chromatographic procedures.

The reactions occurring on the pre-column are as follows:

$$Ph(CH_3)SiCl_2 + Na_2SiF_6 \rightarrow Ph(CH_3)SiF_2 + SiF_4 + 2NaCl$$

$$2PhSiCl_3 + 3Na_2SiF_6 \rightarrow 2PhSiF_3 + 3SiF_4 + 6\,NaCl$$

Using a 10-μl sample and a thermal conductivity detector, phenyl silicon trifluoride can be accurately determined in phenyl silicon dichloride down to 0.01%. The same procedure is applicable to the separation of high boiling chlorosilanes, which are difficult to separate otherwise, and also achieved the separation of various cyanopropylmethylchlorosilanes.

Various workers[474–479] have investigated the pyrolysis of tetraalkylsilanes using gas chromatography combined with mass, infrared and nuclear magnetic resonance spectroscopic techniques. The cleavage of silicon to carbon bonds, especially where at least one of the carbon atoms is attached to a silicon atom which is also part of an aromatic system, has been examined.

The thermal decomposition of arylsilanes has been investigated[365], and the products were identified using retention-time techniques. A graph of log(retention time) against carbon number indicated that the assignments were correct. The compounds studied were phenyltrimethylsilane, benzyltrimethylsilane, *p*-methylbenzyltrimethylsilane, *o*-tolyltrimethylsilane and *p*-tolyltrimethylsilane.

VI. OTHER CHROMATOGRAPHIC TECHNIQUES

a. Thin-layer chromatography. This procedure has been applied[480] to the analysis of silanols, siloxanes and silanes. Mixtures of these substances can be separated on layers of Kieselgel D with benzene, benzene-acetone, light petroleum or light petroleum–benzene as solvent. For the separation of the three classes of compounds, light petroleum–benzene (10:1) was the best solvent; silanols had the lowest R_f values and silanes had the highest.

b. Column chromatography. The column chromatographic separation of pyrolysis products of tetramethylsilane has been studied[481]. The high boiling silicon compounds formed in this pyrolysis could not be separated unchanged by distillation, but were separated on alumina columns into three fractions. The first fraction could be further fractionated on an alumina column with pentane. The second fraction was eluted with benzene and contained strongly fluorescent substances which were characterised and separated by paper chromatography. The third fraction was eluted with benzene–methanol and was further fractionated on alumina columns with similar solvent mixtures. Oligomers of methylphenylsiloxane, dimethylsiloxane, methylhydrosiloxane and methylhydrocyclosiloxane have been separated[482] into fractions of different molecular weight on a column containing Silica gel MSM, activated carbon BAU or quartz sand. The sample was applied to the column in light petroleum solution and eluted with light petroleum–benzene and with benzene. In the various fractions the molecular weights were determined either cryoscopically or ebullioscopically. Quartz sand was the most effective column packing. The greatest molecular weight distribution was observed for dimethylsiloxane, which had a molecular weight range of 600–3200 and an average molecular weight of approximately 1850. The smallest distribution was found for methylhydrocyclosiloxane with a molecular weight range of 600–1200 and an average molecular weight of about 800.

The chromatographic absorption analysis of alkyl and alkyl-aryl siloxanes has been studied[483]. The sample in methanolic solution is applied to a column of activated carbon and the straight-chain components, i.e. $(CH_3)_3SiO[Si(CH_3)(C_6H_5)O]_nSi(CH_3)_3$, are eluted (as a mixture) with methanol, followed by the cyclic species, e.g. $[Si(CH_3)(C_6H_5)O]_n$, which are eluted with benzene. Some separation of the straight-chain components was achieved by successive elution with methanol, ethanol and butanol, and of the cyclic components by successive elution with acetone and benzene.

c. Paper chromatography. In the paper chromatographic separation of pyrolysis products of tetramethylsilane on alumina columns, one of the fractions was subjected to benzene extraction and components of the extract separated by paper chromatography[484].

VII. REFERENCES

1. R. Belcher, D. Gibbons and A. Sykes, *Mikrochim. J.* 76 (1952).
2. A. Sykes, *Mikrochim. Acta*, 1155 (1956).
3. J. A. McHard, P. C. Servais and H. A. Clark, *Anal. Chem.*, **20**, 325 (1945).
4. H. Wilson, *Analyst (London)*, **74**, 243 (1949).
5. M. Jean, *Chem. Anal.*, **37**, 125 (1955).
6. I. A. Voinovitch, *Chem. Anal.*, **40**, 332 (1958).
7. A. Ringbom, P. E. Ahlers and S. Sûtonen, *Anal. Chim. Acta*, **20**, 78 (1959).
8. R. F. Milton, *J. Appl. Chem.*, **1**, 126 (1951).
9. R. F. Milton, *Analyst (London)*, **76**, 431 (1951).
10. L. Halzapfel and G. Gottschalk, *Anal. Chem.*, **142**, 115 (1954).
11. A. J. Christoper and T. R. F. Fennel, *Talanta*, **12**, 1003 (1965).
12. J. H. Wetters and R. C. Smith, *Anal. Chem.*, **41**, 379 (1969).
13. H. J. Horner, in *Treatise on Analytical Chemistry*, Part II, Vol. 12, Wiley–Interscience, New York, 1965, pp. 287–288.
14. T. Urin, T. Wade and A. Hanyo, *J. Chem. Soc. Jpn., Ind. Chem. Sect.*, **60**, 1274 (1957).
15. T. M. Shanina, N. E. Gel'men and L. M. Kiparenko, *Zh. Anal. Khim.*, **20**, 118 (1965).
16. Z. Sir and R. Komers, *Collect. Czech. Chem. Commun.*, **21**, 873 (1956).
17. A. P. Kreshkov, S. V. Syavtsillo and V. T. Shemyatenkova, *Zavod. Lab.*, **22**, 1425 (1956).
18. B. Smith, *Acta Chem. Scand.*, **11**, 558 (1957).
19. L. V. Myshlyaeva, V. V. Krashoschekov and I. V. Sedova, *Ref. Zh. Khim.*, **19GDE** (2), Abstr. No. 2G184 (1965).
20. L. V. Myshlyaeva and I. V. Sedova, *Zavod. Lab.*, **34**, 263 (1968).
21. Ya. I. Chulkov, USSR Patent 121,963 (18/8/59) (1959).
22. A. P. Terent'eva, S. V. Syavtsillo and B. M. Luskina, *Zh. Anal. Khim.*, **16**, 83 (1961).
23. E. I. Fogel'son, *Zavod. Lab.*, **23**, 1427 (1957).
24. A. P. Kreshkov, L. V. Myshlyaeva and V. V. Krasnoshchekov, *Plast. Massy.* (12), 51 (1962).
25. A. P. Kreshkov, L. V. Myshlyaeva, O. B. Khachaturyan and V. V. Krasnoshchekov, *Zh. Anal. Khim.*, **18**, 1375 (1963).
26. L. V. Myshlyaeva and V. V. Krasnoshchekov, *Trudy Mosk. Khim.-Tekhnol. Inst. im D. I. Mendeleeva*, **42**, 178 (1964); *Ref. Zh. Khim.*, **19GDE** (11), Abstr. No. 11G220.
27. A. P. Kreshkov and L. V. Myshlyaeva, *Zavod Lab.*, **29**, 924 (1963).
28. V. Popov, A. Deryabin, L. V. Myshlyaeva and V. V. Krasnoschchekov, *Trudy Mosk. Khim.-Tekhnol. Inst.*, **46**, 7 (1965); *Ref. Zh. Khim.* **19GDE**, 7(5), Abstr. No. 5G214.
29. L. Kalman and A. Vago, *Magyar Kem. Foly.*, **64**, 123 (1958).
30. R. B. Lew and W. Oyung, *Anal. Chem.*, **36**, 1587 (1964).
31. A. P. Kreshkov and N. I. Gludiva, *Trudy Mosk. Khim.-Tekhnol. Inst.*, **18**, 73 (1957); *Ref. Zh. Khim.*, Abstr. No. 1276.
32. M. P. Brown and G. W. A. Fowles, *Anal. Chem.*, **30**, 1689 (1958).
33. R. Reverchon and Y. Legrand, *Chem. Anal.*, **47**, 194 (1965).
34. L. Govett, *Anal. Chim. Acta*, **25**, 69 (1961).
35. A. P. Kreshkov Yu. Ya. Mikhailenko and E. A. Kuchkarev, *Zavod. Lab.*, **30**, 555 (1964).

36. J. Radall and P. D. Hunt, *Anal. Chem.*, **30**, 1280 (1958).
37. F. L. Chau, in *Proceedings of the Society for Analytical Chemistry Conference, Nottingham*, W. Heffer & Sons Ltd., Cambridge, U.K., pp. 89–101, 1962.
38. N. Wanatabe, Y. Yasuda, E. Kato, T. Nakamura, R. Funasaka, K. Shimokana, E. Sato and Y. Ose, *Science of the Total Environment*, **34**, No. 1/2, 169 (1984).
39. Yu. N. Platonov, *Trudy Komiss. Anal.-Khim. Akad. Nauk SSSR*, **13**, 15 (1963); *Ref. Zh. Khim.*, **19GDE** (1), Abstr. No. 19236 (1964).
40. K. A. Andrianov, L. M. Khanonashvili and T. V. Vasil'eva, *Zh. Anal. Khim.*, **16**, 738 (1961).
41. S. Ostrowski, R. Pickós and A. Radecki, *Chemia. Anal.*, **10**, 531 (1965).
42. T. R. F. W. Fennell and J. R. Webb, *Talanta*, **2**, 389 (1959).
43. T. R. F. W. Fennell, M. W. Roberts and J. R. Webb, *Analyst (London)*, **82**, 639 (1957).
44. R. Belcher and A. L. Godbert, *Analyst (London)*, **66**, 194 (1941).
45. T. R. F. W. Fennell and J. R. Webb, *Talanta*, **2**, 105 (1959).
46. H. R. Shell and H. Craig, *Anal. Chem.*, **26**, 996 (1954).
47. S. Harel, E. R. Horman and A. Talmi, *Anal. Chem.*, **27**, 1144 (1955).
48. T. R. F. W. Fennell and J. R. Webb, *Talanta*, **11**, 1323 (1964).
49. O. Schwarzkopf and T. Henlein, *XVth International Congress of Pure and Applied Chemistry (Analytical Chemistry), 1956, Ramos Afono*, Vol. 1, Moita Ltd., Lisbon, 1957, p. 301.
50. T. R. F. W. Fennell, *Chem. Ind.*, 1404 (1955).
51. J. A. Brabson, H. C. Mattraw, G. E. Maxwell, A. Dorrow and M. F. Needham, *Anal. Chem.*, **20**, 504 (1948).
52. S. Greenfield, *Analyst (London)*, **84**, 380 (1959).
53. H. N. Wilson, *Analyst (London)*, **74**, 243 (1949).
54. L. V. Myshlyaeva and T. G. Shatunova, *Trudy Mosk. Khim.-Tekhnol. Inst.*, **48**, 8 (1966); *Ref. Zh. Khim.*, **19GD** (10), Abstr. No. 10G236.
55. Ya. I. Chulkov and L. A. Solov'eva, *Vestn. Tekhnol i Ekon. Inform. Gos. Kom. Sov. Min. SSSR* Po. Khimii (10), 32 (1960); *Ref. Zh. Khim.* **21**, Abstr. No. 21 D40 (1961).
56. A. P. Kreshkov, L. V. Myshlyaeva, E. A. Kucharev and T. G. Shatunova, *Lakokras. Mater. Ikh Primen.* (24), Abstr. No. 24G257 (1966).
57. E. A. Terent'eva, *Zavod. Lab.*, **28**, 807 (1962).
58. E. A. Terent'eva, B. M. Luskina and S. V. Syavtsillo, *Zh. Anal. Khim.*, **16**, 635 (1961).
59. A. P. Terent'eva, B. M. Luskina and S. V. Syavtsillo, *Anal. Abstr.*, **8**, 3309 (1961).
60. B. M. Luskina, A. P. Terent'eva and N. A. Gradskova, *Zh. Anal. Khim.*, **19**, 1251 (1964).
61. B. M. Luskina, A. P. Terent'eva and N. A. Gradskova, *Zh. Anal. Khim.*, **20**, 990 (1965).
62. E. A. Terent'eva and N. N. Smirnova, *Zavod. Lab.*, **32**924 (1966).
63. S. Kohama, *Bull. Chem. Soc. Japan*, **36**, 380 (1963).
64. E. A. Bondarevskaya, V. O. Kuznetskova and S. A. Syavtsillo, *Zh. Anal. Khim.*, **16**, 472 (1961).
65. E. A. Bondarevskaya, A. P. Kreshkov, O. Syavtsillo and V. M. Kuznetskova, *Trudy Komiss. Anal. Khim. Akad. Nauk SSSR*, **13**, 24 (1963); *Ref. Zh. Khim.*, **19GDE** (1), Abstr. No. 1G235 (1964).
66. V. A. Klimova and M. D. Vitalina, *Izv. Akad. Nauk SSSR, Odtel. Khim. Nauk* (12), 2245 (1962).
67. J. A. Magnuson and D. W. Baillargeon, *Anal. Chim. Acta*, **32**, 156 (1965).
68. J. A. Magnuson, *Chem. Anal.*, **53**, 15 (1964).
69. Z. Sir and R. Komers, *Chem. Listy*, **50**, 162 (1956).
70. S. V. Syavtsill, V. T. Shemyatenkova and A. M. Neshumova, *Zavod. Lab.*, **24**, 287 (1958).
71. O. Watt, *Chem. Rev.*, **46**, 317 (1960).
72. L. H. Sommer, E. Dorfman, G. M. Goldberg and F. C. T Whitmore, *Am. Chem. Soc.*, **68**, 488 (1946).
73. F. Hirata and T. Takiguchi, *J. Chem. Soc. Jpn., Ind. Chem. Sect.*, **60**, 1207 (1957).
74. T. A. Khudayakova, E. I. Pavel'eva and R. V. Kozlov, *Zh. Prikl. Khim. (Leningrad)*, **38**, 2002 (1965).
75. Ya. I. Chulkov, USSR Patent No. 118,224 (20/2/59) (1959).
76. L. V. Myshlyaeva, *Trudy Mosk. Khim.-Tekhnol. Inst. im. D. I. Mendeleeva*, **25**, 29, (1957); *Ref. Zh. Khim.*, Abstr. No. 24900 (1958).
77. L. V. Myshlyaeva, Yu. Ya. Mikhailenko, V. V. Krasnoshchekov and E. A. Kuckkarev, *Trudy Khim. Moskov. Khim. Tekhnol Inst.* (44), 139 (1963); *Ref. Zh. Khim.*, **19GDE** (20), Abstr. No. 20G153 (1964).

78. T. Takiguchi, *J. Chem. Soc. Jpn., Ind. Chem. Sect.*, **61**, 1236 (1958).
79. L. V. Myshlyaeva, *Trudy Mosk. Khim. Tekhnol. Inst. im. D. I. Mendeleeva* (25), 29 (1957); *Ref. Zh. Khim.*, Abstr. No. 24900 (1958).
80. T. Takiguchi, *Analyst (London)*, **83**, 482 (1958).
81. T. Takiguchi, *J. Chem. Soc. Jpn., Ind. Chem. Sect.*, **61**, 587 (1958).
82. T. R. F. W. Fennell, *Chem. Ind.*, 1404 (1955).
83. R. Belcher and A. M. G. Macdonald, *Mikrochim. Acta*, 510 (1957).
84. Y. A. Gawargious and A. M. G. Macdonald, *Anal. Chim. Acta*, **27**, 300 (1962).
85. J. Korbl and R. Komers, *Chem. Listy*, **50**, 1120 (1956); *Anal. Abstr.*, **3**, 373, 1384 (1956).
86. K. Z. Uhle, *Anal. Chem.*, **231**, 194 (1967).
87. H. B. Bradley, *Anal. Chem.*, **27**, 2021 (1955).
88. A. Radecki, *Chem. Anal. Warsaw*, **8**, 607 (1959).
89. A. Radecki and R. Pickos, *Roczn. Chem.*, **33**, 57 (1959).
90. V. A. Klimova and E. G. Bereznutskaya, *Zh. Anal. Khim.*, **11**, 292 (1956).
91. V. A. Klimova and E. G. Bereznutskaya, *Zh. Anal. Khim.*, **12**, 424 (1957).
92. H. Lammer, *Chem. Tech. Berlin*, **4**, 491 (1952).
93. W. F. Gilliam, H. A. Lieblafsky and A. F. Winslow, *J. Am. Chem. Soc.*, **76**, 918 (1954).
94. E. W. Balis, H. A. Lieblafsky and L. B. Bronk, *Ind. Eng. Chem. Anal. Ed.*, **17**, 56 (1945).
95. E. G. Rochow and W. F. Gilliam, *J. Am. Chem. Soc.*, **63**, 798 (1941).
96. E. G. Rochow, *J. Am. Chem. Soc.*, **70**, 2170 (1958).
97. T. Aranyi and A. Erdey, *Magy. Kem. Lab.*, **20**, 164 (1965).
98. H. Kautsky, O. Fritz, G. H. R. Siebel and D. Siebel, *Anal. Chem.*, **147**, 327 (1955).
99. V. A. Klimova, M. O. Korshun and E. G. Bereznitskaya, *Zh. Anal. Khim.*, **11**, 223 (1956).
100. V. Terent'eva, S. V. Syavtsillo and B. M. Luskina, *Anal. Khim. Akad. Nauk SSSR*, **13**, 3 (1963); *Ref. Zh. Khim.*, **19GDE** (4), Abstr. No. 4G239 (1964).
101. H. Waledziak, *Chemia Anal.*, **10**, 579 (1965).
102. A. P. Kreshkov, V. T. Shemyatenkova, S. V. Syavtsillo and N. A. Palamarchuk, *Zh. Anal. Khim.*, **15**, 635 (1960).
103. M. G. Voronkov and V. T. Shemyatenkova, *Izv. Akad. Nauk* SSSR (1), 178 (1961).
104. V. M. Krasikova, A. N. Kaganova and V. D. Lobkov, *J. Anal. Chem. USSR (Engl. Transl.)*, **26**, 1458 (1971).
105. V. M. Krasikova, A. N. Kaganova and V. D. Lobkov, *Zh. Anal. Khim.*, **26**, 1635 (1971).
106. D. D. Schleuter and S. Siggia, *Anal. Chem.*, **49**, 2343 (1977).
107. D. D. Schleuter, Ph.D. Dissertation, University of Massachusetts, Amhurst, Massachusetts, 1976.
108. V. M. Krasikova and A. N. Kaganova, *Zh. Anal. Khim.*, **25**, 409 (1970).
109. V. M. Krasikova and A. N. Kaganova, *J. Anal. Chem. USSR (Engl. Transl.)*, **25**, 1212 (1970).
110. G. W. Heylum, R. C. Bajaiski and H. B. Bradlwy, *J. Gas Chrom.*, **2**, 300 (1964).
111. E. R. Bissel, and D. R. Fields, *J. Chromatogr. Sci.*, **10**, 164 (1972).
112. C. L. Hanson and R. O. Smith, *Anal. Chem.*, **44**, 1571 (1972).
113. S. P. Evdokimova, N. A. Isakova and V. F. Evdokimov, *J. Anal. Chem. USSR (Engl. Transl.)*, **26**, 704 (1971).
114. J. Franc and F. Mikes, *Collect. Czech. Chem. Commun.*, **31**, 363 (1966).
115. G. Fritz and H. Burdt, *Anorg. Chem.*, **317**, 35 (1962).
116. A. P. Kreshkov, V. A. Bork and M. I. Aparsheva, *Plast. Massy.* (4), 63 (1965); *Ref. Zh. Khim.*, **19GDE** (21), Abstr. No. 21G196 (1965).
117. A. P. Terent'eva, E. A. Bondarevskaya and T. V. Kirillova, *Zavod. Lab.*, **33**, 156 (1967).
118. N. E. McDougall, *Appl. Spectrosc.*, **19**, 196 (1965).
119. V. A. Bork, A. P. Kreshkov and L. A. Shvyrkova, *USSR Patent.*, 141, 334 (6/10/61) (1961).
120. A. P. Kreshkov, V. A. Bork and L. A. Shvyrkova, *Zh. Anal. Khim.*, **28**, 151 (1962).
121. V. A. Bork and L. A. Shvyrkova, *Trudy Komiss. Anal. Khim. Akad. Nauk SSSR*, **13**, 148 (1963); *Ref. Zh. Khim.*, **19GDE** (22), Abstr. No. 22G159 (1963).
122. O. Cermak and P. Dostal, *Collect. Czech Chem. Commun.*, **28**, 1384 (1963).
123. A. Schwarz, *Z. Anal. Chem.*, **115**, 161 (1939).
124. F. S. Kipping and J. E. Sands, *J. Chem. Soc.*, **119**, 848 (1921).
125. F. S. Malyukova and L. I. Surkova, *Plast. Massy.* (12), 51, (1965), *Ref. Zh. Khim.* 19GD, (10), Abstr. No. 10G237 (1966).

126. A. P. Kreshkov, V. A. Bork and M. Aparsheva, *Zavod. Lab.*, **30**, 1208 (1964).
127. A. P. Kreshkov, V. A. Bork and Aparsheva, *Izv. Vysshucheb Zavod. Khim., Khim. Tekhnol.*, **7** (5), 742 (1964); *Ref. Zh. Khim.*, **19GDE** (11), Abstr. No. 11G212 (1965).
128. Y. Y. Milhailemko, L. P. Seuctskaya and E. G. Kutyrina, *Trudy Komiss. Anal. Khim., Akad. Nauk SSSR*, **13**, 383 (1963). *Ref. Zh. Khim.*, **19GDE** (20), Abstr. No. 20G168 (1963).
129. G. M. Kline, in *Analytical Chemistry of Polymers*, Interscience, New York, 1959, p. 372.
130. C. R. N. Strouts, J. H. Gilfillan and H. N. Wilson, in *Analytical Chemistry—The Working Tools*, Volume 1, Oxford University Press, 1965, p. 417.
131. B. Smith, *Acta. Chem. Scand.*, **11**, 558 (1957).
132. P. Brown and A. L. Smith, *Anal. Chem.*, **30**, 549 (1958).
133. J. S. Fritz and G. H. Schlenk, *Anal. Chem.*, **31**, 1808 (1959).
134. J. A. Magnuson, *Anal. Chem.*, **35**, 1487 (1963).
135. P. Dostal, J. Cermak and P. Novotna, *Collect. Czeck. Chem. Commun.*, **30**, 34 (1965).
136. A. P. Terent'eva, E. A. Bondarevskaya and T. V. Kirillova, *Zh. Anal. Khim.*, **22**, 1242 (1967).
137. A. P. Kreshkov and V. A. Bork, *Zh. Anal. Khim.*, **12**, 764 (1957).
138. A. P. Terent'eva and E. A. Bondarevskaya, *Zh. Anal. Khim.*, **20**, 249 (1965).
139. E. Kreshkov and P. Nessanova, *Zh. Anal. Khim.*, **4**, 220 (1949).
140. S. V. Syavtsillo and E. A. Bondarevskaya, *Zh. Anal. Khim.*, **11**, 613 (1956).
141. V. A. Klimova, K. S. Zabrodina and L. L. Shitikova, *Izv. Akad. Nauk. SSSR, Ser. Khim.* (1), 178 (1965).
142. G. D. Messonova and E. K. Pogosyants, *Zavod Lab.*, **24**, 953 (1958).
143. R. A. Bourrigue, *Chem. Anal.*, **43**, 40 (1954).
144. A. Colsen, *Analyst (London)*, **58**, 594 (1933).
145. W. Kirsten and E. Rogozinsky, *Anal. Abstr.*, **2**, 3080 (1955).
146. V. A. Drozdov and E. G. Vlasova, *Trudy Komiss. Anal. Khim. Akad. Nauk SSSR*, 187 (1963); *Ref. Zh. Khim.-Tekhnol.*, **19GDE** (19), Abstr. No. 19G191 (1963).
147. V. A. Drozdov and E. G. Vlasova, *Trudy Komiss. Anal. Khim. Akad. Nauk SSSR*, **13**, 187 (1963); *Ref. Zh. Khim.*, **19GDE** (19), Abstr. No. 19G191 (1993).
148. A. P. Kreshkov, V. A. Drozdov and E. G. Vlasova, *SSSR, Patent* 137,299 (10/4/61) (1961).
149. F. O. Guenther, *Anal. Chem.*, **30**, 1118 (1958).
150. J. F. Lees and R. T. Lobeck, *Analyst (London)*, **88**, 782 (1963).
151. K. Damm and W. Noll, *Kolloid*, **158**, 97 (1958).
152. J. Haslam and H. A. Willis, in *Identification and Analysis of Plastics*, Van Nostrand, New York, 1965, pp. 256–257.
153. G. M. Kline, in *Analytical Chemistry of Polymers*, Part 1, Vol. 9, *Analysis of Monomers and Polymeric Materials*, Interscience, New York, 1959, p. 373.
154. G. R. Lucas and R. W. Martin, *J. Am. Chem. Soc.*, **74**, 5225 (1952).
155. L. H. Sommer and L. J. Tyler, *J. Am. Chem. Soc.*, **76**, 1030 (1954).
156. K. Damm, D. Belitz and W. Noll, *Angew. Makromol. Chem.*, **76**, 273 (1964).
157. H. Gilman and L. S. Miller, *J. Am. Chem. Soc.*, **73**, 2367 (1951).
158. W. T. Grubb, *J. Am. Chem. Soc.*, **76**, 3408 (1954).
159. R. C. Smith and G. E. Kellum, *Anal. Chem.*, **39**, 338 (1967).
160. G. H. Barnes and N. E. Doughenbough, *Anal. Chem.*, **35**, 1308 (1963).
161. J. S. Fritz and G. H. Schlenk, *Anal. Chem.*, **31**, 1808 (1959).
162. P. Dostal, J. Cermak and P. Novotria, *Collect. Czech. Chem. Commun.*, **30**, 34 (1965).
163. J. A. Magnuson, *Anal. Chem.*, **35**, 1487 (1963).
164. J. A. Magnuson and R. J. Cerri, *Anal. Chem.*, **38**, 1088 (1966).
165. G. E. Kellum and K. L. Uglum, *Anal. Chem.*, **39**, 1623 (1967).
166. F. O. Guenther, *Anal. Chem.*, **30**, 1118 (1958).
167. R. Fuchs, O. Moore, D. Miles and H. J. Gilman, *J. Org. Chem.*, **21**, 1113 (1956).
168. R. Sauer, *J. Am. Chem. Soc.*, **66**, 1707 (1944).
169. J. F. Hyde, *J. Am. Chem. Soc.*, **75**, 2166 (1953).
170. W. Noll, K. Damm and W. Krauss, *Fette U. Seifen Farbe. U. Lack.*, **65**, 17 (1959).
171. A. Berger and J. A. Magnuson, *Anal. Chem.*, **36**, 1156 (1964).
172. K. K. Kunder and M. N. Das, *Anal. Chem.*, **31**, 1358 (1959).
173. A. P. Kreshkov, V. A. Drozdov and E. G. Vlasova, *Izv. Vyssh. Ucheb. Zavedenii i. Khim. Tekhnol.*, **3**, 85 (1960); *Ref. Zh. Khim.*, (18), Abstr. No. 73, 200 (1960).

174. A. P. Kreshkov and V. A. Drozdov, *Dokl. Akad. Nauk, SSSR* **131**, 1345 (1960); *Ref. Zh. Khim.* (18), Abstr. No. 73, 199 (1960).
175. A. P. Kreshkov, *Trudy Mosk. Khim. Tekhnol. Inst. Im D. I. Mendeleeva*, **32**, 333 (1961); *Ref. Zh. Khim.* (8), Abstr. No. 8D211 (1962).
176. A. P. Kreshkov, V. A. Bork and M. I. Aparsheva, *Zh. Anal. Khim.*, **18** (10), 1149 (1963).
177. A. P. Kreshkov, V. A. Drozdov and E. G. Vlasova, *Zavod. Lab.*, **26**, 1080 (1960).
178. A. P. Kreshkov, V. A. Drozdov and E. G. Vlasova, *Izv. Vyssh. Ucheb. Zaved., Khim. Khim. Tekhnol.*, **3**, 80 (1960); *Ref. Zh. Khim.* (18), Abstr. No. 73, 203 (1960).
179. E. G. Kreshkov, V. A. Drozdov and R. R. Tarasyants, *Zavod. Lab.*, **30**, 413 (1964).
180. R. R. Tarasyants, *Trudy Moskov Khim. Tekhnol. Inst.*, **44**, 149 (1963); *Ref. Zh. Khim.*, **19GDE** (20), Abstr. No. 20G155, (1964).
181. V. A. Drozdov and E. G. Vlasova, *Trudy Komiss. Anal. Khim. Akad. Nauk. SSSR*, **13**, 187 (1963); *Ref. Zh. Khim.*, **19GDE** (19), Abstr. No. 19G191 (1963).
182. H. Gilman, R. A. Klein and H. J. Winkler, *J. Org. Chem.*, **26**, 2474 (1961).
183. A. P. Kreshkov, V. A. Drozdov and R. R. Tarasyants, *Plast. Massy*, **4**, 57 (1963).
184. M. F. Borisov, E. Y. Guttsaet, L. M. Shtifman and L. A. Efremova, *Vestn. Tekh. Z. Ekon, Inform. Nanch-Issled. Inst. Tekhe-Ekon. Issled. Gos. Kom. Khim Z. Neft. Prom. Pri. Gosplane, SSSR*, **10**, 51 (1963); *Ref. Zh. Khim.*, **19GDE** (18), Abstr. No. 18G171 (1964).
185. K. K. Popkov and Q. L. Lulchuk, USSR. Pat. 129, 385 (5/6/60) (1960).
186. S. U. Kreingol'd and E. A. Bozbeval'nov, *Trudy Kom. Anal. Khim.*, **16**, 194 (1968).
187. M. Miyamato, *Japan Analyst*, **12**, 233 (1963).
188. A. Schneer, T. Halmos and T. Z. Szekely, *Anal. Chem.*, **182**, 178 (1961).
189. S. Fujiwara and H. Narasaki, *Anal. Chem.*, **36**, 201 (1964).
190. F. D. Krivoruchko, *Zavod Lab.*, **27**, 290 (1961).
191. E. A. Peregud and N. P. Kozlova, *J. Anal. Chem. USSR (Engl. Transl.)*, **9**, 47 (1954).
192. F. D. Krivoruchko, *Zavod Lab.*, **29**, 927 (1963).
193. A. Smith and J. A. McHard, *Anal. Chem.*, **31**, 1174 (1959).
194. G. Munro, J. H. Hunt and L. R. Rowe, *Anal. Chem.*, **51**, 311 (1979).
195. C. A. Burkhard and E. H. Winslow, *J. Am. Chem. Soc.*, **72**, 3276 (1950).
196. C. Eaborn, *J. Chem. Soc.*, 3148 (1953).
197. H. Gilman and G. E. Dunn, *J. Am. Chem. Soc.*, **72**, 2178 (1950).
198. K. Bowden and E. A. Braude, *J. Chem. Soc.*, 1068 (1952).
199. A. Pozefsky and M. E. Grenoble, *Drug Cosmet. Ind.*, **80**, 752 (1957).
200. G. M. Kline (Ed.), in *Chemical Analysis of High Polymers*, Chap. 14, Interscience, New York, 1959.
201. K. P. Bradley and W. J. Potts, *Appl. Spectrosc.*, **12**, 77 (1958).
202. D. N. Ingebrigtson and A. L. Smith, *Anal. Chem.* **26**, 1765 (1954).
203. U. Scheidt, *Appl. Spectrosc.*, **7**, 75 (1953).
204. P. C. Harms, *Anal. Chem.*, **25**, 1141 (1953).
205. N. Wright and M. H. Hunter, *J. Am. Chem. Soc.*, **69**, 803 (1947).
206. A. L. Smith, *Spectrochim. Acta*, **16**, 87 (1960).
207. W. A. Srowacka and T. Burnacka, *Chem. Anal. Warsaw*, **9**, 303 (1964).
208. T. Burnacka and A. Wakroj, *Chemia Anal.*, **10**, 1233 (1965).
209. J. H. Lady, G. M. Bower, R. E. Adams and F. P. Byre, *Anal. Chem.*, **31**, 1100 (1959).
210. A. D. Grant and A. L. Smith, *Anal. Chem.*, **30**, 1016 (1958).
211. W. Fish'l and I. G. Young, *Appl. Spectrosc.*, **10**, 213 (1956).
212. T. Kubota and T. Takamura, *Bull. Chem. Soc. Jpn.*, **33**, 70 (1960).
213. T. Urin and T. Hakamada, *J. Chem. Soc. Jpn., Ind., Chem. Sect.*, **62**, 1421 (1959).
214. T. Urin, S. Tanaka and T. Wada, *J. Chem. Soc. Jpn., Ind. Chem. Sect.*, **62**, 1577 (1959).
215. T. Urin, S. Tanaka and H. Yamamato, *J. Chem. Soc., Ind. Chem. Sect.*, **62**, 1581 (1959).
216. O. Kuratani, *J. Chem. Soc. Jpn., Pure Chem. Sect.*, **73**, 576 (1952).
217. M. Margoshes and V. A. Fassel, *Anal. Chem.*, **27**, 351 (1955).
218. L. A. Efremova and K. K. Popkov, *Zavod Lab.*, **29**, 708 (1963).
219. E. R. Shull, *Anal. Chem.*, **32**, 1627 (1960).
220. P. J. Launer and M. E. Grenoble, *Anal. Chem.*, **32**, 441 (1960).
221. H. Westermark, *Acta Chem. Scand.*, **9**, 947 (1955).
222. E. A. V. Ebsworth, M. Onyszchuki and N. Sheppard, *J. Chem. Soc.*, 1453 (1958).

223. R. West and E. G. Rochow, *J. Org. Chem.*, **18**, 303 (1953).
224. P. Brown and B. L. Smith, *Anal. Chem.*, **30**, 1016 (1958).
225. A. L. Smith and N. C. Angelotti, *Spectrochim. Acta*, **10**, 124 (1972).
226. A. L. Smith, *J. Chem. Phys.*, **21**, 1997 (1953).
227. L. Burnelle and J. S. Duchesne, *Chem. Phys.*, **20**, 1324 (1952).
228. K. Shimazu and H. Murata, *Bull Chem. Soc. Jpn.*, **32**, 46 (1959).
229. M. C. Tobin, *J. Am. Chem. Soc.*, **75**, 1788 (1953).
230. R. Forneris and E. Funck, *Z. Elektrochem.*, **62**, 1130 (1958).
231. H. Kriegsmann, *Z. Elektrochem.*, **61**, 1088 (1957).
232. M. E. Grenoble and P. J. Hanner, *Tenth Annual Symposium on Spectroscopy, Am. Chem. Soc.*, Chicago, 1959.
233. P. Brown and A. L. Smith, *Anal. Chem.*, **30**, 1016 (1958).
234. R. Kniseley, J. A. Fassel and E. Conrod, *Spectrochim. Acta*, **15**, 651 (1959).
235. H. Kriegsmann, *Elektrochem.*, **62**, 1033 (1958).
236. R. N. Dixon and N. Sheppard, *J. Chem. Phys.*, **23**, 215 (1955).
237. C. Newman, J. K. O'Loane and M. K. Wilson, *Chem. Phys.*, **25**, 855 (1956).
238. H. R. Linton and E. R. Nixon, *Spectrochim Acta*, **12**, 41 (1958).
239. D. W. Mayo, H. E. Opitz and J. S. Peake, *J. Chem. Phys.*, **23**, 1344 (1955).
240. H. E. Opitz, J. C. Peake and W. H. Nebergall, *J. Am. Chem. Soc.*, **78**, 292 (1956).
241. A. L. Smith and J. A. McHard, *Anal. Chem.*, **31**, 1174 (1959).
242. L. A. Efremova and K. K. Popkov, *Zavod. Lab.*, **29**, 708 (1963).
243. M. E. Grenoble and P. J. Launer, *Appl. Spectrosc.*, **14**, 85 (1960).
244. W. D. Reents, D. I. Woods and A. M. Mujsce, *Anal. Chem.*, **57**, 104 (1985).
245. P. J. Launer and A. S. Crouse, *Appl. Spectrosc.*, **15**, 118 (1961).
246. H. Z. Kriegsmann, *Anorg. Allg. Chem.*, **299**, 138 (1959).
247. D. C. Smith, J. M. French and J. J. O'Neill, US Naval Research Laboratory Office 9, Technical Services Report No. 2746 (January, 1946).
248. N. Wright and M. J. Hunter, *J. Am. Chem. Soc.*, **69**, 803 (1947).
249. C. W. Young, P. C. Servais, C. C. Currie and M. J. Hunter, *J. Am. Chem. Soc.*, **70**, 3758 (1948).
250. A. L. Smith, *J. Chem. Phys.*, **21**, 1997 (1953).
251. C. W. Young, J. S. Koeler and D. S. McKinney, *J. Am. Chem. Soc.*, **69**, 1410 (1947).
252. H. M. Herchenson, in *Infrared Absorption Spectra-Index (in 1945–1957)*, Academic Press, New York, London, 1959.
253. D. F. Ball, P. L. Goggin, D. C. McKean and L. A. Woodward, *Spectrochim. Acta*, **16**, 11 (1960).
254. S. Kaye and S. Tannbaum, *J. Org. Chem.*, **18**, 1750 (1953).
255. M. C. Harvey, H. W. Negergall and J. S. Peake, *J. Am. Chem. Soc.*, **76**, 4555 (1954).
256. H. Westermark, *Acta Chem. Scand.*, **9**, 947 (1955).
257. L. J. Bellamy, *Infrared Spectra of Complex Organic Molecules* (2nd edn.), Chap. 20, Methuen and Co., London, 1958.
258. G. D. Osbesky and F. F. Bentley, *J. Am. Chem. Soc.*, **79**, 2057 (1957).
259. R. G. Scott and K. C. Frisch, *J. Am. Chem. Soc.*, **73**, 2599 (1951).
260. K. C. Frisch, P. A. Goodwin and R. G. Scott, *J. Am. Chem. Soc.*, **74**, 4584 (1952).
261. D. L. Bailey and A. N. Pines, *Ind. Eng. Chem.*, **46**, 2363 (1954).
262. L. D. Nasiak and H. W. Post, *J. Org. Chem.*, **24**, 489 (1959).
263. L. J. Bellamy, *Infrared Spectra of Complex Organic Molecules* (2nd edn.), Chap. 3, Methuen and Co., London, 1958.
264. J. W. Curry, *J. Am. Chem. Soc.*, **78**, 1686 (1956).
265. M. Kanazachi, *Bull. Chem. Soc. Jpn.*, **26**, 493 (1953).
266. E. R. Schull, R. A. Thursock and C. M. Birdsall, *J. Chem. Phys.*, **24**, 147 (1956).
267. H. Kriegsmann, *Z. Elekrochem.*, **61**, 1088 (1957).
268. M. C. Harvey, H. W. Negergill and J. S. Peake, *J. Am. Chem. Soc.*, **76**, 4555 (1954).
269. J. G. Noltes, M. C. Henry and M. S. Janssen, *Chem. Ind. (London)*, 298 (1959).
270. H. Kriegsmann and K. H. Schowtka, *Z. Phys. Chem. (Leipzig)*, **209**, 261 (1958).
271. L. Spaltier, D. C. Priest and C. W. Harris, *J. Am. Chem. Soc.*, **77**, 6227 (1955).
272. R. D. Kross and V. A. Fassel, *J. Am. Chem. Soc.*, **27**, 5858 (1955).
273. D. H. Whitten, *J. Chem. Soc.*, 1350 (1956).
274. E. N. Beck, W. H. Dandt, H. J. Fletcher, M. H. Hunter and A. J. Barry, *J. Am. Chem. Soc.*, **81**, 1256 (1959).

275. H. A. Clark, A. F. Gordon, C. W. Young and M. J. Hunter, *J. Am. Chem. Soc.*, **73**, 3798 (1951).
276. V. A. Young, R. B. Duvall and N. Wright, *Anal. Chem.*, **23**, 709 (1951).
277. N. B. Colthup, *J. Opt. Soc. Am.*, **40**, 397 (1950).
278. G. R. Wilson, G. H. Hutzel and A. G. Smith, *J. Org. Chem.*, **24**, 381 (1959).
279. J. W. Gilkey and L. J. Tyler, *J. Am. Chem. Soc.*, **73**, 4982 (1951).
280. R. West and E. G. Rochow, *J. Org. Chem.*, **18**, 303 (1953).
281. N. Wright and M. J. Hunter, *J. Am. Chem. Soc.*, **69**, 803 (1947).
282. R. C. Lord, D. W. Robinson and W. C. Schumb, *J. Am. Chem. Soc.*, **78**, 1327 (1956).
283. R. E. Richards and H. W. Thompson, *J. Chem. Soc.*, 124 (1949).
284. C. W. Young, P. C. Servais, C. C. Currie and M. J. Hunter, *J. Am. Chem. Soc.*, **70**, 3758 (1948).
285. H. Kriegsmann, *Z. Anorg. Allg. Chem.*, **298**, 223 (1956).
286. H. Kriegsmann, *Z. Elektrochem.*, **62**, 1033 (1958).
287. A. J. Barry, N. H. Davdt, J. J. Domincone and J. W. Gilkey, *J. Am. Chem. Soc.*, **77**, 4248 (1965).
288. D. C. McKean, *Spectrochim. Acta*, **13**, 38 (1958).
289. C. W. Young, P. C. Servais, C. C. Currie and M. J. Hunter, *J. Am. Chem. Soc.*, **70**, 3758 (1948).
290. *The Analysis of Organosilicon Compounds*, Midland Silicones Ltd., Publication A3-2 7 PP (February 1957).
291. D. Hummel, *Kunststoffe*, **46**, 442 (1956).
292. R. E. Kagarise and L. A. Weinbenger, US Naval Research Laboratory Office of Technical Services, Dept. P13 111438 (May 11th 1954).
293. D. Hummel, *Farbe u. Lack*, **62**, 529 (1956).
294. A. L. Smith, L. H. Brown, L. J. Tyler and M. J. Hunter, *Ind. Eng. Chem.*, **49**, 1903 (1957).
295. T. Yonomoto and K. Senzok, *Bull. Electrotech. Lab. (Tokyo)*, **18**, 428 (1954).
296. W. H. T. Davidson and G. R. Bates, in *Procedures of the Third Rubber Technology Conference*, London, 1954, p. 281.
297. W. Ecknig, M. H. Kriegsmann, H. Rotzsche, H. Jancke, and K. Witke, *Proceedings on Application of Physicochemical Methods of Chemical Analysis, Budapest, Hungary, April 1966*, Abstracts of Plenary Lecture, Volume 111, 476 pp, Section V, Spectrochemical Analysis, 1966, pp. 11–12 (12).
298. M. Kakudo, P. N. Kasai and T. Watase, *J. Chem. Phys.*, **21**, 1894 (1953).
299. W. M. Kakudo, P. N. Kasai and T. Watase, *J. Chem. Phys.*, **21**, 167 (1953).
300. F. Damm and W. Noll, *Kolloid Z.* **158**, 97 (1958).
301. V. I. Kostochkin, M. F. Shostakovshi, V. I. Zil'berbrau and D. A. Kochkin, *Izv. Akad. Nauk SSSR, Ser. Fiz.*, **18**, 726 (1954).
302. H. Z. Kriegsmann, *Anorg. Chem.*, **299**, 78 (1959).
303. Ya. I. Ryskin and M. G. Voronkov, *Zh. Fiz. Khim.*, **30**, 2275 (1956).
304. S. W. Kantor, *J. Am. Chem. Soc.*, **75**, 2712 (1953).
305. Ya. I. Tyskin, *Optika i. Spekroskopiya*, **4**, 532 (1958).
306. H. A. Benesi and A. C. Jones, *J. Phys. Chem.*, **63**, 179 (1959).
307. G. J. Young, *J. Colloid Sci.*, **13**, 67 (1958).
308. M. Folman and D. J. C. Yates, *J. Phys. Chem.*, **63**, 183 (1959).
309. R. S. McDonald, *J. Am. Chem. Soc.*, **79**, 850 (1957).
310. V. I. Lygin, *Vestn. Mosk. Univ., Ser. Mat. Mekh. Astron. Fiz. Khim.*, **13**, 223 (1958).
311. R. S. McDonald, *J. Phys. Chem.*, **62**, 1168 (1958).
312. I. Simon and H. O. McMahon, *J. Chem. Phys.*, **20**, 905 (1952).
313. A. P. Kreshkov, Yu. Ya. Mikhailenko and G. F. Yakimovich, *Zh. Anal. Khim.*, **9**, 208 (1954).
314. R. Okawara, *Bull. Chem. Soc. Jpn.*, **31**, 154 (1958).
315. R. Fornerls and E. Funck, *Z. Elektrochem.*, **62**, 1130 (1958).
316. P. Brown and A. L. Smith, *Anal. Chem.*, **30**, 549 (1958).
317. H. Kriegsmann and K. Light, *Z. Elektrochem.*, **62**, 1163 (1958).
318. A. A. Stuart, C. L. Lau and H. Breederveld, *Recl. Trav. Chem. Pays-Bas*, **74**, 747 (1955).
319. A. P. Kreshkov, Yu. Ya. Mikhailenko and G. F. Yakimovich, *Zh. Fiz. Khim.*, **2**8, 537 (1954).
320. M. Hayashi, *Nippon Kagaku Zasshi*, **79**, 436 (1958).
321. M. O. Bulanin, B. N. Dolgov, T. A. Speranskaya and N. P. Kharitnov, *Zh. Fiz. Khim.*, **31**, 1321 (1957).
322. A. N. Lazarev and M. G. Voronkov, *Optika i. Spekroskopiya*, **4**, 180 (1958).
323. A. N. Lazarev, *Optika i. Spekroskopiya*, **4**, 805 (1958).
324. H. Breederveld and H. I. Watermann, *Recl. Trav. Chem. Pays-Bas*, **73**, 871 (1954).

325. T. Tanaka, *Bull. Chem. Soc. Jpn.*, **31**, 762 (1958).
326. R. Okowaru, *Bull. Chem. Soc. Jpn.*, **31**, 154 (1958).
327. L. J. Bellamy, in *Infrared Spectra of Complex Molecules* (2nd edn.), Methuen and Co., London, 1958, p. 174.
328. F. C. Lanning and M. J. Moore, *J. Org. Chem.*, **23**, 288 (1958).
329. Y. Yamamato, *J. Chem. Soc. Jpn., Ind. Chem. Sect.*, **64**, 1464 (1961).
330. A. L. Smith and N. C. Angelotti, *Spectrochim. Acta*, **15**, 412 (1959).
331. L. Kaplan, *J. Am. Chem. Soc.*, **76**, 5880 (1954).
332. H. Kriegsmann and K. H. Showatka, *Z. Phys. Chem. (Leipzig)*, **209**, 261 (1958).
333. F. A. Anderson and B. Bok, *Acta Chem. Scand.*, **8**, 738 (1954).
334. R. C. Collins and J. R. Nielsen, *J. Chem. Phys.*, **23**, 351 (1955).
335. M. Hayashi, *Nippon Kagaku Zasshi*, **78**, 1472 (1957).
336. E. A. Jones, J. S. Kirby-Smith, P. J. H. Waltz and A. H. Nielson, *J. Chem. Phys.*, **19**, 242 (1951).
337. R. Okawarta, *Bull. Chem. Soc. Jpn.*, **31**, 154 (1958).
338. A. L. Smith, *J. Chem. Phys.*, **21**, 1997 (1953).
339. T. G. Gibian and D. S. McKinney, *J. Am. Chem. Soc.*, **73**, 1431 (1951).
340. R. E. Richards and H. W. Thompson, *J. Chem. Soc.*, 124 (1949).
341. E. A. V. Ebsworth, J. R. Hall, M. J. Mackillop, D. C. McKean, M. Sheppard and L. A. Woodward, *Spectrochim. Acta*, **13**, 202 (1958).
342. H. Kriegsmann and W. Foster, *Z. Anorg. Allg. Chem.*, **298**, 212 (1959).
343. D. W. Robinson, *J. Am. Chem. Soc.*, **80**, 5924 (1958).
344. J. J. McBride and H. C. Beachell, *J. Am. Chem. Soc.*, **74**, 5247 (1952).
345. M. Prober, *J. Am. Chem. Soc.*, **78**, 2274 (1956).
346. H. R. Linton and E. R. Nixon, *J. Chem. Phys.*, **28**, 990 (1958).
347. H. R. Linton and E. R. Nixon, *Spectrochim. Acta*, **10**, 299 (1958).
348. E. A. V. Ebsworth, R. Taylor and L. A. Woodward, *Trans. Faraday Soc.*, **55**, 211 (1959).
349. H. J. Emeleus, A. G. MacDiarmid and A. G. Maddock, *J. Inorg. Nucl. Chem.*, **1**, 194 (1955).
350. H. R. Linton and E. R. Nixon, *J. Chem. Phys.*, **29**, 291 (1958).
351. E. A. Jones, J. S. Kirby-Smith, P. J. H. Woltz and A. H. Nielson, *J. Chem. Phys.*, **19**, 242 (1951).
352. J. Gonbeau and J. Pushing, *Z. Anorg. Allg. Chem.*, **294**, 96 (1958).
353. A. G. MacDiarmid and A. G. Maddock, *J. Inorg. Nucl. Chem.*, **1**, 411 (1955).
354. W. H. Keeber and H. W. Post, *J. Org. Chem.*, **21**, 509 (1956).
355. H. R. Linton and H. R. Nixon, *Spectrochim. Acta*, **15**, 146 (1959).
356. V. A. Zeitler and C. A. Brown, *J. Am. Chem. Soc.*, **79**, 4616 (1957).
357. V. A. Zeitler and C. A. Brown, *J. Phys. Chem.*, **61**, 1174 (1957).
358. W. S. Tatlock and E. G. Rochow, *J. Org. Chem.*, **17**, 1555 (1952).
359. C. S. G. Phillips and P. L. Timms, *Anal. Chem.*, **35**, 505 (1963).
360. P. L. Timms, C. C. Simpson and C. S. G. Phillips, *J. Chem. Soc.*, 1467 (1964).
361. F. Feher and H. Strack, *Naturwissenschaften*, **50**, 570 (1963).
362. F. H. Pollard, G. Nickless and P. C. Uden, *J. Chromatogr.*, **19**, 28 (1965).
363. F. H. Pollard, G. Nickless and P. C. Uden, *J. Chromatogr.*, **14**, 1 (1964).
364. G. Garzo, J. Feketer and M. Blazo, *Acta Chim. Acad. Sci. Hung.*, **51**, 359 (1967).
365. F. H. Pollard, G. Nickless and D. B. Thomas, *J. Chromatogr.*, **22**, 256 (1966).
366. J. A. Semlyn and C. S. G. Phillips, *J. Chromatogr.*, **18**, 1 (1965).
367. J. Bohemen, S. H. Langer, R. H. Perrett and J. J. Purnell, *J. Chem. Soc.*, 2444 (1960).
368. J. A. Semlyn, G. R. Walker, R. C. Blofield and C. S. G. Phillips, *J. Chem. Soc.*, 4949 (1964).
369. L. B. Peetre and B. E. F. Smith, *J. Chromatogr.*, **90**, 41 (1977).
370. L. B. Peetre, O. Ellren and B. E. F. Smith, *J. Chromatogr.*, **88**, 295 (1974).
371. L. B. Peetre, *J. Chromatogr.*, **88**, 311 (1974).
372. A. Di-Lorenzo, *J. Chromatogr.*, **76**, 207 (1973).
373. A. D. M. Hailey and G. Nickless, *J. Chromatogr.*, **49**, 187 (1970).
374. M. Wurst and J. Churacek, *Collect. Czech. Chem. Commun.*, **36**, 3497 (1971).
375. N. S. Nametkin, V. G. Berezkin, N. Y. Vanyukova and V. M. Vodvin, *Nettekhiyma*, **4**, 137; *Ref. Zh. Khim.*, **IGGDE** (24), Abstr. No. 24G220 (1964).
376. A. D. Svegova, L. K. Markov, V. A. Panomarenko and W. D. Zelinsky, *Zh. Anal. Khim.*, **19**, 610 (1960).
377. G. N. Bortnikov, N. S. Vyazankin, N. P. Nikulina and Ya. I. Yashin, *Izv. Akad. Nauk SSSR, Ser. Khim.* (1), 21 (1961).

378. D. Bersadchi and V. Stefan, *Rev. Chem. (Bucharest)*, **15**, 224 (1964).
379. N. A. Palamarchuk, S. V. Syavtsillo, N. M. Turkeltaub and O. Shemyatenkova, *Trudy Komiss. Anal. Khim. Akad. Nauk. SSSR*, **13**, 277 (1963); *Ref. Zh. Khim.*, **19GE** (20), Abstr. No. 20G169 (1963).
380. Y. Abe, *Bunseki Kagaku, Japan Analyst*, **9**, 795 (1960).
381. M. Wurst and Dusek, *Collect. Czech. Chem. Commun.*, **26**, 2022 (1961).
382. G. Fritz and D. Ksinsik, *Z. Anorg. Chem.*, **304**, 241 (1960).
383. H. Rotzsche, *Z. Anorg. Chem.*, **324**, 197 (1963).
384. M. Wurst and R. Dusek, *Collect. Czech. Chem. Commun.*, **27**, 231 (1962).
385. Vront Piscou and O. Parissakis, *J. Chromatogr. Sci.*, **9**, 499 (1971).
386. P. K. Basu, C. J. King and S. Lynn, *J. Chromatogr. Sci.*, **10**, 479 (1972).
387. N. K. Aglinlov, V. V. Luchinkin and G. G. Devyatykh, *Zh. Anal. Khim.*, **23**, 951 (1968).
388. K. Friedrich, *Chem. Ind.*, **2**, 47 (1957).
389. J. Joklik, *Collect. Czech. Chem. Commun.*, **26**, 2079 (1961).
390. K. Kawazumi, S. Lataoka and K. Maruyama, *J. Chem. Soc. Jpn., Ind. Chem. Sect.*, **64**, 784 (1961).
391. N. M. Turkeltaub, N. A. Palamarchuk, V. T. Shemyatenkova and S. V. Syavtsillo, *Plast. Massy*, **4**, 51 (1961); *Ref. Zh. Khim.*, (23), Abstr. No. 23D135 (1961).
392. T. Garzo, F. Till and I. Till, *Magyar Kem Foly.*, **68**, 327 (1962).
393. T. Oiwa, M. Sato, Y. Miyakava and I. Miyazaki, *J. Chem. Soc. Jpn., Pure Chem. Sect.*, **84**, 409 (1963).
394. B. Lengyel, G. Garzo and T. Szekely, *Acta Chim. Acad. Sci. Hung.*, **37**, 37 (1963).
395. H. Rotzsche, *Z. Anorg. Allg. Chem.*, **328**, 79 (1964).
396. G. V. Avdonin, Z. I. Alekseeva, V. N. Detinova, V. D. Merkulov, L. A. Nechaeva, N. A. Palamarchuk, V. E. Syavtsillo-Tronina and S. G. Yagodina, *Plast. Massy*, **3**, 56; *Anal. Abst.*, **15** (6), 3387 (1968).
397. J. Cermak and J. Franc, *Collect. Czech. Chem. Commun.*, **30**, 3278 (1965).
398. E. V. Sivitsova, V. B. Kogan and S. K. Ogorodnikov, *Zh. Prikl. Khim. Leningrad*, **38**, 2609 (1965).
399. A. N. Popov, V. M. Gorbackev and E. I. Torgova, *Izv. Sib. Otdel, Akad. Nauk SSSR (11), Ser. Khim. Nauk (3)*, 17; *Ref. Zh. Khim.*, **19GD** (13), Abstr. No. 13G250 (1967).
400. N. M. Turkel'taub and N. A. Palamarchuk, Trudy Nauchno-Issled. *Geol. Razv. Neft. Inst.*, **64**, 143 (1970); *Anal. Abstr*; **20** (6), 3932 (1971).
401. A. A. Kreshkov, V. A. Drosdov, N. D. Rumyantseva and V. F. Andronov, *Plast. Massy*, **9**, 65 (1971); *Ref. Zh. Khim.*, **19GD** (2), Abstr. No. 2G229 (1972).
402. G. Garzo and F. Till, *Talanta*, **10**, 583 (1963).
403. A. A. Ainshtein and T. I. Shulyat'eva, *Zh. Anal. Khim.*, **27**, 816 (1972); *Anal. Abstr.*, **77** (8), 56183 (1972).
404. K. R. Burson and C. T. Kenner, *Anal. Chem.*, **41**, 870 (1969).
405. J. Frank and M. Wurst, *Gas Kromatografaya, Moskov. Akad. Nauk SSSR*, **289** (1960); *Ref. Zh. Khim.* (13), Abstr. No. 13D222 (1961).
406. J. Frank and M. Wurst, *Collect. Czech. Chem. Commun.*, **25**, 701 (1960).
407. M. Wurst, *Collect. Czech. Chem. Commun.*, **34**, 3297 (1969).
408. M. Wurst, *Collect. Czech. Chem. Commun.*, **30**, 2038 (1965).
409. M. Wurst, *Collect. Czech. Chem. Commun.*, **29**, 1458 (1964).
410. D. Knauz, G. Garzo, P. Gomory and L. Telgdi, *Magyar Kem. Foly.*, **70**, 119 (1964).
411. S. A. Volkov, *Sakodynskii Atomnaya Energuiya*, **17**, 70 (1964).
412. T. Gabor and J. Takacs, *Period. Polytech.*, **10**, 341 (1966).
413. J. H. Taylor, *J. Gas Chromatogr.*, **6**, 557 (1968).
414. A. Kirichenki and B. A. Markov, *Trudy Mosk. Khim.-Technol. Inst.*, **71**, 143 (1972); *Ref. Zh. Chem.*, **19GD** (4), Abstr. No. 4G244 (1973).
415. G. W. Heunish, *Anal. Chem. Acta*, **48**, 405 (1969).
416. C. R. Thrash, *J. Gas Chromatogr.*, **2**, 390 (1964).
417. M. Wurst, *Mikrochim. Acta*, 379 (1966).
418. J. H. Taylor, *J. Gas Chromatogr.*, **6**, 557 (1968).
419. O. Ellren, I. B. Peetre and B. E. F. Smith, *J. Chromatogr.*, **88**, 295 (1974).
420. I. B. Peetre, *J. Chromatogr.*, **90**, 35 (1974).
421. O. Ellren, I. B. Peetre and B. E. F. Smith, *J. Chromatogr.*, **93**, 383 (1974).

422. L. S. Eltren and K. Billehb *J. Chromatogr.*, **30**, 112 (1967).
423. C. B. Moore and H. A. Dewhurst, *J. Org. Chem.*, **36**, 110 (1971).
424. M. Wurst, *Collect. Czech. Chem. Commun.*, **29**, 1458 (1964).
425. M. Wurst, *Anal. Abstr.*, **10**, 2760 (1963).
426. B. M. Luskina, G. N. Turkel'taub and S. V. Syavtsillo, *Zavod. Lab.*, **33**, 1496 (1967).
427. K. Otto and M. Doubek, *Chem. Prumsyl.*, **10**, 476 (1960).
428. H. Rotzche and H. Rosler, Report of Conference on Modern Methods of *Analysis of Organic Compounds*, Munich, October 26th–29th (1960); see also *Anal. Chem.*, **181**, 407 (1961).
429. J. B. Carmichael, D. J. Gordon and C. E. Ferguson, *J. Gas Chromatogr.*, **4**, 347 (1966).
430. N. T. Ivanova, S. V. Syavtsillo and L. D. Prigozhina, *Zavad Lab.*, **39**, 1455 (1973).
431. G. Alexander and G. Garzo, *Chromatographia*, **7**, 190 (1974).
432. G. Alexander and G. Garzo, *Chromatographia*, **4**, 554 (1971).
433. M. Wurst, F. Loember and H. Kelker, *Chromatographia*, **6**, 359 (1973).
434. M. Wurst, F. Loember and H. Kelker, *Collect. Czech. Chem. Commun.*, **36**, 3497 (1971).
435. M. Wurst, *Chemicky Prum.*, **22**, 124 (1972).
436. G. N. Turkel'taub and B. M. Luskina, *J. Anal. Chem. USSR (Engl. Transl.)*, **26**, 2010 (1971); see also *Zh. Anal. Khim.*, **26**, 2243 (1971).
437. F. Sutter and P. Fuchs, *Arzneim-Forsch.*, **22**, 553 (1972).
438. M. Tochner, I. A. Magnuson and L. Z. Soderman, *J. Chromatogr. Sci.*, **7**, 740 (1969).
439. F. H. Pollard, G. Nickless and P. C. Uden , *J. Chromatogr.*, **11**, 312 (1963).
440. G. N. Bortikov, M. N. Bochkarev, N. S. Vyazankin, S. K. R. Ratushnaya and Ya. I. Yashin, *Izv. Akad. Nauk SSSR, Ser. Khim.*, **4**, 851 (1971).
441. E. D. Smith, *J. Chromatogr. Sci.*, **10**, 34 (1972).
442. G. N. Bortnikov, N. S. Vasileiskaya, L. V. Gorbunova, N. P. Nikulina and Ya. I. Yoskin, *Izv. Akad. Nauk SSSR, Ser. Khim.*, **3**, 686 (1970).
443. L. G. Mahone, P. J. Garner, R. R. Buck, T. H. Bane, J. F. Tukera, R. C. Smith and C. L. Frye, *Environmental Toxicology and Chemistry*, **2**, 307 (1983).
444. G. N. Bortnikov, N. S. Vyazankin, E. N. Gladyshev and Ya. I. Yashin, *Izv. Akad. Nauk. SSSR, Ser. Khim.* (7), 1661 (1970).
445. N. S. Vyazankin, G. N. Bortnikov, I. A. Migunova, A. V. Kiselev, Ya. I. Yashin, A. N. Egorochkin and V. F. Mironov, *Izv. Akad. Nauk SSSR, Ser. Khim* (1), 186 (1969).
446. G. N. Bortnikov, A. V. Kiselev, N. S. Yyasankin and Ya. I. Yashin, *Chromatographia*, **4**, 14 (1971).
447. E. V. Sivtsova, *Zh. Prikl. Khim. (Leningrad)*, **45**, 201 (1972).
448. A. D. M. Hanley and G. Nickless, *J. Chromatogr.*, **49**, 180 (1970).
449. J. Franc, *Plast. Liwoty. Kanenk*, **4**, 249 (1967).
450. N. M. Turkel'taub, S. A. Ainshtein and B. V. Kuznetsov, *Khim. Technol. Topl. Masel* (12), 44 (1961); *Ref. Zh. Khim.*, **LL** (ii), Abstr. No. 11D170 (1962).
451. N. M. Turkel'taub, S. A. Shemyatenkova, S. A. Ainshtein and S. V. Syavtsillo, *Trudy Komiss. Anal. Khim. Akad. Nauk SSSR*, **13**, 284 (1963); *Ref. Zh. Khim.*, **19GDE** (20); Abstr. No. 20G167 (1963).
452. G. Garzo and F. Till, *Talanta*, **10**, 583 (1971).
453. M. A. Osman, H. H. Hill, M. W. Holdren and H. H. Wetberg, *Anal. Chem.*, **51**, 1286 (1979).
454. D. Fritz, G. Garzo, T. Szekely and F. Till, *Acta Chim. Acad. Sci. Hung.*, **45**, 301 (1965).
455. G. Garzo and D. Fritz, *Gas Chromatography*, Proceedings of the Sixth International Symposium on Gas Chromatography and Associated Techniques, Rome, Italy, September, 1966.
456. M. Dressler, V. Martinu and J. Janak, *J. Chromatogr.*, **59**, 429 (1971).
457. D. T. Bostick and Y. Talmi, *J. Chromatogr. Sci.*, **15**, 163 (1977).
458. T. G. Crowley, V. A. Fassel and R. N. Kniseley, *Spectrochim. Acta*, **B23**, 771 (1968).
459. D. Fritz, G. Garzo, T. Szekely and F. Till, *Acta Chim. Acad. Sci. Hung.*, **45**, 301 (1965).
460. B. Lengnel, G. Garzo, D. Fritz and F. Till, *J. Chromatogr.*, **24**, 8 (1966).
461. H. H. Hill and W. A. Aue, *J. Chromatogr.*, **122**, 515 (1976).
462. H. H. Hill and W. A. Aue, *J. Chromatogr.*, **140**, 1 (1977).
463. B. D. Quimby, P. C. Uden and R. M. Barnes, *Anal. Chem.*, **50**, 2112 (1978).
464. K. J. Slatkavitz, L. D. Hoey, P. C. Uden and R. M. Barnes, *Anal. Chem.*, **57**, 1846 (1985).
465. M. A. Osman, H. H. Hill, M. W. Holden and H. H. Westbury, *Anal. Chem.*, **51**, 1286 (1979).
466. C. Feldman and D. H. Batistoni, *Anal. Chem.*, **49**, 2215 (1977).
467. J. Franc and J. Dvoracek, *J. Chromatogr.*, **14**, 340 (1964).

468. J. Franc and K. Placek, *J. Chromatogr.*, **48**, 295 (1970).
469. C. A. Burkhaed and F. J. Norton, *Anal. Chem.*, **21**, 304 (1949).
470. H. S. Booth, *Chem. Rev.*, **41**, 97 (1947).
471. W. H. Perlson, T. J. Brice and J. H. Simons, *J. Am. Chem. Soc.*, **67**, 1769 (1945).
472. H. S. Booth and M. L. Freedman, *J. Am. Chem. Soc.*, **72**, 2847 (1950).
473. C. R. Thrash, D. L. Viosinet and K. E. Williams, *J. Gas Chromatogr.*, **3**, 248 (1965).
474. D. F. Helm and E. Mack, *J. Am. Chem. Soc.*, **59**, 60 (1937).
475. C. E. Waring, *Trans. Faraday Soc.*, **36**, 1142 (1940).
476. G. Fritz and B. Raabe, *Z. Anorg. Allg. Chem.*, **229**, 232 (1959).
477. G. Fritz and J. Z. Grobe, *Anorg. Allg. Chem.*, **315**, 157 (1962).
478. G. Fritz, J. Bahl, D. Grobe, D. Aulinger and W. Reerink, *Z. Anorg. Allg. Chem.*, **312**, 301 (1961).
479. G. Fritz, *Fortschr. Chem. Forsch.*, **44**, 459 (1963).
480. K. Uhle, *Z. Chemie*, **7**, 236 (1967).
481. G. Fritz and D. Wick, *Z. Anorg. Allg. Chem.*, **342**, 130 (1966).
482. I. G. Sarishvili, H. V. Sobelovskii, K. P. Grineṿivk and P. Z. Sorobin, *Zh. Prikl. Khim. (Leningrad)*, **40**, 2062 (1967).
483. J. Nagy, O. Brandt and O. Petrik, *Period. Polytech.*, **10**, 443 (1966).
484. G. Fritz and D. Wick, *Z. Anorg. Chem.*, **342**, 130 (1966).

The Chemistry of Organic Silicon Compounds
Edited by S. Patai and Z. Rappoport

CHAPTER 7

Positive and negative ion chemistry of silicon-containing molecules in the gas phase*

HELMUT SCHWARZ
Institut für Organische Chemie der Technischen Universität Berlin, Straße des 17. Juni 135, D-1000 Berlin 12, FRG

*Dedicated to Professor Siegfried Hünig on the occasion of his 65th birthday

I. INTRODUCTION

Organosilicon compounds play an increasingly important role in synthetic organic chemistry as well as in analytical chemistry, and the chemistry of ionic organosilanes forms a subject of current interest[1]. The pronounced effects that silyl groups exert on incipient carbenium ions have been well known for quite some time[2] and have been studied computationally at great length[3]. Noteworthy are, for example, the extremely low reactivity of α-silyl-substituted halogenides towards S_N1 reactions[4], which contrasts with the high reactivity of β-substituted analogues[5]. The chemistry of organosilicon anions is potentially even richer than that of carbanions since not only are silyl anions themselves stable, but silicon stabilizes an adjacent carbanion. Moreover, organosilanes can, in contrast to their carbon analogues, form stable pentavalent silyl anions both in solution and in the gas phase. However, the study of charged organosilanes can be quite difficult if not impossible in solution. This is due to a multitude of problems caused by ion pairing, aggregation, solvent effects, and isomerization and polymerization reactions. Therefore, quite a number of laboratories became interested in the study of the *gas-phase* chemistry of organosilanes, and their results have substantially added to our present knowledge of both the inherent properties of charged organosilicon compounds and of the effects counterions and solvents play in solution.

The techniques employed in the gas-phase studies include classical mass spectrometric techniques, the use of ion cyclotron resonance (ICR) and flowing afterglow (FA) methodologies, and ion beam experiments using tandem mass spectrometry facilities. In this chapter we will not describe any experimental techniques in detail, although reference will be made to them when appropriate[6]. Similarly, we will not focus on the role that the mass spectrometric analysis of silylated compounds plays in elucidating the structures of quite a variety of polar molecules. Treatises of this kind may be found in textbooks on mass spectrometry and in several review articles published on the subject of mass spectrometry of silicon-containing compounds[7]. The number of papers dealing with analytical aspects of the mass spectrometry of silylated polar compounds (which, because of their high volatility and thermal stability, are ideally suited for a gas chromatography–mass spectrometry analysis[7d]) is indeed exceedingly high: from the more than 1500 papers checked dealing with one or other aspect of 'mass spectrometry of silicon-containing compounds' and published between 1974 and December 1986, approximately 80% dealt with analytical applications. These aspects will be completely omitted in the present chapter. We will rather focus on topics which are more directly related to fundamental problems of the gas-phase chemistry of ionic organosilanes. In particular, we will discuss some relevant thermochemical aspects, the proton affinities, gas-phase acidities and the substituent effects (Section II). The well-known propensity of silicon ions to undergo unimolecular rearrangements will be described in great detail in Section III, which will also include a brief discussion of the generation of silicon-containing neutrals having Si–X multiple bonds[8]. Various chemical and physical aspects of ion/molecule reactions of both silicon cations and anions form the subject of Section IV, and in this part emphasis will be given to the fact that a relatively detailed understanding of the chemistry of *isolated* species can be achieved by combining state-of-the-art experiments with high-level *ab initio* molecular orbital (MO) calculations[3h]. The gas-phase chemistry of some doubly-charged organosilicones will be described in Section V.

II. THERMOCHEMISTRY

A. Stabilities of Silicon Hydride Cations

Ionized silanes SiH_n^+ ($n = 1$–5) have been studied both theoretically[9] and experimentally[10], and all species are found to exist in potential minima. For SiH_5^+ (protonated silane) the global minimum corresponds to the C_s structure **1** which, depending on the level of theory used, is predicted to be up to 7 kcal mol^{-1} more stable than the C_{3v} structure **2**. The former can be viewed as a 'side-on' complex of H_2 and SiH_3^+, and the latter as the corresponding 'end-on' complex.

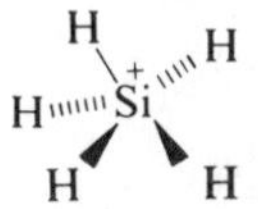

(1)

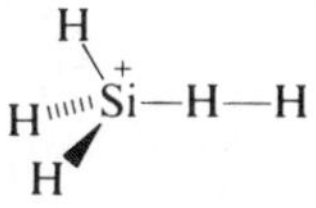

(2)

Experimentally, SiH_5^+ is formed under tandem MS conditions in bimolecular reactions of SiH_4 with NH_2^+, $C_2H_3^+$, $C_2H_6^{+\cdot}$ and $C_3H_8^{+\cdot}$ [11], and the species can also be generated using ICR mass spectrometry by reaction of CH_5^+ with SiH_4[12]. From these proton transfer processes, using thermochemical cycles, the proton affinity (PA) of SiH_4 (defined by the heat of reaction, ΔH_r°, of reaction 1) was determined to be 150–156 kcal mol^{-1}. Thus, SiH_4 is 25–30 kcal mol^{-1} more basic than its carbon analogue CH_4.

$$SiH_5^+ \rightarrow SiH_4 + H^+, \qquad \Delta H_r^\circ \mathrel{\hat{=}} PA \tag{1}$$

The molecular ion of silane, $SiH_4^{+\cdot}$, is known to be much less stable than its carbon counterpart $CH_4^{+\cdot}$. The failures to generate a stable $SiH_4^{+\cdot}$ by electron impact (EI) ionization[13] of SiH_4 could be explained by the calculated[14] small activation energy of 1.6 kcal mol^{-1} needed for the dissociation of $SiH_4^{+\cdot}$ to $SiH_3^+ + H\cdot$[15]. However, Genuit and coworkers[16] reported recently that $SiH_4^{+\cdot}$ can be generated by photoionization mass spectrometry using an ArI resonance lamp as a light source. The rather poor agreement in the literature concerning the ionization threshold[13,17] makes it difficult to determine the heat of formation (ΔH_f°) accurately. MO theory predicts several stable states for $SiH_4^{+\cdot}$, the global minimum corresponding to the C_{3v} structure **3**, followed by the C_{2v} structure **4**. The calculated[9b] ionization energies to generate **3** and **4** from SiH_4 are 0.60 and 0.39 eV smaller than the lowest ionization energy (11.60 eV), determined by photoelectron spectroscopy[17a]. The latter was assigned[17a] to the Jahn–Teller distorted D_{2d} structure **5**.

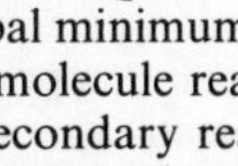

(3)

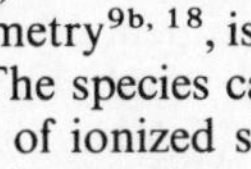

(4)

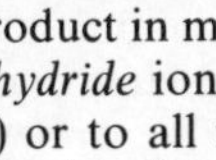

(5)

The most prominent ion in the EI mass spectrum of SiH_4 corresponds to SiH_3^+, which is easily formed (presumably via $SiH_4^{+\cdot}$) if the ionization energy exceeds ca. 12 eV. SiH_3^+, whose global minimum has a D_{3h} symmetry[9b,18], is also the major product in many gas-phase ion/molecule reactions[6,19,20]. The species can be formed by *hydride* ion transfer either to secondary reaction products of ionized silanes (reaction 2) or to all methane primary ions (reactions 3 and 4). From the reactions of $CD_4^{+\cdot}$ and CD_5^+ with SiH_4 and the analysis of the isotopic composition of the silyl ions it was convincingly demonstrated[20] that the latter are formed by *direct hydride* ion transfer from SiH_4 to the ionized hydrocarbon. Deuteriated silane, SiH_4D^+, does not participate in the formation of the silicenium ion SiH_3^+. Similarly, the reactions of $D_2^{+\cdot}$ or D_3^+ with SiH_4 involve exclusively hydride transfer from SiH_4 to generate SiH_3^+ [21]. The fact that SiH_3^+ was the only product of the D_3^+/SiH_4 reaction illustrates the virtual non-existence of SiH_4D^+ as an intermediate. The silicenium ion SiH_3^+ is also generated in many other hydride transfer reactions involving SiH_4 as hydride donor (see Section IV and Ref. 6). Interestingly, SiH_3^+ is also generated by transfer of carbanions R^- ($R=CH_3$ [22], $R=C_2H_5$ [23]) to CH_3^+ (reaction 5) and $C_2H_5^+$ (reaction 6)[24].

$$SiH_2^{+\cdot}+SiH_4 \rightarrow SiH_3^{\cdot}+SiH_3^+ \quad (2)$$

$$CH_n^+ + SiH_4 \rightarrow CH_{n+1} + SiH_3^+ \quad (3)$$

$(n=0\text{–}3)$

$$CH_n^+ + SiH_4 \rightarrow CH_{n-1} + H_2 + SiH_3^+ \quad (4)$$

$(n=4,5)$

$$CH_3^+ + RSiH_3 \rightarrow CH_3R + SiH_3^+ \quad (5)$$

$$C_2H_5^+ + RSiH_3 \rightarrow C_2H_5R + SiH_3^+ \quad (6)$$

Ionized silylene, $SiH_2^{+\cdot}$, prefers according to MO theory[9b] a C_{2v} structure, and the species is formed as a secondary fragment ion under EI of SiH_4 (reactions 7 and 8) with appearance energies of 11.8 eV (for reaction 7) and 16.2 eV (for reaction 8), respectively[13c]. It is also generated by the (formal) transfer of H_2^- from SiH_4 to $C_2H_2^{+\cdot}$ (reaction 9)[25] or $C_2H_4^{+\cdot}$ (reaction 10)[26]. The species seems also to play a role as an intermediate in many other ion/molecule reactions[6].

$$SiH_4 + e^- \rightarrow SiH_2^{+\cdot} + H_2 + 2e \quad (7)$$

$$SiH_4 + e^- \rightarrow SiH_2^{+\cdot} + 2H^\cdot + 2e \quad (8)$$

$$SiH_4 + C_2H_2^{+\cdot} \rightarrow SiH_2^{+\cdot} + C_2H_4 \quad (9)$$

$$SiH_4 + C_2H_4^{+\cdot} \rightarrow SiH_2^{+\cdot} + C_2H_6 \quad (10)$$

For the EI generation of SiH^+ from SiH_4 several appearance energies were reported for reaction 11 (14.5 eV[27], 14.7 eV[13c] and 16.1 eV[13a]), and the large disparity precludes an unambiguous interpretation of the energetics of the reaction. In contrast, the bond dissociation energy for reaction 12 has been determined to be 74.9 kcal mol^{-1} by both spectroscopic studies of silane discharges[28] and by guided beam experiments[29].

$$SiH_4 + e^- \rightarrow SiH^+ + H_2 + H^\cdot + 2e \quad (11)$$

$$SiH^+ \rightarrow Si^{+\cdot}(^2P) + H^\cdot \quad (12)$$

B. Substituted Silicenium Ions

In this sub-section it is not intended to cover the thermochemical aspects of all substituted silicenium ions described in the literature. Rather, we focus on some selected systems which deem to serve as good representatives and the discussion of which may provide an understanding of the substituent effects operative. Some aspects will also be described later in conjunction with ion/molecule reactions of silicon compounds.

While the trimethylsilyl cation Me_3Si^+ has thus far eluded detection in solution, its existence in the gas phase is well established[6]. Under EI trimethylsilyl compounds Me_3SiX (X = alkyl, halogen) form the trimethylsilyl cation, Me_3Si^+ (m/z 73) as one of the most abundant ions. This species is of prime analytical importance, as it generates as a selective low-energy reactant ion in chemical ionization studies very often abundant [M + 73] adduct ions, which can be used to determine the molecular weight of substitutes

M[30]. Me_3Si^+ plays also a decisive role as a central intermediate in many ion/molecule reactions[6] as will be described in detail in Section IV.

Concerning the thermodynamic stability of Me_3Si^+ versus its carbon analogue Me_3C^+, a qualitative conclusion from the early MS work was that both ions should be comparable. This interpretation was aided by theoretical arguments[3c] which state that the more electropositive character of silicon (which would favour Me_3Si^+ over Me_3C^+) should be partly offset by a more favourable hyperconjugative effect in the carbocation case. Recent experimental work[31], using hydride affinities (HA)[32] as a reference reaction for the stability of cations, clearly demonstrates that Me_3Si^+ is 10–20 kcal mol^{-1} *more stable* than Me_3C^+. Direct confirmation of the greater stability of Me_3Si^+ relative to Me_3C^+, at least with respect to hydride transfer, is provided by the observation that reaction 13 under ICR conditions is strictly irreversible.

$$Me_3C^+ + Me_3SiH \rightleftarrows Me_3CH + Me_3Si^+ \qquad (13)$$

Equilibrium measurements involving the transfer of Me_3Si^+ between two bases B^1 and B^2 were used to establish the binding energy of Me_3Si[33]. As Me_3Si^+ can be viewed as a 'hard' acid (according to Pearson's classification of hard and soft acids and bases[34]) the direction of the nucleophilic displacement (reaction 14) gives not only the relative trimethylsilicon affinities of two related bases B^1 and B^2, but also determines a relative hardness scale.

$$Me_3SiB^1 + B^2 \rightleftarrows Me_3SiB^{2+} + B^1 \qquad (14)$$

The binding energies of Me_3Si^+ to oxygen compounds cover the range of 29 kcal mol^{-1} (for H_2O) to 50 kcal mol^{-1} (for $PhCO_2Et$) with typical values for alcohols (MeOH: 38.3 kcal mol^{-1}), ethers (Et_2O: 43.4 kcal mol^{-1}) and ketones (Me_2CO: 43.4 kcal mol^{-1}) in between. It was further observed[33b,d] that there exists an excellent linear correlation between the Me_3Si^+ and H^+ affinities for both oxygen and aromatic bases. Proton affinities are, however, by an order of magnitude larger. It should be mentioned that the adducts formed by reaction of Me_3Ge^+ with bases seem to be more stable than their corresponding Me_3SiB^+ species[33a]. A comparison of the binding energies of Me_3M^+ (M = Si, Sn, C) is revealing. The fact that the binding energies of Me_3Si^+ and Me_3Sn^+ to H_2O (29.0 and 25.7 kcal mol^{-1}) are significantly larger than that of Me_3C^+ (12 kcal mol^{-1} [35]) indicates that silicon is in its behaviour towards oxygen bases closer to the metal (Sn) than to the non-metal (C).

The effects of fluorine on the relative stabilities of silicenium ions $Me_{3-n}F_nSi^+$ ($n = 0$–3) were studied by Murphy and Beauchamp[36a] using ICR. It was observed that the fluoride affinities, $D[R^+ - F^-]$, of the various fluoro- and methyl-substituted silicenium ions followed the order $SiF_3^+ > MeSiF_2{}^+ > Me_2SiF^+ > Me_3Si^+$. Thus, a methyl substituent stabilizes the charge at Si^+ relative to fluorine. Related studies on chlorosilanes are described in Ref. 36b.

The affinity of SiF_3^+ to several bases was found[37] to follow the order: $F_2 < N_2 < HF < SiF_4 < H_2O$. Interestingly, and not yet fully understood, is the finding that the SiF_3^+ affinity to HF (41 kcal mol^{-1})[37] is similar to that of the methyl cation to HF (42 kcal mol^{-1})[37, 38]. The proton affinity of fluorinated silanes $SiH_{4-n}F_n$ ($n = 0, 4$) is, not surprisingly, very much substituent-dependent: while the PA of SiH_4 (150–156 kcal mol^{-1}) is significantly higher than that of CH_4 (126 kcal mol^{-1})[37]. The site of the protonation in SiF_4 is not the silicon but presumably the fluorine lone pair. In trifluorosilanol protonation occurs at the hydroxyl group, thus generating $F_3SiOH_2^+$. The proton affinity of F_3SiOH was determined to be 160 kcal mol^{-1}; this value is significantly smaller than that of silanol, H_3SiOH (198.8 kcal mol^{-1})[30]. The decreases in PA when comparing SiH_4/SiF_4 and H_3SiOH/F_3SiOH are likely to be due to the strong σ-electron withdrawing effects of fluorine ($F \leftarrow Si$).

The directions of proton (deuterium) transfer processes (reaction 15) from the dimethyl-substituted silicenium ion **6** to bases B of known proton affinities were used to determine the heats of formation of neutral 1-methylsilaethene **7** versus neutral dimethylsilylene **8**[40]. For **7** a value of 18 kcal mol^{-1} was obtained, and **8** was determined to be 28 kcal *less* stable. This difference contradicts the experimental findings of other groups[41], who obtained an opposite ordering of stabilities, and is also not in line with theoretical results[42]; the latter predict the two isomers to be of comparable stability.

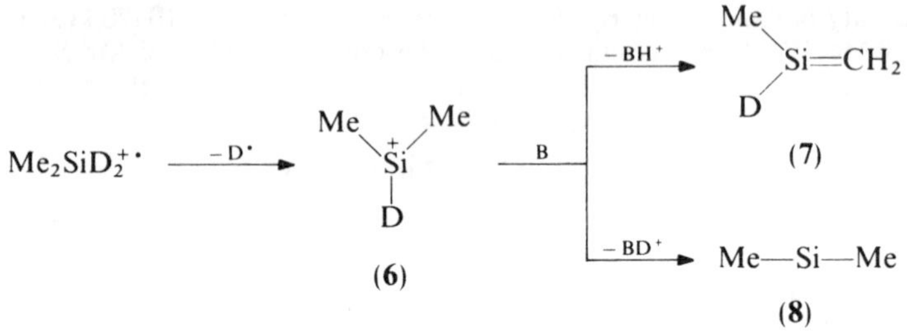

(15)

C. Gas-phase Acidities of Organosilanes

The silyl anion, SiH_3^-, can easily be prepared by proton abstraction from SiH_4 with NH_2^-, and the gas-phase acidity (i.e. the heat of reaction of the process $AH \rightarrow A^- + H^+$) has been determined by the bracketing technique using ICR to be 371 ± 2 kcal mol^{-1}. SiH_4 is thus seen to be about as strong an acid as acetonitrile (372.1 kcal mol^{-1}), HF (371.3 kcal mol^{-1}), dimethyl sulfoxide (372.7 kcal mol^{-1}) and *t*-butanol (373.3 kcal mol^{-1}), and to be significantly more acidic than water (390.8 kcal mol^{-1}). Although proton abstraction from the methyl group of alkylsilanes R_3SiH requires 10–15 kcal mol^{-1} more energy than from the silicon[43], the process to generate the α-silylcarbanions (reaction 16a) is favoured kinetically[44]. As will be discussed in Section IV, both the α-silylcarbanions **9** and the trialkylsilyl anions **10** can be made via specific ion/molecule processes and characterized by reactions with suitable substrates.

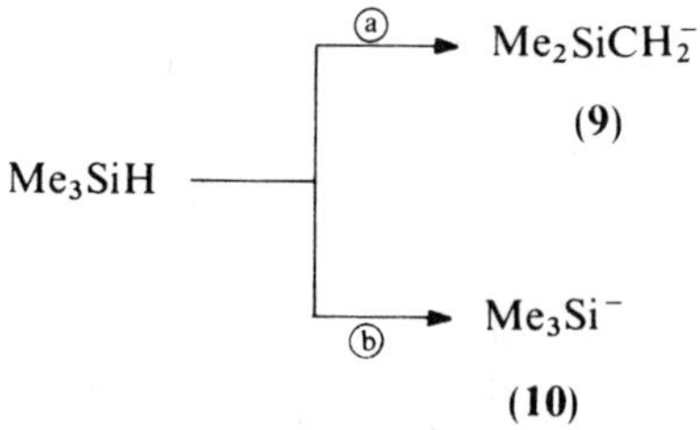

(16)

The enolate of dimethylsilanone was generated independently in two laboratories using different techniques. Squires and coworkers[45] obtained **11** via collision-induced dissociation (CID) of trimethylsiloxide anion, while Braumann and coworkers[46] applied the technique of infrared multi-photon induced loss of CH_4 from Me_3SiO^- (reaction 17). Reaction of **11** with a series of Brønsted acids led to a proton affinity of **11** of 366 kcal mol^{-1} (i.e. the gas-phase acidity of Me_2SiO **12**); thus Me_2SiO is slightly more acidic than acetone (368.8 kcal mol^{-1} [47]).

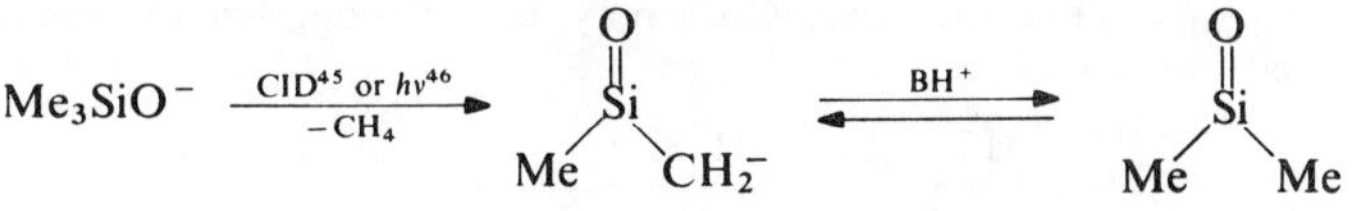

(17)

(11) **(12)**

The gas-phase acidity of dimethylsilene **13** (to yield upon H^+ loss the silaallyl anion **14**) was determined[44] to be 377 kcal mol^{-1}. The comparison of this value with that of isobutene (390 kcal mol^{-1}) indicates a substantial stabilizing effect by silicon, greater than has been observed for the above-mentioned $MeSi(O)CH_2^-$/MeC(O)Me system, but less than in $MeSiCH_2^-$ **15**. The anion of dimethylsilylene **15** [accessible via deprotonation of **8** (reaction 18) with an acidity of 381 kcal mol^{-1}] has been found to be extremely unreactive; this has been explained by *ab initio* calculations in terms of its highly delocalized negative charge[48].

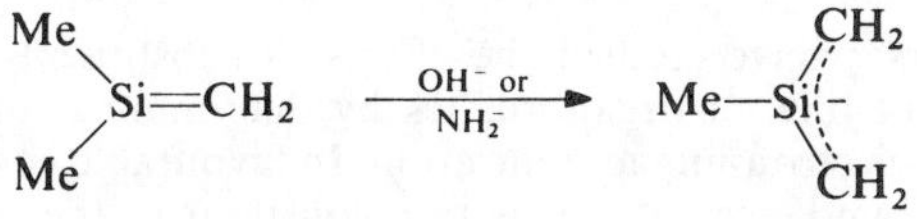

(13) **(14)** (18)

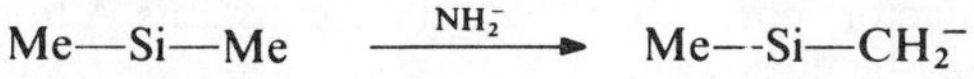

(8) **(15)**

D. Hydride Affinities of Silane and Alkylsilanes

Recent theoretical[49] and experimental[50] studies have converged on the reactions of the potent hydride ion with silane, and in particular on the gas-phase synthesis and characterization of the archetype pentavalent silicon anion, SiH_5^-. *Ab initio* calculations show that H^- should react at the silicon atom of SiH_4 without a barrier to form SiH_5^-, and at the hydrogen periphery to form SiH_3^- and H_2. The calculations further predict that both SiH_5^- and $SiH_3^- \cdots\cdots H_2$ exist in potential wells, but since they are formed in *exothermic* reactions (ΔH_r°: 18–17 kcal mol^{-1}) they will at best be formed as metastable ions which eventually dissociate to SiH_3^- and H_2 (reaction 19). Interestingly, when D^- is the reactant ion, H^- and SiH_2D^- appear in addition to SiH_3^- as products, implying that reversible additions of the nucleophiles to silicon occur during the course of reactive collisions. If one or more hydrogens of SiH_4 are replaced by alkyl groups, the addition of H^- yields *stable* pentavalent silicon anions (reaction 20)[50b]. Presumably, the reaction exothermicity can be dispersed into vibrations and rotations of the alkyl group(s), thereby increasing the lifetime of the complex to an extent that the excess energy can be transferred via collisions to the buffer gas. If D^- is used in reaction 20 the hydrogens and deuterium attached to silicon in $C_5H_{11}SiH_3D^-$ become equivalent, and either may be transferred to CO_2 (reaction 21)[50b]. Similarly, the Et_3SiHD^- product formed by reaction 22 undergoes transfer of either H^- and D^-. For both processes 21 and 22 an identical isotope effect $k_H/k_D = 2.1$ is operative, which may indicate complete randomization of H and D attached to silicon.

$$SiH_4 + H^- \rightarrow [SiH_5^-, H_3Si^- \cdots H_2] \rightarrow SiH_3^- + H_2 \quad (19)$$

$$n\text{-}C_5H_{11}SiH_3 + H^- \rightarrow n\text{-}C_5H_{11}SiH_4^- \quad (20)$$

$$n\text{-}C_5H_{11}SiH_3 + D^- \longrightarrow n\text{-}C_5H_{11}SiH_3D^- \xrightarrow{CO_2} HCO_2^- \text{ and } DCO_2^- \quad (21)$$

$$Et_3SiH + D^- \longrightarrow Et_3SiHD^- \xrightarrow{CO_2} HCO_2^- \text{ and } DCO_2^- \quad (22)$$

E. Substituent Effects on the Stability of Molecular and Fragment Ions

Dube and coworkers[51] have studied the effects of substituents on the stability of molecular and fragment ions of organosilanes by determining ionization/appearance energies (IE/AE) using electron-impact ionization. In favourable cases they were able to correlate the relative abundances of ions with σ-constants of the substituents attached either directly to the silicon atom or to an aryl silicon group. However, in other cases the substantial error bars of the IE/AE energy measurements together with the crude assumptions made for the calculations of ion abundances (QET-type calculations) did not provide any satisfactory correlations with the σ-constants. A brief discussion of some selected examples may be illustrative.

The electron-impact-induced fragmentation of disubstituted benzenes of the formulae $RC_6H_4SiMe_2F$ (R = *p*-NMe_2, *p*-OMe, *p*-, *m*-Me, H, *p*-, *m*-Cl, *p*-, *m*-Br and *p*-NO_2) results mainly in the elimination of a methyl radical from the silyl group[51b, c] (reaction 23). The substituent R exerts a pronounced effect on the relative intensities of **16** and **17** in that electron-releasing substituents increase the abundance of the former and reduce that of the latter, and *vice versa*. Similarly, the activation energy for the reaction **16**→**17**, expressed by the difference AE − IE, is substituent-dependent and large for electron-releasing substituents and small for electron-withdrawing groups. This has been interpreted[51b] as an indication that for R = *p*-NMe_2 and *p*-OMe the charge and spin are mainly located at the substituent R, thereby generating for the silyl substituent an electronic situation comparable to neutral silanes. For the latter, the energy necessary to cleave the Si–C bond is relatively high (for example, dissociation of Me_4Si to $CH_3\cdot$ and $Me_4Si\cdot$ requires 3.7 eV; for R = *p*-NMe_2 the activation energy amounts to 4.0 eV). Conversely, if R corresponds to an electron-withdrawing group (R = *p*-NO_2), the charge distribution in **16** generates an ion, which can be described as an ionized organosilane. For the latter, the dissociation energies are much smaller. For example, the reaction $Me_4Si^{+\cdot} \rightarrow Me_3Si^+ + Me^\cdot$ requires 0.7 eV[17b], and for the reaction **16**→**17** the activation energy shrinks to 0.9 eV if R = *p*-NO_2.

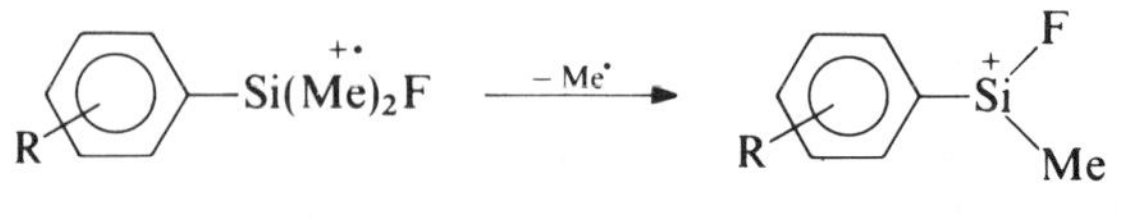

(23)

(16) **(17)**

In a similar fashion substituted phenyldimethylsilanes $RC_6H_4SiHMe_2$ (R = *p*-NMe_2, *p*-OMe, H, *p*-Cl, *m*-Cl, *m*-CF_3) were studied[51a]. It was found that loss of both hydrogen (**18**→**19**) and methyl (**19**→**20**) (reaction 24) were strongly influenced by the nature of R, and the relative abundances of the primary fragments **19** and **20** could be correlated with the σ-constants of R. Furthermore it was observed that, not surprisingly, the abundance of the fragments is inversely proportional to the activation energies E_a of the respective reactions. E_a follows, again, the σ-constants of R, being large for electron-donating groups (*p*-NMe_2: 2.7 eV) and small for electron withdrawing ones (*p*-Cl: 0.1 eV).

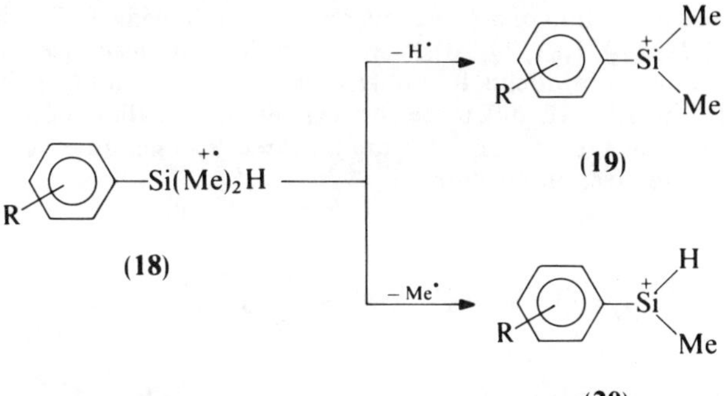

(18) (19) (20) (24)

The electron-impact-induced fragmentation of silyl-substituted phenyldimethylsilanes $Me_2HSiC_6H_4SiX_nMe_{3-n}$ (X = OCH_3, Cl, F; n = 1–3) were also studied with regard to the effect of X on the primary decomposition modes of the molecular ions[51a]. Although there seems to exist a correlation between the electronic properties of X and dissociations involving bonds attached to both silicon atoms, a quantitative analysis could not be provided. Similarly, while the ionization energies show a satisfactory linear dependence on the σ-constants of X, the relatively large error bars prevented the determinations of σ-constants for the SiX_nMe_{3-n} groups.

The decomposition pattern of substituted dimethylmethoxysilane $Me_2Si(OMe)R$ (R = F, Cl, H, allyl, OMe) has been studied in great detail (Ref. 51d and literature cited therein). The principal decomposition route involves the successive elimination of Me˙ and CH_2O (reaction 25). It was observed that the intensity ratio of the ions **23** versus **22** decreased with increasing activation energies E_a (reaction 25). A CNDO/2 analysis further revealed that for the various substituents R the activation energy of the process **22**→**23** could be correlated with the π-bond order of the SiO bond: the higher the latter, the larger E_a and the smaller the propensity of **22** to dissociate to **23**.

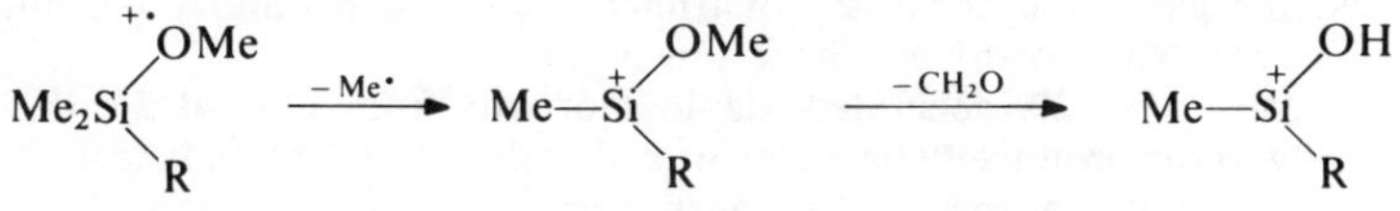

(21) (22) (23) (25)

R	E_a(kcal mol^{-1})	I_{23}/I_{22}
OMe	67	0.82
Me	70	0.60
H	71	0.52
Cl	73	0.39
F	75	0.35

The effects of substituents R on the methyl loss (**24**→**25**) from ionized trimethylaryloxysilanes **24** was also studied by determining the corresponding IE and AE values[51e]. Hammett plots as a function of σ-constants reveal the following: (i) The intentsity ratios Q, where $Q = \lg[(I_{25}/I_{24})/(I_{25\text{-H}}/I_{24\text{-H}})]$, as well as the ionization energies IE of **24** decrease for electron-donating substituents R, and *vice versa.* (ii) The ratio Q is linearly inversely proportional to the AE−IE difference. An explanation of these effects was provided within the frame of the QET theory[51e]. Data for three representative systems with R = *p*-NH_2, H, *p*-NO_2 are given in reaction 26.

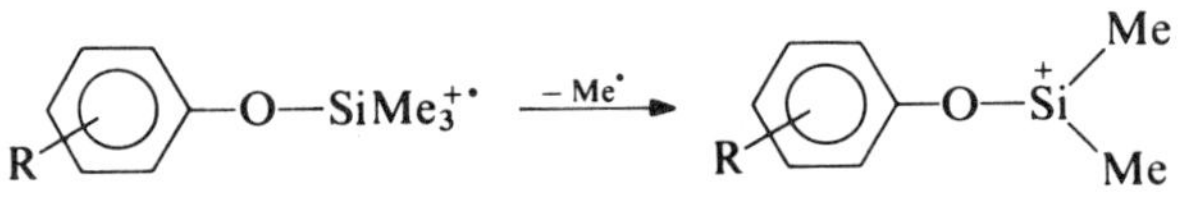

(26)

(24) **(25)**

R	IE(**24**)(kcal mol^{-1})	AE(**25**)(kcal mol^{-1})	Q
p-NH_2	7.4	11.0	−0.4
H	8.4	10.5	0
p-NO_2	9.0	10.6	+0.2

III. REARRANGEMENTS OF IONIZED ORGANOSILANES PRECEDING OR ACCOMPANYING UNIMOLECULAR DISSOCIATIONS

A. Interaction of Silicenium Ions with Remote Electron-rich Centres

Two major types of rearrangements involving silyl centres have been observed. The first involves interaction of a silicenium ion centre, formed by loss of a substituent from silicon in the molecular ion, with distant electron-rich centres in the molecule. The second type, which will be discussed in Section III.B, involves direct transfer of an intact silyl group (such as trimethylsilyl) from one part of the ion to another. Both rearrangement processes may then be followed by unimolecular dissociations.

A variety of interesting, novel rearrangements were reported by Smith and Daves[52], who studied the gas-phase chemistry of trimethylsilylated di- and polyamines using labeled analogues. Noteworthy are the following processes.

The $[M-Me]^+$ ion **27**, generated via loss of Me$^\bullet$ from ionized **26**, decomposes unimolecularly to the product **28** (m/z 160 formed with 48% rel. abundance). This is best explained by assuming a specific hydrogen transfer (**27**→**28**) concomitant with or followed by dissociation (reaction 27).

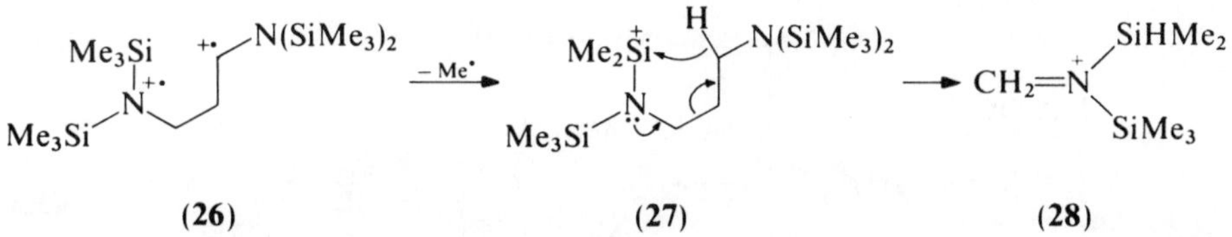

(27)

For **27** ($n=3$) and also for the lower homologue **29** ($n=2$) another interesting pathway concerns the interaction of the silicenium centre with the nitrogen lone pair to generate **30**, which dissociates further to the cyclic product **31** (reaction 28). The chain-length dependence of the formation of **30** as a function of n suggests that the cyclization step is quite sensitive to steric factors.

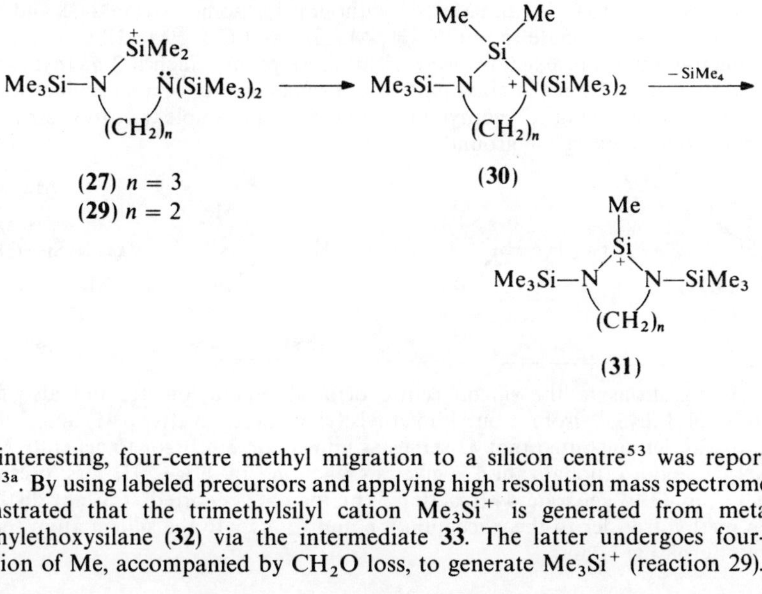

(28)

An interesting, four-centre methyl migration to a silicon centre[53] was reported by Dube[53a]. By using labeled precursors and applying high resolution mass spectrometry he demonstrated that the trimethylsilyl cation Me_3Si^+ is generated from metastable trimethylethoxysilane (**32**) via the intermediate **33**. The latter undergoes four-centre migration of Me, accompanied by CH_2O loss, to generate Me_3Si^+ (reaction 29).

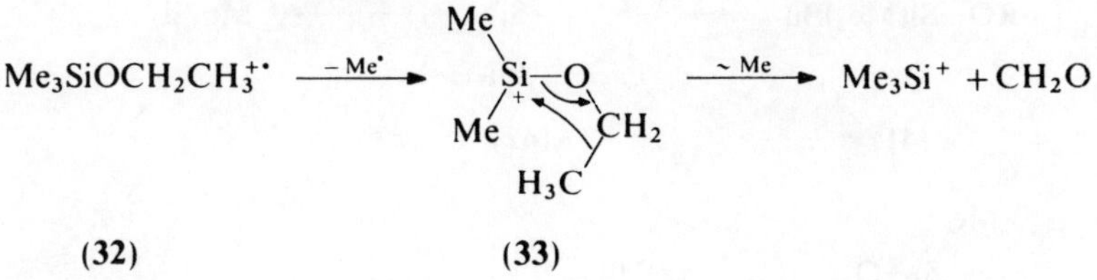

(29)

Interestingly, for the higher homologue trimethyl-*n*-pentyloxysilane **34**, a different pathway (reaction 30) was shown to be operative[54]. Loss of CH_2O from **35** (independently generated) to form **36** is a minor process.

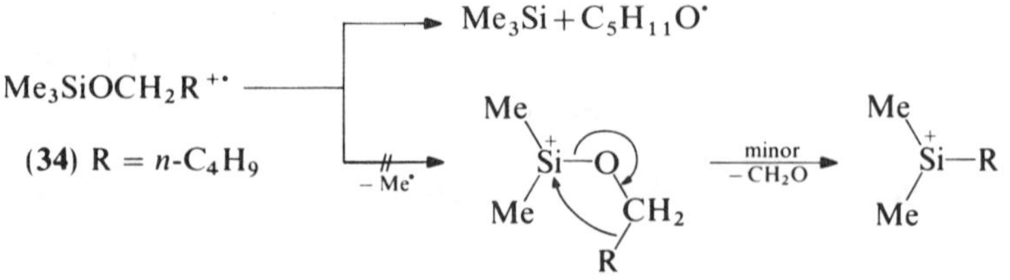

(35) (36)

(30)

The gas-phase chemistry of isomeric $C_5H_{13}Si^+$ ions, including α- and β-silyl-substituted carbocations as well as silicenium ions, was studied[55a] experimentally by using tandem mass spectrometry (MSMS)[56], and the results analyzed in conjunction with *ab initio* MO calculations[55]. It was demonstrated that the bridged ion **37** upon collisional activation dissociates to C_2H_4 and Me_3Si^+ without prior isomerization to **38** and **39**. The fact that **38** and **39** dissociate to both C_2H_4/Me_3Si^+ and C_3H_6/Me_2HSi^+ requires that the fragmentation is preceded by isomerization steps in which 1,2-methyl and 1,2-hydrogen migrations from the alkyl part to the silicon centre of a silicenium ion must occur. Reactions, involving 1,3-methyl migrations via for example transition structure **40** can be ruled out on energetic grounds.

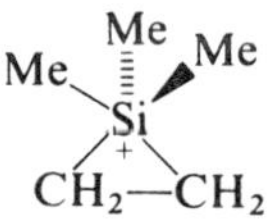

(37)

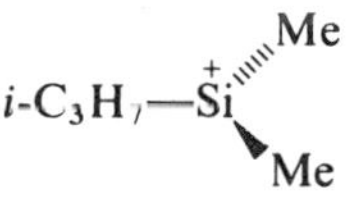

(38)

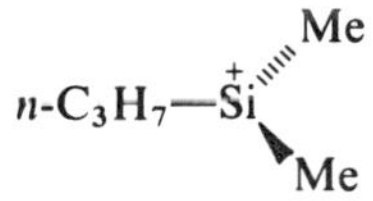

(39)

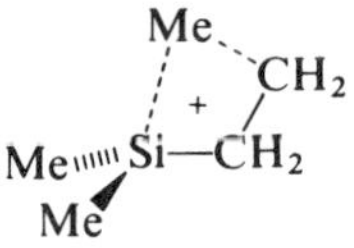

(40)

Methyl migrations to the silicon centre of a silicenium ion account also for the generation of Me_3Si^+ from *t*-butyldimethylsilyl ether derivatives **41** upon electron impact. It was demonstrated that **42** serves as immediate precursor to generate Me_3Si^+ and C_3H_6[57], and a rationalization is given in reaction 31a. For C_3H_6 loss (presumably propene) from **43** to generate **44** (reaction 31b), the data reported[57] do not distinguish between methyl transfer to the silicenium ion and transfer to the silicon atom to which the *t*-butyl group is bound.

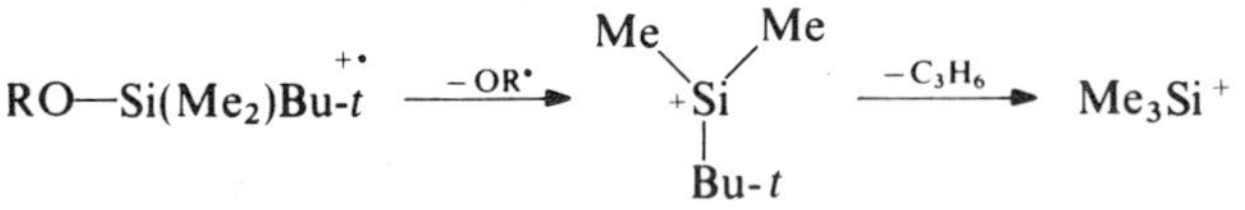

(41) (42)

(31a)

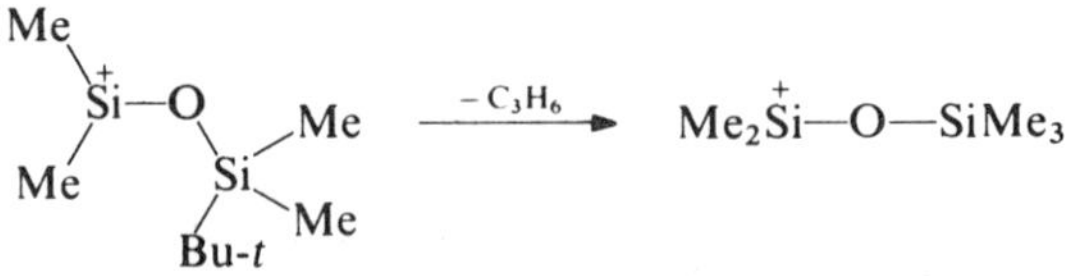

(43) (44)

(31b)

Intramolecular interactions of silicenium ions with aryl rings have been reported repeatedly. The following examples may serve as instructive and representative cases. Species **45**–**47**, generated upon Et˙ loss from the molecular ions of the corresponding ethers, generate in moderate to high yields via loss of CH_2O (from **45**) or C_3H_4O (presumably acrolein from **46** and **47**) the phenyldiethyl silicenium ion **48** (reaction 32)[58].

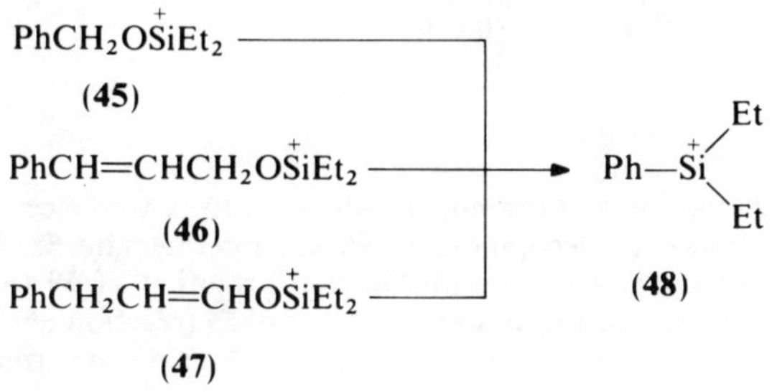

(32)

Interactions of a silyl centre with a remote phenyl group were also studied in great detail by Weber and coworkers[59], and it was demonstrated that the $(M-Me)^+$ ions **49** generated from ionized 1-(trimethylsilyl)-3-phenylpropane ($n=3$, and also some homologues with $n=2,4$) splits off C_nH_{2n} (reaction 33). The abundance of the resulting product ion **50** varies between 3–11% rel. intensity.

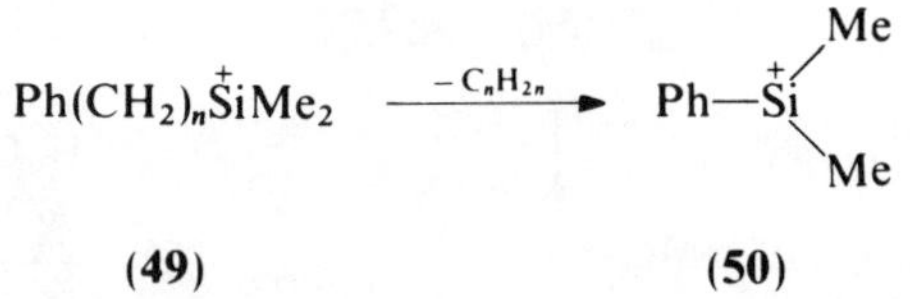

(33)

The base peak in the 70 eV mass spectrum of **51** ($R^1=R^2=Ph$) is due to the ion **53**[60]; the ion is also formed with considerable intensity for $R^1=Me$ (**52**). Although the mechanistic details for the phenyl migration from nitrogen to silicon remain to be elucidated, what is already known is that the substituent R^2 attached to the ring carbon atom of **51** seems to have a pronounced effect on the migratory aptitude of the phenyl group ($R^2=H>Me>Et>Ph$; reaction 34).

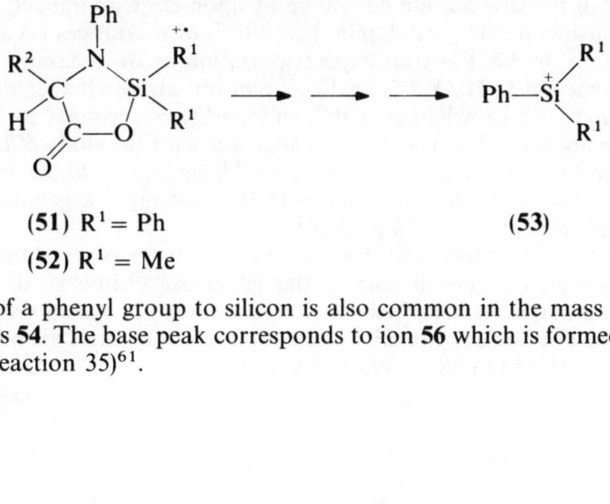

(34)

Migration of a phenyl group to silicon is also common in the mass spectra of α-silyl phenyl ketones **54**. The base peak corresponds to ion **56** which is formed via decarbonylation of **55** (reaction 35)[61].

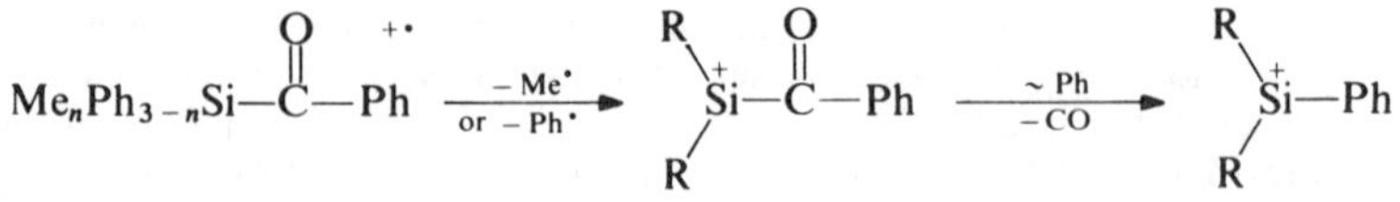

(35)

(54) (55) (56)

(R = Ph or Me)

The migrations of chlorine and bromine to silicon centres were recently reported[62] to occur. Ion **57** was shown to decompose to **59** via intermediate **58**. From the lower homologue **60** loss of C_2H_4O also results in the formation of **59** (reaction 36). The structure of the C_2H_4O also results in the formation of **59** (reaction 36). The structure of the C_2H_4O neutral (acetaldehyde, ethylene oxide, etc.) was not determined.

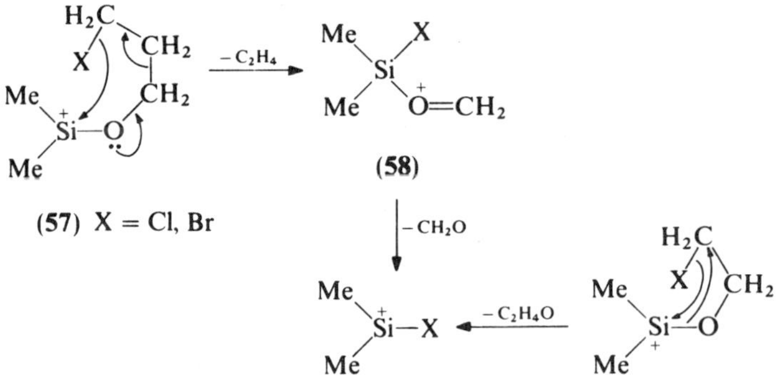

(57) X = Cl, Br (58) (59) (60) X = Cl, Br

(36)

The behaviour of the silyl acetate derivative **61** upon electron impact[63] and thermolysis[64] has met considerable interest. Upon thermolysis **61** rearranges via acyloxy/methyl exchange specifically to **62**. Electron-impact ionization of **61** generates, among other signals, a prominent ion $C_4H_9O_2Si^+$, and the question arose whether this fragment is formed from ionized **62** or whether a different pathway accounts for its formation (reaction 37). The analysis of metastable ions together with the study of labeled precursors suggested that the $[M - Me]^+$ ion **63** generated from ionized **61** serves as immediate precursor. Interaction of the silicenium ion with the carbonyl oxygen accompanied by loss of diphenylcarbene generates **64** ($C_4H_9O_2Si^+$).

Interactions of silicenium ions with the lone pair electrons of the ether oxygen were reported[65] to play also a central role in the gas-phase chemistry of silylated 2,2′-bis(trimethylsilyl)benzhydrols, for which the most significant product ions were shown to be formed from the $[M - Me]^+$ ion **65**. For example, **65** dissociates via **66** and **67** to the alkoxydimethylsilicenium ion **68** (reaction 38).

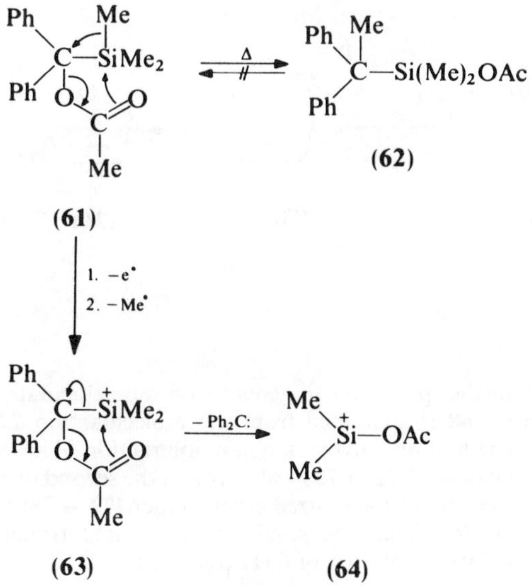

(37)

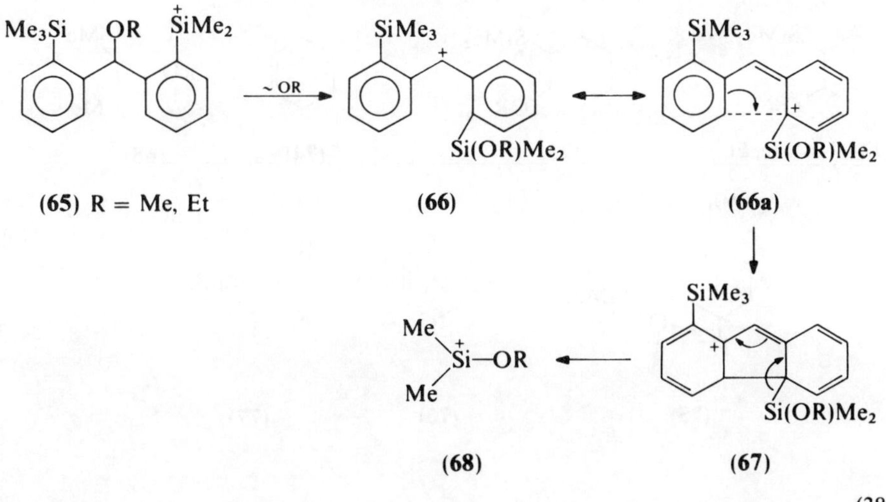

(38)

Similarly, ion **68** is generated via transannular alkoxy transfer to the silicenium ion of **69**[66]. It has not been definitively proven whether **70** is an intermediate or a transition state en route to **68**. The study of the *cis/trans* isomeric pairs of **69** clearly suggests that the formation of **68** is much more enhanced for the *cis* isomer. When starting from *trans*-**69**, some ring-opening processes must be involved. The nature of the neutral formed in the reaction **70** → **68** has yet to be elucidated. It is suggested[66] that starting from *cis*-**69** the bicyclic ether **71** is formed (reaction 39).

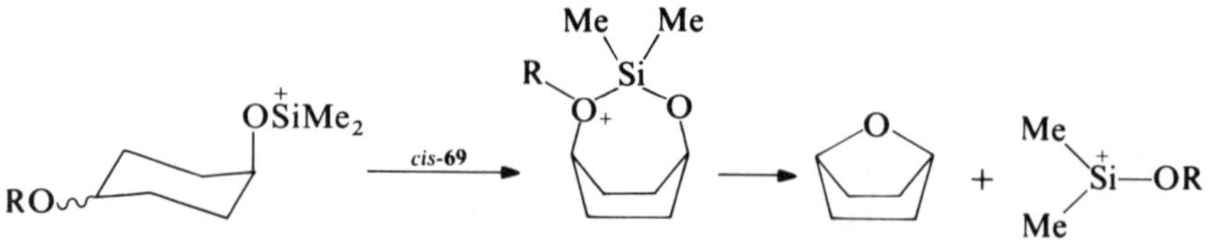

(69) R = Me, Et (70) (71) (68)

(39)

In the same study[66] further pathways to generate **68** were elucidated by using labeled precursors. For example, **68** is generated from the molecular ion **72** via two entirely different mechanisms. The first one involves, again, interaction of an ether oxygen atom with a silicenium ion (reaction 40: **72** → **73** → **68** + **74**). In the second one, in the initial step the intact TMS group migrates to the ionized ether oxygen (**72** → **75**), followed by loss of the cyclohexenyl radical **76**. From the resulting protonated trimethylsilanol **77** the product **68** may then be formed by loss of CH_4 (reaction 40).

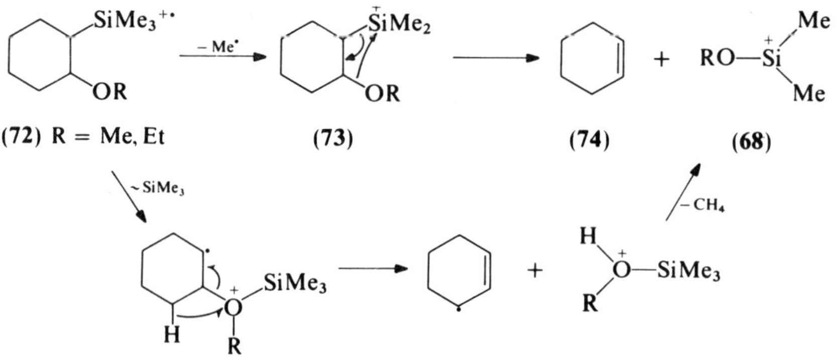

(72) R = Me, Et (73) (74) (68)

(75) (76) (77)

(40)

The bis-trimethylsilyl ethers of stereoisomeric 5,7-undecanediol **78** (as well as the free diols) give upon chemical ionization (protonation with $C_4H_9^+$) MH^+ ions, which show two major processes (reaction 41). One corresponds to a Grob-type fragmentation (β-cleavage) of the ion **79** (formed by loss of Me_3SiOH from protonated **78**) to yield the silylated aldehyde **80**. The second major reaction path involves loss of CH_4 from MH^+, followed by cyclization of the intermediate **81** (siloxy transfer) to **82** and β-cleavage (loss of 1-hexene) to generate **84**[67]. This reaction sequence, in which a centrally located structural unit rather than a peripheral one is eliminated, is rarely observed under chemical ionization and deserves further study.

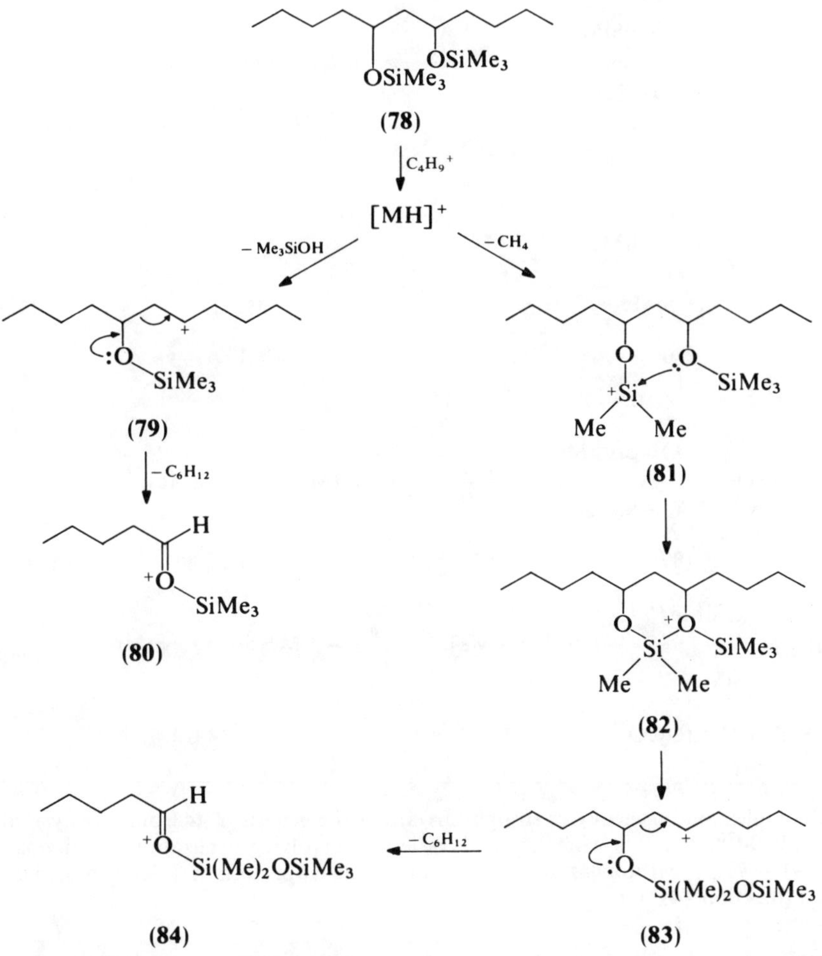

(41)

Quite unusual migrations seem to govern the gas-phase chemistry of $[M - Me]^+$ ions generated from bis-trimethylsilyl derivatives of simple boronic acids[68]. Among them are reversible methyl/phenyl migrations as well as transfer of silyloxy groups to silicenium ions. The so-formed rearrangement products may then serve as precursors to generate the ions at *m/z* 209, 147 and 135, respectively (reaction 42).

The concluding part of this section deals with some structural aspects of the genesis of the ion $Me_3SiOSiMe_2^+$ (**44**) which is ubiquitous in the mass spectra of bis-silylated compounds, and some examples for its formation have already been given in reactions 31 and 42. Ion **84** (reaction 41) upon collisonal activation also dissociates to $Me_3SiOSiMe_2^+$. Formally related with the process **85** → **44** (reaction 42) is the 1,1-elimination of the nitrene RN from the $[M - Me]^+$ ions of the corresponding bis-silylated hydroxylamine derivatives **89** (reaction 43)[69].

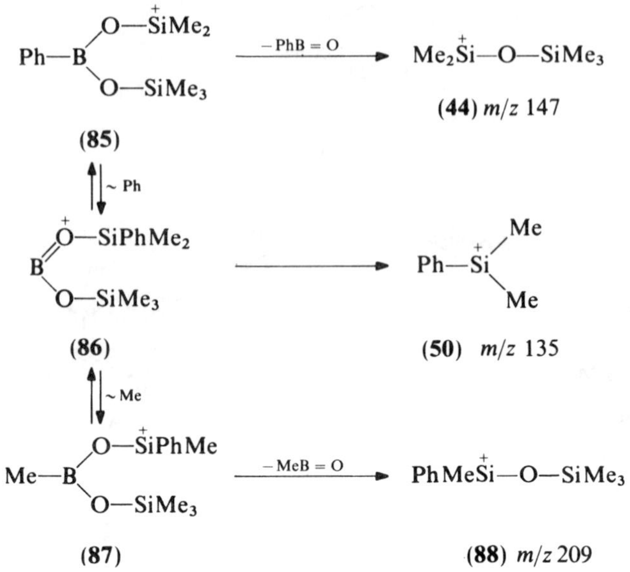

(42)

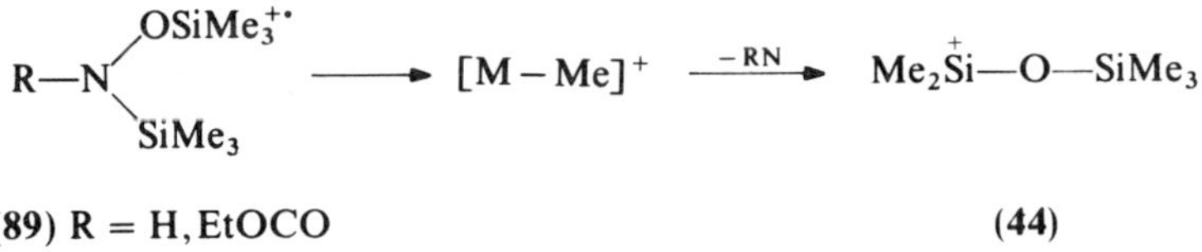

(43)

More complex, but nevertheless straightforward, is the genesis of **44** from bis-silylated amino acids **90**[70]. The following sequence of events involving cyclization via silyloxy transfer (**91** → **92**), decarbonylation (**92** → **93**) and β-cleavage (loss of a Schiff base) was suggested (reaction 44).

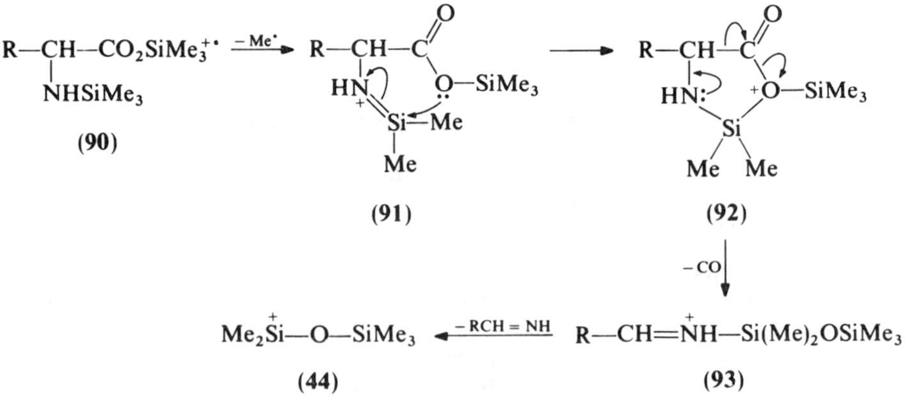

(44)

Stereochemical effects on the formation of **44** from silylated, cyclic diols were repeatedly reported in the literature[66,71,72]. Not surprisingly, it was observed that for those stereoisomers having a *cis*-orientation of the two interacting silyloxy groups the ion **44** was observed with much higher intensity than for those cases in which the two functional groups have a *trans*-orientation relative to each other. As was the case for the formation of the ion **68** (reaction 39), it was suggested that the increase in abundance for the stereoisomers with *cis*-configuration is due to transannular interaction of the silyloxy group with the silicenium ion; for the *trans*-compounds, ring opening preceding dissociation is assumed to account for the formation of ion **44**. It is, again, open to question whether the bicyclic oxonium ions serve as transition structures or intermediates. The stereochemistry of the 'inert' alkyl substituent attached to the ring of the cyclopentane and cyclohexane diols seems to have at best a minor effect on the ease of formation of ion **44**[71,72]. Representative examples for precursors with *cis*-configuration are given in reactions 45 and 46.

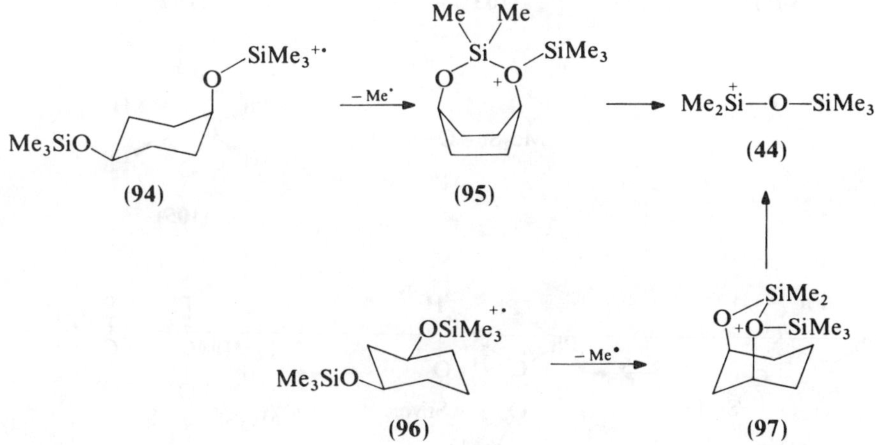

(45)

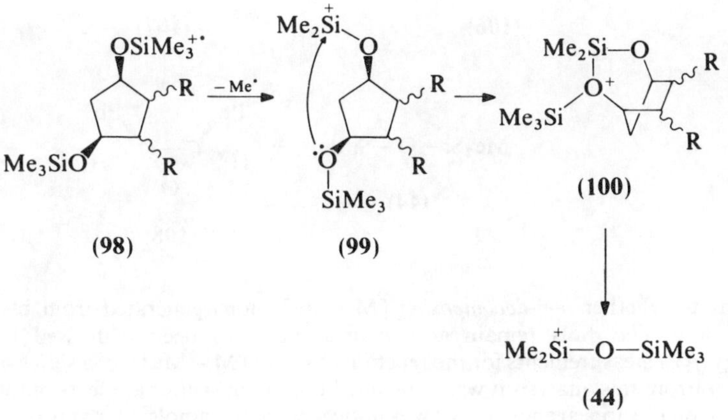

(46)

Arguments derived from conformational analysis were used[73] to account for the fact that the ion **44** is generated in substantially higher abundance when *threo*-1,2-diphenyl-1,2-bis(trimethylsiloxy)ethane (**101**) serves as precursor in comparison with the *erythro*-isomer (**102**). In the former both the cyclization step **103** → **104** and the elimination of *trans*-stilbene oxide (**105**) are favoured on stereochemical grounds (reaction 47). In the *erythro*-precursor **102**, however, the destabilizing interaction of the two phenyl groups retards the formation of **44** (reaction 48).

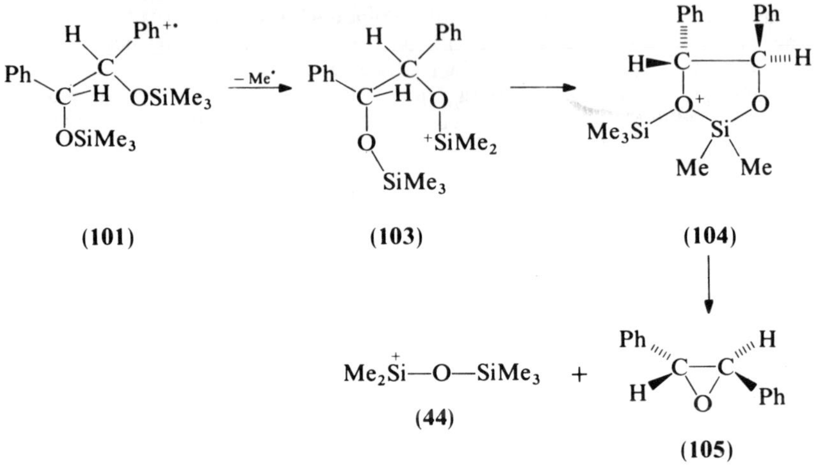

(47)

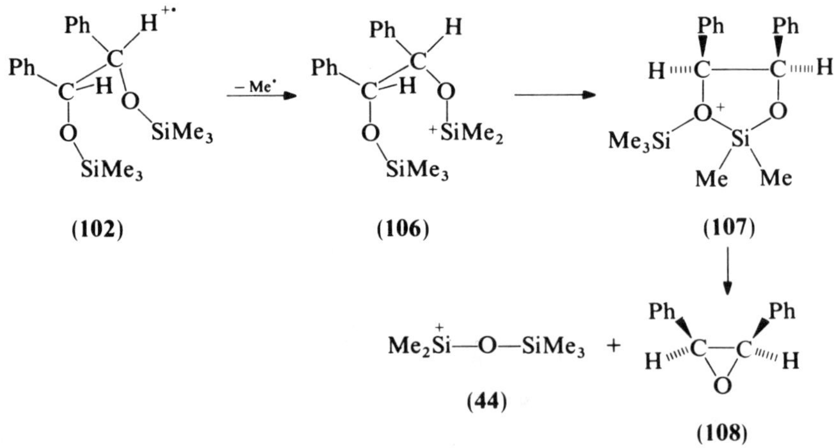

(48)

The question as to whether *non-decomposing* $[M - Me]^+$ ions generated from bis-trimethyl ethers of *acyclic* diols remain acyclic or cyclize has been addressed by appearance energy (AE) measurements for the reactions $M^{+\cdot} \rightarrow [M - Me]^+$ of a series of homologues **109**[74]. From this analysis it was concluded that no measurable effects of the chain length $(CH_2)_n$ on the appearance energy was noticeable. If intramolecular solvation of the silicenium ions by the ether oxygen lone pairs, **110** → **111**, would have occurred, this should have been reflected in the AE data. As this was not observed it was concluded that the non-decomposing $[M - Me]^+$ ions remain acyclic. However, those ions having enough internal energy to eventually decompose to **44** and neutral $C_2H_{2n}O$, (the latter being presumably a cyclic ether **112**) will undergo cyclization. In fact, early studies by McCloskey and coworkers[75] provided evidence that up to 49 membered rings may be involved in the formation of **44** (reaction 49).

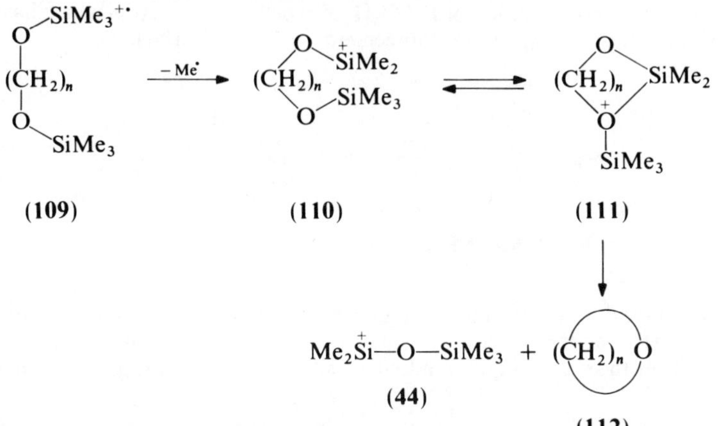

(109) (110) (111) (44) (112)

(49)

Most interestingly, it was also observed that the interaction of silicenium ions with silyloxy groups is even possible in configuratively fixed systems[76]. Thus, the $[M - Me]^+$ ions generated from **113a–e** decompose to **44** (reaction 50). While the high abundance of this fragment is no surprise when X = $(CH_2)_2$ and X = *cis*-CH=CH serve as precursor, the fact that **44** can also be generated from the other precursors requires either stereochemical changes of the multiple bonds upon electron impact (e.g. *trans*-CH=CH → *cis*-CH=CH, *linear* C≡C → *bent* C≡C) or that the Me_3SiO group migrates along the carbon skeleton to the silicenium atom. Although a distinction is not yet possible we should like to draw the reader's attention to the following section, in which ample evidence will be provided for the unusually high migratory aptitude of trimethylsilyl groups.

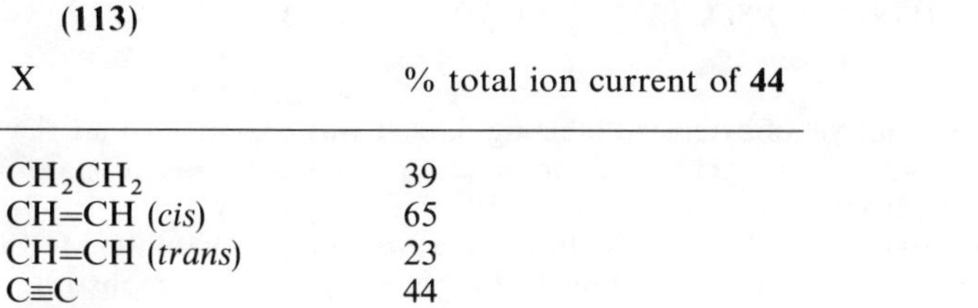

(113) (44)

(50)

113	X	% total ion current of 44
a	CH_2CH_2	39
b	CH=CH (*cis*)	65
c	CH=CH (*trans*)	23
d	C≡C	44
e	$(C≡C)_2$	7

B. Migrations of Silyl Groups to Electron-Deficient Centres

In reaction 40 (preceding section) the rearrangement **72** → **75** can be viewed as a typical example for the migration of a silyl group to an electron-deficient centre. There are indeed numerous papers published on this topic, and in the following we do not present an exhaustive overview but rather focus on a few typical examples. The first one concerns a *McLafferty*-type migration of a Me_3Si group to a positively charged carbonyl oxygen. For example, from the molecular ion of **114** C_2H_4 is expelled in a reaction which has many features typical for the *McLafferty* rearrangement[77, 78] (reaction 51).

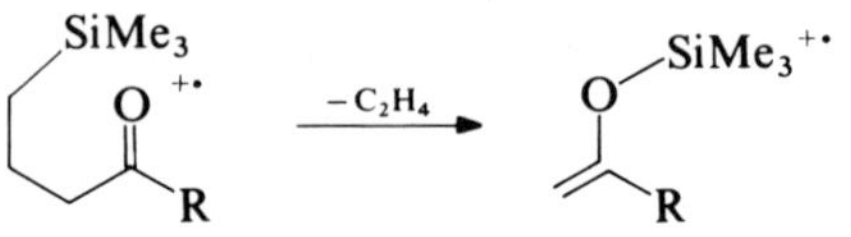

(51)

(**114**) R = OMe, alkyl (**115**)

Weber and coworkers[59] reported the analogous migration of a Me_3Si group to the C=C double bond of **116** (reaction 52), a process which is similar to the well-known[79] transfer of a hydrogen from the γ-carbon of an alkylbenzene to the phenyl ring with concurrent β-cleavage.

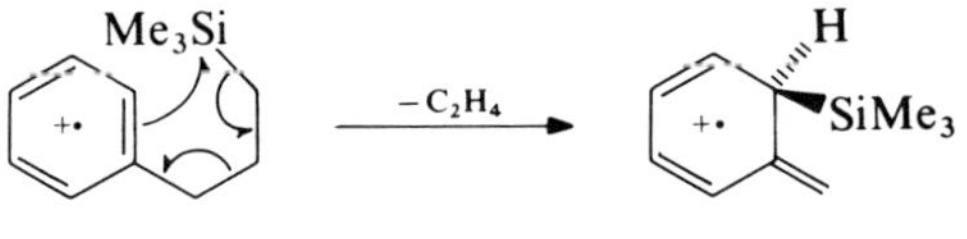

(52)

(**116**) (**117**)

Fragmentation of the molecular ions of **118** (loss of R·) is also associated with migration of the $SiMe_3$ group to generate the non-classical bridged ethylene trimethylsilanium ion **37** (reaction 53)[55, 80].

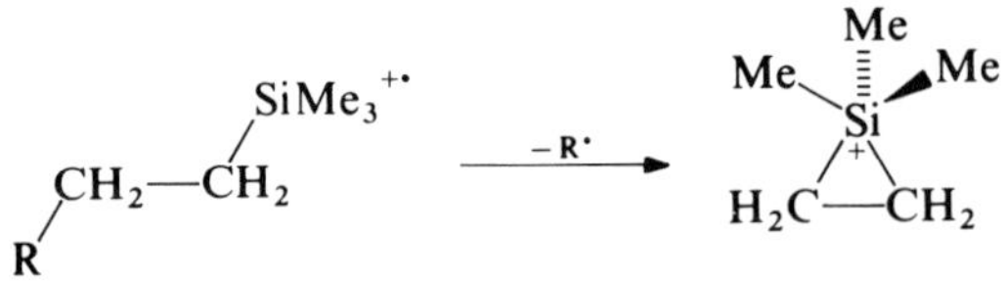

(53)

(**118**) R = PhO, alkyl (**37**)

Based on the analysis of extensive labeling data it was concluded that the relatively abundant $[M-44]^{+\cdot}$ ions generated upon electron-impact ionization of TMS derivatives of 20-hydroxy-5α-pregnan-3-ones (**119**) and 20-hydroxy-4-pregnen-3-ones (**120**) are attributable to the loss of CH_3CHO from C(17) with migration of the Me_3Si group to the charge-retaining fragment. Although the labeling data demonstrate that the CH_3CHO group eliminated contains $C_{(20)}/C_{(21)}$ the detailed mechanism of the rearrangement/dissociation behaviour of **119** and **120** could not be given by the authors[81].

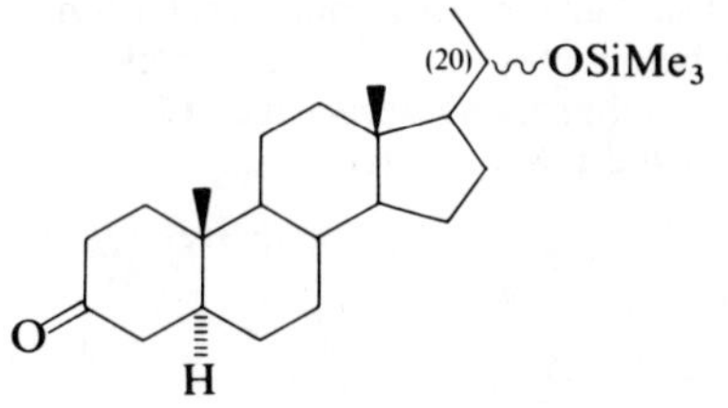

(119)

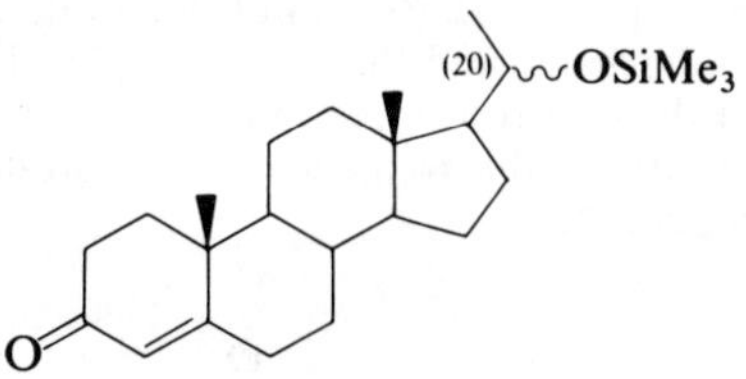

(120)

The electron-impact mass spectra of several ketosilanes can only be explained by invoking considerable rearrangements prior to dissociation. Major pathways very often involve the migration of a silyl group to the ionized carbonyl group following a mechanism being different from the *McLafferty*-type reaction 51. For example, the mass spectra of a series of ionized β-ketosilanes **121** are best explained by assuming that upon electron bombardment the β-ketosilanes undergo rearrangement to a species very similar in structure to (if not identical with) that of the ionized silyl enol ethers **122**. The latter then serve as precursors to generate via loss of $Me_2Si{=}O$ ions **123** and, via elimination of ArC_3H_4 from **124**, the species **125** (reaction 54)[82]. Substitution on the benzene ring shows a significant effect on the formation of ions **123** and **125**, with electron-donating groups favouring the former and electron-withdrawing substituents favouring the latter. The mechanisms for the formation of **123** and **125**, advocated by the authors[82], are given in reaction 54.

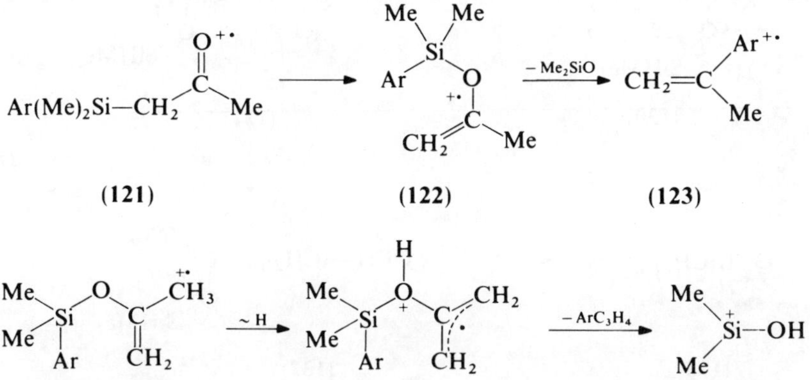

(54)

An interesting rearrangement, paralleling the photochemical behaviour of the neutral, was reported[83] for the molecular ions of tris(trimethylsilyl)arylsilanones **126**. A typical example is given in reaction 55.

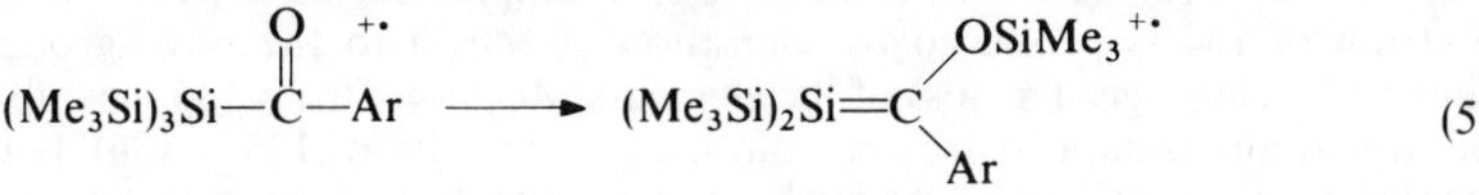

(55)

(126) (127)

Examples for the intramolecular interaction of silyl groups with ionized ester functions are numerous, and the two examples discussed below may suffice to give a glimpse.

In the electron-impact mass spectra of many cyclopropene derivatives **128** an abundant signal is due to the loss of Me_3SiOR from ionized **128** for which reaction 56 has been suggested[84].

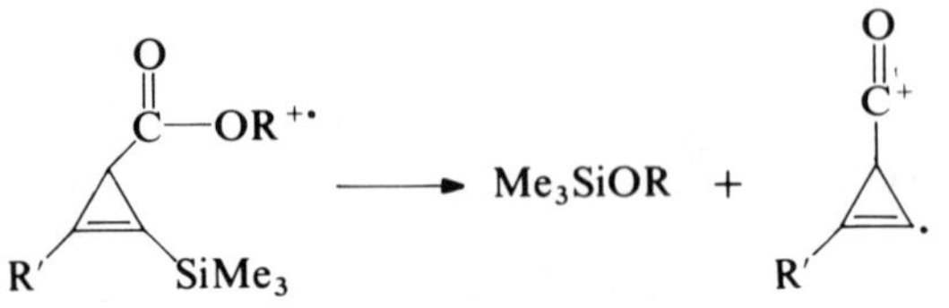

(**128**) R = Me; R' = H, alkyl (**129**)

(56)

Intramolecular interactions between a silyl and an ester group are also characteristic of the mass spectral behaviour of the methyl esters of silylated hydroxycarboxylic acids, e.g. **130** (reaction 57)[85]. Transfer of the silyl function (**130** → **131**) is followed by an α-cleavage (**131** → **132**) and terminated by the expulsion of $Me_2HSiOMe$ (**132** → **133**). A similar reaction sequence is operative for the TMS analogue of **130**[86].

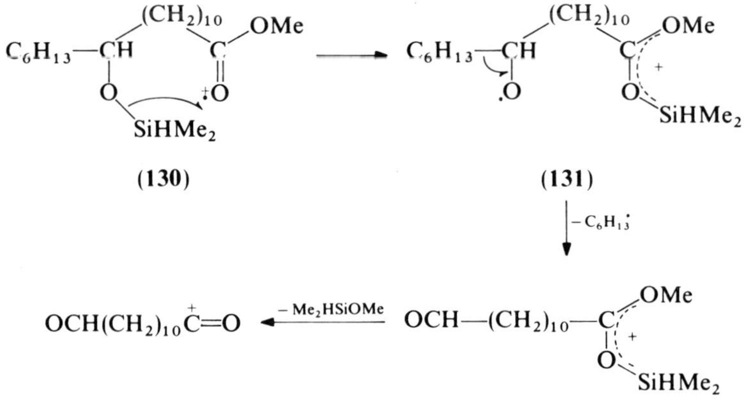

(**130**) (**131**) (**133**) (**132**)

(57)

It should also be mentioned that in the *negative* ion mass spectra of *ortho*-substituted, silylated benzenes many peaks are due to specific rearrangement/dissociation processes involving migrations of the intact Me_3Si group[87]. Most of the mechanistic conclusions are based on the investigation of labeled compounds and comparative studies of the *ortho*-versus *meta*- and *para*-isomers. For example, the base peak (Me_3SiO^-) in the spectrum of **134** is produced by migration of Me_3Si to the nitro group (**134** → **135**), followed by cleavage. The loss of the elements Me_3SiOH from **134** may also proceed via **135**, which abstracts a hydrogen from the *ortho*-position (**135** → **136**). It remains to be established whether Me_3SiOH is eliminated from **136** as an intact moiety or whether $Me_3SiO^{\cdot}$ is formed. On energetic grounds the former is much more likely (reaction 58).

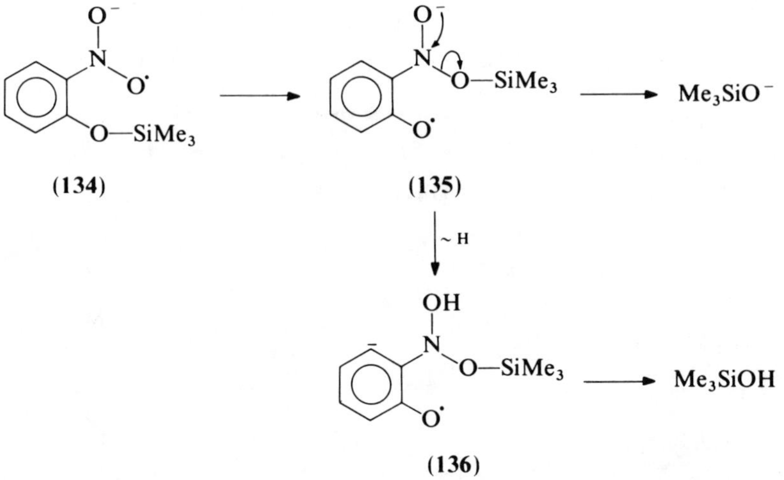

(58)

Similarly, the reaction of the silyl ester **137** commences with a migration of the Me_3Si group (reaction 59: **137** → **138**); the intermediate **138** serves as immediate precursor for the loss of CO_2 and formation of Me_3SiO^-. Elimination of $MeCO_2$ is explained by transfer of a methyl group from **138** to **139** followed by dissociation. Loss of Me_3SiOH may be accounted for by a mechanism comparable to that discussed above for **134** (reaction 59).

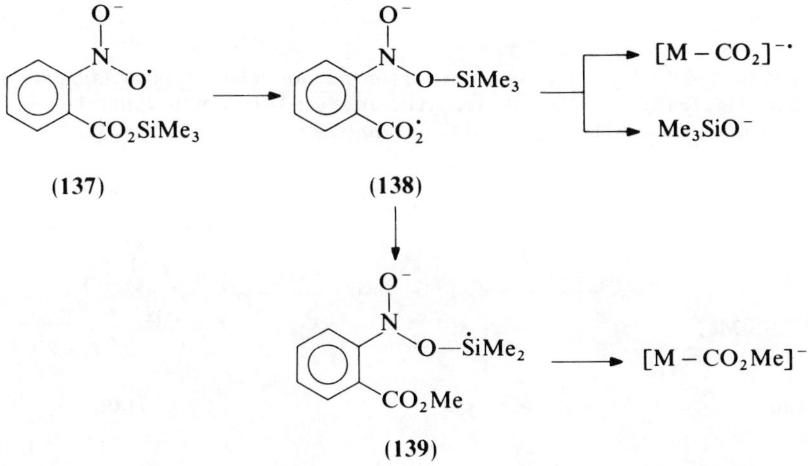

(59)

C. Cyclization Processes Involving Silicenium Ions as Electrophiles

In the previous two sub-sections we have repeatedly referred to cyclic species which, starting from acyclic ions, were formed as *transients* en route to dissociation products (see reactions 28, 38, 39, 41, 44, 45–49). Further examples belonging to this category and emphasizing the role of cyclic intermediates are given in reaction 60, which describes the specific interactions of α/ω-substituents. Ions **142** and **145** are observed as stable ions as well as precursors to generate **59**[88].

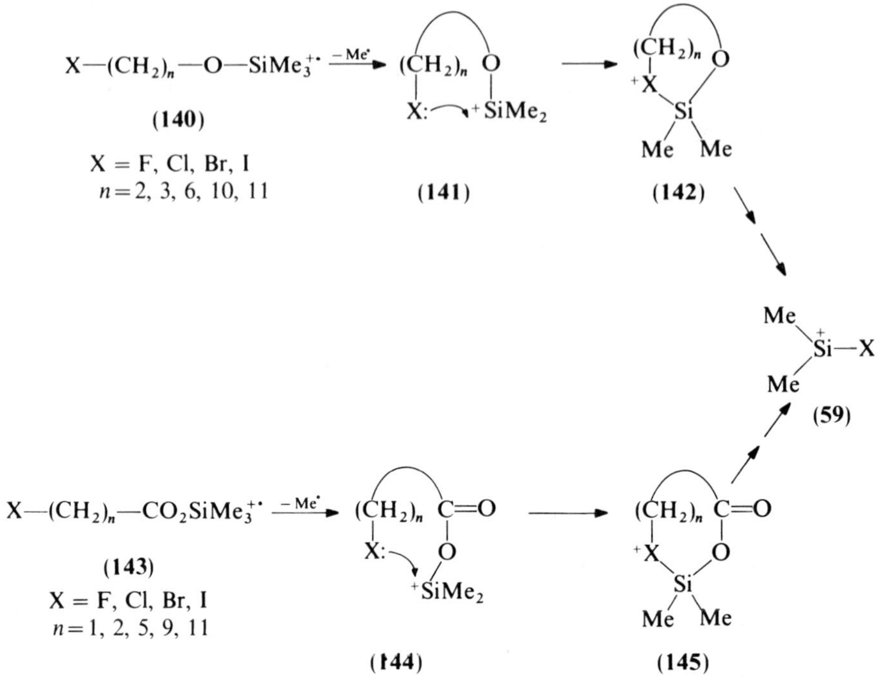

(60)

Similarly, the $[M - CH_3]^+$ ions generated from the molecular ion of silylated ethylmalonic acid **146** are believed[89] to have the cyclic structure **147**; from the latter the loss of CO_2 may proceed via a cycloreversion (**147** → **148**) (reaction 61).

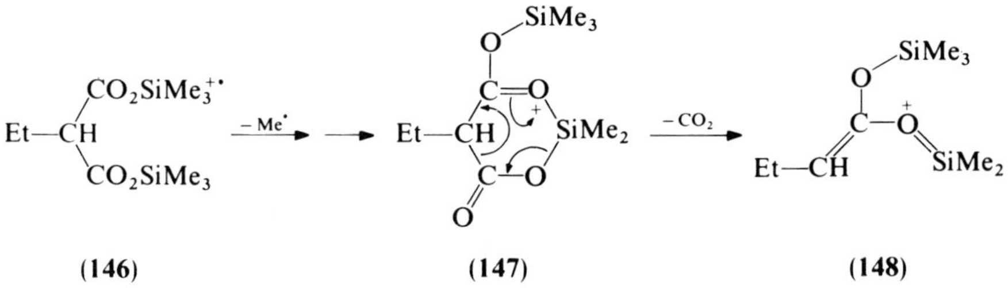

(61)

Cyclic product ions are abundantly generated from appropriately *ortho*-substituted benzene derivatives[90]. With regard to silyl substituents the following examples are worth mentioning. The EI mass spectrum of the TMS ether of 2-methoxyphenol contains strong signals $[M - 30]^{+\cdot}$, and the analyses of labeled precursors, using high-resolution mass spectrometry and linked-scan spectra, are in accord with the mechanism described in reaction 62[91]. For the *meta*- and *para*-isomers the loss of the second methyl group is not observed. In line with MNDO calculations, the most favoured reaction involves the sequence **149** → **150** → **151** → **152**. The alternate pathways are predicted to be much higher in energy. This is shown by the numbers given in parentheses below the structures of reaction 62 which refer to the MNDO calculated heats of formation in kcal mol^{-1} of the respective species.

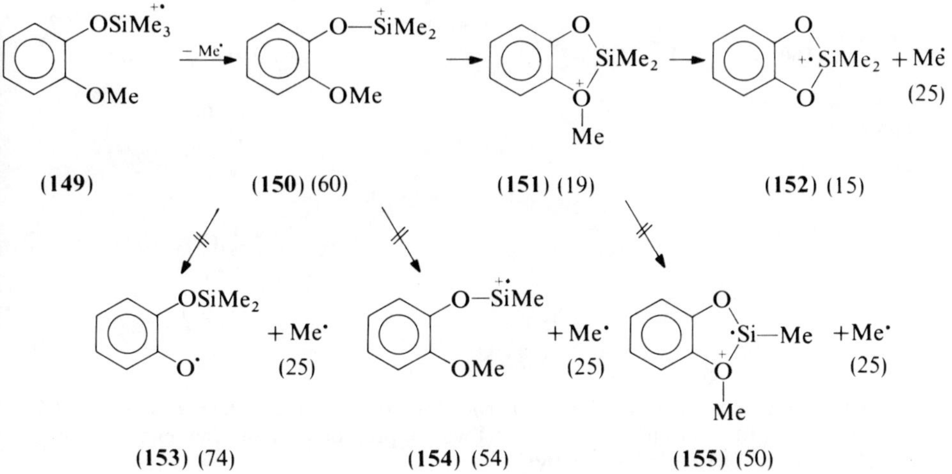

(62)

The ion **152**, which is not only generated from **149** but also from the molecular ion of **156**[51e] (via consecutive losses of Me and NO: reaction 63), dissociates further via the elimination of two methyl radicals. It is believed[51e, 92a] that the species formed have the structures shown in reaction 63. (For an excellent discussion and a comparison with the all-carbon analogue of **158** see Ref. 92b.)

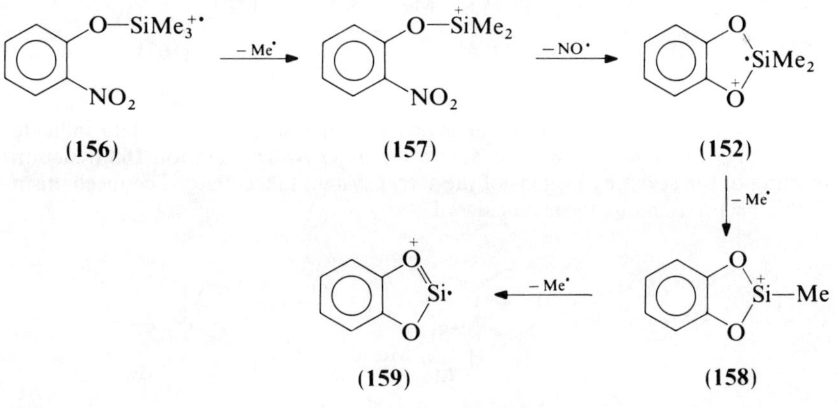

(63)

As demonstrated by reaction 64, cyclic products can also be generated by the interaction of *non*-charged sites. Based on the study of labeled precursors it was shown[92a] that the base peak at $[M-119]^+$ in the EI mass spectra of TMS derivatives of methyl esters of *ortho*-dihydroxy-substituted benzoic acids is due to the successive losses of MeO˙ and Me_4Si. It was argued that the driving force for the reaction **161** → **162** is the relief of steric strain.

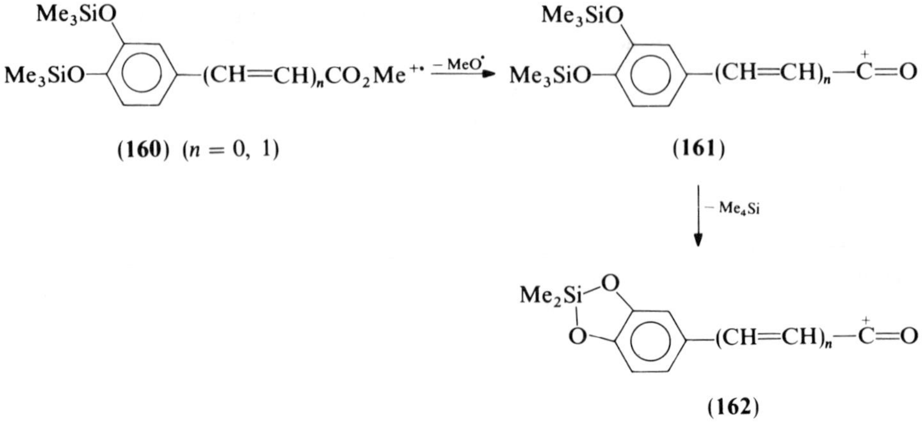

(160) ($n = 0, 1$) (161) (162)

(64)

A cyclization step (reaction 65) is also crucial for the gas-phase chemistry of ion **163**; interaction with the benzene ring yields **164** which, provided the internal energy is high enough, splits off CH_4 to generate **165**[93].

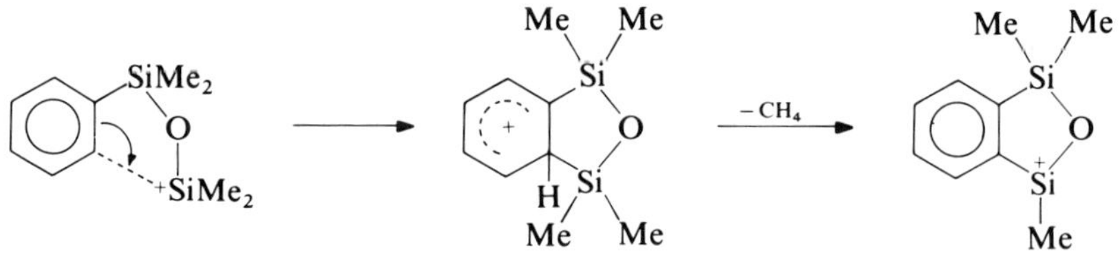

(163) (164) (165)

(65)

Ring closure was also postulated for the styryl system **166**[59]. Labeling data indicate that, prior to CH_4 loss (**167** → **168**) to generate the indenylsilicenium ion **168** (reaction 66), scrambling of the seven hydrogens of the styryl system takes place. The mechanism by which this occurs remains to be elucidated.

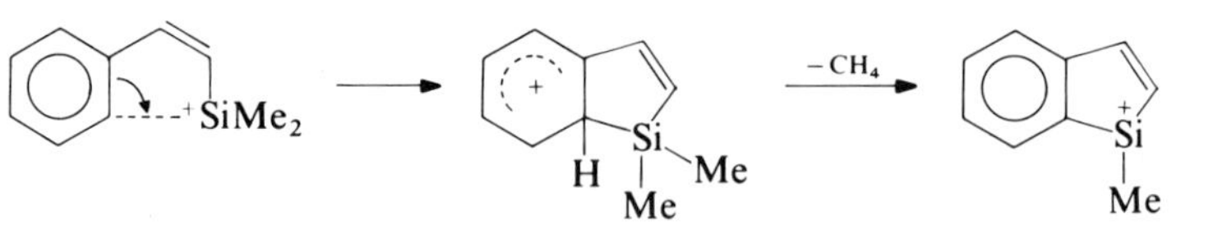

(166) (167) (168)

(66)

The fragmentation pattern of allyloxy-substituted trimethylsilanes was interpreted by invoking first formation, and then interaction of silicenium ions with the C=C double bond (reaction 67)[94]. Again, the loss of RH from **170** is not a site-specific reaction with regard to the hydrogen of RH but seems to be preceded by scrambling processes.

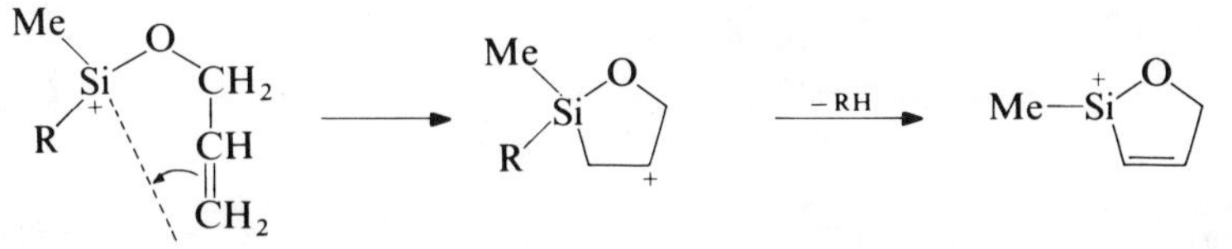

(169) R = Me, H (170) (171) (67)

The mass spectrometric behaviour of trimethylsilyl derivatives of nucleosides, which are extremely suitable for GCMS analysis[95], was studied in great detail by McCloskey and coworkers[96]. As earlier suggested by Westmore and coworkers[97] the extremely high abundances of the $[M - Me]^+$ ions are caused for some TMS derivatives by interaction of the silicenium ions with electron-rich centres, e.g. **172** and **173**.

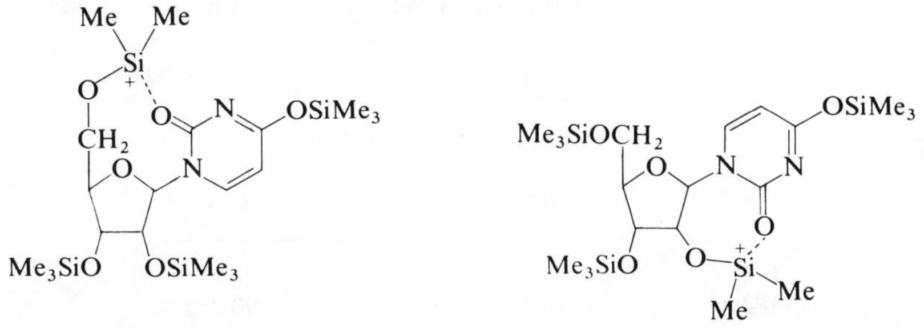

(172) (173)

D. Anchimerically Assisted Dissociations

The cleavage of O–C bonds to generate C-radicals with charge retention at the oxygen atom (**174** → **175**) (reaction 68) is extremely unfavoured on energetic grounds and is observed only under exceptional conditions[98].

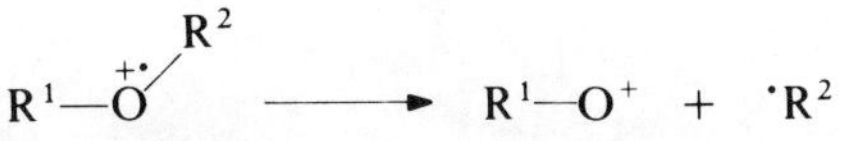

(174) (175) (68)

We have observed[63, 99, 100] that the cleavage of oxygen–carbon bonds can be greatly enhanced if R^1 in **174** contains a substituent which has both high migratory aptitude and affinity to oxygen. Extensive studies reveal that one of the most powerful groups is a silyl function. The overall reaction is given in reaction 69. Noteworthy features are the following: (i) If the migrating group X in **176** corresponds to a carbon or germyl substituent the reaction does not occur. (ii) The product $(M–R)^+$ formed is identical with a species generated by α-cleavage of **179**. (iii) The leaving group R^3 may comprise both stabilized (benzyl, allyl) and non-stabilized carbon radicals (e.g. methyl), and also silyl groups. (iv) The effect Y exerts is the stronger the higher the affinity of Y to the migrating group. Thus if Y = CH_2 the migration of the silyl group associated with cleavage of the Y–R bond does not occur even if R^3 corresponds to a stabilized radical (e.g. benzyl). (v) R^1 and R^2 in **176** can be varied considerably; however, one substituent must correspond to an aromatic system.

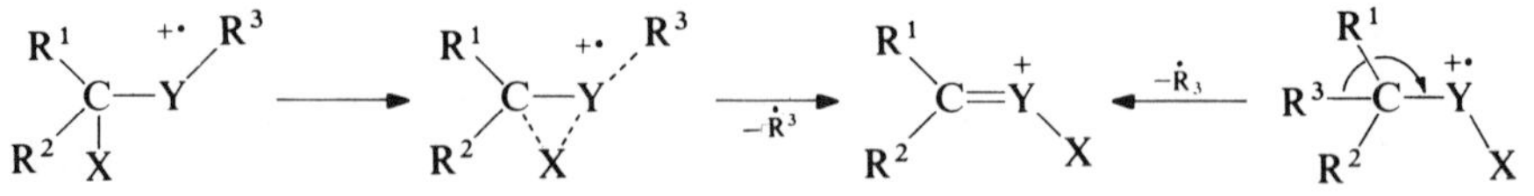

(176) (177) (178) (179)

(69)

Typical examples for an oxygen- and nitrogen-containing substrate are given in reactions 70 and 71.

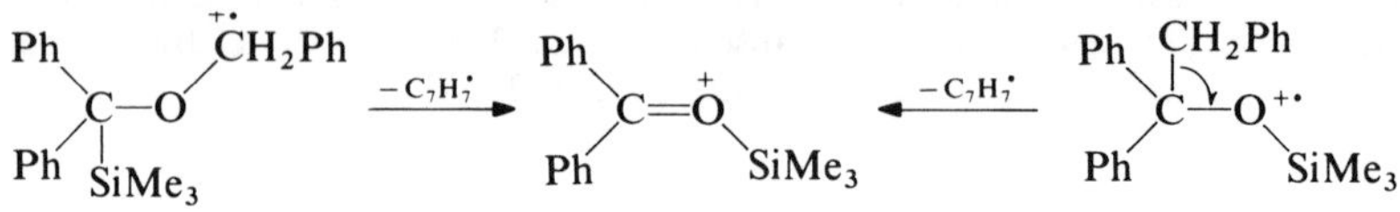

(70)

(180) (181) (182)

Precursor	% total ion current of **181** at 70 eV	12 eV
180	20.7	98.1
182	10.5	90.0

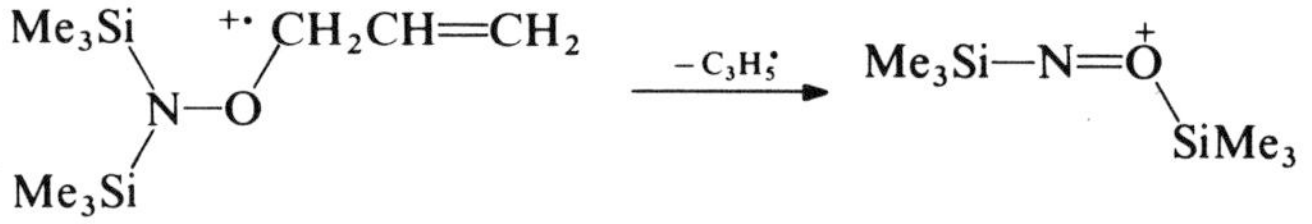

(71)

(183) (184)

	% total ion current of **184**
70 eV	8.8
12 eV	34.5
metastable ions	89.9

We believe that reaction 69 is of quite general applicability and may also account for the generation of ions in many other systems, the genesis of which otherwise would be difficult to explain. For example, the mass spectra of hexamethoxydisilane **185** contains an abundant signal for $[M-Me]^+$[101]; the interaction of a silyl group with the 'incipient' siloxy cation (reaction 72) would certainly lower the activation energy of the reaction. Experiments aimed at testing this tempting hypothesis are under way in our laboratory.

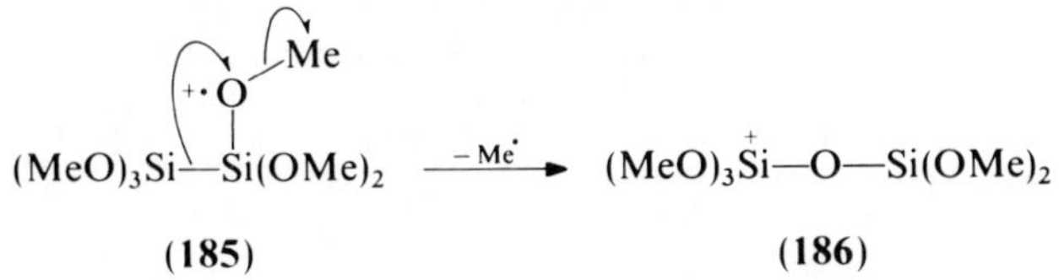

(72)

Similarly, we favour the explanation, given in reaction 73, for the generation of $[M-Me]^+$ ions from ionized methoxymethylsilanes **187**; the alternate mechanism put forward by the authors[102a] (**187** → **188** or **189**) has, unfortunately, not been substantiated by appropriate labeling experiments. Recent experiments[102b] clearly demonstrate that > 99% of metastable **187** ions dissociate via anchimeric assistance to **44**; **188** is formed with < 1%.

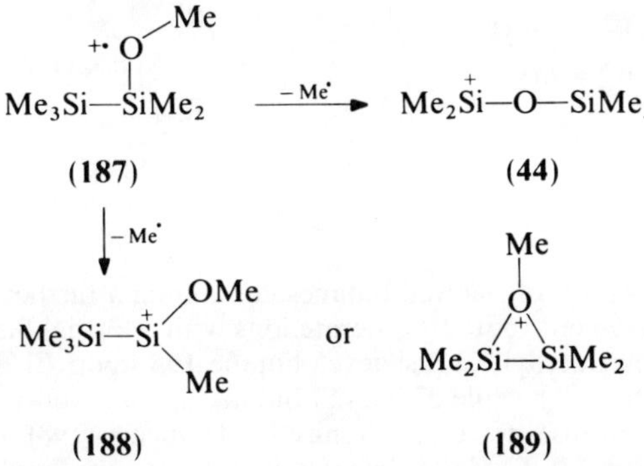

(73)

E. Generation of Fragment Ions Containing Si–X Multiple Bonds

One of the most convenient routes to generate ions (both open and closed shell) containing SiX multiple bonds proceeds via dissociative ionization of silacyclobutane derivatives[103] (reaction 74). Although the mechanistic details of these cyclo-reversions, i.e. concerted versus step-wise, are strictly speaking unknown, most authors seem to favour the concerted variant (**190** → **191**)[83, 103b, 104]. A step-wise process commencing with the formation of silicenium ions (**190** → **192**) followed by β-cleavage (**192** → **191**) was, however, also considered[105] and, as will be demonstrated below, does at least apply for some systems[106].

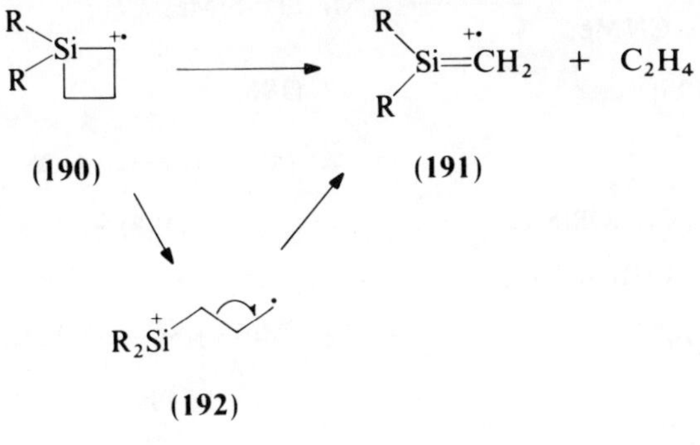

(74)

For 2-substitued silacyclobutanes **193** the product distribution is very much dependent on the substituent R[1 104], and energetic factors (stabilization of the olefin versus the ionized silene) seem to account for this. For example, 2-alkyl-substituted silacyclobutanes **193a** favour the formation of **191**, while 2-aryl-substituted analogues **193b** preferentially decompose to **194** (reaction 75).

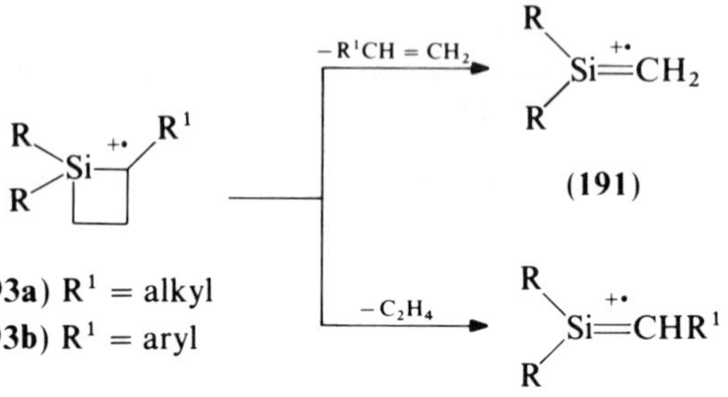

(**193a**) R^1 = alkyl
(**193b**) R^1 = aryl

(**191**)

(**194**)

(75)

Electron-impact ionization of silacyclobutanes containing a further heteroatom X in the ring provides a convenient route to generate ions with a formal SiSi double bond, if X = Si. For example, tetramethyl-1,2-disilacyclobutane **195** upon EI dissociates preferentially to **196**[107] (reaction 76), while disilacyclobutanes having bulky substituents (**197**) depending on R decompose to either ionized silaethene (**198**) or disilene (**199**) derivatives[83, 108] (reaction 77). The latter behaviour seems to parallel closely the thermochemistry of these compounds[83]. Interestingly, species **198** (R = Ph) is identical with intermediate **127** generated from the silyl-substituted alkanone **126**, which, prior to dissociation, rearranges to **127** (reaction 55). In fact, the mass spectral analysis of **127** and **198** (R = Ph) clearly demonstrates that both ions are indeed identical[83]. Tetramethyl disilaethene ions **196** are also produced in the formal *retro*-Diels–Alder reaction of ionized **200** (reaction 78)[109]. Loss of *neutral* **196** to generated **201** is less favoured. Again, reaction 78 has its analogue in the thermal behaviour of neutral **200**[109].

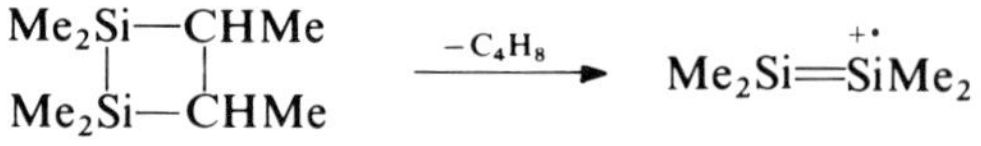

(**195**) (**196**)

(76)

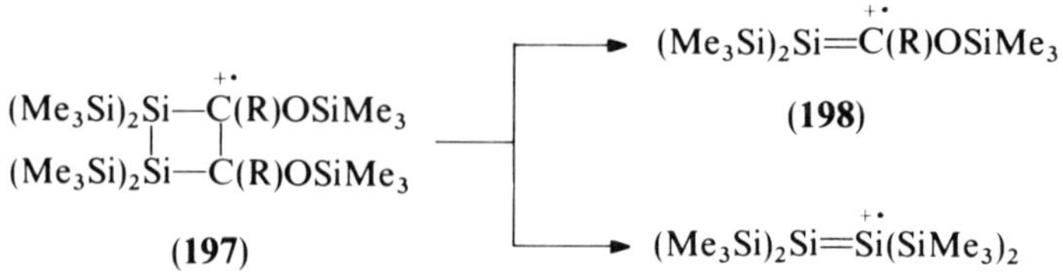

(**197**)

(**198**)

(**199**)

(77)

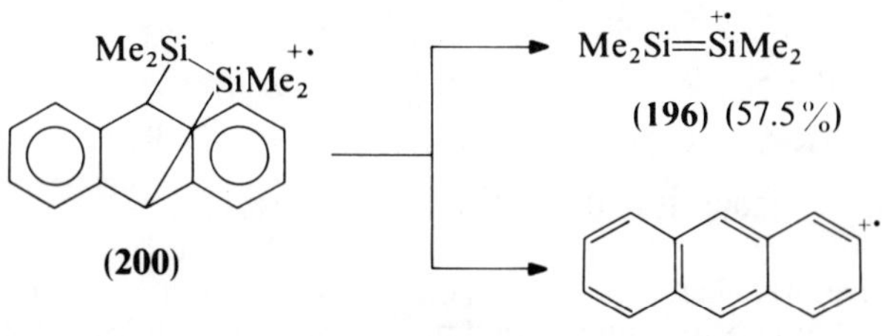

(200) → (196) (57.5%) + (201) (4.4%) (78)

'Cyclo-reversion' of ionized 1,3-disilacyclobutane[110] (**202**) and its ring-methylated derivatives[103b, 111] to generate **203** is not favoured, presumably due to the high heat of formation of **13**. In fact, it was estimated[103b] that the activation energy of reaction 79 is close to 60 kcal mol^{-1}, thus giving enough room for other competing processes.

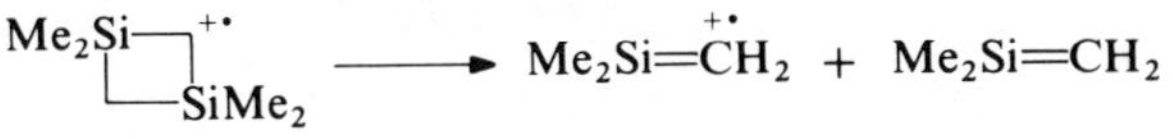

(202) → (203) + (13) (79)

On the other hand, the molecular ions of the 3-silathietane derivatives **204** and **205** do not only dissociate to **203** (from **204**) and **194** ($R = Et$; $R^1 = Me$); more interestingly, the former splits off C_2H_4 to generate at 12 eV ion **208** ($R = Me$) as base peak; similarly, loss of C_4H_8 from **205** generates the higher homologue **208** ($R = Et$) with 80% relative intensity[106]. Obviously, the formation of these ions cannot occur in a one-step process but requires a multi-step sequence. A likely mechanism, having the feature of a metathesis reaction, is suggested in reaction 80.

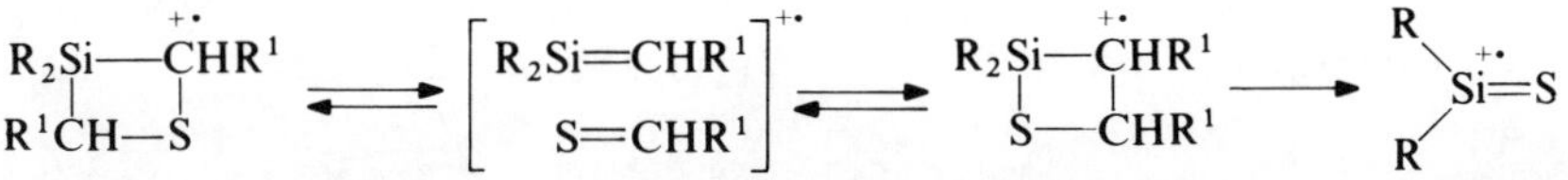

(206) (207) (208) (80)

	R	R^1
(**204**)	Me	H
(**205**)	Et	Me

Another convenient route to generate ionized disilenes takes advantage of 1,2-elimination of stable neutrals Me_3SiX ($X = halogen$) from precursors of the general structure **209**, and it is observed that reaction 81 very often produces **210** as base peaks in the mass spectra of **209**[83, 112].

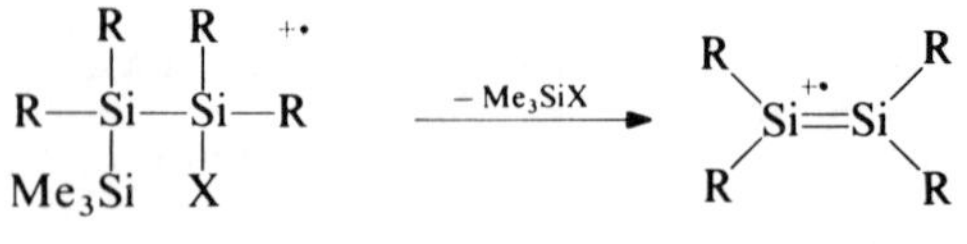

(81)

(209) R = alkyl, $SiMe_3$ **(210)**

Ionized silaarenes, silaallenes and silabutadienes have been suggested to be formed by mechanistically not always fully understood processes. It may suffice to list, out of many examples, the following few cases: HCl loss from ionized halogen-substituted 9,10-dihydro-9-silaanthracenes **211** is believed to yield the silaanthracenium cation radical **212** (reaction 82)[113].

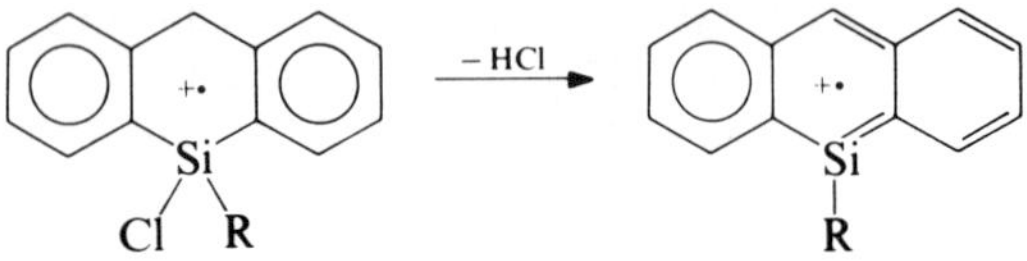

(82)

(211) **(212)**

Ionized 2-silaallene **215** is suggested to be formed via two consecutive cycloreversions starting from the spiro-silane **213** (reaction 83)[114]. Accordingly, loss of C_2H_4 from **216** would generate the silabutadiene derivative **217** (reaction 84)[115].

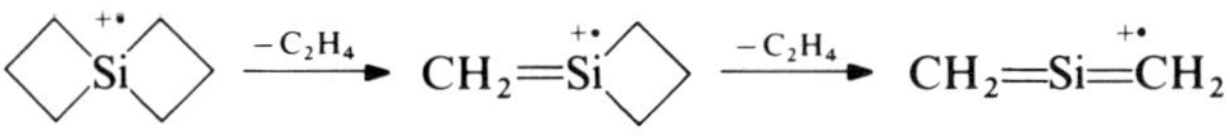

(83)

(213) **(214)** **(215)**

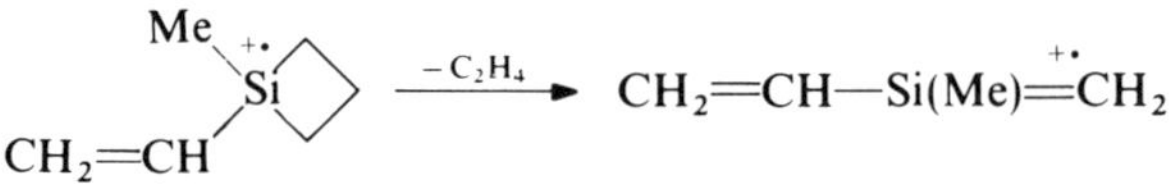

(84)

(216) **(217)**

Closed-shell ions with SiX multiple bonds were repeatedly invoked to explain some unusual features of organosilanes upon electron-impact ionization. Before discussing a few selected examples it is perhaps worth recalling that, according to *ab initio* calculations[41e], some of these silavinyl cation analogues do not exist in potential wells but rather collapse without a barrier into much more stable isomers (for a detailed discussion see Ref. 3h). For example, **218** is predicted to spontaneously rearrange to **219**; the interconversion of **220** to either **219** or **221** would, however, require significant barriers (reaction 85). While not strictly applicable to higher homologues and derivatives of **218**–**221**, the theoretical finding should be kept in mind when discussing the following examples.

$$H_2C{=}\overset{+}{S}i{-}H \longrightarrow Me{-}\overset{+}{S}i \quad H{-}\overset{+}{C}{=}SiH_2 \longleftarrow \overset{+}{C}{-}SiH_3 \qquad (85)$$

(218) **(219)** **(220)** **(221)**

Upon collisional activation from Me_3Si^+ an ion at m/z 43 $(CH_3Si)^+$ is formed, and there are two mechanistic possibilities conceivable for this reaction[116], i.e. consecutive losses of two methyl radicals, and loss of C_2H_4 followed by H_2 loss. In the absence of proper labeling studies it is not possible to distinguish whether H_2 loss from $MeSiH_2^+$ proceeds via a 1,1-elimination to generate **219** or whether a 1,2-elimination is operative. The latter would produce **218** which, of course, would collapse to **219**. Similarly, a 1,1-elimination (involving the methyl group of $MeSiH_2^+$) would result in a yet different m/z 43 product ion, i.e. **220**. While these mechanistic variants (reaction 86) could not be distinguished, the application of tandem mass spectrometry methodologies clearly demonstrated[116] that m/z 43 is formed by consecutive losses of $CH_3^{\cdot}/CH_3^{\cdot}$ *and* C_2H_4/H_2.

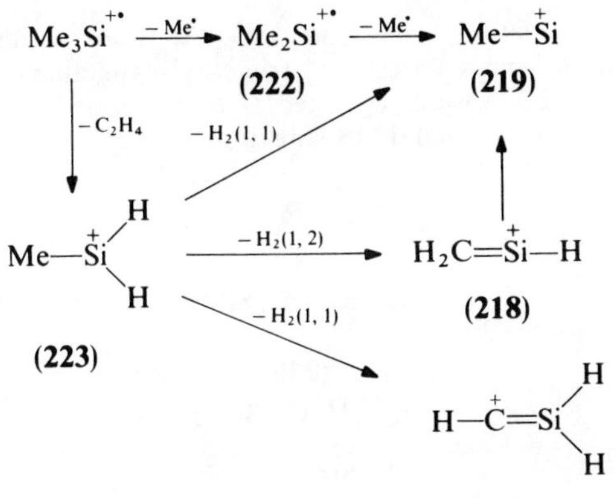

(86)

The substituted silavinyl cation **225** is suggested[61a] to be formed from the molecular ion of **224** via a four-centre elimination process (reaction 87). A re-investigation[61b] revealed that (i) ion **225** is formed to a very minor extent only from **224** and (ii) no metastable ion due to the loss of $(C,H_3,O)^{\cdot}$ from **224** could be detected. Thus, the interesting question as to the nature of $(C,H_3,O)^{\cdot}$, i.e. $CH_3O^{\cdot}$ versus $CH_2OH^{\cdot}$, and the mechanism of the process **224** → **225** remains undecided. Loss of C_2H_4 from the $[M{-}Me]^+$ ion **226** (generated upon electron impact of allylsilanes[117]) is believed to yield **227** (reaction 88). There cannot be any doubt that both processes must follow quite complex pathways.

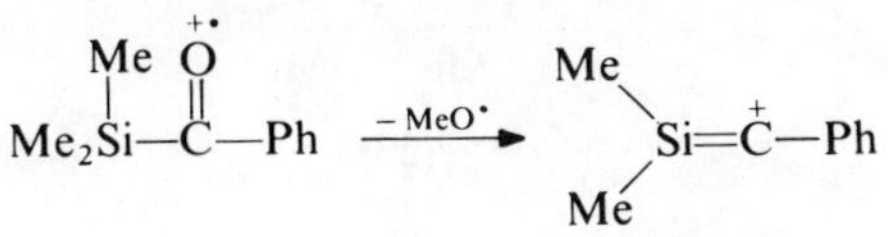

(87)

(224) (225)

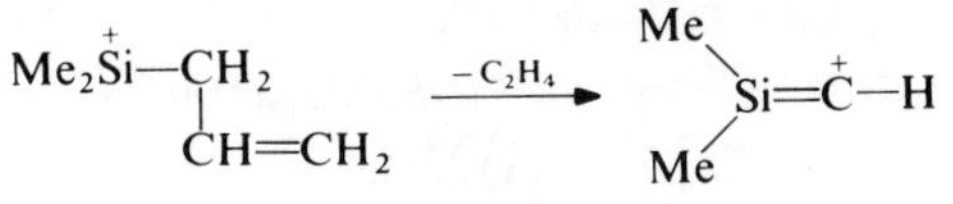

(88)

(226) (227)

Vinyl cation analogues with two silicon atoms were suggested[112] as products formed (in low intensity) upon electron bombardment of linear organopolysilanes. For example, the disilaethene **196** (and **210**: R = Me) undergoes loss of Me· (evidenced by a metastable ion peak) to generate **228** (reaction 89).

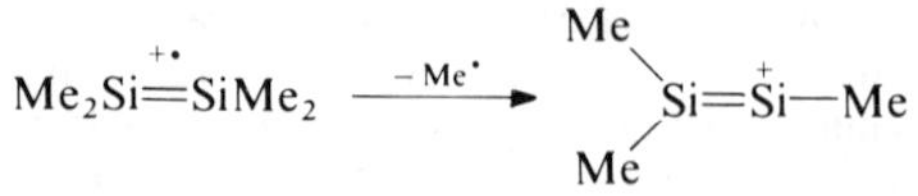

(**196**) (**228**) (89)

The silaallyl cation structures **229**, **230** and **231** were assigned[118,119] to the corresponding ions in the mass spectra of trimethylsilylmethanes, and the silicon-containing cations **232** and **233** were suggested[119] to be formed via complicated pathways (see Section III.F) from ionized TMS-substituted chlorosilanes.

(**229**)
R = H, alkyl, Cl, Me_3Si

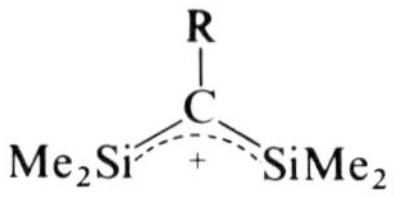

(**230**)
R = H, Cl, $SiMe_3$

(**231**)

$H_2Si{=}CH{-}\overset{+}{Si}{=}CH_2$

(**232**)

$Me_2Si{=}CH{-}\overset{+}{Si}{=}CH_2$

(**233**)

The silicon analogues of the acylium and thioacylium ions are believed to be generated under electron-impact ionization. For example, $CH_3\overset{+}{Si}{=}O$ is suggested to be formed via reactions 90[120] and 91[102], and $CH_3\overset{+}{Si}{=}S$ from **204**[106] (reaction 92).

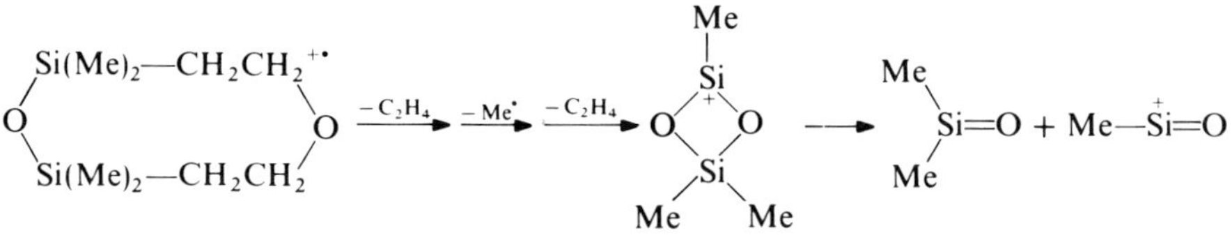

(**234**) (**235**) (**12**) (**236**)

(90)

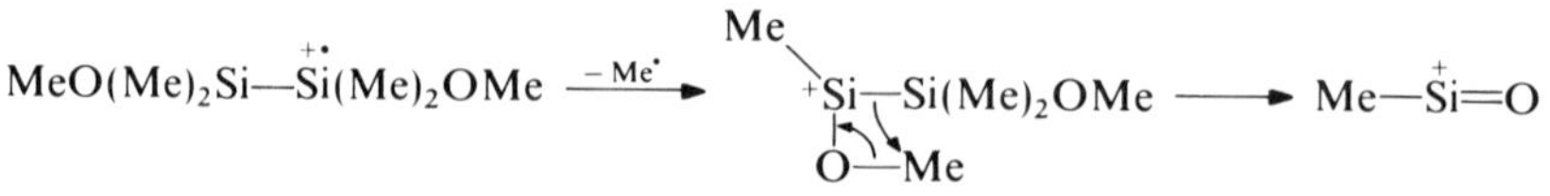

(**237**) (**238**) (**236**)

(91)

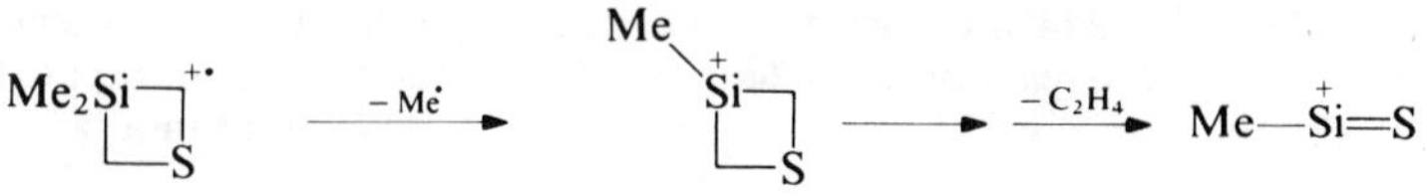

(204) (239) (240)

(92)

F. Si–X Multiple Bond Containing Neutrals Generated in Ionic Dissociation Reactions

'Multiple Bonding to Silicon' is the title of a fascinating review[8] in which the relatively recent chemistry of compounds containing multiple bonds to silicon is exhaustively discussed. In this sub-section we will mention very briefly some ionic systems the decomposition pattern of which is in accord with the assumption that the *neutrals* formed in gas-phase dissociation of ions may contain multiple bonds to silicon. It should be mentioned, however, that the evidence presented is in most cases circumstantial. A few examples have already been mentioned in reactions 54, 79 and 90 of this chapter, and the neutrals discussed were the silicon analogue of acetone, i.e. Me_2SiO (**12**) and 1,1-dimethylsilene $Me_2Si{=}CH_2$ (**13**).

The silicon analogue of acetone, **12** is also suggested to be formed in the dissociation of the oxonium ion **44** to generate Me_3Si^+ (reaction 93)[57, 121]. Similarly, fragment ions generated from disiloxanes[122] and β-ketosilanes[93] are believed to yield **12** when decomposing further.

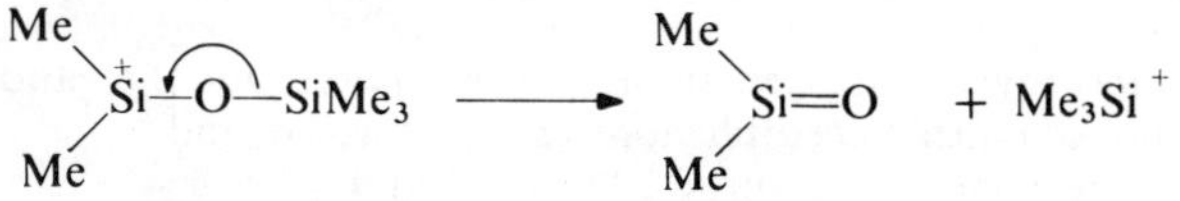

(44) (12)

(93)

The elimination of Me(Et)Si=O (**242**) is suggested to take place from ion **241** (reaction 94); the latter is formed via loss of Me· from **127** (see reaction 55)[83].

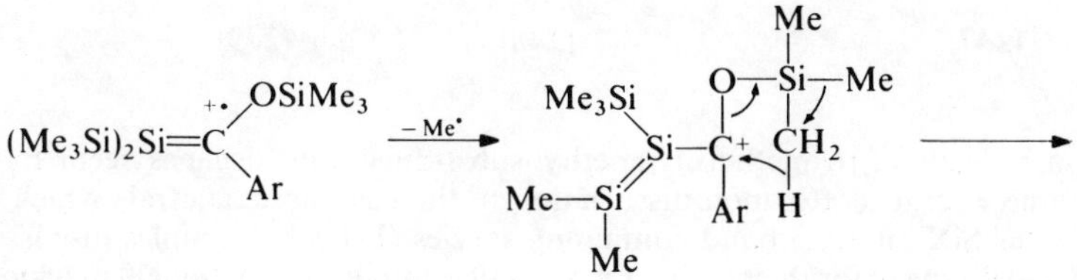

(127) (241)

(94)

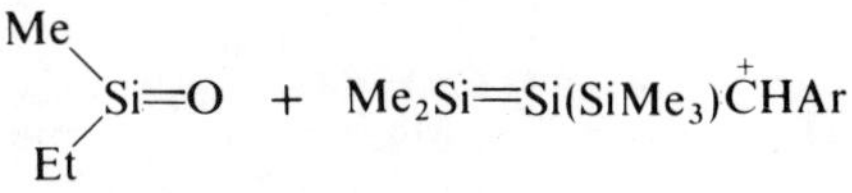

(242) (243)

Loss of neutral Et_2SiO (**244**) is reported[58] to occur when **45** decomposes to generate the $C_7H_7^+$ ion. However, as the reaction **45** → **244** + $C_7H_7^+$ (reaction 95) is not supported by a metastable peak other conceivable routes to generate the $C_7H_7^+$ ion cannot be ruled out.

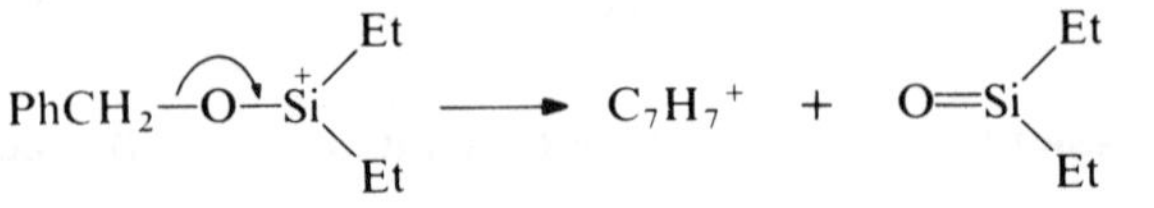

(95)

(**45**) (**244**)

The McLafferty analogous elimination of **245** from **122** was postulated[82] to generate the acetone cation radical **246** (reaction 96).

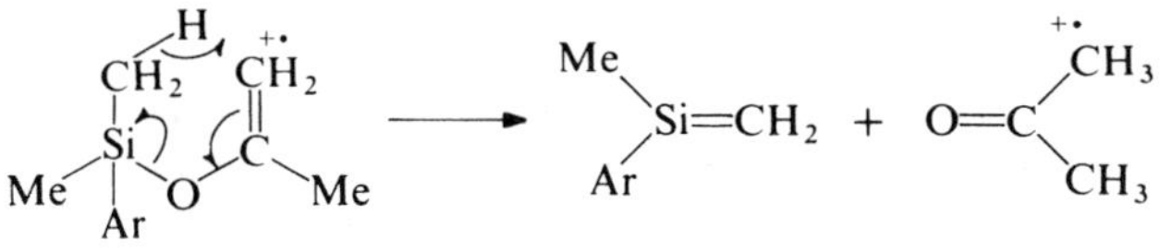

(96)

(**122**) (**245**) (**246**)

Another McLafferty-type hydrogen transfer was reported for the cation radical **247** from which the enol of ionized acetophenone (**249**) is formed, and it was suggested that the neutral may correspond to the silene **248** (reaction 97)[61].

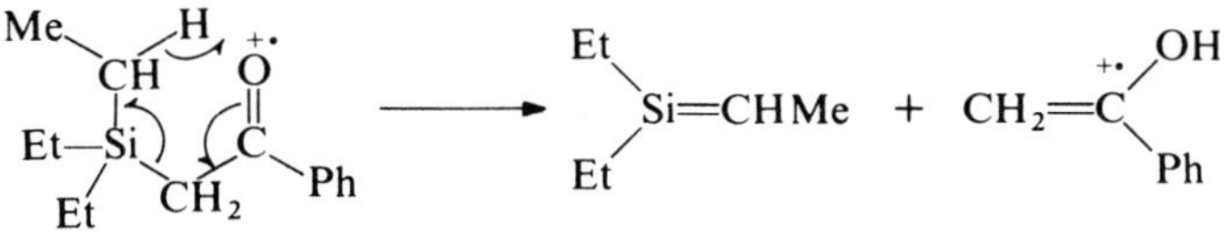

(97)

(**247**) (**248**) (**249**)

The cation radicals of (trimethylsilyl)methyl-substituted chlorosilanes seem to be a convenient source to generate upon dissociation in the gas phase, neutrals which were interpreted[119] as SiX multiple bond containing species. The few examples discussed in reactions 98 and 99 speak for themselves. Of particular interest is reaction 99 in which the generation of silaallene (**256**) is postulated.

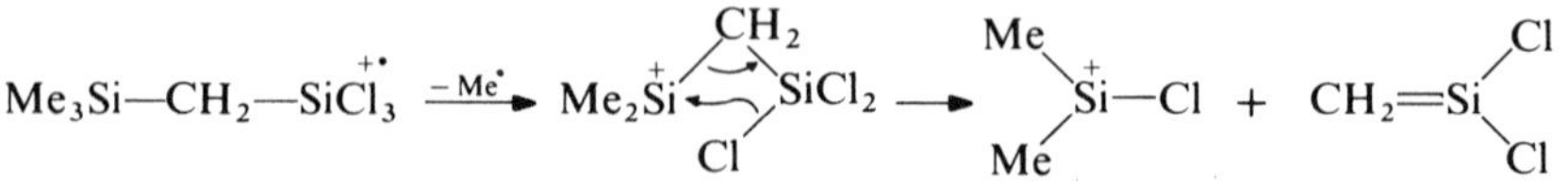

(98)

(**250**) (**251**) (**59**) (**252**)

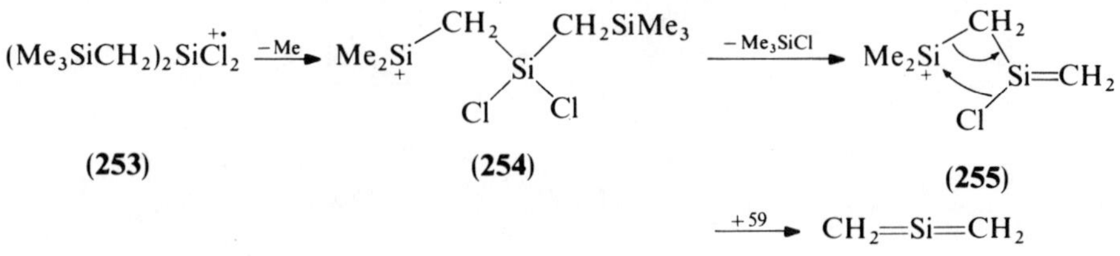

(253) (254) (255) (256)

(99)

G. The Generation of Ionic and Neutral Silylenes via Fragmentation Processes

Divalent silicon species (both neutral and ionic ones) have already been addressed in reactions 15, 18, 63, 86, and also in the section on Thermochemistry. In addition to these the following three examples are illustrative. The extrusion of SiF_2 and $SiCl_2$ from ionized dihalosilanes **257** was reported[123]. Structural requirements for reaction 100 to occur are that R and R′ correspond to either aryl or vinyl groups. If there is one alkyl group present, the reaction is completely suppressed.

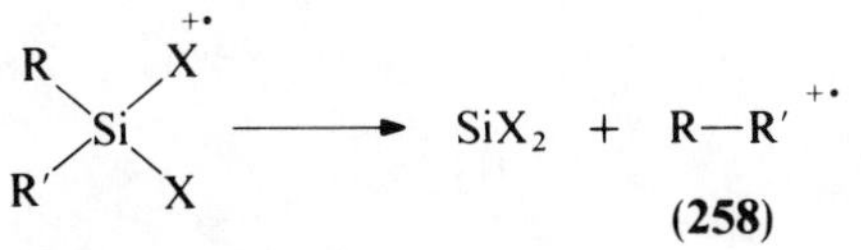

(257) (258) (100)

(X = F, Cl; R, R′ = aryl, vinyl)

The loss of neutral dimethoxysilylene $Si(OMe)_2$ was postulated[101] to connect the $[M - Me]^+$ ion, formed from ionized octamethoxytrisilanes with some daughter ions. No studies were reported which would substantiate this suggestion.

A highly unusual gas-phase reaction was reported[63] for the molecular ion **259**, which undergoes competitive losses of $Si(CD_3)_3Me$ and $SiMe_3CD_3$ in the approximate ratio of 2:1. From the so-formed intermediates, $SiMe_2$ (**8**) and its D_6-labeled analogue $Si(CD_3)_2$ (**8a**) are split off in the ratio 100:1 to eventually generate ionized methylenefluorene. The different rates for the losses of **8**/**8a** were interpreted in terms of stabilities of the intermediates **260** and **261** formed (reaction 101).

IV. ION/MOLECULE REACTIONS OF NEUTRAL AND IONIZED SILICON-CONTAINING MOLECULES

A. Positively Charged Ions

In the preceding sections we have repeatedly addressed ourselves to ion/molecule reactions of silicon-containing species occurring in the gas phase (reactions 2–6, 9, 10, 13–15 for positive ions, and 16–22 for negatively charged ions), and the excellent papers published by Lampe[6], DePuy[44] and Bowie[44, 124] may serve as leading review articles on this important subject. In this section we will focus on a few selected topics, covering quite recent fundamental studies on elementary processes of Si^+ with some simple molecules as well as processes being slightly more complex in nature.

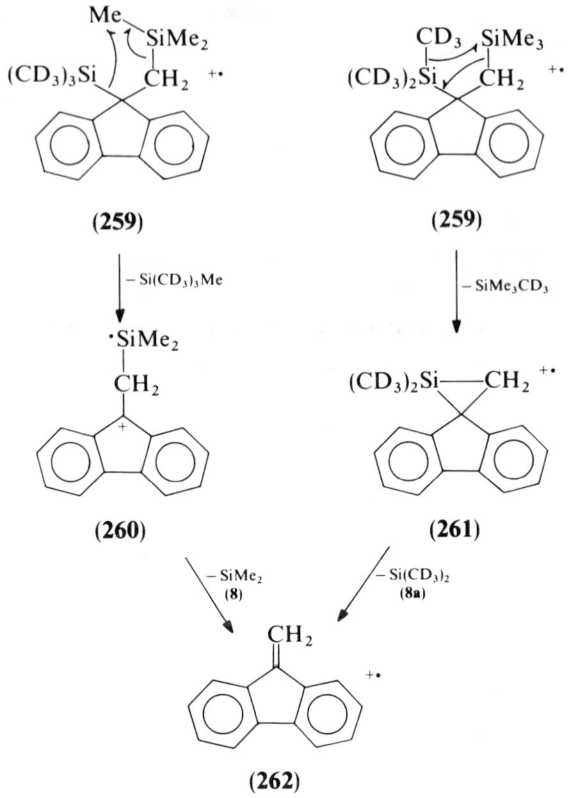

(101)

The reaction of ground state silicon ion (^2P) with silane was recently studied with great detail by Boo and Armentrout[125] using a guided ion beam tandem mass spectrometer. From this study not only a critical re-evaluation of the heats of formation of many $Si_mH_n^+$ ions ($m = 1, 2$; $n = 0, 1, 2, 3$) was possible; more interestingly, quite spectacular processes were uncovered. Among them was an isotope exchange reaction (reaction 102) involving four hydrogen shifts.

$$^{28}Si^+ + {}^{29}SiD_4 \rightarrow {}^{29}Si^+ + {}^{28}SiD_4 \qquad (102)$$

A mechanism which is in accord with the experimental findings, and in particular with the observation that a major branching reaction in the ion/molecule reaction of Si^+/SiH_4 is the formation of $Si_2H_2^+$, is suggested in reaction 103. The first step in the reaction of Si^+ with SiH_4 is presumably oxidative insertion of Si^+ in a SiH bond to form **263**. Ejection of *Si in the isotope exchange reaction 102 requires that **263** undergoes intramolecular hydrogen transfer yielding **265**. Decomposition of the latter can then occur by elimination of SiH_4, the microscopic reversal of the initial insertion step. The isomerization **263**$\rightleftharpoons$**265** almost certainly involves the disilene ion **264** since concerted hydrogen atom transfers, of the kind of dyotropic processes[126], seem unlikely. For the neutral analogue of **263** it has been suggested[127] that the silylsilylene$\rightleftharpoons$disilene rearrangement by H-migration occurs with only a small energy barrier. Dissociation of $Si_2H_4^{+\cdot}$ into $Si_2H_2^{+\cdot}$ and H_2 is the most efficient process at low kinetic energies of Si^+, accounting for over

90% of the products. This product could arise via a 1,1-elimination from **263** (or **265**) to form **266**. An alternative $Si_2H_2^{+\cdot}$ isomer, **267**, could be formed by a 1,1-dehydrogenation of **264**. Alternate pathways involve 1,2-elimination of **264** to form **266**, or of **263** (or **265**) to yield **267**. As these four-centre elimination processes are (formally) symmetry-forbidden they should have relatively high energy barriers, and the authors favour the 1,1-elimination mode. However, a more detailed high level *ab initio* MO calculation would be required to settle this mechanistic dichotomy once and for ever.

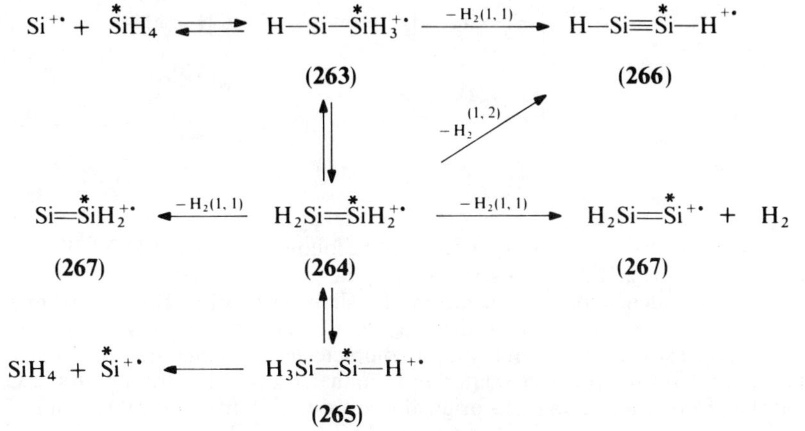

(103)

The reactions of $Si^{+\cdot}$ with H_2O and O_2 and that of $SiO^{+\cdot}$ with H_2 (D_2), which are believed to be involved in the $Si^{+\cdot}$ depletion in the earth's atmosphere above 90 km, were studied in great detail[128]. It was found that reaction 104 accounts for the major loss of $Si^{+\cdot}$ in atmosphere above ca. 90 km, while the association of $Si^{+\cdot}$ with O_2 is the dominant loss process for $Si^{+\cdot}$ below 90 km in the earth's atmosphere. Interestingly, reaction 105 generates initially $SiO_2^{+\cdot}$ in an excited state which undergoes charge transfer with oxygen before relaxation, thus yielding SiO_2. The reactions of $SiO^{+\cdot}$ with H_2 (and D_2) were also found to occur (reaction 106) producing $SiOH^+$. Deprotonation of the latter is suggested to account for the interstellar generation of SiO.

$$Si^{+\cdot} + H_2O \rightarrow SiOH^+ + H^\cdot \qquad (104)$$

$$Si^{+\cdot} + O_2 \rightarrow SiO_2^{+\cdot} \rightarrow SiO_2 \qquad (105)$$

$$SiO^{+\cdot} + H_2 \rightarrow SiOH^+ + H^\cdot \qquad (106)$$

Out of the numerous, elegant ion/molecule reactions of gaseous silicon-containing ions, reported by the Lampe school[6], we should like to confine ourselves to just two examples, i.e. the reaction of SiH_3^+ with ethylene[129] and benzene[130]. Reactions of SiH_3^+ with C_2H_4 ultimately form alkylsilicenium ions, namely $C_2H_5SiH_2^+$, $(C_2H_5)_2SiH^+$ and $(C_2H_5)_3Si^+$. (For a theoretical comparison of the relative stabilities of the various isomers of these species using *ab initio* MO methods, see Ref. 55.) A possible route for the generation of $C_2H_5SiH_2^+$ is suggested in reaction 107. Ion **271** may then serve as a partner to further react with C_2H_4, etc.

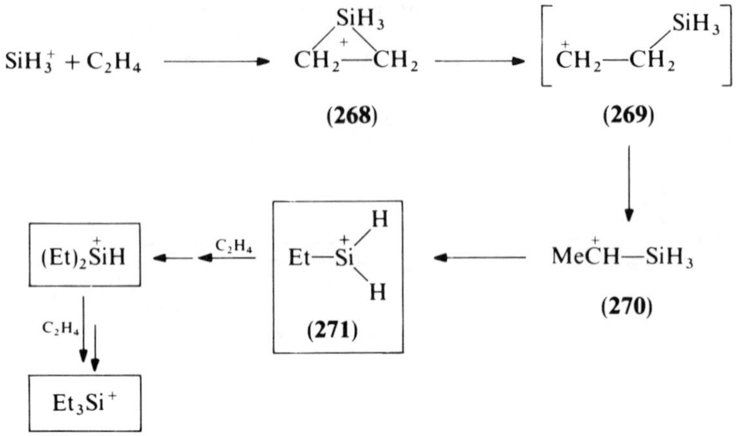

(107)

The ion/molecule reaction of SiD_3^+ with benzene followed by loss of HD (70% of the total H_2 of the total H_2, HD, D_2 losses) from the encounter complex (reaction 108) was interpreted[130] as evidence for the generation of a silabenzyl cation **273**. The latter was also proposed[131] to be the main product in the EI mass spectrum of phenylsilane. However, a re-investigation[132] employing a complete set of deuteriated phenylsilanes leaves no doubt that the hydrogen (deuterium) eliminated originates from both the side-chain and the phenyl ring, and not, as originally suggested[131], from the SiH_3 group. The data[132] which also reveal a kinetic isotope effect $k_H/k_D = 1.63$ [being comparable to that observed for the losses of H/D from ionized toluene (1.58)[133], benzyl cyanide (1.52)[134] and γ-picoline (1.52)[135]] can be interpreted in several ways: (i) H/D loss is preceded by scrambling processes between ring and side-chain hydrogen/deuterium atoms; (ii) the molecular ions of phenylsilane prior to dissociation rearrange to sila-cycloheptatriene analogues, and (iii) ring-opening to silaacetylenes may occur. A distinction is not possible for the time being.

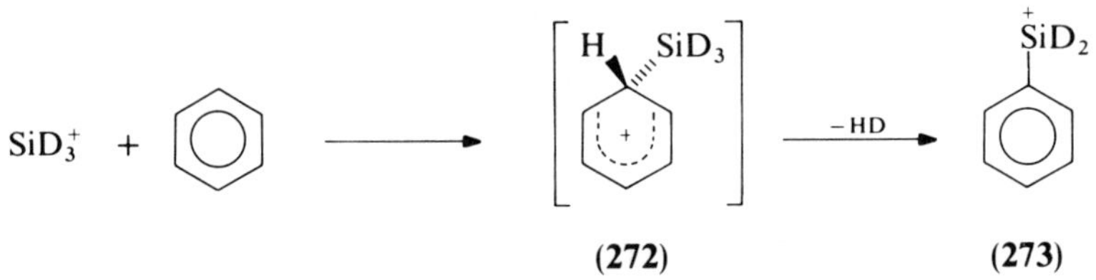

(108)

Detailed studies on the reactivity of the trimethylsilyl cation with nucleophiles were performed by Bowie and coworkers. Out of the many systems investigated the following should be discussed briefly.

Under ICR conditions Me_3Si^+ produces with alcohols ROH a 1:1 adduct[136]. Depending on the nature of the alkyl residue R the adducts fragment by loss of either methane (reaction 109: **274**→**275**) or olefins (**276**→**277**). The study of labeled precursors demonstrated from which site of the adducts **274** and **276** methane and $R'CH{=}CH_2$ are eliminated, and it was observed that both reaction channels are highly specific. Moreover, loss of $R'CH{=}CH_2$ from **276** yields **277** and not **278** as the reaction product undergoes a substitution reaction with ROH to regenerate **276** (by expulsion of H_2O);

this behaviour is typical for protonated trimethylsilanol **277** but not for the silicon-protonated isomer **278**.

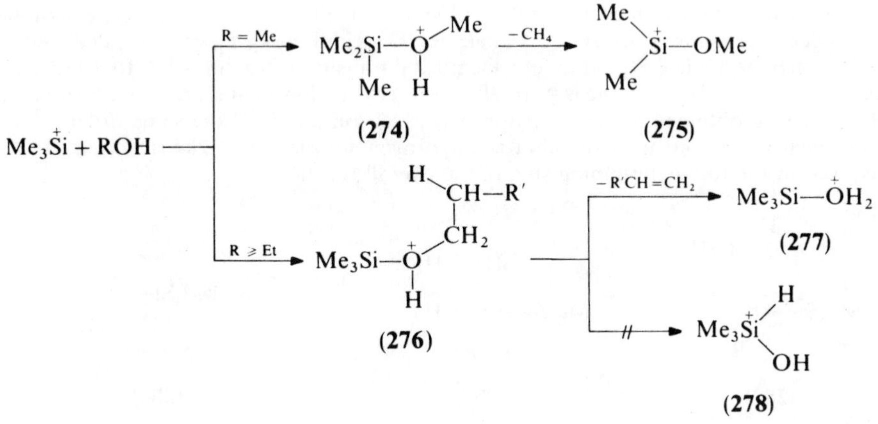

(109)

The 1:1 adduct formed in the nucleophilic attack of a dialkyl ether (ROR′) at the electrophilic silicon centre of Me_3Si^+ also decomposes by elimination of a neutral fragment[137]. When R > Me, the adduct undergoes, via a four-centre mechanism, elimination of an olefin, and when R′ > R the adduct eliminates the larger olefin preferentially. A typical example is given in reaction 110. The 1:1 adduct **279** formed from Me_3Si^+ and ethyl *n*-propyl ether eliminates propene followed by loss of ethylene. The reverse sequence, i.e. elimination of C_2H_4 followed by loss of C_3H_6, is noted, but it is a minor process.

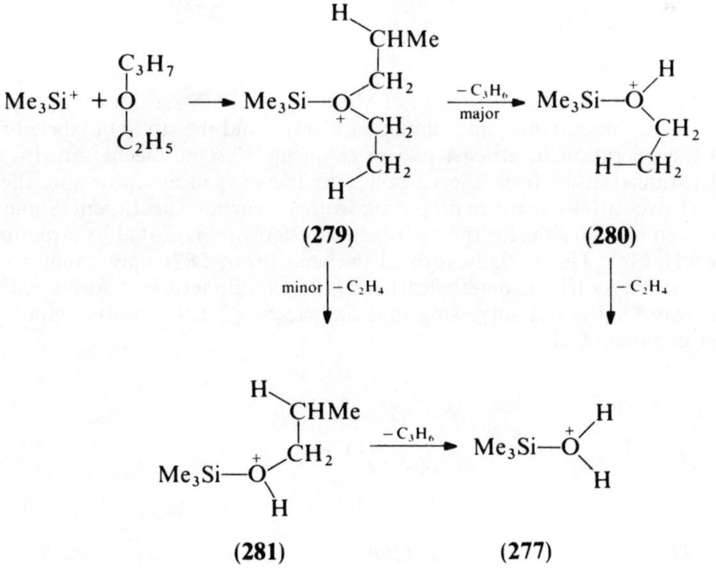

(110)

From the reaction of Me_3Si^+ with labeled ethyl methyl and diethyl ethers[137] it was concluded that the loss of ethylene from the 1:1 adducts is not only site-specific (four-centre elimination) but is also associated with a kinetic isotope effect k_H/k_D which, depending on the internal energy of the adduct ions, varies between 1.39 and 1.8. Isotope effects decrease as the A . . . H . . . A angle decreases and the k_H/k_D value is calculated to be 2.3 when the angle is 90° (as in four-membered transition structures)[138]. In *asymmetrical* systems A ··· H ··· B the isotope effect will be even lower, and thus the upper value of $k_H/k_D = 1.8$ obtained for the reaction of adduct ion **282** is of the same order as the theoretical value, and indicates that the hydrogen transfer **282**→**284** (reaction 111) is involved in the rate-determining step of the overall reaction.

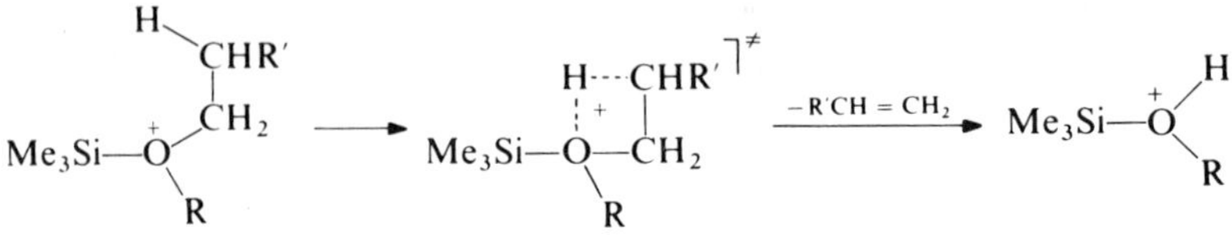

(111)

(282) **(283)** **(284)**

An interesting variant of the elimination process from adducts like **282** was observed when R′ = F. The expected loss of fluoroethylene (**282**→**284**) does *not* take place; rather, loss of Me_3SiF is observed[137]. Although the structure of the ionic fragment for loss of Me_3SiF from **285** could not yet be elucidated, the driving force for the reaction is undoubtedly the strong SiF bond formed. A conceivable process (in which ion/dipole stabilized intermediates could well be involved) is suggested in reaction 112.

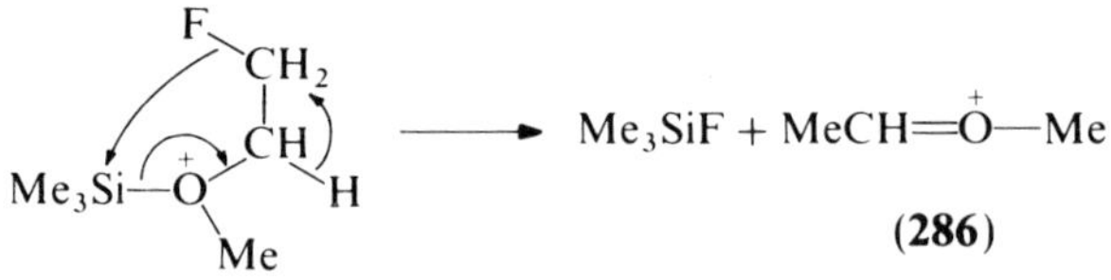

Me_3SiF + MeCH=$\overset{+}{O}$—Me (112)

(285) **(286)**

The 1:1 adduct formed in the reaction of Me_3Si^+ with both 2,2-dimethylpropan-1-ol and 1-methoxy-2,2-dimethylpropane eliminates C_5H_{10}, and the study of labeled precursors reveals the operation of at least two mechanisms[139]. One mechanism involves a formal 1,2-hydride transfer from the carbon α to the oxygen acceptor site; the other mechanism involves a 1,4-hydrogen migration from a γ-carbon substituent. Some plausible yet unproven mechansims for the hydrogen transfers are indicated by structures **287** and **288** (R = H, Me). The initially formed carbene (from **287**) may isomerize to 2-methylbut-2-ene[140]. As trifluoromethylcarbene ($CF_3\ddot{C}H$) undergoes fluorine rearrangement only slowly[141] it is not surprising that the Me_3Si^+/2,2,2-trifluoroethanol adduct **289** does not eliminate C_2F_3H.

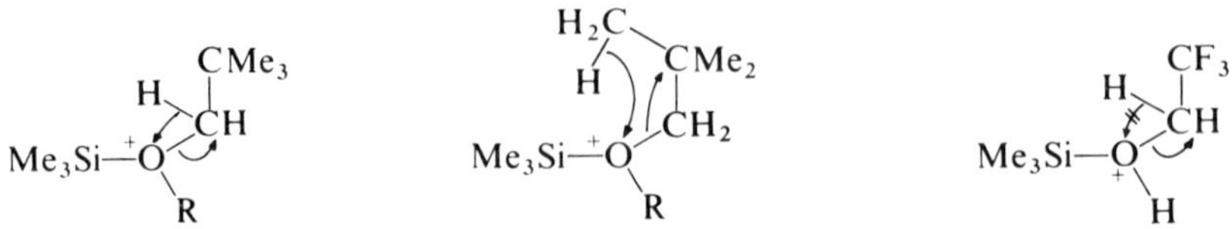

(287) **(288)** **(289)**

Allyldimethylsilyl cations generated via loss of an allyl radical from ionized diallyldimethylsilane, react under ICR conditions with alcohols ROH to produce adducts **290** which undergo a quite remarkable reaction. A 1,5-hydrogen shift, by which the hydrogen migrates from the OH group to the allylic double bond, brings about loss of C_3H_6 to yield the product ion **68** (reaction 113)[142]. The basic pattern which usually governs the gas-phase chemistry of Me_3Si^+ with alcohols (reaction 109) is completely absent, as fragmentation is specifically directed through the allyl group. The corresponding adducts from allyl methyl ether and dialkyl ethers, however, first lose olefin, as do the Me_3Si^+/ether adducts (reaction 110), and then undergo the elimination described in reaction 113.

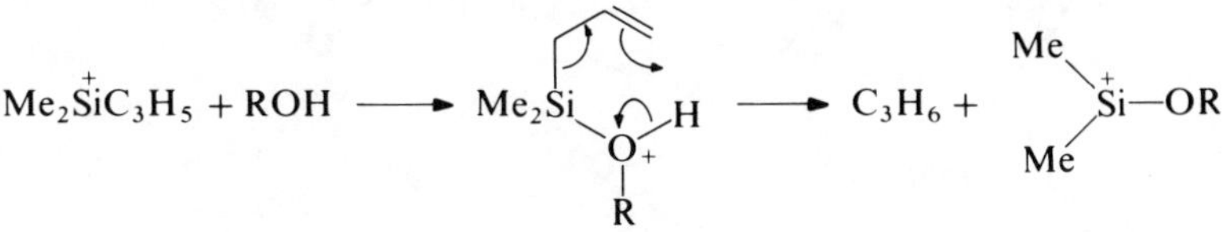

(113)

(290) **(68)**

Characteristic fragmentation pattern were observed for the adducts formed from Me_3Si^+ and alkenyl ethers[143]. In the vinyl ether/Me_3Si^+ system (**291**) a loss of olefin is observed; the vinyl substituent is not involved in the fragmentation (reaction 114).

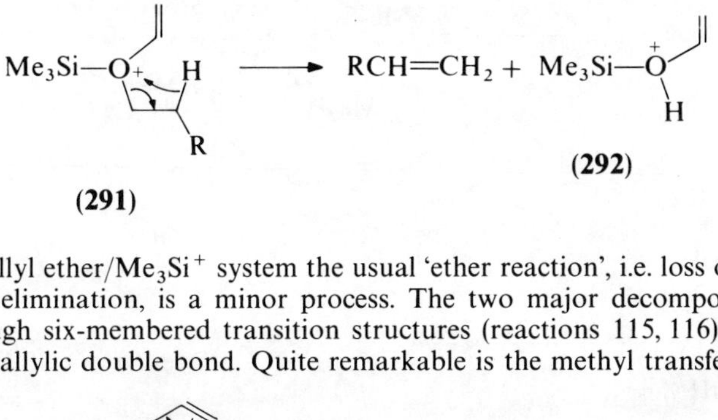

(114)

(291) **(292)**

For the allyl ether/Me_3Si^+ system the usual 'ether reaction', i.e. loss of an olefin via a four-centre elimination, is a minor process. The two major decomposition pathways occur through six-membered transition structures (reactions 115, 116) and, obviously, involve the allylic double bond. Quite remarkable is the methyl transfer **293**→**295**.

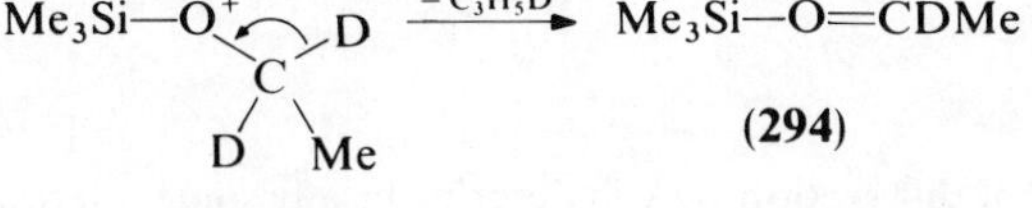

(115)

(294)

(293)

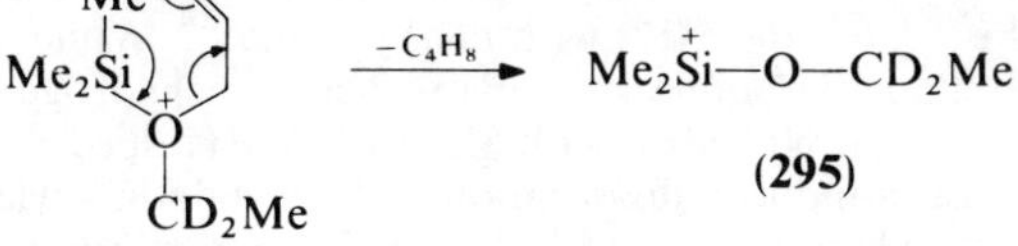

(116)

(295)

(293)

In a combined experimental/*ab initio* MO study, Bowie and coworkers[144] addressed the questions concerning the reaction site of ambident nucleophiles with Me_3Si^+. It was demonstrated that the reaction of Me_3Si^+ with esters, thioesters, thionoesters, anhydrides and thioanhydrides of the general formula A–C(Y)–X–B (A, B: alkyl; B = CO–alkyl; X, Y = O, S) two possible adducts **296** and **297** of comparable energies are formed. From the product observed under ICR conditions it was concluded that structure **297** serves as precursor for dissociation reactions; no firm evidence was available to support the intermediacy of **296** in the decomposition chemistry of the encounter complexes.

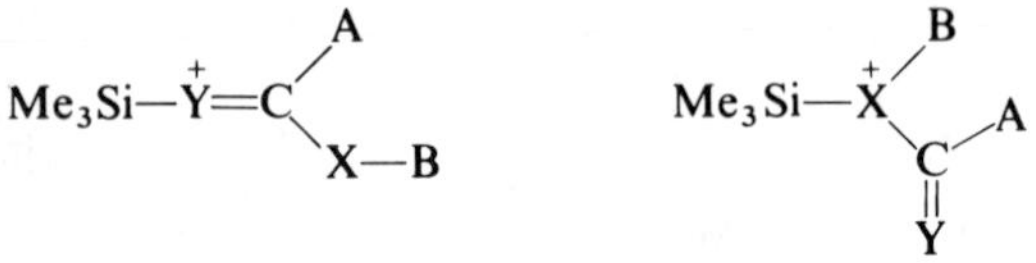

(296) **(297)**

ICR was also used[145] to probe the gas-phase chemistry of ions $Me_2RSiOSiMeR^+$ (R = H, Me) with oxygen-containing compounds. It is no surprise to observe that the chemistry is dictated by the high affinity of the silicenium ion for oxygen, which is manifested by a large variety of addition/elimination processes. A few representative examples are given in reaction 117.

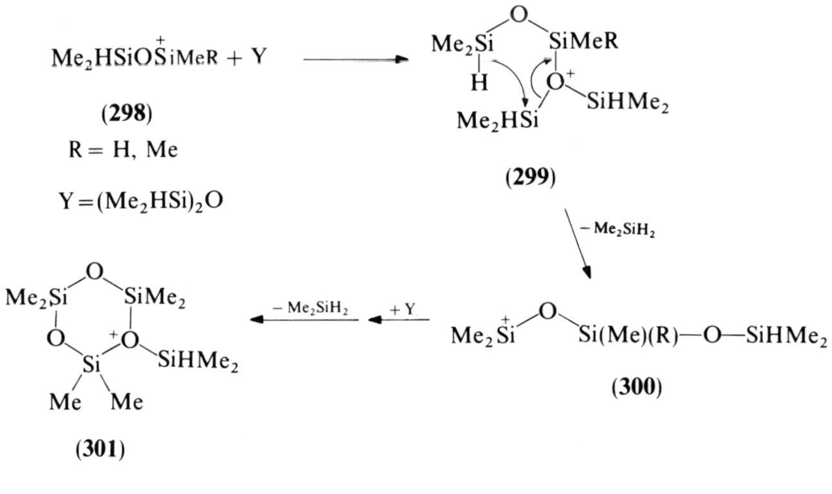

(117)

In the remaining part of this section we will describe briefly some interesting ion/molecule reactions of electrophiles with *neutral* silicon-containing molecules.

Similar processes to those described in reactions 102 and 103 for the insertion of $Si^{+\cdot}$ in a SiH bond were recently described in great detail for the reactions of transition-metal ions (Ti^+, V^+, Cr^+, Fe^+, Co^+ and Ni^+) with methylsilane[146]. While V^+, Cr^+ and Fe^+ are all unreactive with silane, Ti^+ undergoes an exothermic dehydrogenation to generate $Ti{=}SiH_2^+$. Similarly, Co^+ and Ni^+ form with SiH_4 a $Y{=}SiH_2^+$ species, whose structures were confirmed by collision-induced dissociations. The metal ion-silylene bond energies were found to span a range between 61 and 72 kcal mol^{-1}. Correlation between this bond energy and the electronic structures of the metal ion, M^+, supports a bonding scheme in which silylene donates its non-bonding lone pair electrons to an empty 4s orbital of the metal centre. For Co^+ and Ni^+, back-donation of paired 3d electrons from the metal ion into the empty 3p orbital on silicon is suggested to account for the relatively strong bond deduced for some of the metal ions studied. A mechanism for the reactions of M^+ with silane is given in reaction 118.

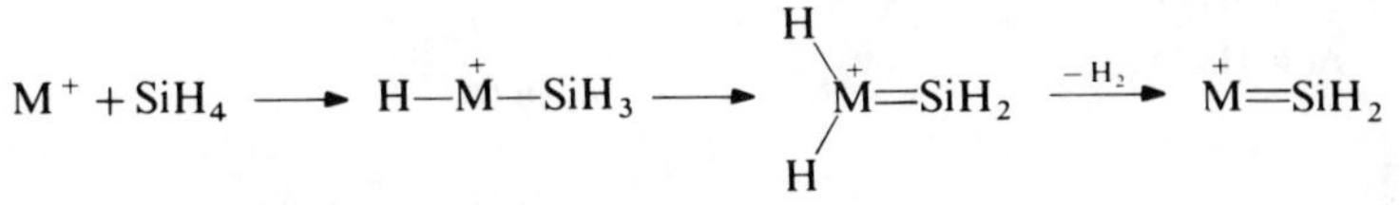

(118)

(302) (303) (304)

Protonation of simple silyl enol ethers, as for example 2-(trimethylsiloxy)propene **305**, has been studied using ICR[147]. From the careful analysis the following conclusions were drawn: (i) Protonation of **305** occurs exclusively at the β-carbon atom; the species formed (**306**) is identical with the adduct between Me_3Si^+ and acetone (reaction 119). (ii) **305** has a proton affinity (PA) which is 16 kcal mol^{-1} higher than that of NH_3. (iii) The reactivity of **306** suggests that most of the positive charge (63%) is located at the Me_3Si group, which is in agreement with *ab initio* calculations.

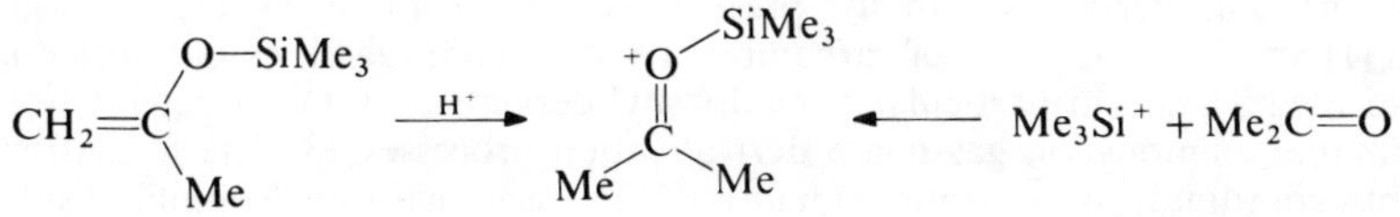

(119)

(305) (306)

The gas-phase acid-induced ring opening of 1,2-epoxy-1-trimethylsilylcyclohexane **307** using H_2O and MeOH as nucleophiles revealed a close parallel to the reaction of **307** in solution. The most relevant results from this interesting study[148] are that, irrespective of the acid A^+ and the nucleophile Nu used, the displacement takes place exclusively at the atom α to silicon with complete inversion of configuration (reaction 120). No evidence for the formation of other conceivable products, which would for example arise from nucleophilic displacement at the Si atom, has been obtained. A 'directing' effect of the trimethylsilyl group (brought about by co-ordination of the nucleophile by the Me_3Si group due to electrostatic interactions) was suggested to be responsible for the high regio- and stereo-chemistry of the reaction.

B. Negatively Charged Ions

Although it is true that there are substantially fewer investigations of the reactions of negative silicon-containing ions than of positive ions, in recent years an opposite trend is clearly discernible in the literature. This is most likely due to an increased interest in the chemistry of silyl-substituted anions as well as to the availability of more sophisticated techniques, in particular ICR and flowing afterglow, which permit the study of these in solution-elusive species. In fact, the quite numerous studies performed over the last 10 years have already led to several review articles on this subject[6, 44, 124, 149]. In the present chapter we already have referred to silicon-containing anions on several occasions, and have in particular discussed the following aspects: hydride affinities of silanes together with the interesting problem of pentavalent SiR_4H^- species (R = H, alkyl), the gas-phase

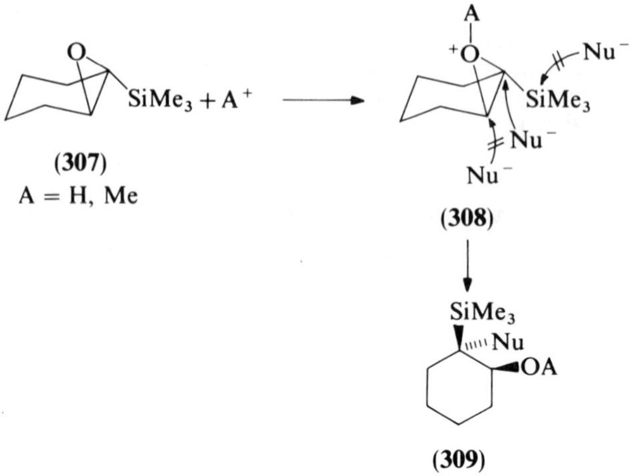

(120)

acidities of organosilanes and the intramolecular migration of the Me_3Si group from an $OSiMe_3$ or CO_2SiMe_3 substituent to an NO_2 function in the radical anions of *ortho*-substituted benzene derivatives. In the present section the following aspects will be discussed: (1) Specific generation of carbanions via gas-phase reactions of fluoride ions with substituted silanes, in particular trimethylsilyl derivatives. (2) Characterization of silicon-containing anions via gas-phase derivatization processes. (3) The formation of stable penta-coordinate silicon intermediates. (4) The gas-phase nucleophilic displacement reactions on silicon.

The specific generation of carbanions in the gas phase by proton abstraction is associated with several problems. In the first place, even simple organic hydrocarbons may have more than one acidic site, and deprotonation may therefore give rise to a mixture of isomers. For example, proton abstraction by OH^- or NH_2^- from propyne **310** generates both the propargyl (**311**) and the methylacetylide anion **312**[150] (reaction 121).

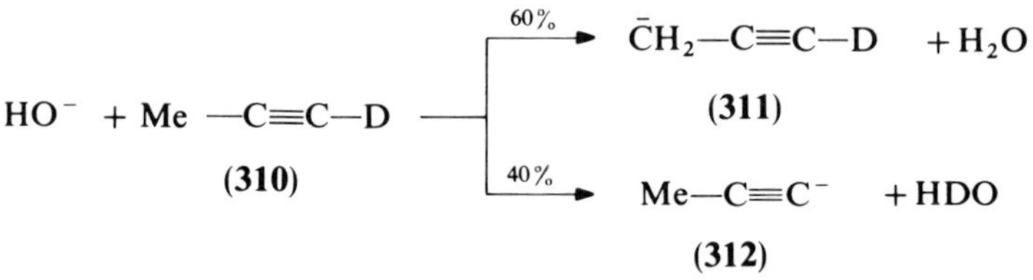

(121)

A second difficulty may result from the fact that the carbanion and the neutral molecule produced in the proton transfer do not always separate immediately[150, 151]. Reprotonation and reabstraction may occur, thereby generating mixtures of isomeric carbanions even if one starts from a symmetrical neutral precursor. A nice example is given in reaction 122.

These and some other problems were overcome by the beautiful method introduced recently by DePuy and coworkers[152], who use the high affinity of fluoride anion to silicon. As described in the general reaction 123 the reaction of trimethylsilyl derivatives with F^- generates efficiently in a regiospecific way carbanions, and the examples given

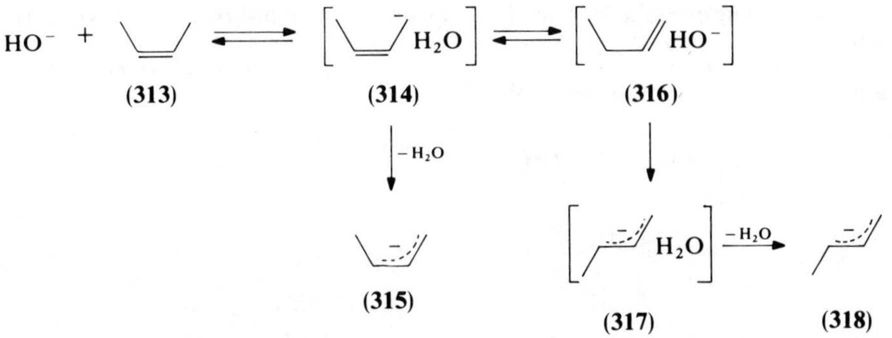

(122)

below may suffice to demonstrate the wide applicability of this new entry to gas-phase anions.

$$\mathrm{R{-}SiMe_3 + F^- \longrightarrow R^- + Me_3SiF} \qquad (123)$$

The driving force for reaction 123 presumably is the formation of the extremely strong silicon–fluoride bond (154 kcal mol^{-1})[8,152] which compensates for the conversions of a high electron affinity radical [EA(F) = 79.5 kcal mol^{-1}][153] to one with low electron affinity [EA(R$^{\cdot}$) < 10–15 kcal mol^{-1} for stabilized radicals like allyl][154]. Carbanions generated via reaction 123 include **311**, **312**, **315** and **318**–**321**. The isomers of $C_3H_5^-$ (i.e. **319**, **322**–**324**) as well as **325** were generated by the sequence described in reaction 124[155], in which NH_2^- was used to replace the Me_3Si function; the resulting carboxylate anion was decarboxylated by collisional activation. The method was also applied to generate in the gas-phase the acyl anion **326** which is not accessible by proton abstraction from acetaldehyde (reactions 125, 126)[156].

$\mathrm{CH_2{=}CH{-}\bar{C}H_2}$ (**319**)

$\mathrm{C_6H_5CH_2^-}$ (**320**)

$\mathrm{R{-}C{\equiv}C^-}$ (**321**) R = Et, n-C_4H_9

$\mathrm{Me{-}CH{=}CH^-}$ (**322**)

$\mathrm{CH_2{=}\bar{C}{-}Me}$ (**323**)

(**324**)

$\mathrm{CH_2{=}CH^-}$ (**325**)

$$\mathrm{R{-}CO_2{-}SiMe_3 \xrightarrow[-\,Me_3SiNH_2]{NH_2^-} R{-}CO_2^- \xrightarrow{-\,CO_2} R^-} \qquad (124)$$

$$\mathrm{Me{-}CHO + OH^- \longrightarrow \bar{C}H_2{-}CHO + H_2O} \qquad (125)$$

$$\mathrm{Me{-}C({=}O){-}SiMe_3 + F^- \longrightarrow Me{-}\bar{C}{=}O + Me_3SiF} \qquad (126)$$

(**326**)

One intriguing extension of reaction 123 involves the competitive displacement of two substituents, and preliminary experiments[152] indicate the preferential displacement of the less basic anion. A typical example is given in reaction 127 in which benzyl anion **327** is displaced in favour of the allyl anion **319**.

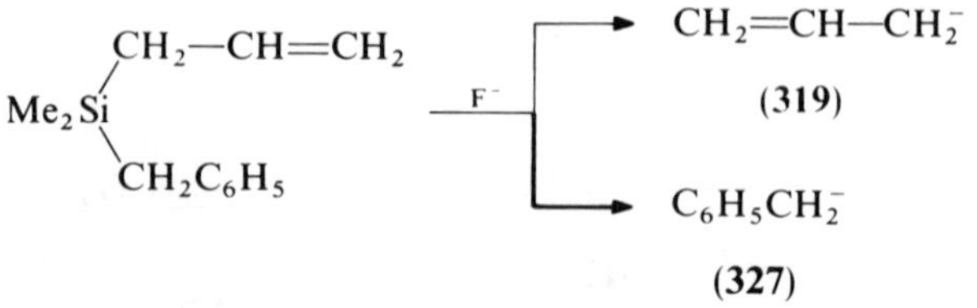

(127)

Several of the anions generated via reaction 123 or other processes are isomeric and it is not easy to distinguish them and to characterize them by using ordinary mass-spectrometric methodologies. DePuy and Damrauer[149] have developed a chemical method to determine the structure of these isomeric anions (including silyl anions, silicon-stabilized carbanions and pentacovalent silicon anions) by using reaction with nitrogen oxide (N_2O) under the conditions of flowing afterglow mass spectrometry. As a case in point we will discuss the problem of distinguishing between silyl anions and silicon-stabilized carbanions.

(Trimethylsilyl)methyl anion **328** reacts with N_2O to generate Me_3SiO^- **329** and the driving force for reaction 128 is obviously the formation of the strong silicon–oxygen bond and the conversion of the highly basic carbanion to the more stable oxide ion **330**.

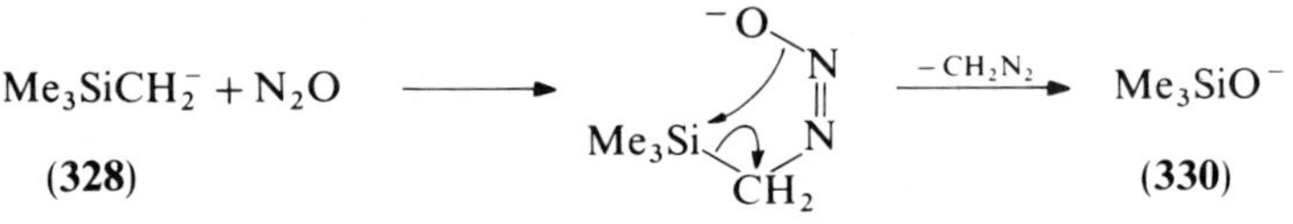

(128)

(329)

In the reaction of the trimethylsilyl anion **10** with N_2O the anion does not suffer from the 'transfer' of a CH_2 group, but the reaction occurs exclusively by oxygen transfer without loss of any substituent from the anionic part (reaction 129).

$$\underset{(\mathbf{10})}{Me_3Si^-} + N_2O \xrightarrow[-N_2]{} \underset{(\mathbf{330})}{Me_3SiO^-} \qquad (129)$$

The latter reaction has been explored for the H_3Si^- anion by using *ab initio* calculations[157], and it has been suggested that the most favoured pathway commences with initial attack of the silicon on nitrogen, followed by loss of N_2 via a four-centre mechanism (reaction 130). This view has been supported experimentally[158]. The fact that almost all gas-phase reactions of N_2O with anions are slow has been explained in terms of charge distribution of the N_2O molecule, in which either end is negatively charged thus repelling the attack of an approaching anion. However, the nitrogen is less negative than the oxygen atom and hence reaction with an anion takes place there.

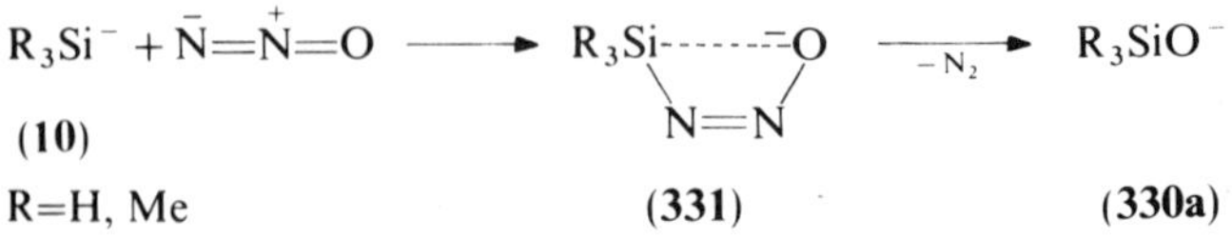

(130)

The distinct reactivity of silyl anions versus carbanions towards N_2O was used[149] to probe the structure(s) of $[M-1]^-$ generated via deprotonation of trimethylsilane with NH_2^-. The fact that the $[M-H]^-$ ion *m/z* 73 upon reaction with N_2O generates an ion at *m/z* 75 and *not* at *m/z* 89 clearly demonstrates that deprotonation has generated ion **9** and not **10** (reaction 131).

$$Me_3SiH + NH_2^- \longrightarrow \begin{cases} Me_2Si(H)CH_2^- \ (\mathbf{9}) \xrightarrow{N_2O} Me_2Si(H)O^- \ (\mathbf{332})\ m/z\ 75 \\ \not\longrightarrow Me_3Si^- \ (\mathbf{10}) \xrightarrow{N_2O} Me_3SiO^- \ (\mathbf{330})\ m/z\ 89 \end{cases} \qquad (131)$$

The high tendency of silicon to form strong bonds to oxygen, nitrogen and sulfur is responsible that silyl anions and α-silylcarbanions have a high tendency to react in the gas phase. In a combined experimental/theoretical study[157] the detailed pathways for the generation of Si–O and Si–S bonds were explored. Although reactions 132a–c appear superficially similar, the authors[157] stress that they occur by fundamentally different mechanisms which are reflected in the extremely different rates of reaction.

$$H_3Si^- \xrightarrow{CO_2} H_3SiO^- + CO \qquad (132a)$$

$$H_3Si^- \xrightarrow{N_2O} H_3SiO^- + N_2 \qquad (132b)$$

$$H_3Si^- \xrightarrow{CS_2} H_3SiS^- + CS \qquad (132c)$$

Reaction of SiH_3^- with CO_2 is rapid (reaction efficiency 0.60). *Ab initio* calculations fail to reveal a low-energy pathway for direct attack on oxygen (with subsequent loss of CO) but do suggest that rearrangement from an initial complex can occur via oxygen addition to silicon (reaction 133).

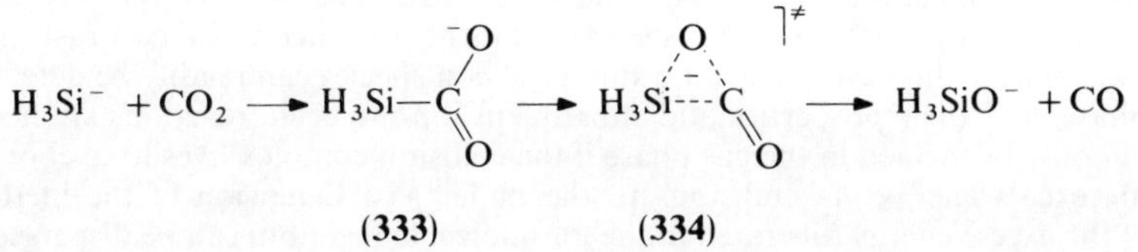

(333) (334) (133)

Reactions of R_3Si^- with N_2O are much slower, and only one in a thousand collisions with N_2O results in a reaction. An explanation for the retardation of the rate has already been provided.

For the reaction of R_3Si^- with CS_2 (and also COS) the MO calculations[157] strongly favour a mechanism involving direct attack on sulfur followed by expulsion of CS (or CO) (reaction 134). This is perhaps due to the better polarisability of sulfur in comparison with oxygen.

$$\underset{(X = S,\ O)}{R_3Si^- + S{=}C{=}X} \longrightarrow \underset{(\mathbf{335})}{\left[R_3Si{-}S{-}\bar{C}{=}X\right]} \xrightarrow{-CX} \underset{(\mathbf{336})}{R_3SiS^-} \qquad (134)$$

Two interesting processes, typical for α-silylcarbanions, are worth mentioning. One concerns the reaction of **328** with carbonyl compounds (for example benzaldehyde), in the course of which **330** is formed. This reaction is the gas-phase analogue of the *Peterson* olefin synthesis (reaction 135)[159].

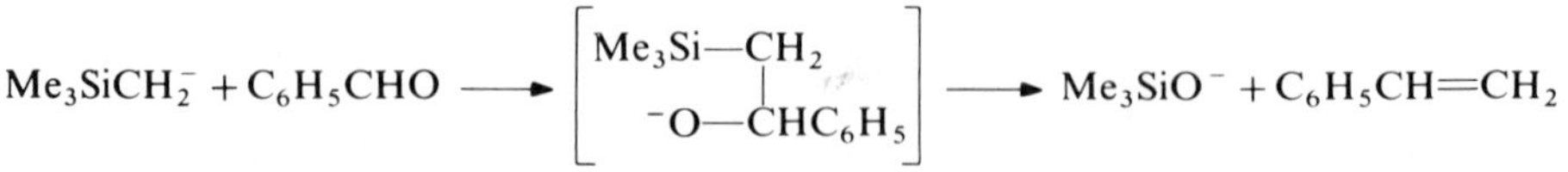

(**328**) (**337**) (**330**)

(135)

The reaction of the enolate of dimethylsilanone **11** with CO_2 can be viewed as a further example for a metathesis process[45]. The ionic product of reaction 136 was suggested to have the structure of the 2-silaacetate anion **339**.

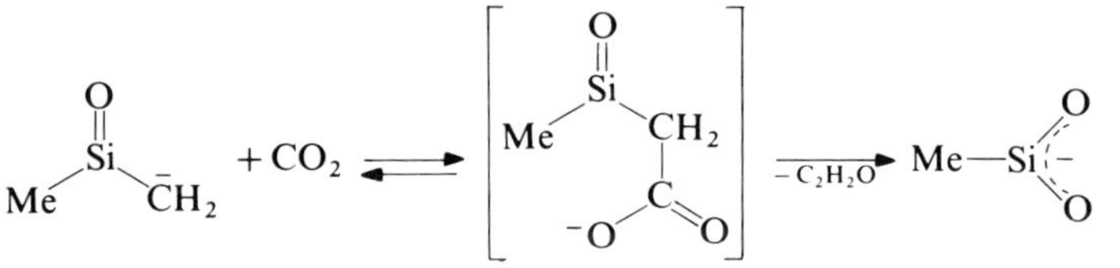

(136)

(**11**) (**338**) (**339**)

Pentacoordinate silicon anions play key roles as reaction intermediates in both the gas phase and in solution. The addition reactions are undoubtedly facilitated by the ease with which the nucleophile adds to silicon and the ability of silicon to be pentacoordinate and together with its substituents to accomodate the negative charge. The role of the nucleophile is apparent in the case of F^-, because of the strength of the silicon–fluorine bond. On the other hand, addition of H^- does lead to stable pentacoordinate silicon adducts only if the substituents are such that charge dispersal is possible. The presence of electronegative groups greatly facilitates attachment to silicon. In contrast to the situation in solution, however, pentacoordinate silicon species can readily be detected in the gas phase, and their properties and subsequent ion/molecule reactions studied. An adduct will only be formed in the gas phase if the collision complex lives long enough to transfer its excess energy via collisions to the buffer gas. Extension of the lifetime is possible if the excess energy (liberated in the ion/molecule reaction) can be dispersed into vibrations and rotations of the substituents; consequently, it is no surprise that Me_4Si, for example, gives a stable adduct with F^- at every collision[159] while no adducts are formed in the reaction with SiH_4[50b]. Similarly, in the reaction of H^- with SiH_4 (reaction 19) no stable SiH_5^- adduct is observed, while replacement of one or more hydrogens by alkyl groups greatly increases the efficiency to form stable pentacoordinate adducts (reaction 20). Once a pentacoordinate silyl anion has been formed and cooled by transferring its excess energy to the buffer gas, its ligand X can be transferred to another molecule in a reaction which is then much less exothermic than that involving X^- itself. For example, the parent SiH_5^- which cannot be generated by direct addition of H^- to SiH_4 is easily accessible by hydride transfer from stabilized hydride adducts to SiH_4 (reaction 137)[50b]. A further example is given in reaction 138. While reaction of alkyltrimethylsilane with F^- results in the formation of an allyl anion **319** (according to the displacement reaction

123), the gentle transfer of F^- from the Me_4SiF^- results in the formation of a stable adduct (reaction 138)[152,160].

$$SiH_4 + C_5H_{11}SiH_4^- \rightarrow SiH_5^- + C_5H_{11}SiH_3 \qquad (137)$$

$$C_3H_5(Me)_3Si + Me_4SiF^- \rightarrow C_3H_5(Me)_3SiF^- + Me_4Si \qquad (138)$$

$$CH_2{=}CHCH_2 \mathrel{\hat{=}} C_3H_5$$

As already mentioned, adduct formation is especially favoured by the presence of electronegative groups attached to silicon. Therefore it is no surprise that in the reactions of F^- with either tetravinylsilane (reaction 139) or tetramethoxysilane (reaction 140) adduct formation is observed. Displacement of either $C_2H_3^-$ or CH_3O^- is not observed. In the case of $(MeO)_4Si$ both the greater strength of the Si–O bond (108 kcal mol^{-1})[161] as compared to the Si–C bond (70–85 kcal mol^{-1}) and the inductive stabilization of the pentacoordinate adduct probably account for this failure[152].

$$(CH_2{=}CH)_4Si + F^- \rightarrow (CH_2{=}CH)_4SiF^- \qquad (139)$$

$$(MeO)_4Si + F^- \rightarrow (MeO)_4SiF^- \qquad (140)$$

Stable adducts can also be generated, even under ICR conditions where the probability to collisionally stabilize the encounter complex is quite remote, if the excess energy is carried away by generating a neutral. For example, CF_3O^- undergoes a net F^- transfer to alkylsilanes[156] to form both stable and decomposing $[M-F]^-$ species. Alkanol-alkoxide negative ions $[R^1O \cdots H \cdots OR^2]^-$ react with alkoxysilanes Me_3SiOR to produce both $[M+R^1O]^-$ and $[M+R^2O]^-$ ions of trigonal bipyramidal geometry[162]. Again, the excess energy liberated in the collison complex is carried off by the neutral alcohol leaving the complex **340** (reaction 141).

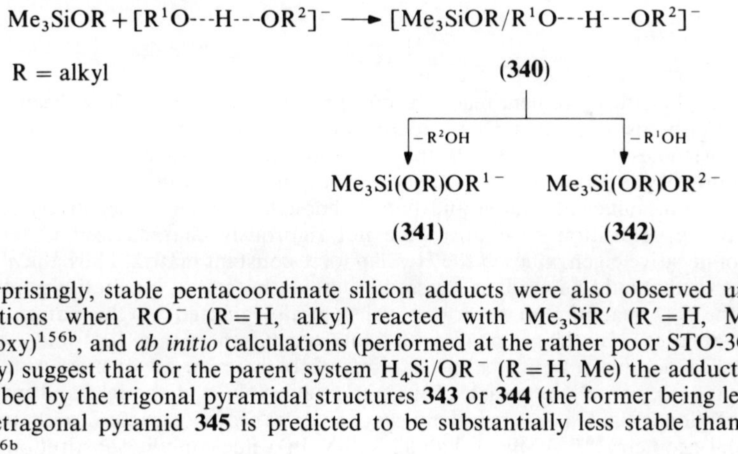

(141)

Surprisingly, stable pentacoordinate silicon adducts were also observed under ICR conditions when RO^- (R = H, alkyl) reacted with Me_3SiR' (R′ = H, Me, vinyl, methoxy)[156b], and *ab initio* calculations (performed at the rather poor STO-3G level of theory) suggest that for the parent system H_4Si/OR^- (R = H, Me) the adducts are best described by the trigonal pyramidal structures **343** or **344** (the former being less stable); the tetragonal pyramid **345** is predicted to be substantially less stable than **343** and **344**[156b].

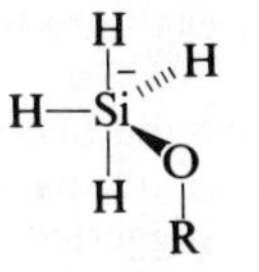

(343)

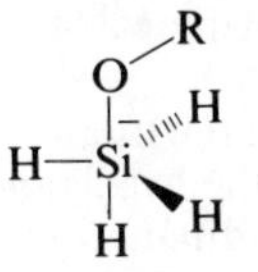

(344)

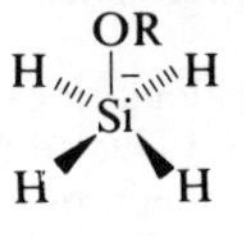

(345)

The problem of apical versus equatorial positions of fluorine in pentacoordinate silicon anions has been studied experimentally by DePuy and coworkers[160]. Reactions of cyclic silanes **346**–**349** with F^- or Me_4SiF^- under flowing afterglow conditions yield stable $[MF]^-$ adducts. The results are consistent with a trigonal bipyramidal geometry of the adduct in which F^-, whose apicophilicity is large[163], occupies the apical position (cf. **350**). In this geometry, the ring strain of the four-membered ring is relieved by allowing the ring-carbon atoms to span one equatorial and one apical position. One might have expected that **348** would have a much higher F^- affinity than either **346** or **349**; it turned out, however, that these F^- affinities are quite close to one another (the difference being probably < 3 kcal mol^{-1}). The explanation favoured by the authors[160] is that fluorine must take in the adduct an equatorial position if both silacyclobutanes are to span equatorial and apical positions (**351**). As mentioned, fluorine prefers strongly an apical position with the consequence that in the $[MF]^-$ adduct of **348** one silacyclobutane ring must occupy an equatorial/equatorial position (cf. **352**). Thus, no particular relief of strain should be expected in the addition of F^- to **348** in comparison to **346**.

(346)

(347)

(348)

(349)

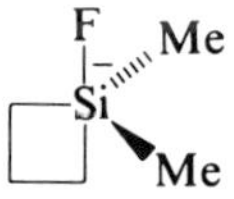

(350)

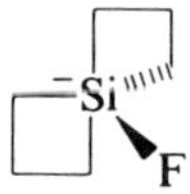

(351)

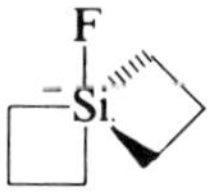

(352)

The nucleophilic displacement reactions on silicon in the gas phase have been studied in great detail by using sophisticated experimental techniques and were complemented by *ab initio* studies[44,156b,164]. While the *ab initio* studies most certainly add to our understanding of gas-phase reactions, a shortcoming in many studies is that (i) the basis sets used are not sufficiently large and flexible enough to describe negatively charged species, and (ii) transition structures were not rigorously characterized in terms of numbers of negative eigenvalues of the Hessian force constant matrix. Thus, the *ab initio* studies reported should be used, as stressed by the authors, as a qualitative guide.

While the S_N2 reaction on carbon is substantially affected by the nature of the solvent[165] and occurs in the gas phase with a rate which is more than 10^{11}-fold as fast as in the most favourable solvent, the mechanisms of the reaction are in most cases completely analogous, namely by attack of the nucleophile (Nu^-) at the rear and inversion of configuration (Walden inversion) by way of a transition structure with trigonal bipyramidal geometry[166]. A silicon halide, R_3SiX, in a nucleophilic substitution might react through a single transition structure **353** (as do carbon analogues) or by way of a pentacoordinate silicon anion intermediate **354**, which decomposes to products (reaction 142).

MNDO calculations[167] and *ab initio* studies[164a] indicate that the one-step reaction (via **353**) is not favoured, even when the five ligands around silicon are space-consuming; instead, the generation of pentacoordinate intermediate **354** is suggested to act as the central link in S_N2 reactions at silicon centres. This is convincingly demonstrated by comparing the rates of gas-phase nucleophilic substitutions at carbon and silicon as a

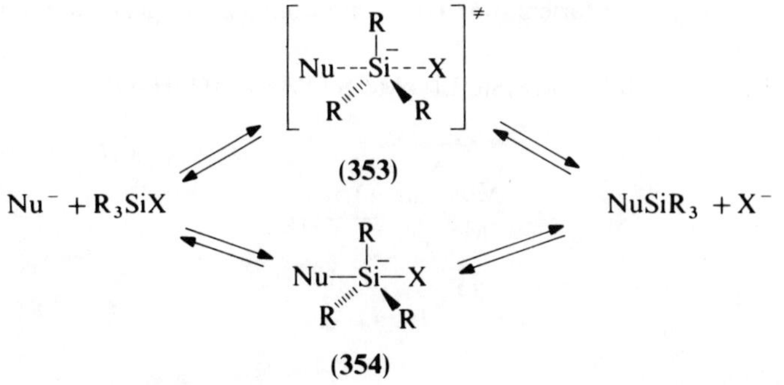

(353) (354) (142)

function of the basicity (nucleophilicity) of the attacking nucleophiles[168]. For methyl iodide there is a gradual reduction in rate as the nucleophile becomes less basic. In distinct contrast, Me_3SiCl reacts at virtually every collision for exothermic reactions, or not at all for endothermic ones. This is precisely the expected behaviour if an intermediate is formed.

Alkoxysilanes, R_3SiOR', in contrast to dialkyl ethers, also undergo nucleophilic displacement reactions in the gas phase (reaction 143)[156b].

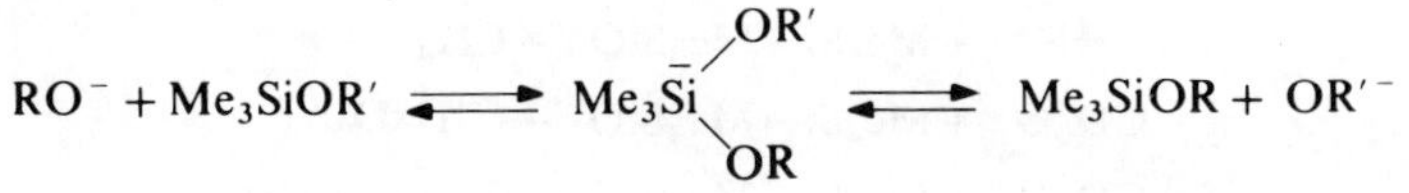

(143)

(355)

When CH_3O^- is used as the base in reaction 143 and R′ corresponds to Me, the study of suitably ^{18}O and ^2H labeled substrates clearly demonstrates that in this particular case, in addition to reaction 143, an S_N2 process at the methyl carbon atom of the methoxysilane takes place (reactions 144, 145). However, all studies performed agree that S_N2 *mechanisms on silicon never resemble that at carbon.*

$$Me^{18}O^- + Me\text{–}O\text{–}SiMe_3 \xrightarrow{S_N2} Me\text{–}^{18}O\text{–}Me + Me_3SiO^- \qquad (144)$$

$$CD_3O^- + Me\text{–}O\text{–}SiMe_3 \xrightarrow{S_N2} CD_3\text{–}O\text{–}Me + Me_3SiO^- \qquad (145)$$

In reaction 141 we have described that stable adducts can be formed, even under ICR conditions, when alkoxide-alcohol clusters are used as reactants[162]. Adduct formation becomes fast, even at low pressure, since the reaction exothermicities are removed by the expulsion of a neutral alcohol molecule. Those adducts which are not efficiently stabilized undergo an ether exchange reaction. When *mixed* clusters such as $EtO^-/HOMe$ are used as reactant ions, an interesting alkoxide exchange reaction is revealed (reaction 146). The surprising aspect of reaction 146 is its high specifity: *the smaller alkoxide ion (in this case* MeO^-*) reacts at silicon while the larger alcohol (EtOH) becomes part of the solvated ion.* Similar reactions occur between alcohol cluster anions and carbon ethers[169], alkoxyboranes[170] and titanium alkoxides[164c]. A preliminary *ab initio* MO analysis suggests that the specifity of reaction 146 may be caused by an initial specific orientation effect: EtO^- bonds to the hydrogen of the ethoxy group of the silane, thus directing methanol towards silicon (**356**). Reports of such orientation effects are quite rare in gas-phase ion chemistry[171].

$$EtO^{-}\text{–}H\text{–}OMe + Me_3SiOEt \rightarrow Me_3SiOMe + EtO\text{–}H\text{–}OEt^{-} \quad (146)$$

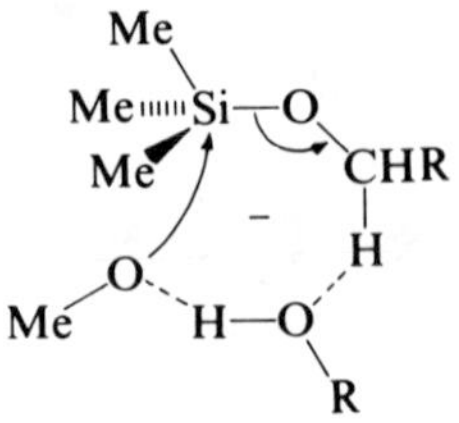

(356)

Depending on the internal energy and the substituents attached to the pentacoordinate silicon adduct anions, not only exchange processes, like reaction 146, or substitutions (reactions 143–145) occur, but alkane elimination is also frequently observed, in particular under ICR conditions. Alkane elimination is favoured if the adduct does not contain a good leaving group (allyl, alkoxide). Three instructive examples are described in reactions 147–149[164b].

$$HO^{-} + Me_4Si \rightarrow Me_3SiO^{-} + CH_4 \quad (147)$$

$$CD_3O^{-} + Me_4Si \rightarrow Me_3SiO^{-} + C_2H_3D_3 \quad (148)$$

$$CD_3O^{-} + Me_4Si \rightarrow Me_2(CD_3O)SiCH_2^{-} + CH_4 \quad (149)$$

When OD^{-} reacts with Me_4Si, the initially formed complex contains insufficient energy to generate the highly basic methyl anion (Me^{-}) as a free species. It is suggested[44, 164b] that Me^{-} begins to depart from the intermediate complex **357** but immediately reacts with the deuterium from the OD group thus generating CH_3D and Me_3SiO^{-}. The mechanism depicted in reaction 150 is substantiated by extensive labeling studies[164b].

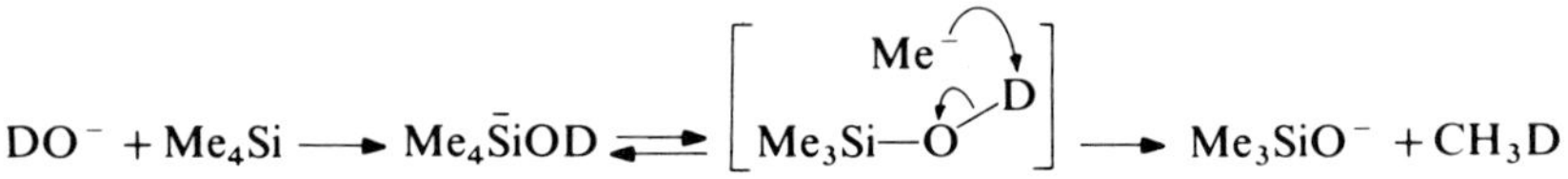

(150)

(357)

Reactions 148 and 149 are thought to proceed analogously; however, the methide ion must find different reaction pathways to escape as methane from the intermediate complex. It does so by proton abstraction from a methyl group (reaction 151: **358**→**359**) or by inducing an S_N2 displacement reaction on the methoxyl group (**358**→**330**)[156b, 164b].

A highly remarkable study, which permits the determination of gas-phase acidities of alkanes by using the properties of alkylsilanes towards nucleophiles like F^{-} or HO^{-}, was recently reported[172]. For example, when dimethyldiethylsilane is allowed to react with HO^{-}, methane and ethane are eliminated in an approximate 2:1 ratio (reaction 152). This result suggests that the methyl anion is easier to expel from the initially formed pentacoordinate adduct **360**, and is therefore a *weaker* base than the ethyl anion, in agreement with theory[173] but opposite to the conclusions one would expect by analogy

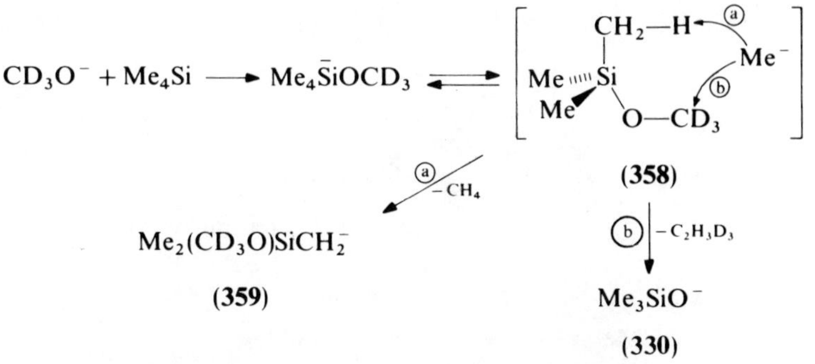

(151)

with the alcohol series (MeO^- is a stronger base than EtO^-)[174]. Although the method used is not particularly sensitive it has the advantage of providing relative gas-phase acidities of saturated alkanes which, otherwise, are not accessible.

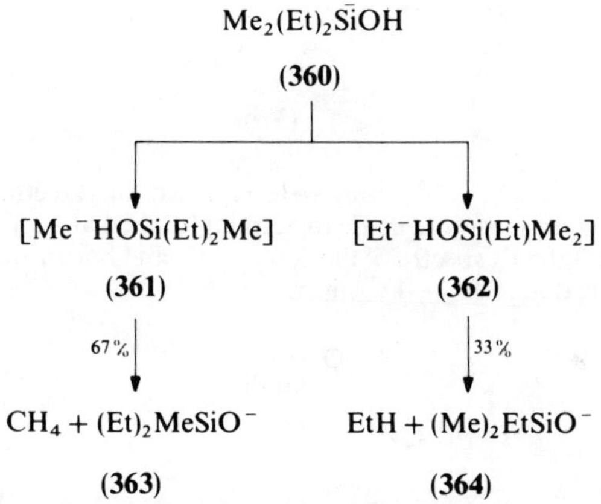

(152)

V. DOUBLY CHARGED IONS

The electropositive nature of silicon is one of the reasons why doubly charged cations are very often encountered in the EI mass spectra of silicon-containing molecules. As the stability of these dications increases with decreased Coulomb repulsion, i.e. with the increased distance between the formally charged centres, it is understandable that the $[M-2Me]^{2+}$ dication is much more pronounced in the EI spectrum of the bis-TMS ether **365** in comparison to **366** (reaction 153)[76].

Similarly, the increase in the relative intensities of the $[M-30]^{2+}$ ions generated from **113b–e** (structures are given in reaction 50) can be explained in a first approximation by the distance criterion[76] (**113b**: 2%; **113c**: 4%; **113d**: 3%; **113e**: 12%). The latter was also used in other cases to explain the abundances of $[M-30]^{2+}$ dications from bis-TMS derivatives[175].

Charge separation has been used to explain the abundancy of doubly charged molecular ions in the EI mass spectrum of **369** (reaction 154)[105].

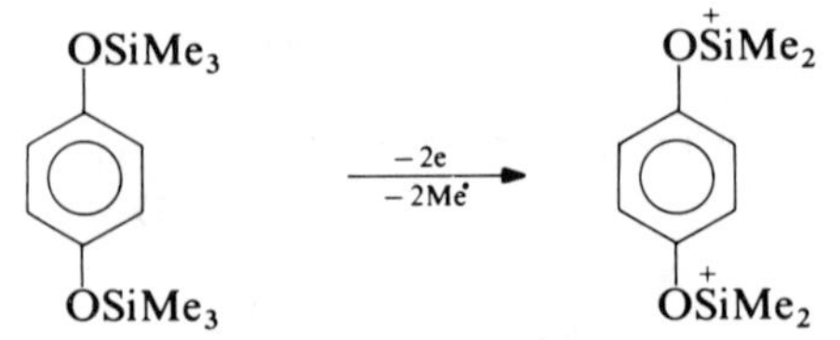

(**365**) (**367**) (30% rel. intensity)

(153)

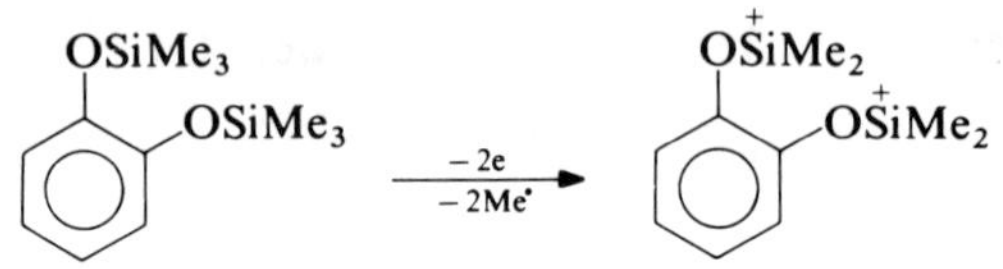

(**366**) (**368**) (1% rel. intensity)

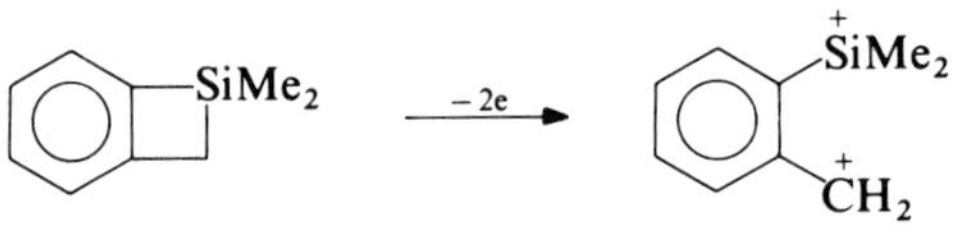

(154)

(**369**) (**370**)

Similarly, doubly charged $[M-1)^{2+}$ ions were reported for the ethers **371** (R = Me, Ph)[176]. No structural suggestions were made to account for the $[M-1]^{2+}$ production. It should be mentioned that the EI spectra of the germanium and boron analogues, **372** and **373**, also contain signals due to $[M-1]^{2+}$ ions.

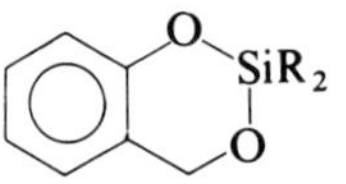

(**371**)

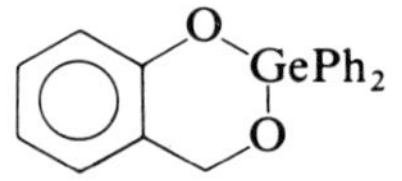

(**372**)

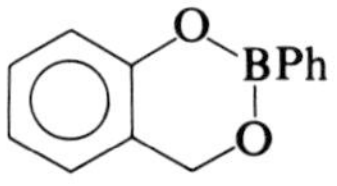

(**373**)

High-level *ab initio* MO calculations [at the MP4/6-31G(d, p)//6-31G(d) level of theory] in conjunction with charge stripping experiments[177, 178] were used to characterize SiH_n^{2+} dications ($n = 1$–5)[178]. SiH_n^{2+} ($n = 1$–3) are observable species. SiH^{2+}, when generated via electron-impact ionization of SiH_4, exists in two stable electronic states (i.e. $X^2\Sigma^+$ and $A^2\Pi$), as predicted by theory and confirmed by experiment. The ionization energies to generate the dications from the corresponding SiH_n^+ monocations were determined, and the following results were obtained: $SiH^+ \rightarrow SiH^{2+\cdot}$ (18.8 eV for the $^1\Sigma^+$ state of SiH^+, and 16.5 eV for the $^3\Pi$ state of SiH^+); $SiH_2^{+\cdot} \rightarrow SiH_2^{2+}$ (16.7 eV); $SiH_3^+ \rightarrow SiH_3^{2+\cdot}$ (15.4 eV). Comparison of the properties of SiH_n^{2+} with the carbon analogues CH_n^{2+} is quite revealing in two respects: (1) Deprotonation energies of the reactions $XH_n^{2+} \rightarrow XH_{n-1}^+ + H^+$ (X = Si, C) is for X = C significantly more exothermic than for the silicon hydrides. (2) The minimum energy structures for SiH_n^{2+} ($n = 4,5$) are best described as complexes between a doubly ionized SiH_{n-2}^{2+} dication and H_2. The latter may complex the dication in an *end-on* (**374**, **376**) or *side-on* fashion (**375**, **377**) [numbers given at the structures refer to 6-31G(d) calculated bond length, in Å]. Both properties are direct consequences of the better capability of silicon to accomodate the positive charge which itself is due to its lower electronegativity and its higher polarisability compared with carbon.

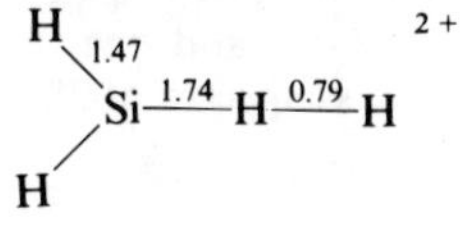

(374) C_{2v}

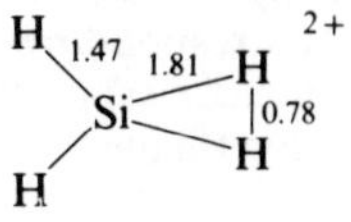

(375) C_{2v}

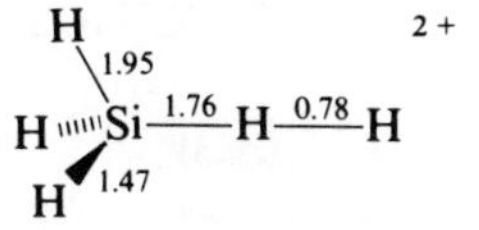

(376) C_s

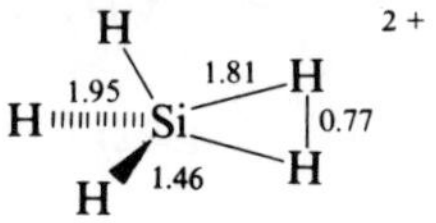

(377) C_1

In a purely theoretical paper, Radom and coworkers[179] studied the geometries and energies of substituted methylene dications HCX^{2+} (X = F, OH, NH_2, CH_3, Cl, SH, PH_2 and, what is of interest in the present context, SiH_3), a number of possible isomeric structures, and transition structures for rearrangement and fragmentation. The results for the $HCSiH_3^{2+}$ dication are rather remarkable. Geometry optimization under C_{3v} symmetry constraint yields a structure **378** with a quite short C–Si bond: 1.83 Å at MP3/6-31G(d, p) compared with 1.88 Å for CH_3SiH_3. This result thus shows a bond shortening, presumably hyperconjugative in orgin, that is not found in the monocation $CH_2SiH_3^+$. **378**, however, has two imaginary frequencies, indicating a second-order saddle point. Rearrangement of **378** along the normal coordinates representing these frequencies leads without a barrier to CH_3SiH^{2+} (**379**), which is predicted to be 106 kcal mol^{-1} more stable than **378**. Thus, $HCSiH_3^{2+}$ is not an equilibrium structure on the $CSiH_4^{2+}$ potential-energy surface and should not be observable. Another low-lying isomer is $SiCH_4^{2+}$ (**380**) which is 3.1 kcal mol^{-1} above **379**, while the silaethene dication **381** is, relative to **379**, somewhat higher in energy by 23 kcal mol^{-1}.

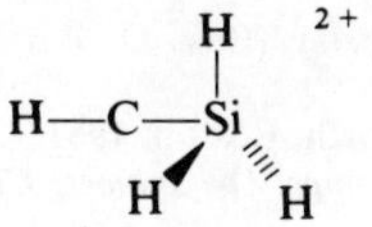

(378) C_{3v}

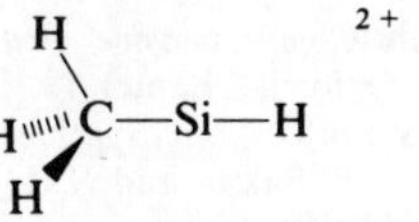

(379) C_{3v}

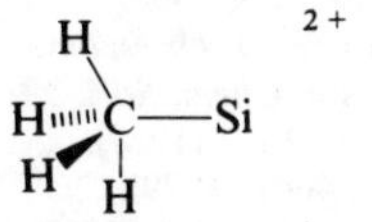

(380) C_{4v}

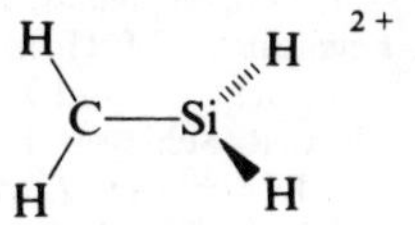

(381) C_{2v}

Gaseous dianions, extremely rarely observed, were described by Bowie and Hale[87] as being formed in some silyl-substituted nitroarenes. Doubly charged molecular ions M^{2-} were indeed observed for **382** and **383**. Species formally corresponding to Me_3SiO^{2-}(!) and NO_2^{2-} were observed in the spectra of **382**, **383** as well as of (*para*-nitrophenoxy)trimethylsilane. The latter compound also gives rise to the formation of **384–386** formally corresponding to Me_3SiO^{2-}, NO_2^{2-} and the structures **384–386** were suggested for the corresponding peaks in the mass spectra of **382**, **383** and of *para*-trimethyl(nitrophenoxy)silane.

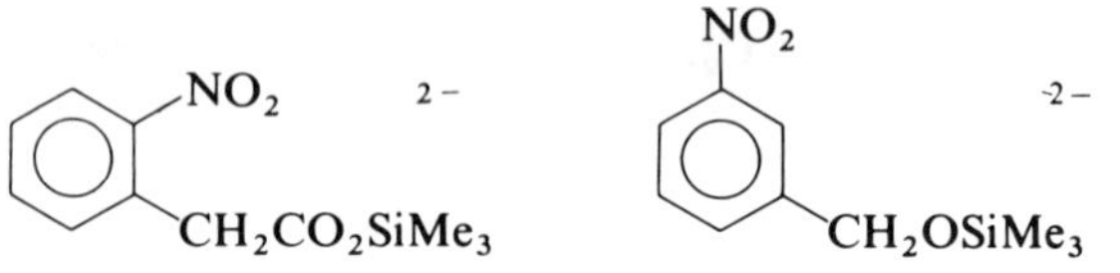

(382) **(383)**

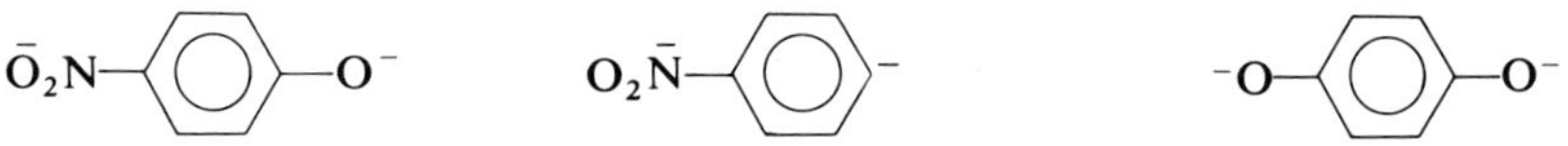

(384) **(385)** **(386)**

VI. ACKNOWLEDGEMENTS

The continuing financial support from the Fonds der Chemischen Industrie, the Gesellschaft von Freunden der Technischen Universität Berlin and the Technische Universität Berlin (Exchange Programme TU Berlin/Hebrew University of Jerusalem) is gratefully acknowledged. I am most grateful to Semiha Akkök and Waltraud Zummack for their literature search and Christine Moser for the efficient typing of the manuscript. The supply of numerous reprints and preprints by Profs. P. B. Armentrout (Berkeley), J. H. Bowie (Adelaide), C. H. DePuy (Boulder), N. M. M. Nibbering (Amsterdam) and Dr. L. Radom (Canberra) has made it possible to include most recent results in the present review.

VII. REFERENCES

1. (a) I. Fleming, in *Comprehensive Organic Chemistry* (Eds. D. Barton and W. D. Ollis), Pergamon Press, Oxford, Chapter 13, 1979.
   (b) E. Colin, *Silicon in Organic Synthesis*, Butterworth, Boston, 1981.
   (c) P. D. Magnus, T. Sakar and S. Djuric, in *Compr. Organomet. Chem.* Pergamon Press, Oxford, Chapter 48 (1982).
   (d) W. P. Weber, *Silicon Reagents for Organic Synthesis*, Springer-Verlag, New York, 1983.
   (e) T. J. Barton, *Pure Appl. Chem.*, **52**, 615 (1980).
   (f) J. A. Gladysz, *Acc. Chem. Res.*, **17**, 326 (1984).
2. C. Eaborn, *Organosilicon Compounds*, Butterworth, Boston, 1960.
3. (a) C. Eaborn, F. Feichtmayr, M. Horn and J. N. Murrel, *J. Organomet. Chem.*, **77**, 39 (1974).
   (b) Y. Apeloig, P. v. R. Schleyer and J. A. Pople, *J. Am. Chem. Soc.*, **99**, 1291 (1977).
   (c) Y. Apeloig and P. v. R. Schleyer, *Tetrahedron Lett.*, 4647 (1977).
   (d) T. Clark and P. v. R. Schleyer, *Tetrahedron Lett.*, 4641 (1979).
   (e) J. B. Lambert and R. B. Finzel, *J. Am. Chem. Soc.*, **104**, 2020 (1982).
   (f) S. G. Wierschke, J. Chandrasekhar and W. L. Jorgensen, *J. Am. Chem. Soc.*, **107**, 1496 (1985).

(g) Y. Apeloig and A. Stanger, *J. Am. Chem. Soc.*, **107**, 2806 (1985).
(h) See also Chapter 2 of this Volume.
4. F. C. Whitmore and L. H. Sommer, *J. Am. Chem. Soc.*, **68**, 481 (1946).
5. (a) L. H. Sommer and G. A. Banghman, *J. Am. Chem. Soc.*, **83**, 3346 (1961).
(b) M. A. Cook, C. Eaborn and D. R. M. Walton, *J. Organomet. Chem.*, **24**, 301 (1970) and **29**, 389 (1971).
(c) F. K. Cartledge and J. P. Jones, *Tetrahedron Lett.*, 2193 (1971).
6. For leading references on various recent MS techniques, see: A. Oppenstein and F. W. Lampe, *Rev. Chem. Intermed.*, **6**, 275 (1986).
7. (a) M. R. Litzow and T. R. Spalding, *Mass Spectrometry of Inorganic and Organometallic Compounds*, Elsevier, Amsterdam, 1973.
(b) V. Y. Orlov, *Russ. Chem. Rev. (Engl. Transl.)*, **42**, 529 (1973).
(c) J. E. Coutant and R. J. Robinson, *Analysis of Silicones* (Ed. A. L. Smith), Chapter 12, Wiley, New York, 1974.
(d) C. F. Poole, in *Handbook of Derivatives for Chromatography* (Eds. K. Blau and G. King), Chapter 4, Heyden, London, 1977.
8. For an excellent overview of this topic see: R. Raabe and J. Michl, *Chem. Rev.*, **85**, 410 (1985). Also see the chapter by these authors in this Volume.
9. (a) P. v. R. Schleyer, Y. Apeloig, D. Arad, B. T. Luke and J. A. Pople, *Chem. Phys. Lett.*, **95**, 477 (1983).
(b) D. Power, P. Brint and T. Spalding, *J. Mol. Struct.*, THEOCHEM, **108**, 81 (1984).
10. See Ref. 6 and literature cited therein.
11. T. M. H. Cheng and F. W. Lampe, *Chem. Phys. Lett.*, **19**, 532 (1973).
12. M. D. Sefcik, J. M. S. Henis and P. P. Gaspar, *J. Chem. Phys.*, **61**, 4329 (1974).
13. (a) F. E. Saalfeld and J. J. Svec, *Inorg. Chem.*, **2**, 46 (1963).
(b) P. Potzinger and F. W. Lampe, *J. Phys. Chem.*, **73**, 3912 (1969).
(c) J. D. Morrison and J. C. Traeger, *Int. J. Mass Spectrom. Ion Phys.*, **11**, 289 (1973).
(d) R. R. Corderman and P. E. Vanier, *J. Appl. Phys.*, **54**, 3987 (1963).
14. M. S. Gordon, *Chem. Phys. Lett.*, **59**, 410 (1978).
15. For a detailed analysis, see Ref. 9b.
16. W. Genuit, A. J. H. Boerboom and T. R. Govers, *Int. J. Mass Spectrom. Ion Processes*, **62**, 341 (1984).
17. (a) A. W. Potts and W. C. Price, *Proc. R. Soc. London, Ser. A*, **326**, 165 (1972).
(b) P. Potzinger, A. Ritter and J. Krause, *Z. Naturforsch.*, **30a**, 347 (1975).
(c) R. Roberge, C. Sandorfy, J. I. Matthews and O. P. Strausz, *J. Chem. Phys.*, **69**, 5105 (1978).
18. A. C. Hopkins and M. H. Lien, *J. Org. Chem.*, **46**, 998 (1981).
19. (a) J. M. S. Henis, G. W. Stewart, M. K. Tripodi and P. P. Gaspar, *J. Chem. Phys.*, **57**, 389 (1972).
(b) T.-Y. Yu, T. M. H. Cheng, V. Kempter and F. W. Lampe, *J. Phys. Chem.*, **76**, 3321 (1972).
20. (a) G. W. Stewart, J. M. S. Henis and P. P. Gaspar, *J. Chem. Phys.*, **57**, 2247 (1972).
(b) T. M. H. Cheng, T.-Y. Yu and F. W. Lampe, *J. Phys. Chem.*, **77**, 2587 (1973).
21. W. N. Allen, T. M. H. Cheng and F. W. Lampe, *J. Chem. Phys.*, **66**, 3371 (1977).
22. (a) G. W. Goodloe, E. R. Haustin and F. W. Lampe, *J. Am. Chem. Soc.*, **101**, 3472 (1979).
(b) J. R. Krause and P. Potzinger, *Int. J. Mass Spectrom. Ion Phys.*, **18**, 303 (1975).
23. G. W. Goodloe and F. W. Lampe, *J. Am. Chem. Soc.*, **101**, 5649 (1979).
24. G. W. Goodloe and F. W. Lampe, *J. Am. Chem. Soc.*, **101**, 6028 (1979).
25. T. M. Mayer and F. W. Lampe, *J. Phys. Chem.*, **78**, 2645 (1974).
26. T. M. Mayer and F. W. Lampe, *J. Phys. Chem.*, **78**, 2433 (1974).
27. H. Neuert and H. Clasen, *Z. Naturforsch.* **7a**, 410 (1952).
28. (a) A. E. Douglas and B. L. Lutz, *Can. J. Phys.*, **48**, 247 (1970).
(b) T. A. Carlson, J. Copley, N. Duric, N. Elander, P. Erman, M. Larsson and M. Lygra, *Astron. Astrophys.*, **83**, 238 (1980).
29. J. L. Elkind and P. B. Armentrout, *J. Phys. Chem.*, **88**, 5454 (1984).
30. (a) T. J. Odiorne, D. J. Harvey and P. Vouros, *J. Phys. Chem.*, **76**, 3217 (1972); *J. Org. Chem.*, **38**, 4274 (1973).
(b) R. N. Stillwell, D. I. Carroll, J. G. Nowlin and E. C. Horning, *Anal. Chem.*, **55**, 1313 (1983).
(c) D. Clemens and B. Munson, *Anal. Chem.*, **57**, 2022 (1985).

31. J. R. Eyler, G. Silverman and M. A. Battiste, *Organometallics*, **1**, 477 (1982).
32. Hydride affinities (HA) are defined as the heterolytic bond dissociation energies $D(R_3M^+ - H^-)$ of the process $R_3M{-}H \rightarrow R_3M^+ + H^-$. By using thermochemical data of R_3MH, H^- and R_3M^+ the hydride affinities are obtained. For Me_3Si^+ the most recent value of its heat of formation is 150.5 kcal mol^{-1} [L. Szepes and T. Baer, *J. Am. Chem. Soc.*, **106**, 273 (1984)]. Earlier values for ΔH_f° (Me_3Si^+) are as follows: 154 kcal mol^{-1} (Ref. 31), 154.8 kcal mol^{-1} [M. K. Murphy and J. L. Beauchamp, *J. Am. Chem. Soc.*, **99**, 2085 (1977)], 160.4 kcal mol^{-1} (Ref. 17b), 171 kcal mol^{-1} [P. Potzinger and F. W. Lampe, *J. Phys. Chem.*, **74**, 719 (1970)] and 176 kcal mol^{-1} [J. M. Gaidis, P. R. Briggs and T. W. Shannon, *J. Phys. Chem.*, **75**, 974 (1971)].
33. (a) V. C. Trenerry, G. Klass, J. H. Bowie and I. A. Blair, *J. Chem. Res. (S)*, 386 (1980).
(b) J. A. Stone and A. C. M. Wojtyniak, 33rd Conference of the American Society of Mass Spectrometry (ASMS), May 26–31, San Diego, Cal., 1985, p. 308.
(c) V. C. Trenerry and J. H. Bowie, *Org. Mass Spectrom.*, **16**, 344 (1981).
(d) A. C. M. Wojtyniak and J. A. Stone, *Int. J. Mass Spectrom. Ion Processes*, **74**, 59 (1986).
34. R. G. Pearson and J. Songstad, *J. Am. Chem. Soc.*, **89**, 1827 (1967).
35. H. Hiraoka and P. Kebarle, *J. Am. Chem. Soc.*, **99**, 360 (1977).
36. (a) M. K. Murphy and J. L. Beauchamp, *J. Am. Chem. Soc.*, **98**, 5781 (1976).
(b) J. R. Krause, P. Potzinger and B. Reimann, *Int. J. Mass Spectrom. Ion Phys.*, **21**, 221 (1976).
37. W. D. Reents, Jr. and A. M. Mujsce, *Int. J. Mass Spectrom. Ion Processes*, **59**, 65 (1984).
38. D. E. Smith and B. Munson, *J. Am. Chem. Soc.*, **100**, 497 (1978).
39. R. Walder and J. L. Franklin, *Int. J. Mass Spectrom. Ion Phys.*, **36**, 85 (1980).
40. C. F. Pau, W. J. Pietro and W. J. Hehre, *J. Am. Chem. Soc.*, **105**, 16 (1983).
41. (a) V. N. Auner and J. Grobe, *Z. Anorg. Allg. Chem.*, **459**, 15 (1979).
(b) T. J. Drahnak, J. Michl and R. West, *J. Am. Chem. Soc.*, **101**, 5427 (1979).
(c) R. T. Conlin and D. L. Wood, *J. Am. Chem. Soc.*, **103**, 1843 (1981).
(d) T. J. Drahnak, J. Michl and R. West, *J. Am. Chem. Soc.*, **103**, 1845 (1981).
(e) See also: G. Maier, G. Mihm, H. P. Reisenauer and D. Littmann, *Chem. Ber.*, **117**, 2369 (1984).
42. (a) Y. Yoshioka and H. F. Schaefer, III, *J. Am. Chem. Soc.*, **103**, 7366 (1981).
(b) H. J. Köhler and H. Lischka, *J. Am. Chem. Soc.*, **104**, 5884 (1982).
43. See Ref. 5 cited in Ref. 44.
44. C. H. DePuy, R. Damrauer, J. H. Bowie and J. C. Sheldon, *Acc. Chem. Res.*, **20**, 127 (1987).
45. S. W. Froelicher, B. S. Freiser and R. R. Squires, *J. Am. Chem. Soc.*, **106**, 6863 (1984).
46. W. Tumas, K. E. Salomon and J. I. Brauman, *J. Am. Chem. Soc.*, **108**, 2541 (1986).
47. J. E. Bartmess and R. T. McIver, in *Gas Phase Ion Chemistry* (Ed. M. T. Bowers), Vol. 2, Chapter 11, Academic Press, New York, 1979.
48. See Ref 43 in Ref. 44.
49. (a) J. C. Sheldon, R. N. Hayes and J. H. Bowie, *J. Am. Chem. Soc.*, **106**, 7711 (1984).
(b) P. Baybutt, *Mol. Phys.*, **29**, 389 (1975).
(c) D. L. Wilhite and L. Spialter, *J. Am. Chem. Soc.*, **95**, 2100 (1973).
(d) F. Keil and R. Ahlrichs, *Chem. Phys.*, **8**, 384 (1975).
(e) Ref. 19 in Ref. 50b.
50. (a) J. D. Payzant, K. Tanaka, L. D. Betowski and D. K. Böhme, *J. Am. Chem. Soc.*, **98**, 894 (1976).
(b) D. J. Hajdasz and R. R. Squires, *J. Am. Chem. Soc.*, **108**, 3139 (1986).
51. (a) G. Dube and V. Chvalovsky, *Coll. Czech. Chem. Comm.*, **39**, 2621, 2641 (1974).
(b) G. Dube and E. Gey, *Org. Mass Spectrom.*, **14**, 17 (1979).
(c) E. Gey and G. Dube, *Int. J. Mass Spectrom. Ion Phys.*, **22**, 103 (1976).
(d) G. Dube, E. Gey and P. Koehler, *Z. Anorg. Allg. Chem.*, **405**, 46 (1974).
(e) G. Dube, *J. prakt. Chem.*, **313**, 905 (1971).
52. R. G. Smith and G. D. Daves, Jr., *J. Org. Chem.*, **43**, 2178 (1978).
53. (a) G. Dube, *Z. Chem.*, **10**, 301 (1970).
(b) See also: R. Pfefferkorn, H.-F. Grützmacher and D. Kuck, *Int. J. Mass Spectrom. Ion Phys.*, **47**, 515 (1983).
54. J. Diekman, J. B. Thomson and C. Djerassi, *J. Org. Chem.*, **32**, 3904 (1967).
55. (a) Y. Apeloig, M. Karni, A. Stanger, H. Schwarz, T. Drewello and G. Czekay, *J. Chem. Soc., Chem. Commun.*, 989 (1987).
(b) D. J. Hajdasz and R. R. Squires, *J. Am. Chem. Soc.*, submitted.

56. F. W. McLafferty (Ed.) *Tandem Mass Spectrometry*, Wiley, New York, 1983.
57. G. Philipou, *Org. Mass Spectrom.*, **12**, 261 (1977).
58. E. I. Mysov, I. S. Akhrem and D. V. Avetisyan, *Org. Mass Spectrom.*, **11**, 358 (1976).
59. W. P. Weber, A. K. Willard and H. G. Boettger, *J. Org. Chem.*, **36**, 1620 (1971).
60. B. Y. K. Ho, L. Spialter and L. D. Smithson, *Org. Mass Spectrom.*, **10**, 617 (1975).
61. (a) B. Y. K. Ho, E. M. Dexheimer, L. Spialter and L. D. Smithson, *Org. Mass Spectrom.*, **14**, 185 (1979).
(b) J. K. Terlouw, W. Zummack and H. Schwarz, unpublished results.
62. S. T. Hill and M. Mokotoff, *J. Org. Chem.*, **49**, 1441 (1984).
63. H. Schwarz, C. Wesdemiotis and M. T. Reetz, *J. Organomet. Chem.*, **161**, 153 (1978).
64. M. T. Reetz and N. Greif, *Angew. Chem., Int. Ed. Engl.*, **16**, 712 (1977).
65. O. S. Akkerman, E. A. I. M. Evers and A. Kerkstra, *Org. Mass. Spectrom.*, **11**, 53 (1976).
66. R. T. Gray, J. Diekman, G. L. Larson, W. K. Musker and C. Djerassi, *Org. Mass Spectrom.*, **3**, 973 (1970).
67. R. Wolfschütz, H. Schwarz, W. Blum and W. J. Richter, *Org. Mass Spectrom.*, **16**, 37 (1981).
68. W. J. A. VandenHeuvel, J. L. Smith and R. A. Firestone, *Org. Mass Spectrom.*, **12**, 724 (1977).
69. H. Schwarz, B. Steiner, G. Zon and Y. H. Chang, *Z. Naturforsch.*, **33b**, 129 (1978).
70. A. M. Lawson, D. B. Ramsden, P. J. Raw and R. Hoffenberg, *Biomed. Mass Spectrom.*, **1**, 374 (1974).
71. M. Maenhaut-Claeys and M. Vandewalle, *Bull. Soc. Chim. Belg.*, **83**, 243 (1974).
72. F. J. Winkler and A. V. Robertson, *Org. Mass Spectrom.*, **10**, 1152 (1975).
73. F. Bohlmann, T. Friedrich, C. Köppel and H. Schwarz, *Org. Mass Spectrom.*, **10**, 158 (1975).
74. C. Köppel, H. Schwarz and F. Bohlmann, *Org. Mass Spectrom.*, **9**, 567 (1974).
75. E. White, V. S. Tsuboyama and J. A. McCloskey, *J. Am. Chem. Soc.*, **93**, 6340 (1970) and references cited therein.
76. H. Schwarz, C. Köppel and F. Bohlmann, *Tetrahedron*, **30**, 689 (1974).
77. W. P. Weber, R. A. Felix and A. K. Willard, *J. Am. Chem. Soc.*, **92**, 1420 (1970).
78. G. Sonneck, K.-G. Baumgarten and D. Habisch, *J. prakt. Chem.*, **322**, 94 (1980).
79. J. D. McCollum and S. Meyerson, *J. Am. Chem. Soc.*, **81**, 4116 (1959).
80. B. Ciommer and H. Schwarz, *J. Organomet. Chem.*, **244**, 319 (1983).
81. A. G. Smith, S. J. Gaskell and C. J. W. Brooks, *Biomed. Mass Spectrom.*, **3**, 161 (1976).
82. R. S.-C. Tsai, G. L. Larson and A. Oliva, *Org. Mass Spectrom.*, **14**, 364 (1979).
83. A. G. Brook, A. G. Harrison and R. K. M. R. Kallury, *Org. Mass Spectrom.*, **17**, 360 (1982).
84. K. Vékey, J. Tamás, G. P. Okonnishnikova, I. E. Dolgy and O. M. Nefedov, *Org. Mass Spectrom.*, **16**, 209 (1981).
85. D. Hunneman and W. J. Richter, *Org. Mass Spectrom.*, **6**, 909 (1972).
86. H. Bosshardt and M. Hesse, *Angew. Chem.*, **86**, 256 (1974).
87. J. H. Bowie and B. A. Hale, *Org. Mass Spectrom.*, **11**, 1105 (1976).
88. E. White and J. A. McCloskey, *J. Org. Chem.*, **35**, 4241 (1970).
89. O. A. Mamer, J. A. Montgomery and C. R. Scriver, *Biomed. Mass Spectrom.*, **8**, 240 (1981).
90. For a review on mass spectrometric *ortho*-effects, see: H. Schwarz, *Top. Curr. Chem.*, **73**, 231 (1978).
91. D. Krauss, H. G. Mainx, B. Tauscher and P. Bischof, *Org. Mass Spectrom.*, **20**, 614 (1985).
92. (a) R. J. Horvat and S. D. Senter, *Org. Mass Spectrom.*, **18**, 413 (1983).
(b) J. G. Liehr and W. J. Richter, *Org. Mass Spectrom.*, **7**, 53 (1973).
93. R. E. Swaim, W. P. Weber, H. G. Boettger, M. Evans and F. M. Bockhoff, *Org. Mass Spectrom.*, **15**, 304 (1980).
94. K. P. Steele and W. P. Weber, *Org. Mass Spectrom.*, **17**, 222 (1982).
95. (a) J. A. McCloskey, A. M. Lawson, K. Tsuboyama, P. M. Krueger and R. N. Stillwell, *J. Am. Chem. Soc.*, **90**, 4182 (1968).
(b) C. Hignite, in *Biochemical Applications of Mass Spectrometry*, (Eds. G. R. Waller and O. C. Dermer) Chapter 5, Wiley, New York, 1980.
96. H. Pang, K. H. Schram, D. L. Smith, S. P. Gupta, L. B. Townsend and J. A. McCloskey, *J. Org. Chem.*, **47**, 3923 (1982).
97. (a) M. A. Quilliam, K. K. Ogilvie, K. L. Sachana and J. B. Westmore, *Org. Mass Spectrom.*, **15**, 207 (1980).
(b) M. A. Quilliam, K. K. Ogilvie and J. B. Westmore, *Org. Mass Spectrom.*, **16**, 129 (1981).
98. (a) H. Schwarz, *Org. Mass Spectrom.*, **15**, 491 (1980).
(b) H. Schwarz, *Nachr. Chem. Techn. Lab.*, **28**, 158 (1980).

(c) H. Schwarz, *Top Curr. Chem.*, **97**, 1 (1981).
(d) H. Schwarz, *Ann. Chim. (Rome)*, 29 (1981).
99. (a) H. Schwarz, M. Kliment, M. T. Reetz and G. Holzmann, *Org. Mass Spectrom.*, **11**, 989 (1976).
(b) H. Schwarz and M. T. Reetz, *Angew. Chem.*, **88**, 726 (1976).
100. B. Ciommer, H. Schwarz, A. Maaroufi, M. T. Reetz and K. Levsen, *Z. Naturforsch.*, **36b**, 771 (1981).
101. F. Höfler, *Monatsh. Chem.*, **106**, 593 (1975).
102. (a) Y. Nakadaira, T. Kobayashi, M. Kira, H. Sakurai, Y. Yamada and Y. Shida, *J. Organomet. Chem.*, **113**, 249 (1976).
(b) T. Weiske, S. Akkök and H. Schwarz, *J. Organomet. Chem.*, in press.
103. For excellent reviews see: (a) Ref. 8, (b) L. E. Gusel'nikov and N. S. Nametkin, *Chem. Rev.*, **79**, 529 (1979).
104. (a) P. Jutzi and P. Langer, *J. Organomet. Chem.*, **139**, 11 (1977).
(b) R. Damrauer, R. A. Davis, M. T. Burke, R. A. Karn and G. T. Goodman, *J. Organomet. Chem.*, **43**, 121 (1972).
105. C. S. Cundy, M. F. Lappert and T. R. Spalding, *J. Chem. Soc., Dalton Trans.*, 558 (1976).
106. L. E. Gusel'nikov, V. V. Volkova, V. G. Zaikin, N. A. Tarasenko, A. A. Tishenkov, N. S. Nametkin, M. G. Voronkov and S. V. Kirpichenko, *J. Organomet. Chem.*, **215**, 9 (1981).
107. D. Seyferth and S. C. Vick, *J. Organomet. Chem.*, **125**, C11 (1977).
108. (a) A. G. Brook and J. W. Harris, *J. Am. Chem. Soc.*, **98**, 3381 (1976).
(b) A. G. Brook, J. W. Harris, J. Lennon and M. El-Sheikh, *J. Am. Chem. Soc.*, **101**, 83 (1979).
109. G. J. D. Peddle, D. N. Roark, A. M. Good and S. G. McGeachin, *J. Am. Chem. Soc.*, **94**, 5837 (1972).
110. (a) G. Fritz and E. Matern, *Z. Anorg. Allg. Chem.*, **426**, 28 (1976).
(b) R. M. Irwin, J. M. Cooke and J. Laane, *J. Am. Chem. Soc.*, **99**, 3273 (1977).
111. R. I. Coulin and P. P. Gaspar, *J. Am. Chem. Soc.*, **98**, 868 (1976).
112. Y. Nakadaira, Y. Kobayashi and H. Sakurai, *J. Organomet. Chem.*, **63**, 79 (1973).
113. (a) P. Jutzi, *J. Organomet. Chem.*, **16**, P17 (1969).
(b) G. L. Van Mouric and F. Bickelhaupt, *Recl. Trav. Chim. Pays-Bas*, **88**, 868 (1969).
(c) P. Jutzi, *Angew. Chem., Int. Ed. Engl.*, **10**, 838 (1971).
114. (a) V. Yu Orlov, L. E. Gusel'nikov and E. S. Finkel'stein, *Izv. Akad. Nauk SSSR, Ser. Khim.*, 1984 (1973).
(b) W. Ando, A. Sekiguchi and T. Migita, *Chem. Lett.*, 779 (1976).
115. See Ref. 513 in Ref. 103b.
116. G. S. Groenewald, M. L. Gross, M. M. Bursey and P. R. Jones, *J. Organomet. Chem.*, **235**, 165 (1982).
117. L. A. Naessens, E. G. Claeys, S. Hoste and G. P. Van der Kelen, *Bull. Soc. Chim. Belg.*, **91**, 507 (1982).
118. D. R. Dimmel, C. A. Wilkie and F. Ramon, *J. Org. Chem.*, **37**, 2665 (1972).
119. K. R. Pope and P. R. Jones, *Organometallics*, **3**, 354 (1984).
120. J. Pola, L. Coak and V. Chvalovsky, *Coll. Czech. Chem. Commun.*, **41**, 368 (1976).
121. J. Diekman, J. B. Thomson and C. Djerassi, *J. Org. Chem.*, **33**, 2270 (1968).
122. R. S.-C. Tsai, G. L. Larson and A. Oliva, *Org. Mass Spectrom.*, **14**, 364 (1979).
123. V. N. Bochkarev, A. N. Polivanov, T. F. Slyusarenko, A. A. Bernadskii, N. N. Silkina and B. N. Klimentov, *Zh. Obshch. Khim.*, **51**, 824 (1981).
124. J. H. Bowie, *Mass Spectrom. Rev.*, **3**, 1 (1984).
125. B. H. Boo and P. B. Armentrout, *J. Am. Chem. Soc.*, **109**, 3549 (1987).
126. For a definition, see: M. T. Reetz, *Tetrahedron*, **29**, 2189 (1973).
127. (a) K. Krogh-Jespersen, *Chem. Phys. Lett.*, **93**, 327 (1982).
(b) P. Ho, M. E. Coltrin, J. S. Binkley and C. F. Melius, *J. Phys. Chem.*, **90**, 3399 (1986).
128. D. W. Fachey, F. C. Fehsenfeld, E. E. Ferguson and L. A. Viehland, *J. Chem. Phys.*, **75**, 669 (1981).
129. T. M. Mayer and F. W. Lampe, *J. Phys. Chem.*, **78**, 2433 (1974).
130. (a) W. N. Allen and F. W. Lampe, *J. Chem. Phys.*, **65**, 3378 (1976).
(b) W. N. Allen and F. W. Lampe, *J. Am. Chem. Soc.*, **99**, 2943 (1977).
131. M. E. Freeburger, B. M. Hughes, G. R. Buell, T. O. Tierman and L. Spialter, *J. Org. Chem.*, **36**, 933 (1971).

132. F. Bohlmann, C. Köppel and H. Schwarz, *Org. Mass Spectrom.*, **9**, 622 (1974).
133. F. Meyer and A. G. Harrison, *J. Am. Chem. Soc.*, **86**, 4757 (1964).
134. N. M. M. Nibbering and Th. J. de Boer, *Tetrahedron*, **24**, 1435 (1968).
135. R. Neeter, N. M. M. Nibbering and Th. J. de Boer, *Org. Mass Spectrom.*, **3**, 597 (1970).
136. I. A. Blair, G. Philipou and J. H. Bowie, *Aust. J. Chem.*, **32**, 59 (1979).
137. V. C. Trenerry, J. H. Bowie and I. A. Blair, *J. Chem. Soc., Perkin Trans. 2*, 1640 (1979).
138. (a) F. H. Westheimer, *Chem. Rev.*, **61**, 265 (1961).
(b) R. A. More O'Ferrall, *J. Chem. Soc. (B)*, 785 (1970).
139. V. C. Trenerry, I. A. Blair and J. H. Bowie, *Aust. J. Chem.*, **33**, 1143 (1980).
140. W. Kirmse and K. Horn, *Chem. Ber.*, **100**, 2698 (1967).
141. (a) R. Fields and R. N. Haszeldine, *Proc. Chem. Soc. (London)*, 22 (1960).
(b) R. A. Mitsch, E. W. Neuvar and P. H. Ogden, *J. Heterocycl. Chem.*, **4**, 389 (1967).
142. I. A. Blair, V. C. Trenerry and J. H. Bowie, *Org. Mass Spectrom.*, **15**, 15 (1980).
143. M. F. Dottore, V. C. Trenerry, D. J. M. Stone and J. H. Bowie, *Org. Mass Spectrom.*, **16**, 339 (1981).
144. V. C. Trenerry, J. C. Sheldon and J. H. Bowie, *Nouv. J. Chim.*, **5**, 193 (1981).
145. C. G. Pitt, M. M. Bursey, D. A. Chatfield and R. S. Greenberg, *J. Organomet. Chem.*, **90**, 269 (1975).
146. H. Kang, D. B. Jacobson, S. K. Shin, J. L. Beauchamp and M. T. Bowers, *J. Am. Chem. Soc.*, **108**, 5668 (1986).
147. M. L. Hendewerk, D. A. Weil, T. L. Stone, M. R. Ellenberger, W. E. Farneth and D. A. Dixon, *J. Am. Chem. Soc.*, **104**, 1794 (1982).
148. P. Crotti, F. Macchia, A. Pizzabiocca, G. Renzi and M. Speranza, *J. Chem. Soc., Chem. Commun.*, 485 (1986).
149. C. H. DePuy and R. Damrauer, *Organometallics*, **3**, 362 (1984).
150. J. H. Stewart, R. H. Shapiro, C. H. DePuy and V. M. Bierbaum, *J. Am. Chem. Soc.*, **99**, 7650 (1977).
151. C. H. DePuy, V. M. Bierbaum, G. K. King and R. H. Shapiro, *J. Am. Chem. Soc.*, **100**, 2921 (1978).
152. C. H. DePuy, V. M. Bierbaum, L. A. Flippin, J. J. Grabowski, G. K. King, R. J. Schmitt and S. A. Sullivan, *J. Am. Chem. Soc.*, **102**, 5012 (1980).
153. J. L. Franklin and P. W. Harland, *Ann. Rev. Phys. Chem.*, **25**, 485 (1974).
154. (a) D. K. Bohme and L. B. Young, *J. Am. Chem. Soc.*, **92**, 3301 (1970).
(b) A. H. Zimmerman and J. I. Brauman, *J. Am. Chem. Soc.*, **99**, 3565 (1977).
155. S. W. Froelicher, B. S. Freiser and R. L. Squires, *J. Am. Chem. Soc.*, **108**, 2853 (1986).
156. (a) C. H. DePuy, V. M. Bierbaum, L. A. Flippin, J. J. Grabowski, G. K. King, R. J. Schmitt and S. A. Sullivan, *J. Am. Chem. Soc.*, **101**, 6443 (1979).
(b) G. Klass, V. C. Trenerry, J. C. Sheldon and J. H. Bowie, *Aust. J. Chem.*, **34**, 519 (1981).
(c) J. C. Sheldon, J. H. Bowie and R. N. Hayes, *Nouv. J. Chim.*, **8**, 79 (1984).
(d) C. H. DePuy, V. M. Bierbaum, R. Damrauer amd J. A. Soderquist, *J. Am. Chem. Soc.*, **107**, 3385 (1985).
157. J. C. Sheldon, J. H. Bowie, C. H. DePuy and R. Damrauer, *J. Am. Chem. Soc.*, **108**, 6794 (1986).
158. S. R. Kass, J. Filley, J. M. van Doren and C. H. DePuy, *J. Am. Chem. Soc.*, **108**, 2849 (1986).
159. R. Damrauer and C. H. DePuy, cited in Ref. 44.
160. S. A. Sullivan, C. H. DePuy and R. Damrauer, *J. Am. Chem. Soc.*, **103**, 480 (1981).
161. J. E. Hukeey, *Inorganic Chemistry: Principles of Structure and Reactivity*, Harper and Row, New York, 1972, p. 699.
162. R. N. Hayes, J. H. Bowie and G. Klass, *J. Chem. Soc., Perkin Trans. 2*, 1167 (1984).
163. J. Emsley and D. Hall, *The Chemistry of Phosphorous*, Wiley, New York, 1976.
164. (a) J. C. Sheldon, R. N. Hayes and J. H. Bowie, *J. Am. Chem. Soc.*, **106**, 7711 (1984).
(b) J. C. Sheldon, R. N. Hayes, J. H. Bowie and C. H. DePuy, *J. Chem. Soc. Perkin Trans. 2* (1987), 275.
(c) H. van der Wel, N. M. M. Nibbering, J. C. Sheldon, R. N. Hayes and J. H. Bowie, *J. Am. Chem. Soc.*, **109**, 5823 (1987).
165. K. Tanaka, G. I. Mackay, J. D. Payzant and D. K. Bohme, *Can. J. Chem.*, **54**, 1643 (1976).
166. W. N. Olmsted and J. I. Brauman, *J. Am. Chem. Soc.*, **99**, 4219 (1977).
167. M. J. S. Dewar and E. Healy, *Organometallics*, **1**, 1705 (1982).
168. R. Damrauer, C. H. DePuy and V. M. Bierbaum, *Organometallics*, **1**, 1553 (1982).

169. R. N. Hayes and J. H. Bowie, *J. Chem. Soc., Perkin Trans. 2*, 567 (1985).
170. R. N. Hayes, J. H. Bowie and J. C. Sheldon, *Organometallics*, **5**, 162 (1986).
171. See also: L. J. de Konig and N. M. M. Nibbering, *J. Am. Chem. Soc.*, **109**, 1715 (1987).
172. C. H. DePuy, V. M. Bierbaum and R. Damrauer, *J. Am. Chem. Soc.*, **106**, 4051 (1984).
173. (a) H. Kollmar, *J. Am. Chem. Soc.*, **100**, 2660 (1978).
(b) G. W. Spitznagel, T. Clark, J. Chandrasekhar and P. v. R. Schleyer, *J. Comput. Chem.*, **3**, 363 (1982).
174. J. I. Brauman and L. K. Blair, *J. Am. Chem. Soc.*, **92**, 5986 (1970).
175. (a) T. Aczel and H. E. Lumpkin, *Anal. Chem.*, **32**, 1819 (1960).
(b) J. L. Beck, W. J. A. VandenHeuvel and J. L. Smith, *Org. Mass Spectrom.*, **4**, 237 (1970).
(c) W. J. A. VandenHeuvel, J. L. Smith and J. L. Beck, *Org. Mass Spectrom.*, **4**, 563 (1970).
(d) J. L. Smith, J. L. Beck and W. J. A. VandenHeuvel, *Org. Mass Spectrom.*, **5**, 473 (1971).
(e) G. G. Smith and C. Djerassi, *Org. Mass Spectrom.*, **5**, 505 (1971).
(f) P. Vouros and K. Biemann, *Org. Mass Spectrom.*, **2**, 375 (1969).
176. R. H. Cragg and M. Nazery, *J. Organomet. Chem.*, **71**, 225 (1974).
177. (a) R. G. Cooks, T. Ast and J. H. Beynon, *Int. J. Mass Spectrom. Ion Phys.*, **11**, 490 (1973).
(b) W. Koch, F. Maquin, D. Stahl and H. Schwarz, *Chimia*, **39**, 376 (1985).
178. (a) W. Koch, G. Frenking and H. Schwarz, *J. Chem. Soc., Chem. Commun.*, 1119 (1985).
(b) W. Koch, G. Frenking, H. Schwarz, F. Maquin and D. Stahl, *J. Chem. Soc., Perkin Trans 2*, 757 (1986).
(c) W. Koch, G. Frenking and C. C. Chang, *J. Chem. Phys.*, **84**, 2703 (1986).
(d) C. J. Porter, C. J. Proctor, T. Ast and J. H. Beynon, *Croat. Chem. Acta*, **54**, 407 (1981).
179. M. W. Wong, B. F. Yates, R. H. Nobes and L. Radom, *J. Am. Chem. Soc.*, **109**, 3181 (1987).

The Chemistry of Organic Silicon Compounds
Edited by S. Patai and Z. Rappoport

CHAPTER **8**

NMR spectroscopy of organosilicon compounds

ELIZABETH A. WILLIAMS
General Electric Co., Corporate Research and Development, PO Box 8, Schenectady, New York 12301, USA

I. INTRODUCTION

Nuclear magnetic resonance (NMR) spectroscopy is a powerful method for molecular structure determination. The early work concentrated on elucidating structures using proton NMR. In the middle 1970s the availability of Fourier transform (FT) NMR spectrometers led to a dramatic increase in the observation of less abundant and lower sensitivity nuclei such as silicon-29 and carbon-13. While ^{1}H NMR remains a valuable tool, there are often clear advantages in using these other nuclei. Carbon-13 NMR, for example, has become one of the most effective methods of determining complex organic structures. There is no rival, however, for the ability of ^{29}Si NMR to detect silicon atoms in different molecular environments. Used together ^{13}C and ^{29}Si NMR often permit quite complex structures of organosilicon compounds to be determined with confidence.

A number of texts exists which thoroughly discuss proton[1,2] and carbon[3] NMR and a detailed discussion of these will not be included here. The aspects of ^{1}H and ^{13}C NMR which specifically relate to the unique features of organosilicon compounds will, however, be discussed. Several review articles on ^{29}Si NMR have also appeared[4–9] and an exhaustive survey of the literature will not be attempted. Instead, a summary of the earlier work on ^{29}Si NMR and a discussion of some of the newer results will be presented.

II. ^{1}H NMR

A. $SiCH_3$

Perhaps the most characteristic resonance of organosilicon compounds is the SiMe resonance which normally occurs at the high field end of the proton chemical shift range between -0.1 and 1.2 ppm. It is the unusually strong shielding of this type of resonance that places it out of the region of most other organic substituents and causes tetramethylsilane (TMS) to be such a convenient reference compound in ^{1}H NMR. The SiMe resonance is usually a singlet and is sensitive to the number and nature of the other substituents on silicon. For example, the chemical shifts of the methyl protons in the series Me_4Si, Me_3SiCl, Me_2SiCl_2, $MeSiCl_3$ are 0, 0.42, 0.79 and 1.14 ppm, respectively[10]. Fortunately, the substituent induced chemical shifts are largely additive. The expected chemical shift, δ_H, can be approximated from equation 1:

$$\delta_H = \Sigma_i A_i \qquad (1)$$

where A_i is the additivity constant and summation is over all the other substituents (L) in the molecule $MeSiL_3$. Additivity constants are collected in Table 1. The data are from Reference 11, which also provides a good review of the proton NMR of organosilicon compounds up to 1974. A note of caution should be interjected at this point. Solvent induced shifts can be substantial in proton NMR and minimization of these effects is essential in order to identify specific compounds or structural groups. The chemical shift of $MeSiCl_3$ in benzene, for example, is 0.57 ppm[12], almost 0.5 ppm shielded relative to $MeSiCl_3$ in CCl_4! Other solvents that might cause substantial medium shifts include polar solvents, solvents with easily polarizable groups such as bromine, and solvents which can engage in specific interactions with the compound to be analyzed. As an example of the last case silanols (SiOH) form good hydrogen bonds with electron pair donor solvents such as dimethyl sulfoxide (DMSO).

B. $SiCH_2$

Another resonance that is often characteristic of organosilanes (because of the strong shielding effect of silicon) is the $SiCH_2R$ resonance which appears between 0.4 and

TABLE 1. Substituent additivity constants for calculating Me shifts of $MeSiL_3$[11]

L	ppm
t-Bu, *i*-Pr, *c*-Hex	−0.05
$(CH_2)_nH$ ($n > 1$)	−0.01
$OSiMe_3$	−0.01
Me	0.00
NR_2	0.00
OR	0.02
H	0.05
$CH_2CH_2CF_3$	0.06
Vinyl	0.07
$ONEt_2$	0.11
F	0.16
NMeAc	0.22
OAc	0.23
SMe	0.24
Ph	0.24
Cl	0.38
Br	0.52
I	0.78

3.8 ppm depending upon the nature of R. If R is an alkyl group, or if the silicon atom is insulated from an electronegative substituent by three methylene groups ($SiCH_2CH_2CH_2X$), the protons of the $SiCH_2$ group can appear as high as 0.4 ppm. In general, any proton on a carbon adjacent to silicon will be shielded relative to its carbon analog. In the case of silylethyl, propyl or longer alkyl chains, these methylene protons are usually shifted out of the range of other organic functional groups and can be identified as proximate to silicon. A tabulation of chemical shift ranges for the $SiCH_2X$ and $Si\underline{CH_2}CH_2X$ protons is shown in Table 2. Additional information is available from the splitting patterns of the $SiCH_2$ protons in $SiCH_2CH_2X$ and $SiCH_2CH_2CH_2X$ compounds. These appear as the AA′ protons in an AA′BB′ or AA′MM′ spin system (depending upon the nature of X). The magnetic nonequivalence arises from the fact that the average coupling constants between an A proton and the two B protons in the different rotamers about the C–C bond need not be equal.[13] The proton spectrum of 3-chloropropyltriethoxysilane is a good example of this pattern and is shown in Figure 1.

TABLE 2. Chemical shift ranges for $SiCH_2$ protons[11]

	Range (ppm)	
X	$SiCH_2CH_2X$	$SiCH_2X$
$CH{=}CH_2$	0.63–1.46	1.42–2.28
CF_3	0.70–1.65	
Ph	0.74–1.60	1.96–3.15
SMe	0.79–1.72	1.67–2.86
OMe	0.80–2.10	2.91–3.52
Cl	1.15–2.00	2.62–3.81
CO_2Me		1.72–2.91

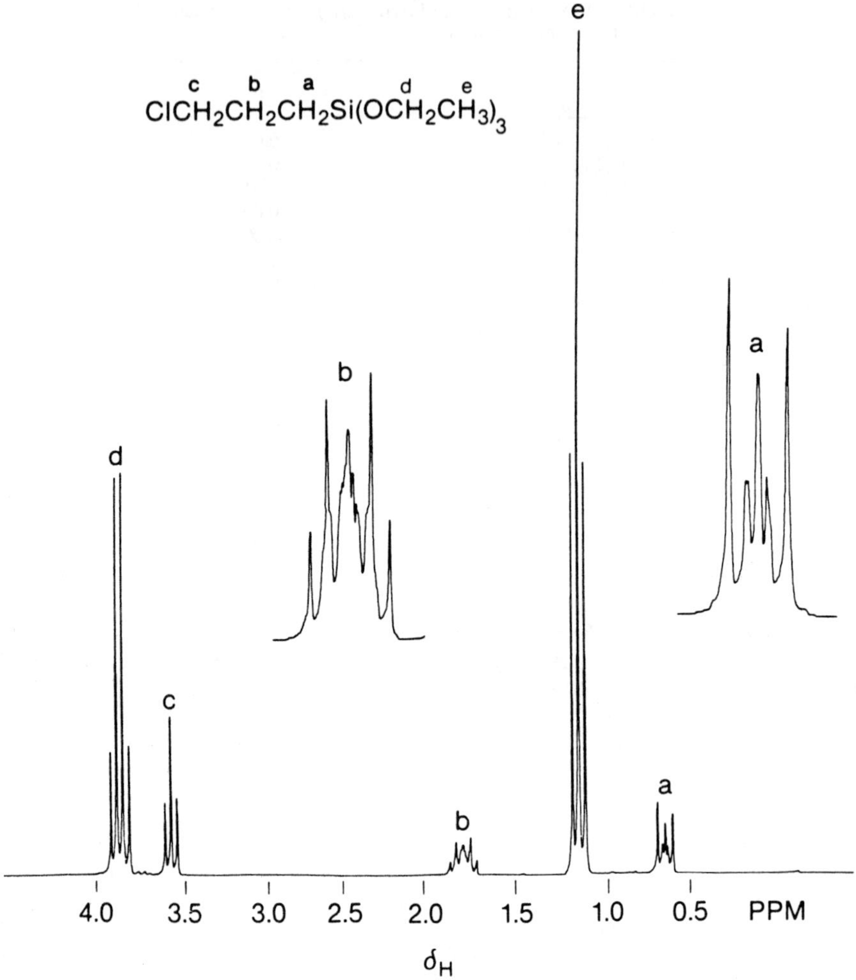

FIGURE 1. Proton NMR spectrum (200 MHz) of chloropropyltriethoxysilane

Note that the $SiCH_2$ protons are shielded relative to the methyl protons of the ethoxy group.

C. SiH

The SiH group is another moiety unique in organosilicon compounds. The range of chemical shifts for this type of proton is from 2.3 ppm for $(Me_3Si)_3SiH$[16] to just below 6 ppm for Cl_3SiH[11]. Even more shielded values were observed for silyl anions[14] with $(H_3Si)_2SiH^-$ appearing at 0.12 ppm. Thus depending upon the other substituents on silicon, the chemical shift of SiH can vary by more than 6 ppm. For common organosilanes (that is, excluding Me_3Si derivatives and silyl anions) the shift range is much smaller

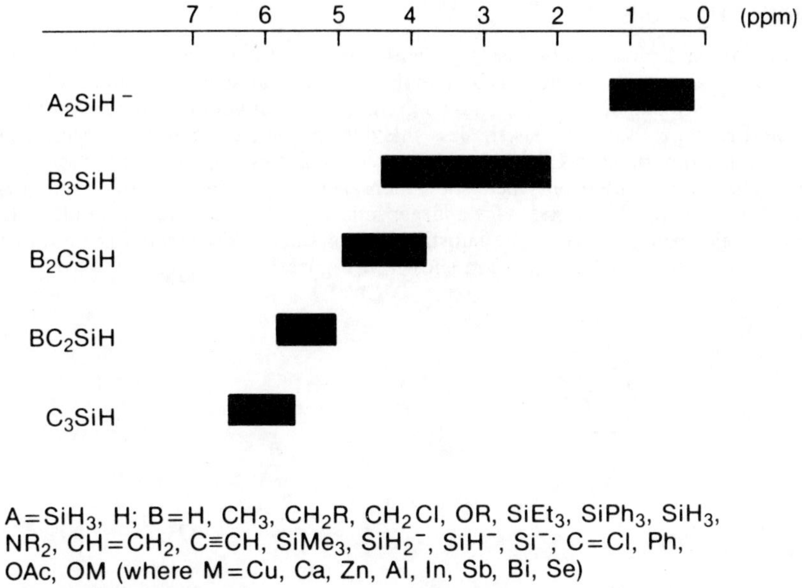

FIGURE 2. Chemical shift ranges for SiH protons[11, 15]

and SiH is observed between 3 and 6 ppm. SiH is usually downfield from CH and is in a region where it can often be differentiated from other groups appearing in the same region of the spectrum, particularly when coupling to other protons is considered. In addition it is frequently possible to observe the ^{29}Si satellites which are present due to the 4.7% natural abundance of this silicon isotope. These coupling constants range from -147 Hz for $(Et_3Si)_3SiH$ to -381 Hz for $HSiF_3$ and show a general trend of increasing J_{SiH} with increasing electronegativity of the substituents on Si[17]. A chart summarizing the shift ranges for different types of SiH protons is shown in Figure 2.

D. SiOH

The chemical shift of the SiOH resonance behaves like other hydroxyl resonances and is dependent on solvent and concentration. Its presence in the 4–7 ppm region of the spectrum may be confirmed by chemical exchange with D_2O. The concentration dependence may be eliminated by the use of a good electron pair donor solvent like DMSO which forms a strong hydrogen bond to the SiOH and eliminates intermolecular association between silanol molecules. This also reduces the rate of proton exchange and has been used to observe spin–spin splitting between the hydroxyl proton and SiH proton (2 Hz) in diphenylsilanol[18], for example. Using this same approach enabled Hampton *et al.*[18] to analyze silanols in the presence of other hydroxyl containing species such as organic alcohols and water.

The use of lanthanide shift reagents for studying and assigning the proton chemical shifts of silanols has also been described[19, 20]. The europium chelate of tris(dipivaloylmethane) [Eu(DPM)$_3$] was found to effect large downfield shifts in the methyl proton resonance of tertiary methylphenylsilanols and smaller changes in the phenyl group protons[20]. No effect on the ^{1}H NMR spectra of alkyl siloxanes was observed.

E. SiCH=CH$_2$ and SiPh

Both vinyl and phenyl substituents in vinyl- and phenylsilanes have proton chemical shifts in the same region as their carbon analogs. In general, spectra of phenylsilanes have two sets of multiplets between 7.1 and 7.8 ppm. The higher field resonance contains both *ortho* and *para* protons whereas the less shielded resonance contains the *meta* protons. This differs from most alkyl monosubstituted benzenes which usually appear as a singlet. The exception is *t*-butylbenzene which also exhibits two sets of resonances, but reversed in intensity. The trend is for a larger separation between the two multiplets with increasing electronegativity of the substituents on silicon. The aromatic multiplets for diphenylsilanediol [$Ph_2Si(OH)_2$] are shown in Figure 3.

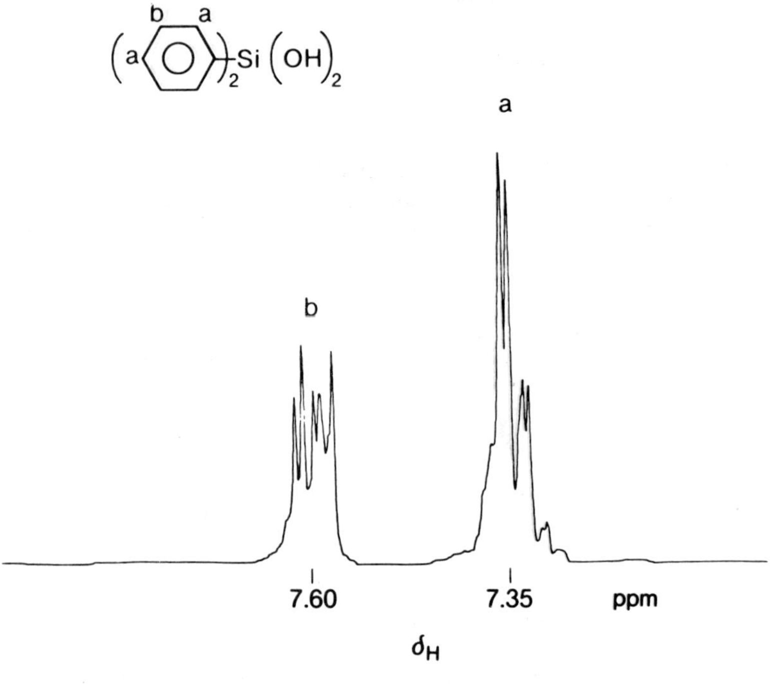

FIGURE 3. Aromatic proton region of the ^{1}H NMR spectrum (200 MHz) of $Ph_2Si(OH)_2$ in DMSO-d$_6$

In summary, the proton chemical shifts for the SiMe protons are generally recognizable by their high field positions relative to the carbon analogs. SiH protons are usually downfield from CH protons and may frequently be identified by spin–spin coupling patterns to other protons or ^{29}Si. The SiOH proton in DMSO is a singlet unless there is a proton directly bound to Si in which case coupling may be seen. Phenyl substituents on silicon exhibit a pair of closely spaced multiplets in the normal aromatic proton region of the spectrum. This spacing increases with electronegative substituents on silicon. This information is summarized in a chart shown in Figure 4.

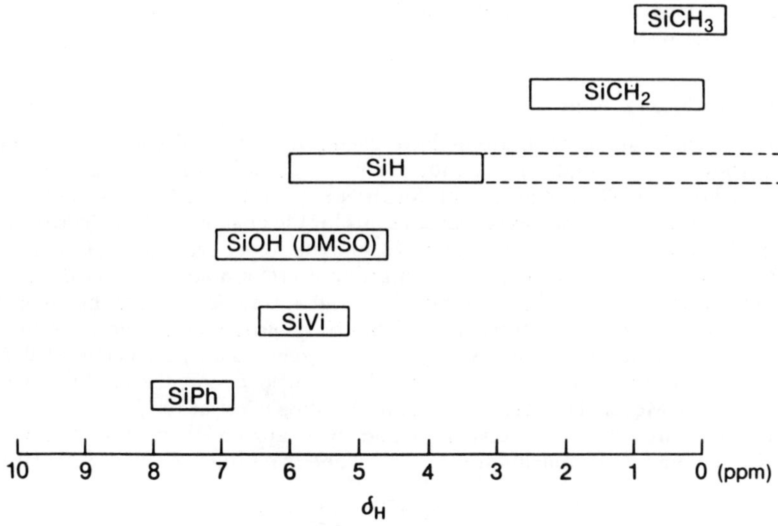

FIGURE 4. General chemical shift ranges for protons in organosilicon compounds. Vi = vinyl

III. ^{13}C NMR

A. $SiCH_3$ and $SiCH_2$

The resonances of alkyl carbon atoms adjacent to silicon are recognized as easily as their attached protons by the unusually shielded position in the ^{13}C NMR spectrum. As in ^{1}H NMR, TMS is used extensively as a chemical shift reference compound due to its position at the high field end of the spectrum. The resonances for $SiCH_3$ carbon atoms occur between -15 and $+20$ ppm relative to TMS. There are differences in the trends of the chemical shifts observed for these carbons relative to the trends observed in the proton NMR of $SiCH_3$ groups, however. Whereas each subsequent replacement of a methyl group with a chlorine atom in the series Me_4Si, Me_3SiCl, Me_2SiCl_2, $MeSiCl_3$ effects a downfield shift of the methyl proton resonance, the carbon chemical shifts are 0, 18, 4 and -16 ppm[21], respectively. This 'sagging pattern' is also observed for silicon chemical shifts and will be discussed in the section on ^{29}Si NMR. It is not observed for the analogous carbon compounds (Me_3CCl, Me_2CCl_2, $MeCCl_3$), however, which exhibit progressive deshielding with each substitution of a chlorine atom for a methyl group[3a].

The methylene carbons adjacent to silicon exhibit chemical shifts between 2 and 35 ppm depending upon substitution on both silicon and carbon. As is the case for ^{1}H NMR, if the carbon atom is insulated from electronegative groups or chain branching (as in $SiCH_2CH_2CH_2X$ compounds) the chemical shift range of the $SiCH_2$ carbon is much smaller and in the 10–20 ppm region. In all cases, however, the $SiCH_2$ resonance is more shielded than the analogous carbon in the CCH_2 analog. For example, the chemical shift of the indicated carbon in $Me_3SiCH_2CH_2CH_2OMe$ is 12.2 ppm whereas the analogous carbon in a similar structure carbon compound $Me_3CCH_2CH_2CH_2OSiMe_3$ is 40.0 ppm[22]. Likewise, the corresponding carbons in $Me_3CCH_2CH_2OH$ and $Me_3SiCH_2CH_2OH$ have chemical shifts of 46.2 and 20.4 ppm, respectively[22]. In both cases the silicon atom has a shielding effect of >20 ppm. Interestingly, in several silacyclopropanes examined by Seyferth and coworkers[23, 24], the carbon atoms in the cyclopropane rings did not exhibit unsually high chemical shifts relative to other cyclopropane derivatives.

B. SiC=C, SiC≡C, and SiPh

Silyl substituted vinyl carbon atoms have chemical shifts in the normal region for carbon–carbon double bonds. The trends are not necessarily the same as for carbon substitution, however. The chemical shifts for three substituted ethylenes are collected in Table 3. Note that the vinyl carbon chemical shifts for the analogous trimethylsilyl and *t*-butyl substituted ethylenes are 138.70 and 148.40 ppm, respectively, for C-1, and 129.57 and 110.40 ppm for C-2[25]. Thus the *t*-butyl group exerts a much stronger deshielding effect on the vinyl carbon to which it is attached, and a likewise stronger shielding effect on the β carbon of the double bond. The trichlorosilyl group affects the chemical shift slightly differently than the trimethylsilyl group as is evidenced by the chemical shifts of trichlorosilylethylene[26]. For this substituent C-2 is deshielded more than C-1 whereas the reverse is true for $Me_3SiCH{=}CH_2$. A strong deshielding effect at the β carbon was also noted for phenyltrimethylsilylacetylene[27] and was attributed to the presence of C-Si (p–d)π bonding with contribution to the ground state from resonance structure **B**:

$$\text{Ph–}\overset{\beta}{\text{C}}{\equiv}\overset{\alpha}{\text{C}}\text{–SiMe}_3 \leftrightarrow \text{Ph–}\overset{+}{\text{C}}{=}\text{C}{=}\overset{-}{\text{Si}}\text{Me}_3$$

(A) **(B)**

For other multiply bonded carbon atoms such as nitriles (C≡N) and isocyanates (N=C=O) virtually no effect was observed in substituting silicon for carbon on the carbon chemical shift of these functional groups. The chemical shifts of the nitrile carbon in Me_3CCN and Me_3SiCN are very close, at 125.1[3b] and 126.9 ppm[28], respectively. The same is true for CH_3CN and SiH_3CN, and SiH_3NCO and CH_3NCO[3b, 28]. In each case the chemical shifts are within 2 ppm.

TABLE 3. Vinyl carbon chemical shifts for some monosubstitued ethylenes[25, 26]

	$RC_{(1)}H{=}C_{(2)}H_2$	
R	C-1 (ppm)	C-2 (ppm)
Me_3Si	138.70	129.57
Cl_3Si	132.5	139.9
Me_3C	148.40	110.40

The use of additive substituent effects to calculate chemical shifts in benzene derivatives has been well documented[3a]. The values obtained for a number of silyl derivatives of benzene are collected in Table 4[29]. The chemical shifts of these compounds are all in the normal range for substituted benzenes (110–150 ppm). The substituent effects differ from those of the analogous carbon containing functional groups and can be differentiated from most other substituents (exceptions are P, Sb, Hg, Bi, Sn, Pb and cationic substituents) by the strong deshielding effect (4–8 ppm) at the *ortho* position.

C. Si–C=O, Si–C$^+$, and Si=C

This unusual set of functional groups has the distinction of the silicon atom exerting a rather dramatic deshielding effect on proximate carbon atoms. In a study of a series of

Table 4. Substituent effects[a] for silyl derivatives of benzene[29]

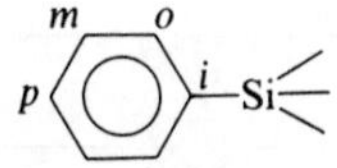

	ipso	*ortho*	*meta*	*para*
SiH_3	−0.5	7.3	−0.4	1.3
SiH_2Me	4.8	6.3	−0.5	1.0
$SiHMe_2$	8.8	5.5	−0.6	0.7
$SiMe_3$	11.6	4.9	−0.7	0.4
SiH_2Ph	2.9	7.1	−0.4	1.3
SiH(Me)Ph	6.7	6.3	−0.5	1.0
$SiMe_2Ph$	9.6	5.6	−0.7	0.6
$SiHPh_2$	4.8	7.3	−0.5	1.3
$Si(Me)Ph_2$	7.6	6.8	−0.7	0.9
$SiPh_3$	5.8	7.9	−0.6	1.1
SiH(Me)OMe	6.7	5.1	−0.7	1.4
$SiMe_2OMe$	8.8	4.8	−0.8	0.9
$SiMe_2OCHMePh$	9.6	4.9	−0.9	0.9
SiMe(Ph)OMe	7.0	5.6	−0.9	1.1
SiMe(Ph)OCHMePh	7.8	5.7	−0.9	1.0
$SiMe(OPh)_2$	4.3	5.4	−0.5	2.1
$SiPh(OPh)_2$	3.8	6.4	−0.5	2.3
SiH(Me)Cl	4.9	5.1	−0.3	2.3
$SiMe_2Cl$	7.7	4.5	−0.4	1.8
SiPh(Me)Cl	5.9	5.5	−0.4	2.0
$SiPh_2Cl$	4.6	6.8	−0.4	2.2
$SiHCl_2$	2.5	4.8	0.0	3.7
$SiMeCl_2$	4.8	4.5	−0.1	3.0
$SiPhCl_2$	3.4	5.5	−0.2	3.2
$SiCl_3$	3.0	4.6	0.1	4.2

[a] Values for $CDCl_3$ solvent; positive values indicate a deshielding effect relative to benzene ($\delta_c = 128.5$ ppm).

trimethylsilyl substituted carbocations Olah and coworkers[30] found that the cationic center of $Ph_2C^+SiMe_3$ had a chemical shift of 259 ppm compared with 229.2 ppm for $Ph_2C^+CH_3$. They also noted a substantial deshielding of the *ipso* carbon (7–10 ppm) of the phenyl ring in the silyl derivative relative to the methyl derivative. Thus in this case the silicon atom affects both the α and β carbons to the cationic center. Similar results were obtained for the (trimethylsilyl)ethynyl substituted cations $Me_3Si\text{–}\overset{\gamma}{C}\equiv\overset{\beta}{C}\text{–}\overset{\alpha}{C}{}^+R_2$, where R=Ph, Me and cyclopropyl. For these cations, however, the silicon atom affected only the β carbon, causing downfield shifts of 15–20 ppm when compared with the analogous *t*-butyl compounds (Me_3Si replaced by Me_3C)[30].

In studies of a number of acylsilanes the carbonyl carbon was found to be deshielded by ~30 ppm compared to the carbon analogues with chemical shifts ranging from 188–246 ppm[31]. The ^{13}C chemical shifts for several of these compounds are collected in Table 5. In the same paper Brook and coworkers reported the chemical shifts for some silaethylenes (one of which had been reported earlier[32]) which also experienced substantial deshielding (>40 ppm) compared with ethylene derivatives of similar substitution. Values for these compounds are shown in Table 6.

In summary, the addition of silicon atoms to organic compounds stretches both ends of the carbon chemical shift range for neutral organic species. From the silylmethyl

TABLE 5. ^{13}C chemical shifts of the carbonyl carbon of some acylsilanes[31]

Compound	$\delta_{C=O}$ (ppm)
$(Me_3Si)_3SiCOCMe_3$	244.6
$(Me_3Si)_3SiCOCEt_3$	246.7
$(Me_3Si)_3SiCOPh$	233.8
$(Me_3Si)_3SiCOC_6H_4OMe$-*p*	230.6
$(Me_3Si)_3SiCOC_6H_4OMe$-*o*	240.0
$(Me_3Si)_2SiCOCF_3$	225.0
$(Me_3Si)_3SiCOOH$	196.8
$(Me_3Si)_3SiCOOSiPh_3$	188.4
$Ph_3SiCOMe$	240.1
$Me_3SiCOMe$	248.7
$Me_3SiCOPh$	233.7
$Et_3SiCOPh$	235.9
$Ph_3SiCOPh$	229.8

TABLE 6. Carbon-13 chemical shifts for silaethylenes[31, 32]

$$(Me_3Si)_2Si{=}C\begin{matrix} OSiMe_3 \\ R \end{matrix}$$

R	δ_c (ppm)
CEt_3	207.3
CMe_3	212.7
1-methyl-1-cyclohexyl	212.9
1-adamantyl[32]	214.2

carbon atoms which appear at the high field end of the spectrum (+20 to −10 ppm) to the highly deshielded carbonyl carbon of the acylsilanes at ~240 ppm, silicon-substituted carbon atoms appear throughout the entire chemical shift range for carbon. A chart summarizing the chemical shifts for distinctive silyl-substituted carbon atoms is shown in Figure 5.

IV. ^{29}Si NMR

The advantage of using ^{29}Si NMR for analyzing organosilicon compounds is that it is often the chemical or structural changes taking place at or near the silicon atom that are under investigation. Since the ^{29}Si chemical shift range is of the same order of magnitude as that for ^{13}C, and substantially larger than that for ^{1}H, the silicon chemical shift is generally a good method for differentiating possible products of a reaction or probing the structure of a molecule. This is particularly true for the polysilanes $[(R_2Si)_n]$ and polysiloxanes $[(R_2SiO)_n]$, where the silicon atoms form the backbone of the polymer chain. In these materials the carbon atoms and protons are on the periphery of the molecule and therefore the ^{13}C and ^{1}H chemical shifts are less sensitive to structural features than the ^{29}Si chemical shifts. Recent improvements in the high resolution NMR of solids has added a new dimension to the study of polysilanes and polysiloxanes, many of which have now been examined in the solid state as well as in solution.

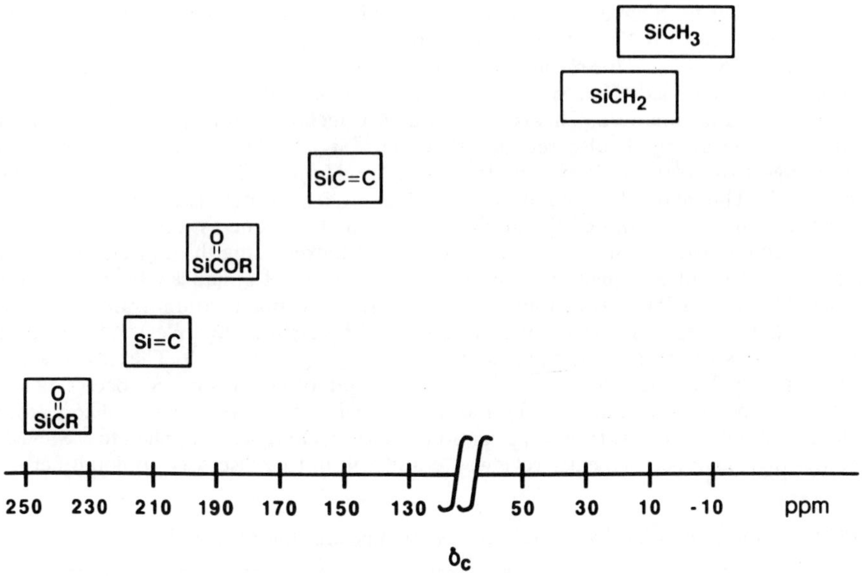

FIGURE 5. Chemical shift ranges for characteristic carbon atoms in organosilanes

Silicon-29 is the only naturally occurring isotope of silicon with nonzero spin ($I = \frac{1}{2}$). Its low natural abundance of 4.7% and low sensitivity relative to ^{1}H places it with carbon-13 (natural abundance 1.1%) in the group of lower sensitivity nuclei. It has the additional problems of long spin–lattice relaxation times (T_1) and negative gyromagnetic ratio (γ)[33]. The latter means that under proton decoupling conditions the nuclear Overhauser effect (NOE) is negative. This can produce reduced signal intensities, null or negative signals depending on the spin–lattice relaxation mechanisms that are operative. As a result, under the normal conditions used to obtain a ^{13}C spectrum (continuous proton decoupling, 1–2 s interval between pulses) one might observe no ^{29}Si signals, even for a concentrated sample. It is possible to use gated decoupling to eliminate the NOE and long delay times between pulses to allow the nuclei to relax, but fortunately there are several alternate solutions to these problems which are more efficient.

The easiest method is to add a small amount (0.01 M) of a paramagnetic relaxation reagent such as $Cr(acac)_3$. This replaces other relaxation mechanisms with the highly efficient electron–nuclear dipole–dipole interaction which reduces T_1 to a few seconds. It also has the effect of removing the NOE since the proton–nuclear dipole–dipole contribution to spin–lattice relaxation is eliminated. The advantage of this technique is that it usually produces quantitative spectra if an adequate pulse interval (>2 s) and gated decoupling (to eliminate any residual NOE) is used. The disadvantages are that some materials may react with the relaxation reagent, and this method affords no enhancement of the signal as is available with some of the other techniques.

The other methods of obtaining silicon-29 NMR spectra take advantage of polarization transfer from the highly abundant, rapidly relaxing protons to effect substantial sensitivity enhancements and allow more rapid acquisition of data. Both selective population transfer (SPT)[34, 35] and J cross polarization (JCP)[36, 37] take advantage of scalar coupling and can produce signals enhanced in the ^{29}Si spectra of up to a factor $|\gamma_H/\gamma_{Si}| = 5$, where γ_H and γ_{Si} are the gyromagnetic ratios for ^{1}H and ^{29}Si, respectively.

These techniques both have limitations which prohibit their general applicability. In SPT spectra one silicon satellite of a proton signal is irradiated prior to application of a nonselective ^{29}Si pulse. This results in population inversion of the ^{1}H–^{29}Si energy levels and produces a coupled ^{29}Si spectrum in which the silicon atom whose coupled protons were irradiated is enhanced. This is not a practical technique for molecules containing multiple silicon atoms. It also requires that the ^{29}Si satellites can be observed in the proton NMR spectrum. JCP requires that both the ^{1}H and ^{29}Si pulses are on resonance and that the Hartman–Hahn condition[38] ($\gamma_H H_H = \gamma_{Si} H_{Si}$) is met. These exacting experimental demands have limited the application of this technique as well.

Substantial sensitivity improvements have been achieved using the nonselective polarization transfer pulse sequences INEPT (insensitive nuclei enhanced by polarization transfer)[39] and DEPT (distortionless enhancement by polarization transfer)[40]. For example, the silanes in Table 7 which were studied by Helmer and West[41, 42] using the INEPT pulse sequences show enhancements factors (E_d) of up to 9. The advantage to both of these pulse sequences is that they may be used for acquiring ^{29}Si spectra as long as there are protons available in the molecule which are coupled to the silicon atoms under investigation. The various polarization transfer techniques as applied to ^{29}Si NMR spectroscopy have been recently reviewed[42] and will not be discussed in detail here.

TABLE 7. Enhancement of ^{29}Si resonances by proton polarization transfer[41, 42]

	E_d (Exp)	δ_{Si} (ppm)	J (Hz)
Me_4Si	9.3	0	6.6
Mc_3SiCl	7.5	29.87	6.8
t-$BuSiCl_3$	3.6	17.75	10.4
Ph_2SiCl_2	3.2	6.47	6.6
$PhSiH_3$	5.7	−59.75	199.9, 6.4 (*ortho*), 1.1 (*meta*)
$(MeO)_4Si$	5.5	−78.25	3.7
$(Me_3\mathbf{Si}OSiMe_2)_2O$	6.5	7.17	6.7 ($^2J_{SiH}$)
$(Me_3SiO\mathbf{Si}Me_2)_2O$	6.2	−21.66	7.3 ($^2J_{SiH}$)
$(Me_3Si)_2C_2$	8.9	−19.13	7.1
$(Me_3Si)_2$	7.3	−19.60	6.5 ($^2J_{SiH}$), 2.7 ($^3H_{SiH}$)
$(Me_2SiO)_5$	7.1	−21.14	7.4
$(EtO)_3SiH$	4.7	−58.82	286.1 ($^1J_{SiH}$), 3.4 ($^3J_{SiH}$)
$(Me_2Si)_6$	7.3	−41.86	6.2 ($^2J_{SiH}$), 3.3 ($^3J_{SiH}$)

The presently known silicon chemical shift range is 825 ppm, although this includes the extreme cases of decamethylsilicocene[43] which is the most shielded resonance reported to date (−577 ppm), and the tetrametallated silicon atoms of μ^3-$Cp(CO)_2FeSiCo_3(CO)_9$ which presently defines the low field (high frequency) end of the spectrum at 248 ppm[44]. Most silicon chemical shifts actually occur in a much smaller range from +50 to −190 ppm relative to Me_4Si (TMS). As in ^{1}H and ^{13}C NMR, TMS is the most commonly used reference compound. All chemical shifts reported here will be given relative to TMS at $\delta_{Si} = 0$ ppm with positive values to high frequency (low field) and negative values to low frequency (high field). Unlike in ^{1}H and ^{13}C NMR, most silicon chemical shifts are upfield from TMS and therefore have negative values. This inconvenience is overridden by the stability, solubility in most organic solvents and relatively short T_1 of TMS, however. One disadvantage of this reference compound is its volatility which makes it unsuitable for use at high temperatures. Many other reference compounds are available

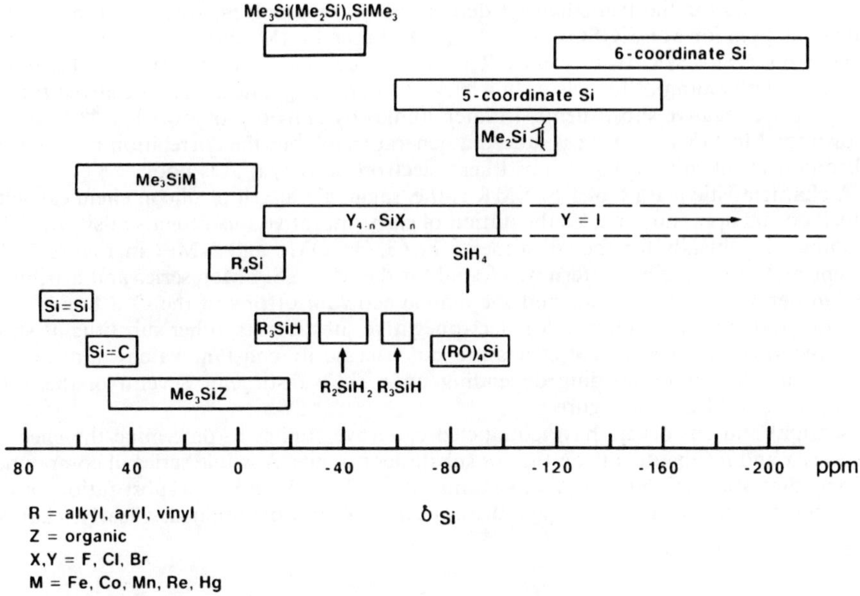

FIGURE 6. Silicon-29 chemical shift ranges for various organosilanes

including $(Me_3Si)_2O$ ($\delta_{Si} = 6.97$ ppm)[5] and the recently suggested tetrakis-(trimethylsilyl)methane ($\delta_{Si} = 3.06$ ppm)[45]. These latter compounds are much less volatile than TMS.

The greatest effect on ^{29}Si chemical shifts occurs with change in coordination number or hybridization of silicon. This is shown in Figure 6 which contains a chart of silicon-29 chemical shift ranges by compound type. In general, four coordinate silicon with organic substituents have chemical shifts in the +45 to −115 ppm region of the spectrum. Five- and six-coordinate silanes are more shielded and appear from −70 to −200 ppm. Silicon atoms in double bonds (Si=C and Si=Si) are deshielded relative to saturated silicon atoms and have chemical shifts in the 35–75 ppm region of the spectrum. This is analogous to ^{13}C NMR in which alkenes are deshielded relative to saturated carbon atoms. Iodo derivatives are unusually shielded as is also observed in ^{13}C NMR. This is where the similarity to ^{13}C NMR ends, however, and the chemical shift trends with substitution of electronegative elements, for example, are quite different for silicon than for carbon. Although it is known in a qualitative sense that substituent electronegativities and silicon coordination number are a major determinant of ^{29}Si chemical shifts, other factors such as steric interactions, variations from tetrahedral bond angles, the possibility of (p-d) π bonding and magnetic anisotropy of neighboring groups must be considered. Discussions of theoretical calculations of ^{29}Si chemical shifts may be found in References 4–9.

A. Methyl Silanes

A large body of data has been collected on substituted methyl silanes, $Me_{4-n}SiX_n$ and $Me_{3-n}X_nSi(CH_2)_mY$. With the exception of coordination number, the greatest effect on ^{29}Si chemical shifts is, of course, caused by substitution changes directly on silicon. The chemical shifts for the trimethylsilyl derivatives, for example, span a range of 65 ppm from 44.6 ppm for $Me_3SiOSO_2CF_3$[46] to −19.58 ppm for $(Me_3Si)_2$[47]. In one of the early papers on ^{29}Si NMR Hunter and Reeves[21] noted that electronegative substituents caused a deshielding of the silicon nucleus and that the greatest effect occurred for the most electronegative substituents. In later studies by Harris and coworkers[48, 49] it was shown for Me_3SiX that there is indeed a general trend, but the correlation between the silicon chemical shift (δ_{Si}) and substituent electronegativity (χ_X) is poor.

A characteristic feature of ^{29}Si NMR is the 'sagging' pattern of silicon chemical shifts which occurs upon multiple α substitution of electronegative substituents at silicon. This is shown graphically for the substituents F, Cl, Br, OMe and NMe_2 in Figure 7. An exception to the sagging pattern was found for the $Me_{4-n}Si(SMe)_n$ series and attributed to van der Waals interactions and the diamagnetic anisotropy of the C–S bonds[50]. In contrast to the results obtained for electronegative substituents, other substituents show unidirectional changes. These effects are not necessarily constant values and may be either shielding or deshielding depending upon the substituent. Several of these are presented graphically in Figure 8.

Schraml and coworkers have conducted extensive studies to determine the effect of various α and β substituents on these α substituent trends. A whole series of compounds of structure $Me_{3-n}X_nSi(CH_2)_mY$ was examined[52–58]. In all cases of β substitution ($m = 1$) the general trend with increasing substitution of X was maintained. Clearly, changes

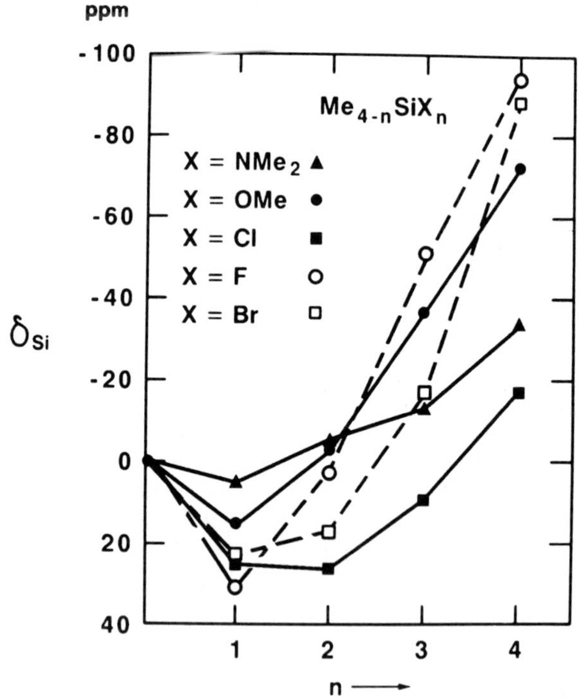

FIGURE 7. ^{29}Si chemical shifts (in ppm relative to TMS) for the methylsilanes $Me_{4-n}SiX_n$ where X is an electronegative substituent. Data are from References 4 and 49

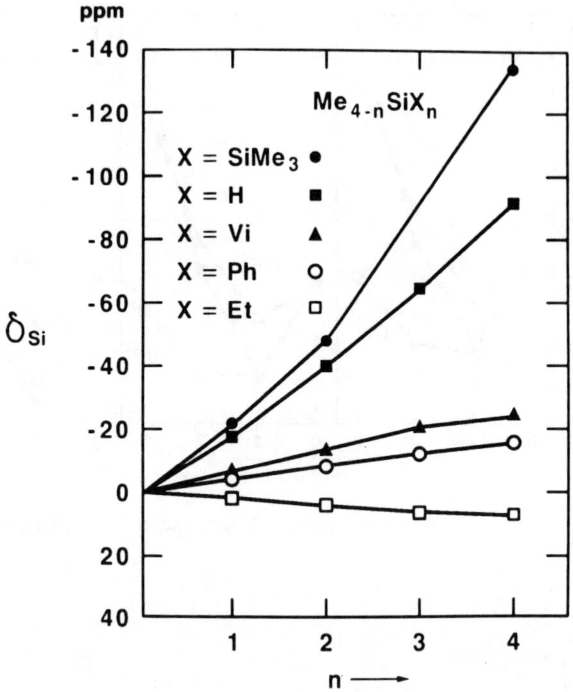

FIGURE 8. ^{29}Si chemical shifts (in ppm relative to TMS) for the methylsilanes $Me_{4-n}SiX_n$. Data are from References 4, 47 and 51

occurring at the more remote β position do not dominate the silicon shielding trends. In general, however, if Y is an electronegative element, the β substituent effect is greater. Similarly, if the substituent $(CH_2)_mY$ was changed from methyl to allyl to vinyl, the shielding trends for the same X substituent remained the same[55]. In all cases, however, the vinyl derivatives were more shielded than the allyl and methyl derivatives. Likewise, the effect of changing Y from Me to CH_2Ph to Ph in the series of compounds $Me_{3-n}X_nSiY$[55] produces virtually no effect on the *trends* of the chemical shift changes although the magnitude may be affected. These trends are shown graphically in Figure 9.

B. Aryl Silanes

One of the most dramatic examples of the influence of other substituents on the direction of chemical shift changes in ^{29}Si NMR is the correlation observed between the silicon chemical shift and Hammett σ constants[59, 60] in the series *p*-$XC_6H_4SiY_3$. The linear correlation between silicon chemical shifts and Hammett σ constants was first noted by Scholl and coworkers[47] for trimethylsilyl substituted benzenes (Y=Me). In subsequent studies[59, 60] it was shown that although a good linear correlation existed between δ_{Si} and σ, the magnitude and signs of the slopes of these lines are not the same. If Y is an electronegative substituent such as F, Cl or OEt, an increase in the electron-withdrawing ability of X produces a shielding effect at silicon. The opposite effect is observed for $Y=CH_3$ and H. The data for Y=OEt and H are shown in Figure 10.

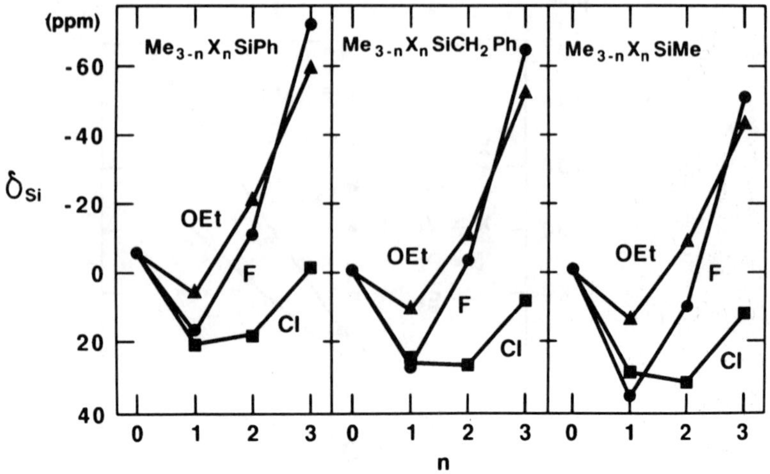

FIGURE 9. Chemical shift trends for $Me_{3-n}X_nSiY$ where Y = Ph, CH_2Ph and Me[55]

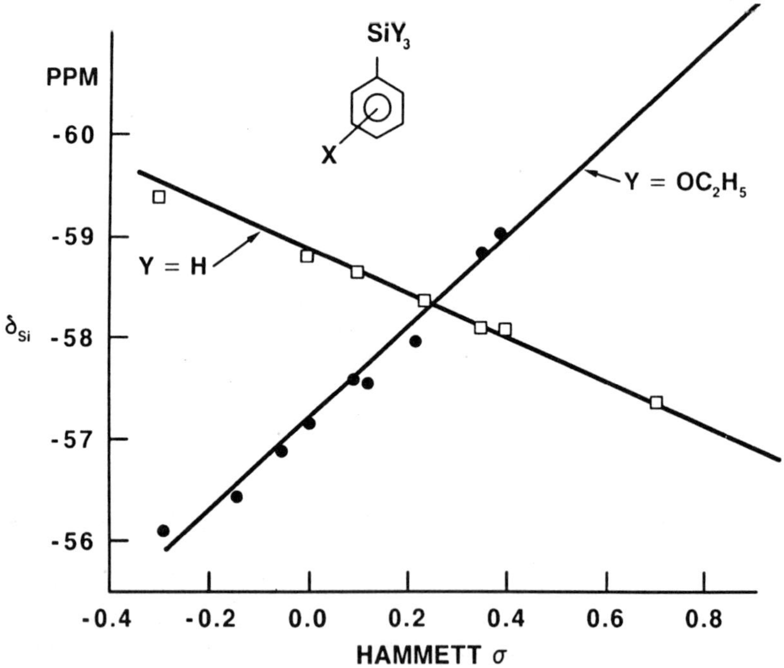

FIGURE 10. Relationship between silicon chemical shift and Hammett σ values for aryltriethoxysilanes and arylsilanes[59, 60]

Similar results were found for a series of alkoxyaminosilanes of general structure $(RO)_nMe_{3-n}SiNHC_6H_4X$-*p*[61] where it was noted that both positive and negative slopes were obtained depending on the number of electronegative atoms attached to silicon. Recently a linear correlation between silicon chemical shifts and the sum of the Taft σ^* constants for the substituents in tetraaryloxysilanes[62], $(p\text{-}XC_6H_4O)_4Si$, was noted.

An oxygen atom between the phenyl ring and the silicon atom has been shown to effectively double the substituent effect in aryloxytrimethylsilanes ($XC_6H_4OSiMe_3$) when compared with aryltrimethylsilanes ($XC_6H_4SiMe_3$)[60, 63] even though the substituent is one bond further removed from the silicon atom. More recently this enhanced sensitivity to substituent effects through the oxygen atom in phenoxysilanes relative to phenylsilanes has been confirmed by Schraml and coworkers[64]. They explored the transmission of the substituent effect of X through the fragment XC_6H_4– (using ^{13}C chemical shifts and CNDO/2 net atomic charges as probes) and found that the results were the same for the aryloxy and aryl derivatives. They concluded that the higher sensitivity of silicon to substituent effects in the aryloxysilanes is due to the silicon atom in the aryloxysilanes having more sensitivity to electron density changes rather than a better transmission of substituent effects through the bonds. This emphasizes the influence the other substituents at silicon have on the magnitude and direction of silicon shieldings effected by one particular substituent.

Recently several dimeric derivatives of catechol have been prepared (**1**–**3**) and their monomer/dimer equilibria examined[65]. It was found that at a temperature of 59 °C **2** and **3** form predominantly monomeric compounds, whereas **1** exists predominantly as the dimer even at this temperature. This was rationalized by considering

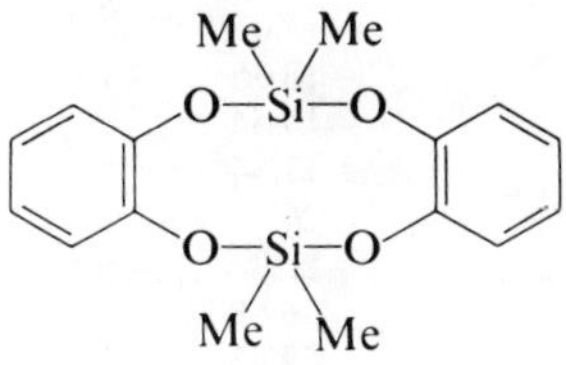

(**1**)

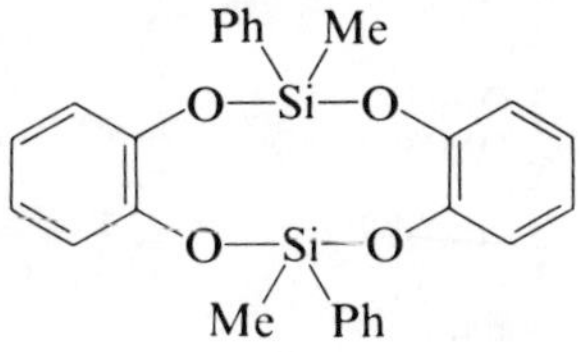

(**2**)

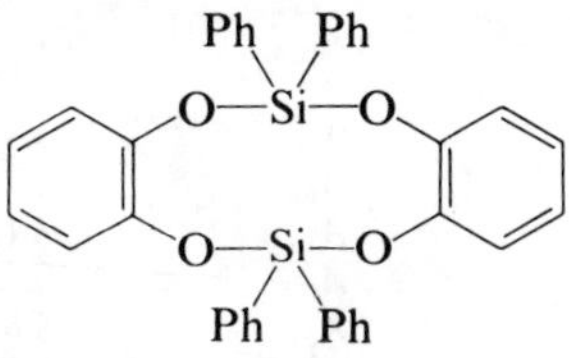

(**3**)

the opposing effects of both ring strain in the five-membered ring of the monomer and the steric interactions between the two substituents on each silicon atom in the dimer. The steric interactions in the dimer would be expected to be greatest in the diphenylsilyl compound, **3**, and indeed this compound dissociates to monomer most easily. The dimethylsilyl compound, **1**, remains dimeric because for this system relief of ring strain dominates. The ^{29}Si chemical shifts for these compounds are shown in Table 8. The normal substituent effect of changing from Me to Ph was observed (about -15 ppm). A deshielding effect of approximately 36 ppm was observed on forming the five-membered ring monomer structure versus the ten-membered ring of the dimer. This deshielding with ring strain is a general feature of ^{29}Si NMR and will be discussed in other sections for cyclosilanes and siloxanes.

C. Halosilanes

A good empirical method for calculating the ^{29}Si chemical shifts in alkylhalosilanes (and other silanes as well) has been recently developed[66] and is based on the method of Grant and Paul[67] for calculating the ^{13}C chemical shifts of haloalkanes and the extension of this method to ^{29}Si NMR by Harris and Kimber[68]. Equation 2 was fitted with 175

TABLE 8. Silicon-29 NMR chemical shifts for catechol derivatives[65]

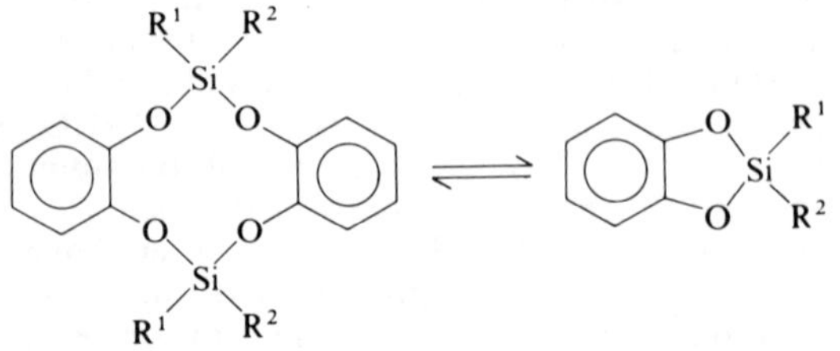

R^1	R^2	δ_{Si} (ppm)	δ_{Si} (ppm)
Me	Me	− 4.96	+32.19
Me	Ph	−19.81, −20.73	+15.48
Ph	Ph	−37.00	−1.37

TABLE 9. Coefficients (ppm) from the multiple linear regression of ^{29}Si chemical shifts[66]

$A_{C(1)}$	18.88	$C_{(C(1), C(1))}$	0.37	$C_{(O, O)}$	−10.21
$C_{C(2)}$	−0.42	$C_{(C(1), C(2))}$	0.44	$C_{(F, F)}$	−23.69
$A_{C(3)}$	−1.21	$C_{(C(1), O)}$	3.31	$C_{(F, Cl)}$	−15.03
$A_{C(4)}$	2.01	$C_{(C(1), Ph)}$	0.01	$C_{(F, Br)}$	−13.35
A_O	13.31	$C_{(C(1), Vi)}$	0.33	$C_{(Cl, Cl)}$	−14.54
A_{Ph}	11.73	$C_{(C(1), F)}$	5.02	$C_{(Cl, Br)}$	−17.54
A_{Vi}	8.12	$C_{(C(1), Cl)}$	2.81	$C_{(Cl, I)}$	−26.73
A_F	23.12	$C_{(C(1), Br)}$	4.04	$C_{(Br, Br)}$	−22.90
A_{Cl}	34.84	$C_{(C(1), I)}$	6.20	$C_{(Br, I)}$	−36.68
A_{Br}	29.78	$C_{(Ph, Ph)}$	2.67	$C_{(I, I)}$	−52.43
A_I	11.11	$C_{(Vi, Vi)}$	2.67		
B	−71.65				

TABLE 10. Calculated and experimental chemical shifts for some halosilanes[66]

	δ_{Si}	
Compound	Calc. (ppm)	Exp. (ppm)
$SiBrI_3$	−276.48	−280.1
$SiBr_2I_2$	−212.73	−212.3
$SiBrICl_2$	−101.04	−98.9
$SiBrCl_3$	−33.58	−34.3
SiH_2Cl_2	−16.51	−11.0
$Si(Ph)MeCl_2$	19.72	17.9
$SiMe_2F_2$	9.10	8.8
$SiCl_2Me_2$	32.83	32.2
$SiF(Bu\text{-}n)_3$	29.32	28.8

pieces of data from 114 compounds using a least-squares multiple regression analysis to give a correlation coefficient of 0.994 and a standard error of 7.67 ppm. The coefficients obtained are shown in Table 9 and the calculated and experimental values for several of the compounds are shown in Table 10. The authors noted that the largest errors were observed for silicon atoms with directly bonded protons suggesting that the diamagnetic shielding term, which is usually ignored in calculations of silicon chemical shifts, may actually be quite important.

$$\delta = B + \sum_i A_i + \sum_{i \neq j} C_{ij} \tag{2}$$

The 'sagging' pattern which is so characteristic of sequential halogen substitution in alkyl and aryl silanes is not observed in the simple mixed silicon halides ($Y_{4-n}SiX_n$). The sagging pattern apparently arises as a result of a substantial electronegativity difference between X and Y which is not present in the mixed halides. The chemical shift trends for the mixed halosilanes are shown in Figure 11[69, 70].

A series of oligomeric perchlorosilanes have been reported[71] from which some trends can be deduced. The data are shown in Table 11. The terminology M=(Si)**Si**Cl_3, D=$(Si)_2$**Si**Cl_2, T=$(Si)_3$**Si**Cl and Q=$(Si)_4$**Si**, which signify monosubstituted, disubstituted, trisubstituted and quarternary silicon atoms, respectively, will be used in this discussion. The Q-type silicon atoms clearly have distinctive chemical shifts of approximately −80 ppm as can be seen from **9** and **11** in Table 11. Although there is only one T-type silicon atom in the table, the chemical shift is sufficiently different from the others to assume that these will also be characteristic (**10**, Si^B). The M and D silicon atoms are not easily differentiated, however, as evidenced by the observation that the chemical shifts of both Si^A and Si^B in **8** are identical. There is a consistent deshielding effect of 2.5–4 ppm

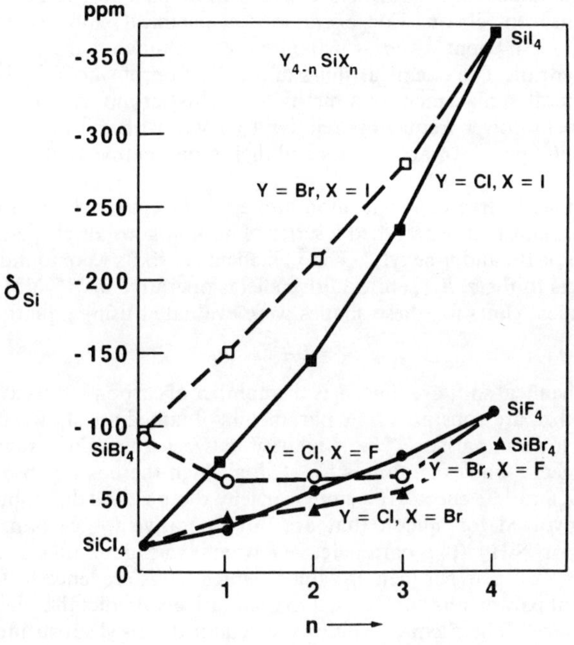

FIGURE 11. Chemical shift trends for mixed silicon halides $Y_{4-n}SiX_n$[69, 70]

TABLE 11. ^{29}Si chemical shifts for oligomeric perchlorosilanes[71]

	δ_{Si}			
Compound	*A*	*B*	*C*	*D*
(**4**) $SiCl_4$	−18.45			
(**5**) $Cl_3SiSiCl_3$	−6.13			
(**6**) $(Cl_3Si^A)_2Si^BCl_2$	−3.70	−7.44		
(**7**) $(Cl_3Si^ASi^BCl_2)_2$	−3.91	−4.15		
(**8**) $(Cl_3Si^ASi^BCl_2)_2Si^CCl_2$	−3.82	−3.82	−1.15	
(**9**) $(Cl_3Si^ASi^BCl_2Si^C(Si^DCl_3)_3$	−3.50	+6.27	−79.48	+4.13
(**10**) $(Cl_3Si^A)_3Si^BCl$	0.0	−31.83		
(**11**) $(Cl_3Si^A)_4Si^B$	+3.95	−80.41		

with each β substituent. Furthermore, the deshielding effect increases slightly with each subsequent replacement of a β chlorine with a $SiCl_3$ substituent, as shown by considering Si^A in compounds **5**, **6**, **10** and **11**.

D. Silicon Hydrides and Alkylsilanes

Almost all silicon hydrides are shielded relative to TMS, and very general chemical shift ranges can be given. Most silicon trihydrides ($RSiH_3$) appear in the region −30 to −70 ppm, silicon dihydrides appear from −20 to −40 ppm, and monohydrides arc in the 0 to −30 ppm region of the ^{29}Si spectrum. Within these ranges there is considerable overlap of the resonances and, as always, consideration must be given to the nature of the other substituents on silicon. For example, the chemical shifts of $(MeO)_3SiH$ and $PhCH_2SiH_3$ are −54.9 ppm[72] and −56.0 ppm[73] even though one is a monohydride and the other a trihydride. The chemical shifts of a collection of silicon hydrides is given in Table 12. In general, replacement of a methyl or phenyl group with a hydrogen effects a shift to higher field (lower frequency) relative to TMS, with the magnitude of the shift between 15 and 30 ppm. Several hydrides and their monomethyl analogues are compared in Table 13.

Alkyl silanes may be treated in a fashion similar to the approach of Grant and Paul for alkanes[67]. Harris and Kimber studied a series of homologous alkyl silanes R_3SiH where R = Me, Et, *n*-Pr, *n*-Bu and *n*-hexyl[68]. The ^{29}Si chemical shifts were found to have similar substituent effects to the α, β, γ, etc. additive effects operative in ^{13}C NMR. Accordingly, the silicon chemical shifts for these silanes were evaluated using equation 3,

$$\delta_{Si} = B + \Sigma_i A_i n_i \tag{3}$$

where B is the chemical shift of silane, n_i is the number of carbon atoms at the ith position from silicon, and A_i are constants. The parameters obtained are shown in Table 14 with the ^{13}C values for comparison. The similarity between the values suggest a common origin for the effects. The exception is for A_α for which there is a substantial difference between the ^{29}Si and ^{13}C effects. It is not clear why these values differ, but there must be other factors involved for silicon that are not operative for carbon. The estimated chemical shift for SiH_4 (the parameter B) which was determined in this work is remarkably close to the experimental value[70] which lends credence to this approach.

The substituent parameters can be used to qualitatively predict the silicon shieldings in other systems as well. The chemical shifts expected for the alkyl substituted silicon atoms of some 1,1,1-dimethylalkyl-2,2,2-triphenyldisilanes were calculated[75] using the trimethyltriphenyldisilane as the base value and the substituent parameters developed by Harris

TABLE 12. ^{29}Si chemical shifts for some silicon hydrides[74]

	Compound	δ_{Si}
SiH	$(Me_3Si)_3$**Si**H	−117.40
	$(H_3Si)_3$**Si**H	−96.0
	Me_3Si**Si**Me_2H	−39.1
	Ph_3SiH	−21.1
	Ph_2MeSiH	−19.5
	$PhMe_2SiH$	−17.15
	$(EtO)_2MeSiH$	−16.1
	n-Pr_3SiH	−8.11
	n-Bu_3SiH	−6.58
	$(PhCH_2)_3SiH$	−3.83
	Et_3SiH	0.15
SiH_2	$(H_3Si)_2$**Si**H_2	−116.5
	$PhMeSiH_2$	−36.80
	Ph_2SiH_2	−34.5
	$(H_3SiCH_2)_2$**Si**H_2	−30.44
	(H_3SiCCl_2)**Si**H_2	−5.09
SiH_3	H_3SiSiH_3	−104.8
	$PhSiH_3$	−59.9
	$PhCH_2SiH_3$	−56.0
	$N(SiH_3)_3$	−39.92
	SiH_4	−91.9
	$KSiH_3$	−165[a]

[a] Reference 14.

TABLE 13. Comparison between silicon hydrides and analogous methyl silanes[74]

	δ_{Si} (ppm)	$\Delta\delta_{(SiH-SiMe)}$
Me_4Si	0	−16.34
Me_3SiH	−16.34	
$(MeO)_3SiMe$	−39.8	−15.1
$(MeO)_3SiH$	−54.9	
$Me_3SiSiMe_3$	−19.78	−19.3
Me_3Si**Si**Me_2H	−39.1	
Cl_3SiMe	12.47	−22.1
Cl_3SiH	−9.6	
Me_2SiH_2	−41.5	−23.7
$MeSiH_3$	−65.2	

TABLE 14. Substituent effects for alkylsilanes[68] and alkanes[67]

Parameter	$^{29}Si^{b}$ (ppm)	$^{13}C^{c}$ (ppm)
B	−91.9	
A_{α}	+25.2	+9.1
$A_{\beta}{}^{a}$	+5.50	+5.75
A_{γ}	−2.75	−2.49
A_{δ}	+0.51	−0.31
4°(1°)	−2.2 ppm	−1.50

[a] Includes effect of chain branching.
[b] Value of parameter used to calculate ^{29}Si chemical shifts from equation 3.
[c] Value of analogous parameter used by Grant and Paul[67] to calculate ^{13}C chemical shifts.

TABLE 15. Calculated[75] and experimental[76] chemical shifts for $Ph_3Si\mathbf{Si}Me_2R$ using additive substituent parameters[68]

R	δ_{Si} (calc)	δ_{Si} (exp)
CH_3		−18.4
n-Pr	−15.7	−17.0
n-Oct	−15.1	−16.8
i-Pr	−9.4	−11.9
t-Bu	−4.1	−8.0

and Kimber[68]. When the calculated shifts are compared with the experimental data (Table 15) it is apparent that the trends are reproduced although the predicted values are slightly lower than the chemical shifts actually observed.

E. Si=C and Si=Si

Within the last eight years two new classes of compounds with a silicon atom double bonded to carbon or another silicon atom have been reported[31, 32, 77, 78]. The silaethylenes (Si=C) and disilenes (Si=Si) behave as would be predicted by analogy with ^{13}C NMR; both types of silicon atoms are strongly deshielded relative to their saturated silicon analogues. The silaethylenes reported to date have chemical shifts in the 40–55 ppm range as shown in Table 16. The other silicon atoms in these compounds do not exhibit unusual chemical shifts and their resonances are found in the same region as those from saturated model compounds. For example, Si^{D} in Table 16 appears in the 12.7–13.6 ppm range which is quite close to the value observed for $MeOSiMe_3$ (13.5 ppm)[21], and Si^{B} and Si^{C} are at approximately −12 ppm whereas $(Me_3\mathbf{Si})_2SiMe_2$ has a chemical shift of −15.9 ppm[76].

The first report of a stable molecule with a silicon–silicon double bond appeared in 1981[78] along with its ^{29}Si chemical shift. Since then the chemical shifts for several disilenes have been reported[79]. All of the compounds are tetrasubstituted with 2,4,6-trimethylphenyl (mesityl) or 2,6-dimethylphenyl substituents on silicon. The chemical shifts for these structurally closely related compounds all fall within 3 ppm (62.5–65.2 ppm). Since chemical shifts in this region are rare, the ^{29}Si chemical shift provides a good method for identifying these species.

TABLE 16. Silicon-29 chemical shifts of some silaethylenes[31]

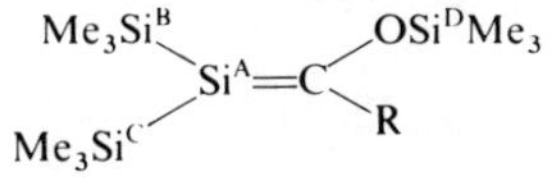

R	A (ppm)	B, C (ppm)	D (ppm)
CMe_3	41.5	−12.1, −12.6	13.4
CEt_3	54.3	−12.4, −12.9	12.7
Adamantyl	41.4	−12.8, −13.4	12.9
1-methyl-1-cyclohexyl	43.5	−12.3, −12.6	13.6

Detailed information about the electronic structure of the silicon–silicon double bond was obtained from the ^{29}Si solid state powder pattern spectrum[80] of tetramesityldisilene. The anisotropy of the chemical shift of Si=Si was found to be comparable to that of C=C in ethylene. In contrast, the anisotropy of Si–Si in the analogous tetramesityldisilane was found to be small which is characteristic of alkanes as well. Based on the similarities of the chemical shift anisotropies it was concluded that the electronic structure of Si=Si closely resembles that of C=C.

F. Cyclic Silanes and Siloxanes

A significant deshielding effect on ^{29}Si chemical shifts occurs with ring strain in cyclic systems. The magnitude of the effect is, of course, dependent upon the type of ring compound involved but the trends for different systems appear to be the same. A number of chemical shifts for cyclic silanes with one silicon atom incorporated into the ring are collected in Table 17[76, 81–83]. In all cases, changing from a six-membered ring to a five-membered ring deshields the silicon atom by 14–20 ppm. A further deshielding of 2 ppm occurs when the silicon atom is in a four-membered ring.

Incorporation of two silicon atoms in a ring decreases the effect of ring size substantially. Table 18 contains the chemical shifts of a number of di- and polysilanes. From the data for the cyclodisilanes it is clear that the effect on the ^{29}Si chemical shift of a decrease in ring size from six to four is 8–10 ppm, unlike the 16–22 ppm for the same change in the cyclic monosilanes. In the cyclic polydimethylsilanes the chemical shift for the five- and six-membered homologues are very close (42.1 and 41.8 ppm[85]) and the strong deshielding effect is only observed on decreasing the ring size to four. Similar results were obtained for the cyclic perchloro- and perbromosilanes[85]. Both the methyl and bromo derivatives are affected strongly by ring size with chemical shift differences between the four- and five-membered rings of 12–14 ppm. The effect is much smaller for the chloro derivatives (<8 ppm).

The chemical shift differences (36 ppm, Table 8) observed for the monomeric and dimeric organosilicon derivatives of catechol were also observed for the analogous derivatives of diols[87] although the magnitude of the effect was less. For compounds **12**–**14** dimerization shifts of 27.0, 6.2, and 6.8 ppm, respectively, were observed on formation of the ten-, twelve- and fourteen-membered rings. Chemical shifts for the monomers are shown with the structures. Compounds **13** and **14** have two chemical shifts due to conformational isomers. The limiting value of the chemical shift as estimated from the open chain analogue $Ph_2Si(OPr\text{-}i)_2$ was not reached until a ring size of 12 atoms was

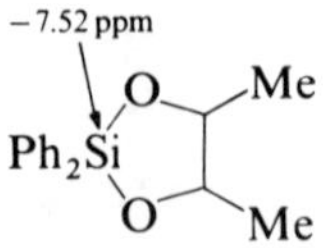

(12)

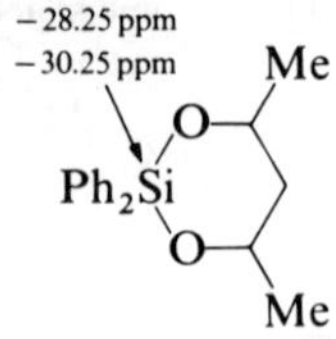

(13)

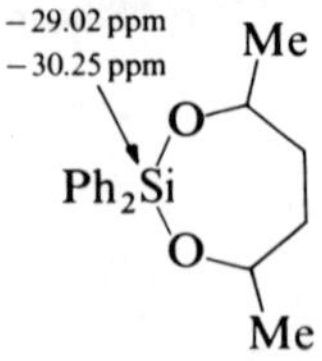

(14)

obtained. The chemical shift relationship with ring strain is consistent with chemical properties[87], such as the markedly decreased stability of the monomeric derivatives 1,2-diols which tend to polymerize.

A similar effect of ring size has been noted for the cyclic siloxanes[88]. The six-membered hexamethyltrisiloxane, $(Me_2SiO)_3$, has a ^{29}Si chemical shift of -9.2 ppm. A shielding

TABLE 17. ^{29}Si chemical shifts of cyclic silanes with one silicon atom incorporated in the ring

Compound	δ_{Si}	Ref.
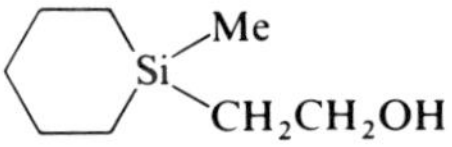	−4.11	81
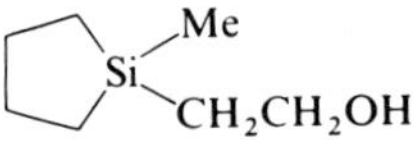	16.33	81
	−1.2	82
	16.77	81
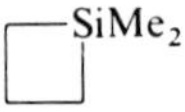	18.90	81
	16.5	83
	17.4	83
	−48.6	76
	−34.8	76

TABLE 18. ^{29}Si chemical shifts of cyclic di- and polysilanes[a]

Compound	δ_{Si}	Ref.
	−22.3	82
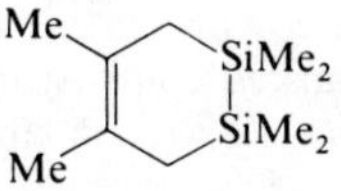	−22.1	82
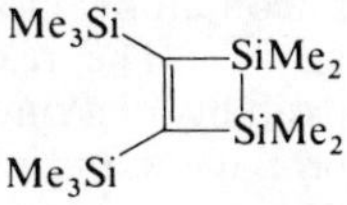	−12.5	82
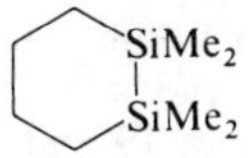	−19.5	82
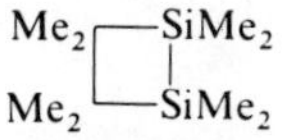	−11.6	82
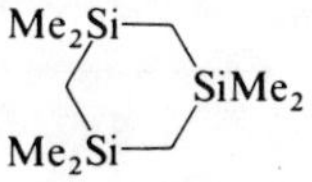	0.3	84
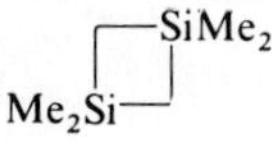	2.73	81
$(Me_2Si)_8$	−39.9	86
$(Me_2Si)_7$	−41.7	86
$(Me_2Si)_6$	−41.81	85
$(Me_2Si)_5$	−42.12	85
$(Me_2Si)_4$	−27.56	85
$(Cl_2Si)_6$	−2.54	87
$(Cl_2Si)_5$	−1.67	87
$(Cl_2Si)_4$	5.86	87
$(Br_2Si)_6$	−25.73	87
$(Br_2Si)_5$	−28.08	87
$(Br_2Si)_4$	−15.97	87

effect of >10 ppm occurs upon increasing the ring size by two atoms; $\delta_{Si} = -19.4$ ppm[5] for $(Me_2SiO)_4$. The ten- and twelve-membered rings of $(Me_2SiO)_5$ and $(Me_2SiO)_6$ have chemical shifts of −21.9 and −22.5 ppm, respectively, which are at the limiting value of the linear dimethylsiloxane repeat unit which has a chemical shift of −21.8 ppm[5].

It may be concluded from the above data that there is a general trend for a deshielding effect with increasing ring strain, but the magnitude of the effect is dependent upon the substituents on the silicon atom and the nature of the ring into which it is incorporated.

As soon as there is appreciable bond angle strain in the silicon bonds a marked effect is noted in the ^{29}Si shielding of the nucleus.

The exceptions to these conclusions include the unusual class of compounds comprised of the silacyclopropanes, disilacyclopropanes and silacyclopropenes. Although cyclic and acyclic tetraalkylsilanes usually appear within a chemical shift range of +5 to −20 ppm, the tetraalkylsilacyclopropanes have chemical shifts in the −50 to −60 ppm region[23, 24, 89]. Analogous results are found in the carbon and proton data for cyclopropane compared with higher ring homologues[90] and are not specific for silicon-containing rings. The disilacyclopropanes prepared by Ishikawa and coworkers[91] have chemical shifts for the ring silicon atoms of −48 to −64 ppm, in marked contrast to the acyclic alkyl and aryl substituted disilanes which have chemical shifts at approximately −20 ppm[76].

Even more unusual are the ^{29}Si chemical shifts for the silicon atoms in silacyclopropene rings which fall into the −87 to −106 ppm range[92, 93]. The reason for the exceptional shielding effect in these rings is not clear. The possibility of a ring current has been considered to explain the unusual shielding in cyclopropane[90] and might also be affecting the chemical shifts in the silacyclopropanes. It has been suggested that the extreme shielding in silacyclopropenes may be due in part to π bonding with involvement of the silicon d-orbitals[92]. The chemical shift data for the silacyclopropanes, disilacyclopropanes and silacylopropenes are collected in Tables 19 and 20.

TABLE 19. Silicon-29 chemical shifts for ring silicon atoms in some representative silacyclopropanes and disilacyclopropanes

Compound	δ_{Si} (ppm)	Ref.
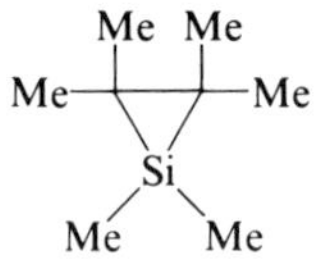	−49.31	23
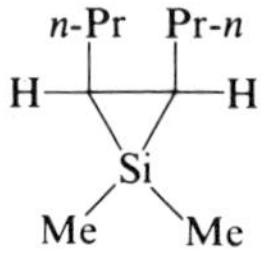	−59.8	89
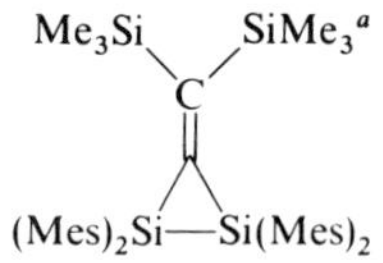	−48.5	91
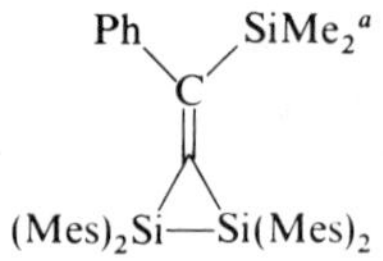	−56.0, −63.3	91

[a] Mes = mesityl.

TABLE 20. Silicon-29 NMR chemical shifts for ring silicon atoms of silacyclopropenes

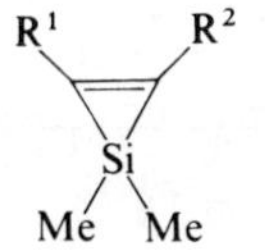

R^1	R^2	δ_{Si} (ppm)	Ref.
Me	*t*-Bu	−87.0	93
Me_3Si	Me_3Si	−106.2	92
Me_2HSi	Me_2HSi	−102.1	93
Me	Me_3Si	−88.6	93
Me_3Si	*t*-Bu	−91.9	93

At the other extreme of the chemical shift range are another class of compounds, the bicyclic silanorbornadienes and silanorbornene derivatives[94]. Silicon-29 NMR data have been reported for eight compounds (Table 21), all of which have resonances at the remarkably high frequency positions of 40–100 ppm from TMS. Although angle strain must certainly be a factor in the strong deshielding of these silicon atoms, other factors are undoubtedly involved. The authors attribute the additional deshielding of these molecules to an enhanced polarization of the ground state arising from σ-π conjugation which leads to a partial positive charge at the silicon atom as shown in Scheme 1.

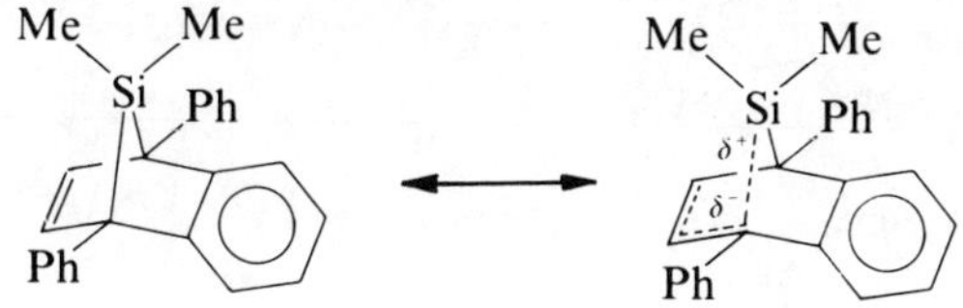

SCHEME 1

G. Five- and Six-coordinate Silicon Compounds

Pronounced shifts to lower frequency occur in the ^{29}Si resonance when silicon compounds with coordination numbers higher than four are obtained. Several studies have now been conducted in which intramolecular coordination of nitrogen to silicon to form a pentacoordinate bond has been shown to produce a strong shielding effect of the silicon chemical shift. Specifically, the silatranes (**15**) are compounds in which a transan-

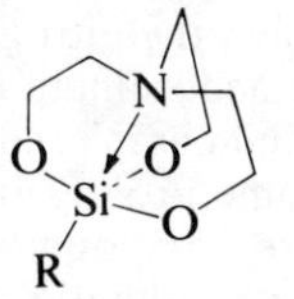

(**15**)

TABLE 21. Silicon-29 chemical shifts for bridge silicon atoms in some bicyclic silanes[94]

Compound	δ_{Si} (ppm)
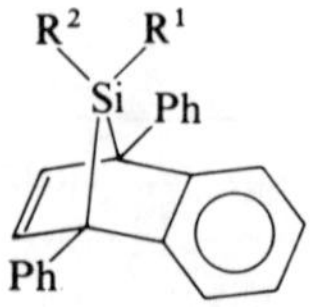	
$R^1 = R^2 = Me$	76.9
$R^1 = Me$, $R^2 = SiMe_3$	82.5
$R^1 = Me$, $R^2 = SiMe_2Ph$	80.2
$R^1 = Me$, $R^2 = SiMe_2SiMe_3$	85.7
$R^1 = R^2 = SiMe_3$	97.7
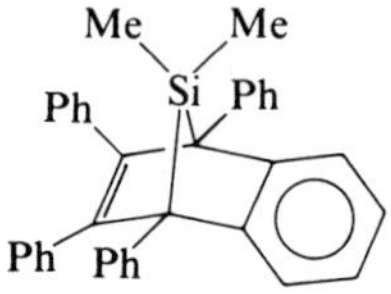	67.7
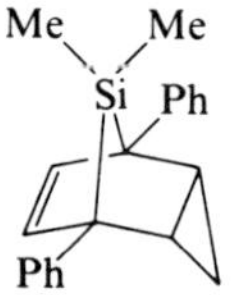	58.5
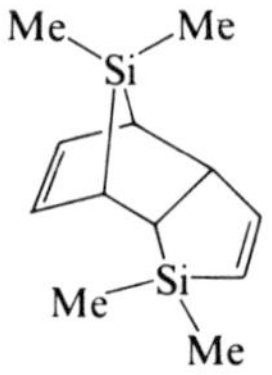	40.8

nular bond from nitrogen to silicon shields the silicon nucleus by approximately 10–30 ppm relative to the analogous triethoxysilanes[95–100]. Silicon-29 chemical shifts were also used to detect the formation of intramolecular coordination in $Ar_2Si(OCH_2CH_2)_2NR$ compounds which can form a transannular N→Si bond only in a boat–boat conformation of the ring[101]. Similar results were obtained for the compounds *o*-$(Me_2NCH_2)C_6H_4SiXYZ$[102] in which the nitrogen can coordinate with the silicon to form a five-membered ring and a pentacoordinate silicon compound. A rapid equilibrium was observed for the uncomplexed and complexed nitrogen species as was evidenced by only one ^{29}Si resonance which became progressively more shielded at lower temperatures as more of the five-coordinate compound was present. From the temperature dependence of the chemical shifts an estimation of the chemical shift changes upon formation of the five-coordinate silicon was made and found to be 20–67 ppm[102].

TABLE 22. Silicon-29 NMR chemical shifts for some five- and six-coordinate organosilicon complexes

Compound		δ_{Si}	Ref.
(16)	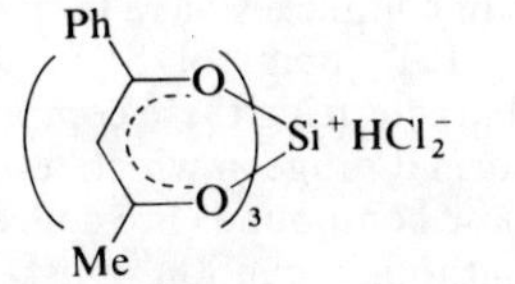	−192.4	105
(17)	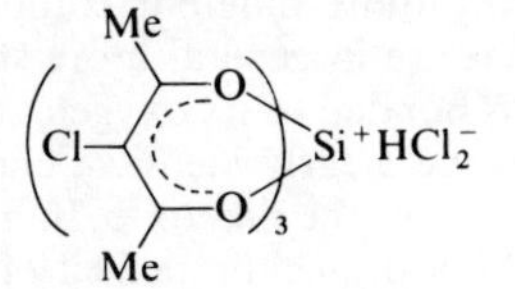	−196.4	105
(18)	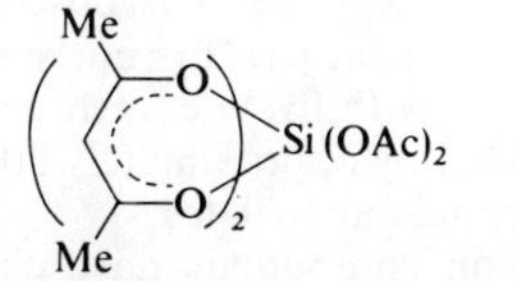	−196.8	105
(19)	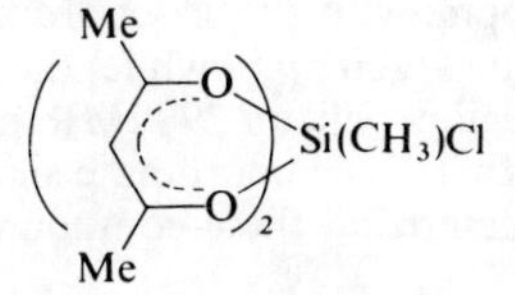	−149.5	105
(20)	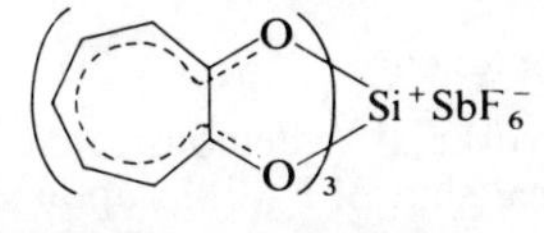	−139.4	105
(21)	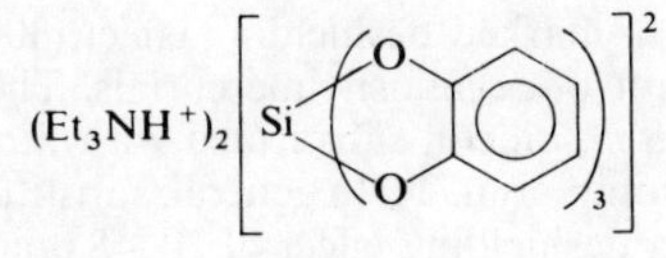	−139.3	105
(22)	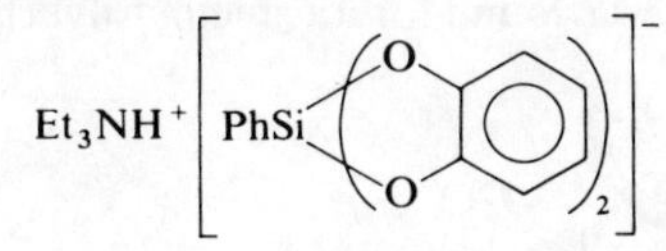	−87.1	105
(23)	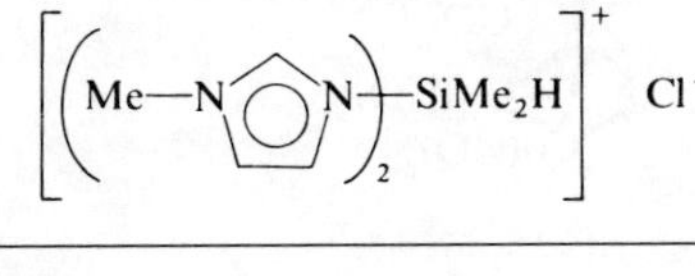	−81	107

Interesting results were obtained for a series of silatranones[103] which showed chemical shifts consistent with five-coordinate silicon for the silatran-3-ones and silatran-3,7-diones. The silicon atom in the silatran-3,7,10-triones, however, is a good electron acceptor and binds electron-donor solvents such as DMSO. For these compounds chemical shifts in the −130 to −150 ppm region, characteristic of six-coordinate silicon, were observed.

A number of penta- and hexacoordinate silicon complexes have been reported[105–107] since the exceptionally shielded resonance (−191.7 ppm) of $SiF_6^{=}$ was reported in 1968[104]. NMR offers an excellent method for characterizing these compounds since their chemical shifts deviate substantially from the normal range in which tetravalent organosilicon compounds occur. Data for several of these compounds are collected in Table 22. From the data several trends are evident, although caution must be taken while attempting to relate information for structurally quite different compounds. The six-coordinate compounds have chemical shifts that are in general more shielded than the five-coordinate compounds. If the silicon atom is bonded to six oxygen atoms as in **16–18** and the chelate ring is derived from a 1,3-diketone, nearly identical chemical shifts are observed regardless of charge type or substitution on the diketone. If, however, the two oxygen atoms are replaced by CH_3 and Cl (**19**), a deshielding effect of almost 50 ppm occurs. The trend for these substituent effects is the same as is observed in the four-coordinate silicon compounds, but the magnitude is larger. It is not surprising that the chelate ring size is also important. The complexes with five-membered chelates, such as tropolone and catechol (**20** and **21**), are deshielded by approximately 50 ppm relative to the similarly substituted complexes with six-membered chelates.

In general the five- and six-coordinate silicon compounds have chemical shifts at unusually low frequency from −50 to −200 ppm, with the six-coordinate compounds appearing at lower frequency from −130 to −200 ppm while the five-coordinate compounds are in the −50 to −150 ppm region. Silicon-29 NMR provides a good method for studying the equilibria between normal four- coordinate silanes and higher-coordinate species in solution as well as characterizing these compounds as penta- or hexacoordinate silicon compounds.

H. Silicon-containing Transition Metal Complexes

An early report by Johannesen and coworkers[104] demonstrated that a marked deshielding effect occurred in the ^{29}Si chemical shift of F_3SiMe upon substitution of a cobalt atom for the methyl group. Ebsworth and coworkers[108] and later studies by Mitchell and Marsmann[109] also found a marked deshielding effect (30–60 ppm) of the mercury atom on the ^{29}Si chemical shift of organosilylmercurials. The most shielded silicon resonance reported to date[44] is for a silicon substituted with three cobalt atoms and one iron atom in the cobalt cluster compound, **24**. In general, substitution of Co, Mn, Fe, Hg and Ru directly on silicon has a deshielding effect of 20–45 ppm[35, 44, 110–112]. The exception to this deshielding effect was found for a trimethylsilyl rhenium complex

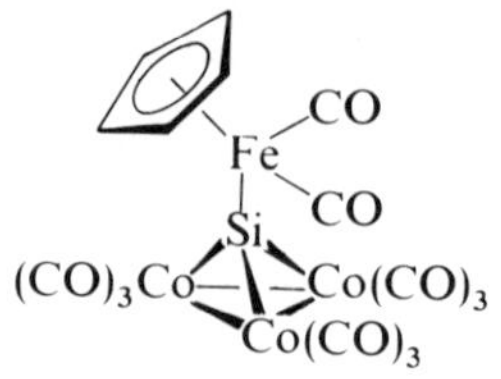

(**24**)

which was shielded relative to TMS[35]. In this study of trimethylsilyl derivatives of $Co(CO)_4$, $Mn(CO)_5$, $Re(CO)_5$ and $(\eta^5\text{-}C_5H_5)Fe(CO)_2$, it was noted that the ^{29}Si chemical shifts paralleled the results obtained previously[113] for ^{119}Sn shifts. Furthermore, it was concluded that π bonding was important in these complexes. The ^{29}Si chemical shifts for several compounds with silicon bonded directly to a transition metal are collected in Table 23. The effect for silicon atoms β to the metal is much less than for the directly bonded silicon atoms, and the γ position is essentially unaffected[112] (see Table 23).

Consideration of the data for a series of $R_3Si(H)Mn(L)(CO)(MeC_5H_4)$ complexes[111] shows that changes in the ligands on the manganese can have a substantial effect on the chemical shift of the silicon atom and that the magnitude of this effect changes with the substituents on silicon. Likewise, it was noted that for the iron complexes $ClMeHSiFe(L)_2(C_5H_5)$, changing from L=CO to $L=PPh_3$ produced a shift to higher field (lower frequency) of 44 ppm[111].

A number of group VI carbonyl silyl substituted carbene and carbyne complexes have been reported[114]. The authors concluded that significant π interactions occur between the metal and the silicon atom either by hyperconjugation or participation of the silicon d orbitals. The chemical shifts for the carbene and carbyne complexes of the same metal were similar. For example, the chemical shfts of $(CO)_5W{=}C(OMe)(SiPh_3)$ and $(C_5H_5)(CO)_2W{\equiv}CSiPh_3$ are −18.3 and −26.7 ppm, respectively.

A number of π-organosilyl transition metal complexes have been studied[110, 115] (Table 24). There is generally a deshielding effect of 6–11 ppm for silicon atoms bonded to a ligand carbon atom directly bonded to a metal, and a very small shielding effect for

TABLE 23. ^{29}Si chemical shifts for some transition metal complexes with silicon directly bonded to the metal

Compound	δ_{Si} (ppm)	Ref.
$Me_3SiRe(CO)_5$	−13.97[a]	35
$Me_3SiMn(CO)_5$	17.95[a]	35
$Me_3SiFe(CO)_2(C_5H_5)$	41.42[a]	35
$Me_3SiCo(CO)_4$	44.42[a]	35
$Me_3Si^BSi^AMe_2Fe(CO)_2(C_5H_5)$	16.95(Si^A), −11.3(Si^B)	112
$Me_3Si^CSi^BMe_2Si^AMe_2Fe(CO)_2(C_5H_5)$	21.22(Si^A), −36.54(Si^B), −15.07(Si^C)	112
$Me_3SiFe(CO)(PPh_3)(C_5H_5)$	34.75	112
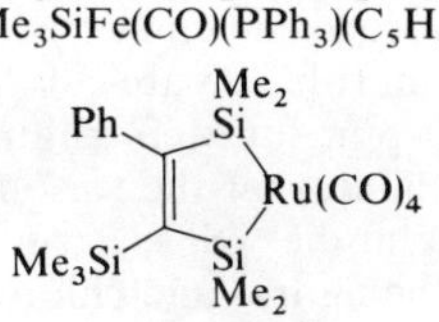	32.40 34.68	110
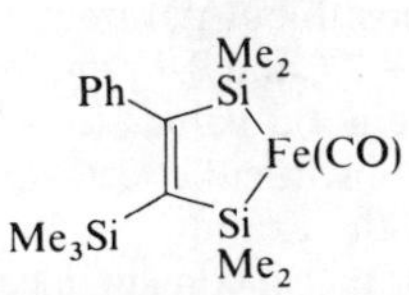	45.39 48.83	110
$Ph_2HSiMn(CO)_2(MeC_5H_4)$	13.5	111
$Ph_2HSiMn(CO)(PMe_3)(MeC_5H_4)$	26.0	111
$Cl_3SiMn(CO)_2(MeC_5H_4)$	54.9	111
$(C_5H_5)(CO)_2FeSiCl_2Co(CO)_4$	101.3	44
$(C_5H_5)(CO)_2FeSiMeCo_2(CO)_7$	83.2	44

[a] Data converted using $\delta_{Si}(Me_3Si)_2O = 6.97$ ppm.

TABLE 24. ^{29}Si chemical shifts for π-organosilyl transition metal complexes[115]

Compound	δ_{Si} Complex	Ligand
$(Me_3SiCH{=}CH_2)Fe(CO)_4^a$	4.43	
$(Me_3Si^BSi^AMe_2CH{=}CH_2)Fe(CO)_4^a$	−11.40(Si^A)	
	−18.58(Si^B)	
$(Me_3SiCH{=}CH_2)_2RhAcac$	−0.62	−7.6
$(Me_3SiCH_2CH{=}CH_2)_2RhAcac$	−1.39	0.39
$(Me_3SiC{\equiv}CSiMe_3)Fe(CO)_4$	−8.54	−19.25
$(Me_3SiC{\equiv}CSiMe_3)Co_2(CO)_6$	1.06	−19.25
$(Me_3Si^BC{\equiv}C{-}C{\equiv}CSi^AMe_3)Co_2(CO)_6$	2.45(Si^A)	−15.99
	−16.91(Si^B)	

^{a} Reference 110.

silicon atoms further removed from the metal upon comparison with the uncomplexed silane[115].

I. Polysilanes

The most extensive chemical shift data are available for the permethylpolysilanes[116]. The data for a series of linear and branched compounds in this series are shown in Table 25. Stanislawski and West[86] used the approach of Grant and Paul[67] to develop substituent parameters to calculate the ^{29}Si chemical shifts and found that the chemical shift of the kth silicon nucleus could be represented by equation 4:

$$\delta_{\mathrm{Di}(k)} = B + \textstyle\sum_l A_l n_{kl} \tag{4}$$

where A_l is the chemical shift parameter for the lth atom from the kth silicon atom and n_{kl} is the number of those atoms present at that position. The parameters were found to be the following: B is a constant (8.5 ppm), $\alpha = -28.5$, $\beta = +3.9$, $\gamma = +1.2$ and $\delta = +0.2$ ppm. The excellent correlation found for the linear polydimethylsilanes did not extend to the cyclic silanes, and the largest deviations were observed for the branched silicon atoms[86]. The chemical shift ranges for the different types of silicon atoms in Table 25 are −9 to −20 ppm for Me_3Si–, −26 to −50 ppm for $Me_2Si{=}$, −79 to −81 ppm for $MeSi{\equiv}$, and −118 to −136 ppm for quarternary silicon atoms. Thus the resonances for the various types of silicon atoms are easily identified since there is no overlap in the ^{29}Si spectrum.

Further studies by Stanislawski and West were conducted on the fluoro- and chloro-terminated linear polydimethylsilanes[117]. Chlorine and fluorine substituent parameters were determined for the α, β and γ positions and found to be +53.0, −2.6 and 0.7 ppm for F, and +42.2, +2.5 and −0.3 ppm for Cl. These parameters gave good values for all chemical shifts with the exception of the dihalodisilanes. From the substituent effects it is apparent that the effect of the halogen is rapidly attenuated down the chain.

An empirical relationship has also been developed by Hahn[118] for the silicon chemical shift of linear and branched polysilanes (Table 26):

$$\delta_{Si} = -96.02 - 2.43a - 3.73a^2 + B_a b + C_a \tag{5}$$

where a is the number of silicon atoms in the α position, b is the number of silicon atoms in the β position, and B_a and C_a are incremental chemical shifts produced by β and γ silicon atoms. As was observed for the polymethylsilanes, little or no overlap between

TABLE 25. ^{29}Si Chemical shifts for permethylated polysilanes[116]

Compound	δ_{Si} A	B	C	D	E	F
$Me_3Si^A–SiMe_3$	−19.7					
$(Me_3Si^A)_2–Si^BMe_2$	−16.1	−48.6				
$(Me_3Si^A–Si^BMe_2)_2$	−15.2	−44.9				
$(Me_3Si^A–Si^BMe_2)_2–Si^CMe_2$	−15.1	−43.6	−40.9			
$(Me_3Si^A–Si^BMe_2–Si^CMe_2)_2$	−15.1	−43.4	−39.5			
$(Me_3Si^A–Si^BMe_2–Si^CMe_2)_2–Si^DMe_2$	−15.0	−43.3	−39.3	−38.2		
$(Me_3Si^A–Si^BMe_2–Si^CMe_2–Si^DMe_2)_2$	−15.0	−43.3	−39.2	−37.9		
$(Me_3Si^A–Si^BMe_2–Si^CMe_2–Si^DMe_2)_2Si^EMe_2$	−15.0	−43.3	−39.1	−37.8	−37.4	
$(Me_3Si^A–Si^BMe_2–Si^CMe_2–Si^DMe_2–Si^EMe_2)_2$	−15.0	−43.3	−39.1	−37.8	−37.6	
$(Me_3Si^A–Si^BMe_2–Si^CMe_2–Si^DMe_2–Si^EMe_2)_2–Si^FMe_2$	−15.0	−43.2	−39.1	−37.8	−37.5	−37.5
$(Me_3Si^A–Si^BMe_2–Si^CMe_2–Si^DMe_2–Si^EMe_2–Si^FMe_2)_2$	−15.0	−43.1	−39.0	−37.6	−37.4	−37.3
$(Me_3Si^A)_3–Si^BMe$	−12.5	−88.1				
$(Me_3Si^A)_4–Si^B$	−9.8	−135.5				
$(Me_3Si^A)_2–MeSi^B–SiMe–(SiMe_3)_2$	−11.7	−81.4				
$(Me_3Si^A)_3–Si^D–Si^CMe_2–Si^BMe_3$	−9.5	−14.7	−39.8	−131.9		
$(Me_3Si^A)_3–Si^D–Si^CMe–(Si^BMe_3)_2$	−9.2	−11.7	−79.8	−129.4		
$(Me_3Si^A)_3–Si^E–Si^CMe_2–Si^DMe–(Si^BMe_3)_2$	−9.5	−11.1	−29.2	−78.8	−126.0	
$(Me_3Si^A)_3–Si^B–Si–(SiMe_3)_3$	−9.5	−130.0				
$(Me_3Si^A)_3–Si^F–Si^CMe_2–Si^DMe_2–Si^EMe–(Si^BMe_3)_2$	−9.4	−11.7	−31.8	−33.2	−81.2	−128.6
$(Me_3Si^A)_3–Si^C–Si^BMe_2–Si–(SiMe_3)_3$	−9.2	−26.0	−118.2			
$(Me_3Si^A)_3–Si^C–Si^BMe_2–SiMe_2–Si–(SiMe_3)_3$	−9.2	−29.0	−126.5			
$[(Me_3Si^A)_3–Si^D–Si^BMe_2–Si^CMe_2]_2$	−9.5	−31.4	−35.7	−128.0		

TABLE 26. ^{29}Si chemical shifts for some polysilanes[118]

Compound	$\delta_{Si}A$	$\delta_{Si}B$	$\delta_{Si}C$	$\delta_{Si}D$	$\delta_{Si}E$
SiH_4					
$(H_3Si)_2$	−103.08				
$(H_3Si^A)_2$–Si^BH_2	−97.98	−115.66			
$(H_3Si^A$–$Si^BH_2)_2$	−97.78	−110.10			
$(H_3Si^A)_3$–Si^BH	−93.62	−136.26			
$(H_3Si^A$–$Si^BH_2)_2$–Si^CH_2	−98.39	−111.54	−107.33		
$(H_3Si^A)_2$–Si^BH–Si^CH_2–Si^DH_3	−94.14	−131.53	−106.92	−98.99	
$(H_3Si^A)_4$–Si^B	−89.60	−165.87			
$(H_3Si^A$–Si^BH_2–$Si^CH_2)_2$	−98.39	−111.28	−107.19		
$(H_3Si^A)_2$–Si^BH–Si^CH_2–Si^DH_2–Si^EH_3	−94.3	−131.6	−103.0	−112.1	−98.2
$(H_3Si^A$–$Si^BH_2)_2$–Si^CH–Si^DH_3	−99.3	−106.8	−127.2	−95.1	
$(H_3Si^A)_2$ – Si^BH_2	−95.62	−127.3			
$(H_3Si^A$–Si^BH_2–$Si^CH_2)_2$–Si^DH_2	−97.70	−110.81	−106.51	−106.73	

resonances of different types exists for the polysilanes and the chemical shift range for SiH_3 is −90 to −103 ppm, for SiH_2 is −103 to −115 ppm, and for SiH is −126 to −136 ppm.

The monosubstituted disilanes, $Me_3Si^ASi^BMe_2X$, have been investigated[119] and found to have substituent effects at Si^B analogous to those observed for the trimethylsilyl compounds, except that the additional silicon atom has a shielding effect on the chemical shift of Si^B. Similar results were observed for Si^B of the isotetrasilanes $(Y_3Si^B)_3Si^BX$ regardless of whether Y = Me, OMe or Cl[120]. The opposite trend was again observed for the case in which silicon is substituted with three electronegative elements as in the tris(trimethylsiloxy)silanes, $(Me_3Si^AO)_3Si^BX$ and electronegative substituents at Si^B were found to produce a shielding effect relative to $X = CH_3$[121]. In all three types of compounds the substituent effects are opposite for Si^A and Si^B.

The effect of increasing chain length and chain branching on silicon chemical shifts of Si^B in $Ph_3Si^ASi^BMe_2R$ (Table 15[76]) is opposite to that observed for $Me_3Si^ASi^BMe_2OR$ and $Me_2(OR)SiSiMe_2OR$[122]. In the first case the effect is to deshield Si^B as was observed for the alkylsilanes, whereas for the last two series of compounds Si^B becomes more shielded with increasing chain length and branching. Very little or no change was observed for Si^A in all cases. Other disilanes which have been studied recently include the series $Si_2Ph_{6-x}Cl_x$, $Si_2Ph_{6-x}H_x$ and $Si_2Cl_{6-x}H_x$[123]. The sagging pattern was observed for the $Si_2Ph_{6-x}Cl_x$ series upon increasing chlorine substitution at the silicon atom under observation. All other series gave monotonic changes, although not necessarily in the same direction. For example, upon increasing chlorine substitution at Si^B of $Ph_{3-x}Si^ACl_xSi^BPh_{3-y}Cl_y$, the chemical shift of Si^A became more shielded for $x = 3$, but became more deshielded for $x = 0$.

Recently, high resolution solution and solid state ^{29}Si NMR studies have been conducted on the polysilanes $(RR'Si)_n$[124–127]. It is obvious that this technique will be a powerful method for determining the microstructure of these polymers. Figure 12 shows a comparison between the ^{29}Si NMR spectrum of poly(methyl-*n*-hexylsilane) and poly(di-*n*-hexylsilane)[124]. In the former case the effects of tacticity can be seen in the fine structure of the spectrum whereas in the latter case for which this is impossible only a single peak is observed.

In the solid state NMR studies of poly(di-*n*-hexylsilane) using both ^{13}C and ^{29}Si NMR, two phases were detected in varying amounts depending upon the temperature[125, 126].

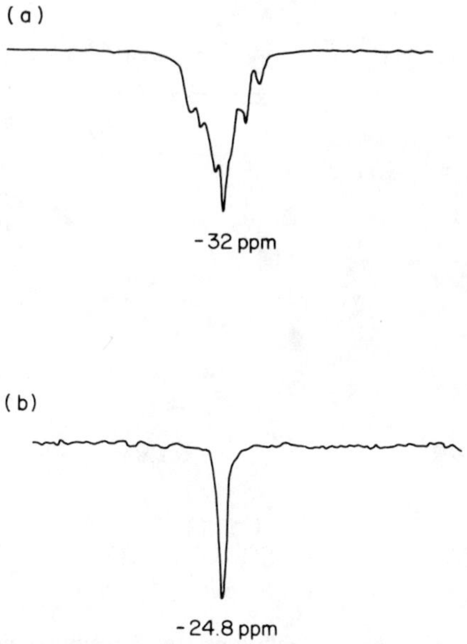

FIGURE 12. ^{29}Si cross polarization/magic angle spinning (CPMAS) spectra of (a) poly(methyl-*n*-hexylsilane) and (b) poly(di-*n*-hexylsilane)[124]

For example, in the ^{29}Si spectrum at low temperatures a single peak was observed which is assumed to contain the silicon backbone in the all *trans* conformations. At higher temperatures two peaks were observed until at 310 K only a single peak at higher field was observed. This higher field resonance arises from the more mobile phase and apparently reflects the insertion of *gauche* conformations in the main and side chains of the polymer[125, 126]. The NMR results were found to be consistent with X-ray and electron diffraction data.

J. Polysiloxanes

The polysiloxanes ('silicones') include the general class of compounds $(R_{3-n}SiO_n)_x$ where $n>0$, $x>2$ and R = alkyl, aryl or H. The terminology used for these compounds incorporates the use of the letters M, D, T and Q which represent $R_3SiO_{0.5}$, $R_2Si(O_{0.5})_2$, $RSi(O_{0.5})_3$ and $Si(O_{0.5})_4$ units, respectively. In addition it is generally accepted that superscripts refer to substituents on silicon other than Me; otherwise Me is understood. For example, 'D' indicates the repeat unit (Me_2SiO), whereas D^{Ph} indicates the repeat unit (MePhSiO) and D^{Ph_2} represents (Ph_2SiO).

The ^{29}Si chemical shift ranges for some structural units in polysiloxanes are shown in Figure 13. Early ^{29}Si NMR studies showed that individual resonance lines can be found for each unique silicon atom of polydimethylsiloxanes up to a ten-unit oligomer[5, 128]. Since then a number of studies of other linear siloxanes have appeared in which tacticity and neighboring group effects on the ^{29}Si chemical shift have been elucidated[74]. The increased spectral resolution relative to ^{1}H and ^{13}C NMR permits the examination of polymer microstructure down to the pentad level in many cases.

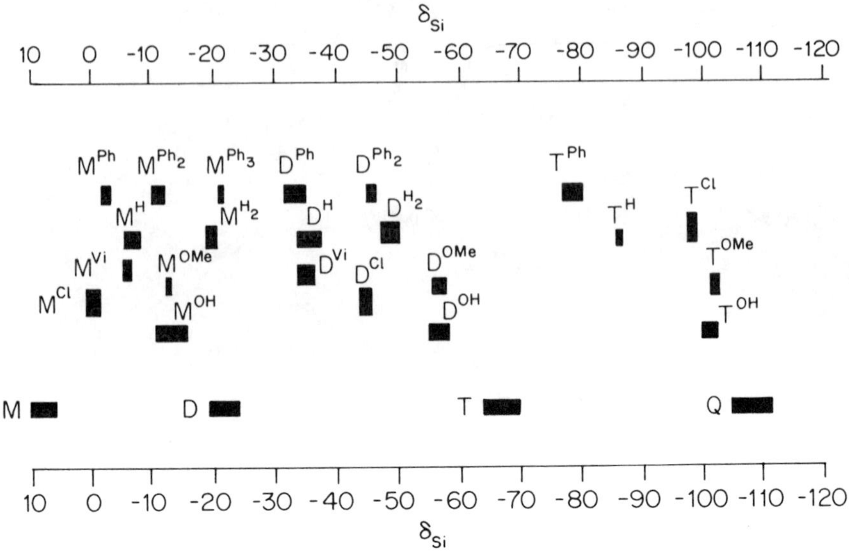

FIGURE 13. Silicon-29 chemical shift ranges for structural units in polysiloxanes

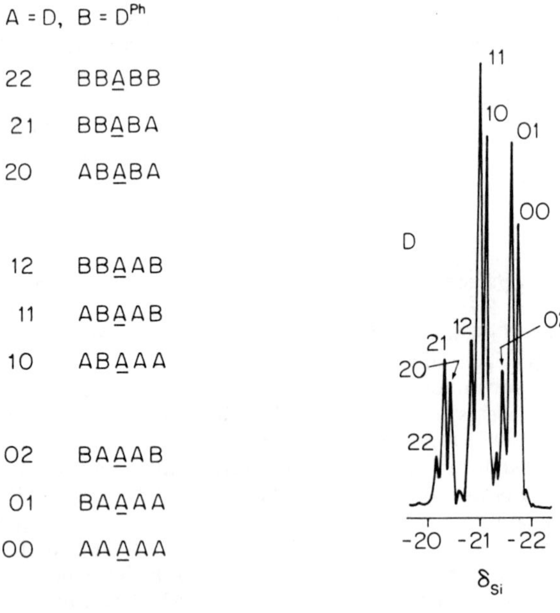

FIGURE 14. D region of the ^{29}Si spectrum of a methylphenylsilicone oil[129]

Some results obtained by Engelhardt and Jancke demonstrate this most clearly. Figure 14 shows the D region of the ^{29}Si spectrum of a methylphenylsilicone oil which contains M, D and D^{Ph} units[129]. The nine different pentad sequences were assigned based on the chemical shifts of model compounds. From the ^{29}Si NMR signal intensities a number of quantitative parameters describing the polymer microstructure can be determined. These include the mole fraction of monomer units, the run number (the average number sequences occurring in a polymer per 100 monomer units), the number average sequence length and the linkage probabilities[129].

More recently ^{29}Si NMR has been used to determine the structure of the various components of silicon resins containing M, D^{OH} and T units[130]. In addition to the expected effects of sequencing on the chemical shifts, diastereotopic M groups were observed in several cases including molecule:

$$\begin{array}{ccccccccc} M & — & T & — & D^{OH} & — & T & — & M \\ & & | & & & & | & & \\ & & M & & & & M & & \end{array}$$

due to the pseudoasymmetric center at D^{OH}. The chemical shifts of the two M groups differed by 0.07 ppm[130].

Silicon-29 NMR has been used to determine chain and block lengths in siloxane oligomers and polymers. The chemical shifts of chain or block terminal silicon atoms differ substantially from the chemical shifts of the silicon atoms in the middle of the siloxane chain (or block, as in a block copolymer)[130–134]. Thus by integrating the areas under the resonances for Si^A and Si^B the average degree of polymerization ($\overline{DP}$) for disiloxanols, for example, can be determined[131] from equation 6:

$$HOSi^AMe_2(OSi^BMe_2)_nOSi^AMe_2OH, \qquad \overline{DP} = 2(B/A) + 2 \tag{6}$$

Using the same approach block lengths in siloxane containing block copolymers have been determined[132, 133]

The chemical shift range for the nitrogen analogue of the siloxanes, the silazanes, is much smaller[135]. The trends of the chemical shifts are generally the same, however, and 'M' groups are deshielded relative to TMS. The 'D' silazane units are more shielded (-6 to -10 ppm), and 'T' groups appear at even higher field[135]. The deshielding effect observed in contracting the ring from the cyclic tetramer to the cyclic trimer in siloxanes is also observed for the silazanes, but the effect is much smaller in keeping with the smaller chemical shift range[136].

K. Spin–Spin Couplings

The coupling of nuclear spins through the bonding framework of the molecule is believed to be dependent upon three factors: an 'orbital' contribution which is a pair of terms arising from the interaction of the field produced by the orbital motion of electrons with the nuclear moments, a 'dipolar' term which arises from the dipole–dipole interaction between nuclear and electron spins, and a 'contact' term which is dependent upon the properties of the electrons at the nucleus[137]. The contact term is usually considered to dominate spin–spin couplings. Indeed, calculations by Beer and Grinter[138] suggested that the orbital and dipolar terms contributed less than 1% to observed $^1J_{SiC}$ coupling. In addition, a number of correlations between various combinations of J_{SiH}, J_{SiC}, J_{SiSi}, J_{CH} and J_{CC} in related systems indicate that the same factors are involved in determining each coupling. This indicates that the contact term is dominant for each and that d-orbital effects for Si are small, at least in the compounds studied. Suggestions have been made, however, that deviations in calculated couplings from experimental values may arise from lack of consideration of d-orbitals[139]. It is clear, however, that the magnitude of the coupling constant is related to the hybridization of the nuclei involved and the electronegativity of the substituents.

Most of the one-bond couplings that have been determined are $^1J_{SiH}$. In general, the magnitude of the coupling is dependent upon the electronegativity of the substituents on silicon with electronegative substituents increasing the absolute magnitude of the coupling and electropositive substituents decreasing it. Not surprisingly the effect is greatest for substituents directly attached to silicon, and β substitution has a much smaller effect. Some values of $^1J_{SiH}$ for silanes are shown in Table 27.

A number of correlations have been drawn between $^1J_{SiH}$ and parameters related to bonding and reactivity. The magnitude of $^1J_{SiH}$ was found to correlate with Hammett σ constants in phenyl and benzyl substituted silanes and disilanes[144, 145], and aryloxy- and (arylthio)dimethylsilanes[146] and with Taft polar substituent constants σ^* in silanes $R^1R^2R^3SiH$ (R = alkyl, Ph, H)[147].

Many one-bond silicon–carbon couplings have also been reported. In general the electronegativity of the substituents on silicon is important. Thus, Harris and Kimber found a linear dependence of the magnitude of $^1J_{SiC}$ and substituent electronegativity in a series of trimethylsilyl compounds, Me_3SiX[48]. The values of $^1J_{SiC}$ were found to correlate well with $^1J_{CC}$ in the analogous *t*-butyl compounds and the authors concluded that the

TABLE 27. Values of ^1J(Si–H) for some silanes

Compound	^1J(Si–H)(Hz)	Reference
$HSiF_3$	−381.7	140
$(EtO)_3SiH$	−287	100
H_2SiF_2	−282	140
H_3SiF	−229	140
H_4Si	−202.5	140
Ph_2SiH_2	−198.2	141
$MeSiH_3$	−194.0	141
Me_2SiH_2	−188.6	141
Me_3SiH	−184.0	141
$(t\text{-Bu})_2SiH_2$	−182.1	141
$(Me_3Si)Me_2SiH$	−173	142
$(Me_3Si)_3SiH$	−155	143

TABLE 28. ^1J(Si–C) for some silanes

Compound	^1J(Si–C) (Hz)	Reference
$CH_2{=}CHSiCl_3$	113	26
$MeSi(OEt)_3$	96.2	155
$MeSiCl_3$	86.6	154
$(Me_3Si)_2\mathbf{Si}{=}\mathbf{C}(OSiMe_3)(Bu\text{-}t)$	83.5	31
$Et_3\mathbf{Si}\text{–}\mathbf{C}{\equiv}CH$	75.0	156
$Me_2Si(OEt)_2$	73.0	155
Me_2SiCl_2	68.3	154
Me_3SiOEt	59.0	48
Me_3SiH	50.8	48
$(Me_3Si)_2\mathbf{SiMe}_2$	37.0	119

TABLE 29. ^1J(Si–Si) for di- and polysilanes

Compound	^1J (Si–Si) (Hz)	Reference
$(Cl_3Si)_2SiCl_2$	186	119
$(MeO)_3SiSiPh_3$	160	119
$[(MeO)_3Si]_3SiOMe$	144.1	122
$[(MeO)_3Si]_3SiH$	132.4	122
$(Cl_3Si)_3SiBr$	130.2	122
$ClMe_2SiSiMe_2(OEt)$	109.8	122
$ClMe_2SiSiMe_2(OPr$-$n)$	107.0	122
$Me_3SiSiMe_2F$	98.7	119
$Me_3SiSiMe_2Cl$	94.0	119
$Me_3SiSiMe_2Ph$	86.1	119
$Ph_3SiSiMe_2(Bu$-$t)$	80.0	76
$(Ph_3Si)_2SiMe_2$	74.5	76
$(Me_3Si)_2SiMe_2$	73.2	119
$(Ph_3Si)_2SiMe$ (Pr-n)	72.2	76
$(Ph_3Si)_2SiEt_2$	70.8	76
$(Me_3Si)_2Si{=}C(OSiMe_3)$ (Bu-t)	70.8	31
$(Me_3Si)_4Si$	52.5	119

contact term is dominant for both couplings. The hybridization of the carbon atom is also important and the absolute magnitude of $^1J_{SiC}$ follows the order $sp > sp^2 > sp^3$ in analogous compounds. Table 28 shows some data for a variety of silanes. The influence of both electronegativity and hybridization are evident in the magnitude of the coupling for $ViSiCl_3$ as opposed to the small value observed for $(Me_3Si)_2SiMe_2$.

One-bond silicon–silicon couplings have been found to correlate with substituent electronegativity and $^1J_{Si^BC}$ in pentamethyldisilanyl compounds $Me_3Si^ASi^BMe_2X$, and $^1J_{SiC}$ in the analogous trimethylsilyl compounds[119]. In the more crowded isotetrasilanes, $(Y_3Si)_3SiX$, the steric effect of X is important as well[120]. In a series of disilanes $Ph_3SiSiMe_2R$, $^1J_{SiSi}$ was found to be dependent upon the inductive effect of the alkyl group R[76]. Representative values for $^1J_{SiSi}$ are collected in Table 29.

The trends in couplings to other nuclei are not as straightforward as $^1J_{SiH}$, $^1J_{SiC}$ and $^1J_{SiSi}$. Substituent effects on $^1J_{SiF}$, for example, vary depending upon the compound examined. For example, $^1J_{SiF}$ is 356 Hz for F_3SiSiH_3[148] and 181 Hz for F_3SiOMe[149] whereas the opposite would be expected from electronegativity considerations. The same trend was observed for the monofluorosilanes $(Me_3Si)_3SiF$ ($^1J_{SiF} = 335$ Hz) and Me_3SiF ($^1J_{SiF} = 274$ Hz)[150]. An extensive compilation of one-bond and long-range couplings involving silicon is available in Reference 8.

L. Spin–Lattice Relaxation

In addition to chemical shifts and coupling constants, information about the structure of organosilicon compounds is often available from spin–lattice relaxation times. Of the four possible relaxation mechanisms; (1) dipole–dipole relaxation (DD), (2) spin–rotation interactions (SR), (3) scalar coupling (SC) and (4) chemical shielding anisotropy (CSA), the dominant relaxation mechanisms for silicon are usually DD and SR. The spin–lattice relaxation times for silicon in organosilicon compounds range from 5–100 s. The longer spin–lattice relaxation times for ^{29}Si as compared with ^{13}C arise from the longer Si–H bond length which decreases the efficiency of the DD relaxation for ^{29}Si. In addition many silicon compounds exist in which silicon is not directly attached to a proton. As a result, the SR mechanism is often more important for ^{29}Si. Studies of the various contributing factors to ^{29}Si spin–lattice relaxation provides detailed information about the motional processes the molecule experiences. For example, in linear polydimethylsiloxanes the M units at the end of the chain exhibit more freedom to rotate about their threefold axis of symmetry whereas the D units are more 'anchored', and to different degrees in the chain[5, 151]. More recently, in studies of silyl transition metal complexes Pannell and Bassindale found a progressive decrease in nuclear Overhauser effect (NOE) (and therefore less DD relaxation) as the silicon atoms move further from the metal atom anchor in $(\eta^5\text{-}C_5H_5)Fe(CO)_2Si^AMe_2Si^BMe_2Si^CMe_3$ $(Si^A \rightarrow Si^C)$[112].

V. SUMMARY

The tremendous growth of ^{29}Si NMR studies, in particular in the past fifteen years, dramatizes the value of NMR for solving structural problems of this class of compounds. Traditional NMR studies have been enhanced by the recent improvements in areas such as high resolution NMR of solids[152] and two-dimensional NMR[153]. The use of these techniques increases the power of the already potent combination of methods available in ^{1}H, ^{13}C and ^{29}Si NMR for the characterization of organosilicon compounds. It is likely that the next fifteen years will witness a continued growth in the number of traditional NMR studies of these compounds as well as in the use of more sophisticated NMR techniques.

VI. REFERENCES

1. F. D. Becker, *High Resolution NMR*, 2nd ed., Academic Press, New York, 1980.
2. J. A. Pople, W. G. Schneider and H. J. Bernstein, *High Resolution Nuclear Magnetic Resonance*, McGraw-Hill, New York, 1959.
3. (a) J. B. Stothers, *Carbon-13 NMR Spectroscopy*, Academic Press, New York, 1972.
   (b) G. C. Levy and G. L. Nelson, *Carbon-13 Nuclear Magnetic Resonance for Organic Chemists*, Wiley, New York, 1972.
4. J. Schraml and J. M. Bellama, in *Determination of Organic Structure by Physical Methods*, Vol. 6 (Eds. F. Nachod, J. J. Zuckerman and E. W. Randall), Academic Press, New York, 1976, pp. 203–269.
5. G. C. Levy and J. D. Cargioli, in *Nuclear Magnetic Resonance Spectroscopy of Nuclei Other than Protons* (Eds. T. Axenrod and G. A. Webb), Chapter 17, Wiley, New York, 1974, pp. 251–274.
6. R. K. Harris, in *NMR and the Periodic Table* (Eds. R. K. Harris and B. E. Mann), Academic Press, London, 1978, pp. 309–340.
7. E. A. Williams and J. D. Cargioli, in *Annual Reports on NMR Spectroscopy*, Vol. 9 (Ed. G. A. Webb), Academic Press, New York, 1979, pp. 221–318.
8. H. Marsmann, in *NMR Basic Principles and Progress*, Vol. 17 (Eds. P. Diehl, E. Fluck and R. Kosfeld), Springer-Verlag, Berlin, 1981, pp. 65–235.
9. E. A. Williams, in *Annual Reports on NMR Spectroscopy*, Vol. 15 (Ed. G. A. Webb), Academic Press, London, 1983, pp. 235–289.
10. J. Rich, private communication, 1987.
11. D. E. Williams, in *Analysis of Silicones* (Ed. A. L. Smith), Chapter 11, Wiley, New York, 1974, pp. 287–323.
12. E. V. van den Berghe and G. P. van der Kelen, *J. Organomet. Chem.*, **59**, 175 (1973).
13. J. W. Emsley, J. Feeney and L. H. Sutcliffe, *High Resolution Nuclear Magnetic Resonance Spectroscopy*, Vol. 1, Pergamon Press, Oxford, 1965, pp. 399–423.
14. H. Bürger, R. Eujen and H. C. Marsmann, *Z. Naturforsch. Teil B*, **29**, 149 (1974).
15. I. E. Saratov, D. A. Khankhodzhaeva and V. O. Reikhsfel'd, *Zh. Obshch. Khim.*, **51**, 111 (1981).
16. H. Bürger and W. Kilian, *J. Organomet. Chem.*, **18**, 299 (1969).
17. Reference 9, p. 275 and references cited therein.

18. J. F. Hampton, C. W. Lacefield and J. F. Hyde, *Inorg. Chem.*, **4**, 1659 (1965).
19. H. Sakurai and J. Hayashi, *J. Organomet. Chem.*, **63**, C7 (1973).
20. A. Lycka, D. Snobl and J. Vencl, *J. Organomet. Chem.*, **149**, C37 (1978).
21. B. K. Hunter and L. W. Reeves, *Can. J. Chem.*, **46**, 1399 (1968).
22. J. Schraml, V. Chvalovský, M. Mägi and E. Lippmaa, *Coll. Czech. Chem. Commun.*, **43**, 3365 (1978).
23. D. Seyferth and D. C. Annarelli, *J. Am. Chem. Soc.*, **97**, 2273 (1975).
24. R. L. Lambert and D. Seyferth, *J. Am. Chem. Soc.*, **94**, 9426 (1972).
25. L. Delmulle and G. P. van der Kelen, *J. Mol. Struct.*, **66**, 309 (1980).
26. G. Miyajima, K. Takahashi and K. Nishimoto, *Org. Magn. Reson.*, **6**, 413 (1974).
27. G. C. Levy, D. M. White and J. D. Cargioli, *J. Magn. Reson.*, **8**, 280 (1971).
28. D. E. J. Arnold, S. Cradock, E. A. V. Ebsworth, J. D. Murdoch, D. W. H. Rankin, D. C. J. Skea, R. K. Harris and B. J. Kimber, *J. Chem. Soc., Dalton Trans.*, 1349 (1981).
29. D. F. Ewing, *Org. Magn. Reson.*, **12**, 499 (1979).
30. G. A. Olah, A. L. Berrier, L. D. Field and G. K. S. Prakash, *J. Am. Chem. Soc.*, **104**, 1349 (1982).
31. A. G. Brook, F. Abdesaken, G. Gutekunst and N. Plavec, *Organometallics*, **1**, 994 (1982).
32. A. G. Brook, F. Abdesaken, B. Gutekunst, G. Gutekunst and R. K. Kallury, *J. Chem. Soc., Chem. Commun.*, 191 (1981).
33. P. C. Lauterbur, in *Determination of Organic Structure by Physical Methods*, Vol. 2 (Eds. E. C. Nachod and W. D. Phillips), Academic Press, New York, 1962.
34. Aa. Linde, H. J. Jakobsen and B. J. Kimber, *J. Am. Chem. Soc.*, **97**, 3219 (1975).
35. S. Li, D. L. Johnson, J. A. Gladysz and K. L. Servis, *J. Organomet. Chem.*, **166**, 317 (1979).
36. R. D. Bertrand, W. B. Moniz, A. N. Garroway and G. C. Chingas, *J. Am. Chem. Soc.*, **100**, 5227 (1978).
37. P. D. Murphy, T. Taki, T. Sogabe, R. Metzler, T. G. Squires and B. C. Gerstein, *J. Am. Chem. Soc.*, **101**, 4055 (1979).
38. S. R. Hartmann and E. L. Hahn, *Phys. Rev.*, **128**, 2042 (1962); A. Pines, M. G. Gibby and J. S. Waugh, *J. Chem. Phys.*, **59**, 569 (1973).
39. G. A. Morris and R. Freeman, *J. Am. Chem. Soc.*, **101**, 760 (1979).
40. D. M. Doddrell, D. T. Pegg and M. R. Bendall, *J. Magn. Reson.*, **48**, 323 (1982); D. T. Pegg, D. M. Doddrell and M. R. Bendall, *J. Chem. Phys.*, **77**, 2745 (1982).
41. B. J. Helmer and R. West, *Organometallics*, **1**, 877 (1982).
42. T. A. Blinka, B. J. Helmer and R. West, *Adv. Organomet. Chem.*, **23**, 193 (1984).
43. P. Jutzi, D. Kanne and C. Krüger, *Angew. Chem.*, **98**, 163 (1986).
44. W. Malisch, H.-U. Wekel, I. Grob and R. H. Kohler, *Z. Naturforsch.*, **37b**, 601 (1982).
45. N. N. Zemlyanskii and O. K. Sokolikova, *Zh. Anal. Khim.*, **36**, 1990 (1981).
46. H. C. Marsmann and H. G. Horn, *Z. Naturforsch.*, **27b**, 1448 (1972).
47. R. G. Scholl, G. E. Maciel and W. K. Musker, *J. Am. Chem. Soc.*, **94**, 6376 (1972).
48. R. K. Harris and B. J. Kimber, *J. Magn. Reson.*, **17**, 174 (1975).
49. R. K. Harris and B. Lemarie, *J. Magn. Reson.*, **23**, 371 (1976).
50. E. V. van den Berghe and G. P. van der Kelen, *J. Organomet. Chem.*, **122**, 329 (1976).
51. K. G. Sharp, P. A. Sutor, E. A. Williams, J. D. Cargioli, T. C. Farrar and K. Ishibitsu, *J. Am. Chem. Soc.*, **98**, 1977 (1976).
52. E. Lippmaa, M. Mägi, G. Engelhardt, H. Jancke, V. Chvalovský and J. Schraml, *Coll. Czech. Chem. Commun.*, **39**, 1041 (1974).
53. J. Schraml, J. Pola, V. Chvalovský, M. Mägi and E. Lippmaa, *J. Organomet. Chem.*, **49**, C19, (1973).
54. J. Schraml, J. Včelák and V. Chvalovský, *Coll. Czech. Chem. Commun.*, **39**, 267 (1974).
55. N. D. Chuy, V. Chvalovský, M. Mägi and E. Lippmaa, *Coll. Czech. Chem. Commun.*, **40**, 875 (1975).
56. J. Schraml, J. Včelák, G. Engelhardt and V. Chvalovský, *Coll. Czech. Chem. Commun.*, **41**, 3758 (1976).
57. J. Schraml, V. Chvalovský, M. Mägi and E. Lippmaa, *Coll. Czech. Chem. Commun.*, **42**, 306 (1977).
58. J. Schraml, N. D. Chuy, V. Chvalovský, M. Mägi and E. Lippmaa, *Org. Magn. Reson.*, **7**, 379 (1975).
59. E. R. Ernst, L. Spialter, G. R. Buell and D. L. Wilhite, *J. Organomet. Chem.*, **59**, C13 (1973).

60. C. R. Ernst, L. Spialter, G. R. Buell and D. L. Wilhite, *J. Am. Chem. Soc.*, **96**, 5375 (1974).
61. J. Pikies, A. Herman and W. Wojnowski, *Z. Anorg. Allg. Chem.*, **498**, 218 (1983).
62. J. Pikies and W. Wojnowski, *Z. Anorg. Allg. Chem.*, **511**, 219 (1984).
63. V. A. Pestunovich, M. E. Larin, E. I. Dubinskaya and M. G. Voronkov, *Dokl. Akad. Nauk SSSR*, **233**, 378 (1977).
64. J. Schraml, R. Ponec, V. Chvalovský, G. Engelhardt, H. Jancke, H. Kriegsmann, M. F. Larin, V. A. Pestunovich and M. G. Voronkov, *J. Organomet. Chem.*, **178**, 55 (1979).
65. R. H. Cragg and R. D. Lane, *J. Organomet. Chem.*, **270**, 25 (1984).
66. D. Cory, A. Wong and W. M. Ritchey, *J. Organomet. Chem.*, **235**, 277 (1982).
67. D. M. Grant and E. G. Paul, *J. Am. Chem. Soc.*, **86**, 2977 (1964).
68. R. K. Harris and B. J. Kimber, *Adv. Mol. Relaxation Processes*, **8**, 15 (1976).
69. G. Engelhardt, R. Radeglia, H. Janke, E. Lippmaa and M. Mägi, *Org. Magn. Reson.*, **5**, 561 (1973).
70. U. Niemann and H. C. Marsmann, *Z. Naturforsch.*, **30b**, 202 (1975).
71. H. C. Marsmann, W. Raml and E. Hengge, *Z. Naturforsch.*, **35b**, 35 (1980).
72. E. A. Williams and J. D. Cargioli, unpublished results.
73. J. Schraml, N. D. Chuy, V. Chvalovský, M. Mägi and E. Lippmaa, *J. Organomet. Chem.*, **51**, C5 (1973).
74. Reference 7 and references cited therein.
75. E. A. Williams, unpublished results.
76. E. A. Williams, J. D. Cargioli and P. E. Donahue, *J. Organomet. Chem.*, **192**, 319 (1980).
77. A. G. Brook, S. C. Nyburg, W. F. Reynolds, Y. C. Poon, Y.-M. Chang, J.-S. Lee and J.-P. Picard, *J. Am. Chem. Soc.*, **101**, 6750 (1979).
78. R. West, M. J. Fink and J. Michl, *Science*, **214**, 1343 (1981).
79. H. B. Yokelson, J. Maxka, D. A. Siegel and R. West, *J. Am. Chem. Soc.*, **108**, 4239 (1986).
80. K. W. Zilm, D. M. Grant, J. Michl, M. J. Fink and R. West, *Organometallics*, **2**, 193 (1983).
81. R. G. Scholl, G. E. Maciel and W. K. Musker, *J. Am. Chem. Soc.*, **94**, 6376 (1972).
82. A. Marchand, P. Gerval, M. Joanny and P. Mazerolles, *J. Organomet. Chem.*, **217**, 19 (1981).
83. M.-L. Filleux-Blanchard, N. D. An and G. Manuel, *Org. Magn. Reson.*, **11**, 150 (1978).
84. G. Fritz and N. Braunagel, *Z. Anorg. Allg. Chem.*, **399**, 280 (1973).
85. D. Kovar, K. Utvary and E. Gengge, *Monatsh. Chem.*, **110**, 1295 (1979).
86. D. A. Stanislawski and R. West, *J. Organomet. Chem.*, **204**, 295 (1981).
87. R. H. Cragg and R. D. Lane, *J. Organomet. Chem.*, **291**, 153 (1985).
88. G. Engelhardt, H. Jancke, M. Mägi, T. Pehk and E. Lippmaa, *J. Organomet. Chem.*, **28**, 293 (1971); D. J. Burton, R. K. Harris, K. Dodgson, C. J. Pellow and J. A. Semlyen, *Polymer Commun.*, **24**, 278 (1983).
89. D. Seyferth and D. C. Annarelli, *J. Organomet. Chem.*, **117**, C51 (1976).
90. J. J. Burke and P. C. Lauterbur, *J. Am. Chem. Soc.*, **86**, 1870 (1964).
91. M. Ishikawa, H. Sugisawa, M. Kumada, T. Higuchi, K. Matsui, K. Hirotsu and J. Iyoda, *Organometallics*, **2**, 174 (1983).
92. D. Seyferth, D. C. Annarelli and S. C. Vick, *J. Am. Chem. Soc.*, **98**, 6382 (1976).
93. D. Seyferth, D. C. Annarelli and S. C. Vick, *J. Am. Chem. Soc.*, **106**, 123 (1984).
94. H. Sakurai, Y. Nakadaira, T. Koyama and H. Sakaba, *Chem. Lett.*, 213 (1983).
95. V. A. Pestunovich, S. N. Tandura, M. G. Voronkov, G. Engelhardt, E. Lippmaa, T. Pehk, V. F. Sidorkin, G. I. Zelchan and V. P. Baryshok, *Dokl. Akad. Nauk SSSR*, **240**, 914 (1978).
96. V. A. Pestunovich, S. N. Tandura, M. G. Voronkov, Y. P. Baryshok, C. I. Zelchan, V. I. Glukhikh, G. Engelhardt and M. Witanowski, *Spectrosc. Lett.*, **11**, 339 (1978).
97. E. E. Liepinsh, I. S. Birgele, I. I. Solomennikova, A. F. Lapsinya, G. I. Zelchan and E. Lukevits, *Zh. Obsch. Khim.*, **50**, 2462 (1980).
98. P. Henscei and H. C. Marsmann, *Acta. Chem. Acad. Sci. Hung.*, **105**, 79 (1980).
99. J. M. Bellama, J. D. Nies and N. Ben-Zvi, *Magn. Reson. Chem.*, **24**, 748 (1986).
100. R. K. Harris, J. Jones and S. Ng. *J. Magn. Reson.*, **30**, 521 (1978).
101. E. Liepins, J. Popelis, I. Birgele, I. Urtane, G. Zelchan and E. Lukevics, *J. Organomet. Chem.*, **201**, 113 (1980).
102. B. J. Helmer, R. West, R. J. P. Corriu, M. Poirier, G. Royo and A. De Saxce, *J. Organomet. Chem.*, **251**, 295 (1983).
103. E. Kupce, E. Liepins, A. Lapsina, G. Zelchan and E. Lukevics, *J. Organomet. Chem.*, **251**, 15 (1983).

104. R. B. Johannesen, F. E. Brinckman and T. D. Coyle, *J. Phys. Chem.*, **72**, 660 (1968).
105. J. A. Cella, J. D. Cargioli and E. A. Williams, *J. Organomet. Chem.*, **186**, 13 (1980).
106. H. C. Marsmann and R. Löwer, *Chem. Z.*, **97**, 660 (1973).
107. A. R. Bassindale and T. Stout, *J. Chem. Soc., Chem. Commun.*, 1387 (1984).
108. S. Cradcock, E. A. V. Ebsworth, N. S. Hosmane and K. M. Mackay, *Angew. Chem.*, **87**, 207 (1975).
109. T. N. Mitchell and H. C. Marsmann, *J. Organomet. Chem.*, **150**, 171 (1978).
110. H. Sakurai, Y. Kamiyama, A. Mikoda, T. Lobayashi, K. Sasaki and Y. Nakadaira, *J. Organomet. Chem.*, **201**, C14 (1980).
111. U. Schubert, G. Scholz, J. Muller, K. Ackermann, B. Worle and R. F. D. Stansfield, *J. Organomet. Chem.*, **306**, 303 (1986).
112. K. H. Pannell and A. R. Bassindale, *J. Organomet. Chem.*, **229**, 1 (1982).
113. D. H. Harris, M. F. Lappert, J. S. Poland and W. McFarland, *J. Chem. Soc. (D)*, 311 (1975).
114. F. H. Köhler, H. Hollfelder and E. O. Fischer, *J. Organomet. Chem.*, **168**, 53 (1979).
115. K. H. Pannell, A. R. Bassindale and J. W. Fitch, *J. Organomet. Chem.*, **209**, C65 (1981).
116. M. Ishikawa, J. Iyoda, H. Ikeda, K. Kotake, T. Hashimoto and M. Kumada, *J. Am. Chem. Soc.*, **103**, 4845 (1981).
117. D. A. Stanislawski and R. West, *J. Organomet. Chem.*, **204**, 307 (1981).
118. J. Hahn, *Z. Naturforsch.*, **35b**, 282 (1980).
119. K. G. Sharp, P. A. Sutor, E. A. Williams, J. D. Cargioli, T. C. Farrar and K. Ishibitsu, *J. Am. Chem. Soc.*, **98**, 1977 (1976).
120. H. C. Marsmann, W. Raml and E. Hengge, *Z. Naturforsch.*, **35b**, 1541 (1980).
121. G. Engelhardt, H. Jancke, R. Radeglia, H. Kriegsmann, M. E. Larin, V. A. Pestunovich, E. I. Dubinskaja and M. G. Voronkov, *Z. Chem.*, **17**, 376 (1977).
122. G. Engelhardt, R. Radeglia, H. Kelling and R. Stendel, *J. Organomet. Chem.*, **212**, 51 (1981).
123. H. Sollradl and E. Hengge, *J. Organomet. Chem.*, **243**, 257 (1983).
124. F. C. Schilling, F. A. Bovey and J. M. Zeigler, *Macromolecules*, **19**, 2309 (1986).
125. G. C. Gobbi, W. W. Fleming, R. Sooriyakumaran and R. D. Miller, *J. Am. Chem. Soc.*, **108**, 5624 (1986).
126. F. G. Schilling, F. A. Bovey, A. J. Lovinger and J. M. Zeigler, *Macromolecules*, **19**, 2663 (1986).
127. W. J. Welsh, K. Beshah, J. L. Ackerman, J. E. Mark, L. D. David and R. West, *Polym. Prepr.*, **24**, 131 (1983).
128. G. Engelhardt, H. Jancke, M. Mägi, T. J. Pehk and E. Lippmaa, *J. Organomet. Chem.*, **28**, 293 (1971).
129. G. Engelhardt and H. Jancke, *Polymer Bull.*, **5**, 577 (1981).
130. R. K. Harris and B. J. Kimber, *Appl. Spectrosc. Rev.*, **10**, 117 (1975).
131. R. W. LaRochelle, J. D. Cargioli and E. A. Williams, *Macromolecules*, **9**, 85 (1976).
132. E. A. Williams, J. D. Cargioli and S. Y. Hobbs, *Macromolecules*, **10**, 782 (1977).
133. R. W. LaRochelle, J. D. Cargioli and E. A. Williams, unpublished results.
134. H. G. Horn and H. C. Marsmann, *Makromol. Chem.*, **162**, 255 (1972).
135. H. Jancke, G. Engelhardt, M. Mägi and E. Lippmaa, *Z. Chem.*, **13**, 435 (1973).
136. B. D. Lavrukhin, B. A. Astapov, A. A. Zhadanov, G. Engelhardt and H. Jancke, *Org. Magn. Reson.*, **18**, 71 (1982).
137. N. F. Ramsey and E. M. Purcell, *Phys. Rev.*, **85**, 143 (1952).
138. M. D. Beer and R. Grinter, *J. Magn. Reson.*, **31**, 187 (1978).
139. K. D. Summerhays and D. A. Deprez, *J. Organomet. Chem.*, **118**, 19 (1976).
140. E. A. V. Ebsworth and J. J. Turner, *J. Chem. Phys.*, **36**, 2628 (1962).
141. C. Shumann and H. Dreeskamp, *J. Magn. Reson.*, **3**, 204 (1970).
142. J. V. Urenovitch and R. West, *J. Organomet. Chem.*, **3**, 138 (1965).
143. H. Bürger and W. Kilian, *J. Organomet. Chem.*, **18**, 299 (1969).
144. F. K. Cartledge and K. H. Riedel, *J. Organomet. Chem.*, **34**, 11 (1972).
145. Y. Nagai, M. Ohtsuki, T. Nakano and H. Watanabe, *J. Organomet. Chem.*, **37**, 81 (1972).
146. H. Watanabe, R. Akaba, T. Iezumi and Y. Nagai, *Bull. Chem. Soc. Jpn.*, **53**, 2981 (1980).
147. Y. Nagai, H. Matsumoto, T. Nakano and H. Watanabe, *Bull. Chem. Soc. Jpn.*, **45**, 2560 (1972).
148. J. E. Drake and N. P. X. Westwood, *J. Chem. Soc. A*, 3300 (1971).
149. W. Airey and G. M. Sheldrick, *J. Inorg. Nucl. Chem.*, **32**, 1827 (1970).
150. H. Bürger and W. Killian, *J. Organomet. Chem.*, **26**, 47 (1971).
151. G. C. Levy, J. D. Cargioli, P. C. Juliano and T. D. Mitchell, *J. Am. Chem. Soc.*, **95**, 3445 (1973).

152. C. A. Fyfe, *Solid State NMR for Chemists*, C.F.C. Press, Guelph, Ontario, Canada, 1983.
153. A. Bax, *Two-Dimensional Nuclear Magnetic Resonance in Liquids*, Delft University Press, Dordrecht, Holland, 1984.
154. H. Dreeskamp and K. Hildenbrand, *Annalen*, 712 (1975).
155. H. Elser and H. Dreeskamp, *Ber. Bunsenges.. Phys. Chem.*, **73**, 619 (1969).
156. K. Kamienska-Trela, *J. Organomet. Chem.*, **159**, 15 (1978).

The Chemistry of Organic Silicon Compounds
Edited by S. Patai and Z. Rappoport

CHAPTER **9**

Photoelectron spectra of silicon compounds

H. BOCK and B. SOLOUKI
Institute of Inorganic Chemistry, Johann Wolfgang Goethe University, Niederurseler Hang, D-6000 Frankfurt (M) 50, West Germany

I. INTRODUCTORY REMARKS: PHOTOELECTRON SPECTROSCOPY TODAY AND SCOPE OF THIS REVIEW[1]*

The early tempestuous years of photoelectron (PE) spectroscopy between the mid-sixties and mid-seventies, which had seen the development of the method[2], the commercial production of PE spectrometers and the assignment of the ionization patterns of thousands of molecules[1-10], have since been superseded by a period of practical application[11-17]. Based on the experience gathered, nowadays reactions of flowing gases—including heterogeneously catalyzed ones[18-21]—are advantageously and 'visually' optimized by the temperature-dependent changes in the continuously recorded real-time PES ionization patterns: the 'molecular fingerprints' of the reactants give way to those of the products. This technique proved to be especially valuable for the generation and unequivocal identification of numerous reactive intermediates[12-16]. With millimole quantities of a suitable precursor and within half a day, the optimum temperature and flow parameters can be determined to prepare the respective short-lived molecule under nearly unimolecular conditions, which in addition diminish consecutive intermolecular reactions like oligomerization or disproportionation. The existence of the reactive species is readily established by comparing its recorded radical cation state sequence with those of chemically related, preferably iso(valence)electronic compounds or with a precalculated one. Often, the optimized conditions allow one to trap the molecule at low temperatures,

* In this chapter, equations, tables, diagrams and schemes are numbered consecutively.

e.g. in a matrix experiment, and to further characterize it by other spectroscopic techniques. The procedure outlined above has been essential in the discovery of many compounds containing silicon centers of coordination numbers 3,2 or 1—from silabenzene via silylenes to phenyl silaisonitriles[22]—and, therefore, its usefulness will be demonstrated first (Section II).

Another application of photoelectron spectroscopy to organosilicon compounds rests on the correlation of ionization and oxidation potentials. Both vertical first electron expulsion in the gas phase (PES time scale 10^{-15} s) and adiabatic single electron transfer in solution (ESR time scale 10^{-6} s) produce the respective radical cation in its ground state either with a 'frozen core' skeleton (IR time scale 10^{-13} s) or with a dynamically induced 'structural change' to allow for an optimum charge delocalization. Based on the validity of approximations concerning the applicability of the Nernst equation, the diffusion to and the absorption at the electrode and, above all, the differences in M and $M^{\cdot\oplus}$ solvation, often linear regressions between first ionization and oxidation potentials result. These allow one to 'predict' single electron transfer reactions in solution based on first vertical ionization potentials, whereas deviations from the established regression line may indicate considerable structural changes. Numerous organosilicon radical cations have been generated this way and characterized by their ESR/ENDOR signal patterns[16, 22]. From the spin and charge distributions, structural changes and molecular dynamics phenomena observed (Section III) much can be learned about the ground state of radical cations.

As concerns the assignment of photoelectron spectra of medium-sized silicon molecules, most of the spectroscopic information from PE band shapes, the advisable radical cation state comparison based on perturbation arguments and suitable methods of calculation as well as bonding models—including knowledge of pitfalls—had already been established in the seventies[1–9]. In Section IV, some prototype cases from 'SiSi/SiSi bond' delocalization via 'hyperconjugative' as well as 'inductive' perturbations of π-systems by silyl substituents to the 'interactions' of silicon centers with adjacent lone pairs are discussed in detail. Although these more qualitative approaches[9, 10] facilitate understanding, shortcomings of MO models neglecting electron correlation[10] and molecular dynamics[10] are obvious. So as not to overstress the simple 'seemingly easy to comprehend' explanation, the body of known photoelectron ionization energies of silicon compounds is presented in tables following typical PE spectra selected for each individual class of molecules (Section V).

The concluding remarks of this report, including also industrially important classes of silicon compounds, point out again that the 'reading' of the 'individual molecular band patterns' and their daily use present no more difficulties than are met in the rationalization of infrared frequencies, NMR couplings or mass spectroscopic fragments. To facilitate understanding, neither special experimental techniques in electron spectroscopy or adjacent fields nor computational studies of the molecules PE spectroscopically investigated are covered extensively and, whenever possible and appropriate, the chemist's point of view is emphasized in the discussions.

We are greatly indebted to Dr. W. Atwell (Dow Corning, Midland, Michigan) for providing the initial profile for the literature search. There are, of course, already numerous valuable reviews available on photoelectron spectroscopy in general and on some of its special aspects. In addition to those already quoted[1–10], there are more books available[23–26], partly dealing with related topics not included here, e.g. ESCA[27, 28] or vacuum UV spectroscopy[29]. A legion of summarizing articles in addition to those already quoted[3, 7–9, 17] has covered general aspects[30–33] or more special topics like the PE spectra of inorganic[32] or organic[33] compounds, or the compounds of certain elements like B(N)[34], P[35], S[36] or F, Cl, Br, I[37] or containing substituent groups like –CN[38], –NCO[39] or –NCS[39].

Summarizing, photoelectron spectroscopy has been developed within the past 20 years into a spectroscopic technique which is particularly valuable for real-time gas analysis in flow systems[11–15, 18–21]. An essential contribution to this status originates from its intimate relationship to chemical bonding, leading to the well-documented assignment of the PE spectroscopic fingerprints of thousands of molecules and allowing their identification and characterization[1–40].

II. REAL-TIME GAS ANALYSIS USING PHOTOELECTRON SPECTROSCOPY: THE DETECTION OF SHORT-LIVED SILICON MOLECULES BY THEIR IONIZATION FINGERPRINTS

To first demonstrate the up-to-date usefulness of photoelectron spectroscopy, its application as a most elegant real-time analysis method in mixtures of flowing gases shall be presented. An outline of the measurement principle as well as of some experimental details (A) will be followed by a detailed discussion of examples of thermally generated unsaturated silicon molecules[40], i.e. those which contain Si centers of low coordination number: for silabenzene and silaethene (B), for silylenes (C) and for phenyl silaisocyanide (D). Assignment of the PE spectra will use the arguments published in the literature and, therefore, these prototype molecules will enlarge the number of cases discussed in detail (cf. Section IV).

A. Principle of PES Measurement and Some Experimental Details

Photoelectron spectroscopy uses the photoionization of a neutral molecule M to its radical cation $M^{\cdot\oplus}$,

$$M + h\nu \rightarrow M^{\cdot\oplus} + e^{\ominus} \qquad (1)$$

to determine the nth vertical ionization potential IE_n^v of M,

$$IE_n^v(M) = h\nu + E_{kin}(e^{\ominus}) \qquad (2)$$

from the difference of the photon energy $h\nu$ and the kinetic energy E_{kin} of the ejected electron. If, as is usual, a helium(I) discharge lamp is chosen as a source of monochromatic photons with $h\nu = 21.21$ eV, all ionization potentials up to 21.21 eV can be measured by counting the emitted electrons of specific kinetic energy[2–4], which are separated when leaving the ionization chamber (Figure 1: H) by passing a cylindrical (Figure 1: I) or spherical analyzer of continuously varying voltage.

For gas-phase reactions monitored by PE spectroscopic real-time analysis in the flow system, the parts of a variable building block apparatus are assembled and directly connected to the spectrometer, which generates the compound stream by its vacuum pump system (Figure 1). Out of numerous possible combinations[12], only a few applications shall be pointed out (cf. Figure 1):

Thermal decomposition of molecules. Compound A is pumped through the heating zone C via bypass E into the spectrometer. Additional options are: after optimization of the decomposition temperature, the pyrolysis product can be isolated in trap E (cf. e.g. References 12, 15, 31, 41–43) or by-products like HCl removed by injecting the stoichiometric amount of, e.g., NH_3 from F, depositing NH_4Cl on the inner wall of the mixing bulb G to record the PE spectrum of the pure pyrolysis product (cf. e.g. References 12, 31, 45–47).

Optimization of heterogeneously catalyzed reactions. Compound A, which can be mixed in B with other components (cf. e.g. References 12, 18, 31) runs over the catalyst heated in C to the optimum temperature determined by continuously recording the PE spectra of the gas mixture (cf. e.g. References 12, 18, 19). Using capillary D, the reaction may be carried out at atmospheric pressure, side-tracking a negligible amount at 10^{-2} mbar pressure for analysis into the PE spectrometer (cf. e.g. References 12, 18).

Detection of novel surface reactions. Compound A runs over a reactive solid heated in C (cf. References 20, 21, 47–49) and the reaction products are either analyzed PE spectroscopically via bypass E or trapped for additional gas-chromatographic determination by switching the stopcocks E (cf. e.g. References 15, 46).

The advantages of the above procedures (as already pointed out in the introduction) are: millimole amounts of the precursor—in the case of heterogeneously catalyzed reactions, therefore, only a few cubic centimeters of the (precious) catalyst—and a couple of hours allow one to optimize temperature and flow conditions in the heated reaction zone (Figure 1) with the gas stream being transported by the vacuum pump system of the PE spectrometer. The spectra can be registered on-line, displaying on a screen continuously the band pattern of the resulting reaction mixture[48, 49]. In most cases, the components can be visually identified by their individual 'molecular ionization finger-

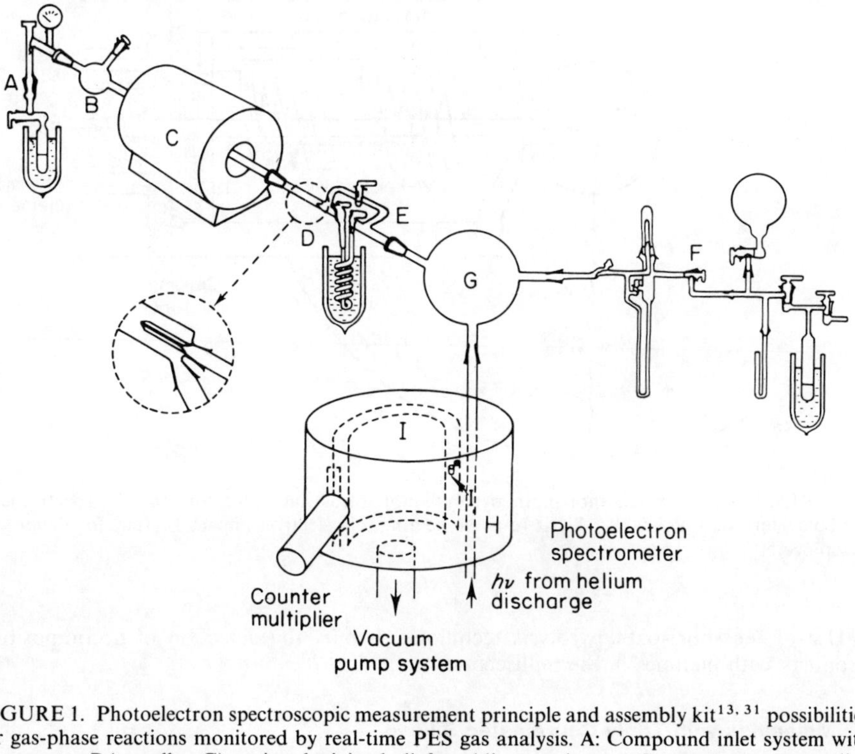

FIGURE 1. Photoelectron spectroscopic measurement principle and assembly kit[13, 31] possibilities for gas-phase reactions monitored by real-time PES gas analysis. A: Compound inlet system with manometer; B (as well as G): optional mixing bulb for adding another gas; C: temperature-controlled oven with quartz tube filled with quartz wool or catalyst; D: optional capillary for branching off a side-stream of reduced pressure to the PE spectrometer while running the gas-phase reaction at atmospheric pressure; E: cooling trap for isolation of compounds after the reaction has been optimized via the bypass; F: gas storage and inlet system, e.g. to add stoichiometric amounts of NH_3 to remove HCl from the pyrolysis gas stream of compound A by depositing NH_4Cl on the inner wall of G. The respective building block apparatus is directly connected to the PE spectrometer containing the ionization chamber H and the analyzer I

prints' and the relative intensity changes allow for quick optimization of the reaction conditions[12, 15].

For the detection of short-lived intermediates the distance between heating zone and target chamber has to be as short as possible. Advantageously, a short-path pyrolysis oven[15, 50] (Figure 2) can be built into the PE spectrometer. The molybdenum tube of the latter is heated resistively and, in addition, also by electron impact (Figure 2: U_1, U_2); the distance between the end of the oven zone and the PES ionization chamber amounts to only 3 cm.

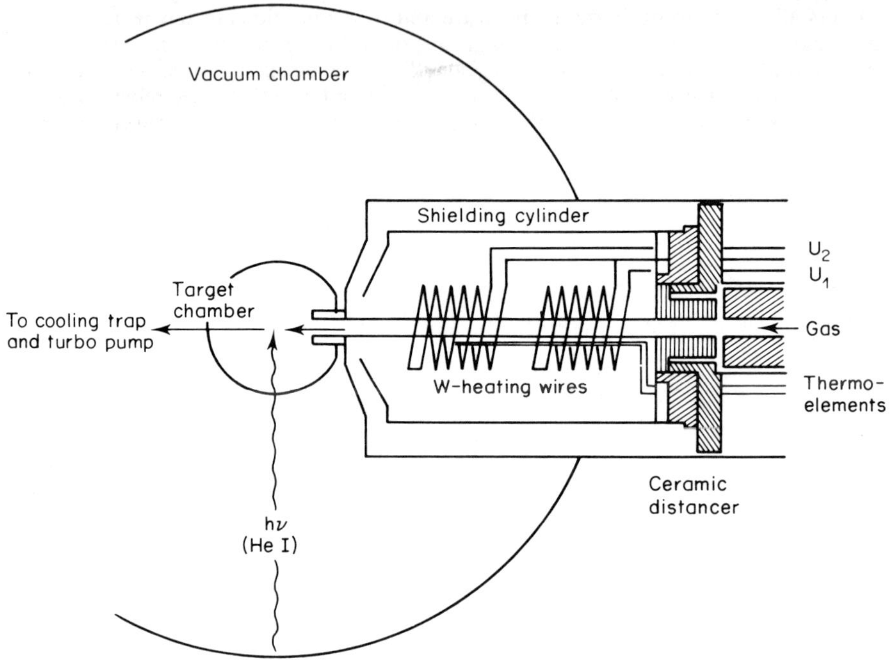

FIGURE 2. Outline of the short-path pyrolysis apparatus integrated into the PE spectrometer Leybold Heraeus UPG 200[12]: U_1 is resistive heating, U_2 electron impact heating; for details see Reference 50

Use of the short-path pyrolysis technique permits the detection of decomposition products with lifetimes in the millisecond range[12, 15, 51].

B. Molecules with Triply Coordinated Silicon

1. PE spectra of silabenzene and silatoluene H_5C_5SiR

The exchange of one CH ring unit by one SiH moiety in benzene, the prototype aromatic compound containing a cyclic 6π electron system, lowers the stability tremendously: silabenzene H_5C_5SiH, a species with SiC multiple bonds, has long eluded conclusive proof of its existence by spectroscopic observation—despite predictions of it being the most stable isomer[52] and having low ionization energies[8]. The first unambiguous PE spectroscopic evidence could be obtained for silatoluene[53] splitting off propene as an advantageous leaving group from a 1-silacyclohexa-2,4-diene precursor and by the analogous generation of silabenzene[54] (equation 3). Still lacking, however, has been an

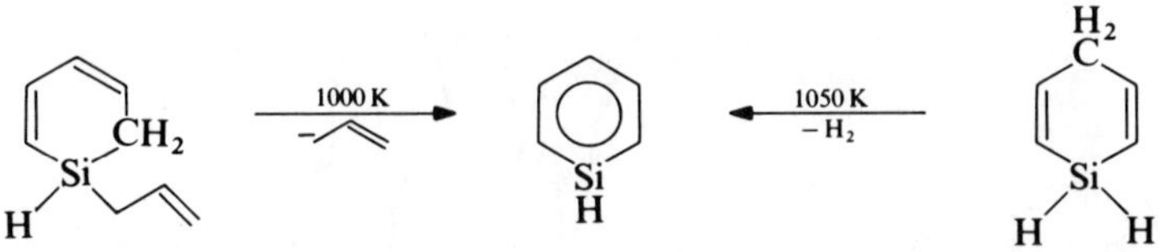

(3)

uncontaminated PE spectrum of H_5C_5SiH, which would allow one to observe all valence ionization states within the He(I) measurement region. On heating 1-sila-2,5-cyclohexadiene in a PE spectroscopically monitored flow system[52], H_2 is expelled as a leaving molecule (equation 3) which, due to its low ionization cross-section, remains invisible under the recorded band pattern (Figure 3: b and c). The ionization pattern of silabenzene is assigned—based on an *ab initio* SCF calculation with a basis set of double-zeta quality, i.e. covering its 50 electrons with 76 groups of Gaussian orbitals[52]—by radical cation state comparison with the iso(valence)electronic benzene (Figure 3: c and d). The 1:1 correspondence between the radical cation state sequences of the two molecules can be illustrated in detail[52]: Comparing the two PE spectra, two dominant features are recognized: the overall shift to lower ionization energies from C_6H_6 to C_5SiH_6 due to the decreasing effective nuclear charge C > Si and the removal of all degeneracies due to lowering of the skeletal symmetry from D_{6h} to C_{2v}. Thus the Jahn/Teller broadening of the first PES band of benzene disappears, the ground state of $C_5SiH_6^{\cdot\oplus}$, $\tilde{X}(^2B_1)$, is lowered by 1.14 eV to 8.11 eV, whereas the following excited π-state, $\tilde{A}(^2A_2)$, is shifted to 9.46 eV, higher by 0.21 eV relative to benzene[5]. The third radical cation state at 11.36 eV, resulting from the first benzene radical cation state, $\tilde{X}(^2E_{2g})$, due to degeneracy split, expectedly also shows a lower ionization energy. Altogether, the 3π-ionization energies of C_5SiH_6 fit perfectly into a perturbation correlation established for group V heterobenzenes[55] at the silicon atom ionization potential, $IE_1(Si) = 8.15$ eV[9, 52] (diagram 4). The regression lines of the correlation diagram exhibit slopes proportional to the HMO coefficients $c_{J\mu}^2$ (shown in squares in diagram 4) at the substitution centers[56], and have served to correctly predict

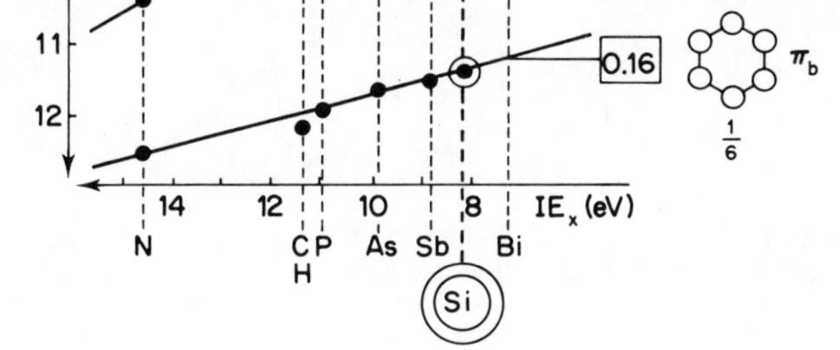

(4)

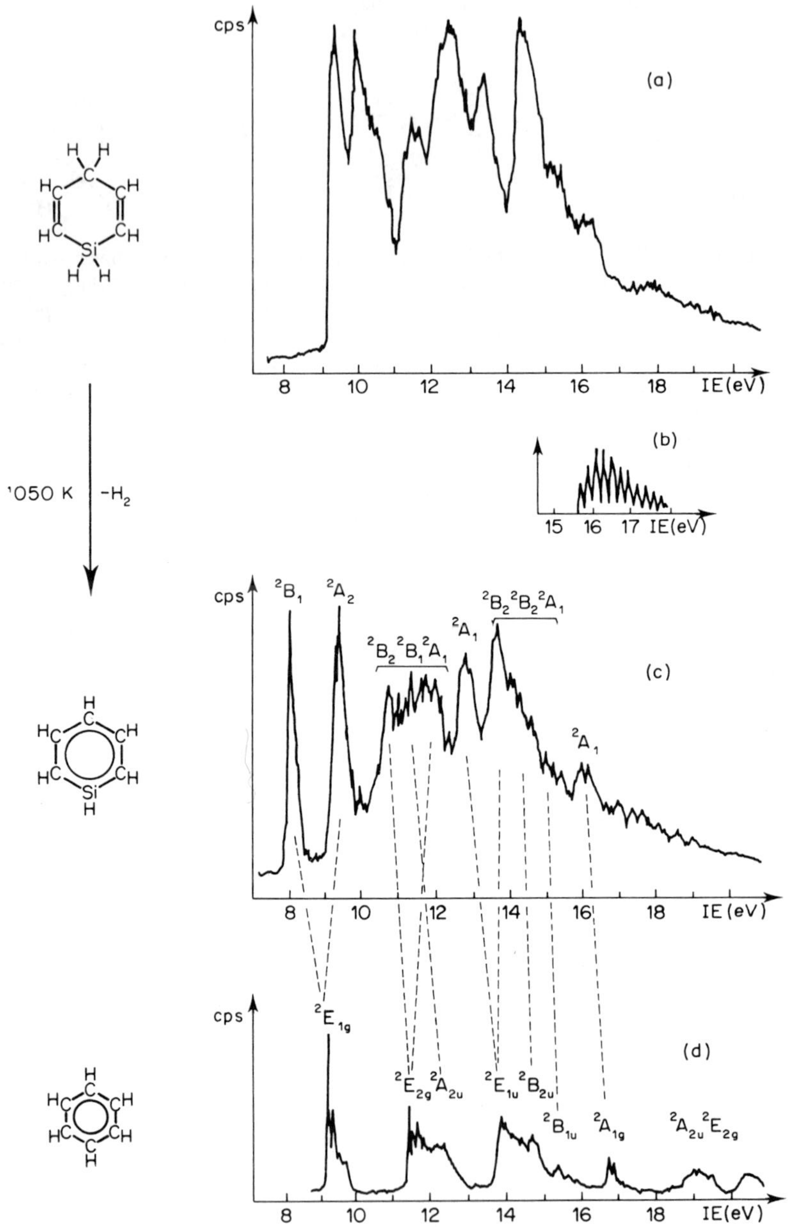

FIGURE 3. He(I) PE spectra of 1-silacyclohexa-2,5-diene, the leaving molecule H_2 and the silabenzene/H_2 mixture at 1050 K as well as the radical cation state comparison with benzene[5] (see text)

the ionization pattern of C_5SiH_6[9], providing a helpful hint in the search for the unstable molecule. Above all, correlation (4) together with the UV spectrum of the matrix-isolated species[12, 54] prove beyond doubt that silabenzene indeed bears the correct name.

2. PE spectrum of silaethene $H_2Si{=}CH_2$

The hunt for this elusive molecule with a half-lifetime in the millisecond range started by calculating a highly correlated wave function, which predicted the following vertical ionization energy differences between the ground state of the neutral molecule $H_2Si{=}CH_2$ and the doublet states of its radical cation[51]:

$H_2Si{=}CH_2^{\cdot\oplus}$	$\tilde{X}(^2B_1)$	$\tilde{A}(^2B_2)$	$\tilde{B}(^2A_1)$	$\tilde{C}(^2B_2)$	$\tilde{D}(^2A_1)$	$\tilde{E}(^2A_1)$	(5)
'IE_n^v'(eV)	8.9 ± 0.1	12.5 ± 0.1	12.9 ± 0.1	14.1 ± 0.1	16.5 ± 0.2	20.4 ± 0.3	

Despite the calculated, rather low first ionization potential, suggesting an isolated, easy-to-detect band in the silaethene PE spectrum, numerous pyrolysis experiments with promising precursors, such as 1,3-disilacyclobutane, yielded no reproducible PE band in the expected region. Eventually, in the thermal retrodiene cleavage of a 5,6-bis(trifluoromethyl) derivative of 2-silabicyclo[2.2.2]octa-5,7-diene, a novel band exhibiting vibrational fine structure appeared within the precalculated region (Figure 4).

The PE spectroscopic evidence for silaethene as the retrodiene dissociation product[51] rests on the coincidence between precalculated and observed ionization energies, on the band intensity increase upon addition of argon to the gas stream, which apparently reduces secondary reactions such as oligomerization, and on the vibrational fine structure of the first PES band, which resembles that of the iso(valence)electronic ethene and precludes with high probability the presence of carbene or silylene isomers. Possibly, the silaethene radical cation $H_2Si{=}CH_2^{\cdot\oplus}$ possesses a twisted ground-state equilibrium structure like $H_2C{=}CH_2^{\cdot\oplus}$ with the two lowest doublet states coupled via the torsional vibration $\tilde{\nu}_6$ (cf. Reference 10).

Silaethene and its deuteriated derivative $D_2Si{=}CH_2$ have been both isolated in a 10 K argon matrix and characterized by IR and UV spectra[51]. 1,1-Dimethylsilaethene[57–59] has been generated by thermolysis of 1,1-dimethyl-1-silacyclobutane; its PE spectroscopically determined first ionization energy, $IE_1^v = 8.0$ eV[58], is lowered by 0.9 eV relative to the parent molecule, expectedly exhibiting a smaller methyl substituent effect than the iso(valence)electronic ethene ($\Delta IE_1 = 1.1$ eV[5]). Numerous calculations have been performed[60] for silaethene, its isomers and derivatives.

C. Molecules with Doubly Coordinated Silicon: PE Spectra of Silylenes SiX_2

Silylenes SiX_2 are an interesting class of compounds[60, 61], attractive from theoretical[60] and spectroscopic[40, 48, 61] as well as from preparative[40, 61] and industrial[62] points of view. Just to select from the vast literature some randomly selected highlights: the parent molecule SiH_2 has a singlet ground state $\tilde{X}(^1A_1)$ in contrast to the triplet $\tilde{X}(^3B_1)$ of the iso(valence)electronic carbene. Some organosilylenes are colored[63], e.g. Me_2Si is yellow. Methyl derivatives can be photochemically and reversibly isomerized to silaethenes like $HSi{-}CH_3 \rightleftharpoons H_2Si{=}CH_2$[60, 64] or, with bulky substituents, dimerize to disilenes $2R_2Si \rightarrow R_2Si{=}SiR_2$[40]. The divalent silicon derivatives insert into numerous main group element bonds[61] and, last but not least, they are of industrial importance, e.g. H_3CSiCl is presumably the intermediate in the Rochow/Müller process for methylchlorosilanes[62]. Although even gas-phase structures have been determined for $SiCl_2$ (equation 8) and $SiBr_2$ by electron diffraction[65], however, PE spectroscopic studies so far are confined to the dihalogen derivatives SiX_2 with X = F[66], Cl[22, 48], Br[67] and I[67] (Figures 5 and 6).

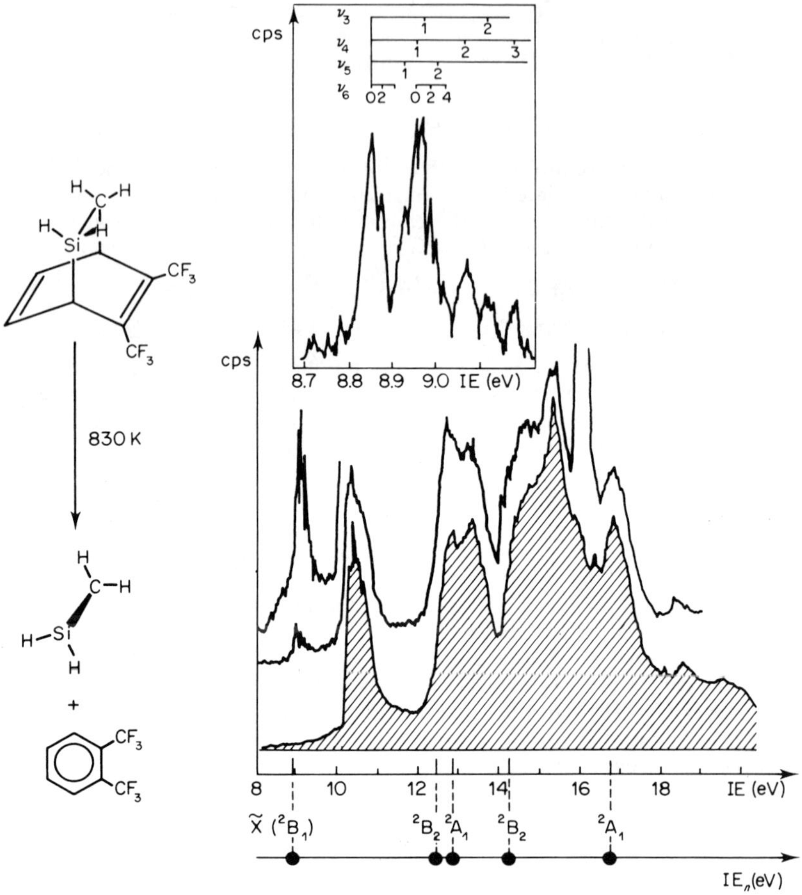

FIGURE 4. HE(I) PE spectrum of the thermal retrodiene dissociation products of 5,6-bis(trifluoromethyl)-2-silabicyclo[2.2.2]octa-5,7-diene at 850 K [hatched: 1,2-bis(trifluoromethyl) benzene] and correlation with precalculated ionization energies (see text). The enlarged insert shows the expanded vibrational fine structure of the ground state of the silaethene radical cation with the following suggested assignments (from top to bottom): $\nu_3(a_1, \delta_{CH_2}) = 1010 \pm 40\ cm^{-1}$, $\nu_4(a_1; \nu_{Si=C} + \delta_{SiH_2}) = 840\ cm^{-1}$, $\nu_5(a_1; \nu_{Si=C} + \delta_{SiH_2}) = 620\ cm^{-1}$ and $\nu_6(a_2; \tau) = 200\ cm^{-1}$

Dichlorsilylene $SiCl_2$ is chosen as a representative example, because it can be generated thermally in many ways, e.g. by passing silicon halides over heated silicon powder (equation 6) or by splitting off naphthalene from the corresponding 3*H*-3-benzosilepine derivative (equation 7)[48]. It also illustrates the use of on-line registration of PE spectra

$$Cl_3SiSiCl_3 \xrightarrow{1450\ K/Si_x} SiCl_4; \quad Cl_3CSiCl_3 \xrightarrow{>1250\ K/Si_x} SiCl_4; \quad SiCl_4 \xrightarrow{1450\ K/Si_x} SiCl_2 \qquad (6)$$

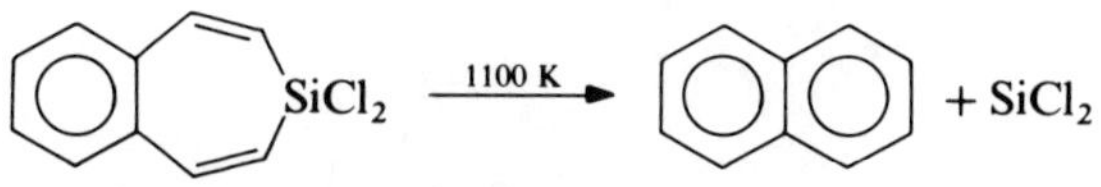

(7)

(Figures 5 and 6) including digital 'spectrum stripping' (Figure 6) as well as various approaches to PE spectroscopic assignment via radical cation state comparison with iso(valence)electronic compounds (equation 8 and Figure 7) or via Koopmans' correlation, $IE_n^v = -\varepsilon_J^{SCF}$, with SCF eigenvalues (scheme 8, cf. Section IV).

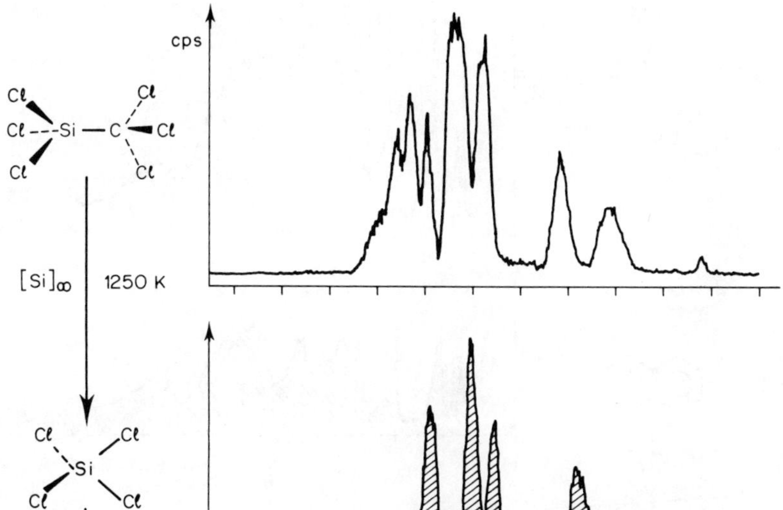

FIGURE 5. He(I) PE spectra of $Cl_3Si–CCl_3$ at 300 K and of the products of its reaction with elemental silicon dispersed in quartz wool at 1250 K yielding $SiCl_4$ (hatched) and at 1450 K yielding dichlorosilylene $SiCl_2$ (dotted)

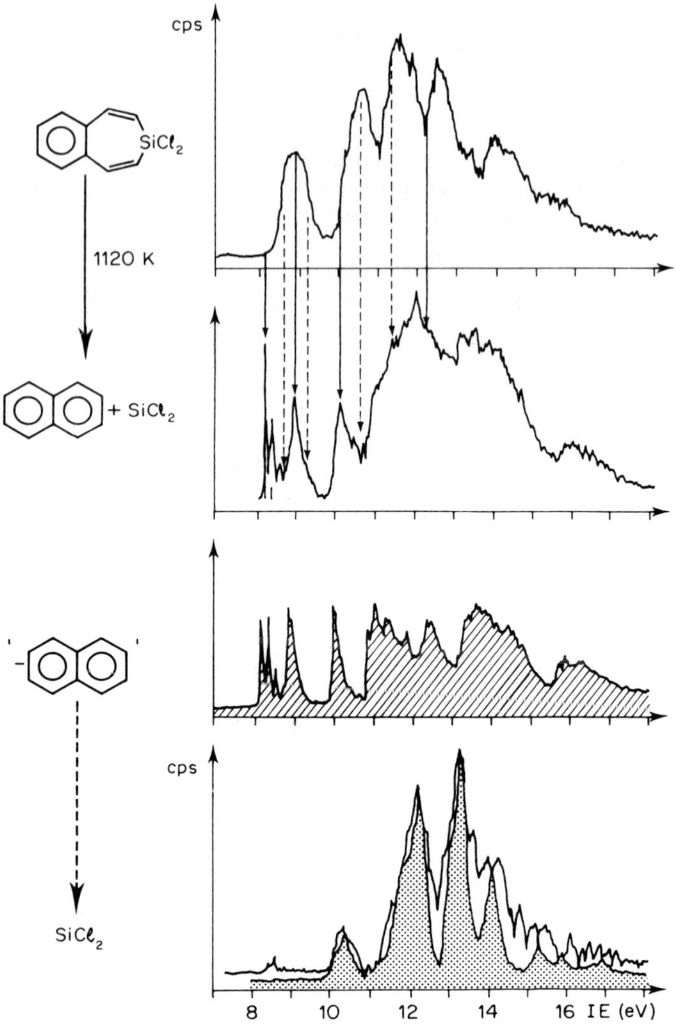

FIGURE 6. He(I) PE spectra of 3,3-dichloro-3*H*-3-benzosilepin at 350 K and of the product mixture after thermolysis at 1120 K showing new bands (→) as well as disappearing ones ---- ►). Digital subtraction of the prerecorded naphthalene PE spectrum[3] (hatched) yields the ionization pattern of $SiCl_2$ (dotted), unequivocally identified by comparison with the band structure resulting from the $SiCl_4 + Si$ reaction (Figure 5)

In detail, for the reaction of trichloromethyltrichlorosilane with elemental silicon[22,48] carried out in the internal heating system (Figure 2), the following changes in the continuously recorded PE spectra are observed while raising the oven temperature (Figure 5): above 1000 K, Cl_3SiCCl_3 begins to decompose, and at 1250 K the known PE spectrum of pure $SiCl_4$ is observed. Transient chlorinated hydrocarbons cannot be detected; possibly, either SiC is formed on the Si surface or $(CCl_2)_\infty$ on cooler parts of the wall. At 1450 K, $SiCl_4$ reacts with Si—as confirmed by independent direct conversion—almost completely to yield $SiCl_2$[22,48]. As another route, the thermolysis of 3,3-dichloro-3*H*-3-benzosilepin has been investigated[48]: at the PE spectroscopically optimized decomposition temperature of 1120 K, the prominent bands of the benzosilepine have largely been replaced by the known[3] ones of naphthalene (Figure 6). Digitally subtracting the PE spectrum of naphthalene from that of the product mixture, the unequivocally identifiable band pattern of $SiCl_2$ emerges from the considerable background noise (Figure 6).

An unequivocal assignment of the $SiCl_2$ ionization pattern, which consists of three single and two double intensity bands, is achieved by comparison of the radical cation states $\tilde{X}$ to $\tilde{E}$ with the ones known for the iso(valence)electronic $SnCl_2$[68], considering as perturbations the opposing influences of decreased effective nuclear charge and reduced bond angle[4] (scheme 8). The assignment is further corroborated by Koopmans' correlation with the eigenvalues $-\varepsilon_J^{MNDO}$ of a geometry optimized MNDO calculation, which satisfactorily reproduces the gas-phase structure of $SiCl_2$ (scheme 8). According to the plausible radical cation state sequence calculated (see scheme 8), the dominant contribution in the ground state of $SiCl_2^{\cdot\oplus}$ should be a largely $3s_{Si}$-type silicon lone pair, which also possibly accounts for a small ionization cross-section[4] and thus would explain the rather low intensity of the PES band at 10.3 eV. The next 4 $M^{\cdot\oplus}$ states are represented by in-plane and out-of-plane chlorine lone pairs, and the sixth at 14.0 eV by an antibonding combination of SiCl bonds.

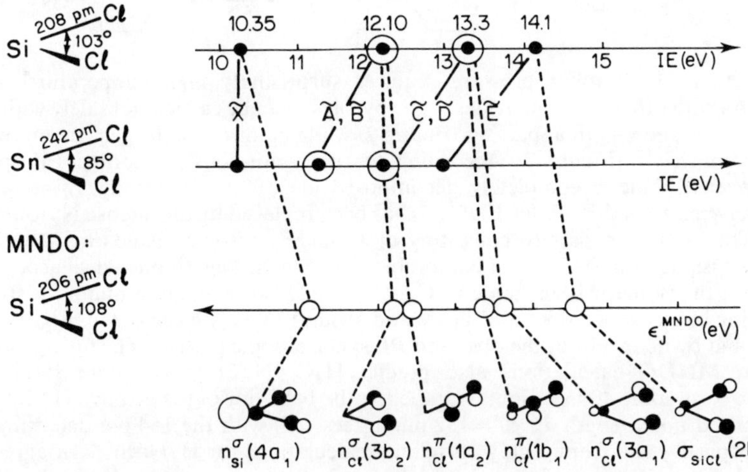

(8)

The geometry-optimized MNDO calculations which almost reproduce the experimental SiCl bond lengths as well as the angle ∢ ClSiCl (scheme 8) suggest, for the ground state of the neutral dichlorosilylene, a charge distribution with a highly positive Si

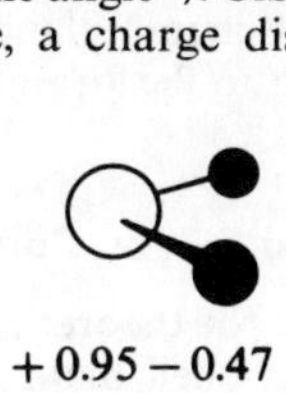

(9)

center,[22] thereby offering an explanation for its insertion activity into many main group element bonds[61].

D. Molecules with Singly Coordinated Silicon: PE Spectrum of Phenyl silaisonitrile H_5C_6–N≡Si

Silicon as a group 4 element is capable of forming compounds with a terminal Si which, in contrast to, for example, CO or isonitriles R–N=C, however, readily oligomerize and therefore have to be generated either under nearly unimolecular conditions in the gas phase or in low-temperature matrices. Along these lines, the matrix photolysis of azidosilane H_3SiN_3 to H–N≡Si had already been reported 20 years ago[69] and its stability relative to the isomer H–Si≡N confirmed by highly correlated quantum-chemical calculations[60, 69]. A literature search for PE spectroscopic studies of molecules containing singly coordinated silicon—which, being performed in gas-flow systems under reduced pressure are well suited for these investigations—yielded only evidence for SiO[13, 70] (cf. Section II.E) and for phenyl silaisonitrile[15, 71].

Triazido(phenyl)silane can be prepared from trichloro(phenyl)silane by $AlCl_3$-catalyzed transazidation with trimethylsilylazide and, after careful decantation from the explosive $AlCl_x(N_3)_{3-x}$ flakes, which precipitated on the walls of the reaction flask, purified by distillation (reaction 10). Its 'controlled explosion' in a nearly unimolecular

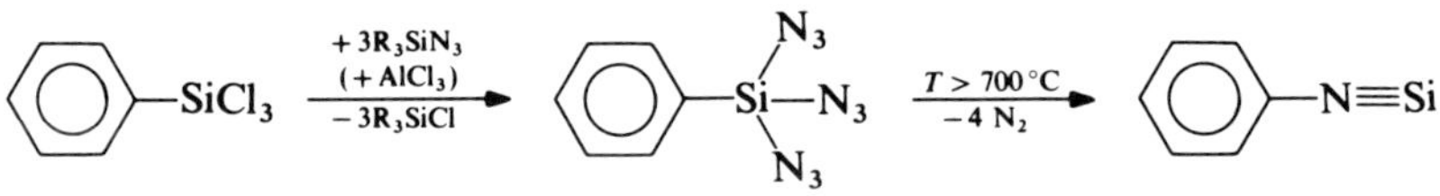

(10)

flow system at 10^{-4} mbar pressure requires surprisingly high temperatures above 1000 K!. As with other covalent azides[15], the pyrolysis can be carried out safely with small amounts in the short-path apparatus (Figure 2) while monitoring the gas stream by PES real-time analysis[12] (Figure 7). According to the continuously recorded PE spectra, triazido(phenyl)silane is completely decomposed at 1100 K: its strong σ_{N_3}-ionization band[15] between 15 and 16 eV (cf. Figure 7) has been replaced by the intense N_2 ionization peaks[5]. The ionization pattern consisting of 11 largely resolved bands between 8 and 14.5 eV is assigned on the basis of comprehensive calculations to phenyl silaisocyanide, whereas for the isomeric silacyanide H_5C_6–Si≡N only two ionization humps of strongly overlapping bands between 9.5 and 11 eV and around 13 eV are suggested (Figure 7: ○), which would by no means fit the observed PE spectroscopic pattern. The fully geometry-optimized MNDO calculations also predict H_5C_6–N≡Si to be more stable than H_5C_6–Si≡N by some 400 kJ/mol(!), paralleling the behavior of the isomers HNSi[69], and the optimized bond length $d^{MNDO}_{N\equiv Si} = 152$ pm agrees well with the 154 pm determined IR spectroscopically and from highly correlated calculations for H–N≡Si[69]. In agreement with the known structure of iso(valence)electronic phenyl isocyanide H_5C_6–N≡C, also for H_5C_6–N≡Si a linear arrangement (C)–N≡Si with NSi triple-bond character is calculated. In the meantime the PE-spectroscopic results[70] have been further supported by the matrix isolation of phenyl silaisocyanide[72].

Altogether, the pyrolysis of triazido(phenyl)silane, driven by the energetically favorable extrusion of four molecules of N_2 leads to the formation of a polar –N≡Si triple bond, with a silicon of coordination number 1.

E. PE Spectra of Other Unsaturated Silicon Compounds

The so-called 'double bond rule', i.e. the theoretically unjustified assumption that main-group elements of periods with outer-shell main quantum numbers $n > 2$ exhibit only

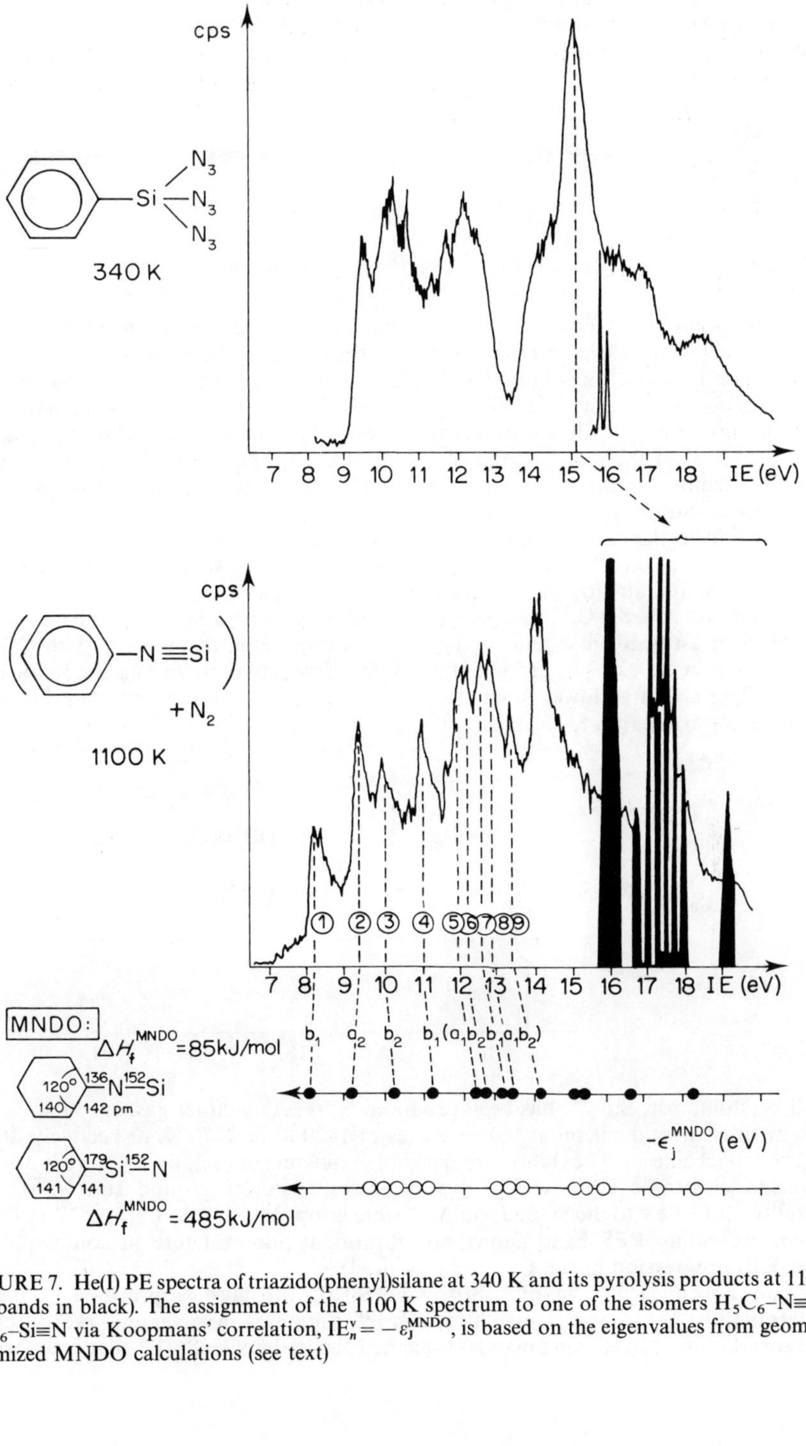

FIGURE 7. He(I) PE spectra of triazido(phenyl)silane at 340 K and its pyrolysis products at 1100 K (N_2 bands in black). The assignment of the 1100 K spectrum to one of the isomers H_5C_6–N≡Si or H_5C_6–Si≡N via Koopmans' correlation, $IE_n^v = -\varepsilon_j^{MNDO}$, is based on the eigenvalues from geometry-optimized MNDO calculations (see text)

limited ability to form $p_\pi p_\pi$-multiple bonds, has been rendered more and more obsolete in recent years. As concerns unsaturated silicon compounds, PE spectroscopic studies under the advantageous and nearly unimolecular gas-phase conditions[12, 15] made possible the detection of small prototype molecules such as those reported in the preceding sections:

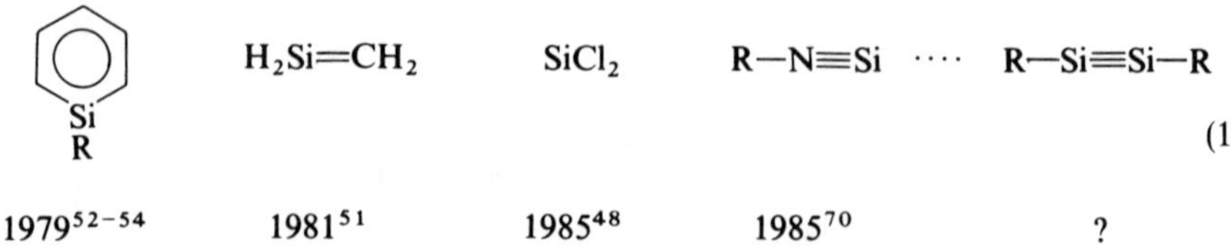

(11)

Again attention shall be focussed on large isolated compounds containing Si=C and Si=Si double bonds[40], which have been kinetically stabilized by being wrapped with bulky hydrocarbon residues. They are, however, beyond the scope of this report, as are other techniques for their generation like photolysis[40], and as are also calculations for numerous interesting small silicon intermediates[60]. The latter, nevertheless, demonstrate future prospects of PE spectroscopically monitored gas-phase thermolysis reactions, for instance, attempts to generate species with Si≡Si triple bonds (scheme 11) from precursors like $(N_3)_2RSi{-}SiR(N_3)_2$[15] (cf. Figure 7).

In addition to the case studies presented in detail (Sections II.B–II.D), the following PES investigations on other unsaturated silicon compounds—listed in order of increasing molecular size and already partly mentioned—have been reported.

Silicon monoxide, Si≡O[70], has been generated by heating a $Si + SiO_2$ mixture and its PE spectrum assigned *inter alia* by the radical cation state comparison with CO and GeO[13] (see scheme 12). The ground state, $\tilde{X}(^2\Sigma^+)$, largely representing the Si lone pair shows a 2.6 eV shift to lower ionization energy relative to CO as expected due to the effective nuclear charges Si < C.

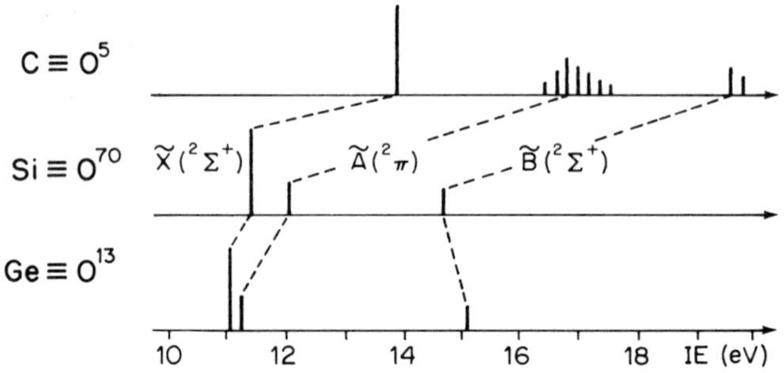

(12)

Silicon difluoride, SiF_2[66], has been produced by reacting either gaseous SiF_4 or solid CaF_2 with elemental silicon at temperatures of 1420 K or 1470 K, respectively. Its PE spectrum (scheme 13) exhibits remarkable differences relative to that of the iso(valence)electronic CF_2, which also possesses a singlet ground state[37]: the first ionization at 11.1 eV to the Si lone pair $M^{\cdot\oplus}$ state is lower by 1.2 eV (CF_2: 12.27 eV)[37] and the corresponding PES band shows no vibrational fine structure in contrast to the 22 peak (!) progression of the CF_2 bending mode[37].

Silicon dibromide and diiodide, $SiBr_2$ and SiI_2[67], can both be prepared by passing the tetrahalides SiX_4 over elemental silicon (cf. Figure 5). A schematic radical cation state correlation of all silicon dihalides (scheme 13) based on the $SiCl_2^{\cdot\oplus}$ sequence [scheme

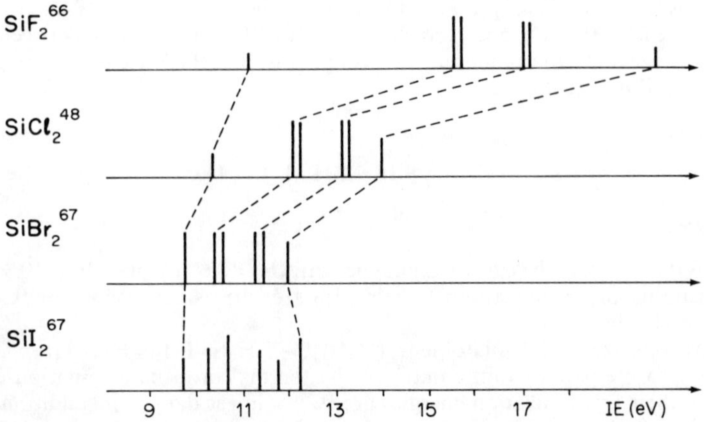

(13)

8: $n_{Si}(4a_1) < n^{\sigma}_{Cl}(3b_2) < n^{\pi}_{Cl}(1a_2) < n^{\pi}_{Cl}(1b_1) < n^{\sigma}_{Cl}(3a_1) < \sigma_{SiCl}(2b_1)$)], reveals the expected shifts to lower ionization energies with decreasing effective nuclear charge of the halide substituents. Especially for SiI_2, the almost equally spaced PE bands demonstrate the dominating influence of spin/orbit coupling[37], which no longer allows a simplified assignment to $M^{\cdot\oplus}$ states with a specific prominent contribution.

Silyl radical, SiH_3[73], has been generated by H abstraction from SiH_4 using F atoms produced in a microwave discharge applied to a F_2/He mixture, and the first band of its PE spectrum recorded in the flow system at 2 cm distance. The extensive vibrational fine structure (scheme 14: A) can be reproduced by a ΔSCF computed Franck/Condon envelope[73] (scheme 14: B). The following results are emphasized[73]: the vertical ionization

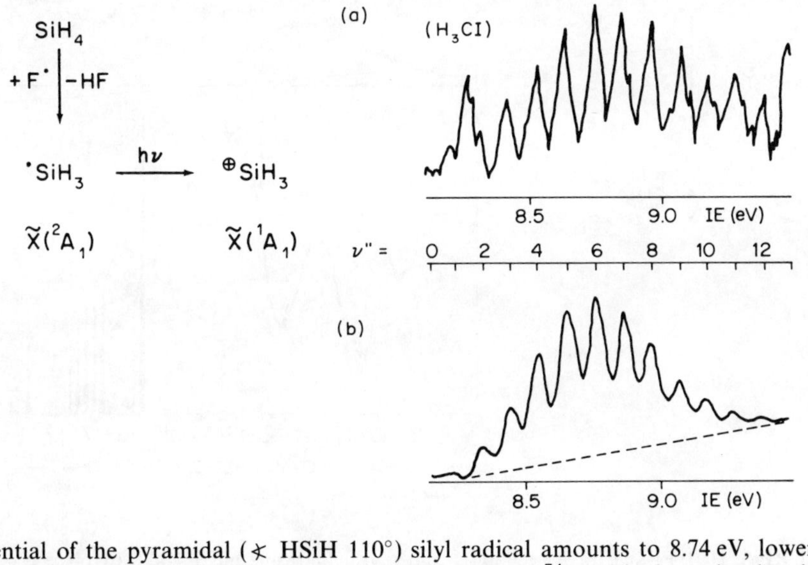

(14)

potential of the pyramidal (∢ HSiH 110°) silyl radical amounts to 8.74 eV, lower by 1.1 eV than that of the planar (!) methyl radical (9.48 eV)[74], and its inversion barrier is calculated to be 20 kJ mol^{-1}. The silyl radical cation produced on vertical ionization (time scale 10^{-15} s) is predicted to relax adiabatically to a planar species (cf. Section III).

Dimethylsilathione, $(H_3C)_2Si{=}S$[75]: On heating the cyclic trimer $(R_2SiS)_3$ in a flow system to at least 500 K, the observed changes in the PES band pattern between 9 eV and 13 eV are assigned to the monomer by Koopmans' correlation based on a geometry-optimized MNDO calculation:

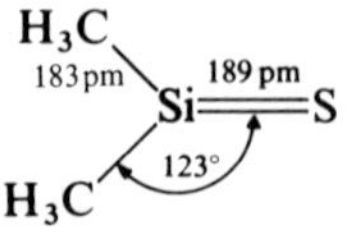

IE^v_n	9.15	10.1	11.05	12.1	(eV)
	n_S	$\pi_{Si=S}$	σ_{SiS}	σ_{SiC}	

(15)

Relative to thioacetone, the carbon analogue with $IE_1 = 8.60$ eV and $IE_2 = 10.3$ eV[77], the characteristic n_S and π ionizations (scheme 15) are surprisingly higher by 0.55 eV and 0.2 eV, respectively.

4-*t*-Butylsilabenzene and -silatoluene, $(H_3C)_3C{-}C_5H_4Si{-}R(R = H, CH_3)$[78]: Generated analogously to the parent compounds[53, 54] by thermal propene elimination from allyl derivatives, their first ionization energies are 0.2 eV lower due to the additional *t*-butyl substituent perturbation.

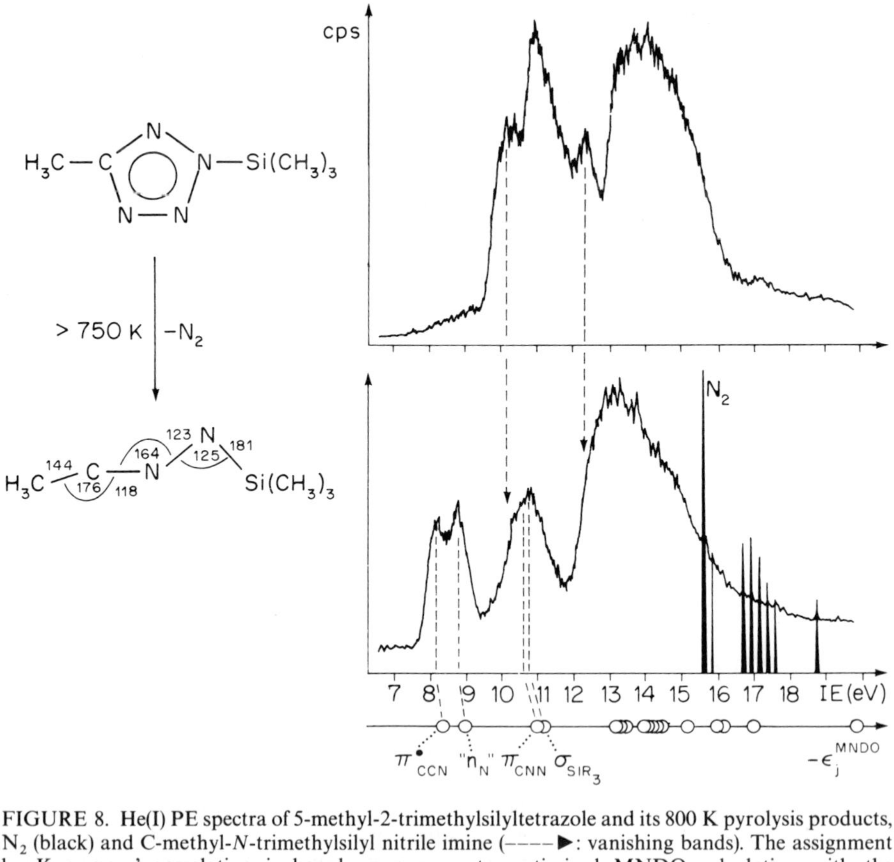

FIGURE 8. He(I) PE spectra of 5-methyl-2-trimethylsilyltetrazole and its 800 K pyrolysis products, N_2 (black) and C-methyl-*N*-trimethylsilyl nitrile imine (----▶: vanishing bands). The assignment by Koopmans' correlation is based on a geometry-optimized MNDO calculation with the approximated bond lengths in pm and angles in deg

N-Trimethylsilylnitrile imines, R–C≡N$^{\oplus}$–N$^{\ominus}$–$SiMe_3$ with R = Me, Ph[79]: In these compounds, Si stabilizes the unsaturated nitrile imine system partly electronically and partly by sterical shielding towards isomerization to carbodiimides or cyclization to indazoles. PE spectroscopic real-time gas analysis has been applied to optimize the conditions for the known[79] vacuum pyrolysis of 5-*R*-2-trimethylsilyltetrazoles to nitrile imines (Figure 8); N_2 is split off quantitatively above 750 K. In the resulting PE spectrum, the three bands in the low-energy region correspond to four ionizations, with the main contributions to the $\tilde{X}$, $\tilde{A}$ and $\tilde{B}$ radical cation states as shown in diagram 16. The

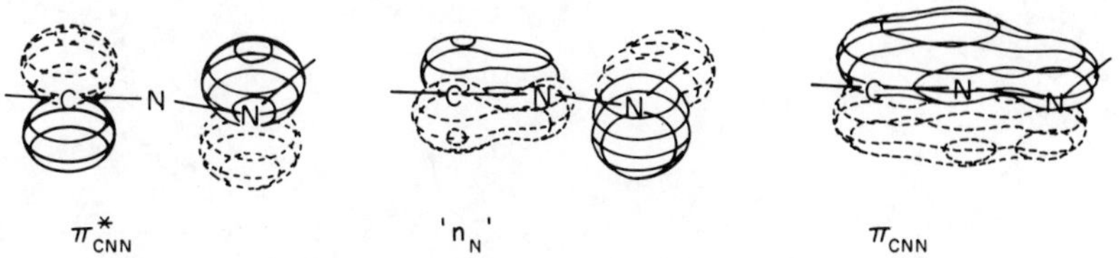

(16)

geometry-optimized MNDO calculation predicts a nitrile imine skeleton, which is planar and bent zigzag-wise (Figure 8), exhibiting a central angle ∢ CNN of about 160° and rather short CN and NN bonds indicating considerable multiple bonding. The calculated charge distribution places a large negative charge on the N center, enhanced relative to the parent HCNNH by the electron-donating R_3Si substituent:

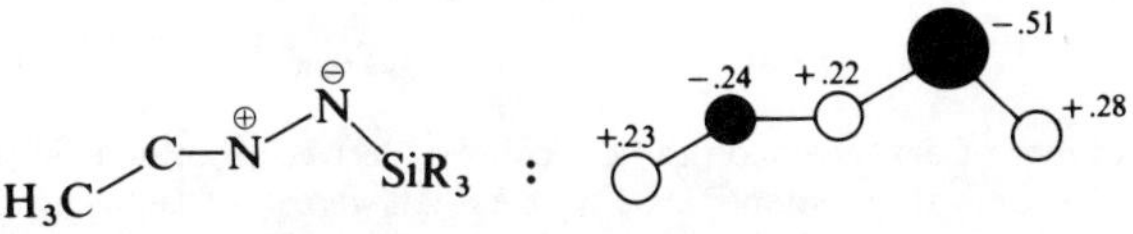

(17)

In closing, it is pointed out that the ionization patterns of nitrile imines (Figure 8) confirm the proposed frontier orbital sequence, which allows their regioselectivity to be rationalized in 1,3-dipolar additions[79].

III. GENERATION OF ORGANOSILICON RADICAL CATIONS IN SOLUTION AS PREDICTED BY FIRST PES IONIZATION ENERGIES AND SOME OF THEIR PROPERTIES

Since 1976[80], a large number of organosilicon radical cations[16] have been generated in $AlCl_3/H_2CCl_2$ solution by single electron oxidation from their neutral precursors (Figure 9).

ESR and ENDOR spectra of organosilicon radical cations yield a wealth of information[16] on their spin distributions, on structural changes during their adiabatic relaxation and on their rigidity or flexibility within the ESR time scale of 10^{-7} s, orders of magnitude longer than the 10^{-15} s required for vertical ionization (Section III.D). Combined with the PES measurements, striking substituent effects of trialkylsilyl groups on π-systems and electron-rich lone-pair centers have been discovered[16], which are nowadays used in preparative organic chemistry. Above all, however, the complementary ESR/ENDOR investigations of organosilicon radical cations in solution provide some insight into their ground-state properties inaccessible by PE spectroscopic measurements.

A. The $AlCl_3/H_2CCl_2$ Oxidation Potential

To study organosilicon radical cations in solution, a general procedure has been developed, indeed a recipe, for their generation based on photoelectron spectroscopy[16]:

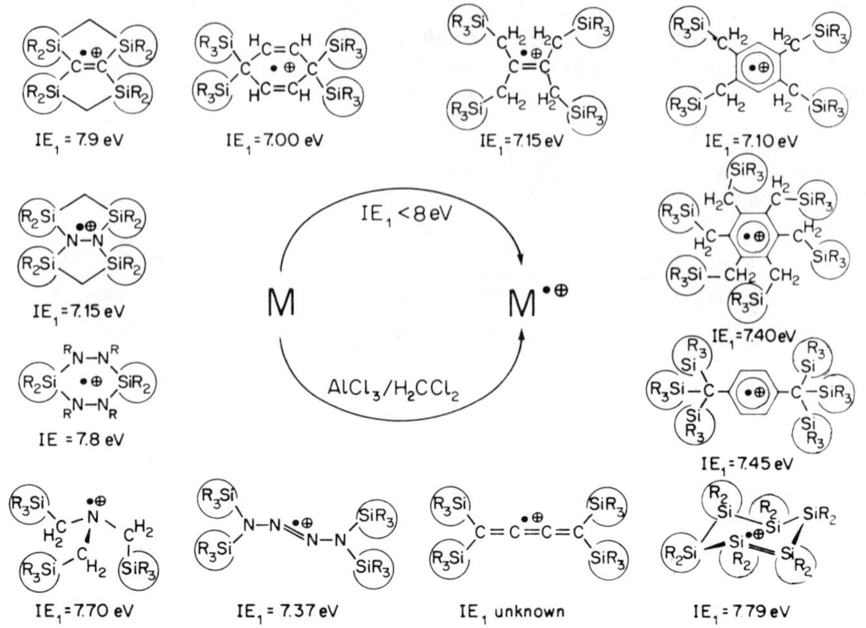

FIGURE 9. Prototypes of organosilicon radical cations, derived from neutral parent compounds exhibiting vertical first ionization potentials below 8 eV. All were oxidized using $AlCl_3/H_2CCl_2$ and characterized by their ESR or ENDOR spectra[16]. Tetrakis(trimethylsilyl)butatriene radical cation is formed on $AlCl_3/H_2CCl_2$ treatment of tetrakis(trimethylsilyl)butyne $(Me_3Si)_3C–C≡C–C(SiMe_3)_3$. The ringed groups represent bulky substituents.

within an hour, the first ionization potential of a promising precursor, predicted either from heuristic correlations of suitable properties of chemically related molecules or by fast semiempirical calculation, can be determined. If the energy needed for electron expulsion in the gas phase is below 8 eV, within another hour the corresponding radical cation can be generated by adding pure $AlCl_3$ to the H_2CCl_2 solution of the neutral molecule and characterized by its unmistakable ESR signal pattern. How does the elegant, selective one-electron oxidation under anhydrous and oxygen-free conditions, using H_2CCl_2 as a seemingly inert solvent (cf. however, equation 18) of advantageously low viscosity,

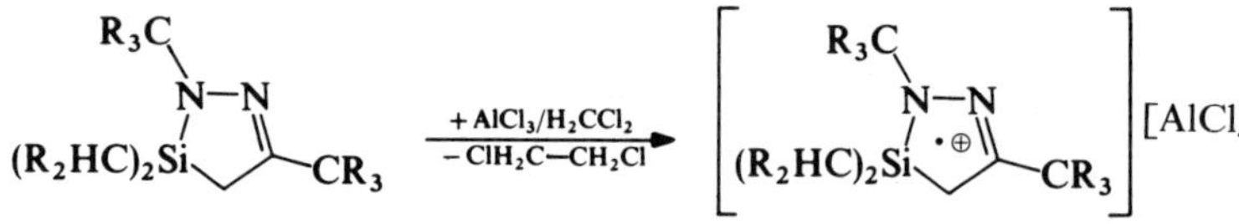

(18)

proceed? Although the actual mechanism is still unknown, several reactions carried out on a preparative scale (see e.g. Reference 81) provide possible clues: the isolated radical ion salt containing the $AlCl_4^{\ominus}$ counteranion has been structurally characterized and 1,2-dichloroethane was identified by GC (equation 18). These observations would support the following assumptions: $AlCl_3$ could subtract $Cl^{\ominus}$ from H_2CCl_2, forming the counter-anion $AlCl_4^{\ominus}$, and the resulting carbonium ion $H_2CCl^{\oplus}$ could, on selective single-electron transfer, yield the radical $H_2CCl^{\cdot}$, which vanishes in consecutive reactions like dimerization to $ClH_2C–CH_2Cl$. An alternative mechanistic proposal[81] could invoke $AlCl_2^{\oplus}$, detected in $AlCl_3/H_2CCl_2$ solutions, e.g., by ^{27}Al–NMR.

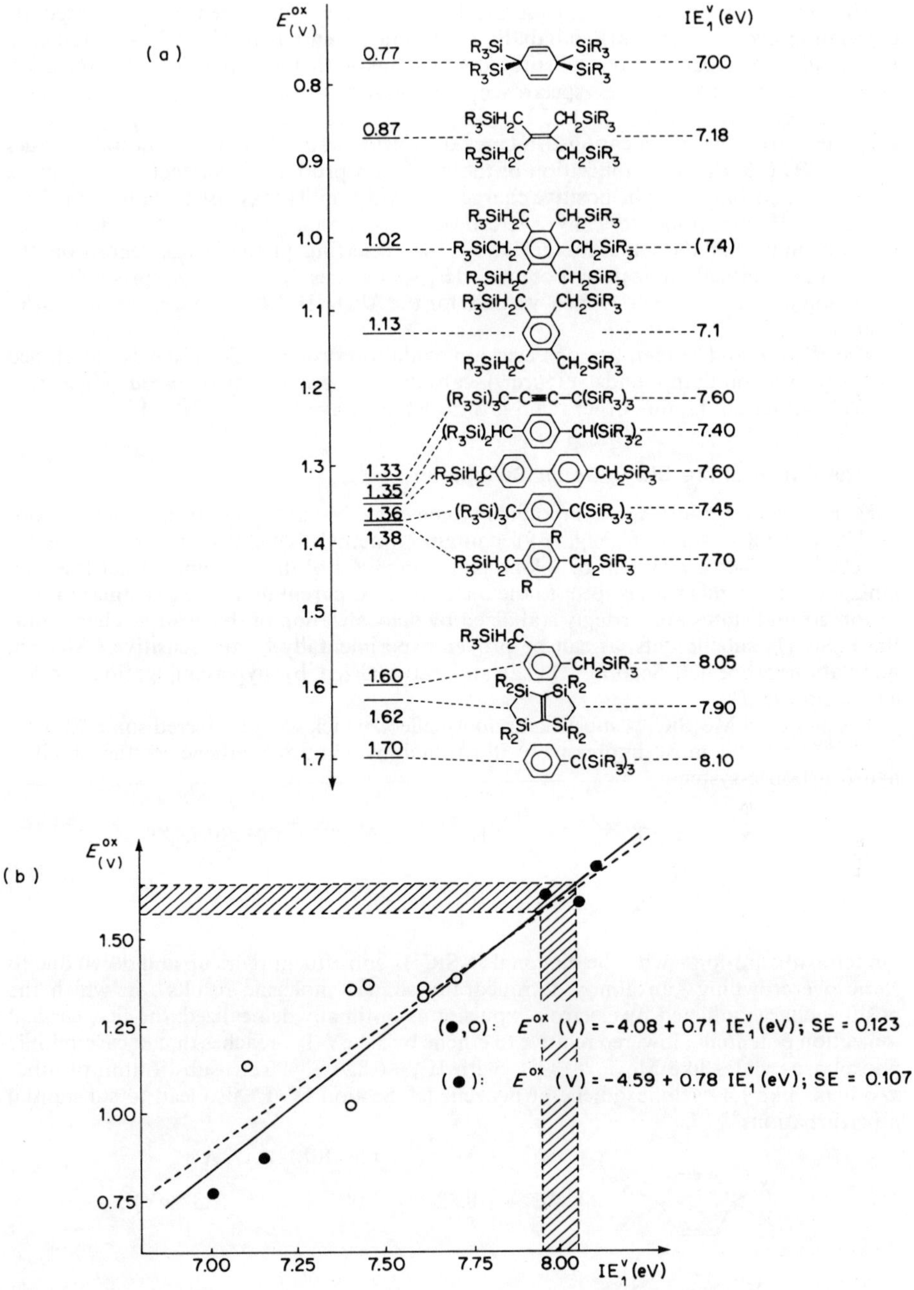

FIGURE 10. (a) Cyclovoltammetric oxidation potentials of selected organosilicon substituted π-systems in H_2CCl_2/0.1 m $R_4N^{\oplus}ClO_4^{\ominus}$ solutions and (b) their correlation with the first vertical PES ionization potentials (circles: sterically overcrowded compounds; hatched area: borderline of oxidizability between $IE_1^v = 7.95$ eV and 8.05 eV, see text).

The oxidation potential of the $AlCl_3/H_2CCl_2$ system has been approximated by correlating cyclovoltammetric adiabatic oxidation potentials in H_2CCl_2 solution of a dozen selected organosilicon substituted π-systems with their first vertical ionization energies determined by PE spectroscopy[81] (Figure 10: a). All but three of these compounds with either $E^{ox} < +1.65$ V or $IE_1^v < 7.95$ eV can be oxidized by $AlCl_3/H_2CCl_2$ treatment to their radical cations. In cases of considerable adiabatic structural changes (Section III.C.3) due to stabilization of the oxidation product by an energetically more favorable distribution of the positive charge, the oxidizability may rise to values of IE_1^v up to 8.25 eV[82]. Omitting sterically overcrowded compounds (Figure 10: b, circles) expectedly improves the standard deviation and, inserting in the linear regression the 'borderline' vertical ionization potential $IE_1^v \sim 8$ eV, results in the surprisingly high oxidation potential $E^{ox} \sim +1.65$ eV vs. SCE for the $AlCl_3/H_2CCl_2$ single electron transfer system.

The efficient and oxygen-free one-electron oxidation system $AlCl_3/H_2CCl_2$, developed for organosilicon compounds (Figure 9), can also be used to advantageously generate radical cations containing other main group elements like B[83], C[84], N[85], P[86] or S[82, 87].

B. The Powerful Me_3SiCH_2 Donor Substituent

Many of the organosilicon compounds, which could be oxidized to their radical cations in $AlCl_3/H_2CCl_2$ solution (Figure 9), contain trimethylsilylmethyl groups Me_3SiCH_2 attached to either a π-system or a lone-pair center. All of them exhibit rather low first ionization potentials with respect to the unsubstituted parent molecules, i.e. their radical cation ground states are strongly stabilized by delocalization of the positive charge into the R_3SiCH_2 substituents, as can be proven experimentally by the sensitive ESR spin population probe (cf. Section III.C.2) and rationalized by hyperconjugation models (cf. Section IV.F).

The powerful Me_3SiCH_2 substituent donor effect, which was discovered some 20 years ago[88, 89], can be most impressively illustrated by choosing ethene as the smallest hydrocarbon π-system:

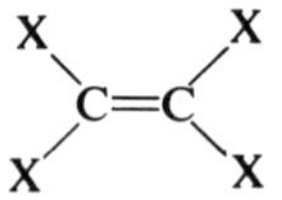

X:	H[5]	CH_3[90]	CH_2SiMe_3[91]
IE_1^v:	10.52	8.27	7.15 eV

(19)

On tetrasubstitution—with the diagonal R_3SiCH_2 substituent pairs up and down due to steric overcrowding—an almost tetrahedrally shaped molecule results[9], in which the positive charge inflicted by electron expulsion is optimally delocalized: the first vertical ionization potential—lowered relative to ethene by 3.4 eV (!)—reaches that of carbanionic phosphorus ylides like $Me_3P^{\oplus}–^{\ominus}CH_2$ with $IE_1^v = 6.82$ eV[92]! Tetrasubstitution of other π-systems like 1,4-cyclohexadiene or benzene (cf. Section IV. F) also lead to substantial π-perturbations[91, 93]:

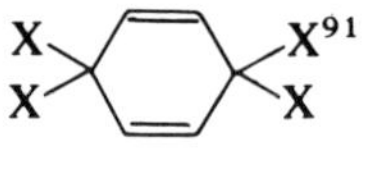

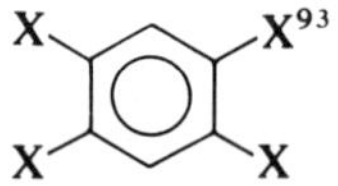

	X:	H	CH_2SiMe_3	ΔIE_1^v
1,4-cyclohexadiene[91]	IE_1^v	8.82	7.00	1.82 (eV)
benzene[93]	IE_1^v	9.25	7.10	2.15 (eV)

(20)

Substituent effects of the same magnitude are observed, if the trimethylsilylmethyl groups are attached to lone-pair centers[102]:

	X:	H	CH_3	CH_2SiMe_3		
SX_2	IE^v_1:	10.43	8.67	8.04	(eV)	(21)
NX_3	IE^v_1:	10.85	8.44	7.66	(eV)	

All the above and numerous other PE spectroscopic results[16] for R_3SiCH_2 groups containing organosilicon compounds demonstrate how strongly radical cation ground states are stabilized due to extensive delocalization of the positive hole generated on ionization or—within a simplifying hyperconjugation model[16, 93]—how powerful donors the R_3SiCH_2 substituents are (Figures 9 and 10).

C. Organosilicon Radical Cations in Solution

Obviously, the close correspondence between the rather instantaneous 'vertical' PES ionization in the gas phase and the considerably slower oxidation in solution, which allows the 'adiabatic' relaxation of the radical cation generated in its solvent cage, are interrelated facets of the same species. Therefore, a short summary on their ESR/ENDOR spectroscopically determined spin populations, their structural changes relative to the neutral precursors and their observable molecular dynamics provides some insight into the reality of molecular radical cation ground states.

1. ^1H and ^{29}Si ESR and ENDOR spectroscopy

The paramagnetic organosilicon species (Figure 9) generated by the single electron transfer to the $AlCl_3/H_2CCl_2$ oxidizing system (Figure 10), on ESR/ENDOR spectroscopic investigation provide much information as demonstrated by the radical cation of 1,4,5,8-tetrakis(trimethylsilyl)-1,2,3,4,5,6,7,8-octahydronaphthalene chosen as an illustrative example (Figure 11). To begin with its ESR spectrum (Figure 11: a)[16], even high resolution can hardly cope with the enormous number of 13875(!) signals of small line widths and greatly varying multiplet intensities; for the center signal (Figure 11: b) alone 925 lines are predicted, of which about 50 become distinguishable at 180 K. ENDOR[22, 94], the NMR measurement 'quantum-jumped' to the ESR mode, reduces the complexity of the $(2n_N I_N+1)$ multiplets for each set of equivalent nuclei N with nuclear spin I_N to a single line pair. Thus in our case (Figure 11: c), only 4 line pairs for the 4 different sets of hydrogens and another one for 1 out of 4 equivalent ^{29}Si nuclei with spin $I_{Si}=1/2$ and observable in natural abundance of only 5.7% (Figure 11: d) remain, with every pair centered around the respective nuclear frequency. All coupling constants and, from triple resonance experiments, all their signs have been determined.

Preceding the first ^{29}Si ENDOR measurement[22, 94], its feasibility had to be estimated: for the correlation rotation time, the most essential condition in the multiparameter optimization experiment, a τ_{rot} range 0.2 to 3 ns[22, 94] was correctly predicted both from molecular and from theoretical considerations.

2. Spin populations and charge distributions

Very interesting information from high-resolution ESR or from ENDOR spectroscopy concerns the distribution of the unpaired electron in the paramagnetic molecular cation: the spin population can be determined experimentally from the electron/nuclear spin coupling, whereas the charge distribution has to be approximated by (open-shell) calculations. As concerns Me_3SiCH_2-substituted π radical cations, the spin population ρ_π

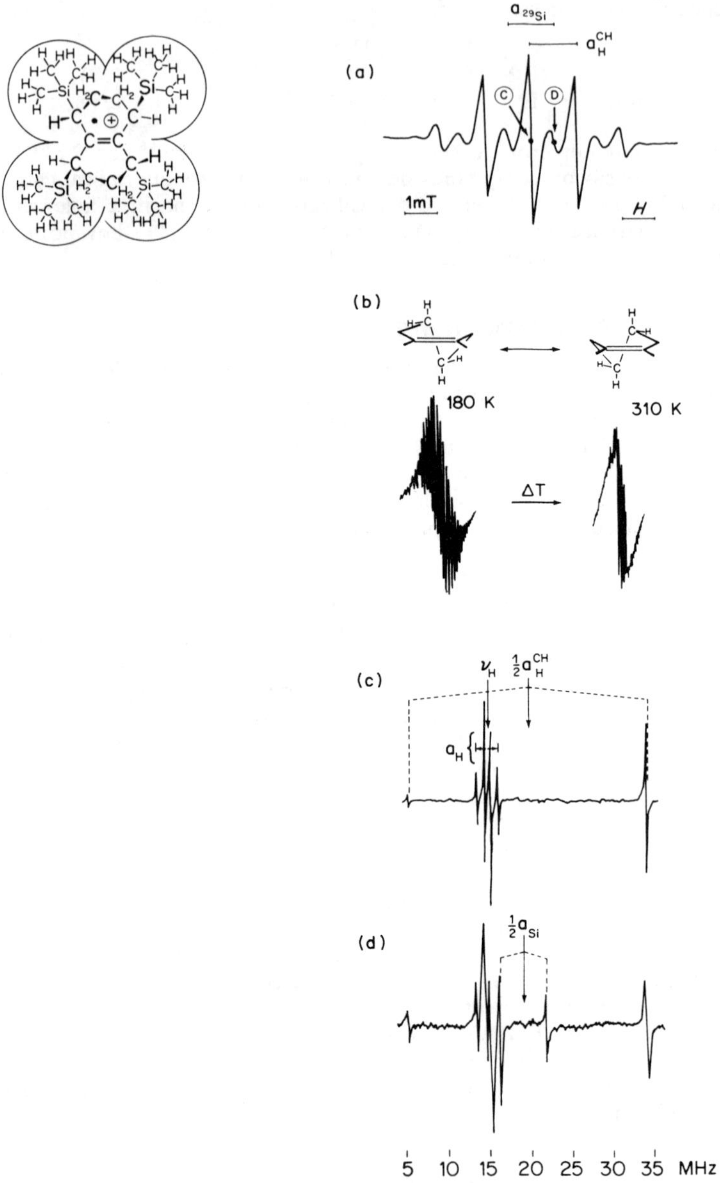

FIGURE 11. ESR and ENDOR spectra of the radical cation of 1,4,5,8-tetrakis(trimethylsilyl)-1,2,3,4,5,6,7,8-octahydronaphthalene[94] generated with $AlCl_3/H_2CCl_2$: (a) ESR spectrum at 200 K with (b) enlarged center line showing the *T*-dependence between 180 K (vibration frozen in) and 310 K (vibration activated); (c) ^1H-ENDOR spectrum at 235 K (all couplings are resolved and all signs determined from TRIPLE resonance measurements) and (d) ^{29}Si-ENDOR spectrum (cf. text)

is transferred by angle-dependent 'σ/π-hyperconjugation', $a_X = (B_0^x + B_2^x \cos^2 \Theta_{CX})$, with all constants B_0^H, B_0^{Si} and B_2^H, B_2^{Si} accessible from correlations of the experimental coupling constants $a_{H, Si}$ vs. $\cos^2\Theta_{CX}$ for series of radical cations with preferred conformations like the following benzene derivatives[95]:

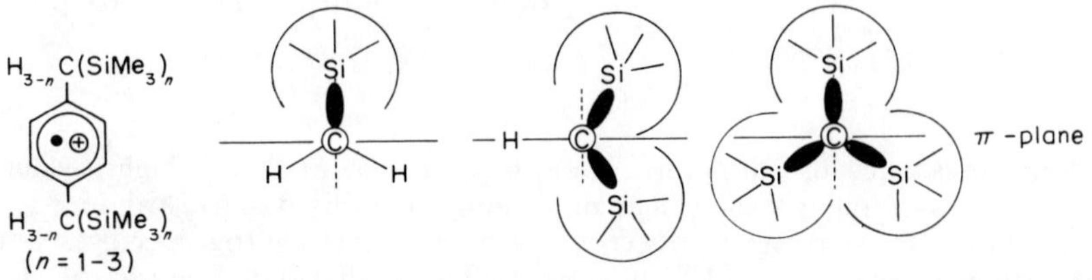

(22)

The conformational assignments from the ESR spectra can be further substantiated by a hyperconjugative perturbation model for the vertical first ionization energies, $IE_1^v = \sum_{CX} d_{CX} (\cos^2 \Theta_{CX})^{95}$, which allows one to both rationalize measured and predict unknown values[95] (cf. also Section IV.F).

From the extensive ESR/ENDOR data base established experimentally [see (6a) to (6z) in Reference 16] the following few prominent features are selected:

(a) The strong delocalizing strength of β-trimethylsilyl groups $CH_{3-n}(SiR_3)_n$ can be sensed by the reduced hydrogen couplings a_H^{Ph} in 1,4-disubstituted benzene radical cations[16]:

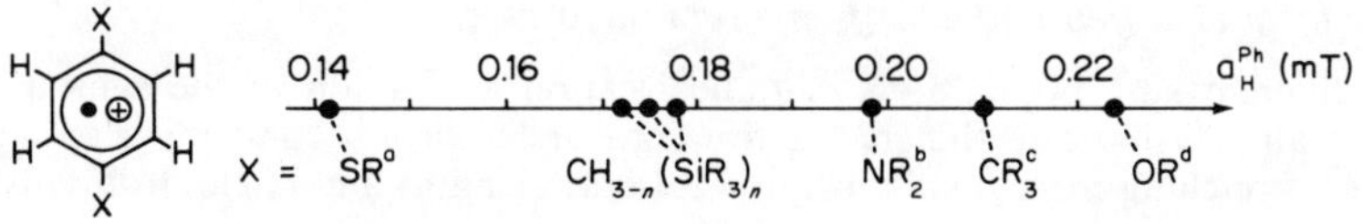

(23)

The effect exceeds that of prototype donors like OR or NR_2 and is surpassed only by alkylthio substituents[16].

(b) In radical ions of octamethyl-2,3,6,8-tetrasilabicyclo[3.3.0]octene (and -octane) derivatives, which are most likely all planar, comparison of $a_{^{29}Si}$ and $a_H^{CH_3}$ coupling constants (see results 24) clearly allows one to read off the central spin populations: in the

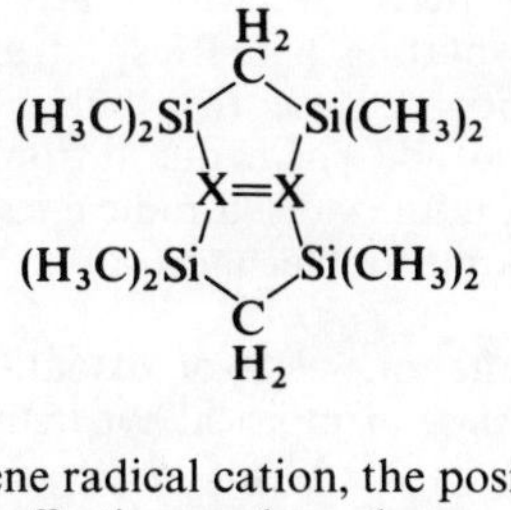

X=X	$C \overset{+}{=} C$[96]	$C \overset{-}{=} C$[96]	$N \overset{+}{=} N$[16]	
$a_{^{29}Si}$	2.271	0.685	0.500	(mT)
a_H	0.062	0.048	0.030	(mT)

(24)

ethene radical cation, the positive hole is largely delocalized at the Si centers due to their low effective nuclear charge, whereas in the radical anion most additional spin density stays in the C=C linkage. The same holds for the iso(valence)electronic hydrazine radical cation, in which formally an *N* lone-pair electron has been expelled to form a 3-electron/2-center π-bond.

(c) Dodecamethylcyclohexasilane, a saturated six-membered ring, which exhibits the rather low first PES ionization energy of only 7.79 eV[97], can also be both oxidized to its radical cation[98] and reduced to its radical anion[98]. The widely differing ESR data of the

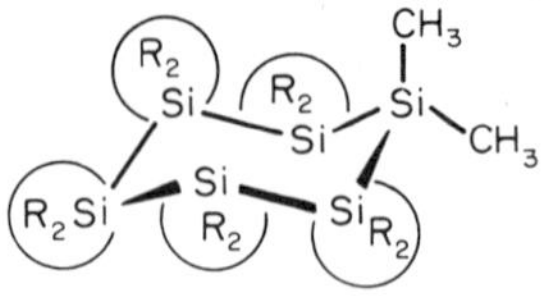

	g	$a_{^{29}Si}$	$a_{^{13}C}$	a_H	
$M^{\cdot\oplus}$	2.0093	2.05	?	0.188 0.094	(25)
$M^{\cdot\ominus}$	2.0028	0.70	1.46	0.049	

two products (see results 25) point to individual species: $M^{\cdot\oplus}$ with high g value and considerable ^{29}Si coupling displays a predominantly Si localized spin distribution and the distinguishable ^{1}H couplings for 2 sets of each 18 methyl hydrogens suggest a chair conformation for the ring. For $M^{\cdot\ominus}$, the g value is close to that of the free electron, the ^{29}Si coupling is small, the ^{13}C coupling dominates and there is only one set of 36 hydrogens, altogether suggesting either a flattened or a rapidly interconverting six-membered ring with most of the spin at the C centers of the equivalent CH_3 groups.

Summarizing, much can be learned about organosilicon radical cations from the ESR/ENDOR data base collected. Their ground state differs completely from that of radical anions of the very same compound (schemes 24 and 25) and their spin population—as well as the charge distribution—can be considerably changed by substituents (scheme 23).

3. *Structural changes and energy hypersurface calculations*

The different spin populations of radical cation and anion of the same precursor molecule already indicate that redox reactions are always accompanied by structural changes[9], stretching from small deviations of bond lengths and angles to totally altered molecular symmetry.

An exemplary case is represented by organosilicon-substituted hydrazine radical cations which, due to their rather low first ionization energies, can be oxidized with the $AlCl_3/H_2CCl_2$ single electron transfer reagent (Figure 9). Starting with the bicyclic tetrasilylated hydrazine, whose flattened structure is reflected by an unusually large n_N/n_N ionization splitting in its PE spectrum[99] (Section V.E), on oxidation the steric constraints will help to achieve a virtually planar structure (Figure 12: $\omega \sim 0°$) of the radical cation ground state. On the other hand, the neutral open-chain derivative with nearly perpendicular $(R_3Si)_2N$ halves as confirmed by minimal n_N/n_N splitting in its PE spectrum[99], on oxidation cannot achieve the advantageous planarity because the bulky R_3Si substituents enforce an intermediate structure (Figure 12: $\omega \sim 45°$) as deduced from the $M^{\cdot\oplus}$ ESR data[99]. Last but not least, for the six-membered tetraazadisila radical cation, ESR coupling proves a half-chair conformation with a slow intramolecular transfer with respect to the ESR time scale[99] (Figure 12).

The structural changes observed ESR spectroscopically after one-electron oxidation of organosilyl-substituted hydrazines[99] can be rationalized by a semiempirical hypersurface for the $N_2H_4^{\cdot\oplus}$ radical cation[100] (Figure 12). Combining its $3 \times 6 - 6 = 12$ degrees of freedom into the 3 angular coordinates ω,α and β[100], the minimum of total energy no longer resembles the *gauche* conformation of neutral N_2H_4 with perpendicular n_N lone pairs, but predicts a fully planar $H_2N^{\cdot\oplus}$–NH_2 with a 3-electron/2-center π-bond. Accompanying the tremendous $C_2 \rightarrow D_{2h}$ symmetry change, the NN bond is predicted to shrink from 145 pm to 128 pm(!), which later on has been confirmed by the X-ray structure of an isolated and exceptionally persistent tetraalkylhydrazine radical cation with an $NN^{\cdot\oplus}$ bond length of 127 pm[101].

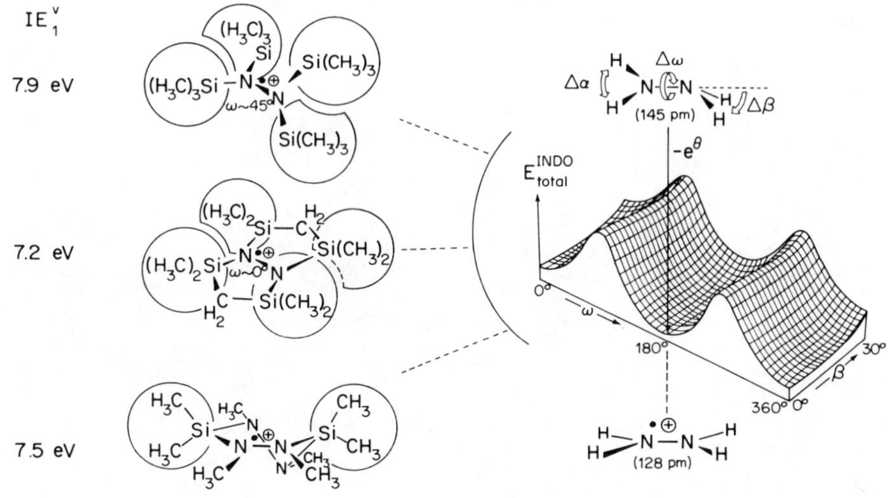

FIGURE 12. Structural changes on $AlCl_3/H_2CCl_2$ oxidation of organosilyl hydrazine derivatives[99] and their rationalization based on an INDO total energy hypersurface[100] for the $N_2H_4^{\cdot\oplus}$ radical cation

4. Radical cation dynamics

Steric overcrowding with bulky Me_3Si substituents considerably increases the persistency of organosilicon radical cations and also enforces restricted rotation, e.g., of R_3SiH_2C groups as confirmed by ESR coupling constant ratios R relative to the corresponding methyl derivatives, ranging from free rotation ($R \sim 1$) to complete fixation ($R \sim 0.5$)[16]:

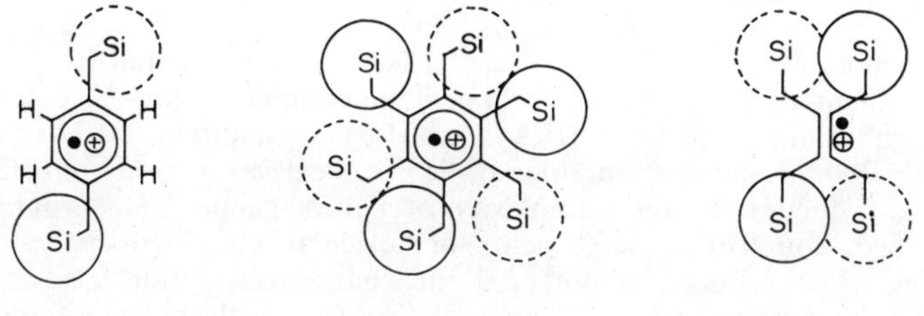

$R = \frac{0.84}{1.41} = 0.60$ $\quad \frac{0.353}{0.645} = 0.55$ $\quad \frac{0.783}{1.66} = 0.47$ (26)

Inside the ESR 'time window' between 10^{-5} to 10^{-8} s, however, numerous dynamic processes for radical cations are feasible. A clearly observable one concerns the ring inversion in the tetrasilyl-substituted octahydronaphthalene radical cation (Figure 11: b). Another representative example is supplied by the tris(trimethylsilylmethyl)amine radical cation (Figure 9), whose temperature-dependent ESR spectrum exhibits a significant line broadening between 180 K and 300 K[102] (scheme 27). The 180 K spectrum shows the three different CH_2 proton coupling constants corresponding to the 'frozen' conformation Γ. On raising the temperature, the six CH_2 hydrogens become equivalent by rapid

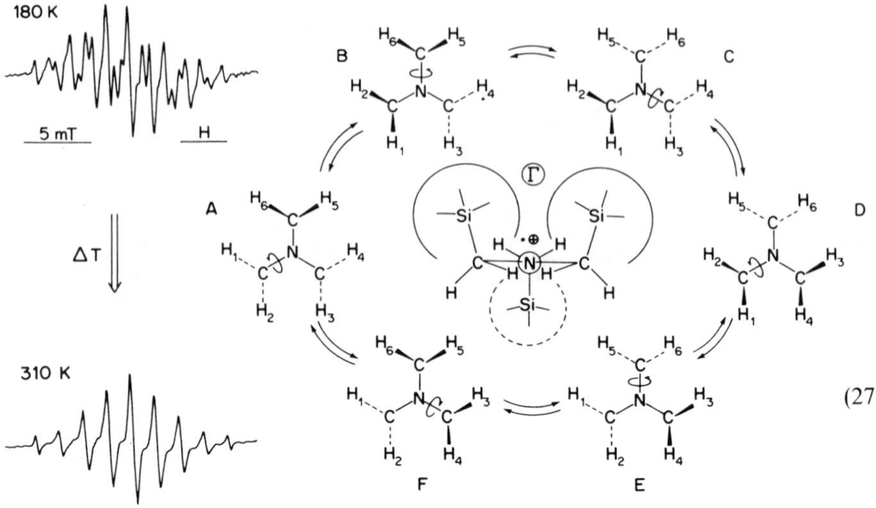

(27)

interconversions [scheme 27: (A)⇄(*F*)], in which each one of the bulky Me_3SiCH_2 groups switches to the other side of the NC_3 plane, offering enough space only for two of the three voluminous substituents.

The fluxtionality of organosilicon radical cations as observed by the ESR/ENDOR probe within its 'time window' shows another essential facet of their ground-state properties: the storage of internal energy in the 3*n*–6 degrees of freedom of nonlinear *n*-atomic molecular ions.

D. Radical Cation States: Vertical and Adiabatic Time Scales

The preceding comments have demonstrated the perpetual usefulness of photoelectron spectroscopy in silicon research: real-time PES gas analysis permits detection of novel short-lived molecules in nearly unimolecular flow systems (Section II) and vertical first ionization potentials allow prediction of precursor compounds for single electron transfer oxidation in solution (Sections III.A and III.B). In addition, a short summary on ESR/ENDOR information about organosilicon radical cations in solution (Section III.C) gave an introduction to the complexity of paramagnetic ionic ground states. An oversimplified, though often useful, picture of molecular states[9] artificially subdivides into topological, structural, nuclear potential and charge distribution facets as well as the skeletal dynamics (diagram 28). Still, among other aspects, the most essential time scale is

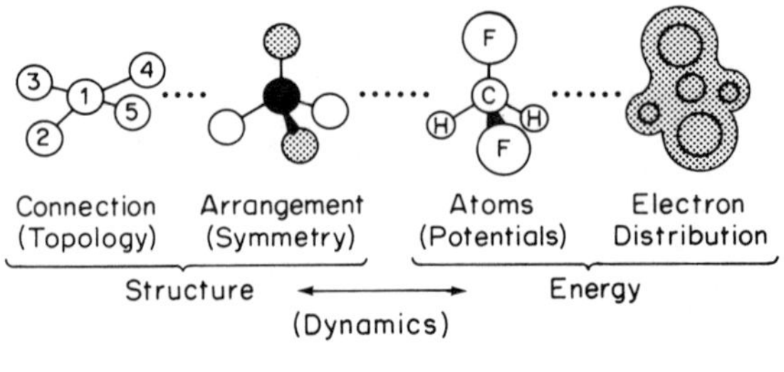

(28)

missing. In PES experiments, the ionizations occur 'vertically', i.e. along the energy scale within about 10^{-15} s generating the molecular radical cation $M^{\cdot\oplus}$ in its various states with a 'frozen core' skeleton. About one order of magnitude later, i.e. at 10^{-14} s, vibrations start and allow for structural adjustment of the positively charged species towards the total energy minimum of the respective $M^{\cdot\oplus}$ state. ESR/ENDOR measurements in a much longer time scale between 10^{-8} and 10^{-5} s therefore look at adiabatically 'relaxed' organosilicon radical cations in their electronic ground states, within solvent cages and accompanied by a counteranion. Both the partly large structural changes (Figure 12) and the activated molecular dynamics (scheme 27) demonstrate that the adiabatic $M^{\cdot\oplus}$ ground states, if at all, can only be rationalized and approximated by total energy hypersurfaces based on reasonable choices of appropriate coordinates out of their $3n$-6 degrees of freedom.

Principles for the interpretation of photoelectron spectra of silicon compounds (Sections II, IV and V) are as follows. Although the rather instantaneous ionization within 10^{-15} s does remove all difficulties concerning $M^{\cdot\oplus}$ structure changes and specific $M^{\cdot\oplus}$ state dynamics, still electronic relaxation is possible and especially the differences in the correlation of n electrons in M and n-1 in $M^{\cdot\oplus}$ should not be neglected. Therefore, all PE spectroscopic information available on individual $M^{\cdot\oplus}$ states (Section IV.B) should predominantly be used for their assignment. Last but not least, orbital models derived from simplified calculations should only be trusted, if their essentials can be substantiated by comparison of 'equivalent states' within series of 'chemically related' molecules[9] (Section IV.A).

The assignment of the PE spectra of silicon compounds, presented in the subsequent Sections IV and V, will whenever possible follow the above guidelines.

IV. RATIONALIZATION OF PHOTOELECTRON SPECTROSCOPIC RADICAL CATION STATE SEQUENCES IN PROTOTYPE SILICON MOLECULES

The assignment of PE spectra, i.e. the rationalization of radical cation state sequences, is not often as straightforward as it might seem at first glance (cf. e.g. References 9, 10, 30, 103) and therefore requires all experimental (Sections IV.A and IV.B) and theoretical (Section IV.C) evidence at hand (Sections IV.D to IV.H).

A. Comparison of Equivalent States of Chemically Related Molecules

PE spectroscopic assignment based on Koopmans' correlation[3–9], $IE_n^v = -\varepsilon_J^{SCF}$, of vertical ionization energies with SCF eigenvalues neglects two opposing effects: the stabilizing electron density redistribution in the vertically generated radical cation and the 'overshooting' difference in correlation energy calculated for the electron expelled[3,4,10,30,103]. Because the two do not usually cancel numerically, so-called 'Koopmans' deviations' result, which are especially large for $M^{\cdot\oplus}$ states with considerable contributions from centers of a highly effective nuclear charge as in fluorine[37] or cyanogen[38] compounds. Therefore, as pointed out repeatedly in the literature, 'one spectrum–one calculation' assignments[103] are rather risky and often meaningless, because even two sets of random numbers, e.g. produced by a chance generator, normally match each other with satisfying correlation coefficients[103]. Additional support for assignments based on Koopmans' correlations may be gained from details of the PE spectroscopic band patterns as discussed in the subsequent section or from comparison of equivalent $M^{\cdot\oplus}$ states of chemically related, preferentially iso(valence)electronic molecules by use of first- and second-order perturbation arguments (cf. e.g. Reference 56). This approach has already been used repeatedly in the preceding sections, for example in the PE spectroscopic assignments of silabenzene by comparison with benzene (Figure 3) and with other heterobenzenes like H_5C_5N or H_5C_5P (scheme 4); of dichlorosilylene $SiCl_2$ by comparison with $SnCl_2$ (scheme 8) or with other dihalosilylenes such as SiF_2 or $SiBr_2$ (scheme 13), or of silicon monoxide SiO by comparison with either CO or GeO (scheme 12).

The perturbation arguments, by which practically every chemist attempts to organize his overview on the otherwise difficult-to-comprehend multitude of chemical compounds and phenomena, may be classified as shown in diagram 29[9]. The distinction between iso- and heteroconjugate or between iso- and extraelectronic $M^{\cdot +}$ state comparison is, of

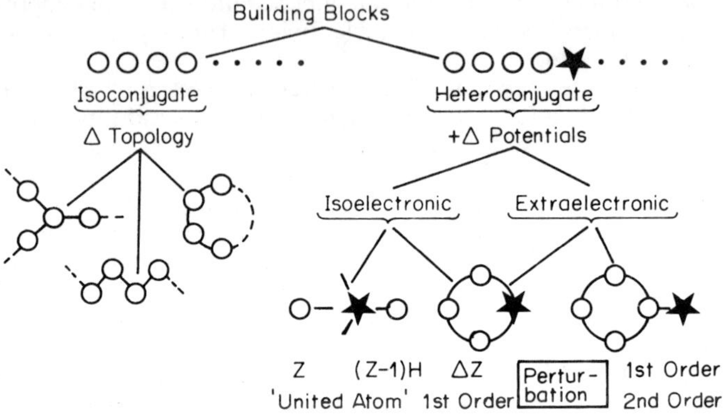

(29)

course, flexible. Often, the extent of perturbation determines the scope and quality of the model chosen, as will be exemplified in more detail for polysilanes (Section IV.D), silylacetylenes (Section IV.E) or R_3SiCH_2-substituted benzenes (Section IV.F).

B. Information from Photoelectron Spectra

For more general information concerning the number of ionization bands within the He(I) measurement region as well as their shapes and intensities displaying vibrational fine structures, spin/orbit coupling or Jahn/Teller distortion of the radical cations generated, reference is made to the many textbooks and summaries on PE spectroscopy already cited[3–9, 24–26, 30, 37]. In the following, this experimental information on individual radical cation states is compared to that available for equivalent ones in chemically related molecules (Section IV.A) and illustrates how the PE spectra, e.g. of the iso(valence)electronic iodo derivatives I-EH_3 with E=C[2, 5, 37] and Si[104] (Figure 13), may be unambiguously assigned even without any calculation:

Number of ionization bands in the He(I) measurements region. According to a useful rule-of-thumb[8, 105] for molecules without centers of extremely low or high effective nuclear charges, He(I) photons usually ionize only $np_{Element}$ and $1s_H$ electrons. Therefore, $(\Sigma np_E + 1s_H)/2$ bands are expected in He(I) PE spectra, e.g. for $ICH_3(5+2+3)/2=5$ bands. With respect to the skeletal C_{3v} symmetry of ICH_3 and the respective irreducible symmetry representations e, a, e for its low-energy $M^{\cdot\oplus}$ states, each two of the ionizations should be (nearly) degenerate.

Spin/orbit splitting[8, 37, 106]. This relativistic effect increases with the atomic number of the respective element: for the 5p-electrons of the iodine atom, it amounts to 0.942 eV and gives rise to two needle-like PES bands in the intensity ratio $2(^2P_{3/2}):1(^2P_{1/2})$. In I–$CH_3$, with one of the 5p-electrons engaged in I–C bonding, the splitting is reduced by one-third to 0.62 eV and the two low-energy needles of now-equal intensity (Figure 13) are therefore assigned to the no-longer-degenerate iodine lone pair $M^{\cdot\oplus}$ states, $\tilde{X}(^2E_{3/2})$ and $\tilde{A}(^2E_{1/2})$,

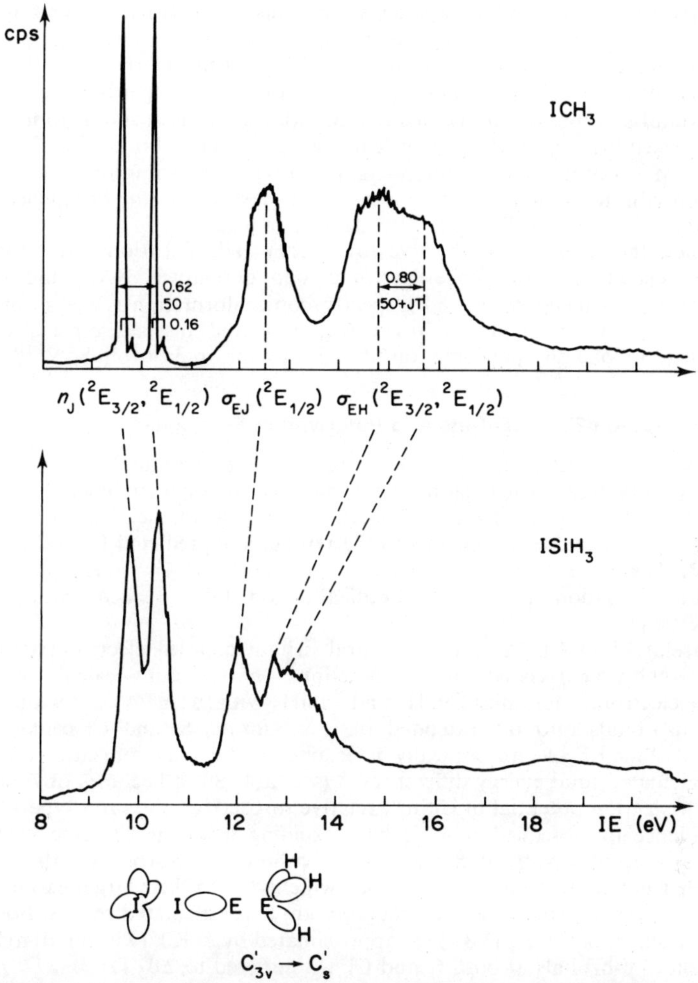

FIGURE 13. He(I) PE spectra of methyl iodide and iodosilane (H. Bock, K. Wittel and G. Becker, unpublished, cf. Reference 104) with radical state assignment within the spin/orbit double group C_{3v} and with dominant $M^{\cdot\oplus}$ state contributions (for details see text)

using the required C_{3v} double-group notation. An analogous assignment holds for the σ_{CH}(e) $M^{\cdot\oplus}$ states (Figure 13), the degeneracy of which is also removed to some extent by spin/orbit interaction.

Jahn/Teller splitting[8,47,107]. The σ_{CH}(e) $M^{\cdot\oplus}$ states of ICH_3 are split predominantly by Jahn/Teller distortion $C_{3v} \rightarrow C_s$, thus explaining the 0.8 eV peak distance of the qualitatively deconvoluted bands which exceeds that due to admixed spin/orbit coupling (Figure 13).

Vibrational fine structures[2–4,9,10,37,107]. In the I–CH_3 PE spectrum (Figure 13), only those accompanying the spin/orbit-coupled n_J ionization needles are clearly visible which, according to their observed frequency $\nu^{\oplus} = 0.16$ eV $\hat{=}$ 1290 cm^{-1}, have been assigned to an increased CH_3 deformation in the respective $M^{\cdot\oplus}$ states.

The He(I) PE spectrum of the iso(valence)electronic $ISiH_3$ (Figure 13) resembles all of the above features and can therefore be compared by the following perturbation arguments: Due to the lower effective nuclear charge of Si < C, the ionization center of gravity shifts from 12.5 eV (C) to 11.6 eV (Si), and due to the longer bond lengths d_{ISi} and d_{SiH} the overall splitting of the five low-energy $M^{\cdot\oplus}$ states is reduced from 5 eV (C) to 3.5 eV (Si). The difference in effective nuclear charge especially shows in the 2 eV lowering of the σ_{SiH} $M^{\cdot\oplus}$ states, which contain a substantial Si contribution. The smaller $\nu^{\oplus}_{SiH}$ deformation vibrational fine structure is no longer resolved in the broadened n_J bands, and also the Jahn/Teller splitting is considerably reduced.

Altogether, the sequence of the five low-energy radical cation states for the two chemically related, i.e. iso(valence)electronic iodo derivatives ICH_3 and $ISiH_3$ can unequivocally be assigned, based on PE spectroscopic information as well as on $M^{\cdot\oplus}$ state comparison, without referring to any calculation, and there is especially no need whatsoever to invoke any participation of fictitious silicon 3d-orbitals[104, 108–110] (!).

C. 'To d or not to d?': Basis Sets and Polarization Functions

Molecules are entities in energy-dependent states (see diagram 28), while orbitals are mathematical functions, which cannot be observables of any experiment[9, 56]. Nevertheless, simple orbital concepts have stimulated whole areas of chemistry from rationalization of the periodic table of elements to the synthesis of predicted Hückel-type organic π-systems[9]. Therefore, how might the controversial[104, 108–110] question of fictitious 3d-orbital participation in ground and radical cation states of silicon-containing molecules be settled?

In a careful PNO–CEPA (pseudo-natural orbital–coupled electron-pair approach) study[111], highly correlated wave functions have been calculated for the iso(valence)electronic molecules $ClCH_3$ and $ClSiH_3$ with (ΔCEPA) and without (ΔSCF) including d-orbitals into the extended basis set for C, Si and Cl centers. From a comparison of the PE spectroscopically determined vertical first ionization energies[5, 104] and the calculated total energy differences Δ (see table 30) it becomes obvious that the increased ionization potential of $ClSiH_3$ relative to $ClCH_3$—the prototype PE spectroscopic evidence for so-called $p_\pi \rightarrow d_\pi$ back-bonding from the chlorine lone pairs to silicon[104]—is reproduced by SCF calculations without any d-orbitals in the basis set[111]. On the other hand, inclusion of d-functions which, due to their larger exponents cover more electron density space, can considerably improve calculated results. For instance, the dipole moment of $ClCH_3$ (1.892 D), approximated by ΔSCF (without d) to be 2.43 D, on inclusion of d-orbitals at both C and Cl, is improved to 2.07 D[111].

M	IE^v_1(exp)	ΔCEPA (with d)	ΔSCF (without d)
$ClCH_3$	11.30 eV	11.09 eV	10.51 eV
$ClSiH_3$	11.61 eV	11.56 eV	11.09 eV

(30)

Summarizing, general agreement has been reached that if d-type polarizing functions are included in extended basis sets to further improve calculations, they have to be included at all essential centers, i.e. also for second-row elements like C, N, O or F[60, 112]. In addition, it is strongly emphasized that different molecular properties, e.g. of carbon vs. silicon or of nitrogen vs. phosphorus compounds, in principle, cannot be traced to any $p_\pi \rightarrow d_\pi$ back-bonding: there are no individual electrons but a total charge distribution (scheme 28) and there are—as pointed out above—no experiments to observe orbitals.

Referring to the much recommended 'comparison of equivalent states of chemically related molecules'[9] (cf. Figures 3 and 13; schemes 4, 8, 12 and 13), also for compounds of other elements like the recently PE spectroscopically characterized $H_3C–P=CH_2$[113], advisable correlation with the ionization patterns of other iso(valence)electronic molecules like $H_3C–N=CH_2$ or $H_3C–C(H)=CH_2$ should always be preferred to fiddling around with nebulous d-orbitals. The answer to the title question is therefore straightforward: 'no d'—except as polarizing functions to improve basis sets for calculations of silicon compounds.

D. Case I: SiSi/SiSi Bond Interactions in Polysilane Radical Cations $Si_nR_m^{\cdot\oplus}$

The band structures of polysilanes Si_nR_m, which are of interest as novel photoconducting materials[114] exhibiting low ionization thresholds and rather small valence bandwidths[115], have recently been both experimentally determined[114, 115] and calculated[115, 116]. The two independent calculations[115, 116] were each parametrized using the ionization energies of polysilanes with up to six silicon centers, which were determined PE spectroscopically almost 20 years ago[97, 117]. To begin with, the considerable difference in effective nuclear charge $Si < C$ will be elucidated by PE spectroscopic comparison of the iso(valence)electronic molecules disilane[117] and ethane[5].

1. PE spectra of disilane and ethane

The ethane radical cation state sequence, $\tilde{X}(^2E_g) < \tilde{A}(^2A_g) < \tilde{B}(^2E_u) < \tilde{C}(^2A_u)$, is well established from its repeatedly recorded PE spectrum[5] (Figure 14). Both $\sigma_{CH}(e)$ $M^{\cdot\oplus}$ states exhibit considerable Jahn/Teller distortions of 0.7 eV and 0.8 eV, respectively.

Comparison with the PE spectrum of the iso(valence)electronic disilane[117] (Figure 14) reveals surprisingly large differences: Judging from the intensities 1 : 2 : 2 of the three low-energy PES bands, the σ_{SiSi} ionization has been shifted by 3 eV(!) relative to the σ_{CC} value. Also, both the $\sigma_{SiH}(e_u)$ and the $3s_{Si}$ dominated $M^{\cdot\oplus}$ states are lower in energy by 2.6 eV and 3.6 eV, respectively. As concerns the σ_{SiH} ionizations, both the e_g/e_u split as well as the $D_{3d} \rightarrow C_{2h}$ Jahn/Teller distortions are smaller than in ethane, reflecting the considerably increased bond distance $d_{SiSi} = 2.35$ pm vs. $d_{CC} = 153$ pm. Altogether, the center of gravity of the six ionizations within the He(I) measurement region is shifted from $\overline{14.9 \text{ eV}}$ for H_3CCH_3[5] by 2.1 eV(!) to $\overline{12.8 \text{ eV}}$ for H_3SiSiH_3[117].

The tremendous lowering of the effective nuclear charge of silicon relative to carbon can be substantiated by the difference in their average atomic ionization energies for all n valence electrons,

$$\overline{IE^{Atom}_{Valence}} = [IE_1(A) + IE_1(A^+) + IE_1(A^{++}) + \cdots + IE_1(A^{+n-1})]/n$$

The experimental quantity $\overline{IE^{Atom}_{Valence}}$ also serves for estimating Slater potentials[9, 56] and is much more reliable than the ambiguously definable and far too imprecise parameter 'electronegativity'[118]. The values of $\overline{IE^{Atom}_{Valence}}$, 25.8 eV for Si and, 11 eV(!) higher, 37.0 eV for C are reflected by a 'united atom' comparison of valence electron ionization energies for third-row isoelectronic species AH_n[119], in which one proton after the other with nuclear charge $Z = 1$ is withdrawn for the preceding central element on its left[9]. In the PE spectroscopic comparison (scheme 31), two regions are easily recognized: the energetically higher $AH_n^{\cdot\oplus}$ states with 3s holes and the band of 3p-type ones, which are to be assigned to the ionization of electron pairs n_A or to the symmetry-adapted bond combinations σ_{AH}. The 3s electrons in the proximity of the nucleus become more readily ionizable with decreasing effective nuclear charge, as represented here by the average ionization energy

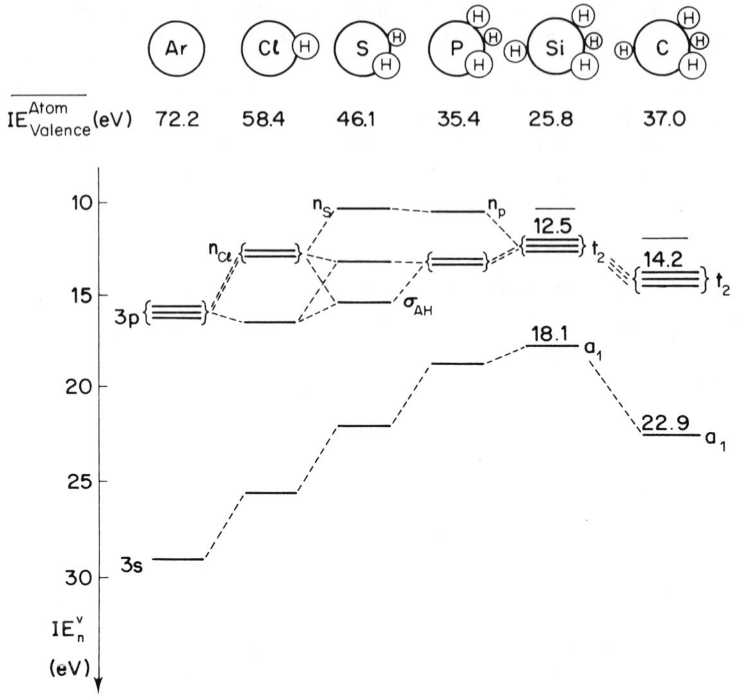

(31)

of the valence electrons of atom A, $\overline{IE^{Atom}_{Valence}}$. The 3p electrons, which are on average farther away from the nucleus, expectedly prove less susceptible to the influence of the core potential and split to characteristic $M^{\cdot\oplus}$ levels of the AH_n molecules. For SiH_4[117, 119] relative to CH_4[5, 119], the center of gravity of the four valence ionization energies is shifted by 2.5 eV(!) in close correspondence to the average shift for six of seven ionizations for Si_2H_6 and C_2H_6 by 2.1 eV (Figure 14).

2. PE spectra of open-chain silanes Si_nH_{2n+2}

Special caution had to be taken to record their PE spectra (Figure 15), because mixtures of silanes Si_nH_m with air, if ignited, explode violently causing harm and damage[117].

As concerns the $Si_nH^{\cdot\oplus}_{2n+2}$ state sequences, their assignment is achieved straightforwardly by PE spectroscopic comparison (Section IV.A). The almost unstructured ionization hills between 11.5 eV and 15 eV are due to σ_{SiH}-dominated $M^{\cdot\oplus}$ states. Presumably, they arise from different conformers of the rather flexible $(SiH_2)_n$ skeletal backbone, as suggested by the low rotational energy barrier of only 4.6 kJ mol^{-1} [117] between the preferred D_{3d} conformations of $H_3Si–SiH_3$[117]. The low-intensity bands around 16 eV (Figure 15) represent $3s_{Si}$ antibonding $M^{\cdot\oplus}$ states with low cross-sections[117] and result from a split around the single $3s_{Si}$ ionization of SiH_4 at 18.1 eV (scheme 31). In the low-energy region between 9 eV and 11 eV, a splitting into $n-1$ $\sigma^{\cdot\oplus}_{SiSi}$ states of the silanes $H(SiH_2)_nH$ becomes either clearly visible or can be deduced after band deconvolution (Figure 15). Relative to the σ_{SiSi} ionization of disilane (Figure 14), the observed $\sigma_{SiSi}/\sigma_{SiSi}$ interaction band patterns in Figure 15 are compressed on their high-energy side, where numerous σ_{SiH} $M^{\cdot\oplus}$ states of identical symmetry will—via antibonding admixtures

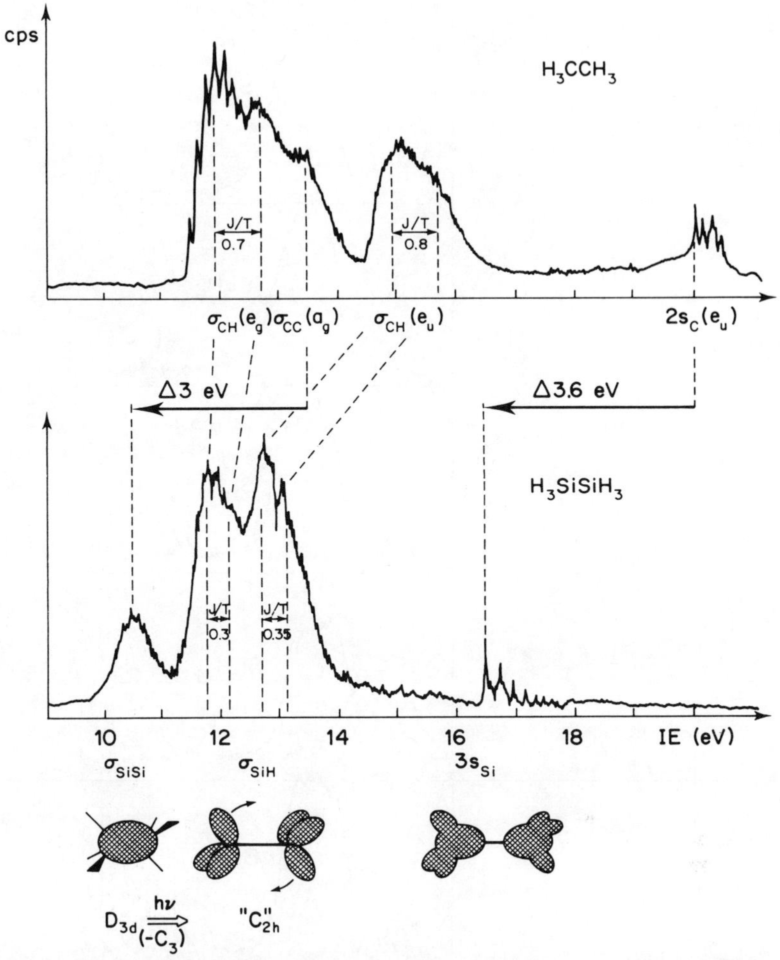

FIGURE 14. He(I) PE spectra of H_3CCH_3 and H_3SiSiH_3, exhibiting Jahn/Teller splitting (J/T) of their $M^{\cdot\oplus}$ (e) states, with qualitative assignment (see text)

within second-order perturbation[56]—shift energetically closest σ_{SiSi} $M^{\cdot\oplus}$ states most strongly[117].

3. Low-energy ionizations of permethyl silanes $Si_n(CH_3)_m$

In contrast to silanes $H(SiH_2)_nH$[117] (Figures 14 and 15), their permethylated open-chain and cyclic derivatives, $H_3C[Si(CH_3)_2]_kCH_3$ and $[Si(CH_3)_2]_r$, exhibit unexpectedly low first ionization energies[97, 98] (Figure 16). For illustration, the following comparisons with permethylated organic π systems are presented: $IE_1 = 8.69$ eV for $Me_3Si–SiMe_3$[97] nearly matches $IE_1 = 8.41$ eV for $Me_2C{=}CMe_2$[90] (cf. scheme 19), or $IE_1 = 7.79$ eV of the

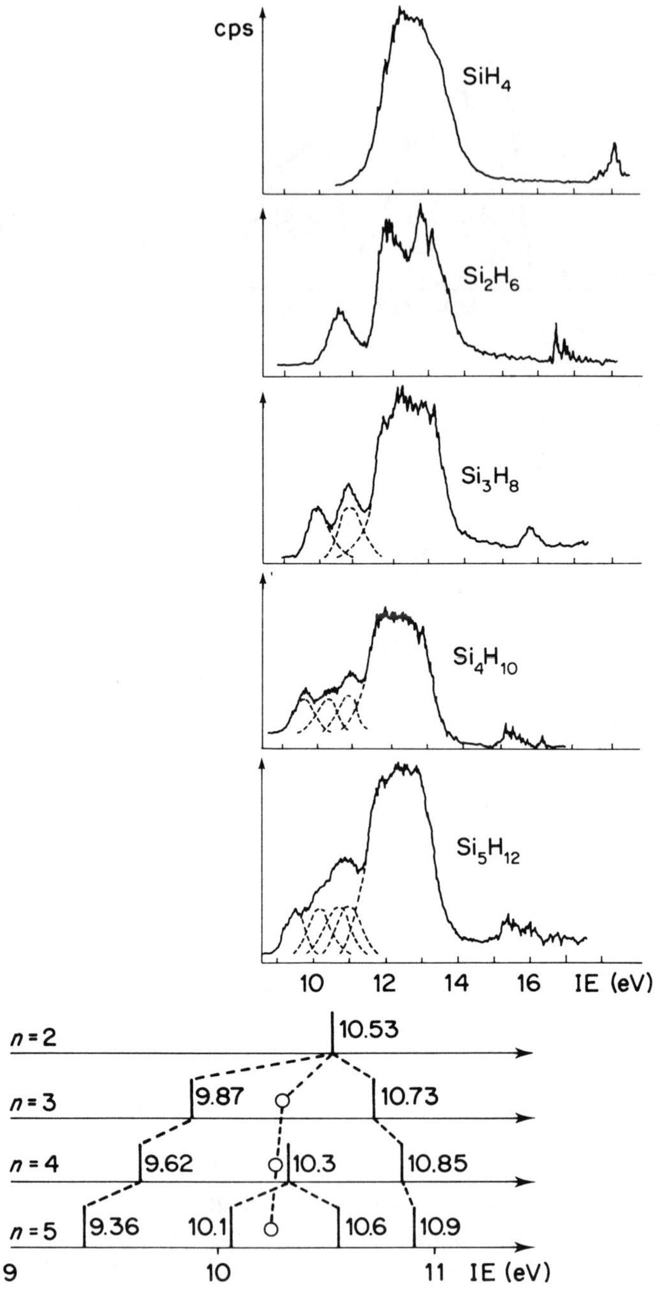

FIGURE 15. He(I) PE spectra of open-chain silanes Si_nH_{2n+2} ($n = 1$ to 5) with enlarged '$\sigma_{SiSi}/\sigma_{SiSi}$' splitting patterns in the 9–11 eV region[117]

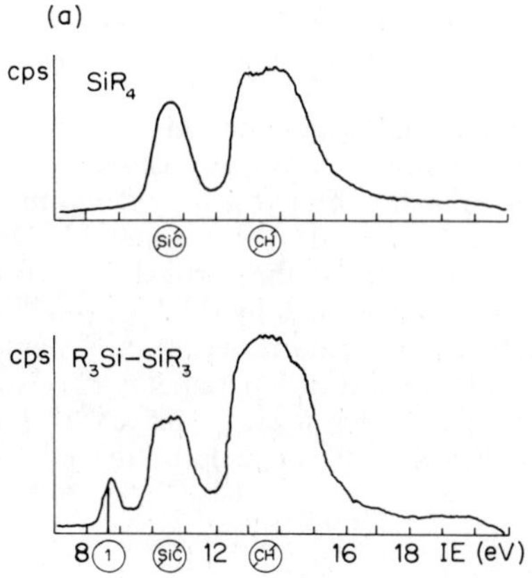

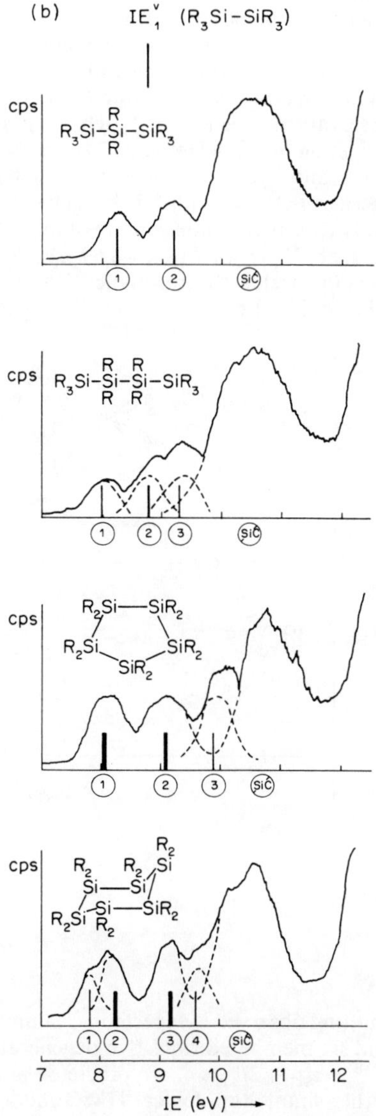

(c)

Compound		①	②	③	④	SiC
$R(SiR_2)_kR$	$k = 1$					10.5
	$k = 2$	8.69				10.4
	$k = 3$	8.19	9.14			10.4
	$k = 4$	7.98	8.76	9.30		10.5
$Si(SiR_3)_4$		8.24	(9.9)			10.4
$(SiR_2)_r$	$r = 5$	7.94	8.91	9.80		10.7
	$r = 6$	7.79	8.16	9.12	9.78	10.4

FIGURE 16. He(I) PE spectra of tetramethylsilane and hexamethyldisilane (a), 7 eV to 12 eV sections for permethylated ($R = CH_3$) polysilanes $R(SiR_2)_{2,3}R$ and $(SiR_2)_{5,6}$ (b) and table of low-energy ionizations (c) of all compounds investigated

puckered cyclohexasilane derivative $Si_6(CH_3)_{12}$[97] equals $IE_1 = 7.85$ eV of planar hexamethylbenzene $C_6(CH_3)_6$[93]. At higher energies two hills of numerous overlapping bands are observed (Figure 16: a): between 10 eV and 11 eV those of $M^{\cdot\oplus}$ states with predominant σ_{SiC} contributions, and between 12 eV and 16 eV both σ_{CH}- and $3s_{Si}$-type states[117]. Most interesting, however, is the low-energy region between 7 eV and 10 eV (Figure 16: b and c), which contains—partly overlapping—as many ionization bands as there are SiSi bonds in the respective polysilane, i.e. k-1 for linear and branched chains or r for r-membered rings. In addition, the energies of these σ_{SiSi}-dominated $M^{\cdot\oplus}$ states split almost symmetrically around the first ionization of Me_3Si–$SiMe_3$ at 8.69 eV[37], an observation which immediately suggests their rationalization by a topology-determined isoconjugate model (scheme 29) like the Hückel molecular orbital (HMO) approach[56] for linear and cyclic hydrocarbon π-systems[8, 9, 117]. Correlation of the vertical ionization energies $IE^v_n(\sigma_{SiSi})$ with Hückel eigenvalue coefficients x_J^{HMO}, defined by $\varepsilon_J^{HMO} = \alpha + x_J^{HMO}\beta$, leads to a nearly linear regression with satisfyingly small standard deviation: it passes through $IE^v_1(R_3Si\text{–}SiR_3) = 8.69$ eV, defined as Coulomb integral parameter α^R_{SiSi} and its slope yields the resonance integral interaction parameter $\beta^R_{SiSi/SiSi} \sim 0.5$ eV[8, 9] (see scheme 32). For all alternate[9] σ^R_{SiSi} systems like $R(SiR_2)_{3,4}R$ or $(SiR_2)_6$, the $\sigma^{\cdot\oplus}_{SiSi/SiSi}$

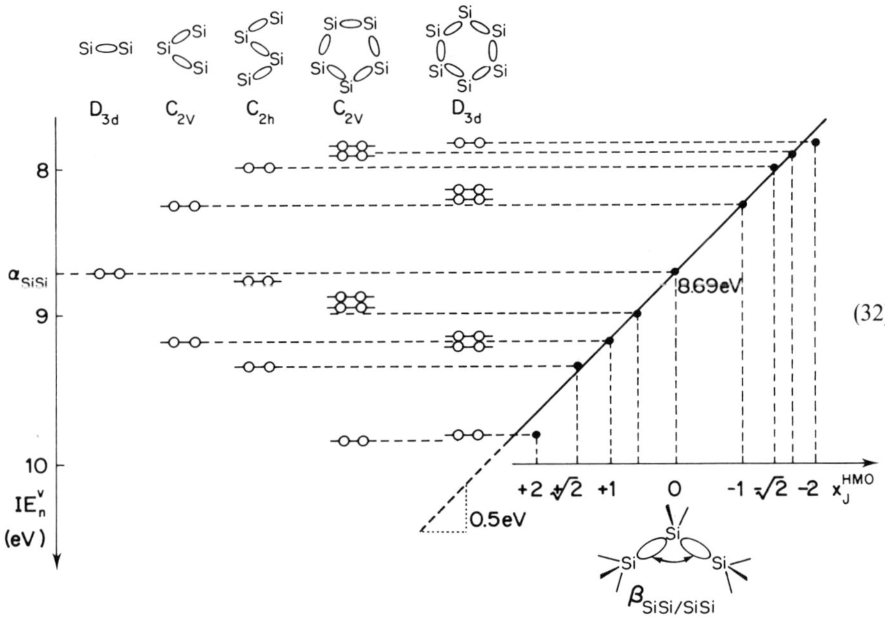

(32)

splitting observed is expectedly[56] almost equidistant from the center α^R_{SiSi}, and for the five- and six-membered rings two degenerate band pairs as well as one identical ionization at $\varepsilon_J^{HMO} = \alpha^R_{SiSi} + 2\beta^R_{SiSi/SiSi}$ are predicted and found, further substantiating application of the rather simple topological HMO model for the saturated σ_{SiSi} systems[8, 9, 97].

There are, however, some restrictive remarks to be added. Slight deviations between observed and predicted values are already apparent from scheme 32, which aggravate with increasing size, e.g. for dodecamethylpentasilane $H_3C[Si(CH_3)_2]_5CH_3$ also a slight compression of the higher-energy σ_{SiSi} ionizations is indicated[117]. The question, why the ionization pattern of silanes are compressed (Figure 15) and differ from the more regular ones of their permethylated derivatives (scheme 32), can be answered from fully-parametrized LCBO–MO (linear combination of bond orbitals–molecular orbitals) models for both H_3Si–SiH_3 and Me_3Si–$SiMe_3$[117]: for the latter, the internal α^R_{SiSi} value is about 3 eV lower than for Si_2H_6 and $\beta^R_{SiSi/SiSi} \sim -1.0$ eV considerably smaller than

$\beta^{H}_{SiSi/SiSi} \sim -1.7$ eV. In between these parameter differences, the borderline for application of the simple σ_{SiSi} HMO model (scheme 32) is crossed.

As pointed out at the beginning of this section, the assigned ionization patterns[97,117] proved to be useful for parametrization of band-structure calculations and for rationalization of the interesting properties of high-molecular-weight polysilanes[114–116]. In conclusion, therefore, diagram 33 presents a comparison of the PE spectrum of Si_5H_{12},

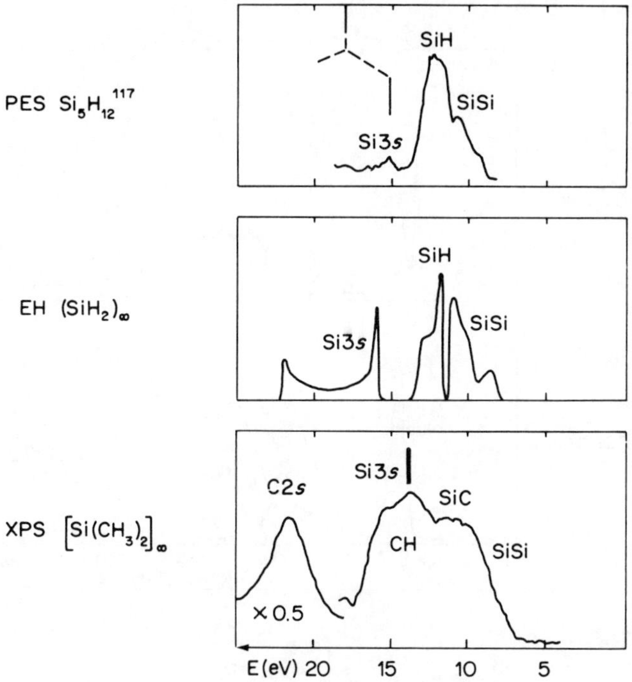

(33)

the $(SiH_2)_\infty$ band structure as calculated by an Extended Hückel (EH) procedure[115] and the experimental band structure of $[Si(CH_3)_2]_\infty$ as determined by X-ray photoelectron spectroscopy XPS[115]. All the features discussed above for both polysilanes Si_nH_m and their methyl derivatives $Si_n(CH_3)_m$ are found again in these interesting novel polymer materials[114–116].

E. Case II: π-Conjugation in Silylacetylene Radical Cation $H_3Si{-}C{\equiv}C{-}H^{\cdot\oplus}$

The delocalization of a positively charged hole generated in organosilicon radical cations vertically by ionization or adiabatically by oxidation (Section III), can be formally accomplished not only via the σ_{SiSi} skeleton (Case I, Section IV.D), but also by π-conjugation into the substituent groups. This $M^{\cdot\oplus}$ state stabilization, usually referred to as 'hyperconjugation' (cf. e.g. Reference 4, 6, 8, 120), can be most clearly exemplified by the PE spectroscopic comparison of acetylene[2,5] with its methyl[2,5,121] and silyl[121] derivatives, $H_3CC{\equiv}CH$ and $H_3SiC{\equiv}CH$, which are both iso(valence)electronic and of identical C_{3v} symmetry (Figure 17).

The assignment of the $M^{\cdot\oplus}$ state sequences advantageously begins with electron counting (Section IV.B); accordingly, the following number of p- (and s-) type ionizations

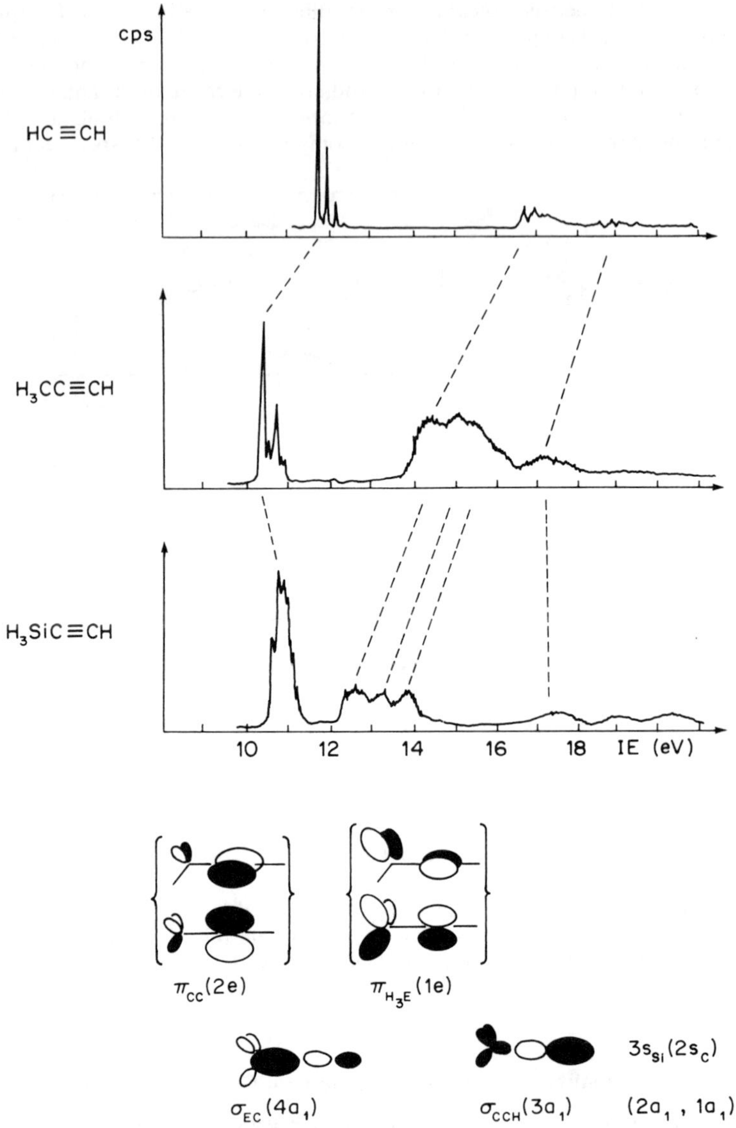

FIGURE 17. He(I) PE spectra of acetylene[2, 5] and its monomethyl[2, 121] and monosilyl[120] derivatives with LCBO–MO assignment (see text) emphasizing hyperconjugative charge delocalization in $\pi_{CC}(e) \pm \pi_{EH_3}(e)$ $M^{\cdot\oplus}$ states (see text)

are expected and found within the He(I) measurement region: 3p (+1s) for HC≡CH, 5p (+1s) for $H_3CC{\equiv}CH$ and, due to the lower effective nuclear charge of Si, 5p (+3s) for $H_3SiC{\equiv}CH$. The pronounced vibrational fine structures of the low-energy PES bands (Figure 17: the sequence of four peaks observed for $HC{\equiv}CH^{\cdot\oplus}$ with distances

$\nu^{\oplus} = 1830\,cm^{-1}$, corresponds[2] to the reduced C≡C stretching vibration $\nu_2 = 1983\,cm^{-1}$ of the neutral molecule HC≡CH)—prove $M^{\cdot\oplus}$ ground states of π symmetry [$D_{\infty h}$: $\tilde{X}(^2\Sigma_u)$ and C_{3v}: $\tilde{X}(^2E)$]. In the H_3C and H_3Si derivatives, they are lowered relative to acetylene with $IE_1^v = 11.4$ eV by 1 eV to 10.4 eV and by 0.7 eV to 10.7 eV[121], respectively, due to stabilizing delocalization of the positive charge into the H_3E substituent group. Within an LBCO–MO approach (Figure 17), this observation can be rationalized by 'hyperconjugative' interaction between the π_{CC} (2e) and π_{EH_3} (1e) components of identical symmetry. Parametrization starts from the mean σ_{EH_3} (e_g and e_u) ionizations of H_3CCH_3 and H_3SiSiH_3 (Figure 14), which provide Coulomb integral values $\alpha^{\pi}_{CH_3} \sim -14.2$ eV and $\alpha^{\pi}_{SiH_3} \sim -12.6$ eV[121]. If inserted together with the (inductively corrected) value $\alpha^{\pi}_{CC} \sim -11.5$ eV for the $\pi_{C\equiv C}$ system into second-order determinants, $\sqrt{(\alpha^{\pi}_{CC} - IE_1)(\alpha^{\pi}_{EH_3} - IE_1)} = \beta^{\pi}_{CC/EH_3}$, the interaction parameters $\beta^{\pi}_{CC/CH_3} \sim -2.1$ eV and $\beta^{\pi}_{CC/SiH_3} \sim -1.2$ eV[121] result, which allow one to correctly predict the three π(e) ionizations of the disubstituted derivatives $H_3CC{\equiv}CCH_3$ and $H_3SiC{\equiv}CSiH_3$[121]. Between the two [$\pi_{CC}(e) \pm \pi_{EH_3}(e)$] $M^{\cdot\oplus}$ states, which each exhibit some Jahn/Teller distortion (cf. section IV.D.1), the $\sigma_{EC}(4_{a_1})$ $M^{\cdot\oplus}$ state is interspersed and at energies above 17 eV there follow the remaining σ-type $M^{\cdot\oplus}$ states of a_1 symmetry with increasing $3s_{Si}$ and $2s_C$ character[121] (Figure 17).

The hyperconjugative stabilization of $M^{\cdot\oplus}$ states has numerous implications and applications[4, 6, 8, 120], of which the following few are emphasized within this context:

(i) Attempts to assign PE spectra of compounds like $H_3SiC{\equiv}CH$ (Figure 17) via Koopmans' correlation, with eigenvalues from semiempirical SCF calculations, often only succeeds if d-orbitals are omitted from the basis set (cf. Section IV.C). For instance, within the CNDO approximation, the exaggerated overlap $S(p_\pi d_\pi) = 0.18$ produces an overpolarized charge distribution for the ground state of the neutral molecule, whereas 'p-only' basis sets satisfactorily reproduce, e.g., dipole moments[121] (see diagram 34).

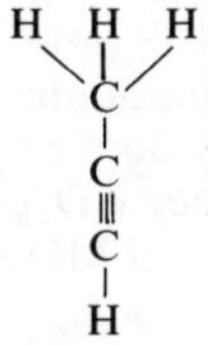

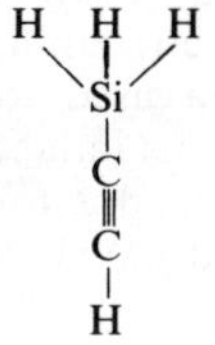

			without	with $3d_{Si}$
H (×3)	$3(+0.038)$	H (×3)	$3(-0.074)$	$3(-0.100)$
C	-0.008	Si	$+0.405$	$+0.416$
C	-0.023	C	-0.169	-0.197
C	-0.182	C	-0.086	$+0.009$
H	$+0.098$	H	$+0.070$	$+0.072$
μ^{calc}	1.013D		0.135D	1.536D
μ^{exp}	0.781D		0.316D	

(34)

(ii) LCBO–MO models like that used for the joint parametrization[121] of the acetylene derivatives $H_3E{-}C{\equiv}CH^{\cdot\oplus}$ and $H_3E{-}C{\equiv}C{-}EH_3^{\cdot\oplus}$ can be extended to include lone-pair/π interactions as in $H_3SiCl^{\cdot\oplus}$ (Section IV.G.1) and even spin/orbit splitting[121] as needed, e.g., to assign the PE spectrum of H_3SiI (Figure 13).

(iii) Complete assignment of the PE spectra of trimethylsilylacetylene and its halogen derivatives $Me_3Si{-}C{\equiv}CX$, X = H, F, Cl, Br, I has been achieved without specifying $\pi_{CC}(e) \pm \pi_{SiC}(e)$ hyperconjugation by using the LCBO–MO parameters $\alpha^{SiR_3}_{SiC} \sim -12.3$ eV and $\beta^{SiR_3}_{SiC/SiC} \sim -1.3$ eV within a second-order perturbation approach[122].

(iv) On the other hand, MO models for substituent effects on π-systems often have been pushed so far as to separate between hyperconjugative, inductive[123] and even fictitious $p_\pi \rightarrow d_\pi$ back-bonding (section IV.C)[124, 125] interactions, i.e. further differentiating between first- and second-order perturbations[56].

Summarizing, MO models for hyperconjugative interactions—especially if parametrized with PE spectroscopic ionization energies—provide a useful rationale to compare equivalent radical cation states of chemically related molecules (section IV.A) including those of silicon compounds. Numerous applications can be found in the literature[2–9] and, as far as silyl-substituted π-systems are concerned, they have not only helped to assign the PE spectra of acteylene derivatives[121] presented (Figure 17), but also those of numerous open-chain (section V.C) and cyclic (sections IV.F and V.D) hydrocarbon π-systems. Being based on Koopmans' correlation ($IE_n^v = -\varepsilon_J^{SCF}$), however, careful handling is recommended[3,4,6,9,10,103] keeping in mind the limitations involved by approximating the reality of open-shell radical cation states within the framework of closed-shell MO calculations.

F. Case III: PE Spectroscopically Parametrized Benzene π-Perturbation by β-Trimethylsilyl Substituents $CH_{3-n}(SiMe_3)_n$

Repeatedly, limitations of simplifying MO models to assign PE spectroscopically recorded ionization patterns and to rationalize even vertically (section III.D) generated radical cation states have been pointed out (cf. e.g. Sections IV.A to IV.C). Organosilicon molecules like those which, due to steric overcrowding, yield persistent radical cations in $AlCl_3/H_2CCl_2$ solution (Figure 9), however, contain such excessive numbers of valence electrons that there is presently no computational chance to calculate any highly correlated wave function. For instance, hexakis(trimethylsilylmethyl)benzene (Figure 9), $C_6(CH_2SiMe_3)_6 \triangleq C_{30}H_{66}Si_6$, contains within itself a total of 210 valence electrons and, according to the counting rule (section IV.B), is supposed to exhibit at least 75(!) ionizations within the He(I) measurement region. Therefore, PE spectra (cf. Figure 18) of such (relatively) rather huge molecules are usually 'crocodile'-shaped due to numerous overlapping σ_{SiC} and σ_{CH} ionizations, i.e. except for some low-energy π-ionization 'forehead humps' they display rather broad and almost contourless 'backs and tails'. Based on the assumptions that the π-type $M^{\cdot\oplus}$ states of these chemically related radical cations are comparable and that most of the numerous individual interactions can be adsorbed in PE spectroscopic parametrization, first[93,95] and second[122] order perturbation models (scheme 29) have been used to rationalize both ionizations and spin populations[16] (scheme 22). As concerns inductively or conjugatively dominated mono-substitutions of the benezene π-system, due to lowering of the symmetry $D_{6h} \rightarrow C_{2v}$, the degeneracy of the $\tilde{X}(^2E_{1g})M^{\cdot\oplus}$ ground state is removed. Within a Hückel MO approach[9]

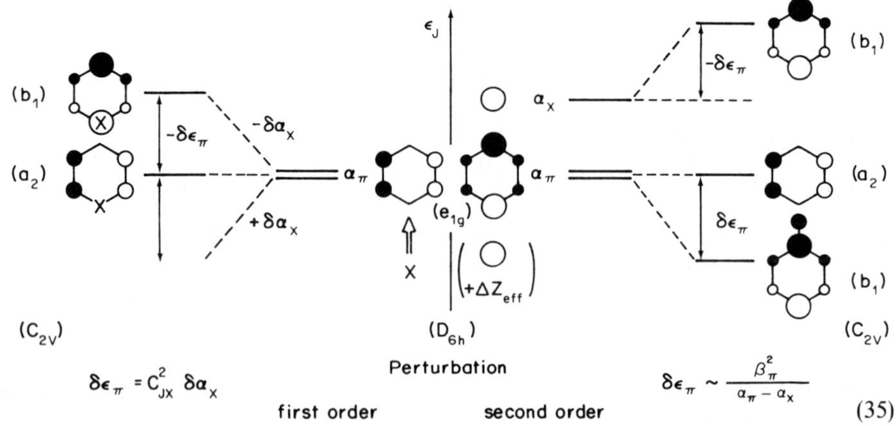

(35)

the first-order perturbation dominated by the potential change $d\alpha_X$ raises or lowers the eigenvalue $\pm\delta\varepsilon_\pi$ depending on the b_1 orbital coefficient at the substitution center (cf. also scheme 4). In contrast, second-order perturbation changes also the b_1 orbital composition by admixture of a substituent orbital of the same b_1 symmetry, with the splitting $\pm\delta\varepsilon_\pi$ being directly proportional to the perturbing interaction squared, $\beta^2_{\pi/X}$, and inversely proportional to the Coulomb energy difference $\alpha_\pi - \alpha_X$ (scheme 35). Contrary to second-order perturbations, those of first order are supposed to be additive (see scheme 36).

$$\delta\epsilon_J^{HMO} = \sum_\mu (c_{J\mu}^{HMO})^2\, \delta\alpha_X$$

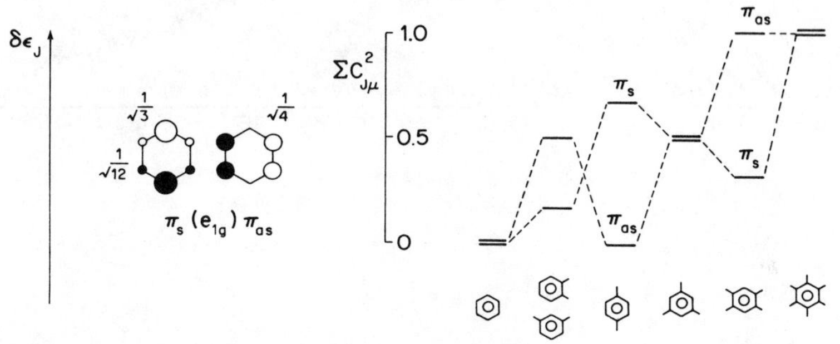

(36)

Depending on the substitution pattern, therefore, also the π_{as} orbital with a nodal plane through centers $\mu = 1,4$ may be raised highest.

Application of first- and second-order perturbation models to organosilicon compounds will be discussed for Me_3SiCH_2[93] and $(Me_3Si)_nCH_{3-n}$[95] substituted benzenes.

1. *π-Ionizations and charge transfer excitations of Me_3SiCH_2-substituted benzenes*

The He(I) PE spectra of Me_3SiCH_2-substituted benzenes exhibit, between 7 eV and 9 eV, two partly overlapping π-ionizations (Figure 18) resulting from the perturbation of the $C_6H_6^{\cdot\oplus}$ ground state, $\tilde{X}(^2E_{1g})$, at 9.24 eV[2-5], whereas the third one, $\tilde{B}(^2A_{2u})$, at

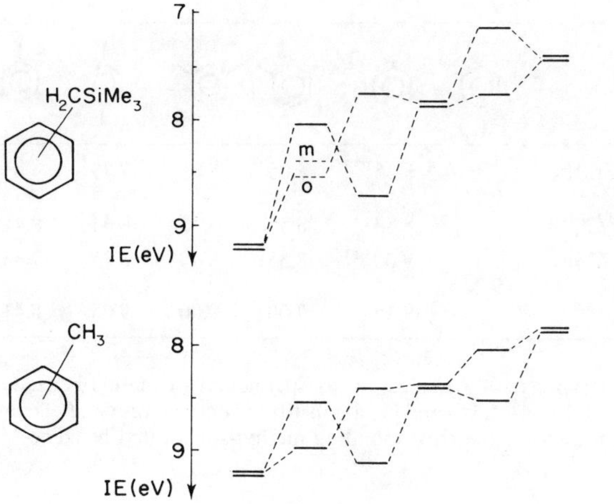

(37)

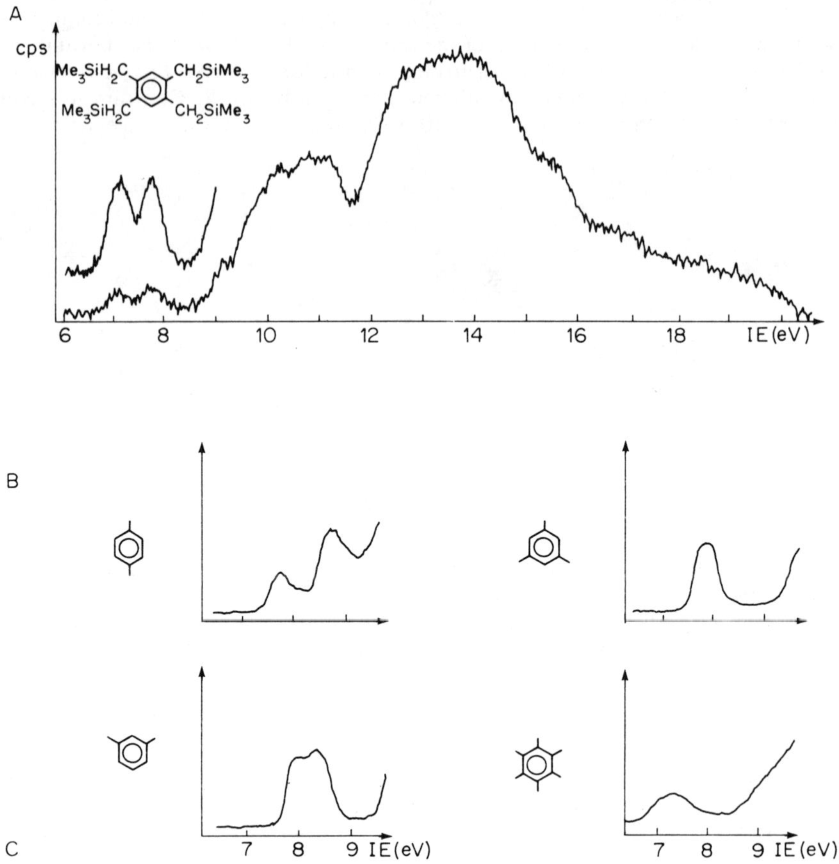

C

Substitution pattern									
IE_1(eV)	CH_2SiMe_3	9.24	8.35[127]	8.05	8.05	7.75	7.85	7.10	7.40
	CH_3[126]		8.84	8.56	8.55	8.44	8.41	8.05	7.85
IE_2(eV)	CH_2SiMe_3	9.24	9.00[127]	8.55	8.40	8.75	7.85	7.75	7.40
	CH_3[126]		9.15	9.00	9.00	9.05	8.41	8.55	7.85

FIGURE 18. He(I) spectra of (A) 2,3,5,6-tetrakis(trimethylsilylmethyl)benzene, (B) extended records 6.5 eV to 9.5 eV for 1,4-, 2,6-, 1,3,5- and hexa-substituted derivatives and (C) comparison of ionization energies $IE_{1,2}$ with those of the corresponding methyl-substituted benzenes[126]

12.4 eV[2−5] is hidden under numerous overlapping π-type PE bands. Their PES data $IE_{1,(2)}(\pi)$ as well as those of the corresponding methyl derivatives[126] (Figure 18: C) can obviously be rationalized by first-order perturbation (cf. schemes 35, 36 and 37). Deviations are observed especially for the tetrakis- and hexakis(trimethylsilylmethyl) derivatives, and originate partly from steric overcrowding (scheme 26) and partly from the extreme perturbation of four or six powerful Me_3SiCH_2 donor groups. These shift the first PE band down to 7.1 eV(!), i.e. passing those of tetracyclic pyrene $C_{16}H_{10}$ at 7.4 eV[93] or *p*-phenylenediamine $H_2NC_6H_4NH_2$ at 7.3 eV[93] close to the ionization region of carbanionic phosphorus ylides $R_3P^{\oplus}-^{\ominus}CH_2$[92] (cf. section III.B). Most striking are the differences $\Delta IE = IE_1(\text{benzene}) - IE_2(\pi_{as})$ observed for mono- and 1,4-disubstituted derivatives of 0.25 eV and 0.5 eV (Figure 18: C), which are predicted to be 0 eV because of the nodal plane through the substitution centers (scheme 35) and therefore suggest introduction of additional perturbation parameters $\Delta\sigma_X$[128]. Least-squares fit yields

$$IE_n^v = 9.25 - [\sum_\mu (c_{J\mu}^{HMO})^2\,\delta\alpha_x + \sum_x \Delta\sigma_x] \qquad \begin{aligned} \delta\alpha_{CH_2SiMe_3} &= 1.33\ \text{eV} \\ \delta\alpha_{CH_3} &= 0.86\ \text{eV} \\ \Delta\sigma_{CH_2SiMe_3} &= 0.24\ \text{eV} \\ \Delta\sigma_{CH_3} &= 0.09\ \text{eV} \end{aligned} \qquad (38)$$

and allows one to reproduce the π_s and π_{as} ionization energies of Me_3SiCH_2-substituted benzenes within the measurement precision of ± 0.2 eV[93]. Also mixed derivatives like 2,3,5,6-tetramethyl-1,4-bis(trimethylsilylmethyl)-benzene are included: predicted are $IE_1(\pi_s) = 7.23$ eV and $IE_2(\pi_{as}) = 7.54$ eV[93], which satisfactorily correlate with the measure-PES data of 7.25 eV and 7.65 eV[93], respectively. This last example further demonstrates the tremendous stabilization of $M^{\cdot\oplus}$ states by β-silyl substituents: the rather low second π_s ionization energy of 2,3,5,6-tetramethylbenzene (Figure 18: C, $IE_1^v = 8.55$ eV) upon the introduction of two R_3SiCH_2 groups in 1- and 4-positions is further lowered by 8.55 eV − 7.25 eV = 1.3 eV(!)[93].

The R_3SiCH_2-substituted benzenes (Figure 18) form, on addition of TCNE in H_2CCl_2 solution, colored donor/acceptor complexes[91, 129]. Their CT excitation energies, $\tilde{\nu}_n^{CT} = a\,IE_n^v(D) + b$, should be linearly correlated to the ionization potentials of the donor under assumptions concerning, e.g., the comparability of vertical and adiabatic data (section III.D) or of the intermolecular complex distances[130]. A general regression for aromatic compounds (scheme 39: ---) also incorporates most of the R_3SiCH_2-substituted benzenes investigated; only the first (scheme 39: ●) and second (scheme 39: ○) CT excitations of tetra- and hexa-substituted derivatives deviate. The reason for both their too high CT excitation energies and their too low molar extinctions[93] is their steric overcrowdedness (scheme 26), which lengthens the intermolecular complex distance to the TCNE acceptor. In turn, deviations from well-established correlations $\tilde{\nu}_n^{CT}/IE_n^v$ as in scheme 39 allow these interesting effects to be detected. As expected, the first-order perturbation (scheme 38) can be empirically adapted to also cover the CT excitations: the parameters $\delta\alpha_{CH_2SiR_3} = 11\,200$ cm^{-1} and $\Delta\sigma_{CH_2SiR_3} = 1000$ cm^{-1} yield a linear regression with a small standard deviation of only 600 $cm^{-1} \mathrel{\hat{=}} 0.08$ eV.

Altogether, first-order perturbation models are well suited even for powerful donors like Me_3SiCH_2 groups to rationalize and, after parametrization, to predict π-ionization energies within series of chemically related compounds. From correlations $\tilde{\nu}_n^{CT}/IE_n^v$ similar to those in scheme 39, values $IE_n^v(\pi)$ can be approximately determined on substituting a PE spectrometer by a test tube containing a TCNE solution in H_2CCl_2. As concerns more general comparisons of, e.g., the almost identical first ionization energies of the tetrahedrally shaped molecules (Figure 9 and structures 40), in which the tetra-substituted π-systems, ethene and benzene, seem to lose their identity, it is pointed out that in both cases the sum of the squared coefficients (scheme 35), $\sum_\mu (c_{J\mu}^{HMO})^2 = 1$ (ethene 2×0.72 and

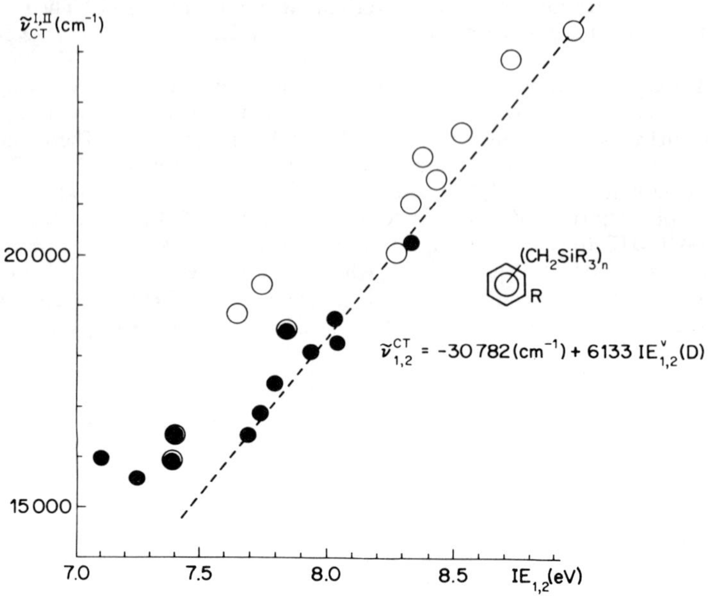

(39)

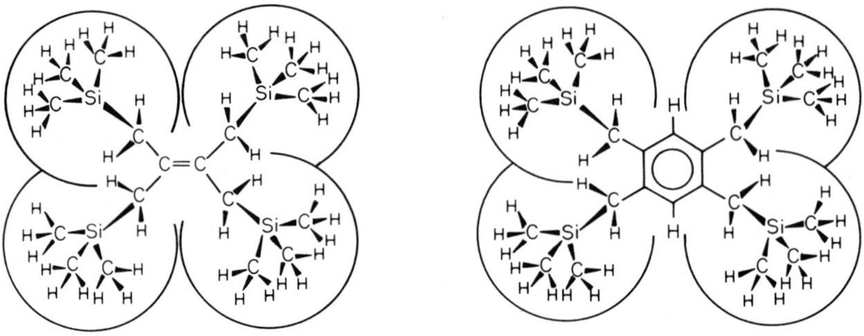

$IE_1 = 7.15$ eV[91] $IE_1 = 7.10$ eV[93]

(40)

benzene $4 \times 0.52)$[56], are the same. On the other hand, the changes in Coulomb potentials, i.e. the parameters $\delta\alpha_{CH_2SiR_3}$ and $\Delta\sigma_{CH_2SiR_3}$ (scheme 38), may not be transferable to different types of compounds. Furthermore, and especially for small systems like ethene, large

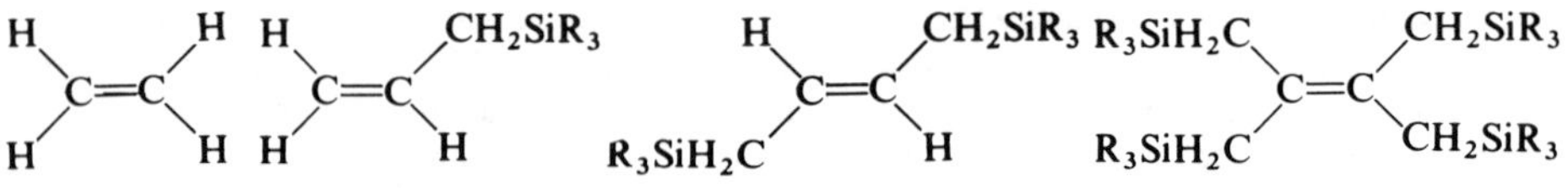

IE_1^v (eV) 10.51[5] 9.0[124] 8.3[91] 7.15[91] (41)

substituent perturbations are no longer additive and exceed the range of applicability of simple first-order models (scheme 35).

2. *Nonadditive effects of substituents* $CH_{3-n}(SiMe_3)_n$

Toluene and *p*-xylene derivatives, in which *n* methyl hydrogens are replaced by $SiMe_3$ groups, show nonadditive differences in their PE spectroscopically determined lowest ionization energies[95] (Table 42). In contrast, the shift in the second π_{as}-type ionizations can be attributed to inductive perturbations $\Delta\sigma_X$ (equation 38) as discussed in the preceding section; especially, the comparable first ionizations for $(Me_3Si)_2CH$ and $(Me_3Si)_3C$ substituted derivatives need further elaboration.

	X^1	X^2	X^3	IE_1^v	IE_2^v
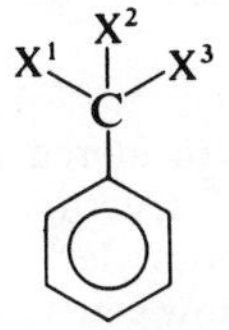	H	H	H	8.84	9.15
	$SiMe_3$	H	H	8.35	9.00
	$SiMe_3$	$SiMe_3$	H	8.10	8.80
	$SiMe_3$	$SiMe_3$	$SiMe_3$	8.10	(<8.9)
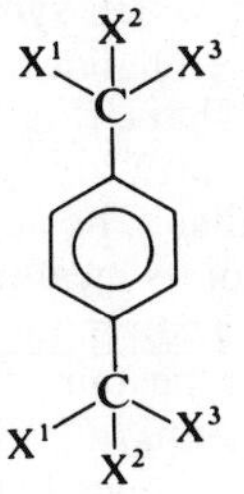	H	H	H	8.44	9.15
	$SiMe_3$	H	H	7.75	8.75
	$SiEt_3$	H	H	7.75	8.80
	$SiMe_3$	$SiMe_3$	H	7.40	8.60
	$SiMe_3$	$SiMe_3$	$SiMe_3$	7.45	(<9.0)

(42)

The radical cations of the R_3Si-substituted xylenes (Table 42) with $IE_1^v < 8$ eV can also be generated in $AlCl_3/H_2CCl_2$ solution[95] (Figure 9). Their ESR spin populations (Section III.C.2) prove rather strong and nearly constant hyperconjugative delocalization from the phenyl rings into the $(R_3Si)_nCH_{3-n}$ substituents in preferred conformations (scheme 22). Furthermore, the first ionization energies of both the trimethylsilylmethyl and the triethylsilylmethyl derivatives, $IE_1^v = 7.75$ eV (scheme 42), are the same. These observations suggest conformation-dependent contributions of the neighboring σ_{CH} and σ_{CS} bonds to the total π-perturbation[95] (scheme 43). From toluene and its threefold R_3Si substituted derivative $(1 \times \cos^2 0° + 2 \times \cos^2 60° = 1.5)$ and the corresponding xylenes

$$D^M = 9.25 - IE_1 = \sum_{CX} d_{CX}^M \langle \cos^2 \Theta \rangle$$

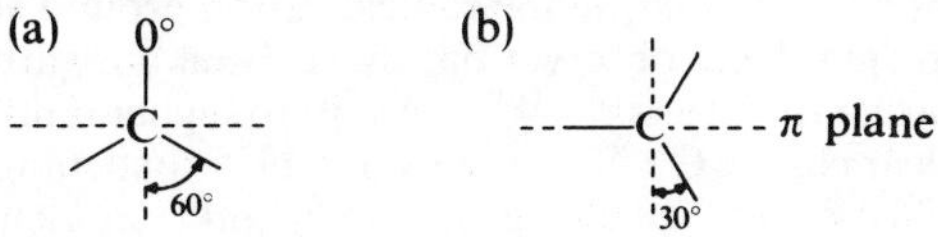

(43)

($\Sigma\langle\cos^2\Theta\rangle = 3.0$), average values $d_{CH} = 0.27$ eV and $d_{CSi} = 0.68$ eV are obtained. Due to identical values $\Sigma\langle\cos^2\Theta\rangle$ for the preferred conformations (scheme 43: b) of $(R_3Si)_2CH$-substituted toluene and xylene derivatives, their first ionization potentials should be the same as those of the $(R_3Si)_3C$-substituted derivatives (scheme 42). For the R_3SiCH_2-substituted benzenes (cf. preceding section), 8.43 eV and 7.60 eV are predicted, which satisfactorily match the PE spectroscopically determined $IE_1^v = 8.35$ eV and 7.75 eV (scheme 42).

In principle, this hyperconjugation model (scheme 43) still represents a first-order perturbation: it is based on the ESR spectroscopically proven, nearly constant phenyl ring spin population in all related radical cations, i.e. constant squared coefficients $c_{J\mu}^2$ (scheme 35), and substitutes the perturbation $\delta\alpha_X$ by the angle-dependent contributions $d_{CX}\langle\cos^2\Theta\rangle$. For a second-order perturbation example, which introduces interactions to additional substituent orbitals (scheme 35), reference is made to the silyl and methyl acetylenes[121] (Section IV.E).

G. Case IV: Lone-pair Ionizations in Silicon Compounds

Heteroelement centers in molecules with so-called 'lone pairs' n_E, i.e. formally containing two valence electrons in only one nuclear potential, are considered to be 'electron rich' and play an essential rôle in determining molecular structures and conformations as suggested, e.g., by the valuable electron-pair repulsion model[131]. The corresponding 'lone pair' ionizations usually occur at relatively low energies[2–9] (cf. scheme 31: HCl, H_2S and PH_3) and generate 'stabilized' radical cation states with considerable n_E contribution. In the PE spectra of all compounds in which symmetry favors localization of the positive charge at the n_E center, as in hydrogen halides[37], lone pair ionizations can be recognized by their almost 'needle-like' band shape (Figure 13): due to only small changes in equilibrium distance on vertical ionization, a high Franck/Condon overlap between the respective potential wells[34] results. If, however, n_E delocalization dominates, 'the term "lone pair" frequently loses much of its meaning and electron pairs in keeping with modern social trends also prefer to live in a commune rather than pairwise isolation'[132]. Therefore the usually, easy to recognize, n_E-type ionizations at low energy have often been used to specify interactions like n_E/n_E, n_E/π or n_E/σ[8] to help in the assignment of PE spectra and to compare equivalent $M^{\cdot\oplus}$ states of chemically related molecules (Section IV.A) and to probe substituent effects.

In the preceding sections the PE spectra of several silicon molecules with more or less pronounced 'lone pair' centers have been discussed, e.g. $SiCl_2$ (Figures 5 and 6, schemes 8 and 13), Si≡O (scheme 12), H_5C_6–N≡Si (Figure 7), $RC{\equiv}\overset{\oplus}{N}{-}\overset{\ominus}{N}SiR_3$ (Figure 8) or $ISiH_3$ (Figure 13). In the following, the n_E-type ionizations of molecules with one adjacent Si center like H_3SiCl, $SiCl_4$ and SiF_4, with the n_E heteroatom embraced by two Si atoms as in $H_3Si{-}O{-}SiH_3$ and with two n_E centers as in the blue azo compound $R_3Si{-}N{=}N{-}SiR_3$, will be presented as characteristic examples of relatively 'electron-rich' silicon compounds.

1. *PE spectra of selected silicon halides: H_3SiCl, SiF_4 and $SiCl_4$*

a. H_3SiCl. The PE spectrum of chlorosilane[104, 109] (Figure 19) has greatly stimulated the question 'as to whether the empty valence-shell d-orbitals in Si or Ge participate significantly in bonding'[104]. Comparing the radical cation ground states of both H_3CCl and H_3SiCl, the following PES observations have been forwarded as evidence for ($n_{Cl} \rightarrow 3d_{Si}$ or $\rightarrow \sigma^*_{SiH_3}$) acceptor effects[104, 109]: the shift to higher ionization energy despite the lower effective nuclear charge C > Si, the considerable broadening of the first PE band (half bandwidths of 0.06 eV and 0.48 eV, respectively) and—on inclusion of H_3SiBr and

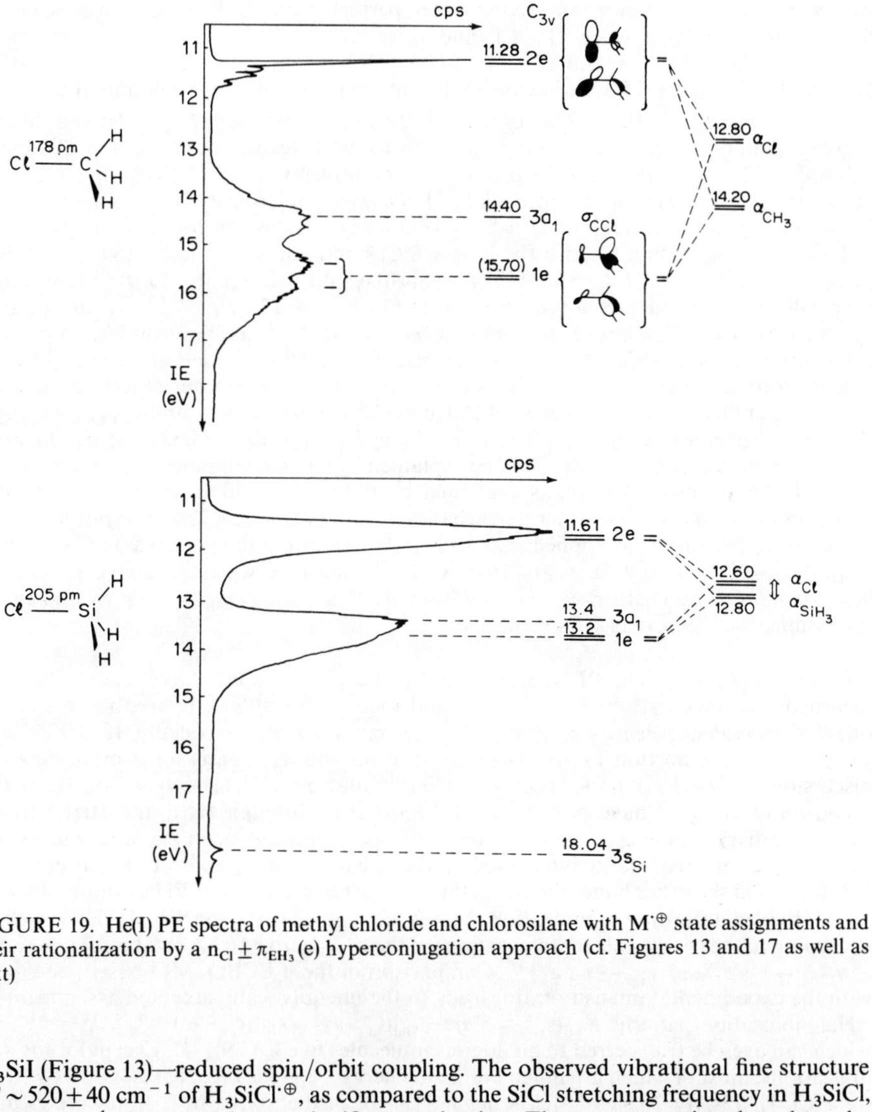

FIGURE 19. He(I) PE spectra of methyl chloride and chlorosilane with $M^{\cdot\oplus}$ state assignments and their rationalization by a $n_{Cl} \pm \pi_{EH_3}$ (e) hyperconjugation approach (cf. Figures 13 and 17 as well as text)

H_3SiI (Figure 13)—reduced spin/orbit coupling. The observed vibrational fine structure $\tilde{\nu}^{\oplus} \sim 520 \pm 40$ cm^{-1} of $H_3SiCl^{\cdot\oplus}$, as compared to the SiCl stretching frequency in H_3SiCl, $\tilde{\nu} = 545$ cm^{-1}, does not show a significant reduction. The argumentation, based on the Franck/Condon overlap of respective potential wells on vertical ionization, although well chosen[104], in principle can neither prove nor disprove d-orbital participation (Section IV.C), and highly correlated wave functions calculated with 'p-only' basis sets[111], for instance, do reproduce the unexpected shift of the first vertical ionization potentials, $IE_1^v(H_3CCl) < IE_1^v(H_3SiCl)$. Therefore the question must rather be directed elsewhere, i.e. 'as to whether there are any other applicable models to explain the PE spectroscopic observations'[8,37]. Referring to both the comparison of equivalent states of related molecules (Section IV.A) and the useful PE spectroscopic parametrization of, e.g., hyperconjugation models (Section IV.E), an interpretation without $p_\pi \rightarrow d_\pi$ back-bonding can be based, for instance, on second-order perturbation as illustrated in scheme 35 for benzene. To begin with H_3CCl, and inserting the $M^{\cdot\oplus}$ state parameters, $\alpha^{\pi}_{Cl} = -12.80$ eV from HCl[2,5] and $\alpha^{\pi}_{CH_3} = -14.20$ eV[121] as the mean e-type ionization of H_3C–CH_3[2,5] (Figure 14 and Section IV.E), into the second-order determinant, $\beta^{\pi}_{Cl/EH_3} = \sqrt{(\alpha_{Cl} - IE_1)(\alpha_{EH_3} - IE_1)}$ yields $\beta^{\pi}_{Cl/CH_3} = -2.1$ eV, identical to β^{π}_{CC/CH_3} derived for the π(e) interaction in methylacetylene (Section IV.E). With respect to the center of gravity, $(\alpha^{\pi}_{Cl} + \alpha^{\pi}_{CH_3})/2$, a symmetrical split into the e-type ionizations at 11.28 eV and 15.70 eV results (Figure 19). The same approach to H_3SiCl, with α^{π}_{Cl} from HCl and $\alpha^{\pi}_{SiH_3} = 12.6$ eV from the e-type ionizations of H_3SiSiH_3 (Figure 14), yields the lower value $\beta^{\pi}_{Cl/SiH_3} = -1.1$ eV, corresponding to both the longer SiCl bond and to the analogous parameter β^{π}_{CC/SiH_3} for H_3Si–C≡CH (Section IV.E). Again the split with respect to the PE spectroscopically determined π(e) ionizations at 11.61 eV and 13.7 eV[104,109] is an (almost) symmetrical one. For better correspondence to the $M^{\cdot\oplus}$ assignments, an inductive interconversion $\alpha^{\pi}_{Cl} \leftrightarrow \alpha^{\pi}_{SiR_3}$ may be considered. Transferring the PE spectroscopic parameters into bonding arguments allows one to rationalize many other experimental data: $\alpha^{\pi}_{CH_3} > \alpha^{\pi}_{SiH_3}$ reflects the difference in effective nuclear charge C > Si, and $\beta^{\pi}_{Cl/CH_3} > \beta^{\pi}_{Cl/SiH_3}$ the reduced overlap due to the different bond lengths $d_{CCl} < d_{SiCl}$ (Figure 19). Broadening of the first band $H_3CCl \rightarrow H_3SiCl$ can be explained by a less-pronounced n_{Cl} contribution to the $H_3SiCl^{\cdot\oplus}$ ground state, as suggested by the smaller difference $\alpha^{\pi}_{SiH_3} - \alpha^{\pi}_{Cl}$ in the denominator of the second-order perturbation term (scheme 35). Last, but not least, the above arguments may be applied, e.g., to H_3SiF, exhibiting the reversed $M^{\cdot\oplus}$ sequence $IE_1(SiH_3) = 12.1$ eV and $IE_3(n_F) = 16.6$ eV[104,109]: starting with the above parameter $\alpha^{\pi}_{SiH_3} = -12.6$ eV and with $\alpha^{\pi}_{F} = -16.1$ eV from HF[37], a value $\beta^{\pi}_{F/SiH_3} = -1.4$ eV and again the symmetrical split of +0.5 eV results.

b. SiF_4 and $SiCl_4$. The PE spectra of both silicon tetrahalides have repeatedly been recorded and assigned[3,4,37,109,133–137] and their $M^{\cdot\oplus}$ states compared to equivalent ones of iso(valence)electronic group 14 molecules $EHal_4$, especially to $CF_4^{\cdot\oplus}$ and $CCl_4^{\cdot\oplus}$[133–138] (cf. Section IV.A). As an illustration and as a basis for a more detailed discussion[138], the He(II) PE spectra of CF_4[134] and SiF_4[134] as well as the He(I) PE spectrum of $SiCl_4$[135] have been selected (Figure 20). Although He(I) and He(II) cross-sections differ[3,5] and cps rates of most PE spectrometers decrease with increasing energies, the relative band intensities of the CF_4 and SiF_4 ionization patterns are 3:3:2:3:1. On the other hand, the altogether 32 valence electrons of $EHal_4$ molecules can be subdivided according to ionization regions $n_{Hal} < \sigma_{EHal} \ll s_{Hal}$ (Section IV.H) as well as according to the irreducible representations of their respective 16 $M^{\cdot\oplus}$ states: $n_{Hal} \rightarrow t_1, t_2$, e, $\sigma_{EHal} \rightarrow t_2, a_1$ and $s_{Hal} \rightarrow t_2, a_1$[138]. Comparison of these LCBO–MO-based sequences with the experimental intensity ratios leads to the unequivocally accepted assignment of $EHal_4$ ionization patterns $n_{Hal}(t_1 > t_2 > e) > \sigma_{Hal}(t_2 > a_1) \gg s_{Hal}(t_2 > a_1)$[2–5,37,109,133–138], which can even be transferred to analogous molecules like $C(CN)_4$[139]. Except for the s_{Hal} ionizations, most of which are much higher in energy—e.g. for HF, $IE^v_4(2s_F) = 39$ eV[37], or for HCl, $IE^v_4(3s_{Cl}) = 25.8$ eV[37] (cf. scheme 31)—all of the above $M^{\cdot\oplus}$ states are clearly recognizable in the PE spectra of CF_4, SiF_4 and $SiCl_4$ (Figure 20), and their individual assignment is readily confirmed by radical cation state comparison (see scheme 44). Under T_d symmetry, the halogen lone pairs n_{Hal} would transform into the n_{Hal}/n_{Hal}-overlap determined order t_1(antibonding) > e(nonbonding) > t_2(bonding)[138]. The t_2 combination is further destabilized by antibonding admixture of $\sigma_{EHal}(t_2)$ components of the same symmetry yielding the $M^{\cdot\oplus}$ state sequence $\tilde{X}(^2T_1) < \tilde{A}(^2T_2) < \tilde{B}(^2E)$ as observed PE spectroscopically. On moving from $CF_4^{\cdot\oplus}$ to $SiF_4^{\cdot\oplus}$, the center of all n_F ionizations

$$\overline{IE^v_n(n_F)} = [3\,IE^v_1(1t_1) + 3\,IE^v_2(3t_2) + 2\,IE^v_3(1e)]/8 = 17.3 \text{ eV}$$

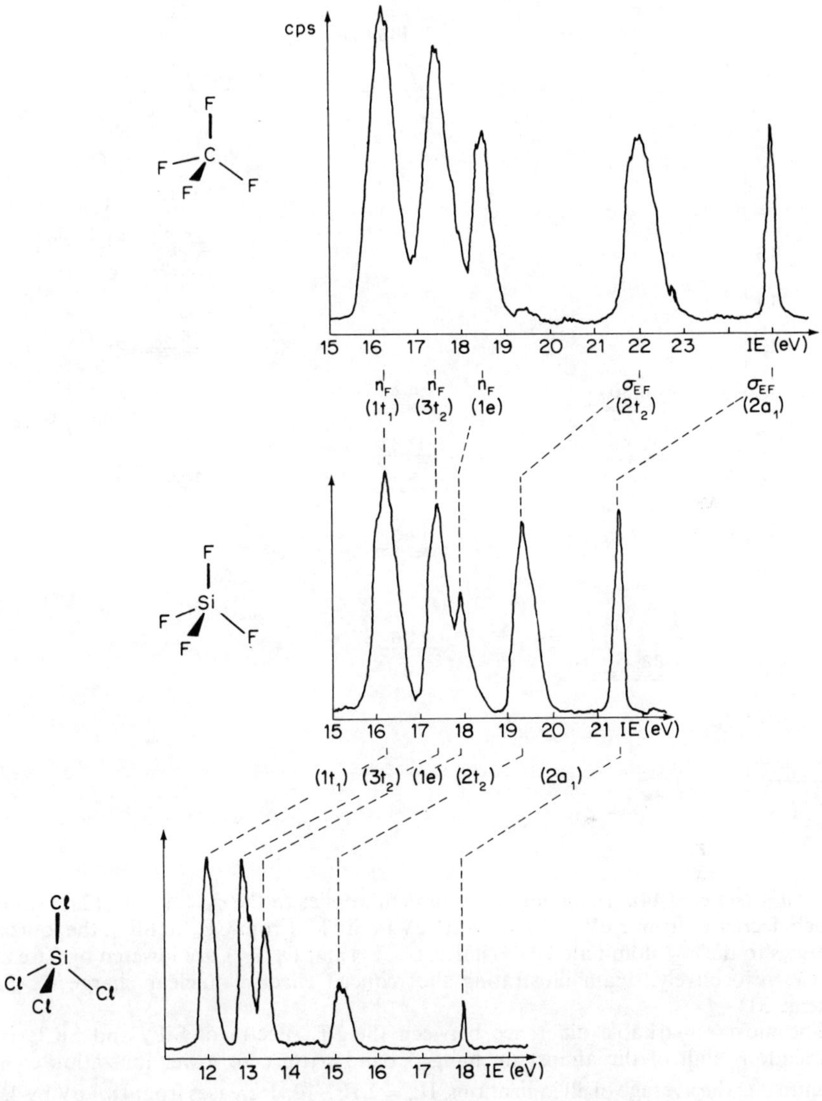

FIGURE 20. He(II) PE spectra of tetrafluoromethane and -silane[134] as well as the He(I) PE spectrum of tetrachlorosilane[135] with $M^{\cdot\oplus}$ state assignments (cf. text)

remains constant, providing some evidence for the dominant fluorine lone-pair character of the $M^{\cdot\oplus}$ states and also for rather small differences in σ_{CF} and σ_{SiF} admixture to the triply degenerate $^2T_2$ state. Compared to HF with $IE_1^v(n_F) = 16.05$ eV[37], the influence of four F ligands with high effective nuclear charge surrounding the E center becomes evident. As concerns the overall n_F splitting, $\Delta IE_{1,3}^v = 2.3$ eV for $CF_4^{\cdot\oplus}$ and 1.6 eV for $SiF_4^{\cdot\oplus}$, it expectedly decreases with decreasing n_F/n_F overlap, i.e. increasing bond lengths

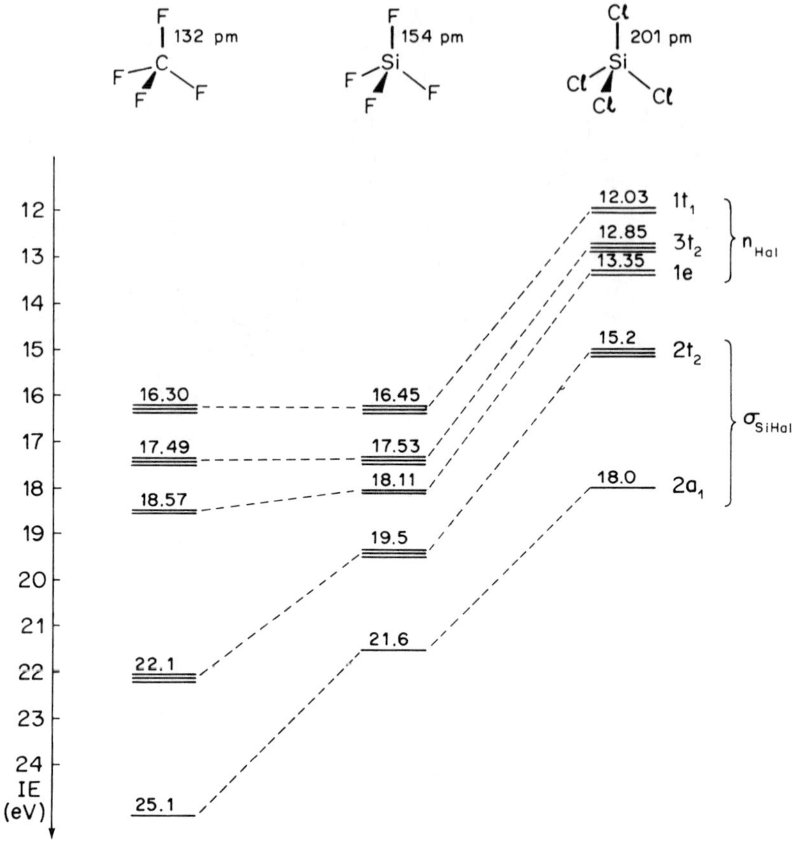

(44)

$d_{CF} < d_{SiF}$ (scheme 44). An analogous argument applies to the $\sigma_{EF}(2t_2)/\sigma_{EF}(2a_1)$ splitting, which decreases from 3 eV in $CF_4^{\cdot\oplus}$ to 2.1 eV in $SiF_4^{\cdot\oplus}$. From CF_4 to SiF_4, the ionization energies to the σ_{EF}-dominated $M^{\cdot\oplus}$ states, $\tilde{C}(^2T_2)$ and $\tilde{D}(^2A_1)$, are lowered by 2.6 eV and 3.5 eV, respectively, again illustrating the reduced effective nuclear charge C > Si (cf. scheme 31).

The most remarkable difference between the PE spectra of SiF_4 and $SiCl_4$ is the tremendous shift of the almost unchanged band pattern to lower ionization energies (Figure 20): the average of all ionizations, $\overline{IE_n^v} = \Sigma IE_n^v/12$, decreases from 18.7 eV by 4.4 eV to 13.8 eV and the center of the halogen lone-pair ionizations, $\overline{IE_n^v(n_{Hal})}$, from 17.3 eV by 4.6 eV to 12.7 eV (scheme 44). Compared to the difference between the first ionizations of HF and HCl[2, 5, 37], $\Delta IE(n_{Hal}) = 16.05\text{ eV} - 12.80\text{ eV} = 3.2\text{ eV}$, a considerably enhanced stabilization of the almost 'ball-shaped' $EHal_4^{\cdot\oplus}$ states relative to the 'bone-shaped' ones of diatomic molecules is suggested. The n_{Cl} lone-pair splitting, $IE_{1,3}^v = 1.3$ eV, reduced relative to SiF_4 by 0.3 eV, again reflects the drop in n_{Cl}/n_{Cl} overlap due to the longer SiCl bonds (scheme 44). The increased difference between the two $M^{\cdot\oplus}$ states of identical t_2 symmetry, $IE_4^v(\sigma_{SiCl}) - IE_2(n_{Cl}) = 2.3$ eV, points to enhanced $n_{Cl}(3t_2) \pm \sigma_{SiCl}(2t_2)$ mixing, and the larger splitting between the two $\sigma_{SiCl}^{\cdot\oplus}$ states, $IE_5^v(2a_1) - IE_4^v(2t_2) = 2.35$ eV, to less polarized $Si^{\delta\oplus}$–$Cl^{\delta\ominus}$ bonds, i.e. to higher electron density around the Si center in $SiCl_4$. A full parametrization of LCBO–MO models with PE spectroscopic ionization energies has been achieved both for chlorohydrocarbons $H_{4-n}CCl_n$[138] and for chlorosilanes $H_{4-n}SiCl_n$[109].

2. PE spectroscopic comparison of $H_3SiOSiH_3$, H_3SiOCH_3 and H_3COCH_3

The measured properties of molecules in their individual states (scheme 28) represent, on comparison, a composite of multiple and usually inseparable influences as illustrated

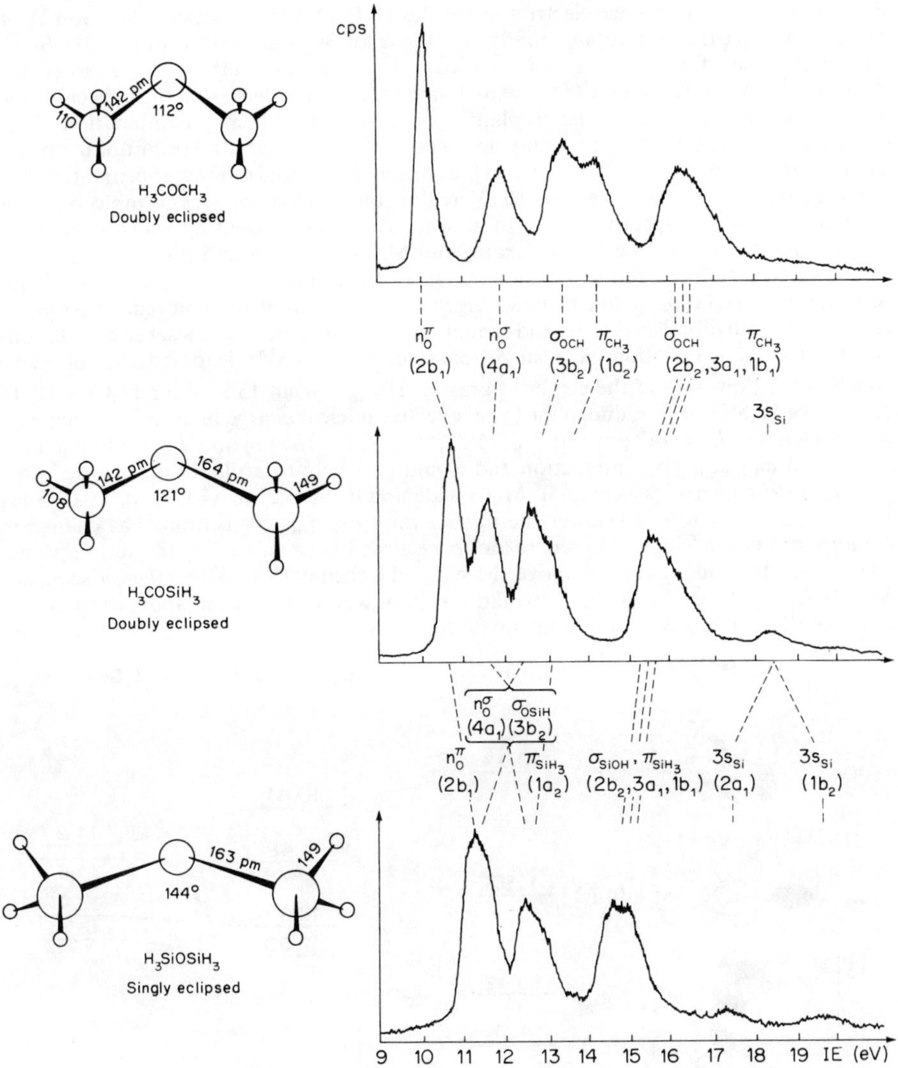

FIGURE 21. He(I) PE spectra of dimethyl ether, methoxysilane and disiloxane[141] with assignment[5,140,141] as well as their gas-phase equilibrium structures[142] (cf. text)

by the preceding discussion of the $EHal_4$ PE spectra. To gain at least some insight into the interplay of several effects in larger molecules, a set of semiempirical calculations, each of which is artificially constrained to varying a single parameter, may help despite the considerable Koopmans' deviations (Section IV.C) encountered. From the numerous literature examples for silicon compounds, PE spectroscopic comparison of disiloxane with dimethyl ether[140, 141] has been selected (Figure 21): their gas-phase equilibrium structures[142] differ not only in bond lengths $d_{OC} < d_{OSi}$ and the considerable increased bond angle $\measuredangle$ COC = 112° to $\measuredangle$ SiOSi = 144°, but also with respect to 'doubly' (four H in front) and 'singly eclipsed' conformations (four H closer to the central oxygen). The He(I) PE spectra of the 20 valence electron-molecules $H_3E–O–EH_3$ expectedly (Section IV.B) exhibit seven partly overlapping bands plus those for $3s_{Si}$ ionizations (Figure 21). Starting with the four lowest $M^{\cdot\oplus}$ states of dimethyl ether, their sequence $\tilde{X}(^2B_1) < \tilde{A}(^2A_1) < \tilde{B}(^2B_2) < \tilde{C}(^2A_2)$ with dominant contributions from the oxygen π- and σ-type lone pairs n_O^π and n_O^σ, the 'in plane' σ_{OCH} bonds and the π_{CH_3} combination of the exclusive a_2 symmetry[143] is generally accepted[5, 140, 141]. In the PE spectra of methoxysilane and disiloxane[140, 141] (Figure 21), additional $3s_{Si}$ ionizations appear at higher energies, whereas in the 10 eV to 16 eV region the number of recognizable bands is reduced to 4 or 3, respectively, due to additional overlap caused by shifts in opposite directions (scheme 45). Thus for disiloxane, four $M^{\cdot\oplus}$ states are found within only a 1.5 eV distance and therefore definite assignment, especially for the n_O^σ and σ_{OSiH} ones (45: ?), via Koopmans' correlation with calculated eigenvalues[141] cannot be achieved. Attempts to separate the individual electronic and structural effects (Figure 21) by a set of calculations constrained to a variation of a single parameter in each[141] leads to the following conclusions: Lowering of the center of gravity, $\overline{IE}_{1\text{ to }7}$, from 13.7 eV for H_3COCH_3 to 12.8 eV for $H_3SiOSiH_3$, is due to the lower effective nuclear charge of Si < C. The increase in bond lengths $d_{CO} = 142$ pm and $d_{CH} = 109$ pm to $d_{SiO} = 163$ pm and $d_{SiH} = 149$ pm leads to reduced $n_O^\pi \pm \pi_{SiH_3}$ (b_1) interaction and would thus help to explain the increase in the first ionization energy (scheme 45). Angle widening from 112° to 144° in $H_3Si–O–SiH_3$ should leave n_O^π unaffected, lower the n_O^σ and raise the σ_{OEH} ionization. The change in conformation from 'singly eclipsed' to 'doubly eclipsed' is calculated[141] to exert opposing effects and shift the σ_{OSiH} level above the n_O^σ level (scheme 45: ?). Altogether, a sequence $\tilde{X}(n_O^\pi) < \tilde{A}(\sigma_{OSiH}) < \tilde{B}(n_O^\sigma) < \tilde{C}(\pi_{SiH_3})$ would result, however, within the above-emphasized limits of the numerous assumptions involved.

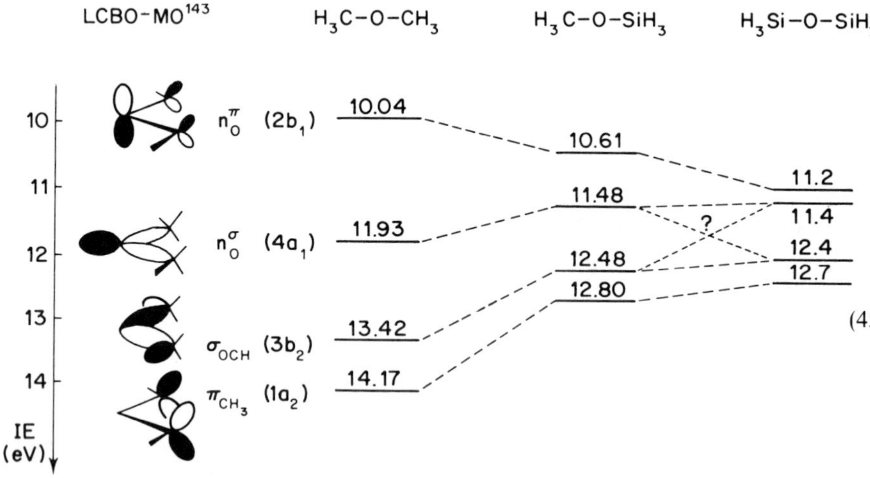

(45)

As concerns the conformational change between the neutral molecules H_3C–O–CH_3 and H_3Si–O–SiH_3 as detected by electron diffraction[142], the 'doubly eclipsed' arrangement with four hydrogens closer to the oxygen lone pairs presumably helps to alleviate the considerable electron density accumulated in this region of disiloxane due to the tremendous difference in effective nuclear charge between O and Si. This point of view is further supported by comparison with the iso(valence)electronic sulfides H_3C–S–CH_3 and H_3Si–S–SiH_3, which exhibit negligible changes in the central angle ($\sphericalangle$CSC = 99°, $\sphericalangle$SiSSi = 97°)[144] and the same 'singly eclipsed' conformations: in their PE spectra[140, 145], the first four ionization bands remain separated. In addition, it shall be pointed out that $N(SiH_3)_3$, with a large difference in effective nuclear charge N $\gg$ Si, is planar while $P(SiH_3)_3$, with much smaller Si $\rightarrow$ P electron-density polarization, exhibits a rather normal structure with an angle SiPSi of only 97° (cf. Reference 35).

3. PE and UV spectra of blue R_3Si–N=N–SiR_3

The surprising blue color of bis(trimethylsilyl)diimine[146], due to an n $\rightarrow \pi^*$ excitation[147] near the borderline between visible and infrared spectral regions, has stimulated investigation of its molecular properties[149]:

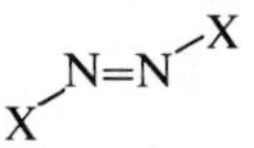

	X = $SiMe_3$	POR_2	C_6H_5	CMe_3	H	F	(46)
$\tilde{\nu}_m^{n \rightarrow \pi^*}$ (cm^{-1})	12750	18000	22600	27200	28600	(45000)	

Relative to *trans*-diimine, the n $\rightarrow \pi^*$ excitation of the trimethylsilyl derivative is lowered by 16000 $cm^{-1} \triangleq$ 2 eV, whereas the iso(valence)electronic di-*t*-butyl derivative exhibits a shift of only 1400 $cm^{-1} \triangleq$ 0.17 eV[149] (Figure 22).

The PE spectra of the 60 valence electron molecules Me_3E–N=N–EMe_3 (Figure 22), due to numerous overlapping bands in both the σ_{SiC} (or σ_{CC}) and σ_{CH} ionization regions, are expectedly 'crocodile-shaped' (cf. Section IV.F). Nevertheless, the first vertical ionization potentials, $IE_1^v(n_N^+)$ = 7.1 eV for E = Si and 8.2 eV for E = C[149], can be precisely determined from the isolated low-energy bands. As concerns the assignment, one best starts with the PE spectrum of azomethane H_3CN=NCH_3, which contains only 24 valence electrons and shows three well-separated ionizations between 8.9 eV and 12.3 eV[150] (see scheme 47): n_N^-/n_N^+ splits with a subsequent 'through bond' interaction between the $n_N^+(a_g)$ and $\sigma_{NN}(a_g)$ components of identical symmetry (scheme 47), while both the $\pi_{NN}(a_u)$ and $n_N^-(b_u)$ levels remain essentially unperturbed. For larger molecules

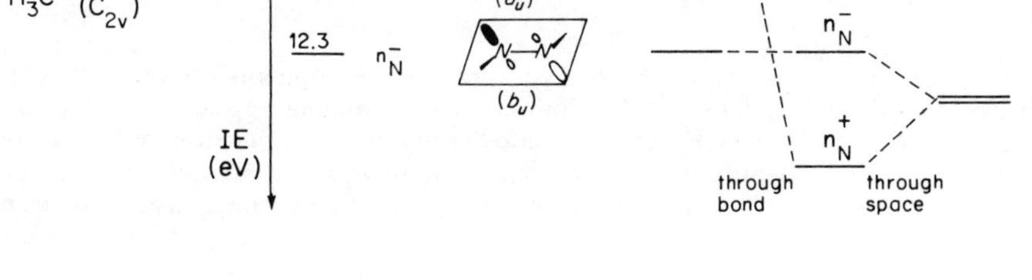

(47)

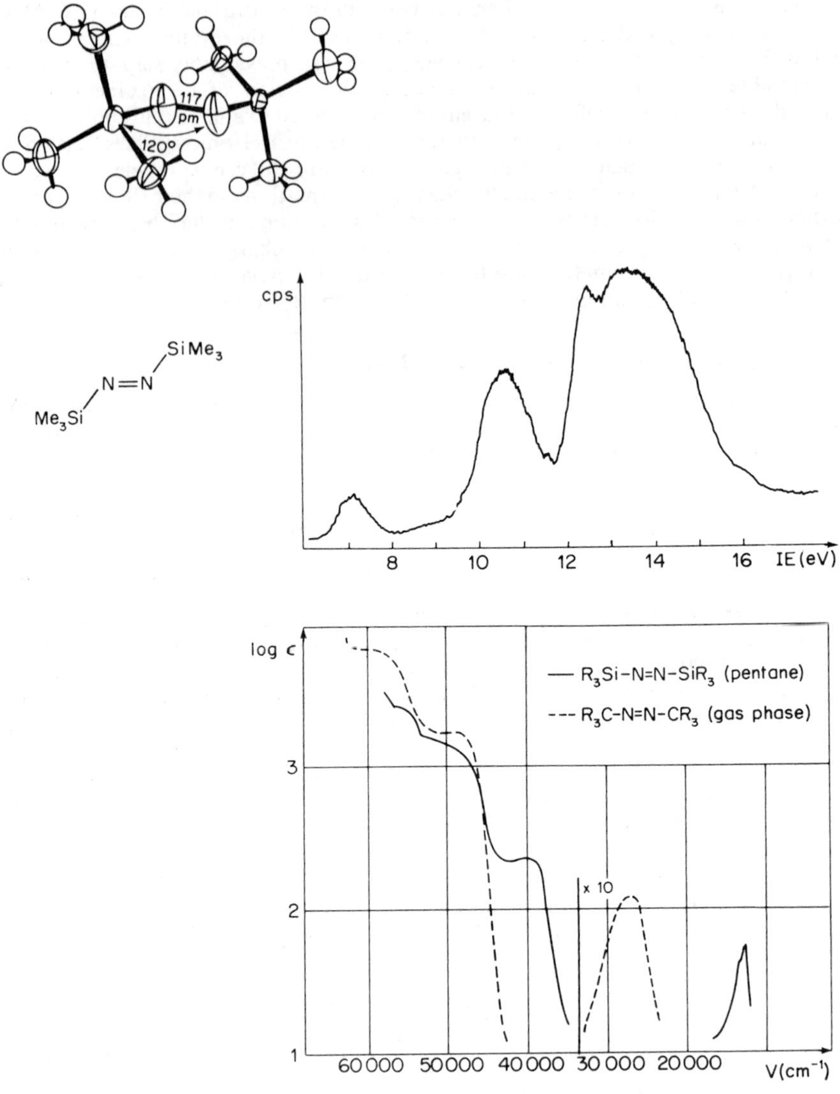

FIGURE 22. Bis(trimethylsilyl)diimine: solid-state structure[146], gas-phase PES ionization pattern[149] and UV/visible spectrum[149] in pentane solution compared to the gas-phase one of the iso(valence)electronic *t*-butyl derivative[149] (cf. text)

like the diimine derivatives $Me_3E{-}N{=}N{-}EMe_3$, however, Koopmans' assignments based on molecular orbital calculations suffer from severe shortcomings repeatedly pointed out (cf. Sections IV.A, IV.C or IV.F): the $M^{\cdot\oplus}$ state density in some ionization regions is too high for any reliable identification of adjacent regions based on eigenvalues, e.g. from semiempirical procedures[9]. Although the overall reproduction of the ionization patterns seems to be satisfactory[149] (scheme 48), the individual molecular orbitals of large and low-symmetry molecules tend to become rather uncharacteristic due to extensive mixing of symmetry-equivalent components: thus the seventh and eighth occupied orbitals calculated for $Me_3Si{-}N{=}N{-}SiMe_3$, although appearing to represent its π_{NN} and n_N^- $M^{\cdot\oplus}$ states (scheme 48), only contain about 50% and 30% contributions from the N centers (scheme 49)[149]. On the other hand, the almost steady rise of the first ionization energy from the

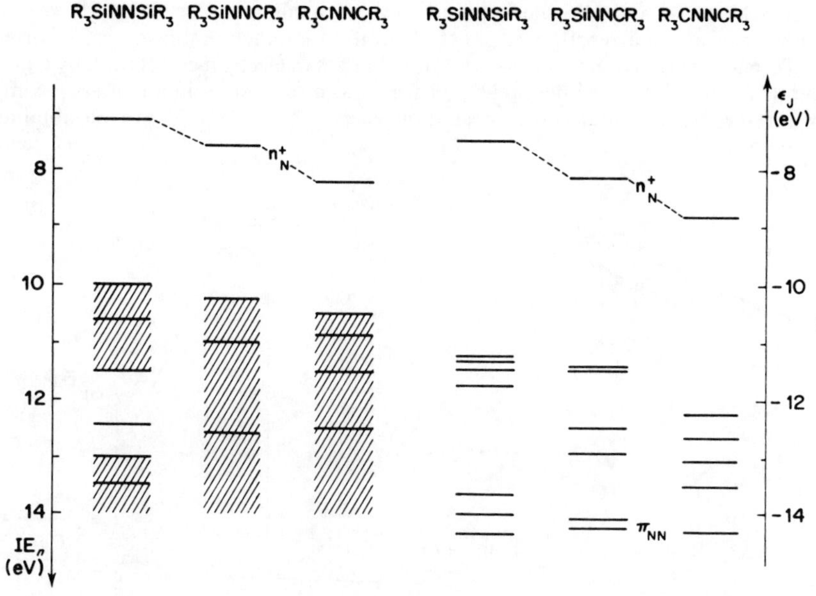

(48)

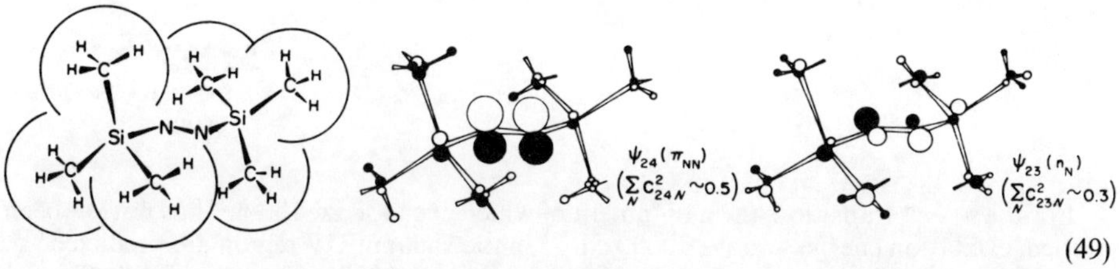

(49)

Si/Si via the mixed Si/C to the C/C substituted diimine derivatives (scheme 48) may be correlated with the calculated charge densities at the N centers of the neutral molecules[149]

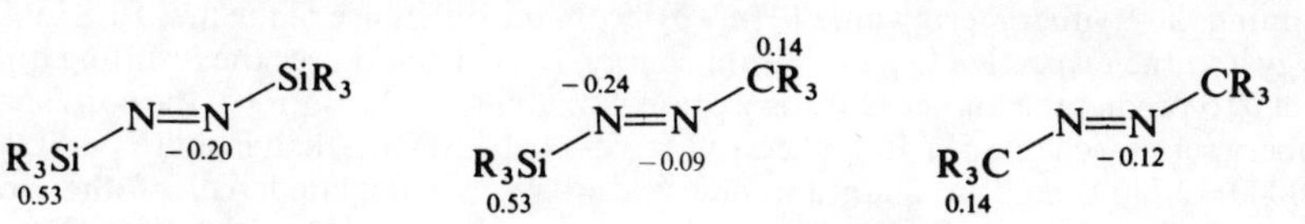

(50)

and again confirms the 'inductive' electron donation Si→N due to the large difference in effective nuclear charge Si ≪ N. In addition, the large polarization calculated (scheme 50) provides a rationale for the structure of $Me_3Si{-}N{=}N{-}SiMe_3$[146] (Figure 22), which exhibits only slightly increased angles ∢ SiNN 120° relative to the literature values for azoalkanes, ∢ CNN 106° to 113°, but a rather short NN bond length of only 117 pm—half way between those of carbon azo compounds with $d_{NN} = 123$ pm to 127 pm and the nitrogen molecule with $d_{NN} = 111$ pm!

In elaborating the origin of the surprising blue color of bis(trimethylsilyl)diimine, tentative literature speculations[149] concerning a linear SiNNSi skeletal bone or $3d_{Si}$-stabilization of the π^* state (cf. Section IV.C) can be discarded. Simple semiempirical Koopmans' correlations, $IE_n^v = -\varepsilon_J^{SCF}$, satisfactorily reproduce the lowest n_N^+ ionization energy (schemes 48 and 51), whereas the π^* level is calculated to be 9.6 eV above (scheme 51). On inclusion of a zeroth order of configuration interaction, however, i.e. correcting the difference in eigenvalues $\varepsilon(\pi^*) - \varepsilon(n_N^+)$ for changes in electron correlation by Coulomb integral (J) and exchange integral (K) terms, a more realistic estimate of approximately 1 eV for the experimental $n \to \pi^*$ excitation energy $\tilde{\nu}_m^{n \to \pi^*} = 1.5$ eV can be obtained[149] (scheme 51).

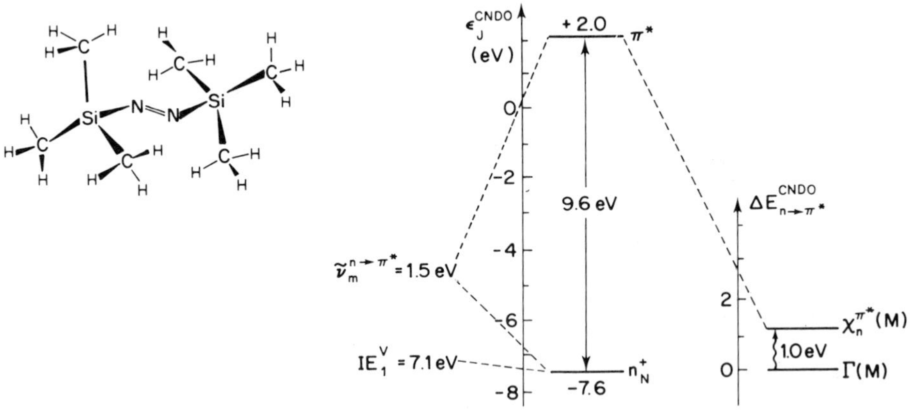

$$\Delta E_{n\to\pi^*}^{CNDO} = \underbrace{\epsilon_{\pi^*} - \epsilon_n - J_{n\pi^*}} + 2K_{n\pi^*}$$
$$= 9.6 - 8.6 + 2 \times 0$$
$$= 1.0 \text{ eV}$$

For the $\pi \to \pi^*$ transition, the assignment of which even for azoalkanes has not yet been settled, excitation energies above 70000 cm^{-1} in the vacuum UV region are predicted[148]. The UV bands observed for $Me_3Si{-}N{=}N{-}SiMe_3$ at 40000, (49000) and 52000 cm^{-1} (Figure 22), in close analogy to those of azomethane[148] at 54000 cm^{-1} and 63500 cm^{-1}, may be members of a $n_N^+ \to$ np Rydberg series[149]. The evaluation procedure[148] starts by obtaining the Rydberg term value $R/(n-\delta)^2$ from the difference of the first PES ionization energy and the respective UV excitation energy, $IE_1^v - \tilde{\nu}_n$, and uses the resulting quantum defect δ to predict the higher UV absorption maxima $\tilde{\nu}_{n+1}, \tilde{\nu}_{n+2}, \cdots$ of the corresponding Rydberg series (scheme 52). If applied to $Me_3Si{-}N{=}N{-}SiMe_3$, then from $IE_1^v - \tilde{\nu}_n = 57300 - 40000 = 17300$ cm^{-1} a quantum defect $\delta = 0.48$ is calculated, i.e. of the order of magnitude found for np-Rydberg series in other compounds[148], which is in accord with the above $n_N^+ \to 3p$ excitation at 40000 cm^{-1} and allows one to predict the higher $n_N^+ \to 4p$ and $n_N^+ \to 5p$ Rydberg transitions at 49000 cm^{-1} as well as 52000 cm^{-1}—in satisfactory agreement with the UV adsorption curve recorded for bis(trimethylsilyl)diimine (Figure 22).

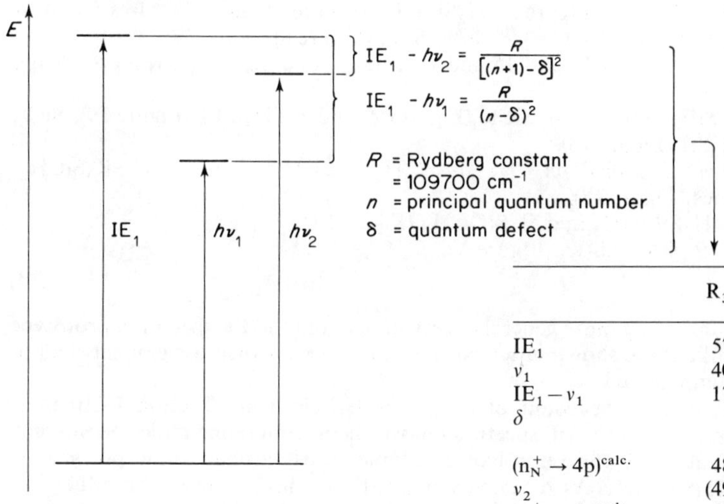

(52)

	$R_3Si–N{=}N–SiR_3$	
IE_1	57300	cm^{-1}
ν_1	40000	cm^{-1}
$IE_1-\nu_1$	17300	cm^{-1}
δ	0.48	
$(n_N^+ \rightarrow 4p)^{calc.}$	48500	cm^{-1}
ν_2	(49000)	cm^{-1}
$(n_N^+ \rightarrow 5p)^{calc.}$	51900	cm^{-1}
ν_3	52000	cm^{-1}

A summary of facts and discussions indicates that the blue bis(trimethylsilyl)diimine is unique not only because of the tremendous $n_N^+ \rightarrow \pi^*$ shift relative to azoalkanes, but also because of its extremely short NN bond length and its rather low first ionization energy[149]. All of these molecular property changes can be rationalized by a strong electron donation from the R_3Si substituents to the N=N bond according to the difference in effective nuclear charge Si $\ll$ N. In addition, there is evidence for low-lying np Rydberg states[149], which can play a rôle in photochemistry[148]. For example[149], 1,2,3-trisilacycloheptanes on irradiation with a low-pressure mercury lamp expel almost quantitatively dialkylsilylenes[151] (reaction 53). For this reaction, a three-membered ring transition state

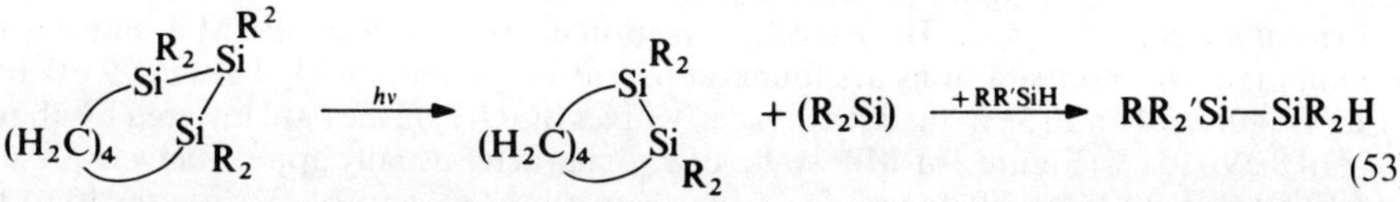

(53)

has been suggested[151]. Substituting both the first ionization potential of octamethyltrisilane[97], $IE_1^v = 8.19$ eV at 66000 cm^{-1}, and the mean first UV excitation energy of trisilacycloheptanes[151], $\tilde{\nu} \sim 46000$ cm^{-1}, into the Rydberg formula (scheme 52) yields a term value $R/(n-\delta)^2 \sim 20000$ cm^{-1}, which could correspond to a $4p_{Si}$ Rydberg state[148]. Under local C_{2v} symmetry, only the two $\sigma_{SiSiSi} \rightarrow 4p_{Si}$ transitions in the SiSiSi plane would be allowed, fitting nicely into the suggested three-membered ring transition-state model. Accordingly, also in the photochemical synthesis of disilenes $R_2Si{=}SiR_2$ from trisilanes $R_3Si(SiR_2)SiR_3$[152] analogous to equation 53, $4p_{Si}$ Rydberg states may be essential.

H. PE Spectra of Prototype Silicon Molecules: Information and Application

In the preceding discussions on information from and assignment of PE spectra (Sections II.B–II.E, IV.B and IV.D–IV.G), those of the following silicon-containing molecules have been presented:

Intermediates: Silabenzene (Figure 3), $H_2Si{=}CH_2$ (Figure 4), $SiCl_2$ (Figures 5 and 6), $H_5C_6{-}N{\equiv}Si$ (Figure 7) and $H_3CC{\equiv}N^{\delta\oplus}{-}N^{\delta\ominus}{-}SiR_3$ (Figure 8).

Polysilanes: Si_nH_m (Figures 14 and 15 and scheme 33) and $Si_n(CH_3)_m$ (Figure 16 and scheme 32).

Silicon halides: SiF_4 (Figure 20), $SiCl_2$ (Figures 5 and 6), H_3SiCl (Figure 19), $SiCl_4$ (Figure 20) and H_3SiI (Figure 13). (54)

Silicon-substituted hydrocarbon π-systems: $H_3SiC{\equiv}CH$ (Figure 17), R_3SiCH_2-substituted benzenes (Figure 18).

SiO derivatives: $H_3SiOSiH_3$ and H_3SiOCH_3 (Figure 21).

SiN derivatives: $R_3SiN{=}NSiR_3$ (Figure 22).

(54)

In conclusion, the following more general observations from the PE spectra of prototype silicon molecules like those shown in scheme (54) and rules for their assignment shall be summarized and emphasized:

Number of ionizations. The count of $np_{Element} + 1s_H$ electrons (Section IV.B) must include also $3s_{Si}$, because in the PE spectra of most silicon-containing molecules (except, e.g., SiF_4) additional $3s_{Si}/2$ ionization bands become clearly visible, e.g. in polysilanes, H_3SiX, $SiCl_4$ or $H_3SiOSiH_3$. As repeatedly pointed out, this is due to the rather low effective nuclear charge of silicon centers, best illustrated by the $M^{\cdot\oplus}$ state comparison of the iso(valence)electronic series Ar to SiH_4 based on a 'united atom' approach (scheme 31). Calculated charge distributions for the neutral molecules (e.g. schemes 9, 17, 34 or 50) reproduce the electron donation from Si to adjacent centers of higher effective nuclear charge.

PE spectroscopic information. Si prototype molecules with up to 12 atoms (Figure 3) and/or 32 valence electrons (Figure 20) can show well-resolved band patterns. On the other hand, even in small molecules (Figure 21) and especially in all those which contain numerous $Si(CH_3)_n$ subunits (Figures 16, 18 or 22) PE bands will overlap: sometimes 'crocodile-shaped' contours result and only the isolated low-energy 'forehead-humps' can be assigned via Koopmans' correlation with calculated eigenvalues or by $M^{\cdot\oplus}$ state comparison. Compared to carbon analogues, vibrational fine structures are usually less pronounced (cf. e.g. Figure 17 or 19) and spin/orbit coupling (Figure 13) as well as Jahn/Teller splittings (Figure 14) are smaller.

Typical ionization regions. In general, ionization energies to generate $M^{\cdot\oplus}$ states with predominant σ_{SiSi} contributions are found in parent polysilanes Si_nH_m between 9 eV and 11 eV (Figures 14 and 15); in their methyl derivatives $Si_n(CH_3)_m$ they are lowered by about 1 eV to 8 eV–10 eV (Figure 16). $M^{\cdot\oplus}$ states of σ_{SiC} character usually appear between 10 eV and 12 eV (Figure 16 or 22), those of σ_{SiH} type between 12 eV and 14 eV (Figures 13 to 16 and 17 as well as scheme 33) and the σ_{CH} ones between 12 eV and 16 eV (Figure 13, 16 or 22). Last but not least, $3s_{Si}$-type ionizations are often observed between 16 eV and 20 eV (Figures 14, 15, 17, 19 and 20, and also scheme 33). Other PE bands, which originate, e.g., from π-systems (Figure 3, 9 or 18), are shifted to lower energies relative to the parent compounds, especially by R_3SiCH_2 donor substituents (Sections III.B. and IV.F), whereas n_N-type ionizations from silyl-substituted heterocenters often exhibit higher energies (Figure 13, 19 or 21). Finally, the extreme stabilization of $M^{\cdot\oplus}$ states by several R_3SiCH_2 substituent groups shall be pointed out again (cf. e.g. Figures 9 and 18 or schemes 19–21 and 40). With all the above ionization regions frequently covered, most silicon-containing molecules only exhibit small 'observation windows', in which substituent perturbations (like in scheme 29 or 35) are difficult to detect PE spectroscopically.

Assignment of PE spectra. Although for small Si prototype molecules, their $M^{\cdot\oplus}$ state sequences may be reliably calculated using highly correlated wave functions (Section

IV.C), for those with more than about 3 heavy atoms, there is only the recommendation to compare equivalent $M^{\cdot\oplus}$ states of chemically related, preferentially iso(valence)electronic molecules (Section IV.A). This approach has been used throughout the preceding discussions, and especially for those PE spectra, for which the density of $M^{\cdot\oplus}$ states in certain regions are too high to permit Koopmans' correlations, $IE_n^v = -\varepsilon_J^{SCF}$, with semiempirical as well as *ab initio* eigenvalues, because both do not account for $M^{\cdot\oplus}$ electron redistribution or electron correlation. On the other hand, first- or second-order LBCO–MO perturbation models can be readily parametrized using the 'experimental' PES ionization energies (cf. e.g. Sections IV.D, IV.E, IV.F or IV.G.1). In any case, there is no need to include fictitious $3d_{Si}$-orbitals in discussions of $M^{\cdot\oplus}$ state sequences of Si-containing molecules other than as polarization functions (Section IV.C).

In addition to providing valuable information on the radical cation states of silicon molecules and, via Koopmans' correlations, also on their electronic structure in the neutral ground state, numerous applications of PE spectra and/or individual ionization energies have been reported in the literature[2–31]. The following examples were discussed in the preceding sections:

(a) PE spectroscopic ionization 'fingerprints' can advantageously be used to optimize gas-phase reactions[12] and especially to characterize short-lived reactive intermediates (cf. Section II).

(b) Substituent effects on π-systems or heteroatom centers with electron pairs are often readily determined and parametrized (cf. e.g. Section IV.F).

(c) First vertical ionization energies are linearly correlated to adiabatic one-electron transfer oxidation potentials, if the structures of the radical cations generated either do not change considerably or change in the same way (cf. Section III.A.). For organosilicon compounds, selective one-electron oxidation to their radical cations in water- and oxygen-free $AlCl_3/H_2CCl_2$ solution (Section III.A) can be predicted by $IE_1^v < 8$ eV (cf. e.g. Figure 9).

(d) The same applies to charge-transfer excitation energies of donor molecule complexes with an identical acceptor; deviations from the resulting linear regressions provide hints on changes in the intermolecular interactions (Section IV.F.1).

(e) The PE spectroscopic ionization patterns of linear polysilanes, Si_nH_{2n+2} and $Si_n(CH_3)_{2n+2}$, for higher *n*, approach the band structures of the corresponding polymers (scheme 33) and their vertical ionization energies are useful in parametrizing the respective calculations (cf. Section IV.D).

(f) Rydberg states of polyatomic molecules, which may play an essential rôle in photochemical reactions, especially of polysilanes, can be detected and assigned by determining the quantum defect δ in the value of the term $R/(n-\delta)^2$ from the difference $IE_1^v - h\nu$ between the respective UV excitation and the vertical first ionization energies (Section IV.G.3).

There are many more applications of PE spectroscopic band patterns and vertical ionization energies reported in the literature[2–31]. However, some correlations with other molecular properties in other molecular states are based on assumptions[9] which neglect the reality of the $M^{\cdot\oplus}$ states observed and should therefore be viewed with care. To quote again one facet of many models, inclusion or omission of $3d_{Si}$-orbital functions can neither prove nor disprove their influence on the bonding in silicon-containing molecules (Section IV.C). On the other hand and in conclusion, it is emphasized how much one can learn about vertically (Sections II and IV) and adiabatically (Section III) generated radical cations and the different time scales involved (Section III.D) from investigations including PE spectroscopy.

Based on the preceding detailed discussions (Sections II to IV), the PE spectroscopic publications concerning silicon-containing molecules will be reviewed.

V. IONIZATION POTENTIALS OF SILICON-CONTAINING MOLECULES

A. Data Compilation and Selection of Prototype PE Spectra

In the preceding sections, almost 40 PE spectra of selected silicon-containing molecules have been displayed in the figures summarised in scheme 54. For $M^{\cdot\oplus}$ state comparison, usually the ionization patterns of related and especially of iso(valence)electronic carbon species have also been shown. For the subsequent general overview, the PE spectroscopic investigations as retrieved by the literature search[153] are classified according to the following types:

saturated molecules Si_nR_m (Section V.B),
Silicon-substituted open-chain (Section V.C) and cyclic (Section V.D) hydrocarbon π-systems, (55)
silyl derivatives of elements from groups 15 (Section V.E), 16 (Section V.F) and 17 (Section V.G) of the periodic table, and
miscellaneous silicon-containing compounds (Section V.H).

With few exceptions, every silicon-containing molecule will be mentioned only once in the above order of Sections V.B to V.H, and within them according to further specification of the corresponding compounds. Referring to the preceding detailed discussions (Sections II to IV), the assignment by the respective investigators cited will not be commented on. Consequently, this also concerns the large number of publications, in which various properties of silicon-containing molecules have been calculated at various levels of sophistication, and no comparison will be made with the $M^{\cdot\oplus}$ state sequences of chemically related molecules containing elements other than silicon. From the often rather contourless (Section IV.H) PE spectra, only 'characteristic' vertical ionization energies are selected; for all others reference is made to the relevant publications. While noting again the 38 PE spectra already reproduced (Figures 3–8 and 13–22), only few additional ones illustrating special facets of $M^{\cdot\oplus}$ states are shown.

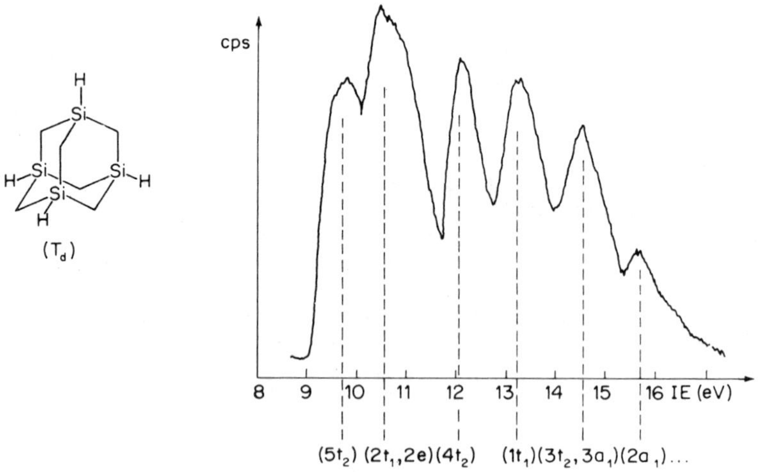

FIGURE 23. He(I) PE spectrum of tetrahedral silamantane with assignment[162]

B. Saturated Molecules Si_nR_m

The saturated silicon-containing molecules Si_nR_m are subdivided according to the number of Si centers, $n=1$ and $n>1$, and to substituents R=H and R=alkyl.

SiR_4. The PE spectrum of SiH_4[117,119,154,155] is reproduced in Figure 15, and that of $Si(CH_3)_4$[97,136,156] in Figure 16. In addition, the vibrational fine structure of the low-energy PE band of the radical $SiH_3^{\cdot}$[73,157] is shown in scheme 14. PE spectra are also reported for all compounds listed in Table 56. As concerns their first ionization energies, H_3C substitution of SiH_4 lowers the values by 1 eV[158] [Figure 15: $IE_1^v(Si(CH_3)_4) = 10.6$ eV], methyl to ethyl exchange by 0.6 eV and ring closure to four- and five-membered saturated rings by 1.6 eV(!) or by 0.8 eV[159]. There is a special $\sigma_{SiC/SiC}$ interaction in (C_{2h}?) 1,2-bis(trimethylsilyl)ethane[160], whereas the n_O ionization of cyclic ketones remains constant despite varying ring size, i.e. changing distance to the SiR_2 link[169]. The PE spectrum of the 56 valence-electron molecule silamantane $(HSi)_4(CH_2)_6$ (Figure 23) is nevertheless well-resolved due to the T_d symmetry of the latter. Both donor ($X=CH_3$) and acceptor (X=F) substituents considerably affect the first skeletal ionization (Table 56: $\Delta IE_1^v = 1.8$ eV).

M ($R=CH_3$)		References	IE_1^v (eV)	
R_nSiH_{4-n},	$n=1$	158	11.61	($4a_1$)
	2		11.2	($4b_1$)
	3		10.7	(5e)
$CH_3CH_2SiH_3$		158	10.95	(10a′)
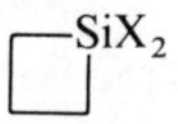	X=H	159	10.05	(a_1)
	CH_3		9.40	(a_1)
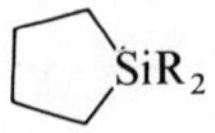			9.75	(a_1)
$R_3Si(CH_2)_nSiR_3$				
	$n=1$	160	9.5	(σ_{SiC})
	2		8.78	(σ_{SiC})
	3		9.41	(σ_{SiC})
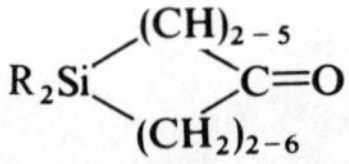		161	9.2	(n_O)
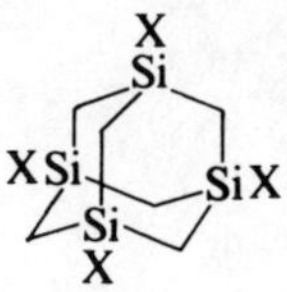	$X=CH_3$	162	8.85	
	H		9.7	
	Cl		10.41	
	F		10.61	
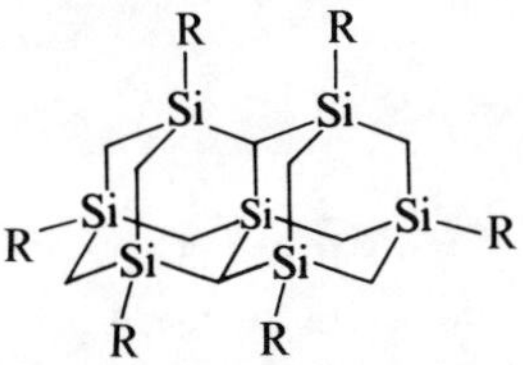			9.11	

(56)

Si_nH_m. The PE spectra of Si_2H_6[117] and of the higher linear polysilanes Si_nH_{2n+2}[117] with $n=3,4,5$ are shown in Figures 14 and 15. Reference is also made to the calculated band structure[114–116] of the polymer $(SiH_2)_\infty$ shown in scheme 33.

Si_nR_m. The PE spectra of linear $Si_n(CH_3)_{2n+2}$[97] ($n=2$[97, 157], 3 and 4) as well as of cyclic $Si_n(CH_3)_{2n}$[97] ($n=5,6$) permethylated polysilanes are shown in Figure 16. Other PE spectroscopic investigations concern the di- and polysilane derivatives listed in Tables 57 and 58. The $M^{\cdot\oplus}$ states with main contributions from SiSi bonds can vary their respective ionization energies tremendously: relative to $IE_1^v=10.53$ eV of H_3SiSiH_3[117] (Figure 14), in hexamethyldisilane with $IE_1^v=8.69$ eV[97] (Figure 16) it is already lowered by 1.8 eV (Section IV.D). Table 57, however, shows that it may be as low as 7.5 eV on phosphorus ylide substitution[165], i.e. adjacent to a 'carbon lone pair'[92] and as high as 13.2 eV in F_3Si–SiF_3[167]—a range of almost 6 eV! On the other side, neither *t*-butyl replacement[163] of H_3C groups nor vinyl[164] or phenyl[127] π-conjugation cause any larger perturbation. Relative to hexamethyl disilane[97] (Figure 16), the first vertical ionization energy of the tin derivative R_3Sn–SiR_3 is only lowered by 0.4 eV[166].

M ($R=CH_3$)	References	IE_n^v (eV)	
R_3C–R_2Si–SiR_2–CR_3	163	8.52	(σ_{SiSi})
$H_2C=CH$–R_2Si–SiR_3	164	8.56	(σ_{SiSi})
$H_2C=CH$–R_2Si–SiR_2–$HC=CH_2$	164	8.63	(σ_{SiSi})
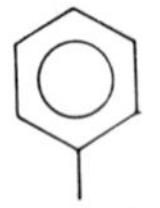 R_2Si–SiR_3	127	8.35	(σ_{SiSi})
$R_3P=CH$–R_2Si–SiR_3	165	6.85	(n_c^-)
		8.63	(σ_{SiSi})
$R_3P=CH$–R_2Si–SiR_2–$N=PR_3$	165	6.18	(n_c^-)
		7.44	(σ_{SiSi})
		7.84	(n_n)
R_2Si–SiR_2 / $R_3P=C$ $C=PR_3$ / SiR_2 (ring)	165	6.11/6.43	(n_c^-)
		7.56	(σ_{SiSi})
R_3Sn–SiR_3	166	8.32	(σ_{SnSi})
Cl_3Si–$SiCl_3$	48, 164	10.91	($3a_{1g}$)
F_3Si–SiF_3	167	13.20	($3a_{1g}$)

(57)

M ($R=CH_3$)	References	IE_n^v (eV)
$R-(SiR_2)_5-R$	97	7.7–9.6 (σ_{SiSi})
$Si(SiR_3)_4$	97	8.24 (t_1)
		(9.9) (a_1)
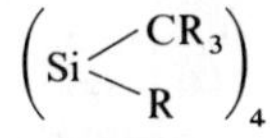	163	7.42, 7.85, 8.15 (σ_{SiSi})
$(Si_5R_9)-SiR_3$	98	7.8–10.0 (σ_{SiSi})
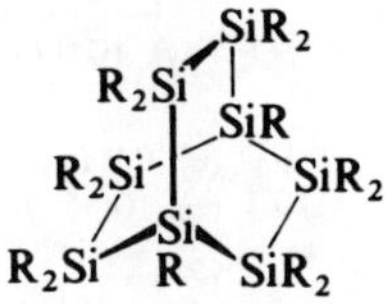	98	7.8–10.0 (σ_{SiSi})

(58)

The ionization patterns of permethylated polysilanes have been thoroughly discussed using a topological SiSi/SiSi bond interaction model (Section IV.D.3). Chain-lengthening from $n=4$ to 5[97] lowers the value only by $\Delta IE_1^v = -0.15$ eV[97] (Figure 16 and Table 58), further confirming the developing band structure (scheme 33); in contrast, branching raises the value by $\Delta IE_1^v = +0.24$ eV. A special case is the sterically overcrowded, 'butterfly-shaped' tetrakis(*t*-butyl)-substituted cyclotetrasilane, exhibiting its $M^{\cdot\oplus}$ ground state at 7.42 eV (Table 58). Larger systems like the five-membered ring isomer Si_6R_{12} or the norbornane-like Si_8R_{14} both have their $\sigma_{SiSi}^{\oplus}$ states in an identical ionization region between 7.8 eV and 10 eV[98] (Table 58).

C. Silicon-substituted Open-chain π-Systems

Numerous PE spectroscopic investigations have dealt with organosilicon derivatives of ethene (Tables 59 and 60), cyclopentadiene (Table 61) and acetylene (Table 62), partly arranged in rings with intermittent conjugational barriers. Also covered within this subsection will be all compounds with carbon to heteroatom π-bonds, in which the silyl substituent is connected to the carbon center (Table 63).

Monosilyl-substituted ethenes. Special interest has been focused on effects of α and β silyl substituents in terms of 'through bond' and 'through space' interactions[6, 8] and also on special arrangements of π-ligands, e.g. for spiro conjugation or transannular effects (Table 59). Whereas monosilyl ethene[125] exhibits a higher first ionization energy than the iso(valence)electronic propene ($IE_1^v = 10.03$ eV[5]), on substitution with CH_2SiH_3[125] it is lowered beyond that of the iso(valence)electronic butene-1 ($IE_1^v = 9.72$ eV[125]). The difference between each of the two silyl derivatives, which amounts to 0.88 eV for SiH_3[125] and 0.8 eV for $Si(CH_3)_3$[168], illustrates the donor properties of silyl groups. Substituents $-SiCl_3$ and $-Si(OCH_3)_3$ increase the ionization energy (Table 59). Fixation of the $Si(CH_3)_2$ unit in a five-membered ring obviously allows a better hyperconjugative $M^{\cdot\oplus}$ stabilization (Section IV.F) and further lowers $IE_1^v(\pi)$ by 0.4 eV (Table 59). The interaction between several vinyl groups at one Si center remains small[171] and a transannular interaction in 1-silapentene-3 derivatives has been excluded by several authors (see e.g. References 172 and 173).

M (R = CH_3)		References	IE^v_1 (eV)	
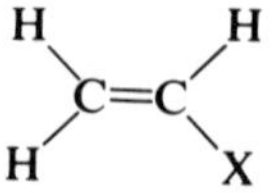	X = SiH_3	125	10.37	(2a″)
	CH_2SiH_3	125, 168	9.49	(3a″)
	SiR_3	168, 169	9.86	(π_1)
	CH_2SiR_3	168, 169	9.1	($\pi + CH_2Si$)
	$SiCl_3$	169	10.7	(π)
	$Si(OR)_3$	169	11.0	(π)
		170	9.27	(π)
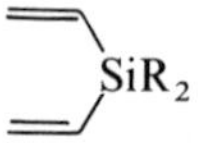		171	9.8 (b_1), 10.0 (a_2)	
$Si(HC{=}CH_2)_4$		171	9.7 (e), 9.9 (a_2); 10.1 (e), 10.35 (b_1)	
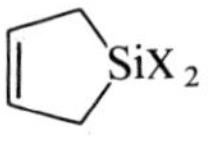	X = H	170, 171	9.21	(π)
	= CH_3	170, 173	9.1	(π)
	= CH_3 | $HC{=}CH_2$	170	9.05; 9.7	(π); (π)
	= C_6H_5	170	9.1	(π)
	= OCH_3	164	9.59	(π)
	= Cl	172	9.63	(π)
	= F	172	9.62	(π)

(59)

Polysilyl ethenes. Polysilyl-substituted ethene derivatives have been mainly investigated by PE spectroscopy to detect those with $IE^v_1 \leqslant 8$ eV, which can be oxidized to their radical cations adiabatically in solution using the selective one-electron transfer reagent $AlCl_3/H_2CCl_2$ (Section III). All organosilyl ethene derivatives with $IE^v_1 < 8$ eV, as expected, could be oxidized in solution (Figure 10). On fourfold β-trimethylsilyl substitution, the ionization potentials of $H_2C{=}CH_2$ at 10.52 eV[2, 5] or that of $(H_3C)_2C{=}C(CH_3)_2$ at 8.41 eV[91] are lowered by well over 3 eV or 1.2 eV, respectively, and that of hexadiene-2,4 at 8.82 eV[91] by almost 1.9 eV—tremendous effects, which further illustrate the powerful donor properties of CH_2SiR_3 substituents (Sections III.B and IV.F).

Silyl-substituted cyclopentadienes. For silylated butadienes, no entries were found in the literature other than first adiabatic ionization energies determined by mass spectroscopy, ranging for the disilyl-substituted isomers $C_4H_2(SiR_3)_2$ between 8.43 and

M (R = CH_3)		References	IE^v_n (eV)	
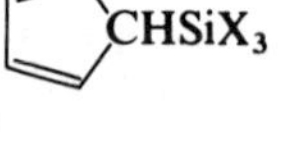	X = F	176	9.1, 10.9	π(a′, a″)
	Cl		9.0, 10.6	
	H		8.7, 10.2	
	CH_3	91	8.3, 9.1	$\pi(a_2, b_1)$
		91	8.05, 9.1	

(61)

(60)

M ($R = CH_3$)		References	IE^v_n (eV)
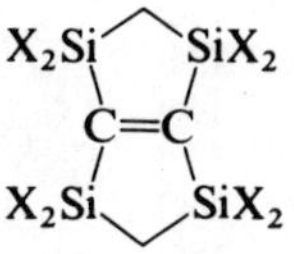	X = F H CH_3	96	10.16 8.93 7.98 } (b_{1g})
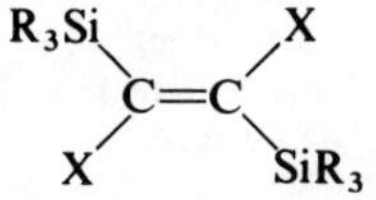	X = H Cl Br	174	9.19 8.96 8.83 } (π)
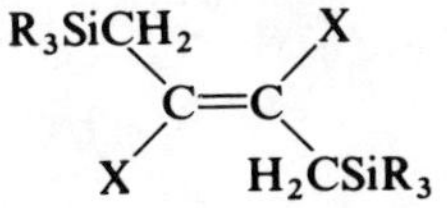	X = H CH_3 CH_2SiR_3	91	8.30 7.70 7.15 } (π)
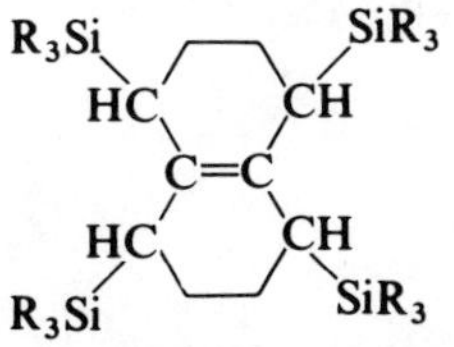		91	6.98 (π)
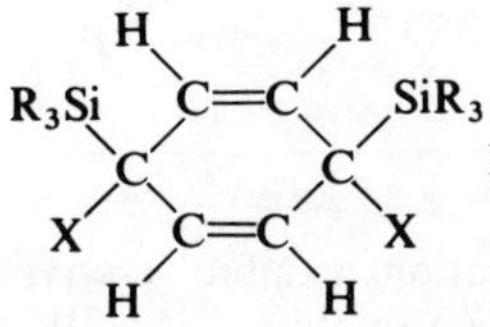	X = H SiR_3	91	7.70, 9.30 7.00, 9.05 } (π)

8.65 eV[175], and for cyclopentadienes (Table 61). Silyl-substituted cyclopentadienes are fluxional[176], nevertheless the PE spectra are well-resolved up to 11 eV. Because the $M^{\cdot\oplus}$ ground state, $\tilde{X}(^2A_2)$, supposedly has a nodal plane passing through the substitution center and the $\pi(b_1)$ ionization is more strongly perturbed by β silyl groups[91] relative to the unsubstituted parent system C_5H_6[176] with $IE^v_1(a_2) = 8.6$ eV and $IE^v_2(b_1) = 10.7$ eV, the shifts amount to 0.3 eV or 0.5 eV for $\pi(a_2)$ and to 1.6 eV for $\pi(b_1)$[91].

Silyl-substituted acetylenes. The 'hyperconjugative' interaction between the C≡C moiety and SiH_3 as well as CH_3 groups of local C_{3v} symmetry has been discussed extensively as one of the best examples of how to parametrize LCBO–MO models by PE spectroscopic $M^{\cdot\oplus}$ state energies (Section IV.E). The ionization patterns of the following mono- and disilyl acetylene derivatives are also reported in the literature[153] (cf. Table 62).

Disilylacetylene $H_3Si–C≡C–SiH_3$ exhibits a first veritcal ionization potential, which is lowered by almost 1 eV relative to the parent HC≡CH ($IE^v_1 = 11.40$ eV[2,5]), but raised by 0.85 eV with respect to the iso(valence)electronic dimethyl derivative ($IE^v_1 = 9.61$ eV[121]). Halogen substitution of both $H_3Si–C≡CH$ ($IE^v_1 = 10.7$ eV[121]) and $R_3Si–C≡CH$ ionizations are moderate[177,178] and chain lengthening does not exert a substantial influence[179] (Table 62). Whereas the nine-membered ring triacetylene (3,3,6,6,9,9-hexamethyl-

M (R = CH_3)		References	IE_n^v (eV)
$H_3Si–C≡C–X$,	X = SiH_3	121, 177	10.46 (3e)
	Cl	177	10.22 (3e)
	Br	177	10.17 (3e)
$R_3Si–C≡C–X$,	X = H	122	10.2 (7e)
	F		10.2 (8e)
	Cl		9.7 (9e)
	Br		9.6 (12e)
	I		9.4 (15e)
	C_6H_5	178	8.37 (b_1), 9.28 (a_2)
$R_3Si–(C≡C)_n–SiR_3$,	$n = 1$	178	8.85 (e)
	2	179	8.8 (e)
	3		8.7 (e)
$(R_3Si)_3C–C≡C–C(SiR_3)_3$		178	7.60 (e)
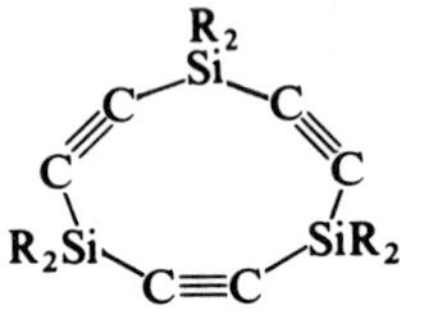		180	9.3–10.0 ($2a_2''$, $4a_1'$, $5e'$)
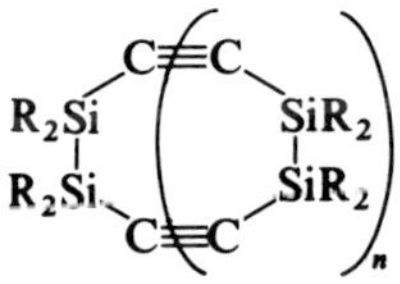	$n = 1$	180	8.13 ($5a_g$), 9.05 ($2b_{2g}$)
	2		8.45 ($5a_1'$), 9.5 ($7e'$, $2a_2''$)

(62)

3,6,9-trisilanona-1,4,7-triyne) according to MNDO calculations exhibits a π-type first ionization[180], the 8- and 12-membered rings each with two or three $R_2Si–SiR_2$ units possess a σ-type $M^{\cdot\oplus}$ ground state with predominant SiSi contributions[180] and, formally, would therefore have to be added to Table 57.

The final remark concerns the PE spectrum of $(R_3Si)_3C–C≡C–C(SiR_3)_3$, a 'Guiness Book of Records' acetylene, which contains 156 valence electrons[178] (Figure 24). In this

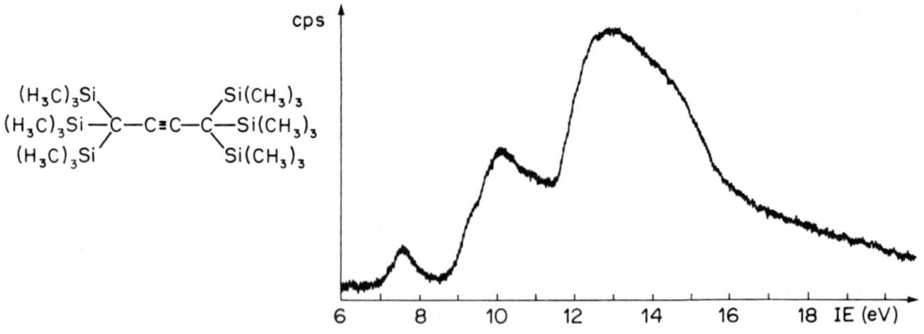

FIGURE 24. He(I) PE spectrum of 1,1,1,4,4,4-hexakis(trimethylsilyl)-2-butyne with IE_1^v = 7.60 eV[178]

'ball-shaped' molecule, six bulky trimethylsilyl groups provide complete steric shielding for the little C≡C 'bone', and its ionization energy is lowered relative to acetylene by 3.8 eV(!)[178]. Because $IE_1^v < 8$ eV, it can be oxidized by $AlCl_3/H_2CCl_2$ and under R_3Si group transfer forms tetrakis(trimethylsilyl)butatriene radical cation $(R_3Si)_2C{=}C{=}C{=}C(SiR_3)_2^{\cdot\oplus}$ (Figure 9).

Silyl derivatives of carbon to heteroatom π-bonds. The compounds in Table 63 qualify for the title definition. Trimethylsilylmethylidine phosphane $R_3Si{-}C{\equiv}P$ has been prepared by energetically advantageous thermal R_3SiCl elimination from the disilylated precursor[50] (Figure 25) in the short-pathway pyrolysis apparatus (Figure 2). Its ionization pattern extracted by digital substraction of on-line recorded PE spectra[48] yields $IE_1^v = 9.9$ eV, higher by 0.2 eV relative to the iso(valence)electronic $R_3C{-}C{\equiv}P$, a liquid,

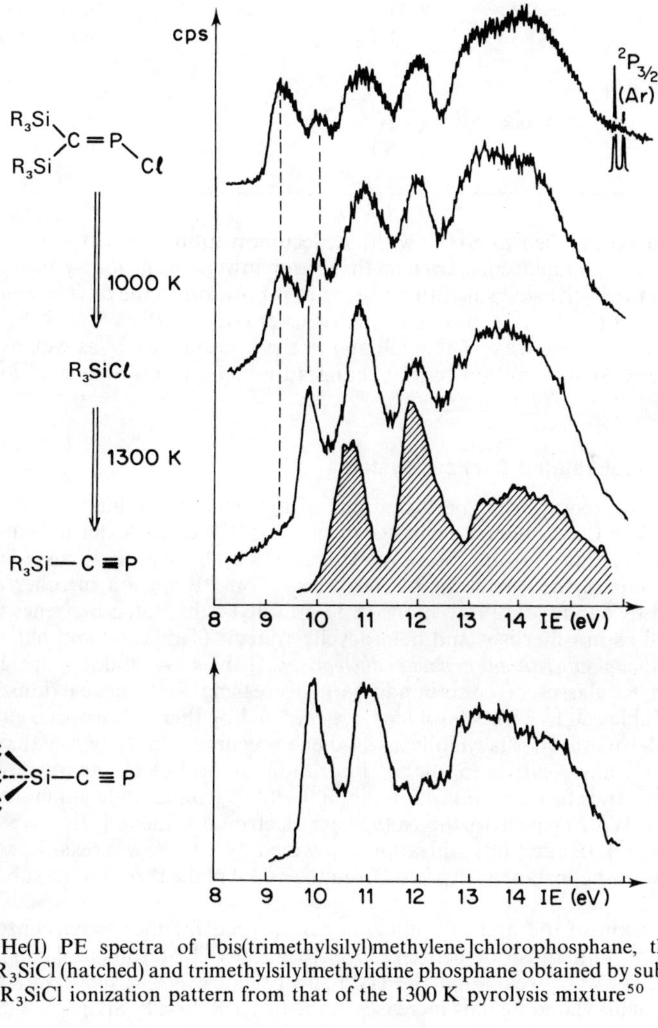

FIGURE 25. He(I) PE spectra of [bis(trimethylsilyl)methylene]chlorophosphane, the thermal elimination of R_3SiCl (hatched) and trimethylsilylmethylidine phosphane obtained by subtraction of the digitalized R_3SiCl ionization pattern from that of the 1300 K pyrolysis mixture[50]

M (R=CH_3)		References	IE^v_n (eV)
$R_3Si—C{\equiv}P$		50	9.90 (e)
$R_3Si(R)C{=}O$		181	8.60 (n_O)
$R_3Si(RH_2COOC)C{=}N_2$		182	8.58 (π)
$R_3Si(X)C{=}PR_3$	X=H	165	6.81
	SiR_3		6.92 (n_C^-)
$R_3Si(X)C{=}AsR_3$	X=H	183	6.56
	SiR_3		6.66 (n_C^-)

(63)

which can be distilled at 61 °C without decomposition[50]. Starting from acetone (IE^v_1 =9.70 eV[5]), R_3Si replacement of a methyl group lowers the n_O lone-pair ionization due to σ_{SiC} admixture—the silyl substituent being in β-position to the oxygen center (cf. Section IV.F)—by 1.1 eV[181]. For diazoacetic ethyl ester (IE^v_1=9.18 eV[182]), $R_3Si \rightarrow H$ exchange lowers IE^v_1 only by 0.6 eV (Table 63) and in the phosphorus[165] as well as in the arsenic ylides, the R_3Si substitution hardly changes the '$n_C^{\ominus}$ carbon lone pair' ionization at all [cf. e.g. $IE^v_1(R_3P{=}CH_2)$=6.87 eV[92]].

D. Silicon-substituted Cyclic π-Systems

First- and second-order perturbations of the benzene system (scheme 35) either by 'internal' Si → C ring center exchange (Section B.II.1) or by 'external' substitution with, e.g., R_3SiCH_2 donor groups (Sections III.B and IV.F) have been discussed in detail. In the following tables, the numerous publications[153] on PE spectra of other derivatives are organized by subdividing into mono- and polysilyl-substituted benzenes (Tables 64 and 66) as well as in other iso- and heterocyclic systems (Tables 67 and 68).

Monosilicon-substituted benzene derivatives. Tables 64 and 66 are both arranged according to classes of compounds with decreasing first vertical ionization energies (cf. also Tables 60 to 63). Silatoluene[53], generated by thermal propene elimination from the 1-allyl-1-methyl-1-silacyclohexa-2,4-diene precursor similarly to equation 3, exhibits a −0.4 eV shift relative to the π_s ionization at 8.11 eV[52] recorded for the parent silabenzene, thus further confirming the $\tilde{X}(^2B_1)M^{\cdot\oplus}$ ground-state assignment of the latter (Figure 3). With respect to the iso(valence)electronic toluene [$IE_{1,2}$=8.82 eV (π_s) and 9.36 eV $(\pi_{as})^2$], the first (π_s) ionization is lowered by 1.1 eV, whereas the second (π_{as}) one remains almost constant as expected from the nodal plane through the substitution center (scheme 35).

Comparison of the first two ionizations recorded for monosilyl benzene derivatives (scheme 65) with those of benzene [$\pi(e_{1g})$=9.25 eV[2]] or toluene [$\pi_s(a')$=8.82 eV and $\pi_{as}(a'')$=9.36 eV[2]] reveals that the H_3Si substituent exerts a small acceptor effect. The donation of alkylated ligands increases in the order $R_3Si < R_3SiCH_2 \sim R_3SiR_2Si$ (scheme

M ($R=CH_3$)		References	IE^v_n (eV)
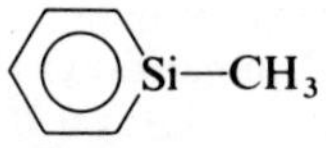		53	7.7 (π_s), 9.15 (π_{as})
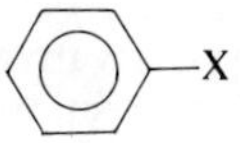	$X=SiH_3$	184	9.18 (π_s), 9.43 (π_{as})
	SiR_3	2, 185, 186	9.05 (π_s), 9.30 (π_{as})
	CH_2SiR_3	127, 185	8.42 (π_s), 9.08 (π_{as})
	SiR_2SiR_3	127	8.35 (σ_{SiSi}), 9.07 (π_{as})
	SiF_3	187	9.85 (π_{as}), 10.23 (π_s)
	SiF_2R		9.55 (π_s), 9.75 (π_{as})
	$SiCl_3$		9.55 (π_s), 9.85 (π_{as})
	$SiCl_2R$		9.52 (π_s), 9.75 (π_{as})
	$SiClR_2$		9.3 (π_s), 9.7 (π_{as})
$(C_6H_5)_nSiH_{4-n}$	$n=2$	188	9.23 (π)
	3		9.13 (π)
	4	189	8.96 (π)
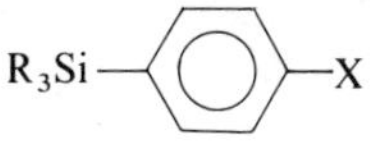	$X=NO_2$	190	9.80 (n_s)
	Cl		9.03 (n_s)
	F		9.0 (n_s)
	Br		8.67 (n_s)
	OCH_3		8.47 (n_s)
	SCH_3	191	7.93 (n_s)
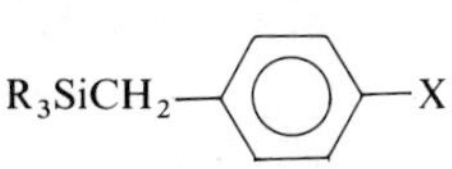	$X=SCH_3$	191	7.72 (n_s)
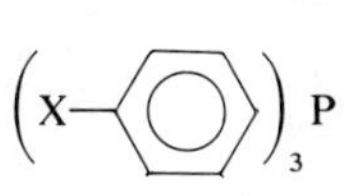	$X=H$	192	7.92 (n_p); 9.2 (π)
	SiR_3		7.67 (n_p); 8.84, 9.02 (π)

(64)

65). As concerns the disilane derivative, the symmetry-governed $\pi_s + \sigma_{SiSi}$ splitting[127] not only yields a third ionization at 10.03 eV, but both centers of gravity of the starting levels, $(8.69 + 2 \times 9.25)/3 = 9.1$ eV, as well as the resulting ones, $(8.35 + 9.07 + 10.03)/3 = 9.1$ eV, are almost identical and thus exclude any other dominant contribution to these $\pi^{\cdot\oplus}$ states. On the other hand, SiF_3 as well as $SiCl_3$ act as strong acceptors (Table 64). The polyphenylsilanes $(C_6H_5)_nSiH_{4-n}$ neither display ΔIE^v_n shifts relative to the monophenylsilane nor substantial splittings like in the analogous carbon compounds[188, 189]: the broad low-energy PE band, under which many π-ionizations overlap, indicates

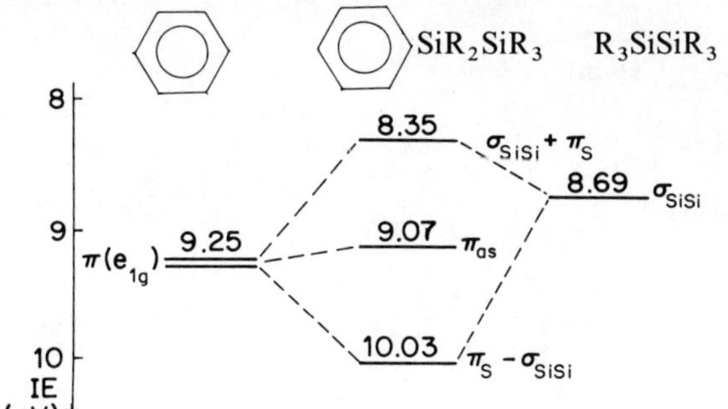

(65)

negligible phenyl/phenyl interactions. The most effective π_s-perturbation (scheme 35) in $R_3SiC_6H_4X$ derivatives causes shifts from 9.80 eV (NO_2)[190] to 7.93 eV (SCH_3)[191], still lowered by 0.2 eV on exchange of $R_3Si \rightarrow R_3SiCH_2$ groups. The identical values for X = H and X = F are expected by the 'perfluoro effect'[30, 37], due to which σ-acceptor and π-donor interactions of F substituents with planar π-systems often cancel numerically. For the finally listed triphenylphosphanes (Table 64), the n_P ionization energy sinks by 0.25 eV on threefold *para*-SiR_3 substitution[192].

Polysilyl-substituted benzenes. The PE spectra of prototype benzene derivatives with several powerful R_3SiCH_2 donor substituents have been discussed before (Sections III.B and IV.F). Most of the additional compounds listed in table 66 were also synthesized to detect those with $IE_1^v < 8$ eV, from which radical cations can be prepared adiabatically in solution using the selective one-electron oxidizing reagent $AlCl_3/H_2CCl_2$ (Figures 9, 10 and 11; cf. also Table 60).

M ($R = CH_3$)		References	IE_n^v (eV)	
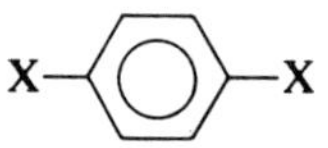	$X = SiR_3$ CH_2SiR_3	193 93	8.7 (π_s) 7.86 (π_s)	9.0 (π_{as}) 8.76 (π_{as})
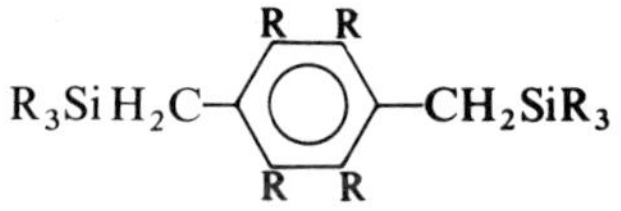		93	7.25 (π_{as}),	7.65 (π_s)
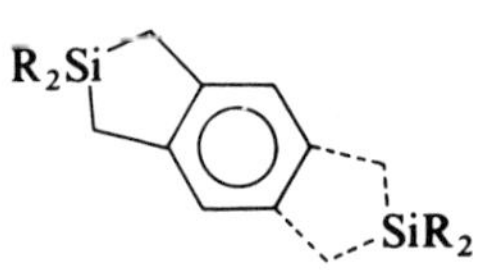		194 93	8.41 (π_{as}) 7.80 (π_{as}),	 8.30 (π_s)
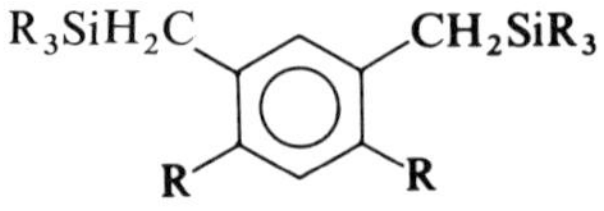		93	7.95 8.45	7.70 (π_s) 8.35 (π_{as})
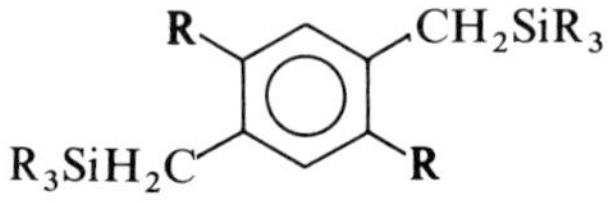				
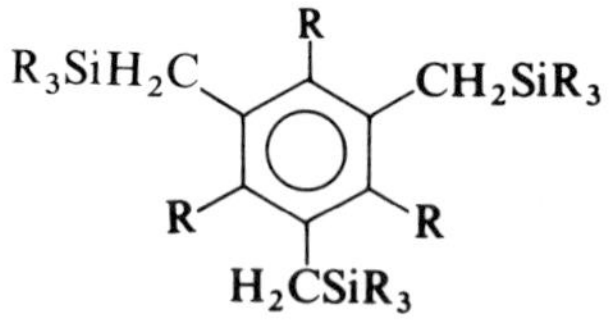		93	7.40 (e)	
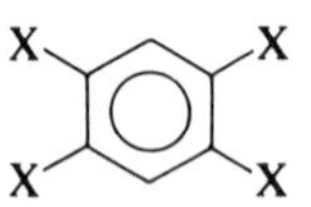	$X = SiR_3$ CH_2SiR_3	195 93	8.30 (π_{as}) 7.10 (π_{as}),	9.0 (π_s) 7.75 (π_s)

(66)

1,4-Disubstituted benzenes with large π-coefficients $(c_{J\mu}^{HMO})^2 = 0.33$ at each substitution center emphasize the different donor properties of α and β SiR_3 groups: relative to 1,4-di(*t*-butyl)benzene with lowest ionizations at 8.40 eV (π_s) and 8.90 eV (π_{as}), the π-perturbation by R_3Si ligands is somewhat smaller, whereas R_3SiCH_2 substituents push the π_s-ionization by −0.54 eV down to 7.86 eV (Table 66). Additional fourfold methyl substitution changes the $M^{\cdot\oplus}$ state sequence and lowers the value further to 7.25 eV. Five-membered ring formation hardly changes much. Altogether, the PE data of mixed H_2C/R_3SiCH_2-substituted compounds can be rationalized—including changing $M^{\cdot\oplus}$ state sequences—by the perturbation model introduced in scheme 38. Finally, the 1,2,3,5-tetrasubstituted derivatives reconfirm the different perturbations already observed with 1,4-disubstituted derivatives: Relative to benzene $[IE_1^v(^2E_{1g}) = 9.25\ eV$[2,5]], the π_{as} ionization is shifted −0.95 eV by 4 R_3Si, but −2.1 eV(!) by 4 R_3SiCH_2 groups (cf. scheme 40).

Silicon-substituted iso- and heterocyclic π-systems. The PE spectroscopic investigations detected by a literature search[153] are listed in Tables 67 and 68. 3,3-Dichloro-3*H*-3-benzosilepin can be used as $SiCl_2$ precursor, which thermally splits off naphthalene (Figure 6). For 1,4-bis(trimethylsilyl)naphthalene, in which a second ring is added to the benzene, relative to the ionizations of the parent $C_{10}H_8$ at 8.15 eV and 8.88 eV[193], rather small changes are caused by $2R_3Si$ groups in 1,4-positions. 1,4-Bis(trimethylsilylmethyl)biphenyl with $IE_1 \leq 8$ eV can be oxidized by $AlCl_3/H_2CCl_2$ to its radical cation[196]. Spiroconjugation between Si-connected fluorenyl moieties remains small.

(67)

M ($R = CH_3$)	References	IE_n^v (eV)
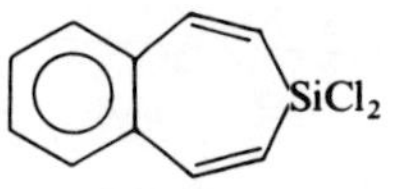	48	8.8 (π)
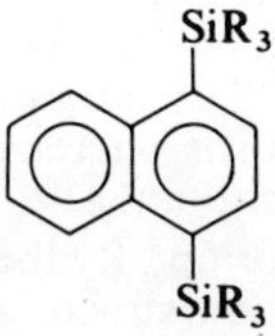	193	7.85, 8.6 (π)
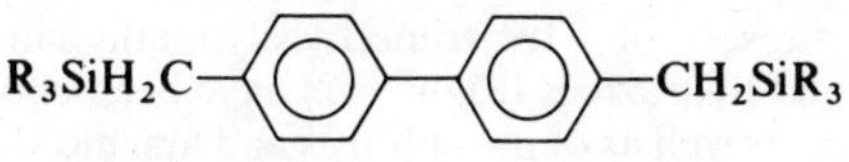	196	7.60 (π)
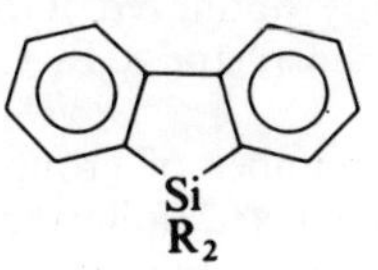 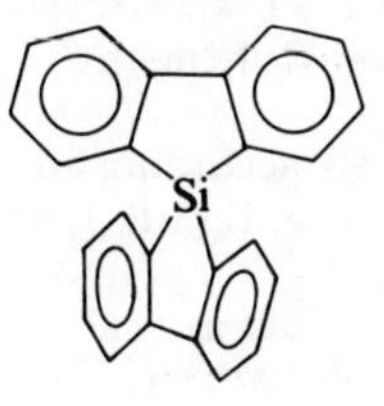	197	7.9 → 7.85, 8.0; 8.8 → 8.85, 9.1

M (R = CH_3)		References	IE_n^v (eV)
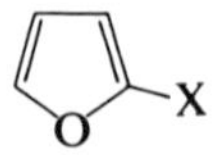	X = SiH_3 $SiHR_2$ SiR_3	197	8.92, 10.40 8.62, 10.22 8.53, 10.07 } (π)
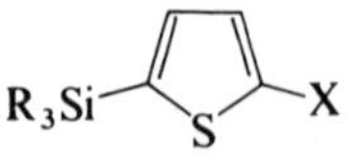	X = H SiR_3	198 199	8.68, 9.22 8.20, 8.81 } (π)
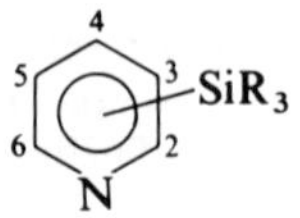	2 4 2, 6	200	8.90, 9.30, 8.50 } (n_N), 9.30, 9.40, 8.95 } (π)
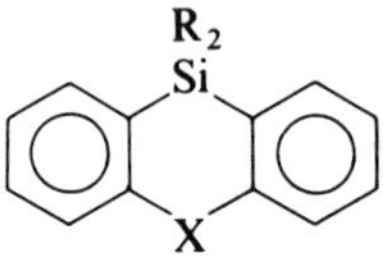	X = O S NR	201	7.92 7.81 7.17 } (π_7)

(68)

The PE spectra of silicon-substituted furans, thiophens, pyridines and 9,10-hetero-anthracenes confirm the expected substituent perturbations; for additional derivatives see the literature cited. The $M^{\cdot\oplus}$ state sequence for R_3Si-substituted pyridine derivatives, $\tilde{X}(n_N) < \tilde{A}(\pi)$, has some implication also for charge-transfer excitations of pyridine/acceptor complexes[200].

E. Silyl Derivatives of Group 15 Element Compounds

A variety of silyl-substituted nitrogen compounds as well as some derivatives of the higher group 15 elements P, As and Sb have been investigated PE spectroscopically[153]. Examples already discussed within the context of previous sections comprise phenyl silaisocyanide H_5C_6–N≡Si and its thermal generation by N_2 split-off from triazidophenyl-silane $H_5C_6Si(N_3)_3$ (Figure 7), the pyrolysis of 5-methyl-2-trimethylsilyltetrazole to the polar *N*-trimethylsilyl nitrile imine $H_3C–C≡\overset{\oplus}{N}–\overset{\ominus}{N}–SiR_3$ (Figure 8) or the PE and UV spectra of the blue bis(trimethylsilyl)diimine including evidence for np_{Si} Rydberg excitations (Section IV.G.3). To illustrate generation, ESR characterization, structural changes and molecular dynamics of silyl-substituted radical cations generated adiabatically in $AlCl_3/H_2CCl_2$ solution (Figure 9), those of tris(trimethylsilylmethyl)amine $^{\cdot\oplus}N(CH_2SiR_3)_3$ (scheme 27) as well as silylated hydrazines (Figure 12) have been chosen (Section III). Silyl derivatives of diazomethane as well as of phosphorus and arsenic ylides have already been listed in Table 63 and the PE spectroscopically optimized R_3SiCl elimination from $(R_3Si)_2C$=PCl to yield trimethylsilylmethylidine phosphane $R_3Si–C≡P$ is displayed in Figure 25.

The subsequent overview will be organized according to the number of group 15 element centers in open-chain silyl derivatives, i.e. $E_x(SiR_3)_n$ with $x = 1$ or 2, followed by silyl-substituted heterocycles and miscellaneous compounds including pseudohalogen derivatives $R_3Si–X$ with X = NCO, NCS and NSO as well as metal silylamides.

Silyl amines and analogous P, As and Sb derivatives. The α and β silyl-substituted ammonia derivatives are listed in Table 69 in order of decreasing first ionization energies;

M ($R = CH_3$)	References	$IE_n(n_N)$ (eV)
$(H_3Si)_3N$	202	9.7
$F_3Si{-}NR_2$	203	9.60
$Cl_3Si{-}NR_2$	204	9.30
$(H_3Si)_2NR$	202	9.2
$R_3SiH_2C{-}NH_2$	102	9.07
$(R_3Si)_2NH$	165, 205	8.66
$(R_3Si)_3N$	165	8.60
$R_3SiHRC{-}NH{-}R$	102	8.55
$H_3Si{-}NR_2$	202	8.5
$R_3SiH_2C{-}NH{-}CH_2R$	102	8.46
$R_3Si{-}NH{-}CR_3$	206	8.41
$(R_3SiH_2C)_2NH$	102	8.36
$(R_3Si)_2NR$	207	8.21
$R_3SiH_2C{-}NR_2$	102	8.20
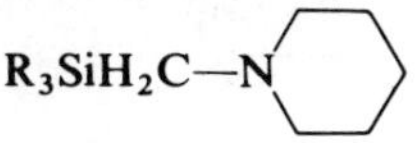	102	8.18
$R_3Si{-}NR_2$	207	8.03
$R_3SiH_2C{-}N(CH_2R)_2$	102	7.93
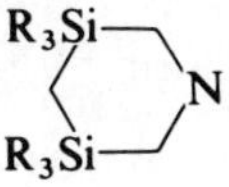	102	7.90
$(R_3SiH_2C)_2NR$	102	7.86
$(R_3SiH_2C)_2NCH_2R$	102	7.86
$(R_3SiH_2C)_3N$	102	7.66
$Cl_2Si(NR_2)_2$	204	8.81
$Si(NR_2)_4$	204, 208	8.39 (. . . 8.52)
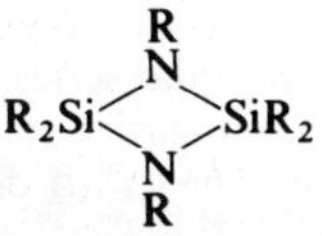	205	8.22 (b_{3u}) 8.55 (b_{2g})

(69)

the resulting radical-cation ground states are each assigned n_N referring to their predominant nitrogen lone-pair component in the distribution of the positive charge. The first ionization energies $IE_1(n_N)$ stretching from 9.7 eV for $N(SiH_3)_3$ over a remarkable 2 eV range down to 7.66 eV for $N(CH_2SiR_3)_3$, however, demonstrate a strong substituent dependence on the charge delocalization. With respect to the parent systems NH_3 with $IE_1(n_N) = 10.85$ eV[5] or $N(CH_3)_3$ with $IE_1(n_N) = 8.44$ eV[5], no additivity is observed in series like $(R_3Si)_nNR'_{3-n}$ with both R and R′ either H or CH_3 (scheme 70: -----), although for the silylated derivatives $(H_3Si)_nN(CH_3)_{3-n}$ with $n \geqslant 1$ an almost linear increase results (scheme 70: ———). Within the series of trimethylsilylmethyl-substituted amines, the first ionization energies decrease in the substituent sequence $R_3SiCH_2 > RH_2C > H_3C > H$, each additively (scheme 70: ———) and the individual substituent perturbations can

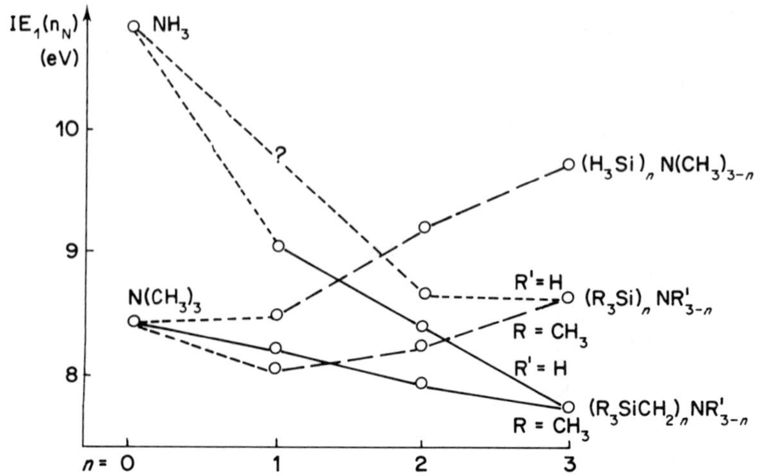

(70)

be parametrized by a least-squares fit to the experimental data in Table 69 by the expression[102]

$$\mathrm{IE}_1^{\mathrm{calc}}\ (\mathrm{eV}) = \overline{\mathrm{IE}}_R - \sum_R \Delta\mathrm{IE}_R = 9.71 - 0.68\,\mathrm{n}_{\mathrm{CH_2SiR_3}} - 0.50\,\mathrm{n}_{\mathrm{CH_2CH_3}} - 0.43\,\mathrm{n}_{\mathrm{CH_3}} \quad (71)$$

yielding a linear regression $\mathrm{IE}^{\mathrm{calc}}$ (eV) = 0.329 + 0.961 $\mathrm{IE}^{\mathrm{obs}}$ with a standard deviation of only 0.07 eV[102]. The seemingly, partly irregular substituent dependence of the lone-pair ionization energies of silylated amines can be rationalized by referring to known structural differences[209, 210]: whereas NH_3, $N(CH_3)_3$ and presumably also all R_3SiCH_2 derivatives are trigonal pyramidal, for $N(SiH_3)_3$, $H_3CN(SiH_3)_2$ as well as for $N(SiR_3)_3$ planar NX_3 skeletons have been determined. The increase of the $\Delta\mathrm{IE}_1(\mathrm{n_N})$ values within the planar or nearly planar series of silyl amines $(R_3Si)_nN(CH_3)_{3-n}$ from $n=1$ to $n=3$ (scheme 70:———) may then be discussed in terms of decreasing 'hyperconjugation' contributions $CH_3 > SiR_3 > SiH_3$ and the decrease in the pyramidal series of silylamines $(R_3SiCH_2)_nNR'_{3-n}$ in terms of increasing 'hyperconjugative' contributions $CH_3 < CH_2SiR_3$. As expected, the end point is reached by 7.66 eV (!) determined for

M (R = CH_3)	References	$\mathrm{IE}_1(\mathrm{n_E})$ (eV)
$F_3Si{-}PH_2$	203	11.06
$F_3Si{-}AsH_2$	203	10.90
$H_3Si{-}PH_2$	202	9.9
$F_3Si{-}AsR_2$	203	9.4
$(H_3Si)_3P$	202	9.3
$(H_3Si)_3As$	202	9.3
$Cl_3Si{-}AsR_2$	203	9.2
$Cl_3Si{-}PR_2$	203	9.1
$R_3Si{-}PH_2$	211	9.1
$(R_3Si)_2PH$	211	8.35
$(R_3Si)_3As$	212	(8.3)
$(R_3Si)_3P$	211, 212	7.9
$(R_3Si)_3Sb$	212	7.6

(72)

$N(CH_2SiR_3)_3$, further substantiating the tremendous electron-donor effect of trimethylsilylmethyl groups as discussed repeatedly for other classes of compounds (cf. e.g. Section III.B or Section IV.F).

Silyl derivatives of PH_3, AsH_3 and SbH_3 have been less systematically investigated PE spectroscopically (Table 72) than those of NH_3 (Table 69). Again, F_3Si groups are strong acceptor substituents: relative to $H_3Si{-}PH_2$[202], the n_p-type ionization is increased in $F_3Si{-}PH_2$[207] by almost 1.2 eV (Table 72). Threefold SiH_3 substitution to $(H_3Si)_3P$ lowers the IE_1 value by only 0.6 eV to 9.3 eV[202], whereas the value $IE_1(n_P) = 7.9$ eV observed for $(R_3Si)_3P$ demonstrates a much larger effect of more bulky ligands on the $M^{\cdot\oplus}$ charge delocalization. Differences between iso(valence)electronic P and As derivatives are rather small.

(73)

ⓝ	M (R = CH_3)		References	IE_n^v (n_E) (eV)
②	$R_3Si{-}N{=}N{-}X$	X = SiR_3	149, 213	7.1 (n_N^+)
		= CR_3	149	7.6 (n_N^+)
		= C_6H_5	149	7.85 (n_N^+)
		= CH_3	213	8.02
	$(RO)_3Si{-}N{=}N{-}CR_3$		213	7.88
	$R_3Si{-}N{=}C{=}N{-}X$	= SiR_3	214	9.1
		= CR_3		8.9
	$R_3Si{-}N{=}PR_3$		165	8.30 (n_N, π_{PN})
	$R_3P{=}CH{-}R_2Si{-}R_2Si{-}N{=}PR_3$		165	6.18 (π_{PC})
	$(R_3Si)_2N{-}N(SiR_3)_2$		99	7.95 (n_N)
	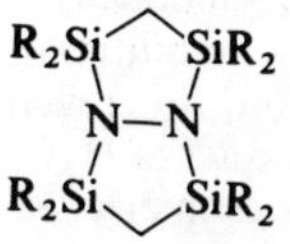		99	7.15 (n_N)
	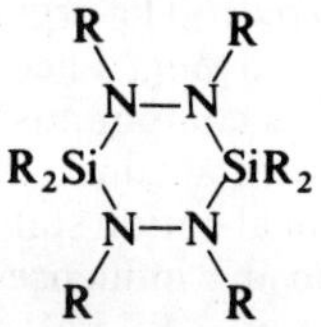		99	7.5 . . . 7.9 (n_N)
	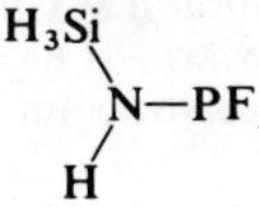		215	11.0 (n_P)
③	$X_3Si{-}NNN$	X = H	216	10.3 ($\pi_{N_3}^*$); 11.0 ($\sigma_{N_3}^*$)
		= CH_3	15, 22, 216	9.6 ($\pi_{N_3}^*$); 10.3 ($\sigma_{N_3}^*$)
		= C_6H_5	217	9.2 ($\pi_{C_6H_5}$)
	$R_2Si\ (NNN)_2$		217	10.00; 10.35 ($\pi_{N_3}^*$)
	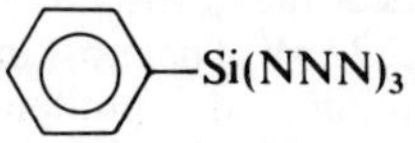		217	9.3 ($\pi_{C_6H_5}$)
④	$(R_3Si)_2N{-}N{=}N{-}N(SiR_3)_2$		213	7.4 (n_N)

Silyl derivatives of open-chain compounds with ≥ 2 N or P centers. According to the literature search[153] the molecules in Table 73 qualify for this subsection, i.e. silyl-substituted azo, carbodiimide, phosphorus imide, hydrazine, azide and tetrazene compounds. Most of them are of considerable size: for example, tetrakis(trimethylsilyl)hydrazine $C_{12}H_{36}N_2Si_4$ contains 110 valence electrons—74 of n_{PE} and 1_{SH} type[8]— and therefore most of the 37 PES bands[8] expected in the He(I) measurement range do overlap (cf. e.g. Section IV.F). Nevertheless, valuable information can be deduced from isolated low-energy ionizations: to continue with $(R_3Si)_2N$–$N(SiR_3)_2$, if the observed splitting between the n_N^- and n_N^+ nitrogen lone-pair dominated $M^{\cdot\oplus}$ states is inserted into the empirical relation $\Delta IE_{1,2}(n_N/n_N) = A \cos\omega + B$, which has been parametrized using structurally fixed derivatives like five- or six-membered rings (Table 73)[99], a dihedral angle $\omega \sim 86°$ is estimated between the twisted $(R_3Si)_2N$ molecular halves[99]. This value almost coincides with the experimentally determined $\omega = 82.5°$ for $(H_3Si)_2N$–$N(SiH_3)_2$[218]. Referring in addition to the adiabatic generation of radical cations in $AlCl_3/H_2CCl_2$ solution (Section III), due to their low $IE_1 \leq 8$ eV all hydrazine derivatives (Table 73) could be oxidized and the tremendous structural changes predicted by MNDO hypersurface calculations (Figure 12) verified by ESR[16, 99].

For the amazingly blue R_3Si–N=N–SiR_3, the low first ionization energy has been used to assign np_{Si} Rydberg states in its UV spectrum[149] (cf. Section IV.G.3). Another more general remark concerns the unusual thermal stability of trimethylsilyl azide, which has been repeatedly and erroneously attributed to fictitious ($p_N \rightarrow d_{Si}$) back-bonding[219] (cf. Section IV.C). In fact, due to the lower effective Si nuclear charge, the R_3Si group acts predominantly as σ-donor towards the N_3 subunit [cf. $(H_3C)_3C$–N_3: IE_1 $(\pi^*_{N_3}) = 9.4$ eV and $IE_2(\sigma^*_{N_3}) = 10.7$ eV[15]], and thus increases the barrier for the kinetically important $R_3SiN \cdots NN$ extrusion. In addition, the relative thermodynamic stabilities of the azide and the imine, resulting on decomposition from 1,2-R migration, are reversed[15]. As pointed out already in Section II.C, H_5C_6–N≡Si, the pyrolysis product of H_5C_6–$Si(N_3)_3$, is also stabilized by about 400 kJ mol^{-1} relative to the isomeric silacyanide H_5C_6–Si≡N[15].

Silyl-substituted group 15 heterocycles and miscellaneous compounds. The organosilicon derivatives of the heterocyclic ring systems listed in Table 74 are arranged according to characteristic links from B to O. All of them are individual classes of compounds lacking a common denominator and therefore provide a chance to warn against unjustified overgeneralizations concerning substituent effects in different radical-cation states. So far, R_3Si groups at N centers have exerted, e.g., the following opposing influences on first vertical (n_N) ionization energies: that of planar $N(SiR_3)_3$ is higher by $+0.15$ eV than that of pyramidal $N(CH_3)_3$ (Table 70), whereas those of the isostructural azo compounds R_3Si–N=N–SiR_3 (Table 73) and H_3C–N=N–CH_3 (scheme 47) differ by -1.88 eV in the opposite direction. The following selected comments should therefore be complemented from the literature cited in Table 74.

The permethylated boron heterocycles $R_5B_2N_3$ and $R_4B_2N_2S$ are ionized at 7.47 eV and 8.00 eV, respectively, and thus the R_3Si groups exert no, or almost negligible, effects[220]. The opposite is found for the extremely electron-rich 1,4-diazine derivatives: compared to the iso(valence)electronic cyclohexadiene (Table 60: $IE_1 = 7.70$ eV), the first ionization is lowered by 1.54 eV and, relative to the mono aza ring system (Table 74: $IE_1^v = 8.80$ eV), even by 2.66 eV. Additional permethylation reduces the $n_N^+(b_{3u})/n_N^-(b_{1g})$ split, presumably due to steric overcrowding[221], from 3.34 eV to 2.22 eV. The most interesting PE spectroscopic result from the careful investigation[222] of mono and di(phthalocyanine) bicapped linear (OSiO) chains is the π-splitting of 0.3 eV. The R_3Si-substituted SiN_4 heterocycles containing N_4 or N_5 subunits and exhibiting $IE_1^v < 8$ eV have, as expected, been oxidized to their radical cations in $AlCl_3/H_2CCl_2$ solution[223] (cf. Section III). The

(E) M(R=CH_3)	References	IE_n^v (eV)
(B) 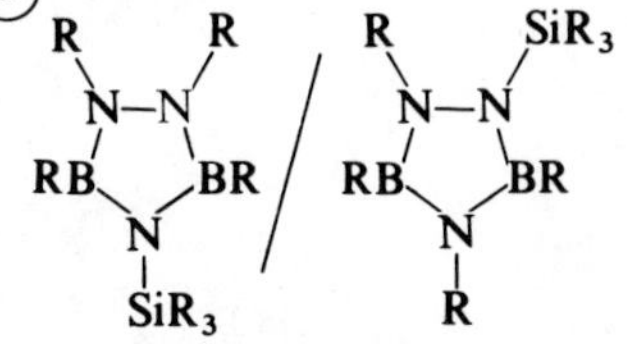	220	7.48/7.56 (π_3, a_2)
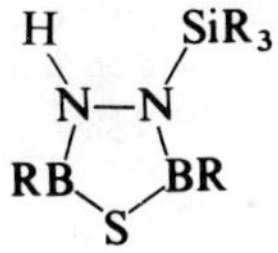	220	8.25 (π_3, a_2)
(C) 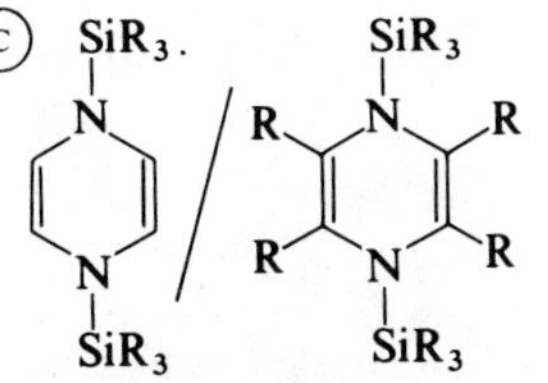	221	6.16/6.48 (π, b_{3u}) 9.5 /8.10 (π, b_{2g}) 9.5 /8.7 (π, b_{1g})
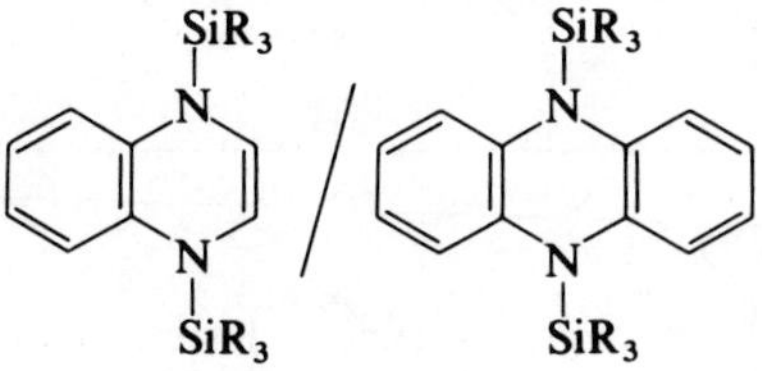	221	6.50/6.5 (π, b_1/b_{3u}) 8.25/7.9 (π, a_2/b_{1g}) 9.00/8.9 (π, b_1/a_u)
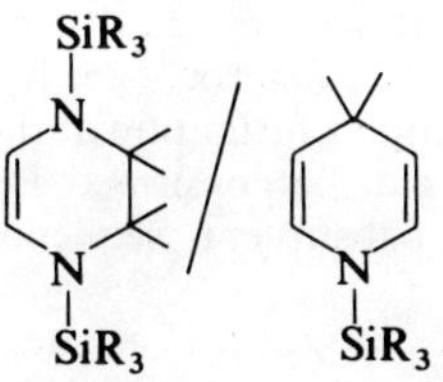	221	6.60/7.30 (π, b_1) 8.80/9.55 (π, a_2)
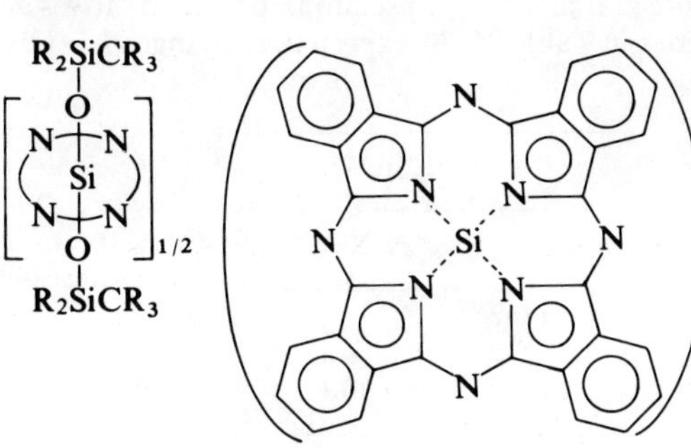	222	6.46/6.11 (π, $2a_{1u}$) /6.41

(74)

(E) M(R=CH_3)		References	IE^v_n (eV)
(N) 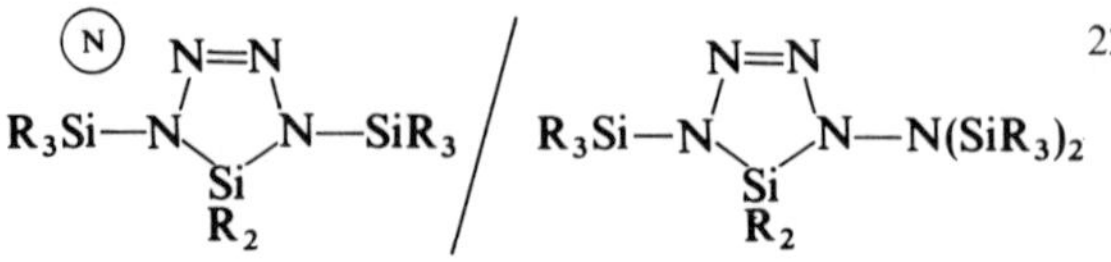		223	7.97/7.65 (n_N)
(P) 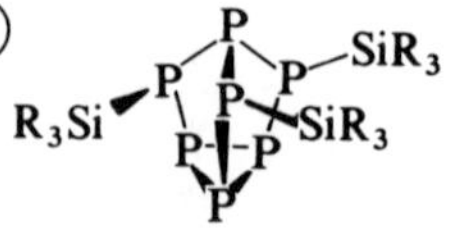		211	7.58, 8.18, 8.56, 9.38 (n_P)
		224	8.0, 8.5, 9.1 (n_P)
(O) 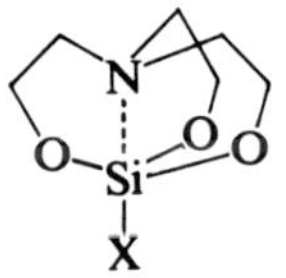	X=H	225	10.1 ($n_N \rightarrow Si$)
	CH_3		8.7
	OCH_2R		10.6
	[$N(CH_2CH_2OH)_3$]		[8.7]

(74)

PE spectrum of the tris(trimethylsilyl)-substituted P_7 main group element cluster (Figure 26) can be tentatively assigned[211], despite its 110 valence electrons, by $M^{\cdot\oplus}$ state comparison with P_4S_3—iso(valence)electronic to the P_7H_3 parent system—guided by MNDO calculations based on the known structure[226]. The $P_4(SiR_2)_3$ heterocluster has been compared to $P_7(CH_3)_3$[224], and the differences in carbonyl complex formation at the basal P_3 unit rationalized. The N · · · Si bonding in silatranes can be confirmed PE spectroscopically[225], with the interaction increasing in the substituent sequence $CH_3 < H < OCH_2R$ (Table 74).

To conclude the PE spectroscopic discussion of silyl derivatives of group 15 element compounds as retrieved from the literature search[153], their pseudohalogen derivatives are listed in Table 75 and the metal silylamides in Table 76. As expected, exchange H → CH_3

M	References	IE^v_1 (eV) X=H	IE^v_1 (eV) X=CH_3
X_3Si–NCO	216	11.10	10.3
X_3Si–NCS	216	9.54	9.3
X_3Si–NSO	227	11.5	10.3

(75)

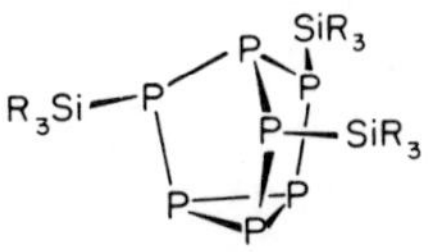

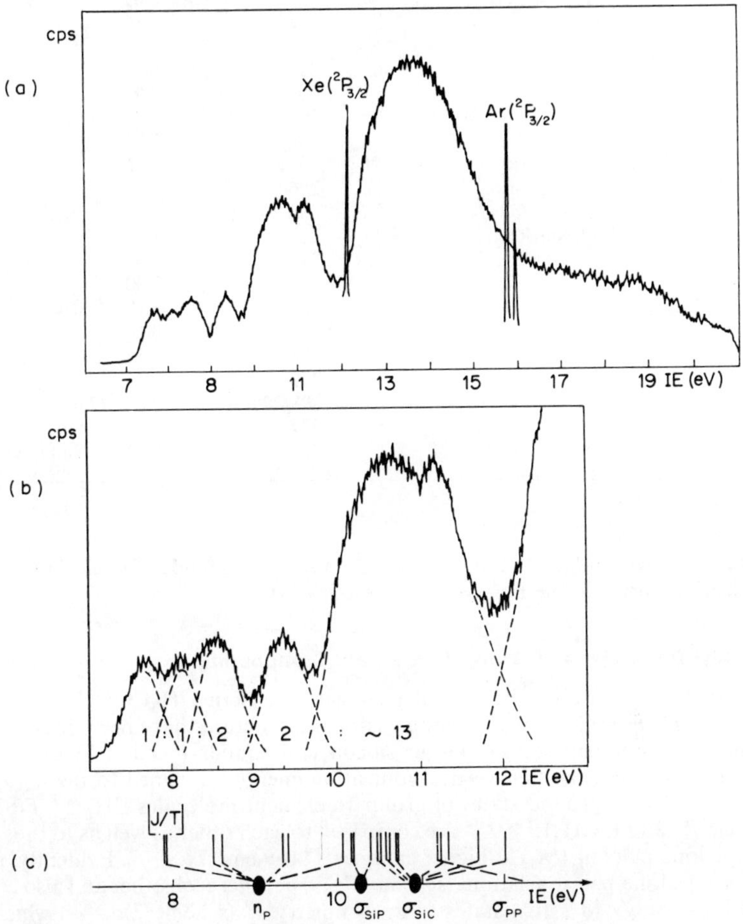

FIGURE 26. He(I) PE spectrum of tris(trimethylsilyl)-heptaphosphanortricyclene (a) with expanded scale 7–13 eV (b) showing relative intensities of deconvoluted bands and tentative assignment (c) based on $M^{\cdot\oplus}$ state comparison with iso(valence)electronic P_4S_3 as well as Koopmans' correlation with MNDO eigenvalues[211]

stabilizes the $M^{\cdot\oplus}$ ground state of linear H_3SiNCX and twice bent H_3SiNSO molecules; for bent $X_3Si{-}N_3$ (Table 73) the first ionization energy is lowered from 10.3 eV to 9.6 eV. Relative to those of the parent amines, the first ionizations $IE_1(n_N)$ of most metal amides (Table 76) are smaller. Comparison of their PES data, however, are hampered by different structures, such as the Li compound being a dimer and some of the trivalent amides either

M (R=CH$_3$)		References	IE_1^v (n_N) (eV)
Met–N(SiR$_3$)$_2$	Met=Li	228	8.46 (dimer)
	Na	228	8.1
Met(N(SiR$_3$)$_2$)$_2$	Met=Mg	228	8.87
	Zn	206	8.50
	Hg	206	8.33
	Co	228	8.08
	Pb	206	7.92
	Sn	206	7.75
	Ge	206	7.71
Met(N(SiR$_3$)(CR$_3$))$_2$	Met=Pb	206	7.26
	Sn	206	7.26
	Ge	206	7.24
Met(N(SiR$_3$)$_2$)$_3$	Met=Al	228	8.73
	Ga	229	8.39
	In	229	8.30
	Ce	228	8.19 (7.07d)
	Sc	229	8.18
	Eu	228	8.14
	Ti	229	8.14
	Cr	228/229	8.8 (7.6 d)
	Fe	229	8.7 (7.9 d)
	U	228	8.04 (5.96 f)

(76)

planar or pyramidal[228]. For Cr, Fe, Ce and U trisamides (Table 76), d- and f-type ionizations precede the n_N ones (cf. Section V.H).

F. Silyl Derivatives of Group 16 Element Compounds

The PE spectra of the iso(valence)electronic series H_3COCH_3, H_3SiOCH_3 and $H_3SiOSiH_3$ as well as their differing gas-phase structures have been discussed within the context of lone-pair ionizations in silicon compounds (Section IV.G.2). For a more general overview, the first vertical ionization energies assigned to the $n_N(b_2)$ lone-pair-dominated $M^{\cdot\oplus}$ ground states of group 16 element molecules EH_2[119], $E(SiH_3)_2$[140, 141] (Table 78) and $E(CH_3)_2$[135, 134] are compared to each other as well as to those of the EH_2 s-type lone pairs n_E^s ($^1a_1$) at higher energies[119] (scheme 77: ---). Evidently, the slopes of the π-type lone-pair ionizations (scheme 77: —) parallel each other and also correlate with the high-energy $M^{\cdot\oplus}$ (n_E^s) states of a_1 symmetry[119] (scheme 77: ---) which reflect the effective nuclear charges of the respective elements O, S, Se and Te[9]. As concerns the disilyl derivatives, as expected, no deviations due to fictitious $n_E^\pi \rightarrow d_{Si}$ back-bonding (Section IV.C) are observed. The compound sequence $EH_2 < E(SiH_3)_2 < E(CH_3)_2$ can be rationalized by increased stabilization of the $M^{\cdot\oplus}$ ground states due to increased $\oplus$charge delocalization. The same approach holds for other substituent effects listed in Table 78: $H_3Si \rightarrow R_3Si$ exchange lowers the first ionization potential and, especially, β SiR_3 groups exert tremendous shifts down to 8.03 eV for $S(CH_2SiR_3)_2$[102] or to 7.66 eV for $(R_3Si)_3C$–S–R[232] (Table 78). Towards π-systems, $OSiR_3$ substituents[194] are weaker donors than R_3SiCH_2 groups[93] (Section III.B) but stronger than CH_3[93] (Figure 18). Unusually large is the effect of a R_3SiCH_2S substituent towards the benzene π-system, shifting the first ionization potential to 7.81 eV[191] (Table 78), i.e. almost matching that of a dimethylamino group [$IE_1(H_5C_6–N(CH_3)_2) = 7.65$ eV[35]]. In conclusion, it might be

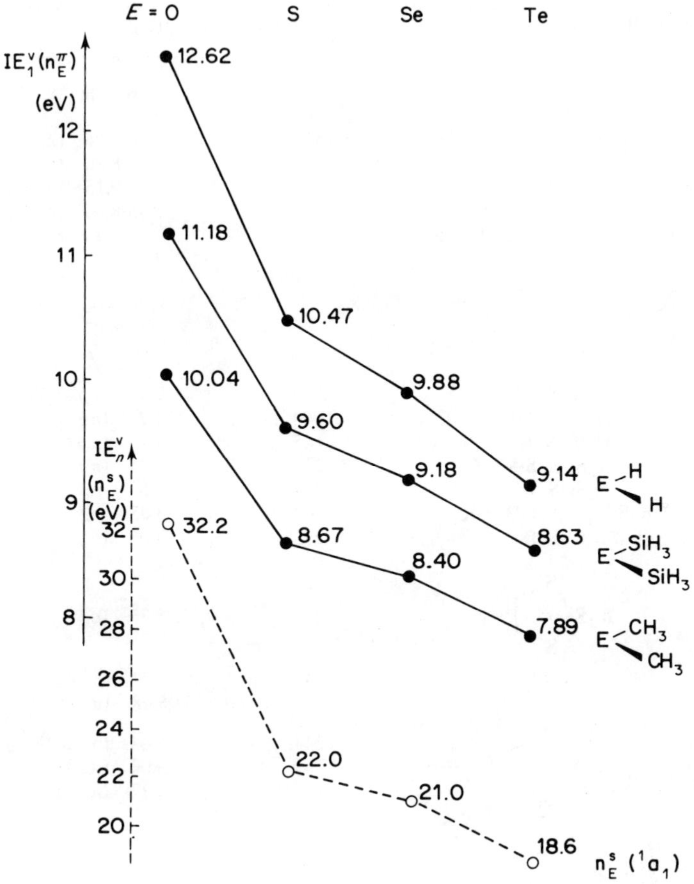

(77)

Ⓔ	M (R = CH_3)	References	IE_n^v (eV)	
Ⓞ	Cl_3Si–O–CH_2R	204	11.30	(n_O)
	H_3Si–O–SiH_3	140, 141	11.18	(n_O)
	H_3Si–O–R	141	10.61	(n_O)
	R_3Si–O–SiR_3	165	9.88	(n_O)
	$Cl_2Si(OCH_2R)_2$	204	10.78	(n_O)
	$Si(OR)_2$	159	9.5	(n_O)
	$ClSi(OCH_2R)_3$	204	10.52	(n_O)
	$Si(OCH_2R)_4$	204, 225	9.77	(n_O)

(78)

Ⓔ M (R = CH_3)	References	IE_n^v(eV)
$R_3SiO{-}N{=}N{-}OSiR_3$	213	9.6 (π_{NO})
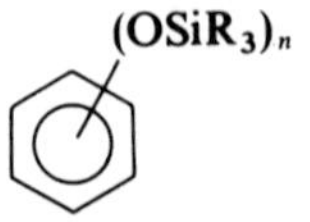 1, 2	194	8.30/8.95 (π_{as}/π_s)
1, 3		8.33/8.80 (π_{as}/π_s)
1, 4		8.00/9.15 (π_s/π_{as})
1, 2, 4, 5		7.45/8.40 (π_{as}/π_s)
1, 2, 3, 4, 5, 6		7.40 (π_s/π_{as})
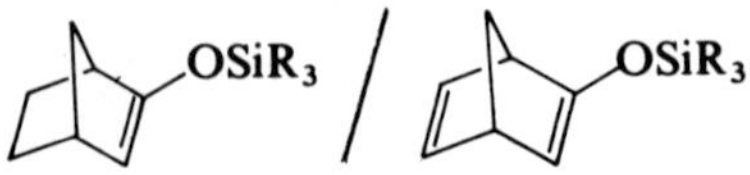		8.09/8.06 (π) 9.07
Ⓢ $H_3Si{-}S{-}H$	140	9.97 (n_s)
$H_3Si{-}S{-}SiH_3$	140, 145	9.7 (n_s)
$H_3Si{-}S{-}R$	145	9.1 (n_s)
$R_3Si{-}S{-}SiR_3$	231	8.74 (n_s)
$R_3Si{-}S{-}R$	231	8.69 (n_s)
$R_3SiH_2C{-}S{-}CH_2SiR_3$	102, 232	8.03 (n_s)
$(R_3Si)_3C{-}S{-}R$	232	7.66 (n_s)
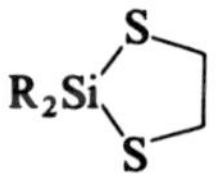	233	$\overline{8.95}$ (n_s)
$R_2Si(SR)_2$	233	8.72 (n_s^+) 8.98 (n_s^-)
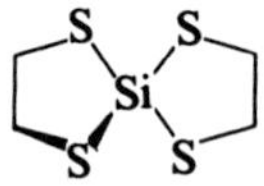	234	8.64 (n_s, a_2) 9.00 (n_s, b_2) 9.37 (n_s, e)
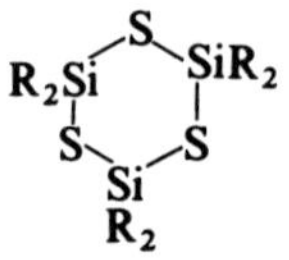	75	8.4 9.0
$(R_3Si)_2N{-}S{-}S{-}N(SiR_3)_2$	232	8.25 ($n_s - n_N$) 8.65 ($n_s + n_N$)
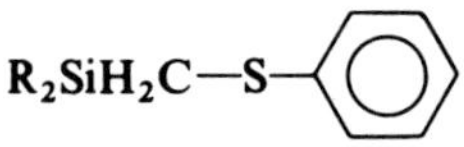	191	7.81 ($n_s - \pi_s$)
Ⓢⓔ $H_3Si{-}Se{-}SiH_3$	140	9.18 (n_{Se})
Ⓣⓔ $H_3Si{-}Te{-}SiH_3$	140	8.63 (n_{Te})

(78)

pointed out that two R_3Si ligands stabilize hyponitrous acid HON=NOH to an extent which allows vaporization of the respective diester without decomposition to record its PE spectrum[213].

G. Silyl and Organo Silyl Halides

Throughout this review again and again, PE spectra of silicon and organosilicon halides have been presented, assigned and discussed. Thus it has been demonstrated how the gas-phase preparation of $SiCl_2$ can be optimized PE spectroscopically from both $Cl_3SiSiCl_3$ and Si via $SiCl_4$ (Figure 5) or from dichlorobenzsilepin (Figure 6), and how its ionization pattern (scheme 8) is assigned by $M^{\cdot\oplus}$ state comparison, e.g. with SiF_2, $SiBr_2$ and SiI_2 (scheme 13). The PE spectra of $ISiH_3$ and ICH_3 have been selected (Figure 13) to illustrate accessible information on radical cation states like vibrational fine structures, Jahn/Teller splitting or spin/orbit coupling. Those of the iso(valence)electronic molecules $ClSiH_3$ and $ClCH_3$ served to introduce hyperconjugative perturbation models and to

M ($R=CH_3$)	References	IE_n^v (eV)
H_3SiF	104, 109, 236	12.6 (π_{SiH}, 2e), 16.2 (n_F^π, 1e)
H_3SiCl	104, 109	11.6 (n_{Cl}, 2e), 13.7 (π_{SiH}, 1e)
H_3SiBr	104, 109	10.9, 11.1 (n_{Br}^π {$2e_{3/2}$, $2e_{1/2}$}), 13.3 (π_{SiH}, 1e)
H_3SiI	104	9.8, 10.3 (n_I^π {$2e_{3/2}$, $2e_{1/2}$}), 12.8 (π_{SiH}, 1e)
F_3SiH	167, 236	14.5 (σ_{SiH}, $3a_1$), 15.9, 16.4 (n_F^π {$1a_2$, 3e})
F_3SiCH_3	167	14.8 (σ_{CH}, 4e), 15.6, 16.3 (n_F^π {$1a_2$, 3e})
F_3SiCl	167	13.4 (n_{Cl}^π, 4e), 16.4, 16.7 (n_F^π {$1a_2$, 3e}) (79)
SiI_4	67, 237	9.4 −11.2 (n_I)
Si_2Br_6	67	10.1 (σ_{SiSi}), 10.6−12.2 (n_{Br})
Si_4Br_8	67	8.8, 10.0 (σ_{SiSi}), 10.3−12.0 (n_{Br})
R_3SiF	158	11.0 (σ_{SiC}), 15.5−16.0 (n_F)
R_3SiCl	37, 163, 204, 238	10.5, 10.80 {(σ_{SiC}), 11.9, 12.2 {(n_{Cl})
R_3SiBr	239	10.2, 10.4 {(n_{Br}), 11.4 (σ_{SiBr})
	159	9.9, 10.3 {(σ_{Ring}), 11.8 (n_{Cl})
$(R_3C)R_2SiCl$	163	9.8, 10.7 {(σ_{SiC}), 12.1 (n_{Cl})
R_3SiH_2CCl	240	10.2 (σ_{SiC}), 11.0 (n_{Cl})

clearly explain their parametrization using vertical ionization energies (Figure 19), and those of SiF_4, $SiCl_4$ and CF_4 (Figure 20) to fully discuss the T_d symmetry-adapted $M^{\cdot\oplus}$ state lone-pair n_{Hal} and σ_{SiHal} bond combinations (scheme 44). The real-time PES gas analysis in a flow system allowed one both to control the thermal elimination of R_3SiCl from $(R_3Si)_2C{=}PCl$ and to identify the short-lived $R_3Si{-}C{\equiv}P$ formed (Figure 25). In addition, numerous Si halogen derivatives and their characteristic ionization energies have already been included in Tables 56, 57, 59–62, 64, 67, 69, 72 and 78. Last but not least, we note the review on PE spectra of halogen compounds within this same series[37], which includes a chapter on halogen derivatives of silicon, germanium and tin covering the literature up to 1982; only some references[153] are compiled in Table 79.

Referring to the radical cation states of HCl presented in scheme 31, singly coordinated halogen centers formally possess one lone pair of s-character—ionized at least for F and Cl outside the He(I) measurement region >21.21 eV[37]—and two of p-type n_{Hal} at lower energy, subject to spin/orbit coupling on electron expulsion (Figure 13). The latter are delocalized, e.g. in $H_3SiCl^{\cdot\oplus}$ (Figure 19)—as can be rationalized by hyperconjugative interaction without any necessity to include fictitious d-orbitals (Section IV.C). Their relative position on the $M^{\cdot\oplus}$ energy scale depends on the effective nuclear charges and thus, for H_3SiF of C_{3v} symmetry (Table 79), the reversed sequence π_{SiH} (2e) < n_F (1e) is observed PE spectroscopically. From the numerous polyhalogen compounds Hal_2SiR_2 or Hal_3SiR investigated (cf. e.g. References 37, 104, 158, 204, 238), the derivatives F_3SiR have been selected, exhibiting n_F ionizations between 15.6 eV and 16.7 eV (Table 79) and demonstrating the F_3Si acceptor effect by increasing, e.g., the n_{Cl} ionization relative to H_3SiCl by 1.88 eV[167] (Table 79). In addition to the SiX_4 (Figure 20) and X_3SiSiX_3 (Table 79) molecules already mentioned[37], also PE spectra of, e.g., SiI_4 or $Br_3SiSiBr_3$ have been recorded, which no longer can be properly assigned neglecting spin/orbit coupling[237]. Octabromocyclotetrasilane is presently studied with respect to its thermal decomposition[67], which might yield $Br_2Si{=}SiBr_2$. As concerns the organosilicon halides, both trimethylsilyl fluoride and chloride do possess $M^{\cdot\oplus}$ ground states with predominant σ_{SiC} contribution (Table 79) and, for instance, the n_{Cl} ionization remains almost constant at 12 eV despite cyclization to a four-membered ring[159] or bulky R_3C substituents[240]. As discussed over and over again (cf. e.g. Section III.B or Section IV.F), shifting the R_3Si group into β-position by inserting an H_2C link generates the powerful electron-donor substituent R_3SiCH_2 and the n_{Cl} ionizations are lowered by 1 eV to 11 eV[220] (cf. scheme 21).

H. Metallorganic and Miscellaneous Silicon-containing Compounds

Several molecules, which likewise could have been included in Table 80, have already been listed elsewhere: $R_3Si{-}SnR_3$ as a disilane homologue in Table 57, the phthalocyanine complexes due to their central $R_3Si{-}O(Si{-}O)_{1,2}{-}SiR_3$ chains as silyl ethers in Table 74 or the various metal silylamides $Met(N(SiR_3)_2)_n$ as N derivatives in Table 76. The compounds in Table 80 are now commented on.

B_5H_9 with $n=5$ centers and $n+1=6$ skeletal electrons is a nidoborane and its $M^{\cdot\oplus}$ ground state at 10.53 eV has been assigned $\sigma(B_5)$ (4e). All three different positions 1,2 and μ can be substituted by SiH_3 groups; the (relatively) largest lowering of the first ionization energy is observed for $(\mu\text{-}SiH_3)B_5H_8$[241] (Table 80).

Four different classes of R_3SiCH_2-substituted metal compounds have been investigated PE spectroscopically (Table 80). The twofold coordinated derivatives $Met(CH_2SiR_3)_2$ with Met = Ge, Sn and Pb exhibit well-separated bands at low energy[206] and their $M^{\cdot\oplus}$ ground states are assigned by comparison with atomic ionization energies (Ge: 7.5 eV, Sn: 7.0 eV and Pb: 6.8 eV[214]) to dominant contributions of the respective 'lone pairs n_{Met}'. For $Cd(CH_2SiR_3)_2$ the d^9 hole states $4d_{5/2}$ and $4d_{3/2}$ are found at 17.5 eV and 18.2 eV[242] as

M (R=CH$_3$)		References	IE^v_n (eV)
Ⓑ 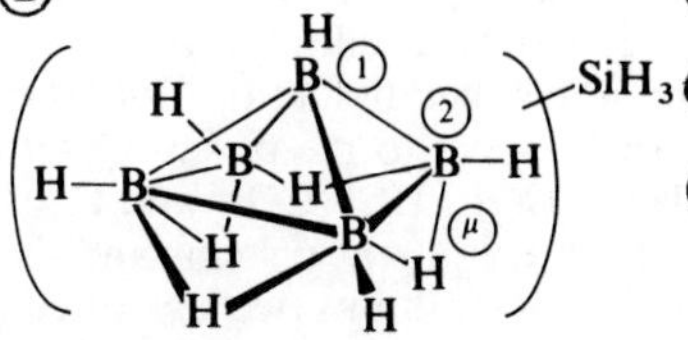		① 241	10.40 } (4e)
		②	10.42
		(μ)	10.17
		(B$_5$H$_9$)	(10.53)
Ⓒ Met(CH$_2$SiR$_3$)$_2$	Met=Ge	206	7.75 } (n$_{Met}$) 8.87 } (σ_{MetC})
	Sn	206	7.42 (n$_{Met}$) 8.33 (σ_{MetC})
	Pb	206	7.25 (a$_1$) 7.98 (b$_2$)
	Cd	242	8.8 (σ_u)
	Hg	206	8.12 (σ_{HgC})
Met(CH$_2$SiR$_3$)$_4$	Sn	240	8.71 (σ_{SnC}, t$_2$)
	Pb	240	8.14/8.86 (σ_{PbC}, t$_2$)
	Ti	243	8.58 } (σ_{MetC})
	Zr	243	8.64 (σ_{MetC})
	Hf	243	8.58
	Cr	240	7.26 (3d$_{cr}$), 8.69(σ_{CrC}, t$_2$)
	ReO	244	8.00 (5d$_{Re}$), 8.46 (σ_{ReC}, e)
(R$_3$SiCH$_2$)$_3$Mo≡Mo(CH$_2$SiR$_3$)$_3$		245	7.5 { (π_{MoMo}, 5e) (σ_{MoC}, 3a$_{2u}$)
(η^5-C$_5$H$_5$)$_2$V–CH$_2$SiR$_3$		246	6.32 } 6.85 } (3d$_V$), 7.90 (σ_{VC})
(η^4-(7-R$_3$Si)C$_7$H$_7$)Fe(CO)$_3$		247	7.96 (3d$_{Fe}$)
(η^8-(R$_3$Si)C$_8$H$_7$)$_2$U		248	6.05 (5f$_U$)
(Met) X$_3$SiMn(CO)$_5$	X=H	249	8.99/9.38 (3d$_{Mn}$)
	CH$_3$	249	9.0/9.3 (3d$_{Mn}$)
	F	249	9.8 (3d$_{Mn}$), 10.4 (σ_{MnSi})
X$_3$SiCo(CO)$_4$	X=H	249	8.85/9.9 (3d$_{Co}$)
	Cl	250	9.11/10.32 (3d$_{Co}$)
H$_3$SiRe(CO)$_5$		249	9.1/9.6 (5d$_{Re}$)
(Si) 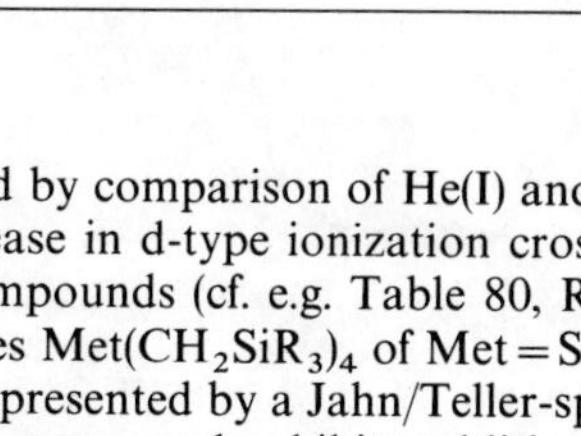 (C$_{2v}$?)		251	6.7 (π, 3e$_{1g}$?) 7.5 (n$_{Si}$, 4a$_{1g}$?) 8.1 (π, 3e$_{1u}$?)

(80)

confirmed by comparison of He(I) and He(II) spectral intensities: the decrease in p-type and increase in d-type ionization cross sections are used for numerous assignments of metal compounds (cf. e.g. Table 80, References 245–248, 251). For the tetrasubstituted derivatives Met(CH$_2$SiR$_3$)$_4$ of Met=Sn, Pb, Ti, Hf and Zr, as expected, the M$^{\cdot\oplus}$ ground state is represented by a Jahn/Teller-split (? cf. Figures 13 and 14) σ_{MetC}(t$_2$) combination. The lead compound exhibits additional spin/orbit coupling (cf. Section IV.B) and the Cr(d^2)[240] and ReO(d^1)[244] centered ones, low-energy 3d$_{Cr}$ or 5d$_{Re}$ ionizations. The first

PES band of the Mo≡Mo triply bonded compound[245] presumably contains both π_{MoMo} and σ_{MoC} ionizations[245] and in the η^5-C_5H_5 vanadium derivative[241], the $3d_V$ electrons are ionized first.

The silyl-substituted η^4-cycloheptatriene and η^8-cyclooctatetraene complexes of Fe and U have been investigated to obtain PE spectroscopic information concerning substituent effects on either fluxional behavior or bonding: their first ionization potentials are lowered by 0.2 eV[247] or remain constant[248] with respect to the parent systems.

The PE spectra of silylated carbonyl hydrides $HMn(CO)_5$, $HCo(CO)_4$ and $HRe(CO)_5$ have been assigned by $M^{\cdot\oplus}$ state comparison with both the parent systems and the CH_3 and GeH_3 derivatives: except for the F_3Si compound, which exhibits two separated bands at low energy, the σ_{MetSi} ionizations are presumably hidden under the second $3d_{Met}$ or $5d_{Met}$ PES bands[249].

The last topic to be mentioned in this review is a most unusual and surprising one: reaction of bis(pentamethylcyclopentadienyl)silicon dichloride $(R_5C_5)_2SiCl_2$ with sodium naphthalide yields decamethyl silicocen $(R_5C_5)_2Si$[251], the first silicon sandwich with tenfold coordinated Si(!). The crystals contain both coplanar (D_{5d}) and bent (C_{2v}) conformations[251]. Meanwhile the novel compound has been evaporated and its PE spectrum recorded[252] (Figure 26). At low ionization energies, three PES bands in the approximate intensity ratio 2:1:2 are recorded which, for a D_{5d} isomer, could be assigned as denoted in Table 80[252]. The global energy minimum for $(R_5C_5)_2Si$, however, is calculated for the C_{2v} silylene[252] (cf. Section II.C), another fascinating and, until now, unknown molecule of the silicon world.

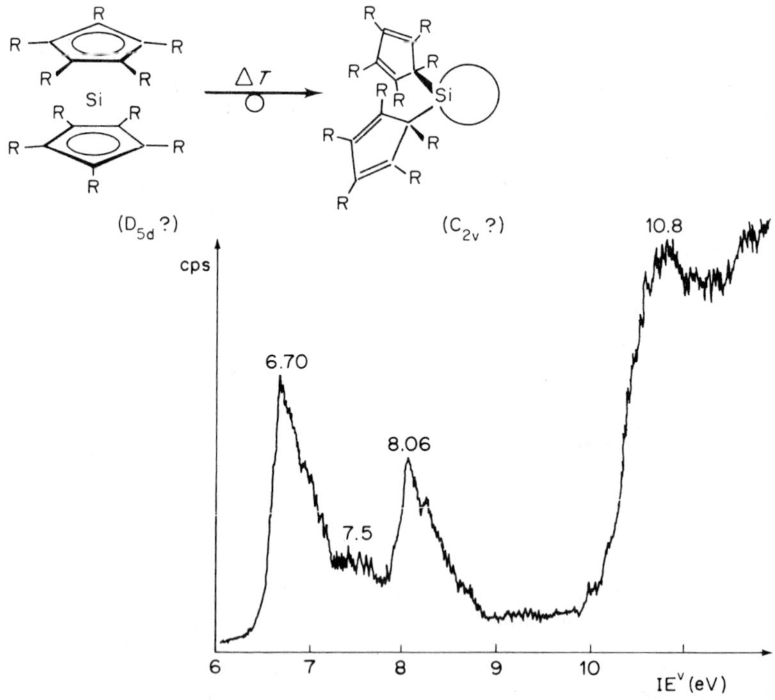

FIGURE 27. He(I) PE spectrum (6eV to 12eV) of bis(pentamethylcyclopentadienyl)silicon[252] of unknown gas-phase structure (cf. text)

VI. SUMMARY: PHOTOELECTRON SPECTRA OF SILICON COMPOUNDS AND THEIR USEFUL APPLICATION

It has been both an effort and a challenge to summarize and again discuss the ionization 'fingerprints' of silicon compounds as supplied by the literature search profile[153], twenty years after their first photoelectron spectra were recorded[154] (cf. Section I). Nowadays, about 500 are known and several thousand radical-cation states of silicon-containing molecules are assigned based on PE spectroscopic information, on comparison of equivalent $M^{\cdot\oplus}$ states of chemically related molecules and on Koopmans' correlation with SCF eigenvalues calculated at various levels of sophistication (cf. Section IV. A, B, C and H).

Radical cations are generated vertically on photoionization, i.e. within 10^{-15} of a second, and both their ground state and excited states are observed simultaneously by PE spectroscopy, all of them in a 'frozen core' mode, because molecular dynamics is activated only later after about 10^{-13} of a second (cf. Section III.D). Nevertheless, neither the change in electron correlation nor the electronic relaxation accompanying the vertical electron expulsion can be neglected—they will indispensibly emerge as 'Koopmans defects' in $IE_n^v/\varepsilon_J^{SCF}$ correlations and are always traceable to oversimplifying molecular-orbital approximations (cf. Section IV.C). On the other hand, much can be learned from just looking at PE spectra, for example at the manifold of different band shapes[3,4] such as vibrational fine structures, spin/orbit coupling or Jahn/Teller broadening (cf. e.g. Section IV.B). The first point to be made concerning the useful application of PE spectra of silicon compounds, therefore, is that we should learn as much as possible about the reality of their molecular radical cation states.

The experience gathered from vertically excited—and, therefore, less complicated—radical cation states, especially of organosilicon compounds, provides easier access to adiabatically generated radical cations in solution (Section III): structural changes accompany every electron transfer and molecular dynamics are activated at least at about room temperature. Ionization energies are not only experimental data easily accessible by PE spectroscopy, but useful in numerous aspects: thus substituent effects can be elegantly determined from the perturbation of suitable parent systems—usually from the literature or with little effort (cf. e.g. Section III.B and Section IV.F). Topological consequences for chains, rings and clusters may be anticipated and band structures for solids like polysilanes approximated (Sections IV.D.2 and 3). Oxidation potentials may be correlated (Figure 10), charge transfer excitations predicted (scheme 39), UV spectra screened for Rydberg states (scheme 52), structural distortions rationalized (scheme 43) and so on and so on—the second answer to be given to the title question as to how PE spectra of silicon compounds may be usefully applied.

In addition to knowledge accumulated by and large on the individual silicon compounds and their respective molecular states—from structures and their energy dependence to potentials and charges—as building blocks also for the preparatively oriented chemist, their PE spectroscopic ionization patterns can be successfully applied for identification especially of short-lived molecules, such as H_5C_5SiH (Figure 3), $H_2C{=}SiH_2$ (Figure 4), $SiCl_2$ (Figures 5 and 6), SiO (scheme 12), SiH_3 (scheme 14), $RC{-}N{=}N{-}SiR_3$ (Figure 8) or $R_3Si{-}C{\equiv}P$ (Figure 25). It is the flow system under reduced pressure, i.e. the design of a PE spectrometer (Figures 1 and 2) coupled with the real-time PES gas analysis[12], which in addition allows the optimization of the respective reaction conditions (Section II), as is convincingly demonstrated by the safe thermal decomposition of $H_5C_6Si(N_3)_3$ into 4 N_2 and $H_5C_6{-}N{\equiv}Si$ (Figure 7), the first organosilicon molecule with triply bonded, singly coordinated Si. Also, heterogeneously catalyzed gas reactions can be advantageously optimized using PES real-time analysis[12,18,19] and, due to only millimole quantities of compounds needed to investigate the temperature dependence over several hundred degrees in 5 to 6 hours, are best suited to screen catalysts for a given reaction[21] or, vice versa, to test the effectiveness of a given catalyst for other reactions[20]. There are limits for reaction mixtures of compounds with overlapping PES bands[12] but, e.g., for those of chlorosilanes H_2SiCl_2 and $HSiCl_3$, the potential applicability of real-time PES gas analysis has been illustrated by computer simulation[12].

This chapter has attempted to present how the useful application of photoelectron spectroscopy also for silicon compounds ranges from gaining more reliable insight into their bonding in terms of molecular states via correlations with other molecular properties, to their identification by the individual ionization 'fingerprints' in real-time gas analysis. The techniques allow one to generate and characterize short-lived intermediates as well as to optimize the conditions of both homogeneous and heterogeneously catalyzed gas reactions.

VII. REFERENCES

1. Part 114 of *Photoelectron Spectra and Molecular Properties*; also part 70 of *Gasphase Reactions*; for parts 113 or 69, respectively, cf. H. Bock, J. Dörr, M. Haun, B. Solouki, S. Aygen, M. Bankmann, O. Breuer, R. Dammel, T. Hirabayashi, D. Jaculi, J. Mintzer, S. Mohmand, H. Müller, P. Rosmus, B. Roth, J. Wittmann and H.-P. Wolf, *J. Mol. Struct.*, in press.
2. D. W. Turner, C. Baker, A. D. Baker and C. R. Brundle, *Molecular Photoelectron Spectroscopy*, Wiley–Interscience, London, 1970, contains a summary, illustrated with hundreds of He(I) photoelectron spectra of prototype molecules.
3. The five volumes of the handbook: A. D. Baker and C. R. Brundle (Eds.), *Electron Spectroscopy: Theory, Techniques and Applications,* Academic Press, London, 1977–84, contain a comprehensive view of the topic, including X-ray photoelectron spectroscopy, in individual contributions written by experts.
4. J. W. Rabalais, *Principles of Ultraviolet Photoelectron Spectroscopy*, Wiley, New York, 1977, is mentioned among the numerous monographs because of its literature review and the lucid presentation of difficult topics, such as photoionization cross-sections.
5. K. Kimura, S. Katsumata, Y. Achiba, T. Yamazaki and S. Iwata, *Handbook of HeI Photoelectron Spectra of Fundamental Organic Molecules*, Halstead Press, New York, 1981, is the standard compendium containing several hundred PE spectra assigned by *ab initio* calculations and many references.
6. In *Special Reports on 'Electronic Structure and Magnetism of Inorganic Compounds'* (Ed. P. Day), Chemical Society, London, comprehensive literature reviews on technical details of the instrumentation as well as on the PE spectra of small molecules are provided by A. F. Orchard, S. Evans and A. Hamnett in Vols. 1 (1970) to 3 (1972), and by R. G. Edgell and A. W. Potts in Vols. 6 (1980) and 7 (1982).
7. E. Heilbronner has made fundamental contributions to the assignment of photoelectron spectra of organic compounds. Summaries are found in: *Pure Appl. Chem.*, **7**, 9 (1971); R. Daudel and B. Pullmann, Eds., *The World of Quantum Chemistry*, D. Reidel, Dordrecht, 1974, p. 211; Reference 3 (together with J. P. Maier); *Helv. Chim. Acta*, **60**, 2248 (1977); *Phys. Scr.*, **16**, 202 (1977) (together with T. B. Jones, E. Kloster-Jensen and J. P. Maier).
8. H. Bock and B. G. Ramsey, *Angew. Chem.*, **85**, 773 (1973); *Angew. Chem., Int. Ed. Engl.*, **12**, 734 (1973), report on the interpretation of the PE spectra of main group element compounds by comparison of equivalent radical cation states of chemically related molecules, based on simple MO methods.
9. H. Bock, *Angew. Chem.*, **89**, 631 (1977); *Angew. Chem., Int. Ed. Engl.*, **16**, 613 (1977), discusses advantages and limits of the rationalization of molecular states by molecular orbital models, including numerous examples of radical cation states from photoelectron spectra.
10. H. Köppel, L. S. Cederbaum, W. Domcke and S. S. Shaik, *Angew. Chem.*, **95**, 221 (1983); *Angew. Chem., Int. Ed. Engl.*, **22**, 210 (1983), discuss 'Non Born/Oppenheimer effects' in radical cations. See also L. S. Cederbaum, W. Domcke, J. Schirmer and W. v. Niessen, *Adv. Chem. Phys.*, **5**, 115 (1986) on 'Correlation effects in the ionization of molecules: breakdown of the molecular orbital picture'.

11. On future trends of PE spectroscopy see, for example, the report by H. Bock, B. Solouki, S. Aygen, G. Bert, T. Hirabayashi, S. Mohmand, P. Rosmus and J. Wittmann, *Nachr. Chem. Tech. Lab.*, **26**, 634 (1978) or *J. Mol. Struct.*, **60**, 31 (1980).
12. H. Bock and B. Solouki, *Angew. Chem.*, **93**, 425 (1981); *Angew. Chem., Int. Ed. Engl.*, **20**, 427 (1981), report on application of PE spectroscopy to determine thermal decomposition routes to detect unstable intermediates and to optimize heterogeneously catalyzed reactions (142 references).
13. J. M. Dyke, N. Jonathan and A. Morris, *Int. Rev. Phys. Chem.*, **2**, 3 (1982), summarize some 'Recent progress in the study of transient species with vacuum ultraviolet photoelectron spectroscopy'.
14. H. W. Kroto, *Chem. Soc. Rev.*, **11**, 435 (1982), reports on some studies of transient molecules in the gas phase.
15. H. Bock and R. Dammel, *Angew. Chem.*, **99**, 518 (1987); *Angew. Chem., Int. Ed. Engl.*, **26**, 487 (1987), demonstrate the usefulness of PE spectroscopy in studies of gas-phase pyrolyses of azides and report on information obtained concerning the pathways of thermal fragmentation.
16. H. Bock and W. Kaim, *Acc. Chem. Res.*, **15**, 9 (1982) review ESR/ENDOR information on organosilicon radical cations in solution, generated by single electron transfer as predicted by PE spectroscopic first ionization potentials.
17. R. S. Brown and F. S. Jörgensen, in Reference 3, Vol. 5, pp. 1–122, review 'Conformational analysis by photoelectron spectroscopy'.
18. H. Bock, B. Solouki, J. Wittmann and H.-J. Arpe, *Angew. Chem.*, **90**, 933 (1978); *Angew. Chem., Int. Ed. Engl.*, **17**, 986 (1978); cf. also *Chem. Ber.*, **115**, 2326 (1982).
19. H. Bock, J. Mintzer, J. Wittmann and J. Russow, *Angew. Chem.*, **92**, 136 (1980); *Angew. Chem., Int. Ed. Engl.*, **19**, 147 (1980); cf. also *Chem. Ber.*, **115**, 2346 (1982).
20. H. Bock and H. P. Wolf, *Angew. Chem.*, **97**, 411 (1985); *Angew. Chem., Int. Ed. Engl.*, **24**, 418 (1985). Cf. also H. Bock, G. Tschmutowa and H. P. Wolf, *J. Chem. Soc., Chem. Commun.*, 1068 (1986).
21. H. Bock and O. Breuer, *Angew. Chem.*, **99**, 492 (1987); *Angew. Chem., Int. Ed. Engl.*, **26**, 461 (1987).
22. Cf. e.g. H. Bock, 'Reaction pathways of carbon and of organosilicon compounds as viewed by a preparative chemist', *L'actualite chimique*, 33 (1986/3). See also H. Bock, B. Solouki, P. Rosmus, R. Dammel, P. Hänel, B. Hierholzer, U. Lechner-Knoblauch and H. P. Wolf, 'Recent investigations on short-lived organosilicon molecules and molecular ions', in *Organosilicon and Bioorganosilicon Chemistry* (Ed. H. Sakurai), E. Horwood Ltd., Chichester, 1985, pp. 45–73.
23. T. A. Carlson, *Photoelectron and Auger Spectroscopy*, Plenum Press, New York, 1975.
24. J. H. D. Eland, *Photoelectron Spectroscopy*, Butterworth, London, 1974, surveys the experimental information available from PE spectroscopic investigation of radical cation states.
25. A. D. Baker and D. Betteridge, *Photoelectron Spectroscopy—Chemical and Analytical Aspects*, Pergamon Press, Oxford, 1972.
26. R. E. Ballard, *Photoelectron Spectroscopy and Molecular Orbital Theory*, Adam Hilger, Bristol, 1978, focuses on a few selected examples of small molecules to introduce the reader to PE spectroscopic assignment.
27. K. Siegbahn, C. Nordling, A. Fahlmann, R. Nordberg, K. Hamrin, J. Hedman, G. Johannson, T. Bergmark, S. E. Karlsson, I. Lindgren and B. Lindberg, *Nova Acta R. Soc. Sci. Ups.*, Ser. IV, **20** (1967).
28. K. Siegbahn, G. Johannson, J. Hedman, P. F. Heden, K. Hamrin, U. Gelius, T. Bergmark, L. O. Werme, R. Manne and Y. Bear, *ESCA Applied to Free Molecules*, North-Holland, Amsterdam, 1969.
29. M. B. Robin, *Higher Excited States of Polyatomic Molecules*, Vols. I and II, Academic Press, New York, 1974 and 1975.
30. C. R. Brundle and M. B. Robin, in *Determination of Organic Structure by Physical Methods*, Vol. 3 (Eds. F. C. Nachod and J. J. Zuckerman), Academic Press, New York, 1971, p. 1.
31. H. Bock, in *Scattering Techniques* (Ed. P. Day), D. Reidel, Dordrecht, 1981.
32. R. L. DeKock and D. R. Lloyd, in *Advances in Inorganic Chemistry and Radiochemistry*, Vol. 16 (Ed. H. J. Emeleus and A. G. Sharpe), Academic Press, London, 1974, p. 66.
33. J. P. Maier, *J. Chem. Soc., Ann. Rep. B*, 75 (1974).

34. H. Bock, in Gmelins *Handbuch der Anorganischen Chemie*, Vol. 8/22, Part 4, Springer-Verlag, Berlin, 1975, p. 170.
35. H. Bock, *Pure Appl. Chem.*, **44**, 343 (1975).
36. R. Gleiter and J. Spanget-Larsen, *Top. Curr. Chem.*, **86**, 139 (1979).
37. K. Wittel and H. Bock, in *The Chemistry of Functional Groups*, Supplement D (Eds. S. Patai and Z. Rappoport), Wiley–Interscience, Chichester, 1983, pp. 1499–1603; 406 references.
38. H. Stafast and H. Bock, in *The Chemistry of Functional Groups*, Supplement C (Eds. S. Patai and Z. Rappoport), Wiley–Interscience, Chichester, 1982, pp. 137–185; 113 references.
39. K. Wittel, J. K. Meeks and S. P. McGlynn, in *The Chemistry of Functional Groups: Cyanates and their Thio Derivatives* (Ed. S. Patai), Vol. 1, Wiley–Interscience, Chichester, 1977, p. 1.
40. For reviews covering predominantly those unsaturated silicon compounds, which are kinetically stabilized by bulky inert substituents, cf. e.g. G. Raabe and J. Michl, *Chem. Rev.*, **85**, 419 (1985) or G. Müller, *Nachr. Chem. Tech. Lab.*, **34**, 778 (1986) and the numerous references given.
41. H. Bock and R. Dammel, *Z. Naturforsch.*, **42b**, 301 (1987).
42. H. Bock, R. Dammel and D. Jaculi, *J. Am. Chem. Soc.*, **108**, 7844 (1986).
43. H. Bock, S. Mohmand, T. Hirabayashi, G. Maier and H. P. Reisenauer, *Chem. Ber.*, **116**, 273 (1982).
44. B. Solouki, P. Rosmus and H. Bock, *J. Am. Chem. Soc.*, **98**, 6054 (1976); cf. also H. Bock, Chem. Rundsch. **34/29**, 3 (1981).
45. H. Bock, T. Hirabayashi, S. Mohmand and B. Solouki, *Angew. Chem.*, **89**, 106 (1977); *Angew. Chem., Int. Ed. Engl.*, **16**, 105 (1977).
46. H. Bock and M. Bankmann, *Angew. Chem.*, **98**, 288 (1986); *Angew. Chem., Int. Ed. Engl.*, **25**, 265 (1986).
47. H. Bock, T. Hirabayashi and S. Mohmand, *Chem. Ber.*, **114**, 2595 (1981).
48. H. Bock, B. Solouki and G. Maier, *Angew. Chem.*, **97**, 205 (1985); *Angew. Chem., Int. Ed. Engl.*, **24**, 205 (1985).
49. M. Binnewies, B. Solouki, H. Bock, R. Becherer and R. Ahlrichs, *Angew. Chem.*, **96**, 704 (1984); *Angew. Chem., Int. Ed. Engl.*, **23**, 731 (1984).
50. B. Solouki, H. Bock, R. Appel, A. Westerhaus, G. Becker and G. Uhl, *Chem. Ber.*, **115**, 3747 (1982).
51. P. Rosmus, H. Bock, B. Solouki, G. Maier and G. Mihm, *Angew. Chem.*, **93**, 616 (1981); *Angew. Chem., Int. Ed. Engl.*, **20**, 598 (1981). For the matrix isolation of silaethene see G. Maier, G. Mihm and H. P. Reisenauer, *Angew. Chem.*, **93**, 615 (1981); *Angew. Chem., Int. Ed. Engl.*, **20**, 597 (1981).
52. H. Bock, P. Rosmus, B. Solouki and G. Maier, *J. Organomet. Chem.*, **271**, 145 (1984). For calculations see also M. J. S. Dewar, D. H. Lo and C. A. Ramsden. *J. Am. Chem. Soc.*, **97**, 1311 (1975) or H. B. Schlegel, B. Coleman and M. Jones Jr., *J. Am. Chem. Soc.*, **100**, 6499 (1978).
53. H. Bock, A. Bowling, B. Solouki, T. J. Barton and G. T. Burns, *J. Am. Chem. Soc.*, **102**, 429 (1980).
54. B. Solouki, P. Rosmus, H. Bock and G. Maier, *Angew. Chem.*, **92**, 56 (1980); *Angew. Chem., Int. Ed. Engl.*, **19**, 51 (1980). For the matrix isolation of silabenzene see G. Maier, G. Mihm and H. P. Reissenauer, *Angew. Chem.*, **92**, 58 (1980); *Angew. Chem., Int. Ed. Engl.*, **19**, 52 (1980).
55. C. Batich, E. Heilbronner, V. Hornung, A. J. Ashe III, D. T. Clark, U. T. Cobley, D. Kilcast and I. Scanlan, *J. Am. Chem. Soc.*, **95**, 928 (1973). For C_5BiH_6 cf. J. Bastide, E. Heilbronner, J. P. Maier and A. J. Ashe III, *Tetrahedron Lett.*, 411 (1976).
56. Cf. e.g. E. Heilbronner and H. Bock, *The HMO Model and its Application*, Vols. 1–3, Wiley–Interscience, London, 1975.
57. T. Koenig and W. McKenna, *J. Am. Chem. Soc.*, **103**, 1212 (1981).
58. J. M. Dyke, G. D. Josland, R. A. Lewis and A. Morris, *J. Phys. Chem.*, **86**, 2913 (1982).
59. The gas-phase electron diffraction structure has been determined by P. G. Mahaffy, R. Gutowsky and L. K. Montgomery, *J. Am. Chem. Soc.*, **102**, 2854 (1980); for a corrected Si=C bond length cf. Y. Yoshioka, J. D. Goddard and H. F. Schaefer III, *J. Am. Chem. Soc.*, **103**, 2452 (1981).
60. A most valuable summary of both theoretical and experimental work on existing as well as still elusive small silicon intermediates containing up to 5 atoms and their comparison to the corresponding carbon analogues is provided by B. T. Luke, J. A. Pople, M.-B. Krogh-Jespersen,

Y. Apeloig, J. Chandrasekhar and P. v. R. Schleyer, *J. Am. Chem. Soc.*, **108**, 260 and 270 (1986) with well over 300 references.

61. Cf. e.g. P. Gaspar, in *Reactive Intermediates* (Eds. M. Jones jr. and R. A. Moss), Wiley–Interscience, New York, Vol. 1 (1978), p. 229; Vol. 2 (1981), p. 335 and Vol. 3 (1985), p. 333, and references therein.
62. Cf. the summary of earlier literature by W. M. Atwell and D. R. Weyenburg, *Angew. Chem.*, **81**, 485 (1969), *Angew. Chem., Int. Ed. Engl.*, **8**, 469 (1969). For recent surface investigations on Cu-doped silica cf. W. F. Banholzer, N. Lewis and W. Ward, *J. Catal.*, **101**, 405 (1986).
63. R. S. Grey and H. F. Schaefer III, *J. Am. Chem. Soc.*, **108**, 5804 (1986).
64. Cf. e.g. H. P. Reisenauer, G. Mihm and G. Maier, *Angew. Chem.*, **94**, 864 (1982); *Angew. Chem., Int. Ed. Engl.*, **21**, 854 (1982).
65. I. Hargittai, G. Schultz, J. Tremmel, N. D. Kagramanov, A. K. Maltsev and O. M. Nefedov, *J. Am. Chem. Soc.*, **105**, 2895 (1983).
66. T. P. Fehlner and D. W. Turner, *Inorg. Chem.*, **13**, 754 (1974), as well as N. P. C. Westwood, *Chem. Phys. Lett.*, **25**, 588 (1974).
67. H. Bock, M. Kremer and B. Solouki, unpublished results. Cf. M. Kremer, Master thesis, University of Frankfurt, 1986.
68. S. Evans and A. F. Orchard, *J. Electron Spectrosc. Relat. Phenom.*, **6**, 207 (1975) or D. H. Harris, M. F. Lappert, J. B. Pedley and G. J. Sharp, *J. Chem. Soc., Dalton Trans.*, 945 (1976).
69. J. F. Ogilvie and S. Cradock [*Chem. Commun.*, 364 (1966)] derived the structural data ($d_{NSi} = 154$ pm, $d_{HN} = 100$ pm and $\measuredangle$HNSi = 180°) from IR spectroscopic measurements. J. N. Murrell, H. W. Kroto and M. F. Guest [*J. Chem. Soc., Chem. Commun.*, 619 (1977) and *Astrophys. J.*, **219**, 886 (1978)] calculated HNSi to be 372 kJ mol^{-1} more stable than the isomeric HSiN and R. Preuss, R. J. Buenker and S. D. Peyer-Imhoff [*J. Mol. Struct.*, **49**, 171 (1978)] predicted a barrier of 55 kJ mol^{-1} for the rearrangement HSiN$\rightarrow$HNSi from CI calculations.
70. E. A. Colbourn, J. M. Dyke, E. P. F. Lee, A. Morris and I. R. Trickle, *Mol. Phys.*, **35**, 873 (1978); see also Reference 13.
71. H. Bock and R. Dammel, *Angew. Chem.*, **97**, 128 (1985); *Angew. Chem, Int. Ed. Engl.*, **24**, 111 (1985).
72. Cf. G. Gross and J. Michl, *Chem. Eng. News*, **63**, No. 12, p. 31 (1985).
73. J. M. Dyke, N. Jonathan, A. Morris, A. Ridha and M. J. Winter, *Chem. Phys.*, **81**, 481 (1983).
74. Cf. e.g. F. A. Houle and J. L. Beauchamp, *J. Am. Chem. Soc.*, **101**, 4070 (1979) and references cited therein.
75. C. Guimon, G. Pfister-Guillonzo, H. Lavayssiere, G. Dousse, J. Barreau and J. Stage, *J. Organomet. Chem.*, **249**, C17 (1983).
76. Selenoformaldehyde, H_2C=Se, is best generated by the analogous pyrolysis of its trimer; cf. H. Bock, S. Aygen, P. Rosmus, B. Solouki und E. Weissflog, *Chem. Ber.*, **117**, 187 (1984).
77. H. Bock, T. Hirabayashi and S. Mohmand, *Chem. Ber.*, **115**, 492 (1982).
78. G. Märkl, D. Rudnick, R. Schulz and A. Schweig, *Angew. Chem.*, **94**, 211 (1982); *Angew. Chem., Int. Ed. Engl.*, **21**, 221 (1982); Suppl. 523.
79. H. Bock, R. Dammel, S. Fischer and C. Wentrup, *Tetrahedron Lett.*, **28**, 617 (1987).
80. H. Bock, G. Brähler, G. Fritz and E. Matern, *Angew. Chem.*, **88**, 765 (1976); *Angew. Chem., Int. Ed. Engl.*, **15**, 669 (1976) as well as H. Bock, W. Kaim and J. W. Connolly, *Angew. Chem.*, **88**, 766 (1976); *Angew. Chem., Int. Ed. Engl.*, **15**, 700 (1976).
81. H. Bock and U. Lechner-Knoblauch, *J. Organomet. Chem.*, **294**, 295 (1985).
82. Cf. e.g. H. Bock, U. Stein and A. Semkow, *Chem. Ber.*, **113**, 3208 (1980).
83. Cf. e.g. H. Nöth, W. Winterstein, W. Kaim and H. Bock, *Chem. Ber.*, **112**, 2494 (1979).
84. Cf. e.g. H. Bock, B. Roth and G. Maier, *Chem. Ber.*, **117**, 172 (1984).
85. Cf. e.g. H. Bock, W. Kaim, A. M. Semkow and H. Nöth, *Angew. Chem.*, **90**, 308 (1978); *Angew. Chem., Int. Ed. Engl.*, **17**, 286 (1978); H. Bock, W. Kaim, M. Kira, L. Rene and H.-G. Viehe, *Z. Naturforsch.*, **39b**, 763 (1984) or H. Bock, P. Hänel, W. Kaim and U. Lechner-Knoblauch, *Tetrahedron Lett.*, **26**, 5115 (1985).
86. Cf. e.g. W. Kaim, H. Bock and H. Nöth, *Chem. Ber.*, **111**, 3276 (1978).
87. Cf. e.g. H. Bock and G. Brähler, *Chem. Ber.*, **112**, 3081 (1979); H. Bock, G. Brähler, U. Henkel, R. Schlecker and D. Seebach, *Chem. Ber.*, **113**, 289 (1980); H. Bock, G. Brähler, D. Dauplaise and

J. Meinwald, *Chem. Ber.*, **114**, 2622 (1981); H. Bock, W. Schulz and U. Stein, *Chem. Ber.*, **114**, 2632 (1981); H. Bock, U. Stein and P. Rittmeyer, *Angew. Chem.*, **94**, 540 (1981); *Angew. Chem., Int. Ed. Engl.*, **21**, 533 (1982); H. Bock, B. Roth and R. Schumaker, *Phosphorus and Sulfur*, **21**, 79 (1984) or H. Bock and P. Rittmeyer, *Chem. Ber.*, **119**, 3766 (1986).

88. H. Bock and H. Alt, *Angew. Chem.*, **79**, 934 (1967); *Angew. Chem., Int. Ed. Engl.*, **6**, 943 (1967). Cf. also *J. Am. Chem. Soc.*, **92**, 1569 (1970).
89. H. Bock and H. Seidl, *J. Organomet. Chem.*, **13**, 91 (1968). Cf. also *Chem. Ber.*, **101**, 2815 (1968) or *J. Chem. Soc. (B)*, 1158 (1968).
90. G. Bieri, F. Burger, E. Heilbronner and J. P. Maier, *Helv. Chim. Acta*, **60**, 2223 (1977) and references cited therein.
91. H. Bock and W. Kaim, *J. Am. Chem. Soc.*, **102**, 4429 (1980); cf. also *Tetrahedron Lett.*, **27**, 2343 (1977) and references cited therein.
92. K. A. Ostoja Starzewski and H. Bock, *J. Am. Chem. Soc.*, **98**, 8486 (1976).
93. H. Bock and W. Kaim, *Chem. Ber.*, **111**, 3552 (1978).
94. H. Bock, B. Hierholzer, H. Kurreck and W. Lubitz, *Angew. Chem.*, **95**, 817 (1983); *Angew. Chem., Int. Ed. Engl.*, **22**, 787 (1983) and, especially, the details in *Angew. Chem. Suppl.*, 1088 (1983).
95. H. Bock, W. Kaim and H. E. Rohwer, *Chem. Ber.*, **111**, 3573 (1978).
96. H. Bock, G. Brähler, G. Fritz and E. Matern, *Angew. Chem.*, **88**, 765 (1976); *Angew. Chem., Int. Ed. Engl.*, **15**, 699 (1976) as well as *Z. Anorg. Allg. Chem.*, **439**, 173 (1978).
97. H. Bock and W. Enßlin, *Angew. Chem.*, **83**, 435 (1971); *Angew. Chem., Int. Ed. Engl.*, **10**, 404 (1971). For a recent study of $SiMe_4$ cf. J. E. Bice, K. H. Tan, G. M. Bancroft and J. S. Tse, *J. Am. Chem. Soc.*, **26**, 4106 (1987) and references cited therein.
98. H. Bock, W. Kaim, M. Kira and R. West, *J. Am. Chem. Soc.*, **101**, 7767 (1979); cf. also B. Hierholzer, Ph.D. Thesis, University of Frankfurt, 1988 and E. Carberry, R. West and G. E. Glass, *J. Am. Chem. Soc.*, **91**, 5446 (1969).
99. H. Bock, W. Kaim, H. Nöth and A. Semkow, *J. Am. Chem. Soc.*, **102**, 4421 (1980); cf. also *Angew. Chem.*, **90**, 308 (1978); *Angew. Chem., Int. Ed. Engl.*, **17**, 286 (1978).
100. H. Bock, R. Dammel and B. Roth, Molecular state fingerprints and semiempirical hypersurface calculations, in *Inorganic Rings and Clusters* (Ed. A. H. Cowley), ACS Symp. Ser., **232**, 139 (1983).
101. S. F. Nelson, W. C. Hollinsed, C. R. Kessel and J. C. Calabrese, *J. Am. Chem. Soc.*, **100**, 7876 (1978).
102. H. Bock, W. Kaim, M. Kira, H. Osawa and H. Sakurai, *J. Organomet. Chem.*, **164**, 295 (1979).
103. E. Heilbronner, Organic chemical photoelectron spectroscopy, in *Molecular Spectroscopy* (Ed. A. R. West), Heyden Sons Ltd., London, 1977, pp. 422–451.
104. S. Cradock and R. A. Whiteford, *Trans. Faraday Soc.*, **67**, 3425 (1971); see also the discussion in Reference 37. Cf. S. Craddock and E. A. V. Ebsworth, *J. Chem. Soc., Chem. Commun.*, 57 (1971). The PE spectra of silyl halides recorded in Frankfurt (cf. Figures 13 and 19), therefore, have not been published.
105. Cf. also K. Kimura, S. Katsumata, T. Yamazaki and H. Wakabayashi, *J. Electron Spectrosc. Relat. Phenom.*, **6**, 41 (1975) and references cited therein.
106. For an easy-to-comprehend introduction cf. also F. Brogli and E. Heilbronner, *Helv. Chim. Acta*, **54**, 1423 (1971) or K. Wittel, H. Bock and R. Manne, *Tetrahedron*, **30**, 651 (1974).
107. Cf. also E. Heilbronner, K. A. Muszkat and J. Schäublin, *Helv. Chim. Acta*, **54**, 58 (1971).
108. Cf. e.g. H. Kwart and K. G. King, d-Orbitals in the chemistry of silicon, phosphorus and sulfur, in *Reactivity and Structure Concepts in Organic Chemistry*, Vol. 3, Springer-Verlag, Heidelberg 1977 or E. A. V. Ebsworth, in *Organometallic Compounds of Group IV Elements* (Ed. A. G. McDiarmid), M. Dekker, New York, 1968, p. 1.
109. D. C. Frost, F. G. Herring, A. Katrib, R. A. N. McLean, J. E. Drake and N. P. C. Westwood, *Can. J. Chem.*, **49**, 4033 (1971) as well as *Chem. Phys. Lett.* **10**, 347 (1971).
110. Cf. also e.g. D. Armstrong, J. Jamieson and P. Perkins, *Theor. Chim. Acta*, **25**, 396 (1972) or G. DeAlti, P. DeCleva and A. Sgamellotti, *Chem. Phys.*, **35**, 283 (1978).
111. A. M. Semkow, P. Rosmus, H. Bock and P. Botschwina, *Chem. Phys.*, **40**, 377 (1979) and references cited therein; e.g. for H_3SiCl. J. M. Howell and J. R. van Wazer, *J. Am. Chem. Soc.*, **96**, 3064 (1974). For CH_3SiF, cf. e.g. A. Breeze, G. A. D. Collins and D. W. J. Cruickshank, *J. Chem. Soc., Chem. Commun.*, 445 (1971).
112. For another review (cf. Reference 60) of calculations for molecules containing third-row

elements, cf. M. W. Schmidt, P. N. Truong and M. S. Gordon, *J. Am. Chem. Soc.*, **109**, 5217 (1987), with 115 references.
113. H. Bock and M. Bankmann, *Angew. Chem.*, **98**, 287 (1986); *Angew. Chem., Int. Ed. Engl.*, **25**, 265 (1986).
114. Cf. e.g. G. Loubriel and J. M. Zeigler, *Phys. Rev. B*, **33**, 4203 (1986) or L. A. Harrah and J. M. Zeigler, *Macromolecules*, **20**, 601 (1987) and references cited therein.
115. K. Seki, T. Mori, H. Inokuchi and K. Murano, *Bull. Chem. Soc. Jpn.*, in press.
116. K. Takeda, H. Teramae and N. Matsumoto, *J. Am. Chem. Soc.*, **108**, 8186 (1986) or K. Takedo, N. Matsumoto and M. Fukuchi, *Phys. Rev. B*, **30**, 5871 (1984).
117. H. Bock, W. Enßlin, F. Feher and R. Freund, *J. Am. Chem. Soc.*, **98**, 668 (1976) and references cited therein. For attempted PE spectra simulation cf. W. Enßlin, H. Bergmann and S. Elbel, *J. Chem. Soc., Faraday Trans. 2*, 913 (1975).
118. The electronegativity is defined by Pauling and Mulliken as an energy, and by others as a force or as a dimensionless ratio. Cf. numerous textbooks, e.g. Holleman–Wiberg, *Lehrbuch der Anorganischen Chemie*, Editions 91–100, de Gruyter, Berlin, 1985, p. 152. The entire range of chemistry is embraced by the ridiculously small difference of only 3.54 units (Fr 0.86 to Ne 4.40) and the variation of the central atom in various substituent groups like C in CH_3, CN or CF_3 is not accounted for at all; cf. e.g. J. Hinze, *Fortschr. Chem. Forsch.*, **9**, 448 (1968).
119. A. W. Potts and W. C. Price, *Proc. R. Soc. London, Ser. A*, **326**, 165, 181 (1972); cf. also H. J. Lempka, T. R. Passmore and W. C. Price, *Proc. R. Soc. London, Ser. A*, **304**, 53 (1968).
120. Cf. e.g. the review by C. G. Pitt, *J. Organomet.*, **61**, 49 (1973) and references cited therein.
121. W. Enßlin, H. Bock and G. Becker, *J. Am. Chem. Soc.*, **96**, 2757 (1974) and references cited therein. For HC≡CH and H_3C–C≡CH see also References 2 and 5.
122. G. Bieri, F. Brogli, E. Heilbronner and E. Kloster-Jensen, *J. Electron Spectrosc. Relat. Phenom.*, **1**, 67 (1972/73). For the phenyl derivative cf. S. Elbel, K. Lienert, A. Krebs and H. tom Dieck, *Liebigs. Ann. Chem.*, 1785 (1981). For the phenyl derivative cf. S. Elbel, K. Lienert, A. Krebs and H. tom Dieck, *Liebigs Ann. Chem.*, 1785 (1981).
123. Cf. e.g. C. Müller, A. Schweig and H. Vermeer, *J. Am. Chem. Soc.*, **100**, 8056 (1978) as well as the numerous references to their own work.
124. Cf. e.g. U. Weidner and A. Schweig, *J. Organomet. Chem.*, **39**, 261 (1972). For vinylsilane, the ridiculous Koopmans' deviation $\varepsilon_J^{CNDO} - IE_1^v = 4.9$ eV (!) results on including $3d_{Si}$ orbitals into the basis set. Modified CNDO/2 yields the much smaller, through still considerable, deviation of 1.8 eV (see Reference 125).
125. P. Mollère, H. Bock, G. Becker and G. Fritz, *J. Organomet. Chem.*, **46**, 89 (1972).
126. Cf. e.g. L. L. Miller, G. D. Nordblom and E. A. Mayer, *J. Org. Chem.*, **37**, 916 (1972).
127. C. G. Pitt and H. Bock, *J. Chem. Soc., Chem. Commun.*, 28 (1972).
128. M. Klessinger, *Angew. Chem.*, **84**, 545 (1972); *Angew. Chem., Int. Ed. Engl.*, **11**, 525 (1972).
129. Cf. H. Bock and H. Alt, *J. Am. Chem. Soc.*, **92**, 1569 (1970) or H. Bock, H. Alt and H. Seidl, *Chem. Ber.*, **101**, 2815 (1968).
130. Cf. e.g. B. G. Briegleb, *Elektronen-Donator-Acceptor-Komplexe*, Springer-Verlag, Berlin, 1961, or B. R. Foster, *Organic Charge-Transfer Complexes*, Academic Press, New York, 1969, and references cited in both.
131. Cf. e.g. R. J. Gillespie, *Angew. Chem.*, **79**, 885 (1967); *Angew. Chem., Int. Ed. Engl.*, **6**, 819 (1967) or *Molecular Geometry*, van Nostrand Reinhold Co., London, 1972, and references cited therein.
132. E. Haselbach, E. Heilbronner, A. Mannschreck and W. Seitz, *Angew. Chem.*, **82**, 880 (1970); *Angew. Chem., Int. Ed. Engl.*, **9**, 902 (1970).
133. P. J. Bassett and D. R. Lloyd, *J. Chem. Soc. (A)*, 641 (1971).
134. D. R. Lloyd and P. J. Roberts, *J. Electron. Spectrosc. Relat. Phenom.*, **7**, 325 (1975).
135. J. C. Green, M. L. H. Green, P. J. Joachim, A. F. Orchard and D. W. Turner, *Phil. Trans. R. Soc. London, Ser. A*, **268**, 111 (1970).
136. A. E. Jonas, G. K. Schweitzer, F. A. Grimm and T. A. Carlson, *J. Electron Spectrosc. Relat. Phenom.*, **1**, 29 (1972/73) and references cited therein.
137. B. W. Yates, K. H. Tan, G. M. Bancroft, L. L. Coatsworth and J. S. Tse, *J. Chem. Phys.*, **83**, 4906 (1985).
138. For more general reviews on PE spectra of $EHal_4$ molecules see References 3 and 37, and especially A. W. Potts, H. J. Lempka, D. G. Steets and W. C. Price, *Phil. Trans. R. Soc. London, Ser. A*, **268**, 59 (1970) or R. N. Dixon, J. N. Murrell and B. Narayan, *Mol. Phys.*, **20**, 611 (1971).

139. H. Bock, E. Haselbach, E. Maier and H. Stafast, *Helv. Chim. Acta*, **59**, 1035 (1976); see also Reference 38.
140. S. Cradock and R. Whiteford, *J. Chem. Soc., Faraday Trans. 2*, 281 (1972).
141. H. Bock, P. Mollère, G. Becker and G. Fritz, *J. Organomet. Chem.*, **61**, 113 (1973).
142. A. Almenningen, O. Bastiansen, V. Ewing, K. Hedberg and T. Traetteberg, *Acta Chem. Scand.*, **17**, 2455 (1963). For H_3COCH_3 cf. K. Kimura and M. Kubo, *Nature*, **39**, 533 (1959) and for H_3COSiH_3 cf. C. Glidewell, D. W. H. Rankin, A. G. Robiette, G. M. Sheldrick, B. Beagley and J. M. Freemann, *J. Mol. Struct.*, **5**, 417 (1970) with the 'doubly eclipsed' conformation being calculated to have the total energy minimum (see Reference 141).
143. For visualized molecular orbital diagrams cf. W. C. Jorgensen and L. Salem, *The Organic Chemist's Book of Orbitals*, Academic Press, New York, 1973.
144. For H_3CSCH_3 cf. L. Pierce and M. Hayashi, *J. Chem. Phys.*, **35**, 479 (1961) and for $H_3SiSSiH_3$ cf. A. Almenningen, K. Hedberg and R. Seip, *Acta Chem. Scand.*, **17**, 2264 (1963).
145. P. Mollère, H. Bock, G. Becker and G. Fritz, *J. Organomet. Chem.*, **61**, 127 (1973).
146. Synthesis: N. Wiberg, W. Joo and W. Uhlenbrock, *Angew. Chem.*, **80**, 661 (1968); *Angew. Chem., Int. Ed. Engl.*, **7**, 640 (1968). Structure: M. Veith and H. Bärnighausen, *Acta Crystallogr., Sect. B*, **30**, 1806 (1974).
147. For reviews on color and constitution of azo compounds see H. Bock, *Angew. Chem.*, **77**, 469 (1965); *Angew. Chem., Int. Ed. Engl.*, **4**, 457 (1965) or Reference 148, Vol. II, pp. 68ff.
148. M. B. Robin, *Higher Excited States of Polyatomic Molecules*, Academic Press, New York, 1975.
149. H. Bock, K. Wittel, M. Veith and N. Wiberg, *J. Am. Chem. Soc.*, **98**, 109 (1976) and reference cited therein.
150. E. Haselbach and E. Heilbronner, *Helv. Chim. Acta*, **53**, 684 (1970). For 'through space' and 'through bond' interactions see also References 7 and 8 or R. Hoffmann, *Acc. Chem. Res.*, **4**, 1 (1971).
151. H. Sakurai, Y. Kobayashi and Y. Nakadeira, *J. Am. Chem. Soc.*, **93**, 5272 (1971) as well as B. G. Ramsey, *J. Organomet. Chem.*, **67**, C67 (1974).
152. Cf. e. g. the summary by R. West, *Angew. Chem.*, **99**, 1231 (1987); *Angew. Chem., Int. Ed. Engl.*, **26**, 1201 (1987).
153. A literature-search profile for a complete retrieval of all PES ionization energies ever recorded for silicon-containing molecules is difficult to design. The excerpt from the DOW CORNING data file, generously provided by Dr. W. Atwell, comprised 330 entries, which were compared to the card index collected over the years at our institute, and in a few cases complemented. References to all PES investigations of Si compounds not included will be appreciated.
154. B. P. Pullen, T. A. Carlson, W. E. Moddemann, G. K. Schweitzer, W. E. Bull and F. A. Grimm, *J. Chem. Phys.*, **53**, 768 (1970).
155. T. Heinis, K. Boerlin and M. Jungen, *Chem. Phys. Lett.*, **110**, 429 (1984).
156. Cf. also S. Evans, J. C. Green, P. J. Joachim, A. F. Orchard, D. W. Turner and J. P. Maier, *J. Chem. Soc., Faraday Trans. 2*, 905 (1972), as well as comments by R. Boschi, M. F. Lappert, J. B. Redley, W. Schmidt and B. T. Wilkins, *J. Organomet. Chem.*, **50**, 69 (1973).
157. M. R. Nimlos and G. B. Ellison, *J. Am. Chem. Soc.*, **108**, 6522 (1986).
158. L. Szepes, G. Naray-Szabo, P. P. Colonna and G. Distefano, *J. Organomet. Chem.*, **117**, 141 (1976). Cf. also R. Roberge, C. Sandorfy, J. I. Matthews and O. P. Strauß, *J. Chem. Phys.*, **69**, 5105 (1978) and references cited therein.
159. C. S. Cundy, M. F. Lappert, J. B. Pedley, W. Schmidt and B. T. Wilkens, *J. Organomet. Chem.*, **51**, 99 (1973). Cf. also P. D. Mollère and K. N. Houk, *J. Chem. Soc.*, **99**, 3226 (1977).
160. A. Hosomi and T. G. Traylor, *J. Am. Chem. Soc.*, **97**, 3682 (1975).
161. G. Guimon, G. Pfister-Guillozzo, A. Faucher, P. Mazerolles and Y. Limouzin, *J. Organomet. Chem.*, **131**, 365 (1977).
162. W. Schmidt, B. T. Wilkins, G. Fritz and R. Huber, *J. Organomet. Chem.*, **59**, 109 (1973) and W. Schmidt, *Tetrahedron*, **29**, 2129 (1973).
163. T. F. Block, M. Biernbaum and R. West, *J. Organomet.*, **131**, 199 (1977).
164. V. I. Khvostenko, B. G. Zykov, V. P. Yuriev, V. F. Mironov, G. I. Kovelzon, A. A. Panasenko, V. D. Sheludyakow and I. A. Gailyunas, *J. Organomet. Chem.*, **218**, 155 (1981).
165. K. A. Ostoja-Starzewski, H. tom Dieck and H. Bock, *J. Organomet. Chem.*, **65**, 311 (1974).
166. L. Szepes, T. Koranyi, G. Naray-Szabo, A. Modelli and G. Distefano, *J. Organomet. Chem.*, **217**, 35 (1981).

167. S. Cradock, E. A. V. Ebsworth and R. A. Whiteford, *J. Chem. Soc., Dalton Trans.*, 2401 (1973).
168. Cf. Reference 124, also U. Weidner and A. Schweig, *Angew. Chem.*, **84**, 167 (1972); *Angew. Chem., Int. Ed. Engl.*, **11**, 146 (1972), also *J. Organomet.*, **37**, C29 (1972) and also A. Schweig, U. Weidner and G. Manuel, *J. Organomet. Chem.*, **67**, C4 (1974).
169. V. P. Yuriev, A. A. Panasenko, V. I. Khvostenko, L. M. Khalilov, Y. B. Yasman, M. M. Timoshenko, Y. V. Chizkov, B. G. Zykov, I. I. Furlei and S. R. Rafikov, *J. Organomet. Chem.*, **166**, 169 (1979).
170. C. Guimon, G. Pfister-Guillonzo, G. Manuel and P. Mazerolles, *J. Organomet. Chem.*, **149**, 149 (1978).
171. U. Weidner and A. Schweig, *Angew. Chem.*, **84**, 550, 551 (1972); *Angew. Chem., Int. Ed. Engl.*, **11**, 536, 537 (1972), and also R. K. Hill and D. A. Cullison, *J. Am. Chem. Soc.*, **95**, 5426 (1973) and J. Berger and W. Grahn, *Tetrahedron Lett.*, 557 (1973).
172. I. Bertoti, S. Cradock, E. A. V. Ebsworth and R. A. Whiteford, *J. Chem. Soc., Dalton Trans.*, 937 (1976).
173. A. Schweig, U. Weidner and G. Manuel, *Angew. Chem.*, **84**, 899 (1972); *Angew. Chem., Int. Ed. Engl.*, **11**, 837 (1972); cf. also H. Schmidt and A. Schweig, *Tetrahedron Lett.*, 1437 (1973).
174. W. Enßlin and H. H. Schmidtke, *Inorg. Chim. Acta*, **24**, 159 (1977).
175. H. Bock and H. Seidl, *J. Am. Chem. Soc.*, **90**, 5694 (1968).
176. S. Cradock, R. H. Findlay and M. H. Palmer, *J. Chem. Soc., Dalton Trans.*, 1650 (1974) as well as S. Cradock, E. A. V. Ebsworth, H. Moretto and D. W. H. Rankin, *J. Chem. Soc., Dalton Trans.*, 390 (1975).
177. S. Cradock, E. A. V. Ebsworth and A. R. Green, *J. Chem. Soc., Dalton Trans.*, 759 (1978).
178. W. Kaim and H. Bock, *J. Organomet. Chem.*, **164**, 281 (1979).
179. E. Haselbach, U. Klemm, U. Buser, R. Gschwind, M. Jungen, E. Kloster-Jensen, J. P. Maier, O. Marthaler, H. Christen and P. Baertschi, *Helv. Chim. Acta*, **64**, 823 (1981).
180. R. Gleiter, W. Schäfer and H. Sakurai, *J. Am. Chem. Soc.*, **107**, 3046 (1985).
181. B. G. Ramsey, A. Brook. A. R. Bassindale and H. Bock, *J. Organomet. Chem.*, **74**, C41 (1974). Cf. also R. S. Brown, *Can. J. Chem.*, **53**, 2446 (1975).
182. A. Fadini, E. Glozbach, P. Krommes and J. Lorberth, *J. Organomet. Chem.*, **149**, 297 (1978).
183. K.-H. A. Ostoja-Starzewski, W. Richter and H. Schmidbaur, *Chem. Ber.*, **109**, 473 (1976).
184. R. A. McLean, *Can. J. Chem.*, **51**, 2089 (1973).
185. P. K. Bischof, M. J. S. Dewar, D. W. Goodman and T. B. Jones, *J. Organomet. Chem.*, **82**, 89 (1974).
186. T. Veszpremi, Y. Harada, K. Ohno and H. Mutoh, *J. Organomet. Chem.*, **244**, 115 (1983). For assignment cf. T. Veszpremi and J. Nagy, *J. Organomet. Chem.*, **255**, 41 (1983).
187. T. Veszpremi, Y. Harada, K. Ohno and H. Mutoh, *J. Organomet. Chem.*, **266**, 9 (1984).
188. G. Distefano, S. Pignataro, L. Szepes and J. Borossay, *J. Organomet. Chem.*, **104**, 173 (1976).
189. I. Novak and A. W. Potts, *J. Organomet. Chem.*, **262**, 17 (1984).
190. Y. Limouzin and J. C. Maire, *J. Organomet. Chem.*, **92**, 169 (1975) and **105**, 179 (1976).
191. F. P. Colonna, R. Danieli, G. Distefano and A. Ricci, *J. Chem. Soc., Perkin. Trans. 2*, 306 (1976).
192. M. A. Weiner, M. Lattmann and S. O. Grim, *J. Org. Chem.*, **40**, 1292 (1975).
193. W. Kaim, H. Tesmann and H. Bock, *Chem. Ber.*, **113**, 3221 (1980).
194. W. Kaim, *J. Organomet. Chem.*, **282**, 1 (1985).
195. H. Bock and W. Kaim, *Z. Anorg. Allg. Chem.*, **459**, 103 (1979).
196. W. Kaim and H. Bock, *Chem. Ber.*, **111**, 3585 (1978).
197. B. G. Zykov, N. P. Erchak, V. I. Khvostenko, E. Lukevits, V. F. Matorykina and N. L. Asfarandiarov, *J. Organomet. Chem.*, **253**, 301 (1983).
198. T. Veszpremi, Y. Harada, K. Ohno and H. Mutoh, *J. Organomet. Chem.*, **252**, 121 (1983).
199. H. Bock and B. Roth, *Phosphorus and Sulfur*, **14**, 211 (1983).
200. E. Heilbronner, V. Hornung, H. Bock and H. Alt, *Angew. Chem.*, **81**, 537 (1969); *Angew. Chem., Int. Ed. Engl.*, **8**, 524 (1969) E. Heilbronner, V. Hornung, F. H. Pinkerton and S. F. Thames, *Helv. Chim. Acta*, **55**, 289 (1972).
201. A. Ricci, D. Pietropaolo, G. Distefano, D. Macciantelli and F. P. Colonna, *J. Chem. Soc. Perkin Trans. 2*, 689 (1977).
202. S. Cradock, E. A. V. Ebsworth, W. J. Savage and R. A. Whiteford, *J. Chem. Soc., Faraday Trans. 2*, 68, 934 (1972).

203. R. Demuth, *Z. Naturforsch, B*, **32**, 1252 (1977).
204. M. C. Green, M. F. Lappert, J. B. Pedley, W. Schmidt and B. T. Wilkins, *J. Organomet. Chem.*, **31**, C55 (1971).
205. T. Veszpremi, L. Bihatsi, Y. Harada, K. Ohno and H. Mutoh, *J. Organomet. Chem.*, **280**, 39 (1985).
206. D. H. Harris, M. F. Lappert, J. B. Pedley and G. J. Sharp, *J. Chem. Soc., Dalton Trans.*, 945 (1976).
207. D. H. Aue, H. M. Webb and M. T. Bowers, *J. Am. Chem . Soc.*, **97**, 4136 (1975).
208. S. G. Gibbins, M. F. Lappert, J. B. Pedley and G. J. Sharp, *J. Chem. Soc., Dalton Trans.*, 72 (1975).
209. Cf. P. Livant, M. L. McKee and S. D. Worley, *Inorg. Chem.*, **22**, 895 (1983) and all the structure references given therein.
210. Cf. also the review on shapes of simple silyl compounds by E. A. V. Ebsworth, *Acc. Chem. Res.*, **20**, 295 (1987).
211. H. Bock, B. Solouki, G. Fritz and W. Hoelderich, *Z. Anorg. Allg. Chem.*, **458**, 53 (1979) and references cited therein.
212. W.-W. Dumont, H. J. Breunig, H. Schumann, H. Götz, H. Juds and F. Marschner, *J. Organomet. Chem.*, **96**, 49 (1975). For $(R_3Si)_3$ As no PE spectrum is given; the one for $(R_3Si)_3P$ has to be rescaled[211].
213. J. Kroner, W. Schneid, N. Wiberg, B. Wrackmeyer and G. Ziegleder, *J. Chem. Soc., Faraday Trans. 2*, 1909 (1978).
214. H. Bock and U. Krynitz, unpublished results; cf. U. Krynitz, Ph.D. Thesis, Univ. Munich, 1969.
215. S. Cradock and D. W. H. Rankin, *J. Chem. Soc., Faraday Trans.* 2, 940 (1972).
216. S. Cradock, E. A. V. Ebsworth and J. D. Murdoch, *J. Chem. Soc., Faraday Trans. 2*, 67, 86 (1972).
217. H. Bock and R. Dammel, unpublished results; cf. R. Dammel, Ph.D. Thesis, Univ. Frankfurt, 1986.
218. C. Glidewell, D. W. Rankin, A. G. Robiette and G. M. Sheldrick, *J. Chem. Soc., (A)*, 318 (1970).
219. Cf. e.g. W. T. Reichle, *Inorg, Chem.*, **3**, 402 (1964); R. West and J. S. Thayer, *J. Am. Chem. Soc.*, **84**, 1763 (1962) or S. Cradock and E. A. V. Ebsworth, *J. Chem. Soc. (A)*, 1420 (1968) and references cited therein.
220. J. Kroner, D. Nölle H. Nöth and W. Winterstein, *Chem. Ber.*, **108**, 3830 (1975).
221. W. Kaim. *J. Am. Chem. Soc.*, **105**, 707 (1983).
222. E. Ciliberto, K. A. Doris, W. J. Pietro, G. M. Reisner, D. E. Ellis, I. Fragala, F. H. Herbstein, M. A. Ratner and T. J. Marks, *J. Am. Chem. Soc.*, **106**, 7748 (1984).
223. H. Bock, W. Kaim, N. Wiberg and G. Ziegleder, *Chem. Ber.*, **111**, 3150 (1978) and references cited therein.
224. R. Gleiter, M. C. Boehm, M. Eckert-Maksić, W. Schäfer, M. Baudler, Y. Aktalay, G. Fritz and K.-D. Hoppe, *Chem. Ber.*, **116**, 2972 (1983).
225. S. Cradock, E. A. V. Ebsworth and I. B. Muiry, *J. Chem. Soc., Dalton Trans.*, 25 (1975).
226. W. Hönle and H. G. von Schnering, *Z. Anorg. Allg. Chem.*, **440**, 171 (1978).
227. S. Cradock, E. A. V. Ebsworth, G. D. Meikle and D. W. H. Rankin, *J. Chem. Soc., Dalton Trans.*, 805 (1975).
228. J. C. Green, M. Peyne, E. A. Seddon and R. A. Andersen, *J. Chem. Soc., Dalton Trans.*, 887 (1982) and references cited therein.
229. M. F. Lappert, J. B. Pedley, G. J. Sharp and D. C. Bradley, *J. Chem. Soc., Dalton Trans.*, 1737 (1976).
230. K. N. Houk, N. G. Rondom, M. N. Paddon-Row, C. W. Jefford, P. T. Huy, P. D. Burrow and K. D. Jordan, *J. Am. Chem. Soc.*, **105**, 5563 (1983).
231. G. Distefano, A. Ricci, F. P. Colonna, D. Pietropaolo and S. Pignataro, *J. Organomet. Chem.*, **78**, 93 (1974).
232. H. Bock and U. Stein, unpublished results, cf. U. Stein, Ph.D. Thesis, Univ. Frankfurt, 1980.
233. F. Bernardi, G. Distefano, A. Modelli, D. Pietropaolo and A. Ricci, *J. Organomet. Chem.*, **128**, 331 (1977).
234. E. Andoni, M. Bossa, C. Cauletti, C. Furlani and A. Palma, *J. Organomet. Chem.*, **244**, 343 (1983).
235. Cf. G. Tschmutowa and H. Bock, *Z. Naturforsch.*, **31b**, 1611 (1976) and references cited therein.
236. G. DeAlti, P. Decleva and A. Sgamellotti, *Chem. Phys.*, **35**, 283 (1978) review and assign the F_nSiH_{4-n} PE spectra based on MS-X_α calculations (cf Reference 110).

237. I. Novak and H. Vancik, *J. Electron Spectrosc. Relat. Phenom.*, **24**, 289 (1981).
238. R. Roberge, C. Sandorfy and O. P. Strausz, *Theor. Chim. Acta*, **52**, 171 (1979).
239. A. Flamini, E. Semprini, F. Stefani, S. Sorriso and G. Cardaci, *J. Chem. Soc., Dalton Trans.*, 731 (1976).
240. S. Evans, J. C. Green and S. E. Jackson, *J. Chem. Soc., Faraday Trans. 2*, 191 (1973).
241. J. A. Ulman and T. P. Fehlner, *J. Am. Chem. Soc.*, **98**, 1119 (1976).
242. D. K. Creber and G. M. Bancroft, *Inorg. Chem.*, **19**, 643 (1980).
243. M. F. Lappert, J. B. Pedley and G. Sharp, *J. Organomet. Chem.*, **66**, 271 (1974).
244. J. C. Green, D. R. Lloyd, L. Galyer, K. Mertis and G. Wilkinson, *J. Chem. Soc., Dalton Trans.*, 1403 (1978).
245. B. E. Bursten, F. A. Cotton, J. C. Green, E. A. Seddon and G. G. Stanley, *J. Am. Chem. Soc.*, **102**, 4579 (1980).
246. J. C. Green, M. P. Payne and J. H. Teuben, *Organometallics*, **2**, 203 (1983).
247. I. L. Fragala, J. Takats and M. A. Zerbo, *Organometallics*, **2**, 1502 (1983).
248. G. Bruno, E. Ciliberto, R. D. Fischer, I. Fragala and A. W. Spiegl, *Organometallics*, **1**, 1060 (1982).
249. S. Cradock, E. A. V. Ebsworth and A. Robertson, *J. Chem. Soc., Dalton Trans.*, 22 (1973); *Chem. Phys. Lett.*, **30**, 413 (1975).
250. J. N. Louwen, R. R. Andrea, D. J. Stufkens and A. Oskam, *Z. Naturforsch., B*, **38**, 194 (1983).
251. P. Jutzi, D. Kaml and C. Krüger, *Angew. Chem.*, **98**, 163 (1986); *Angew. Chem., Int. Ed. Engl.*, **25**, 164 (1986).
252. R. Gleiter and P. Jutzi, unpublished results; R. Gleiter, personal communication, Univ. Heidelberg.

The Chemistry of Organic Silicon Compounds
Edited by S. Patai and Z. Rappoport

CHAPTER **10**

General synthetic pathways to organosilicon compounds

LEONHARD BIRKOFER and OSKAR STUHL†
Institute of Organic Chemistry, University of Düsseldorf, D-4000 Düsseldorf 1, West Germany

† Present address:
R. J. Reynolds Tobacco GmbH, P.O. Box 100885, D-5000 Cologne 1, West Germany

I. ABBREVIATIONS

Ac = acetyl
Ar = aryl
Bu = butyl
m-CPBA = *m*-chloroperbenzoic acid
DDQU = 2,3-dichloro-5,6-dicyanobenzoquinone
DCB = dicyclohexyl borane
DIBAL = diisobutylaluminum hydride
DME = dimethoxyethane
DMF = dimethyl formamide
Et = ethyl
Hex = hexyl
HMPA = hexamethylphosphoric triamide
LDA = lithium diisopropylanilide
LDMC = lithium dimethyl cuprate(I)
LiTMP = lithium 2,2,6,6-tetramethylpiperidide
M = metal
Me = methyl
Mes = 2,4,6-trimethylphenyl-('mesityl')
Naph = naphthyl
NBS = *N*-bromosuccinimide
PE = petroleum ether
Ph = phenyl
Pr = propyl

Py	= pyridine	TMS	= trimethylsilyl
RT	= room temperature	TMSA	= trimethylazidosilane
TES	= triethylsilyl	Tos	= toluenesulfonyl('tosyl')
THF	= tetrahydrofuran	triflate	= trifluoromethylsulfonate
TMEDA	= tetramethylethylenediamine		

II. GENERAL INTRODUCTION

Over the last three decades the importance of organosilicon chemistry has greatly increased. While at the beginning the main aim of research was the synthesis of Si–C compounds, later research activities have expanded and Si–N, Si–O as well as Si compounds with other group VI elements have attracted much attention. Besides the synthetic aspects, organosilicon chemistry has become a major tool in organic chemistry for the preparation of silylated synthons as reactive intermediates in organic synthesis and for an ever-increasing number of technical applications.

In this chapter we discuss the major reaction pathways to synthesize Si–C, Si–N, Si–O and Si–S(Se,Te) compounds. Our prime aim is to summarize the main reactions of interest to the synthetic chemist in laboratory or pilot-scale preparations of the various classes of organosilicon compounds. Hence, we emphasise the experimental side and not theoretical or mechanistic details which have already been widely discussed in the literature cited. Furthermore, we wanted to avoid extensive overlap with other chapters.

Wherever feasible, this review article covers only literature published during the last ten years, until late 1986; for earlier publications it may be advisable to refer also to previously published review papers[1–20].

Since this review covers a heterogeneous scientific area, we chose to start the respective sections with a brief theoretical introduction—where needed—instead of one comprehensive introduction at the beginning.

III. ALKYLSILANES

A. Introduction

Common synthetic pathways to obtain mono-, di-, tri- and tetraalkylsilanes are reactions of the respective halogenosilanes with lithiated alkyl compounds, with alkylmagnesium halides (Grignard reactions) and with alkyl halides in a Wurtz–Fittig-type reaction, as well as the addition of silanes to alkenes, i.e. hydrosilylation.

The first three methods are especially suitable for the synthesis of symmetrical alkylsilanes. Hydrosilylation offers a wide variety of applications which give rise to a large number of non-symmetrical organosilanes to be discussed in a separate section (*vide infra*).

B. Symmetrical Alkylsilanes

In this section we discuss the synthesis of tri- and tetraorganylsilanes.* A routine method to obtain tetraorganylsilanes is the reaction of lithiated alkyl compounds with silicon halides which affords the desired organylsilanes (equation 1a) with yields of up to 97%—illustrated by the synthesis of tetrabutylsilane (**3**) (equation 1b)[21]. Alternatively,

* In this review the term "organyl" is used to describe *any* organic group (alkyl, aryl, heteroaryl). While this is not an accepted usage in the English chemical literature, it is a needed term and there is no generally used simple alternative for it.

tetraalkoxylsilane (**4**)[22] or silane (**5**)[23], respectively, can be directly converted into tetraorganylsilanes by employing BuLi (equations 2a and 3a, respectively) which is illustrated by representative examples (equations 2b and 3b, respectively).

$$SiX_4 + 4\ RM \xrightarrow[-4MX]{} SiR_4 \qquad (1a)$$

$$\underset{(\mathbf{1})}{SiCl_4} + 4\ \underset{(\mathbf{2})}{BuLi} \xrightarrow[-4LiCl]{} \underset{(\mathbf{3})}{SiBu_4} \qquad (1b)$$

$$Si(OR')_4 + 4\ RM \xrightarrow[-4MOR']{} SiR_4 \qquad (2a)$$

$$\underset{(\mathbf{4})}{Si(OEt)_4} + 4\ \underset{(\mathbf{2})}{BuLi} \xrightarrow[-4LiOEt]{} \mathbf{3} \qquad (2b)$$

$$\underset{(\mathbf{5})}{SiH_4} + 4\ RM \xrightarrow[-4MH]{} SiR_4 \qquad (3a)$$

$$\mathbf{5} + 4\ EtLi \xrightarrow[-4LiH]{} \underset{(\mathbf{6})}{SiEt_4} \qquad (3b)$$

Steric effects (e.g. bulkiness) are less important in reactions employing lithiated alkyl reagents compared to Grignard-type approaches. While it is feasible to obtain triisopropylsilane (**9**) by the reaction of isopropyllithium (**8**) and trichlorosilane (**7**) (equation 4)[24], the corresponding Grignard reaction only yields diisopropylsilane.

$$\underset{(\mathbf{7})}{HSiCl_3} + 3\ \underset{(\mathbf{8})}{\text{iso-PrLi}} \xrightarrow[-3LiCl]{} \underset{(\mathbf{9})}{HSi(\text{iso-Pr})_3} \qquad (4)$$

The homologous tetraisopropylsilane (**11**) cannot be accomplished by means of **8** from tetrachlorosilane (**1**) (analogously to equation 1) owing to the bulkiness of the isopropyl moiety. In this case the route via the more reactive fluoro(triisopropyl)silane (**10**) plus **8** is more preferable and forms **11** at a yield of 77% in a non-polar medium (i.e. petroleum

$$\underset{(\mathbf{10})}{FSi(\text{iso-Pr})_3} + \mathbf{8} \xrightarrow[-LiF]{} \underset{(\mathbf{11})}{Si(\text{iso-Pr})_4} \qquad (5)$$

ether) during 6 d under reflux (equation 5)[24a]. Similarly, the reaction of $SiCl_4$ with cyclohexyllithium (**12**) leads to chloro(tricyclohexyl)silane (**13**)[25] (equation 6). However, tetracyclohexylsilane (**15**) can be obtained analogously to equation 1—when the more

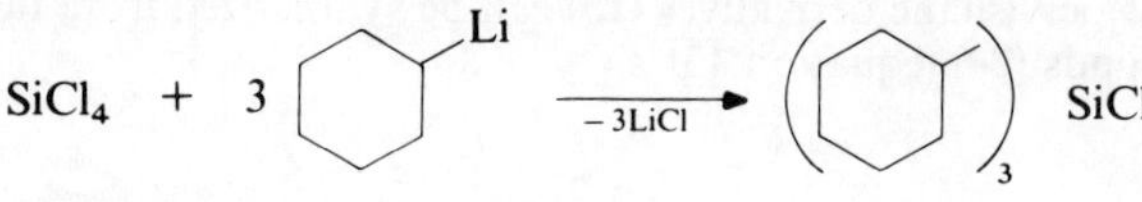

(6)

(**12**) (**13**)

reactive tetrafluorosilane (**14**) is applied in ligroin at 55–75 °C (equation 7)[26]. Generally speaking, this latter type of reaction allows a large number of tetraorganylsilanes to be synthesized with bulky moieties.

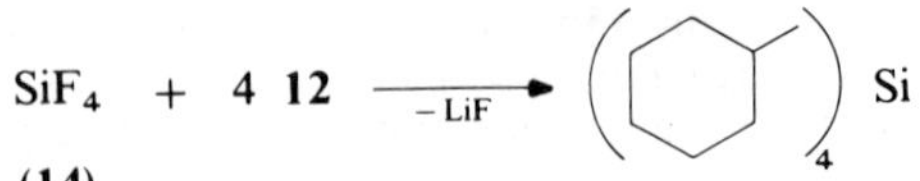

(14) (15) (7)

C. Non-symmetrical Alkylsilanes

An alternative approach—in addition to those outlined above—to synthesize alkylsilanes is the reaction of chlorotrimethylsilane (**16**) with olefins and magnesium in the presence of HMPA[27]. For instance, octene-1 (**17**) can be converted by the aforementioned set of reagents in a pressure vessel at 200 °C/3d into the desired octyltrimethylsilane (**18**) in 50% yield (equation 8)[27]. At lower temperatures (3d/reflux) partial allyl rearrangement occurs and, besides **18**, a monosilylated olefin (**19**) is formed in a low yield (20%) which cannot be improved, even after addition of catalysts such as $FeCl_3$ or $TiCl_4$ (equation 9)[27]. Variously substituted alkylsilanes can be formed from trichloroalkylsilanes or from chlorotrialkylsilanes, respectively, by means of the Grignard reaction (equation 10)[28].

$$\underset{(\mathbf{16})}{Me_3SiCl} + \underset{(\mathbf{17})}{Me(CH_2)_5CH{=}CH_2} \xrightarrow{\text{Mg/HMPA}} \underset{(\mathbf{18})}{Me_3Si(CH_2)_7Me} \tag{8}$$

$$\mathbf{16} + \mathbf{17} \xrightarrow{\text{Mg/HMPA/FeCl}_3\text{ or TiCl}_4} \mathbf{18} + \underset{(\mathbf{19})}{Me(CH_2)_4CH{=}CHCH_2SiMe_3} \tag{9}$$

$$\begin{aligned} RSiCl_3 + 3R'MgBr &\xrightarrow[-3MgBrCl]{} RSiR_3' \\ R_3SiCl + R'MgBr &\xrightarrow[-MgBrCl]{} R_3SiR' \end{aligned} \tag{10}$$

In a similar vein silylated cycloalkanes **21** can be obtained from the corresponding halogenocycloalkanes **20** with Me_3SiCl and Mg in THF as solvent. However, the larger the ring size, the smaller the yield (equation 11)[29].

$$\underset{(\mathbf{20})}{(CH_2)_n\ CHCl} \xrightarrow[-MgCl_2]{Me_3SiCl/Mg/THF} \underset{(\mathbf{21})}{(CH_2)_n\ CHSiMe_3} \tag{11}$$

In a related manner 1-TMS-adamantane (**23**) can be accomplished by the employment of Me_3SiCl and sodium in HMPA with 1, 3-dichloroadamantane (**22**) which, in a first step, affords dehalogenation and, subsequently, silylation[30] (equation 12). Analogously, arylbicyclo[2.2.2]octylsilane derivatives (**25**) can be synthesized from the corresponding lithiated compounds (**24**) (equation 13)[31].

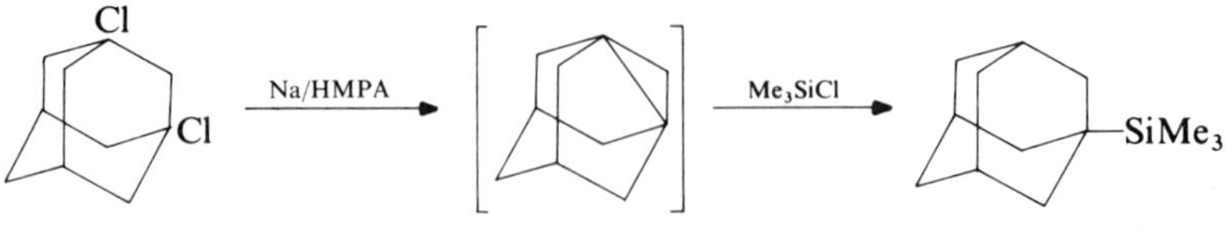

(22) (23) (12)

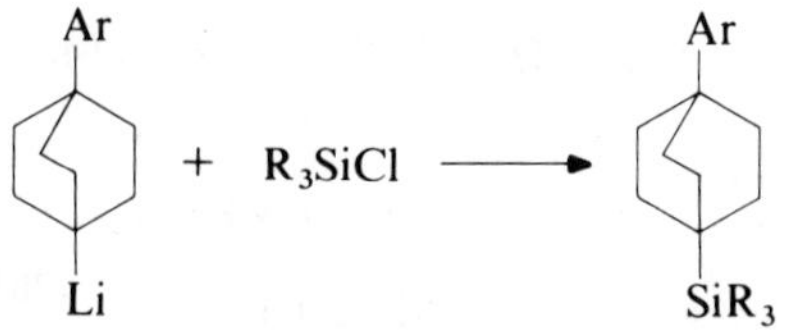

(13)

(24) **(25)**

Ar = p-FC_6H_4, m-FC_6H_4, Ph
R = Cl, H, Me, Et

By slightly modifying the afore-mentioned procedure it is feasible to accomplish the synthesis of cyanomethyltrimethylsilane (**27**) from Me_3SiCl, bromoacetonitrile (**26**) and Zn in THF/benzene (equation 14)[32] (yield: 81%).

$$\underset{\textbf{(26)}}{BrCH_2CN} \xrightarrow{Me_3SiCl/Zn/THF/C_6H_6} \underset{\textbf{(27)}}{Me_3SiCH_2CN} \qquad (14)$$

D. Hydrosilylation

The addition of silanes across multiple bonds occurs in the presence of catalysts, mostly complexes of VIII B group elements (e.g. Co, Ni, Pd, Pt) (equation 15). Concerning the course of reaction it can be generally said that, predominantly, (a) the hydrosilylation of simple alkenes and alkynes places the silicon atom at the less substituted carbon atom; and (b) via catalysts and reaction conditions a stereospecific course of reaction can be accomplished. Furthermore, as a very positive side-effect, asymmetrical hydrosilylation can be realized if chiral catalysts are employed[33a, 33b, 33c]. For further details we recommend comprehensive review papers on this subject[33a–c].

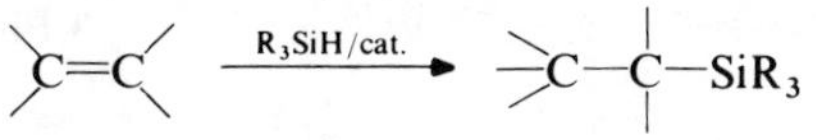

(15)

In the following we will discuss some representative examples which illustrate the great variety of applications.

The hydrosilylation of 1-hexene (**28**) with triethoxysilane (**29**) in the presence of a poly-γ-mercaptopropylsiloxane platinum complex leads to hexyltriethoxysilane (**30**) (equation 16), whereas the hydrosilylation of acetylene (**31**) can afford both the mono- and disilylated product, i.e. vinyltriethoxysilane (**32**) and 1,2-bis(triethoxysilyl)ethane (**33**) respectively (equation 17)[34].

$$\underset{\textbf{(28)}}{CH_3(CH_2)_3CH{=}CH_2} + \underset{\textbf{(29)}}{HSi(OEt)_3} \xrightarrow{\text{cat.}} \underset{\textbf{(30)}}{CH_3(CH_2)_5Si(OEt)_3} \qquad (16)$$

$$\underset{\textbf{(31)}}{HC{\equiv}CH} \xrightarrow{\textbf{29}/\text{cat.}} \underset{\textbf{(32)}}{CH_2{=}CHSi(OEt)_3} + \underset{\textbf{(33)}}{(EtO)_3SiCH_2CH_2Si(OEt)_3} \qquad (17)$$

As a catalyst, dichloro-bis(2,4,6-trimethylpyridine)platinum(II) is very suitable for the hydrosilylation of **28** or styrene (**36**) with chlorodimethylsilane (**34**), leading in the case of

$$\underset{}{\mathbf{28}} + \underset{(\mathbf{34})}{ClMe_2SiH} \xrightarrow{\text{cat.}} \underset{(\mathbf{35})}{CH_3(CH_2)_5SiMe_2Cl} \qquad (18)$$

$$\underset{(\mathbf{36})}{PhCH{=}CH_2} + \mathbf{34} \xrightarrow{\text{cat.}} \underset{(\mathbf{37})}{PhCH(SiMe_2Cl)Me} + \underset{(\mathbf{38})}{PhCH_2CH_2SiMe_2Cl} \qquad (19)$$

36 to silylation of either the α- or the β-carbon atom (equations 18 and 19)[35]. The hydrosilylation of bicyclo[2.2.1]heptene (**39**) with $HSiR_2R'$ in the presence of a platinum complex with silyl ligands as catalyst gives rise to 1-silylbicyclo[2.2.1]heptane (**40**) in 76–85% yield (equation 20)[36]. The hydrosilylation of phenylacetylene (**41**) by means of

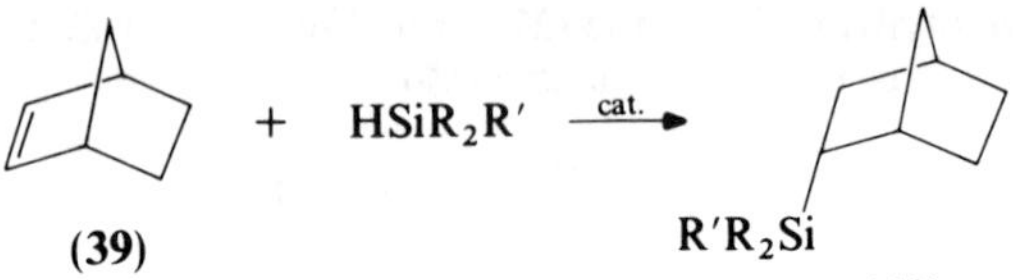

(**39**) (**40**) (20)

	yield
R = Me, R′ = Cl	85%
R = R′ = Et	76%

$HSiR_2Me$ affords both α- and β-silylated products, the latter being formed in higher yield (equation 21). However, the analogous treatment of styrene (**36**) exclusively leads to the silylation of the β-carbon atom (equation 22)[37]. The hydrosilylation of β-pinene (**45**) with

$$\underset{(\mathbf{41})}{PhC{\equiv}CH} + HSiMeR_2 \xrightarrow{(Ph_3P)_3RhCl/HMPA} \underset{(\mathbf{42})}{PhCH{=}CHSiMeR_2} + \underset{(\mathbf{43})}{PhCH(SiMeR_2)CH_3} \qquad (21)$$

R=Et, *i*-Pr, *i*-Bu

$$\underset{(\mathbf{36})}{PhCH{=}CH_2} + HSiEtMe_2 \xrightarrow{(Ph_3P)_3RhCl/HMPA} \underset{(\mathbf{44})}{PhCH_2CH_2SiEtMe_2} \qquad (22)$$

di- and trichlorosilane using Ni-acetylacetonate as a catalyst causes the formation of two distinct products (equation 23)[38].

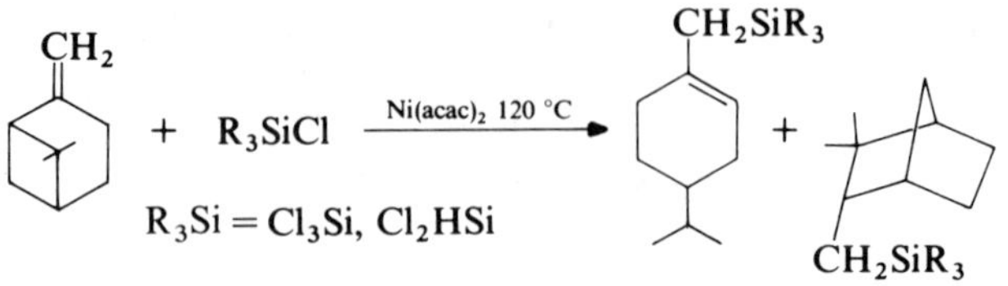

(**45**) (**46**) (**47**) (23)

E. Organyldisilanes

For the synthesis of organyldisilanes, two major synthetic pathways are common: (a) reaction of halogenotriorganylsilanes with alkali metals, and (b) reaction of halogenosilanes with metallated organylsilanes (Scheme 1). The syntheses and applications are treated in another chapter of this volume.

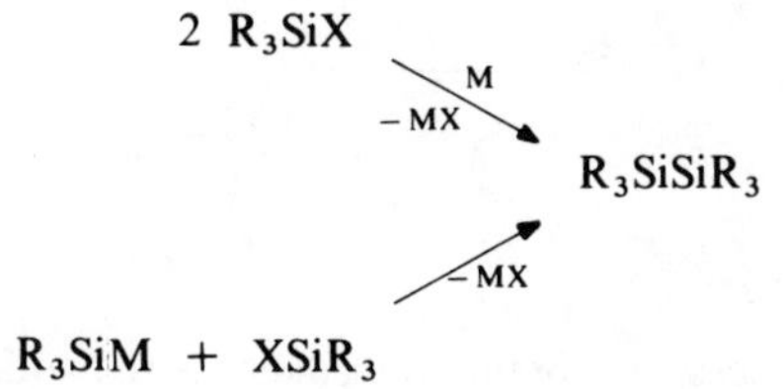

SCHEME 1

IV. VINYLSILANES

A. Introduction

Vinylsilanes have been widely recognized as facile intermediates for many sorts of applications[39]. While vinylsilanes undergo electrophilic substitution at the carbon atom containing the silyl group, the double-bond geometry is quite frequently retained. As a consequence vinylsilanes are well suited as precursors to a variety of reaction products with geometrically defined double bonds. Also feasible is an addition–elimination sequence, however here the reaction conditions have an influence on the stereochemistry of the reaction products.

The main synthetic pathways to vinylsilanes are: (a) direct processes in the gas phase, (b) Grignard reactions, (c) metallation reactions (e.g. with lithiated reagents), (d) hydrosilylation of alkynes and related addition reactions, (e) hydrogenation of alkynylsilanes, and (f) elimination reactions of alkylsilanes. Methods (a)–(e) are mostly used; method (f), however, has become less common nowadays.

B. Linear Vinylsilanes

On an industrial scale, the production of vinylsilanes (**49**) from vinyl chloride (**48**) and a chlorosilane (**7**) as a direct process can only be carried out at high temperatures (equation 24)[40]. The concomitant formation of by-products due to side-reactions and/or

$$\underset{(\mathbf{48})}{CH_2{=}CHCl} + \underset{(\mathbf{7})}{HSiCl_3} \xrightarrow[-HCl]{450-500\,^{\circ}C/40s} \underset{(\mathbf{49})}{CH_2{=}CHSiCl_3} \qquad (24)$$

decomposition is disadvantageous. Additionally, compounds like 1,1,2-trichloroethene (**50**), which contain an activated double bond, can undergo this type of reaction (equation 25)[41], although at temperatures above 600 °C decomposition reactions increasingly occur. At lower temperatures one has to choose carefully the reaction conditions in order

$$\underset{(\mathbf{50})}{Cl_2C{=}CHCl} + \underset{(\mathbf{51})}{HSiRCl_2} \xrightarrow[-HCl]{600-700\,^{\circ}C} \underset{(\mathbf{52})}{Cl_2C{=}CHSiRCl_2} \qquad (25)$$

not to allow addition reactions to become dominant. This is illustrated by the example in equation 26 where, under conditions of a radical substitution (catalyst: di-*t*-butylperoxide), the yield is 12% for **54** (substitution) and 5% for **55** (addition)[42].

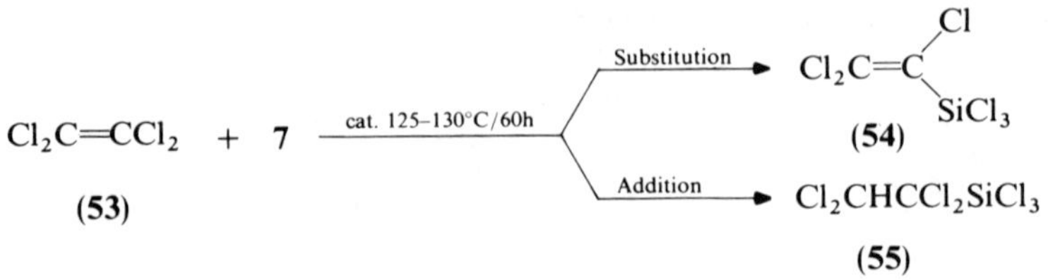

(26)

A direct industrial process to obtain vinylsilanes is the addition of silanes (e.g. **51**) to acetylene and its derivatives which has to be conducted at higher temperatures in a pressure vessel (equation 27)[43]. A major challenge is to determine the correct reaction period with an optimal reagent:substrate ratio in order to avoid further reaction to the corresponding disilylalkanes.

$$\underset{(\mathbf{31})}{HC{\equiv}CH} + \mathbf{51} \rightarrow \underset{(\mathbf{56})}{CH_2{=}CHSiRCl_2} \qquad (27)$$

If alkynes (**57**) react with dichlorosilane (**58**) mainly three types of products are formed (**59**–**61**) (equation 28)[44] which underlines the necessity to apply optimal reaction conditions to accomplish the desired reaction pathway to **59**.

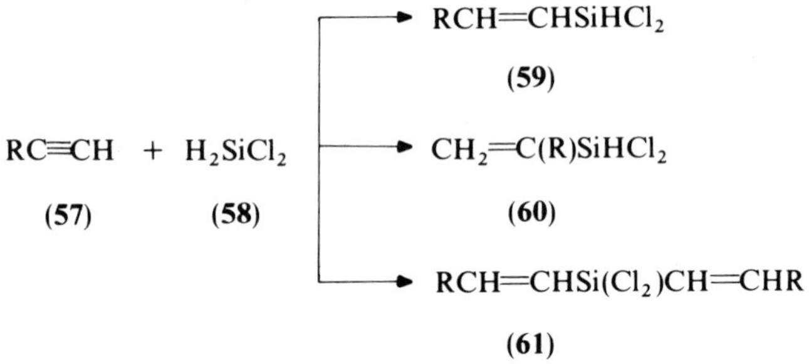

(28)

In the laboratory the classical method to synthesize vinylsilanes is the Grignard method by which, e.g., tetravinylsilane (**63**) can be obtained from tetrachlorosilane and vinylmagnesium chloride (**62**) (equation 29)[45]. Sometimes the inverse approach, i.e.

$$4\,\underset{(\mathbf{62})}{CH_2{=}CHMgCl} + SiCl_4 \xrightarrow[-4MgCl_2]{} \underset{(\mathbf{63})}{(CH_2{=}CH)_4Si} \qquad (29)$$

alkylation of chlorovinylsilanes, is more advisable as can be shown for the synthesis of trimethylvinylsilane (**65**) from trichlorovinylsilane (**49**) (equation 30)[46].

$$\mathbf{49} + 3\,\underset{(\mathbf{64})}{MeMgBr} \rightarrow \underset{(\mathbf{65})}{H_2C{=}CHSiMe_3} \qquad (30)$$

The treatment of a dialkylamino ether (**67**) with 1-trimethylsilylvinylmagnesium bromide (**66**) gives rise to α-dialkylaminomethylvinyltrimethylsilane (**68**) (equation 31)[47]

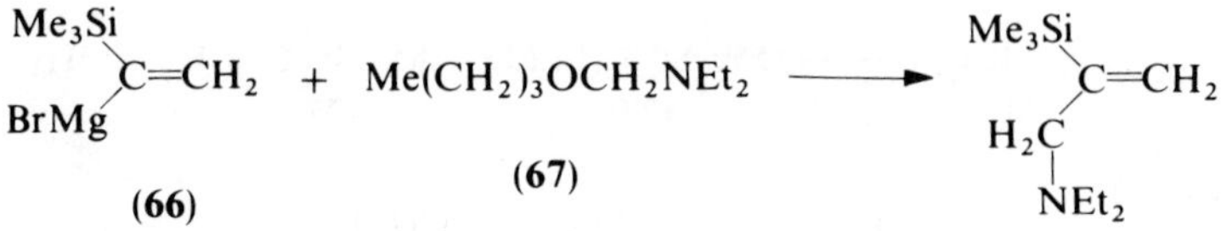
(31)

(**66**) (**67**) (**68**)

(yield: 70%). In some cases, e.g. when bulky moieties are involved, the application of catalysts can be useful. An example is the synthesis 2-TMS-dec-1-ene (**71**) from α-bromovinyltrimethylsilane (**69**) via Grignard reaction in presence of tetrakis(triphenylphosphine)palladium(0) (equation 32)[48]. In a similar vein, Grignard

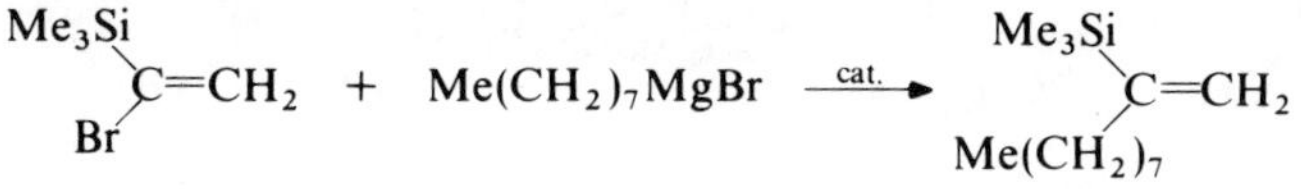
(32)

(**69**) (**70**) (**71**)

reagents from heterocycles (**72**) react with α-bromovinyltrimethylsilane (**69**) in the presence of $PdCl_2$ as a catalyst to form 1-heteroarylvinyltrimethylsilane (**73**) (equation 33)[49]; the yield is 81%.

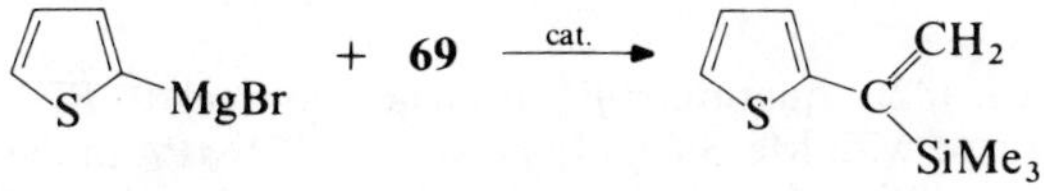
(33)

(**72**) (**73**)

Another classical approach to vinylsilanes is a Wurtz reaction with vinyl chloride (**48**) and tetrachlorosilane and sodium (equation 34)[50]. Similarly, this type of reaction can

$$4\ \mathbf{48} + SiCl_4 \xrightarrow[-\ 8NaCl]{8Na} \mathbf{63} \qquad (34)$$

also be applied to a variety of vinyl derivatives. The reaction of *trans*-1,2-dipropyl-1-halogenoethene (**74**) with Me_3SiCl and sodium in ether is stereospecific and yields, under retention of configuration, *trans*-1,2-dipropyl-1-TMS-ethene (**75**)[51]. When comparing the homologous vinyl halides, vinyl iodide was shown to be the most effective stereospecific substrate (equation 35)[51]. Using lithium, 1-chloro-1,1-bis(TMS)-ethane (**76**) can be

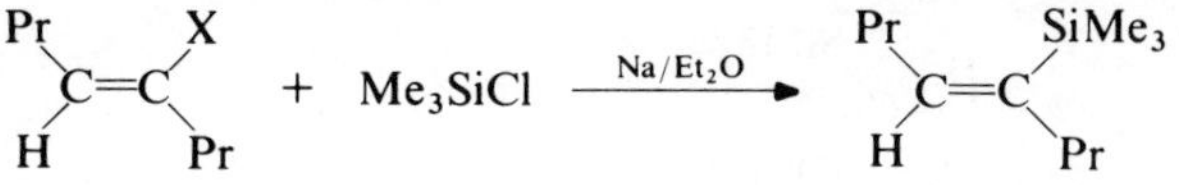
(35)

(**74**) (**75**)

X = Cl, Br, I

converted into the corresponding lithiated product which, under elimination of LiH, yields 1,1-bis(TMS)-ethene (**78**) (yield: 55%) (equation 36)[52]. The by-product, 1,1-bis(TMS)-ethane (**79**), is formed from unreacted **76** with the generated LiH (equation 37)[52].

$$\underset{(\mathbf{76})}{(Me_3Si)_2C(Cl)Me} \xrightarrow{Li} \underset{(\mathbf{77})}{(Me_3Si)_2C(Li)Me} \rightarrow \underset{(\mathbf{78})}{(Me_3Si)_2C{=}CH_2} + LiH \qquad (36)$$

$$\mathbf{76} + LiH \xrightarrow[-LiCl]{} \underset{(\mathbf{79})}{(Me_3Si)_2CHMe} \qquad (37)$$

The treatment of 1,5-dichloro-2-methylpent-1-ene (**80**) with butyllithium and, subsequently, with Me_3SiCl at -110 °C gives access to the 1-TMS-derivative (**81**) (equation 38)[53]. *Trans-β*-TMS-styrene (**84**) is available by the reaction of bis(TMS)-methyllithium

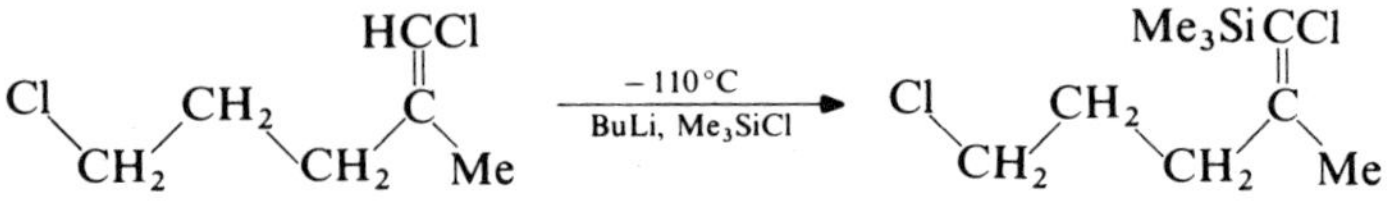

(**80**) (**81**) (38)

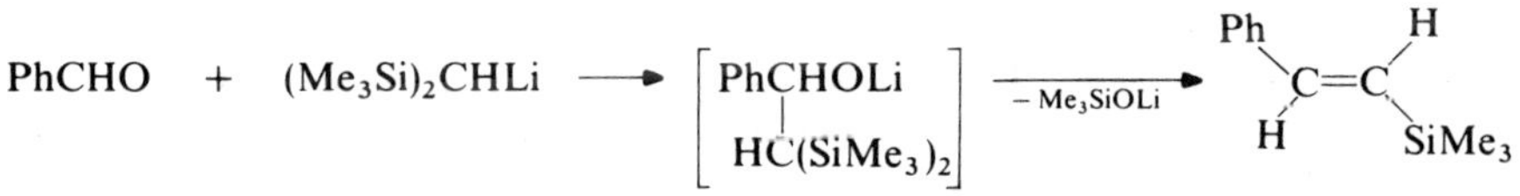

(**82**) (**83**) (**84**) (39)

(**83**) with benzaldehyde (**82**) (equation 39)[54]. A facile approach to *E*-1,2-bis(TMS)-ethene (**85**) is the treatment of **65** with Me_3SiCl, Mg powder and HMPA in the presence of $FeCl_3$ as a catalyst (equation 40)[55]. However, *E*-1,2-dialkyl-1-TMS-ethene, e.g. (**89**), is formed

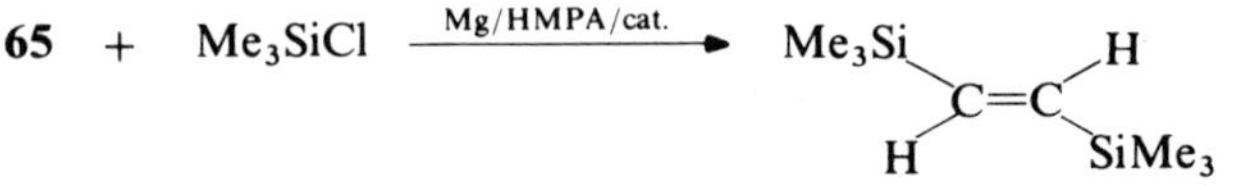

(40)

(**85**)

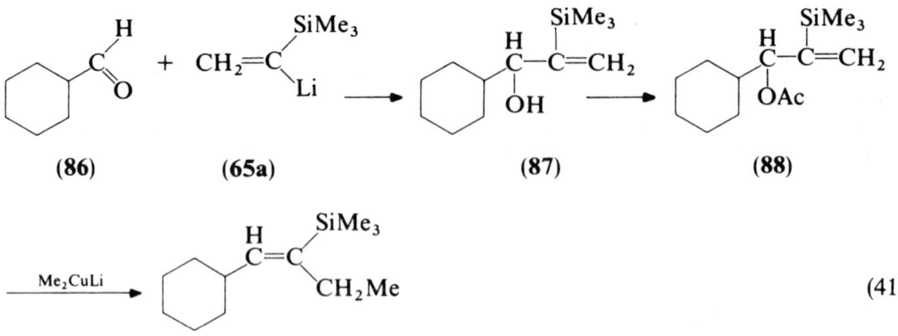

(**86**) (**65a**) (**87**) (**88**)

(**89**) (41)

by the reaction of α-lithiated vinyltrimethylsilane (**65a**) with cyclohexane carbaldehyde (**86**) which, in the first step, yields a 2-TMS-alcohol (**87**) which in turn is converted into the corresponding acetate (**88**) and, finally, treatment with a dialkyl copper lithium reagent via S_N2' coupling gives rise to **89** (94% optical purity) (equation 41)[56].

An alternative synthetic pathway to vinylsilanes starts from ketones (**90**), via their sulfonylhydrazones (**91**) to form with alkyllithium the corresponding vinylcarbanions which are then converted with Me_3SiCl into the desired vinylsilanes (**92**) (equation 42)[57-61].

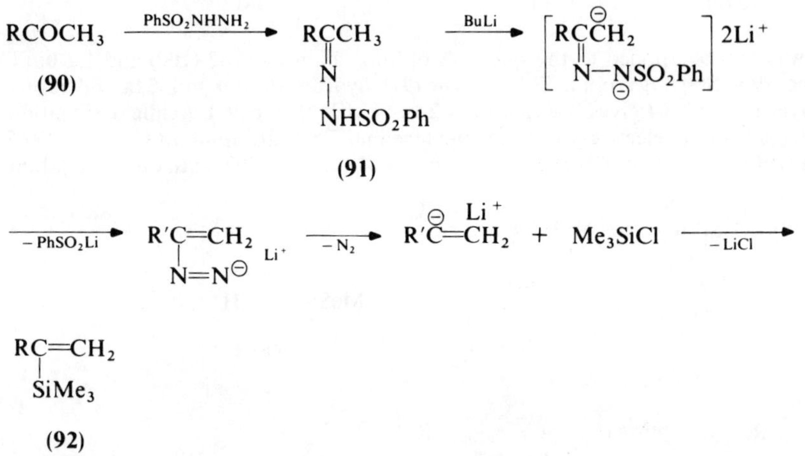

(42)

Another very useful route to vinylsilanes is hydrosilylation of alkynes by means of chlorosilanes, trialkoxysilanes, methyldichloro- or dialkoxysilanes, respectively, as catalyzed by ruthenium-, rhodium- or platinum complexes (equation 43a)[62]. A well-known example is the reaction of **31** with **7** in the presence of a ruthenium complex (equation 43b)[62].

$$\underset{(\mathbf{31})}{HC{\equiv}CH} + HSiR_3 \xrightarrow{\text{cat.}} H_2C{=}CHSiR_3 \qquad (43a)$$

$$\mathbf{31} + \underset{(\mathbf{7})}{HSiCl_3} \xrightarrow{(Ph_3P)_2RuCl} \underset{(\mathbf{49})}{CH_2{=}CHSiCl_3} \qquad (43b)$$

E-1-TMS-alkenes (**95**) can be achieved via the hydrosilylation of 1-alkynes, e.g. (**93**) with **7** or **51a** in the presence of H_2PtCl_6 and subsequent methylation by means of **64**

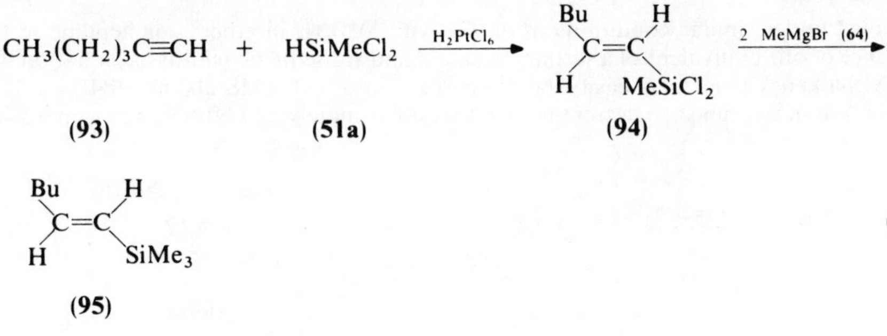

(44)

(equation 44)[63]. Analogously, *E*-1,2-bis(TMS)-ethene (**85**) can be obtained with a yield as high as 85% from TMS-acetylene (**57a**) plus dimethylchlorosilane (**34**), catalyzed by H_2PtCl_6 and subsequently methylated by MeMgI (equation 45)[64]. The same reaction

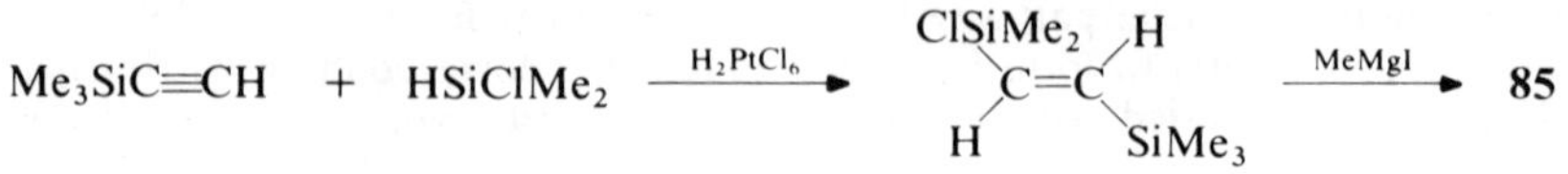

(**57a**) (**34**) (**96**) (45)

pathway can be applied to the synthesis of long-chain mono-2-(**100**) and 1,2-bis(TMS)-alkenes (**99**). Starting from a TMS-alkyne (**97**), hydrosilylation with **51a** and subsequent methylation with **64** gives rise to an *E*-1,2-bis(TMS)-alkene (**99**). Acidic desilylation with acetic acid/water selectively removes the terminal TMS-grouping to form a 1-TMS-alk-1-ene (**100**) (equation 46)[65]. The reaction of 1-TMS-alkynes (**97**) with diisobutylaluminum

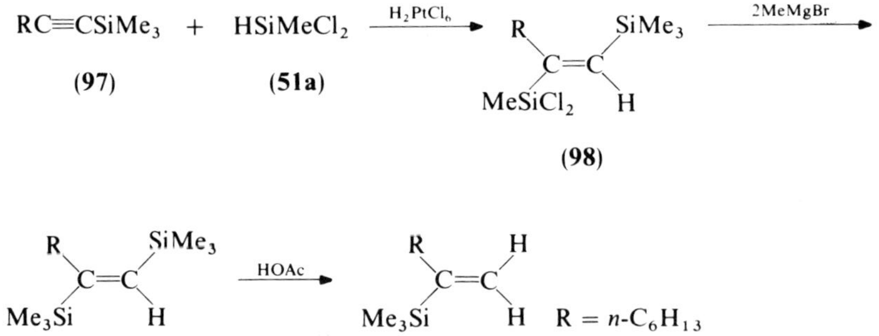

(**97**) (**51a**) (**98**)

(**99**) (**100**) $R = n\text{-}C_6H_{13}$ (46)

hydride (DIBAL) in heptane as a solvent and subsequent protolysis affords *E*-1-TMS-alkenes (**84**) (equation 47)[66]. However, the enantiomeric *Z*-1-TMS-alkenes (**84a**) can be

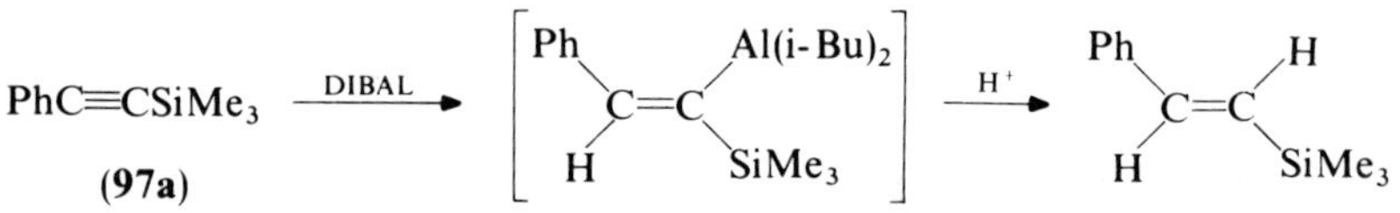

(**97a**) (**84**) (47)

obtained under similar conditions from **97a** with DIBAL in ether[67] or heptane in the presence of one equivalent of a tertiary amine[66] and subsequent acidolysis (equation 48). *Z*-TMS-alkenes (**84a**) can easily be isomerized into *E*-1-TMS-alkenes (**84**) via UV-irradiation in pyridine or via treatment with approximately 5% NBS[67]. The application

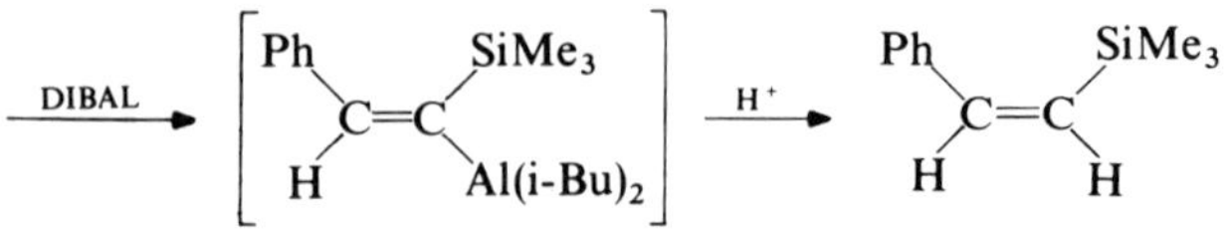

(**97a**) (**84a**) (48)

of tris-TMS-aluminium to alkynes (**101**) affords vinylsilanes (**102a/b**) in the *trans*-form exclusively (equation 49)[68].

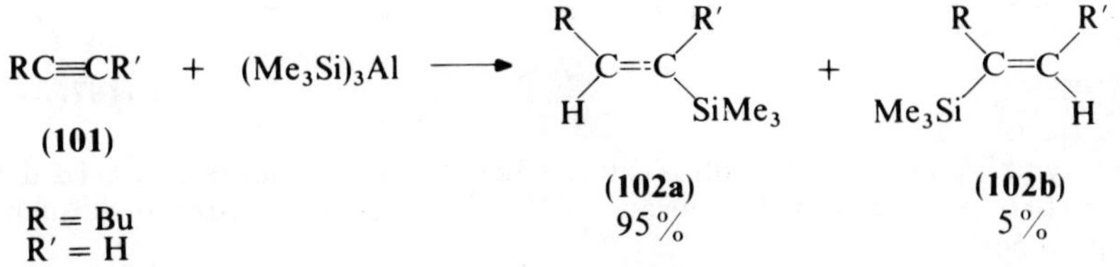

(101) R = Bu, R′ = H; (102a) 95%; (102b) 5% (49)

The addition of HBr to alkynylsilane (**57a**) catalyzed by peroxide leads to *E*-2-bromo-1-vinyltrimethylsilane (**103**) (equation 50)[69]. An alternative route to *Z*-1-TMS-alk-1-enes

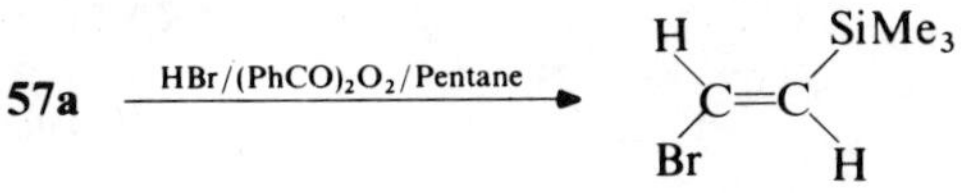

(50)

(103)

(**84a**) is the catalytic hydrogenation of 1-TMS-alkynes (**97a**) by means of Raney-nickel[70] or a Ni-boride catalyst[71] or palladium[71] (equation 51). However, the hydrogenation

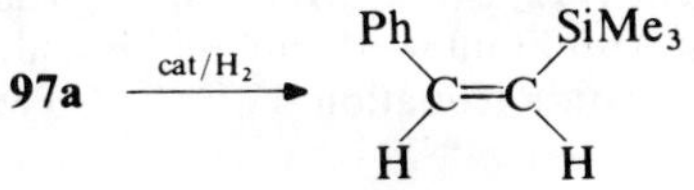

(51)

(84a)

conditions can have an effect on the course of the reaction. Thus, when 4,6-bis(TMS)-hex-5-yn-2-one (**104**) is treated with H_2/Raney-nickel, the corresponding hex-5-en-2-one (**105**) is formed; however $Pd/BaSO_4$ as the catalyst affords the hex-4-en-2-one (**106**) after a displacement of a proton took place (equation 52)[72].

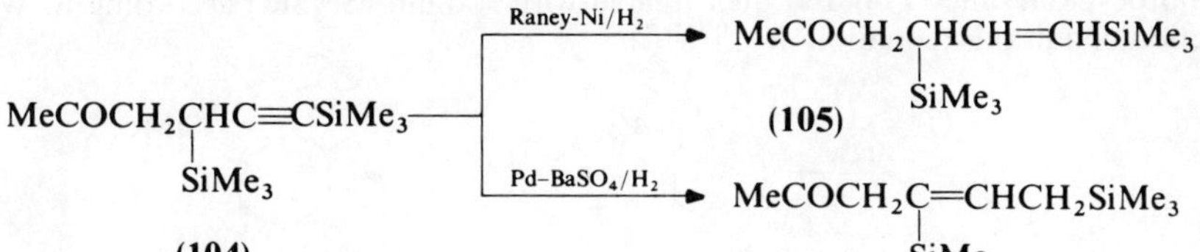

(52)

(104) **(106)**

An additional hydrogenation-type procedure is hydroboration of 1-TMS-alkynes, e.g. (**97b**), by means of dicyclohexylborane (DCB) in TMF and, consecutively, acidolysis by means of acetic acid/H_2O_2, to yield *Z*-1-TMS-alk-1-ene (**107**) (equation 53)[73].

Finally, two substitution reactions are outlined briefly since they resemble the previously described hydrosilylation reactions. In the presence of tetrakis

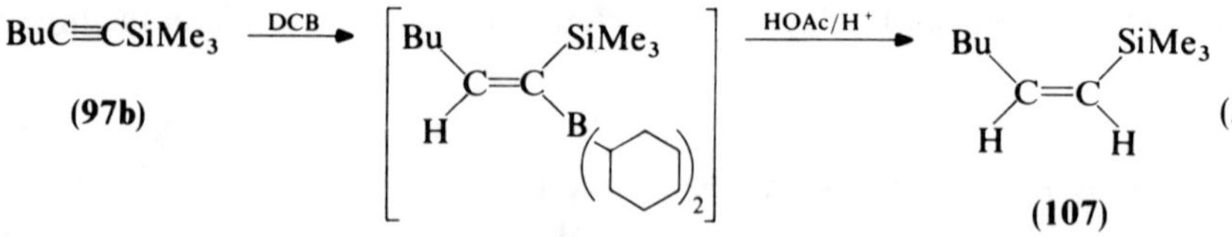

(97b) (107) (53)

(triphenylphosphine)palladium(0) as catalyst, both vinyl chloride (**48**) and *E*-1,2-dichloroethene (**109**) can be silylated by means of 1,1,2,2-tetrachloro-1,2-dimethyldisilane (**108**) (equation 54)[74].

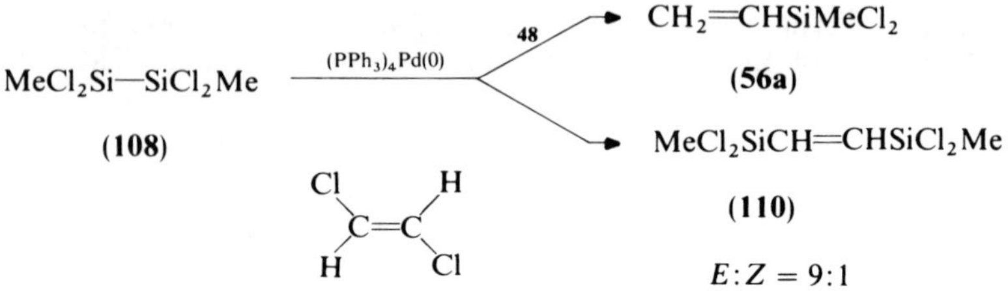

(108) (56a) (110) (109) (54)

C. Cyclic Vinylsilanes

1-TMS-cyclohexa-1,4-diene (**112**) can be synthesized (yield: 85%) from 1,3-bis(TMS)-benzene (**111**) via treatment with lithium/ethanol and liquid ammonia at −70 °C (Birch reaction) under partial desilylation (equation 55)[75, 76]. 1-TMS-cyclohex-1-ene (**115**) can

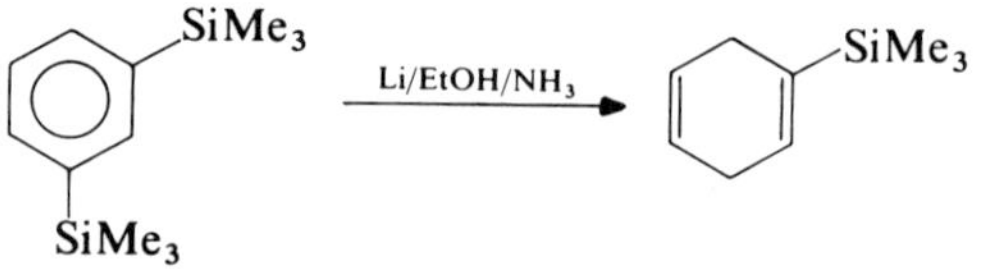

(111) (112) (55)

be formed from cyclohexanone (**113**), which is converted by PCl_5 into the corresponding 1-chlorocyclohexane (**114**) and then treated with sodium/Me_3SiCl according to Wurtz reaction conditions (equation 56)[54, 77, 78].

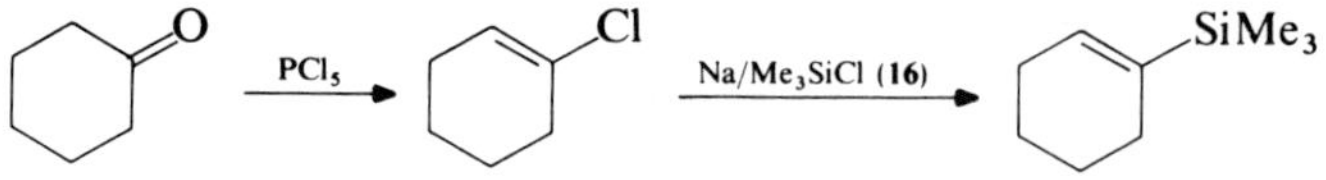

(113) (114) (115) (56)

V. ALLYLSILANES

A. Introduction

Allylsilanes, especially allyltrimethylsilane, are useful organosilicon compounds of great versatility and well-suited for a large number of applications in organic syntheses.

In general, one can say that allylsilanes are configurationally stable and do not undergo allylic rearrangement at ordinary temperatures.

The main synthetic routes to allylsilanes are Grignard reactions, reductive silylation, and Wittig-type reactions with halogenosilanes and hydrosilylation of dienes.

B. Linear Allylsilanes

The Grignard reaction is an appropriate way to synthesize allylalkylsilanes, shown in equation 57 for the series of allylmethylsilanes[79, 80]. Besides **116**, the chloride **116a** can be

$$\underset{\textbf{(116)}}{CH_2{=}CHCH_2MgBr} \begin{cases} \xrightarrow{Me_3SiCl} CH_2{=}CHCH_2SiMe_3 \quad \textbf{(119)} \\ \xrightarrow{Me_2SiCl_2\,(117)} (CH_2{=}CHCH_2)_2SiMe_2 \quad \textbf{(120)} \\ \xrightarrow{MeSiCl_3\,(118)} (CH_2{=}CHCH_2)_3SiMe \quad \textbf{(121)} \end{cases} \tag{57}$$

$$\underset{\textbf{(116a)}}{CH_2{=}CHCH_2MgCl} + SiCl_4 \longrightarrow \underset{\textbf{(122)}}{(CH_2{=}CHCH_2)_4Si} \tag{58}$$

used to obtain tetraallylsilane (**122**)[81, 82] (equation 58). However, when the silane has bulky substituents, only one allylic moiety can be introduced by the Grignard method. Thus, by means of (**116a**) it is also feasible to prepare allyldiphenylsilane (**124**) from diphenylsilane (**123**) as starting material (equation 59)[83]. Alternatively, allylsilanes can be

$$\textbf{116a} + \underset{\textbf{(123)}}{Ph_2SiH_2} \longrightarrow \underset{\textbf{(124)}}{CH_2{=}CHCH_2SiHPh_2} \tag{59}$$

obtained by employment of reagents such as TMS-methylmagnesium chloride (**126**) with esters, forming in the first step a bis-β-silylated alcohol which, under elimination of silanol, yields the allylsilane (**127**)[84] (equation 60).

$$\underset{\textbf{(125)}}{RCOOEt} + \underset{\textbf{(126)}}{2Me_3SiCH_2MgCl} \longrightarrow RC(OH)(CH_2SiMe_3)CH_2SiMe_3 \xrightarrow{-Me_3SiOH} \underset{\textbf{(127)}}{RC(CH_2SiMe_3){=}CH_2} \tag{60}$$

$R = H, Ph, PhCH_2CH_2$

Alkenes can be converted via reductive silylation to allylsilanes by the system chloroalkylsilane/Mg/HMPA in the presence of $FeCl_3$ and/or $TiCl_4$ as catalysts, involving an allylic rearrangement; by-products are some hydrogenated compounds

$$\underset{\textbf{(128)}}{RCH_2CH{=}CHR'} + \textbf{16} \xrightarrow[-MgCl_2]{Mg/HMPA/FeCl_3/TiCl_4} \underset{\textbf{(129)}}{RCH{=}CHCHR'SiMe_3} + \underset{\textbf{(130)}}{R(CH_2)_2CHR'SiMe_3} + \underset{\textbf{(131)}}{Me_3SiSiMe_3} \tag{61}$$

(equation 61)[85]. Applying these conditions of the reductive silylation to a butadiene derivative (**132**) yields the corresponding allylsilane (**133**)[86] (equation 62). However, if 1,3-

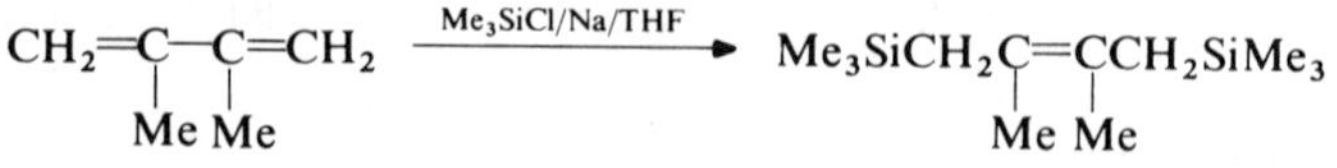

(62)

(**132**) (**133**)

butadiene (**134**) is treated with Mg and then with dichlorodimethylsilane (**117**), under cyclization 1,1-dimethyl-1-silacyclopent-3-ene (**135**) is formed (equation 63)[87].

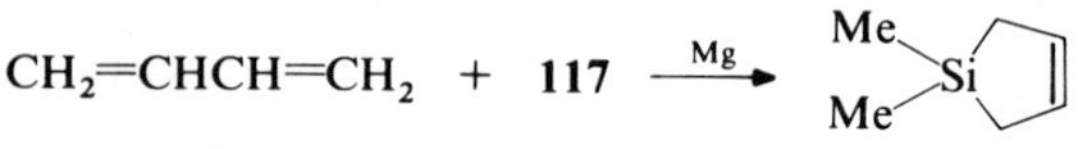

(63)

(**134**) (**135**)

Allyl methyl ether (**136**) can be converted into an allylsilane (**137**) by means of the reagent combination Me_3SiCl/Na (equation 64)[88]. In several cases the application of

(64)

(**136**) (**137**)

metallated silyl reagents in the presence of copper(I) salts is preferable, for instance, for the silylation of 1-phenyl-3-chloroprop-1-ene (**138**) which forms the two regioisomers (**139a/b**) in the ratio 94:6 (equation 65)[89].

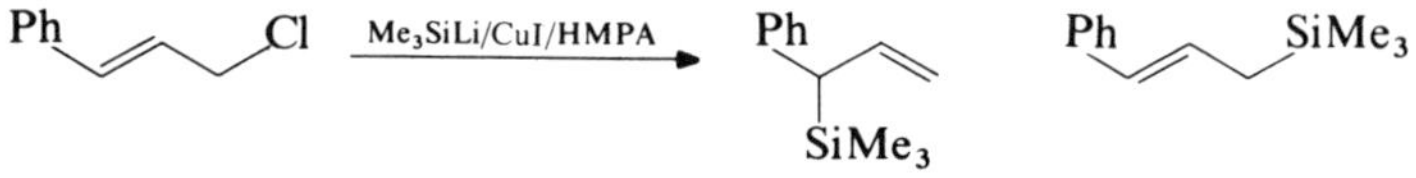

(65)

(**138**) (**139a**) (**139b**)

Non-symmetrical secondary allyl acetates and urethanes react in a regiospecific manner with lithium(dimethylphenylsilyl)cuprate to afford allylsilanes. Thus, *E*-pent-3-en-2-ylacetate (**140**) is converted into *E*-2-dimethylphenylsilylpent-3-ene (**141**) (yield: 70%), by the silylating agent in THF/ether/pentane at 0 °C (equation 66)[90].

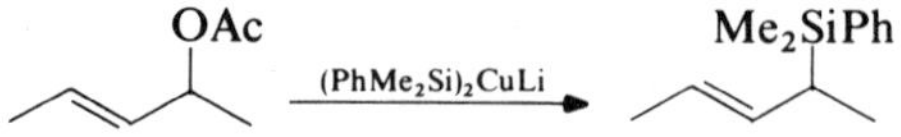

(66)

(**140**) (**141**)

Another way to synthesize allyltriorganylsilanes is the reaction of (2-triorganylsilylethylidene)phosphoranes with carbonyl compounds according to the Wittig reaction. For instance, heptanal (**143**) plus 2-TMS-ethylidenephosphorane (**142**) form (non-2-en-1-yl)trimethylsilane (**144**) (equation 67)[91, 92].

$$\underset{(\mathbf{142})}{Me_3SiCH_2CH{=}PR_3} + \underset{(\mathbf{143})}{OCH(CH_2)_5CH_3} \xrightarrow[-R_3PO]{} \underset{(\mathbf{144})}{Me_3SiCH_2CH{=}CH(CH_2)_5CH_3} \quad (67)$$

Dienes can be converted into allylsilanes via hydrosilylation in the presence of catalysts, e.g., butadiene (**134**) when treated with trichlorosilane/tetrakis-(triphenylphosphine)palladium(0) gives rise to but-2-en-1-yl-trichlorosilane (**145**)

$$\mathbf{134} + HSiCl_3 \xrightarrow{cat/5h/100\,°C} \underset{(\mathbf{145})}{CH_3CH{=}CHCH_2SiCl_3} \quad (68)$$

(yield: 85%) (equation 68)[93]. If the same palladium catalyst is applied to the hydrosilylation of a 1,3-diene such as **134**, a Z-1,4-disilylated product (**146**) can be obtained with

134 + $Cl_3SiSiCl_3$ → 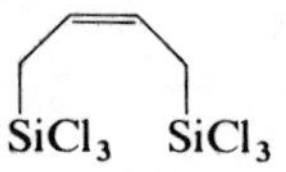 (69)

(**146**)

hexachlorodisilane (equation 69)[94]. However, the hydrosilylation of **134** with trimethylsilane (**147**) in the presence of $Fe(CO)_5$ as catalyst leads to three silylated regioisomers (**148a–c**) (equation 70)[95].

134 + Me_3SiH (**147**) $\xrightarrow{Fe(CO)_5/h\nu}$ 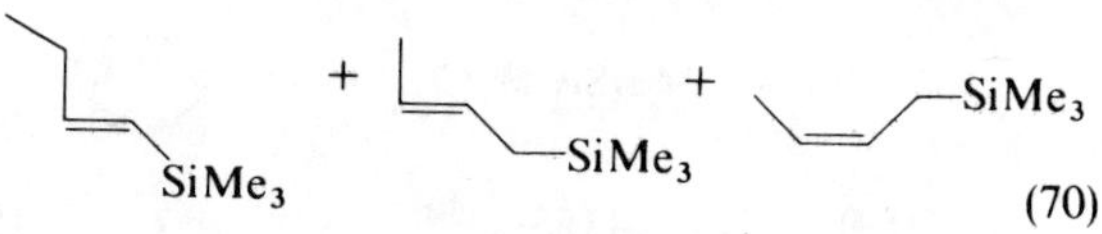 (70)

(**148a, b, c**)

The hydrosilylation of β-pinene (**45**) with **7** and 2,2′-azobisisobutyronitrile (catalysis via free radicals) affords 4-isopropyl-1-trichlorosilylmethylcyclohexene (**46**) (equation 71)[96].

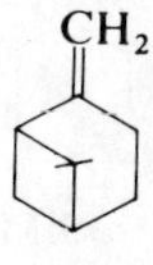 + **7** $\xrightarrow{cat.}$ 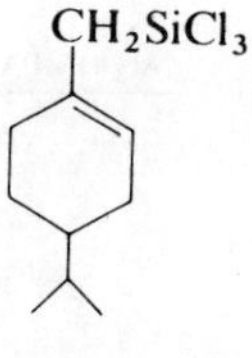 (71)

(**45**) (**46**)

Finally, the reaction of the optically active (+) (1R, 2S, 3S)-di-μ-chloro-bis(1-methyl-3-phenyl-π-allyl)dipalladium (**149**) with 1,1,1-trichloro-2,2,2-trimethyldisilane (**150**) in the

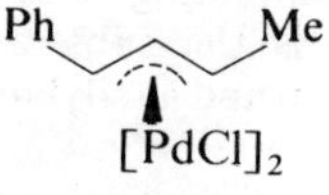 $\xrightarrow{Cl_3SiSiMe_3(\mathbf{150})/cat}$ 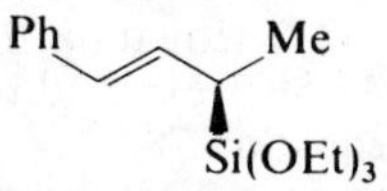 (72)

(**149**) (**151**)

presence of triphenylphosphine, using THF as solvent, leads after alcoholysis (ethanol/triethylamine) with high yield (80%) to the optically active *E*-1-phenyl-3-triethoxysilyl-but-1-ene (**151**) (equation 72)[97].

C. Cyclic Allylsilanes

Metallation reactions/reductive silylation of cyclic allyl halides are very common for the synthesis of cyclic allylsilanes. Therefore, a routine procedure is the treatment of cyclic allyl halides (**152**) with chlorotrimethylsilane/Mg in THF. The yield is dependent on the respective ring size (equation 73)[98].

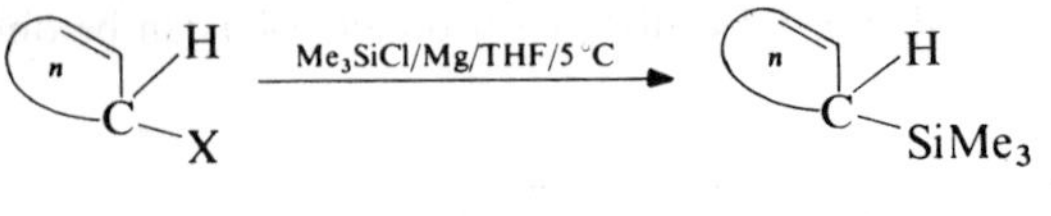

(**152**) (**153**)

n	% yield
5	94
6	64
7	32
8	22

(73)

If cyclopentadiene (**154**) is treated with sodium and then with chloromethyltrimethylsilane (**155**), 2- and 3-TMS-methylcyclopentadiene (**156, 157**) are formed (equation 74)[99].

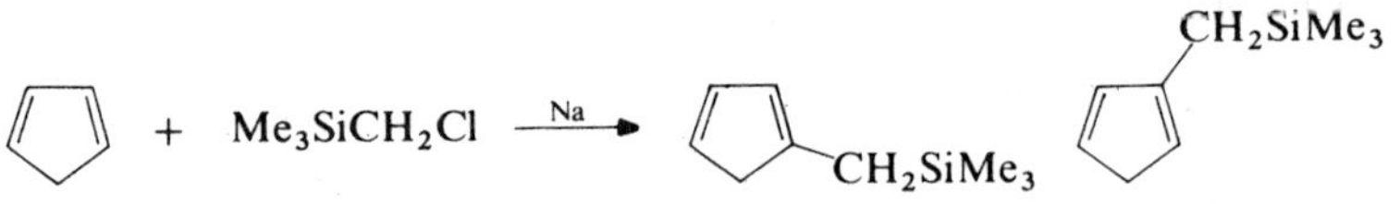

(74)

(**154**) (**155**) (**156**) (**157**)

The reductive silylation of 1,3-cyclohexadiene (**158**) with chlorotrimethylsilane/Mg/HMPA affords bis (TMS)-cyclohexenes (**159a–c**) (equation 75)[100]. When reductive

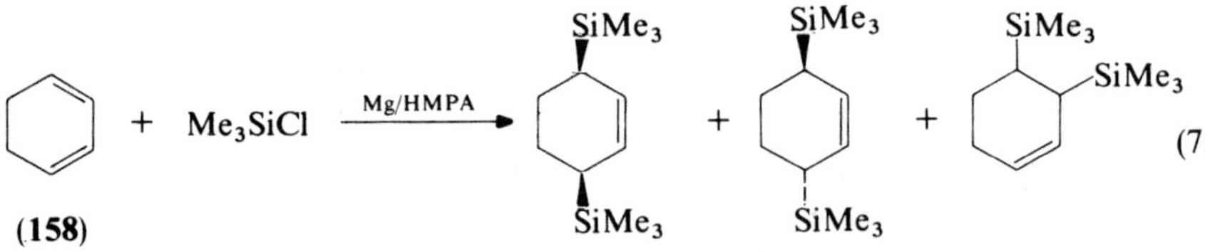

(75)

(**158**) (**159a, b, c**)

silylation with $Me_3SiCl/Li/THF$ is applied to aromatic substrates such as benzene (**160**), *p*-xylene (**161**) and *m*-xylene (**162**)[101] at low temperatures, the corresponding bis(TMS)-cyclohexadiene derivatives (**163**, **164**, **165**) are formed (equation 76)[102]. The reductive silylation of *p*-bis(TMS)-benzene (**166**) leads to the corresponding 1,1,4,4-tetrakis(TMS)-cyclohexadiene (**167**) (equation 77)[86]. If indole (**168**) is the substrate for reductive silylation with Me_3SiCl (excess)/Li/THF, both *N*-silylation and C-silylation of the benzo-moiety occurs (equation 78)[103].

In a related manner, TMS-benzocyclopropane (**172**) can be synthesized from benzocyclopropane (**171**) via lithiation with subsequent silylation (equation 79)[104]. Under

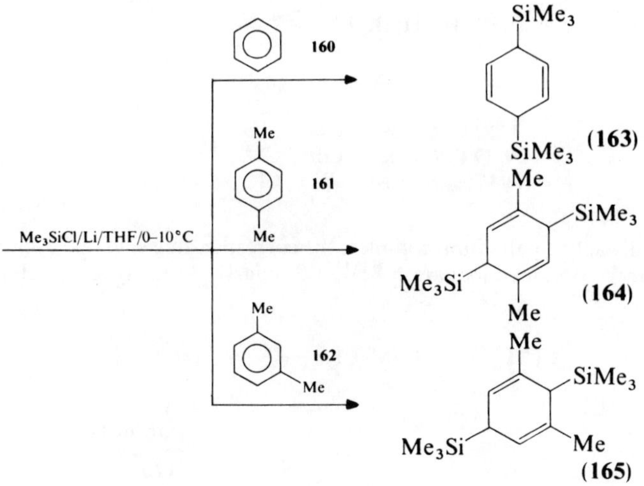

(76)

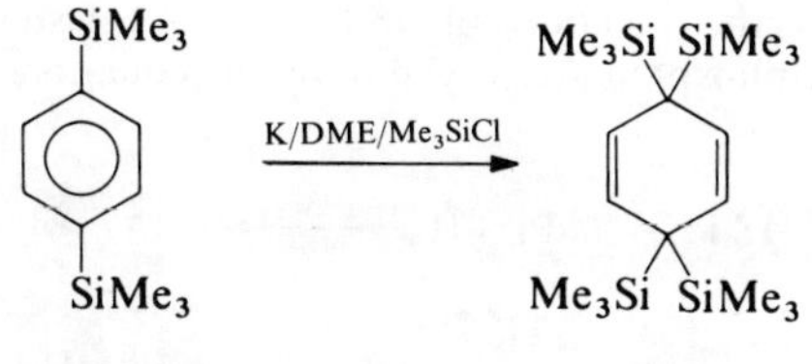

(166) (167)

(77)

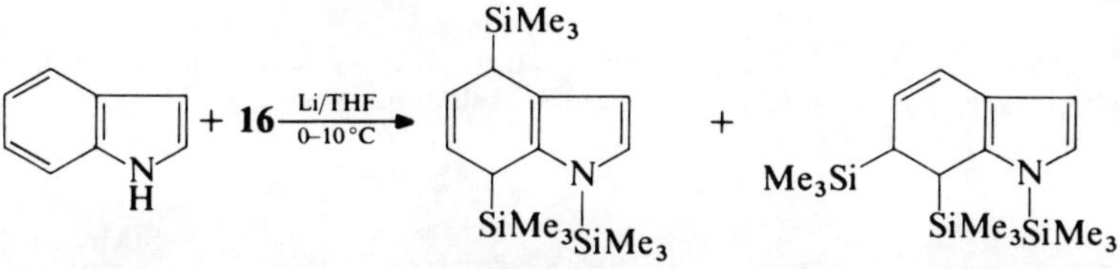

(168) (169) (170)

(78)

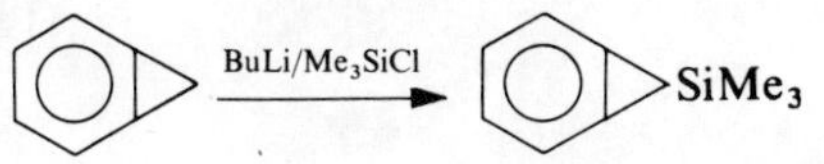

(171) (172)

(79)

hydrosilylation-type conditions the synthesis of a large variety of cyclic allylsilanes can be accomplished.

In the presence of $Co_2(CO)_8$ as catalyst, cyclopentadiene (**154**) can add triethoxysilane (**29**) or chlorodimethylsilane (**34**) (equation 80)[105], whereas via catalysis by $Pd(PPh_3)_4$ the reaction of **154** with trichlorosilane (**7**) is feasible (equation 80)[106]. In these cases the yields are in the range of 91–93%. If the hydrosilylation of **154** is performed via catalysis

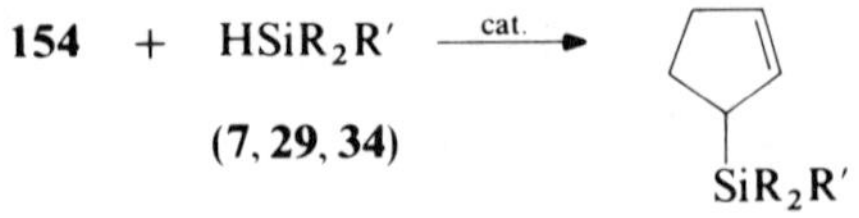

(80)

7/173: R = R′ = Cl
29/174: R = R′ = OEt
34/175: R = Me; R′ = Cl

(173–175)

by the optically active palladium complex, the corresponding silylated cyclopentene (**176**) in *S*-form can be obtained (equation 81)[107]. Similarly, 3-(phenylsilyl)cyclopentene (**178**)

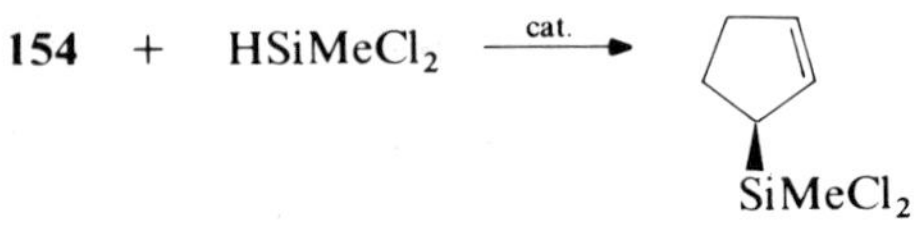

(81)

(176)

can be obtained from the reaction of **154** with phenylsilane (**177**) catalyzed by bis(acac)nickel/triphenylphosphine/triethylaluminum (equation 82)[108]. Hydrosilylation

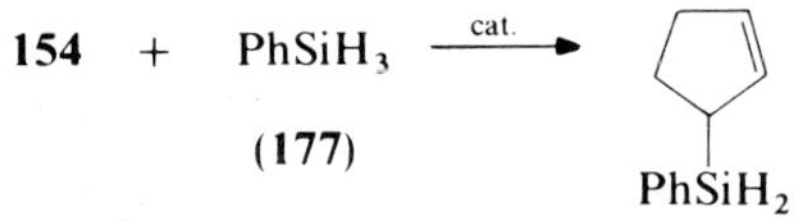

(82)

(178)

(related) procedures are also applicable to larger ring systems, e.g. 1,3-cyclooctadiene (**179**), which gives rise to 3-TMS-cyclooctene (**180**) (equation 83)[109].

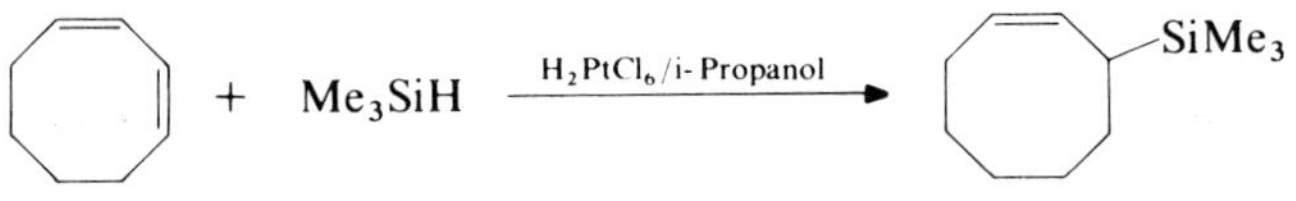

(83)

(179) **(180)**

VI. ALKYNYLSILANES

A. Introduction

Alkynylsilanes—reactive precursors and versatile synthons for various reactions—can be obtained mainly by silylation techniques of metallated alkynes (including Grignard derivatives), or by elimination reactions of halogenated alkanes and alkenes in the presence of a silylating agent. The main classes of alkynylsilanes will be discussed as well as some derivatives which may be of further interest to organic synthesis.

B. Ethynylsilanes

1. Monosilylethynes

A common approach is the reaction of halogenoalkylsilanes with alkali metal (lithium, sodium)[110, 111] or magnesium (Grignard)[112] derivatives of acetylene to form silylated acetylenes which can be illustrated by the synthesis of TMS-acetylenes (**182**/**57a**) (equation 84). The reaction can be slightly modified by using bis(TMS)-sulfate (**184**) as silylating agent (equation 85)[113].

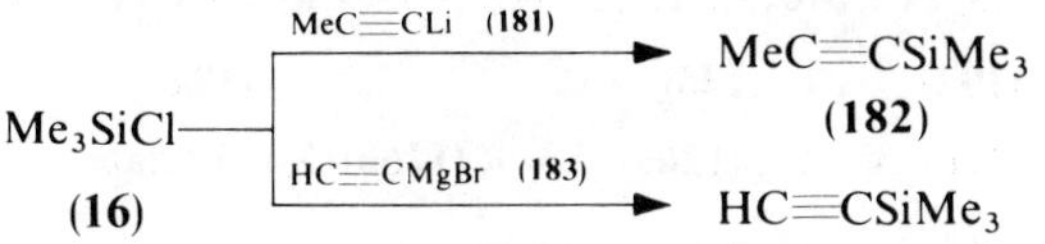

(**16**) (**182**) (**57a**) (84)

$$(Me_3Si)_2SO_4 + NaC{\equiv}CH \rightarrow \mathbf{57a} \qquad (85)$$

(**184**) (**185**)

An alternative method is the treatment of 1,1,2-trichloroethene (**50**) with lithium to obtain **57a** (equation 86)[114]. Another option is to apply methyllithium to halogenated alkanes or alkenes with subsequent employment of bromo- or fluorosilanes, in order to obtain silylated halogenoacetylenes (equation 87)[115].

$$ClCH{=}CCl_2 + 4\,Li + \mathbf{16} \xrightarrow[-4\,LiCl]{THF,\,-78\,^{\circ}C} \mathbf{57a} \qquad (86)$$

(**50**)

$$ClHC{=}CHCl \xrightarrow[-LiCl]{MeLi/Et_2O} LiC{\equiv}CCl \xrightarrow{FSiH_3(\mathbf{186})} H_3SiC{\equiv}CCl \qquad (87)$$

(**109**) (**187**)

2. Bisilylethynes

The classical route to bisilylacetylenes is the reaction of the di-Grignard derivative of acetylene (**188**) (equation 88)[116]. Also, the reaction of tetrachloroethene (**53**) with lithium and Me_3SiCl in THF at $-78\,^{\circ}C$ gives rise to **189** (equation 89)[114]. Alternatively, the application of methyllithium to halogenated alkanes or alkenes with subsequent employment of bromo- or fluorosilanes affords bis-silylated acetylenes (equation 90)[115].

$$BrMgC{\equiv}CMgBr \xrightarrow{Me_3SiCl} Me_3SiC{\equiv}CSiMe_3 \qquad (88)$$

(**188**) (**189**)

$$Cl_2C{=}CCl_2 + 6\,Li \xrightarrow{2Me_3SiCl/THF,\,-78\,^{\circ}C} \mathbf{189} \qquad (89)$$

(**53**)

$$Br_2HCCHBr_2 \xrightarrow[-LiBr]{MeLi/Et_2O} LiC{\equiv}CLi + 2\,BrSiH_3 \longrightarrow H_3SiC{\equiv}CSiH_3 \qquad (90)$$

(**190**) (**191**) (**192**)

3. *Propargylsilanes*

A convenient route is the reaction of alkynyllithium with TMS-methyl halides or TMS-triflate (equation 91)[116a, 117]. Starting from 3-TMS-prop-1-ynes (**195**) one can synthesize 1,3-bis(TMS)-prop-1-ynes (**196**) via metallation with butyllithium and subsequent silylation (equation 92)[118]. Alternatively, **196** (R = H) is the reaction product of the silylation of prop-1-yn-3-yl ethers (**197**); by-products are silylated propene (**198**) and propane (**199**) (equation 93)[119].

$$\underset{(\mathbf{193})}{RC{\equiv}CLi} + \underset{(\mathbf{126})}{Me_3SiCH_2X} \xrightarrow{BuLi/Hexan/THF} \underset{(\mathbf{194})}{RC{\equiv}CCH_2SiMe_3} \qquad (91)$$

(**126**) X = Cl; (**126a**) X = I; (**126b**) X = Triflate

$$\underset{(\mathbf{195})}{HC{\equiv}CCH(R)SiMe_3} \xrightarrow{BuLi/16-30\,^{\circ}C\ Me_3SiCl} \underset{(\mathbf{196})}{Me_3SiC{\equiv}CCH(R)SiMe_3} \qquad (92)$$

$$\underset{(\mathbf{197})}{HC{\equiv}CCH_2OR} + Me_3SiCl \xrightarrow{Li/THF/5\,^{\circ}C} \mathbf{196} + \underset{(\mathbf{198})}{Me_3SiCH{=}C(SiMe_3)CH_2SiMe_3} + \underset{(\mathbf{199})}{(Me_3Si)_2CH{-}CH(SiMe_3)CH_2SiMe_3} \qquad (93)$$

When allene (**200**) is treated with two equivalents of butyllithium and of Me_3SiCl **196** (R = H) is formed (equation 94)[119a]. If TMS-acetylene (**57a**) reacts with ethylmagnesium bromide in the presence of *N,N*-dimethyl formamide, 3-TMS-propynal (**201**) can be isolated (69% yield) which can form 2-TMS-ethynyl-1,3-dioxane (**202**) with 1,3-propanediol in 81% yield (equation 95)[120].

$$\underset{(\mathbf{200})}{CH_2{=}C{=}CH_2} \xrightarrow{BuLi/THF/-50\,^{\circ}C} CH_2{=}C{=}\overset{\ominus}{C}HLi^+ \xrightarrow{BuLi/THF} Li^+\overset{\ominus}{C}H_2-C{\equiv}\overset{\ominus}{C}Li^+ + 2\,Me_3SiCl \longrightarrow \mathbf{196} \qquad (94)$$

$$\underset{(\mathbf{57a})}{Me_3SiC{\equiv}CH} \xrightarrow{EtMgBr/HCONMe_2} \underset{(\mathbf{201})}{Me_3SiC{\equiv}CCHO} \xrightarrow{HO(CH_2)_3OH} \qquad (95)$$

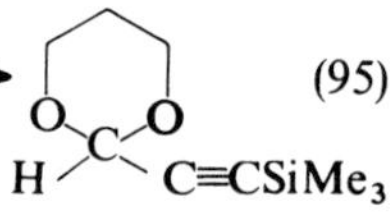

(**202**)

4. *Variously substituted ethynylsilanes*

Ethyl 3-TMS-2-propynoate (**204**) is formed by the reaction of TMS-acetylene (**57a**) with butyllithium and subsequent treatment with triethoxycarbenium tetrafluoroborate and hydrolysis (equation 96)[121]. If bis(TMS)-acetylene (**189**) is employed with acyl halides or acid anhydrides in the presence of a Lewis acid (e.g. $AlCl_3$) as a catalyst, then

$$\mathbf{57a} \xrightarrow{\text{BuLi}} Me_3SiC{\equiv}CLi \xrightarrow[-\text{LiBF}_4]{(EtO)_3CBF_4} Me_3SiC{\equiv}CC(OEt)_3 \xrightarrow{H_2O} \underset{(\mathbf{204})}{Me_3SiC{\equiv}CCOOEt} \qquad (96)$$

alkynonylsilanes are formed. 'Application of acetyl-(**205a**), propanoyl-(**205b**), *p*-nitrobenzoyl-(**205d**), chloroacetyl chloride (**205c**), respectively, affords the 1-TMS-derivative of but-1-yn-3-one (**206a**), pent-1-yn-3-one (**206b**), 3-(*p*-nitrophenyl)prop-1-yn-3-one (**206d**), 4-chlorobut-1-yn-3-one (**206c**) in high yield (90%) (equation 97)[122].

$$\mathbf{189} + \underset{(\mathbf{205a\text{–}d})}{RCOCl} \xrightarrow{AlCl_3} \underset{(\mathbf{206a\text{–}d})}{Me_3SiC{\equiv}CCOR} \qquad (97)$$

205a/206a $R = Me$
205b/206b $R = Et$
205c/206c $R = CH_2Cl$
205d/206d $R = p\text{-}C_6H_4NO_2$

C. Polyynylsilanes and their Derivatives

By means of the CuCl–TMEDA complex as a catalyst one can synthesize 1,4-bis(TMS)-1,3-butadiyne (**207**) from TMS-acetylene (**57a**) in the presence of oxygen (equation 98)[123]. Alternatively, **207** and the corresponding TES-compound (**210**), respectively, can be formed by the reaction of **16** or the corresponding TES-Cl (**209**) with the di-Grignard derivative of 1,3-butadiyne (**208**) (equation 99)[124, 125].

$$\mathbf{57a} \xrightarrow{[CuCl\cdot TMEDA]I,\ O_2} \underset{(\mathbf{207})}{Me_3SiC{\equiv}C\text{–}C{\equiv}CSiMe_3} \qquad (98)$$

$$\underset{(\mathbf{208})}{BrMgC{\equiv}C\text{–}C{\equiv}CMgBr} + \underset{(\mathbf{16},\ \mathbf{209})}{2\,R_3SiCl} \rightarrow \underset{(\mathbf{207/210})}{R_3SiC{\equiv}C\text{–}C{\equiv}CSiR_3} \qquad (99)$$

16/207: $R = Me$
209/210: $R = Et$

If 1-alkylalk-7-yn-3-ols are acetylated with acetic anhydride and then treated with dimethyllithium cuprate via S_N2 replacement, the corresponding 1-TMS-alkynes (e.g. **212**) can be isolated (equation 100)[126].

$$\underset{(\mathbf{211})}{Me_3SiC{\equiv}CCH(OH)(CH_2)_4CH_3} \xrightarrow{Ac_2O} Me_3SiC{\equiv}CCH(OAc)(CH_2)_4CH_3$$

$$\xrightarrow{Me_2CuLi} \underset{(\mathbf{212})}{Me_3SiC{\equiv}CCH(CH_3)(CH_2)_4CH_3} \qquad (100)$$

Oct-1-yn-3-ol (**213**) can be converted at very low temperature into the corresponding 1-TMS-oct-1-yn-3-ol (**214**) leaving the configuration unaffected (equation 101)[127]. Consecutively, **214** can be converted via hydroboration with DCB and subsequent oxidation into the corresponding carboxylic acid.

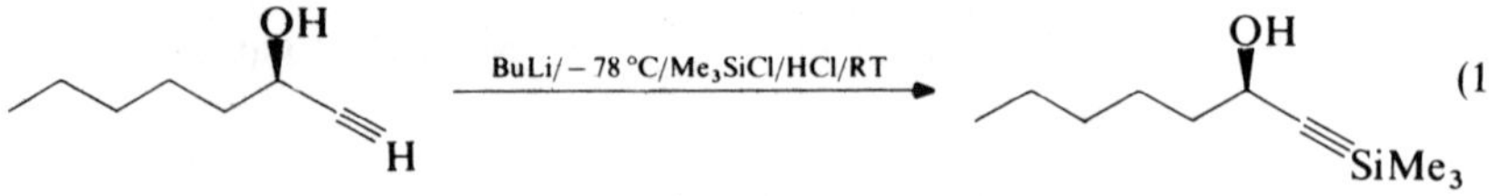

(213) (214) (101)

Finally, silylated penta-3,4-dien-1-ynes (**216**) are synthesized from 1,5,5,5-tetralithiopenta-1,3-diyne (**215**) with Me_3SiCl in THF at very low temperature (equation 102)[128].

$$\underset{(\mathbf{215})}{LiC{\equiv}C{-}C{\equiv}CCLi_3} \rightleftharpoons Li_2C{=}C{=}C(Li)C{\equiv}CLi \xrightarrow{Me_3SiCl/O\ 78\,^{\circ}C} \underset{(\mathbf{216})}{(Me_3Si)_2C{=}C{=}C(SiMe_3){-}C{\equiv}CSiMe_3} \qquad (102)$$

VII. ARYLSILANES

A. Introduction

Arylsilanes have been synthesized since the beginning of the systematical investigation of organosilicon chemistry. Owing to their extraordinary properties, arylsilanes have long been substances of interest for, e.g., technical applications. For the synthesis of arylsilanes several pathways are of interest using Grignard reagents as well as other metallated reagents (such as Wurtz-type and reductive silylation-type reactions). In the following we will briefly discuss the synthesis of various classes of arylsilanes.

B. Phenylsilanes

The classical reaction—already applied by Dilthey[129]—is the Grignard reaction. Depending on the stoichiometric ratio, mono-, di-, and triphenylsilanes can be synthesized from tetrachlorosilane and phenylmagnesium bromide (**217**) (equation 103)[129].

$$SiCl_4 \xrightarrow{PhMgBr\ (\mathbf{217})} PhSiCl_3\ (\mathbf{218})$$
$$SiCl_4 \xrightarrow{2\ \mathbf{217}} Ph_2SiCl_2\ (\mathbf{219})$$
$$SiCl_4 \xrightarrow{3\ \mathbf{217}} Ph_3SiCl\ (\mathbf{220}) \qquad (103)$$
$$SiCl_4 \xrightarrow{4\ \mathbf{217}/160\text{–}180\,^{\circ}} Ph_4Si\ (\mathbf{221})$$

Tetraphenylsilane (**221**) can only be formed at higher temperatures[130] since under ordinary reaction conditions the steric hindrance is an impairing factor.

As silicon components for the Grignard reaction, chlorosilanes are used predominantly since these are most widely available and since bromosilanes are less reactive, require higher temperatures and deliver lower yields. Fluorosilanes are the most reactive halogenosilanes which are applied to the synthesis of silanes with especially bulky substituents[131] (*vide infra*).

Alternative approaches to synthesize **221** include the reaction of phenyllithium (**222**) with $SiCl_4$ (yield: 99%) (equation 104)[132], or the reaction of **222** with triphenylsilane (**223**) (equation 105)[133]. A facile way for preparing e.g. triphenylsilane (**223**) is to react diphenylsilane (**123**) with **217** (equation 106)[134]. Analogously to the Wurtz synthesis, **221** can be obtained by the reaction of chlorobenzene, gaseous SiF_4 and sodium at 150 °C according to a US patent[135].

$$\underset{(\mathbf{222})}{4\,\mathrm{PhLi}} + \mathrm{SiCl_4} \xrightarrow[-\mathrm{LiCl}]{} \mathbf{221} \tag{104}$$

$$\mathbf{222} + \underset{(\mathbf{223})}{\mathrm{Ph_3SiH}} \xrightarrow[-\mathrm{LiH}]{} \mathbf{221} \tag{105}$$

$$\underset{(\mathbf{123})}{\mathrm{Ph_2SiH_2}} + \mathbf{217} \xrightarrow[-\mathrm{MgBrH}]{\mathrm{THF}} \mathbf{223} \tag{106}$$

A Wurtz synthesis with halogenobenzene (**224**) and Me_3SiCl affords the respective phenyltrimethylsilane (**225**) (equation 107a). This route can also be applied to halogenobenzenes with substituents in *ortho* position as demonstrated by equation 107b[136],

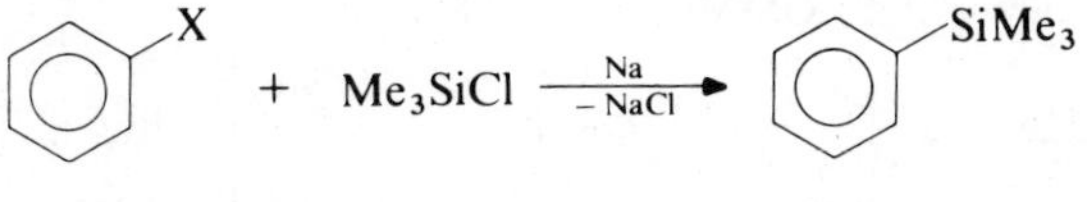

(107a)

(**224**) (**225**)

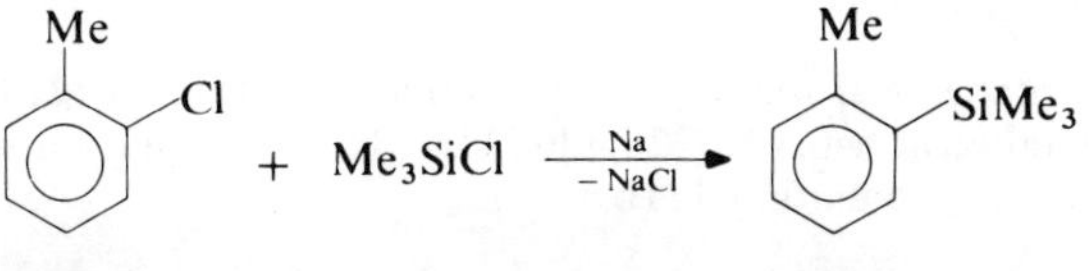

(107b)

(**226**) (**227**)

for the synthesis of *o*-TMS-toluene (**227**). A similar approach uses the combination Mg/HMPA, instead of sodium, for '*in situ* syntheses' to obtain trialkylaryl- or trialkylheteroarylsilanes from chlorotrialkylsilanes and aryl or heteroaryl halides, respectively (equation 108). For an overview see the work of Effenberger and Häbich[137].

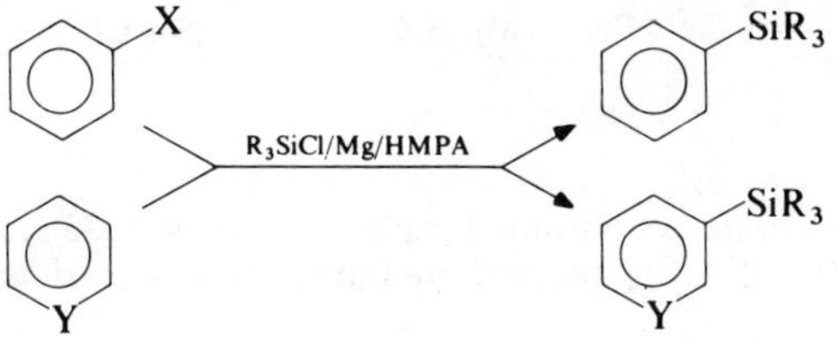

(108)

A related method is lithiation of an aryl halide with subsequent treatment by means of chloroalkylsilanes to achieve alkyl aryl silanes as can be shown by the synthesis of bis[3,4-dimethylphenyl]dimethylsilane (**229**) (equation 109)[138].

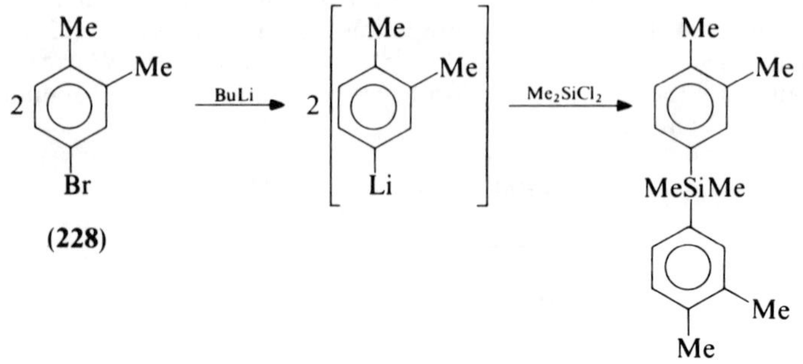

(**228**) (109)

(**229**)

A scheme that affords *o*-hydroxyphenyltrialkylsilanes is the conversion of trialkyl-(*o*-bromophenoxy)silane (**230**) by means of *t*-butyllithium via an anionic TMS-rearrangement to 2-hydroxyphenyltrimethylsilane (**231**) (equation 110)[139].

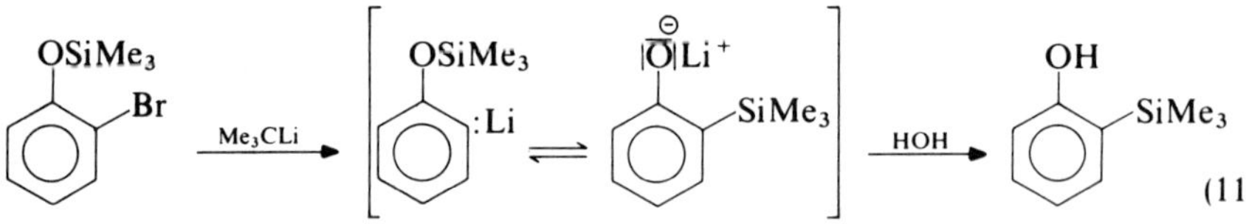

(110)

(**230**) (**231**)

Finally, an electrochemical method for trimethylphenylsilanes (**225**) has been established[140] using an halogenobenzene (**224**) plus Mg/Me_3SiCl (equation 111a); the assumed mechanism is outlined in equation 111b.

$$\underset{(\mathbf{224})}{\text{PhX}} \xrightarrow{e^-/Me_3SiCl/Mg} \mathbf{225} \tag{111a}$$

$$\begin{aligned} &\text{Mg} \longrightarrow \text{Mg}^{2+} + 2e \\ &\text{PhX} + 2e^- \longrightarrow \text{Ph}^- + \text{X}^- \\ &\text{Ph}^- + Me_3SiCl \longrightarrow \mathbf{225} + Cl^- \end{aligned} \tag{111b}$$

C. Biphenylsilanes

A well-established route to tetrakis(4-biphenylyl)silane (**233**) is the reaction of 4-bromobiphenyl (**232**) with $SiCl_4$ and sodium (yield: 90%) according to Wurtz synthesis (equation 112)[141].

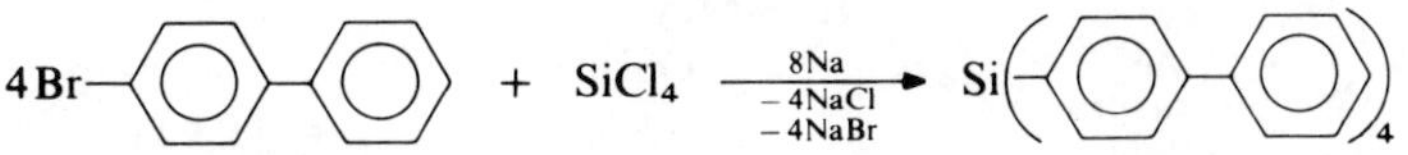

(232) (233) (112)

D. Polycyclic Silanes

1. Naphthylsilanes

An established pathway to naphthylsilanes is the Grignard reaction which, in this case, requires fluorosilanes as silicon components, as can be shown by the synthesis of bis (1-naphthyl)difluorosilane (**234**) (equation 113a)[131]. However, for the synthesis of tetrakis (2-naphthyl)silane (**235**) it is preferable to use lithiated naphthalene (equation 113b)[142]. In

$$2\ \text{1-NaphMgBr} + SiF_4 \xrightarrow[-2MgBrF]{} (\text{1-Naph})_2SiF_2 \qquad (113a)$$

(234)

$$SiCl_4 \xrightarrow{\text{2-NaphLi}} (\text{2-Naph})_4Si \qquad (113b)$$

(235)

a slight modification of equation 113b, (8-iodonaphth-1-yl)trimethylsilane (**238**) and bis(8-iodonaphth-1-yl)methylsilane (**239**) can be obtained (equation 114)[143].

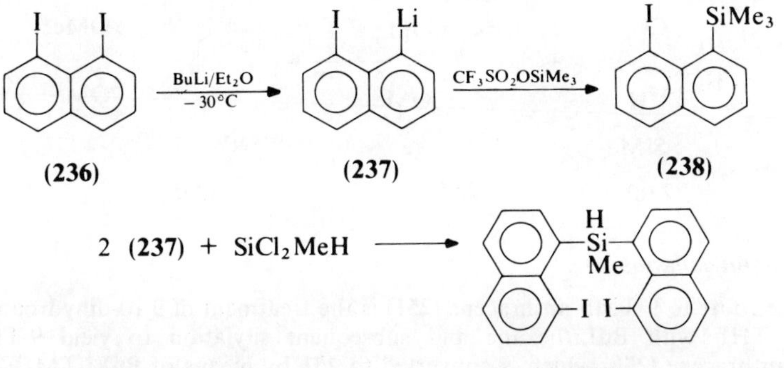

(236) (237) (238)

(239) (114)

A very common route to naphthyltrialkylsilanes is reductive silylation of e.g. 1-(**240**) or 2-trimethylsiloxynaphthalene (**242**), or of 1,5-bis(trimethylsiloxy)naphthalene (**244**) which can all be converted into the corresponding TMS-naphthalene compounds **241**, **243** and **245**, respectively, by means of Me_3SiCl/Na (equation 115)[144, 145]. A reactive intermediate

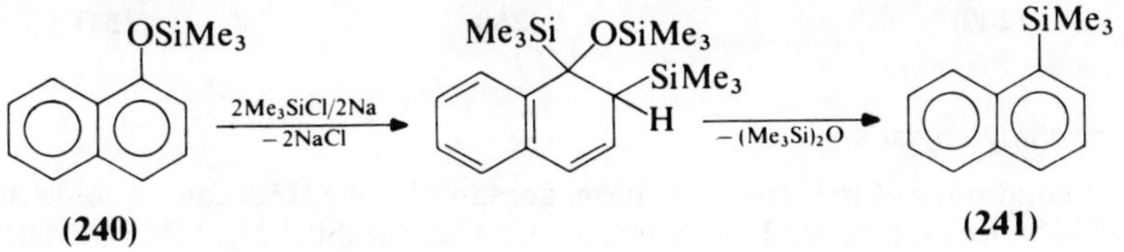

(240) (241)

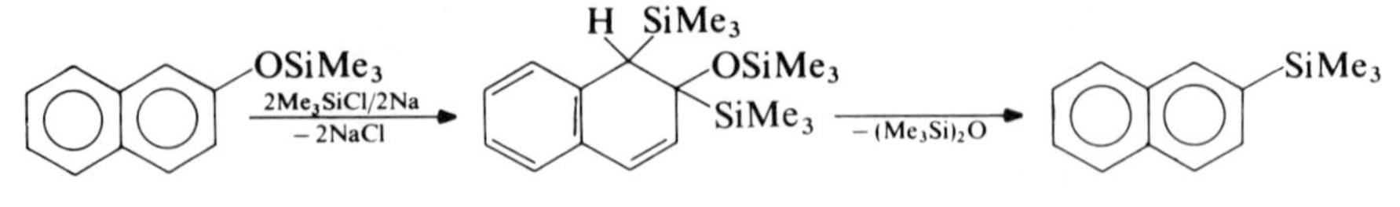

(242) (243)

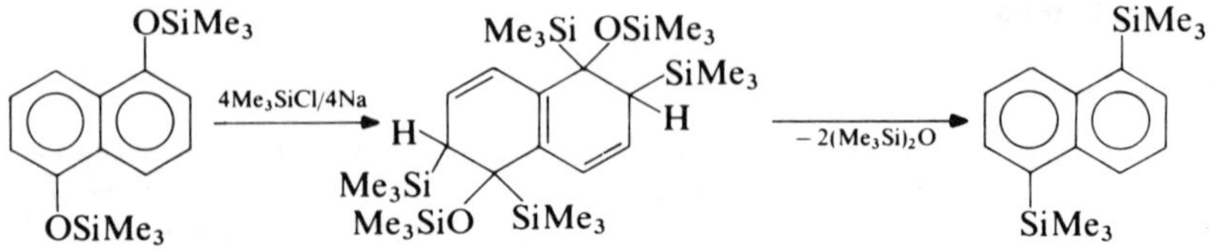

(244) (245) (115)

is most likely being formed in the first step of the reaction scheme which, under elimination of hexamethyldisiloxane, gives rise to the various silylated naphthalene compounds.

An alternative reaction pathway is the direct synthesis of silylated naphthalenes, e.g. via a Diels–Alder reaction. Thus, the addition of 1,4-bis(TMS)-butadiene (**246**) to 2,3-dimethoxy-*p*-benzoquinone (**247**) affords 2,3-dimethoxy-5,8-bis(TMS)-1,4-naphthoquinone (**248**) (equation 116)[146].

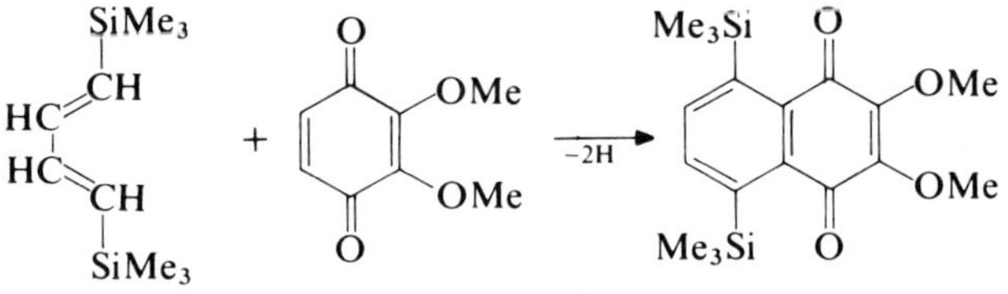

(246) (247) (248) (116)

2. Anthracenylsilanes

A facile route to 9-TMS-anthracene (**251**) is the treatment of 9,10-dihydroanthracene (**249**) in THF with BuLi/hexane and subsequent silylation to yield 9-TMS-9,10-dihydroanthracene (**250**) which is converted to **251** by means of BuLi/TMEDA in the presence of $CdCl_2$ (equation 117)[147].

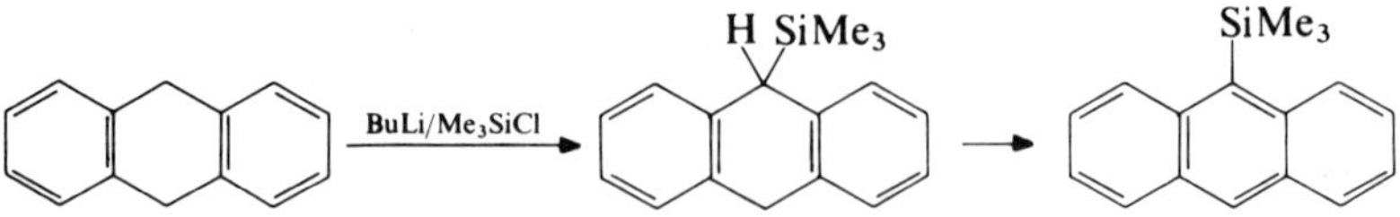

(249) (250) (251) (117)

3. Acenaphthylenylsilanes

Under conditions of reductive silylation, acenaphthylene (**252**) can be converted into 1,2-bis(TMS)-acenaphthene (**253**) by means of either $Me_3SiCl/Mg/HMPA$ (yield: 81%)

or $Me_3SiCl/Li/THF$ (yield: 21%). Subsequent treatment of **253** using BuLi/TMEDA/$CdCl_2$ gives rise to 1,2-bis(TMS)-acenaphthylene (**254**), or **253** can be partially desilylated to 1-TMS-acenaphthylene (**255**) by DDQU or chloranil[148]. Alternatively, **255** is formed via the silylation of 1-bromoacenaphthylene (**256**) (equation 118)[148].

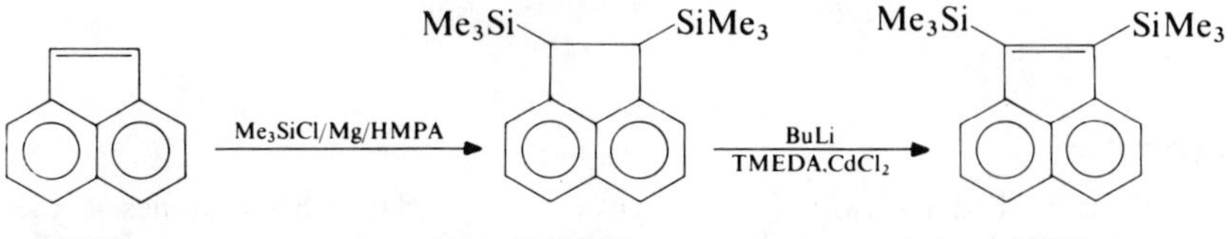

(252) **(253)** **(254)** (118)

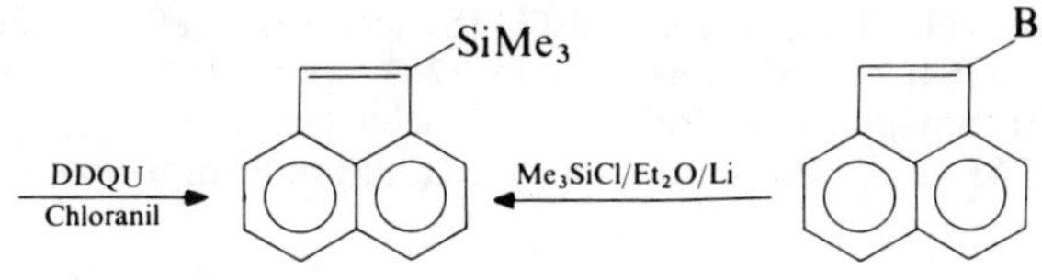

(253) **(255)** **(256)**

VIII. HETEROARYLSILANES

A. Introduction

Heteroarylsilanes have been widely appreciated as facile intermediates for a large number of applications in organic synthesis. The main synthetic pathways to heteroarylsilanes are silylation procedures by means of metallated reagents, including Grignard reagents (although the latter method is no longer very common), and direct syntheses such as 1,3-dipolarophilic cycloaddition. In the following we will outline the syntheses of several main classes of heteroarylsilanes.

B. Furans

Silylated furan derivatives (**258**) with a silyl moiety at the carbon atom adjacent to the oxygen atom can be obtained by the silylation of furan compounds (**257**) (equation 119)[149, 150].

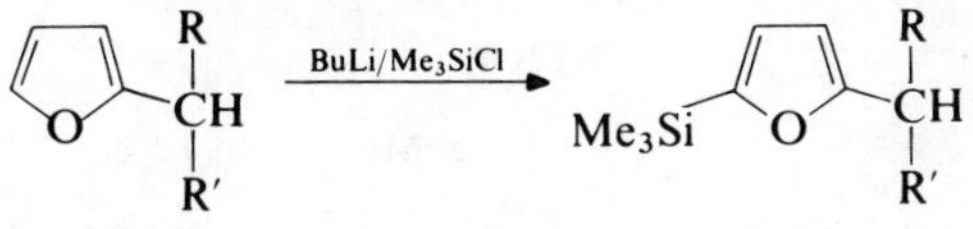

(119)

(257a,b) **(258a,b)**

257a/258a: R = R′ = OEt
257b/258b: R= OEt, R′ = H

C. Thiophenes

3-TMS-thiophene (**260**) can be formed from 3-bromothiophene (**259**) (equation 120)[137].

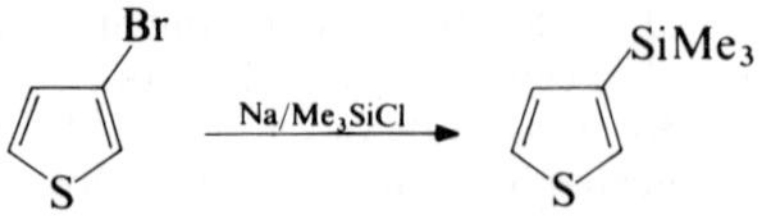

(120)

(259) **(260)**

D. Pyrazoles

A well-established pathway for the synthesis of silylated heteroarenes is the 1,3-dipolarophilic cycloaddition of alkynylsilanes (**97a**/**189**) with diazo compounds (**261**) or (**262**). Thus, TMS-phenylacetylene (**97a**) and diazomethane (**261**) form 3-TMS-4-phenylpyrazole (**263**), whereas bis(TMS)-acetylene (**189**) and **261** afford 3,4-bis-(TMS)-pyrazole (**264**). Ethyl diazoacetate (**262**) plus **189** give rise to 3,4-bis(TMS)-5-ethoxycarbonylpyrazole (**265**) in good yields (equation 121)[151]. The sterospecific formation of **263** via 1,3-dipolarophilic cycloaddition from **97a** and **261** can be considered as

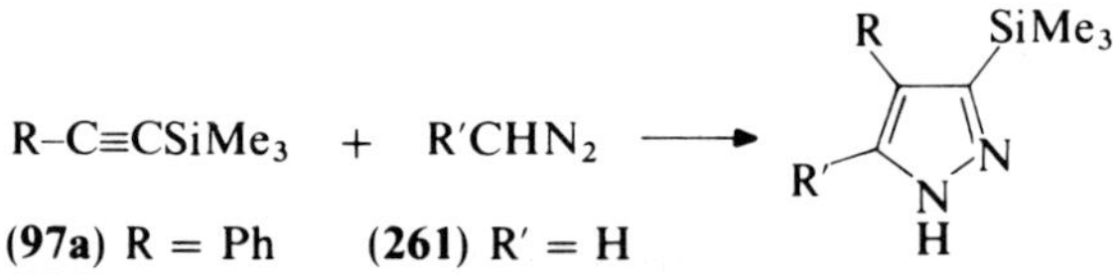

(121)

(**97a**) R = Ph (**261**) R′ = H
(**189**) R = $SiMe_3$ (**262**) R′ = CO_2Et

(**263**) R = Ph; R′ = H
(**264**) R = $SiMe_3$; R′ = H
(**265**) R = $SiMe_3$; R′ = CO_2Et

evidence for a directive influence of the TMS grouping, because only one of the two possible reaction products is usually favoured. Thus, when **261** and **262** are applied to TMS-acetylene (**57a**) only one isomer is formed, 3-TMS-pyrazole (**266**) and 3-TMS-5-ethoxycarbonylpyrazole (**267**), respectively (equation 121a)[152]. In addition, the directive

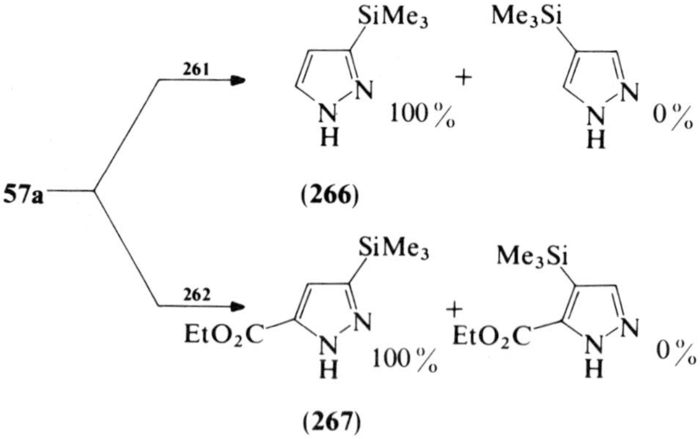

(121a)

influence of the TMS moiety is more pronounced than that of the acetyl grouping, as can be shown when **261** is used with TMS-acetylacetylene (**268**). Both isomers result, **269** being the main product (70%) (equation 122)[153].

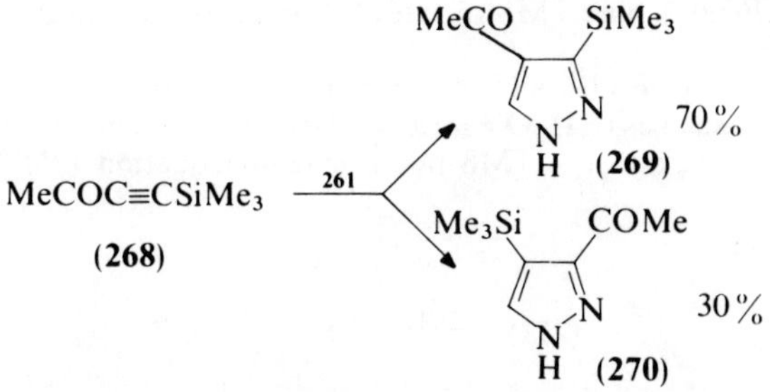

(122)

An easy procedure to obtain 3-TMS-pyrazoles is the selective desilylation of 3,4-bis(TMS)-pyrazole (**271**), e.g., by means of bromine at 0 °C the corresponding 4-bromo-3-TMS-pyrazole (**272**) can be isolated; substitution of the remaining TMS moiety by means of bromine only takes place at 80 °C[151]. An alternative approach is protolytic desilylation of **271**, which in the first step selectively forms **266** and only at 100 °C leads to the cleavage of the remaining TMS grouping[152]; in addition, the application of nitrating acid leads to replacement of the silyl moiety at the 4-position, exclusively, which is most prone

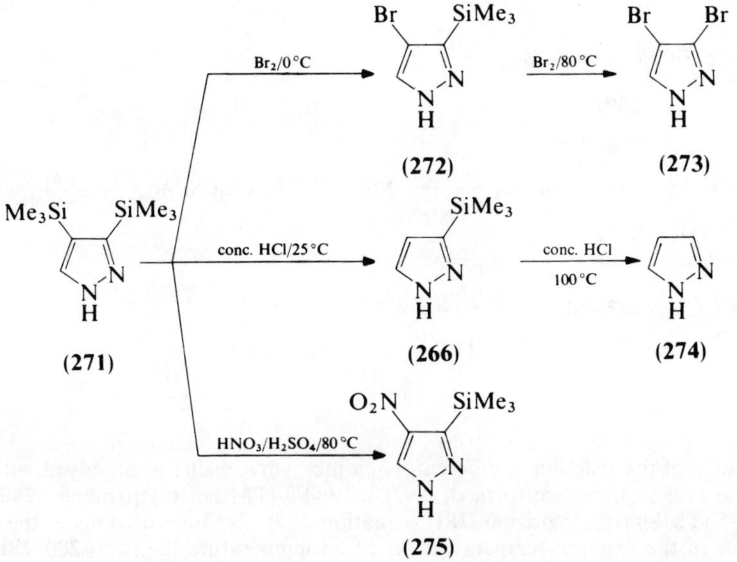

(123)

to electrophilic replacements (equation 123)[152]. The reaction of 3,4-bis(TMS)-pyrazole (**271**) with concentrated nitric acid leads to 3,4-bis(TMS)-pyrazolium nitrate (**276**).

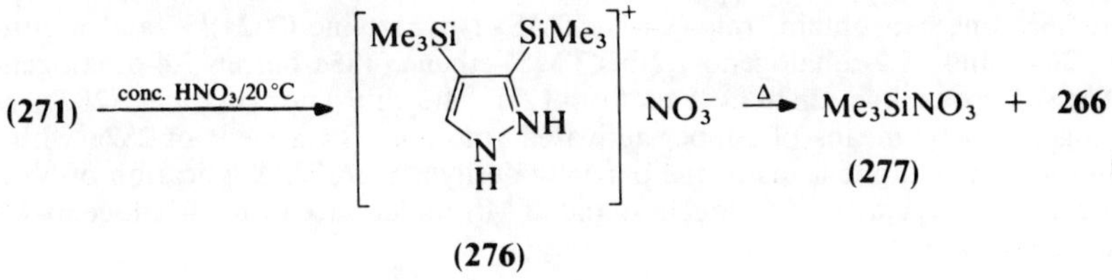

(124)

Thermolysis of **276** gives pure TMS-nitrate (**277**) in nearly quantitative yield, besides **266** (equation 124)[153].

In contrast to these results, silyl moieties in 3-positions are removed by common nucleophilic agents such as $C_2H_5O^{\ominus}$ since the 3-positions are more prone to nucleophilic attacks, thereby giving rise to 4-TMS-pyrazole (**278**) (equation 125)[152].

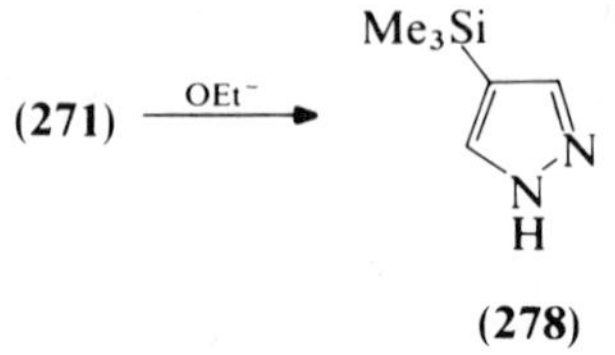

(125)

3-TMS methylpyrazole (**279**) is formed by the reaction of 1,4-bis(TMS)-buta-1,3-diyne (**207**) with anhydrous hydrazine in abs. ethanol[154]. The reaction is initiated by a nucleophilic attack of hydrazine at carbon-1 of **207**, forming a hydrazone via rearrangement and, after partial desilylation and ring closure, **279** can be isolated (equation 126).

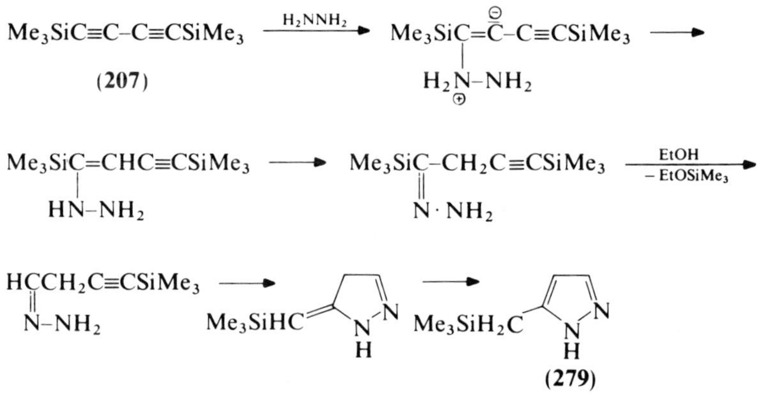

(126)

The course of the reaction is different when methylhydrazine is employed with **207**, in this case two isomers are formed, i.e. 1-methyl-3-(TMS)methylpyrazole (**280**) and 1-methyl-5-(TMS)methylpyrazole (**281**) (equation 127)[154]. Quite obvious is the directive influence of the reaction temperature: at room temperature the ratio **280**:**281** is 1:10, whereas at 70–80 °C the ratio is 4:1. The course of reaction is directed by the nucleophilic attack of either nitrogen atom of methyl hydrazine, thereby leading to **280** or **281**, respectively.

In a similar vein, via 1,3-dipolarophilic cycloaddition of **261** with *trans*-bis(TMS)-ethene (**85**) one can obtain *trans*-3,4-bis(TMS)-1-pyrazoline (**282a**)[155] and accordingly with **261** and 1,2-dihalogeno-1,2-bis(TMS)-ethene (**85a/b**)*trans*-3,4-dihalogeno-3,4-bis(TMS)-1-pyrazoline (**282b/c**) (equation 128). The dibromopyrazoline (**282c**) can be dehalogenated by means of copper-activated zinc dust. Treatment of **282a** with tetrachloro-*o*-benzoquinone leads to the partially desilylated **283a**. Application of NaOH to **282b/c** effects cleavage of halogeno- and TMS-moiety to form 3-halogeno-4-TMS-pyrazole (**283b/c**).

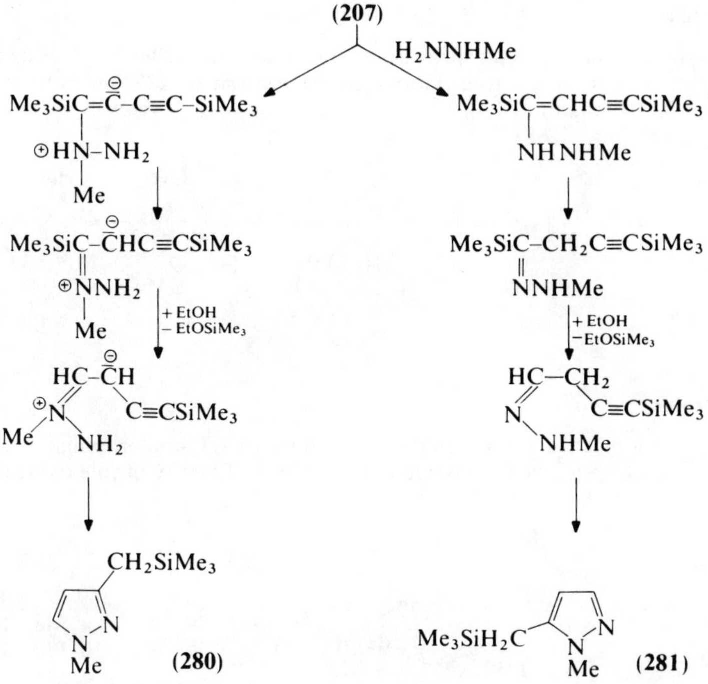

(127)

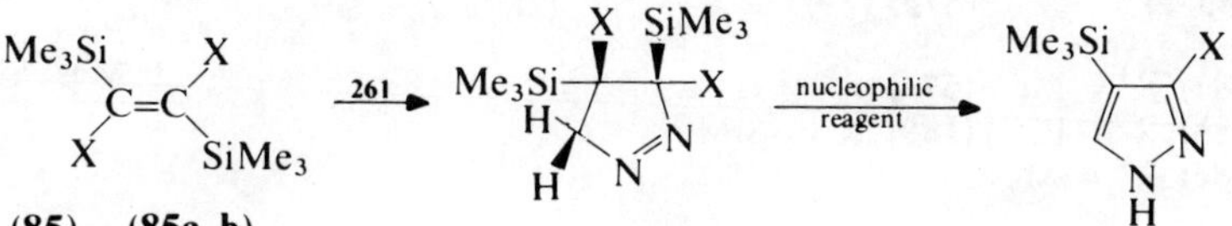

(128)

(85), **(85a, b)**

(282a–c)

(283a–c)

85, 282a/283a: X = H
85a, 282b/283b: X = Cl
85b, 282c/283c: X = Br

282c $\xrightarrow{\text{Zn/Cu}}$ **271**

E. Triazoles

The synthesis of TMS-triazoles can be accomplished by the reaction of 2,2-dimethyl-2-silahex-3-yn-5-one (**268**) with azides (**284**), e.g. hydrogen azide (**284a**) or phenyl azide (**284b**) (equation 129)[154].

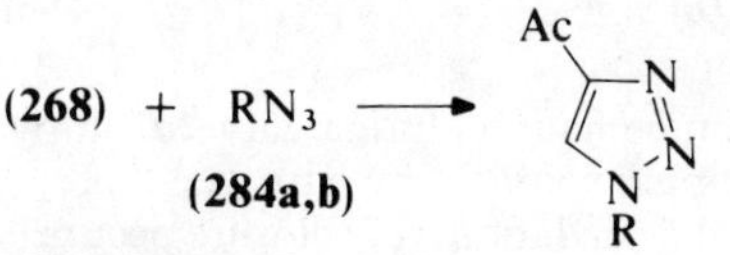

(129)

(284a,b)

284a/285a: R = H
284b/285b: R = Ph

(285a, b)

F. Oxazoles

A facile route to silyloxazoles starts from 4-methyloxazole (**286**), involves its treatment with butyllithium, subsequent silylation and cyclization to **287** under heating in the presence of KOH (equation 130)[156].

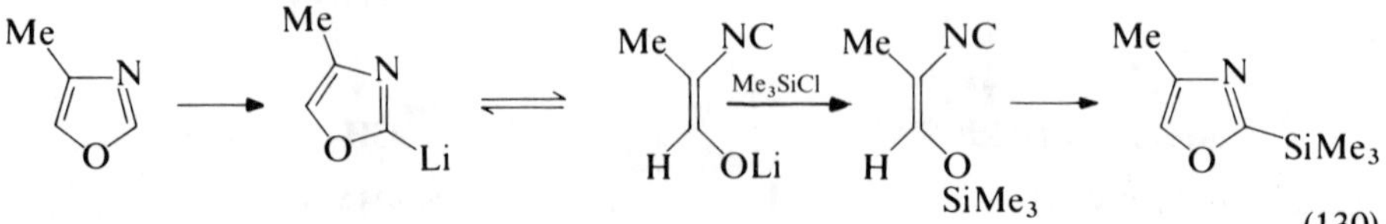

(286) **(287)** (130)

G. Isoxazoles

A very elegant reaction is the synthesis by means of a 1,3-dipolarophilic cycloaddition of nitrile oxides with alkynylsilanes (equation 131)[157]. If mesityl nitrile oxide (**288c**) and

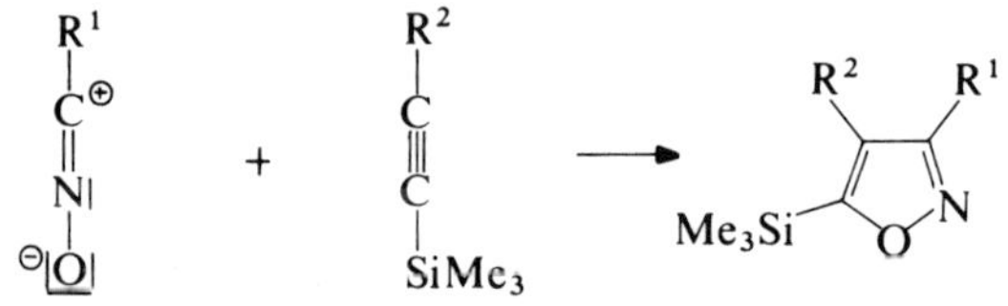

(288a–c) **(57a, 189)** **(289a–f)**

(**288a**) R^1 = Me (**57a**) R^2 = H
(**288b**) R^1 = Ph (**189**) R^2 = $SiMe_3$
(**288c**) R^1 = Mes

289	R^1	R^2
a	Me	H
b	Ph	H
c	Mes	H
d	Me	$SiMe_3$
e	Ph	$SiMe_3$
f	Mes	$SiMe_3$

(131)

3,3-dimethyl-3-silapenta-1,4-diyne (**290a**) or 3,3,4,4-tetramethyl-3,4-disilahexa-1,5-diyne (**290b**), respectively, are taken instead, one is able to synthesize the dimeric bis(3-mesityl-5-isoxazolyl)silanes **291a** and **291b**, respectively (equation 132)[157]. Additionally, the

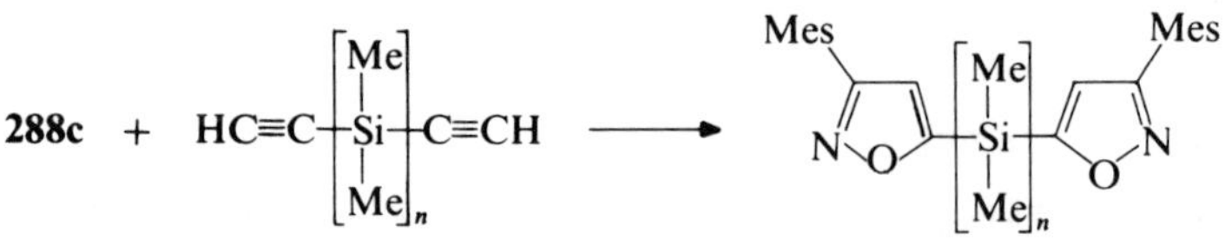

(132)

(290a–b)
(**a**) $n = 1$
(**b**) $n = 2$

(291a–b)
(**a**) $n = 1$
(**b**) $n = 2$

employment of hydroxylamine hydrochloride with **207** in pyridine as a solvent leads to **289a**. In this case nucleophilic attack by nitrogen takes places at carbon-2 of the butadiyne and, after partial desilylation, ring closure occurs (equation 133a)[154]. Alternatively, if **268** is taken instead to react with hydroxylamine hydrochloride, the reaction leads to **289a**, after cyclization from the oxime **292**, shown to be a relatively stable intermediate (equation 133b)[154]. In an analogous vein, 5-(TMS)methylisoxazole (**293**) can be synthesized from **207** and hydroxylamine hydrochloride in a polar medium (NaOAc/EtOH) (equation 133c)[154].

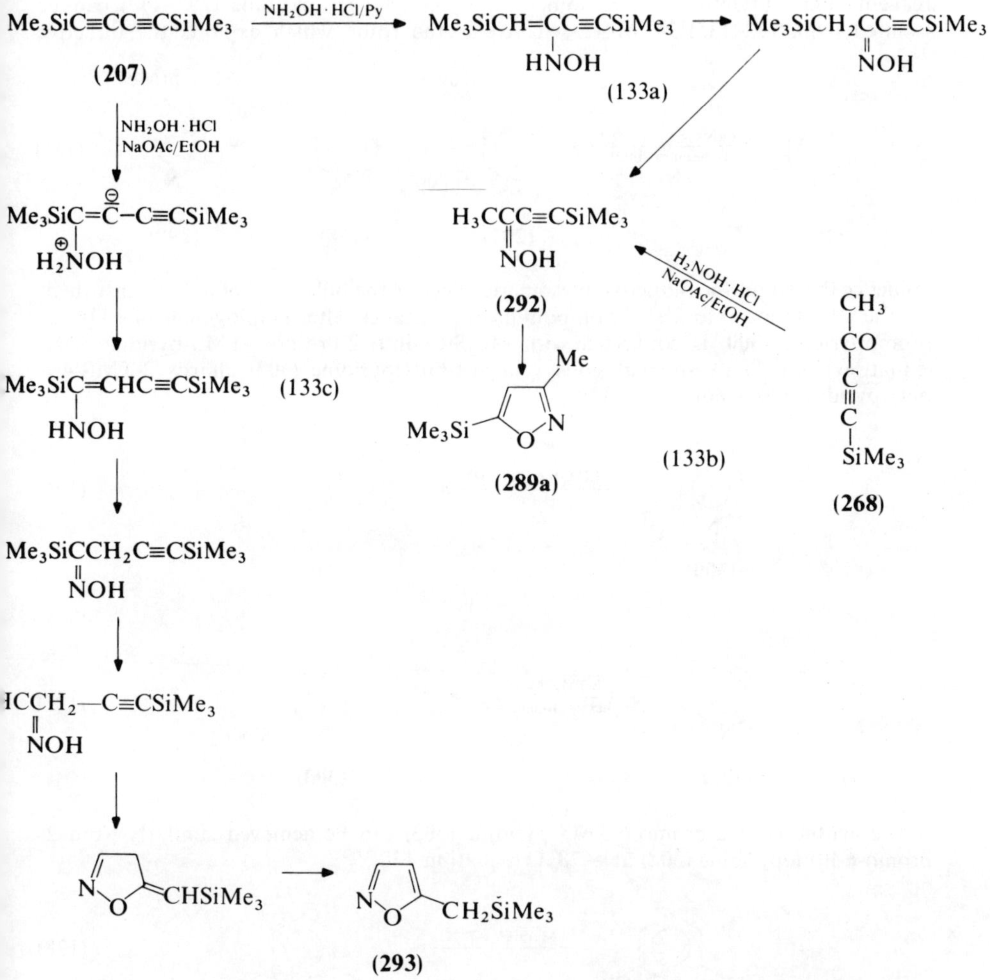

(**207**) (133a) (**292**) (**289a**) (133b) (**268**) (133c) (**293**)

Finally, a very helpful synthetic pathway to silylisoxazoles is the 1,3-dipolarophilic cycloaddition of TMS-nitrile oxide (**294**) with alkenes such as styrene which gives rise to, e.g., 3-TMS-4,5-dihydro-5-phenylisoxazole (**295**) (equation 134)[158].

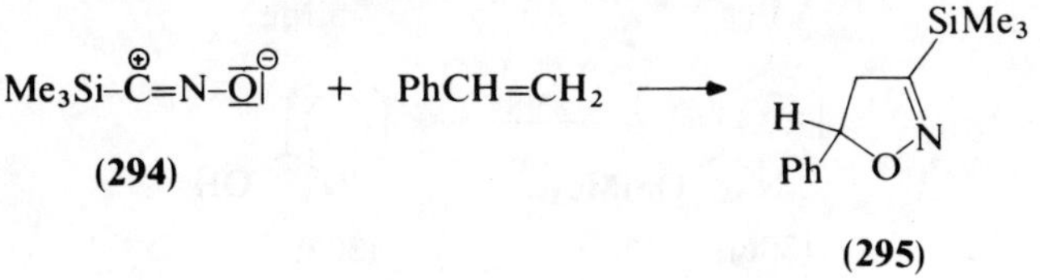

(**294**) (**295**) (134)

H. Pyridines

When an attempt is made to synthesize a silylpyridine in a direct approach from pyridine (**296**) with a silyl component, e.g. Me_3SiCl in the presence of electrophilic reagents like butyllithium, the isomers 2-,3- and 4-TMS-pyridine (**297**–**299**) can be isolated (equation 135)[159]. More selective is the route which exploits the directive

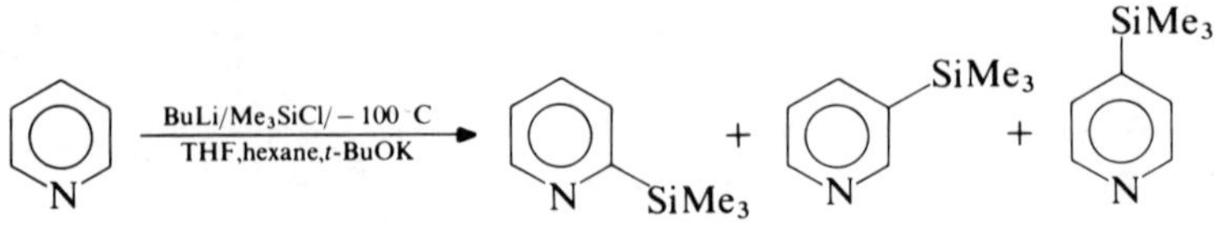

(135)

(**296**) (**297**) (**298**) (**299**)

influence (based on the inductive/mesomeric effect or the bulkiness) of a first substituent for the conversion into a silyl compound. For instance, after employment of LDA, 2-bromopyridine (**300**) is converted with Me_3SiCl into 2-bromo-5-TMS-pyridine (**301**) (equation 136)[160]. In an analogous vein, 4-*t*-butylpyridine (**302**) affords 2-TMS-4-*t*-butylpyridine (**303**) (equation 137)[161].

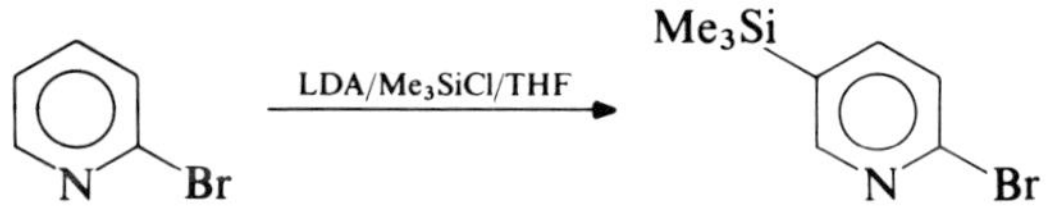

(136)

(**300**) (**301**)

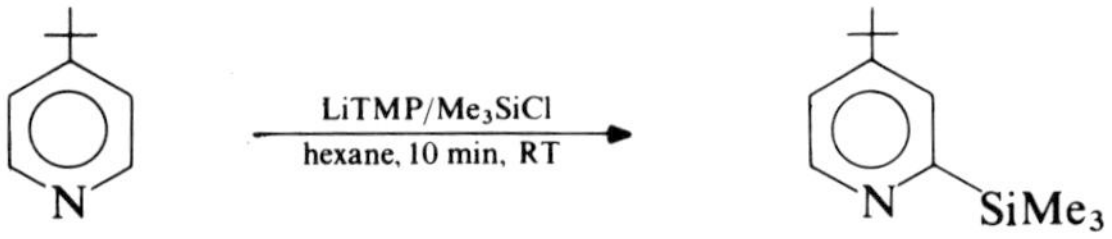

(137)

(**302**) (**303**)

The synthesis of 2-bromo-6-TMS-pyridine (**305**) can be achieved similarly from 2-bromo-6-lithiopyridine (**304**) at $-78\,^{\circ}C$ (equation 138)[162].

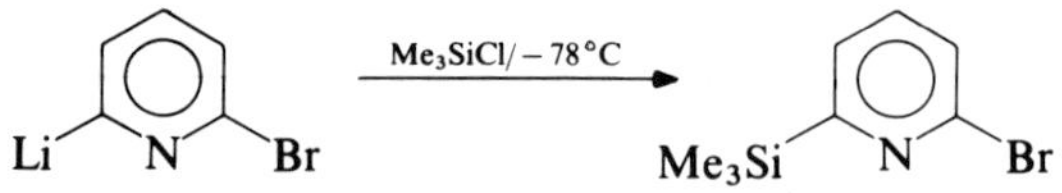

(138)

(**304**) (**305**)

(139)

(**306**) (**307**)

An alternative approach to silylpyridines is the treatment of bromosiloxypyridines with a potent metallating agent such as *t*-BuLi which, after an anionic rearrangement of the silyl moiety, leads to the formation of a hydroxy-TMS-pyridine, e.g. **307** (equation 139)[139].

I. Pyridazines

In a different pattern, by using silylalkynes, silyl-substituted pyridazines are obtainable from the tetrazine derivative **308** in a Diels–Alder-type reaction. Thus 4-TMS-(**309a**) and 4,5-bis(TMS)-3,6-bis(methoxycarbonyl)pyridazine (**309b**) can be isolated in very high yields. The products are inert against acid-catalyzed desilylation (equation 140)[163].

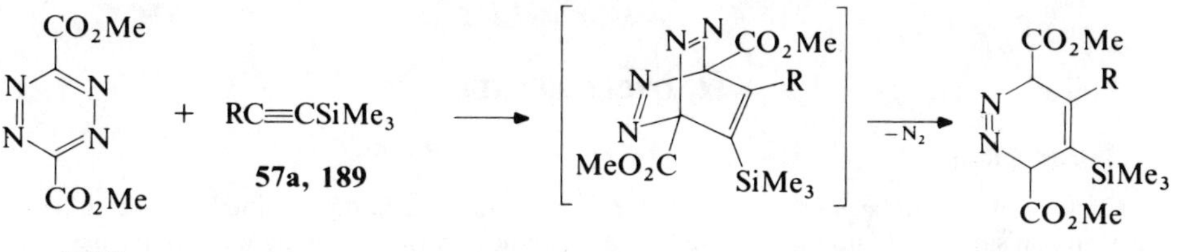

(308) **57a, 189** (309a, b)

(**a**) R = H
(**b**) R = $SiMe_3$

(140)

Depending on the alkyne component, the course of reaction can be quite different. Whereas **268** in the keto form reacts analogously to equation 140, the enol form **268a** leads to the corresponding TMS-alkynylpyridazine (**311**) (equation 141)[164].

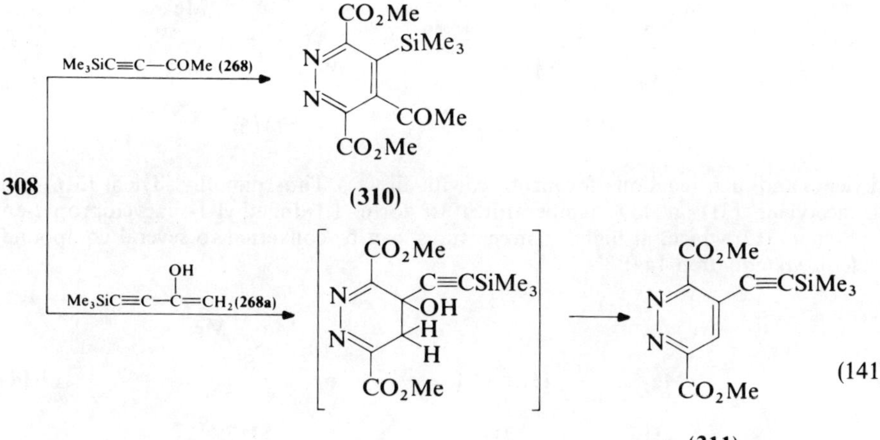

(310) (311)

(141)

J. Quinolizines

By means of a sophisticated Diels–Alder-type reaction, silylquinolizine derivatives can be synthesized, starting from silylpyridines[165]. Thus, 4-TMS-pyridine (**299**) affords with two equivalents of dimethyl but-2-yn-1,4-dioate (**312**) in good yield 8-TMS-4*H*-1,2,3,4-tetrakis(methoxycarbonyl)quinolizine (**313**) (equation 142)[166].

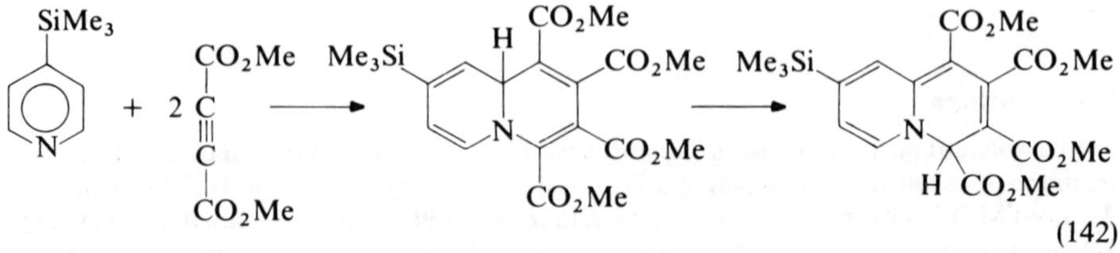

(142)

(**299**) (**312**) (**313**)

IX. CYCLOSILANES

A. Introduction

Cyclosilanes can be synthesized according to the methods outlined above in the sections on saturated and unsaturated organylsilanes. In the following we will discuss the various classes of cyclic silanes in the order of increasing ring size.

B. Three-membered Rings

The classical approach to achieve a silirane is the treatment of 2,4-dibromo-2,3,3,4-tetramethyl-3-silapentane (**314**) with Mg in THF to form 1,1,2,2,3,3-hexamethylsilirane (**315**) (equation 143)[167]. More recent syntheses involve, e.g., the *in situ* formation of

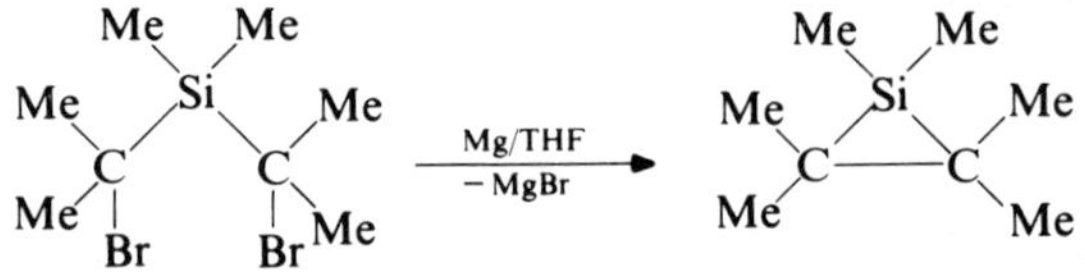

(143)

(**314**) (**315**)

silylenes and their reactions, for instance, with alkynes. Thus, dimethylsilylene (**316**) adds to acetylene (**31**) at low temperatures to form 1,1-dimethyl-1-silacycloprop-2-ene (-silirene) (**317**) which, at higher temperatures, can be converted to several compounds (*vide infra*) (equation 144)[168].

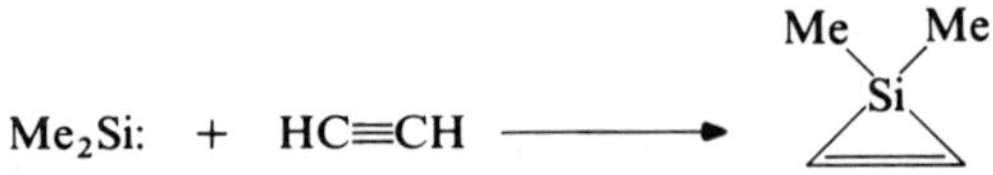

(144)

(**316**) (**31**) (**317**)

In a slightly modified manner, tris(TMS)-phenylsilane (**318**) is converted into phenyl-TMS-silylene under conditions of photolysis, which can add to allyl ethyl ether to give rise to 1-phenyl-1-TMS-2-ethoxymethyl-1-silacyclopropane (-silirane) (**319**) (equation 145)[169].

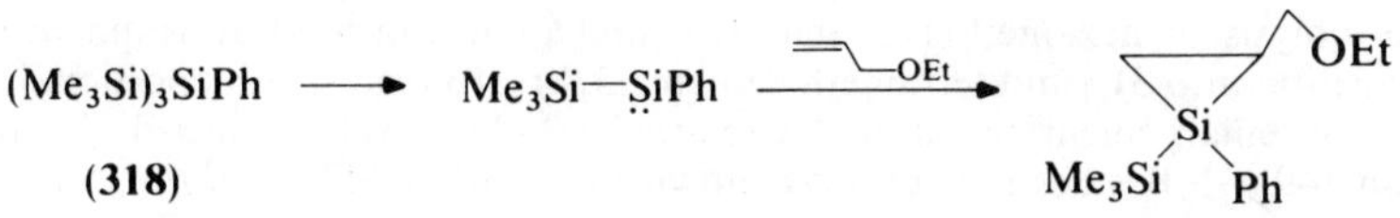

(318) (319) (145)

Stable disilacyclopropanes (disiliranes) are obtainable by means of the addition of dimesitylsilylene to 1,1-dimesityl-3,3-bis(TMS)-1-silapropadiene (**320a**) and 1,1-dimesityl-3-phenyl-3-TMS-1-silapropadiene (**320b**), respectively, to yield the corresponding 1,1,2,2-tetramesityl-3-[bis(TMS)methylene]-(**321a**) and 1,1,2,2-tetramesityl-3-[phenyl-(TMS)-methylene]1,2-disilacyclopropane (-disilirane) (**321b**) (equation 146)[170].

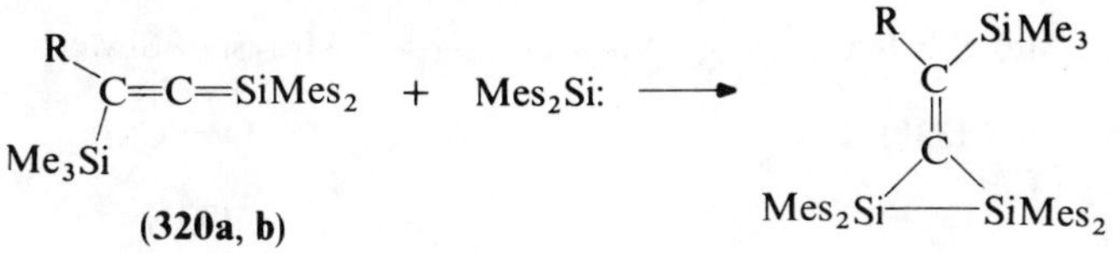

(320a, b) (321a, b) (146)

320a/321a : R = $SiMe_3$
320b/321b : R = Ph

In a related manner, disilacyclopropane (disilirane) derivatives can be synthesized by the reaction of disilene (**322**) with other reactive compounds. Thus, via diazomethane one can add a methylene component to **322** to form the corresponding 1,2-disilacyclopropanes (**323**). These are versatile intermediates which give rise to a large variety of compounds (*vide infra*). One type of application, the conversion of **323** with *m*-CPBA to a 1,3-disila-2-oxacyclobutane (disilaoxetane) derivative (**324**), is shown in equation 147[171].

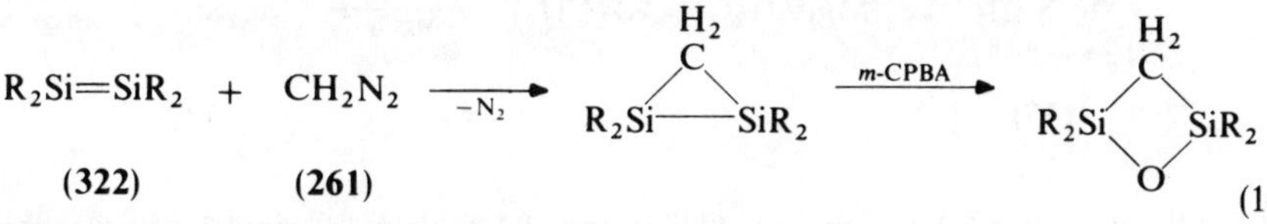

(322) (261) (147)

R = 2,6-dimethylphenyl (323) (324)

C. Four-membered Rings

A direct approach to silacyclobutane (siletane) derivatives is the reaction of methylsilenes with other unsaturated compounds. Thus, the silaethylene (silene) (**326**), formed from pivaloyl-tris(TMS)-silane (**325**) via thermolysis (10 h/140 °C), reacts with 1-phenylprop-1-yne to yield the silacyclobutene (silete) derivative **327** (equation 148)[172].

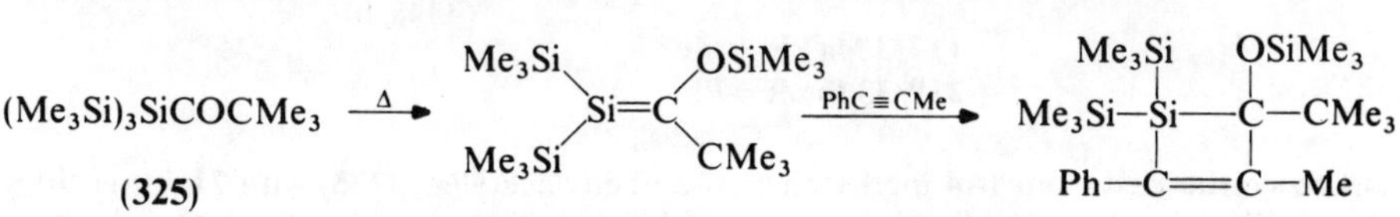

(325) (326) (327) (148)

The thermolysis of hexamethylsilirane (**315**) under inert (argon) atmosphere leads to decomposition into **316** and tetramethylethene (**328**). **316** reacts with additional **315** and induces ring enlargement to afford octamethyl-1,2,-disilacyclobutane(-disiletane) (**329**) (equation 149)[173]. Besides the route as shown in equation 147, the formation of a 1,2-

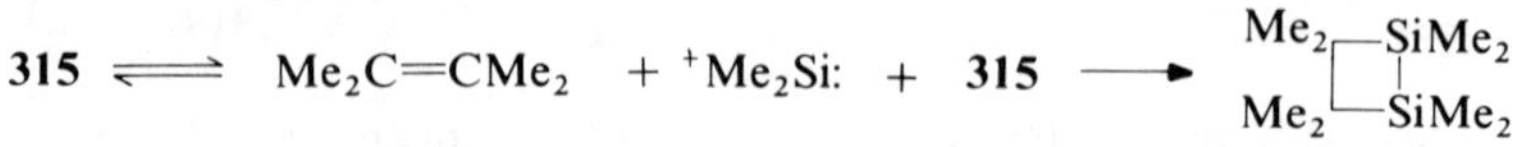

(328) (316) (329) (149)

disilaoxetane (**331**) can be accomplished by the reaction of a disilene, e.g. **330**, with a ketone (equation 150)[174].

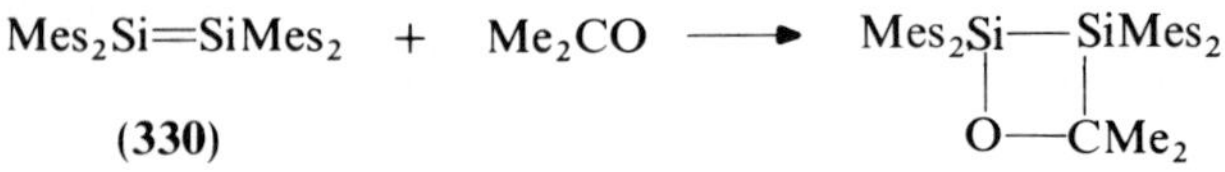

(330) (331) (150)

D. Five-membered Rings

A standard synthesis shall first be discussed. Dialkyldichlorosilanes and but-2-en-1-ylmagnesium (the latter having been obtained by reduction of 1,3-butadiene with Mg) yield silacyclopentenes; e.g., **117** plus but-2-en-1-ylmagnesium afford 1,1-dimethyl-1-silacyclopent-3-en (-3-silolene) (**332**) (equation 151)[175]. In a related manner, 1,3,5-hexatriene (**333**) and **117** or **219**—in the presence of Mg and catalytic amounts of CuCl, in

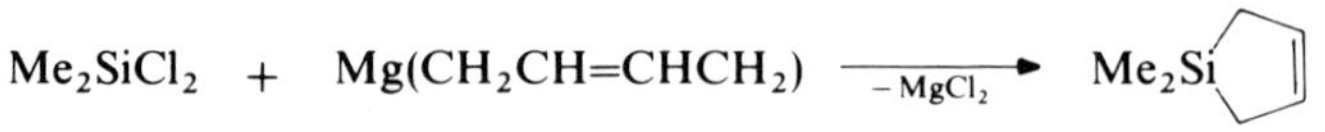

(117) (332) (151)

THF—directly afford 1,1-dimethyl- (**334a**) and 1,1-diphenyl-2-vinyl-1-silacyclopent-3-ene (-3-silolene) (**334b**), respectively (equation 152)[176]. In addition, 2,5-bis(4-methylphenyl)-1,1,3,4-tetraphenyl-2-silacyclopenta-2,4-diene (-tetraphenylsilole) (**337**)

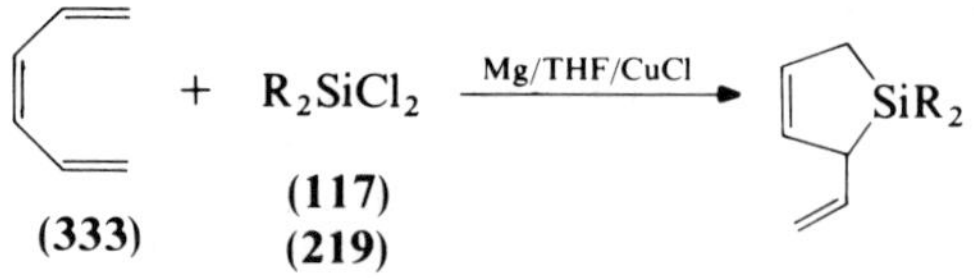

(333) (117) (219) (334a, b) (152)

117/334a: R = Me
219/334b: R = Ph

can be synthesized from 1-(4-methylphenyl)-2-phenylacetylene (**335**) with **219** after lithiation with subsequent cyclization (equation 153)[177]. The reaction of 1,2-dimethoxytetramethyldisilane (**338**) with two equivalents of styrene (**36**) or methylstyrene (**339**) in

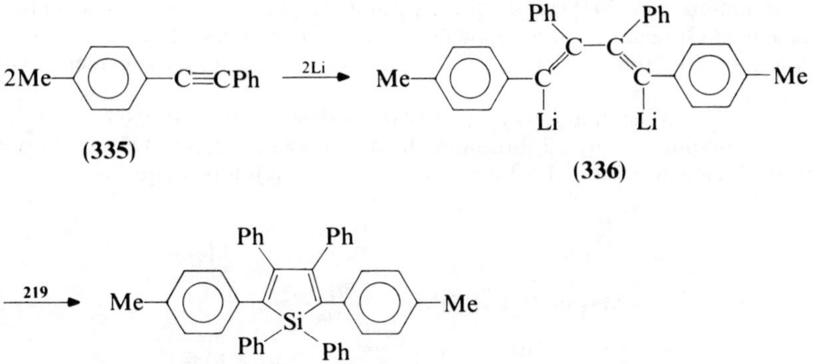

(335) (336)

(337)

(153)

presence of a catalyst leads to 1,1-dimethyl- (**340a**) and 1,1,2,4-tetramethyl-2,4-diphenyl-1-silacyclopentane (-silolane) (**340b**), respectively, which are formed as mixtures of *cis/trans*-isomers (equation 154)[178].

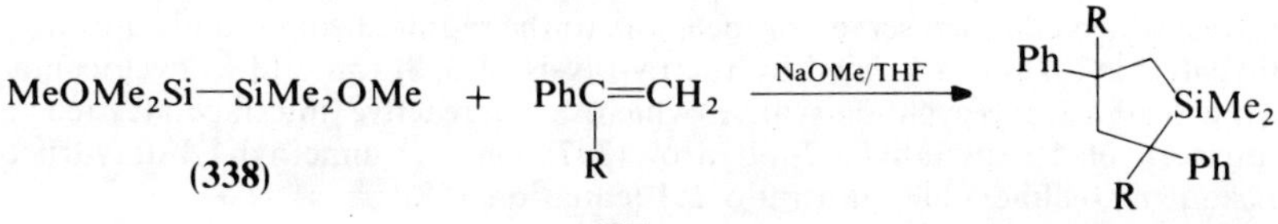

(338) (36) (339)

36/340a: R = H (**340a, b**)
339/340b: R = Me

(154)

A convenient reaction pathway to silafluorenes is the pyrolysis of, e.g., dimethyldiphenylsilane (**341**) in a pressure vessel which gives rise to 5,5-dimethyl-5-silafluorene (-dibenzosilole) (**342**) in 25% yield (equation 155)[179].

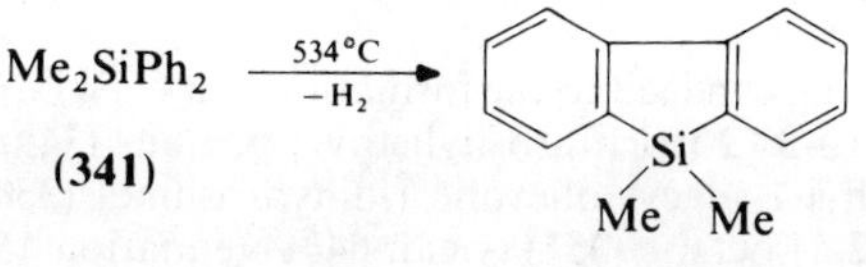

(341) (342)

(155)

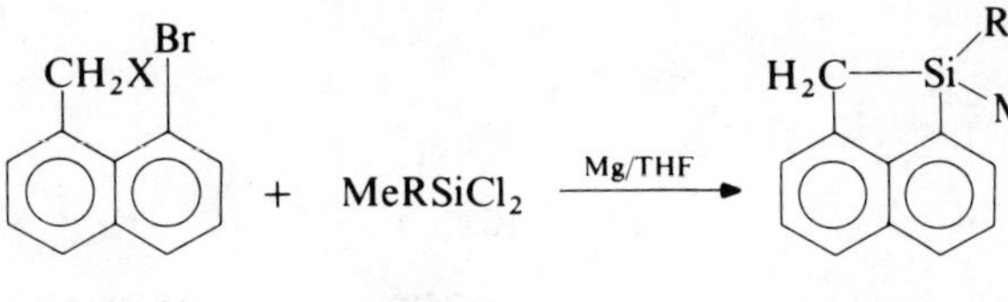

(156)

(**343a, b**) (**51**)(**117**) (**344a, b**)

			yield %
(**343a**) X = Cl	(**51**) R = H	(**344a**) R = H	64
(**343b**) X = Br	(**51**) R = H	(**344a**) R = H	54
(**343a**) X = Cl	(**117**) R = Me	(**344b**) R = Me	71

1-Silaacenaphthenes (**344**) can be directly obtained when 1-bromo-8-halogenomethylnaphthalene (**343**) reacts with alkyldichlorosilanes in the presence of magnesium. The application of tri- and tetrachlorosilane predominantly affords polymers (equation 156)[180].

Finally, via a cyclization reaction, disilacyclopentane derivatives can be formed. Thus, 1-(bromodimethylsilyl)-2-(bromomethyldimethylsilyl)ethane (**345**) in the presence of lithium is converted to 1,1,3,3-tetramethyl-1,3-disilacyclopentane (-disilolane) (**346**) (equation 157)[181].

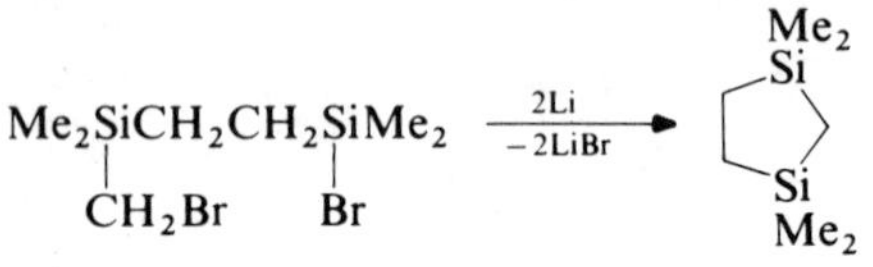

(157)

(**345**) (**346**)

E. Six-membered Rings

A facile route to silacyclohexane compounds involves the formation of silacyclohexene derivatives, which can then serve as precursors to the required compounds. For instance, dimethylsilylene (**316**) (e.g formed by the pyrolysis of **338**) can add to cyclopentadiene (**154**) and lead to a bicyclo-derivative which, as a reactive intermediate, rearranges to a mixture of 1,1-dimethyl-1,2-dihydro- (**347**) and 1,1-dimethyl-1,4-dihydro-1-silacyclohexadiene (-siline) (**348**) at a ratio 2:1 (equation 158)[182].

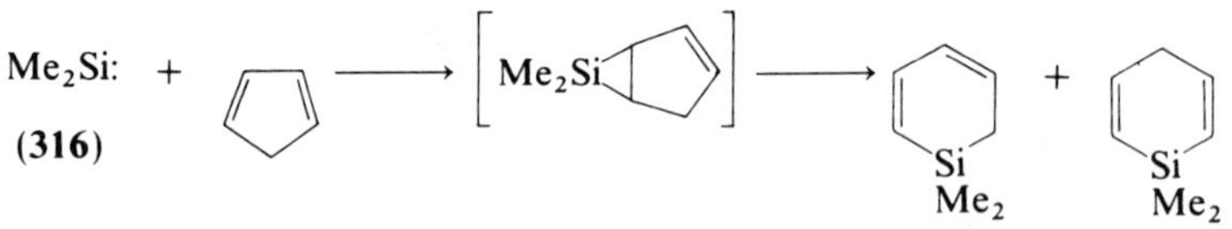

(158)

(**316**) (**154**) (**347**) (**348**)

Intramolecular coupling can lead to the formation of a C–Si bond, so that cyclic silanes result. Thus, 1,5-dichloro-3-[2-(trichlorosilyl)ethyl]-pentane (**349**) with Mg in ether yield both 1,1-dichloro-4-ethyl-1-silacyclohexane (-dihydrosiline) (**350**) (yield: 21%) and 1-chloro-1-silabicyclo[2.2.2]-octane (**351**) (yield: 64%) (equation 159)[183].

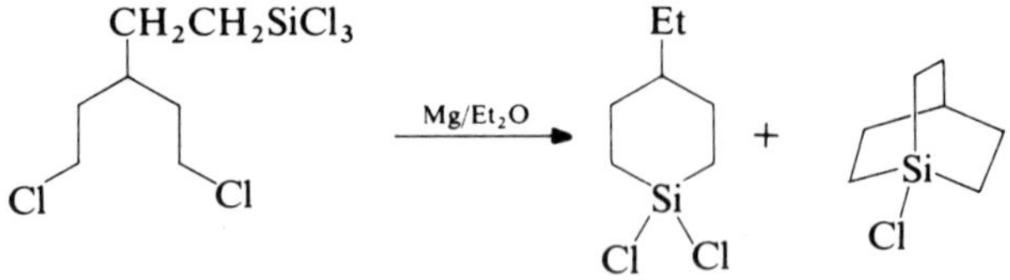

(159)

(**349**) (**350**) (**351**)

The addition of dichlorosilane to 1,4-cyclohexadiene (**352**) in the presence of hexachloroplatinic acid as catalyst unexpectedly affords 3-dichlorosilylcyclohexene (**353**) in 96% yield[184]. This, on heating in the presence of the catalyst, rearranges into 7,7-

dichloro-7-silanorbornane (**354**), the reduction of which to 7-silanorbornane (**355**) can be accomplished by means of $LiAlH_4$ (equation 160)[184].

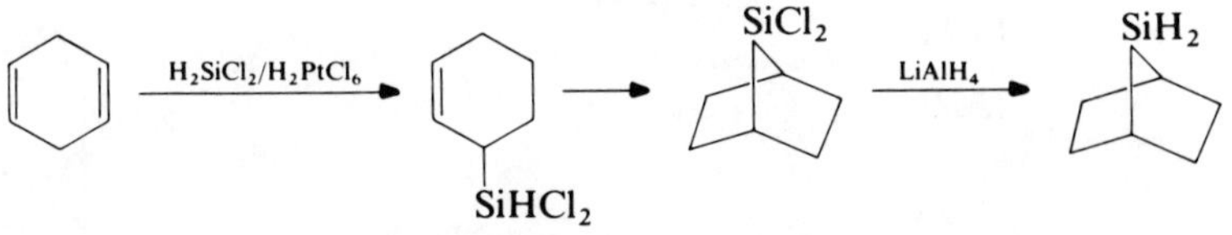

(160)

(352) **(353)** **(354)** **(355)**

9-Silaanthracene derivatives can be obtained by means of metallated precursor compounds. For instance, the bis-Grignard compound of bis(*o*-chlorophenyl)methane (**356**) forms with chloromethylmethyldichlorosilane (**357**) 9-chloromethyl-9-methyl-9,10-dihydro-9-silaanthracene (**358**) (equations 161)[185]. In a related manner, the dilithio

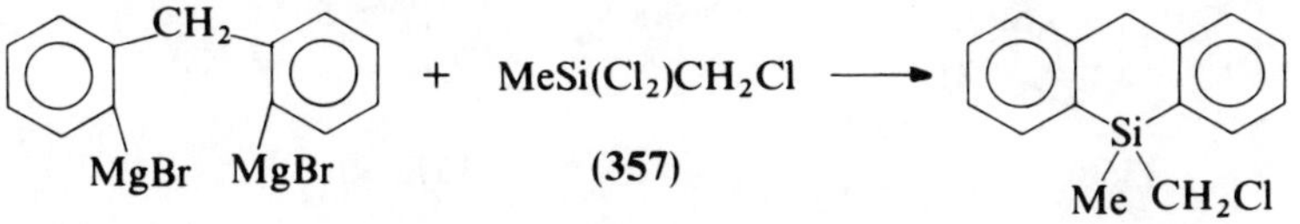

(161)

(356) **(357)** **(358)**

compound of bis(2-bromophenyl)phenylmethane (**360**) can react with dimethyldichlorosilane (**117**) to yield 9,9-dimethyl-10-phenyl-9,10-dihydro-9-silaanthracene (**361**) (equation 162)[186]. An extension of this approach is the synthesis of 2,2-dimethyl-2,3-dihydro-2-silaphenalene (**363**) from 1,8-bis(chloromethyl)naphthalene (**362**) (equation 163)[187]. This

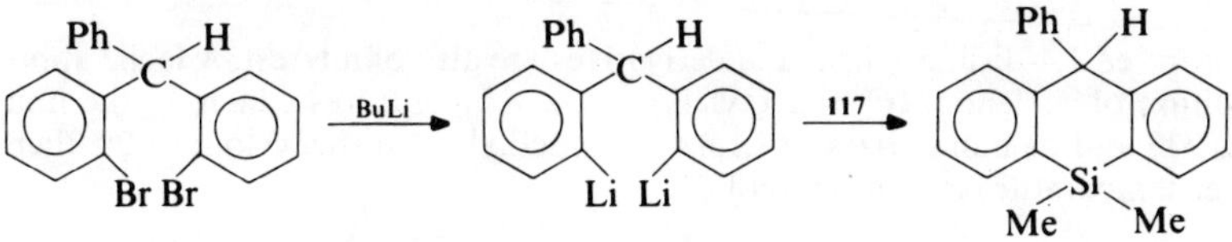

(162)

(359) **(360)** **(361)**

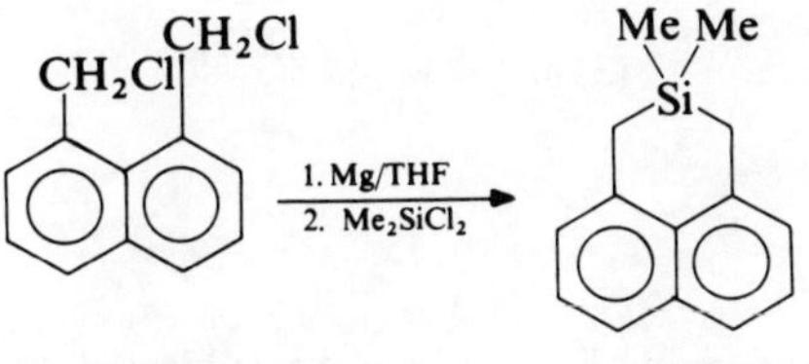

(163)

(362) **(363)**

synthetic pathway is also applicable to prepare sila-heteroaryl compounds, e.g., the *o,o'*-dilithiated derivative of diphenyl ether (**364**) reacts with phenylsilane to give rise to 9-phenyl-9-silaxanthene (**365**) (equation 164)[188]. Alkylating and phenylating reactions at the silicon atom are also feasible, e.g., by means of Grignard reagents according to the

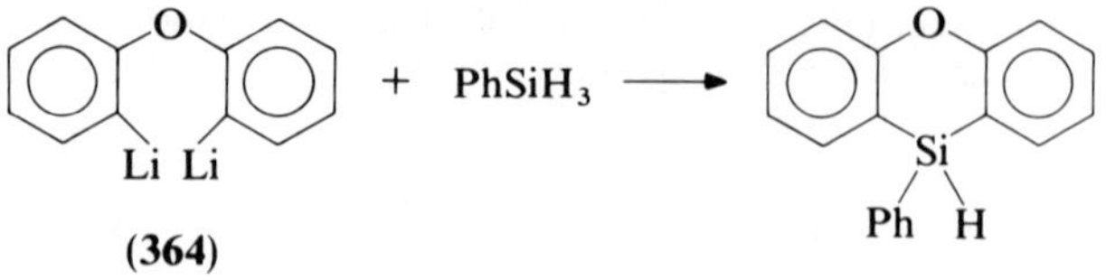

(364) (365) (164)

general reaction methods to synthesize, for instance, alkylsilanes. Thus, optically active 2-naphth-1-yl-2-silatetralin (**366**) can react with various Grignard reagents—in the presence of bis(triphenylphosphine)nickel(0) as catalyst—with retention of configuration to yield 2-naphth-1-yl-2-organyl-2-silatetralin derivatives (**367a**–d) (equation 165)[189].

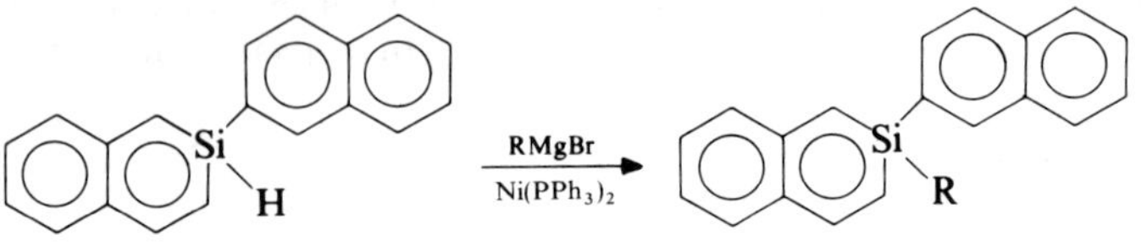

(366) (367a–d) (165)

R	Optical purity%	$[\alpha]_D^{\circ}$	No.
Me	95	+78	**367a**
Vinyl	96	+96	**367b**
Allyl	98	+97	**367c**
Ph	95	+122	**367d**

Unsaturated 1,4-disilacyclohexane derivatives are also of interest. A facile approach is the addition of silylene (**316**) to acetylene at low temperatures which, in the first stage, leads to **317** and then dimerizes to 1,1,4,4-tetramethyl-1,4-disilacyclohexa-2,5-diene (**368**) at higher temperature (equation 166)[168].

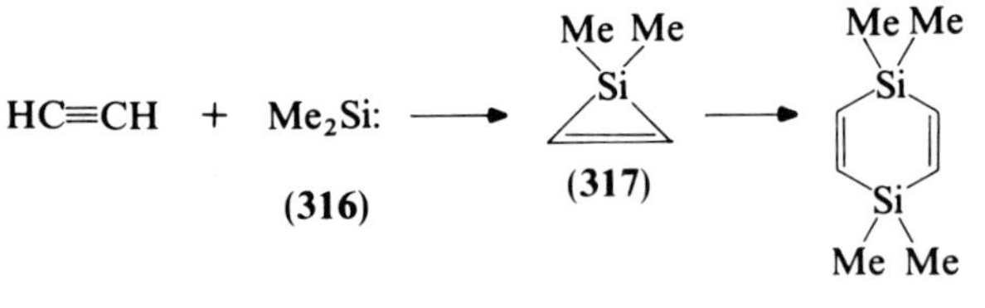

(316) (317) (368) (166)

In addition, cyclotrisilane derivatives can be subjected to photolysis to generate organylsilylenes. For instance, the photolysis of hexakis-*t*-butylcyclotrisilane (**369**) in the presence of substituted acetylenes, e.g. (**370**) affords 1,1,4,4-tetrakis-*t*-butyl-2,3,5,6-tetraorganyl-1,4-disilacyclohexa-2,5-diene (**371**) (equation 167)[190]. Alternatively, thermolysis of disilanes in the presence of alkynes is a viable route to disilacyclohexane derivatives. Thus, thermolysis of 1,2-dimethoxy-1,1,2,2-tetramethyldisilane (**338**) in the presence of diphenylacetylene (**370**) in a pressure vessel—via reactive intermediates and,

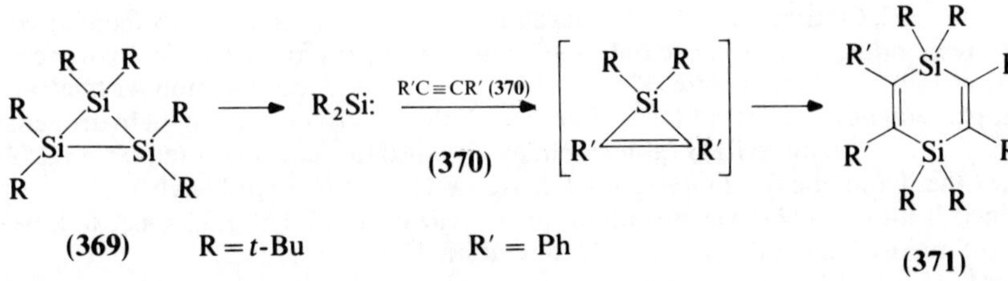

(369) R = *t*-Bu (370) R′ = Ph (371) (167)

finally, dimerization—yields 1,1,4,4-tetramethyl-2,3,5,6-tetraphenyl-1,4-disilacyclohexa-2,5-diene (-1,4-dihydro-1,4-disilenine) (**372**) (equation 168)[191]. The same reaction and reaction product (**372**) as outlined by equation 168 occurs when Me_2Si: is generated from 3,3-dimethylbenzosilepin (cf. *vide infra*) in the presence of **370**[192].

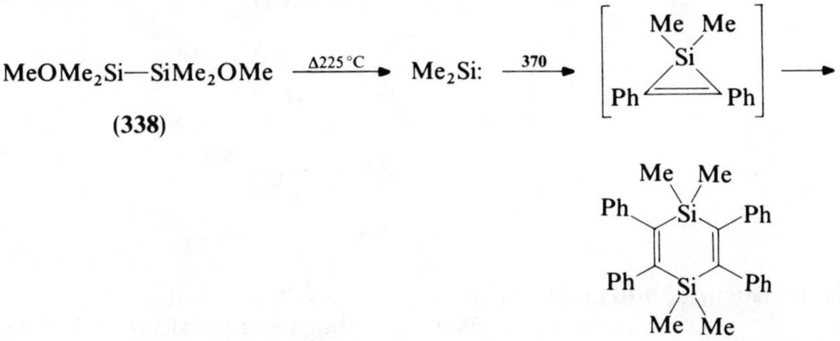

(338) (372) (168)

Another class of compounds worth mentioning are 1,2-disilacyclohexane derivatives, which can be obtained via cyclization of 1,4-dihalogenoalkanes with dihalogenodisilanes. For instance, 1,4-dibromo-but-2-ene (**373**) and 1,2-dichloro-1,1,2,2-tetramethyldisilane

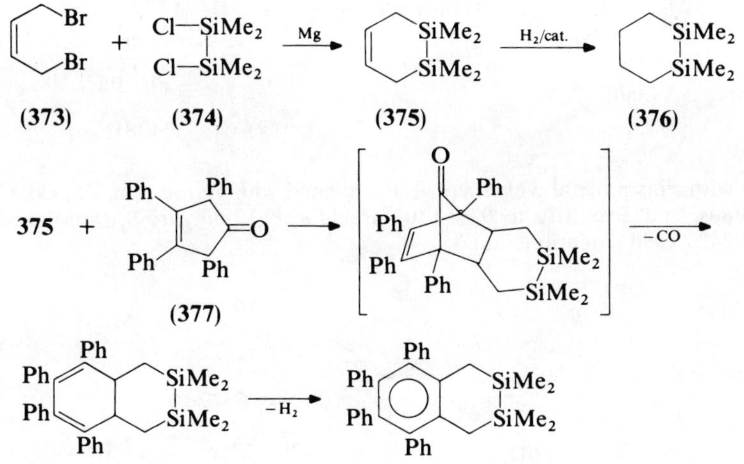

(373) (374) (375) (376) (377) (378) (169)

(**374**) form 1,1,2,2-tetramethyl-1,2-disilacyclohex-4-ene (**375**), which can be converted into the corresponding 1,2-disilacyclohexane (**376**) by means of catalytic hydrogenation (equation 169)[193]. Furthermore, **375** can undergo a Diels–Alder reaction with tetraphenylcyclopentadienone (**377**) and via decarbonylation and subsequent dehydrogenation affords 2,2,3,3-tetramethyl-5,6,7,8-tetraphenyl-2,3-disilatetralin (**378**) (equation 169)[193]. On the other hand, the thermolysis of 1,1,2,2-tetrakis(2,6-dimethylphenyl)-1,2-disilacyclopropane(-disilirane) (**323**) via intramolecular cyclization affords 1,3,3-tris(2,6-dimethylphenyl)-8-methyl-1,3-disilatetralin (**379**) (equation 170)[171].

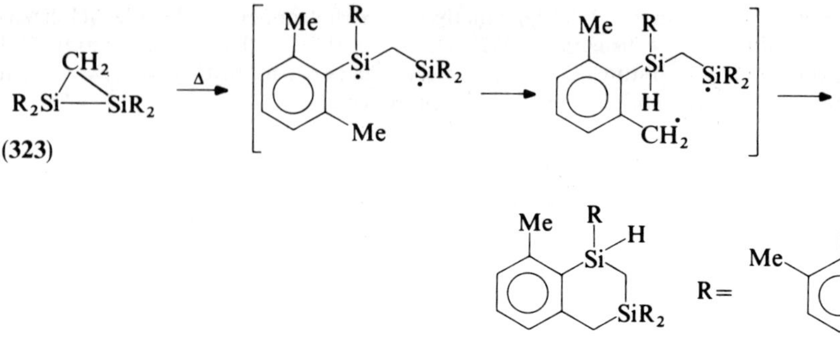

(**323**) (**379**) (170)

For the formation of 9,10-disilaanthracene derivatives there are two versatile approaches. In the first, *o*-dilithiobenzene (**380**) reacts with two equivalents of a silane, e.g. diphenyldichlorosilane (**219**), to form 9,9,10,10-tetraphenyl-9,10-dihydro-9,10-disilaanthracene (**381**) (equation 171)[194]. In the second, *o*-dichlorobenzene (**382**) reacts

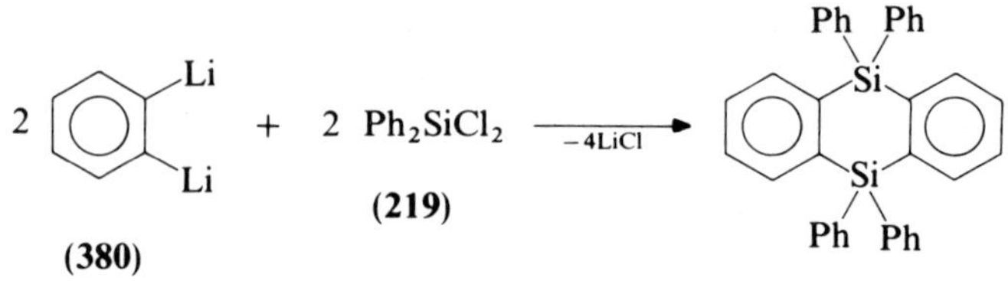

(**380**) (**219**) (**381**) (171)

directly with silicon metal which has been activated with various Cu, Zu, Cd, Co or Ni compounds, and gives rise to 9,9,10,10-tetrachloro-9,10-dihydro-9,10-disilaanthracene (**383**) in 84% yield (equation 172)[195].

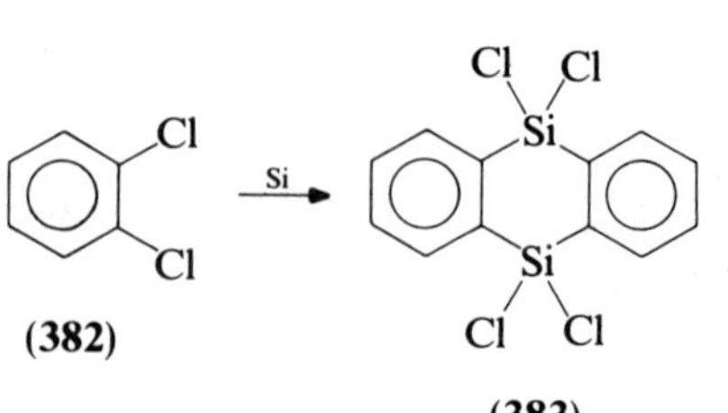

(**382**) (**383**) (172)

The reaction of 2,2′-dichlorobiphenyl (**384**) with butyllithium and subsequent cyclization with 1,2-dichloro-1,1,2,2-tetramethyldisilane (**374**) affords 9,9,10,10-tetramethyl-9,10-dihydro-9,10-disilaphenanthrene (**385**) (equation 173)[196].

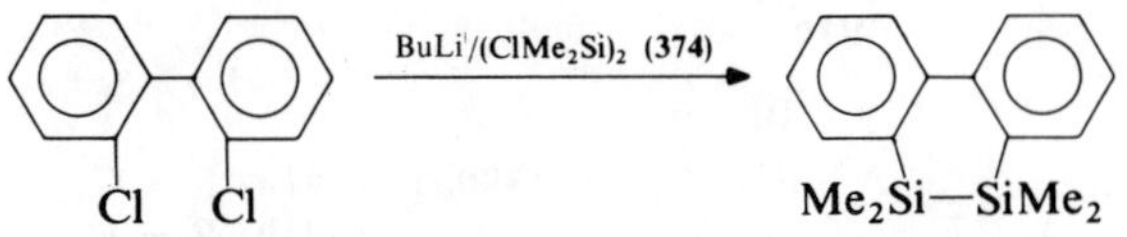

(173)

(384) **(385)**

Finally, in a one-step reaction of bis-*t*-butylsilylene (**369a**) with nitriles, the 3,3,6,6-tetrakis-*t*-butyl-2,5-diorganyl-3,6-dihydro-3,6-disilapyrazines (**387**) can be formed (equation 174)[190] (cf. *vide supra*: equations 166–168).

369a + 2R′CN → 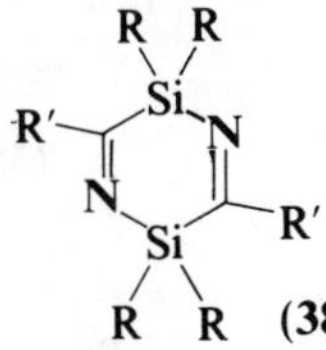

(174)

(386a, b) **(387a, b)**

(386a) R′ = Me **(387a)** R = *t*-Bu; R′ = Me
(386b) R′ = Ph **(387b)** R = *t*-Bu; R′ = Ph

F. Seven-membered Rings

The best known compounds of this class are benzosilepines. The classical reaction is the cyclization of the di-Grignard compound of 1,2-bis(*β*-bromoethyl)benzene (**388**) by means of substituted dichlorosilanes (**1/58/117/219**) to afford 3-substituted 1,2,4,5-tetrahydro-3*H*-3-benzosilepines (**389a–d**) (equation 175)[197]. **389a** and **d** can easily be brominated by means of NBS to yield the corresponding 1,5-dibromo compounds

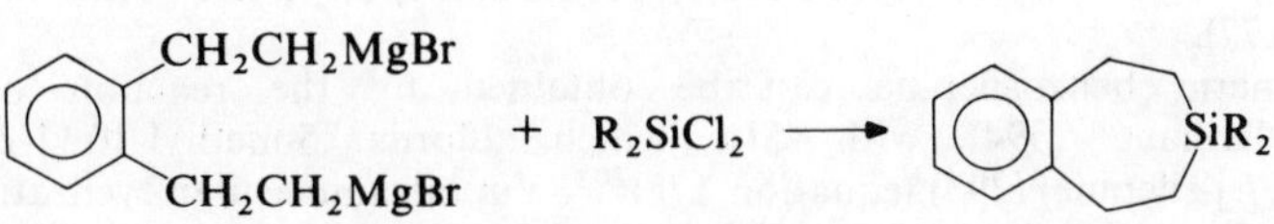

(175)

(388) **(1), (58), (117), (219)** **(389a–d)**

1/389a : R = Cl **58/389b** : R = H
117/389c : R = Me **219/389d** : R = Ph

390a, b. Via dehydrobromination by either thermolysis (in the case of **390a**)[200] or by means of 1,5-diazabicyclo[4.3.0]non-5-ene (in the case of **390b**)[198] the 3,3-disubstituted 3*H*-3-benzosilepines **391a** or **391d**, respectively, are obtained. Further treatment of **391a** with either $LiAlH_4/LiAlH[(CH_3)_3CO]_3$ or MeLi yields the 3*H*-3-benzosilepine derivatives **391b** and **391c**, respectively (equation 176)[198a].

Starting from **388** plus either **389a**[199] or **391a**[200], respectively, one can synthesize the corresponding spirosilepines, i.e. 3,3′-spiro-bi[1,2,4,5-tetrahydro-3*H*-3-benzosilepine]

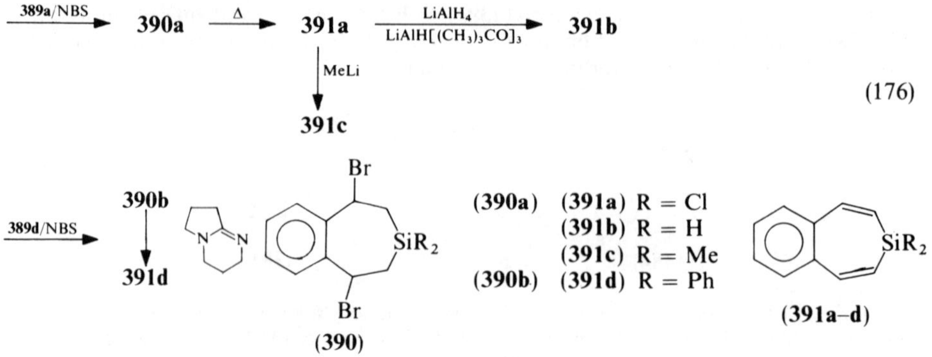

(176)

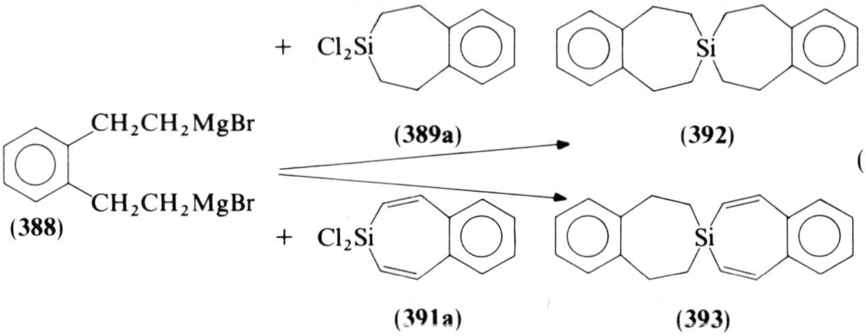

(177)

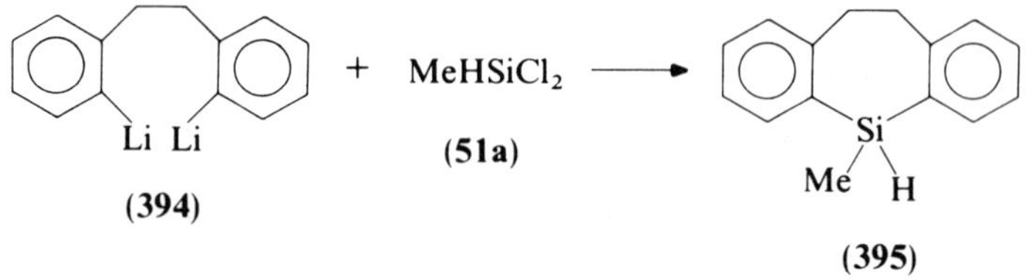

(178)

(**392**) and spiro[(1,2,4,5-tetrahydro-3*H*-3-benzosilepine)-3,3′-3′*H*-3′-benzosilepine] (**393**) (equation 177).

An isomeric benzosilepine can be obtained by the reaction of 1,2-bis(2-lithiophenyl)ethane (**394**) with **51a** which affords 5-methyl-10,11-dihydro-5*H*-(dibenzo[*b*,*f*]-silepine) (**395**) (equation 178)[201]. Via intramolecular cyclization of [4-(2-chlorophenyl)butyl]diphenylsilane (**396**) in the presence of Na, suspended in toluene, the synthesis of 2:3-benzo-1,1-diphenyl-1-silacyclohept-2-ene (also named 1,1-diphenyl-6,7,8,9-tetrahydro-5*H*-(benzo[*b*]-silepine) (**397**) can be achieved at 56% yield (equation 179)[202].

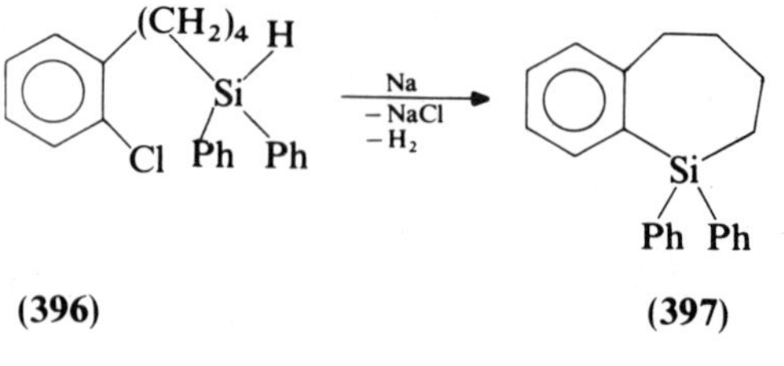

(179)

Referring to equation 169, 1,1,2,2-tetramethyl-1,2-disilacyclohex-4-ene (**375**) reacts with oxygen under ring extension to form 2,2,7,7-tetramethyl-1-oxa-2,7-disilacyclohept-4-ene (**398**) (equation 180)[193].

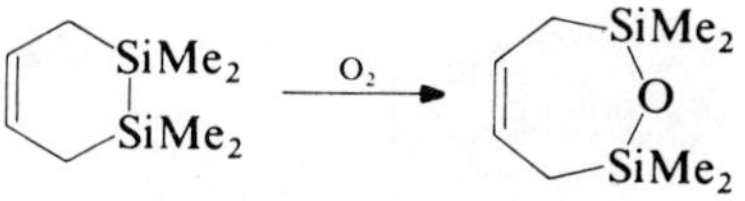

(**375**) (**398**)

(180)

G. Bicyclic Silanes

When 1,2-bis(bromomethyl)cyclobutane (**399**) plus dimethyldichlorosilane are concomitantly treated with Mg in THF in the presence of catalytic amounts of CuCl, 3,3-dimethyl-3-silabicyclo[3.2.0]heptane (**400**) is formed (equation 181)[203]. Furthermore,

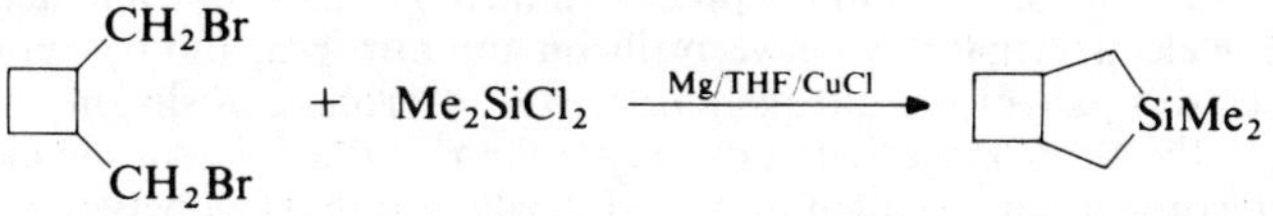

(**399**) (**400**)

(181)

heating (100 °C/8 h) of [2-(cyclohexen-3-yl)ethyl)]dichlorosilane (**401**) in the presence of H_2PtCl_6 leads to the *cis/trans* isomers (4:1) of 7,7-dichloro-7-silabicyclo[4.3.0]nonane (**402a/b**) (equation 182)[184]. Finally, it should be remembered that in equation 158 the

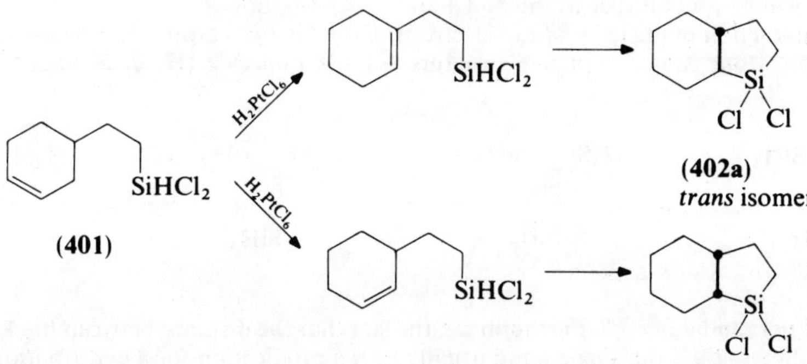

(**401**)

(**402a**)
trans isomer

(**402b**)
cis isomer

(182)

formation of 1-chloro-1-silabicyclo[2.2.2]octane (**351**)—as a major product—was outlined.

H. Silacyclophanes

In a one-step reaction 3,3,4,4,7,7,8,8-octamethyl-3,4,7,8-tetrasilacycloocta-1,5-diyne (**403**) and 2,3-dimethyl-1,3-butadiene afford 1,1,2,2,5,6,9,9,10,10,13,14-dodecamethyl-1,2,9,10-tetrasila[2.2]orthocyclophane (**404**) (equation 183)[204].

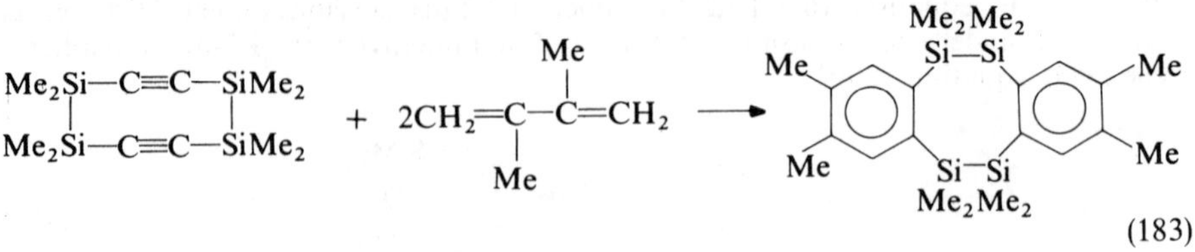

(403) (404) (183)

X. SILICON–NITROGEN COMPOUNDS

A. Introduction

The properties of the Si–N bond are predetermined by two important facts, namely, by the difference in electronegativity between silicon and nitrogen, and by optional ractions of the lone electron pair of the nitrogen atom with d-orbitals of silicon.

According to the electronegativity table of Pauling[205] the difference in electronegativity between silicon and nitrogen amounts to 1.2, which is midway between the difference of the Si–C bond (0.7) and that of the Si–O bond (1.7), and corresponds to that of the Si–Cl bond. In addition, the ionic character of compounds with Si–N bond(s), which amounts to 30%, is again midway between that of the Si–C bond (12%) and that of the Si–O bond (50%). The substantial difference in electronegativity in the Si–N bond has consequences for the chemical properties of such compounds concerning their thermal stability and also their marked sensitivity of solvolysis, which is not so pronounced towards water—in contrast to the Si–Cl and Si–O–Me bonds.

The interaction of the lone pair of electrons of the nitrogen atom with the d-orbitals of the silicon atom causes a planar structure for the molecule $(H_3Si)_3N$ instead of the

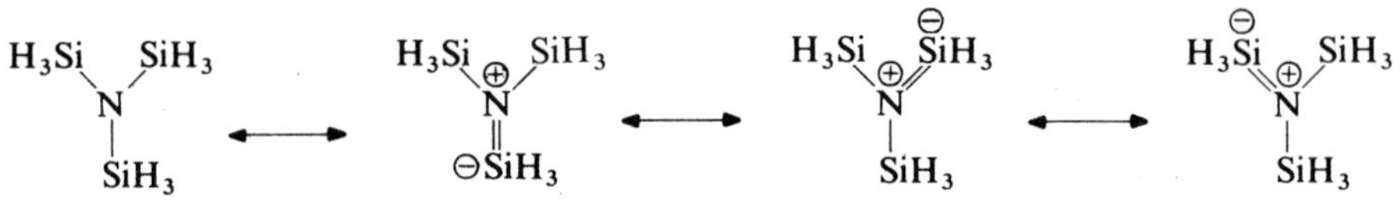

expected pyramidal one[206]. Furthermore, the fact that the distance between the Si and N atom is shorter than the single bond usually is, is an indication for a contribution of the lone pair of electrons to the π–bond. On the other hand, this decreases the polarity of the Si–N bond so that protonation as well as electrophilic attacks are impeded. Also, the basicity of silylated amines compared with the N–H and N–C bond of other, e.g. methylated, amines is thereby reduced.

The most important compounds with Si–N bonds are aminosilanes (**I**), organylsilazanes (**II**) and cyclosilazanes (**III**), in which silicon atoms are linked to NH or NR moieties. In the following the synthesis of these and further classes of compounds with Si–N bonds will be discussed.

$$R_nSi(NR')_{4-n} \qquad (R_3Si)_nNR'_{3-n} \qquad (R_2SiN)_n$$

(**I**) (**II**) (**III**)

B. Aminosilanes

1. Monoaminosilanes

The synthesis of aminosilanes from halogenosilanes and amines can usually be accomplished in an inert solvent like diethyl ether, benzene or carbon tetrachloride. Mostly, the reaction is carried out at room temperature and, if necessary, with cooling; an alternative approach is to conduct the reaction in liquid ammonia.

The reaction of Me_3SiCl with ammonia affords with 60% yield exclusively hexamethyldisilazane (**405**). As an intermediate, TMS-amine is formed and is rapidly condensed to give rise to **405**, a compound which does not undergo further reactions with an excess of the silane (equation 184)[207].

$$2\,Me_3SiCl \xrightarrow[-NH_4Cl]{NH_3} \underset{\textbf{(405)}}{Me_3SiNHSiMe_3} \qquad (184)$$

The stability of triorganylsilanes increases with increasing size of the organyl grouping, and from chlorotriethylsilane and ammonia the reaction product is triethylaminosilane (**406**) whereas the corresponding disilazane is only present in minute amounts (equation 185)[208]. With larger organyl groupings in the silane the corresponding aminosilane is formed exclusively, e.g. tripropylaminosilane from tripropylchlorosilane and ammonia[209].

$$Et_3SiCl \xrightarrow[-NH_4Cl]{NH_3} \underset{\textbf{(406)}}{Et_3SiNH_2} \qquad (185)$$

Trialkoxychlorosilanes (**407**) with ammonia form the corresponding trialkoxyaminosilanes (**408**) (equation 186)[210]. In quantitative yield one can obtain trisilylamine (**410**) from chlorosilane (**409**) plus ammonia (equation 187)[211].

$$\underset{\textbf{(407)}}{(RO)_3SiCl} \xrightarrow[-NH_4Cl]{NH_3} \underset{\textbf{(408)}}{(RO)_3SiNH_2} \qquad (186)$$

$$\underset{\textbf{(409)}}{3\,H_3SiCl} \xrightarrow[-NH_4Cl]{NH_3} \underset{\textbf{(410)}}{(H_3Si)_3N} \qquad (187)$$

An alternative route to aminosilanes is the reaction of halogenosilanes with metallated amides and amines. Thus, one can synthesize, e.g. *t*-butyl-*t*-butylaminofluorophenylsilane (**413**) in 91% yield from phenyl-*t*-butyldifluorosilane (**411**) and lithio-*t*-butylamide (**412**) (equation 188)[212]. Analogously *t*-butyl-(1-methylprop-1-yl)-phenylaminofluorosilane (**416**) can be obtained from lithiated aniline **415** with the respective fluorosilane **414** (equation 189)[213].

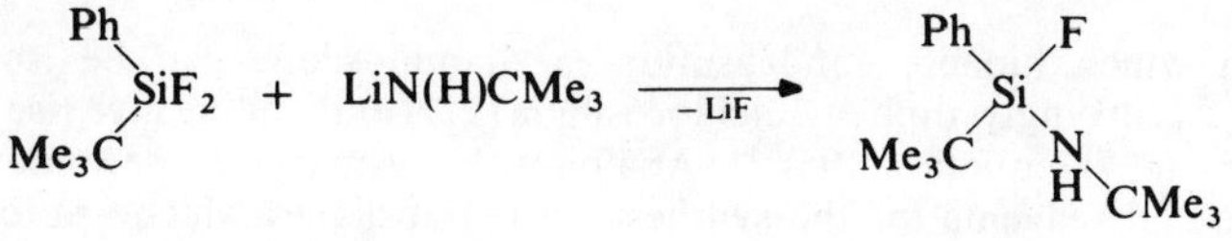

(188)

(411) **(412)** **(413)**

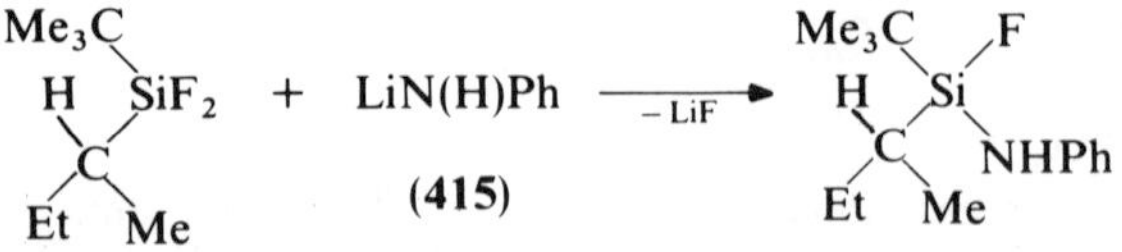

(189)

(414) **(415)** **(416)**

In some cases, the metallated primary or secondary amino component can be formed as an intermediate in the presence of catalytic amounts of metal hydrides (e.g. NaH) when using organyl-*H*-silanes (**418**) to prepare organylaminoorganylsilanes (**419**) (equation 190)[214]; the catalyst is recovered after the reaction.

$$\underset{\textbf{(417)}}{R'_2NH} \xrightarrow{\text{NaH}} [R'_2NNa] \xrightarrow[-\text{NaH}]{R_3SiH\ \textbf{(418)}} \underset{\textbf{(419)}}{R'_2NSiR_3} \qquad (190)$$

The reverse approach is also feasible, i.e., the reaction of amines with metallated organylsilanes. An example is the synthesis of butylaminotriphenylsilane (**422**) from lithiated triphenylsilane (**420**) with an excess of butylamine (**421**) in THF (equation 191)[215].

$$\underset{\textbf{(420)}}{Ph_3SiLi} + \underset{\textbf{(421)}}{BuNH_2} \rightarrow \underset{\textbf{(422)}}{Ph_3SiN(H)Bu} \qquad (191)$$

Finally, enamines and silanes in the presence of a catalyst such as $PdCl_2$ afford aminosilanes (equation 192)[216]. This reaction is an analogy to hydrosilylation (*vide supra*).

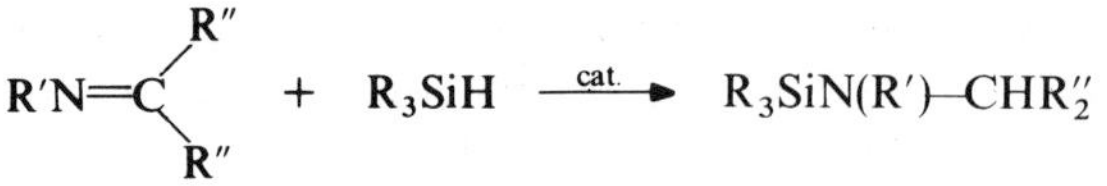

(192)

(423) **(424)**

2. Diaminosilanes

If mono- and diorganylsilanes react with ammonia, normally linear and cyclic polysilazanes result, since diaminosilanes are not stable and form condensation products (equation 193)[217]. However, if at least one bulky organyl substituent is attached to the

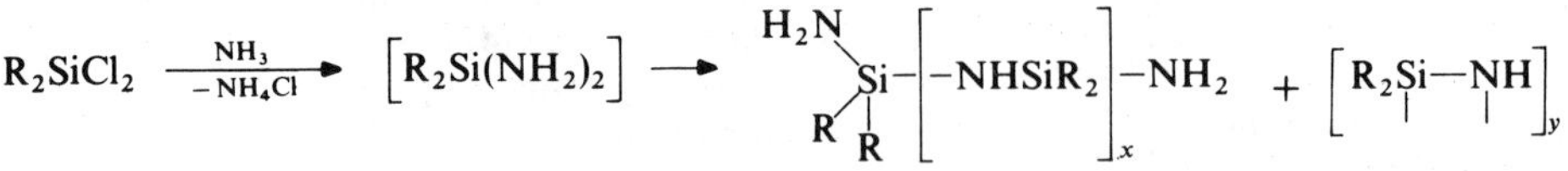

(193)

silicon atom which hinders condensation, a diaminosilane can be isolated (equation 194)[218]. Accordingly, diphenyldichlorosilane (**219**) and aniline give rise to dianilinodiphenylsilane (**427**) (equation 195)[218]. Analogously, primary and secondary amines or amides are useful reagents for the synthesis of aminosilanes which are formed via an equilibrium reaction, such as that for monoamines (equation 196). The course of reaction is strongly influenced by the number and type of organyl moieties at the nitrogen atom. In general, the tendency of amines, compared with the corresponding reactions with

$$\underset{\textbf{(425a, b)}}{RR'SiCl_2} \xrightarrow[-NH_4Cl]{NH_3} \underset{\textbf{(426a, b)}}{RR'Si(NH_2)_2} \tag{194}$$

425a/426a: R = Me, R′ = *t*-Bu
425b/426b: R = Me, R′ = *t*-Bu

$$\underset{\textbf{(219)}}{Ph_2SiCl_2} \xrightarrow[-PhNH_2\cdot HCl]{PhNH_2/PE} \underset{\textbf{(427)}}{Ph_2Si(NHPh)_2} \tag{195}$$

$$R_3SiX + 2\,HNR'_2 \rightleftharpoons R_3SiNR'_2 + R'_2NH\cdot HX \tag{196}$$

ammonia, is less pronounced to form disilazanes from halogenoorganylsilanes; only methylamine gives rise to a disilazane (*vide infra*).

C. Silazanes

1. Linear silazanes

The standard approach is the reaction of two equivalents of triorganylhalogenosilane in non-polar solvents (e.g. petroleum ether) with ammonia or a primary amine, such as methylamine (**428**)[219]. This yields the corresponding silazanes **405** and **429**, respectively (equation 197). **429** is also the reaction product if **428** is treated with the silylating agent

$$2\,Me_3SiCl \xrightarrow{NH_3} \underset{\textbf{(405)}}{Me_3SiNHSiMe_3}$$

$$2\,Me_3SiCl \xrightarrow{MeNH_2\ \textbf{(428)}} \underset{\textbf{(429)}}{Me_3SiN(Me)SiMe_3\ (85\%)} \tag{197}$$

N-methyl-*N*-TMS-acetamide (**430**) (equation 198)[220]. Furthermore, an alternative approach is the application of *N*-metallated aminosilanes which can afford halogenosilazanes, e.g. the silazane-type [(dialkylfluorosilyl)-alkylamino]alkyldifluorosilane (**431**) is formed in 95% yield (equation 199)[221]. Linear polysilazanes are also formed as condensation products of diaminosilanes[217] (cf. equation 193).

$$\underset{\textbf{(428)}}{MeNH_2} + 2\,\underset{\textbf{(430)}}{MeCON(Me)SiMe_3} \xrightarrow[-AcNHMe]{} \textbf{429} \tag{198}$$

$$R^1R^2FSiN(Li)R^3 + F_3SiR^4 \rightarrow \underset{\textbf{(431)}}{R^1R^2FSi{-}N(R^3)SiR^4F_2} \tag{199}$$

2. Cyclic silazanes

Cyclic silazanes are the reaction products when diorganyldihalogenosilanes are employed with ammonia or primary amines, especially when the reaction takes place at a high dilution ratio. For example, both hexamethylcyclotrisilazane (**432**) and octamethylcyclotetrasilazane (**433**) are formed when dimethyldichlorosilane reacts with ammonia (equation 200)[212]. Cyclic polysilazanes are also formed via the condensation of diaminosilanes[217] (cf. equation 193).

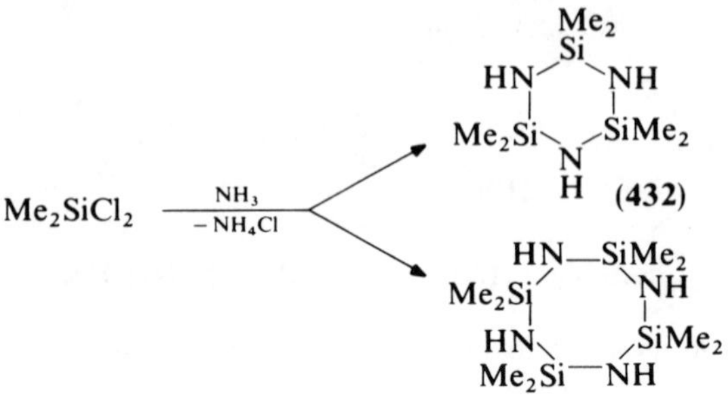

(**433**) (200)

D. Amidosilanes

The NH_2 grouping of amides can react with, e.g., Me_3SiCl to afford an amidosilane. Thus, acetamide (**434**) forms acetamidosilane (**435**) with one equivalent of **16** in the presence of an HCl acceptor (e.g. triethylamine), in 90% yield (equation 201)[222]. Alternatively, from trifluoroacetic anhydride (**436**) and heptamethyldisilazane (**429**) one can obtain (*N*-methyltrifluoroacetamido)trimethylsilane (**437**) in 95% yield (equation 202)[219]. However, if acetamide (**434**) is employed with an excess of Me_3SiCl (e.g. 2.5–3.0 equivalents), bis(TMS)-acetamide (**438**) can be isolated (equation 203)[223]. Analogously, the corresponding bis(TMS)-acrylamide can be synthesized from acrylamide.

$$\underset{(\mathbf{434})}{MeCONH_2} + \underset{\mathbf{16}}{Me_3SiCl} \xrightarrow{Et_3N/benzene} \underset{(\mathbf{435})}{MeCONHSiMe_3} \tag{201}$$

$$\mathbf{429} + \underset{(\mathbf{436})}{(CF_3CO)_2O} \rightarrow \underset{(\mathbf{437})}{CF_3CON(Me)SiMe_3} \tag{202}$$

$$\mathbf{434} + 2\,Me_3SiCl \rightarrow \underset{(\mathbf{438})}{MeCON(SiMe_3)_2} \rightleftharpoons MeC(OSiMe_3){=}N{-}SiMe_3 \tag{203}$$

Organylsilylurea derivatives are obtainable via addition of isocyanates to organylaminosilanes. Thus, *N*,*N*-bis(TMS)-aniline (**439**) with an excess of phenylisocyanate (**440**) yields 1,3-bis(TMS)-1,3-diphenyl urea (**441**) (equation 204)[224]. *N*-organylsilylthiourea

$$\underset{(\mathbf{439})}{(Me_3Si)_2NPh} + \underset{(\mathbf{440})}{PhN{=}C{=}O} \longrightarrow \underset{(\mathbf{441})}{Me_3SiN(Ph){-}CO{-}N(Ph){-}SiMe_3} \tag{204}$$

derivatives are afforded by the reaction of aminotriorganylsilanes with methylisothiocyanates. For instance, *N*,*N*-diorganyl-*N*′-organyl-*N*′-TMS-thiourea is formed by means of organylaminotriorganylsilane and organylisothiocyanate (equation 205)[225].

$$R^1_3SiNR^2_2 + R^3N{=}C{=}S \rightarrow R^1_3SiN(R^3)C({=}S)NR^2_2 \qquad (205)$$

E. *N*-Organylsilyl-amidines and -enamines

The hydrosilylation of carbodiimides with organyl-*H*-silanes in the presence of a catalyst such as $PdCl_2$ or tris(triphenylphosphine)rhodium(I) chloride can give rise to *N*-organylsilylformamidines, e.g., from diisopropylcarbodiimide (**443**) and triethylsilane (**442**) in the presence of $PdCl_2$ *N*,*N*′-diisopropyl-*N*-triethylsilylformamidine (**444**) is formed (equation 206)[226]. Also lithium diisopropylamide reacts with *N*-phenylimines

$$\underset{\textbf{(442, 9)}}{R_3SiH} + \underset{\textbf{(443)}}{R'N{=}C{=}NR'} \xrightarrow{PdCl_2/140\text{–}200\,^\circ C} \underset{\textbf{(444)}}{R_3SiN(R')CH{=}NR'} \qquad (206)$$

(442) R = Et R′ = *i*-Pr

(**445**) to yield lithiated *N*-phenylenamines (**446**) which, for their part, with Me_3SiCl deliver *N*-phenyl-*N*-TMS-enamines (**447**) (equation 207)[227].

$$\underset{\textbf{(445)}}{PhN{=}C(R)CH_2R} \xrightarrow{(i\text{-}Pr)_2NLi} \underset{\textbf{(446)}}{PhN(Li){-}C(R){=}CHR} \xrightarrow{Me_3SiCl} \underset{\textbf{(447)}}{Me_3SiN(Ph){-}C(R){=}CHR} \qquad (207)$$

F. *N*-Organylsilyl-imines and -hydroxylamines

For the synthesis of tris (TMS)-ketimine (**448**), acetonitrile (**386a**) is lithiated with *t*-butyllithium and then converted into **448**; as a by-product [bis(TMS)amino]-TMS-acetylene (**449**) occurs (equation 208)[228]. In addition, tris(TMS)-ketimine (**448**) can be

$$\underset{\textbf{(386a)}}{MeCN} \xrightarrow{Me_3CLi} \left[Li_2C{=}C{=}NH \rightleftharpoons LiC{\equiv}C{-}NHLi\right]$$

$$\xrightarrow{Me_3SiCl} \underset{\textbf{(448)}}{(Me_3Si)_2C{=}C{=}NSiMe_3} \qquad \underset{\textbf{(449)}}{Me_3SiC{\equiv}C{-}N(SiMe_3)_2} \qquad (208)$$

obtained by treatment of TMS-acetonitrile (**27**) with butyllithium and Me_3SiCl (equation 209)[228a].

$$Me_3SiCH_2CN \xrightarrow{BuLi} Me_3SiCH(Li){-}CN \xrightarrow{Me_3SiCl} \qquad (209)$$

(**27**)

$$(Me_3Si)_2CHCN \rightleftharpoons (Me_3Si)_2C{=}C{=}NH \xrightarrow{BuLi/Me_3SiCl} \textbf{448}$$

If benzonitrile (**386b**) is lithiated with phenyllithium, a *N*-lithiated imine is formed which, with halogenoorganylsilanes, yields *N*-silylated ketimines, e.g., diphenylmethylenaminotrimethylsilane (**450**) (equation 210)[229].

$$PhCN \xrightarrow{PhLi} [Ph_2C{=}NLi] \xrightarrow[-\,LiCl]{Me_3SiCl} Me_3SiN{=}CPh_2 \qquad (210)$$

(**386b**) (**450**)

The phosphorus analogues, *N*-TMS-triorganylphosphimines (**453**), can be synthesized via the reaction of triorganylphosphine (**451**) with TMS-azide (**452**) by heating several

$$R_3P + Me_3SiN_3 \xrightarrow[-N_2]{\Delta} R_3P{=}NSiMe_3 \qquad (211)$$

(**451**) (**452**) (**453**)

hours (equation 211)[230]. While it is expected that the reaction of alkylaminotriorganylsilanes (**454**) with cyanogen chloride will afford a mixture of triorganylsilylorganylcarbodiimides (**455**) and triorganylsilylorganylcyanamides (**456**), in the case of alkylaminotrimethylsilanes the formation of the carbodiimides (**455**)—probably for steric reasons—is preferred (equation 212)[231].

$$3RNHSiMe_3 \xrightarrow{ClCN} 2RN{=}C{=}NSiMe_3 \leftrightarrows 2RN(CN)SiMe_3 \qquad (212)$$

(**454**) (**455**) (**456**)

Indirect routes[232] can be employed to synthesize *N*,*N*-bis(TMS)-carbodiimide (**458**) (a) by applying silver imidazole to *N*,*N*-bis(TMS)-thiourea (**457**) (equation 213), (b) by reaction of Me_3SiCl with silver cyanamide (**459**) (equation 214), (c) by reaction of Me_3SiCl with hydrogen cyanamide (**460**) in the presence of a tertiary amine (equation 215) and (d) from cyanogen bromide (**461**) and hexamethyldisilazane (**405**) (equation 216).

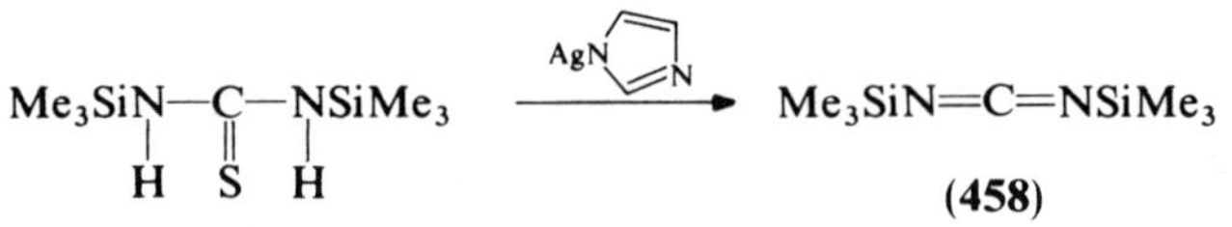

(**457**) (**458**) (213)

$$Me_3SiCl + Ag_2NCN \rightarrow \textbf{458} \qquad (214)$$

(**459**)

$$Me_3SiCl + H_2NCN \xrightarrow[-Et_3N \cdot HCl]{Et_3N} \mathbf{458} \quad (215)$$

(460)

$$BrCN + Me_3SiNHSiMe_3 \xrightarrow[-Me_3SiNH_2]{-Me_3SiBr} \mathbf{458} \quad (216)$$

(461) **(405)**

N,O-bis(triorganylsilyl)hydroxylamines (**462**) can be obtained from the application of hydroxylamine hydrochloride to hexaorganyldisilazane (equation 217)[233, 234]. Furthermore, tris(TMS)-hydroxylamine (**464**) is formed by the reaction of diethylaminotrimethylsilane (**463**) with hydroxylammonium sulfate (equation 218)[235].

$$R_3SiNHSiR_3 + [NH_3OH]^+Cl^- \xrightarrow{-NH_4Cl} R_3SiNHOSiR_3 \quad (217)$$

(462)

$$5Me_3SiNEt_2 + [NH_3OH]^+_2 SO_4^{--} \longrightarrow (Me_3Si)_2NOSiMe_3 \quad (218)$$

(463) **(464)**

G. *N*-Organylsilyl-hydrazines and -azenes

The reaction of halogenotriorganylsilanes with hydrazine affords *N,N'*-disubstituted hydrazine compounds, exclusively, whereas monosubstituted hydrazines normally react with the silane only at the non-substituted *N* atom (equation 219)[236]. If hydrazine (as

$$R_3SiX \xrightarrow{H_2NNH_2} R_3SiNHNHSiR_3 \;\mathbf{(465)}$$

$$R_3SiX \xrightarrow{H_2NNHPh} R_3SiNHNHPh \;\mathbf{(466)} \quad (219)$$

mono-hydrate) is treated with diethylaminotrimethylsilane (**463**) at elevated temperatures under inert atmosphere, tris(TMS)-hydrazine (**467**) can be isolated (equation 220)[235].

$$H_2NNH_2 + 3Me_3SiNEt_2 \xrightarrow[-HNEt_2]{} (Me_3Si)_2NNHSiMe_3 \quad (220)$$

(463) **(467)**

The homologous tetrakis(TMS)-hydrazine (**469**) can be synthesized from lithiated tris(TMS)-hydrazine (**468**) (equation 221)[237]. A viable route to 1,1,2-trimethyl-2-TMS-hydrazine (**471**) is the lithiation of 1,1,2-trimethylhydrazine (**470**) and subsequent conversion into **471** (equation 222)[238]. When lithiated tris(TMS)-hydrazine (**468**) is employed

$$(Me_3Si)_2NN(Li)SiMe_3 \xrightarrow[-LiCl]{Me_3SiCl} (Me_3Si)_2NN(SiMe_3)_2 \quad (221)$$

(468) **(469)**

$$\underset{(\mathbf{470})}{Me_2NNHMe} \xrightarrow{BuLi} Me_2NN(Li)Me \xrightarrow{Me_3SiCl} \underset{(\mathbf{471})}{Me_2NN(Me)SiMe_3} \qquad (222)$$

with tosylazide, bis(TMS)-diimine (**472**), also called bis(TMS)-diazene, is isolated (equation 223)[237]. The thermolysis of bis(TMS)-diimine (**472**) gives rise to 1,1,4,4-tetrakis(TMS)-tetrazene (**473**) (equation 224)[239].

$$\mathbf{468} \xrightarrow[-N_2]{TosN_3} \underset{(\mathbf{472})}{Me_3SiN{=}NSiMe_3} + TosN(Li)SiMe_3 \qquad (223)$$

$$2\ \mathbf{472} \xrightarrow{\Delta} 2\left[(Me_3Si)_2\overset{\oplus}{N}{=}\overset{\ominus}{N}\right] \longrightarrow \underset{(\mathbf{473})}{(Me_3Si)_2NN{=}NN(SiMe_3)_2} \qquad (224)$$

H. *N*-Silylated Amino Acids

The treatment of an amino acid *per se* by means of halogenoorganylsilanes cannot be accomplished, since the formation of dipolar ions $^{\oplus}NH_3CHRCOO^{\ominus}$ impedes any reaction with halogenated silanes. Hence, a viable approach is to esterify the amino acid first to **474** and then convert it in the presence of a HCl acceptor into the corresponding *N*-TMS-amino acid (**475**) (equation 225)[240, 241]. Another method is to employ the sodium salts of amino acids (**476**) which yields *N*-TMS-amino acid TMS-ester (**477**) (equation 226)[242].

$$\underset{(\mathbf{474})}{H_2NCH(R)CO_2Et} \xrightarrow{Me_3SiCl/Et_3N} \underset{(\mathbf{475})}{Me_3SiNHCHRCO_2Et} \qquad (225)$$

$$\underset{(\mathbf{476})}{H_2NCH(R)CO_2Na} \xrightarrow{Me_3SiCl/Et_3N} \underset{(\mathbf{477})}{Me_3SiNHCH(R)CO_2SiMe_3} \qquad (226)$$

A preceding esterification or salt-forming reaction can be avoided when silylating agents such as hexamethyldisilazane (**405**) or triethylaminosilane (**406**) are used. In this route the amino as well as the carboxyl groups are concomitantly silylated so that **477**-type compounds can be isolated (equation 227)[243].

$$H_2NCH(R)CO_2H + \mathbf{405} \xrightarrow[-NH_3]{} \mathbf{477} \qquad (227)$$

In presence of $PdCl_2$ the synthesis of *N*-TES-amino acid TES-esters (**478**) is feasible when amino acids are treated with either triethylsilane (**442**) or **406** (equation 228)[244]. For

$$H_2NCH(R)CO_2H + \mathbf{442} \xrightarrow{PdCl_2} \underset{(\mathbf{478})}{Et_3SiNHCH(R)CO_2SiEt_3} \qquad (228)$$

analytical purposes (e.g. via GC) it is sufficient to heat small amounts (20–50 mg) of the amino acid(s) containing a sample with bis(TMS)-acetamide (**438**) to carry out silylation[239].

I. Common Silylation Procedures

The most widely used methods of silylation of compounds with an amino group are the use of hexamethyldisilazane (**405**) or bis(TMS)-acetamide (**438**). Thus, diamines like trimethylene- (**479a**) or tetramethylenediamine (**479b**), respectively, or arylamines such as *o*-diaminobenzene (**481**), can be converted into the corresponding bis(TMS)-derivatives in yields of up to 95% by means of **405** (equations 229 and 230)[245].

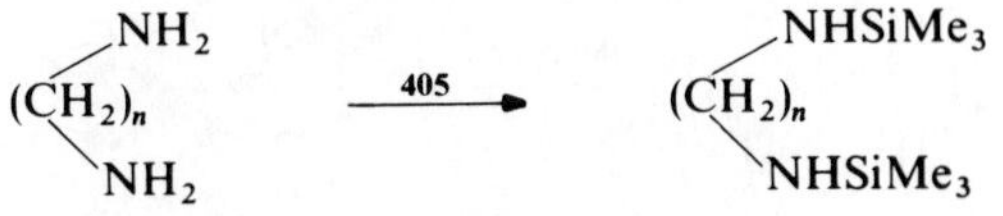

(229)

(479a, b) (**a**) $n = 3$
(**b**) $n = 4$

(480a, b)

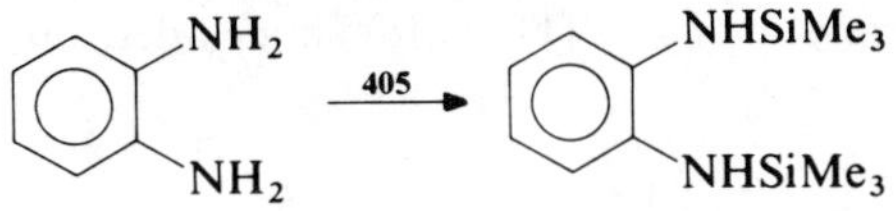

(230)

(481) **(482)**

Analogously, 4,5-diaminouracil (**483**) and **405** give rise to 2,4,5,6- tetrakis(TMS)-4,5-diaminouracil (**484**) in 90% yield including both *N*- and *O*-silylation (equation 231)[245].

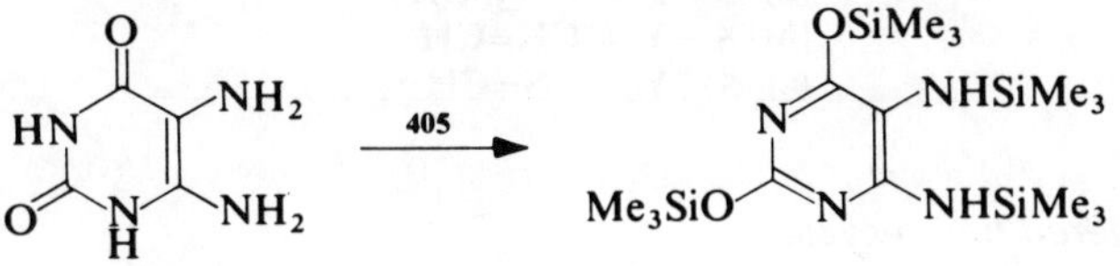

(231)

(483) **(484)**

J. Silylation of *N*-heterocycles

The reactions of silyl compounds with *N*-heterocycles are either substitutions or reductive silylation-type reactions at the N-atom(s) of the ring system. Suitable silylating agents are **16**, **405** or **406**. Due to the pronounced reactivity of the heterocyclic silyl derivatives and the tendency towards rapid cleavage of the silyl moiety, these substances are predominantly utilized as facile intermediates.

1. Five-membered heterocycles

By varying the afore-mentioned silylating agents according to the specific case, *N*-heterocycles can be converted into the corresponding silyl derivatives in high yields e.g. pyrazole (equation 232)[246]. Analogously, the NH-functions of 2,2′-bis(2-imidazoline) (**487**) can easily be silylated (equations 233)[247]. In the same manner, 2-aminothiazoline

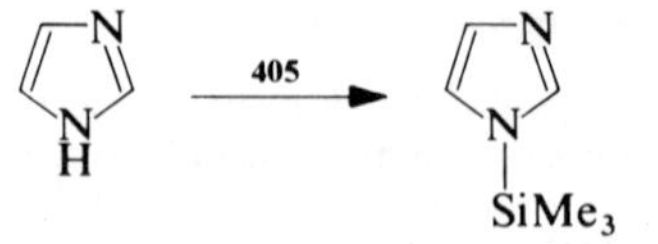

(485) (486) (232)

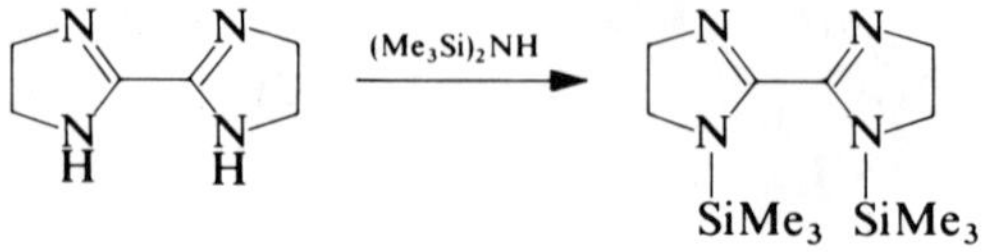

(487) (488) (233)

(**489a**), 2-aminothiazole (**489b**) and 2-aminothiadiazole (**489c**), respectively, can be converted into the corresponding 2-[bis(TMS)-amino]-derivatives **490a**–**490c** (equation 234)[248].

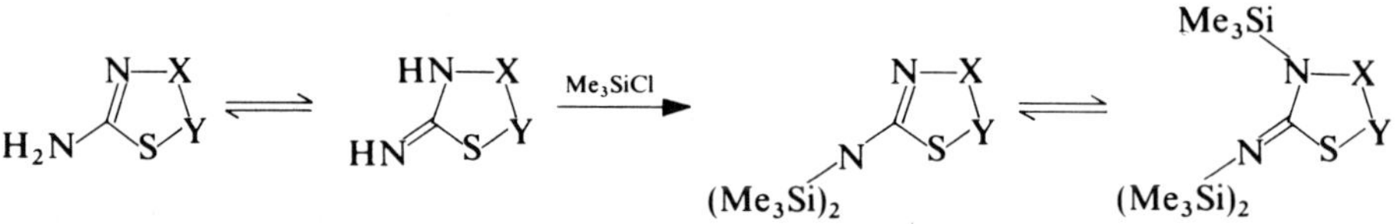

(**489a–c**) (**490a–c**) (234)

(**a**) $X-Y = CH_2–CH_2$
(**b**) $X-Y = CH{=}CH$
(**c**) $X-Y = -N{=}CH$

2. Six-membered heterocycles

Lactames and dioxopiperazines, when treated with Me_3SiCl in the presence of a tertiary amine as HCl acceptor, afford the corresponding TMS derivatives. Thus, the treatment of 2,5-dioxopiperazine (**491**) leads to 1,4-bis(TMS)-2,5-dioxopiperazine (**492**) (equation 235)[249]. The same applies to the silylation of triazines, e.g. **493**, which can be

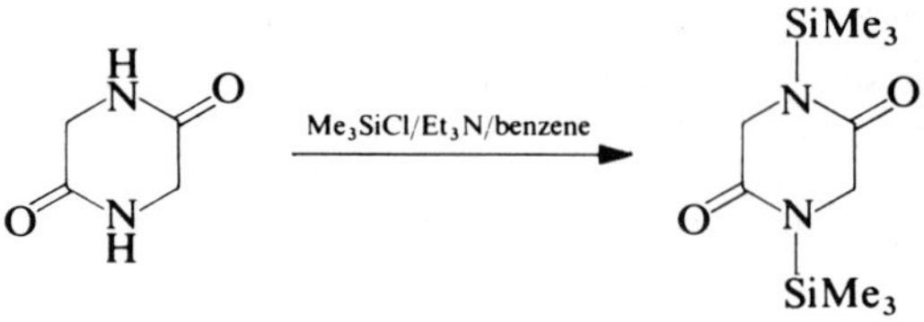

(**491**) (**492**) (235)

converted into the corresponding *N*-TMS-triazines (**494**) (equation 236)[250]. Heteroaryl compounds such as pyrazine (**495**)[251], phenazine (**497**)[252] or acridine (**499**) can be converted via reductive silylation into the bis(TMS)-derivatives (equation 237). In addition, bis(TMS)-mercury (**502**) has been shown to be an excellent reagent for the

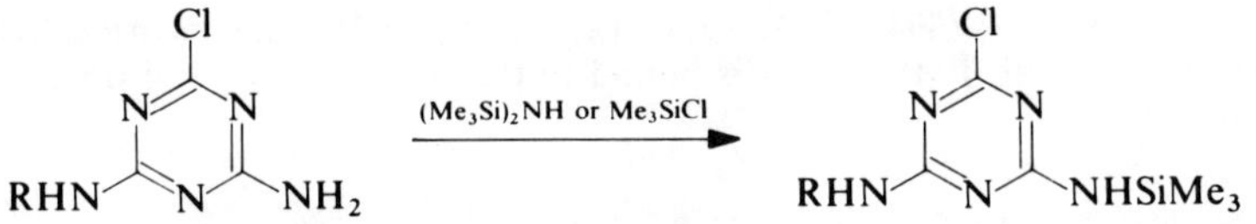

(236)

(493) **(494)**

R = Et, Pr, *i*-Pr, *t*-Bu, EtCHMe

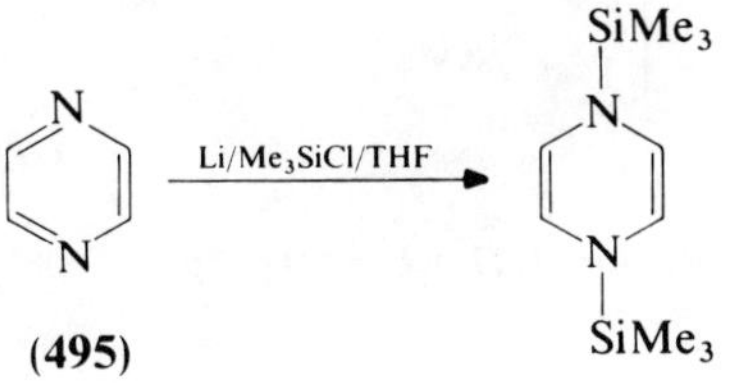

(495) **(496)** (237)

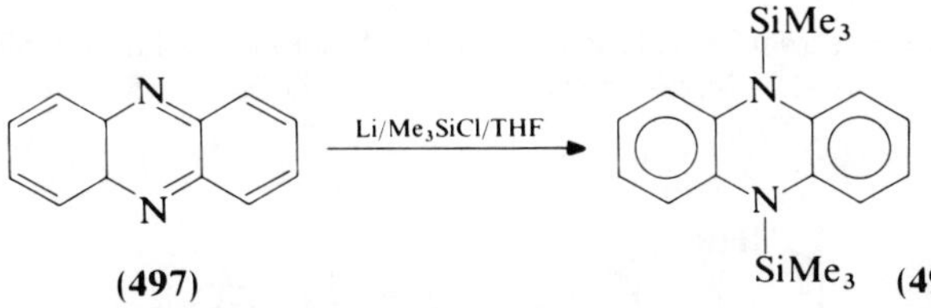

(497) **(498)**

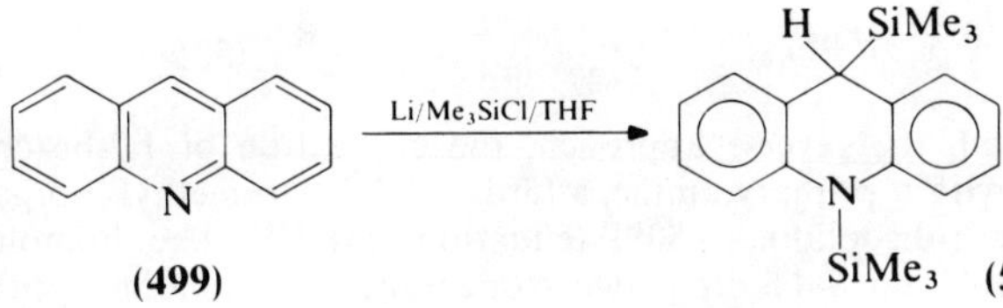

(499) **(500)**

reductive dimerization of 4-phenylpyridine (**501**) during which 1,1′,2,2′-tetrahydro-4,4′-diphenyl-1,1′-bis(TMS)-2,2′-bipyridine (**503**) can be isolated in quantitative yield (equation 238)[253]. Alternatively, one can obtain N-silylated heterocycles by direct syntheses, e.g. via 1,3-dipolarophilic cycloaddition of TMS-azide (**452**) to acetylenes which leads to

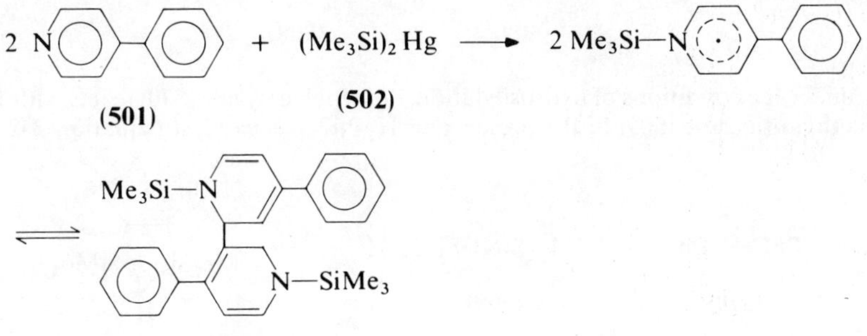

(501) **(502)** **(503)** (238)

silylated 1,2,3-triazoles (**504**)[254]. It could be shown by IR- and NMR-spectroscopic investigations that the TMS moiety is bound to the central N atom of the triazole ring (equation 239).

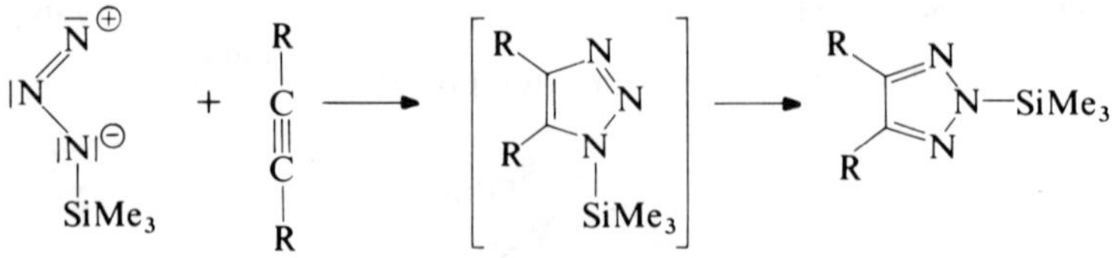

(239)

(**452**) (**504**)

R = Me, Pr, Bu, $(CH_2)_2C{\equiv}CH$, Ph

K. Nitrogen-containing Cyclosilanes

The application of tetrachlorosilane to *N,N'*-dialkylethylenediamine (**505**) gives rise to 1,3-dialkyl-2,2-dichloro-1,3-diaza-2-silaycyclopentane (-silolidine) (**506**) (equation 240)[255].

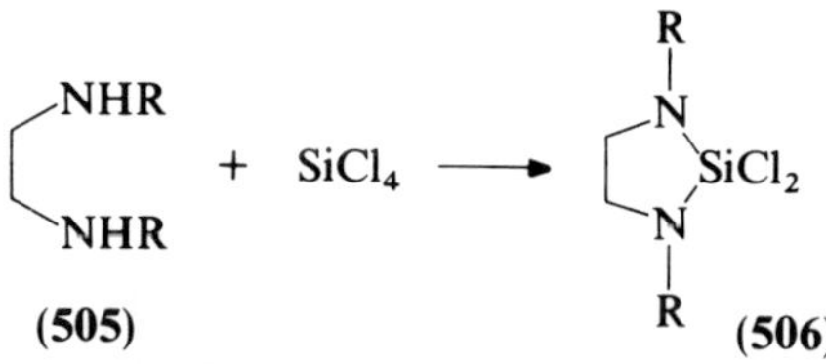

(240)

(**505**) (**506**)

On the other hand, the reverse approach, the cyclization of 1,2-bis(chlorodimethylsilyl)ethane (**507**) with a primary amine, affords 1,1,3,3-tetramethyl-2-organyl-1,3-disila-2-azacyclopentane (-disilolidine) (**508**) (equation 241)[256]. The homologous 1,1,3,3-tetramethyl-1,3-disila-2-aza-4,5-diphenylcyclopentane (-disilolidine) (**510**) is formed

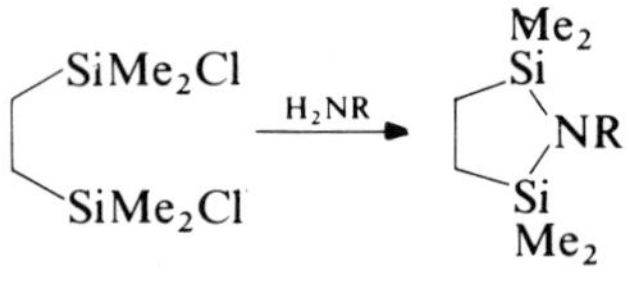

(241)

(**507**) (**508**)

when, under the conditions of hydrosilylation, diphenylacetylene (**370**) reacts with 1,1,3,3-tetramethyldisilazane (**509**) in the presence of H_2PtCl_6 as catalyst (equation 242)[257].

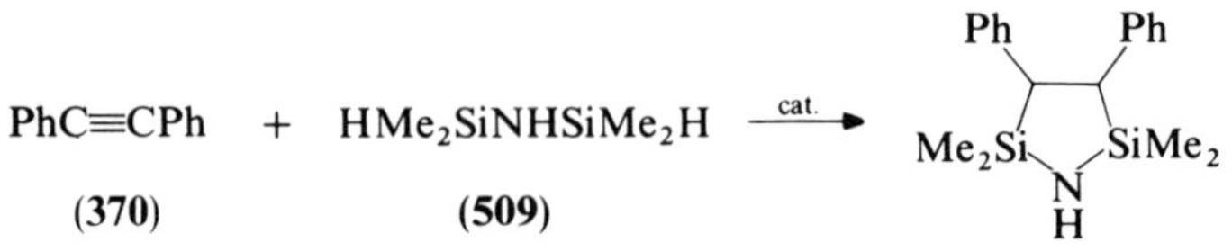

(242)

(**370**) (**509**) (**510**)

L. Azidosilanes

Organylazidosilanes (general formula: $R_nSi(N_3)_{4-n}$; R = Me, Ph; $n = 3,2,1$) are thermally stable derivatives of hydrogen azide. This is most likely due to an enhanced mesomeric stabilization by formation of a d_π–p_π bond between N and Si atoms.

Especially, trimethylazidosilane (TMSA) (**452**) is an excellent reagent for preparative purposes, since it can be applied to many reactions which otherwise require the application of the explosive hydrogen azide. A number of synthetic methods for the preparation of azidosilanes are known. Starting materials are mostly metal azides, which are applied to halogenosilanes in THF; via this approach TMSA (**452**) is obtained in 81% yield (equation 243)[230, 258] from Me_3SiCl.

$$\underset{}{Me_3SiCl} + \underset{\textbf{(511)}}{NaN_3} \xrightarrow{THF} \underset{\textbf{(452)}}{Me_3SiN_3} \tag{243}$$

In a eutectic mixture of fused LiCl/KCl or in $ZnCl_2$/KCl, sodium azide and Me_3SiCl yield TMSA (**452**) in 73% yield. By means of the same methodology dimethyldiazidosilane (**512**) can be synthesized (equation 244)[259]. Analogously, from methyltrichlorosilane a

$$\underset{\textbf{(117)}}{Me_2SiCl_2} + 2\,NaN_3 \xrightarrow[-NaCl]{250\,°C} \underset{\textbf{(512)}}{Me_2Si(N_3)_2} \tag{244}$$

mixture of methyltriazidosilane (**513**), methyldiazidochlorosilane (**514**) and methylazidodichlorosilane (**515**) is obtained (equation 245)[259]. Furthermore, a mixture of sodium azide/aluminum chloride in THF with Me_3SiCl gives rise to **452** in 90% yield[260].

$$MeSiCl_3 + \textbf{511} \longrightarrow \underset{\textbf{(513)}}{MeSi(N_3)_3} + \underset{\textbf{(514)}}{MeSi(N_3)_2Cl} + \underset{\textbf{(515)}}{MeSiN_3Cl_2} \tag{245}$$

Similarly, triphenylchlorosilane and diphenyldichlorosilane can react with lithium azide to afford the corresponding azidosilanes (**516**) and (**517**), respectively (equations 246a and 246b)[261]

$$Ph_3SiCl \xrightarrow[-LiCl]{LiN_3/THF} \underset{\textbf{(516)}}{Ph_3SiN_3} \tag{246a}$$

$$Ph_2SiCl_2 \xrightarrow{LiN_3/Py} \underset{\textbf{(517)}}{Ph_2Si(N_3)_2} \tag{246b}$$

A halogen-azide replacement easily occurs when the formed, lower boiling chlorosilane (e.g. Me_3SiCl) can be removed immediately from the reaction mixture by distillation.

Thus, from the reaction of Me_3SiN_3 with chlorodiphenylmethylsilane (**518**) in the presence of small amounts of $AlCl_3$, Me_3SiCl continuously distills from the reaction mixture and **519** can be isolated (equation 247)[262]. Via this reaction pathway one can also

$$\underset{(\mathbf{518})}{MePh_2SiCl} + Me_3SiN_3 \xrightarrow[-Me_3SiCl]{AlCl_3} \underset{(\mathbf{519})}{MePh_2SiN_3} \qquad (247)$$

$$PhSiCl_3 + Me_3SiN_3 \xrightarrow[-Me_3SiCl]{AlCl_3} \underset{(\mathbf{520})}{PhSi(N_3)_3} \qquad (248)$$

synthesize triazidosilanes, e.g. triazidophenylsilane (**520**) (equation 248)[263]. It is hypothesized that $AlCl_3$ may to some extent function as azide transferring agent by intermediately forming aluminum azide. Me_3SiN_3 is also obtained by the reaction of hexamethyldisilazane (**405**) with hydrogen azide (equation 249)[230]. Since these azidosilanes are very sensitive to moisture, synthesis, general handling and further reactions have to be carried out in total exclusion of moisture. Otherwise, due to hydrolysis, hydrogen azide can be formed which can cause explosions.

$$\mathbf{405} \xrightarrow{HN_3} Me_3SiN_3 \qquad (249)$$

M. Organyl-pseudohalogenosilanes

Another class of compounds featuring Si–N bonds are organyl-pseudohalogenosilanes, i.e. (1) cyano-/isocyanosilanes: $R_nSi(CN)_{4-n}/R_nSi(NC)_{4-n}$; (2) isocyanatosilanes: $R_nSi(NCO)_{4-n}$; and (3) isothiocyanatosilanes: $R_nSi(NCS)_{4-n}$ ($n = 1,2,3$). The synthesis by means of fused salts[259] is also applicable to obtain this class of compounds.

Cyano- and isocyanosilanes form an equilibrium which is temperature-dependent: at room temperature the cyano structure is preferred, at higher temperatures the isocyano structure[264]. Cyanotrimethylsilane (**521**) can be obtained by the reaction of hydrogen cyanide with Me_3SiCl/triethylamine in diethyl ether[265] (equation 250), or from sodium

$$Me_3SiCl \xrightarrow{HCN/Et_3N} \underset{(\mathbf{521})}{Me_3SiCN} \qquad (250)$$

cyanide with Me_3SiCl/*N*-methylpyrrolidone[266, 267]. In addition, the application of potassium cyanide to bis(TMS)-sulfate is an alternative approach (equation 251)[268].

$$(Me_3Si)_2SO_4 \xrightarrow{KCN} \mathbf{521} \qquad (251)$$

Quite common too is the treatment of alkyl- or arylhalogenosilanes with silver cyanide to synthesize the corresponding cyanosilanes (equation 252)[269]. Finally, **521** can also be

$$Me_2SiCl_2 \xrightarrow[-AgCl]{AgCN} \underset{(\mathbf{522})}{Me_2Si(CN)_2}$$

$$Ph_3SiCl \xrightarrow{AgCN} \underset{(\mathbf{523})}{Ph_3SiCN} \qquad (252)$$

obtained by the reaction of hexamethyldisilazane (**405**) with hydrogen cyanide (equation 253)[270].

$$\mathbf{405} \xrightarrow{\text{HCN}} \mathbf{521} \qquad (253)$$

Isocyanatotrimethylsilane (**524**) is formed via a direct synthesis from Me_3SiCl and hydrogen isocyanate in diethyl ether in the presence of a HCl acceptor (e.g. trimethylamine) (equation 254)[271]. In the fused state chlorotriphenylsilane reacts with urea or

$$Me_3SiCl \xrightarrow{HNCO/Me_3N} \underset{(\mathbf{524})}{Me_3SiN{=}C{=}O} \qquad (254)$$

thiourea and isocyanato- (**525**) and isothiocyanatotriphenylsilane (**526**) respectively, are formed (equation 255)[272, 273]. Analogously to equation 251, bis(TMS)-sulfate (**184**) and

$$Ph_3SiCl \xrightarrow{H_2NCONH_2} \underset{(\mathbf{525})}{Ph_3Si{-}N{=}C{=}O}$$

$$Ph_3SiCl \xrightarrow{H_2NCSNH_2} \underset{(\mathbf{526})}{Ph_3Si{-}N{=}C{=}S} \qquad (255)$$

$$\mathbf{184} + H_2NCONH_2 \longrightarrow \mathbf{524} \qquad (256)$$

urea form **524** (equation 256)[274]. If silver isocyanate is applied to TMS 4-bromobutanoate, **524** is the reaction product (equation 257)[275] in 74% yield. Alternatively, hexamethyldisilazane (**405**) gives rise to **524** when treated with phosgene (equation 258)[276].

$$Br(CH_2)_3CO_2SiMe_3 + AgNCO \xrightarrow{-AgBr} \mathbf{524} \qquad (257)$$

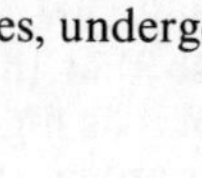

$$\mathbf{405} + COCl_2 \xrightarrow[-NH_4Cl]{-Me_3SiCl} \mathbf{524} \qquad (258)$$

If silver cyanate is applied to dimethyl[tris(TMS)methyl]iodosilane (**527**) at first the corresponding cyanate (**528**) results which, at higher temperatures, undergoes rearrangement to yield the isocyanate **529** (equation 259)[277].

$$\underset{(\mathbf{527})}{(Me_3Si)_3CSiMe_2I} \xrightarrow{AgNCO/CH_2Cl_2} \underset{(\mathbf{528})}{(Me_3Si)_3CSiMe_2OCN} \xrightarrow{150\,^{\circ}C} \underset{(\mathbf{529})}{(Me_3Si)_3CSiMe_2NCO} \qquad (259)$$

XI. SILICON–OXYGEN COMPOUNDS

A. Introduction

The properties of the Si–O bond significantly differ from those of the Si–C and Si–N bonds. Due to the substantial difference in electronegativity between silicon (1.8) and oxygen (3.5) of 1.7 units, the Si–O bond has a more pronounced polarity than the analogous C–O bond (1.0). The ionic character of the Si–O bond (50%) is very high compared with the Si–C (12%) or Si–N bond[205]. The consequence of this is that Si–O compounds are relatively stable against homolytic cleavage such as thermolysis, but very sensitive towards heterolytic attack like, e.g., solvolysis.

Classification and structure of Si–O compounds

According to the substituents at the oxygen atom of the Si–O bond, various types of compounds have to be distinguished:

1. Silanols and their condensation products, the siloxanes

Si–OH and –Si–O–Si–

2. Organyloxysilanes

–Si–OR, R = alkyl, aryl

3. Acyloxysilanes (also named silyl esters)

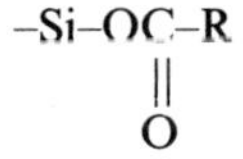

4. Silyl enol ethers

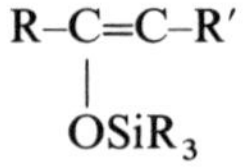

5. Cyclic silicon–oxygen compounds (e.g. silaoxa-heterocycles)

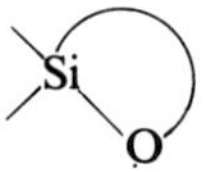

B. Silanols

Organylsilanols are highly polar, form hydrogen bonds and undergo association[278–282], so that their boiling points are relatively high, e.g., the boiling point of trimethylsilanol is as high as the one of hexamethyldisiloxane (~100 °C). Organylsilanediols and -triols are mostly solids. Owing to their polarity and their hydrogen bonds, for instance, diethyl- and diisopropyl-silanedioles are water soluble. However, water solubility decreases with increasing chain length of the alkyl group. From the homologous series of arylsilanols, only the triols are moderately soluble in water.

Compared with alcohols, the hydrogen of the hydroxyl group of silanols is more acidic and more reactive. Thus, trimethylsilanol is more acidic and shows a higher tendency towards forming hydrogen bonding than *t*-butanol[283]. Furthermore, triphenylsilanol turns out to be more acidic than trialkylsilanol, but is still less acidic than phenol[284].

Catalyzed by acids or bases, organylsilanols undergo condensation to form hexaorganyldisiloxanes (equation 259a):

$$2\ R_3SiOH \xrightarrow[-H_2O]{} R_3SiOSiR_3 \qquad (259a)$$

On the whole, the lower the number of hydroxyl groups and the larger and bulkier the organyl moieties at the silicon atom which protect the hydroxyl group, the less is the probability that silanols undergo condensation. In addition, the pH of the medium has an effect: in the presence of strong acids (which, e.g., are formed during hydrolysis of halogenosilanes) silanols easily undergo condensation. Since alkali can also have an enhancing effect, one has to carry out all reactions in a neutral medium in order to obtain the unstable silanols in good yield.

Thus, for the preparation of the sensitive trimethylsilanol, a solution of fluorotrimethylsilane (**530**) in diethyl ether is titrated with sodium hydroxide under cooling and stirring, so that the reaction mixture is maintained at a neutral pH (phenolphthalein as indicator); trimethylsilanol (**531**) is dissolved in the ether layer and can be isolated in 70% yield (equation 260)[285].

$$\underset{(\mathbf{530})}{Me_3SiF} \xrightarrow[-NaF]{NaOH} \underset{(\mathbf{531})}{Me_3SiOH} \qquad (260)$$

An alternative route to **531** which bypasses the above-mentioned problems is the hydrolytic cleavage of *N*-TMS-acetamide (**435**) in ice water (73% yield) (equation 261)[286]. Furthermore, the hydrolysis of hexamethyldisilazane (**405**) by means of 1*N* hydrochloric acid which neutralizes generated ammonia gives rise to **531** (equation 262)[287].

$$\underset{(\mathbf{435})}{MeCONHSiMe_3} \xrightarrow[-AcNH_2]{H_2O} \mathbf{531} \qquad (261)$$

$$\mathbf{405} \xrightarrow[-NH_3]{H_2O} 2\ \mathbf{531} \qquad (262)$$

More stable is triethylsilanol (**533**) which can be prepared by hydrolysis of chlorotriethylsilane. If only water is employed, a mixture of **533** and hexaethyldisiloxane is formed; however, with a dilute solution of alkali (NaOH or $NaHCO_3$) in ether, **533** can be isolated in 95% yield[285]. **533** can also be obtained by the hydrolysis of hexaethyldisilathiane (**532**) in dioxane at room temperature (equation 263)[288]. Organyl-*H*-silanes, when treated with anhydrous alkali hydroxides in alcoholic or non-polar solution, give rise to organylsilanolates which can be converted into organylsilanols by means of an excess of either alcohol or water. An example illustrating this is the conversion of optically active organyl-*H*-silanes into optically active organylsilanols. For instance, (−)methylnaphth-1-ylphenylsilane (**534**) affords (−)methylnaphth-1-ylphenylsilanol (**535**) (equation 264)[289] quantitatively.

$$\underset{(\mathbf{532})}{(Et_3Si)_2S} \xrightarrow[-H_2S]{H_2O} \underset{(\mathbf{533})}{2\ Et_3SiOH} \qquad (263)$$

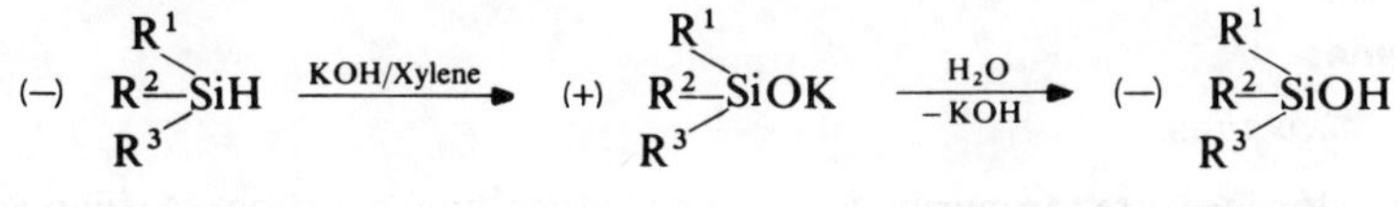

(264)

(**524**) (**535**)

$R^1 = Me,\ R^2 = 1\text{-Naph},\ R^3 = Ph$

A dialkyl-*H*-silanol like diisopropyl-*H*-silanol (**537**) can be obtained in 85% yield by the reaction of trichlorosilane with isopropylmagnesium chloride and subsequent hydrolysis of the formed chlorosilane **536** (equation 265)[290].

$$HSiCl_3 \xrightarrow[-MgCl_2]{2Me_2CHMgCl} \underset{(\mathbf{536})}{(Me_2CH)_2SiHCl} \xrightarrow[-HCl]{H_2O} \underset{(\mathbf{537})}{(Me_2CH)_2SiHOH} \qquad (265)$$

Several silanediols can be synthesized by hydrolysis of diorganyldihalogenosilanes with dilute alkali solutions (NaOH or $NaHCO_3$). Thus, diethylsilanediol (**539**) is obtained from diethyldichlorosilane (**538**), and diphenyldichlorosilane affords diphenylsilanediol (**540**) (equations 266 and 267)[291]. An alternative is the hydrolysis of diorganyldialkoxysilanes (equation 268)[292, 293].

$$\underset{(\mathbf{538})}{Et_2SiCl_2} \xrightarrow[-NaCl]{H_2O/NaOH/Et_2O} \underset{(\mathbf{539})}{Et_2Si(OH)_2} \qquad (266)$$

$$Ph_2SiCl_2 \xrightarrow{H_2O/(NH_4)_2CO_3} \underset{(\mathbf{540})}{Ph_2Si(OH)_2} \qquad (267)$$

$$\underset{(\mathbf{541})}{Me_2Si(OR)_2} \xrightarrow[-ROH]{H_2O} \underset{(\mathbf{542})}{Me_2Si(OH)_2} \qquad (268)$$

Phenylsilanetriol (**544**) is the reaction product of the hydrolysis of phenyltrimethoxysilane (**543**) by means of a dilute solution of acetic acid[294] or, alternatively, of the hydrolysis of trichlorophenylsilane with aniline in aqueous solution[295] (equation 269).

$$\underset{(\mathbf{543})}{PhSi(OMe)_3} \xrightarrow{H_2O/AcOH} \underset{(\mathbf{544})}{PhSi(OH)_3} \xleftarrow[-PhNH_2\cdot HCl]{H_2O/PhNH_2} PhSiCl_3 \qquad (269)$$

Finally, hydroxyaryldihydrosilanols (**546**) are obtainable from lithiated lithium phenolate (**545**) and dichlorosilane (equation 270)[296].

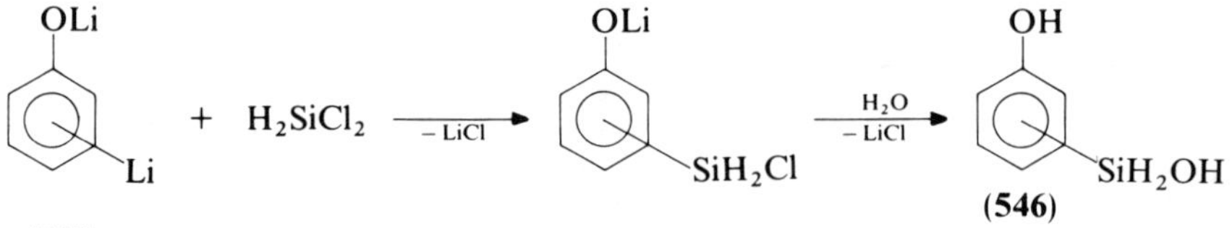

(**545**) (**546**) (270)

C. Siloxanes

1. *Linear siloxanes*

Linear siloxanes are primarily formed by either condensation of silanols or by hydrolysis of organylsilanes with functional (e.g. halogeno, amino, etc.) substituents (equation 271).

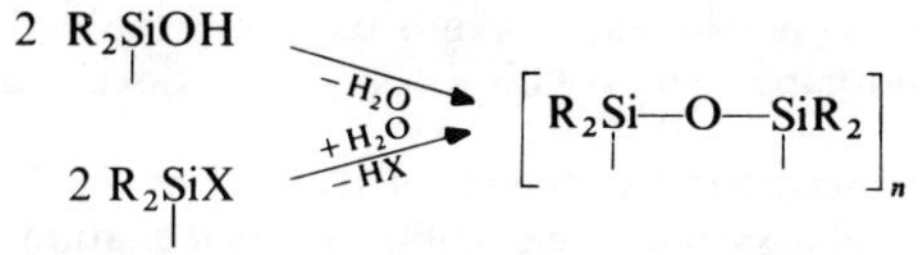

(271a)

2. Cyclic siloxanes

Cyclic siloxanes are primarily synthesized by either condensation of silanediols or by hydrolysis of organylsilanes with functional (e.g. halogeno, amino, etc.) substituents (equation 272).

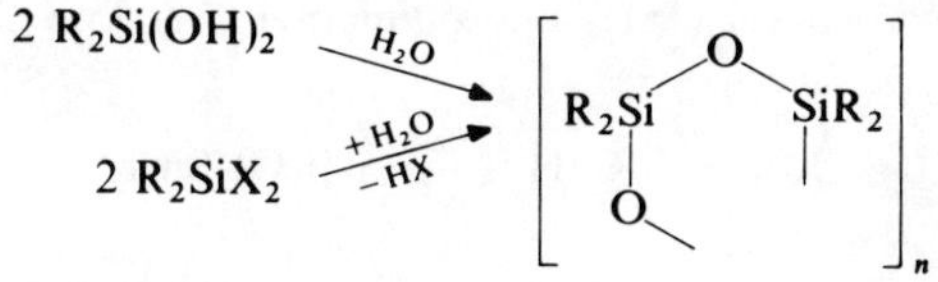

(271b)

A comprehensive review of this important class of compounds is given in this volume in the chapter by Kendrick.

D. Organyloxysilanes

Organyloxy-organylsilanes have the general structure

$$R_{4-n}Si(OR')_n \qquad n = 1,2,3$$

Organyloxy-organylsilanes are sensitive to moisture and are suitable substrates for reactions involving organometallic reagents.

1. Alkoxyorganylsilanes

Alkoxyorganylsilanes can be synthesized from tetraalkoxy-, halogenoorganyl-, pseudohalogenoorganyl-, organyl-*H*-silanes as well as organosilicon–nitrogen compounds. In terms of resistance to hydrolysis, alkoxyarylsilanes are more stable than alkoxy-alkylsilanes.

When halogenoorganylsilanes are treated with alcohol, the hydrogen halide has to be either continuously removed from the reaction medium or neutralized (equation 272).

$$R_{4-n}SiX_n + nR'OH \xrightarrow[-nBHX]{nB} R_{4-n}Si(OR')_n \qquad n = 1,2,3 \qquad (272)$$

For instance, in one technical process alcohol vapour is pressed from the bottom, into reaction column, while the chloroorganylsilane enters at the top whereas the developing hydrogen chloride is removed at the top of the column[297]. In another process, the reaction time is only 30 s[298] and the hydrogen chloride is neutralized with ammonia, which is being continuously fed into the reaction system[299, 300].

The reaction of, e.g., chlorotris(2-methylphenyl)silane (**547**) with various C_1–C_4 alcohols gives rise to alkoxytriarylsilanes (**548a–d**) (equation 273)[301,302]. If a large excess of

$$o\text{-}Tol_3SiCl \xrightarrow[-HCl]{ROH} o\text{-}Tol_3SiOR \qquad (273)$$

(**547**) (**548a–d**)

(a) R = Me (b) R = Et (c) R = Pr (d) R = Bu

Tol = MeC_6H_4

alcohol is employed, it is not urgent to remove or trap the hydrogen halide. Further reactions of halogenoorganylsilanes with alcohols are discussed comprehensively by Pawlenko[302].

If less than the stoichiometrically equivalent amount of alcohol is employed, partial alcoholysis to halogenoalkoxysilanes, e.g. (**549**), occurs (equation 274)[303].

$$Me_2SiCl_2 \xrightarrow[-HCl]{\text{butanol}} \underset{\textbf{(549)}}{Me_2SiClOBu} \qquad (274)$$

The reaction of tetrachlorosilane with tertiary alcohols in the presence of pyridine leads to dialkoxydichlorosilanes (**550**) (equation 275)[304]. If Me_2SiCl_2 reacts with a long-chain alcohol, e.g. iso-octanol (**551**), chlorodimethyl-*i*-octyloxysilane (**552**) is formed (equation 276)[305].

$$SiCl_4 + 2\ ROH \xrightarrow[-py\cdot HCl]{py} \underset{\textbf{(550)}}{(RO)_2SiCl_2} \qquad (275)$$

$$Me_2SiCl_2 + \underset{\textbf{(551)}}{i\text{-}C_8H_{17}OH} \longrightarrow \underset{\textbf{(552)}}{Me_2Si(Cl)OC_8H_{14}\text{-}i} \qquad (276)$$

An elegant procedure to synthesize alkoxysilanes is the reaction of alcohols with TMS-triflate[306]. If 'perfluorinated resinsulfonic acid' is treated with TMS-triflate, binding of TMS-triflate to the resin occurs. The resulting product, named 'Nafion-TMS', is a potent silylating agent for alcohols and phenols which have to be dissolved in dichloromethane. The yield is dependent on the reaction time. Thus, ethanol in 12 h gives rise to ethoxytrimethylsilane in quantitative yield.

In another route, Me_2SiCl_2 and dimethoxydimethylsilane (**553**) in the presence of $AlCl_3$ as a catalyst yield methoxychlorodimethylsilane (**554**) (equation 277)[307].

$$Me_2SiCl_2 + \underset{\textbf{(553)}}{Me_2Si(OMe)_2} \xrightarrow{AlCl_3} \underset{\textbf{(554)}}{Me_2Si(Cl)OMe} \qquad (277)$$

Alkoxychlorosilanes can be used for the synthesis of silanes with non-identical alkoxy groups (equations 278)[307]. Instead of alcohols, sodium alcoholate can be employed with halogenosilanes (equation 279)[308]. Generally speaking, alcoholates are by far more reactive than alcohols towards organylsilanes in the presence of an amine, to give rise to organylsilanes with bulky alkoxy moieties. Thus, the reaction of tetrachlorosilane with *t*-butanol in the presence of amine alone affords bis-*t*-butoxydichlorosilane, while the use of sodium *t*-butoxide leads to tetrakis-*t*-butoxysilane[309].

$$\textbf{554} + Me_3COH \xrightarrow[-Et_3N\cdot HCl]{Et_3N/\text{pentane}} \underset{\textbf{(555)}}{Me_2Si(OMe)OCMe_3} \qquad (278)$$

$$RSiH_2X + EtONa \xrightarrow[-NaX]{} RSiH_2OEt \qquad (279)$$

$$R = \text{alkyl, aryl}$$

Furthermore, alkoxyorganylsilanes can also be synthesized from a variety of silicon–nitrogen compounds, e.g. hexamethyldisilazane (**405**), or other aminosilanes, in good yield. Thus far, this approach is not so well established since the starting material itself has to be synthesized. However, the great advantage of aminosilanes is that ammonia is the only by-product of the reaction and side-reactions which are common with halogenated silanes do not occur. When the reaction rate is slow, acid catalysis has an enhancing effect; acid catalysts used are HCl or catalytic amounts of chlorosilane (which both immediately form NH_4Cl, an effective catalyst) (equation 280)[310]. The same approach is applicable to the synthesis of [penta-2,4-diyne-1-yloxy]trimethylsilane[311], [3-cyanocyclobut-1-yloxy]trimethylsilane[312] and [4-oxohex-3-yloxy]trimethylsilane[313].

$$\tfrac{1}{2}\,\mathbf{405} + CH_3(CH_2)_6OH \xrightarrow[-NH_3]{} Me_3SiO(CH_2)_6CH_3 \qquad (280)$$

The solvolytic cleavage of Si–H bonds by means of alcohols and phenols proceeds at a lower rate than with Si–N bonds; therefore, it is preferable to apply a catalyst. For instance, the reaction of diethylsilane (**557**) with methanol leads to diethyldimethoxysilane (**558**) when catalyzed by tris[triphenylphosphine]rhodium(I) chloride (equation 281)[314].

$$\underset{\mathbf{(557)}}{Et_2SiH_2} \xrightarrow{MeOH/[Ph_3P]_3RhCl} \underset{\mathbf{(558)}}{Et_2Si(OMe)_2} \qquad (281)$$

From benzophenone (**560**) and phenylsilane (**559**), mono- (**561**) and bis(diphenylmethyl)phenylsilane (**562**) can be obtained (equation 282)[315]. The asymmetric dimethyl(2,2-dimethyl-1-phenylprop-1-yloxy)phenylsilane (**563**) can be synthesized in the presence of a cationic rhodium complex (equation 283)[316]. An alkoxy replacement can be used for the synthesis of alkoxysilanes with non-identical alkoxy moieties, e.g. for diethoxymethoxysilane (**565**) (equation 284)[317].

$$\underset{\mathbf{(559)}}{PhSiH_3} + \underset{\mathbf{(560)}}{Ph_2CO} \longrightarrow \underset{\mathbf{(561)}}{PhSi(H_2)OCHPh_2} + \underset{\mathbf{(562)}}{PhSi(H)(OCHPh_2)_2} \qquad (282)$$

$$Me_3CCOPh + Me_2PhSiH \xrightarrow{cat.} \underset{\mathbf{(563)}}{Me_3CCPhHOSiMe_2Ph} \qquad (283)$$

$$\underset{\mathbf{(564)}}{(Me_3O)_3SiH} + EtOH \longrightarrow \underset{\mathbf{(565)}}{(EtO)_2Si(H)OMe} \qquad (284)$$

An alternative route to alkoxyorganylsilanes is the application of organometallic compounds. Tetraalkoxysilanes with Grignard reagents yield mono, di-regular print or trialkoxysilanes, depending on the silane: Grignard ratio (equation 285)[318–321].

$$Si(OR^1)_4 + nR^2MgHal \xrightarrow[nR'OMgHal]{} R^2_nSi(OR^1)_{4-n} \qquad (285)$$

An additional route to alkoxyarylsilanes is the reaction of tetraalkoxysilanes with organometallic compounds, e.g., tetraethoxysilane plus α-lithiated naphthalene gives rise to a variety of ethoxynaphth-1-ylsilanes (**566**–**568**) (equation 286)[322]. The reaction of

$$Si(OEt)_4 \xrightarrow{RLi} \underset{\mathbf{(566)}}{RSi(OEt)_3}$$

$$Si(OEt)_4 \xrightarrow{2\,RLi} \underset{\mathbf{(567)}}{R_2Si(OEt)_2} \qquad (286)$$

$$Si(OEt)_4 \xrightarrow{3\,RLi} \underset{\mathbf{(568)}}{R_3Si(OEt)}$$

R = 1-Naph

tetrabutoxysilane with triethylaluminum etherate leads to a mixture of ethyltributoxy (35%), diethyldibutoxy (45%) and triethylbutoxysilanes (20%)[323].

An interesting reaction is the addition of 1-phenyl-2-TMS-acetylene to carbonyl compounds, e.g. cyclohexanone (**113**) in the presence of catalytic amounts of tetrabutylammoniumfluoride to form, for instance, [1-(phenylethynyl)cyclohex-1-yloxy]trimethylsilane (**569**) (equation 287)[324]. However, if 4-*t*-butylcyclohexanone (**570**) reacts

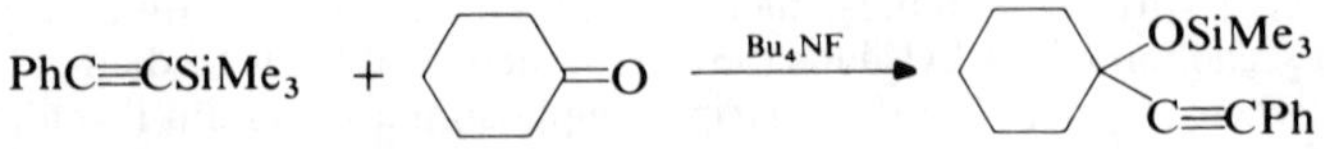

(**113**) (**569**) (287)

with a trialkylsilane, e.g. triethylsilane (**442**), in the presence of a catalyst such as $(Ph_3P)_3RhCl$, then [4-*t*-butylcyclohex-1-yloxy]triethylsilane (**571**) can be isolated (equation 288)[325].

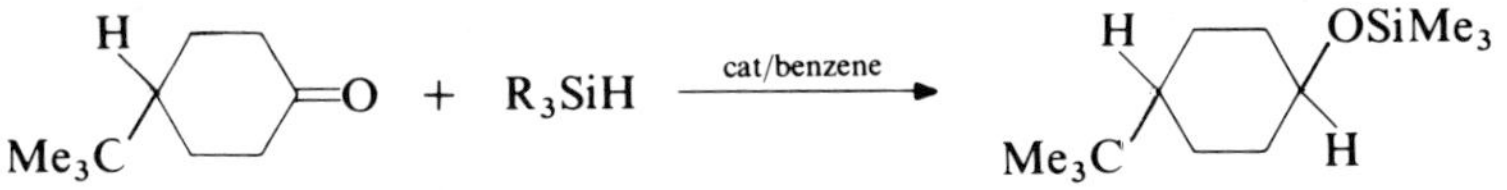

(**570**) (**571**) (288)

The addition of Me_3SiN_3 to aliphatic aldehydes (**572a–f**) in the presence of zinc chloride leads to α-(trimethylsiloxy)alkyl azides (**573a–f**) which, by thermolysis, give the corresponding *N*-TMS amides (**574a–f**) via rearrangement (equation 289)[326]. Me_3SiN_3

$$Me_3SiN_3 + RCHO \xrightarrow{ZnCl_2} RCH(N_3)OSiMe_3 \xrightarrow[-N_2]{\Delta} RCON(H)SiMe_3 \qquad (289)$$

(**572a–f**) (**573a–f**) (**574a–f**)

572	573	574	R
a	**a**	**a**	$Me(CH_2)_2$
b	**b**	**b**	$Me(CH_2)_3$
c	**c**	**c**	$Me(CH_2)_4$
d	**d**	**d**	Me_2CH
e	**e**	**e**	Me_3C
f	**f**	**f**	Cl_2C

forms with cyclohexane epoxide (**575**) or with styrene epoxide (**577**) (*trans*-2-azidocyclohex-1-yloxy)trimethylsilane (**576**) and 2-azido-1-phenyleth-1-yloxytrimethylsilane (**578**), respectively (equations 290 and 291)[326]. Otherwise, ring-opening

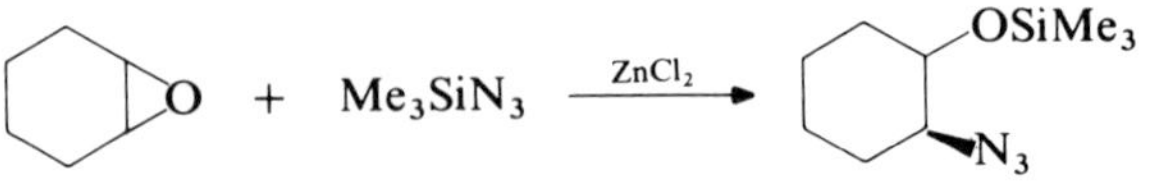

(**575**) (**576**) (290)

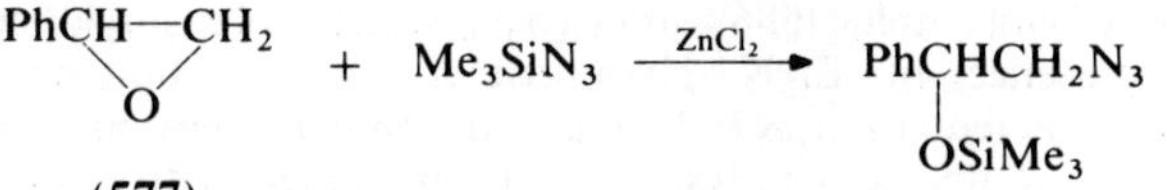

(577) (578) (291)

reactions of epoxides, such as cyclopentane epoxide (**579**), can be accomplished by means of TMS-triflate to yield, for example, (cyclopent-1-ene-3-yloxy)trimethylsilane (**580**) (equation 292)[327].

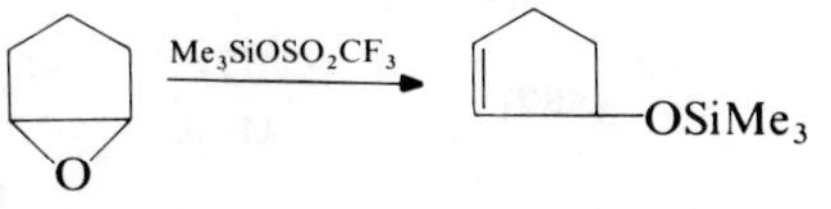

(579) (580) (292)

Another way to synthesize alkoxyorganylsilanes is the reaction of halogenosilanes with alkyl orthoformate **29** (equations 293)[328] in the presence of a Lewis acid.

$$Me_2SiCl_2 + 2HC(OEt)_3 \xrightarrow[-HCO_2Et,\ -EtCl]{AlCl_3} Me_2Si(OEt)_2 \qquad (293)$$

(29) (581)

Finally, an alternative approach to alkoxyorganylsilanes is the treatment of chloroorganylsilane with carboxylic acids or their esters and sodium in HMPA/THF. Thus, from acetic acid, Me_3SiCl and sodium 1,1-bis(TMS)-1-(TMS-oxy-)ethane (**582**) is obtained, whereas Me_3SiCl and β-halogenoesters (**583**) yield (1-alkoxycyclopropan-1-yloxy)trimethylsilanes (**584**) (equations 294 and 295)[313].

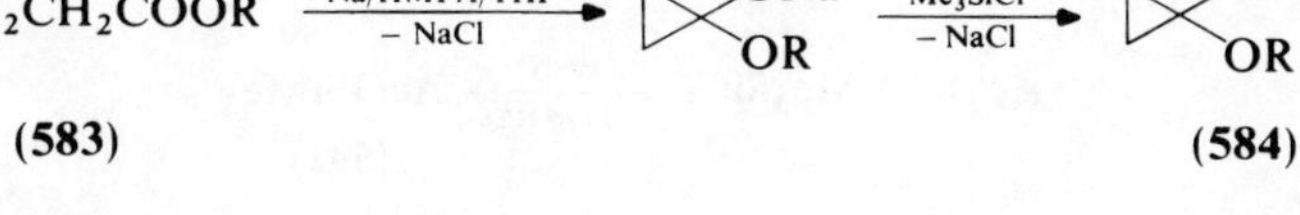

(582) (294)

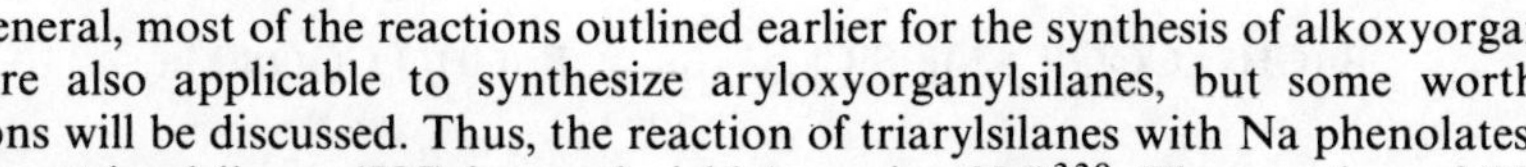

(583) (584) (295)

2. Aryloxyorganylsilanes

In general, most of the reactions outlined earlier for the synthesis of alkoxyorganylsilanes are also applicable to synthesize aryloxyorganylsilanes, but some worthwhile additions will be discussed. Thus, the reaction of triarylsilanes with Na phenolates leads to aryloxytriarylsilanes (**585**) in good yield (equation 296)[329]. The reaction of 'Nafion-TMS' (*vide supra*) with phenols in the presence of Et_3N affords phenoxytrimethylsilane in 86% yield after a reaction period of 18h[306].

$$Ar_3SiH + PhONa \rightarrow Ar_3SiOPh \qquad (296)$$

(585)

The treatment of glutarimide (**586**) with two equivalents of *t*-butyldimethylsilyl triflate (**587**) in the presence of Et_3N gives rise to 2,6-bis(*t*-butyldimethylsiloxy)-3,4-dihydropyridine (**588**) (equation 297)[330]. Quinone (**589**) and hexamethyldisilazane (**405**) in the presence of iodine as catalyst form 1,4-bis(TMS)-hydroquinone (**590**) (equation 298)[331].

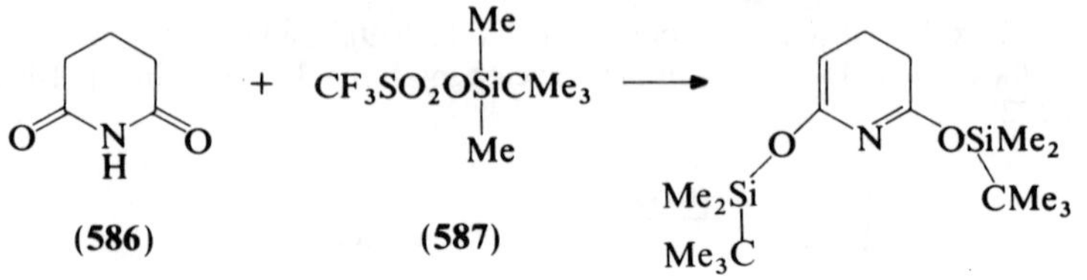

(**586**) (**587**) (**588**) (297)

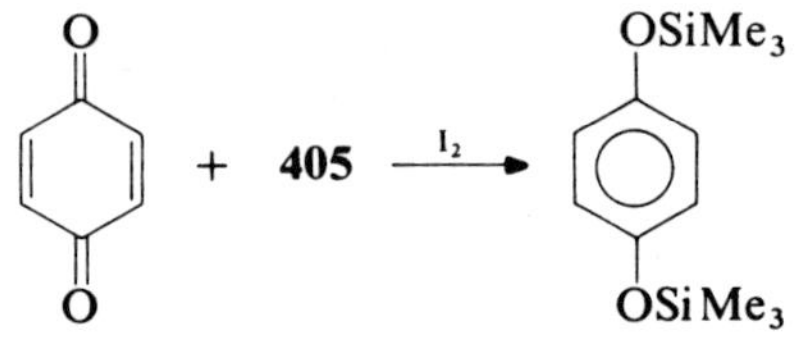

(**589**) (**590**) (298)

3. *Acyloxysilanes*

In this section only acyloxyorganylsilanes deriving from carboxylic acids will be discussed. These compounds are very sensitive to solvolysis by means of water or alcohols; they are well soluble in inert organic solvents, and can be distilled without decomposition. For the synthesis of acyloxyorganylsilanes, halogenosilanes are treated with carboxylic acids in the presence of amines (e.g. diethyl-, triethylamine, pyridine, etc.) as acceptors for the generated hydrogen halide (equation 299)[332]. Besides, acetoxytrimethylsilane (**591**)[333] as well as α- (**592**) and ω-bromoacyloxytrimethylsilanes (**593**) can be obtained similarly (equations 300 and 301)[334].

$$\mathrm{AcOH + Me_3SiCl \xrightarrow[-Et_3NHCl]{Et_3N} AcOSiMe_3} \quad \mathbf{(591)} \tag{299}$$

$$\mathrm{RCBrHCOOR + Me_3SiCl \xrightarrow[-Et_3N\cdot HCl]{Et_3N} RCBrHCOOSiMe_3} \quad \mathbf{(592)} \qquad \mathrm{R = H, Me, Et} \tag{300}$$

$$\mathrm{Br(CH_2)_{\mathit{n}}COOR + Me_3SiCl \xrightarrow[-Et_3N\cdot HCl]{Et_3N} Br(CH_2)_{\mathit{n}}COOSiMe_3} \quad \mathbf{(593)} \tag{301}$$

Acid anhydrides can also be used with chlorosilanes (**594**). If acetic anhydride is used, the produced acetyl chloride is removed by distillation. Via this approach, for instance, chloromethyldiacetoxymethylsilane (**595**) is obtainable (equation 302)[335]. Alternatively, Na or Ag carboxylates can react with chlorosilanes in inert solvents like diethyl ether, petroleum ether, benzene, etc.[336].

$$\underset{\mathbf{(594)}}{Ac_2O + Cl_2SiMeCH_2Cl} \xrightarrow{-AcCl} \underset{\mathbf{(595)}}{MeSi(OAc)_2(CH_2Cl)} \qquad (302)$$

In the presence of catalysts such as tris[triphenylphosphine]rhodium(I) chloride[337], colloidal Ni[338], $NiCl_2$[339], a mixture of aluminum/iodine[339] or Hg(II)acetate[339], trialkyl-*H*-silanes **147**, **442** can be employed with carboxylic acids (**596**) (equation 303).

$$R_3SiH + R'COOH \xrightarrow[-H_2]{cat.} R'COOSiR_3 \qquad (303)$$

$$\mathbf{(597)}$$

$$\mathbf{(147)}\ R=Me \qquad \mathbf{(596)}$$
$$\mathbf{(442)}\ R=Et \qquad R'=Et$$

Another approach is the application of silanols as silyl component to acid anhydrides and chlorides, as shown for the synthesis of acetoxytriethylsilane (**598**) and TES-chloroformate (**599**), respectively (equation 304)[340]. Transesterification may afford certain acyloxyorganylsilanes; the produced carboxylic acid has to be removed, e.g. by

$$Et_3SiOH \xrightarrow{Ac_2O} \underset{\mathbf{(598)}}{MeCOOSiEt_3}$$
$$Et_3SiOH \xrightarrow{COCl_2} \underset{\mathbf{(599)}}{ClCOOSiEt_3} \qquad (304)$$

distillation (equation 305)[341]. Organylsilanthiols (**601**) with carboxylic acids also yield acyloxyorganylsilanes (**602**) (equation 306)[342].

$$\mathbf{598} + Cl_3CCOOH \xrightarrow[-AcOH]{} \underset{\mathbf{(600)}}{Cl_3CCOOSiEt_3} \qquad (305)$$

$$\underset{\mathbf{(601)}}{R_3SiSH} + R'COOH \xrightarrow[-H_2S]{} \underset{\mathbf{(602)}}{R'COOSiR_3} \qquad (306)$$
$$R=Ph \qquad R'=Me,\ F_3C$$

According to expectation Si–N bonds can be cleaved by carboxylic acids to form Si–O bonds[315]. Thus, bis(TMS)-carbodiimide (**458**) and ethyl acetate (**603**) form **591** (equation 307)[343].

$$\underset{\mathbf{(458)}}{Me_3SiN{=}C{=}NSiMe_3} + \underset{\mathbf{(603)}}{2AcOEt} \xrightarrow[-Et_2NCN]{} \underset{\mathbf{(591)}}{2AcOSiMe_3} \qquad (307)$$

TMS-TMS-formate (**604**) is obtainable via the reaction of bis(TMS)-mercury (**502**) with carbon dioxide (equation 308)[344]. From aminotriethylsilane and CO_2 the corresponding carbamate (**605**) is formed in 76% yield[345] whereas Me_3SiCl and ammonium

$$\underset{\mathbf{(502)}}{(Me_3Si)_2Hg} \xrightarrow{CO_2/HMPA} \underset{\mathbf{(604)}}{Me_3SiCOOSiMe_3} \qquad (308)$$

carbamate (**606**) give rise to the corresponding TMS carbamate (**607**)[346] (equations 309 and 310).

$$Et_3SiNH_2 \xrightarrow{CO_2} \underset{\textbf{(605)}}{H_2NCOOSiEt_3} \tag{309}$$

$$\underset{\textbf{(606)}}{H_2NCOONH_4} + Me_3SiCl \xrightarrow[NH_4Cl]{THF/0\,^{\circ}C} \underset{\textbf{(607)}}{H_2NCOOSiMe_3} \tag{310}$$

E. Silyl Enol Ethers

The so-called 'silyl enol ethers' (enoxyorganylsilanes) are important synthones, e.g, for regiospecific preparation of enolates, aldol condensation, synthesis of α-substituted carbonyl derivatives and for thermal or photochemical cycloaddition. For the preparation of silyl enol ethers the corresponding aldehydes and ketones first have to be enolized and then treated with silylating agents in the presence of a base. Thus, from butanal (**608**) and Me_3SiCl, *cis/trans*-1-trimethylsiloxybut-1-ene (**609**) is obtained (equation 311)[347], while 1-trimethylsiloxy-1-phenylethene (**610**) is the product from acetophenone (**90a**) (equation 312)[347].

$$\underset{\textbf{(608)}}{MeCH_2CH_2CHO} + Me_3SiCl \xrightarrow{Et_3N/DMF} \underset{\textbf{(609)}}{MeCH_2CH{=}CHOSiMe_3} \tag{311}$$

$$\underset{\textbf{(90a)}}{PhCOCH_3} + Me_3SiCl \xrightarrow{Et_3N/DMF} \underset{\textbf{(610)}}{Ph{-}C(OSiMe_3){=}CH_2} \tag{312}$$

This approach is also applicable to cyclic ketones. Thus 2-methylcyclohexanone (**113a**) gives a mixture of 1-methyl-2-trimethylsiloxycyclohex-1-ene (**611**) (77% yield) and 6-methyl-1-trimethylsiloxycyclohex-1-ene (**612**) (22% yield) (equation 313)[347]. However, if

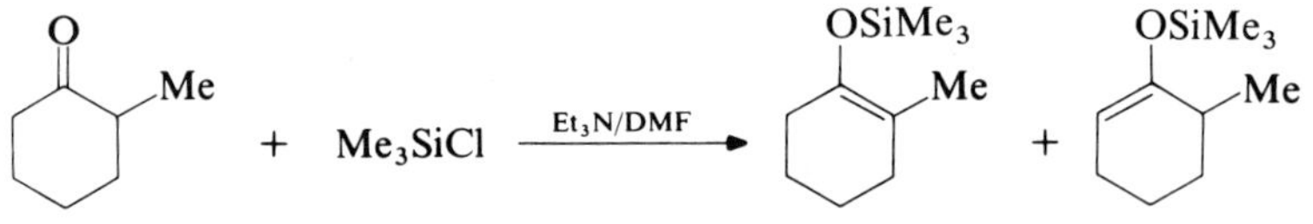

(313)

(**113a**) (**611**) (**612**)

the enolate formation is carried out separately with LDA, in DME, subsequent silylation gives **612** in 74% yield, **613** in 6% and **611** only in 1% yield[347]. Using potassium hydride in dioxane with **113a**, **611** can be obtained in 34% yield, **612** in 59% yield and *cis/trans*-1-methyl-2-trimethylsiloxycyclohexane (**613**) in 1.2% yield (equation 314)[348].

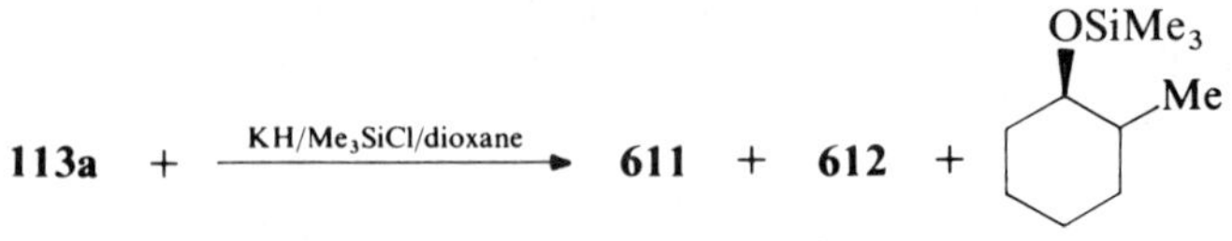

(314)

(**613**)

If enolizable ketones are used, isomeric reaction products may result; e.g., from butan-2-one both 2-trimethylsiloxy-but-2-ene (**614**) and but-1-ene (**615**) are formed (equation 315)[349] in a ratio of 2:1.

$$\text{MeCOEt} \xrightarrow{Me_3SiCl/NaI/Et_3N} \underset{(\mathbf{614})}{\text{Me—}\underset{\text{OSiMe}_3}{\text{C}}\text{=CHMe}} + \underset{(\mathbf{615})}{\text{CH}_2\text{=}\underset{\text{OSiMe}_3}{\text{C}}\text{—CH}_2\text{Me}} \qquad (315)$$

An α,α-dibromoketone such as **616**, when treated at first with LDMC and then silylated, yields under debromination the α-methylated trimethylsiloxy compounds (**612**) (equation 316)[350].

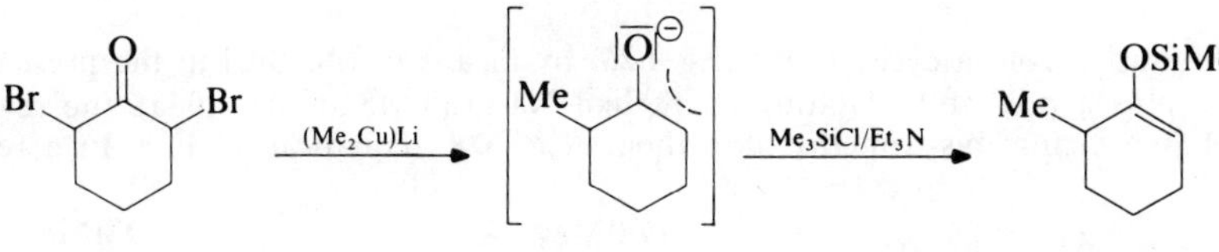

(**616**) (**612**) (316)

By means of the combination hexamethyldisilazane (**405**)/Me_3SiI in pentane solution, 3-oxo-1-phenylbut-1-ene (**617**) or methyl 5-oxo-5-phenylpentanoate (**619**) give rise to *trans*-4-phenyl-2-trimethylsiloxybuta-1,3-diene (**618**) and methyl 5-phenyl-5-trimethylsil-oxypent-4-enoate (**620**), respectively (equations 317 and 318)[351].

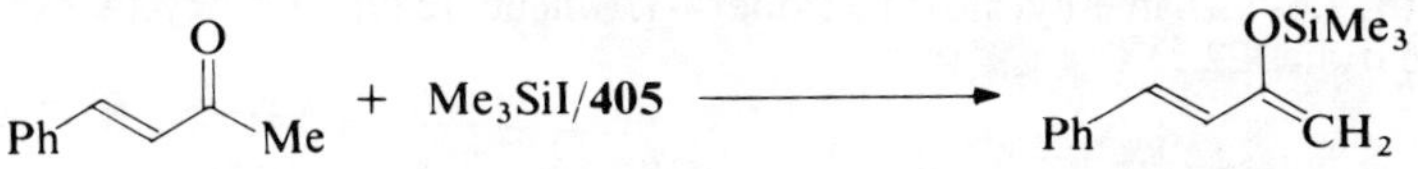

(**617**) (**618**) (317)

$$\underset{(\mathbf{619})}{\text{PhCOCH}_2\text{CH}_2\text{CH}_2\text{COOMe}} \xrightarrow{Me_3SiI/405} \underset{(\mathbf{620})}{\text{Ph}\underset{\text{OSiMe}_3}{\text{C}}\text{=CHCH}_2\text{CH}_2\text{COOMe}} \qquad (318)$$

Depending on the reaction conditions, derivatives of dicarboxylic acids lead to different types of products. For instance, if diethyl succinate (**621**) is treated with LDA

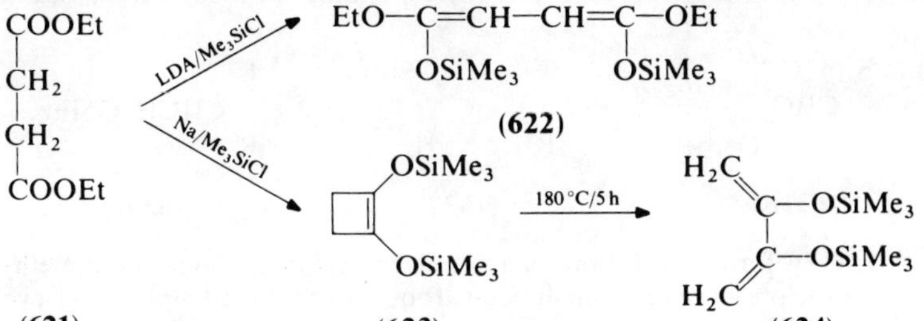

(**621**) (**623**) (**624**) (319)

and Me_3SiCl then 1,4-diethoxy-1,4-bis(trimethylsiloxy)-1,3-butadiene (**622**)[352] can be isolated; however, using Na and **621** in the first step, 1,2-bis(trimethylsiloxy)cyclobut-1-ene (**623**)[353] is formed and can be thermolyzed to 2,3-bis(trimethylsiloxy)buta-1,3-diene (**624**)[354] (equation 319). Alternatively, **624** can be obtained from 2,3-butane-dione (**625**) by means of TMS-triflate in the presence of triethylamine (equation 320)[355].

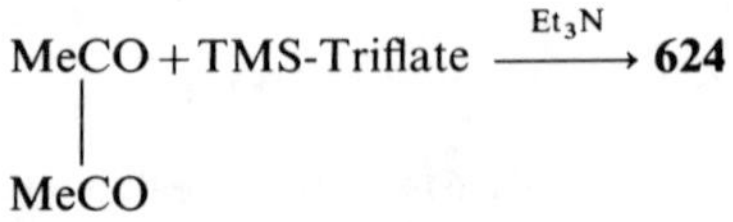

(320)

(**625**)

Analogously, from a cyclic dione (e.g. **626**) by means of Me_3SiCl in the presence of pyridine plus a powerful silylating agent like Na-bis(TMS)amide (**405a**), the reaction product is a vicinal bis-trimethylsiloxydiene (e.g. **628**) (equation 321)[356]. In a related

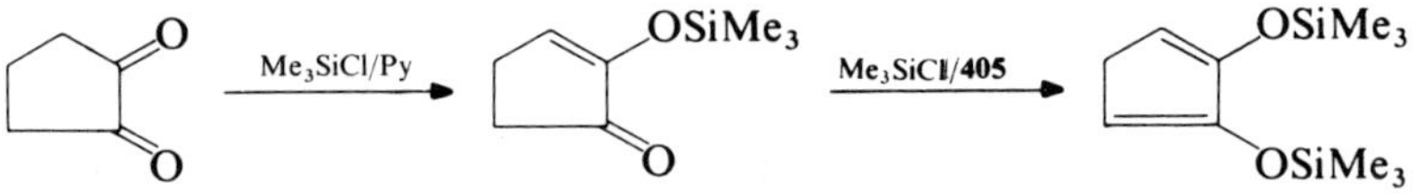

(321)

(**626**) (**627**) (**628**)

manner, benzoquinone can react with the combination TMS-triflate/triethylamine to yield both 1,4-bis(trimethylsiloxy)cyclohexa-1,4-diene (**629**) and -cyclohexa-1,3-diene (**630**) (equation 322)[355].

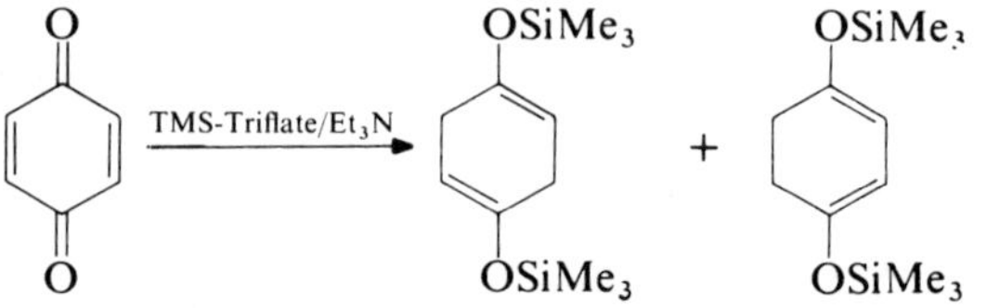

(322)

(**629**) (**630**)

Alternatively, triorganyl-*H*-silanes can be added across the carbonyl function under hydrosilylation-type conditions, e.g., triethylsilane (**442**) reacts with aldehydes in the presence of a Ni catalyst to yield the corresponding TES-enol ether **632** (equation

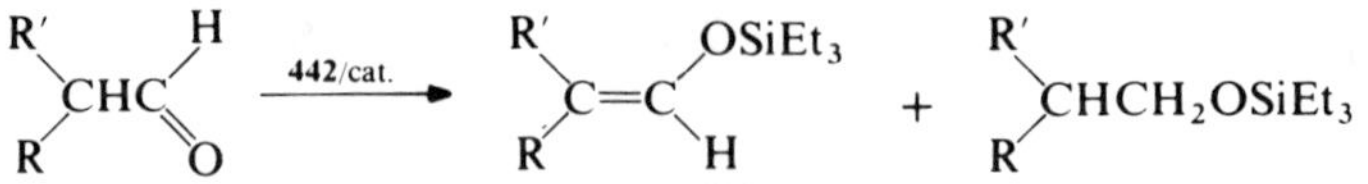

(323)

(**631**) (**632**) (**633**)

323)[357, 358]; **633** is formed as by-product. Similarly, cyclohexanone and dimethylphenylsilane (**634**) in the presence of dicobalt octacarbonyl form dimethylphenyl (1-cyclohexen-1-yloxy)silane (**635**) and dimethylphenylcyclohexyloxysilane (**636**) (equation 324)[359],

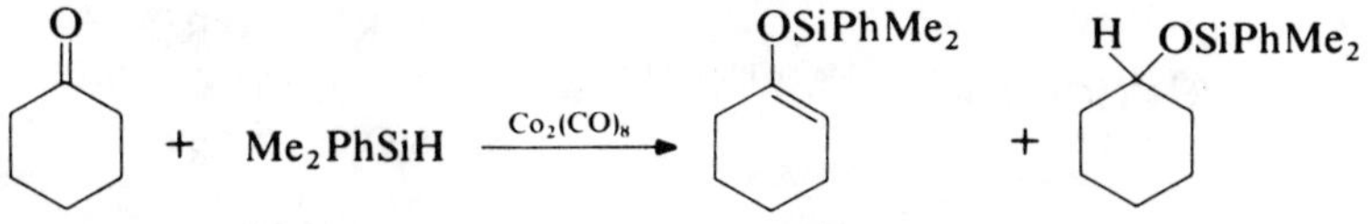

(324)

(634) **(635)** **(636)**

$$MeCOCHMe_2 + \mathbf{634} \xrightarrow{Co_2(CO)_8} Me_2PhSi{-}O{-}C(Me){=}CMe_2$$

(637) **(638)**

and dimethyl[3-methylbut-2-en-2-yloxy]phenylsilane (**638**) is obtained under the same conditions from 3-methylbutan-2-one (**637**)[360].

The reaction of alkenes with trialkylsilanes (**639**) and carbon monoxide in the presence of Co, Ru and Rh catalysts leads to silyl enol ethers which are extended by one carbon atom. Since this reaction is also applicable to cycloalkenes, one can synthesize diethylmethylsiloxymethylene cyclohexane (**640**) in 89% yield (equation 325)[354].

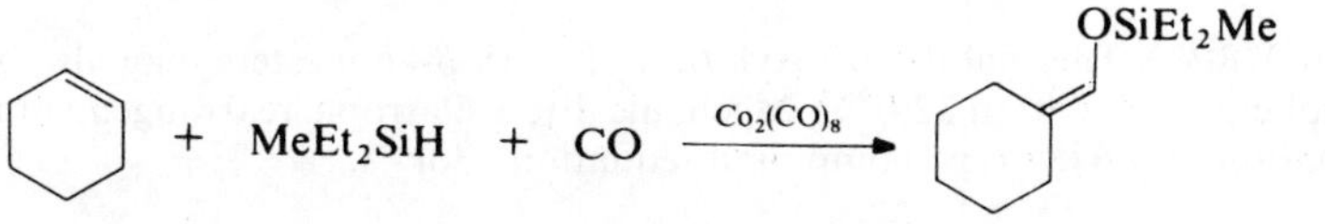

(325)

(639) **(640)**

Another approach to silyl enol ethers is the acidic ring opening of α,β-epoxysilanes (to form diols) with subsequent silylation (equation 326)[361]. Also, the application of Me_3SiCl to ketones in the presence of lithium sulfide and triethylamine yields silyl enol ethers

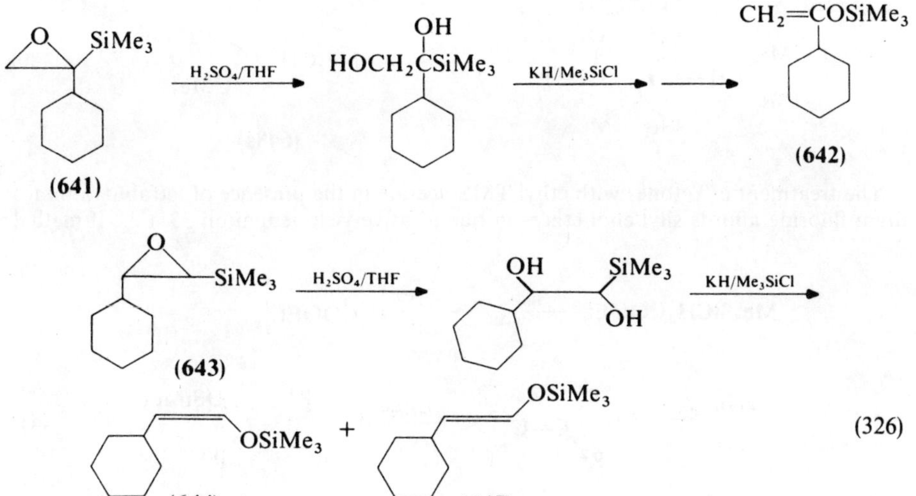

(641) **(642)** **(643)** **(644)** **(645)**

(326)

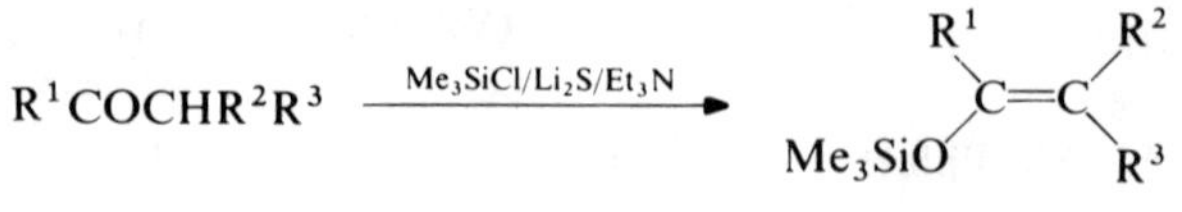

(327)

(646)

(equation 327)[362]. Silylated allylic alcohols can be rearranged to silyl enol ethers in the presence of a ruthenium catalyst at elevated temperatures over a period of 10–20 h

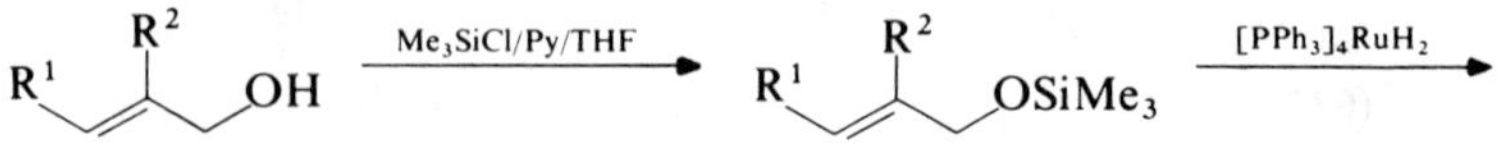

$R^1 = H, Me, Ph, R^2 = H, Me$

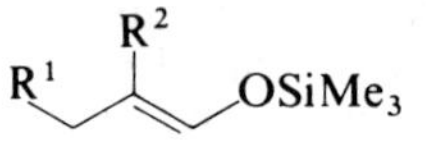

(328)

(647)

(equation 328)[363]. Thermal decarboxylation of TMS β-keto esters may also give rise to TMS-enol ethers (equation 329)[364]. Mechanically, a silatropic rearrangement may cause the formation of a **648a**-type compound (equation 330)[365].

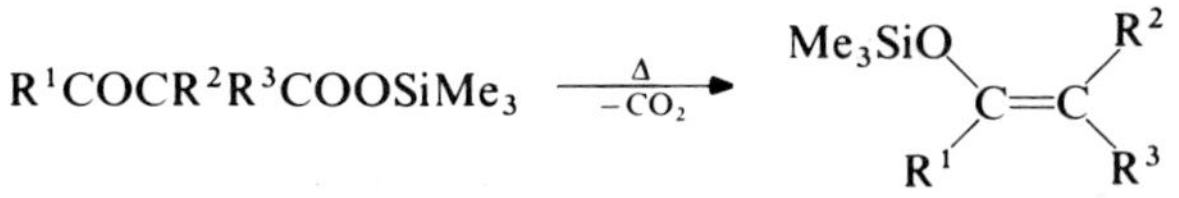

(329)

$R^1 = Ph, i\text{-}Pr$, c-Hex
$R^2 = R^3 = Me$, c-Hex

(648)

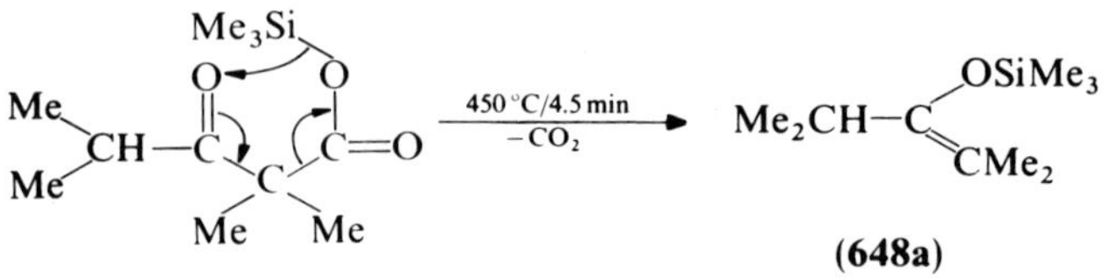

(330)

(648a)

The treatment of ketones with ethyl TMS-acetate in the presence of tetrabutylammonium fluoride affords silyl enol ethers in quantitative yield (equation 331)[366]. If methyl

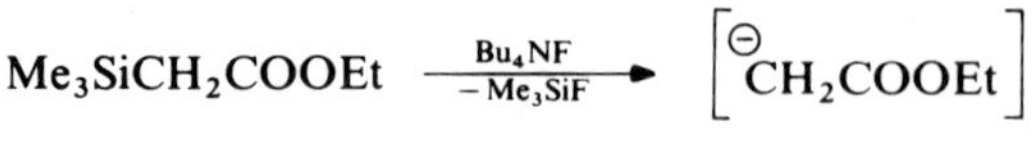

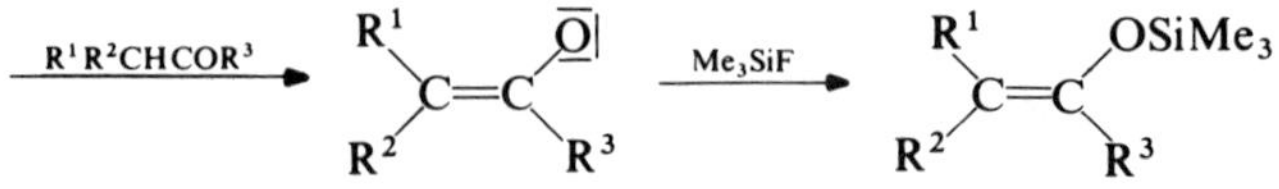

(331)

(649)

acetate (**650**) is first treated with LDA, it can yield 1-methoxy-1-trimethylsiloxyethene (**651**) (equation 332)[367]. If ethyl TMS-acetate is treated according to equation 332,

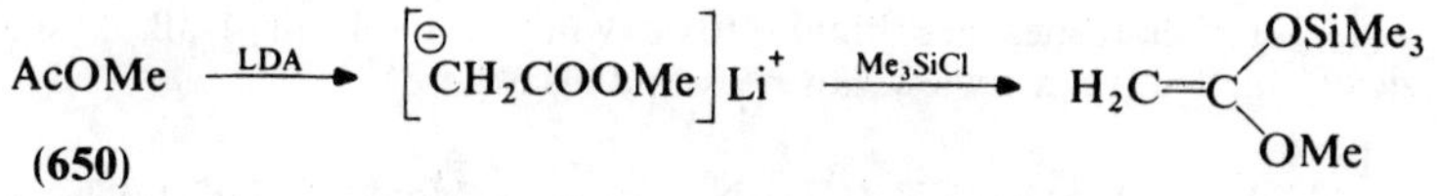

(650) (651) (332)

1-ethoxy-1-trimethylsiloxy-2-TMS-ethene can be isolated[368]. Analogously, using a carboxylic acid, one can obtain bis(trimethylsiloxy)ethene (**652**) (equation 333)[369].

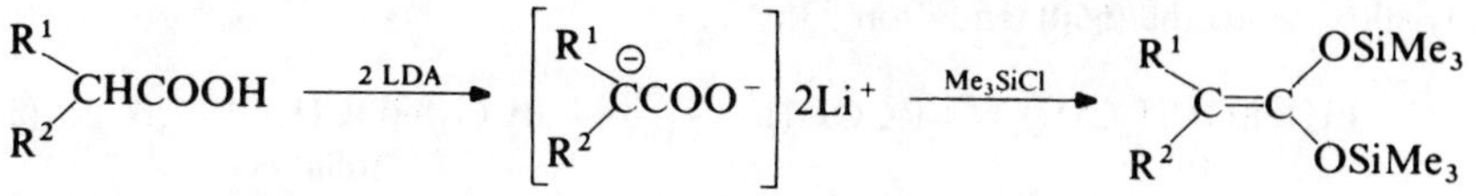

(652) (333)

α,β-Unsaturated esters can be converted via hydrosilylation into the corresponding trimethylsiloxy compounds (equation 334)[370].

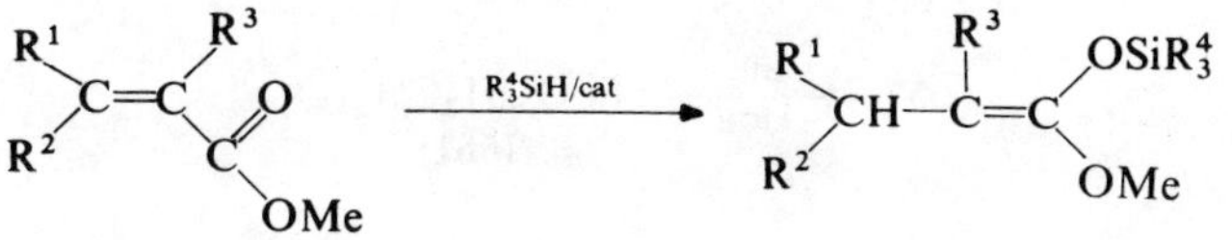

(334)

(653)

R^1	R^2	R^3	R^4	Yields
Me	H	H	Pr	73.5
Et	H	H	Et	75.4
H	H	H	Et	31.6

F. Cyclic Silicon–Oxygen Compounds

1. *Silatranes*

1-Organyl-2,8,9-trioxy-5-aza-1-silabicyclo[3.3.3]undecanes (**655**), so-called silatranes, are synthesized by alcoholysis of trialkoxysilanes with trialkanol amines (**654**) (equation 335)[371, 372]. X-ray analyses showed that the lone pair of electrons of the nitrogen atom

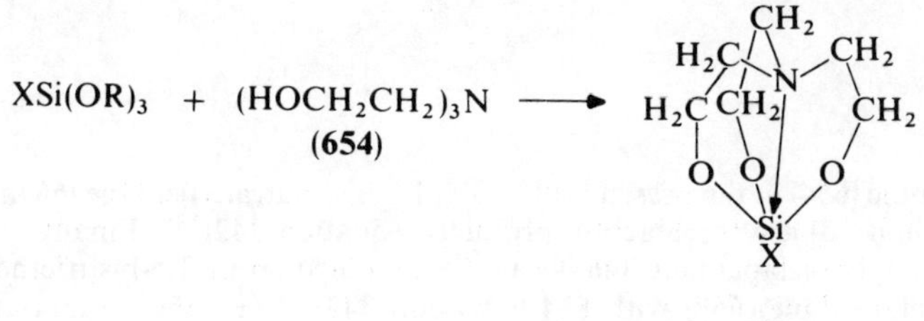

(335)

(655)

forms a coordinate bond with the d_π-orbitals of the silicon atom[373, 374]. Another pathway to silatranes is the reaction of polysiloxanes (**656**) with **654** as catalyzed by potassium hydroxide (equation 336)[375]. An important property of silatranes is their physiological activity, e.g., 1-arylsilatranes are highly toxic while 1-alkyl- or 1-alkoxysilanes are physiologically ineffective; a review can be found elsewhere[376].

$$\frac{1}{n}(\mathrm{MeSiO_{1.5}})_n + (\mathrm{HOCH_2CH_2})_3\mathrm{N} \xrightarrow{-\mathrm{H_2O}} \mathrm{MeSi(OCH_2CH_2)_3N} \qquad (336)$$

(**656**) (**654**) (**655a**)

1-Acyloxysilatranes (**658**) can be obtained from 1-ethoxysilatrane (**657**) with carboxylic acids in the presence of the corresponding acid anhydride (equation 337)[377]. Transetherification of **657** by means of triphenylsilanol (**659**) gives rise to 1-(triphenylsiloxy)silatrane (**660**) (equation 338)[377].

$$\mathrm{EtOSi(OCH_2CH_2)_3N} + \mathrm{RCOOH} \xrightarrow[-\mathrm{EtOH}]{(\mathrm{RCO})_2\mathrm{O}} \mathrm{RCO_2Si(OCH_2CH_2)_3N} \qquad (337)$$

(**657**) (**658**)

$$\mathbf{657} + \mathrm{Ph_3SiOH} \longrightarrow \mathrm{Ph_3SiOSi(OCH_2CH_2)_3N} \qquad (338)$$

(**659**) (**660**)

1-Fluorosilatrane (**661**) is obtainable from **657** with hydrogen fluoride (equation 339)[377], and other halogenosilatranes (**663**) by the reaction of 1-hydrosilatrane (**662**) with

$$\mathbf{657} \xrightarrow[-\mathrm{EtOH}]{+\mathrm{HF}} \mathrm{FSi(OCH_2CH_2)_3N} \qquad (339)$$

(**661**)

$$\mathrm{HSi(OCH_2CH_2)_3N} \xrightarrow[\mathrm{HX}]{\mathrm{X_2}} \mathrm{XSi(OCH_2CH_2)_3N} \qquad (340)$$

(**662**) (**663**)

X = Cl, Br, I

the corresponding halogen (equation 340)[378]. 1-Organyl-3-oxosilatranes (**666**) are formed by the reaction of organyltrimethoxysilane (**664**) with [bis-(hydroxyethyl)amino]acetic acid (**665**) (equation 341)[379]. If 1,2-bis(triethoxysilyl)alkanes (**667**) are employed with

$$\mathrm{RSi(OMe)_3} + \mathrm{(HOCH_2CH_2)_2NCH_2COOH} \longrightarrow$$

(**664**) (**665**)

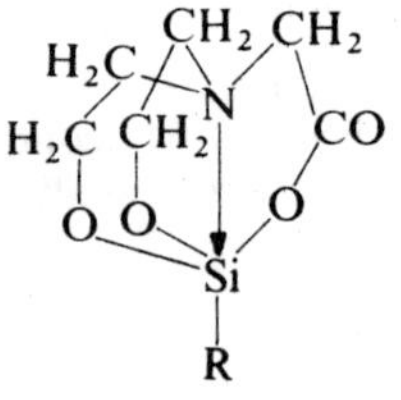

(**666**) (341)

triethanolamine (**654**) in the presence of KOH, 1,2-bis(silatranyl)ethane (**667a**, $n = 1$) and -hexane (**667b**, $n = 3$) are the reaction products (equation 342)[380]. Finally, 3,3-dimethyl-1,5-bis(silatranyl)-3-silapentane (**669**) can be obtained from 1,5-bis(triethoxysilyl)-3,3-dimethyl-3-silapentane (**668**) with **654** (equation 343)[380]. Further compounds are discussed by Voronkov and coworkers[381].

$$(EtO)_3Si(CH_2CH_2)_nSi(OEt)_3 + 2(HOCH_2CH_2)_3N \xrightarrow{KOH}$$

(667) **(654)**

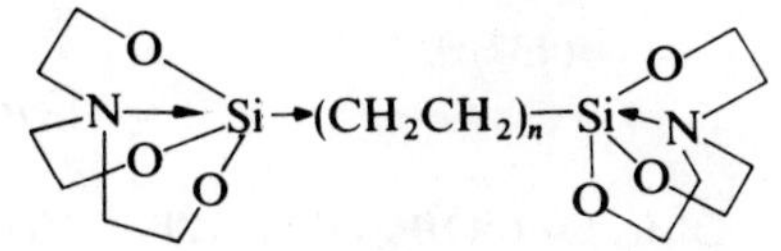

(342)

(667a, b)

(a) $n = 1$
(b) $n = 3$

$$(EtO)_3SiCH_2CH_2{-}SiMe_2{-}CH_2CH_2Si(OEt)_3 + 2\ \mathbf{654} \xrightarrow{KOH}$$

(668)

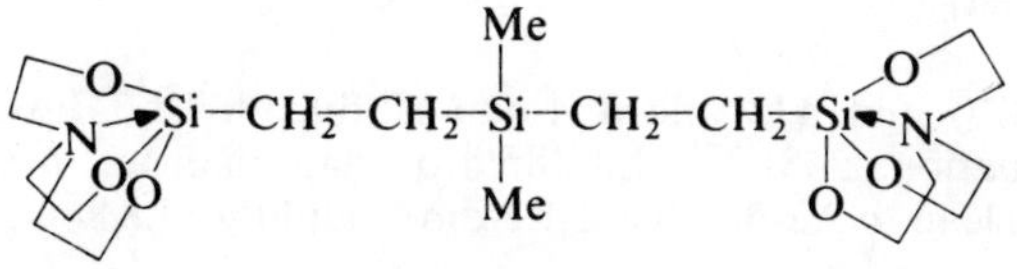

(343)

(669)

2. *Silaoxacycloalkanes*

Aldehydes and ketones with 1,1-dimethyl-2,3-bis(TMS)-silirene (**670**) give silaoxacyclopentene derivatives (**671**) via an insertion reaction (equation 344)[382]. If dimethylsilylene

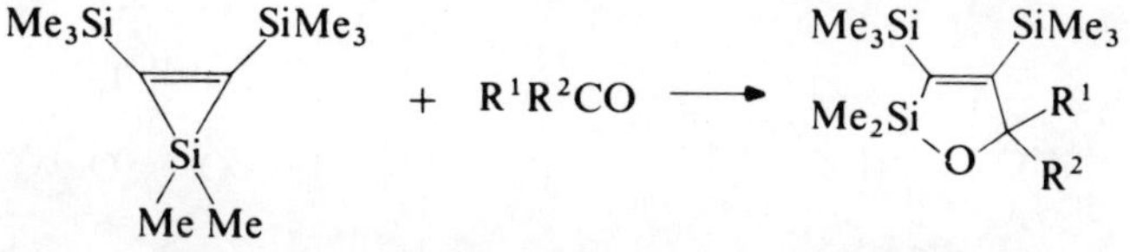

(344)

(670) **(671)**

reacts with oxetane (**672**) in toluene solution, ring expansion yields 2,2-dimethyl-1-oxa-2-silacyclopentane (**673**), together with allyloxydimethylsilane (**674**) (equation 345)[383, 384].

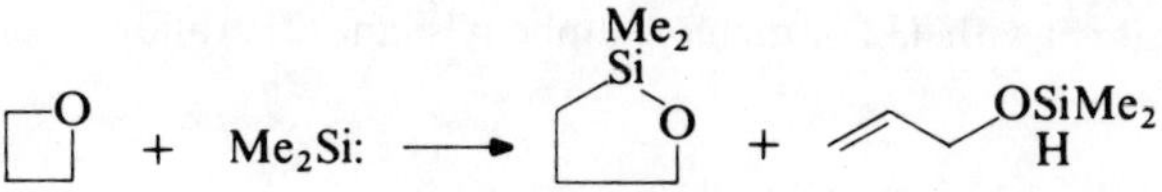

(345)

(672) **(673)** **(674)**

Vicinal diols (**675**) can react with hexamethylcyclotrisilazane (**432**) to form 1,3-dioxa-2-silacyclopentane derivatives (**676**) (equation 346)[385]. Analogously, *o*-dihydroxybenzene

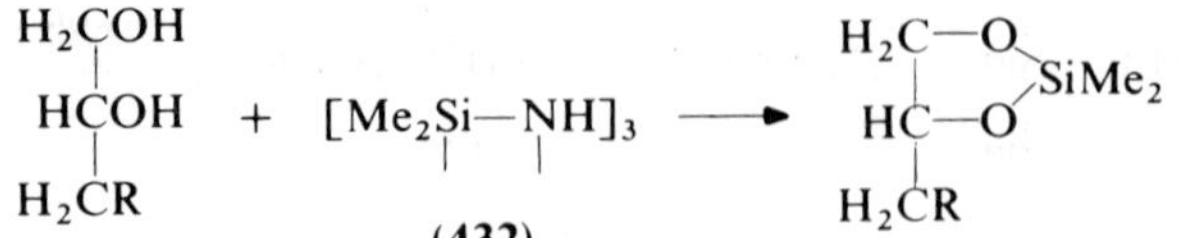

(675) (432) (676) (346)

R = Cl, Br, I, OMe, OEt, OPr, OBu, OPh, OCH_2Ph

(**675a**) and dichlorodimethylsilane yield a benzo-1,3-dioxa-2-silacyclopentane derivative (**676a**) (equation 347)[386].

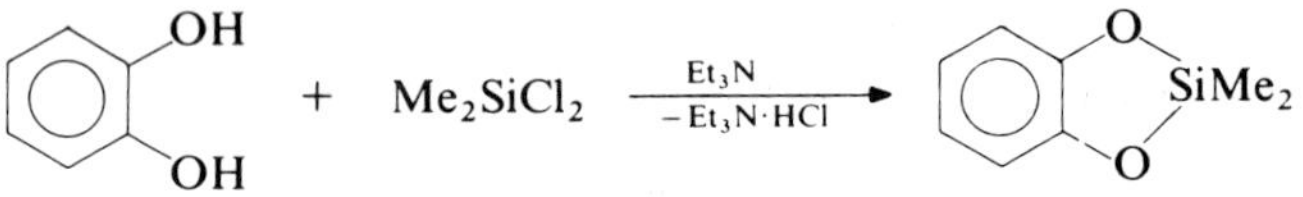

(675a) (676a) (347)

Salicylic acid (**677**) and **432** afford 4-oxo-2,2-dimethyl-1,3-dioxa-2-silabenzocyclohexane (**678**) (equation 348)[387]. 1,8-dihydroxynaphthalene (**679**) and di-*t*-butyldichlorosilane (**680**) lead to 2,2-di-*t*-butyl-1,3-dioxanaphtho[1.8.8*a*–*e*.*f*]-2*H*-2-siline (**681**)

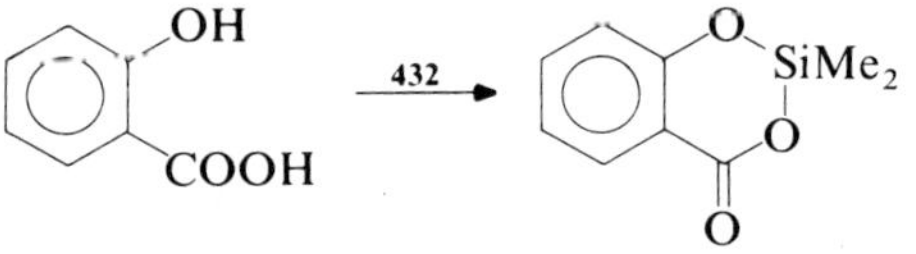

(677) (678) (348)

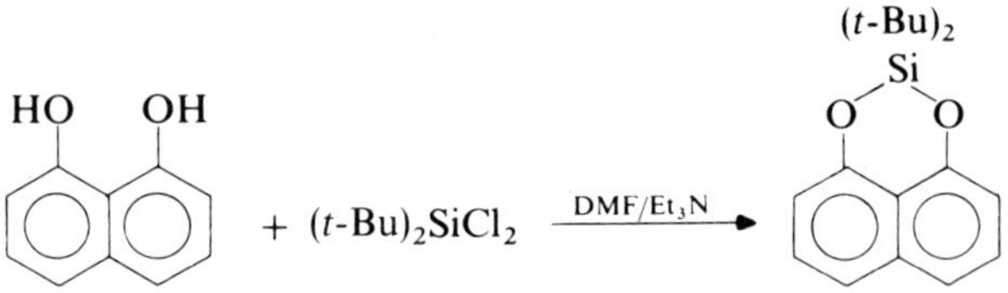

(679) (680) (681) (349)

(equation 349)[388]. Cyanosilylation of a 1,3-diketone leads to the isomeric derivatives of 3,5-dioxa-4-silacyclohex-1-ene (**683, 684**) (equation 350)[389]. 1,2-bis(hydroxymethyl)benzene (**685**) with **432** or dichlorodiphenylsilane (**219**) affords 3,3-dimethyl- (**686**)

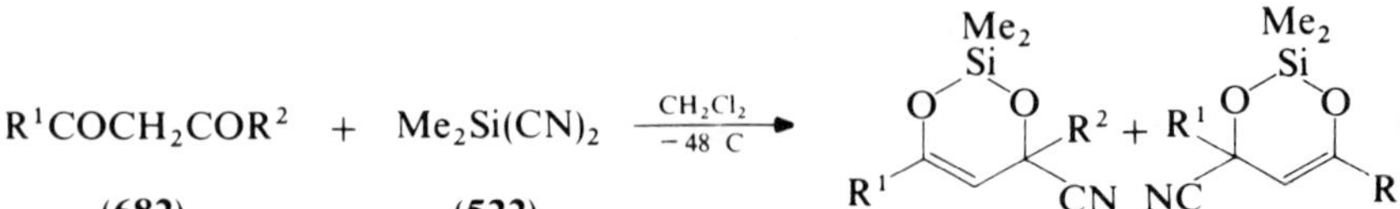

(682) (522) (683) (684) (350)

and 3,3-diphenyl-2,4-dioxa-1,2,4,5-tetrahydro-3H-3-benzosilepine (**687**), respectively (equation 351)[390]. If **685** is treated with tetramethoxysilane (**688**), 3,3′-spiro-bi[2,4-dioxa-1,2,4,5-tetrahydro-3H-3-benzosilepine] (**689**) results (equation 352)[390]. By means of **432**

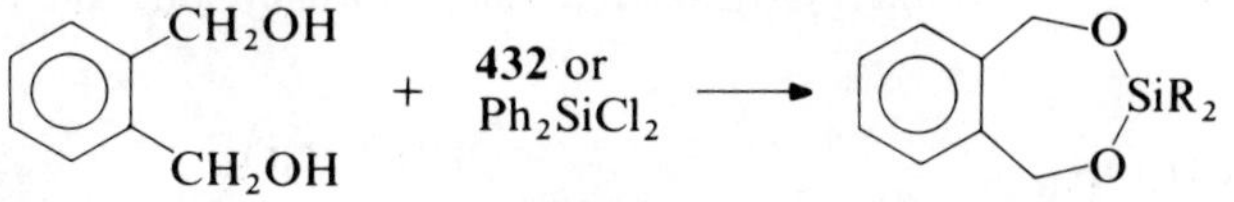

(**219**)

(**685**) (**686**) R = Me, (**687**) R = Ph

(351)

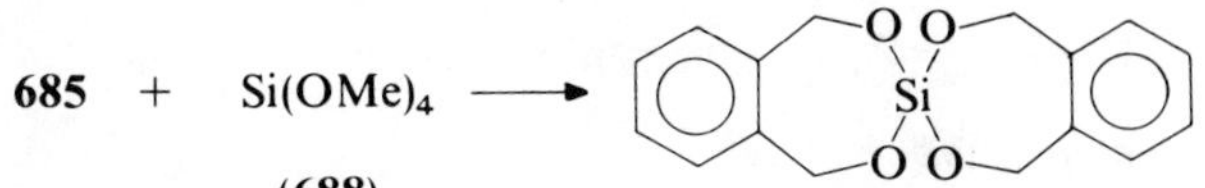

(**688**) (**689**)

(352)

or **219** and 1,2-bis(hydroxymethyl)cyclohexane (**690**) the corresponding 3,3-diorganyl-2,4-dioxaperhydro-3-benzosilepines **691** and **692** are formed (equation 353)[391].

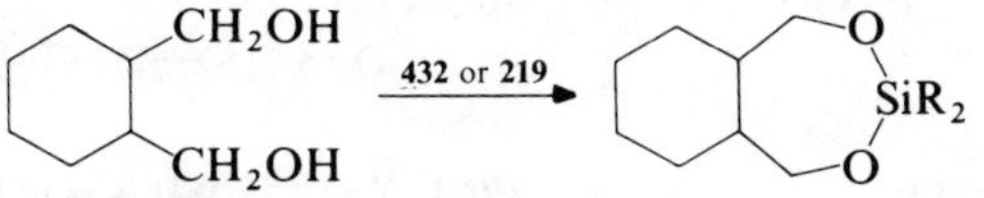

(**690**) (**691**) R = Me, (**692**) R = Ph

(353)

Under analogous conditions, 1,8-bis(hydroxymethyl)naphthalene (**693**) gives rise to 3,3-diorganyl-2,4-dioxa-1,2,4,5-tetrahydronaphtho[1,8,8a–f,g]-3H-3-silocins (**694, 695**) (equation 354)[391], while from 2,2′-bis(hydroxymethyl)diphenyl (**696**) the corresponding

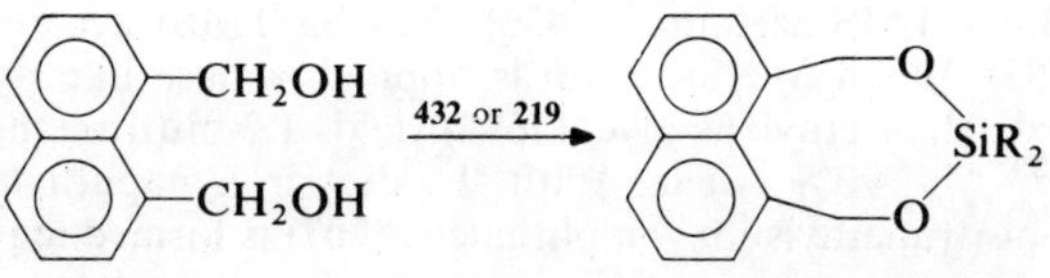

(**693**) (**694**) R = Me, (**695**) R = Ph

(354)

3,3-diorganyl-2,4-dioxa-1,2,4,5-tetrahydrodibenzo[f,h]-3H-3-silonins (**697**, **698**) can be isolated (equation 355)[391].

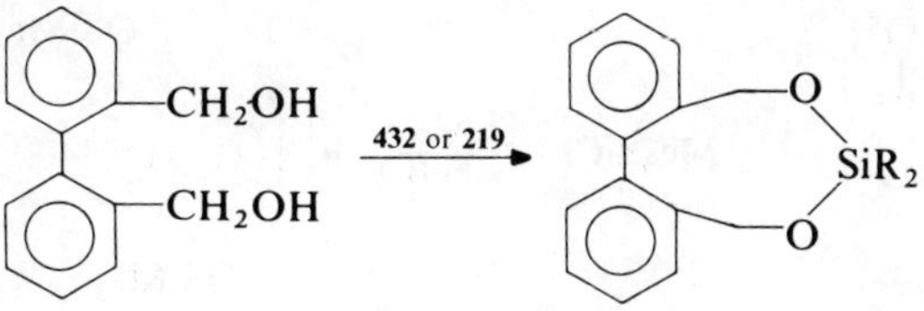

(**696**) (**697**) R = Me, (**698**) R = Ph

(355)

3. *Disilatetroxacycloalkanes*

The reaction of halogenosilanes with polyhydric alcohols leads quite often to oligomers. Thus, the reaction of ethylene glycol (**699**) with dichlorodimethysilane (**117**) affords the ten-membered 2,2,7,7-tetramethyl-1,3,6,8-tetroxa-2,7-disilacyclodecane (**700**) (equation 356)[392].

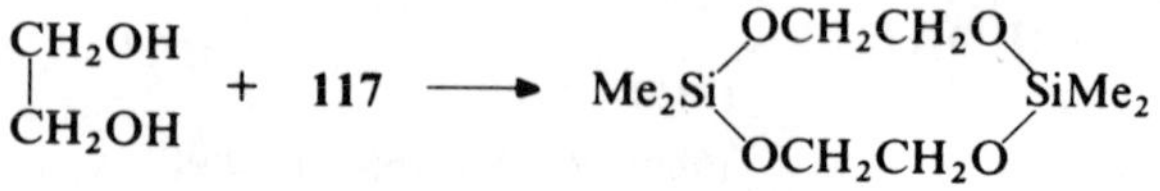

(356)

(**699**) (**700**)

4. *Silaoxaparacyclophanes*

The reaction of 4,4′-dihydroxydiphenyl sulfide (**701**) and -sulfone (**702**) with **432** yields the corresponding silaoxaparacyclophanes (**703**) and (**704**), respectively (equation 357)[393].

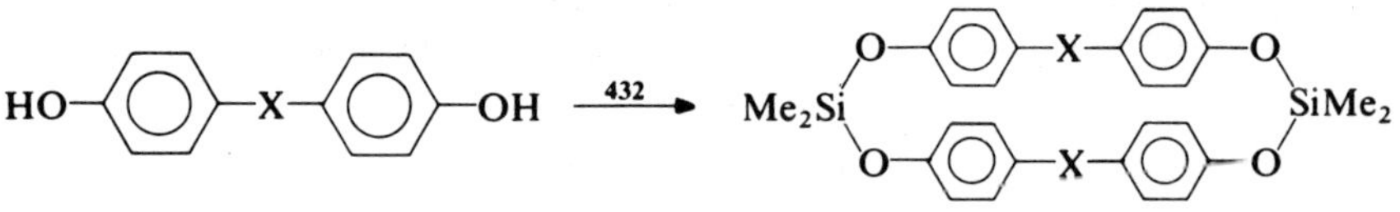

(357)

(**701**) X = S, (**702**) X = SO_2 (**703**) X = S, (**704**) X = SO_2

G. Silylation Techniques

For the silylation of polyols, halogenosilanes are rarely applied in order to prevent side-reactions so a variety of other silylating agents are in use. These include hexamethyldisilazane (**405**), *N*-TMS-acetamide (**435**), *N,O*-bis(TMS) acetamide (**438**) and TMS-diethylamine (**463**). When Me_3SiCl (**16**) is applied, a base like pyridine as HCl acceptor has to be used. Thus, ethylene glycol (**699**) yields 1,2-bis(trimethylsiloxy)ethane (**705**) (equation 358)[394] in 90% yield. With 1,5-dihydroxynaphthalene (**706**) and Me_3SiCl/pyridine 1,5-bis(trimethylsiloxy)naphthalene (**707**) is formed (equation 359)[395],

$$\mathbf{699} + Me_3SiCl \xrightarrow[-Py.HCl]{Py} Me_3SiOCH_2CH_2OSiMe_3 \quad (358)$$

(**705**)

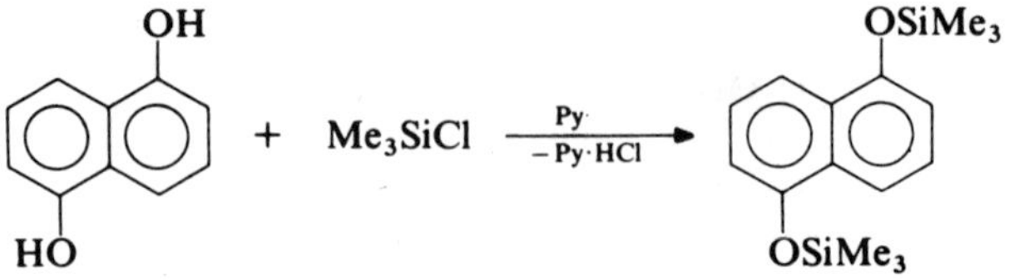

(359)

(**706**) (**707**)

whereas 1-hydroxynaphthalene (**708**) plus Me_2SiCl_2/pyridine yields dimethyl bis(1-naphthoxy)silane (**709**) (equation 360)[395].

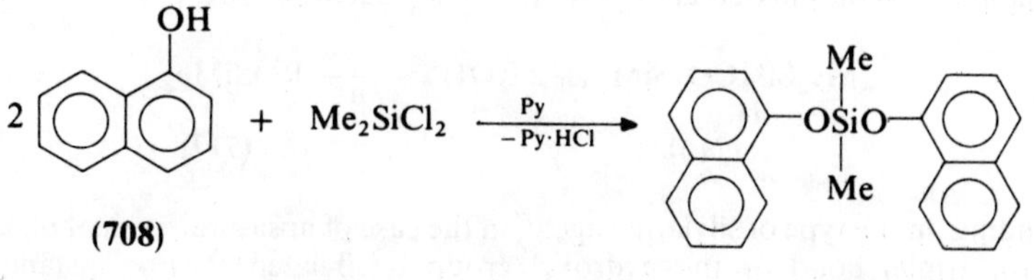

(708) (709) (360)

435 is an excellent silylating agent for carbohydrates such as glucose which, in pyridine solution, forms the tetrakis-TMS-derivative (**711**), but in molten state, the pentakis-TMS-derivative (**712**) (equation 361)[396]. With vicinal diols like **675**, **405** forms the corresponding bis (TMS) compounds (**713**) (equation 362)[385, 397]. For cycloaliphatic diols, such as

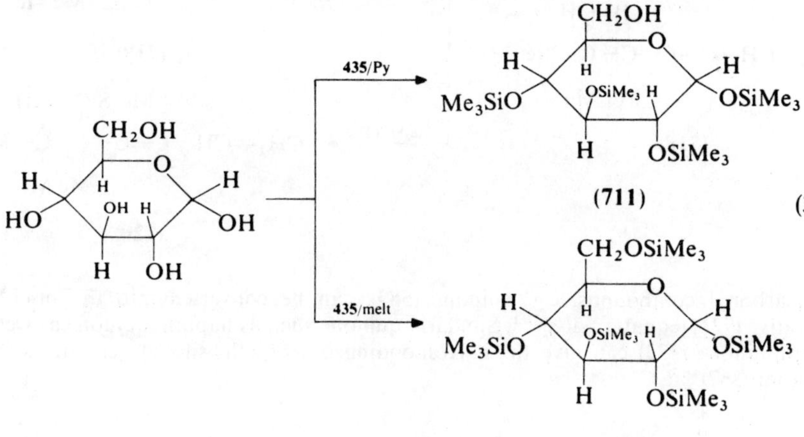

(711) (712) (361)

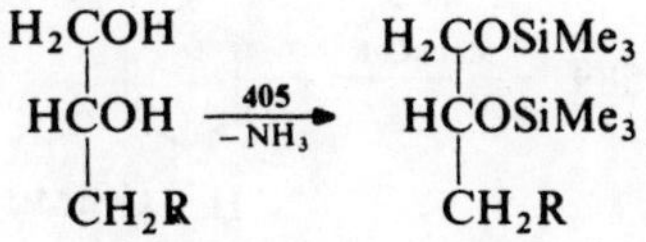

(675) (713) (362)

1,2-dihydroxycyclohexane (**714**), bis(TMS)-acetamide (**438**) is well suited to form the corresponding 1,2-bis(trimethylsiloxy)cyclohexane (**715**) (equation 363)[398].

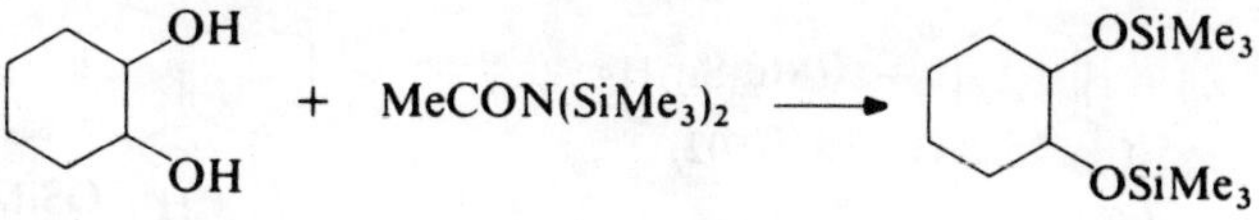

(714) (438) (715) (363)

Another facile silylating reagent for hydroxyl groups of alcohols, phenols and carboxylic acids is *N,O*-bis(TMS)-carbamate (**716**) which gives yields between 85–95%, and only ammonia and carbon dioxide are the by-products (equation 364)[399].

$$\underset{\textbf{(716)}}{Me_3Si\underset{H}{N}CO_2SiMe_3} + 2\,ROH \xrightarrow[-CO_2]{-NH_3} \underset{\textbf{(717)}}{ROSiMe_3} \qquad (364)$$

Depending on the type of silylating agent, in the case of unsaturated alcohols either the double (or triple) bond or the hydroxyl group is silylated[398]. For instance, if 2,3-dihydroxy-2,3-dimethylhept-6-en-4-yne (**718**) is treated with triethylsilane, then C-silylation to **719** occurs, whereas the combination Me_3SiCl/pyridine leads to *O*-silylation (**720**) (equation 365)[400].

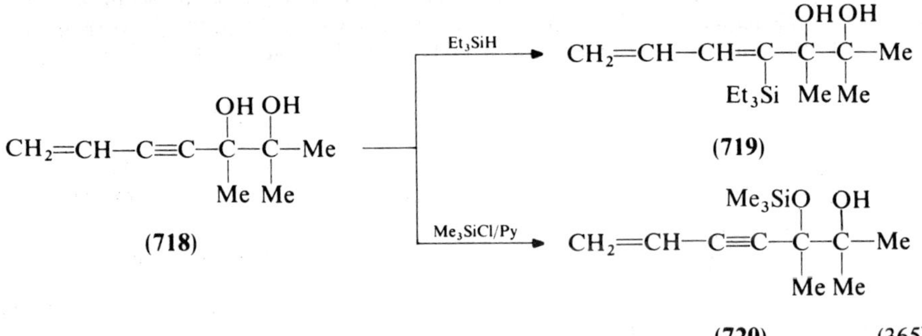

(365)

Dicarbonyl compounds, e.g. quinone (**589**), can be converted into the bis(TMS) derivative (**721**) (equation 366)[401]. Similarly quinone such as naphthoquinone as well as anthraquinone (**722**) can give the corresponding bis(trimethylsiloxy) derivatives (**723**) (equation 367)[402].

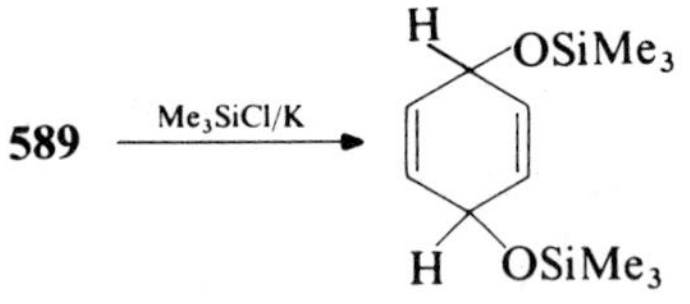

(721) (366)

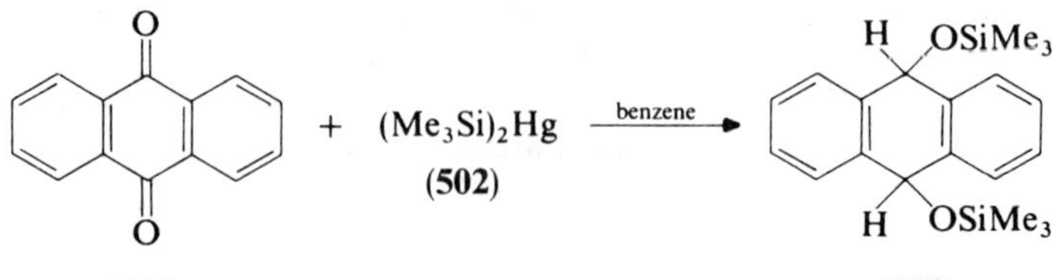

(722) (502) (723) (367)

XII. SULPHUR-, SELENIUM- AND TELLURIUM-CONTAINING ORGANOSILICON COMPOUNDS

A. Introduction

In this section the synthesis of compounds containing the –C–Si–Y– (Y = S, Se, Te) moiety is discussed. Literature published after 1980 predominates. For earlier literature it may be advisable to refer to previously published reviews[403–406].

B. Organosilicon–Sulphur Compounds

1. *Organylsilanethiols*

For the synthesis of organylsilanethiols R_3SiSH (**725**) a common approach is to employ chloro- or bromosilanes (**724**) with hydrogen sulphide in the presence of tertiary amines as hydrogen halide acceptor[407–411] (equation 368). Without an HX acceptor no reaction occurs; even $SiCl_4$ reacts with H_2S only above 500 °C to yield trichlorosilanethiol[412] (**726**) (equation 369).

$$\underset{(\mathbf{724})}{R_3SiX} + H_2S \xrightarrow[{[R'_3NH]^+X^-}]{R'_3N} \underset{(\mathbf{725})}{R_3SiSH} \qquad X = Cl,\ Br \tag{368}$$

$$SiCl_4 + H_2S \xrightarrow{500\,^\circ C} HCl + \underset{(\mathbf{726})}{Cl_3SiH} \tag{369}$$

An alternative synthesis of (**725**) is the reaction of aminosilanes with hydrogen sulphide[413, 414] (equation 370) or that of trialkylsilanes with hydrogen sulphide catalyzed by Pd/Al_2O_3 (equation 371)[415]. Arylsilanethiols (**728**) are formed when a triarylsilane is treated with sulphur in an inert solvent (e.g. in decalin under reflux) (equation 372)[416, 417]

$$\begin{gathered} R_3SiNH_2 + H_2S \xrightarrow[-NH_3]{} R_3SiSH \\ \underset{(\mathbf{727})}{Me_3SiNHPh} + H_2S \rightarrow \underset{(\mathbf{725a})}{Me_3SiSH} + PhNH_2 \end{gathered} \tag{370}$$

$$R_3SiH + H_2S \xrightarrow[-H_2]{Pd/Al_2O_3} \mathbf{725} \tag{371}$$

$$Ar_3SiH + S \longrightarrow \underset{(\mathbf{728})}{Ar_3SiSH} \tag{372}$$

In contrast to the afore-mentioned types of compounds, unsubstituted silanethiol (**730**) can be obtained either by treatment of disilathiane (**729**) (for preparation of **729**; see Section XII, B.2 below) with H_2S[418] (equation 373) or via the reaction of chlorosilane (**731**) with alkali- or magnesium hydrogen sulphides (equation 374)[419].

$$\underset{(\mathbf{729})}{H_3SiSSiH_3} + H_2S \longrightarrow \underset{(\mathbf{730})}{2\ H_3SiSH} \tag{373}$$

$$\underset{(\mathbf{731})}{H_3SiCl} + NaSH \xrightarrow[-NaCl]{} \mathbf{730} \qquad (374)$$

2. Disilathianes

a. Linear disilathianes. Hexaorganodisilathianes $R_3SiSSiR_3$ (**733**) can be obtained via the reaction of sodium silanethiolates with chlorosilanes[418, 420, 421] (equation 375) or via

$$\underset{(\mathbf{732})}{R_3SiSNa} + R_3SiX \xrightarrow[NaX]{} \underset{(\mathbf{733})}{R_3SiSSiR_3} \qquad (375)$$

the reaction of trialkylaminosilane (**734**) or alkoxydialkylchlorosilane (**735**) with hydrogen sulphide[422, 423] (equations 376a and b). Also, the direct application of hydrogen

$$3\ \underset{(\mathbf{734})}{R_3SiNH_2} + 2\ H_2S \xrightarrow[-3NH_3]{} \mathbf{733} \qquad (376a)$$

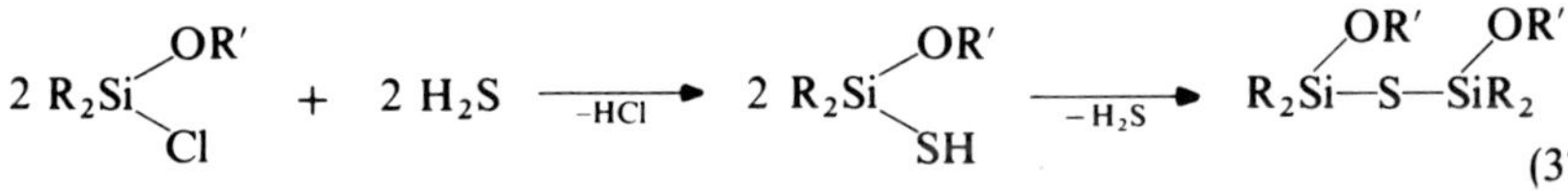

(376b)

(**735**) (**736**) (**737**)

sulphide to a solution of (1-triorganylsiloxyethylideneamino)triorganylsilane (**738**) in petroleum ether leads to (**733**)[424] (equation 377). Furthermore, **739** can be obtained by

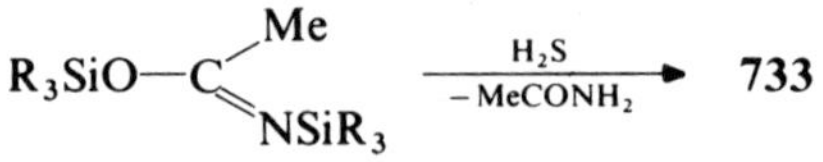

(377)

(**738**)

means of hydrogen sulphide and hexamethyldisilazane (**405**) using imidazole as a catalyst[425]. Mechanistically, it can be expected that as a reactive intermediate, *N*-trimethylsilylimidazole is being formed, which reacts with **739** by addition of H_2S (equation 378).

(**274**) (**739**) (378)

Another route for the preparation of **739** is the reaction of chloromethylsilane, with either Na_2S[426], NH_4HS[427] or Li_2S[419] (equation 379). By means of mercury[428] or silver

$$2\ Me_3SiCl + Na_2S \longrightarrow \underset{(\mathbf{739})}{Me_3SiSSiMe_3} \qquad (379)$$

sulphide[429], iodotriorganylsilanes can be converted into the corresponding disilathianes **733** (equation 380).

$$R_3SiI + HgS(Ag_2S) \xrightarrow[(-AgI)]{-HgI_2} \mathbf{733} \qquad (380)$$

Non-symmetrically substituted hexaorganyldisilathianes can be synthesized from halotriorganylsilanes and sodium triorganylsilanethiolate (equation 381)[420]. A convenient way to accomplish the synthesis of symmetrical disilathianes (**733**) with bulky

$$Me_3SiCl + Ph_3SiSNa \xrightarrow[-NaCl]{} Me_3SiSSiPh_3 \qquad (381)$$

moieties is the reaction of the corresponding tetraorganylsilane (**740**) with sulphur, leading in the first step to the corresponding alkylthiosilane (**741**) which, at higher temperatures, forms **733**[430] (equations 382 and 383).

$$\underset{(\mathbf{740})}{R_4Si} + S \longrightarrow \underset{(\mathbf{741})}{R_3SiSR} \qquad (382)$$

$$2\ \mathbf{741} \longrightarrow \underset{(\mathbf{733})}{R_3SiSSiR_3} \qquad (383)$$

$R = n\text{-Bu}$

b. Cyclic silathianes. If dihalogenodialkylsilanes (**742**) are treated with H_2S, an equilibrium of a six-membered hexaalkylcyclotrisilathiane (**743**) and a four-membered te-

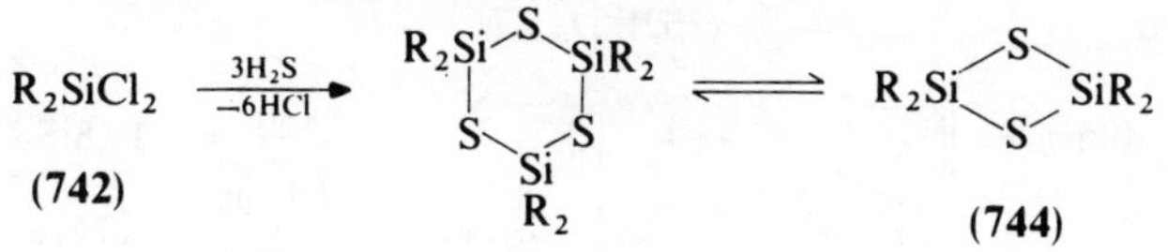

(384)

(**742**) (**743**) (**744**)

traalkylcyclodisilathiane (**744**)[431] results (equation 384). 2,2,6,6-Tetraalkyl-1-thia-2,6-disilacyclohexane (**746**) is formed from 1,3-bis(halogenodialkylsilyl)propane (**745**) with H_2S in pyridine[432] (equation 385). The reaction of dichlorophenylmethylsilane (**747**) with

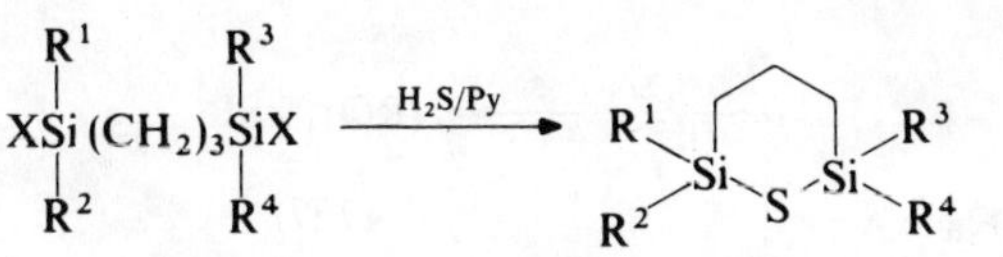

(385)

(**745**) X = Cl, Br (**746**)

1,2-ethanedithiol (**748**) in an inert solvent (e.g. benzene) in the presence of triethylamine gives rise to 2-phenyl-2-methyl-1,3-dithia-2-silacyclopentane (**749**)[433] (equation 386).

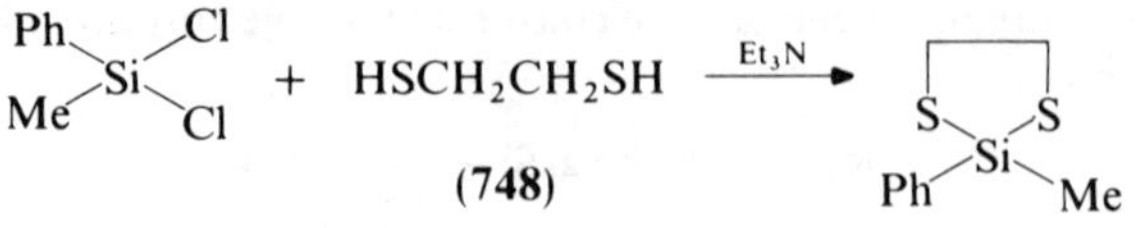

(747) (748) (749) (386)

c. *Bis(triorganylsilyl)di- and polysulphanes.* This class of compounds (**751**) can be easily synthesized by treating sodium triorganylsilanethiolate with the corresponding dichloro-di- and-polysulphanes (**750**) (equation 387)[434].

$$2\,R_3SiSNa + ClS_nCl \xrightarrow[-2NaCl]{} R_3SiS_nSiR_3 \qquad (387)$$

(**750**) (**751**)

3. *Alkylthioorganylsilanes*

Triorganylsilanes in a one-step reaction with a thiol in the presence of a rhodium catalyst yield alkylthiosilanes (**741a**)[435] (equation 388). Halosilanes and thiols in the

$$R^1_3SiH + R^1SH \xrightarrow[-H_2]{[Ph_3P]_3RhCl} R^2_3SiSR^2 \qquad (388)$$

(**752**) (**741a**)

presence of an amine also yield **741**-type compounds[410] (equation 389). **741a** is also obtained when hexamethyldisilazane (**405**) reacts with a thiol using imidazole as a catalyst (which probably forms a reactive intermediate) (equation 390)[425].

$$R_3SiX + R'SH \xrightarrow[{[R^3_3NH]^+X^-}]{R^3_3N} R_3SiSR' \qquad (389)$$

(**752**)

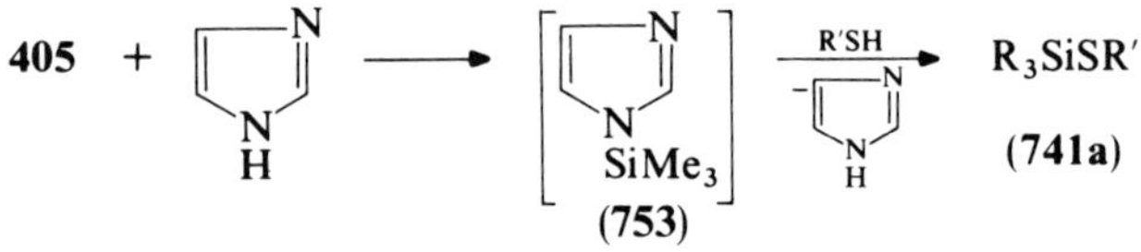

(390)

4. *Alkylthioalkoxysilanes*

The formation of alkylthioalkoxysilanes (**757/758**) is achieved via the conversion of halogenotrialkoxysilanes (**755**) or dihalogenodialkoxysilanes (**756**) with sodium alkylthiolate (**754**)[433] (equation 391).

$$R'SNa\ (\mathbf{754}) \xrightarrow[THF/5\,h,\ \Delta]{(RO)_3SiX\ (\mathbf{755})} (RO)_3SiSR'\ (\mathbf{757}) \qquad (391)$$

$$R'SNa\ (\mathbf{754}) \xrightarrow[THF/5\,h,\ \Delta]{(RO)_2SiX_2\ (\mathbf{756})} (RO)_2Si(SR')_2\ (\mathbf{758})$$

X = Cl, R = *i*-Pr, R′ = *t*-Bu

5. *Tetrakis(alkylthio)silanes*

The reaction of a thiol with tetrachlorosilane yields tetrakis(alkylthio)silanes (**759**)[433] (equation 392).

$$RSH + SiCl_4 \xrightarrow[-[Et_3NH]^+Cl^-]{Et_3N} \underset{\substack{\mathbf{(759)} \\ R = Et}}{Si(SR)_4} \qquad (392)$$

6. *Alkenylthioorganylsilanes*

A common reaction pathway to alkenylthioorganylsilanes (**759a**) is the radical addition of a trialkylsilanethiol to an alkyne[436] (equation 393).

$$R_3SiSH + HC{\equiv}CR' \longrightarrow \underset{\mathbf{(759a)}}{R_3SiCH{=}CHR'} \qquad (393)$$

7. *Alkylthiotriarylsilanes*

A direct route to alkylthiotriarylsilanes (**761**) is the treatment of a lithiated triarylsilane (**760**) with an alkyl halide in the presence of sulphur[437] (equation 394) or of sodium triarylsilanethiolate (**762**) with an alkyl halide (equation 395).

$$\underset{\mathbf{(760)}}{Ar_3SiLi} + RX + S \xrightarrow[-LiX]{} \underset{\mathbf{(761)}}{Ar_3SiSR} \qquad (394)$$

$$\underset{\mathbf{(762)}}{Ar_3SiSNa} + RX \xrightarrow[-NaX]{} \underset{\mathbf{(761)}}{Ar_3SiSR} \qquad (395)$$

8. *Acylthioorganylsilanes and related compounds*

Acylthiotriorganylsilanes (**764**) can be synthesized from alkali triorganylsilanethiolates (**763**) with acyl chlorides[437] (equation 396). In contrast, if chlorotriorganylsilanes are

$$\underset{\mathbf{(763)}}{R_3SiS^{\ominus}\,M^+} + R'COCl \xrightarrow[-MCl]{} \underset{\mathbf{(764)}}{R_3SiSCOR'} \qquad (396)$$

employed with ammonium dithiocarboxylates (**765**), thioacylthiotriorganylsilanes (**766**) can be isolated[438] (equation 397). Of some interest for the synthetic chemist may be

$$R_3SiCl + \underset{\mathbf{(765)}}{\left[R'C(=S){-}S\right]^- NH_4^+} \xrightarrow[X\,=\,Cl]{-NH_4^+X^-} \underset{\mathbf{(766)}}{R_3SiC(=S){-}R'} \qquad (397)$$

aminothiocarbonylthioorganylsilanes (**768**), which are the reaction products when dialkylaminotriorganylsilanes (**767**) are treated with carbon disulfide in an insertion-type reaction[439] (equation 398).

$$R^1_3SiNR^2_2 + CS_2 \longrightarrow R^1_3SiS{-}\underset{\underset{S}{\|}}{C}{-}NR_2 \qquad (398)$$

(767) **(768)**

$R^1 = Me, R^2 = Et$

C. Organosilicon–Selenium Compounds

Relatively few reactions and few compounds of this type have been reported so far. Some representative examples are shown below.

1. Linear compounds

In a one-step reaction triethylsilane and selenium form triethylsilaneselenol **(769)**[440] (equation 399). The related hexamethyldisilaselenane **(770)** can be obtained from chlorotrimethylsilane and disodium selenide[441] (equation 400). Alkylselenoorganylsilanes **(772)** are formed when halogenoorganylsilanes react with methylselenol **(771)** in the presence of tertiary amines[422] (equation 401) or via cleavage of a Si–N bond by **771** (equation 402)[422].

$$Et_3SiH + Se \xrightarrow{215\,°C} Et_3SiSeH \qquad (399)$$

(769)

$$2\,Me_3SiCl + Na_2Se \xrightarrow[-2NaCl]{} Me_3SiSeSiMe_3 \qquad (400)$$

(770)

$$R_3SiX + MeSeH \xrightarrow[R_3NH^+X^-]{R'_3N} R_3SiSeMe \qquad (401)$$

(771) **(772)**

$$R_3SiNH_2 + \mathbf{771} \longrightarrow \mathbf{772} \qquad (402)$$

(734)

$R = Me$

2. Cyclic compounds

Cyclic silicon–selenium compounds can be prepared by an insertion-type reaction of a cyclic organosilicon compound with selenium; e.g., 2,2-dibutyl-1-selena-2-silacyclopentane **(774)** is the product of the reaction of 1,1-dibutylsilacyclobutane **(773)** with selenium (equation 403)[443].

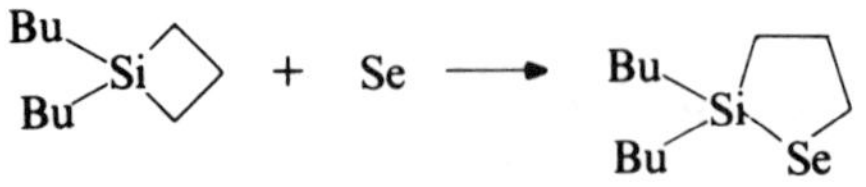

(403)

(773) **(774)**

D. Organosilicon–Tellurium Compounds

When trialkylsilanes are heated with elementary tellurium, hexalkyldisilatelluranes **(775)**[440] are formed (equation 404). An alternative route to **775** is the reaction of dilithium

$$2\,R_3SiH + Te \xrightarrow[-H_2]{280\,^\circ C/22\,h} (R_3Si)_2Te \quad (775) \tag{404}$$

$$2\,R_3SiX + Li_2Te \xrightarrow[-2\,LiX]{} 775 \tag{405}$$

(776)

$$(R_3Si)_2Te \xrightarrow{CF_3COOH/20\,^\circ C} R_3SiTeH \tag{406}$$

(775) (776)

R = Et

telluride (**776**) with halogenotriorganylsilanes[419] (equation 405). A trialkylsilanetellurol (**776**) is obtained when a strong acid, such as trifluoroacetic acid, is employed with hexaalkyldisilatellurane (**775**)[440] (equation 406). Phenyltrimethylsilyl telluride (**777**) can be synthesized when phenyltellurium magnesium bromide is treated with a halogenotrimethylsilane; however, not only **777** is formed but also **775** and 1-trialkylsiloxy-4-phenyltellurabutane (**778**)[444] (equation 407).

$$PhTeMgBr + Me_3SiCl \longrightarrow \underset{(777)}{Me_3SiTePh} + \underset{(778)}{Me_3SiO(CH_2)_4TePh} + 775 \tag{407}$$

XIII. ACKNOWLEDGEMENT

The authors wish to thank Mrs. Vera Foremny and Ms. Annemarie Wurm for their excellent assistance in the preparation of the manuscript.

XIV. REFERENCES

1. C. Eaborn, *Organosilicon Compounds*, Butterworths, London, 1960; Academic Press, New York, 1960.
2. L. Birkofer and A. Ritter, *Angew. Chem.*, **77**, 414 (1965); *Angew. Chem., Int. Ed. Engl.*, **4**, 417 (1965).
3. L. H. Sommer, *Stereochemistry, Mechanism and Silicon*, McGraw-Hill, New York, 1965.
4. A. D. Petrov, B. F. Mironov, V. A. Ponomarenko and E. A. Chernyshev, *Synthesis of Organosilicon Monomers*, Heywood, London, 1964; Consultant Bureau, New York, 1964.
5. K. Rühlmann, *Z. Chem.*, **5**, 130 (1965).
6. V. Bažant, V. Chvalovsky and J. Rathouský (Eds.), *Organosilicon Compounds*, Academic Press, New York, 1965.
7. V. Bažant, V. Chvalovský and J. Rathouský (Eds.), *Handbook of Organosilicon Compounds*, Marcel Dekker, New York, 1973.
8. W. Noll (Ed.), *Chemistry and Technology of Silicones*, 2nd edn., Verlag Chemie, Weinheim, 1967; Academic Press, New York, 1968.
9. A. E. Pierce, *Silylation of Organic Compounds*, Pierce Chemical Co, Rockford Ill., 1968.
10. J. F. Klebe, *Acc. Chem. Res.*, **3**, 299 (1970).
11. S. S. Washburne, *J. Organomet. Chem.*, **83**, 155 (1974).
12. I. Fleming, *Chem. Ind. (London)*, 449 (1975).
13. S. S. Washburne, *J. Organomet. Chem.*, **123**, 1 (1976).
14. (a) D. Seyferth and R. B. King (Eds.), *Organometallic Chemistry Reviews*, Annual Surveys, Vols. 1–11, 13, Elsevier, Amsterdam, 1976–1981.
    (b) R. B. King and J. P. Oliver (Eds.), *Organometallic Chemistry Reviews*, Annual Surveys, Vols. 14, 17, Elsevier, Amsterdam, 1984, 1985.

15. C. Eaborn and R. W. Bott, in *Organometallic Compounds of the Group IV Elements*, Vol. 1, Part. I (Ed. A. G. MacDiarmid), Marcel Dekker, New York, 1968, pp. 105–536.
16. E. W. Colvin, *Chem. Soc. Rev.*, **7**, 15 (1978).
17. S. Pawlenko, in *Methoden der organischen Chemie* (*Houben-Weyl*), Vol. 13, Part 5 (Eds. E. Müller and O. Bayer), Georg Thieme Verlag, Stuttgart, 1980, pp. 1–502.
18. L. Birkofer and O. Stuhl, in *Topics in Current Chemistry*, Vol. 88 (Ed. F. L. Boschke), Springer-Verlag, Berlin, Heidelberg, New York, 1980, pp. 33–88.
19. R. Calas, *J. Organomet. Chem.*, **200**, 11 (1980).
20. (a) J. Y. Corey, *J. Organomet. Chem.*, **313**, 1 (1986).
(b) G. L. Larson, *J. Organomet. Chem.*, **313**, 141 (1986).
21. H. Gilman and R. N. Clark, *J. Am. Chem. Soc.*, **69**, 967 (1947).
22. H. Gilman and R. N. Clark, *J. Am. Chem. Soc.*, **68**, 1675 (1946).
23. J. S. Peake, W. H. Nebergall and Y. T. Chen, *J. Am. Chem. Soc.*, **74**, 1526 (1952).
24. H. Gilman and R. N. Clark, *J. Am. Chem. Soc.*, **69**, 1499 (1947).
(a) M. Weidenbruch, W. Schiffer, G. Hägele and W. Peters, *J. Organomet. Chem.*, **90**, 145 (1975).
25. W. H. Nebergall and O. H. Johnson, *J. Am. Chem. Soc.*, **71**, 4022 (1949).
26. A. D. Petrov and T. J. Chernysheva, *Zh. Obshch. Khim.*, **24**, 1189 (1954); *Chem. Abstr.*, **49**, 12276 (1955).
27. J. Dunoguès, J. P. Pillot, N. Duffaut, N. Ardoin and R. Calas, *J. Organomet. Chem.*, **82**, 217 (1974).
28. F. C. Whitmore, L. H. Sommer, P. A. DiGiorgio, W. A. Strong, R. E. van Strien, D. L. Bailey, H. K. Hall, E. W. Pietrusza and G. T. Kerr, *J. Am. Chem. Soc.*, **68**, 475 (1946).
29. J. M. Reuter, A. Sinha and R. G. Salomon, *J. Org. Chem.*, **43**, 2438 (1978).
30. T. Sasaki, K. Shimizu and M. Ohno, *Synth. Commun.*, **14**, 853 (1984).
31. W. Adcock, G. L. Aldous and W. Kitching, *J. Organomet. Chem.*, **202**, 385 (1980).
32. I. Matsuda, S. Murata and Y. Ishii, *J. Chem. Soc., Perkin Trans. 1*, 26 (1979).
33. (a) E. Lukevics, Z. V. Belyakova, M. G. Pomerantseva and M. G. Voronkov, *J. Organomet. Chem. Library*, **5**, 1 (1977).
(b) J. L. Speier, *Adv. Organomet. Chem.*, **17**, 407 (1979).
(c) G. K. I. Magomedov, K. A. Andrianov, O. V. Shkolonik and B. A. Izmailov, *J. Organomet. Chem.*, **149**, 29 (1978).
34. L.-Z. Wang and Y.-Y. Jiang, *J. Organomet. Chem.*, **251**, 39 (1983).
35. R. Rumin, *J. Organomet. Chem.*, **247**, 351 (1983).
36. M. Green, J. L. Spencer, F. G. A. Stone and C. A. Tsipis, *J. Chem. Soc., Dalton Trans.*, 1519 (1977).
37. G. Kuncova and V. Chvalovský, *Collect. Czech. Chem. Commun.*, **45**, 2240 (1980).
38. V. V. Kaverin, I. M. Salimgareeva, I. N. Brui, F. F. Kayumov, A. V. Kuchin, A. Yu. Spivak and V. P. Yur'ev, *Izv. Akad. Nauk SSSR, Ser. Khim.*, 2657 (1980); *Chem. Abstr.*, **95**, 7464 z (1981).
39. D. J. Ager, *Chem. Soc. Rev.*, **11**, 493 (1982).
40. DOS 2210189 (1973), Union Carbide Corp. (Inv. Y. P. Mui and E. W. Bennet); *Chem. Abstr.*, **79**, 146 646 (1973).
41. E. A. Chernyshev, V. F. Mironov and A. D. Petrov, *Izv. Akad. Nauk SSSR, Otdel Khim. Nauk*, 2147 (1960); *Chem. Abstr.*, **55**, 14345 (1961).
42. E. T. McBee, C. W. Roberts and G. W. R. Puerckhauer, *J. Am. Chem. Soc.*, **79**, 2326 (1957).
43. M. F. Shostakovskii and D. A. Kochkin, *Izv. Akad. Nauk SSSR, Otdel. Khim. Nauk*, 1150 (1956); *Chem. Abstr.*, **51**, 4935 (1957).
44. R. A. Benkeser and D. F. Ehler, *J. Organomet. Chem.*, **69**, 193 (1974).
45. S. D. Rosenberg, J. J. Walburn, T. D. Stankovich, A. E. Balint and H. E. Ramsden, *J. Org. Chem.*, **22**, 1200 (1957).
46. R. Nagel and H. W. Post, *J. Org. Chem.*, **17**, 1379 (1952).
47. G. Courtois and P. Miginiac, *Bull. Soc. Chim. Fr. II*, 395 (1982).
48. C. Huynh and G. Linstrumelle, *Tetrahedron Lett.*, 1073 (1979).
49. A. Minato, K. Suzuki, K. Tamao and M. Kumada, *Tetrahedron Lett.*, **25**, 83 (1984).
50. A. D. Petrov, V. F. Mironov and V. G. Glukhovtsev, *Izv. Akad. Nauk SSSR, Otdel. Khim. Nauk, Nauk*, 461 (1956); *Chem. Abstr.*, **50**, 16663 (1956).
51. P. F. Hudrlik A. K. Kulkarni, S. Jain and A. M. Hudrlik, *Tetrahedron*, **39**, 877 (1983).
52. G. Fritz and W. Himmel, *Z. Anorg. Allg. Chem.*, **448**, 40 (1979).

53. R. Münstedt, U. Wannagat and D. Wrobel, *J. Organomet. Chem.*, **264**, 135 (1984).
54. I. Fleming and A. Pearce, *J. Chem. Soc., Perkin Trans. 1*, 2485 (1980).
55. J. Dunoguès, J. P. Pillot, N. Duffaut and R. Calas, *C.R. Acad. Sci. Paris, Ser. C.*, 467 (1974).
56. W. Mychajlowskij and T. H. Chan, *Tetrahedron Lett.*, 4439 (1976).
57. T. H. Chan, A. Baldassar re and D. Massuda, *Synthesis*, 801 (1976).
58. R. T. Taylor, C. R. Degenhardt, W. P. Melega and L. A. Paquette, *Tetrahedron Lett.*, 159 (1977).
59. A. R. Chamberlin, Y. E. Stemke and F. T. Bond, *J. Org. Chem.*, **43**, 147 (1978).
60. W. E. Fristad, T. R. Bailey and L. A. Paquette, *J. Org. Chem.*, **43**, 1620 (1978).
61. L. A. Paquette, W. E. Fristad, D. S. Dime and T. R. Bailey, *J. Org. Chem.*, **45**, 3017 (1980).
62. H. Watanabe, M. Asami and Y. Nagai, *J. Organomet. Chem.*, **195**, 363 (1980).
63. R. A. Benkeser, M. L. Burrous, L. E. Nelson and J. V. Swisher, *J. Am. Chem. Soc.*, **83**, 4385 (1961).
64. L. Birkofer and T. Kühn, *Chem. Ber.*, **111**, 3119 (1978).
65. P. F. Hudrlik, R. H. Schwartz and J. C. Hogan, *J. Org. Chem.*, **44**, 155 (1979).
66. J. J. Eisch and M. W. Foxton, *J. Org. Chem.*, **36**, 3520 (1971).
67. G. Zweifel and H. P. On, *Synthesis*, 803 (1980).
68. L. Rösch, G. Altnau and G. Jas, *Chem. Ztg.*, **106**, 441 (1982).
69. R. K. Boeckman, Jr. and K. J. Bruza, *J. Org. Chem.*, **44**, 4781 (1979).
70. W. P. Weber, A. K. Willard and H. G. Boettger, *J. Org. Chem.*, **36**, 1620 (1971).
71. R. B. Miller and G. McGarvey, *J. Org. Chem.*, **43**, 4424 (1978).
72. K. Atsumi and I. Kuwajima, *Tetrahedron Lett.*, 2001 (1977).
73. R. B. Miller and T. Reichenbach, *Tetrahedron Lett.*, 543 (1974).
74. H. Matsumoto, S. Nagashima, T. Kato and Y. Nagai, *Angew. Chem.*, **90**, 288 (1978); *Angew. Chem., Int. Ed. Engl.*, **17**, 279 (1978).
75. C. Eaborn, R. A. Jackson and R. Pearce, *J. Chem. Soc., Perkin Trans. 1*, 2055 (1974).
76. C. Eaborn, R. A. Jackson and R. Pearce, *J. Chem. Soc., Perkin Trans. 1*, 470 (1975).
77. T. H. Chan, A. Baldassarre and D. Massuda, *Synthesis*, 801 (1976).
78. G. Nagendrappa, *Synthesis*, 704 (1980).
79. L. H. Sommer, L. J. Tyler and F. C. Whitmore, *J. Am. Chem. Soc.*, **70**, 2872 (1948).
80. A. D. Petrov and V. F. Mironov, *Dokl. Akad. Nauk SSSR.*, **80**, 761 (1951); *Chem. Abstr.*, **46**, 11102 (1952).
81. S. O'Brien, M. Fishwick, B. McDermott, M. G. H. Wallbridge and G. A. Wright, *Inorg. Synth.*, **13**, 73 (1972).
82. E. W. Abel and R. J. Rowley, *J. Organomet. Chem.*, **84**, 199 (1975).
83. H. Gilman and E. A. Zuech, *J. Am. Chem. Soc.*, **81**, 5925 (1959).
84. I. Fleming and A. Pearce, *J. Chem. Soc., Perkin Trans. 1*, 251 (1981).
85. J. Dunoguès, J. P. Pillot, N. Duffaut, N. Ardoin and R. Calas, *J. Organomet. Chem.*, **82**, 217 (1974).
86. H. Bock and W. Kaim, *Tetrahedron Lett.*, 2343 (1977); *J. Am. Chem. Soc.*, **102**, 4429 (1980).
87. W. Y. Richter, *Synthesis*, 1102 (1982).
88. D. Tzeng and W. P. Weber, *J. Org. Chem.*, **46**, 265 (1981).
89. J. G. Smith, N. R. Quinn and M. Viswanathan, *Synth. Commun.*, **13**, 773 (1983).
90. I. Fleming and A. P. Thomas, *J. Chem. Soc., Chem. Commun.*, 411 (1985); *J. Chem. Soc., Chem. Commun.*, 1456 (1986).
91. I. Fleming and I. Paterson, *Synthesis*, 446 (1979).
92. D. Seyferth, K. R. Wursthorn and R. E. Mammarella, *J. Org. Chem.*, **42**, 3104 (1977).
93. Y. Tsuij, M. Hara and K. Ohno, *Tetrahedron*, **30**, 2143 (1974).
94. H. Matsumoto, K. Shono, A. Wada, I. Matsubara, H. Watanabe and Y. Nagai, *J. Organomet. Chem.*, **199**, 185 (1980).
95. I. Fischler and F. W. Grevels, *J. Organomet. Chem.*, **204**, 181 (1981).
96. J. P. Pillot, G. Déléris, J. Dunoguès and R. Calas, *J. Org. Chem.*, **44**, 3397 (1979).
97. T. Hayashi, A. Yamamoto, T. Iwata and Y. Ito, *J. Chem. Soc., Chem. Commun.*, 398 (1987).
98. J. M. Reuter, A. Sinha and R. G. Salomon, *J. Org. Chem.*, **43**, 2438 (1978).
99. V. A. Mironov, M. E. Dolgaya, S. A. Yankovskii and E. N. Pukhova, *Izv. Vyssh. Uchebn. Zaved., Khim. Khim. Tekhnol.*, **26**, 923 (1983); *Chem. Abstr.*, **100**, 34601 v (1984).
100. G. Wickham and W. Kitching, *Organometallics*, **2**, 541 (1983).
101. M. Laguerre, J. Dunoguès, R. Calas and N. Duffaut, *J. Organomet. Chem.*, **112**, 49 (1976).
102. M. Laguerre, J. Dunoguès and R. Calas, *Tetrahedron Lett.*, **21**, 831 (1980).

103. C. Biran, B. Efendene and J. Dunoguès, *J. Organomet. Chem.*, **253**, C 13 (1983).
104. B. Halton, C. J. Randall and P. J. Stang, *J. Am. Chem. Soc.*, **106**, 6108 (1984).
105. G. K. I. Magomedov, K. A. Andrianov, O. V. Shkolnik, B. A. Izmailov and V. N. Kalinin, *J. Organomet. Chem.*, **149**, 29 (1978).
106. J. Hetflejš and V. Vaisarova, *Czech. Patent 171.581; Chem. Abstr.*, **89**, 24529 t (1978).
107. T. Hayashi, K. Kabeta, T. Yamamoto, K. Tamao and M. Kumada, *Tetrahedron Lett.*, **24**, 5661 (1983).
108. V. P. Yur'év, I. M. Salimgareeva, O. Zh. Zhebarov and G. A. Tolstikov, *Zh. Obshch. Khim.*, **45**, 2568 (1975); *Chem. Abstr.*, **84**, 59640 (1976).
109. K. Yamamoto and M. Kumada, *J. Organomet. Chem.*, **13**, 131 (1968).
110. C. Eaborn and D. R. M. Walton, *J. Organomet. Chem.*, **4**, 217 (1965).
111. E. J. Corey, J. A. Katzenellenbogen, N. W. Gilman, S. A. Roman and B. W. Erickson, *J. Am. Chem. Soc.*, **90**, 5618 (1968).
112. U. Krüerke, *J. Organomet. Chem.*, **21**, 83 (1970).
113. M. F. Shostakovskii, N. V. Komarov and O. G. Yarosh, *Zh. Prikl. Khim.*, **38**, 435 (1965); *Chem. Abstr.*, **62**, 13172 (1965).
114. R. West and L. C. Quass, *J. Organomet. Chem.*, **18**, 55 (1969).
115. S. Cradock, E. A. V. Ebsworth and A. R. Green, *J. Chem. Soc., Dalton Trans.*, 759 (1978).
116. K. C. Frisch and R. B. Young, *J. Am. Chem. Soc.*, **74**, 4853 (1952).
(a) S. K. Chiu and P. E. Peterson, *Tetrahedron Lett.*, **21**, 4047 (1980).
117. J. Pornet, N. B. Kolani, D. Mesnard, L. Miginiac and K. Jaworski, *J. Organomet. Chem.*, **236**, 177 (1982).
118. J. Pornet, D. Mesnard and L. Miginiac, *Tetrahedron Lett.*, **23**, 4083 (1982).
119. M. Laguerre, J. Dunoguès and R. Calas, *J. Organomet. Chem.*, **145**, C 34 (1978).
(a) F. Jaffe, *J. Organomet. Chem.*, **23**, 53 (1970).
120. K. J. H. Kruithof, R. F. Schmitz and G. W. Klumpp, *Tetrahedron*, **39**, 3073 (1983).
121. G. Boche and J. Bigalke, *Tetrahedron Lett.*, **25**, 955 (1984).
122. L. Birkofer, A. Ritter and H. Uhlenbrauck, *Chem. Ber.*, **96**, 3280 (1963).
123. D. R. M. Walton and F. Waugh, *J. Organomet. Chem.*, **37**, 45 (1972).
124. I. A. Shikhiev, M. F. Shostakovskii and L. A. Kayutenko, *Dokl. Akad. Nauk Azerbaidzhan SSR*, **15**, 21 (1959); *Chem. Abstr.*, **53**, 15957 (1959).
125. Le Quan Minh, J. C. Billiotte and P. Cadiot, *C.R. Acad. Sci. Paris*, **251**, 730 (1960); *Chem. Abstr.*, **55**, 5389 i (1961).
126. R. S. Brinkmeyer and T. L. MacDonald, *J. Chem. Soc., Chem. Commun.*, 876 (1978).
127. T. L. MacDonald, D. R. Reagan and R. S. Brinkmeyer, *J. Org. Chem.*, **45**, 4740 (1980).
128. T. L. Chwang and R. West, *J. Am. Chem. Soc.*, **95**, 3324 (1973).
129. W. Dilthey and F. Eduardoff, *Ber. Dtsch. Chem.*, **37**, 1139 (1904).
130. N. W. Cusa and F. S. Kipping, *J. Chem. Soc. (London)*, 1040 (1933).
131. A. D. Petrov and T. I. Chernysheva, *Izv. Akad. Nauk SSSR, Otdel. Khim. Nauk*, 820 (1951); *Chem. Abstr.*, **46**, 7556 (1952).
132. H. Gilman and R. N. Clark, *J. Am. Chem. Soc.*, **68**, 1675 (1946).
133. W. H. Nebergall, *J. Am. Chem. Soc.*, **72**, 4702 (1950).
134. H. Gilman and E. A. Zuech, *J. Am. Chem. Soc.*, **79**, 4560 (1957).
135. R. H. Wiley, *US. Patent 2238 669* (Du Pont) (1941); *Chem. Abstr.*, **35**, 4782 (1941).
136. H. Freiser, M. V. Eagle and J. Speier, *J. Am. Chem. Soc.*, **75**, 2821 (1953).
137. F. Effenberger and D. Häbich, *Liebigs Ann. Chem.*, 842 (1979).
138. J. R. Pratt and S. F. Thames, *J. Org. Chem.*, **38**, 4271 (1973).
139. G. Simchen and J. Pfletschinger, *Angew. Chem.*, **88**, 444 (1976); *Angew. Chem., Int. Ed. Engl.*, **15**, 428 (1976).
140. P. Pons, C. Biran, M. Bordeau, J. Dunoguès, S. Sibille and J. Perichon, *J. Organomet. Chem.*, **321**, C 27 (1987).
141. L. Spialter, D. C. Priest and C. W. Harris, *J. Am. Chem. Soc.*, **77**, 6227 (1955).
142. H. Gilman and C. C. Brannen, *J. Am. Chem. Soc.*, **72**, 4280 (1950).
143. H. E. Katz, *Organometallics*, **5**, 2308 (1986).
144. L. Birkofer and N. Ramadan, *J. Organomet. Chem.*, **44**, C 41 (1972).
145. L. Birkofer and N. Ramadan, *J. Organomet. Chem.*, **92**, C 41 (1975).
146. L. Birkofer and V. Foremny, *Z. Chem.*, **23**, 250 (1983).
147. Hee Cho and R. G. Harvey, *J. Org. Chem.*, **40**, 3097 (1975).

148. M. Laguerre, G. Félix, J. Dunoguès and R. Calas, *J. Org. Chem.*, **44**, 4275 (1979).
149. E. Lukevics, N. P. Erchak, I. Castro, S. Roite, I. Mazeika, A. P. Gaukhman and J. Popelis, *Zh. Obshch. Khim.*, **54**, 1315 (1984); *Chem. Abstr.*, **101**, 171334 v (1984).
150. A. V. Anisimov, G. N. Murina, L. V. Mozhaeva, N. B. Kazennova and E. A. Viktorova, *Khim. Geterotsikli. Soedin*, 744 (1984); *Chem. Abstr.*, **101**, 230265 y (1984).
151. L. Birkofer and M. Franz, *Chem. Ber.*, **100**, 2681 (1967).
152. L. Birkofer and M. Franz, *Chem. Ber.*, **105**, 1759 (1972).
153. L. Birkofer and M. Franz, *Chem. Ber.*, **105**, 470 (1972).
154. L. Birkofer and K. Richtzenhain, *Chem. Ber.*, **112**, 2829 (1979).
155. L. Birkofer and T. Kühn, *Chem. Ber.*, **114**, 2293 (1981).
156. A. Dondoni, T. Dall'Occo, G. Fantin, M. Fogagnolo, A. Medici and P. Pedrini, *J. Chem. Soc., Chem. Commun.*, 258 (1984).
157. L. Birkofer and R. Stilke, *Chem. Ber.*, **107**, 3717 (1974).
158. F. De Sarlo, A. Brandi, A. Goti, A. Guarna and P. Rovero, *Heterocycles*, **20**, 511 (1983).
159. J. Verbeek, A. V. E. George, R. L. P. de Jong and L. Brandsma, *J. Chem. Soc., Chem. Commun.*, 257 (1984).
160. F. Marsais, B. Laperdrix, T. Guengoer, M. Mallet and G. Queguiner, *J. Chem. Res. Synop.*, 278 (1982).
161. St. L. Taylor, D. Y. Lee and J. C. Martin, *J. Org. Chem.*, **48**, 4156 (1983).
162. P. Jutzi and O. Lorey, *J. Organomet. Chem.*, **104**, 153 (1976).
163. L. Birkofer and R. Stilke, *J. Organomet. Chem.*, **74**, C 1 (1974).
164. L. Birkofer and E. Hänsel, *Chem Ber.*, **114**, 3154 (1981).
165. L. Birkofer and B. Wahle, *Chem. Ber.*, **116**, 2564 (1983).
166. D. G. Anderson, M. A. M. Bradney and D. E. Webster, *J. Chem. Soc. (B)*, 450 (1968).
167. D. Seyferth and D. C. Annarelli, *J. Am. Chem. Soc.*, **97**, 2273 (1975).
168. E. A. Halevi and R. West, *J. Organomet. Chem.*, **240**, 129 (1982).
169. M. Ishikawa, K. Nakagawa and M. Kumada, *J. Organomet. Chem.*, **214**, 277 (1981).
170. M. Ishikawa, H. Sugisawa, M. Kumada, T. Higuchi, K. Matsui, K. Hirotsu and J. Iyoda, *Organometallics*, **2**, 174 (1983).
171. S. Masamune, S. Murakami, H. Tobita and D. J. Williams, *J. Am. Chem. Soc.*, **105**, 7776 (1983).
172. A. G. Brook, J. W. Harris, J. Lennon and M. ElSheikh, *J. Am. Chem. Soc.*, **101**, 83 (1979).
173. D. Seyferth, D. C. Annarelli, S. C. Vick and D. P. Duncan, *J. Organomet. Chem.*, **201**, 179 (1980); D. Seyferth, E. W. Goldman and J. Escudié, *J. Organomet. Chem.*, **271**, 337 (1984).
174. M. J. Fink, D. J. De Young and R. West, *J. Am. Chem. Soc.*, **105**, 1070 (1983).
175. W. Y. Richter, *Synthesis*, 1102 (1982).
176. L. Birkofer and D. Prüstel, *J. Organomet. Chem.*, **60**, C 1 (1973).
177. J. C. Brunet and C. Lemahieu, *J. Organomet. Chem.*, **49**, 381 (1973).
178. H. Watanabe, J. Inose, T. Muraoka, M. Saito and Y. Nagai, *J. Organomet. Chem.*, **244**, 329 (1985).
179. R. W. Coutant and A. Levy, *J. Organomet. Chem.*, **10**, 175 (1967).
180. J. S. Kiely and P. Boudjouk, *J. Organomet. Chem.*, **182**, 173 (1979).
181. G. Fritz and P. Schober, *Z. Anorg. Chem.*, **372**, 21 (1970).
182. R.-J. Hwang, R. T. Conlin and P. P. Gaspar, *J. Organomet. Chem.*, **94**, C 38 (1975).
183. R. F. Cunico and F. Drone, *J. Organomet. Chem.*, **150**, 179 (1978).
184. R. A. Benkeser, E. C. Mozdzen, W. C. Muench, R. T. Roche and M. P. Siklosi, *J. Org. Chem.*, **44**, 1370 (1979).
185. J. Y. Corey, E. A. Francis, M. S. Bursten and J. C. Kunz, *J. Organomet. Chem.*, **210**, 149 (1981).
186. F. Bickelhaupt and G. L. van Mourik, *J. Organomet. Chem.*, **67**, 389 (1974).
187. L. M. Engelhardt, R. I. Papasergio, C. L. Raston and A. H. White, *J. Chem. Soc., Dalton Trans.*, 311 (1984); cf. R. I. Papasergio and C. L. Raston, *J. Chem. Soc., Chem. Commun.*, 1023 (1982).
188. H. Gilman and W. J. Trepka, *J. Org. Chem.*, **27**, 1418 (1962).
189. R. J. P. Corriu and J. P. Masse, *Chem. Commun.*, 213 (1970).
190. M. Weidenbruch, A. Schäfer, K. Peters and H. G. von Schnering, *J. Organomet. Chem.*, **314**, 25 (1986).
191. W. H. Atwell and D. R. Weyenberg, *J. Am. Chem. Soc.*, **90**, 3438 (1968).
192. L. Birkofer, *J. Organomet. Chem. Library*, **9**, 295 (1980).
193. L. Birkofer and W. Weniger, *Chem. Ber.*, **106**, 3595 (1973).
194. H. Gilman and E. A. Zuech, *J. Am. Chem. Soc.*, **82**, 3605 (1960).

195. E. A. Chernyshev, N. G. Komalenkova, L. N. Shamshin, R. A. Turetskaya and M. A. Luzganova, *Zh. Obshch. Khim.*, **50**, 1790 (1980). *Chem. Abstr.*, **93**, 239510 (1980).
196. Hitachi Ltd., *Jpn. Kokai Tokkyokoho (Jap. Patent) 5821, 684; Chem. Abstr.*, **98**, 198431 (1983).
197. L. Birkofer and E. Krämer, *Chem. Ber.*, **102**, 427 (1969).
198. L. Birkofer and H. Haddad, *Chem. Ber.*, **102**, 432 (1969).
198. a. L. Birkofer and H. Haddad, *Chem. Ber.*, **105**, 2101 (1972).
199. L. Birkofer and H. Haddad, *J. Organomet. Chem.*, **164**, C 17 (1979).
200. L. Birkofer, H. Haddad and H. Zamarlik, *J. Organomet. Chem.*, **25**, C 57 (1970).
201. J. Y. Corey and R. Z. Farrell, *J. Organomet. Chem.*, **153**, 15 (1978).
202. H. Gilman and O. L. Marrs, *J. Org. Chem.*, **29**, 3175 (1964).
203. L. Birkofer and D. Prüstel, *J. Organomet. Chem.*, **32**, C 75 (1970).
204. H. Sakurai, Y. Nakadaira, A. Hosomi and Y. Eriyama, *Chem. Lett.*, 1971 (1982).
205. L. Pauling, *The Nature of the Chemical Bond*, Cornell University Press, Ithaca, 1978.
206. K. Hedberg, *J. Am. Chem. Soc.*, **77**, 6491 (1955).
207. R. O. Sauer, *J. Am. Chem. Soc.*, **66**, 1707 (1944).
208. R. O. Sauer and R. H. Hasek, *J. Am. Chem. Soc.*, **68**, 241 (1946).
209. E. Larsson and R. Mårin, *Acta Chem. Scand.*, **5**, 1173 (1951).
210. R. Schwarz and F. Weigel, *Z. Anorg. Allg. Chem.*, **268**, 291 (1952).
211. C. A. Burkhard, *J. Am. Chem. Soc.*, **71**, 963 (1949).
212. S. D. Brewer and C. P. Haber, *J. Am. Chem. Soc.*, **70**, 3888 (1948).
213. U. Klingebiel, D. Bentmann and A. Meller, *Z. Anorg. Allg. Chem.*, **441**, 118 (1978).
214. H. Gilman, B. Hofferth, H. W. Melvin and G. E. Dunn, *J. Am. Chem. Soc.*, **72**, 5767 (1950).
215. H. Gilman and G. D. Lichtenwalter, *J. Am. Chem. Soc.*, **82**, 3319 (1960).
216. Y. Nagai, I. Ojima and T. Ogure, *Jap. Patent 7601693* (Sagami Chemical Research Center) (1976); *Chem. Abstr.*, **85**, 46173 (1976).
217. W. De Man, J. van Steenis and H. I. Waterman, *Rec. Trav. Chim. Pays-Bas*, **67**, 864 (1948).
218. L. H. Sommer and L. J. Tyler, *J. Am. Chem. Soc.*, **76**, 1030 (1954).
219. L. Birkofer and G. Schmidtberg, *Chem. Ber.*, **104**, 3831 (1971).
220. L. Birkofer and D. Brokmeier, *Tetrahedron Lett.*, 1325 (1968).
221. U. Klingebiel, J. Neemann and A. Meller, *Z. Anorg. Allg. Chem.*, **462**, 207 (1980).
222. L. Birkofer, A. Ritter and H. Dickopp, *Chem. Ber.*, **96**, 1473 (1963).
223. L. Birkofer, A. Ritter and W. Gieβler, *Angew. Chem.*, **75**, 93 (1963).
224. J. F. Klebe, J. B. Bush Jr and J. E. Lyons, *J. Chem. Soc.*, **86**, 4400 (1964).
225. G. Oertel, H. Malz and H. Holtschmidt, *Chem. Ber.*, **97**, 891 (1964).
226. I. Ojima and S. Inaba, *J. Organomet. Chem.*, **72**, C 11 (1974).
227. H. Ahlbrecht and D. Liesching, *Synthesis*, 746 (1976).
228. G. A. Gornowicz and R. West, *J. Am. Chem. Soc.*, **93**, 1714 (1971).
(a) S. Inoue and Y. Sato, *Organometallics*, **5**, 1197 (1986).
229. L.-H. Chan and E. G. Rochow, *J. Organomet. Chem.*, **9**, 231 (1967).
230. L. Birkofer, A. Ritter and P. Richter, *Chem. Ber.*, **96**, 2750 (1963); L. Birkofer and S. M. Kim, *Chem. Ber.*, **97**, 2100 (1964).
231. I. Ruppert, *Angew. Chem.*, **89**, 336 (1977); *Angew. Chem., Int. Ed. Engl.*, **16**, 311 (1977).
232. L. Birkofer, A. Ritter and P. Richter, *Tetrahedron Lett.*, 195 (1962).
233. J. P. Llonch, F. Duboudin, G. Vinçon, F. Dabescat and E. Frainnet, *J. Organomet. Chem.*, **52**, C 24 (1973).
234. R. West and P. Boudjouk, *J. Am. Chem. Soc.*, **95**, 3987 (1973).
235. K. Witke, P. Reich and H. Kriegsmann, *J. Organomet. Chem.*, **15**, 37 (1968).
236. U. Wannagat and W. Liehr, *Angew. Chem.*, **69**, 783 (1957); *Z. Anorg. Allg. Chem.*, **297**, 129 (1958).
237. N. Wiberg, W.-Ch. Joo and W. Uhlenbrock, *Angew. Chem.*, **80**, 661 (1968); *Angew. Chem., Int. Ed. Engl.*, **7**, 640 (1968).
238. E. M. R. Kiremire, G. J. Leigh, J. R. Dilworth and R. A. Henderson, *Inorg. Chim. Acta*, **83**, L 83 (1984).
239. N. Wiberg and W. Uhlenbrock, *Angew. Chem.*, **82**, 47 (1970); *Angew. Chem., Int. Ed. Engl.*, **9**, 70 (1970).
240. L. Birkofer and A. Ritter, *Angew. Chem.*, **68**, 461 (1956).
241. L. Birkofer and A. Ritter, *Liebigs Ann. Chem.*, **612**, 22 (1958).
242. K. Rühlmann, *J. Prakt. Chem.*, **9**, 86 (1959).

243. L. Birkofer and A. Ritter, *Chem. Ber.*, **93**, 424 (1960).
244. L. Birkofer, E. Bierwirth and A. Ritter, *Chem. Ber.*, **94**, 821 (1961).
245. L. Birkofer, H. P. Kühlthau and A. Ritter, *Chem. Ber.*, **93**, 2811 (1960).
246. L. Birkofer, P. Richter and A. Ritter, *Chem. Ber.*, **93**, 2804 (1960).
247. E. Duranti and C. Balsamini, *Synthesis*, 815 (1974).
248. H. R. Kricheldorf, *Liebigs Ann. Chem.*, **745**, 81 (1971).
249. K. Rühlmann and B. Rupprich, *Liebigs Ann. Chem.*, **686**, 226 (1965).
250. D. Berrer and C. Vogel, *Dos 2208329* (Agripat, S.A.) (1972); *Chem. Abstr.*, **77**, 152236 (1972).
251. R. A. Sulzbach and A. F. M. Iqbal, *Angew. Chem.*, **83**, 145 (1971); *Angew. Chem., Int. Ed. Engl.*, **10**, 127 (1971).
252. L. Birkofer and N. Ramadan, *Chem. Ber.*, **108**, 3105 (1975).
253. W. P. Neumann and K. Reuter, *Chem. Ber.*, **112**, 936 (1979).
254. L. Birkofer and P. Wegner, *Chem. Ber.*, **99**, 2512 (1966); L. Birkofer, A. Ritter and P. Richter, *Chem. Ber.*, **96**, 2750 (1963).
255. G. Eisele, H. Autzen and U. Wannagat, *Monatsh. Chem.*, **109**, 1267 (1978).
256. S. Djuric. J. Venit and P. Magnus, *Tetrahedron Lett.*, **22**, 1787 (1981).
257. I. A. Vostokov and Yu. I. Dergunov, *Khim. Elementoorg. Soedin.*, **63**, (1981); *Chem. Abstr.*, **97**, 216302a (1982).
258. L. Birkofer and P. Wegner, *Org. Synth.*, **50**, 107 (1970).
259. W. Sundermeyer, *Z. Anorg. Allg. Chem.*, **313**, 290 (1961); *Chem. Ber.*, **96**, 1293 (1963).
260. R. West and J. S. Thayer, *J. Am. Chem. Soc.*, **84**, 1763 (1962).
261. N. Wiberg, F. Raschig and R. Sustmann, *Angew. Chem.*, **74**, 388, 7176 (1962).
262. K. Rühlmann, A. Reiche and M. Becker, *Chem. Ber.*, **98**, 1814 (1965).
263. W. Sundermeyer, *Angew. Chem.*, **74**, 875 (1962).
264. T. A. Bither, W. H. Knoth, R. V. Lindsey jr. and W. H. Sharkey, *J. Am. Chem. Soc.*, **80**, 4151 (1958).
265. B. Uznanski and W. J. Stec, *Synthesis*, 154 (1978).
266. S. Hünig and G. Wehner, *Synthesis*, 522 (1979).
267. T. Hiiro and H. Sakurai, *Jap. Patent. (Kokai) 78 132 525* (Kawaken Fine Chemical Co.) (1977/78); *Chem. Abstr.*, **90**, 168721 (1979).
268. J.-P. Picard, A. Aziz-Elyusufi, R. Calas, J. Dunoguès and N. Duffaut, *Organometallics*, **3**, 1660 (1984).
269. J. J. McBride jr. and H. C. Beachell, *J. Am. Chem. Soc.*, **74**, 5247 (1952); J. J. McBride jr., *J. Org. Chem.*, **24**, 2029 (1959).
270. T. A. Bither, W. H. Knoth, R. V. Lindsey jr. and W. H. Sharkey, *J. Am. Chem. Soc.*, **80**, 4158 (1958).
271. P. R. Steyermark, *J. Org. Chem.*, **28**, 586 (1963).
272. H. Gilman, B. Hofferth and H. W. Melvin, *J. Am. Chem. Soc.*, **72**, 3045 (1950).
273. J. Goubeau and J. Reyhing, *Z. Anorg. Allg. Chem.*, **294**, 96 (1950).
274. M. G. Voronkov, V. K. Roman and E. A. Maletina, *Synthesis*, 277 (1982).
275. L. Birkofer, A. Ritter and J. Schramm, *Chem. Ber.*, **95**, 426 (1962).
276. J. Pump and U. Wannagat, *Monatsh. Chem.*, **93**, 352 (1962).
277. C. Eaborn, P. D. Lickiss, G. Marquina-Chidsey and E. Y. Thorli, *J. Chem. Soc., Chem. Commun.*, 1326 (1982).
278. W. T. Grubb and R. C. Osthoff, *J. Am. Chem. Soc.*, **75**, 2230 (1953).
279. H. Kriegsmann, *Z. Anorg. Allg. Chem.*, **294**, 113 (1958).
280. R. West and R. H. Baney, *J. Am. Chem. Soc.*, **81**, 6145 (1959).
281. C. G. Barraclough, D. C. Bradley, J. Lewis and I. M. Thomas, *J. Chem. Soc.*, 2601 (1961).
282. K. Licht and H. Kriegsmann, *Z. Anorg. Allg. Chem.*, **323**, 190, 239, 252 (1963).
283. H. Schmidbaur, *Chem. Ber.*, **97**, 830 (1964).
284. R. West and R. H. Baney, *J. Inorg. Nucl. Chem.*, **7**, 297 (1958).
285. L. H. Sommer, E. W. Pietrusza and F. C. Whitmore, *J. Am. Chem. Soc.*, **68**, 2282 (1946).
286. L. Birkofer, A. Ritter and H. Dickopp, *Chem. Ber.*, **96**, 1473 (1963).
287. R. O. Sauer, *J. Am. Chem. Soc.*, **66**, 1707 (1944).
288. C. Eaborn, *J. Chem. Soc.*, 3077 (1950).
289. L. H. Sommer, W. D. Korte and C. L. Frye, *J. Am. Chem. Soc.*, **94**, 3463 (1972).
290. H. Gilman and R. N. Clark, *J. Am. Chem. Soc.*, **69**, 1499 (1947).
291. T. Takiguchi, *J. Org. Chem.*, **24**, 989 (1959).

292. S. W. Kantor, *J. Am. Chem. Soc.*, **75**, 2712 (1953).
293. J. F. Hyde, *J. Am. Chem. Soc.*, **75**, 2166 (1953).
294. L. J. Tyler, *J. Am. Chem. Soc.*, **77**, 770 (1955).
295. T. Takiguchi, *J. Am. Chem. Soc.*, **81**, 2359 (1959).
296. M. Seki, H. Ito and I. Ohara, *Jap Patent (Kokai) 7488840* (Oji Paper Co.) (1974); *Chem. Abstr.*, **82**, 171175 (1975).
297. H. A. Clark, *US Patent 2553845* (Dow Corning Corp.) (1951); *Chem. Abstr.*, **46**, 130 (1952).
298. F. Augst and F. Sufrage, *DOS 2033373* (VEB Chemiekombinat Bitterfeld), (1971); *Chem. Abstr.*, **74**, 125839 (1971).
299. J. L. Speier, *US Patent 2626272* (Dow Corning Corp.), (1953); *Chem. Abstr.*, **48**, 1418 (1954).
300. R. Kleinstück, *DOS 2521399* (Bayer A. G.), (1976); *Chem. Abstr.*, **86**, 72863 (1977).
301. H. Gilman and G. N. R. Smart, *J. Org. Chem.*, **19**, 441 (1954).
302. S. Pawlenko, in *Methoden der organischen Chemie (Houben-Weyl)*, Vol. 13, Part 5 (Eds. E. Müller and O. Bayer), Georg Thieme Verlag, Stuttgart, 1980, p. 190.
303. R. O. Sauer, *J. Am. Chem. Soc.*, **68**, 138 (1946).
304. G. Schott, L. Engelbrecht and H. J. Holdt, *Z. Anorg. Allg. Chem.*, **459**, 177 (1979).
305. M. V. Sobolevskii, S. I. Kleshchevnikova, E. N. Lebedev and E. A. Abramova, *Khim. Prom.-st.*, 181 (1978); *Chem. Abstr.*, **89**, 129578 m (1978).
306. S. Murata and R. Noyori, *Tetrahedron Lett.*, **21**, 767 (1980).
307. S. P. Hopper, M. J. Tremelling and E. W. Goldman, *J. Organomet. Chem.*, **156**, 331 (1978).
308. W. S. Miller, J. S. Peake and W. H. Nebergall, *J. Am. Chem. Soc.*, **79**, 5604 (1957).
309. N. S. Nametkin, A. V. Topchiev and L. S. Povarov, *Dokl. Akad. Nauk SSSR*, **117**, 245 (1957); *Chem. Abstr.*, **52**, 8943 (1958).
310. S. H. Langer, S. Connell and I. Wender, *J. Org. Chem.*, **23**, 50 (1958).
311. B. F. Coles and D. R. M. Walton, *Synthesis*, 390 (1975).
312. P. Ykman and H. K. Hall jr., *J. Organomet. Chem.*, **116**, 153 (1976).
313. K. Rühlmann, *Synthesis*, 236 (1971).
314. Y. Nagai, I. Ojima and T. Kogure, *Jap. Patent 7442615* (Sagami Chemical Research Center) (1974); *Chem. Abstr.*, **81**, 105690 (1974).
315. H. Gilman and D. Wittenberg, *J. Org. Chem.*, **23**, 501 (1958).
316. T. Hayashi, K. Yamamoto, K. Kasuga, H. Omizu and M. Kumada, *J. Organomet. Chem.*, **113**, 127 (1976).
317. T. Suzuki, N. Imaki, K. Nakanome and T. Yamaura, *Jpn. Kokai Tokkyo Koho*, **80**, 33, 457 (1980); *Chem. Abstr.* **93**, 167629 u (1980).
318. R. R. McGregor and E. L. Warrick, *US. Patent 2442053* (Corning Glass Works) (1948); *Chem. Abstr.*, **42**, 7786 (1948).
319. A. V. Topchiev and N. S. Nametkin, *Dokl. Akad. Nauk SSSR*, **80**, 897 (1951); *Chem. Abstr.*, **46**, 7534 (1952).
320. V. Cappucio, R. Pisani and I. Dukli, *Ital. Patent 470505* (Montecatini Societa Generale) (1952); *Chem. Abstr.*, **48**, 13717 (1954).
321. H. W. Post and C. H. Hofrichter, *J. Org. Chem.*, **4**, 363 (1939); **5**, 443 (1940).
322. H. Gilman and C. G. Brannen, *J. Am. Chem. Soc.*, **73**, 4640 (1951).
323. H. Jenker, *DOS 974764* (Kali Chemie AG.) (1953); *Chem. Abstr.*, **56**, 4795 (1962).
324. E. Nakamura and I. Kuwajima, *Angew. Chem.*, **88**, 539 (1976); *Angew. Chem., Int. Ed. Engl.*, **15**, 498 (1976).
325. M. F. Semmelhack and R. N. Misra, *J. Org. Chem.*, **47**, 2469 (1982).
326. L. Birkofer and W. Kaiser, *Liebigs Ann. Chem.*, 266 (1975).
327. S. Murata, M. Suzuki and R. Noyori, *Bull. Chem. Soc. Jap.*, **55**, 247 (1982).
328. A. Gancberg and J. Vandevelde, *Ind. Chem. Belge*, **6**, 591 (1964).
329. H. Gilman and G. N. R. Smart, *J. Org. Chem.*, **19**, 441 (1954).
330. W. Ried and U. Reiher, *Chem. Ber.*, **120**, 657 (1987).
331. M. Matsumoto, S. Koike, I. Matsubara, T. Nakano and Y. Nagai, *Chem. Lett.*, 533 (1982).
332. D. N. Andreev and E. V. Kukharskaya, *Zh. Obshch. Khim.*, **30**, 2782 (1960); *Chem. Abstr.*, **55**, 15332 (1961).
333. Y. Etienne, *C.R. Acad. Sci. Paris*, **235**, 966 (1952).
334. L. Birkofer, A. Ritter and J. Schramm, *Chem. Ber.*, **95**, 426 (1962).
335. M. Jakoubková, P. Reich, Ž. Papouskova, P. Novák, J. Pola and V. Chvalovsky, *Collect. Czech. Chem. Commun.*, **38**, 3471 (1972).

336. H. H. Anderson and H. Fischer, *J. Org. Chem.*, **19**, 1296 (1954).
337. Y. Nagai, I. Ojima and S. Inaba, *Jap. Patent 74110634* (Sagami Chemical Research Center), (1974); *Chem. Abstr.*, **83**, 206415 (1975).
338. N. F. Orlov, R. A. Bogatkin, Z. I. Sergeeva and M. G. Voronkov, *Zh. Obshch. Khim.*, **32**, 650 (1962); *Chem. Abstr.*, **57**, 15141 (1962); *Zh. Obshch, Khim.*, **32**, 2561 (1962); *Chem. Abstr.*, **58**, 9114 (1963).
339. B. N. Dolgov, N. P. Kharitonov and M. G. Voronkov, *Zh. Obshch. Khim.*, **24**, 861 (1954); *Chem. Abstr.*, **49**, 8094 (1955).
340. V. F. Mironov, V. D. Sheludyakov, G. D. Khatuntsev and V. L. Kozlikov, *USSR. Patent 330 173* (1972); *Chem. Abstr.*, **77**, 48623 (1972).
341. H. H. Anderson and G. M. Stanislow, *J. Org. Chem.*, **18**, 1716 (1953).
342. B. Becker and W. Wojnowski, *Z. Anorg. Allg. Chem.*, **422**, 89 (1976).
343. J. E. Dracke, R. T. Hemmings and H. E. Henderson, *Inorg. Nucl. Chem. Lett.*, **12**, 563 (1976).
344. G. Neumann and W. P. Neumann, *J. Organomet. Chem.*, **42**, 277 (1972).
345. L. Birkofer and P. Sommer, *J. Organomet. Chem.*, **99**, C 1 (1975).
346. L. Birkofer and P. Sommer, *J. Organomet. Chem.*, **35**, C 15 (1972).
347. H. O. House, L. J. Czuba, M. Gall and H. D. Olmstead, *J. Org. Chem.*, **34**, 2324 (1969).
348. P. F. Hudrlik and J. M. Takacs, *J. Org. Chem.*, **43**, 3861 (1978).
349. P. Cazeau, F. Moulines, O. Laporte and F. Duboudin, *J. Organomet. Chem.*, **201**, C 9 (1980).
350. G. H. Posner, J. J. Sterling, C. E. Whitten, C. M. Lentz and D. J. Brunelle, *J. Am. Chem. Soc.*, **97**, 107 (1975).
351. R. D. Miller and D. R. McKean, *Synthesis*, 730 (1979).
352. N. R. Long. and M. W. Rathke, *Synth. Commun.*, **11**, 687 (1981).
353. K. Rühlmann, *Synthesis*, 236 (1971).
354. J. M. Denis and J. M. Conia, *Tetrahedron Lett.*, 4593 (1972).
355. G. Simchen and W. Kober, *Synthesis*, 259 (1976).
356. M. T. Reetz and G. Neumeier, *Chem. Ber.*, **112**, 2209 (1979).
357. E. Frainnet and R. Bourhis, *J. Organomet. Chem.*, **93**, 309 (1975).
358. Yu. I. Baukov and I. F. Lutsenko, *Organomet. Chem. Rev. A*, **6**, 355 (1970).
359. H. Sakurai, K. Miyoshi and Y. Nakadaira, *Tetrahedron Lett.*, 2671 (1977).
360. H. Sakurai and Y. Nakahira, *Jap. Patent (Kokai) 7916429* (Ajinomoto Co. Inc.) (1977/1979); *Chem. Abstr.*, **91**, 57175 (1979).
361. P. F. Hudrlik, R. H. Schwartz and A. K. Kulkarni, *Tetrahedron Lett.*, 2233 (1979).
362. G. A. Ohla, B. G. B. Gupta, S. C. Narang and R. Malhotra, *J. Org. Chem.*, **44**, 4272 (1979).
363. H. Suzuki, Y. Koyama, Y. Moro-Oka and T. Ikawa, *Tetrahedron Lett.*, 1415 (1979).
364. Y.-N. Kuo, J. A. Yahner and C. Ainsworth, *J. Am. Chem. Soc.*, **93**, 6321 (1971).
365. R. M. Coates, L. O. Sandefur and R. D. Smillie, *J. Am. Chem. Soc.*, **97**, 1619 (1975).
366. E. Nakamura, T. Murofushi, M. Shimizu and I. Kuwajima, *J. Am. Chem. Soc.*, **98**, 2346 (1976).
367. M. W. Rathke and D. F. Sullivan, *Synth. Commun.*, **3**, 67 (1973).
368. I. Matsuda, *J. Organomet. Chem.*, **321**, 307 (1987).
369. C. Ainsworth, C. Chen and Y. N. Kuo, *J. Organomet. Chem.*, **46**, 59, 73 (1972).
370. E. Yoshii, Y. Kobayashi, T. Koizumi and T. Oribe, *Chem. Pharm. Bull. Jpn.*, **22**, 2767 (1974).
371. A. B. Firestone, *US Patent 2.953 545* (Westinghouse Electric Corp.), (1960); *Chem. Abstr.*, **55**, 4045 i (1961).
372. C. L. Frye, G. E. Vogel and J. A. Hall, *J. Am. Chem. Soc.*, **83**, 996 (1961).
373. J. W. Turley and F. B. Boer, *J. Am. Chem. Soc.*, **91**, 4129 (1969).
374. F. B. Boer and J. W. Turley, *J. Am. Chem. Soc.*, **91**, 4134 (1969).
375. M. G. Voronkov and G. Zelcans, *Khim. Geterotsikl. Soedin., Akad. Nauk Latv. SSSR*, 51 (1965); *Chem. Abstr.*, **63**, 5670 (1965).
376. M. G. Voronkov, in *Topics in Current Chemistry*, Vol. 84 (Ed. F. L. Boschke), Springer Verlag, Berlin–Heidelberg–New York, 1980, pp. 77–135.
377. C. L. Frye, G. A. Vincent and W. A. Finzel, *J. Am. Chem. Soc.*, **93**, 6805 (1971).
378. M. G. Voronkov, G. I. Zelcans, G. F. Tsybulya and P. G. Volfson, *USSR Patent 299, 510* (1971); *Chem. Abstr.*, **75**, 63955 (1977).
379. E. Popowski, M. Michalik and H. Kelling, *J. Organomet. Chem.*, **88**, 157 (1975).
380. L. Birkofer and K. Grafen, *J. Organomet. Chem.*, **299**, 143 (1986).
381. M. G. Voronkov, V. M. Dyakov and S. V. Kirpichenko, *J. Organomet. Chem.*, **233**, 1 (1982).
382. D. Seyferth, S. C. Vick and M. L. Shannon, *Organometallics*, **3**, 1897 (1984).

383. T. Y. Y. Gu and W. P. Weber, *J. Am. Chem. Soc.*, **102**, 1641 (1980).
384. D. Tzeng and W. P. Weber, *J. Org. Chem.*, **47**, 1976 (1982).
385. L. Birkofer and O. Stuhl, *J. Organomet. Chem.*, **187**, 21 (1980).
386. M. Jaćović, *Z. Anorg. Allg. Chem.*, **288**, 324 (1956).
387. R. H. Cragg and R. D. Lane, *J. Organomet. Chem.*, **212**, 301 (1981).
388. B. M. Trost and C. G. Caldwell, *Tetrahedron Lett.*, **22**, 4999 (1981).
389. I. Ryu, S. Murai, A. Shinonaga, T. Horriike and N. Sonoda, *J. Org. Chem.*, **43**, 780 (1978).
390. L. Birkofer and O. Stuhl, *J. Organomet. Chem.*, **164**, C 1 (1979).
391. L. Birkofer, W. Grüner and O. Stuhl, *J. Organomet. Chem.*, **194**, 159 (1980).
392. R. H. Krieble and C. A. Burkhard, *J. Am. Chem. Soc.*, **69**, 2689 (1947).
393. L. Birkofer and O. Stuhl, *J. Organomet. Chem.*, **177**, C 16 (1979).
394. M. M. Sprung and L. S. Nelson, *J. Org. Chem.*, **20**, 1750 (1955).
395. R. C. Finch and H. W. Post, *J. Org. Chem.*, **24**, 969 (1959).
396. L. Birkofer, A. Ritter and F. Bentz, *Chem. Ber.*, **97**, 2196 (1964).
397. A. Klemer, E. Buhe and R. Kutz, *Liebigs Ann. Chem.*, **739**, 185 (1970).
398. Z. Lasocki, *Synth. React. Inorg. Met.-Org. Chem.*, **3**, 29 (1973).
399. L. Birkofer and R. Sommer, *J. Organomet. Chem.*, **99**, C 1 (1975).
400. A. I. Nogaideli and N. A. Rtvetliavshili, *USSR Tr. Tbilis. Univ. (5)*, 101 (1972); *Chem. Abstr.*, **79**, 115673 (1973).
401. T. Murakawa, K. Fujii, S. Murai and S. Tsutsumi, *Bull. Chem. Soc. Jpn.*, **45**, 2520 (1972); *Chem. Abstr.*, **77**, 126740 (1972).
402. A. G. Beaumont, C. Eaborn and R. A. Jackson, *J. Chem., Soc. (B)*, 1624 (1970).
403. A. Hass, *Angew. Chem.*, **77**, 1066 (1965); A. Haas, in *Sulfur in Organic and Inorganic Chemistry*, Vol. 1 (Ed. A. Senning), Marcel Dekker, New York, 1971, pp. 3 ff.
404. E. W. Abel and D. A. Armitage, *Adv. Organomet. Chem.*, **5**, 1 (1967).
405. S. N. Borisov, M. G. Voronkov and E. Ya. Lukevits, *Organosilicon Derivatives of Phosphorous and Sulfur*, Plenum Press, New York–London, 1971.
406. D. Brandes, *J. Organomet. Chem. Library*, **7**, 257 (1979).
407. V. Etienne, *Bull. Soc. Chim. France*, 791 (1953).
408. F. Fehér and H. Goller, *Z. Naturforsch.*, **22b**, 1223 (1967).
409. E. P. Lebedev, D. V. Fridland and V. O. Reikhsfel'd, *Zh. Obshch. Khim.*, **44**, 294 (1974). *Chem. Abstr.*, **80**, 121048 (1974).
410. A. V. Topchiev, N. S. Nametkin and L. S. Povarov, *Issled Obl. Kremniiorg. Soedin., Akad. Nauk SSSR, Inst. Neftekhim. Sinteza, Sb. Statei*, 130 (1962); *Chem. Abstr.* **59**, 6430 f (1963).
411. T. Sakakibara, T. Hirabayashi and Y. Ishii, *J. Organomet. Chem.*, **46**, 231 (1972).
412. W. C. Schumb and W. J. Bernard, *J. Am. Chem. Soc.*, **77**, 862 (1955).
413. E. Larsson and R. C. I. Mårin, *Acta Chem. Scand.*, **5**, 964 (1951).
414. C. G. Pitt, *J. Am. Chem. Soc.*, **91**, 6613 (1969).
415. L. H. Sommer and J. D. Citron, *J. Org. Chem.*, **32**, 2470 (1967).
416. R. Calas, N. Duffaut, B. Martel and C. Paris, *Bull. Soc. Chim. France*, 886 (1961).
417. L. Birkofer, A. Ritter and H. Goller, *Chem. Ber.*, **96**, 3289 (1963).
418. A. G. MacDiarmid, *Quart. Rev.*, **10**, 208 (1956).
419. G. Champetier, Y. Etienne and R. Kullmann, *C.R. Acad. Sci. Paris*, **234**, 1985 (1952).
420. E. Karakoyunlu and D. Brandes, *Chem. Ztg.*, **100**, 441 (1976).
421. E. N. Gladyshev, V. S. Andreevichev, N. S. Vyazankin and G. A. Razuvaev, *Zh. Obshch. Khim.*, **40**, 939 (1970).
422. E. Larsson and R. E. I. Mårin, *Swed. Patent 138357* (Uddeholms Aktiebolag) (1952); *Chem. Abstr.*, **48**, 2761 (1954).
423. E. P. Lebedev, D. V. Fridland and V. O. Reikhsfel'd, *Zh. Obshch. Khim.*, **44**, 2784 (1974); *Chem. Abstr.*, **82**, 73080 (1975).
424. D.-H. Nguyen, Yu. I. Baukov and I. F. Lutsenko, *Zh. Obshch. Khim.*, **39**, 922 (1969); *Chem. Abstr.*, **71**, 50138 (1969).
425. D. N. Harpp and K. Steliou, *Synthesis*, 721 (1976); E. Louis and G. Urry, *Inorg. Chem.*, **7**, 1253 (1968).
426. E. W. Abel, *J. Chem. Soc.*, 4933 (1961).
427. E. P. Lebedev, V. A. Barburina and V. O. Reikhsfel'd, *Zh. Obshch. Khim.*, **45**, 348 (1975); *Chem. Abstr.*, **82**, 156422 (1975).

428. E. A. V. Ebsworth, R. Taylor and L. A. Woodward, *Trans. Farady Soc.*, **55**, 211 (1959).
429. H. H. Anderson and H. Fischer, *J. Org. Chem.*, **19**, 1296 (1954).
430. M. Schmidt and H. Schumann, *Z. Anorg. Allg. Chem.*, **325**, 130 (1963).
431. D. L. Mayfield, R. A. Flath and R. L. Best, *J. Org. Chem.*, **29**, 2444 (1964).
432. M. Hong, Y. Zhao and X. Zhou, *Gaodeng Xuexiao Huaxue Huebao*, **4**, 735 (1983); *Chem. Abstr.*, **100**, 209948 t (1984).
433. W. Wojnowski and J. Pikies, *Z. Anorg. Allg. Chem.*, **508**, 201 (1984).
434. F. Fehér and G. Goller, *Z. Naturforsch.*, **22b**, 1224 (1967).
435. I. Ojima, M. Nihonyanagi and Y. Nagai *J. Organomet. Chem.*, **50**, C 26 (1973); I. Ojima and Y. Nagai, *J. Organomet. Chem.*, **57**, C 42 (1973).
436. M. G. Voronkov, R. G. Mirskov, O. S. Stankevich, S. P. Sitnikova and E. O. Tsetlina, *Dokl. Akad. Nauk SSSR*, **221**, 347 (1975). *Chem. Abstr.*, **83**, 10315 (1975).
437. H. Gilman and G. D. Lichtenwalter, *J. Org. Chem.*, **25**, 1064 (1960).
438. S. Kato, M. Mizuta and Y. Ishii, *J. Organomet. Chem.*, **55**, 121 (1973).
439. H. Breederveld, *Recl. Trav. Chim. Pays-Bas.*, **81**, 276 (1962).
440. M. N. Bochkarev, L. P. Sanina and N. S. Vyazankin, *Zh. Obshch. Khim.*, **39**, 135 (1969); *Chem. Abstr.*, **70**, 96876 (1969).
441. M. Schmidt and H. Ruf, *Z. Anorg. Chem.*, **321**, 270 (1963).
442. J. W. Anderson, G. K. Barker, J. E. Drake and M. Rodger, *J. Chem. Soc., Dalton Trans.*, 1716 (1973).
443. J. Dubac and P. Mazerolles, *C.R. Acad. Sci. Paris*, **267**, 411 (1968).
444. C. H. W. Jones and R. D. Sharma, *J. Organomet. Chem.*, **268**, 113 (1984).

The Chemistry of Organic Silicon Compounds
Edited by S. Patai and Z. Rappoport

CHAPTER **11**

Recent synthetic applications of organosilanes

GERALD L. LARSON
Petrarch Systems, Bartram Road, Bristol, PA 19007, USA

I. INTRODUCTION

The unabated growth of applications of organosilanes in the resolution of synthetic problems has continued. This includes new applications and modifications of older reactions and innovations. A single chapter covering all aspects of this burgeoning field is essentially an impossible task. Consequently, the coverage has been limited to selected articles from the last two years making this somewhat of a progress report.

A number of reviews have been written on various topics of organosilanes in synthesis. Some of these are listed along with their topics[1].

II. TRIMETHYLCHLOROSILANE

Long considered a useful reagent for silylations of active hydrogen species, for trapping of organometallic reagents and as a useful reagent for the synthesis of other more useful organosilanes, trimethylchlorosilane (TMSCl) has recently found use due to its 'low' reactivity with organometallic reagents. For example, Cooke, Jr.[2] has found that esters are directly converted to ketones via their reaction with an organolithium reagent when the reaction is carried out in the presence of trimethylchlorosilane, which serves to trap the intermediate lithiated ketal (equation 1). One does find the formation of the tertiary alcohol and some attack at silicon in this procedure. The employment of Grignard reagents gives only the tertiary alcohol, perhaps due to a lower reactivity of the halomagnesium alkoxide with the chlorosilane.

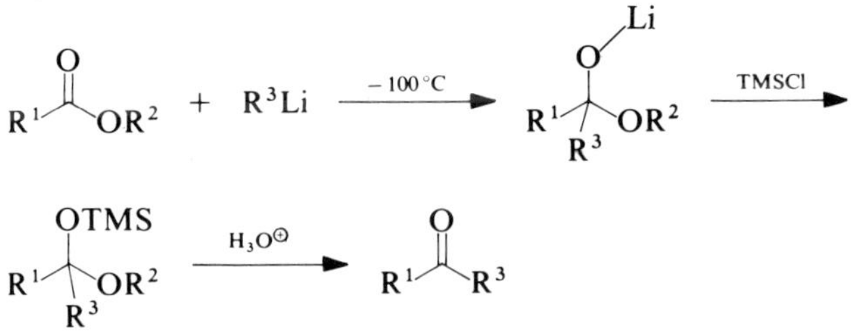

(1)

A number of groups have seized on the Corey and Boaz[3] observation that trimethylchlorosilane enhances the addition of organocopper species to α,β-unsaturated carbonyl compounds. The reaction goes with the stoichiometric organocopper reagents or with half an equivalent of a dialkylorganocuprate[4] (equation 2). The reaction has been shown to proceed with a Grignard reagent in the presence of a catalytic amount of Cu(I) as well[5] (equation 3).

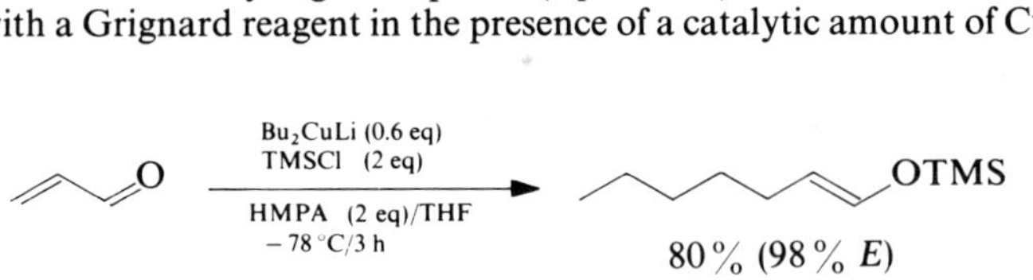

(2)

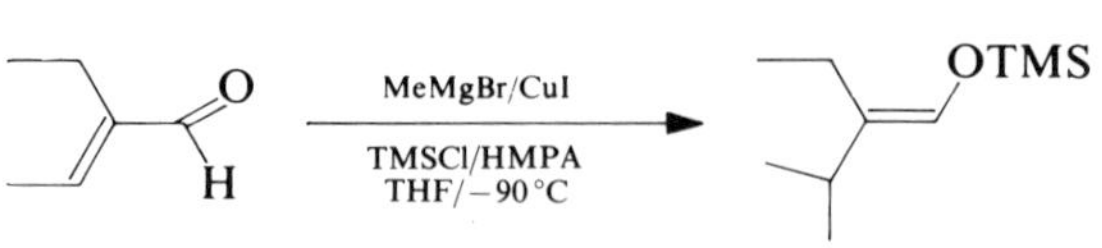

(3)

Most recently, the combination of trimethylchlorosilane and TMEDA was shown to greatly enhance the addition of stoichiometric organocopper reagents to enones[6] (equation 4).

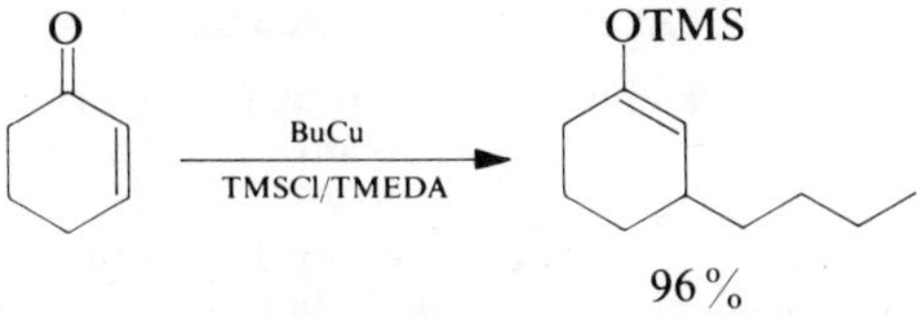

(4)

(without TMSCl 70%; without TMEDA 6%)

The effect of trimethylchlorosilane on the conjugate addition of organocuprates has been investigated with α,β-unsaturated esters and amides[7] in addition to the already discussed α,β-unsaturated aldehydes and enones.

Cyclopropyl carbinols were treated with trimethylchlorosilane to give the chloride. Direct substitution is the reaction pathway when the carbinol is secondary and ring-opened product if the carbinol is tertiary[8]. The stability of the carbocationic intermediate seems to be important in determining the product ratio. Secondary cyclopropyl carbinols give only direct substitution with trimethylchlorosilane alone, a mixture in the presence of lithium bromide and only ring-opened product in the presence of lithium iodide (equation 5). The increased amount of ring-opened material in the presence of bromide and iodide is presumably due to the stronger Lewis acids trimethylbromo- and trimethyliodosilane in these latter two reactions.

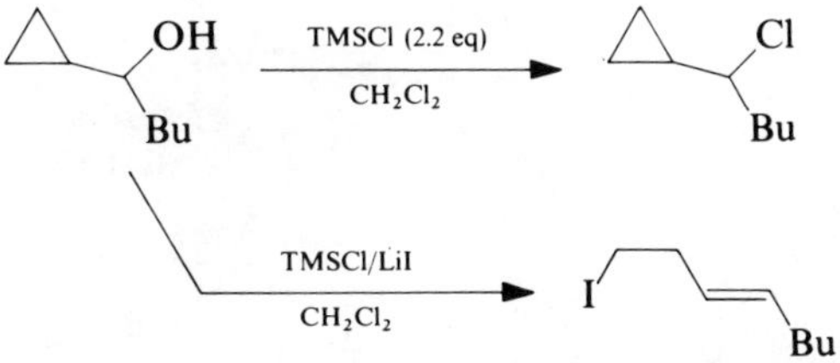

(5)

Trimethylbromosilane has been used to deesterify phosphate esters. Two such examples are shown here: Chouinard and Bartlett's conversion of shikimic acid to 5-enolpyruvylshikimate 3-phosphate[9] (equation 6) and the Degenhardt and Burdsall preparation of ethylenylidenebis(phosphonic acid)[10] (equation 7).

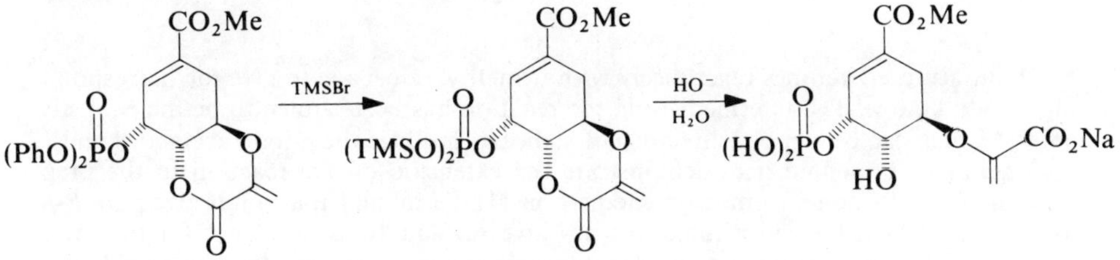

(6)

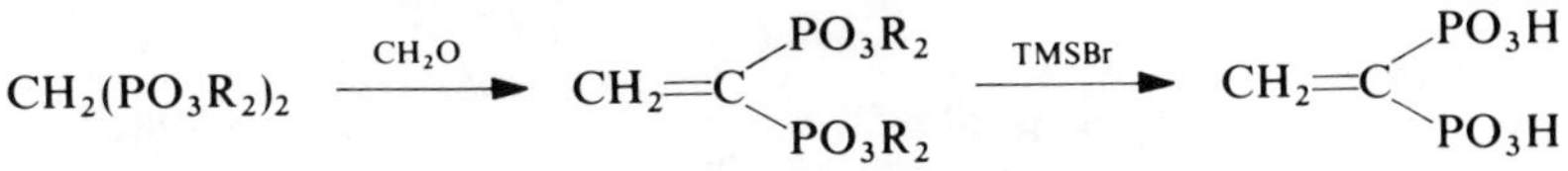

(R = various alkyl groups)

(7)

III. TRIMETHYLIODOSILANE

Although somewhat overshadowed by its cousin, trimethylsilyl trifluoromethanesulfonate (see below), as a Lewis acid catalyst, trimethyliodosilane (TMSI) has found use in this regard as seen by the examples given below. It was used in the key rearrangement employed in the synthesis of hirsutene[11] (equation 8). The α-iodoester **1** was cyclized in good yield with trimethyliodosilane[12] (equation 9). In this latter cyclization it is argued that the α-iodo ester is first converted to the silyl ketene acetal, which then cyclizes with the silylated enone. It was also employed by Hudlicky and coworkers[13] in the rearrangement of vinylcyclopropanes (equation 10).

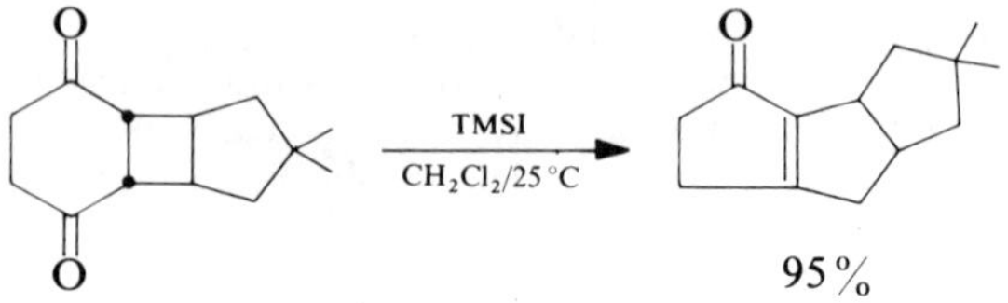

(8)

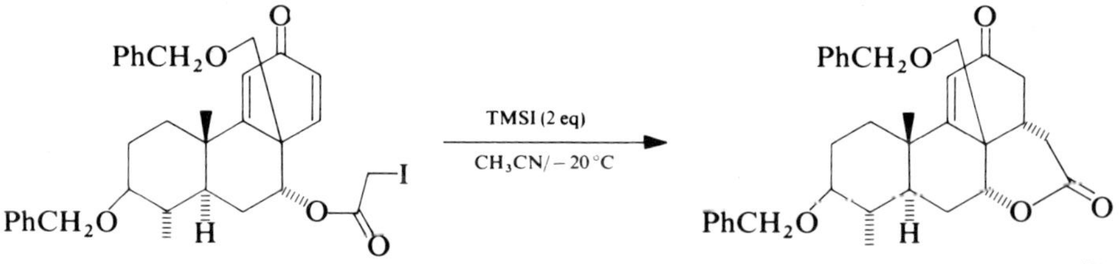

(1) (9)

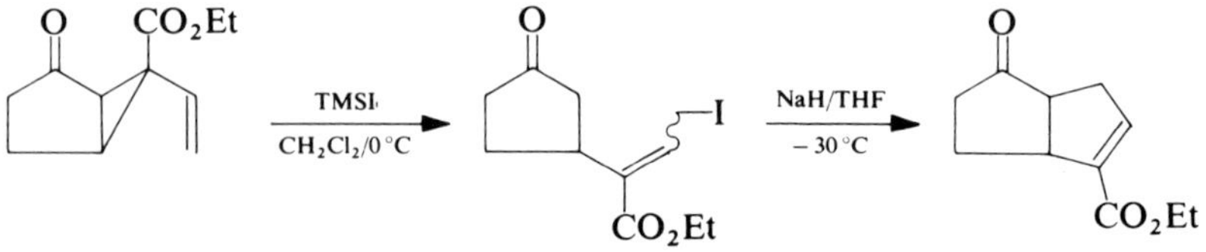

(10)

α,β-unsaturated ketones react nicely with trimethyliodosilane to give the corresponding β-iodo ketone. The intermediate in the reaction has been shown to be the enol silyl ether[14] which can be trapped directly with a diol to produce the β-iodo acetal or ketal[15]. The examples presented here demonstrate the extension of this reaction to the preparation of β-iodoenones from ynones (equation 11). Terminal triple bonds give pure *E*-β-iodo enones[16, 17], but substituted alkynes give *E,Z* mixtures. Evidence for the intermediacy of the enol silyl ether comes from trapping of this intermediate with an aldehyde in a Mukaiyama reaction[18] (equation 12).

(11)

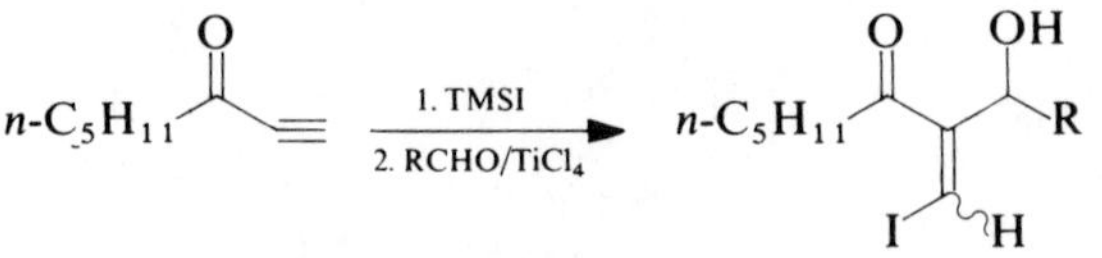

(12)

Grieco and coworkers[19] employed the trimethyliodosilane/hexamethyldisilazane (HMDS) combination, first introduced by Miller and McKean[20], to prepare dienol silyl ether **2** (equation 13). This enol silyl ether was converted via the Rubottom reaction to the sensitive α-hydroxy enone A ring of Quassimarin, **3**.

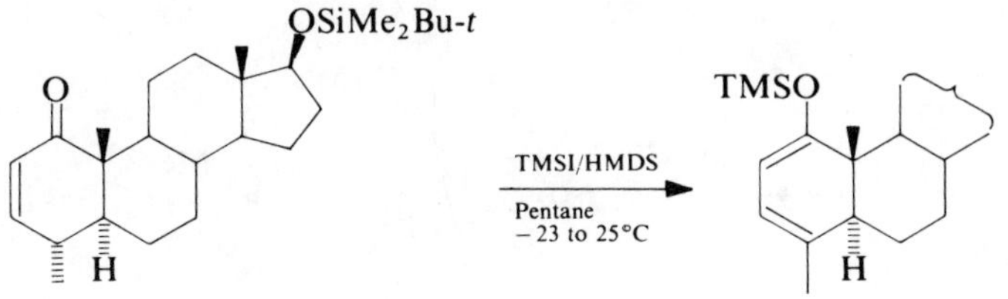

(13)

(**2**)

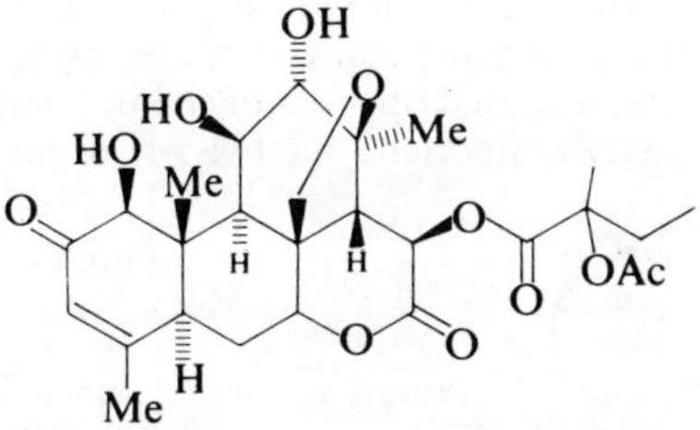

(**3**)

It has been known for some time that trimethyliodosilane is an excellent reagent for the dealkylation of ethers and the saponification of esters. Two recent examples show that this can be done with enol methyl ethers[21] (equation 14) and carbamates[22] (equation 15) as well.

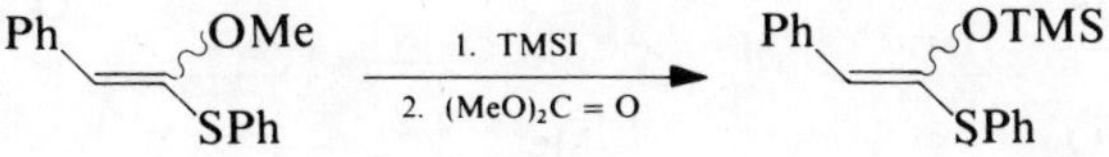

(14)

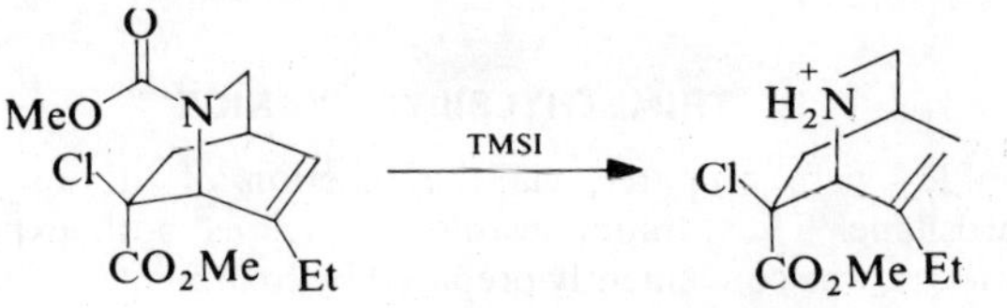

(15)

In related reactions the trimethyliodosilane opening of ketal **4** was shown to be highly stereospecific, with the equatorial oxygen reacting with the silicon moiety as shown [23] (equation 16).

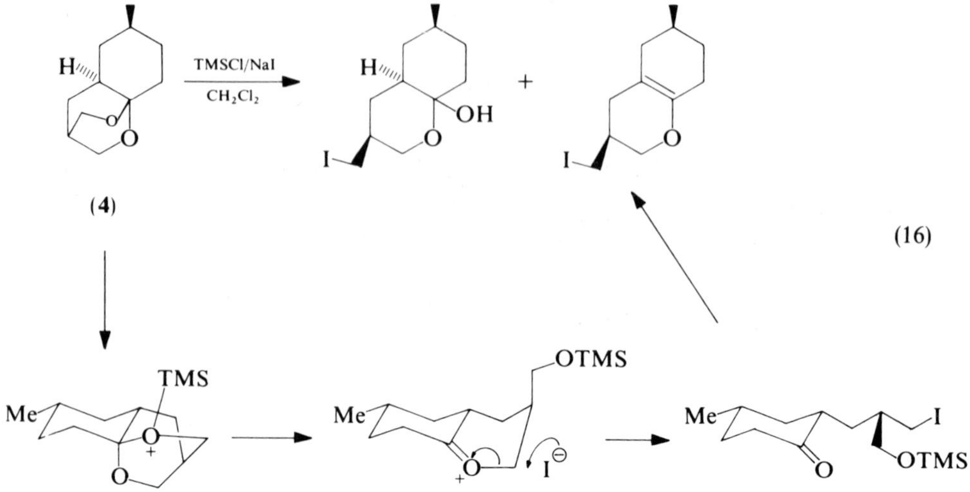

(4)

(16)

Epoxide **5** was cleanly and stereoselectively opened with *tert*-butyldimethyliodosilane, generated *in situ*[24]. Unfortunately, a fair amount of the hydrolyzed material **6** was also formed (equation 17). Compound **5** was used to prepare some 3,14-dihydroxytrichothecenes. The trimethyliodosilane-mediated opening of enantiomerically pure **7** was the key step in the preparation of singly antipodal 2-methyl-3-hydroxy ester[25] (eqation 18).

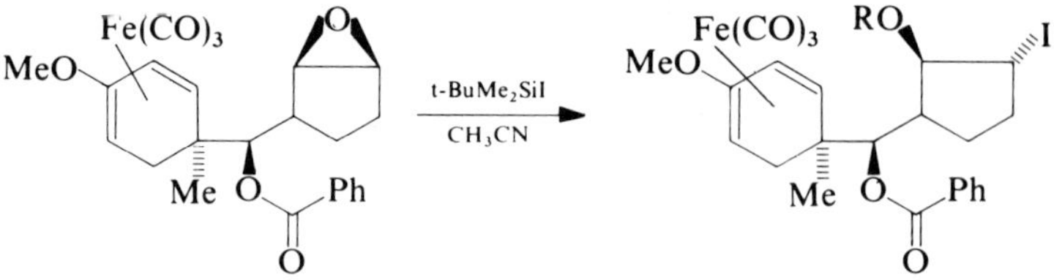

(17)

(**5**)

R = t-$BuMe_2Si$
R = H (**6**)

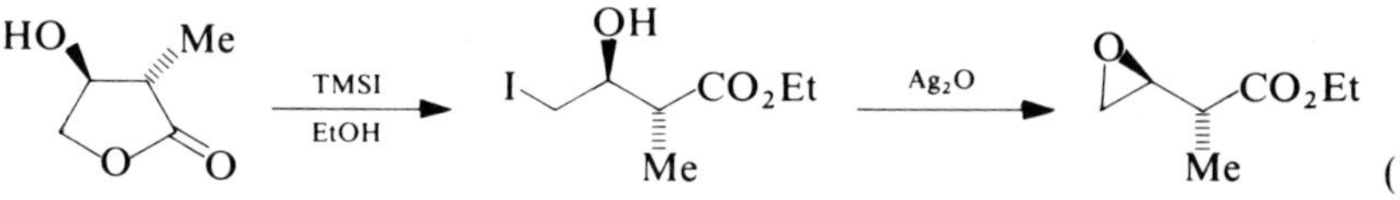

(**7**) (18)

IV. TRIMETHYLSILYL CYANIDE

Several silyl cyanides were prepared via the reaction of lithium cyanide with the appropriate chlorosilane. The lithium cyanide, which is both expensive and has a tendency to polymerize, was conveniently prepared by treatment of trimethylsilyl cyanide with *n*-butyllithium (equation 19). Using 2,6-di-*tert*-butylphenol as a test substrate, the authors found that although trimethylsilyl cyanide is less reactive than bistrimethylsilyl acetamide (BSA) as a silylating agent, it is more reactive than trimethylsilyl triflate[26].

$$R_3SiCl + LiCN \longrightarrow R_3SiCN + LiCl \qquad (19)$$

$$TMSCN \xrightarrow{BuLi} LiCN$$

In an extension of the addition of trimethylsilyl cyanide to aldehydes and ketones, it was found that the combination of trimethylsilyl cyanide and lithium cyanide in tetrahydrofuran adds to these systems in high yield[27]. The lithium ion must serve as a catalyst in this reaction, eliminating the need for the zinc chloride that is ordinarily used for this purpose (equation 20).

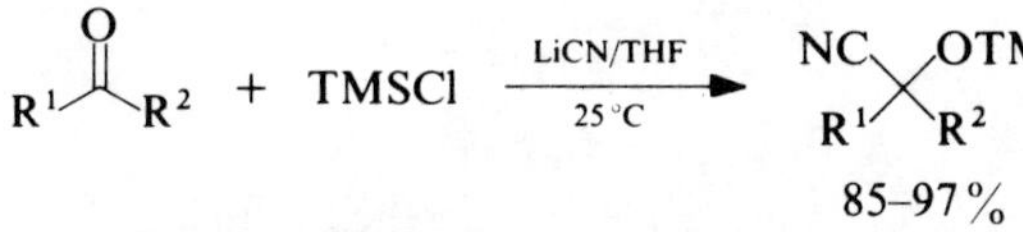

(20)

A novel entry into these important intermediates is the 'anionic' addition of the trimethylsilyl group, generated from **8**, to acyl cyanides[28]. The final product is the result of an umpolung effect of the addition when the adduct, an α,β-unsaturated acyl anion equivalent, adds to another molecule of acyl cyanide (equation 21).

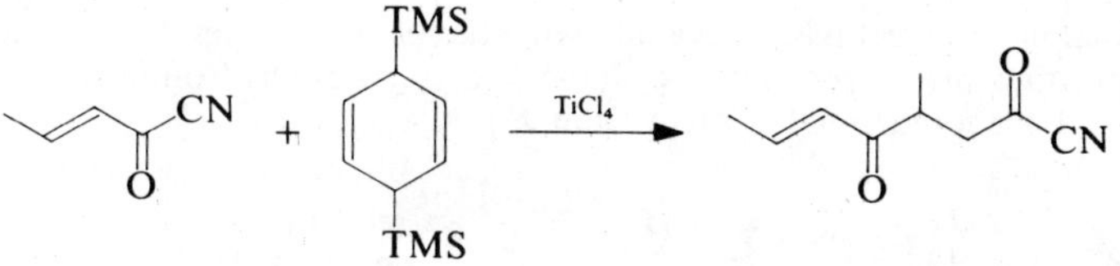

(**8**)

(21)

The addition of trimethylsilyl cyanide to aldehydes was carried out by Reetz and coworkers[29] who employed the optically active chloroborane **9** as the Lewis acid catalyst. The enantiomeric excesses were found to be between 12 and 16% (equation 22).

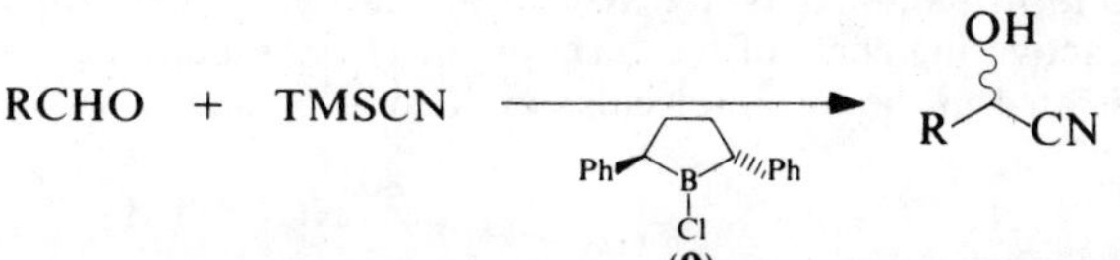

(**9**)

(22)

Trimethylsilyl cyanide adds to tropones under the catalysis of zinc iodide to give 1-trimethylsiloxy-2-cyanotropylidene or cyanotropone derivatives[30] (equation 23).

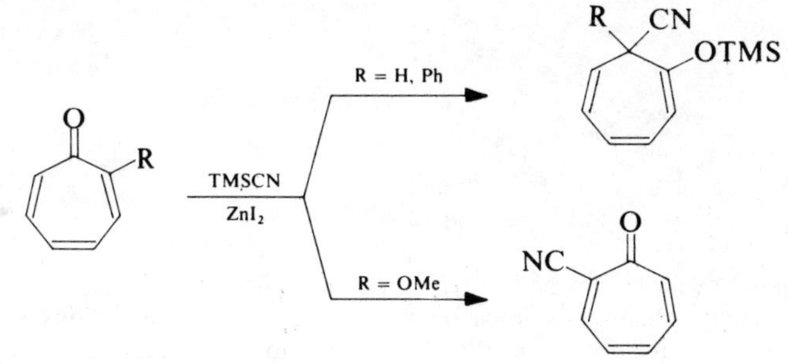

(23)

A study of the mechanism of the addition of trimethylsilyl cyanide to epoxide **10** is consistent with the generation of a carbenium ion[31] (equation 24). The structure of the principal product **11** was confirmed by an X-ray crystal structure. *Tert*-butyldimethylsilyl cyanide opens epoxides to give the *tert*-butyldimethylsilyl protected β-hydroxy isocyanides[32] (equation 24a).

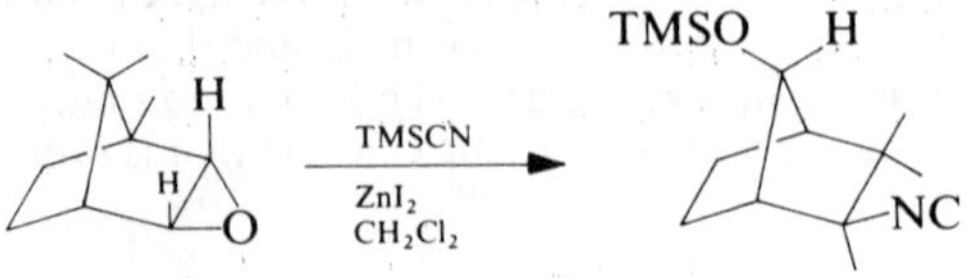

+ 8 other products (24)

(10) **(11)** 72%

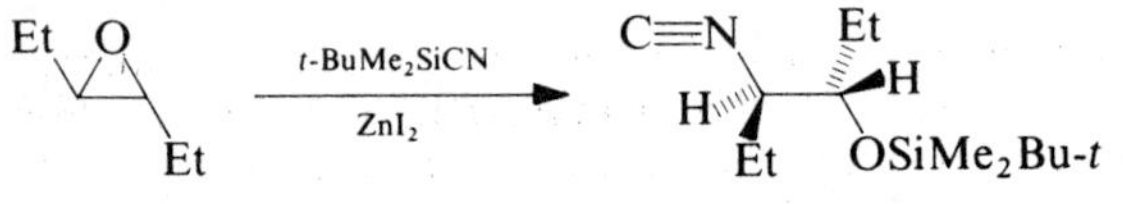

(24a)

74%

The reaction of trimethylsilyl cyanide with the *cis* ene diones **12**, produced photochemically from furans, provides 2-cyano-5-hydroxy-2,5-dihydrofurans, which can be oxidized to 4-cyanobutenolides[33] (equation 25).

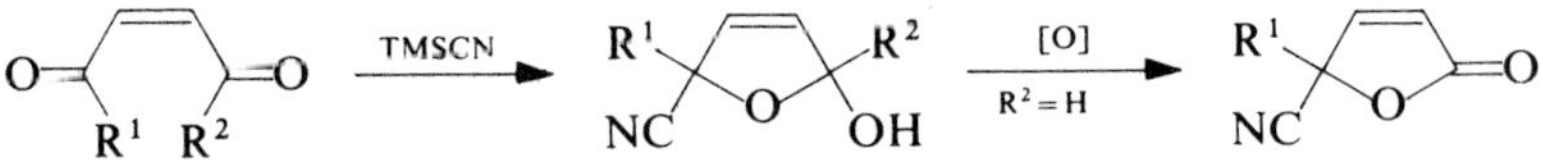

(25)

(12)

In an interesting application it was found that the addition of trimethylsilyl cyanide to formylfurans to form **13** not only allowed for alkylation of the formyl group, but also removed the deactivating effect of the carbonyl in Diels–Alder reactions of the furan[34]. This is demonstrated in the intramolecular cycloaddition shown in equation 26.

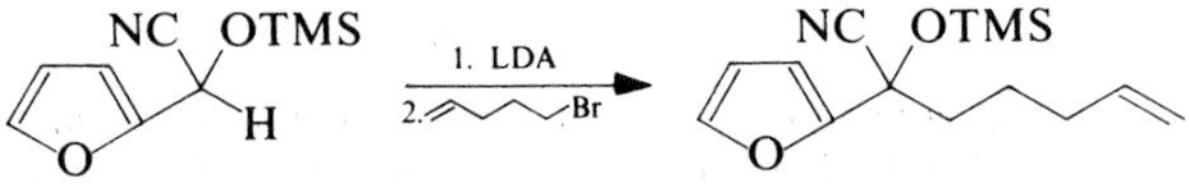

(13)

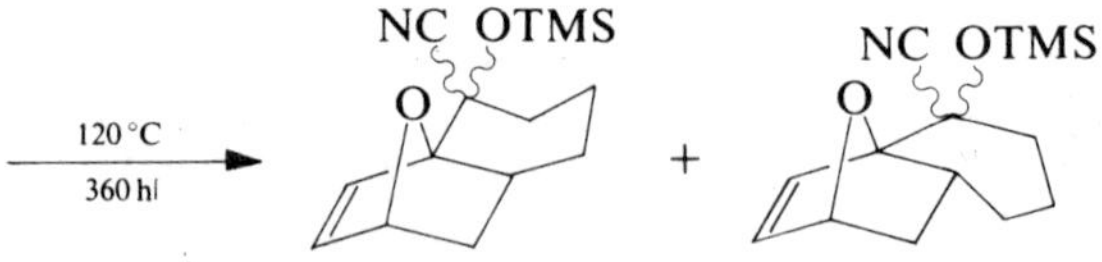

(26)

88 : 12

Trimethylsilyl cyanide can be used to introduce the nitrile group in a substitution reaction, especially when a stable carbenium ion can be formed. Thus, chloride **14** is converted stereospecifically to nitrile **15**[35] (equation 27). Benzyl or β,γ-allyl nitro compounds are substituted with trimethylsilyl cyanide[36] (equation 28). In the case of the allyl systems both possible regioisomers are obtained. Carbamate **16** was reacted with trimethylsilyl cyanide in the presence of triethylamine to produce the acyl cyanide **17** as part of a synthesis of 6-α-substituted penicillins[37] (equation 29).

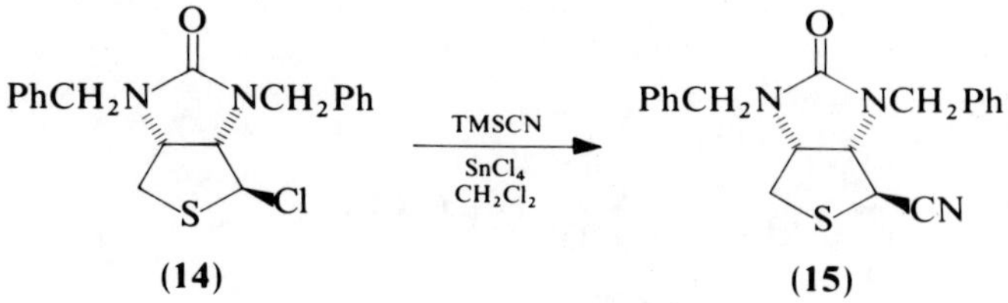

(27)

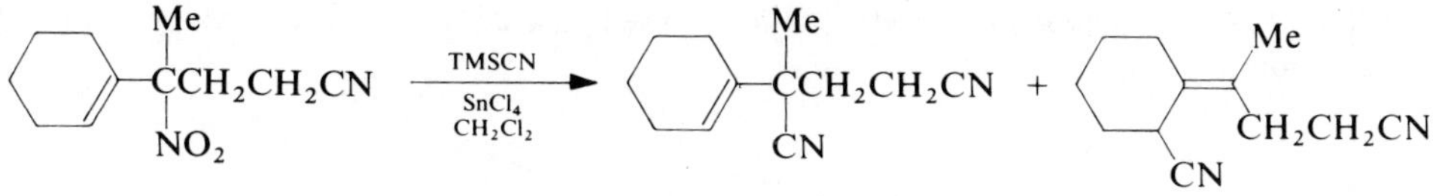

(28)

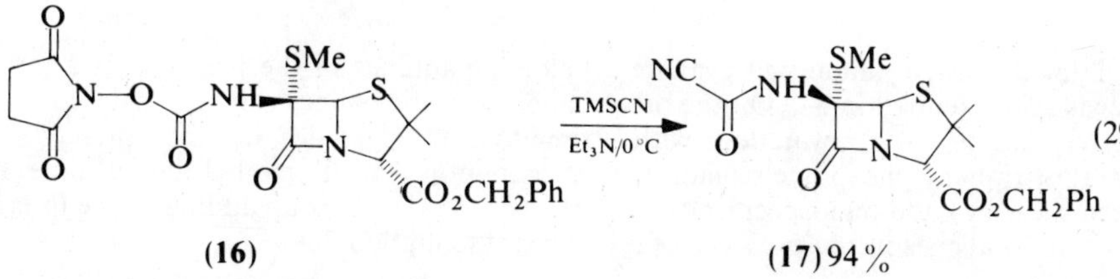

(29)

Trimethylsilyl cyanide converts, 1,2,4,5-tetrazines to 4-aminopyrazoles in refluxing toluene[38] (equation 30).

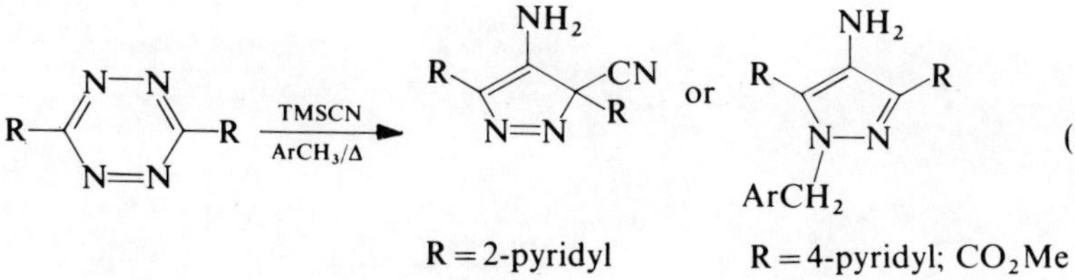

(30)

The silylated cyanohydrins, formed by the addition of trimethylsilyl cyanide to ketones or aldehydes, were reacted with Grignard reagents to give, upon hydrolysis of the intermediate, α-hydroxy ketones **18** in good yields[39] (equation 31). A Reformatsky approach was used to react ester enolates with these same intermediates[40]. In this case the reaction also takes place at the nitrile group, but hydrolysis brings about lactonization to the tetronic acids (equation 32).

It has been shown that olefins can be hydrocyanated with trimethylsilyl cyanide and bis(cyclopentadienyl)hydridochlorozirconium(IV). The reaction proceeds somewhat

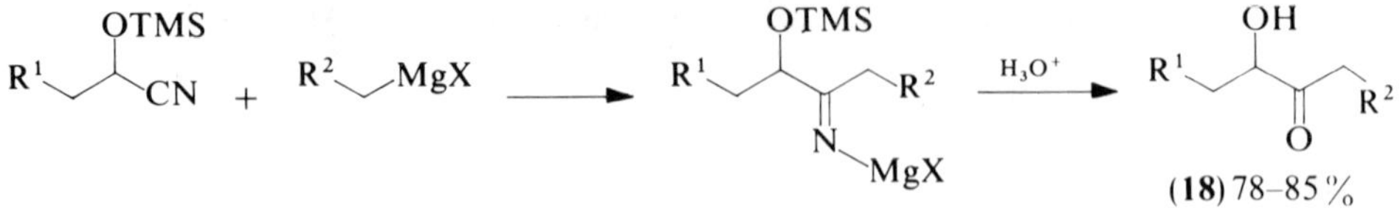

(18) 78–85%

(31)

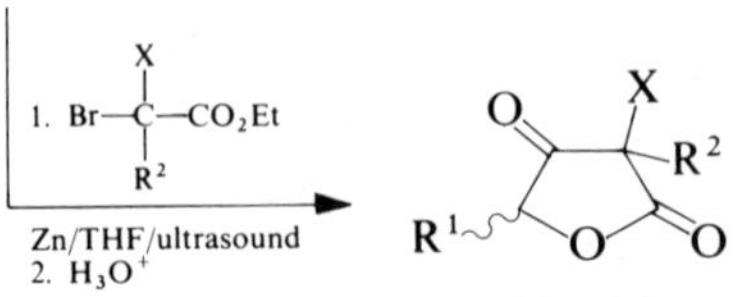

48–69%

(X = F, CF_3) (32)

better, however, when *tert*-butyl cyanide is used instead of trimethylsilyl cyanide[1] (equation 33).

45%

(33)

Trimethylsilyl cyanide will cyanate 1-haloadamantanes in the presence of a strong Lewis acid[42] (equation 34). On the other hand, aromatic halides can be converted to the corresponding aryl cyanides with trimethylsilyl cyanide in the presence of tetrakis(triphenylphosphine)palladium(0)[43] (equation 35). Trimethylsilyl cyanide has been shown to add regiospecifically, with the trimethylsilyl group adding to the internal carbon, to allenes in the presence of a catalyst[44] (equation 36).

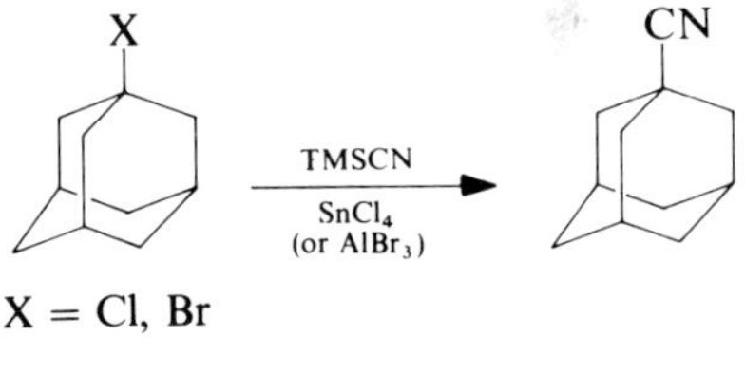

X = Cl, Br

(34)

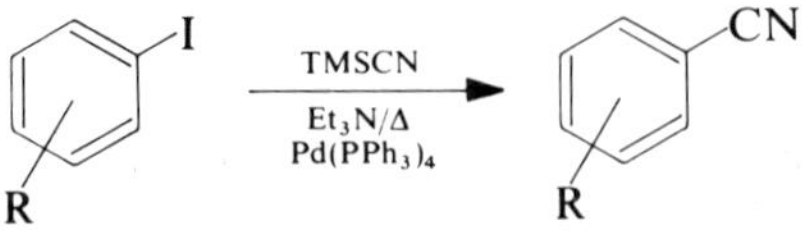

(35)

(36)

(catalyst = $PdCl_2$, $NiCl_2$, $PdBr_2$ or DIBAH)

Trimethylsilyl cyanide will cycloadd to acetylenes. Both trimethylsilylacetylenes[45] (equation 37) and arylacetylenes[46](equation 38) have been investigated.

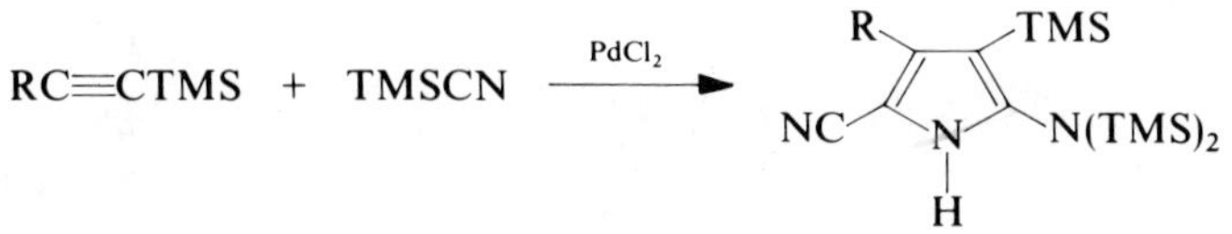

(37)

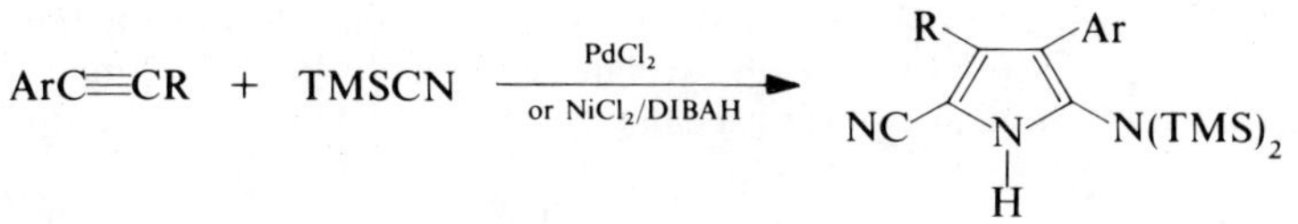

(38)

V. TRIMETHYLSILYL TRIFLUOROMETHANESULFONATE (TRIMETHYLSILYL TRIFLATE)

Trimethylsilyl triflate (TMSOTf) has grown in popularity as a selective, relatively easily handled, soluble Lewis acid and has been applied as such in several instances. Many of these are often in conjunction with other silylated reagents, making for a dual silicon-mediated transformation. Several representative examples of trimethylsilyl triflate as a Lewis acid are given.

Bis-trimethylsilylated α-hydroxy acids can be reacted with pivaldehyde with trimethylsilyl triflate catalysis to give 1,3-dioxolane-4-ones[47] (equation 39). In a similar vein the bis-trimethylsilylated enantiomerically enriched (88% ee) β-hydroxy acid **19** can be condensed with aldehydes in the presence of trimethylsilyl triflate to produce the 1,3-dioxane-4-ones[48] (equation 40). These were in turn reacted with organocuprates to provide optically active secondary alcohols.

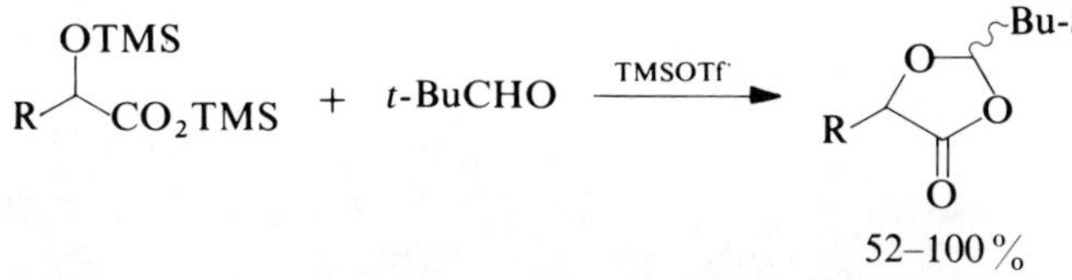

(39)

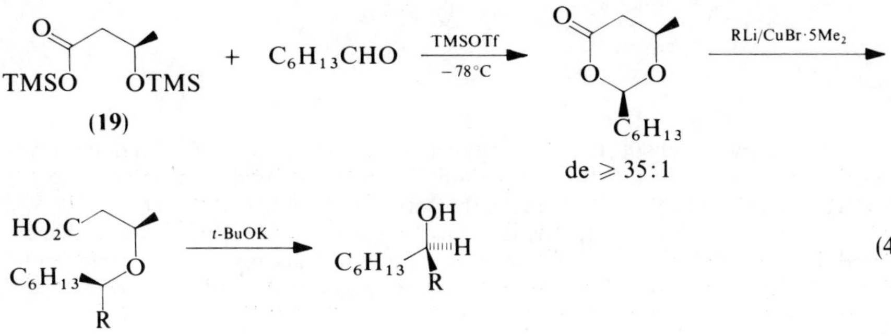

(40)

Imines can be prepared in excellent yields from the trimethylsilyl triflate catalyzed reaction of bis-trimethylsilylated primary amines and ketones or aldehydes[49] (equation 41).

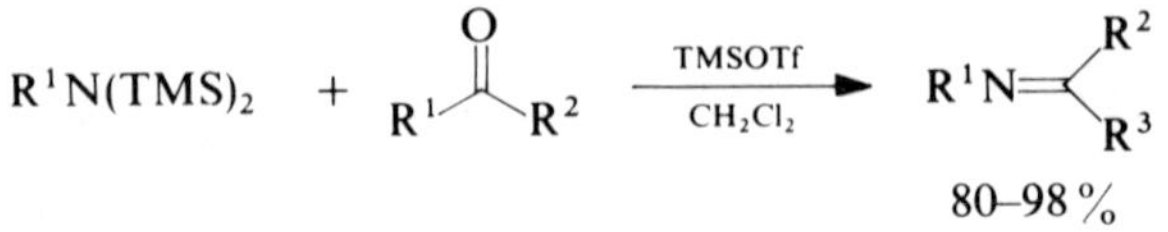

(41)

The reaction of acetals and ketals with enol silyl ethers, usually catalyzed with the more traditional Lewis acids such as boron trifluoride etherate or titanium tetrachloride, can be carried out with trimethylsilyl triflate catalysis[50]. In the example shown in equation 42 the resulting allylsilane was annulated with titanium tetrachloride.

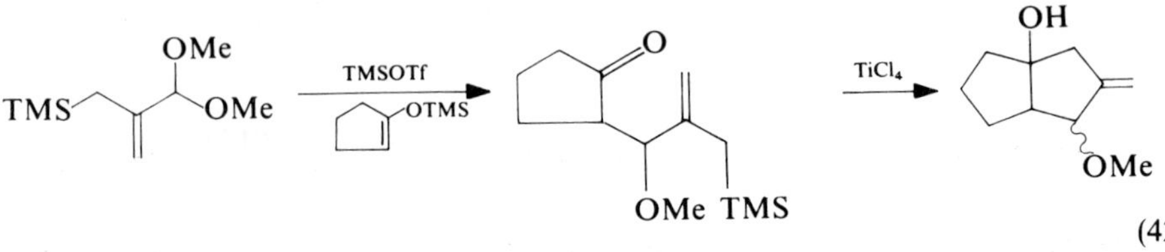

(42)

Acetals and ketals of ethylene glycol are readily formed from bis-trimethylsilyl ethylene glycol, the aldehyde or ketone and trimethylsilyl triflate catalysis[51] (equation 43). The reaction is quite selective for ketones over aldehydes (equation 44).

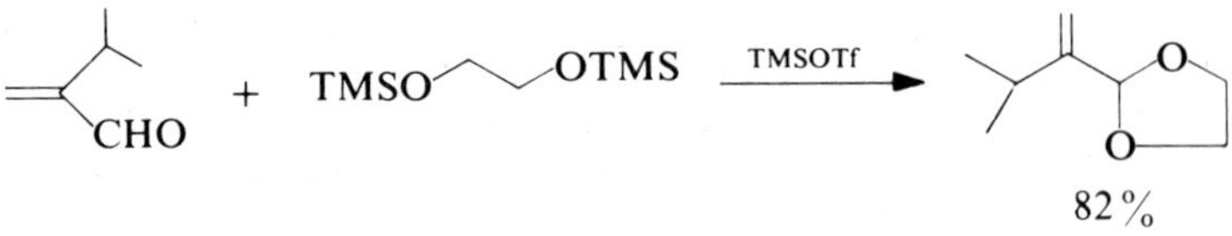

(43)

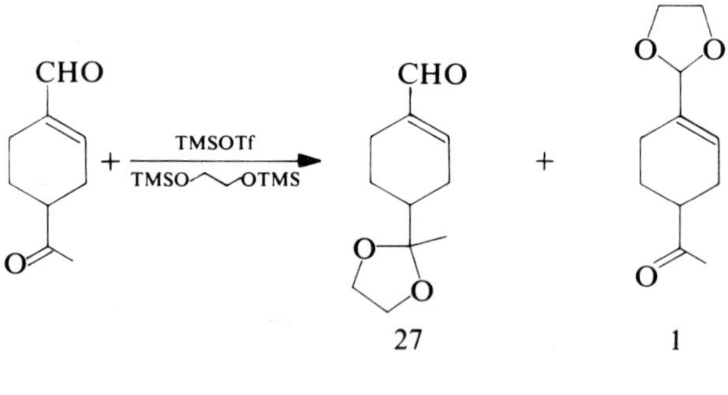

(44)

Optically active 2-substituted-1,3-propanediol derivatives were prepared as shown in equation 45. The bis-trimethylsilylated 1,3-diols were condensed with *l*-menthone in the presence of trimethylsilyl triflate to give the corresponding diastereomeric ketals, which were separated and subjected to further transformations. For example, the cleavage was carried out via a Mukaiyama reaction to generate the α-alkoxy ketone, which could be protected and then cleaved to the optically active, unsymmetrical, monoprotected 1,3-diol[52].

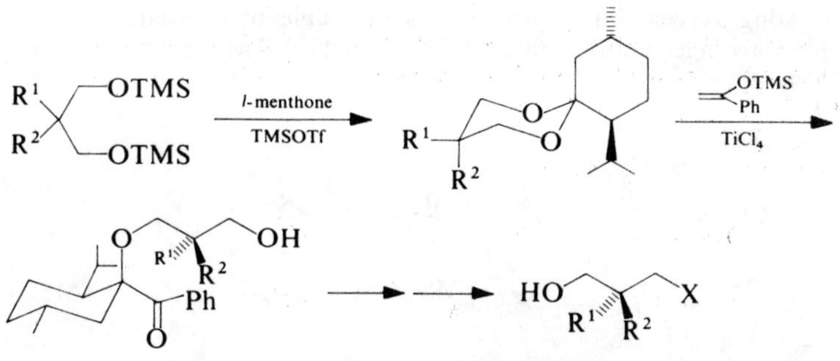

(45)

X = OTHP, OCPh$_3$, SPh

Trimethylsilyl triflate was found to be superior to silver triflate in the glycosidation of 4-demethoxyanthracyclinones[53] (equation 46).

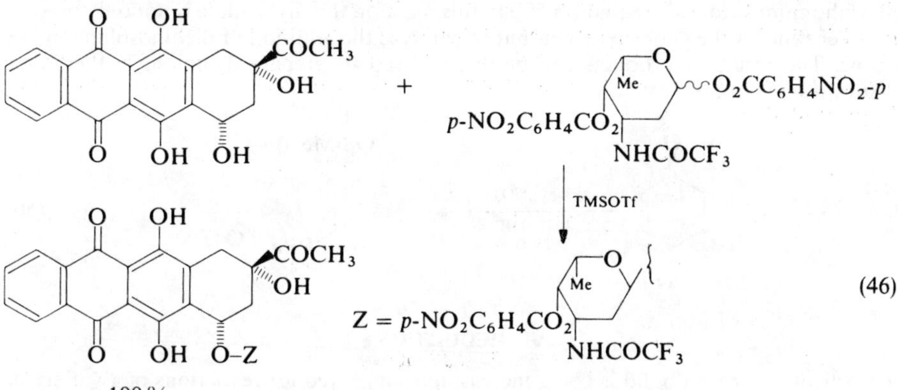

(46)

~100%

Trimethylsilyl triflate was used to bring about the cyclopentannulation shown in equation 47[54]. It was used to simultaneously catalyze the phenylselenylation of the double bond and cyclization of vinylsilane **20**[55] (equation 48).

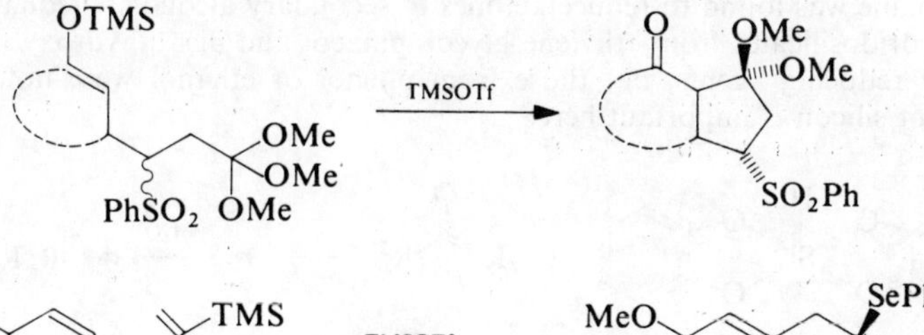

(47)

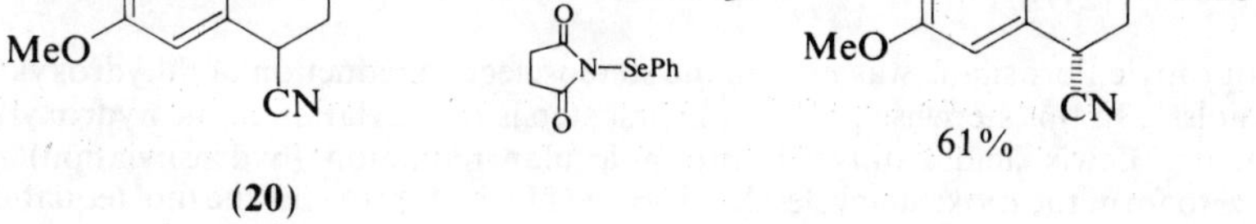

(48)

(20)

An interesting oxygen ylide is proposed as a possible intermediate in the [2,3] Wittig sigmatropic rearrangement illustrated in equation 49[56]. The reaction gives high erythro and Z selectivity.

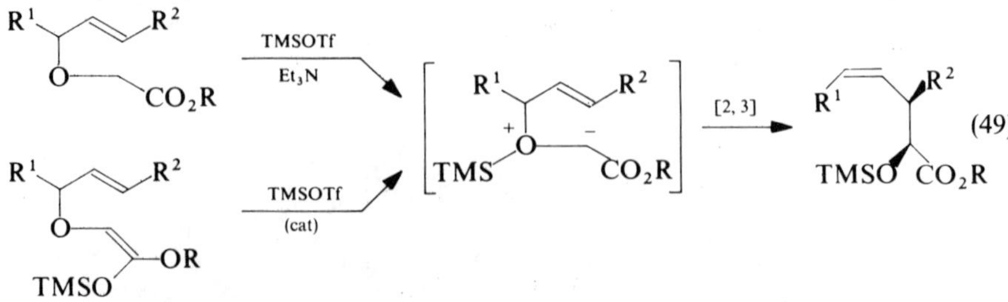

(49)

Enones can be reacted with *tert*-butyldimethylsilyl triflate and triphenylphosphine to produce the regiospecifically generated enol *tert*-butyldimethylsilyl ether of β-triphenylphosphonium ketones[57] (equation 50). In this reaction the silyl triflate is not only present in order to form the enol silyl ether, but to catalyze the addition of the phosphine to the enone. The resultant products can be further used to prepare dienol silyl ethers and β-substituted enones.

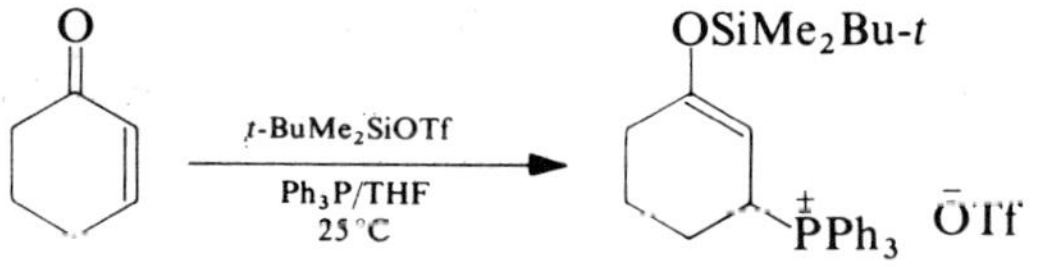

(50)

VI. REDUCTIONS

The silicon hydrogen bond is being increasingly employed for reductions of a variety of functional groups. In many cases the reduction procedures represent major improvements in terms of ease of reaction and workup over existing methodology.

Ketones are reduced by silanes, particularly in the presence of a good silicophile such as fluoride ion. This reaction has recently been shown to occur via single electron transfer[58]. The interesting silicohydride **21** formed *in situ* from dilithiocatechol and trichlorosilane was found to reduce ketones to secondary alcohols[59] (equation 51). The related hydridosilicates from ethylene glycol, pinacol and *o,o'*-dihydroxybiphenyl were also good reducing agents, but those from phenol or ethanol were not. Clearly the pentavalent silicon is important here.

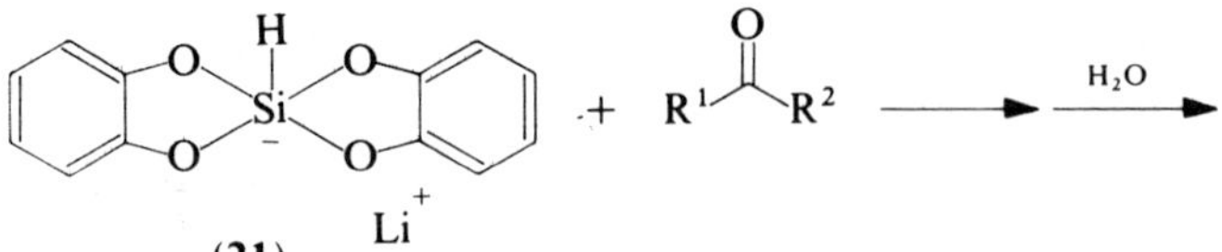

(51)

(**21**)

Diisopropylchlorosilane was used in the stereoselective reduction of β-hydroxyketones to 1,3-diols[60]. In this stepwise process the first step is the silylation of the hydroxyl group and then a Lewis acid catalyzed intramolecular reduction (hydrosilylation) of the carbonyl to form the dioxasilacycles **22**, cleavage of which provides the diol (equation 52).

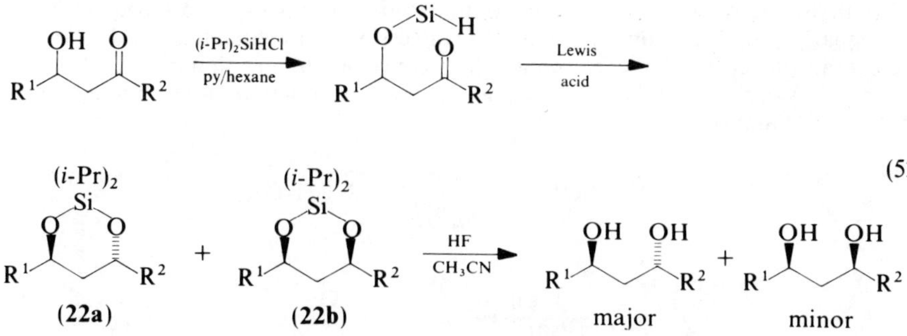

(52)

The combination of triethylsilane and trifluoromethanesulfonic acid reduces aryl ketones to the methylene derivative[61] (equation 53).

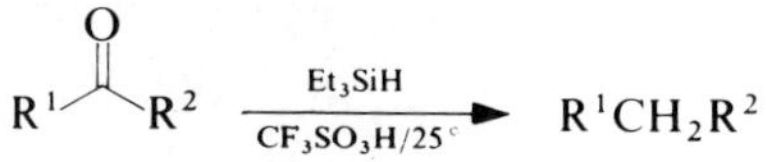

(R^1 and/or R^2 = aryl) (53)

Enones can be reduced to the saturated ketones with triethylsilane and Wilkinson's catalysis[62] (equation 54). Interestingly, the same product was prepared via a Tiffeneau–Demjanov ring expansion wherein trimethylsilyl cyanide was used (equation 55). The selective reduction of the double bond of enones can also be carried out with diphenylsilane in the presence of palladium(0) or palladium(II) and zinc chloride[63] (equation 56), or more effectively with phenylsilane and molybdenum hexacarbonyl[64] (equation 57). This latter reagent was also used to reduce the double bond of α,β-unsaturated esters, amides and nitriles.

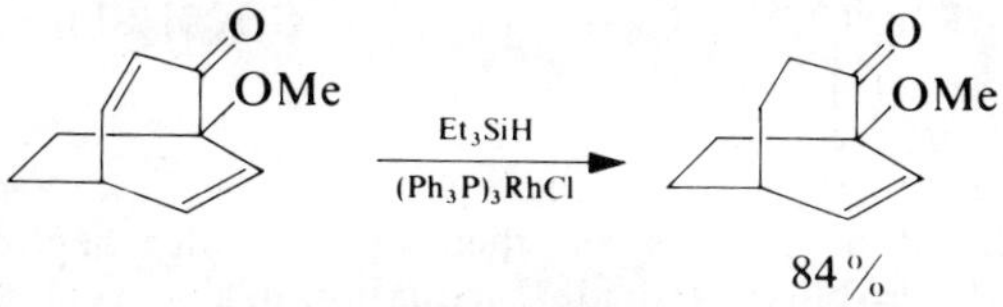

(54)

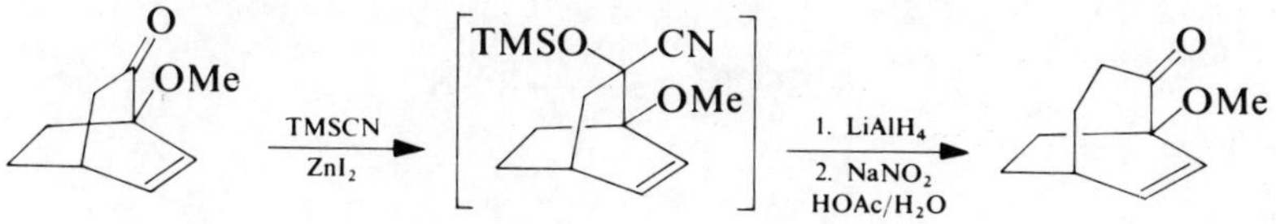

(55)

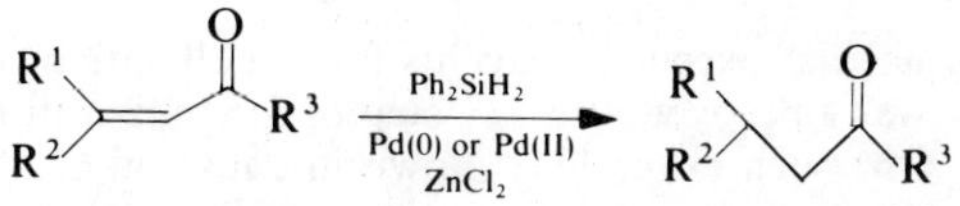

(56)

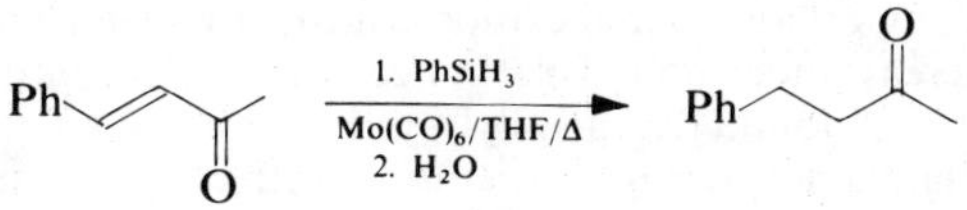

(57)

Triethylsilane reduces ketals to ethers in excellent yields in the presence of Nafion-H[65](equation 58). When one considers the conversion of a ketone to a ketal to an ether, the overall transformation corresponds to a reductive etherification of the ketone. This general class of reduction has also been applied to the reduction of the anomeric carbon of sugars[66](equation 59).

$$R^1CH(OR^2)_2 \xrightarrow[\text{Nafion-H, } CH_2Cl_2/\Delta]{Et_3SiH} R^1CH_2OR^2 \qquad (58)$$

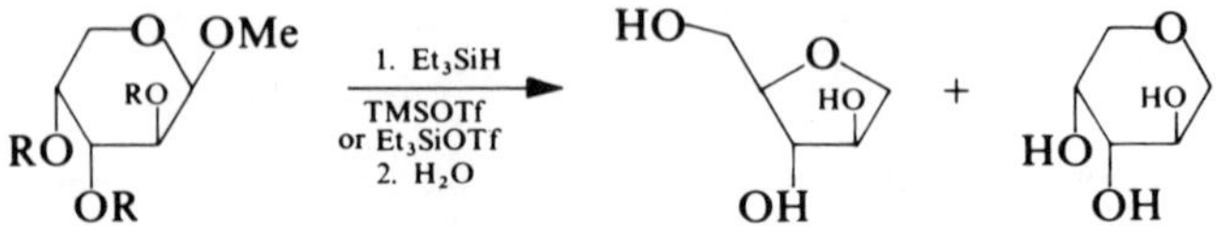

(59)

R = TMS	major	minor
R = Et_3Si	minor	major

Triethylsilane in the presence of a Lewis acid catalyst has been found to reduce α-nitrosulfides to thio ethers[67] (equation 60). The β-nitrosulfides are reduced with rearrangement (equation 61). Benzyl nitro systems are also reduced under these conditions. Tri-*n*-butyltin hydride is still the best reagent for the reduction of alkyl nitro compounds, but in these instances it would reduce the thiophenoxy group as well.

$$RCH(SPh)NO_2 \xrightarrow[SnCl_4]{Et_3SiH} RCH_2SPh \qquad (60)$$

$$R^1R^2C(SPh)CH_2NO_2 \xrightarrow[SnCl_4]{Et_3SiH} R^1R^2CHCH_2SPh \qquad (61)$$

Alcohols are reduced to alkanes via their esters under neutral conditions with triphenylsilane and di-*tert*-butyl peroxide[68] (equation 62).

$$R^1CO_2R^2 \xrightarrow[(t\text{-}BuO)_2/\Delta]{Ph_3SiH} R^2\text{-}H \qquad (62)$$

VII. ALLYLSILANES

Allylsilanes have surfaced as excellent reagents for the allylation of a variety of electrophiles. This reaction was employed as a key step in the preparation of pyranonaphthoquinone antibiotics. One such example is shown in equation 63[69].

The diastereoselectivity of the allylation of α-benzyloxy aldehyde **23** was studied[70] (equation 64). It was found that under chelation control a very high percentage of the syn adduct is formed whereas under non-chelation control the anti product is favored.

Taddei and coworkers[71] found that an allylsilane containing an optically active ligand could be used to allylate aldehydes in an enantioselective manner (equation 65). Of the

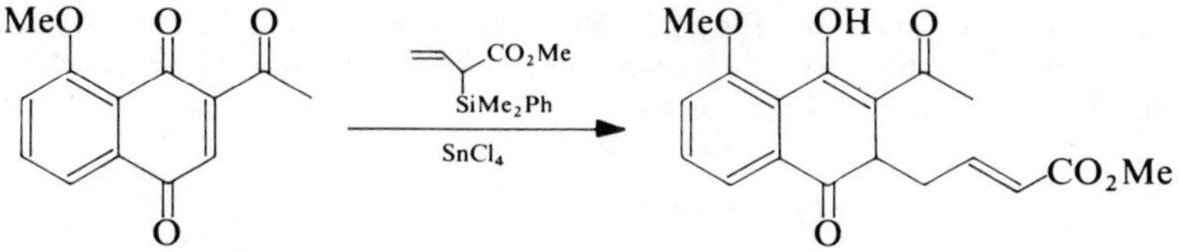

(63)

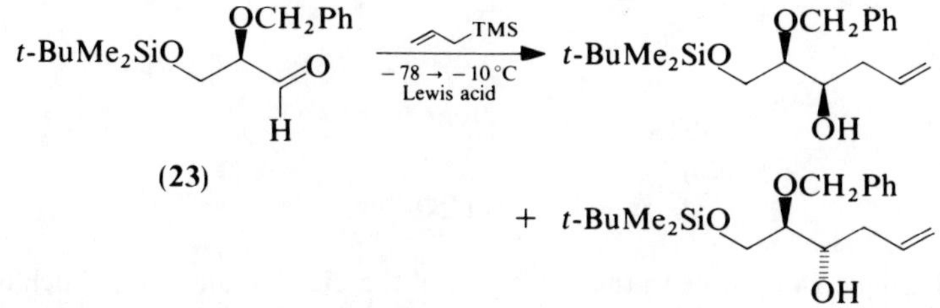

(23)

(64)

$TiCl_4$ or $SnCl_4$	>98%	<2%
$BF_3{\cdot}OEt_2$	19%	81%

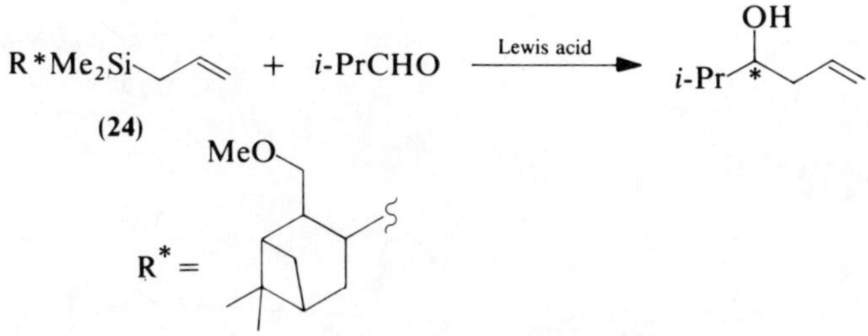

(65)

(24)

variety of Lewis acids investigated for this transformation zinc iodide (59% ee) and titanium tetrachloride (46% ee) gave the best results.

Taddei and coworkers[72] also found that allyltrimethylsilane would react with aldehydes to form 4-halotetrahydropyrans in good to excellent yields (equation 66). Interestingly, the reaction gives stereospecifically the all *cis* product.

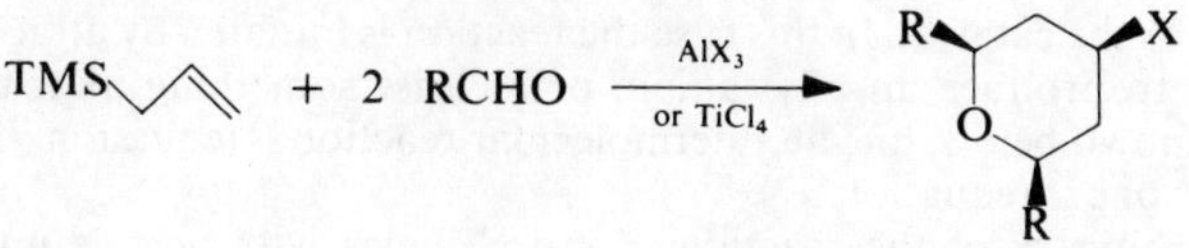

(66)

The allylation of aldehydes has also been done in an intramolecular fashion to produce ring systems. An intramolecular allylation was the key step in the formation of the

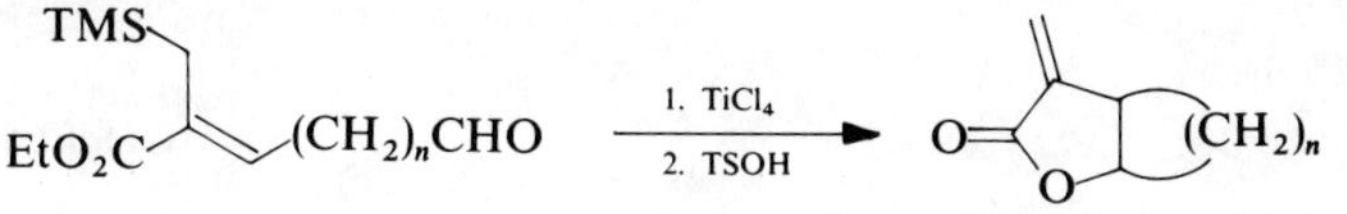

(67)

(**25a**) $n = 3$ 83%

(**25b**) $n = 4$ 90%

bicyclic α-methylene-γ-lactones **25a** and **25b**[73] (equation 67). Allylsilanes of the general structure **26** were cyclized to the exo-methylenecyclopentanes **27** in good yields by Molander and Andrews[74] (equation 68).

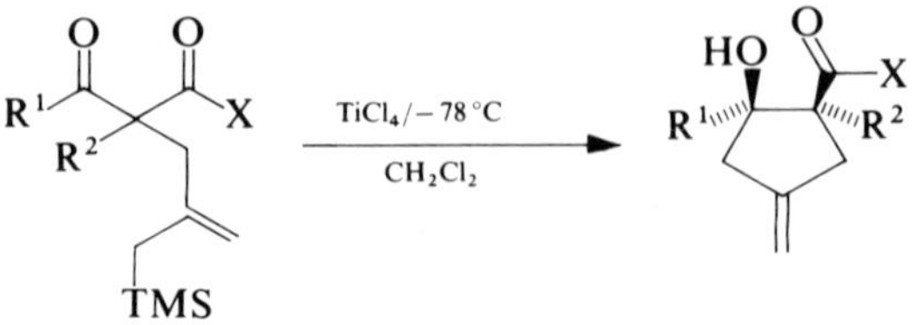

(26) **(27)**

(X = OEt, NR$_2$) (68)

The allylsilane **28** cyclizes to the δ position of the dienone under ethyldichloroaluminum catalysis[75] (equation 69). On the other hand, the corresponding propargylsilane **29** gives only trace amounts of product resulting from attack at the δ position and **30** as the major product (equation 70).

(69)

(28)

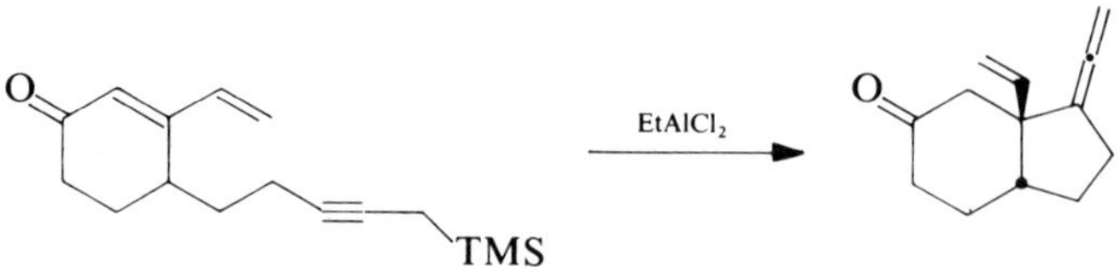

(70)

(29) **(30)**

The allylation of electrophiles with allylsilanes can also be carried out in the presence of fluoride ion as the catalyst. In this case the reaction is initiated by attack of the fluoride ion on silicon to produce an allyl anion, or at least something akin to it. Two such examples are shown below, one an intermolecular reaction[76] (equation 71) and the other intramolecular one[77] (equation 72).

The stereochemistry of the reaction of crotylsilanes with acetals under Lewis acid catalysis was studied by Sakurai and coworkers[78] (equation 73). It was found that with

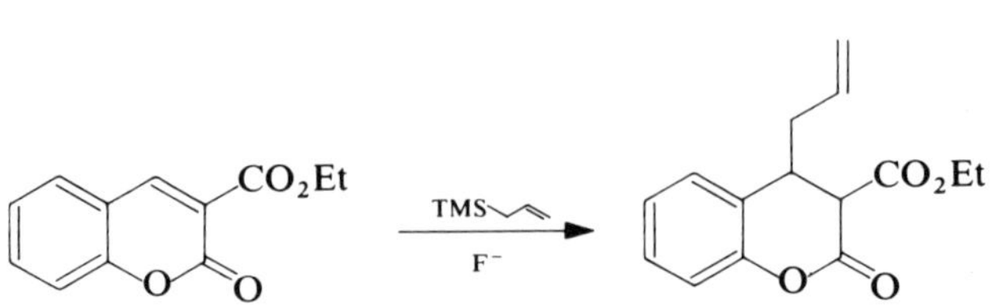

91% (71)

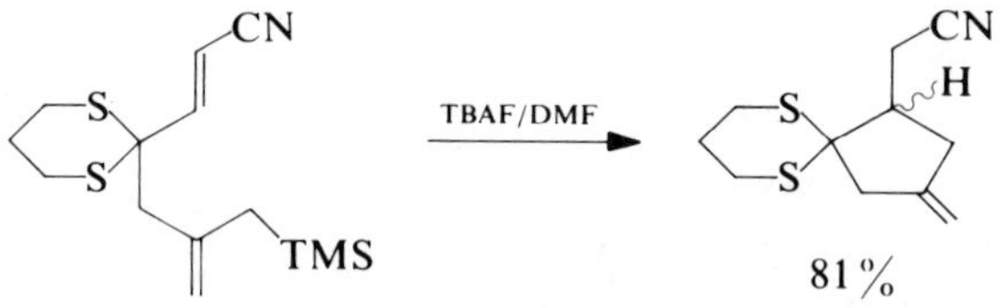

(72)

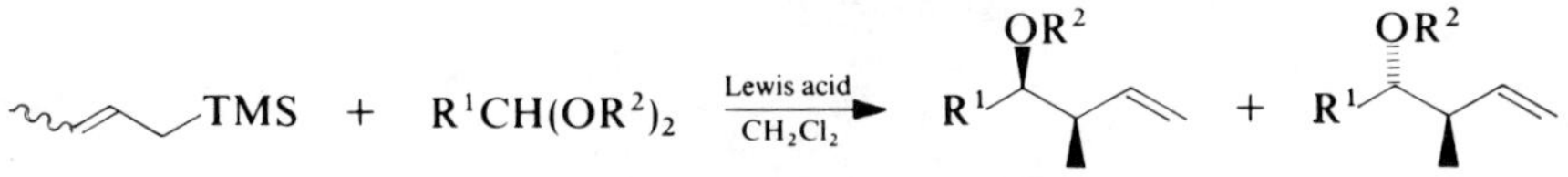

(73)

the acetals of alkyl aldehydes the *E* crotylsilane gives more *syn* adduct and that the *Z* gives more of the *anti* adduct with the exception of the acetal of pivaldehyde. On the other hand, with aryl aldehydes the *Z* crotylsilane gives more of the *anti* adduct with the *E* crotylsilane giving more of the *syn* product. The *Z* and *E* crotylsilanes were also reacted with activated glycal derivatives[79] (equation 74). Here again the *E* crotylsilane favors the formation of the *syn* product and the *Z* crotylsilane the *anti* product.

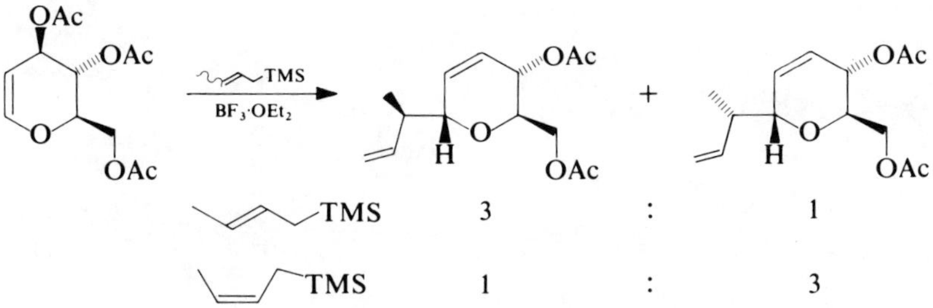

(74)

Allylsilane **31** was employed in fine fashion in the biomimetic polyene cyclization as a key step in a route to the 21-hydroxyprogesterone system[80] (equation 75).

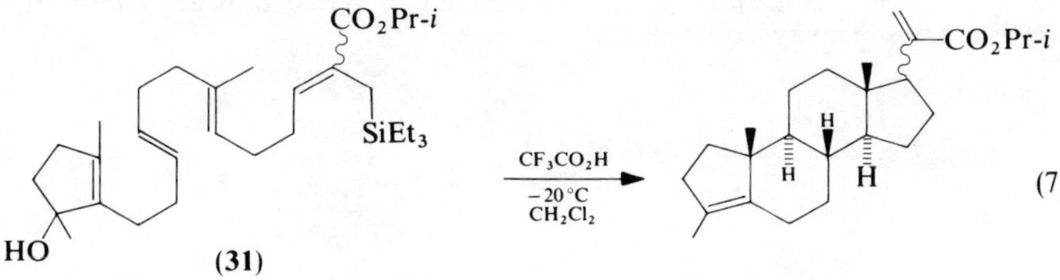

(75)

The allylsilane **32** was used to prepare dienones via the tin tetrachloride catalyzed condensation with acetals or aldehydes followed by an elimination reaction[81] (equation 76). Interestingly, when the condensation is catalyzed with fluoride ion only the α-adduct is observed (equation 77).

The reaction of allylsilanes with nitriles leads to β,γ-unsaturated ketones after hydrolysis[82] (equation 78).

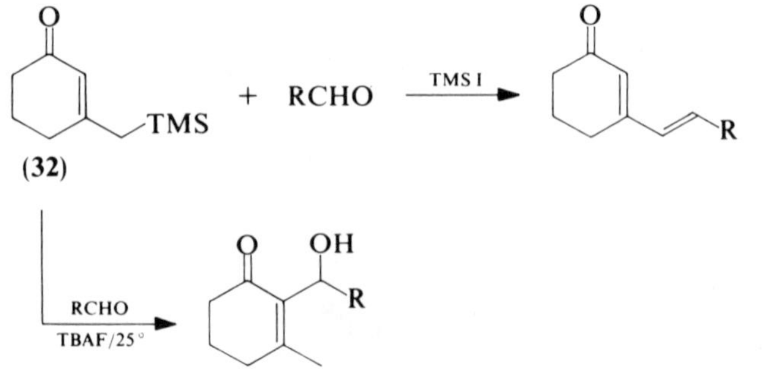

(32)

(76)

(77)

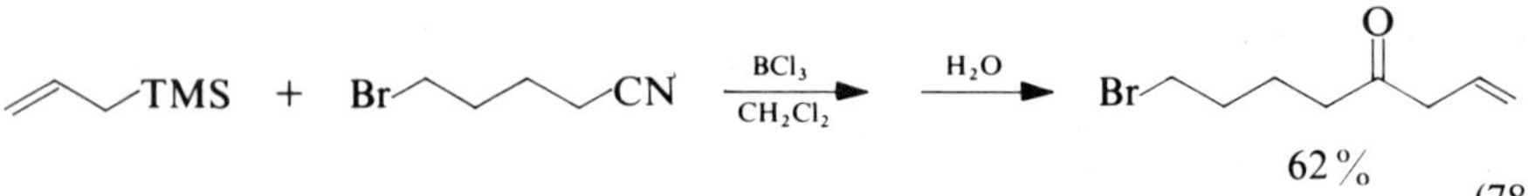

(78)

It has been shown by Yoshida and coworkers[83] that allylsilanes react electrochemically with alcohols to provide allyl ethers (equation 79). Unfortunately, a mixture of regioisomers is obtained from the reaction.

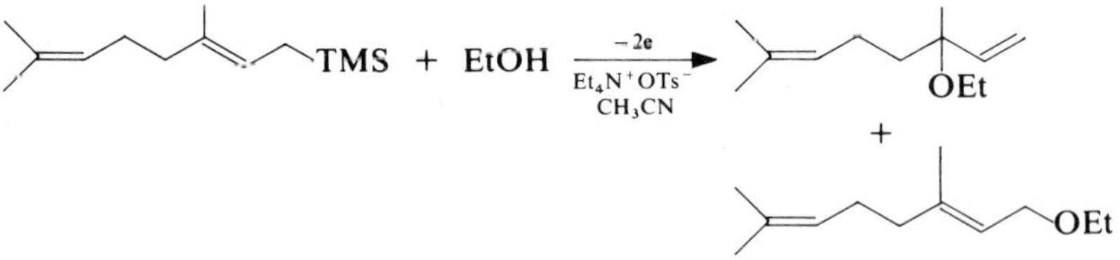

(79)

The lithium reagent produced from the deprotonation of allylsilanes has in the past proven somewhat capricious in its α versus γ reactivity. It has now been demonstrated by Ikeda and Yamamoto[84] that treatment of the lithium reagent with titanium tetraisopropoxide and then an aldehyde gives clean α attack with subsequent Peterson elimination leading to *Z* (*syn* elimination) and *E* (*anti* elimination) terminal dienes with excellent (>50:1) stereoselectivity (equation 80).

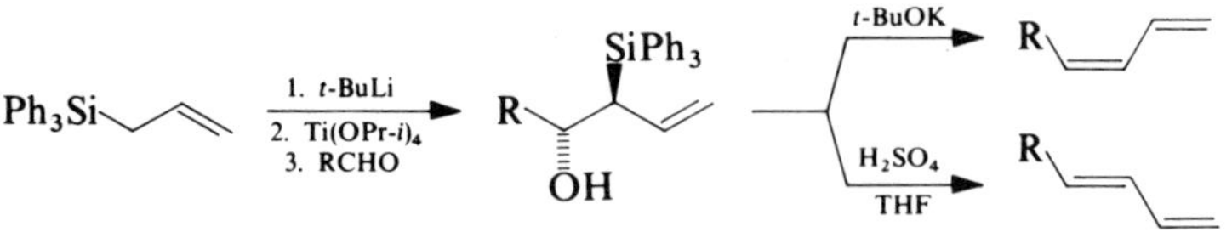

(80)

The lithium reagent produced from (diisopropylamino)allyldimethylsilane was treated sequentially with zinc chloride and an aldehyde and then subjected to hydrogen peroxide oxidation to produce erythro diols of the type **33** in good to excellent yields[85] (equation 81). The presence of the hydrolyzable amino group on the silicon allows the oxidation step to take place.

The acetoxyallylsilane **34** and similar systems react with palladium(0) or palladium(II) to form the trimethylenemethane palladium complexes, which have been employed by

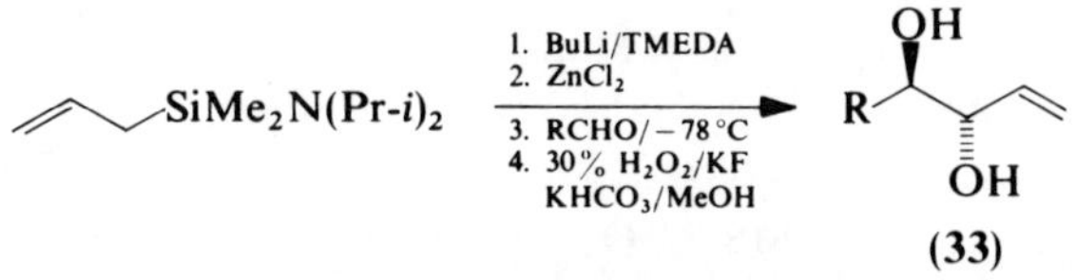

(81)

the Trost group in the formation of a multitude of cyclopentyl systems. This subject has been reviewed by Trost[86]. Most recently it has been found that the [3+2] cycloaddition of the above-mentioned complexes parallel those of the Diels–Alder reaction with the same acceptors[87]. One example is shown (equation 82). The [3+2] cycloaddition of the trimethylenemethane palladium complex and bicycloheptadienes followed by flash vacuum pyrolysis leads to 4-methylenecyclopentenes[88] (equation 83). In the presence of tri-*n*-butyltin acetate as a cocatalyst the reagent will add to aldehydes to give 3-methylene tetrahydrofurans[89] (equation 84). A related reagent **35** has been prepared and reacted with dipolarophiles[90] (equation 85). This reagent, however, gives the carboxylated cyclopentyl system. It is argued that the elimination of methyltrimethylsilylcarbonate liberates carbon dioxide and trimethylmethoxysilane and that the carbon dioxide is trapped by the anion **36** prior to the cycloaddition to give the observed product. Jones and Kemmitt[91] have found that similar systems will react with imines to give 3-methylene pyrrolidines (equation 86).

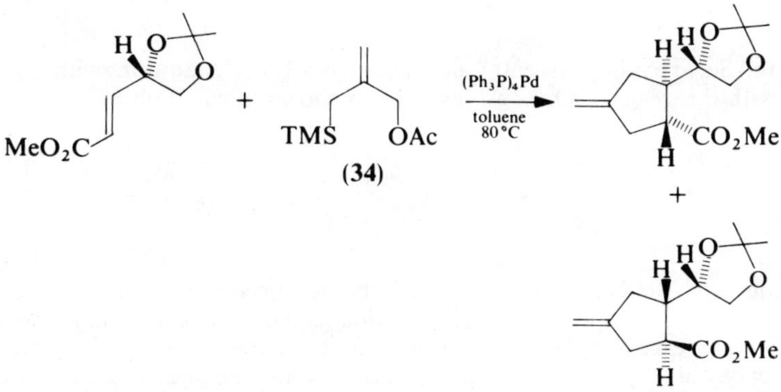

(82)

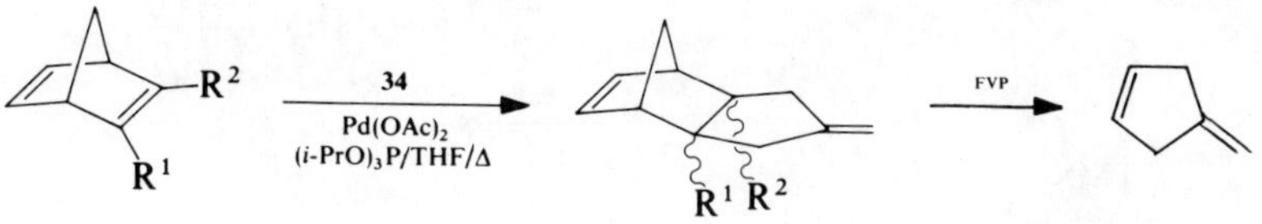

(83)

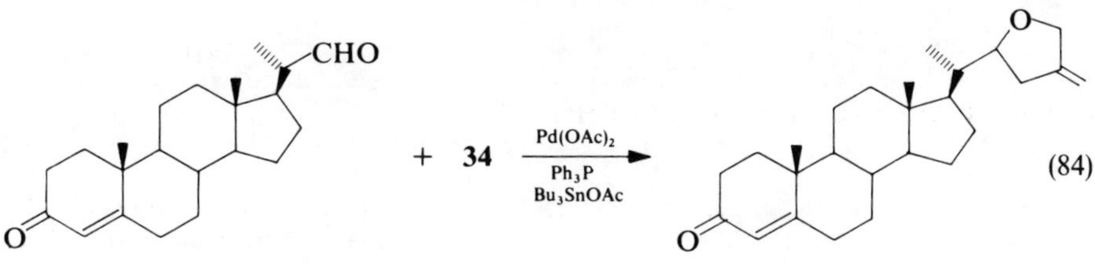

(84)

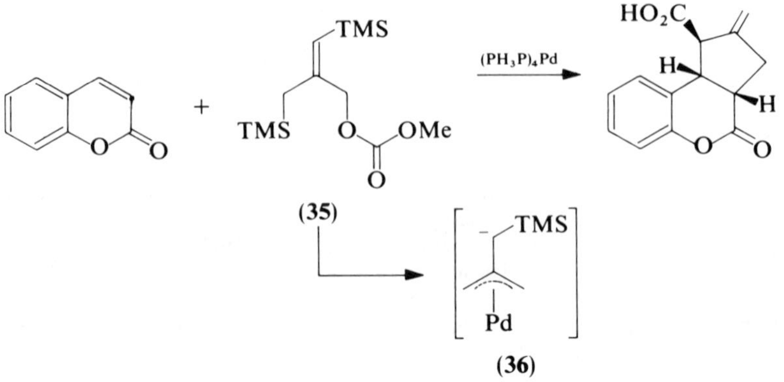

(85)

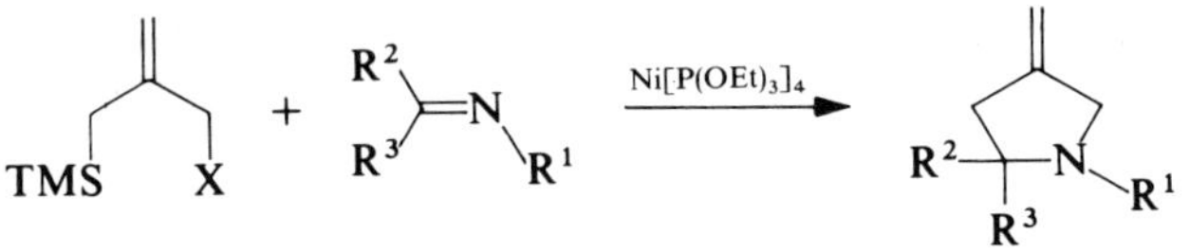

(86)

$X = OSO_2Me, Cl, OAc$

Trost and Seoane[92] have recently shown that [6+3] cycloadditions with tropones are possible with this reagent. One such example is shown in equation 87[92].

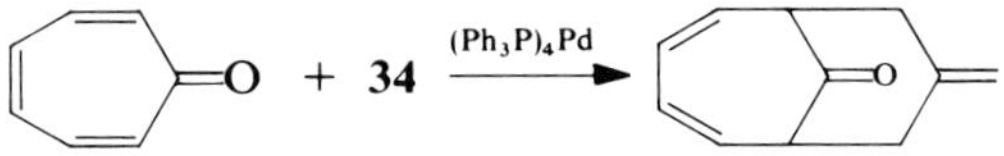

(87)

Molander and Shubert[93] have reported the addition of iodoallylsilane **37** to 1,2-diketones under the influence of tin(II) fluoride (equation 88). The diastereoselectivity of the reaction is excellent. This same reagent/catalyst system brought about the cycloaddition to epoxy aldehyde **38**[94] (equation 89). In this case the diastereoselectivity was not nearly so high in all systems investigated.

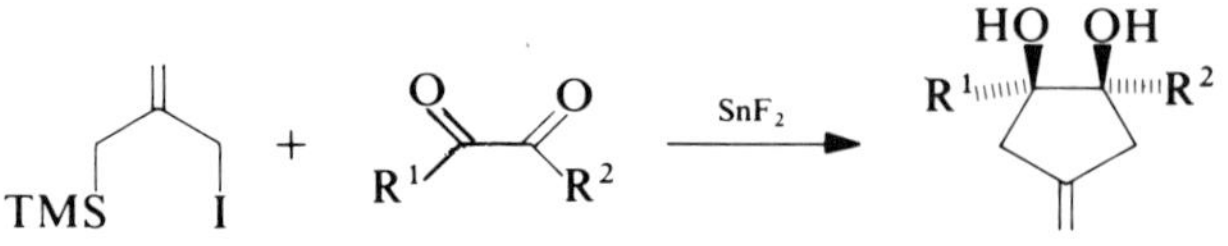

(88)

(37)

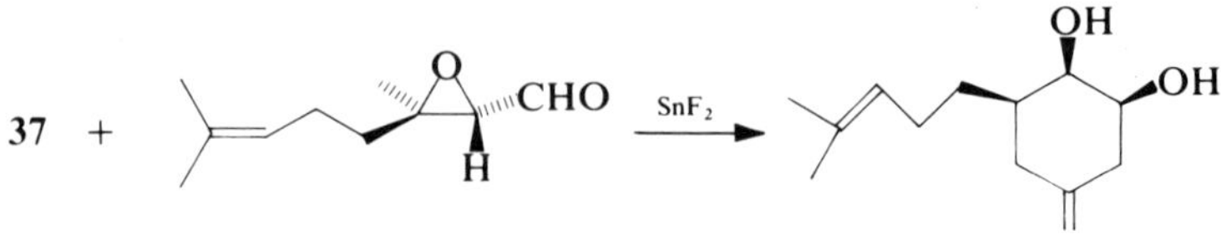

(89)

(38) major diastereomer

The aminoallylsilane **39** was converted by Grieco and Fobare[95] to the Mannich base, which cyclized under the same conditions to the vinyl pyrrolidine **40** (equation 90). In a similar reaction the aminoallylsilanes **41** and **42** were converted to their cyclic amines[96] (equation 91).

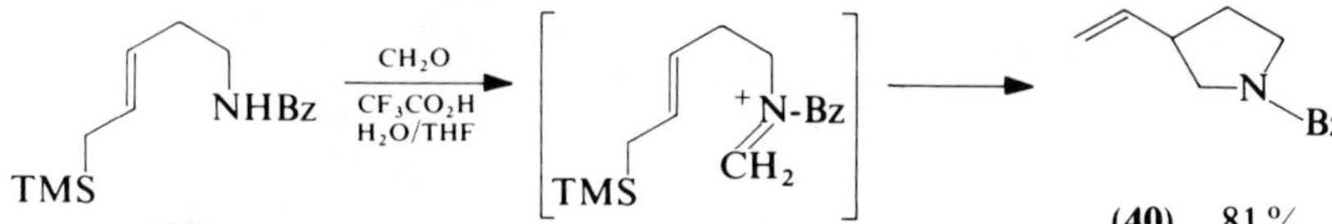

(**39**) (**40**) 81% (90)

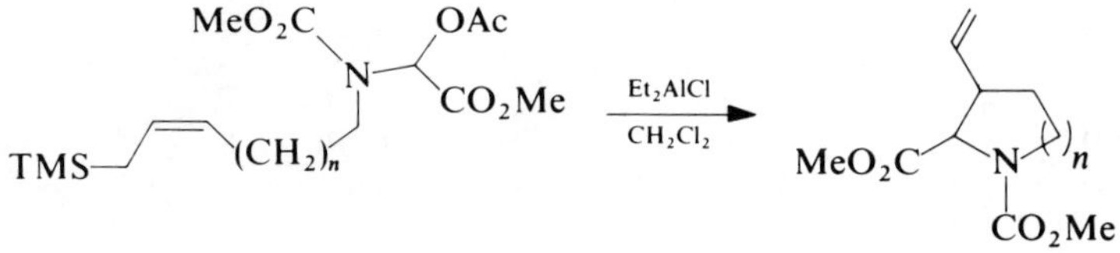

(91)

(**41**) $n = 1$ $n = 1$ 78%

(**42**) $n = 2$ $n = 2$ 75%

VIII. ETHYNYLSILANES

The acetylenic silanes have often been employed in order to protect a terminal acetylene during a reaction sequence. They have also found use in directing the regioselectivity of a reaction as, for example, in the hydroalumination or hydroboration reactions. The few examples given here illustrate the protective aspect of the acetylenic silanes and also some of the possibilities. Dicarboxylic acids are nicely cyclized to the anhydrides by (trimethylsilyl)ethoxyacetylene, a stable alternative to the explosive parent ethoxyacetylene[97] (equation 92). In some pretty work by Diercks and Vollhardt[98] trimethylsilylacetylene, an extremely useful acetylene derivative, was coupled with hexabromobenzene in the presence of palladium(II) catalyst to give hexa(trimethylsilylethynyl)benzene, which could be desilylated to hexaethynylbenzene in 28% overall yield (equation 93). This material was then cyclized with 3 equivalents of bis(trimethylsilyl)acetylene, another highly useful acetylene derivative, and cyclopentadienylcobalt dicarbonyl to produce **43**, which was desilylated to **44**. Compound **44** (also **43**), due to the possibility of forming a cyclobutadiene in one of the two Kekule structures, is a bond-fixed cyclohexatriene. The same system, **44**, was prepared from trimethylsilyl-1,3-butadiyne, a more stable form of butadiyne (equation 94).

Trimethylsilylacetylene can be used to form the monolithium or magnesium reagent allowing chemistry of lithioacetylene or halomagnesioacetylene without complications from the dimetallo derivatives. Two examples are given here. Trimethylsilylethynyllithium was added to azetidinone **45** to give, after allylation and desilylation, enyne **46**, which

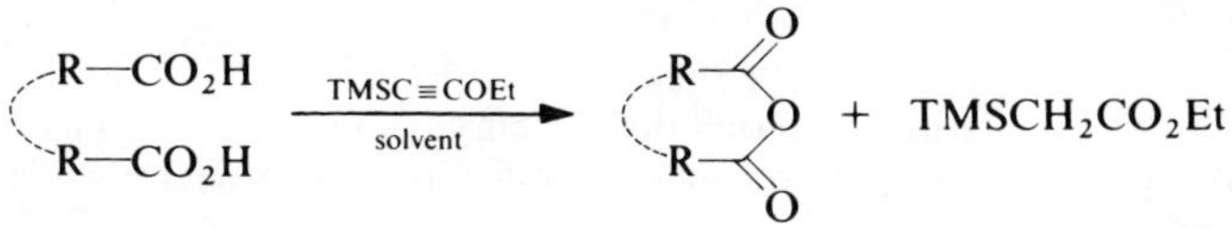

(92)

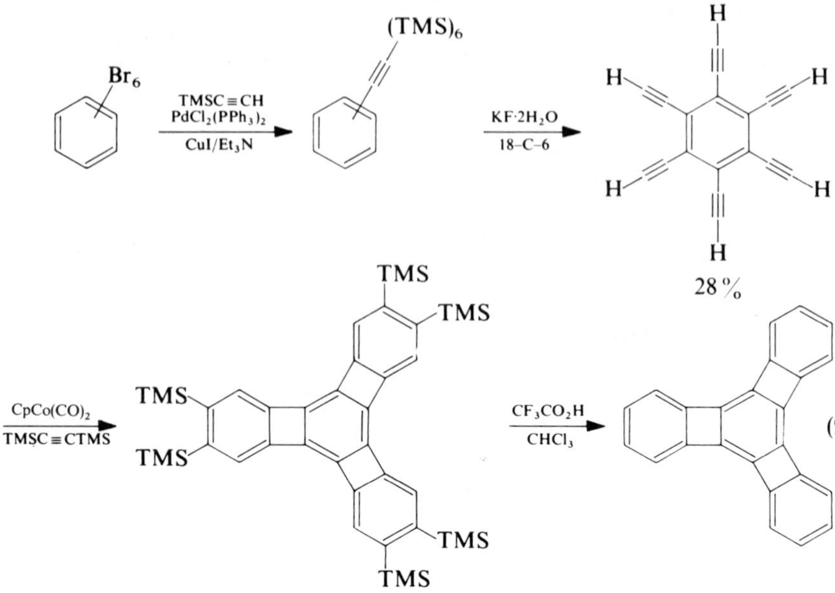

(43) **(44)**

(93)

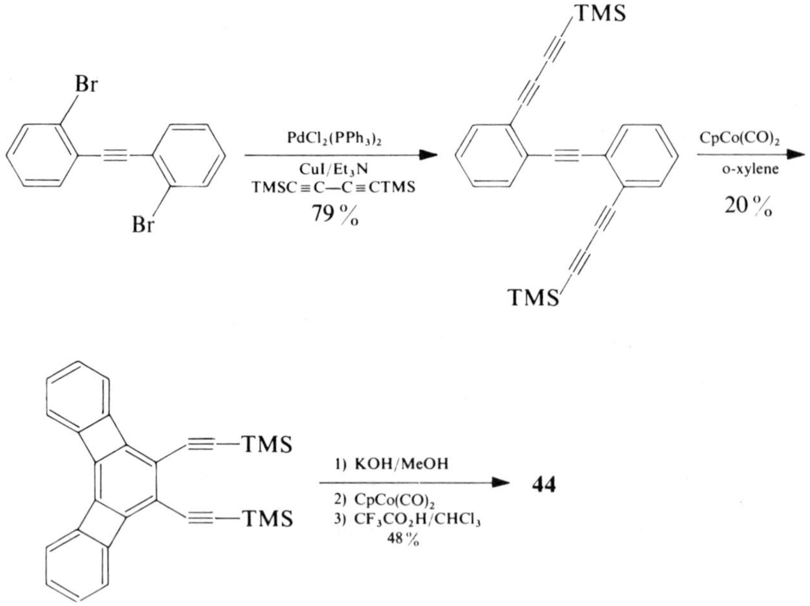

(94)

was cyclized to **47**[99] (equation 95). Trimethylsilylethynylmagnesium bromide was reacted with the optically pure sulfinate **48** to give optically active (*R*) sulfoxide **49** in 80% yield[100] (equation 96).

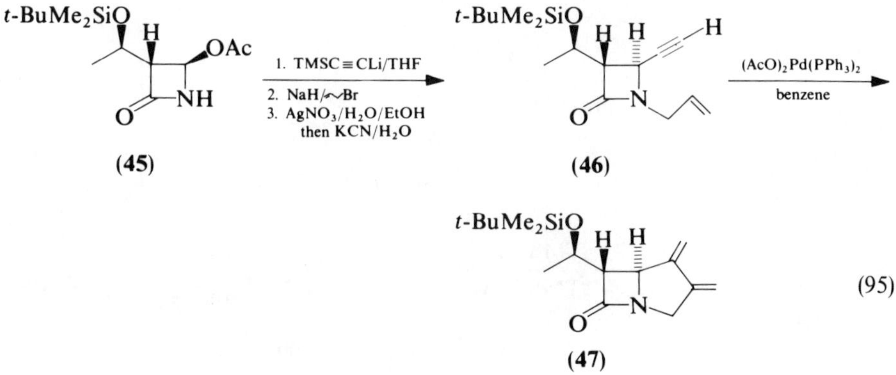

(45) (46) (47)

(95)

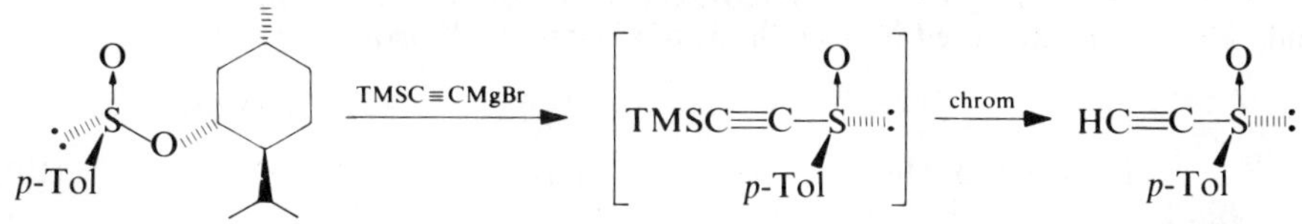

(48) (49) (*R*)

(96)

It has been found that oxidation of ethynylsilanes with osmium tetroxide leads to α-keto esters in moderate yield[101] (equation 97). It has also been reported that ethynylsilanes react with alkyl halides under the influence of aluminum chloride to give the corresponding disubstituted acetylene[102] (equation 98). The alkyl halide must be tertiary or adamantyl.

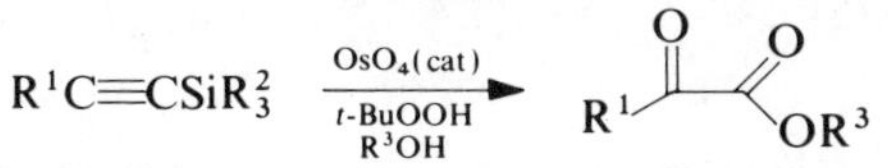

(97)

$$R^1C{\equiv}CTMS + R^2X \xrightarrow[CH_2Cl_2]{AlCl_3} R^1C{\equiv}CR^2 \qquad (98)$$

IX. α-SILYL ORGANOMETALLICS IN THE OLEFINATION REACTION

The use of α-silyl organometallic reagents, principally lithium and Grignard reagents, to convert aldehydes and ketones to olefins is an extremely useful reaction in organic chemistry. Its many applications have their origin in the work of Peterson and the reaction is most often referred to as the Peterson olefination reaction. The reaction has been reviewed[103]. Some new additions to the applications of the Peterson-type olefination of carbonyls are given here. It was found by Johnson and Tait[104] that although trimethylsilylmethyllithium reacts with enolizable ketones to produce large amounts of the deprotonated material, when converted to its cerium reagent it gives excellent yields of the addition product (equation 99). Halton, Stang and coworkers[105] have prepared cycloproparenes utilizing the Peterson olefination concept (equation 100).

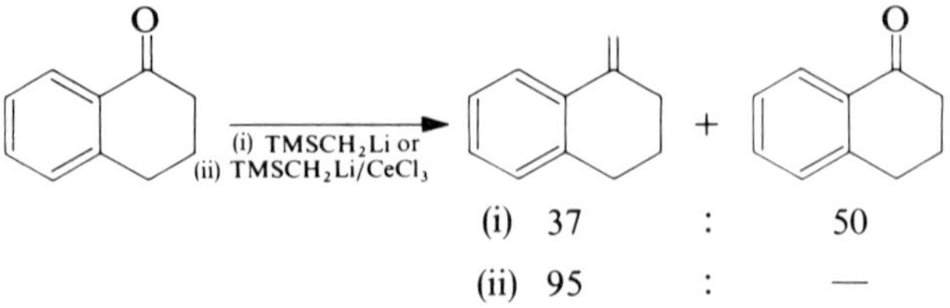

(99)

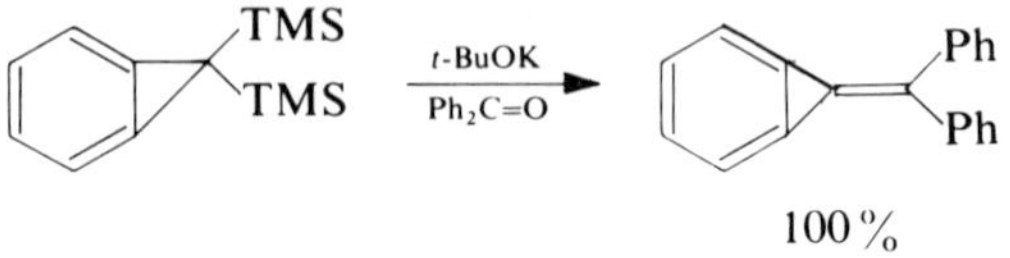

(100)

The dienoate **52**, prepared in a stereoselective fashion from the α-silyl-α-lithio ester **50** and aldehyde **51**, was used in a synthesis of aklavinone[106] (equation 101).

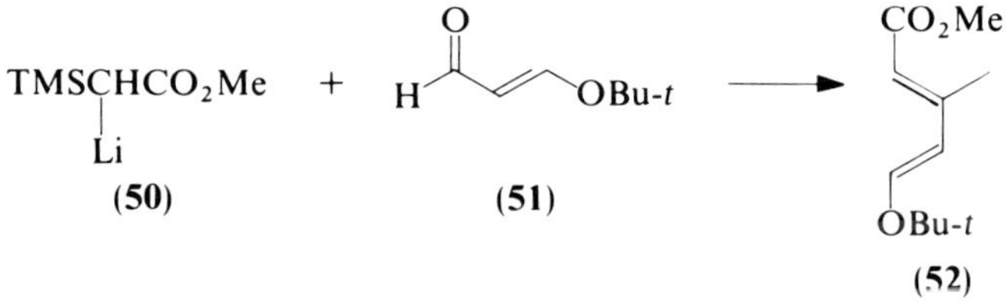

(101)

Vinylphosphonates can be prepared from the lithium reagent of α-silyl phosphonates[107] (equation 102).

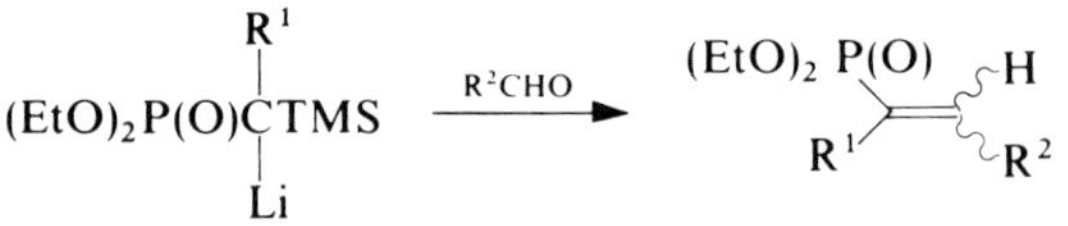

(102)

Ketene *O,S*-acetal derivatives can be readily prepared from the reaction of [methoxy(phenylthio)(trimethylsilyl)methyl]lithium with aldehydes and ketones[108] (equation 103). It is interesting to point out that the methyl enol ether was converted to the enol silyl ether by reaction with trimethyliodosilane. This was then hydrolyzed to the thio ester **53** (equation 104).

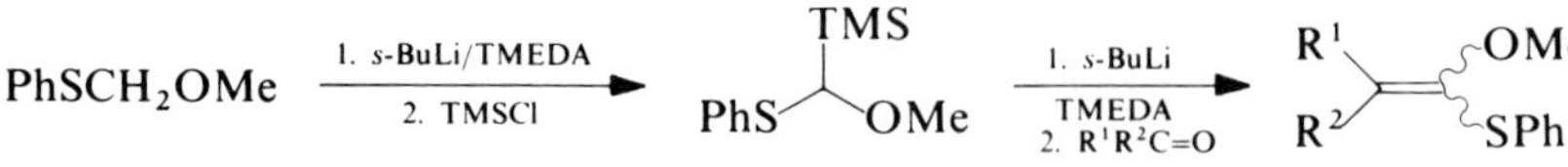

(103)

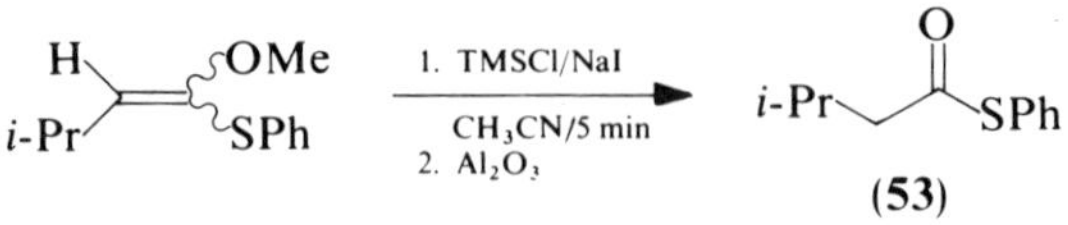

(104)

The lithium reagent of **54** was reacted with aldehydes and ketones to produce the vinylphosphonates **55**, which were then oxidized and deprotected to provide a convenient route to α-hydroxy acids[109] (equation 105).

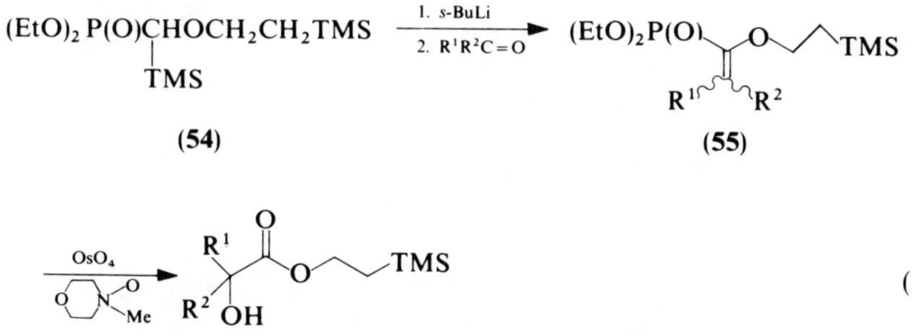

(54) (55)

(105)

Although not a Peterson approach in the pure sense, the β-hydroxysilanes created from the reaction of vinylmagnesium bromide and β-ketosilanes are eliminated to produce 2-substituted 1,3-dienes[110] (equation 106). The β-ketosilanes are prepared from the reaction of α-silyllithium reagents or trimethylsilylmethylmagnesium bromide as illustrated in equations 107 and 108.

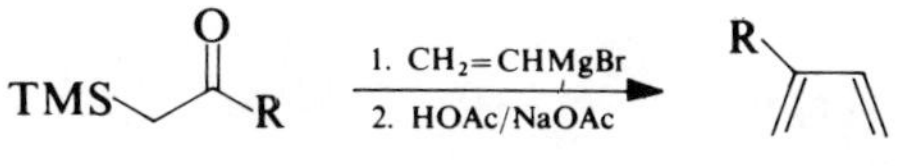

(106)

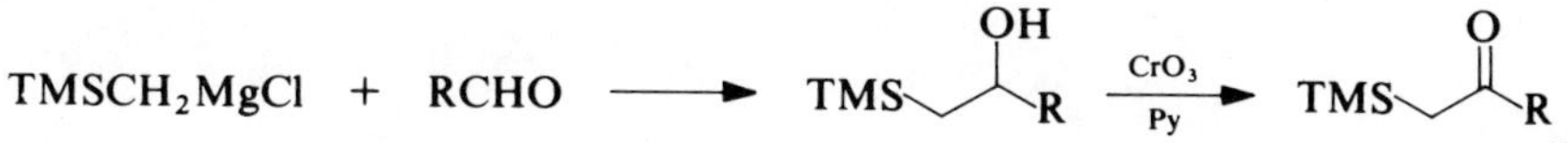

(107)

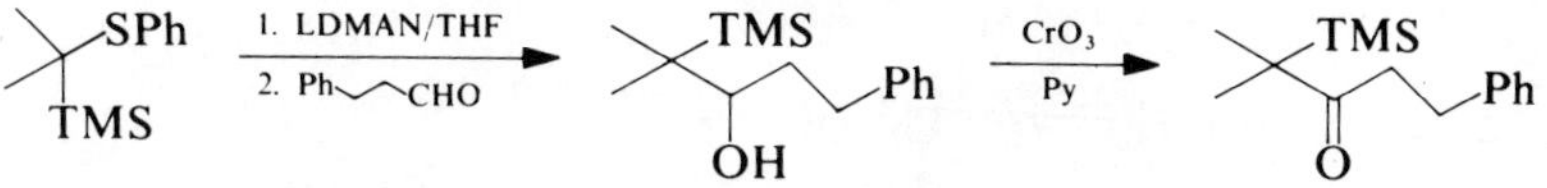

(108)

X. SILICON AND THE FLUORIDE ION

The very strong silicon–fluorine bond has been utilized in many applications of organosilanes in organic synthesis. This allows for the chemoselective interaction of a silicon group in a molecule even in the presence of a host of other functional groups. Advantage has been taken of this property of silicon in the selective deprotection of silylated acids and alcohols. More recently, however, it has become especially useful in the preparation of certain high-energy species. This allows for the chemoselective generation of such a species in the presence of a potential substrate. In this section several such examples will be given.

The reaction of tetra-*n*-butylammonium fluoride (TBAF) (available as a THF solution although it always contains some water) with benzylsilanes in the presence of an aldehyde or ketone provided homobenzyl alcohols[111] (equation 109). It is presumed that the reaction proceeds through the tetra-*n*-butylammonium salt of the benzyl anion. This

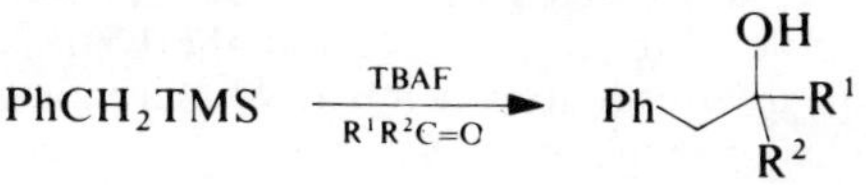

(109)

even allowed for the preparation of the 'anion' **56**, which contains both a chloro and a nitro group. The direct preparation of this 'anion' by conventional means is clearly not possible[112] (equation 110).

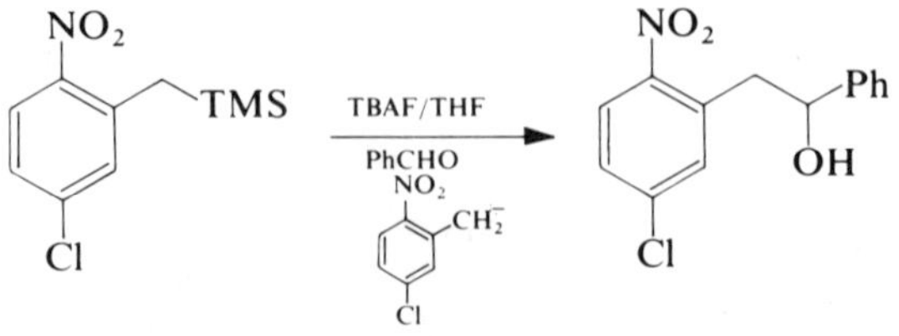

(56) (110)

The aminomethylation of aldehydes proved possible employing **57**, fluoride ion and the aldehyde in a single flask[113] (equations 111 and 112). Other N-protected α-amino benzylsilanes were also used.

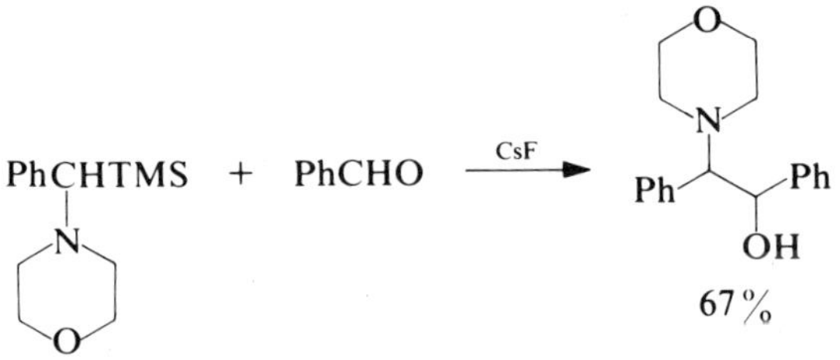

67% (111)

(57)

57% (112)

The α-bromosilane **58** was used to generate the benzyl anion, which then formed the carbene resulting in cyclization to the aminofuran **59**. Intermediate **59** could then be trapped with dienophiles providing the dihydronaphthalenes **60**[114] (equation 113).

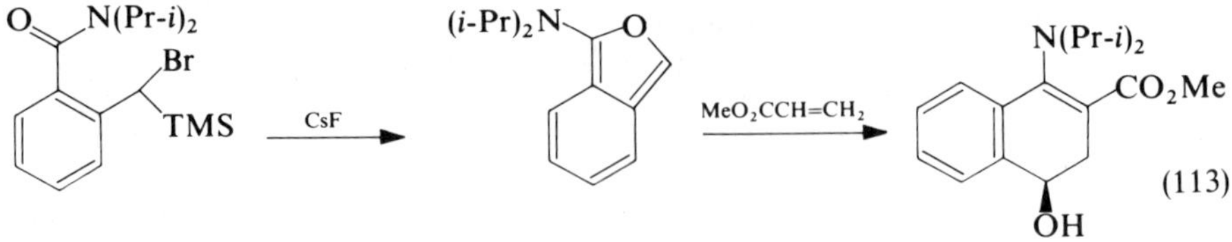

(113)

(58) **(59)** **(60)** 44%

Trimethylsilyltriazole **61** was reacted with aldehydes and ketones to give the corresponding β-hydroxytriazole[116] (equation 114), but the bistrimethylsilyl derivative **62** produces a Peterson-type intermediate, which reacts to give the α,β-unsaturated triazole in good to excellent yields[115] (equation 115).

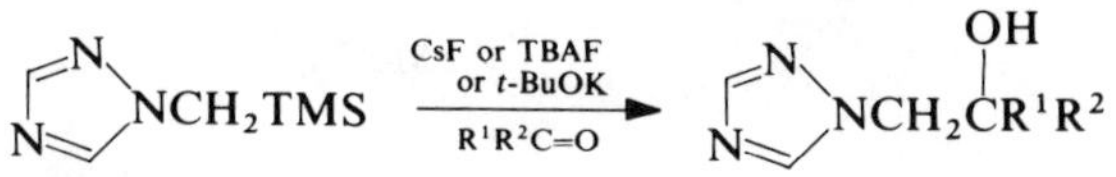

(114)

(61)

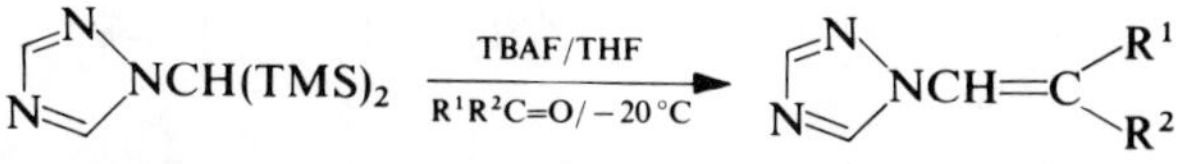

(115)

(62)

In similar chemistry a series of benzylsilanes were employed in the preparation of β-hydroxypyrazoles, imidazoles, tetrazoles and triazoles[116] (equation 116).

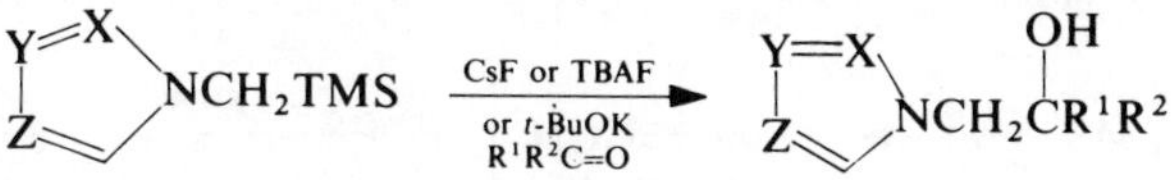

(116)

(X = Y = Z = CH; X = N, Y = Z = CH; X = Y = CH, Z = N, X = Y = Z = N; X = Z = N, Y = CH)

The protiodesilylation of an α-trimethylsilyl thioketone *S*-oxide with TBAF (the proton presumably comes from the water present) gives the aryl thioaldehye *S*-oxide in excellent yield[117] (equation 117).

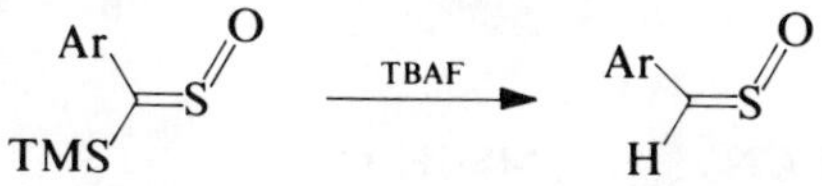

(117)

It has long been known that *N,O*-bis(trimethylsilyl)amides are thermally converted to nitriles. This reaction does not work in all cases, however. It has now been found that a number of Lewis acid catalysts or TBAF will bring about the elimination[118] (equation 118).

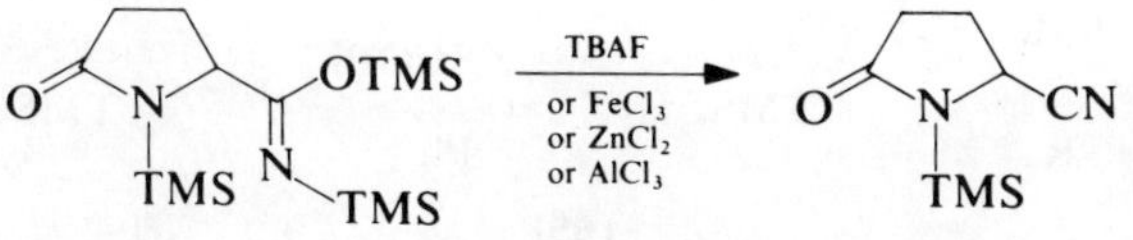

(118)

Undoubtedly, the single largest synthetic application of fluoride ion and silicon is in the generation of azomethine ylides and related intermediates. The use of organosilanes in the generation of sulfonium ylides has been reviewed[119]. The pioneering and extensive work of Padwa and his group have shown that a variety of syntheses can be accomplished via the *in situ* generation of azomethine ylides from *N*-[trimethylsilylmethyl] methylamino ethers and fluoride ion[120] (equation 119). It had been shown earlier that this reaction could be catalyzed by trifluoroacetic acid[121].

Acetonitrile ylide was generated from α-silyl imine **63** and silver fluoride. This same ylide could be generated from the reaction of methylene with acetonitrile. The chemoselective silicon–fluoride ion route seems best, however, offering the advantages of readily available, easily handled reagents. The so-generated ylide was cycloadded to several

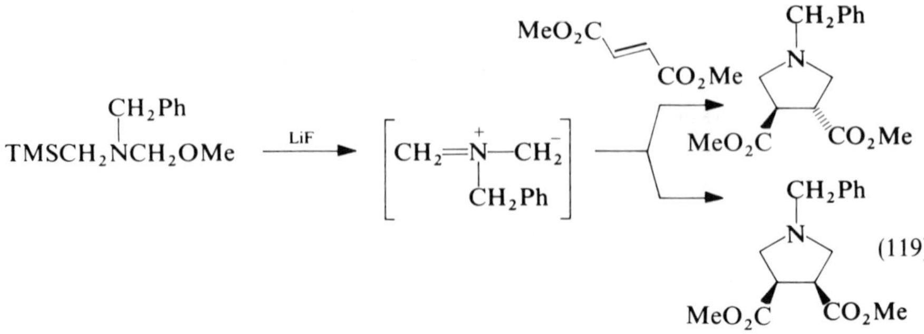

(119)

dipolarophiles[122] (equations 120 and 121). The requisite **63** was readily prepared in high yield according to equation 122.

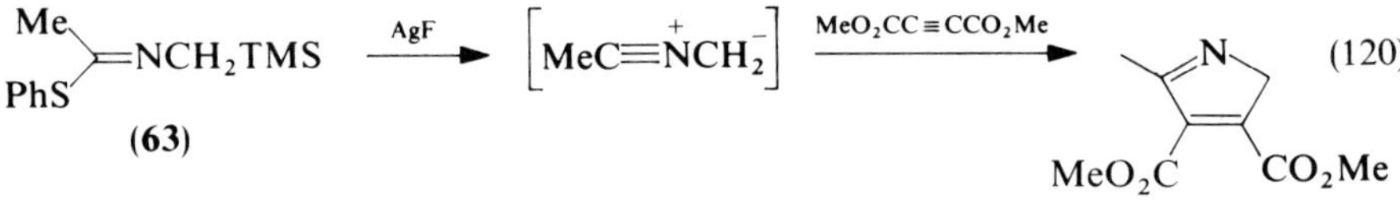

(**63**) (120)

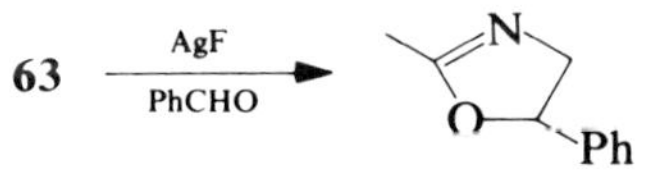

(121)

$$CH_3CN \; + \; TMSCH_2OTf \xrightarrow{PhSH} \mathbf{63} \qquad (122)$$

α-Trimethylsilyl imines **64**, **65** and **66** were used in conjunction with fluoride ion to generate the azomethine ylides **67**, **68** and **69**, respectively. All of the so-generated azomethine ylides were cycloadded to 1,3-dipolarophiles[123] (equations 123 and 124).

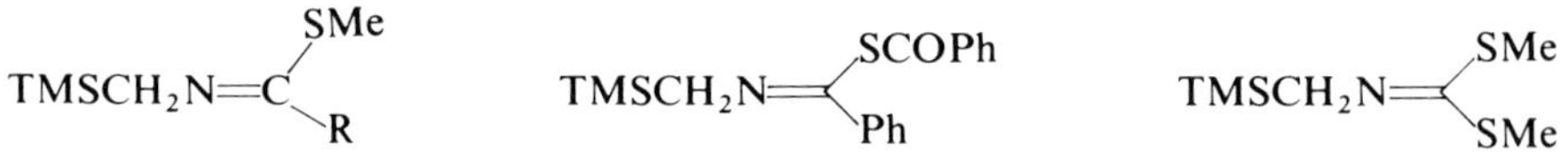

(**64**) (**65**) (**66**)

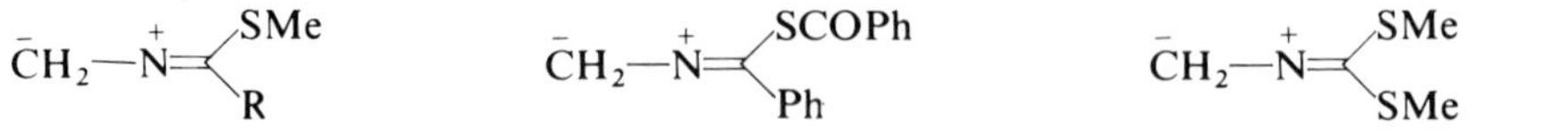

(**67**) (**68**) (**69**)

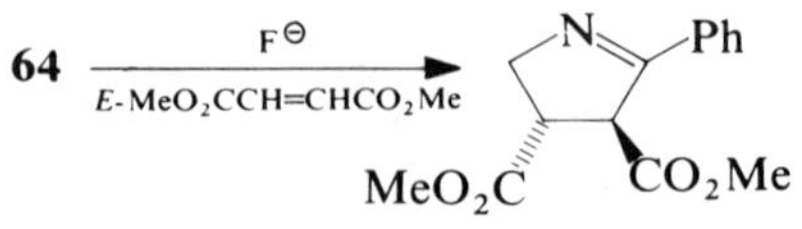

(123)

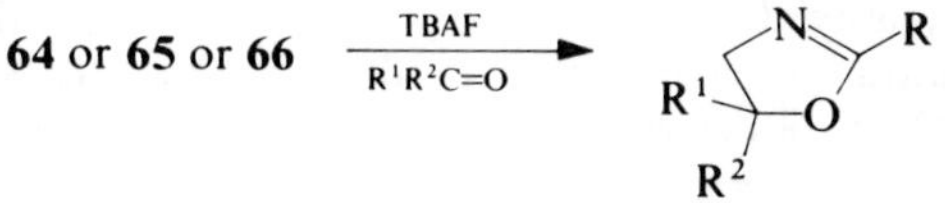 (R = Ph, SMe, Et, *i*-Pr) (124)

A general route to N-protonated azomethine ylides from *N*-(trimethylsilylmethyl)amidines and thioamides and their cycloadditions to olefins was reported. This is illustrated in a general way in equation 125[124].

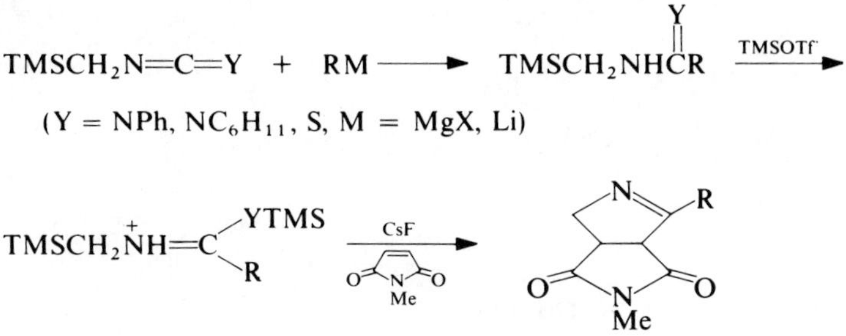

(125)

A precursor to the simplest nitrile ylide **71** was generated by the action of silver fluoride on **70** (equation 126). The cycloaddition of **71** with methyl propiolate is shown[125] (equation 127).

$$\underset{\mathbf{(70)}}{TMSCH_2N(CH_2Ph)CH(CN)_2} \xrightarrow{AgF} \underset{\mathbf{(71)}}{\left[\overset{+}{C}H_2{-}\overset{+}{N}(CH_2Ph){=}CHCN\right]} \qquad (126)$$

 (127)

A number of applications of the above-presented approach have been made in the synthesis of bicyclic amines. It has been used in the preparation of pyrrolo[1,2-*a*]indoles starting from *N*-trimethylsilylmethyl indoles, which are reacted with silver fluoride in an oxidation–reduction reaction to provide the desired products in good to excellent yields[126] (equation 128). This same approach was applied to the preparation of some

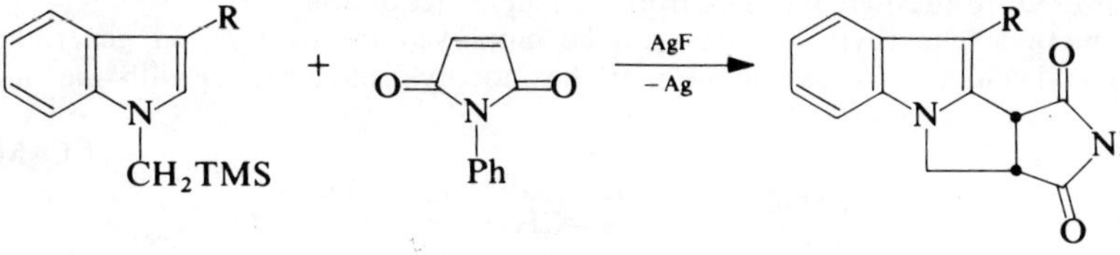 (128)

(R = H, Me, CHO, CH_2CN)

nonstabilized 2-methylthiazolium methylides, which undergo cycloadditions with dipolarophiles[127] (equation 129). An intramolecular version of this reaction has been employed in an approach to the synthesis of the erythrinane skeleton[128] (equation 130).

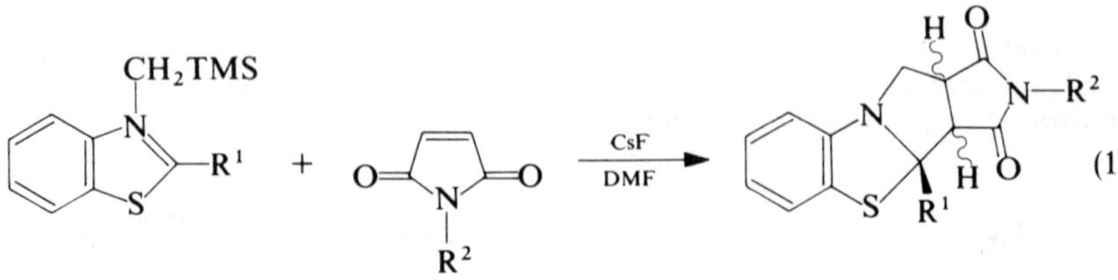

(129)

(R^1 = H, Me; R^2 = Me, *p*-Tol)

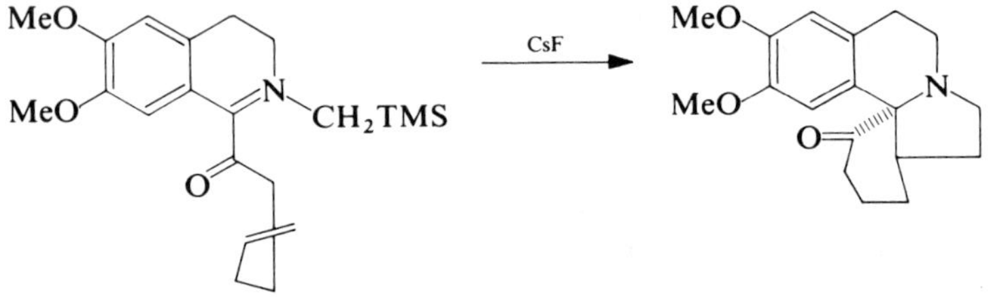

(130)

The amine **72** was used in concert with benzoyl fluoride to generate *N*-benzoylazomethine ylide, which was cycloadded directly to 1,3-dipolarophiles leading to 3-substituted or 3,4-disubstituted N-benzoylpyrrolidines in high yields[129] (equation 131). This also represents an extremely efficient way to generate an *N*-protected azomethine ylide in the presence of a dipolarophile from readily available, stable starting materials.

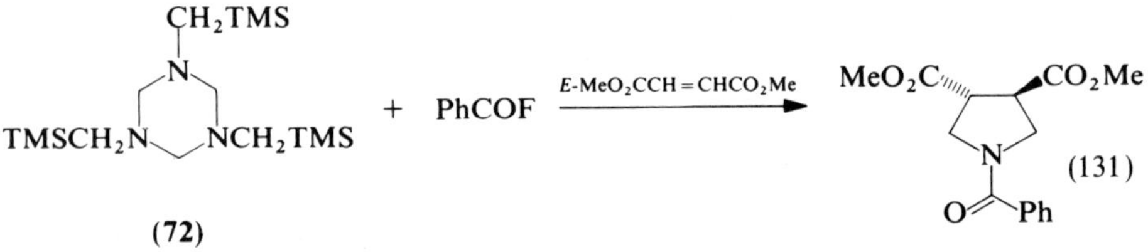

(131)

(72)

Chloromethyl(trimethylsilylmethyl)sulfide, prepared from mercaptomethyltrimethylsilane and paraformaldehyde, reacts with cesium fluoride to generate the thiocarbonyl ylide **73**. Here again the reactive intermediate can be generated in the presence of an olefin leading to tetrahydrothiophenes (equation 132) or dimethyl acetylene dicarboxylate (DMAD) to produce the dihydrothiophene[130] (equation 133).

Bis(trimethylsilylmethyl)sulfoxides can be employed in the thermal generation of thiocarbonyl ylides with loss of the inert hexamethyldisiloxane. As with the fluoride

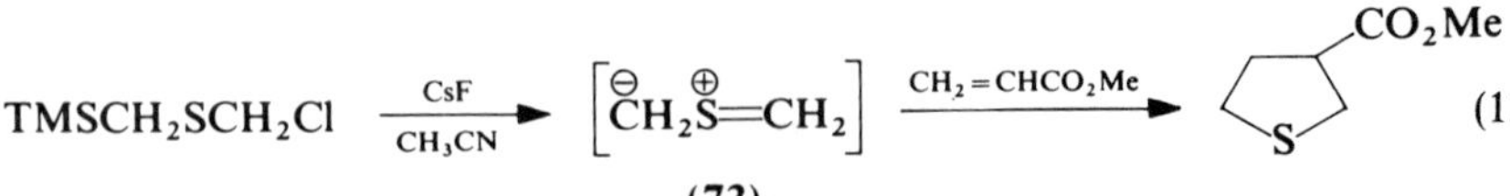

(132)

(73)

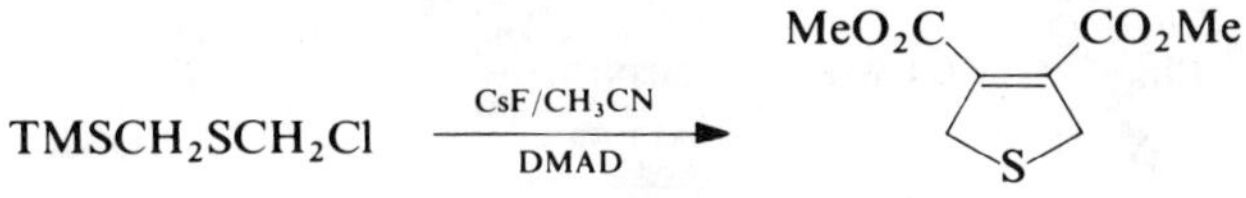

(133)

approaches mentioned above, this procedure allows for the generation of the reactive intermediate in the presence of a dipolarophile[131] (equation 134).

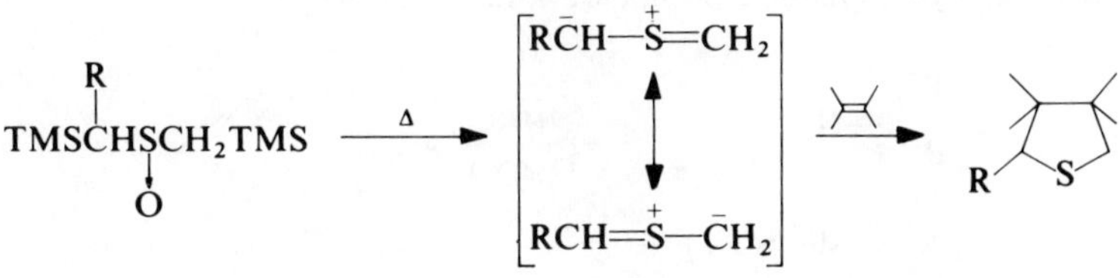

(134)

XI. MISCELLANEOUS

A. Trimethylsilyl Azide

Trimethylsilyl azide ($TMSN_3$) is a more easily handled, stable alternative to hydrazoic acid. It has found many uses in organic synthetic applications. Most recently it has been found to readily convert hindered alkyl halides to alkyl azides in the presence of a Lewis acid. Thus, tertiary alkyl azides can be formed from tertiary alkyl chlorides, trimethylsilyl azide and tin tetrachloride. In this work the azides were converted to the benzamides[132] (equation 135). Similar conditions were used by Olah and coworkers[133] to prepare secondary and tertiary cyclic azides in good yields (equation 136).

$$R^1R^2R^3C{-}Cl + TMSN_3 \xrightarrow[15\text{–}20\%,\ 6d]{SnCl_4} R^1R^2R^3C{-}N_3 \xrightarrow[\substack{3.\ PhCOCl\\ NaOH}]{\substack{1.\ (EtO)_3P\\ 2.\ HClg}} R^1R^2R^3C{-}NHBz \quad (135)$$

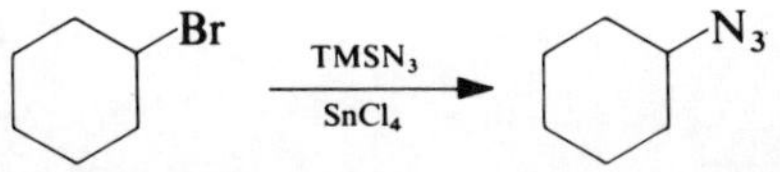

(136)

The opening of epoxides with trimethylsilyl azide provides an excellent route into β-azido alcohols. The reaction is catalyzed by Lewis acids and the azide resulting from attack at the carbon atom best able to tolerate a positive charge is observed. The regioselectivity is greater than 92% in these ring openings. Two examples are given in equations 137[134] and 138[135].

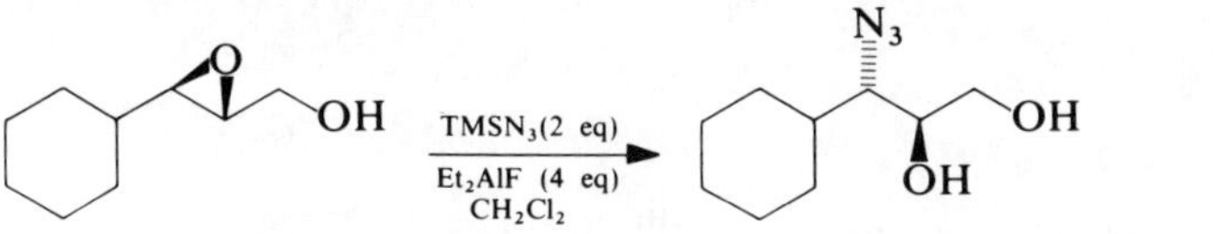

(137)

85% (2% other regioisomers)

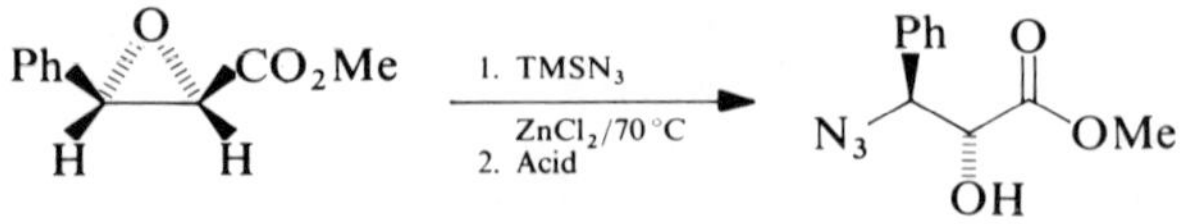

(138)

The sulfonate esters of hydroxylamines react with trimethylsilyl azide in acetonitrile to give the corresponding tetrazoles in moderate to excellent yields[136] (equation 139). The benzoates of the hydroxylamine do not work nearly as well.

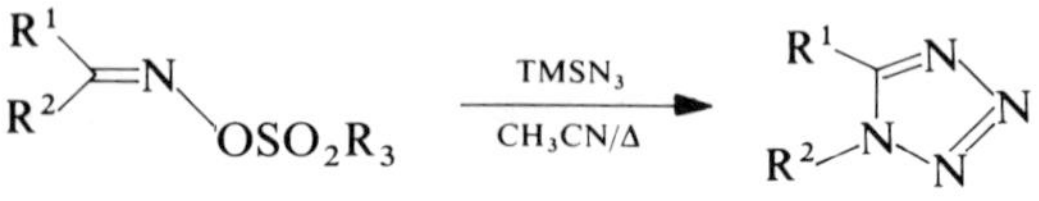

(139)

(R^3 = Me, *p*-Tol)

B. Mercaptosilanes

Thioalkoxysilanes have been used to prepare 1-thioglycosides[137] (equation 140). Clearly, in this reaction the trimethylsilyl group is activating the acetate of the anomeric carbon followed by attack on the resultant carbenium ion by the thiomethoxy group. Allyl sulfides are readily prepared from allyl carbonates, thioalkoxytrimethylsilanes and palladium(0) catalysis[138] (equation 141). Under these same conditions thioalkoxytrimethylsilanes will open vinyl epoxides in an S_N2' process[138] (equation 142). It turns out that bis(trimethylsilyl)sulfide reacts with two equivalents of methyllithium in tetrahydrofuran to give tetramethylsilane and an active, soluble form of lithium sulfide. The thus-generated lithium sulfide reacts nicely with α,ω-dihalides to give the cyclic thio ethers[139] (equation 143).

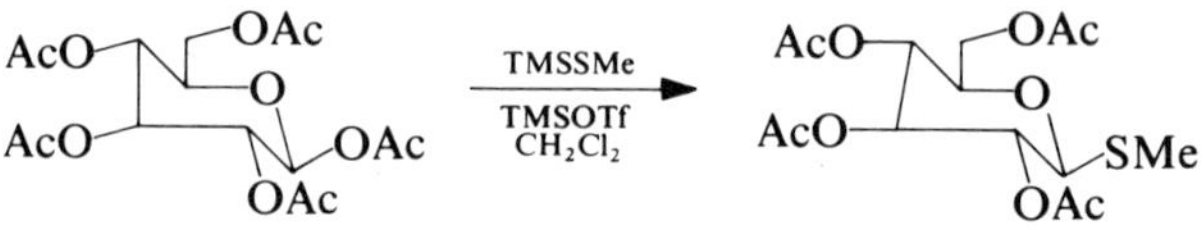

(140)

84%

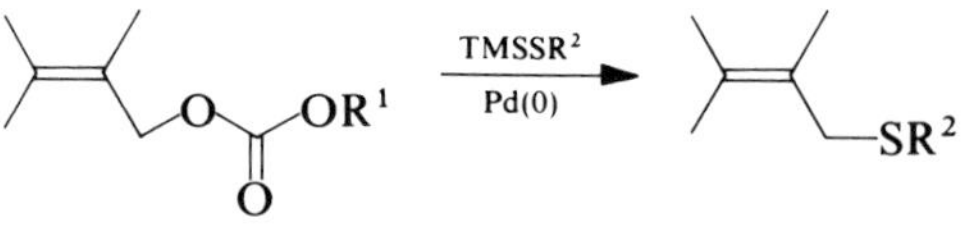

(141)

(142)

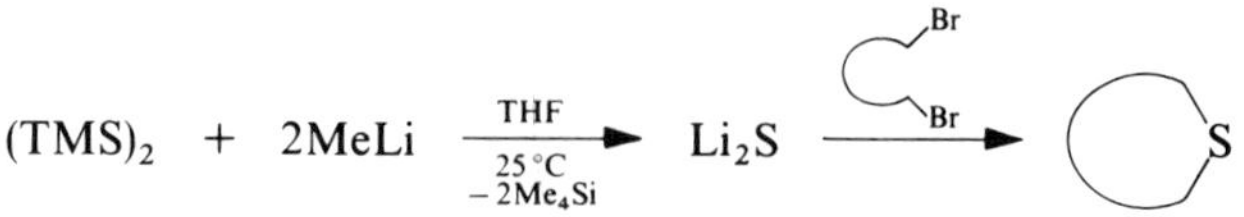

(143)

C. Amino- and Phosphinosilanes

Bis(trimethylsilyl)acetamide(BSA) or better diethylaminotrimethylsilane were used to *N*-silylate sulfoximine **74** allowing it to be deprotonated and then reacted with an electrophile. Deprotection resulted in the substituted sulfoximine in good overall yields[140] (equation 144).

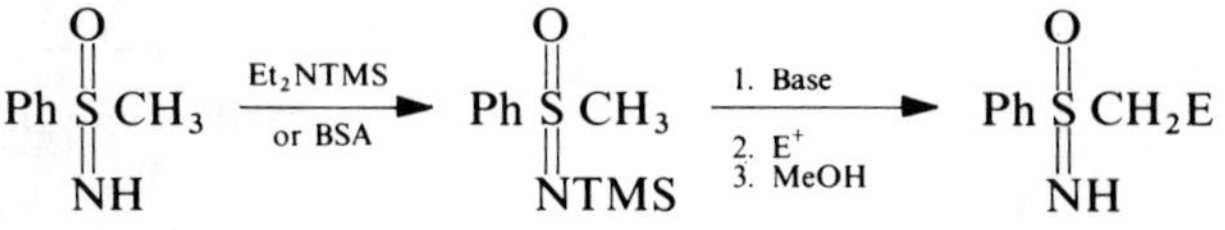

(74) (144)

Diphenylphoshinotrimethylsilane and diphenylphosphinotrimethyltin can be used to phosphinylate aromatic rings under palladium(II) catalysis[141] (equation 145). This opens up a new synthetic approach to arylphosphines not available from the more traditional route of adding an arylmetallic reagent to a chlorophosphine.

$$\mathrm{TMSPPh_2 + ArI \xrightarrow{PdCl_2(PPh_3)_2} Ph_2P{-}Ar} \qquad (145)$$

D. Silicon as a Hydroxyl Equivalent

It is, of course, well known that the boron–carbon bond can be oxidized to the corresponding alcohol in high yield. It has now been amply shown that the silicon–carbon bond can be oxidized to alcohols under conditions that are remarkably similar to those for the organoborane oxidations. This permits the use of organosilanes as a disguised hydroxyl group, the advantage of which is that the silicon–carbon bond is much less reactive than the boron–carbon bond which, although stable to water, reacts with oxygen. The one important requirement for this important transformation is that the silicon group have at least one fluorine atom, or leaving group, attached. It happens, however, that this can be a phenyl group because it can be removed electrophilically to produce an oxidizable silicon. This is nicely demonstrated by the use of diastereoselectively prepared β-phenyldimethylsilyl esters, which can be oxidized to the β-hydroxy esters[142] (equation 146). A further application of this chemistry resulted in a formal synthesis of thienamycin[143] (equation 147). The lithium enolate of ethyl β-phenyldimethylsilylbutyrate was condensed with imines to give the azetidinones, which could be oxidized[144] (equation 148). The use of the β-phenyldimethylsilyl ester not only permits the eventual introduction of the hydroxyl group, but also provides modest to excellent *cis* stereoselectivity at positions 3 and 4. The use of the enolate of ethyl butyrate alone gave low (15–25%) yields of a mixture of *cis* and *trans* isomers.

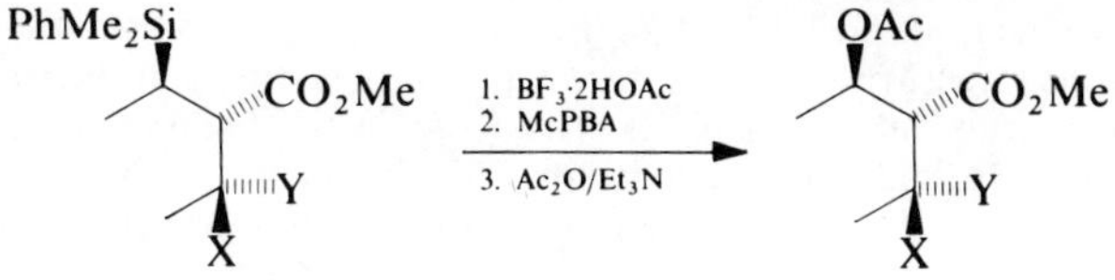

(146)

(X = OAc, Y = H)
(X = H, Y = OAc)

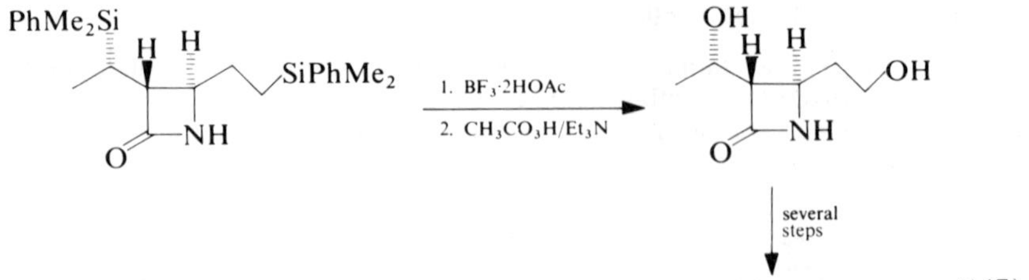

(147)

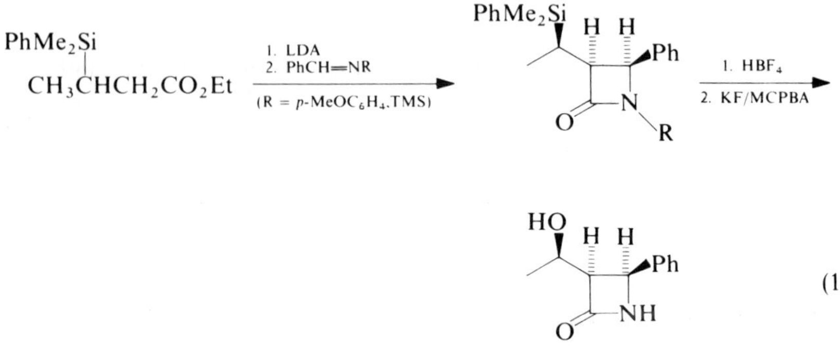

(148)

The oxidation of alkylsilyl alkoxides has been extended to the preparation of α-hydroxy ketones from alkenylsilanes. Thus, 5-decenyltrimethylsilane was treated with *meta*-chloroperbenzoic acid and the resulting α-silyl epoxide oxidized with hydrogen peroxide in the presence of fluoride ion to give 6-hydroxy-5-decanone in 88% yield[145] (equation 149).

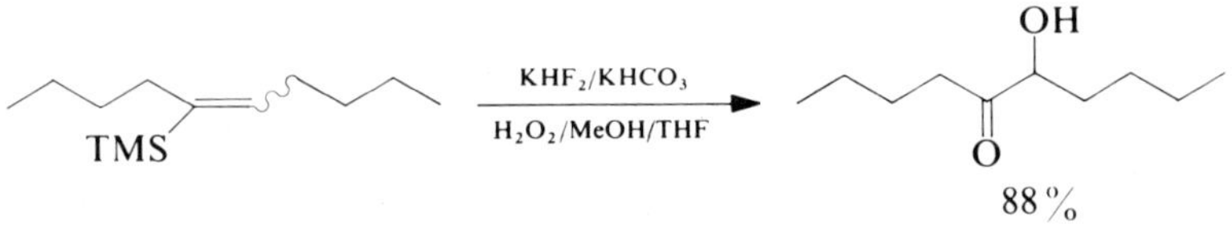

(149)

The oxidation of the silicon–carbon bond to alcohols has been extended to the regioselective preparation of diols from unsaturated alcohols. In this sequence tetramethyldisilazane is reacted with the alcohol group and then the double bond is intramolecularly hydrosilylated. Subsequent oxidation-hydrolysis provides the diol in good yield. A reasonably thorough study of this diol synthesis has been made[146]. The general reaction is shown in equation 150 and an indication of the regioselectivity of the reaction as a function of the structure of the unsaturated alcohol is shown in the various structures below. Homoallylic and allylic alcohols tend to form the five-membered silacycle and consequently the 1,3-diols upon oxidation. This is illustrated in equations 151 and 152[147]. The stereoselectivity of the reaction with acyclic homoallylic alcohols favors the *syn* product.

In a related approach bromomethyldimethylchlorosilane was used to silylate the hydroxyl group of propargyl alcohols and the resultant bromomethylsilyl ether reacted

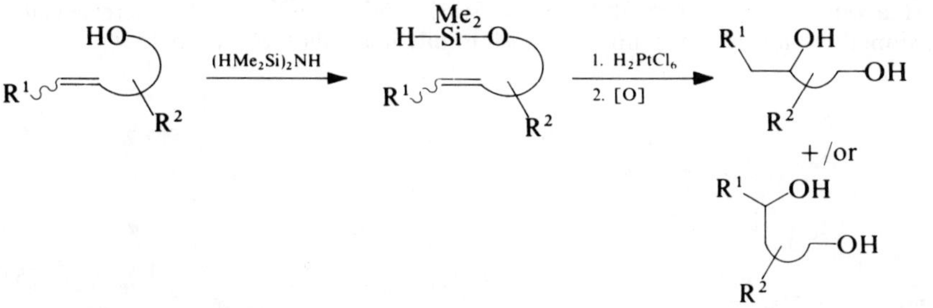

(150)

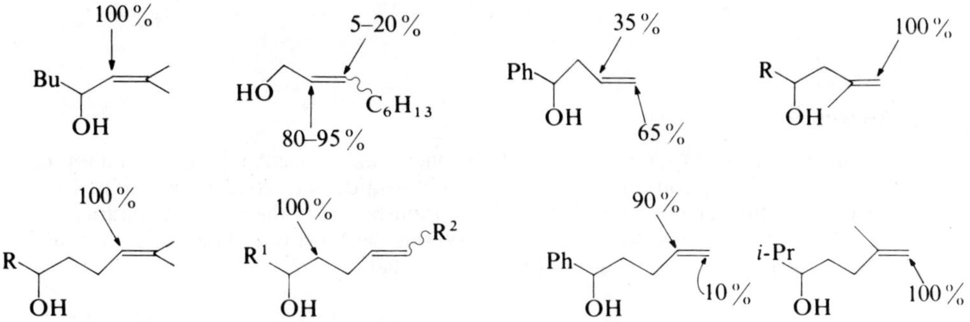

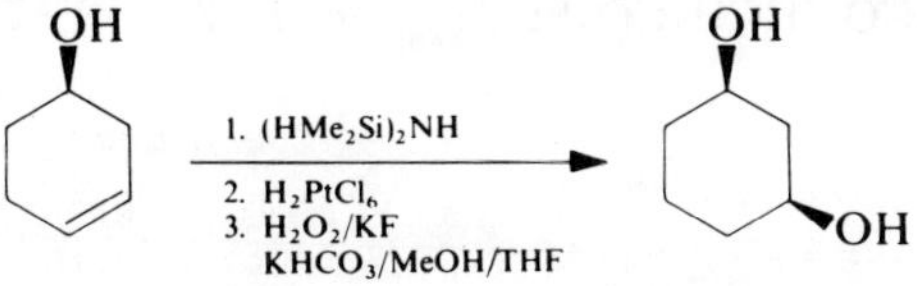

(151)

(152)

with triphenyltin hydride to give an *exo* methylene silacycle. Oxidation of this intermediate provided the interesting diol **75**[148] (equation 153). Treatment of the silacyclic intermediate with methyllithium in ether provides the allylsilane **76** (equation 154). A

(153)

(75)

(154)

(76)

similar reaction sequence was utilized by Stork and Sofia[149] in the stereospecific reductive methylation of the double bond of an allylic alcohol (equation 155).

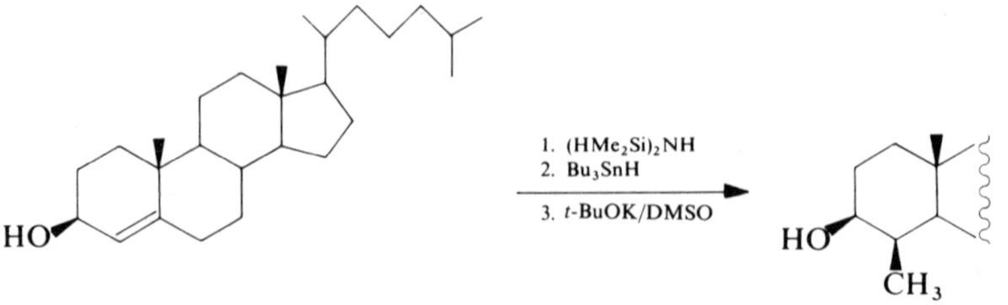

(155)

E. Arylsilanes

Wilson and Jacob[150] have employed the facile and selective iododesilylation of arylsilanes in the preparation of the iodophenyldipeptide derivative **77** (equation 156). Unfortunately, the penicillin derivative **78** decomposed under the reaction conditions.

The bis(trimethylsilyl)benzene derivative **79** was selectively acylated and then desilylated to give 2,6-dimethoxyacetophenone[151] (equation 157).

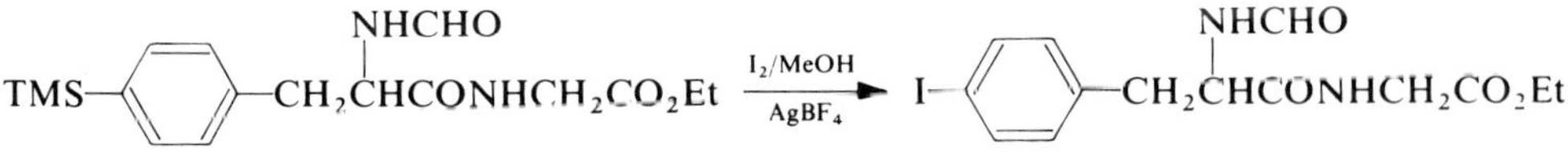

54.5%
(**77**)

(156)

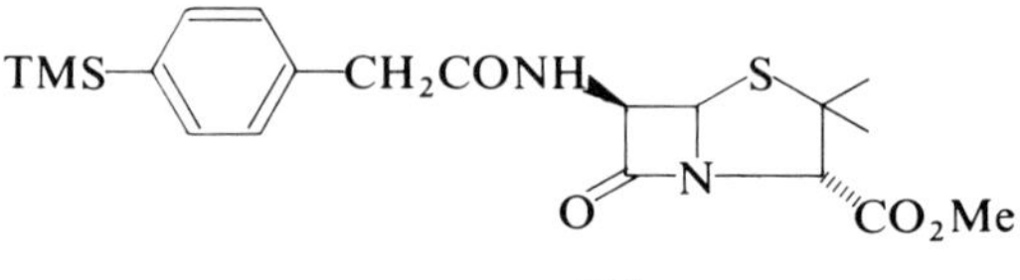

(**78**)

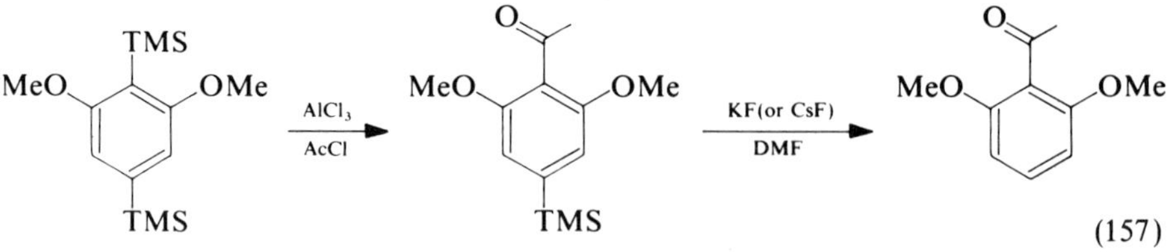

(**79**)

(157)

A variety of useful electrophilic substitutions of 3-trimethylsilylindoles were reported by Majchrzak and Simchen[152]. Of particular interest is the ability to bring about the 1,4-addition to methyl acrylate, acrolein and enones (equations 158 and 159).

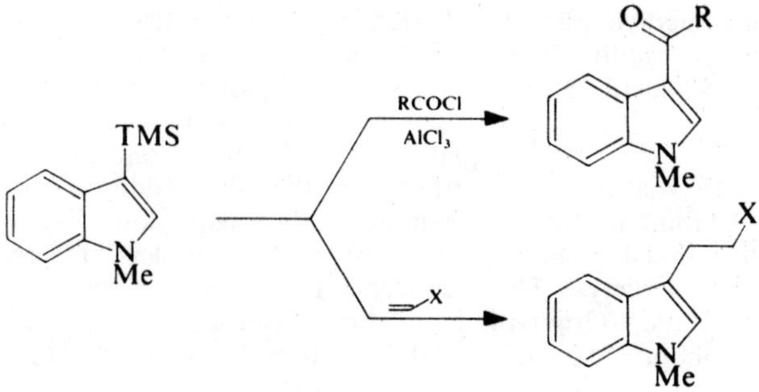

(158)

(159)

(X = CN, CO_2Me, Ac)

F. α-Silyl Esters

The synthetic utility of α-silyl esters has been amply demonstrated by several examples. The basis for this chemistry is the observation that ester lithium enolates can be directly C-silylated with methyldiphenylchlorosilane, a reagent which avoids the more common O-silylation[153]. The α-silyl-γ-valerolactone **80** was converted in two steps and high yield to racemic ancepsenolide by condensation of its lithium enolate with decane 1,10-dicarboxaldehyde followed by isomerization to the endocyclic double bonds of the natural product[154] (equation 160). Treatment of the α-silyl-γ-butyrolactone **81** or **80** with a Grignard reagent followed by pyridinium chlorochromate (PCC) oxidation provides 4-oxo aldehydes and 1,4-diketones, respectively[155] (equation 161).

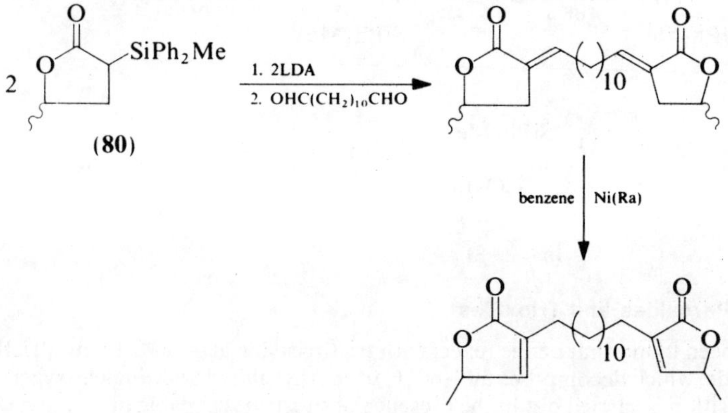

(**80**)

(160)

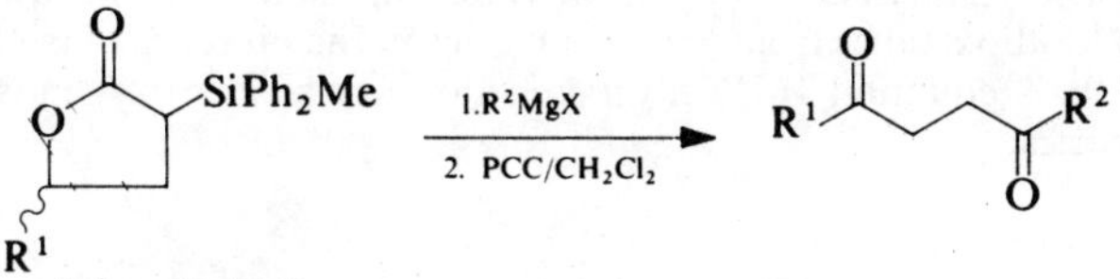

(161)

R[1] = H (**81**)
R[1] = Me (**80**)

Treatment of α-(methyldiphenylsilyl) esters of the general structure **82** with a Grignard reagent (but not an organolithium reagent) results in the formation of the α-silyl ketone, which can be desilylated to the ketone with potassium fluoride (equation 162) or thermally converted to the enol silyl ether, providing a general route to the regiospecific preparation of enol silyl ethers[156] (equation 163). That the second equivalent of Grignard reagent enolizes the α-silyl ketone formed from the initial addition was verified by the high-yield methylation of the intermediate enolate (equation 164). It is interesting and useful to find that the enolization occurs only on the less-hindered α-carbon not bearing the silyl moiety. This conversion of esters to ketones via the methyldiphenylsilylation–Grignard–desilylation sequence has been employed in the preparation of monoprotected 1,4- and 1,5-diketones[157] (equation 165).

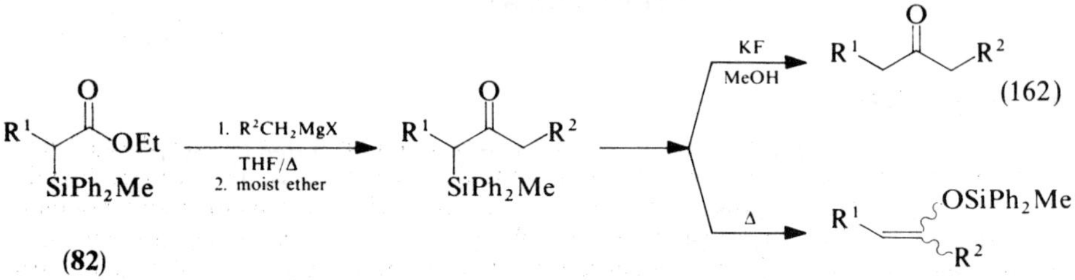

(82)

(162)

(163)

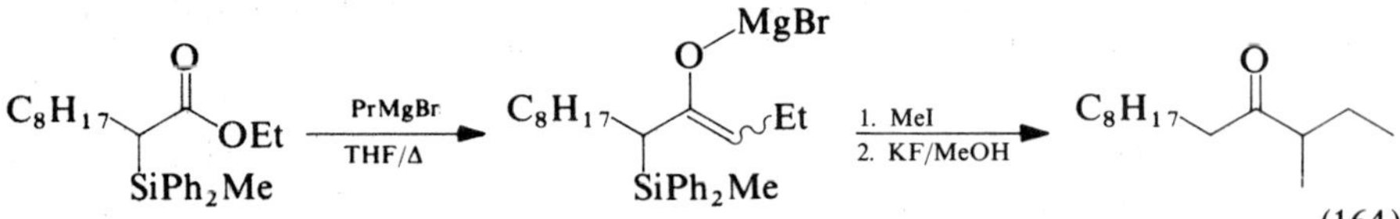

(164)

(165)

($n = 2$, 3)

G. Silyl Peroxides and Trioxides

It has been found that ozone reacts with triethylsilane at −78 °C to give triethylsilylhydrotrioxide, which decomposes at −60 °C to triethylsilanol and singlet oxygen. When the decomposition is carried out in the presence of substrates capable of reacting with singlet oxygen, one obtains the products in excellent yield. Anthracene, for example, reacts to form the adduct in 92% yield[158] (equation 166). This same reagent, triethylsilylhydrotrioxide, forms 1,2-dioxetanes from vinyl aromatics in the presence of triphenylphosphite ozonide (TPPO)[159] (equation 167). In this reaction free singlet oxygen is not the major oxygenating species.

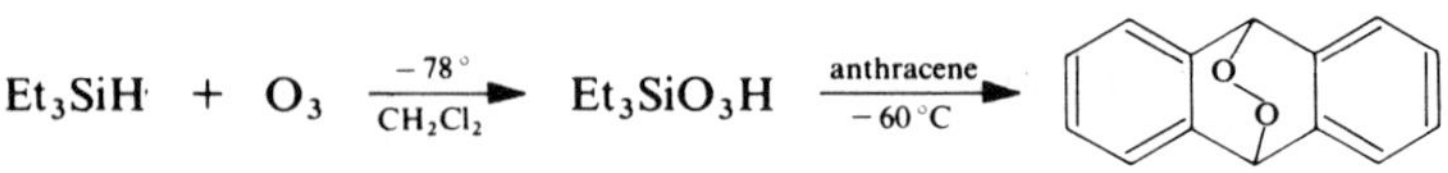

(166)

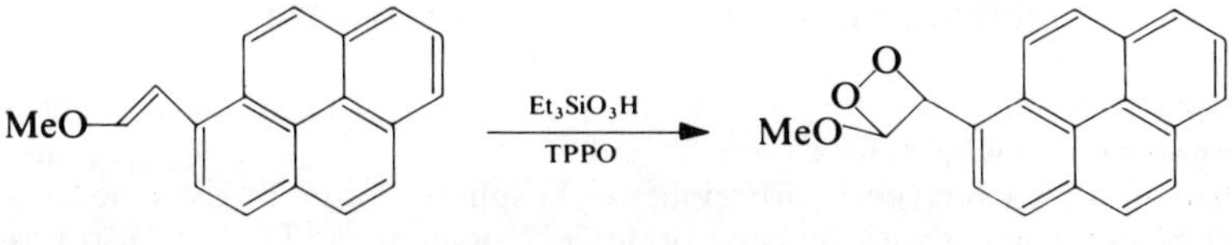
(167)

Bis(trimethylsilyl)peroxide was reacted with 3-lithiofuran and 3-lithiothiophene to give the 3-trimethylsiloxy derivatives in good yield. These derivatives proved to be good dienes for cycloadditions[160] (equation 168). This same silyl peroxide was employed in the oxidation of phosphites to phosphates in high yield[161] (equation 169).

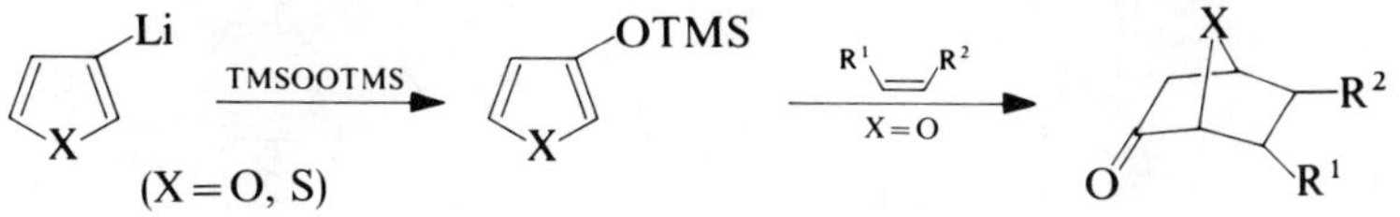
(168)

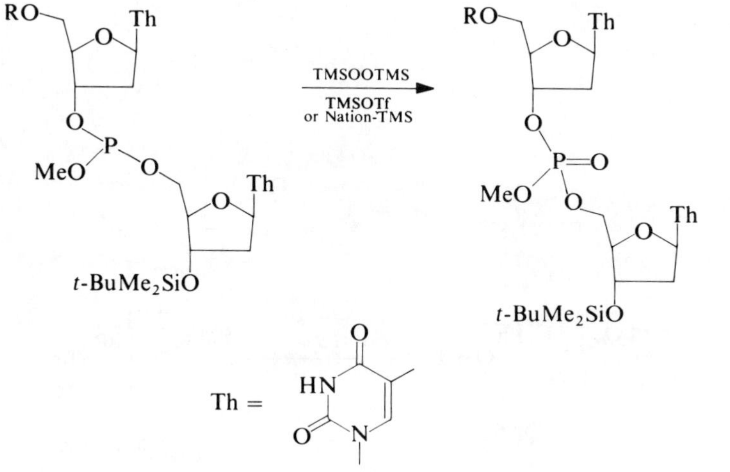
(169)

H. Other

Lithiotrimethylsilylmethyl carbonate, formed from the reaction of lithium trimethylsilylmethoxide and carbon dioxide, can be deprotonated and the resultant dianion reacted with esters, amides, or nitriles to give, after hydrolysis, α-hydroxy ketones in good yields[162] (equation 170). The dianion thus represents a methanol dianion equivalent.

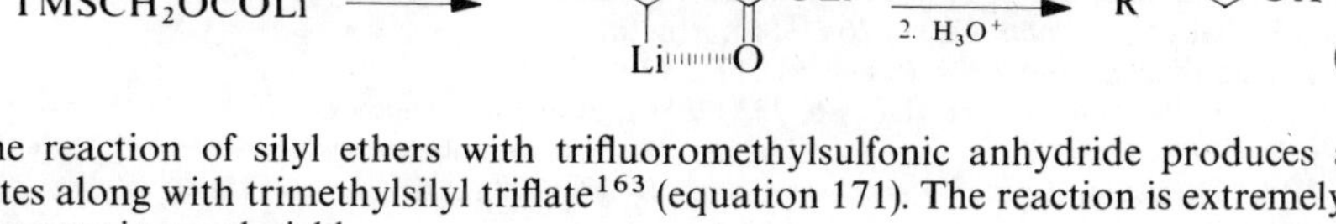
(170)

The reaction of silyl ethers with trifluoromethylsulfonic anhydride produces alkyl triflates along with trimethylsilyl triflate[163] (equation 171). The reaction is extremely fast and occurs in good yield.

$$ROTMS + (CF_3SO_2)_2O \xrightarrow[0°C/15\,min]{CH_2Cl_2} ROTf + TMSOTf \qquad (171)$$

Acylsulfenyl halides are formed by the reaction of *N*-bromosuccinimide or *tert*-butyl hypochlorite with thiocarboxylic acid *O*-trimethylsilyl esters[164] (equation 172). Treatment of silyl carboxylates with triphenylphosphine dibromide gives the corresponding acyl bromide along with the silyl bromide[165] (equation 173). It is also possible to convert silyl ethers to the alkyl bromide with this reagent[165] or with a mixture of carbon tetrabromide and triphenylphosphine[166]. The mild nature and selectivity of the reaction is underscored by the conversion of the trimethylsiloxy β-lactam **83** to the bromide **84**[165] (equation 174).

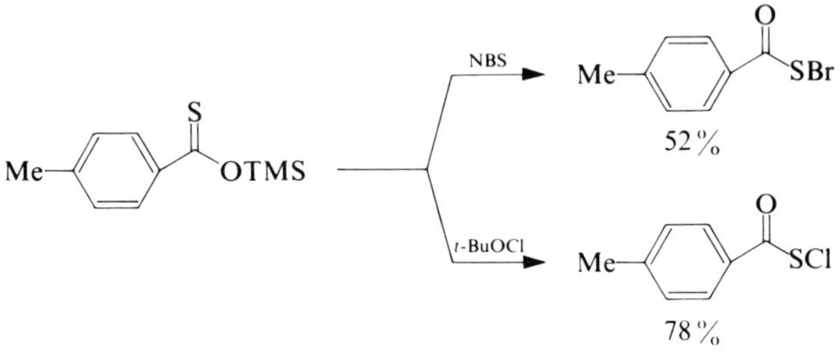

(172)

$$R^1CO_2SiR^2_3 \xrightarrow{Ph_3PBr_2} R^1\overset{O}{\overset{\|}{C}}Br + R^2_3SiBr \qquad (173)$$

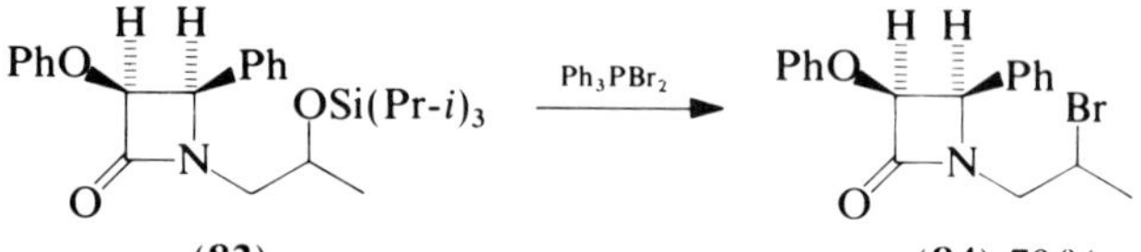

(**83**) (**84**) 70% (174)

XII. REFERENCES

1. (a) E. Colvin, *Silicon in Organic Chemistry*, Butterworths, Boston, 1981.
   (b) W. P. Weber, *Silicon Reagents for Organic Synthesis*, Springer-Verlag, New York, 1983.
   (c) I. Fleming, *Organic silicon chemistry*, in *Comprehensive Organic Chemistry* (Eds. D. Barton and W. D. Ollis), Pergamon Press, New York, 1979.
   (d) I Fleming, *Actual. Chim.*, 41 (1986); enol silyl ethers.
   (e) M. Lalonde and T. H. Chan, *Synthesis*, 817 (1985); protecting groups.
   (f) H. Sakurai, *Actual. Chim.*, 26 (1986); general.
   (g) H. Sakurai, *Pure Appl. Chem.*, **57**, 1759 (1985); carbon–carbon bond formation.
   (h) L. A. Paquette, *Chem. Rev.*, **86**, 733 (1986); cyclopropylsilanes.
   (i) L. E. Overman, *Lect. Heterocycl. Chem.*, **8**, 59 (1985); heterocyclizations with vinylsilanes.
   (j) A. Dondoni, *Lect. Heterocycl. Chem.*, **8**, 13 (1985); silylthiazoles and silyloxazoles in synthesis.
2. M. P. Cooke, Jr., *J. Org. Chem.*, **51**, 951 (1986).

3. E. J. Corey and N. W. Boaz, *Tetrahedron Lett.*, **26**, 6019 (1985).
4. E. Nakamura, S. Matsuzawa, Y. Horiguchi and I. Kuwajima, *Tetrahedron Lett.*, **27**, 4029 (1986).
5. Y. Horiguchi, S. Matsuzawa, E. Nakamura and I. Kuwajima, *Tetrahedron Lett.*, **27**, 4025 (1986).
6. C. R. Johnson and T. J. Marren, *Tetrahedron Lett.*, **28**, 27 (1987).
7. A. Alexakis, J. Berlan and Y. Besace, *Tetrahedron Lett.*, **27**, 1047 (1986).
8. G. Blame, G. Fournet and J. Gore, *Tetrahedron Lett.*, **27**, 1907 (1986).
9. P. M. Chouinard and P. A. Bartlett, *J. Org. Chem.*, **51**, 75 (1986).
10. C. R. Degenhardt and D. C. Burdsall, *J. Org. Chem.*, **51**, 3488 (1986).
11. M. Iyoda, T. Kushida and S. Kitami, *J. Chem. Soc., Chem. Commun.*, 1049 (1986).
12. A. S. Demir, R. S. Gross, N. K. Dunlap, A. Bashir–Hashemi and D. S. Watt, *Tetrahedron Lett.*, **27**, 5567 (1986).
13. A. Fleming, G. Sinai–Zingde, M. Natchus and T. Hudlicky, *Tetrahedron Lett.*, **28**, 167 (1987).
14. G. L. Larson and R. Klesse, *J. Org. Chem.*, **50**, 3627 (1985).
15. R. D. Miller and D. R. McKean, *Tetrahedron Lett.*, 2305 (1979).
16. S. H. Cheon, W. J. Christ, L. D. Hawkins, H. Jin, Y. Kishi and M. Taniguchi, *Tetrahedron Lett.*, **27**, 4759 (1986).
17. M. Taniguchi, S. Kobayashi, M. Nakagawa, T. Hino and Y. Kishi, *Tetrahedron Lett.*, **27**, 4763 (1986).
18. M. Taniguchi, T. Hino and Y. Kishi, *Tetrahedron Lett.*, **27**, 4767 (1986).
19. R. F, Spohn, P. A. Grieco and R. P. Nargund, *Tetrahedron Lett.*, **28**, 2491 (1987).
20. R. D. Miller and D. R. McKean, *Synthesis*, 730 (1979).
21. S. Hackett and T. Livinghouse, *J. Org. Chem.*, **51**, 879 (1986).
22. S. Raucher, B. L. Bray and R. F. Lawrence, *J. Am. Chem. Soc.*, **109**, 442 (1987).
23. S. L. Schreiber and B. Hulin, *Tetrahedron Lett.*, **27**, 4561 (1986).
24. A. J. Pearson and Y. S. Chen, *J. Org. Chem.*, **51**, 1939 (1986).
25. M. Larcheveque and S. Henrot, *Tetrahedron Lett.*, **28**, 1781 (1987).
26. K. Mai and G. Patil, *J. Org. Chem.*, **51**, 3545 (1986).
27. R. Yonada, K. Santo, S. Harusawa and T. Kurihara, *Synthesis*, 1054 (1986).
28. M. Santelli, D. El Abed and A. Jellal, *J. Org. Chem.*, **51**, 1199 (1986).
29. M. T. Reetz, F. Kunisch and P. Heitman, *Tetrahedron Lett.*, **27**, 4721 (1986).
30. K. Saito and H. Kojima, *Bull. Chem. Soc. Jpn.*, **58**, 1918 (1985).
31. P. G. Gassman, K. Okuma, A. Lindbeck and R. Allen, *Tetrahedron Lett.*, **27**, 6307 (1986).
32. P. G. Gassman and L. M. Haberman, *J. Org. Chem.*, **51**, 5010 (1986).
33. I. Saito, Y.-H. Kuo and T. Matsuura, *Tetrahedron Lett.*, **27**, 2757 (1986).
34. K. Fischer and S. Huenig, *J. Org. Chem.*, **52**, 564 (1987).
35. H. A. Bates and S. B. Rosenblum, *J. Org. Chem.*, **51**, 3447 (1986).
36. H. Miyake and K. Yamamura, *Tetrahedron Lett.*, **27**, 3025 (1986).
37. A. V. Stachulski, *J. Chem. Soc., Chem. Commun.*, 401 (1986).
38. M. Takahashi and H. Kikuchi, *Tetrahedron Lett.*, **28**, 2139 (1987).
39. M. Gill, M. J. Kiefel and D. A. Lally, *Tetrahedron Lett.*, **27**, 1933 (1986).
40. T. Kitazume, *Synthesis*, 855 (1986).
41. S. L. Buchwald and S. J. LaMaire, *Tetrahedron Lett.*, **28**, 295 (1987).
42. G. A. Olah, O. Farooq and G. K. S. Prakash, *Synthesis*, 1140 (1985).
43. N. Chatani and T. Hanafusa, *J. Org. Chem.*, **51**, 4714 (1986).
44. N. Chatani, T. Takeyasu and T. Hanafusa, *Tetrahedron Lett.*, **27**, 1841 (1986).
45. T. Kusumoto, H. Tamejiro and O. Koreharu, *Tetrahedron Lett.*, **27**, 4197 (1986).
46. N. Chatani and T. Hanafusa, *Tetrahedron Lett.*, **27**, 4201 (1986).
47. T. R. Hoye, B. H. Peterson and J. D. Miller, *J. Org. Chem.*, **52**, 1351 (1987).
48. S. L. Schreiber and J. Reagen, *Tetrahedron Lett.*, **27**, 2945 (1986).
49. T. Morimoto and M. Sekiya, *Chem. Lett.*, 1371 (1985).
50. T. V. Lee, K. A. Richardson and D. A. Taylor, *Tetrahedron Lett.*, **27**, 5021 (1986).
51. J. R. Hwu, L.-C. Leu, J. A. Robl, D. A. Anderson and J. M. Wetzel, *J. Org. Chem.*, **52**, 188 (1987).
52. T. Harada, T. Hayashiya, I. Wada, N. Iwaake and A. Oku, *J. Am. Chem. Soc.*, **109**, 527 (1986).
53. Y. Kimura, M. Suzuki, T. Matsumoto, R. Abe and S. Terashima, *Bull. Chem. Soc. Jpn.*, **59**, 423 (1986).

54. S. De Lombaert, I. Nemery, B. Roekens, J. C. Carretero, T. Kimmel and L. Ghosez, *Tetrahedron Lett.*, **27**, 5099 (1986).
55. E. D. Edstrom and T. Livinghouse, *Tetrahedron Lett.*, **27**, 3483 (1986).
56. K. Mikami, O. Takahashi, T. Tabei and T. Nakai, *Tetrahedron Lett.*, **27**, 4511 (1986).
57. A. P. Kozikowski and S. H. Jung, *J. Org. Chem.*, **51**, 3400 (1986).
58. D. Yang and D. D. Tanner, *J. Org. Chem.*, **51**, 2267 (1986).
59. M. Kira, K. Sato and H. Sajurai, *J. Org. Chem.*, **52**, 948 (1987).
60. S. Anwar and A. P. Davis, *J. Chem. Soc., Chem. Commun.*, 831 (1986).
61. G. A. Olah, M. Arvanaghi and L. Ohannesian, *Synthesis*, 770 (1986).
62. T. Uyehara, Y. Kabasawa, T. Furuta and T. Kato, *Bull. Chem. Soc. Jpn.*, **59**, 539 (1986).
63. E. Keinan and N. Greenspoon, *J. Am. Chem. Soc.*, **108**, 7314 (1986).
64. E. Keinan and D. Perez, *J. Org. Chem.*, **52**, 2576 (1987).
65. G. A. Olah, T. Yamato, P. S. Iyer and G. K. S. Prakash, *J. Org. Chem.*, **51**, 2826 (1986).
66. J. A. Bennek and G. R. Gray, *J. Org. Chem.*, **52**, 892 (1987).
67. N. Ono, T. Hashimoto, T. X. Jun and A. Kaji, *Tetrahedron Lett.*, **28**, 2277 (1987).
68. H. Sano, M. Ogata, T. Mayako and T. Migita, *Chem. Lett.*, 77 (1986).
69. H. Uno, *J. Org. Chem.*, **51**, 350 (1986).
70. M. T. Reetz and K. Kesseler, *J. Org. Chem.*, **50**, 5434 (1985).
71. L. Coppi, A. Mordini and M. Taddei, *Tetrahedron Lett.*, **28**, 969 (1987).
72. L. Coppi, A. Ricci and M. Taddei, *Tetrahedron Lett.*, **28**, 973 (1987).
73. K. Nishitani and K. Yamakawa, *Tetrahedron Lett.*, **28**, 655 (1987).
74. G. A. Molander and S. W. Andrews, *Tetrahedron Lett.*, **27**, 3115 (1986).
75. D. Schinzer, J. Steffen and S. Solyom, *J. Chem. Soc., Chem. Commun.*, 829 (1986).
76. G. Majetich, A. Casares, D. Chapman and M. Behnke, *J. Org. Chem.*, **51**, 1745 (1986).
77. G. Majetich, R. W. Desmond, Jr. and J. J. Soria, *J. Org. Chem.*, **51**, 1753 (1986).
78. A. Hosomi, M. Ando and H. Sakurai, *Chem. Lett.*, 365 (1986).
79. S. J. Danishefsky, S. DeNinno and P. Lartey, *J. Am. Chem. Soc.*, **109**, 2082 (1987).
80. W. S. Johnson, C. Newton and S. D. Lindell, *Tetrahedron Lett.*, **27**, 6027 (1986).
81. Y. Hatanaka and I. Kuwajima, *J. Org. Chem.*, **51**, 1932 (1986).
82. H. Hamana and T. Sugasawa, *Chem. Lett.*, 921 (1985).
83. J. Yoshida, T. Murata and S. Isao, *Tetrahedron Lett.*, **27**, 3373 (1986).
84. Y. Ikeda and H. Yamamoto, *Bull. Chem. Soc. Jpn.*, **59**, 657 (1986).
85. K. Tamao, E. Nakajo and Y. Ito, *J. Org. Chem.*, **52**, 957 (1987).
86. B. M. Trost, *Angew. Chem.*, **98**, 1 (1986).
87. B. M. Trost and S. M. Mignani, *Tetrahedron Lett.*, **27**, 4137 (1986).
88. B. M. Trost, J. M. Balkovec and S. R. Angle, *Tetrahedron Lett.*, **27**, 1445 (1986).
89. B. M. Trost and S. A. King, *Tetrahedron Lett.*, **27**, 5971 (1986).
90. B. M. Trost, S. M. Mignani and T. N. Nanninga, *J. Am. Chem. Soc.*, **108**, 6051 (1986).
91. M. D. Jones and R. D. W. Kemmitt, *J. Chem. Soc., Chem. Commun.*, 1201 (1986).
92. B. M. Trost and P. R. Seoane, *J. Am. Chem. Soc.*, **109**, 615 (1987).
93. G. A. Molander and D. C. Shubert, *J. Am. Chem. Soc.*, **108**, 4683 (1986).
94. G, A. Molander and D. C. Shubert, *J. Am. Chem. Soc.*, **109**, 576 (1987).
95. P. A. Grieco and W. F. Fobare, *Tetrahedron Lett.*, **27**, 5067 (1986).
96. H. H. Mooiweer, H. Hiemstra, H. P. Fortgens and W. N. Speckamp, *Tetrahedron Lett.*, **28**, 3285 (1987).
97. Y. Kita, S. Akai, N. Ajimura, M. Yoshigi, T. Tsugoshi, H. Yasuda and Y. Tamura, *J. Org. Chem.*, **51**, 4150 (1986).
98. R. Diercks and K. P. C. Vollhardt, *J. Am. Chem. Soc.*, **108**, 3150 (1986).
99. B. M. Trost and S.-F. Chan, *J. Am. Chem. Soc.*, **108**, 6053 (1986).
100. H. Kosugi, M. Kitaoka, K. Tagami and H. Uda, *Chem. Lett.*, 805 (1986).
101. P. C. Page and S. Rosenthal, *Tetrahedron Lett.*, **27**, 1947 (1986).
102. G. Capozzi, R. Ottana, G. Romeo and F. Marcuzzi, *Gazz. Chim. Ital.*, **115**, 311 (1985).
103. D. J. Ager, *Synthesis*, 384 (1984).
104. C. R. Johnson and B. D. Tait, *J. Org. Chem.*, **52**, 281 (1987).
105. B. Halton, C. J. Randall, G. J. Gainsford and P. J. Stang, *J. Am. Chem. Soc.*, **108**, 5949 (1986).
106. A. Guingant and J. d'Angelo, *Tetrahedron Lett.*, **27**, 3729 (1986).
107. E. E. Aboujaoude, S. Lietje, N. Collignon, M. P. Teulade and P. Savignac, *Synthesis*, 934 (1986).

108. S. Hackett and T. Livinghouse, *J. Org. Chem.*, **51**, 879 (1986).
109. J. Binder and E. Zbiral, *Tetrahedron Lett.*, **27**, 5829 (1986).
110. P. A. Brown, R. V. Bonnert, P. R. Jenkins and M. R. Selim, *Tetrahedron Lett.*, **28**, 693 (1987).
111. B. Beenetau, M. Bordeau and J. Dunogues, *Bull. Soc. Chim. Fr.*, 90 (1985).
112. G. Bartol, M. Bosco, R. Dalpozzo and P. E. Todesco, *J. Org. Chem.*, **51**, 3694 (1986).
113. O. Tsuge, J. Tanaka and S. Kanemasa, *Bull. Chem. Soc. Jpn.*, **58**, 1991 (1985).
114. C. W. Chen and P. Beak, *J. Org. Chem.*, **51**, 3325 (1986).
115. S. Shimizu and M. Ogata, *J. Org. Chem.*, **52**, 2314 (1987).
116. S. Shimizu and M. Ogata, *J. Org. Chem.*, **51**, 3897 (1986).
117. B. F. Bonini, G. Mazzanti, P. Zani, G. Maccagnani, G. Barbaro, A. Battaglia and P. Giorgianni, *J. Chem. Soc., Chem. Commun.*, 964 (1986).
118. B. Rigo, C. Lespagnol and M. Pauly, *Tetrahedron Lett.*, **27**, 347 (1986).
119. E. Vedejs, *Chem. Rev.*, **86**, 941 (1986).
120. A. Padwa and W. Dent, *J. Org. Chem.*, **52**, 235 (1987).
121. Y, Terao, H. Kotaki, N. Imai and K. Achiwa, *Chem. Pharm. Bull.*, **33**, 2762 (1985).
122. A. Padwa, J. R. Gasdaska, M. Tomas, N. J. Turro, Y. Cha and I. R. Gould, *J. Am. Chem. Soc.*, **108**, 6739 (1986).
123. O. Tsuge, S. Kanemasa, T. Yamada and K. Matsuda, *J. Org. Chem.*, **52**, 2523 (1987).
124. O. Tsuge, S. Kanemasa and K. Matsuda, *J. Org. Chem.*, **51**, 1997 (1986).
125. A. Padwa, P. Eisenbarth, M. K. Venkataramanan and G. S. K. Wong, *J. Org. Chem.*, **52**, 2427 (1987).
126. A. Padwa and J. R. Gasdaska, *J. Am. Chem. Soc.*, **108**, 1104 (1986).
127. O. Tsuge, S. Kanemasa and S. Kuraoka, *Bull. Chem. Soc. Jpn.*, **58**, 1570 (1985).
128. M. Westling, R. Smith and T. Livinghouse, *J. Org. Chem.*, **51**, 1159 (1986).
129. T. Morimoto, Y. Nezu and K. Achiwa, *Chem. Pharm. Bull.*, **33**, 4596 (1985).
130. A. Hosomi, Y. Matsuyama and H. Sakurai, *J. Chem. Soc., Chem. Commun.*, 1073 (1986).
131. M. Aono, C. Hyodo, Y. Terao and K. Achiwa, *Tetrahedron Lett.*, **27**, 4039 (1986).
132. A. Koziara, K. Osowska-Pacewicka, S. Zawadzki and A. Zwierzak, *Synthesis*, 487 (1987).
133. G. K. S. Prakash, M. A. Stephenson, J. G. Shih and G. A. Olah, *J. Org. Chem.*, **51**, 3215 (1986).
134. (a) K. Maruoka, H. Sano and H. Yamamoto, *Chem. Lett.*, 599 (1985).
(b) D. Shinou and M. Emziane, *Tetrahedron Lett.*, **27**, 4423 (1986).
135. J.-N. Denis, A. E. Greene, A. A. Serra and M.-J. Luche, *J. Org. Chem.*, **51**, 46 (1986).
136. K. Nishiyama and I. Miyata, *Bull. Chem. Soc. Jpn.*, **58**, 2419 (1985).
137. V. Pozsgay and H. J. Jennings, *Tetrahedron Lett.*, **28**, 1375 (1987).
138. B. M. Trost and T. S. Scanlon, *Tetrahedron Lett.*, **27**, 4141 (1986).
139. K. Steliou, P. Salama and J. Corriveau, *J. Org. Chem.*, **50**, 4969 (1985).
140. K.-J. Hwang, *J. Org. Chem.*, **51**, 99 (1986).
141. S. E. Tinney and J. K. Stille, *J. Org. Chem.*, **52**, 748 (1987).
142. I. Fleming and J. D. Kilburn, *J. Chem. Soc., Chem. Commun.*, 305 (1986).
143. I. Fleming and J. D. Kilburn, *J. Chem. Soc., Chem. Commun.*, 1198 (1986).
144. D. A. Burnett, J. C. Galucci and D. J. Hart, *J. Org. Chem.*, **50**, 5120 (1985).
145. K. Tamao and K. Maeda, *Tetrahedron Lett.*, **27**, 65 (1986).
146. K. Tamao, T. Tanaka, T. Nakajima, R. Sumiya and Y. Ito, *Tetrahedron Lett.*, **27**, 3377 (1986).
147. K. Tamao, T. Nakajima, R. Sumiya, H. Arai, N. Higuchi and Y. Ito, *J. Am. Chem. Soc.*, **108**, 6090 (1986).
148. E. Magnol and M. Malacria, *Tetrahedron Lett.*, **27**, 2255 (1986).
149. G. Stork and M. J. Sofia, *J. Am. Chem. Soc.*, **108**, 6826 (1986).
150. S. R. Wilson and L. A. Jacob, *J. Org. Chem.*, **51**, 4833 (1986).
151. R. Calas, *C. R. Acad. Sci. Paris, Ser. 2,* **301**, 1289 (1985).
152. M. W. Majchrzak and G. Simchen, *Synthesis*, 956 (1986).
153. G. L. Larson and L. M. Fuentes, *J. Am. Chem. Soc.*, **103**, 2418 (1981).
154. G. L. Larson and R. Betancourt de Perez, *J. Org. Chem.*, **50**, 5257 (1985).
155. R. Betancourt de Perez, L. M. Fuentes, G. L. Larson, C. L. Barnes and M. J. Heeg, *J. Org. Chem.*, **51**, 2039 (1986).
156. G. L. Larson, D. Hernandez, I. Montes de Lopez-Cepero and L. E. Torres, *J. Org. Chem.*, **50**, 5260 (1985).
157. (a) I. Montes de Lopez-Cepero, A. Santiago and G. L. Larson, *Synth. Commun.*, **16**, 705 (1986).
(b) A. Santiago, I. Montes de Lopez-Cepero and G. L. Larson, *Synth. Commun.*, **16**, 697 (1986).

158. E. J. Corey, M. M. Mehrotra and A. U. Khan, *J. Am. Chem. Soc.*, **108**, 2472 (1986).
159. G. H. Posner, M. Weitzberg, W. M. Nelson, B. L. Murr and H. H. Seliger, *J. Am. Chem. Soc.*, **109**, 278 (1987).
160. L. Camici, A. Ricci and M. Taddei, *Tetrahedron Lett.*, **27**, 5155 (1986).
161. Y. Hayakawa, M. Uchiyama and R. Noyori, *Tetrahedron Lett.*, **27**, 4191 (1986).
162. A. R. Katritzky and S. Sengupta, *Tetrahedron Lett.*, **28**, 1847 (1987).
163. C. Aubert and J. P. Begue, *Synthesis*, 759 (1985).
164. T. Murai, S. Oida, S. Min and S. Kato, *Tetrahedron Lett.*, **27**, 4593 (1986).
165. J. M. Aizpurua, F. P. Cossio and C. Palomo, *J. Org. Chem.*, **51**, 4941 (1986).
166. H. Mattes and C. Benezra, *Tetrahedron Lett.*, **28**, 1697 (1987).

The Chemistry of Organic Silicon Compounds
Edited by S. Patai and Z. Rappoport

CHAPTER **12**

Acidity, basicity and complex formation of organosilicon compounds

ALAN R. BASSINDALE and PETER G. TAYLOR
PORCG, Department of Chemistry, The Open University, Milton Keynes, MK7 6AA, England

I. INTRODUCTION

A. Acidity and Basicity

The related concepts of acidity and basicity are fundamental to studies of chemical behaviour. The Brønsted acidity of a protonic acid HA is usually expressed in terms of pK_a, as defined by the negative of the logarithm of the equilibrium constant for reaction 1.

$$HA \rightleftharpoons A^- + H^+ \qquad (1)$$

The Brønsted basicity of a base B can be expressed as either the pK_b for the reaction

$$B + H^+ \rightleftharpoons BH^+ \qquad (2)$$

or, as is increasingly common practice[1], as the pK_a of the conjugate acid, BH^+, through the equilibrium

$$BH^+ \rightleftharpoons B + H^+ \qquad (3)$$

The standard state for pK_a measurement is infinite dilution in water at 25°. This has several disadvantages, particularly for those interested in silicon chemistry; solubility in water is often limited, and many silyl derivatives of organic compounds are hydrolytically unstable. Solvation effects in water are very strong indeed, and can mask small changes in 'intrinsic' acidity or basicity. Additionally, the autoprotolysis constant of water precludes direct measurement of very strong and very weak acids and bases. Dimethyl sulphoxide has been suggested as a suitable alternative solvent for setting standard acidities and basicities[1], but this is unlikely to supplant the aqueous basis of pK_a measurements and, furthermore, labile silicon compounds, such as halosilanes, can react with DMSO.

There is a strong body of opinion that gas phase studies of acid–base behaviour through

$$HA_{(g)} \rightleftharpoons H^+_{(g)} + A^-_{(g)} \qquad (4)$$

give the best indication of 'intrinsic' acidity and basicity[2], and the use of modern mass spectroscopic methods is becoming commonplace in such studies.

Lewis acidity and basicity usually refers to the propensity to form complexes with donor or acceptor molecules,

$$A + X \rightleftharpoons AX \qquad (5)$$

for the Lewis acidity of A and

$$B{:} + X \rightleftharpoons BX \qquad (6)$$

for the Lewis basicity of B. There is no single scale equivalent to pK_a for measuring Lewis acidity or basicity, but there are several empirical correlations that refer to the donor or acceptor behaviour. The Gutmann Donor Number[3] is defined as $-\Delta H/\text{kcal mol}^{-1}$ for the interaction of a Lewis base with $SbCl_5$ in a dilute solution in CH_2Cl_2[4,5], and is considered a useful indicator of Lewis basicity. Taft and coworkers[6,7] through a multiparameter analysis of linear solvation energy relationships defined a hydrogen bond acceptor basicity, β, which is measure of a solvent's ability to donate an electron pair in a solvent–solute hydrogen bond. Similar parameters are available for Lewis acids[3,6,7]. A review by Jensen in 1978 discusses definitions of Lewis acidity and basicity[8].

B. The Measurement of Acidity and Basicity of Silicon Compounds

The majority of measurements of acidity and basicity of silicon compounds have been concerned with the influence of silicon on a neighbouring functional group as, for example, the acidity of silanols, R_3SiOH and the basicity of aminosilanes, R_3SiNR_2 and $R_3SiCH_2NR_2$. The SiH bond is polarized Si^+H^- and direct measurement of the acidity of silanes in solution does not appear to have been accomplished. Accordingly, this review concentrates on the effects of R_3Si on neighbouring functional groups.

In this section we review briefly the methods that have been used in measuring acidity and basicity. Detailed references will be found in subsequent sections, where the 'acidity' and 'basicity' of various organosilicon functional compounds are discussed. Compared with the wealth of data on organic compounds, the literature on organosilicon compounds is quite limited. The approach we have taken is to discuss the effect of silyl substitution on Lewis and Brønsted behaviour and hydrogen-bonding together in the same sections. This seems appropriate for several reasons. Many of the studies were not concerned with the measurement of pK_a but were designed to evaluate the relative effect of silyl substitution. In general, the Lewis and Brønsted behaviours are parallel, justifying a common discussion, and separate sections for each type of measurement would unnecessarily obscure the essential conclusions. The chapter is organized so that similar functional groups are examined together and functional groups with silicon directly bonded are treated separately from those with more remote silicon substitution.

The measurement of acidity and basicity of silicon compounds follows quite well-defined phases. The first studies were conventional acid–base titrations in which pK_a values were determined. The solvents varied, but were usually aqueous or water/alcohol mixtures. These studies were limited by solubility and resistance to attack by protic solvents.

Following the pK_a measurements, considerable efforts were expended on evaluating the interaction of silicon compounds with donor or acceptor solvents. The measurement of relative acidity or basicity followed similar lines. For acidity, the interaction of a dilute solution of the silicon compounds with a hydrogen bond acceptor, such as ether or THF, was monitored. The most common approach was to measure the difference in infrared stretching frequency between free X–H and hydrogen-bonded X–H. The difference was taken to be proportional to the enthalpy of hydrogen bonding, according to Badger and Bauer[9], and Drago[10]. For basicity a similar method was applied, but a reference acid, usually $CHCl_3$ or $CDCl_3$, was monitored for changes in C–H or C–D stretching frequency. The Badger–Bauer methodology has been strongly criticized[11] as being highly inaccurate in evaluating $-\Delta H$ for H-bond formation. Despite criticisms the method is often considered reliable for a closely related series of compounds. The infrared stretching frequency differences were not the only parameters used to estimate the enthalpy of hydrogen bonding with donor or acceptor solvents. Thermodynamic measurements by calorimetry were also carried out, and NMR spectroscopy was applied but found rather imprecise.

At the same time as more organic chemists were using spectroscopic measurements to assess acidity and basicity, inorganic methodology was also being applied. In particular, silyl substituted bases were reacted with Lewis acids and thermodynamic parameters were assessed for complex formation. In many cases direct measurement of heats of formation or stability constants proved difficult to obtain, but relative Lewis basicities were measured.

Currently, effort is being expended in two directions. Gas-phase measurements by high-pressure mass spectroscopy, ion cyclotron resonance spectroscopy and flowing afterglow methods have elucidated 'intrinsic' acidities and basicities, and the results from the various techniques appear to be relatively consistent[2]. The gas-phase acidities and proton affinities are free from solvent perturbations making the data accessible to modelling by calculation, which is the other current area of interest. The calculation of proton affinities and gas-phase acidities is becoming highly sophisticated as, for accurate correlations with experiment, *ab initio* methods with extended basis sets are essential.

A discussion of the effects of silicon substitution on acidity and basicity of various functional groups follows. Despite the variety of methods applied to the problem there is a clear concensus that two features dominate in determing the acidity and basicity of functional silicon compounds. The first is the π-type electron withdrawal by R_3Si in R_3Si–XH which lowers basicity and increases acidity, relative to the carbon analogues. The second is the strongly electron-releasing effect of R_3SiCH_2 in R_3SiCH_2XH which increases basicity and decreases acidity relative to the carbon analogues.

II. ACIDITY

A. R_3Si Directly Bonded to the Acidic Functional Group

1. *The effect of R_3Si on the acidity of R_3SiXH (X=O, S, NR, PR, CR_2)*

Silanols, R_3SiOH, are considerably more acidic than the corresponding alcohols, R_3COH. Sommer and coworkers[12], in 1946, found that trimethylsilanol was converted to its sodium salt in NaOH solution, and commented on the increased reactivity of the hydroxyl proton in silanols compared to alcohols. West and Baney[13] subsequently showed that triphenylsilanol is sufficiently acidic to be titrated in pyridine, whereas triphenylcarbinol is not. West[14, 15] was the first to propose that the enhanced acidity of silanols was due to (p-d)π bonding, in which one of the oxygen lone pairs overlapped with the vacant silicon 3d orbitals. This model has remained popular as it is conceptually simple and provides a satisfactory explanation for the delocalisation of electrons α to silicon.

In addition to the titration data, infrared spectroscopic studies were particularly common in the 1960s for evaluating the acidity and basicity of silicon compounds. West[15] used the Badger–Bauer relationship[9] in examining the shift of the OH stretching band on hydrogen bonding to ether and mesitylene by silanols and carbinols. The silanols gave band shifts nearly twice as large as the carbinols, and phenyl rings on silicon or carbon markedly increased the band shifts and, by implication, the acidity. West[16] argued that the 'acidity' measured by these infrared studies was a Lewis acidity, as hydrogen bonding is analogous to other Lewis acid–base interactions. The acidity order, arylsilanols > alkylsilanols > arylcarbinols > alkylcarbinols, was established[15]. Steric effects were said to be unimportant in determining the infrared band shifts $\Delta\nu$. Arylsilanols appeared to be about as acidic as phenols from these studies. Further work by West and coworkers[16] using a similar infrared method[17] suggested that the group IV Ph_3MOH acidity decreased in the order $Ph_3SiOH > Ph_3GeOH \geqslant Ph_3COH \geqslant Ph_3SnOH$, Ph_3PbOH. Efficient (p-d)π bonding to silicon was postulated to explain the high acidity of Ph_3SiOH compared with the other group IV hydroxy compounds. Matwiyoff and Drago[18], in connection with an evaluation of the Allred and Rochow electronegativity scale[19], repeated West's studies[16] on the group IV hydroxy compounds and extended the study to include tetramethylene sulphoxide, TMSO, as hydrogen bond acceptor for Ph_3MOH (M=C, Si, Ge). The conclusions were similar to those of West concerning (p-d)π bonding.

Steward and Fussaro[20] measured pK_a values for some R_3SiOH compounds in DMSO using tetramethylammonium hydroxide as base. The results are given in Table 1. The pK_a of Ph_3GeOH was determined as 17.06. The order of acidity for Ph_3MOH (M=C, Si, Ge) is in agreement with the relative acidity determined by hydrogen bonding studies[16]. However, the enhanced acidity of $Ph_3SiSiPh_2OH$ was not reflected in a hydrogen bonding study. An explanation was advanced for the pronounced acidity enhancement of $Ph_3SiSiPh_2OH$ involving an interaction between an oxygen non-bonding electron pair

TABLE 1. pK_a values of some silanols in DMSO[20]

Compound	pK_a (σ)[a]
Ph_3COH	16.97 (0.24)[b]
Ph_3SiOH	16.57 (0.11)[b]
	16.63 (0.07)[c]
Ph_3SiCPh_2OH	14.33 (0.13)[c]
$Ph_3SiSiPh_2OH$	13.87 (0.13)[c]
	13.99 (0.03)[b]

[a] Standard deviation obtained from a number of determinations.
[b] Indicator 4-nitrodiphenylamine, pK_a 14.3 ± 0.3.
[c] Indicator 2,4-dinitrodiphenylamine, pK_a 12.4 ± 0.3.

and the lowest unoccupied MO of the SiSi system. This α-effect was designated $n_0 \rightarrow LUMO_{SiSi}$. The LUMO could be either σ^* or π^*.

An NMR study[21] showed that the ^{29}Si NMR resonance in silanols was dependent on the basicity of the solvent. A linear correlation between $\delta^{29}Si$ and the Gutmann donor number[3, 4, 22] was observed.

The observation of enhanced acidity for R_3SiXH compared with R_3CXH compounds appears to be general. Salinger and West[23] applied the Badger–Bauer[9] infrared band shift method to silanethiols and the order of acidity obtained was $Ph_3SiSH \gg PhSH > Me_3SiSH > Ph_3CSH > Me_3CSH$. Relative Brønsted acidities were also determined using a titrimetric method with tetrabutylammonium hydroxide as titrant in pyridine solution. The half-neutralization potential was taken as a measure of the equilibrium acid strength. The results together with some potentiometric titrations in 50% dioxane–water using sodium hydroxide as base are given in Table 2. The scale of Brønsted acidities as determined by this method was

$$Ph_3SiSH > PhSH > Ph_3CSH > PhOH > Ph_3SiOH > Me_3SiSH, Me_3CSH$$

These results were said to be consistent with efficient (p-d)π bonding in the silanethiols, as in the silanols. The high Brønsted acidity of silanethiols was postulated to be due to increased π-bonding in the silanethiolate anions relative to the parent silanethiols.

TABLE 2. Half-neutralization potentials for some silanethiols and silanols[23]

	$E_{1/2}$ (V)	
Compounds	Bu_4NOH/pyridine	$NaOH/H_2O$/dioxane
PhCOOH		0.10
CH_3COOH		0.09
Ph_3SiSH	−0.21	0.09
PhSH	−0.34	[a]
Ph_3CSH	−0.56	[a]
PhOH	−0.68	
Ph_3SiOH	−0.80	
Me_3SiSH	[a]	
Me_3CSH	[a]	

[a] Too weak an acid to titrate.

Silyl phosphines, $(H_3Si)_2PH$ and $(H_3Si)PH_2$ are also stronger acids than phosphine, as shown by chemical equilibrium and displacement studies by Sheldrick and coworkers[24, 25]. The order of Brønsted acidity was found to be $(H_3Si)_2PH > H_3SiPH_2 > PH_3 > CH_3PH_2 > (CH_3)_2PH$. No explanation for the observed order was proposed, but recently[26] Glidewell and Thompson carried out *ab-initio* molecular orbital calculation on these and related compounds. All calculations were single-configuration SCF calculations using the GAUSSIAN 80 program system. The 3-21G basis set without d functions was used for most calculations. The calculations covered proton affinities, defined as $-\Delta E^{\ominus}$, for the reaction shown in equation 7 and acidities, defined as $+\Delta E^{\ominus}$, for the reaction shown in equation 8.

$$B + H^+ \longrightarrow BH^+ \tag{7}$$

$$AH \longrightarrow A^- + H^+ \tag{8}$$

The calculated gas-phase acidities, converted to kcal mol^{-1}, are shown in Table 3. For phosphine derivatives the calculations are in good agreement with the experimental observations with the acidity falling in the order $(H_3Si)_2PH > H_3SiPH_2 > H_3CPHSiH_3 > PH_3 > CH_3PH_2 > (CH_3)_2PH$.

Similar calculations on ammonia derivatives again reproduced the acid strengthening property of α-silicon groups. In this case methyl substitution for H increased the Brønsted acidity, whereas in the phosphine case methyl substitution had an acid weakening effect. Consequently a slightly different order of acidity for ammonia derivatives was observed, $(H_3Si)_2NH > H_3SiNHCH_3 > H_3SiNH_2 > (CH_3)_2NH > CH_3NH_2 > NH_3$. It can be seen that the silyl group is significantly more acidity-enhancing than methyl. Glidewell and Thompson repeated the calculations for the SiH_3PH_2 and SiH_3NH_2 including d functions[26], and concluded that, although the absolute energies were lower with the inclusion of d functions, the 3-21G basis set without d functions did accurately reflect the trend in acidity and basicity for these compounds.

In addition to the amino and phosphine derivatives Glidewell and Thompson[26] calculated the carbon acidity of $(H_3Si)_2CH_2$ and $(H_3Si)_3CH$, and again silyl substitution increased the gas-phase acidity by about 20 kcal mol^{-1} (cf. *ca.* 10 kcal mol^{-1} for the analogous phosphines and 33 kcal mol^{-1} for the amines).

TABLE 3. Calculated gas-phase acidities for some H_3SiXH compounds[26]

1. *Phosphine derivatives*				
Compound	PH_3	H_3SiPH_2	$(H_3Si)_2PH$	$H_3CPHSiH_3$
Acidity (kcal mol^{-1})	372	356	346	363
Compound		H_3CPH_2	$(H_3C)_2PH$	
Acidity (kcal mol^{-1})		379	385	
2. *Ammonia derivatives*				
Compound	NH_3	H_3SiNH_2	$(H_3Si)_2N$	H_3CNSiH_3
Acidity (kcal mol^{-1})	462	418.4	385	418.1
Compound		H_3CNH_2	$(H_3C)_2NH$	
Acidity (kcal mol^{-1})		449	439	
3. *Methane derivatives (loss of C–H in each case)*				
Compound	CH_4^a	$H_3SiCH_3^a$,	$(H_3Si)_2CH_2$	$(H_3Si)_3CH$
Acidity (kcal mol^{-1})	446	416	397	377

[a] From Reference 27, *ab initio* calculation double-ζ basis set.

These results can be compared with the calculations of Hopkinson and Lien[27] on CH_4 and CH_3SiH_3 which complete the set. Although the calculations used different basis sets, both sets of results given in Table 3 are without d-orbital functions, and the trend of successive silyl stabilization of α-carbanions is well reproduced.

In conclusion, silicon adjacent to an X–H functional group is demonstrably stabilizing to the R_3SiX^- anion resulting in enhanced acidity. The question of the origin of this stabilization remains open; d-orbital participation by silicon is an attractive explanation because of its simplicity but *ab initio* calculations without d functions also satisfactorily reproduce the trends in acidity.

2. The effect of R_3Si on the acidity of $R_3SiCOOH$

The α-silyl-substituted carboxylic acids $R_3SiCOOH$ are, rather surprisingly, more acidic than their carbon analogues. This was first reported by Steward and coworkers[28] who measured ionisation constants in ethanol–water mixtures by a potentiometric technique. The results are given in Table 4. The enhanced acidity of the α-silyl carboxylic acids was considered to be in accord with (p–d)π bonding rather than the inductive effect. As silicon is inductively electron releasing, an acid-weakening effect would be expected on these grounds. Steric effects arising from solvent exclusion in the vicinity of the anion were considered but rejected on the basis of comparisons between Ph_3CCOOH and $Ph_3GeCOOH$ in ethanol–water and DMSO. The same carboxylic acids were reexamined by Steward and coworkers[29] giving essentially identical results, and $Me_3SiCOOH$ was

TABLE 4. Ionization constants of R_3MCOOH in ethanol–water mixtures

Compound	$pK_a(\sigma)^a$ 45% w/w ethanol	$pK_a(\sigma)$ 76% w/w ethanol
CH_3COOH	5.64(0.06)[b] 5.61(0.01)[c]	6.78(0.02)[b] 6.57(0.02)[c]
$(CH_3)_3CCOOH$	6.41(0.08)[c]	7.96(0.05)[c]
$(CH_3)_3SiCOOH$	6.60(0.04)[c]	7.74(0.02)[c]
$(CH_3)_2PhCCOOH$	6.00(0.03)[b] 6.05(0.01)[c]	7.45(0.02)[b] 7.45(0.04)[c]
$(CH_3)_2PhSiCOOH$	5.96(0.10)[b] <6.06[c]	7.28(0.05)[c]
CH_3Ph_2CCOOH	5.77(0.04)[b] 5.81(0.01)[c]	7.20(0.03)[b] 7.15(0.09)[c]
$CH_3Ph_2SiCOOH$	5.66(0.05)[b] <6.0[c]	7.03(0.09)[b] 7.04(0.05)[c]
Ph_3CCOOH	— —	6.70(0.03)[b] 6.78(0.02)[c]
$Ph_3SiCOOH$	— —	6.23(0.09)[b] 6.10(0.10)[c]

[a] Standard deviation obtained from a number of determinations.
[b] Reference 28.
[c] Reference 29.

also measured. Acidity was found to increase consistently with phenyl substitution at silicon. The (p–d)π bonding argument was elaborated. Spectroscopic studies by Steward and coworkers[30] suggested that little π(p–d) bonding was evident in the ground state of the acid form.

Ramsey[31] challenged the view of Steward by pointing out that the HOMO of the carboxylic π-system is nodal at carbon. He suggested that 1,3(p–d)π bonding between Si and oxygen in the anion could be important. This question was taken up by Kuznesof[32] who carried out CNINDO molecular orbital calculations on $H_3SiCOOH$, H_3SiCOO^- and CH_3COOH and CH_3COO^-. The gas-phase acidities thus calculated gave the correct acidity order with $H_3SiCOOH$ being more acidic than CH_3COOH. The inclusion of d orbitals on silicon had a negligible effect on the acidity of $H_3SiCOOH$. In evaluating the importance of 1,3(p–d)π interactions in the anion it was noted that the 1,3Si(d_{yz})–O(p_y) bond order was 0.119, almost identical to the 1,2Si(d_{yz})–C(p_y) bond order of 0.118, despite the very small overlap integral for the 1,3 interaction. This was not considered to be sufficiently large to be a major factor in anion stabilization. The 'anomalous' acidity of α-silylcarboxylic acids was attributed to the larger size of silicon, relative to carbon, and its greater polarizability, facilitating stabilization of the negative charge in the anion.

Steward[33] examined the effect of β-silicon substitution on α-silylcarboxylic acids by measuring the pK_a of $(Me_3Si)_3SiCOOH$ as 9.33 and $Ph_3SiSiPh_2COOH$ as 7.95. Comparison with the data in Table 4 shows that β-silicon reduces the acidity of α-silylcarboxylic acids. These results ruled out extensive extended d-orbital participation, as was observed for the alcohols[20]. It was suggested that steric hindrance to solvation could be responsible for the acid weakening effect of the β-silyl substitution on α-silylcarboxylic acids.

Although α-silylcarboxylic acids are generally stronger acids than their carbon analogues it cannot be stated with confidence that the phenomenon is well understood. Several alternative explanations are available[28,31–33].

B. The Effect of Silicon Remote from the Acidic Functional Group

1. R_3SiCH_2OH and R_3SiCH_2COOH compounds

No quantitative data are available for the acidity of R_3SiCR_2OH derivatives or their homologues. Brook and Pannell[34] observed a linear relationship between $\Sigma\sigma^*$, the sum of the Taft polar substituent constants, and the chemical shift of the OH proton in DMSO of α-silylcarbinols Ph_3SiCR_2OH. The OH resonance frequency was as expected for the relative acidity where the substituent effects are uncomplicated by resonance effects. Caution was, however, urged in the use of such correlations, particularly for unrelated compounds. Peddle[35] carried out a similar study, extending the range of compounds to $R_3SiCH_2CH_2OH$ derivatives, but no firm conclusions can be drawn from the study.

The data for carboxylic acid derivatives are better established. Sommer[36] showed that β-silicon groups are acid-weakening for R_3SiCH_2COOH derivatives. The data are given in Table 5, and were determined by a potentiometric method, presumably in aqueous solution. Eaborn, Bott and coworkers[37] repeated the titration of Me_3SiCH_2COOH in water and obtained the same value as Sommer[36]. It was concluded[36] that inductive electron release by silicon decreased the acid strength and that steric hindrance to solvation was also important[37].

2. Aromatic carboxylic acids and phenols containing R_3Si

The effect of *m*- and *p*-Me_3Si groups on the acidity of benzoic acids has been studied by several groups[38–42]. Apparent ionization constants for these compounds are given in

TABLE 5. pK_a data for R_3SiCH_2COOH compounds calculated from apparent dissociation constants

Compound	pK_a	Reference
Me_3SiCH_2COOH	5.22	36
	5.22	37
$Me_3SiOSiMe_2CH_2COOH$	5.22	36
$PhMe_2SiCH_2COOH$	5.26	36
$Me_3SiCH_2CH_2COOH$	4.91	36
Me_3CCH_2COOH	5.00	36
CH_3CH_2COOH	4.87	36

Table 6. Much of this work was aimed at evaluating the extent of (p–d)π bonding to silicon in aromatic systems. The *m*-Me_3Si group should exert a mainly inductive releasing electronic effect and therefore be acid weakening. That was the general concensus of opinion until the work of Zuckerman and coworkers[42], who found *m*-$Me_3SiC_6H_4COOH$ to be stronger than both benzoic acid and *p*-$Me_3SiC_6H_4COOH$. The *p*-Me_3Si group was postulated[39–41] to have an electron-releasing inductive effect and an electron-withdrawing resonance effect which were finely balanced resulting in *p*-Me_3Si being either very slightly activating or very slightly deactivating. Zuckerman and coworkers[42] strongly opposed this view for carboxylic acids, based on measurements of $\Delta G^{\ominus}$, $\Delta H^{\ominus}$ and $\Delta S^{\ominus}$ for the ionization of silyl carboxylic acids. They found that the *m*- and *p*-Me_3Si and Me_3C substituted benzoic acids had large enthalpy values and small $\Delta S^{\ominus}$ values as shown in Table 7. On the basis of these data and other arguments, (p–d)π bonding was rejected and the acidity of the silyl-substituted acids was claimed to follow from the inductive electron-releasing properties of Me_3Si and the hydrophobicity of the Me_3Si group preventing highly ordered solvent molecules efficiently solvating the anions.

Chatt and Williams[43] found that successive phenyl substitution in *p*-$Ph_{3-n}Me_nC_6H_4COOH$ increased the acidity. This was interpreted as increasingly efficient (p–d)π bonding, slightly offset by the change in inductive effect of phenyl relative to methyl. Solvent effects were also considered important[41,43]. Other explanations, such as that of Zuckerman[42], are presumably also applicable.

Benkeser and Krysiak[40] measured the apparent ionization constant of *m*- and *p*-trimethylsilylphenols in 30% v/v ethanol–water solution by potentiometric titration. Phenol was found to have a pK_a in this medium of 10.85, *m*-trimethylsilylphenol 11.01 and *p*-trimethylsilylphenol 10.64. Therefore the *p*-Me_3Si group is slightly acid-enhancing and the *m*-Me_3Si group acid-weakening. The view was taken that the *p*-Me_3Si group was conjugated with the phenoxy group to some extent. Based on NMR studies of the OH proton in DMSO solution Zuckerman and Fenton[44] suggested that there is a larger (p–d)π interaction in trimethylsilyl phenols than in the corresponding benzoic acids.

It is apparent that the effect of trimethylsilyl substitution on the acidity of phenols and carboxylic acids has not been established to universal satisfaction. The concensus opinion, if not the most recent, seems to be that through (p–d)π interactions, *p*-Me_3Si groups enhance acidity, but the question has apparently not been discussed since 1970.

III. BASICITY

A. The Basicity of Functional Groups Directly Bonded to Silicon, R_3SiX ($X=NR_2$, OR, SR, PR_2)

It is clearly established that silyl groups reduce the basicity of heteroatoms to which they are directly bonded. The first indication that silylamines are weak bases appears to

TABLE 6. Apparent ionization constants for silyl-substitued benzoic acids at 25°

$K_a \times 10^6$					
m-$Me_3SiC_6H_4COOH$	p-$Me_3SiC_6H_4COOH$	C_6H_5COOH	Solvent	Method	Reference
0.96	1.41	1.78	$EtOH/H_2O$ 1:1 v/v	potentiometric titration	38
1.00	1.59	1.59	$EtOH/H_2O$ 1:1 v/v	potentiometric titration	39
57.5	53.7	63.1	H_2O	spectrophotometric	40
—	1.11	1.05	60.1% $EtOH/H_2O$ (w/w)	differential potentiometric titration	41
81.4	63.4	62.8	H_2O	spectrophotometric	42

TABLE 7. Thermodynamic data for the ionization of some benzoic acid derivatives at 298.15 K[42]

Acid	$\Delta G^{\ominus}$ (cal mol^{-1})	$\Delta H^{\ominus}$ (cal mol^{-1})	$\Delta S^{\ominus}$ (cal mol^{-1} deg^{-1})
C_6H_5COOH	5735	150	−18.7
p-MeC_6H_4COOH	5952	1100	−16.3
p-$Me_3CC_6H_4COOH$	5988	2900	−10.4
p-$Me_3SiC_6H_4COOH$	5719	3060	−8.9
m-MeC_6H_4COOH	5825	1200	−15.3
m-$Me_3CC_6H_4COOH$	5729	2470	−10.9
m-$Me_3SiC_6H_4COOH$	5579	2320	−10.9

be the observation by Emeleus and Miller[45] that although Me_2NSiH_3 formed an unstable adduct with H_3SiCl, $MeN(SiH_3)_2$ and $N(SiH_3)_3$ formed no adduct. Subsequently trisilylamine was shown to form no adduct with Me_3B[46] and the order of basicity towards Me_3B was shown to be[47] $Me_3N > Me_2NSiH_3 > MeN(SiH_3)_3$. This order of basicity was rationalised in terms of resonance involving (p–d)π bonds and supported by the, then unpublished, report[48] that $N(SiH_3)_3$ has a planar Si_3N skeleton as obtained by gas-phase electron diffraction. The resonance energy in Me_2NSiH_3 was estimated[47] as about 9 kcal mol^{-1}. Ebsworth[49] considered how much extra energy must be derived from π-bonding to stabilise planar molecules such as $(H_3Si)_3N$ and other silylamines with widened angles. The minimum energy needed was estimated as 2–3 kcal mol^{-1}/Si–N bond, but a more realistic estimate seemed to be 10–15 kcal mol^{-1}/Si–N bond. MO calculations[50] using the Pariser–Parr–Pople SCF method gave an order of magnitude for the π-bond energy in trisilylamine as 16 kcal mol^{-1}/Si–N bond but this was suggested to be possibly too high by a factor of up to two. An *ab initio* SCF–LCAO–MO study[51] on H_3SiNH_2 found it to be non-planar and d functions on silicon did not affect the inversion barrier or geometry. The HOMO was found to be a π-type MO but it was not decided whether this was a result of (p–d)π bonding or polarization.

Aylett and coworkers[52–55] examined stability and chemistry of the adducts of a number of aminosilanes with Lewis acids. Disilazane is a very weak base[53] forming no complex with Me_3B. $PhN(SiH_3)_2$ formed no complex with Me_3B or H_3SiI and the geometry about the nitrogen atom was tentatively suggested to be planar[54].

The first study on trimethylsilylamine derivatives was by Jarvie[56] who measured the enthalpy of mixing of silylamines with chloroform. The enthalpy of mixing was taken as an indication of base strength of the amines and, for one series, fell in the order $Me_3SiNHMe > Me_3SiNHEt > Me_3SiNPr^i > Me_3SiNHBu^t$. It was suggested that the electron donor power of the silylamines decreased as the bulk of the groups on silicon or nitrogen increased. This was disputed by Abel and coworkers[57] who observed the infrared shifts and relative intensities of the C–D band in $CDCl_3$ on mixing with aminosilanes. A large number of aminosilanes were observed and $\Delta\nu$ fell in the order $Me_3SiNHBu^t > Me_3SiNHPr^i > Me_3SiNHEt$ in accord with the expected inductive order. It was observed that for highly hindered compounds the ratio of the areas of the unshifted C–D absorption to the CD–N absorption became rather small, suggesting that steric effects affected the equilibrium constant, but not necessarily the basicity. Thus it seemed probable that Jarvie's results were complicated by steric effects and the true order of basicity is based on the inductive effect of the organic group rather than its bulk. Abel[57] also observed that N-phenyl groups reduced the basicity, as did the substitution of H or R by $SiMe_3$. $(Me_3Si)_3N$ was not a strong enough base to cause any shift in the frequency of the C–D band in $CDCl_3$. Other studies on the aminosilanes, relating to (p-d)π bonding and basicity, include NMR measurements of ^{15}N–H coupling constants[58, 59] which concluded that $PhNHSiMe_3$ was pyramidal and that there was no stereochemical evidence for (p-d)π bonding. A normal mode analysis of Me_3SiNMe_2 measured the dihedral angle between the two SiNC planes to be 144° and concluded that (p-d)π bonding was important[60].

Quantitative pK_a measurements are generally not possible for aminosilanes because of their ready hydrolysis, however some data are available for highly hindered compounds as given in Table 8. It is difficult to understand the significance of the pK_a values for these compounds since data are not available for the analogous carbon compounds (which are in several cases unknown). However, it can be seen from Table 8 that all of the hindered aminosilanes are at least 4.7 pK_a units less basic than Bu^tNH_2 and are of similar basicity to aromatic amines. Eaborn[63] suggested that steric inhibition to solvation of the protonated species may lower the basicity, and that a simpler silylamine such as Me_3SiNH_2 could be significantly more basic.

TABLE 8. pK_a values for the conjugate acids of highly hindered aminosilanes in *ca.* 40/60 v/v H_2O/THF at 293 K

Compound	pK_a	Reference
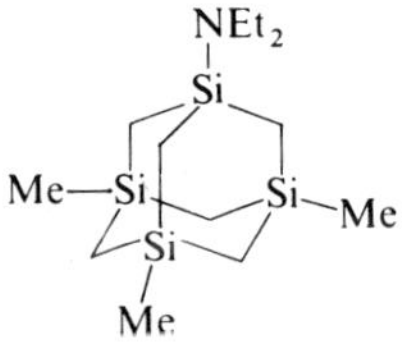	5.7	61, 62
$Bu^t_3SiNH_2$	5.8	61, 62
$Bu^t_3SiNHMe$	4.6	61, 62
$Bu^t_3SiNMe_2$	3.2	61, 62
$(Me_3Si)_3CSiMe_2NH_2$	4.1	63
m-$MeC_6H_4NH_2$	3.75	63
p-$MeC_6H_4NH_2$	4.05	63
Bu^tNH_2	10.5	63

Gas-phase proton affinity data are available[64] for $^tBuMe_2SiNMe_2$ and $^tBuMe_2CNMe_2$. It was shown, by measurement of equilibrium constants for proton-transfer reactions between the amines and reference bases by pulsed ion cylcotron resonance mass spectrometry, that $^tBuMe_2SiNMe_2$ with a proton affinity of 225.6 kcal mol^{-1} is 5 kcal mol^{-1} less basic than $^tBuMe_2CNMe_2$. An explanation was proposed for the reduced basicity of the silylamine involving overlap between a low-lying LUMO of the silicon group of π^* symmetry and the nitrogen lone-pair. It did not prove possible, by this technique, to measure the proton affinities of Me_3Si derivatives as transamination reactions prevented equilibrium from being attained (equations 9 and 10).

$$Me_3SiNMe_2 + BH^+ \rightleftharpoons Me_3Si\overset{+}{N}Me_2H + B \quad (9)$$

$$Me_3Si\overset{+}{N}Me_2H + NMe_3 \rightarrow Me_3Si\overset{+}{N}Me_3 + HNMe_2 \quad (10)$$

Glidewell and Thompson[26] calculated the gas-phase proton affinity of H_3SiNH_2 to be about 9 kcal mol^{-1} smaller than that of H_3CNH_2. Data for some calculated proton affinities of silylamines and phosphines are given in Table 9. Substitution of successive methyl groups into ammonia caused an increase in proton affinity, as predicted by simple

TABLE 9. Calculated gas-phase proton affinities for some silylamines and phosphines[a]

1. *Ammonia derivatives*					
Compound	NH_3	H_3SiNH_2	$(H_3Si)_2NH$	$(H_3Si)_3N$	CH_3NHSiH_3
Gas-phase proton affinity (kcal mol^{-1})	227.2	228.3	225.7	220.4	235.9
		H_3CNH_2 237.2	$(H_3C)_2NH$ 244.4	$(H_3C)_3N$ 248.8	
2. *Phosphine derivatives*					
Compound	PH_3	H_3SiPH_2	$(H_3Si)_2PH$	$(H_3Si)_3P$	CH_3PHSiH_3
Gas-phase proton affinity (kcal mol^{-1})	184.4	199.4	209.6	219.4	209.2
		H_3CPH_2 198.1	$(H_3C)_2PH$ 210.1	$(H_3C)_3P$ 219.33	

[a] Data from Reference 26 *ab initio* calculation, double-ζ basis set.

inductive effects, whereas substitutions of silyl groups initially caused an increase in proton affinity but subsequent substitution caused a decrease in basicity. Unexpectedly, silyl and methyl groups both increased the basicity of phosphines to an almost identical extent. In a later study Hendewerk and coworkers[65] calculated the proton affinities of H_3SiXH (X = PH, NH, S, O) and compared these with the methyl analogues. The results, using a double-ζ basis set with polarization functions on the heavy atoms, corrected for zero point energy, and the conversion of ΔE to ΔH are given in Table 10. The silyl group was, in this case, found to lower the proton affinity, relative to methyl, by 4.5–6.9 kcal mol^{-1} for N, P and S. By use of cyclotron resonance, operating in a drift cell mode, these authors were able to estimate the proton affinity of Me_3SiNMe_2 at 5.6 kcal mol^{-1} less than Me_3CNMe_2, consistent with the calculated value for the H_3Si and CH_3 difference of 6.9 kcal mol^{-1}. There is little experimental data available for silylphosphines, but $(H_3Si)_3P$ is known, and fails to form complexes with BF_3[66] and $Fe(CO)_5$ or $CH_3Mn(CO)_5$[67]. The Si–P–Si bond angle in $(H_3Si)_3P$ has been reported[68] as 96.5° which seems to rule out (p-d)π bonding, but to account for the weakly nucleophilic character of silylphosphines, Ebsworth[67] suggested that the dz_2 orbital on silicon (the z axis is defined along the Si–P bond direction) could overlap with a phosphorus lone-pair of predominantly s character, i.e. (s—d)σ-bonding.

TABLE 10. Absolute proton affinities of some silyl compounds

Compound	Absolute proton affinity[a] (kcal mol^{-1})	Experimental[b]
CH_3NH_2	219.1	214.8
H_3SiNH_2	212.2	—
CH_3PH_2	207.6	205.5
H_3SiPH_2	203.4	—
CH_3OH	181.9	185.6
H_3SiOH	181.9	—
CH_3SH	184.2	189.3
H_3SiSH	179.7	—

[a] From Reference 65.
[b] From Reference 69, but see the note in Reference 28 of Reference 65.

The basicity of silicon oxygen compounds shows strong parallels with that of silicon nitrogen compounds. Sternbach and MacDiarmid[70] found that H_3SiOCH_3 and $(H_3Si)_2O$ were weaker bases than $(CH_3)_2O$ towards diborane, and Emeleus[71] was unable to form complexes of disiloxane with boron halides, hydrogen iodide or methyl iodide. West and coworkers[15, 16, 72], using phenol or pyrrole as the hydrogen donor, measured relative basicities (suggested to be Lewis acidities[16]) by comparing the shifts in OH frequencies in the infrared spectra. The solution basicity, by this method, was found to decline in the order $R_3COCR_3 > R_3COSiR_3 \gg R_3SiOSiR_3$. Silanols were found to be almost as basic as carbinols[15] and it was initially suggested that only one of the oxygen lone-pairs of electrons was involved in π-bonding and that π-bonding had little effect on basicity. Comparison of Ph_3SiOH with Ph_3GeOH suggested that the basicity of Ph_3SiOH was considerably reduced and the conclusions were modified accordingly. Siloxanes were found to be much weaker bases than alkoxysilanes and (p-d)π bonding was postulated to be stronger in the former[72]. Similar conclusions were reached by other groups[73–75], using the same Badger–Bauer methodology[9]. Abel[73], using $CDCl_3$ as the reference acid, found that $(Me_3Si)_2O$ and $(Me_3Si)_2S$ were significantly less basic than their carbon analogues. For the series Me_3SiXR (X=O, S; R=CH_3, C_2H_5, ${}^iC_3H_7$ and ${}^tC_4H_9$) the basicity of the oxygen series increased monotonically as R changed from CH_3 to ${}^tC_4H_9$, whereas the basicity of the sulphur series was relatively unaffected by the nature of R. The polarizability of sulphur, compared with oxygen, was suggested to allow a possible compensating increase in π-bonding, neutralizing the inductive increase in basicity. A study involving IR and NMR measurements on a number of alkoxysilanes[76] reported that the basicity of these compounds was not linearly dependent on the Taft σ^* constants; the α-effect was also discussed. The ^1H NMR shifts of $CHCl_3$ mixed with siloxanes have also been correlated with basicity, as measured by infrared shifts[77]; reduced basicity on substitution by silicon was again found, and arguments were advanced for some type of extra bonding in SiOSi systems compared with COC systems. Rather remarkably for the time (1961) (p-d)π bonding was viewed with scepticism and the author called for detailed spectral measurement and correlation with orbital theory. This was answered, in part, by Bock and coworkers[78] in 1973 with a photoelectron spectral study of methyl and silyl ethers complemented with CNDO/2–SCF–MO calculations. It was concluded that d-orbitals may well be important, but on going from methyl ether to methoxysilane to disiloxane there are at least three other perturbations to the ether system: effective nuclear charge differences between C and Si, widening of the bond angle at oxygen and possibly conformational changes. West[79] obtained the only quantitative thermodynamic data on solution basicity by examining phenol hydrogen bonding to siloxanes and alkoxysilanes, and used Bock's molecular orbital model to rationalise the increased basicity of siloxanes as the SiOSi bond angle is compressed. According to that model, on compression of the SiOSi angle, the highest SiOSi σ-bonding orbital is destabilized, probably accounting for the higher basicity of strained siloxanes, rather than reduced π-bonding. Some of the thermodynamic data obtained by West are given in Table 11. The measured value of ΔH was taken as a measure of basicity and the results were in general consistent with the earlier qualitative studies. The basicity for the isostructural series falls in the order $R_3COCR_3 > R_3COSiR_3 > R_3CSCR_3 > R_3SiOSiR_3 \gg R_3SiSSiR_3$. Structural effects were found to affect basicity markedly, with basicity decreasing with decreasing alkyl chain branching. The order of basicity from ΔH, $Me_3COCMe_3 > Me_3SiOCMe_3 > Me_3COEt > Me_3SiOEt$, Bu_2O was different from that obtained from infrared spectroscopy, namely $Me_3COCMe_3 > Me_3COEt > Bu_2O > Me_3SiOEt > Me_3SiOCMe_3$. This was taken as an example of the breakdown of the Badger–Bauer relationship[81] for alkoxysilanes and ethers taken together in the same series. These authors[79] found (p-d)π bonding from oxygen to silicon to be the most plausible explanation for reduced basicity of silicon oxygen compounds, but hyperconjugation[82] could not easily be ruled out.

Pitt and coworkers[83] using ion cyclotron resonance spectroscopy obtained the following order for gas-phase proton affinities $(Me_3C)_2O \gg Me_3COCHMe_2 \gg (Me_2CH)_2O \gg Me_3COMe \gg (Me_3Si)_2O \gg (Me_2SiH)_2O$, Me_3SiCH_2OMe, $> Me_3SiOMe > Me_3SiOH$, Et_2O, $PhOMe > Et_2O$. The silyl ethers were found to be substantially more basic than Me_2O, EtOH, MeOH and H_2O, and the substituent effect was $Me_3C > Me_2CH > Me_3Si > Me_2SiH > Et > Me > H >$. Comparisons were made with the solution basicities as measured by IR spectroscopy. Alkyl ethers and alcohols have essentially the same order of basicity in both the gas phase and in solution. The silyl ethers and anisole change significantly with the strength of the reference acid. In the gas phase $(Me_3Si)_2O$ is a stronger base than ether, which is not the case in solution. It was postulated that polarizability of the base becomes important with the strong reference acid in the gas phase. In solution the lone-pair delocalization of siloxanes and silyl ethers, by (p-d)π bonding or hyperconjugation, reduces the basicity. In the gas phase the stronger acid was said to polarize the total electron distribution of the base. It was also noted that despite

TABLE 11. Thermodynamic constants for hydrogen bonding of phenol to silicon and carbon compounds

Compound	$-\Delta H$ (kcal mol^{-1})	$-\Delta G$ (kcal mol^{-1}) 301 K	$-\Delta S$ (cal deg^{-1} mol^{-1})	Reference
Me_3COCMe_3	7.31	0.745	22.0	80
$Me_3SiOCMe_3$	6.89	0.148	22.6	79
$Me_3COCH_2CH_3$	6.52	1.400	17.2	80
$Me_3SiOCH_2CH_3$	5.76	0.876	16.4	79
$Me_3SiOSiMe_3$	3.82	−0.849	15.7	79
$(Me_2SiO)_3$	4.48	−0.001	15.0	79
$(Me_2SiO)_4$	3.65	−0.399	13.6	79
$(Me_2SiO)_5$	3.52	−0.341	13.0	79
Me_3CSCMe_3	4.87	0.423	14.9	80
$Me_3SiSCMe_3$	1.93	0.821	3.7	79
$Me_3SiSSiMe_3$	0.82	1.472	−2.2	79

additional polarization, silyl ethers were still found to be less basic than their isostructural ethers suggesting that some type of π bonding is still important. The gas-phase proton affinity of trimethylsilanol has been measured by high-pressure mass spectroscopy from the enthalpy change for the reaction[84],

$$Me_3Si^+ + H_2O \rightleftharpoons Me_3SiOH_2^+ \tag{11}$$

The proton affinity of Me_3SiOH was 183.7 kcal mol^{-1} by this method, and Me_3COH had a value of 193.7 kcal mol^{-1}. This lowering of proton affinity of the silicon compound by 10 kcal mol^{-1} is consistent with the general order found by Pitt[83], although Hendewerk[65] found that, by contrast to amines, phosphines and thiols, the calculated proton affinities for H_3SiOH and H_3COH were identical at 181.9 kcal mol^{-1}. Glidewell and Thompson[26], on the other hand, calculated the proton affinities of Me_2O and $(H_3Si)_2O$ as 213.1 kcal mol^{-1} and 201.3 kcal mol^{-1} respectively, a reduction in basicity of 11.8 kcal mol^{-1} for the silicon compound compared with the carbon compound.

In summary, solution measurements by infrared spectroscopy, NMR methods, gas-phase proton measurements and *ab initio* calculations are in remarkable agreement that the basicity of a heteroatom adjacent to silicon is reduced. The most widespread explanation is that electron density is transferred from the heteroatom to silicon in a manner that can most readily be modelled and understood by (p-d)π bonding.

B. The Effect of Silicon Remote from the Basic Centre

1. The basicity of $R_3Si(CH_2)_nXR$ (X=NR, O)

In contrast to silicon directly bonded to the basic centre, silylamines separated by one or more CH_2 groups from the nitrogen atom exhibit enhanced basicity. In 1951 two groups independently published basic dissociation constants for silyl-substituted amines[85,86]; the results are given in Table 12. The Me_3SiCH_2 group is strongly electron releasing and both investigators found that $Me_3SiCH_2NH_2$ was more basic than propylamine, and $(Me_3SiCH_2)_2NH$ was found to be particularly basic. As expected, alkyl groups on nitrogen increased the basicity and the aniline derivative was only feebly basic[85]. The basicity on increasing the number of methylene groups between N and Si falls in the order

$$Me_3SiCH_2CH_2NH_2 > Me_3SiCH_2NH_2 > Me_3SiCH_2CH_2CH_2NH_2 > Me_3CCH_2NH_2$$

Chvalovsky[87], using the Badger–Bauer method, found a different order and extended the number of methylene groups to four. The order obtained by the infrared method, with $\Delta\nu$ for the CD bond in $CDCl_3$ given in parentheses, was $Me_3SiCH_2CH_2NH_2$ (46) > $Me_3Si(CH_2)_4NH_2$ (43) ~ $Me_3Si(CH_2)_3NH_2$ (42) > $Me_3SiCH_2NH_2$ (36). Butylamine gave a $\Delta\nu$ of 43 cm^{-1}. It was suggested that a hyperconjugative effect might prevail over the inductive one of Me_3SiCH_2 causing a relative lowering of basicity, although no indication was given of how this effect might operate. The Me_3SiCH_2 group is generally considered to be electron releasing by hyperconjugation in the sense[88].

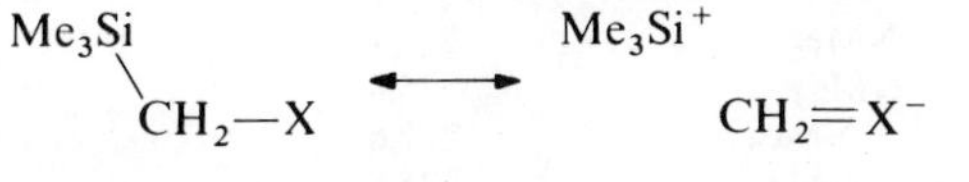

(12)

There was general agreement that no intramolecular coordination occurred between nitrogen and silicon in these $Me_3Si(CH_2)_nNH_2$ compounds, as that would be expected to reduce the basicity and no evidence was found for that effect. It is probable that solvation

TABLE 12. Dissociation constants for silylamines in aqueous solution

Compound	$K_b \times 10^4$	Method	Reference
$Me_3SiCH_2NH_2$	9.6	*a*	85
	8.5	*b*	85
	9.1	*c*	86
$C_6H_5SiMe_2CH_2NH_2$	2.8	*a*	85
	2.3	*b*	85
$Me_3SiCH_2NHC_6H_{11}$-c	17	*a*	85
	9.1	*b*	85
$Me_3SiCH_2NHPr^i$	12	*a*	85
$(Me_3SiCH_2)_2NH$	14	*a*	85
	4–10	*b*	85
	25	*c*	86
$Me_3SiCH_2CH_2NH_2$	9.7	*c*	86
$Me_3SiCH_2CH_2CH_2NH_2$	5.6	*c*	86
$Me_3CCH_2NH_2$	1.6	*c*	86
$C_3H_7NH_2$	4.9	*c*	86
	3.9	*a*	85

[a] Titration of amine salt with NaOH.
[b] 'Extrapolation method' from p*K* vs ionic strength, extrapolated to infinite dilution.
[c] Potentiometric titration.

is important in determining the solution basicities of silicon-substituted amines as gas-phase proton affinities show only a very small variation with *n* for $Me_3Si(CH_2)_nNMe_2$[64] as shown in Table 13. Both Me_3SiCH_2 and $Me_3SiCH_2CH_2$ increase the basicity of the amine by 1.3 kcal mol^{-1} compared with the carbon analogue.

There have been a number of infrared, Badger–Bauer studies on the basicities of ethers, $R_3Si(CH_2)_nOR$[90–93], and much of the evidence, together with unpublished data, has been summarized[94]. The Me_3SiCH_2 group, as would be expected, appears to be electron releasing and enhances the basicity of ethers. Both Ph_3SiCH_2 and H_3SiCH_2 are said to be electron withdrawing, reducing basicity[94], through the 'α-effect' in which electron density is transferred from a heteroatom to silicon in a R_3SiCH_2X system[95–97]. Bellama and Gerchman[90] measured the shift in the methanol OH infrared frequency on addition

TABLE 13. Gas-phase proton affinities for some silyl substituted amines, measured by pulsed ion cyclotron resonance[64]

Compound	Proton affinity (kcal mol^{-1})[a]
$Me_3SiCH_2NMe_2$	227.1
$Me_3CCH_2NMe_2$	225.8
$Me_3Si(CH_2)_2NMe_2$	227.6
$Me_3C(CH_2)_2NMe_2$	225.8
$Me_3Si(CH_2)_3NMe_2$	227.6

[a] Based on a proton affinity of NH_3 of 201.0 ± 2 kcal mol^{-1} (taken from Reference 89).

to $H_3M(CH_2)_nOCH_3$ ($n=1$–4). The results suggest that H_3SiCH_2 is significantly electron withdrawing, $H_3SiCH_2CH_2$ moderately so, and as n increases beyond two the basicity of the silicon compounds approach those of the carbon analogues. It would be interesting to see if these trends are confirmed by quantitative measurements in solution or in the gas phase.

2. *The effect of Me_3Si on the basicity of anilines*

There do not appear to have been extensive studies on the effect of *m*- and *p*-Me_3Si groups on the basicity of aromatic phenols, ethers, thiols or thioethers, although the substituent effect of R_3Si groups on aromatic systems has been well studied[88].

Hammett[98] reported that *m*-Me_3Si increased the basicity of aniline whereas *p*-Me_3Si slightly decreased the basicity. These results were similar to those of Benkeser[99] who reported pK_a values in 50% ethanol/water at 25°, of 4.35 for dimethylaniline, 4.41 for the *m*-Me_3Si derivative and 3.98 for the *p*-Me_3Si derivative. These data are consistent with cross-conjugation between Me_3Si and NMe_2 in the *para* position. This conjugation is not possible in the *meta* position and the dominant effect is the inductive electron supply of Me_3Si.

IV. COMPLEXES OF SILICON COMPOUNDS

A. Introduction

Numerous penta[100–109,118] and hexacoordinate[110–114] complexes of silicon are known, and some examples of current interest are shown in structures **1**, **2**, **3** and **4**. These

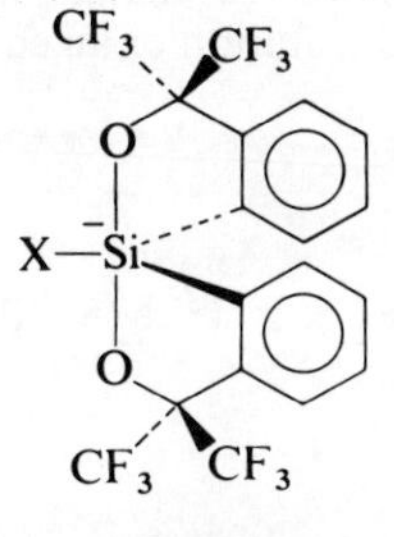

(1) (Ref. 109)

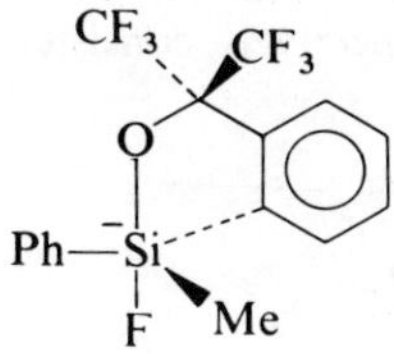

(2) (Ref. 108)

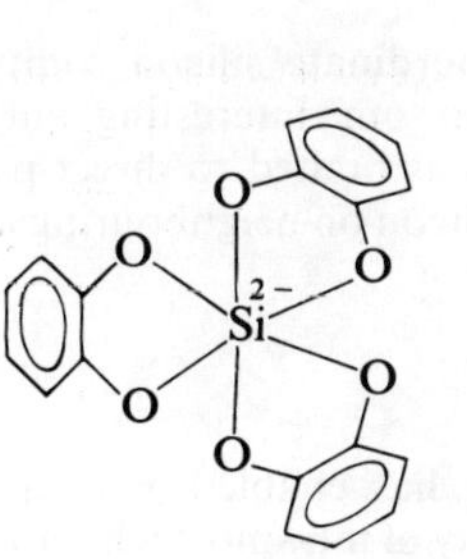

(3) (Ref. 114)

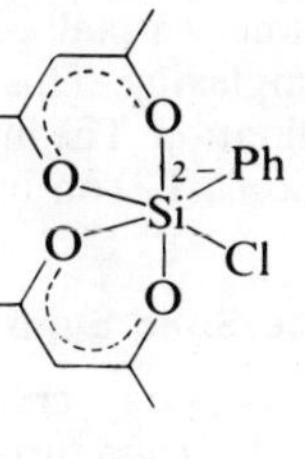

(4) (Ref. 112)

complexes are generally known as hypervalent or extracoordinate silicon compounds and are reviewed elsewhere in this volume[115]. Hypervalent silicon compounds usually have three or more electronegative ligands around Si. There are also a considerable number of silicon species known in which the presence of an intramolecular coordinating ligand perturbs the molecule so that the geometry around the silicon atom tends towards trigonal bipyramidal or possibly square pyramidal. Examples of this type of intramolecular complexing include compounds **5**[103] and **6**[107].

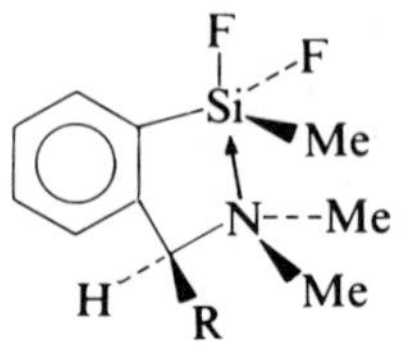

(5)

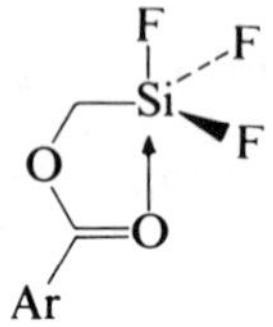

(6)

Corriu and coworkers[103] have used the induced diastereotopicity in **5** to study the dynamic behaviour by NMR spectroscopy. Two different dynamic processes were observed, one with $\Delta G^{\neq}$, 9.4 kcal mol^{-1} attributable to pseudorotation at silicon, and one with $\Delta G^{\neq}$, 11.8 kcal mol^{-1} attributable to Si–N bond breaking. Klebe and Hensen[116] studied the dynamics of compound **7** by ^{19}F NMR and stated that the energy barrier for fluorine equilibration of 7.5 kcal mol^{-1} was best accounted for by reversible dissociation of the N–Si bond. By contrast Damrauer and Danahey[117] recently extended and repeated ^{19}F NMR studies initially carried out by Klanberg and Muetterties[118], and found that **8** and **9** undergo intramolecular fluorine exchange with $\Delta G^{\neq} = 11.7$ and 9.9 kcal mol^{-1} respectively, consistent with pseudorotation or an equivalent mechanism.

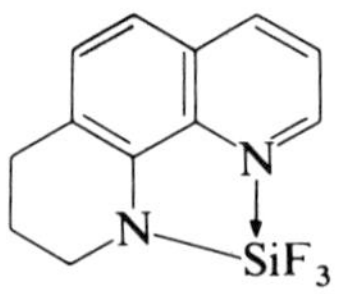

(7)

$Ph_2SiF_3^-$

(8)

$PhMeSiF_3^-$

(9)

The question of pseudorotation is covered in more detail in the chapter on mechanisms of silicon reactions[119].

As the chemistry of hypervalent or extracoordinate silicon compounds is covered elsewhere in this volume, we shall concentrate on some interesting, but less well surveyed aspects of silicon complexing. This brief survey is limited to direct participation of the silicon atom in coordination. The influence of silicon on neighbouring atom coordination is mentioned in Section III.A on basicity.

B. Complexing in the Solid State

In organic chemistry X-ray crystallography has enabled a fuller understanding of reaction mechanisms at carbon through the study of intramolecular atomic distances and deviations from tetrahedral geometry[120]. Although not developed to the same extent there are a number of X-ray crystallographic studies on organosilicon compounds in which inter- and intra-molecular interactions have been revealed. These studies can therefore provide models for inter- or intra-molecular complexing, and it is in that light that we shall examine them.

An early X-ray study on a molecule that can exhibit intramolecular coordination was reported by Voronkov and coworkers[121] for 1-chlorosilatrane **10**. The SiCl bond length

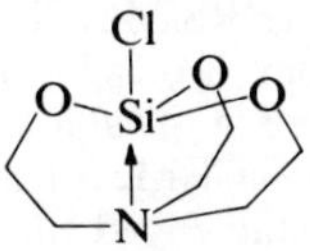

(10)

was 215 pm compared with 209 pm for Me_3SiCl. The SiO_3 fragment is almost planar and a linear correlation of SiN bond lengths in silatranes with inductive σ constants of substituents at silicon was reported. In the silatranes the geometry of the molecule is such that the nitrogen and chlorine, or other ligand on silicon, are constrained to apical positions in the developing trigonal bipyramid. This can be conceived as a model for the intermediate in nucleophilic substitutions at silicon taking place with inversion of configuration. Silicon is also known to undergo substitution with retention of configuration, necessitating an intermediate with one electronegative ligand equatorial and the other axial in the trigonal bipyramid. For intramolecular complexes where there is a choice of geometry for two electronegative ligands, the available evidence from X-ray crystallography indicates that they are more likely to be found in apical positions. The heterocyclic compound **11** has been studied by X-ray crystallography[122]. The geometry

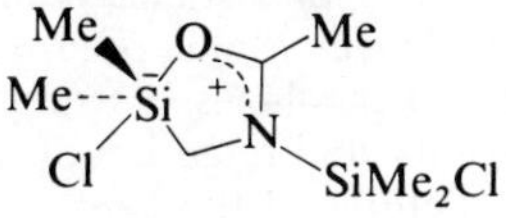

(11)

of the molecule indicated five-coordination at silicon with distorted trigonal bipyramidal form. The oxygen and chlorine atoms occupy apical sites and the O–Si distance is 191.8 pm, and the SiCl bond length for the five-coordinate silicon fragments is 234.8 pm, compared with 205.0 pm for the four-coordinate Si–Cl bond. The sum of the van der Waals radii for Si and Cl is 380 pm, so although the SiCl bond is lengthened considerably on extra coordination it can still be classified as a covalent bond[123].

In recent X-ray studies compounds of type **12** were examined[124, 125]. The silicon atom in each of the complexes of type **12**, $RSiMe_2Cl$, $RSiMeCl_2$ and $RSiCl_3$, is of trigonal

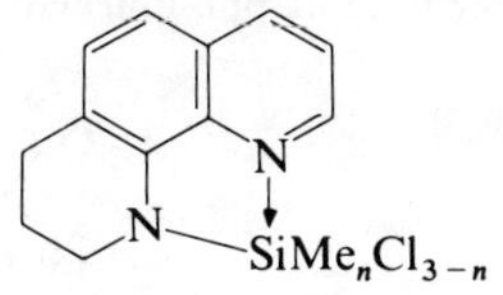

(12)

bipyramidal geometry. There is no intramolecular coordination for the $RSiMe_3$ derivative **12**. The coordinating nitrogen of the pyridine-type ring occupies one apical position with a chlorine atom the other. The saturated-ring nitrogen occupies an equatorial position. The apical SiCl interatomic distance falls in the order $RSiMe_2Cl > RSiMeCl_2 > RSiCl_3$ with the respective values being 226.9, 220.7 and 215.0 pm. The pyridine nitrogen–silicon distance follows the same trend but not in such a dramatic fashion. Si–N_{py} distances are $RSiMe_2Cl$, 202.9; $RSiMeCl_2$, 202.7; and $RSiCl_3$, 198.4 pm. These bond lengths are some 14–18% longer than the equatorial SiN bonds. Surprisingly the sensitivity to moisture of the chlorosilanes of type **12** was reported[124] to decrease in the order $RSiClMe_2 > RSiCl_2Me > RSiCl_3$ which is the inverse of the expected order. The assumption was challenged that the X-ray structures of these compounds can be associated with species along the S_N2 pathway on the grounds that, although the apical SiCl bond length increases in the order $RSiClMe_2 > RSiCl_2Me > RSiCl_3$, the 'entering' nitrogen–silicon distance also increases in this order. This is perhaps an oversimplified view as other factors, such as the influence of the equatorial substituents and the 'lateness' of the proposed transition state, need to be considered. When there is no electronegative 'leaving' group as in compounds **13** and **14** there is still significant N-silicon bonding, but as expected for the silicon bearing no

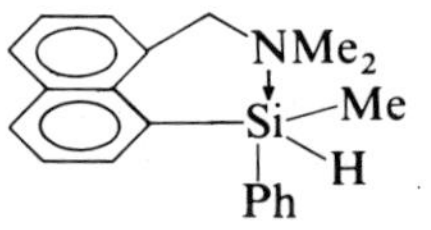

(13)

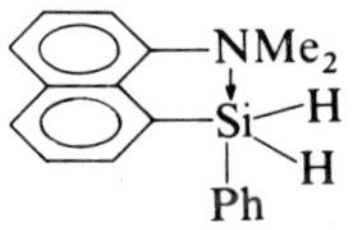

(14)

strongly electronegative ligands, the NSi distance is long, at 226 pm for **13**[126]. Interestingly, in both **13** and **14** the hydrogen atoms occupy equatorial sites and the phenyl group is apical. Substitution of SiH in silanes is strongly favoured over SiMe or SiPh substitution by nucleophilic reagents, and retention of configuration is the predominant stereochemical pathway. The structures of the compounds **13** and **14** is evidence that apical entry, probably followed by pseudorotation, is the most efficient pathway for retentive mechanisms[126]. Recent calculation by Dieters and Holmes support this view[128,129], and the proposal[127] that equatorial attack of nucleophile was favoured has been withdrawn by Corriu following his studies on compounds **13** and **14**[126].

There are many examples of intramolecular rearrangements involving migration of silicon groups between electronegative atoms, and in examples that have been studied retention of configuration at silicon is predominant[130]. X-Ray crystallographic studies can help to elucidate the factors involved in these migrations through studies of intramolecular coordination. Colvin, Seebach, Dunitz and coworkers[131] studied the silyl nitronates **15** and **16**, which in solution undergo rapid 1, 3 migration of the silicon group. In these molecules both intra- and inter-molecular interactions with the silicon can be conceived. The crystal structure revealed that for **15** the SiO bonded distance is 171.6 pm and the non-bonded SiO distance is 283.2 pm, markedly shorter than the sum of the Si

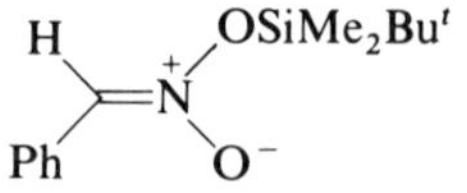

(15)

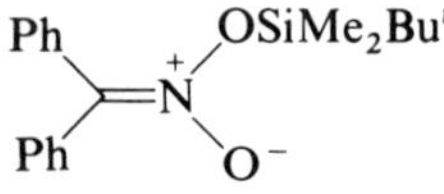

(16)

and O van der Waals radii (362 pm) which was taken to indicate a specific attraction between the non-bonded Si and O atoms. It was suggested that the structure correspon-ded to a slight displacement of the molecule along the reaction path towards the transition state for O–Si--O switching. The results for **16** were very similar and the same conclusion was drawn. Neither **15** nor **16** had any unusually short intermolecular contacts. Barrow, Ebsworth and coworkers have determined the X-ray crystal structure of a series of silicon compounds and analysed the data for abnormally short intra- and inter-molecular contacts to silicon. By contrast to the Bu^tSiMe_2 nitronate derivatives[131], silyl acetate[132], $H_3SiOOCCH_3$ and silyl monothioacetate[133], $H_3SiOCSCH_3$ do exhibit abnormally short SiO or SiS intermolecular contacts. For $H_3SiOOCCH_3$ the heavy atom skeleton is almost planar with SiO and C=O bonds *cis* to one another[132]. The intramolecular non-bonded Si---O distance is 283.2 pm which is 80 pm less than the sum of the van der Waals radii, but methyl acetate was examined in the same study and the intramolecular non-bonded C--O distance is 261.1 pm which is 60 pm less than the sum of the van der Waals radii of C and O. However, there are no short intermolecular contacts in methyl acetate whereas there is a particularly short non-bonded intermolecu-lar Si--O distance of 272.1 pm. The O–Si--O angle is 173.1° corresponding to two apical oxygen atoms in a developing trigonal bipyramid, and this is illustrated in Figure 1. Similar results were obtained for the monothioester[133] and the short intermolecular Si–S interaction is illustrated in Figure 2. We may conclude from the information so far that silicon can undergo specific, directional, weak interactions with neighbouring electronegative elements. When two electronegative groups are present a diapical arra-ngement is favoured[124,125] but an apical/equatorial arrangement can occur if the geometry of the molecule so demands[131]. The preference for a diapical arrangement of electronegative groups is perhaps best illustrated by the silyl acetate[132] and silyl monothioacetate[133] where both inter- and intra-molecular coordination is observed, but the intermolecular Si–O or Si--S distance is shorter than the corresponding intra-molecular separation.

Barrow, Ebsworth and Harding[134] used the correlation technique developed by Burgi[135] to analyse a series of crystallographic studies, carried out by various groups. Briefly, this method is concerned with an analysis of structural parameters in the solid-state complexes as the geometry changes from tetrahedral to bipyramidal along an S_N2-like reaction profile. Figure 3 illustrates the parameters used; the N--Si distance

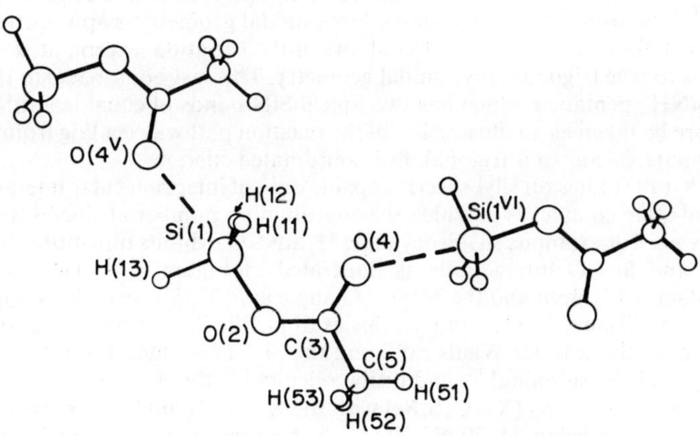

FIGURE 1. View of silyl acetate molecules in the crystal to highlight the O(4^V) · · · Si intermolecu-lar contacts. Reproduced by permission of the Royal Society of Chemistry.

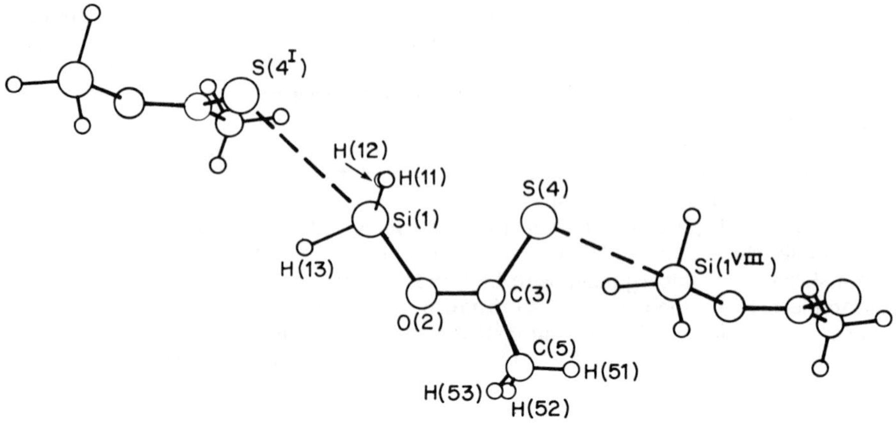

FIGURE 2. Crystalline silyl thioacetate: view of reference molecule and two neighbouring molecules showing labelling scheme and intermolecular Si · · · S contacts. Reproduced by permission of the Royal Society of Chemistry.

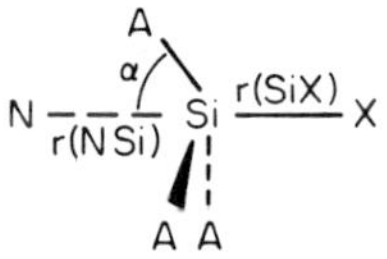

FIGURE 3. Parameters used in the correlation analysis of nucleophilic attack at silicon in an S_N2-like process

r(N--Si) becomes shorter as the intermediate is approached; the SiX bond distance *r*(SiX) becomes extended, and the N--Si–A angle, α, tends to 90°. Figure 4 shows the scatter plot for a series of silatranes, and other compounds with a long Si--N interaction[134]. It can be seen that as the trigonal bipyramidal geometry is approached, *r*(Si--N) decreases significantly whereas *r*(SiX) shows little variation except at $\alpha = 90°$ which corresponds to true trigonal bipyramidal geometry. This last point refers to the structure of the H_3SiNH_2 pentamer which has two apical SiN bonds of equal length[136]. Figure 4 can therefore be taken as an illustration of the reaction pathway leading from tetrahedral four-coordinate silicon to a trigonal, five-coordinated species.

Further X-ray studies on silyl species capable only of intermolecular interactions show similar solid-state complexes in which the coordination number at silicon is extended to five or even six. For example, silyl isocyanate H_3SiNCO exhibits directional intermolecular Si--N and Si--O interactions as illustrated in Figure 5[134]. The intermolecular Si--O contact is 331 ppm and the N–Si--O angle is 76.7°. The N--Si–N angle is 173.8° and the N--Si distance is 331.1 pm. In this example the Si--O contact is some 9% less than the sum of the van der Waals radii and the Si--N contact is about 11% less. For disiloxane[137], disilyl sulphide[138], and disilyl selenide[138] the X-ray structures all showed directional Si--X contacts (X = O, S, Se) with the X–Si--X angles close to 180°, and the long Si--X contact being 35–50 pm less than the sum of the van der Waals radii. By contrast the planar trisilylamine shows no tendency to form long, directional Si--N bonds[138], consistent with its inability to form complexes with Me_3B^{46}.

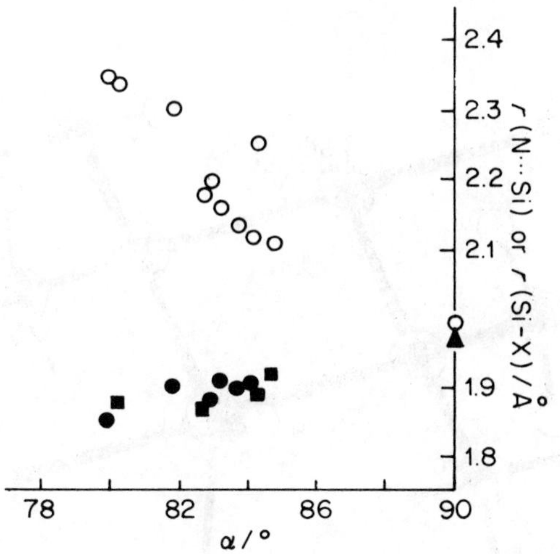

FIGURE 4. Scatter plot of $r(N \cdots Si)$, open circles and $r(Si–X)$ [X = C(sp^2) (●), C(sp^3) (■) or N (▲)] against α. Reproduced by permission of the Royal Society of Chemistry.

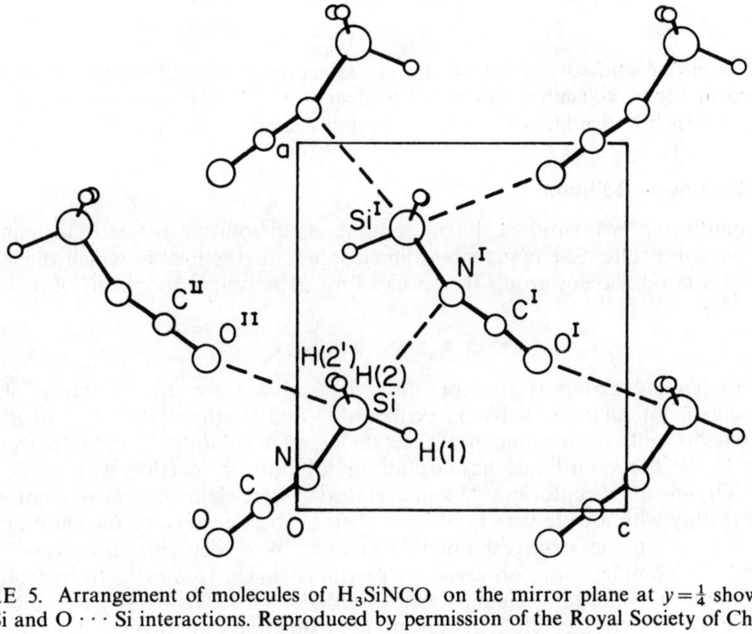

FIGURE 5. Arrangement of molecules of H_3SiNCO on the mirror plane at $y = \frac{1}{4}$ showing the $N \cdots Si$ and $O \cdots Si$ interactions. Reproduced by permission of the Royal Society of Chemistry.

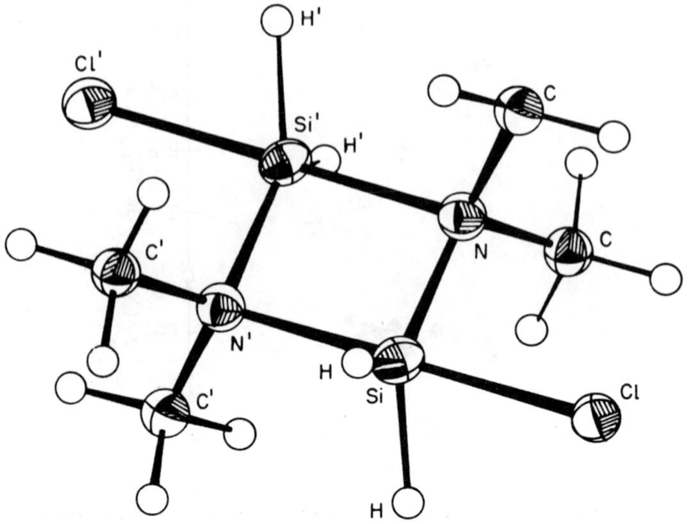

FIGURE 6. Structure of chlorosilyl-*N*,*N*-dimethylamine in the crystal at 116 K. The Si_2N_2Cl group is almost planar: the dimer contains a center of inversion. Bond lengths (pm) and angles (deg): Si–Cl 223.1(6), Si–N 181.4(13), Si–H 149(1), N–C 149.8(18), Si–N 205.4(3), Cl–Si–N 96.1(5), Si–N–C 112.8(9), C–N–C 108.8(14), N–Si–N: 83.0(5), Si–N–Si 97.0(6). Reproduced by permission of VCH Verlagsgesellschaft from *Angewandte Chemie*

Chlorosilyl-*N*,*N*-dimethylamine in the crystal consists of well separated dimers, as illustrated in Figure 6, each dimer being regarded as a 'frozen' intermediate in an S_N2 reaction in which N displaces Cl to form the ionic form of the dimer[139].

C. Complexing in Solution

The complexing behaviour of silicon compounds in solution is less well understood than in the solid state. For many silane–nucleophile interactions in which the silane is bonded to a good leaving group the species formed is simply the result of a substitution[140–142],

$$Nu + R_3SiX = R_3Si\overset{+}{N}uX^- \qquad (13)$$

Where no leaving group is present, infrared studies show the existence of weak solvent–silicon interactions in trialkoxysilanes[143] and triethylsilane[144]. Trimethylsilyl acetate was shown to have a unidentate acetate ligand in solution[145] and tetraacetatosilane formed the hexacoordinate hexaacetate in solution on reaction with acetic anhydride[146]. Organosilicon chlorides[147] and acetates[148] were claimed to form complexes in solution readily with a wide variety of Lewis bases. These results for the chlorides have been challenged in the strongest possible manner by Yoder and coworkers[149] who attributed the 'complexation' observed by thermochemical studies[147] to hydrolysis by trace amounts of water, and claimed that essentially no Lewis acidity is exhibited by Me_3SiCl towards pyridine, triphenyl phosphine oxide and dimethyl sulphoxide.

V. REFERENCES

1. R. Stewart, *The Proton: Applications to Organic Chemistry*, Academic Press, Orlando, USA, 1985.
2. R. W. Taft, in *Progress in Physical Organic Chemistry*, Vol. 14 (Ed. R. W. Taft), Wiley, New York, 1983, p. 247.
3. V. Gutmann, *Coordination Chemistry in Non-Aqueous Solutions*, Springer-Verlag, Vienna, 1969.
4. V. Gutmann, *Coord. Chem. Rev.*, **18**, 225 (1976).
5. V. Gutmann and E. Wychera, *Inorg. Nucl. Chem. Lett.*, **2**, 257 (1966).
6. R. W. Taft, T. Gramstad and M. J. Kamlet, *J. Org. Chem.*, **47**, 4557 (1982).
7. M. J. Kamlet, J.-L. M. Abboud, M. Abraham and R. W. Taft, *J. Org. Chem.*, **48**, 2877 (1983).
8. W. B. Jensen, *Chem. Rev.*, **78**, 1 (1978).
9. R. M. Badger and S. H. Bauer, *J. Chem. Phys.*, **5**, 839 (1937).
10. R. S. Drago and T. D. Epley, *J. Am. Chem. Soc.*, **91**, 2883 (1969) and references cited therein.
11. E. M. Arnett, L. Joris, E. Mitchell, T. S. S. R. Murty, T. M. Gorrie and P. v. R. Schleyer, *J. Am. Chem. Soc.*, **92**, 2365 (1970).
12. L. H. Sommer, E. W. Pietrusza and F. C. Whitmore, *J. Am. Chem. Soc.*, **68**, 2282 (1946).
13. R. West and R. H. Baney, *J. Inorg. Nucl. Chem.*, **7**, 297 (1958).
14. R. H. Baney, K. J. Lake, R. West and L. S. Whatley, *Chem. Ind.*, 1129 (1959).
15. R. West and R. H. Baney, *J. Am. Chem. Soc.*, **81**, 6145 (1959).
16. R. West, R. H. Baney and D. L. Powell, *J. Am. Chem. Soc.*, **82**, 6269 (1960).
17. L. P. Kuhn, *J. Am. Chem. Soc.*, **74**, 2492 (1952).
18. N. A. Matwiyoff and R. S. Drago, *J. Organomet. Chem.*, **3**, 393 (1965).
19. A. L. Allred and E. G. Rochow, *J. Inorg. Nucl. Chem.*, **5**, 269 (1958) and **20**, 167 (1961).
20. O. W. Steward and D. R. Fussaro, *J. Organomet. Chem.*, **120**, C28 (1977).
21. E. A. Williams, J. D. Cargioli and R. W. Larochelle, *J. Organomet. Chem.*, **108**, 153 (1976).
22. V. Gutmann, *Chem. Ind.*, 102 (1971).
23. R. M. Salinger and R. West, *J. Organomet. Chem.*, **11**, 631 (1968).
24. K. D. Crosbie, C. Glidewell and G. M. Sheldrick, *J. Chem. Soc. (A)*, 1861 (1969).
25. C. Glidewell and G. M. Sheldrick, *J. Chem. Soc. (A)*, 350 (1969).
26. C. Glidewell and C. Thompson, *J. Comput. Chem.*, **3**, 495 (1982).
27. A. C. Hopkinson and M. H. Lien, *J. Org. Chem.*, **46**, 998 (1981).
28. O. W. Steward, H. W. Irwin, R. A. Gartska and J. O. Frohlinger, *J. Chem. Soc. (A)*, 3119 (1968).
29. O. W. Steward, J. E. Dziedzic, J. S. Johnson and J. O. Frohlinger, *J. Org. Chem.*, **36**, 3480 (1971).
30. O. W. Steward, J. E. Dziedzic and J. S. Johnson, *J. Org. Chem.*, **36**, 3475 (1971).
31. B. G. Ramsey, *Spectroscopy in Inorganic Chemistry*, Vol. II (Eds. C. N. R. Rao and J. R. Ferraro), Academic Press, New York, 1971, p. 249.
32. P. M. Kunznesof, F. B. T. Pessini and R. E. Bruns, *J. Am. Chem. Soc.*, **94**, 9087 (1972).
33. O. W. Steward, G. L. Heider and J. S. Johnson, *J. Organomet. Chem.*, **168**, 33 (1979).
34. A. G. Brook and K. H. Pannell, *J. Organomet. Chem.*, **8**, 179 (1967).
35. G. J. Peddle, R. J. Woznow and S. G. McGeachin, *J. Organomet. Chem.*, **17**, 331 (1969).
36. L. H. Sommer, J. E. Gold, C. M. Goldberg and N. S. Marans, *J. Am. Chem. Soc.*, **71**, 1509 (1949).
37. R. W. Bott, C. Eaborn, K. C. Pande and T. W. Swaddle, *J. Chem. Soc.*, 1217 (1962).
38. J. D. Roberts, E. A. McElhill and R. Armstrong, *J. Am. Chem. Soc.*, **71**, 2923 (1949).
39. J. D. Roberts and C. M. Regan, *J. Am. Chem. Soc.*, **75**, 4102 (1953).
40. R. A. Benkeser and H. Krysiak, *J. Am. Chem. Soc.*, **75**, 2421 (1953).
41. J. Chatt and A. A. Williams, *J. Chem. Soc.*, 4403 (1954).
42. J. M. Wilson, A. G. Briggs, J. E. Sawbridge, P. Tickle and J. J. Zuckerman, *J. Chem. Soc. (A)*, 1024 (1970).
43. J. Chatt and A. A. Williams, *J. Chem. Soc.*, 688 (1956).
44. J. J. Zuckerman and D. E. Fenton, *Inorg. Chem.*, **7**, 1323 (1968).
45. H. J. Emeleus and N. Miller, *J. Chem. Soc.*, 819 (1939).
46. A. B. Burg and E. S. Kulijan, *J. Am. Chem. Soc.*, **72**, 3103 (1950).
47. S. Sujishi and S. Witz, *J. Am. Chem. Soc.*, **76**, 4631 (1954).
48. K. Hedberg, *J. Am. Chem. Soc.*, **77**, 6491 (1955).
49. E. A. V. Ebsworth, *The Physical Basis of the Chemistry of Group IV Elements* in *Organometallic Compounds of the Group IV Elements*, Vol. 1 (Ed. A. G. MacDiarmid), Dekker, New York, 1968.

50. P. G. Perkins, *J. Chem. Soc., Chem. Commun.*, 268 (1967).
51. J. M. Lehn and B. Munsch, *J. Chem. Soc., Chem. Commun.*, 994 (1970).
52. B. J. Aylett and L. K. Peterson, *J. Chem. Soc.*, 4043 (1965).
53. B. J. Aylett and M. J. Hakim, *J. Chem. Soc. (A)*, 639 (1969).
54. B. J. Aylett and M. J. Hakim, *J. Chem. Soc. (A)*, 800 (1969).
55. B. J. Aylett and M. J. Hakim, *J. Chem. Soc. (A)*, 1788 (1969).
56. A. W. Jarvie and D. Lewis, *J. Chem. Soc.*, 1073 (1963).
57. E. W. Abel, D. A. Armitage and G. R. Willey, *Trans. Faraday Soc.*, **60**, 1257 (1964).
58. E. W. Randall, J. J. Ellner and J. J. Zuckerman, *J. Am. Chem. Soc.*, **88**, 622 (1966).
59. E. W. Randall and J. J. Zuckerman, *J. Am. Chem. Soc.*, **90**, 3167 (1968).
60. A. Marchand and M.-T. Forel, *J. Organomet. Chem.*, **156**, 341 (1978).
61. G. D. Homer and L. H. Sommer, *J. Organomet. Chem.*, **67**, C10 (1974).
62. P. M. Nowakowski and L. H. Sommer, *J. Organomet. Chem.*, **178**, 95 (1979).
63. C. Eaborn and F. M. S Mahmoud, *J. Organomet. Chem.*, **220**, 139 (1981).
64. K. J. Shea, R. Gobeille, J. Bramblett and E. Thompson, *J. Am. Chem. Soc.*, **100**, 1611 (1978).
65. M. Hendewerk, R. Frey and D. A. Dixon, *J. Phys. Chem.*, **87**, 2026 (1983).
66. A. G. MacDiarmid, *Angew. Chem., Int. Engl., Ed.* **5**, 418 (1966).
67. E. A. V. Ebsworth, C. Glidewell and G. M. Sheldrick, *J. Chem. Soc. (A)*, 352 (1969).
68. B. Beagley, A. G. Robiette and G. M. Sheldrick, *J. Chem. Soc. (A)*, 3002 (1968).
69. S. G. Lias, D. M. Shold and P. Ansloos, *J. Am. Chem. Soc.*, **102**, 2540 (1980).
70. B. Sternbach and A. G. MacDiarmid, *J. Am. Chem. Soc.*, **83**, 3384 (1961).
71. H. J. Emeleus and M. Onyszchuck, *J. Chem. Soc.*, 604 (1958).
72. R. West, L. S. Whatley and K. J. Lake, *J. Am. Chem. Soc.*, **83**, 761 (1961).
73. E. W. Abel, D. A. Armitage and D. B. Brady, *Trans. Faraday Soc.*, **62**, 3459 (1966).
74. K. Ulbrich, M. Jakoubkova and V. Chvalovsky, *Coll. Czech. Chem. Commun.*, **33**, 1693 (1968).
75. S. G. Shevchenko, E. L. Brodskaya. Y. L. Frolov, A. M. Sklyanova, R. G. Mirsov and M. G. Voronkov, *Zh. Obshch. Khim.*, **44**, 1925 (1974).
76. J. Pola, Z. Papouskova and V. Chvalovsky, *Coll. Czech. Chem. Commun.*, **41**, 239 (1976)
77. C. M. Huggins, *J. Phys. Chem.*, **65** 1881 (1961).
78. H. Bock, P. Mollere, G. Becker and G. Fritz, *J. Organomet. Chem.*, **61**, 113 (1973).
79. R. West, L. S. Wilson and D. L. Powell, *J. Organomet. Chem.*, **178**, 5 (1979).
80. R. West, D. L. Powell, M. K. T. Lee and L. S. Whatley, *J. Am. Chem. Soc.*, **86**, 3227 (1964).
81. R. West, D. L. Powell, L. S. Whatley, M. K. T. Lee and P. v. R. Schleyer, *J. Am. Chem. Soc.*, **84**, 3221 (1962).
82. C. G. Pitt, *J. Organomet. Chem.*, **61**, 49 (1973).
83. C. G. Pitt, M. M. Bursey and D. A. Chatfield, *J. Chem. Soc., Perkin Trans. 2*, 434 (1976).
84. J. A. Stone, A. C. M. Wojtyniak and W. Wytenburg, *Can. J. Chem.*, **64**, 575 (1985).
85. J. E. Noll, B. F. Daubert and J. L. Speier, *J. Am. Chem. Soc.*, **73**, 3871 (1951).
86. L. H. Sommer and J. Rockett, *J. Am. Chem. Soc.*, **73**, 5130 (1951).
87. F. Fialova, V. Bazant and V. Chvalovsky, *Coll. Czech. Chem. Commun.*, **38**, 3837 (1973).
88. A. R. Bassindale and P. G. Taylor, Chapter 14 in Part 2 of this Volume.
89. J. F. Wolf, R. H. Staley, I. Koppel, M. Taagepera, R. T. McIver Jr., J. L. Beauchamp and R. W. Taft, *J. Am. Chem. Soc.*, **99**, 5417 (1977).
90. J. M. Bellama, and L. L. Gerchman, *Inorg. Chem.*, **14**, 1618 (1975).
91. J. Pola and V. Chvalovsky, *Coll. Czech. Chem. Commun.*, **41**, 581 (1976).
92. C. H. Van Dyke and J. T. Wang, *Inorg. Chem.*, **6**, 1741 (1967).
93. J. M. Bellama and J. B. Davison, *Inorg. Chem.*, **14**, 3118 (1975).
94. J. Pola and V. Chvalovsky, *Coll. Czech. Chem., Commun.*, **45**, 854 (1980).
95. M. G. Voronkov, V. P. Feshin, V. P. Mironov, S. A. Mikhailants and T. K. Gar, *Zh. Obshch. Khim.*, **41**, 2211 (1971).
96. M. G. Voronkov, V. P. Feshin, L. S. Romaneko, J. Pola and V. Chvalovsky, *Coll. Czech. Chem. Commun.*, **41**, 2718 (1976).
97. J. Pola and V. Chvalovsky, *Coll. Czech. Chem. Commun.*, **42**, 3581 (1977).
98. L. P. Hammett, *Physical Organic Chemistry*, McGraw-Hill, New York, 1940, p. 184, cited in Ref. 99.
99. R. A. Benkeser and H. R. Krysiak, *J. Am. Chem. Soc.*, **75**, 2421 (1953).

100. R. R. Holmes, R. O. Day, J. J. Harland, A. C. Sau and J. M. Holmes, *Organometallics*, **3**, 341 (1984).
101. R. R. Holmes, R. O. Day, J. J. Harland and J. M. Holmes, *Organometallics*, **3**, 347 (1984).
102. A. R. Bassindale and T. Stout, *J. Chem. Soc., Chem. Commun.*, 1388 (1984).
103. R. J. P. Corriu, A. Kpoton, M. Poirier, G. Royo and J. Y. Corey, *J. Organomet. Chem.*, **277**, C25 (1984).
104. R. J. P. Corriu, M. Poirier and G. Royo, *J. Organomet. Chem.*, **233**, 165 (1982).
105. C. Breliere, F. Carre, R. J. P. Corriu, A. De Saxce, M. Poirier and G. Royo, *J. Organomet. Chem.*, **205**, Cl (1981).
106. C. L. Frye, *J. Am. Chem. Soc.*, **92**, 1205 (1970).
107. M. G. Voronkov, Yu. L. Frolow, V. M. D'Yakov, N. N. Chipanina, L. I. Gubanova, G. A. Gabrilova, L. V. Klyba and T. N. Aksamentova, *J. Organomet. Chem.*, **201**, 165 (1980).
108. W. G. Farnham and R. L. Harlow, *J. Am. Chem. Soc.*, **103**, 4608 (1981).
109. W. H. Stevenson and J. C. Martin, *J. Am. Chem. Soc.*, **104**, 309 (1982).
110. R. West, *J. Am. Chem. Soc.*, **80**, 3246 (1958).
111. J. J. Flynn and F. P. Boer, *J. Am. Chem. Soc.*, **91**, 5756 (1969).
112. N. Serpone and K. A. Hersh, *J. Organomet. Chem.*, **84**, 177 (1975).
113. A. Boudin, G. Cerveau, C. Chuit, R. J. P. Corriu and C. Reye, *Angew. Chem., Int. Ed. Engl.*, **25**, 473 (1986).
114. A. Boudin, G. Cerveau, C. Chuit, R. J. P. Corriu and C. Reye, *Angew. Chem., Int. Ed. Engl.*, **25**, 474 (1986).
115. R. J. P. Corriu and C. Young, Hypervalent silicon compounds (a chapter in this book).
116. G. Klebe and K. Hensen, *J. Chem. Soc. Dalton Trans.*, 5 (1985).
117. R. Damrauer and S. E. Danahey, *Organometallics*, **5**, 1490 (1986).
118. F. Klanberg and E. L. Muetterties, *Inorg. Chem.*, **7**, 155 (1968).
119. A. R. Bassindale and P. G. Taylor, Chapter 13 in this Volume.
120. H.-B. Burgi and J. D. Dunitz, *Acc. Chem. Res.*, **16**, 161 (1983).
121. A. Kemme, J. Bleidelis, V. A. Pestunovich, V. P. Baryzhok and M. G. Voronkov, *Dokl. Akad. Nauk SSSR*, **243**, 668 (1978).
122. K. D. Onan, A. T. McPhail, C. H. Yoder and R. W. Hillyard Jr., *J. Chem. Soc., Chem. Commun.*, 209 (1978).
123. C. H. Yoder, C. M. Ryan, G. F. Martin and P. S. Ho, *J. Organomet. Chem.*, **190**, 1 (1980).
124. G. Klebe, J. W. Bats and K. Henson, *J. Chem. Soc., Dalton. Trans.*, 1 (1985).
125. G. Klebe, J. W. Bats and K. Henson, *Z. Naturforsch., B*, **38**, 825 (1983).
126. C. Breliere, F. Carre, R. J. P. Corriu, M. Poirier and G. Royo, *Organometallics*, **5**, 388 (1986).
127. R. J. P. Corriu, C. Guerin and J. J. E. Moreau, *Top. Stereochem.*, **15**, 43 (1984) and references cited therein.
128. J. A. Dieters and R. R. Holmes, *J. Am. Chem. Soc.*, **109**, 1686 (1987).
129. J. A. Dieters and R. R. Holmes, *J. Am. Chem. Soc.*, **109**, 1692 (1987).
130. A. R. Bassindale and A. G. Brook, in *Rearrangements in Ground and Excited States*, Vol. 1 (Ed. P. de Mayo), Academic Press, New York, 1980.
131. E. W. Colvin, A. K. Beck, B. Bastini, D. Seebach, Y. Kai and J. D. Dunitz, *Helv. Chim. Acta*, **63**, 697 (1980).
132. M. J. Barrow, S. Cradock, E. A. V. Ebsworth and D. W. H. Rankin, *J. Chem. Soc., Dalton Trans.*, 1988 (1981).
133. M. J. Barrow, E. A. V. Ebsworth, C. M. Huntley and D. W. H. Rankin, *J. Chem. Soc., Dalton Trans.*, 1131 (1982).
134. M. J. Barrow, E. A. V. Ebsworth and M. M. Harding, *J. Chem. Soc., Dalton Trans.*, 1838 (1980).
135. H.-B. Burgi, *Inorg. Chem.*, **12**, 2321 (1973).
136. R. Rudman, W. C. Hamilton, S. Novick and T. D. Goldfarb, *J. Am. Chem. Soc.*, **89**, 5157 (1967).
137. M. J. Barrow, E. A. V. Ebsworth and M. M. Harding, *Acta Crystallogr., Sect. B*, **35**, 2093 (1979).
138. M. J. Barrow and E. A. V. Ebsworth, *J. Chem. Soc., Dalton Trans.*, 563 (1984).
139. D. G. Anderson, A. J. Blake, S. Cradock, E. A. V. Ebsworth, D. W. H. Rankin and A. J. Welch, *Angew. Chem., Int. Ed. Engl.*, **25**, 1 (1986).
140. I. R. Beattie and E. W. Parrett, *J. Chem. Soc. (A)*, 1784 (1966).
141. H. J. Campbell-Ferguson and E. A. V. Ebsworth, *J. Chem. Soc. (A)*, 705 (1967).

142. A. R. Bassindale and T. Stout, *Tetrahedron Lett.*, **26**, 3403 (1985).
143. B. Marciniec and H. Gulinska, *J. Organomet. Chem.*, **146**, 1 (1978).
144. S. Ya Khorshev, V. L. Tsvetkova and A. N. Egorochkin, *J. Organomet. Chem.*, **264**, 169 (1984).
145. N. W. Alcock, V. M. Tracy and T. C. Waddington, *J. Chem. Soc., Dalton Trans.*, 2243 (1976).
146. N. W. Alcock, V. M. Tracy and T. C. Waddington, *J. Chem. Soc., Dalton. Trans.*, 2238 (1976).
147. D. P. Graddon and B. A. Rana, *J. Organomet. Chem.*, **140**, 21 (1977).
148. D. P. Graddon and B. A. Rana, *J. Organomet. Chem.*, **136**, 315 (1977).
149. J. N. Spencer, S. W. Barton, B. M. Calder, C. D. Corsico, L. E. Harrison, M. E. Mankuta and C. H. Yoder, *Organometallics*, **4**, 394 (1985).

The Chemistry of Organic Silicon Compounds
Edited by S. Patai and Z. Rappoport

CHAPTER **13**

Reaction mechanisms of nucleophilic attack at silicon

ALAN R. BASSINDALE and PETER G. TAYLOR
PORCG, Department of Chemistry, The Open University, Milton Keynes MK7 6AA, England

I. INTRODUCTION

The mechanisms of reactions involving silicon have proved to be a most fruitful area of research for over forty years. Silicon engages in nucleophilic and radical reactions covering almost all known reaction mechanisms.

Reactions of organosilicon compounds described as 'electrophilic' do not generally take place with electrophilic attack at silicon. Such reactions, involving cleavage of Si–X bonds under electrophilic conditions usually consist of electrophilic activation of a leaving group, with concerted or subsequent nucleophilic attack at silicon[1a]. Reactions such as aromatic substitution of arylsilanes, and many transformations of allyl and vinyl silanes fall into this class and are covered in the chapter on Activating and Directive Effects of Silicon[1b].

Radical and gas phase reactions, and the siliconium ion question are covered elsewhere in this volume.

Therefore in this review we have chosen to limit the scope to nucleophilic substitutions at silicon. This is an area of both commercial and academic interest. The commercial interest lies largely in the hydrolysis reactions of functional silicon compounds to produce silicones. The academic interest, in addition to the hydrolysis reactions, stems from the stereochemical outcome of reaction of chiral silanes, in which either retention or inversion of configuration may result. The question of the nature of the intermediates and/or transition states in nucleophilic substitutions is one that has exercised many groups over the years. Our knowledge of silicon reaction mechanisms is steadily increasing, and the contributions of Sommer[1c, 2] Prince[3] and Corriu[4–6] to nucleophilic substitutions are noteworthy. It will become obvious to the reader, however, that this topic is by no means closed and remains the subject of lively debate and some controversy.

II. THE MECHANISM OF NUCLEOPHILIC SUBSTITUTION AT SILICON

A. Introduction

The mechanism of nucleophilic substitution at silicon has been reviewed numerous times in recent years. The early work, up to 1964, has been reviewed by Sommer[1c]. Further work, up to 1972, has been reviewed by Sommer[2] and Prince[3]. More recently Corriu has published a number of reviews on this topic.[4–6] The purpose of this review is twofold: to discuss more recent work, and to compare mechanistic evidence obtained from a range of techniques.

We start by examining the direct nucleophilic substitution of a leaving group by a nucleophile, and in the next part of this section, C, we discuss nucleophilic catalysis of substitution and racemisation.

B. Direct Nucleophilic Substitution of a Leaving Group by a Nucleophile

1. Stereochemistry of substitution

We begin by examining direct exchange of a leaving group X by a nucleophile Y. Much of the discussion in this area has centred on whether the five-coordinate species **1** exists as

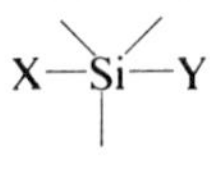

(1)

an activated complex or as an intermediate, that is, mechanism A, an elementary reaction, *versus* mechanism B.

Mechanism A  (1)

Mechanism B 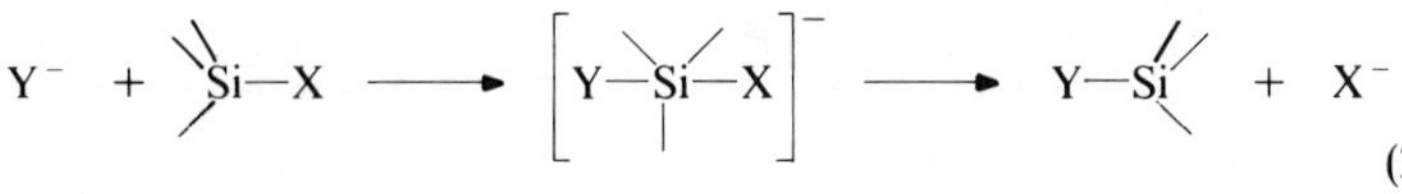 (2)

As in substitution at carbon, stereochemical and kinetic studies provide the key to differentiating between these two possibilities. The stereochemical data are examined first. Substitution at silicon leads to both retention and inversion of configuration and the stereochemical outcome depends upon the nature of the leaving group, the nucleophile, solvent, complexing agents, and whether or not the silicon is part of a ring.

a. The effect of nucleophile and leaving group. Table 1 shows some typical stereochemical data as a function of the leaving group. Sommer[2] has pointed out that good leaving groups of low basicity ($pK_a < 6$, e.g. Br and Cl) tend to give inversion of configuration while poor leaving groups of high basicity ($pK_a > 10$, e.g. H and RO) usually (but not always) result in retention of configuration. Thus he argued that it was the basicity, not the electronegativity, that is the critical factor in determining stereochemical outcome. Although the fluoride leaving group ($pK_a \sim 4$) should fall in the category of a good leaving group, retention is observed in some cases making this a borderline group. Similarly a thiol leaving group ($pK_a \sim 10$) should be a poor leaving group yet inversion is often observed. Comparison of the SMe and OPh leaving groups which have similar pK_a values show the former to have a greater preference for inversion[2]. Both Sommer[2] and Corriu[5] suggest that bond polarization and ease of Si–X stretching during attack by the nucleophile may be important factors. Thus with Si–Cl, Si–Br and Si–SMe bonds which are highly polarizable the leaving groups are displaced with inversion whereas with the far less polarizable SiF bond the leaving group is displaced with retention. Corriu has also suggested[5] an empirical relationship between the observed stereochemistry and the ability of leaving groups to be displaced:

$$\mathrm{Br} \sim \mathrm{Cl} > \mathrm{SR} \sim \mathrm{F} > \mathrm{OMe} > \mathrm{H} \quad \text{ability to be displaced}$$

$$\text{Inversion} \longrightarrow \text{retention} \quad \text{stereochemistry}$$

This order was devised from the observation of preferential loss of one group in bifunctional silanes, or by competition studies. Sommer[18, 21] has suggested that inversion occurs when the countercation is able to assist the departure of the leaving group. With poor leaving groups such as RO^- it is difficult to displace the leaving group as the free ion, as suggested by its high basicity. Coupled with this, elements such as oxygen have free electron pairs that can be donated to the electrophilic part of the substituting group and in so doing any π-bonding between the oxygen and silicon is reduced. Calculations[2] show that with organometallic reagents such as those shown in series 2 and 3 of Table 1 the difference in bond energies between the Si–C bond and the Si–F bond needs to be offset by the concerted formation of a metal fluorine bond which has a high heat of formation. Thus for poor leaving groups Sommer has proposed a cyclic mechanism involving electrophilic assistance.[2] Whilst a four-membered ring was proposed he did not discount the possibility of a six-membered ring (see **2**). This species could be tetragonal pyramidal with X and Y basal, or trigonal bipyramidal with Y apical and X equatorial. In

TABLE 1. Stereochemical data for the substitution of 1-NaphPhMeSiX as a function of the leaving group X

Substrate	Series 1		Series 2		Series 3		Series 4	
	Predominant stereochemistry with potassium hydroxide or alkoxide	Ref.	Predominant stereochemistry with alkyl lithium reagent	Ref.	Predominant stereochemistry with alkyl Grignard reagent	Ref.	Predominant stereochemistry with $LiAlH_4$	Ref.
1-NaphPhMeSiBr	Inv	7	Inv	14	Inv	14	Inv	7, 11, 14
1-NaphPhMeSiCl	Inv	7, 8	Inv	10, 15, 16	Inv[a]	19	Inv	11, 20
1-NaphPhMeSiOCOR	Inv	9	—	—	—	—	Inv	9
1-NaphPhMeSiSR	—	—	Ret	14, 17	Inv	14	Inv	14, 17
1-NaphPhMeSiF	Ret	10	Ret	10, 15, 16	Inv[a]	19	Inv	20
1-NaphPhMeSiOR	Ret	8, 11, 12	Ret	15, 16, 18	Ret[a]	19	Ret	8, 11, 20
1-NaphPhMeSiH	Ret	8, 13	Ret	15, 16, 18	—	—	—	—

[a] In these cases the substrate was the vinyl derivative 1-NaphPh$\langle$Si / X.

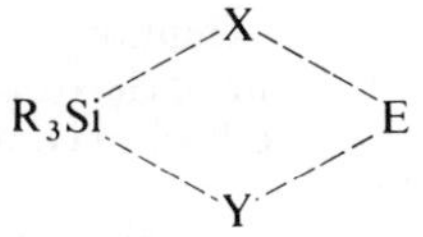
(2)

this latter case pseudorotation may occur to permit apical exit of X with overall retention.

Evidence for such electrophilic assistance comes from solvent and countercation effects. The stereochemistry of substitution of alkoxy, mercapto and fluorosilanes with alkoxide ion changes dramatically as the composition of the solvent is varied.[2] As the percentage of alcohol in the benzene solution increases, there is a move from retention of stereochemistry towards inversion. Similarly, alcohols which have an increased capacity for hydrogen bonding are able to promote inversion more than those that are less able to hydrogen bond. Sommer suggested that in benzene, electrophilic assistance from the countercation is required to enable the leaving group to be expelled (see **3**). However, as

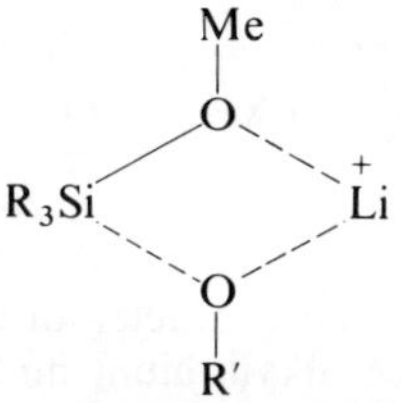
(3)

the alcoholic content of the solution increases the solvent provides sufficient electrophilic assistance for removal of the poor leaving group through hydrogen bonding, thus allowing inversion to proceed. Obviously the proportion of inversion would depend upon the hydrogen bonding ability of the solvent, as was observed (see **4**).

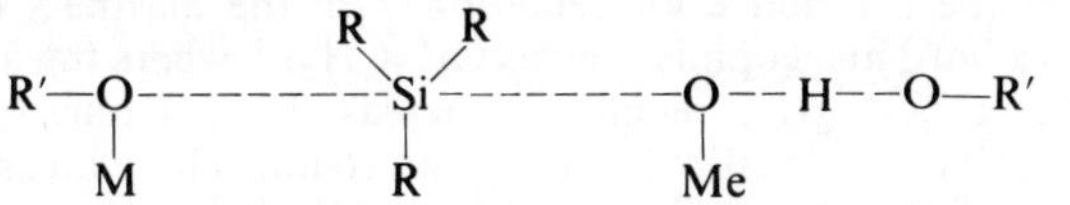
(**4**)

When the countercation is varied from lithium to sodium to potassium, the proportion of inversion increases. Here the lithium–oxygen bond is relatively covalent and thus favours a retentive mechanism. However, the more ionic oxygen–sodium bond allows the components to function more independently such that the RO^- may attack the back face of the silicon tetrahedron while electrophilic assistance by M^+ of the leaving group from the front side aids inversion of configuration. This is further supported by studies on the influence of added salts. Addition of $MgBr_2$ to a Grignard reagent or Li^+ to an alkyl lithium reagent results in a change from retention to inversion[22,23]. Thus complexation of the added metal ion with the leaving group enhances the leaving-group ability and the free nucleophile or ion pair can then react with inversion. These conclusions have been criticised by Corriu, who found that complexation of the lithium cation of alkyl lithium reagent using TMEDA or kyrptofix 211 led to increased retention with increased rate[23]. He argued that if species **3** were important in determining the streochemical outcome, then removal of the counterion by complexation should favour inversion of configuration, which was not observed with alkyl lithium reagents. This shift towards inversion is observed however when the counterion of allyllithium is complexed[24, 25]. Nucleophilic solvents such as THF or DME promote retention when compared to ether, again accompanied by an increase in rate[19, 22]. Again, if electrophilic assistance were important the use of a more nucleophilic solvent should lead to complexation of the countercation with an increase in the proportion of inversion.

The stereochemistry of substitution is also greatly dependent on the nature of the nucleophile. Table 2 shows how the stereochemical outcome varies as a function of the nucleophile for a range of leaving groups including the borderline fluorosilanes and thiosilanes. After carrying out numerous studies using a range of leaving groups and nucleophiles, Sommer[2] suggested that localised carbanions such as alkyl lithium reagents exist as tight ion pairs where the localized negative charge makes the countercation a poor electrophile. Thus electrophilic assistance leads to predominant retention of configuration, where concerted formation of the Si–C and Li–F bonds are mutually helpful in the breaking of the Si–F and Li–C bonds. However, delocalized carbanions such as benzyl lithium will have a more ionic carbon–lithium bond and the ion pair will be looser. The lithium is therefore more electrophilic and can provide electrophilic assistance with predominant inversion (see **5**).

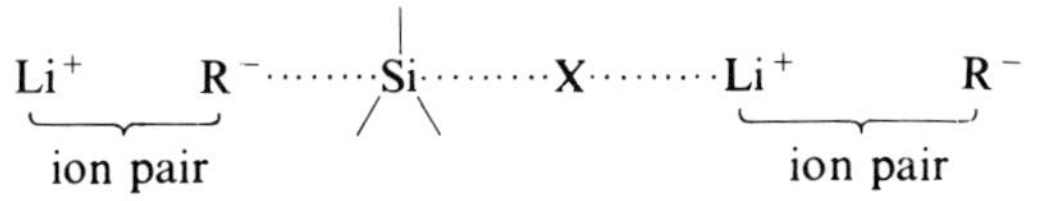

(5)

Sommer[2] emphasized that high ionic character of the C–Li bond in itself is not sufficient to promote inversion since alkyllithium and alkynyllithium reagents have similar stereochemical outcomes. Both high ionic character and charge delocalization are required.

Corriu[5, 6] has pointed out a parallel between the stereochemistry at silicon and the regioselectivity of attack at α-enones. Nucleophiles that favour retention at silicon also favour 1,2-addition to α-enones whereas those that react with inversion give rise to 1,4-addition.

Corriu has related the preference for retention with the hardness of the attacking nucleophile[5, 6]. Thus a hard nucleophile such as CH_3CH_2Li, where the negative charge is concentrated on the carbon, gives retention whereas the soft nucleophile $PhCH_2Li$, where the negative charge is delocalized, leads to inversion. These ideas have received a theoretical justification from the work of Anh and Minot[26], who carried out perturbational studies on substitution at silicon with either retention or inversion. Following previous treatments of the Walden inversion[27, 28] they proposed a frontier orbital interaction between the nucleophile's HOMO and the substrate's LUMO as shown in Figure 1. A frontside attack, giving retention, corresponds to an attack on the large lobe on silicon, but this will be opposed by an unfavourable out-of-phase overlap between the

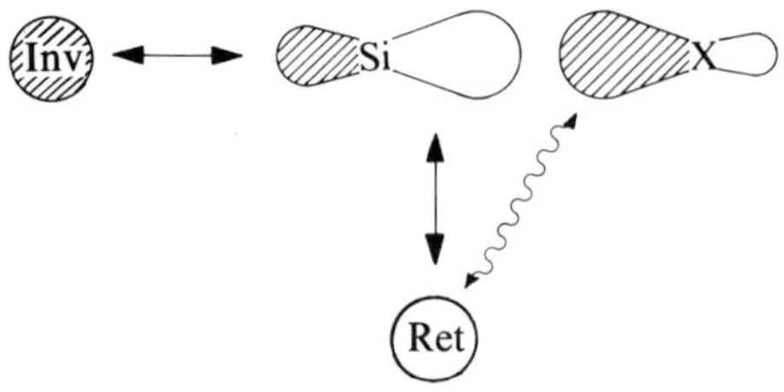

FIGURE 1. Frontier orbital interaction in substitution at silicon

TABLE 2. Stereochemical data for the substitution of 1-NaphPhMeSiX as a function of the nucleophile

Nucleophile	Stereochemical outcome with 1-NaphPhR^1SiOR^2	Ref.	Stereochemical outcome with 1-NaphPhR^1SiF	Ref.	Stereochemical outcome with 1-NaphPhR^1SiSR^2	Ref.
RLi (R = benzyl)	Inv	15, 18,	Inv	15, 10,	Inv	14, 17
RMgBr (R = benzyl)	Inv	18	Inv	18	Inv	14
$LiAlH_4$	Ret	8, 11, 12	Inv	26	Inv	14, 17
RLi (R = alkyl)	Ret	14, 16, 18	Ret	15, 16, 10	Ret	14, 17

nucleophile and the leaving group. However, backside attack with inversion corresponds to an attack on the small lobe on Si with no unfavourable interaction. Thus the probability of observing retention depends upon the balance between the favourable interaction between the nucleophile and the reaction centre and the unfavourable interaction between the nucleophile and the leaving group. The authors propose that retention is observed for silicon and not for carbon for a number of reasons. Firstly, the Si–X bond will be longer, diminishing the unfavourable interaction between X and the nucleophile for frontside attack. Secondly, the valence orbitals are 3s and 3p, which are more diffuse and overlap better with the nucleophile at larger distances. Finally, compared to carbon, the silicon has an increased contribution to the substrate LUMO whilst that of the leaving group is diminished. Using this model it is possible to account for the dependence of the stereochemical outcome on the hardness of the nucleophile. A hard reagent is usually small, with contracted valence orbitals, so long-range overlap with the leaving group will be negligible and frontside attack, leading to retention of configuration, is favoured. A soft nucleophile, however, usually has diffuse valence orbitals and the unfavourable interaction with the leaving group will be large and inversion becomes the favoured pathway. However, this is not the only controlling factor. A change in the hardness of a nucleophile implies a modification of its energy level. It has been shown that in aprotic solvents the HOMO of a hard anion lies at a higher energy than that of a soft anion, but this is reversed in protic solvents[29]. Thus in aprotic solvents a hard nucleophile has a high-energy HOMO and the frontier orbital interaction is predominant leading to retention. A soft anion has a low-energy HOMO and the importance of the HOMO–LUMO interaction is decreased and that of the nucleophile HOMO–substrate superjacent MO is increased. As shown in Figure 2, the larger lobe of the silicon superjacent orbital points to the rear, favouring inversion of configuration. Using similar arguments it is possible to rationalize the influence of the leaving group. On replacement of fluorine for chlorine as the leaving group, the increase in electronegativity lowers the energy of the leaving group atomic orbital thus decreasing its contribution to the LUMO, favouring retention. At the same time the valence orbitals of the leaving group are contracted, reducing the unfavourable interaction with the nucleophile favouring retention. However, the Si–X bond also shortens which will increase this unfavourable interaction. Whilst it is difficult to decide *a priori* which factors predominate, calculations indicated that bond shortening does not outweigh the other effects. Thus replacing the leaving group by a higher element from the same group of the periodic table will increase the proportion of retention. Comparison between groups is however more difficult. When hydrogen is the leaving group the Si–H bond is short and its electronegativity is quite low, favouring inversion. However, retention is usually observed because the repulsion between the 1s orbital of the leaving group and the nucleophile will be very small.

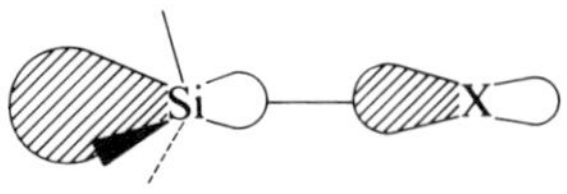

FIGURE 2. Structure of the substrate superjacent MO

Corriu has used this hard/soft nucleophile concept to explain the solvent and counter-cation effects[5,6]. He suggests that in benzene the RO^-M^+ species have a localized negative charge on the oxygen atom and thus are hard and react with retention. On increasing the alcohol content the charge on the anion is dispersed by hydrogen bonding, producing a softer species that reacts with inversion, as observed (see **6**). Similarly with

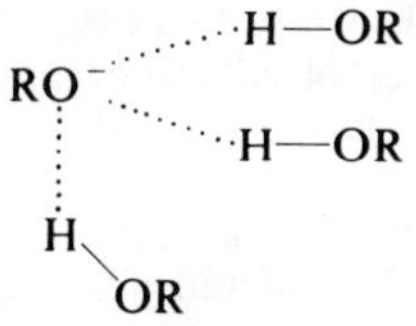
(6)

organometallic nucleophiles, increasing the nucleophilic character of the solvent will enhance the complexation of the cation. This leaves a more naked carbanion, which is harder and therefore predominantly reacts with retention (see **7**). Finally, the organolithium RLi will be a harder reagent than the analogous Grignard reagent RMgX leading to more retention.

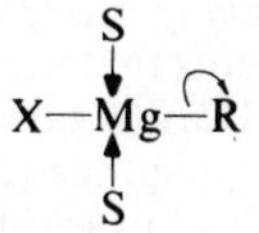
(7)

Lithium aluminium hydride generally reacts with inversion of configuration, in accord with the ability of hydride ion to act as a soft nucleophile. On the other hand, *i*-Bu_2AlH in hexane reacts with retention. Sommer[2] suggests that this is a result of a very strong electrophilic centre and a weaker nucleophilic centre proceeding with electrophilic assistance (see **8**). Corriu[5, 6] has pointed out that the stereochemistry of reaction at silicon

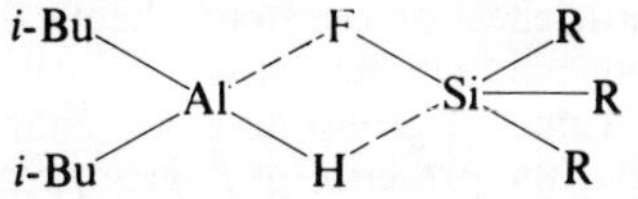
(**8**)

is matched by the regioselectivity of $LiAlH_4$ and *i*-Bu_2AlH reduction of α-enones (1,4- and 1,2-addition, respectively). If hydride ion from $LiAlH_4$ is acting as a soft nucleophile, the corollary of this is that *i*-Bu_2AlH behaves as a hard nucleophile. The addition of ether to *i*-Bu_2AlH[2] results in a move from retention to inversion in line with a move from 1,2- to 1,4-addition for α-enones. Sommer suggests that this is because the reducing H is more hydridic and nucleophilic in the etherate complex resembling that in $LiAlH_4$. Thus the strong electrophilic centre and weak nucleophilic centre have disappeared and there is no need to proceed via retentive electrophilic assistance. In terms of the hard/soft approach Corriu[4, 5] proposes that as the *i*-Bu_2AlH becomes solvated the Al–H hydrogen becomes more hydridic as the negative charge is localised on the hydrogen (see **9**). This is the same

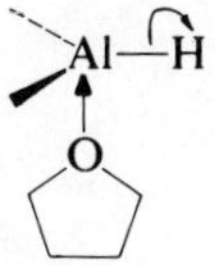
(**9**)

type of argument as that applied to the effect of nucleophilic solvents on alkyllithium and Grignard reagents, **7**. However, in this case Corriu suggests that localization leads to a softer nucleophile. It was proposed that in *i*-Bu_2AlH the nucleophilic electrons are in the bond which is contracted. Displacement of the electrons onto the diffuse hydrogen valence orbitals gives a softer anion. This singular behaviour of alanes is also observed when LiBr is added to $LiAlH_4$ in ether[30]. In this case retention is favoured, the opposite trend to that observed for the addition of salts to organometallics. Here it was proposed that interactions between the Li^+ and the Al–H bonds prevent localization of the negative charge on hydrogen.

Evidence for electrophilic assistance has also been obtained from studies on bifunctional organosilanes, **10**. In general, the preference for cleavage with X or Y as the leaving

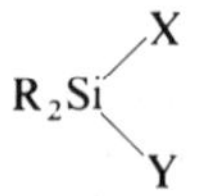

(10)

group depends upon leaving-group ability, as discussed earlier. However, the F and OR leaving groups have sufficiently close leaving-group ability to allow other factors to be revealed. Corriu maintains that soft reagents that lead to inversion cleave only the better leaving group, namely fluorine[5, 6, 31, 32]. Hard organolithium reagents substitute the two leaving groups concurrently with predominant loss of the better leaving group, fluoride ion. In contrast, hard Grignard reagents lead to selective cleavage of the alkoxy group. It was proposed that the Li^+ cation has no great affinity for either group whereas the magnesium atom has a great affinity towards oxygen. Thus, the OR group complexed with magnesium is the better leaving group. Corriu pointed out that in these cases the electrophilic assistance only determines which group is cleaved and does not affect the stereochemistry.

b. The effect of substrate structure. Sommer suggests that the structure of the alkyl or aryl groups on the silicon has little effect on the stereochemical outcome of substitution[2]. For example, despite the difference in the electron-withdrawing ability, the relacement of a pentafluorophenyl group by a phenyl group does not change the stereochemistry to any significant extent[2]. Similarly, the presence of R groups of moderate to large steric requirement leaves the stereochemistry unchanged[2]. There is, however, a significant change when silicon is made part of a strained-ring structure. Providing the leaving group is exocyclic there is a move to retention as the angle strain of the ring at silicon increases, as shown in Table 3. Nevertheless, the general trends relating to the nucleophiles and leaving groups observed with acyclic systems still apply to these cyclic systems. For example, a chlorine leaving group gives more inversion than the fluorine leaving group.

The opposite stereochemical trend is observed with endocyclic leaving groups. In this case angle strain favours inversion, as shown in Table 4. Nevertheless, the same trends observed with acyclic systems are still dominant. For example, the SR group gives more inversion than the OR group. With substitution at phosphorous these trends have been explained by the need for small rings to be apical/equatorial (90° bond angle), **11**, rather than diequatorial (120° bond angle), **12**, in a pentacoordinate intermediate.

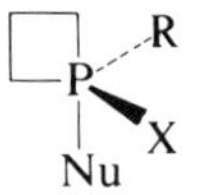

(11)

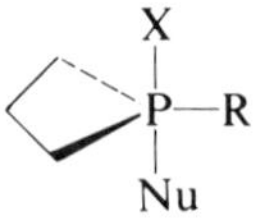

(12)

A similar argument can apply to silicon. With exocyclic leaving groups the requirements for small rings to be axial/equatorial and axial entry of the nucleophile give an intermediate **13** which results in retention, as is observed experimentally. When the leaving group is endocyclic, the small ring provides extra incentive for the leaving group to be axial. With axial entry this leads to preferred inversion **14**, as observed.

TABLE 3. Stereochemical outcome of substitution with $LiAlH_4$ on cyclic silicon compounds with the leaving group exocyclic

Compound	Angle	X=Cl	X=F	Ref.
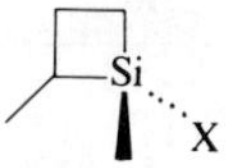	90°	Ret	Ret	33, 34
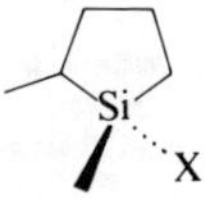	93–96°	Inv	racemic	35, 36
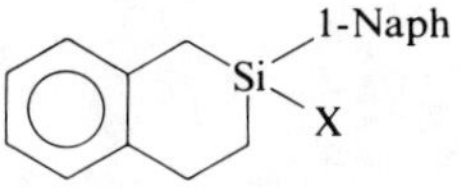	105°	Inv	racemic	37
R_3SiX	109°	Inv	Inv	2

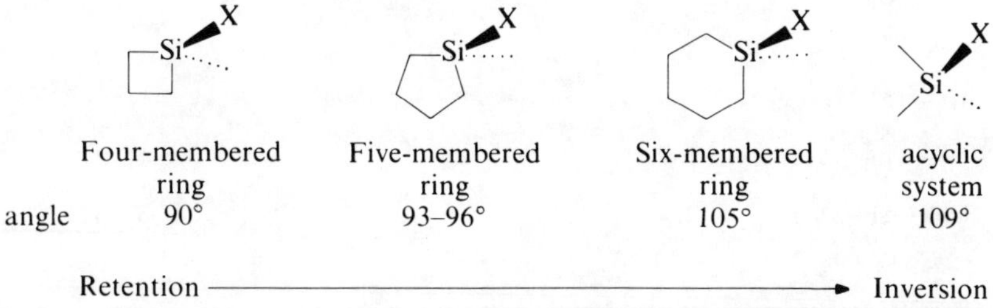

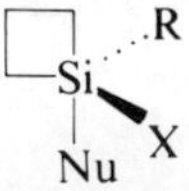
(13)

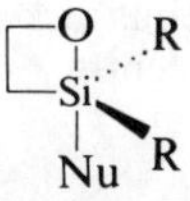
(14)

An alternative explantion has been forwarded by Anh and Minot[26]. For a tetracoordinated silicon atom, **15**, if the R^1SiR^2 angle is decreased (by including R^1, Si and R^2 in a

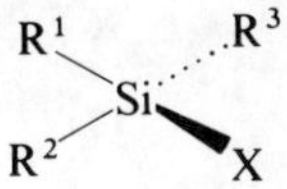
(**15**)

strained ring, for example) the R^3SiX angle becomes more open. This results in the Si–X bond having more s character, increasing the favourable overlap between the reaction centre and incoming nucleophile. Thus retention is favoured. If Si and X are both included in a strained ring, the Si–X bond contains less s character, decreasing the overlap between the reaction centre and the nucleophile favouring inversion, accounting also for

TABLE 4. Stereochemical outcome of substitution with various nucleophiles on cyclic silicon compounds with the leaving group endocyclic

Compound	Nucleophile			Ref.
	$LiAlH_4$	p-$AnCH_2Li$	$CH_2{=}CHCH_2Li$	
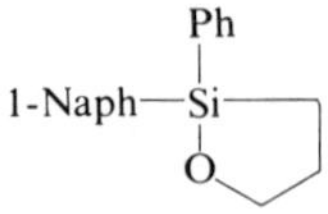	Ret	Inv	Inv	5, 38, 39
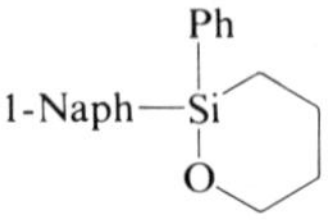	Ret	Ret	Ret	40
1-Naph—Si(Ph)(Me)—OMe	Ret	Ret	Ret	2
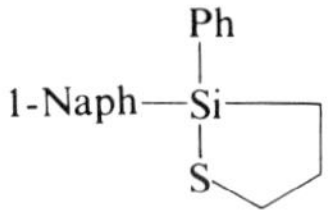	Inv	—	—	41
1-Naph—Si(Ph)(Me)—SMe	Inv	—	—	17

the increase in rate on going from **16** to **18**[5]. This order reflects the increase in angle strain owing to the bi- and tricyclic system; thus the SiX bond in **18** has more s character than in **16**. The rate of substitution is thus increased because of the increased overlap between the nucleophile and the reaction centre.

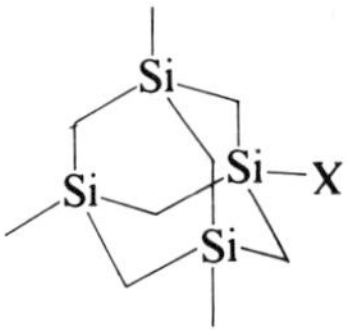

(16)

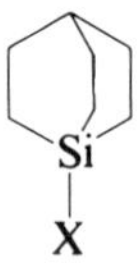

(17)

(18)

Increasing rate of substitution ⟶

2. *Kinetics of substitution*

Having reviewed the stereochemistry of direct substitution we now turn our attention to kinetic aspects of substitution. However, in this section we shall concentrate on observations that we feel require special comment.

Kinetic data from solvent/nucleophile-catalysed substitutions will be discussed in a later section.

Schowen and Swain[42] observed a change in kinetic order for the hydrolysis of aryloxytriphenylsilanes in 60% acetonitrile–40% water. At low base concentrations the reaction was second order in hydroxide ion, but as the base concentration was increased the reaction became first order. The authors suggested that this behaviour was consistent with the scheme of equation 3, which yields the kinetic law of equation 4 upon application of the steady-state approximation. At low concentration of HO^-,

$$HO^- + Ph_3SiOAr \underset{k_{-1}}{\overset{k_1}{\rightleftharpoons}} Ph_3\overset{-}{Si}(OH)(OAr) \xrightarrow[[OH]]{k_2} \text{Products} \qquad (3)$$

$$k_{obs} = k_1k_2[HO^-]^2/(k_{-1} + k_2[HO^-]) \qquad (4)$$

$k_2[HO^-] \ll k_{-1}$ and the reaction will be second order in HO^-. When the concentration of HO^- is high, $k_2[HO^-] \gg k_{-1}$ and equation 4 reduces to first order in HO^-. They were able to demonstrate this change in a rate-limiting step using a plot of $k_{obs}/[HO^-]$ versus $[HO^-]$. This is shown in Figure 3, which is linear at low concentrations but levels off at higher concentrations, as would be expected for a change from second to first order. In contrast, the hydrolysis of methoxytriphenylsilane is second order in hydroxide ion throughout the range studied. They suggest that because of the poor leaving-group ability of MeO^-, k_2 will be smaller and thus the change in rate-limiting step (when $k_2[OH^-] \gg k_1$) will occur at higher base concentrations. It is worth noting that a plot of log k against log $[OH^-]$ gives perfect straight lines of slopes 1.50 (c.c. 0.998) and 1.67 [ignoring first point as in Figure 1 of Reference 42 (c.c. 0.995)] for p-$CH_3OC_6H_4OSiPh_3$ and $CH_3OSi(C_6H_5)_3$, respectively. No indication of any curvature arising from a change in order (and hence slope) from 2 to 1 is observed. Subsequently, Eaborn[43] suggested that the behaviour observed by Swain was due to an acidic (or acid-generating) impurity in the acetonitrile which reduced the base concentration in the medium. On removal of this acidic impurity Eaborn obtained a satisfactory first-order

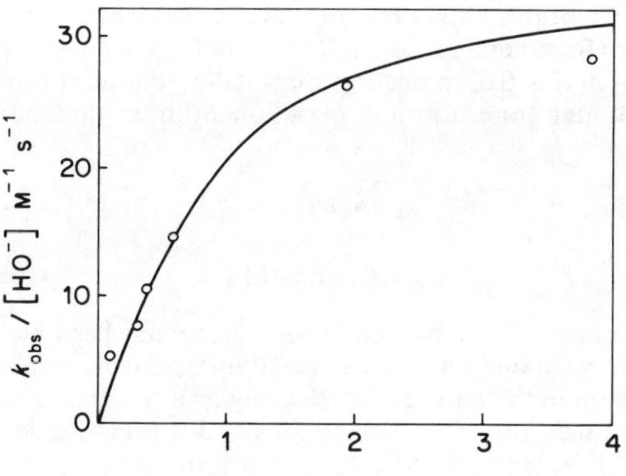

FIGURE 3. Plot of $k_{obs}/[HO^-]$ versus $[HO^-]$ for hydrolysis of p-methoxyphenoxytriphenylsilane

dependence of the observed rate constant on the base concentration for both substrates over the range examined by Swain. This is consistent with (i) a concerted S_N2-type mechanism, (ii) a mechanism involving a five-coordinate intermediate where loss of the alkoxide is not catalysed by HO^-, or (iii) a mechanism where such a step is not kinetically significant. A log/log plot of Eaborn's data gives a straight line of slopes 1.03 (c.c. 1.000) and 0.979 (c.c. 0.998) for p-$CH_3OC_6H_4OSiPh_3$ and CH_3OSiPh_3, respectively.

Swain was able to demonstrate that the methanolysis of methoxyphenoxymethylsilane was catalysed by both formate ion and formic acid in combination, suggesting a concerted formation and fission of entering and leaving groups. When the free energy for methanolysis of aryloxytriphenylsilanes (equation 5) is plotted against the free energy of neutralisation of the corresponding phenols (equation 6) as a function of substituent, curvature is observed[44]. This could arise from a change in mechanism from a concerted mechanism with good leaving groups, to a two-step mechanism involving a pentacoordinate intermediate, with poor leaving groups. However, the authors favour an explanation in terms of variation in the Si–O dπ-pπ bonding in the activated complex.

$$B + CH_3OH + Ph_3SiOAr \rightarrow CH_3OSiPh_3 + ArO^- + BH^+ \tag{5}$$

$$B + HOAr \rightarrow BH^+ + {}^-OAr \tag{6}$$

Grant and Prince examined the chloride exchange reactions of triphenylchlorosilane with [^{36}Cl]-labelled tetrabutylammonium chloride in benzene[45]. In this case a plot of log k against log [Salt] give a slope of 1.6 (c.c. 0.997). This was interpreted in two ways: firstly in terms of two competing pathways, one proceeding via an MCl ion pair and the other via an MCl dimer (equation 7). Whilst the data fit this expression reasonably well, no curvature of the log/log plot from first order (slope 1) to second order (slope 2) was observed. The second interpretation depended upon the predominant form of MCl being the dimer $(MCl)_2$ yet the reaction proceeding via the minority trimer $(MCl)_3$ (equation 8). This latter proposal conflicts with the work of Pettit and Bruckenstein[46], who suggest that the major species present in such solution is the trimer $(MCl)_3$. Prince's data may, however, have been complicated by the presence of a secondary salt effect leading to an enhanced order of reaction as was observed for the exchange of trityl chloride[45]. Prince also measured the rate of exchange and racemization of 1-naphthylphenylmethylchlorosilane[45] and found that exchange occurred about 100 times faster than racemization. This implies that substitution of the good leaving group Cl^- by the nucleophile Cl^- occurs preferentially with retention. This is the opposite to that which would be expected from the extensive work of Sommer and Corriu. The order for exchange was found to be 1.5 yet that for racemization was 0.5, indicating that at low concentrations of salt inversion predominates, yet at high concentrations of salt retention predominates, which is difficult to reconcile.

$$k_{obs} = k_1[MCl] + k_2[MCl]^2 \tag{7}$$

$$3(MCl)_2 \rightleftharpoons 2(MCl)_3 \tag{8}$$

The dependence of rate of reaction on leaving group has been measured by a number of workers. Sommer[1c, 2] found the order of reactivities shown in Table 5. Clearly the rate of reaction depends upon the nature of the leaving group with the rate decreasing as the basicity of the leaving group increases. Corriu, however, found only a very small variation of rate with leaving group[47, 48], as shown in Table 6. The results fall into two categories, those leading to inversion or retention with a rate ratio of about one, and those of the F and OMe groups leading to inversion with a rate ratio of $>10^3$. Corriu points out that the rate ratios approximating to one show that the slow step does not

TABLE 5. Rate constants for direct nucleophilic substitution at silicon

	Base-catalysed solvolysis of Me_3SiX ($l\ mol^{-1}\ min^{-1}$)	Hydrolysis of Ph_3SiX (min^{-1})
X=H	10^{-2}	—
X=OMe	1	—
X=OPh	10^2	—
X=F	too fast to measure	10^{-6} (37 M aqueous acetone)
X=Cl	too fast to measure	10^{-2} (4 M aqueous acetone)
X=Br	too fast to measure	

TABLE 6. Relative rates of nucleophilic substitutions at silicon

Leaving groups for comparison	Substrate	Nucleophile	Solvent	$k_1:k_2$
(i) Reactions occurring with retention				
F:OMe	I	EtMgBr	Et_2O	1.5–2.5
	I	PhMgBr	Et_2O	23
	I	MeMgBr	THF	1–1.5
	I	allylLi	Et_2O	4–5
	II	BuLi	Et_2O	50
F:Cl	I	PhMgBr	Et_2O	20–40
OMe:H	II	PhLi	Et_2O	1.5–2
	II	BuLi	Et_2O	5–6
H:D	II	PhLi	Et_2O	1.3
	II	BuLi	Et_2O	1.0
(ii) Reactions occurring with inversion				
F:OMe	I	allyl MgBr	Et_2O	$>10^5$
	I	$PhCH_2MgCl$	Et_2O	$2\times10^3 < k_F/k_{OMe} < 5\times10^3$
	II	allyl MgBr	Et_2O	$>10^3$
F:Cl	I	CH_3MgBr	Et_2O	1
	I	allyl MgBr	Et_2O	0.9
	I	crotyl MgBr	Et_2O	1
	I	$PhCH_2MgCl$	Et_2O	1.6
	III	CH_3MgBr	Et_2O	0.32
	III	allyl MgBr	Et_2O	0.5
	III	$PhCH_2MgCl$	Et_2O	0.7
	III	crotyl MgBr	Et_2O	0.5
	III	propyl MgBr	Et_2O	1.7

I = 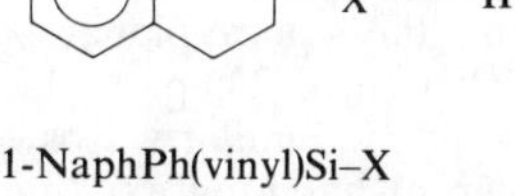

II = 1-NaphPhEtSi–X

III = 1-NaphPh(vinyl)Si–X

involve Si–X bond cleavage. Thus a mechansm involving a single elementary step should be eliminated. These results agree with the slow formation of a pentacoordinate intermediate ($k_2 > k_{-1}$) with little stretching of the Si–X bond. The small kinetic isotope effect confirms this proposal. These results seem to be in conflict with the key role played by the leaving group in determining the stereochemical outcome. As mentioned earlier it is the C–X bond polarizability and ease of Si–X stretching which determines the nature of the reaction.

$$R_3SiX + Nu \underset{k_{-1}}{\overset{k_1}{\rightleftharpoons}} R_3SiXNu \xrightarrow{k_2} \text{products}$$

In general, the reactions of 1-NaphPhMeSiCl with RLi give high stereospecificity and excellent yields of 1-NaphPhMeSiR. These findings suggest that if a two-step mechanism such as that shown above were operative, k_2 must be greater than k_{-1}. If k_{-1} were greater than k_2, then breakup to products would be slower than return to starting materials. It therefore seems unreasonable for the intermediate to return many times to reactant involving Si–C bond rupture without extensive exchange of the other organic groups and formation of optically inactive mixed products.

3. *The geometry of the intermediate/transition state*

Having discussed the stereochemistry and kinetics of substitution it is useful to examine the geometry of the activated state.

a. Inversion. The simplest model for the transition state and/or intermediate leading to inversion is a distorted trigonal bipyramid (TBP) with the entering and leaving group both apical (see **19**). Since the stereochemical outcome depends upon the basicity of

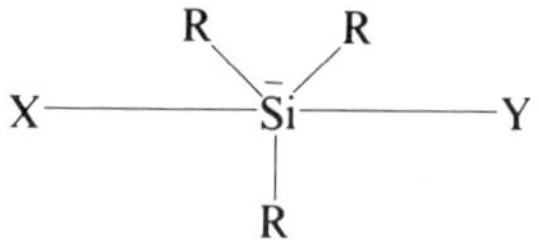

(**19**)

the leaving group, i.e. its ability to stabilize a negative charge, rather than on electronegativity, Sommer[2] has suggested that there is considerable ionic stretching in the transition state, placing negative charge on the leaving group X^-. If the apical Si–X and Si–Y bonds are long and weak, five equivalent $3sp^3d$ bonds cannot be involved. A more likely explanation is three equatorial $3sp^2$ hybrid bonds with the R groups, which are short and strong, and two axial $3p_zd_{z^2}$-orbitals with the entering and leaving groups. This latter situation gives rise to a three-centre four-electron hypervalent bond. Since two of the electrons are bonding and two non-bonding, the bonds will be longer[49]. This follows the general trend observed in other 10-X-5 compounds[50]. The X-ray structures of a number of 10-Si-5 siliconates have been determined[51, 52]. Martin[52] found that the apical Si–O bond length is increased significantly when compared to the corresponding four-coordinate system, whereas with the equatorial Si–C bonds the increase is only modest. It is interesting to note that the X-ray results of these siliconates, and those of others, show that pentacoordinate silicon derivatives are non-rigid and reveal progressive distortions between the trigonal bipyramid and the square-planar structure. The X-ray crystal structure of **20** has an Si–Cl bond length of 235 pm, which is much longer than the conventional distance of about 200 pm[53]. Similarly Ebsworth[54] has shown that the chlorosilyl-*N*,*N*-dimethylamine dimer **21** has long axial-type Si–N bonds and shorter

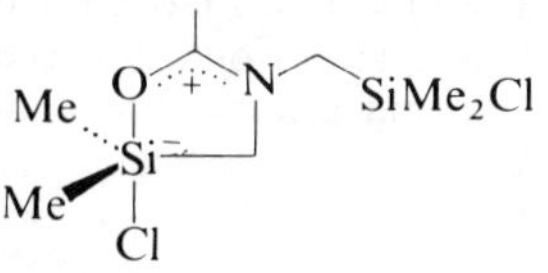
(**20**)

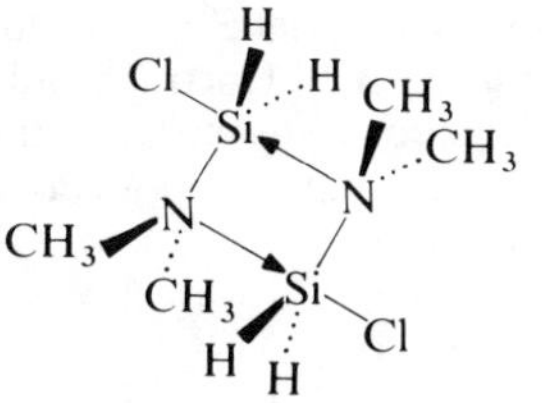
(**21**)

equatorial-type Si–N bonds. Molecular-orbital calculations at the STO-3G level by Deiters and Holmes[55] also show that as the nucleophile approaches in an apical position, the *trans* axial bond to the leaving group increases in length.

There is little conclusive evidence as to whether a pentacoordinate intermediate with a finite lifetime is involved in invertive processes. The fact that carbon nucleophiles do not lead to exchange with other carbon substituents suggests that if such an intermediate is formed it is not formed reversibly, unless its lifetime is too short to allow axial–equatorial exchange via pseudorotation. Sommer[2] suggests that if a pentacoordinate intermediate is formed, it is likely to be only a small minimum in the reaction coordinate diagram. Gas-phase studies of the reaction of trimethylchlorosilane with various nucleophiles using the flowing afterglow technique[56] show that methyl iodide and trimethylchlorosilane have a very different response of rate of reaction to changes in the basicity of the nucleophile. The authors suggest that the difference is so drastic that different mechanisms must be operating in the two systems and that substitution at silicon proceeds via a pentacoordinate intermediate.

b. Retention. Retention requires the nucleophile to attack from the same side as the leaving group and this can give rise to two different trigonal pyramidal activated states, **23** and **24**. **23** is the result of attack of the nucleophile at the R^1–X edge of the tetrahedron

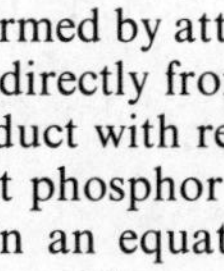
(**22**)

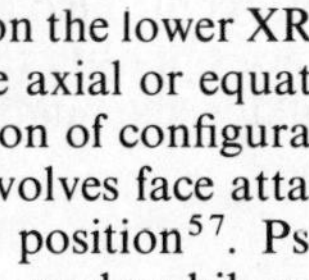
(**23**)

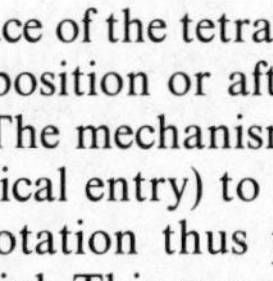
(**24**)

22 whereas **24** is formed by attack on the lower XR^2R^3 face of the tetrahedron **22**. Loss of the leaving group directly from the axial or equatorial position or after pseudorotation then gives the product with retention of configuration. The mechanism generally accepted for retention at phosphorus involves face attack (apical entry) to give **24**, where the leaving group is in an equatorial position[57]. Pseudorotation thus places the leaving group in the axial position and the nucleophile equatorial. This amounts to axial entry and axial departure, as may be expected from the principle of microscopic reversibility. However, as pointed out by Corriu[58], for asymmetric reactions where the entering and leaving groups are different, the reaction pathway does not necessarily have to be symmetrical. In general, for phosphorus, the first trigonal bipyramid **24** has the groups so arranged as to give the most stable intermediate on the basis of their relative apicophilicity, which in turn depends upon their electronegativity.

For silicon, the preference for apical positions is not simply related to electronegativity but depends on the polarizability or the ability of the Si–X bond to be stretched. The experimental order is Br, Cl, OAc > F, SR > H, OR[59] and OR > C_{sp^2} > H[60]. (This order has been questioned by Klebe[61], who found that fluorine was more apicophilic than chlorine.) With this in mind Corriu[62] had argued that such a mechanism of substitution cannot operate for silicon, since retention of monofunctional silanes would proceed via a higher-energy trigonal bipyramid with the apicophilic leaving group equatorial. For some time Corriu favoured edge attack, which places the leaving group apical (equation 9).

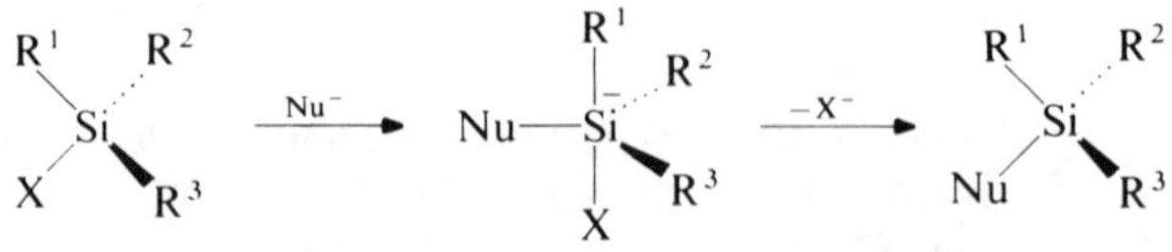
(9)

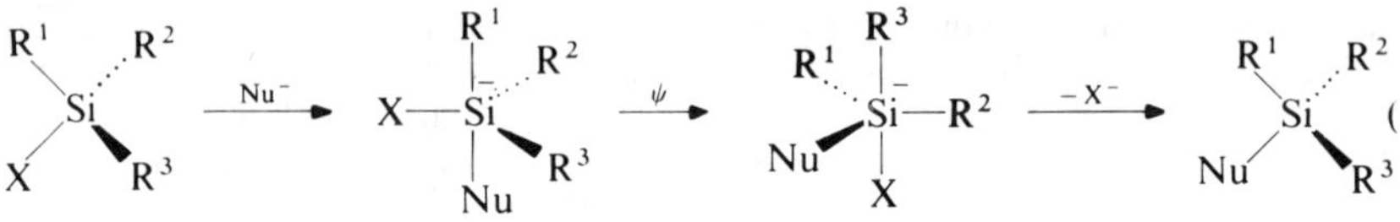
(10)

In support of this he argued that, with the alternative mechanism (equation 10), it is difficult to explain how disubstituted silanes such as 1-Naph SiFOMen (Men = menthoxide) lose both groups with retention[32,4,5]. Apical entry keeping one group apical would lead to **25** or **26**, where subsequent loss of the apical leaving group would lead to

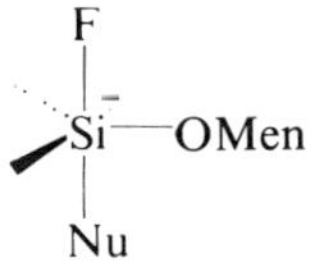
(**25**)

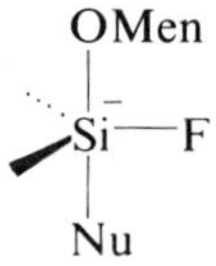
(**26**)

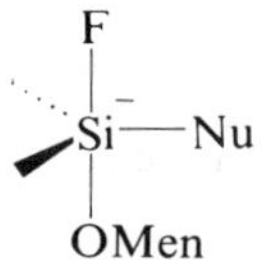
(**27**)

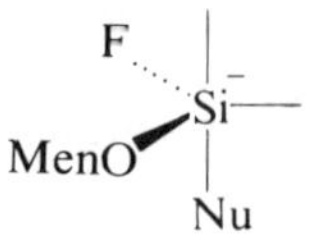
(**28**)

inversion. Only edge attack leading to **27** or face attack to **28** can account for the observed stereochemistry, and he argued that **28** will have too high an energy since both apicophilic groups are equatorial. However, Holmes[55] has pointed out that in related phosphorus reactions the use of bulky non-departing groups causes a change from inversion to retention. In these kinetically controlled reactions steric hindrance forces the nucleophile to attack the tetrahedral face opposite the largest ligand such that the two apicophilic groups X and Y end up equatorial (equation 11). Holmes further argues that

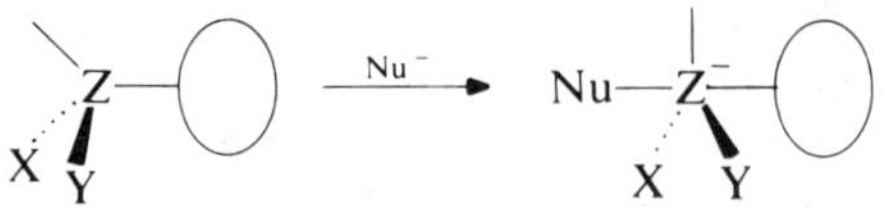
(11)

the naphthyl group may be bulky enough to achieve this. Corriu also quotes the work of Martin[52] in support of edge attack. The siliconates, **29**, are obtained by nucleophilic attack on the corresponding spirosilanes. The product has the two oxygen groups apical, with the entering nucleophile equatorial. Martin has shown that pseudorotation is easier in siliconates than in the isostructural isoelectronic phosphorane, **30** (for inversion $\Delta G^{\neq} = 26.0\,\text{kcal mol}^{-1}$ for **29**, Nu = Ph, and $\Delta G^{\neq} = 28.3\,\text{kcal mol}^{-1}$ for **30**). Thus fast axial

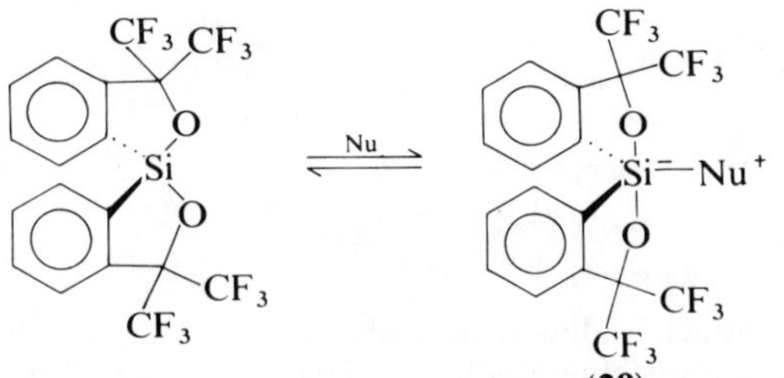

(29)

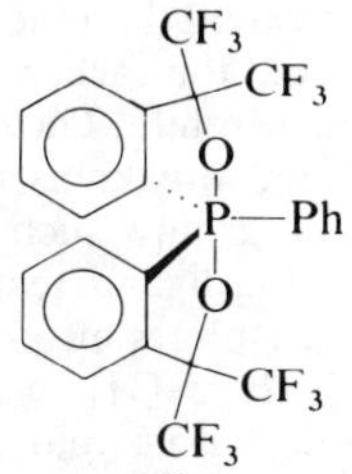

(30)

attack leading to a higher-energy isomer followed by pseudorotation to the lower-energy form **29** may account for these observations[63]. Minot[64] has performed calculations of the HOMO of a pentacoordinate D_{3h} intermediate (SiX_5^-) which suggests that the bond lengths in the equatorial and axial positions of SiH_5^- are nearly equal. He also points out that a soft reagent in the apical position is able to stabilize a negative charge, weakening the *trans* axial bond but strengthening the equatorial bonds. However, a hard reagent in the equatorial position weakens the perpendicular apical bonds whilst strengthening the *trans* equatorial bonds.

Recent molecular-orbital calculations by Deiters and Holmes[55] suggest that edge attack is very unlikely. They performed *ab initio* calculations at the STO-3G level for the reaction 12 using a number of different reaction pathways. The structures at points along the reaction coordinate were minimised using molecular mechanics programs. The four

$$SiH_3X + H^- \longrightarrow (HSiH_3X)^- \longrightarrow SiH_3H + X^- \quad \text{when } X = Cl, F \qquad (12)$$

pathways were:

(i) H^- attack at the HHH face (axial attack leading to inversion);
(ii) H^- attack at the HHX face (axial attack leading to retention);
(iii) H^- attack at the H–X edge (equatorial attack leading to retention);
(iv) H^- attack at the H–H edge (equatorial attack leading to inversion).

They found that edge attack had a higher energy than face attack. In fact, for edge attack, as the reaction progresses the incoming nucleophile slips over to a lower-energy tetrahedral face and follows the course of axial entry. Calculations also suggest that using the square-pyramidal planar form **31** as the transition state, the pseudorotation (equation 13) has a relatively low energy, such that axial entry, pseudorotation, axial exit is a viable pathway.

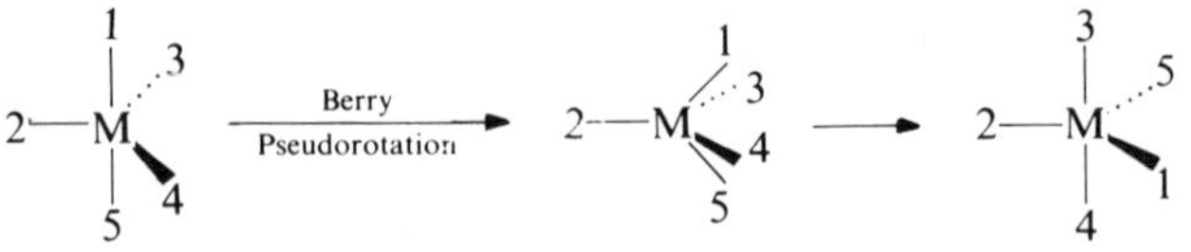

(13)

(31)

Furthermore, face attack leading to inversion and face attack leading to retention were similar in energy for fluorine, but for chlorine inversion was preferred. This again agrees with the experimental stereochemical behaviour of fluorine and chlorine. The results also show the importance of Si–X bond stretching. The Si–Cl bond stretched under the influence of the nucleophile to a larger extent than the Si–F bond and, for axial attack, the stretching is greater when Si–X was axial than when equatorial. The stretching was also dependent upon the nature of the nucleophile. They propose that with a good leaving group such as chlorine, the Si–Cl bond breaks as the Si–Nu bond forms, leading to inversion. However, with poor leaving groups such as fluorine or hydrogen, a tightly bound intermediate is favoured allowing ligand rearrangement, so that inversion and retention routes have similar energies. Holmes also found that the calculated apicophilicity of the leaving group ($Cl^- > SH^- > F^- > OH^- > H^-$) correlated with the experimentally determined tendency towards inversion; that is, groups with low apicophilicity tended towards retention. Furthermore, these apicophilicities were found to be dependent upon the nucleophile. Thus soft nucleophiles increase the apicophilicity of the leaving group and hence favour inversion processes.

Recent X-ray crystallographic studies[60] on **32** show the Si–H bonds to be both equatorial. Corriu[60] points out that these results show that equatorial attack certainly is

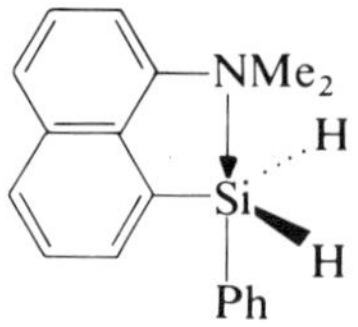

(32)

not the most favourable process and that nucleophilic displacement of the hydrogen with retention takes place by apical attack of the nucleophile with the Si–H bond equatorial.

The mechanism of retention, involving axial attack, and axial exit by definition involves an intermediate of sufficient lifetime to allow pseudorotation. Since carbon nucleophiles do not lead to exchange with other substituents in a retentive process where R^2 could also be lost, the energy difference between the activated complex for equation 14 and the

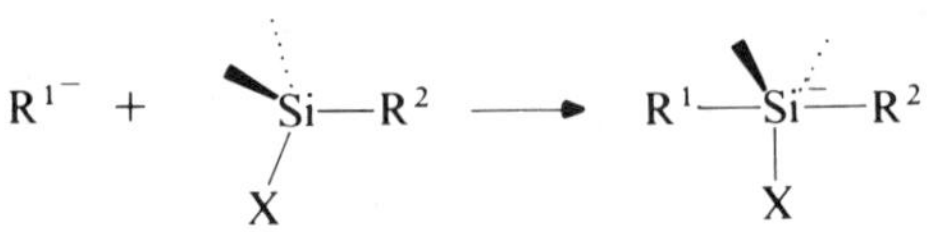

(14)

intermediate must be greater than that of pseudorotation. Recently, a number of interesting studies have been performed to investigate pseudorotation in pentacoordinate silicon species. We shall now review the available data and examine what evidence is provided as to whether pseudorotation is involved in the retentive process.

4. Pseudorotation in pentacoordinate silicon species

The pioneering work of Klanberg and Muetterties on pentafluorosilicates SiF_5^-, $RSiF_4^-$ and $R_2SiF_3^-$ suggested the possibility of pseudorotation at silicon[65]. Recently, there have been a number of studies on pseudorotation of pentacoordinate silicon species.

Corriu has performed a number of dynamic NMR studies in systems of the type **33**. Exchange of A, B and C lead to line broadening of atoms or groups that make up A,

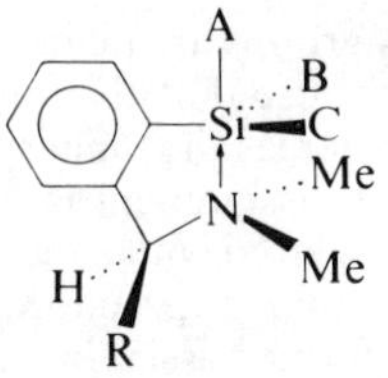

(**33**)

B and C. When A, B and C are different, the two methyl groups on the nitrogen are diastereotopic such that this ligand exchange also leads to broadening of these methyls as well as the benzylic protons (R = H). For example, at low temperature when no ligand exchange occurs, the ^{19}F and ^{1}H NMR data for **34** show the fluorine atoms, the methyl

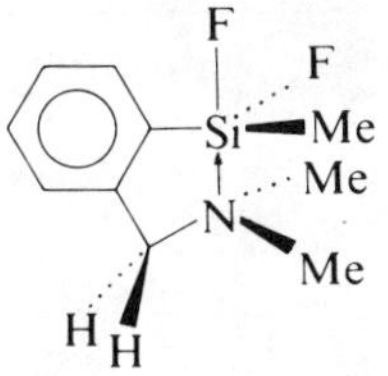

(**34**)

groups (of NMe_2) and the benzylic protons to be non-equivalent. However, on warming these groups become equivalent.

Ligand exchange could occur via pseudorotation or a dissociative process involving cleavage of the Si–N bond (equation 15). The use of compounds of the type **35** allows

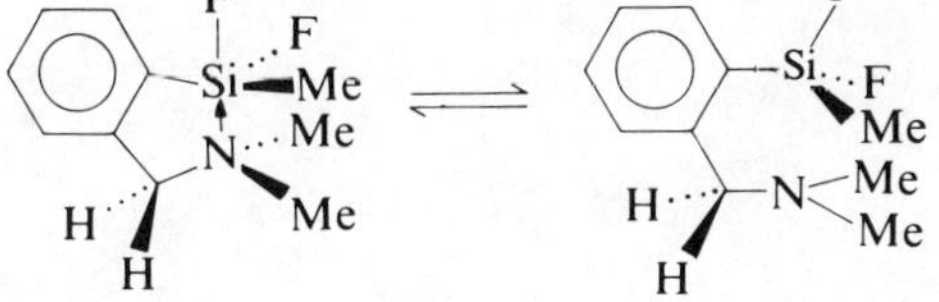

(15)

distinction between these two mechanisms[66]. The two methyl groups are diastereotopic owing to the chirality of the benzyl group. Pseudorotation will cause equivalence of the fluorines but not the two methyl groups. However, a dissociative pathway causes the fluorines and the methyl groups to be made equivalent. Variable-temperature studies using ^{19}F NMR show an equivalence mechanism with $\Delta G^{\neq}$ of 13.1 kcal mol^{-1}, whereas

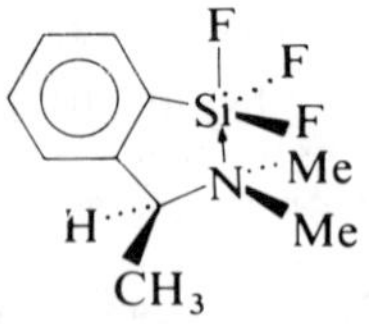
(35)

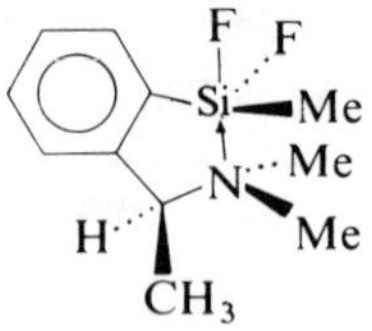
(36)

the ¹H NMR shows that the NMe_2 groups are made equivalent via a process with a ΔG‡ of 15.8 kcal mol^{-1}. The authors propose that the lower-energy pathway shown by 19F NMR arises from pseudorotation whereas the higher-energy process shown by ¹H NMR is due to a dissociative process. This assignment is consistent with studies on **36**. Since one of the fluorine groups has been replaced with a methyl, the ¹H NMR studies show both processes, one with a $\Delta G^{\neq}$ of 9.4 kcal mol^{-1} (pseudorotation) and the other with a $\Delta G^{\neq}$ of 11.8 kcal mol^{-1} (dissociative). The former assignment agrees with the $\Delta G^{\neq}$ of 9.3 kcal mol^{-1} obtained using ^{19}F NMR. In all these cases Si–F coupling is retained, showing

TABLE 7. Energy barriers to ligand exchange

Compound	$\Delta G^{\neq}$ (kcal mol^{-1})	$\Delta H^{\neq}$ (kcal-mol^{-1})	$\Delta S^{\neq}$ (eu)	Proposed mechanism	Ref.
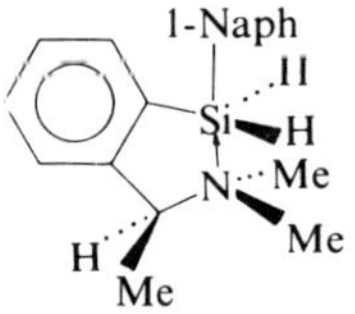	9.7	—	—	dissociative	68
	>10	—	—	pseudorotation	
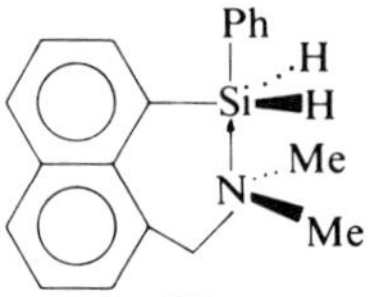 (37)	10.2	—	—	dissociative	68
	>10	—	—	pseudorotation	
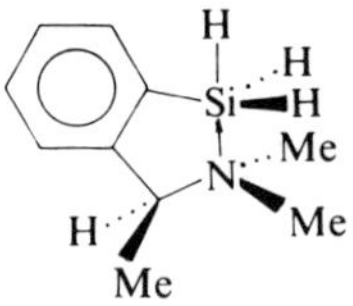	10	—	—	dissociative	68
	≪7	—	—	pseudorotation	
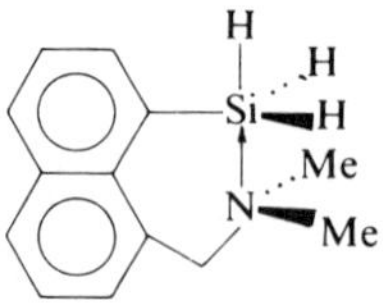 (38)	10.5	—	—	dissociative	68
	≪7	—	—	pseudorotation	

TABLE 7. (*continued*)

Compound	$\Delta G^{\neq}$ (kcal mol^{-1})	$\Delta H^{\neq}$ (kcal-mol^{-1})	$\Delta S^{\neq}$ (eu)	Proposed mechanism	Ref.
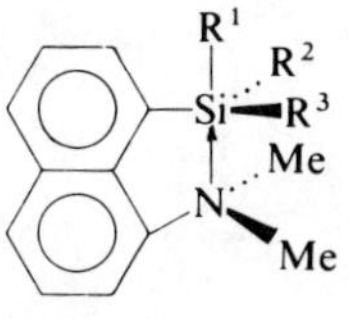					
$SiR^1R^2R^3$ =					
SiPhMeH	22	—	—	dissociative	69
	>22	—	—	pseudorotation	69
SiPhMeCl	20	—	—	dissociative	69
	>20	—	—	pseudorotation	69
SiPhMeF	23	—	—	dissociative	69
	>23	—	—	pseudorotation	69
SiPhMe(OMe)	22	—	—	dissociative	69
	>22	—	—	pseudorotation	69
SiPhHCl	>22	—	—	dissociative	69
	>22	—	—	pseudorotation	69
SiPhHBr	>21	—	—	dissociative	69
	>21	—	—	pseudorotation	69
SiMeHCl	22	—	—	dissociative	69
	>22	—	—	pseudorotation	69
$SiPhCl_2$	11	—	—	pseudorotation	69
$SiPhF_2$	12	—	—	pseudorotation	69
$SiPh(OMe)_2$	9	—	—	pseudorotation	69
$SiMeCl_2$	9	—	—	pseudorotation	69
$Si(OMe)_3$	<7	—	—	pseudorotation	69
$Si(OEt)_3$	<7	—	—	pseudorotation	69
SiF_3	12	—	—	pseudorotation	69
$Ph_2SiF_3^-$	11.7	11.2	2.5	pseudorotation	70
$PhMeSiF_3^-$	9.9	9.4	−4.4	pseudorotation	70
$PhSiF_4^-$	<7	—	—	pseudorotation	70
$F_2SiPh_3^-$	>?	—	—	pseudorotation	70
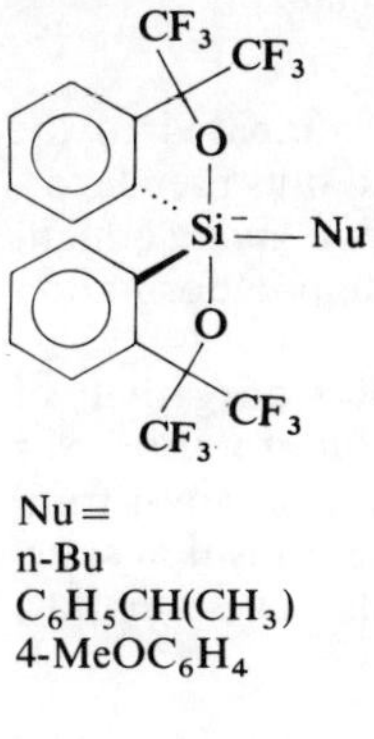					
Nu =					
n-Bu	28.6	—	—	pseudorotation	52
$C_6H_5CH(CH_3)$	29.0	—	—	pseudorotation	52
4-$MeOC_6H_4$	26.6	—	—	pseudorotation	52

TABLE 7. (*continued*)

Compound	$\Delta G^{\neq}$ (kcal mol^{-1})	$\Delta H^{\neq}$ (kcal-mol^{-1})	$\Delta S^{\neq}$ (eu)	Proposed mechanism	Ref.
C_6H_5	26.0	—	—	pseudorotation	52
3-$CF_3C_6H_4$	25.6	—	—	pseudorotation	52
3,5-$(CF_3)_2C_6H_3$	25.5	—	—	pseudorotation	52
C_6F_5	21.9	23.9	4.5	pseudorotation	52
C_6H_5O	20.4	—	—	pseudorotation	52
F	16.6	14.1	−8	pseudorotation	67
NC	16.8	—	—	pseudorotation	52
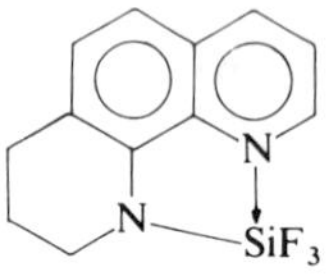 **(39)**	—	9.9	—	dissociative	61

that intermolecular fluorine exchange does not occur and the ^{29}Si NMR is consistent with a pentacoordinate species.

Barriers to ligand exchange have been measured in a number of cases and both pseudorotation and dissociative mechanisms have been observed, as shown in Table 7.

Damrauer and Danahey[70] examined pseudorotation in compounds of the type $[Ph_nSiF_{5-n}]^-$. Early studies by Muetterties concluded that no SiF or FF couplings could be observed above $-100\,^{\circ}$C, implying an intramolecular process at low temperatures and a dissociative process at higher temperatures. However, Damrauer and Danahey found that both FF and SiF couplings could be observed above $-100\,^{\circ}$C and ^{29}Si–^{19}F coupling could be observed at room temperature in the ^{29}Si NMR. A plot of $\log(k/T)$ versus $1/T$ is linear over the 110 °C temperature range studied, indicating no change in mechanism.

The barrier to pseudorotation seems to depend on the number of electronegative groups present. Corriu found that the barrier increased in the order

monofunctional	bifunctional	trifunctional
$\Delta G^{\neq} > 20$	$\Delta G^{\neq} \sim 9$–12	$\Delta G^{\neq} < 7$

The larger $\Delta G^{\neq}$ for the phenyldihydro compound **37** (see Table 7) compared to the trihydro compound **38** was ascribed to the phenyl being apical in **37** and thus pseudorotation requiring the less apicophilic hydrogen to go into the axial position giving a high-energy barrier. In **38** one hydrogen is always axial, and thus the barrier to pseudorotation is less[68].

Martin[52] found that the barrier for pseudorotation decreased with electronegativity of the group Y in **40**. He suggested that the transition state for the pseudorotation was similar to **41**, and hence there is an increase in the negative charge on Y as it moves from the equatorial position in the ground state to the apical position in the transition state. The use of Y groups that are more electronegative stabilizes the transition state, thus reducing $\Delta G^{\neq}$.

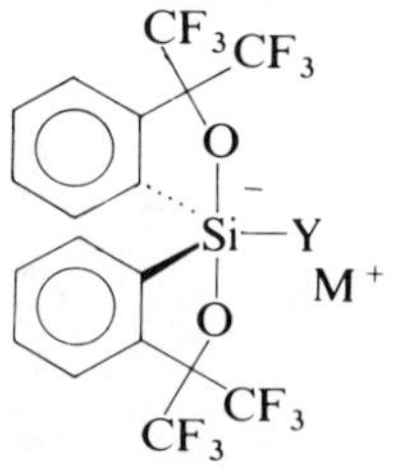

(40)

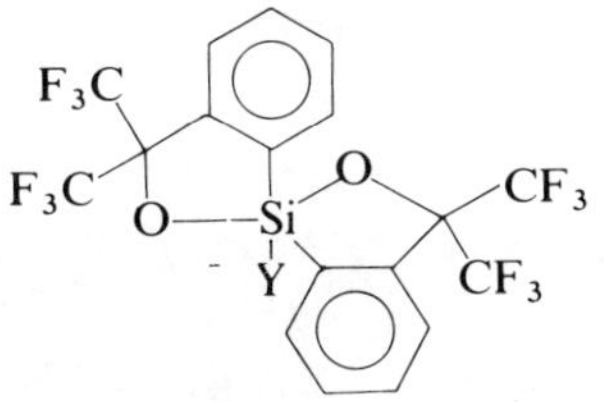

(41)

The energy barrier for a dissociative process depends upon the strength of the Si–X bond. For example, Klebe[61] proposes that in **39** a dissociative process with a low $\Delta H^{\neq}$ operates because the dative Si–N bond is weak.

These results indicate that pseudorotation is to be expected with silicon, and indeed Martin[52] has shown that it has a lower barrier than the corresponding phosphorus compounds. Thus axial entry, pseudorotation and axial exit which is generally accepted for phosphorus is a viable mechanism for retention at silicon. It must be remembered that the pseudorotation process proposed in this retentive mechanism will have a much lower energy barrier than many of those shown in Table 7. If this were not the case retentive mechanisms would not be stereoprecise and racemization via many pseudorotations would compete. Thus the pentacoordinate species **42** is formed in a higher-energy state with the apicophilic leaving group equatorial. One pseudorotation leading to **43** will not therefore have a high barrier. Racemization however requires a higher-energy intermediate of the form **44** with both apicophilic groups equatorial.

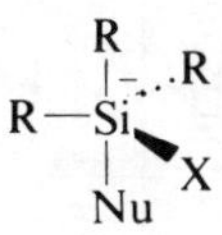

(42)

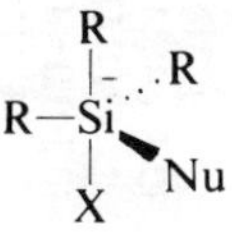

(43)

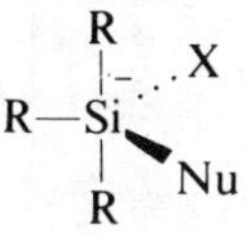

(44)

C. Nucleophile-catalysed Substitution at Silicon and Racemization

Nucleophilic substitution of silicon has frequently been found to be catalysed by the presence of other nucleophiles[71–80]. Eaborn and Modena[74a] found that hydrolysis and methanolysis of triisopropylsilyl chloride in aprotic media had an order in nucleophile of greater than one. Similarly Prince[73] found an order of 4.2 with respect to water for the hydrolysis of Ph_3GeCl in ether-acetone mixture. The hydrolysis and alcoholysis of chlorosilanes in aprotic solvents has been shown to be catalysed by the addition of chloride ions[74b] with the rate equation 16. Similarly Corriu[76–78] and Frye[79] have found

$$\text{rate} = k[R_3SiCl][ROH][Cl^-] \tag{16}$$

that the hydrolysis of a range of chlorosilanes is first order in nucleophilic catalysts such as pyridines, HMPT, DMSO and DMF as shown in Table 8. The reaction between bifunctional silanes and water, alcohols, thiols and amines is also catalysed by nucleophilic agents[72]. The stereochemistry of this type of reaction has been shown to be

TABLE 8. Kinetic orders for the nucleophile-catalysed hydrolysis of chlorosilanes

Substrate	Reactant	Nucleophile (and concentration range)	Solvent	Order	c.c.	Ref.
Ph_3SiCl	H_2O	HMPT ($0.5–3 \times 10^{-3}$M)	Anisole	1.05[a]	0.984	78
				1.19[b]	0.997	78
Ph_3SiCl	H_2O	DMSO ($0.3–1.4 \times 10^{-2}$ M)	Anisole	0.78[c]	0.999	78
Ph(1-Naph)MeSiCl	H_2O	DMF ($0.3–3.0 \times 10^{-2}$ M)	Anisole	0.64	0.990	78
Ph_3SiCl	H_2O	HMPT ($0.5–2.0 \times 10^{-3}$ M)	THF	0.81	0.990	78
Ph_3SiCl	H_2O	Cl^- (0.27–0.9 M)	dioxan	0.92[d]	0.985	74b
				0.76[e]	0.997	

[a] $[Ph_3SiCl] = 7.4 \times 10^{-3}$ M.
[b] $[Ph_3SiCl] = 9 \times 10^{-3}$ M.
[c] $[Ph_3SiCl] = 3.3 \times 10^{-3}$ M.
[d] $[H_2O] = 0.25$ M.
[e] $[H_2O] = 0.5$ M.

retentive, even with chlorosilanes where direct nucleophilic substitution usually gives inversion[76, 77]. These types of reactions are particularly important because of their use in forming acyloxy, alkoxy and amino silanes as protecting groups and synthetic reagents, as well as in the production of silicones.

A reaction related to the nucleophile-catalysed hydrolysis of silanes is the nucleophile-catalysed racemization of silanes. Corriu[81–86], Cartledge[87, 88] and Chojnowski[89] have examined the racemization of a range of halosilanes using nucleophiles such as HMPA, DMF, DMSO and Ph_3PO. The order with respect to nucleophile was found to vary from 1 to 3, as shown in Table 9. As well as displaying high order with respect to the nucleophile, these reactions are characterized by entropies of activation which are often large and negative and enthalpies of activation which are small and sometimes negative, as shown in Table 10. The relationship to nucleophilic-catalysed hydrolysis is further demonstrated by the similar activation parameters obtained for these reactions as illustrated by Table 11.

Racemization is slowed down as the steric crowding around the silicon increases[85] (Table 12). A similar trend is observed for hydrolysis[74b] of halosilanes. Whilst these two reactions are generally similar, different behaviour is sometimes observed. An examination of the electronic effect of phenyl substituents in Ar(1-Naph)(OMen)SiCl, Ar(1-Naph)EtSiCl[84] and α-ArFcBrSiCl[85] show that methyl and methoxy substituents have little effect on the rate of racemization whereas the CF_3 group effectively stops racemization. In contrast electron donation to silicon reduces the rate of the hydrolysis [74a].

In view of the overall similarity between racemization and hydrolysis, it is useful to discuss the two mechanisms together. Modena and Allen propose that hydrolysis and methanolysis of silyl chlorides involves base catalysis as shown in **45**[74b]. They found that

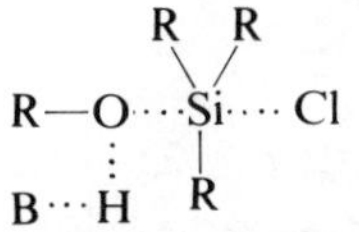

(45)

the rate decreased by a factor of two on going from H_2O to D_2O indicating that proton transfer is important in the transition state.

These conclusions have been criticised by Corriu[76–78] and more recently by Frye[79]. Such a mechanism does not explain the dramatic change in stereochemistry from inversion to retention that occurs when nucleophiles are added. The striking similarity between the rate equations and activation parameters for nucleophile-catalysed hydrolysis and racemization indicates a common mechanism; however, there is no pathway analogous to that of Modena that can be proposed for racemization. Frye[79] has shown that for a series of nucleophiles the rate of hydrolysis depends not on the basicity but the nucleophilicity of the nucleophile, as shown in Table 13. For example, pyridine and quinuclidine have virtually identicial k_1 values but very different base strengths, also the rate is greatly enhanced for the 'super-nucleophile'[91] dimethylaminopyridine. The nucleophile-catalysed alcoholysis of the bifunctional silane (1-Naph)PhSiHF eventually gives as the only monosubstitution product (1-Naph)PhSiFOR (equation 17).[2] Corriu maintains that if the mechanism of Modena operated, only the Si–F bond would be substituted as seen for other direct nucleophilic substitutions. The alternative mechanism that he proposes does not, however, explain why Si–H should be more labile than Si–F. Indeed, at the beginning of the reaction the amount of alkoxysilane, arising from Si–F cleavage, is substantial. However, this disappears as the reaction progresses such that only the fluoroalkoxysilane remains. This indicates that the Si–F bond is in fact more labile than the Si–H bond and can be interpreted in terms of fast reversible equilibrium

TABLE 9. Kinetic orders for the nucleophile-catalysed racemization of halosilanes[a]

Substrate	Nucleophile (and concentration range)	Solvent	Order	c.c	Ref.
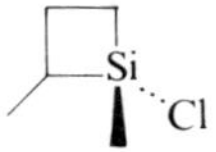	HMPA (2×10^{-5}–2×10^{-3} M)	CCl_4	1.24	0.999	88
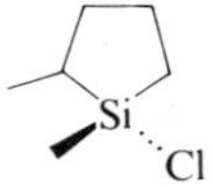	HMPA (3×10^{-4}–5×10^{-3} M)	CCl_4	1.55	0.998	88
1-Naph[i]PrFcSiCl	HMPA (1.14×10^{-2}–2.85×10^{-2} M)	benzene	2.31	0.996	88
1-NaphPhFcSiCl	HMPA (0.59×10^{-2}–2.85×10^{-2} M)	benzene	1.94	0.998	85
1-Naph(4-An)FcSiCl	HMPA (0.57×10^{-2}–7.85×10^{-2} M)	benzene	1.87	0.998	85
t-BuPh(1-Naph)SiBr	HMPA (1×10^{-1}–3.5×10^{-1} M)	CCl_4	2.19	1.000	83
EtPh(1-Naph)SiCl	DMF (0.8×10^{-1}–5×10^{-1} M)	CCl_4	2.05	0.996	82
(OMen)Ph(1-Naph)SiCl	HMPA (0.44×10^{-1}–2.49×10^{-1} M)	CCl_4	2.24	0.998	82
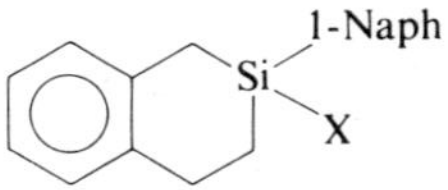	DMF (0.7×10^{-1}–2.5×10^{-1} M)	toluene	1.94	0.995	82

(PhCHMe)$SiMe_2Cl$	NMI (0.1–0.6 M)	silane	3.37	—	90
(PhCHMe)$SiMe_2Cl$	HMPA (0.6–0.4 M)	silane	2.33	—	90
(PhCHMe)$SiMe_2Br$	DMPU (0.01–0.145 M)	silane	0.86	—	90
	(0.008–0.13 M)	1:1 silane/CH_2Cl_2	1.17	—	90
(PhCHMe)$SiMe_2Br$	DMEU (0.09–0.45 M)	silane	1.70	—	90
(PhCHMe)$SiMe_2Br$	DMF (0.12–0.36 M)	silane	2.35	—	90
(PhCHMe)$SiMe_2Br$	NMP (0.2–0.4 M)	silane	2.02	—	90
(PhCHMe)$SiMe_2Br$	NMI (5.6×10^{-2}–9.6×10^{-2} M)	1:1 silane/CH_2Cl_2	1.64	—	90
(PhCHMe)$SiMe_2Br$	NMPO (0.015–0.058 M)	1:1 silane/CH_2Cl_2	1.60	—	90
(PhCHMe)$SiMe_2Br$	HMPA (0.005–0.016 M)	1:4 silane/CH_2Cl_2	1.48	—	90

[a] HMPA = hexamethylphosphortriamide; NMI = *N*-methylimidazole; DMPU = dimethylpropyleneurea; DMEU = dimethylethyleneurea; NMP = *N*-methylpyrolidinone; NMPO = 1-methyl-2-pyridone.

TABLE 10. Activation parameters for nucleophile-catalysed racemization

Substrate	Nucleophile	Solvent	E_a (kcal mol^{-1})	$\Delta H^{\neq}$ (kcal mol^{-1})	$\Delta S^{\neq}$ (cal mol^{-1} deg^{-1})	$\Delta S^{\neq}$ (kcal mol^{-1})	Ref.
t-BuPh(1-Naph)SiBr	HMPA	CCl_4	3.1	2.5	−56.5	16.8	83
i-PrPh(1-Naph)SiCl	HMPA	CCl_4	3.75	3.15	−55	19.6	82
EtPh(1-Naph)SiCl	HMPA	CCl_4	0.98	0.39	−57	17.3	82
EtPh(1-Naph)SiCl	DMF	CCl_4	0	0	≈ −70	20.6	82
(OMen)(*p*-An)(1-Naph)SiCl	HMPA	CCl_4	1.10	0.5	−62	19	82
	HMPA	CCl_4	11	10.5	−41	—	88
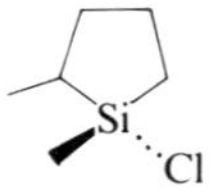	HMPA	CCl_4	6.2	5.5	−55	—	88

TABLE 11. Activation parameters for nucleophile-catalysed hydrolysis

Substrate	Catalyst	Solvent	$\Delta H^{\neq}$ (kcal mol^{-1})	$\Delta S^{\neq}$ (cal mol^{-1} deg^{-1})	Ref.
Ph_3SiCl	HMPA	Anisole	−3.4	−56	76
MeNaphPhSiCl	DMSO	Anisole	1.4	−43	76
MeNaphPhSiCl	DMF	Anisole	2.6	−40	76
(OMen)NaphPhSiCl	HMPA	Anisole	−2.9	−46	76

TABLE 12. Rate constants for the racemization of various chlorosilanes in HMPA

Substrate	(1-Naph) FcMeSiCl	(1-Naph) FcEtSiCl	(1-Naph) Fc^{i}PrSiCl	(1-Naph) *t*-BuFcSiCl	(1-Naph) Ph^{i}PrSiCl
rate constant (mol^{-2} l^{2} min^{-1})	instant racemization	instant racemization	0.58	2.8×10^{-3}	4.5

TABLE 13. The variation of the rate of alcoholysis of Ph_2SiCl_2 with different nucleophile catalysts

Nucleophile	pK_B	$k_1$rel[a]	$k_2$rel[b]
Et_3N	3.36	0.06	<0.25
quinuclidine	3.42	90	1
4-dimethylaminopyridine	4.3	190,000	13,300
2.6-lutidine	7.2	9	4
3,5-lutidine	7.8	300	100
pyridine	8.8	100	100
N,*N*-dimethylaniline	9.6	0.8	0.5

[a] Relative rate of hydrolysis of the first step $Ph_2SiCl_2 + ROH \longrightarrow Ph_2SiClOR + HCl$.
[b] Relative rate of hydrolysis of the second step $Ph_2SiClOR + ROH \longrightarrow Ph_2Si(OR)_2 + HCl$.

formation of the alkoxysilane in conjunction with a slower irreversible formation of the fluoroalkoxysilane.

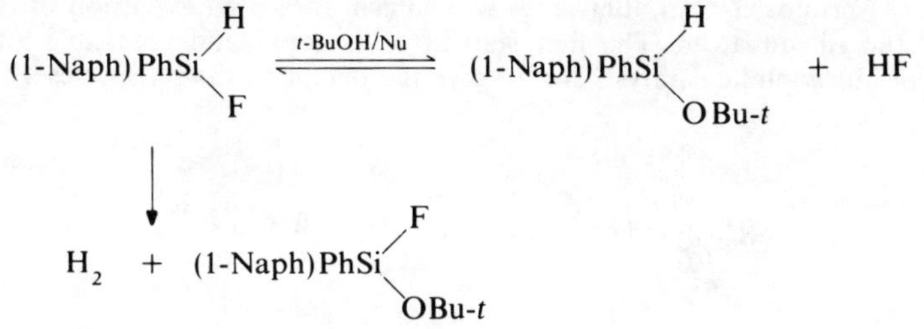

(17)

Sommer[92, 93] has proposed that the methanol-catalysed racemization of 1-napthylphenylmethylfluorosilane involves formation of a pentacoordinate intermediate followed by pseudorotation (equation 18). Three successive pseudorotations are required for

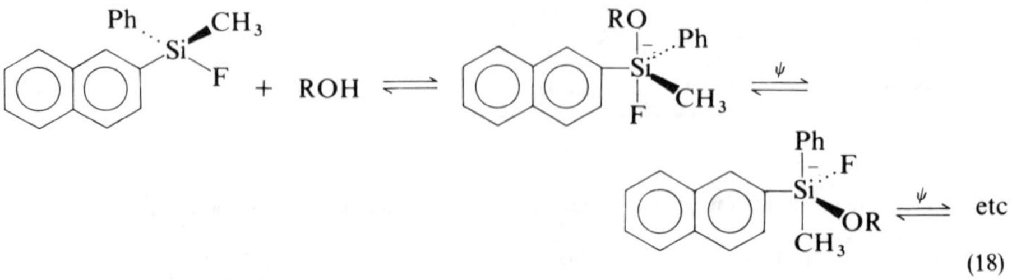

(18)

racemization, followed by loss of the solvent from the axial position. The details of such pseudorotations have been dealt with earlier. Kumada[94] has proposed a similar mechanism for the isomerisation of *trans*-diiodosilacyclohexanes.

However, it has been pointed out[81–83] that the high order with respect to nucleophile, as given in Table 9, is not compatible with such a mechanism. Indeed, Sommer reported the reaction to be third order in nucleophile but rationalized this in terms of ion-pair formation and solvation effects. Corriu has also pointed out that the energy of activation of racemization (Table 10) is smaller than that expected for pseudorotation (Table 7)[82].

Martin[52] has shown that racemization by pseudorotation is important for **46**. This reaction has a first-order dependence on the nucleophile and pseudorotation has been

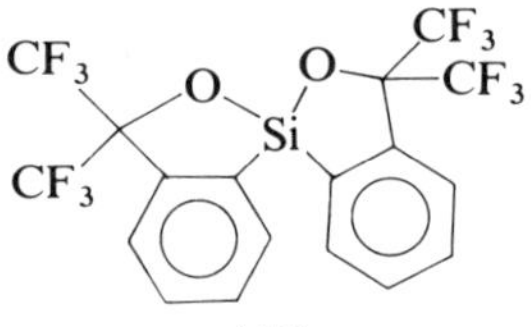

(**46**)

shown to occur for a series of siliconates **40**. The racemization of **46** is hindered by bulky nucleophiles and has a moderately negative entropy of activation in accord with an ordered intermediate/transition state like **40**. Indeed, at low temperature the five-coordinate species can be observed with some nucleophiles. A mechanism involving heterolysis of the Si–O bond can be ruled out on the basis of substituent effects of the ligand and the lack of a solvent effect. The intermediacy of a six-coordinate species can be ruled out on the basis of the negligible effect of different nucleophile concentrations on the energy barrier to inversion at low temperatures.

Corriu has proposed[72] an alternative mechanism involving extension of the coordination of the silicon atom. The first step involves rapid and reversible attack by a molecule of nucleophilic catalyst, Nu, to give the pentacoordinate species **47** (equation

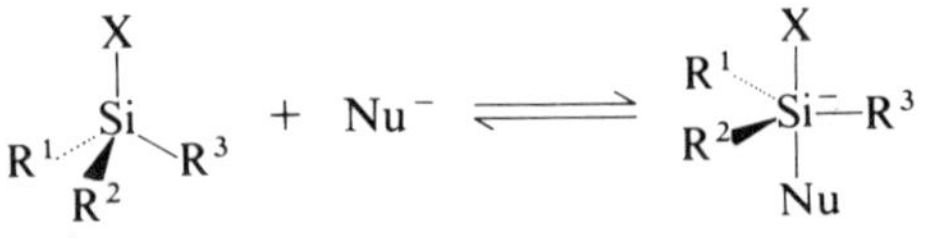

(**47**) (19)

19). The slow step in the process is then the attack by a second nucleophile at an R–X edge to give a hexacoordinate intermediate or transition state. The second nucleophile is the alcohol in the case of alcoholysis or the nucleophilic catalyst in the case of racemization (equations 20 and 21). For alcoholysis, collapse of this activated state then

racemization

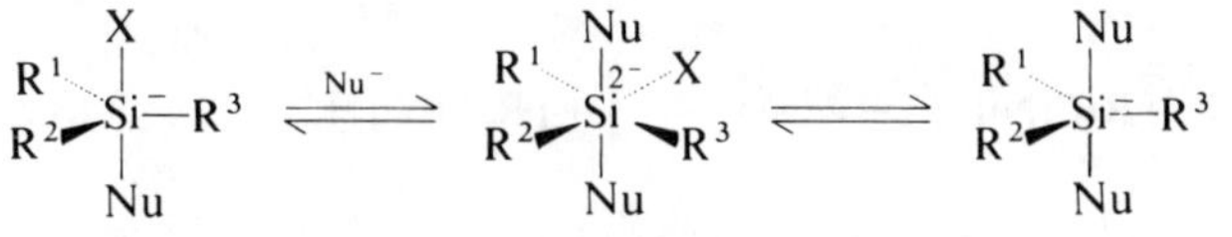

(47) (48) (49)

intermediate or transition state (20)

alcoholysis

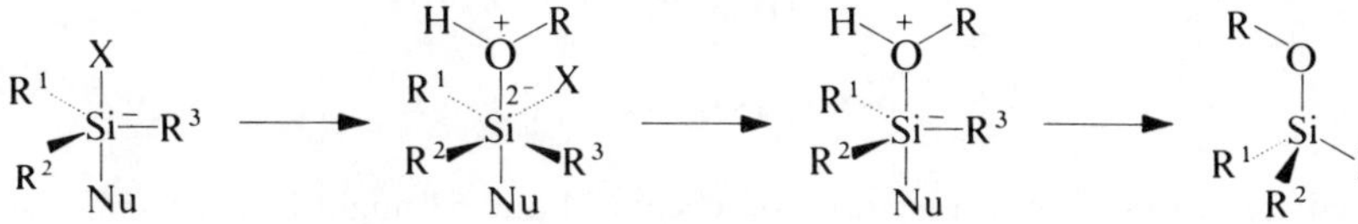

(47) (50) (51) (21)

intermediate or transition state

leads to the substitution product with retention. For racemization, since **48** and **49** both have a plane of symmetry, the reversible formation of such species leads to inversion and retention. Whether racemization involves solely the intermediate **48** or whether this is a transition state leading to **49** as the only intermediate is undecided, and evidence in favour of the hexacoordinate silicon[72,4] and the pentacoordinate silicon has been presented[88].

Evidence in favour of this mechanism involving extra-coordinate intermediates comes from a variety of sources. As well as explaining the stereochemistry, this mechanism is also in agreement with the kinetics. Corriu has found that racemization is second order with respect to the nucleophile (see entries 3 to 9 in Table 9) whereas alcoholysis is first order in the alcohol and nucleophile. Both of these rate equations are in agreement with the pre-equilibrium formation of an intermediate followed by rate-limiting attack by a second nucleophile. The small values obtained for $\Delta H^{\neq}$ and the large negative values of the entropy of activation show the reaction to be entropy controlled with a highly organized transition state, as would be expected for a hexacoordinate transition state/intermediate. The retardation of racemization by a CF_3 substituent on the phenyl ring of Ar(1-Naph)EtSiCl and Ar(1-Naph)(OMe)SiCl has been explained in terms of the CF_3 group attenuating the polarizability (lengthening) of the Si–Cl bond making the silicon in **47** less electrophilic[84].

Cartledge[87,88] has used halosilacyclobutanes and pentanes as probes to distinguish racemization via **48** from racemization via **49**. He observed rapid halide exchange between halosilanes in the presence of nucleophiles (equation 22) and proposes that this occurs via ionic intermediates of the form shown in reaction 23. If halide exchange and racemization, which occur alongside each other, proceed via similar intermediates, this would suggest **49** to be the key intermediate in racemization. Corriu[72], however, prefers the hexacoordinate complex, explaining halide exchange between (1-Naph)PhSiHF and Me_3SiCl as proceeding via **54** with concerted transfer of chlorine and fluorine. If **49** is the

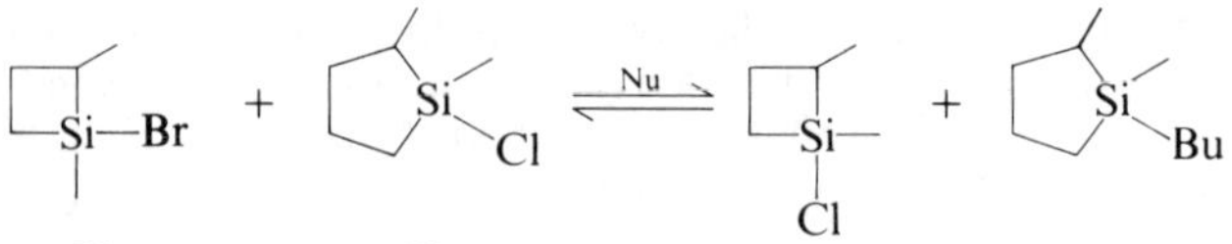 (22)

(52) (53)

$$X^-[R_3SiNu_2]^+ + R^1_3SiY \longrightarrow [R_3SiNu_2]^+\ Y^- + R^1_3SiX \tag{23}$$

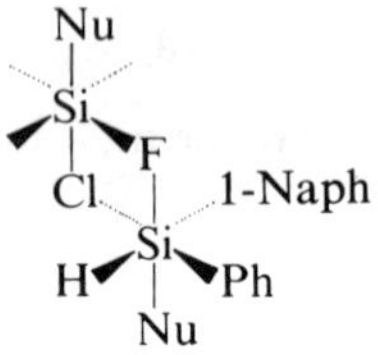

(54)

intermediate through which racemization occurs **52** and **53** might be expected to show decreased rates of isomerization relative to acyclic or large-ring chlorosilanes, since the C–Si–C angle in **49** is about 120° but in **52**, it is 80°, and in **53**, 96°. However, if **48** is the intermediate, greater rates should be obtained since the C–Si–C angle is about 90°.

The details of the plots of log k_{obs} vs log [Nu] are given in Table 9, where k_{obs} is the first-order rate constant with respect to substrate. A plot of k_{obs}/[Nu] vs [Nu] for **53** is a straight line with a positive intercept on the *y*-axis. Cartledge interpreted this in terms of competing first and second-order processes (equation 24). For the four-membered ring compound **52** the plot of k_{obs}/[Nu] vs [Nu] is curved. This was interpreted in terms of a reaction that is second order at low concentration of the nucleophile but first order at high concentrations. This agrees with the rate equation 26 obtained using the steady-state approximation in equation 25. However, no such change in the slope from 2 to 1 is observed in the plot of log k_{obs} vs log [Nu].

$$k_{obs} = k_a[\mathrm{Nu}] + k_b[\mathrm{Nu}]^2 \tag{24}$$

$$\mathrm{Nu} + R_3SiCl \underset{k_{-1}}{\overset{k_1}{\rightleftharpoons}} R_3SiCl(Nu) \xrightarrow[\mathrm{Nu}]{k_2} R_3SiCl(Nu)_2 \tag{25}$$

$$k_{obs} = \frac{k_1 k_2[\mathrm{Nu}]^2}{k_{-1} + k_2[\mathrm{Nu}]} \tag{26}$$

The first-order process in equation 24 for **53** was interpreted as a pseudorotation mechanism. It was proposed that the second-order pathway proceeds through penta-coordinate species as in equation 27. The four-membered ring compound was 800 times faster than the five-membered ring compound which, in turn, was about 700 times as reactive as the corresponding acyclic compound. Since the ring is diequatorial in **55**, it is

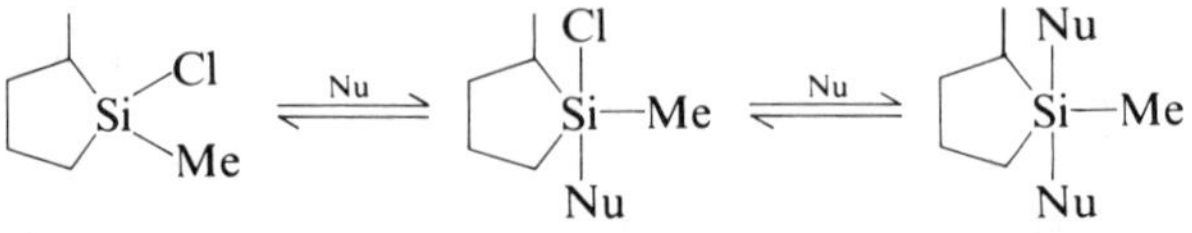 (27)

(55)

difficult to account for the faster rate of **53** compared to acyclic compounds. For **52**, the mechanism shown in equation 28 was proposed. Here, formation of the 90° axial/equatorial angle in the trigonal bipyramid leads to rate acceleration. The kinetic analysis also suggests that the second step is accelerated by ring strain even more than the initial formation of a pentacoordinate intermediate. Compound **56** with both apicophilic groups

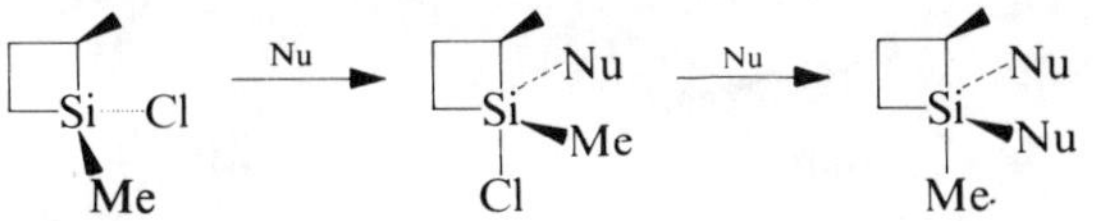

(28)

(56)

equatorial is not, however, likely to be the most stable form of this intermediate. Frye has criticized this type of mechanism[79] and stated:

'In our opinion, however, his (Corriu's) conclusions that initially formed pentacoordinate silanes then react with solvolysing agents via hexacoordinate transition states or intermediates are untenable since they essentially beg the question of the marked rate enhancements; i.e., the silicon center of a pentacoordinate adduct should be markedly less susceptible to attack by a nucleophilic solvolysing agent than the original tetracoordinate silane for at least two very good reasons: 1. Attack at the silicon will be encumbered by five ligands instead of only four. 2. The extra-coordinate silicon will bear a formal negative charge because of the formation of an essentially dative bond with the nucleophile.

Thus, both of the above considerations would be expected to sharply retard subsequent attack by any other nucleophile. The well-known enhanced solvolytic stability of the alkoxy- and halo-silatranes[9] strongly attests to the validity of this assertion'.

This latter criticism deserves further discussion. The uncatalysed and catalysed mechanisms of hydrolysis are shown in equations 29 and 30 respectively. For Corriu's

$$\underset{\textbf{(57)}}{ROH + R'_3Si\text{–}X} \xrightarrow{k_1} ROSiR'_3 + HX \tag{29}$$

$$Nu + R_3SiX \underset{}{\overset{K}{\rightleftharpoons}} \underset{\textbf{58}}{R_3SiXNu} \xrightarrow[ROH]{k_2} ROSiR_3 + HX \tag{30}$$

mechanism to operate, not only must the pentacoordinate species **58** be more electrophilic than the tetracoordinate starting materials, **57**, but the concentration of the intermediate, albeit small, should be reasonable such that $Kk_2\,[Nu] > k_1$. Since the equilibrium formation of the pentacoordinate species is likely to be well to the left, the nucleophilic enhancement of the pentacoordinate species compared to the tetracoordinate species must be substantial. It is therefore unlikely that the chloride-ion-catalysed hydrolysis of chlorosilanes[74] proceeds via such a mechanism because the concentration of the species **59** would be too small.

(59)

In favour of this mechanism Corriu has found that **60**, which exists predominantly in the pentacoordinate form, reacts with alcohols and acids to give mono or disubstitution

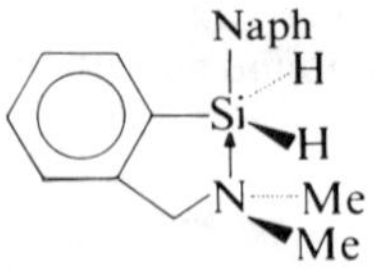

(**60**)

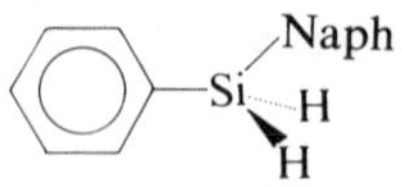

(**61**)

products whereas the corresponding four-coordinate species **61** does not react at all[96]. Martin has examined the reactivity of pentacoordinate species of the form **40**[52]. Treatment of **46** with a large excess of a range of nucleophiles gave only the pentacoordinate product with no evidence of a hexacoordinate intermediate or transition state. For example, racemization of **40** has been shown to proceed via pseudorotation rather than the hexacoordinate species[52].

Nu = pyridine, $Bu_4N^+F^-$, $Ph_4N^+OH^-$, $C_6H_5^-Li^+$

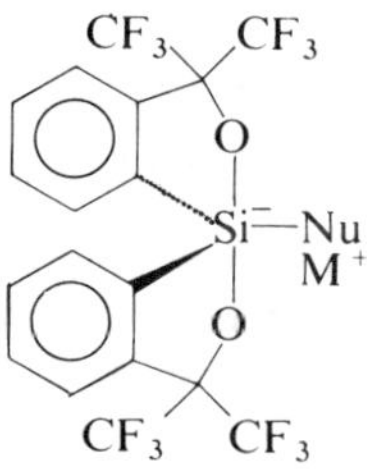

(**40**)

Martin, however, points out that many other silanes readily form hexacoordinate species and indeed, with SiF_4, some nucleophiles form only this type of compound[97]. He argues that **46** is reluctant to form hexacoordinate species because **40** has two electronegative groups apical and three electropositive groups equatorial. Hexacoordinate species have optimum stabilization when all ligands are electronegative, for example, with four such groups **62** can be prepared although it is rather unstable[98].

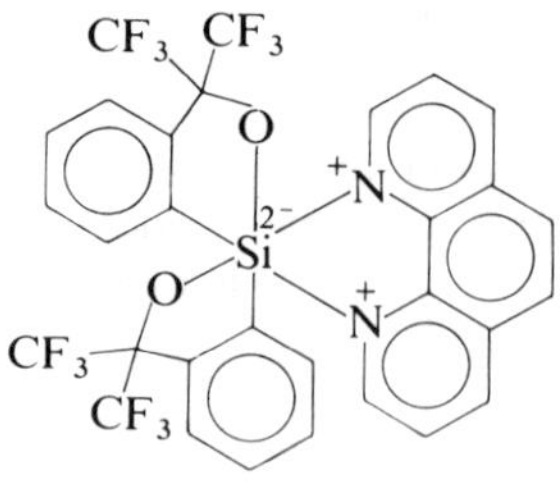

(**62**)

Recently, Corriu[99,100] investigated the reactions of nucleophiles with compounds of the form **63** which contain four electronegative groups. Reactions with $LiAlH_4$, i-Bu_2AlH

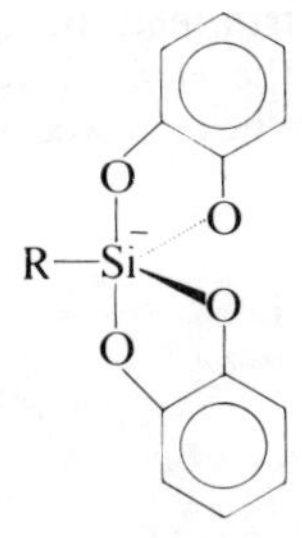

(63)

or Et_3Al/NaH give $RSiH_3$, but Grignard reagents and organolithiums are more selective, and the products depend upon the amount of organometallic reagent (equations 31 and

$$\mathbf{63} + 2\ R'Li \xrightarrow[LiAlH_4]{} RR'_2SiH \tag{31}$$

$$\mathbf{63} + \text{excess } R'MgBr \longrightarrow RSiR'_3 \tag{32}$$

32). Comparison of the reactivity of **63** with that of **64** indicates the pentacoordinate species to be more reactive towards nucleophiles than the four-coordinate species. Even

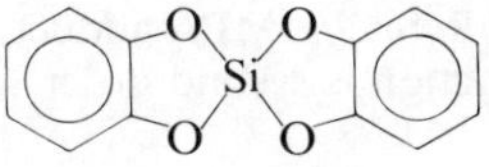

(64)

the six-coordinate species **65** has been shown to be reactive towards nucleophiles (equation 33).

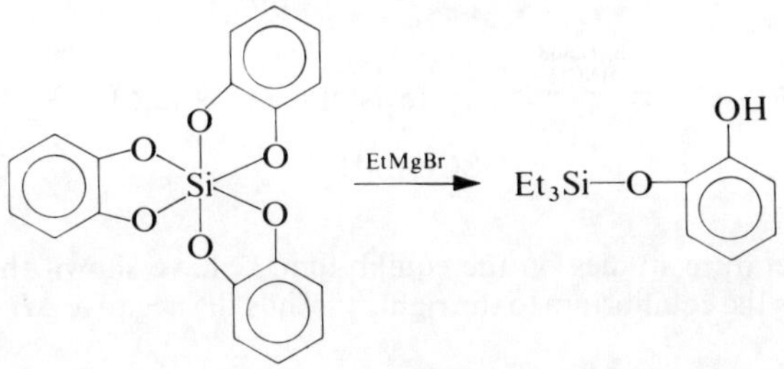 (33)

Further support for the enhanced electrophilicity of silicon in five-coordinate species compared to four-coordinate species comes from X-ray fluorescence spectra, which reveal a higher positive charge on the silicon atom of silatranes $XSi(OCH_2CH_2)_3N$ compared to triethoxy silanes $XSi(OCH_2CH_3)_3$ which is in contradiction to the widely accepted ideas[101]. Molecular-orbital calculations by Baybutt[102] show that the positive charge on silicon increases by 0.16–0.20 *e* on increasing the coordination from 4 to 5 (equation 34, X=H). Voronkov[103] estimates a larger change for X=H, 0.303 *e*, which decreases for X=F(+0.128 *e*) and X=Cl(+0.112 *e*).

$$XSiH_3 + H^- \longrightarrow [X-Si(H_3)-H]^- \tag{34}$$

Chojnowski[104] has proposed a fourth mechanism for nucleophile-catalysed racemization and substitution. He isolated the HMPA adducts of trimethylbromosilane and trimethyliodosilane. Conductivity studies showed them to be ionic with the proposed

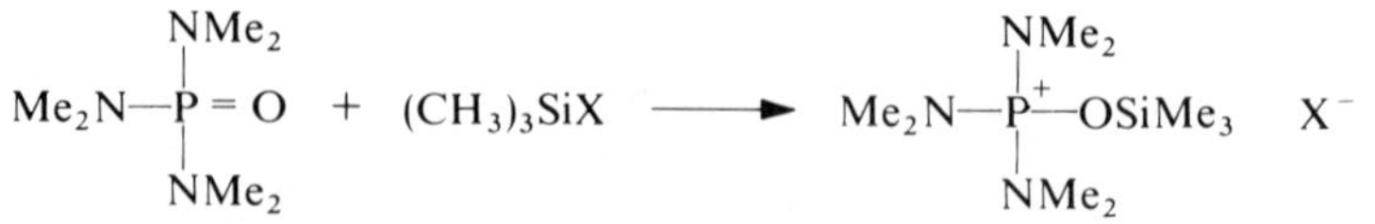

(35)

(65)

structure **65** (equation 35). A similar result was obtained by Beattie from trimethylphosphine oxide and trimethyliodosilane[105] and similarly by Ebsworth[16]. Chojnowski proposed that HMPA-catalysed substitution of halogen at silicon may involve the transient formation of a phosphonium cation containing a four-coordinate silicon. The retention of configuration observed in hydrolysis would then be the result of two consecutive inversions (equation 36). Since the first step is a pre-equilibrium, the rate would be first

$$R_3SiCl \underset{}{\overset{\text{inversion}}{\underset{\text{HMPA}}{\rightleftharpoons}}} R_3SiO{-}\overset{+}{P}(NMe_2)_3 \quad Cl^- \xrightarrow[\text{R'OH}]{\text{inversion}} R_3SiOR' \qquad (36)$$

order with respect to the concentration of HMPA and ROH in agreement with established rate equations. Racemization involves rate-limiting invertive attack of a second HMPA molecule on the R_3SiCl/HMPA adduct, **66** (Scheme 1). Again, if the first step is a pre-equilibrium the reaction is second order with respect to the nucleophile.

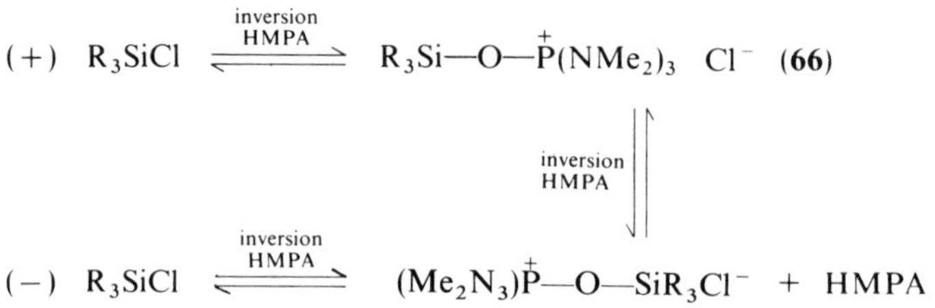

SCHEME 1

Variable-temperature studies on the equilibrium 37 have shown that decreasing the temperature shifts the equilibrium to the right[107]. Thus the negative ΔH [-10 kcal mol^{-1}

$$R_3SiX + Nu \overset{K}{\rightleftharpoons} [R_3SiNu]^+ + X^- \qquad (37)$$

for $(CH_3)_3SiCl$ and trimethylsilylimidazole] for this reaction reduces the $\Delta H^{\neq}$ obtained for the overall reaction (equation 38) leading to the low values give in Tables 10 and 11.

$$R_3SiX + Nu \overset{k}{\rightleftharpoons} [R_3SiNu]^+ + X^- \xrightarrow[\substack{Nu \\ (inv)}]{k_2} [NuSiR_3]^+ X^- \qquad (38)$$

$$k_{obs} = Kk_2[Nu]^2$$

The high negative entropies of activation can be explained in terms of the highly ordered arrangement of solvent molecules around the ionic transition state. As an indication of this, ΔS for reaction 37 (X=Cl, Nu=trimethylsilylimidazole) was found to be −41 eu[107].

Bassindale[107] has examined the ^{29}Si, ^{13}C and ^{1}H NMR spectra for a range of 1:1 mixtures of Me_3SiX with DMF in CD_2Cl_2 and found four-coordinate adducts resulting from O-silylation to be formed when X=I, OSO_2CF_3 and Br. No such adducts were formed with the weak nucleophile CD_3CN.

Similarly, a wide range of nucleophiles have been shown to form four-coordinate adducts with $Me_3SiOSO_2CF_3$ (Nucleophile=pyridines, 2-pyridone, pyridine–*N*–oxide, phosphine oxides, ureas, imidazoles)[108]. Competition studies between Me_3SiX and *N*-trimethylsilylimidazole led to the order of adduct formation[109,110].

$$X = ClO_4 > CF_3SO_3 > I > Br \gg Cl \gg OAc$$

A similar order has been observed using conductiometric titrations[107]. For X = I and OSO_2CF_3, the equilibrium lies substantially on the side of adduct formation. For X = Br the equilibrium constant is smaller. However, for X = Cl the equilibrium lies on the side of reactants. Competition studies using different nucleophiles led to the order *N*-methylimidazole > 4-*N*,*N*-dimethylaminopyridine > HMPA > 1,2-dimethylimidazole > *N*-methyl-2-pyridone > pyridine-*N*-oxide > triphenylphosphine oxide > dimethylpropyleneurea > 2,4-dimethylpyridine > *N*-methyl-2-pyridone > DMF > 3,5-dimethylpyridine > dimethylethyleneurea > pyridine > trimethylamine. This order follows the nucleophilicity as suggested by the Kamlet-Taft β constants[111] and k_{rel} found by Frye for the alcoholysis of Ph_2SiCl_2[79], clearly demonstrating the relationship between the ability to form adducts and the ability to catalyse nucleophilic substitution. The X-ray crystal structure of the salt **67** has been obtained showing it to be four-coordinate[112]. Whilst most of these complexes are easily solvolysed, Frye and Klosowski have prepared the solvolytically stable tetrasilaadamantyl derivative **68**[113]. Recent studies[109] have shown that, as expected, salts **69** are highly susceptible to nucleophilic attack.

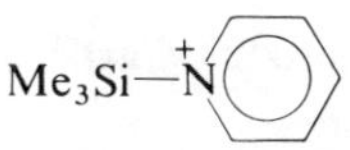

(67)

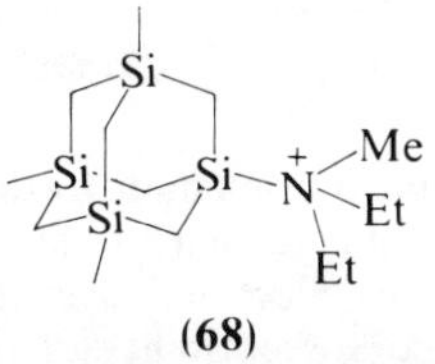

(68)

(69)

Thus the formation of nucleophilic adducts of electrophilic silanes is clearly established. Corriu[114] has re-examined the work of Chojnowski. He found that triorganohalogermanes and stannanes, which are more easily racemized than triorganohalosilanes, do not give ionic adducts with HMPT, indicating that between the elements racemization is not directly related to adduct formation. He also found that the crystalline adducts of Ph_3SiCl with HMPT and Me_3SiBr with DMF do not conduct electricity and thus suggested they were not ionic. Bassindale[107, 108], however, attributes this to tight ion pairing in the non-polar CD_2Cl_2. The ^{1}H chemical shift of the H2 of imidazole in salts of the form **69** was found to be highly variable depending upon the solvent and counterion, and this was attributed to different extents of ion pairing. In more polar solvents such as CD_3CN, the salts are less ion paired and the chemical-shift variation is smaller. The chemical shifts are in agreement with the order for ion pairing.

$$Cl > Br > I > OSO_2CF_3 > OClO_3$$

Corriu[114] suggests that Chojnowski's mechanism may only operate with very labile Si–Br and Si–I bonds and good nucleophiles. It should be noted that the difficulty in observing ionic adducts of the form $R_3Si\overset{+}{Nu}\ X^-$ does not in itself disprove this mechanism. Such adducts could be reactive intermediates formed in low concentrations.

Corriu also criticizes this mechanism as it cannot account for the nucleophile- induced epimerization of chlorocyclobutanes[114]. As discussed earlier, in strained systems retentive substitution predominates, such that two consecutive such substitutions by HMPA would only lead to retention. However, whilst retention is the major pathway, inversion may still occur as a minor route. Whereas retention would not reveal itself, one invertive substitution in the equilibrium 39 would lead to racemization. Such a pathway is

$$R_3SiX + Nu \rightleftharpoons R_3SiNu^+ + X^- \qquad (39)$$

doubtful because the reaction would be first order in nucleophile and it is hard to explain why the cyclobutane **70** would be so much more reactive than the corresponding acyclic species. A more plausible possibility is pseudorotation of the intermediate formed in the second step of mechanism 40. Retentive attack on **71** would normally lead to **72**.

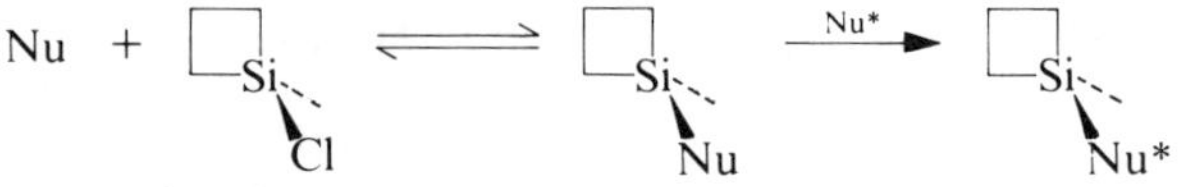

(40)

(**70**) (**71**)

Pseudorotation is then followed by loss of Nu to give back a similar molecule **71**. However, if **72** underwent pseudorotation to **73** racemization would occur. Whilst this

(**72**) (**73**)

would be a minor route, this is the same intermediate as that proposed by Cartledge[87, 88] (equation 28), and thus both mechanisms have a similar standing.

Recently, Chojnowski[89] examined the effect of dielectric constant on the racemization of (1-Naph)PhMeSiCl. A pronounced susceptibility of the reaction to changes in the polarity of the medium was observed. This would be consistent with mechanisms involving intermediates of the form **65** or **49** but not **48**. He also examined the rate of transilylation (equation 41), which will involve concurrent racemization of the chiral

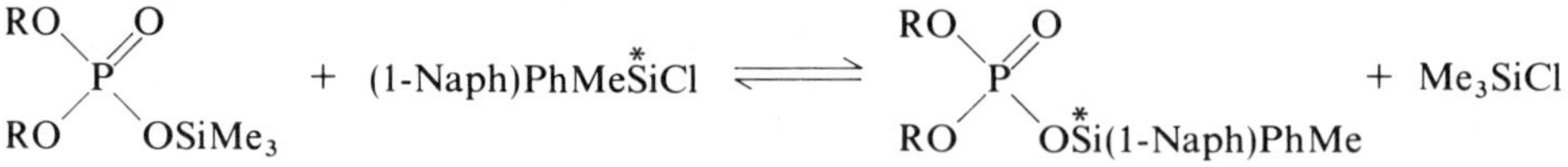

(41)

silane. The proposed mechanism is shown in Scheme 2. Equilibrium formation of the intermediate **74** requires loss of silicon in the back step of reaction one to be faster than substitution by a second phosphate molecule in the second step (i.e. $k_2[(RO)_2P(O)OSiMe_3] \ll k_{-1}$). Since k_3 involves a similar loss of silicon, it is likely that

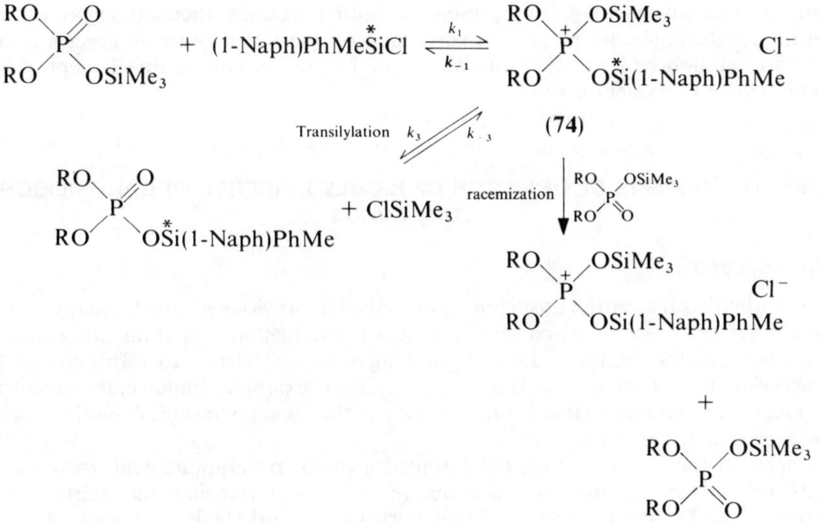

SCHEME 2

$k_3 \gg k_2[(RO)_2P(O)OSiMe_3]$. Hence if this mechanism were operating, transilylation should be faster than racemization. In agreement with this, the racemization via step 2 was shown to be over two orders of magnitude slower than transilylation via step 3. In fact a faster racemization via intermolecular silyl group migration was shown to occur with most substrates (equation 42).

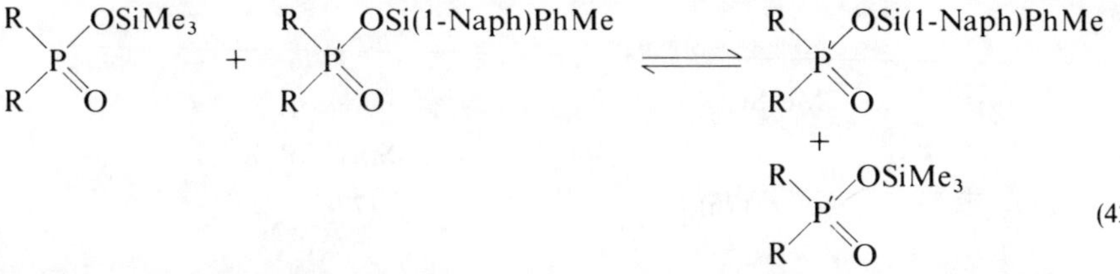

(42)

Chojnowski proposes that these results are consistent with the double-displacement mechanism but not a mechanism involving a pentacoordinate species 75. He argues that

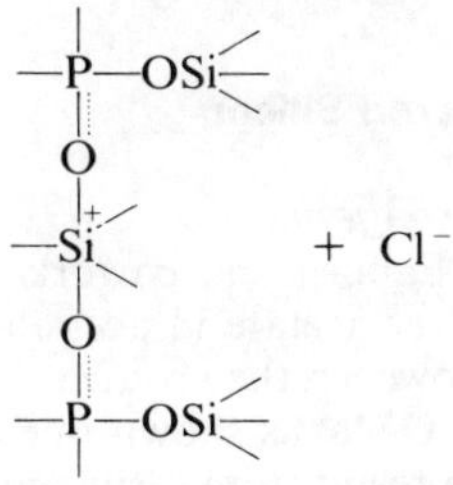

(75)

if this type of intermediate was present in both processes, then attack at the four-coordinate silicon should be faster than attack on the five-coordinate species in order that transilylation be faster than racemization. This conflicts with the concept of 'extra reactivity' of hypervalent centres.

III. MECHANISMS OF REACTION OF HIGHLY HINDERED ORGANOSILICON COMPOUNDS

A. Introduction

Nucleophilic substitution at silicon, as described in previous sections, usually proceeds through five- or possibly six-coordinated silicon intermediates or transition states. For silicon bearing the common alkyl or aryl ligands, the barrier to formation of such extracoordinated intermediates is quite low. In these examples bimolecular substitution is so facile that unimolecular substitution, even in the most favourable cases, is so slow by comparison as to be unobservable.

Eaborn, in the last few years, has initiated a major programme concerned with the effects on silicon reaction mechanisms of severely increasing the steric bulk at silicon[115, 116]. During the course of this work Dewar and Healy[117] suggested that the major factor in the difference between silicon and carbon is not the availability of d orbitals on silicon, but more simply the small size of the carbon atom, preventing ready formation of five-coordinated intermediates. A consequence of this argument might be that silicon, when particularly encumbered with bulky ligands, could become carbon-like in its chemistry[145]. Whereas this has been proved to some extent, the chemistry of silicon still provides us with some surprises.

The starting point for Eaborn's work is the tris(trimethylsilyl)methyl ligand, **76**, now abbreviated to 'trisyl' or 'Tsi'. The trisyl ligand, or some variation thereof, is attached to a

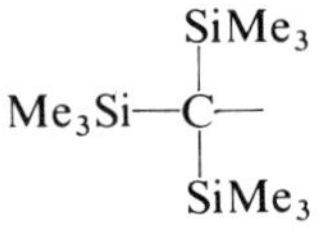

(**76**)

$(Me_3Si)_3CSiR_2X$

(**77**)

functional silicon centre, SiR_2X, to give compounds **77** in which transformations at silicon may be studied. In the following sections we briefly review the variety of reactions at highly hindered silicon, including bimolecular and unimolecular substitution, rearrangements and fragmentation reactions.

B. Substitutions at Highly Hindered Silicon

1. Reactions with electrophilic reagents

a. Trisyl derivatives. In 1978, Eaborn and coworkers[118] observed that reactions of $TsiSiMe_2I$ and $TsiSiH_2I$ with silver acetate in acetic acid gave $TsiSiMe_2OCOMe$ and $TsiSiH_2OCOMe$, respectively. However, the product of the similar reaction of $TsiSiPh_2I$ was $(Me_3Si)_2(SiPh_2Me)SiMe_2OCOMe$, as shown in equation 43. This rearranged product was tentatively suggested to result from a siliconium ion intermediate. It has now been established that an intermediate cation is involved in such reactions, but not of the

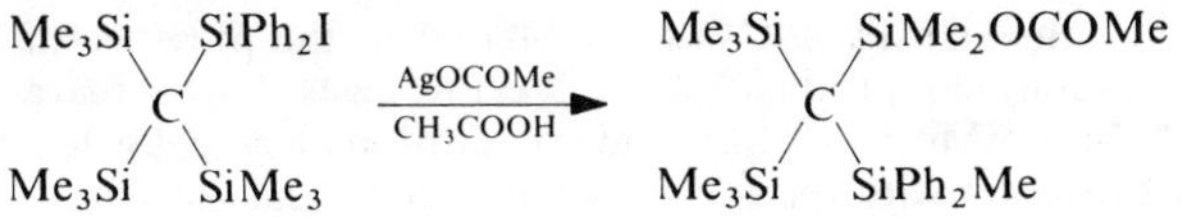
(43)

conventional type. Bridged siliconium ions, **78**, best explain the many metal ion-assisted reactions of trisyl compounds.

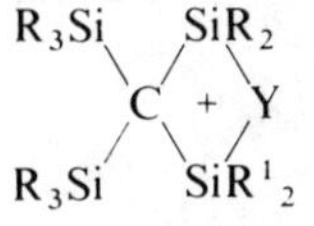

(**78**)

Consistent with bridged cationic intermediates, there are 1,3-migrations of methyl groups in many reactions of TsiSiRRX compounds with silver(I) or mercury(II) salts in acid or alcoholic solvents[119, 120]. A typical reaction[120] is shown in equation 44. From

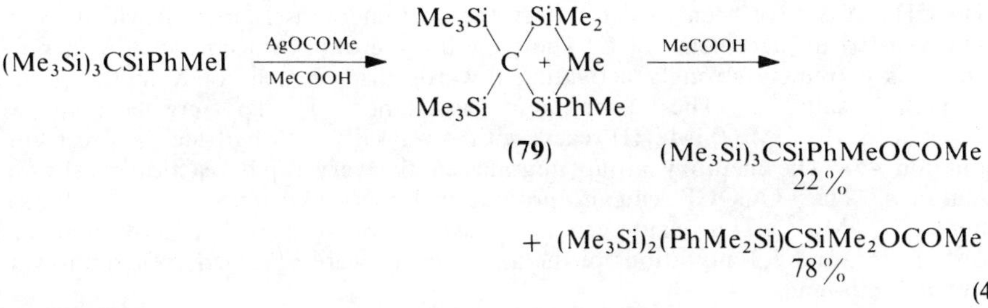
(44)

this and other observations[119, 120] it appears that the bridged ion **79** can be attacked at either silicon atom by the nucleophilic solvent, but the less sterically hindered site is preferentially substituted. This accounts for the observation that TsiSiPh$_2$I undergoes reaction with metal salts with exclusive rearrangement[120] whereas no rearrangement is observed with TsiSiH$_2$I. Other evidence for an intermediary cationic species includes the characterization of SiONO$_2$ products in addition to the alkoxides SiOR when trisyl iodides are treated with ROH/AgNO$_3$ mixtures[120]. The same product distribution is obtained from reaction of either TsiSiMe$_2$I or TsiSiMe$_2$Br with AgNO$_3$ in MeOH, pointing to a common intermediate.[120]

In these metal-assisted solvolyses there is a large rate variation with X in SiX; the order of ease of displacement is $I > Br \gg Cl$[120].

Reactions of TsiSiR$_2$I and TsiSiR$_2$H with I$_2$ and ICl also proceed through bridged siliconium ions[121, 122]. For example, TsiSiPh$_2$I on reaction with one equivalent of ICl in CCl$_4$ gave exclusively the rearranged product, (Me$_3$Si)$_2$C(SiMe$_2$Cl)(SiPh$_2$Me)[121]. The same product was obtained by reaction of TsiSiPh$_2$H with a 2.2 molar equivalent of ICl[121]. Similarly[122], TsiSiPh$_2$I with ICl in MeOH gave (Me$_3$Si)$_2$C(SiPh$_2$Me)(SiMe$_2$OMe) and the rearranged chloride, (Me$_3$Si)$_2$C-(SiMe$_2$Cl)(SiPh$_2$Me). These reactions, and those with iodine, are consistent with methyl-bridged, ion-paired cationic intermediates formed by abstraction of I$^-$ by I$_2$ or ICl.

A reaction that points out clearly the analogy between highly hindered silicon components and carbon compounds is the oxidatively assisted nucleophilic substitution of $TsiSiR_2I$ (and t-Bu_3SiI), by m-CPBA in MeOH in the presence of LiCl[123]. In this reaction the exclusive product for R = Ph was the rearranged chloride $(Me_3Si)_2C(SiPh_2Me)(SiMe_2Cl)$. Thus, under conditions where, for less hindered silicon compounds, S_N2 reaction with the solvent would occur, reactions analogous to those of alkyl iodides, involving cationic intermediates, are predominant[123].

b. Anchimeric assistance. The cation **79** contains a bridging methyl group, but it has been shown subsequently that many other groups are more effective at bridging than methyl. Unpublished calculations[124] for the equilibrium in equation 45 are reported[125]

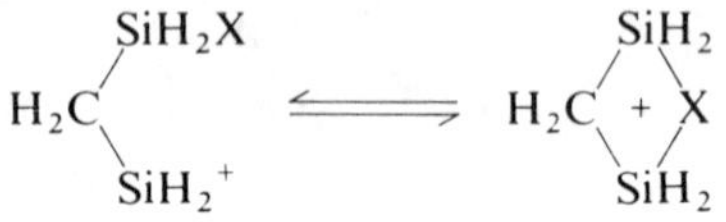

(45)

to show that the bridged ion is of similar energy to the open ion for X=Me, but becomes some 63 kcal mol^{-1} more stable for X = OH. The order of stabilization of the bridged ion is said to be X = OH > F > Cl > Me.

Compounds of the type $(Me_3Si)_2C(SiMe_2Y)SiMe_2X$ (where X = H, Cl, Br, I; Y = OMe, $CH{=}CH_2$, N_3) have been used to study the anchimeric assistance provided by Y in metal-assisted displacements of X. The γ-methoxy group, in accordance with prediction[125], is extremely strongly activating towards displacement of X in the presence of metal salts[126]. The anchimeric assistance is so great that even $(Me_3Si)_2C(SiMe_2OMe)(SiMe_2H)$ reacts with silver salts with hydride displacement[127] (equation 46). The methoxy group migrates in the very rapid reaction[126] shown in equation 47. The γ-$OCOCF_3$ can also provide anchimeric assistance, but not to the same extent as γ-OMe[127]. The γ-azido group, N_3, is also able to provide effective anchimeric assistance[128], and 1,3-migrations of the azido group were observed with suitably substituted compounds.

$$(Me_3Si)_2C(SiMe_2OMe)(SiMe_2H) \xrightarrow{AgOCOCF_3} (Me_3Si)_2C(SiMe_2OMe)(SiMe_2OCOCF_3) \quad (46)$$

$$(Me_3Si)_2C(SiMe_2OMe)SiPh_2Br \xrightarrow[EtOH]{AgClO_4} (Me_3Si)_2C(SiMe_2OEt)(SiPh_2OMe) \quad (47)$$

Similar effects were observed for the isomeric γ-chlorides, $(Me_3Si)_2C(SiEt_2Cl)(SiMe_2I)$ and $(Me_3Si)_2C(SiMe_2Cl)(SiEt_2I)$, on treatment with $AgBF_4$ in CH_2Cl_2[128]. If the ion **80** was fully formed, and had sufficient lifetime, both compounds

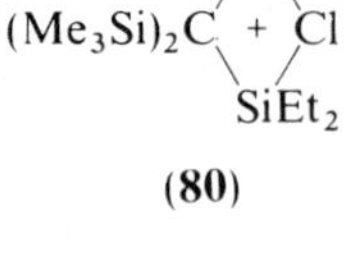

(80)

would give the same mixture of 1,3-chlorofluorosilanes. This was not the case, but the ratios were not dramatically different. In each example it was observed that there was rather more of the unrearranged product than might have been expected on simple steric grounds. It was suggested that liberation of BF_4^- close to the silicon from which I^- was abstracted could have affected the product balance[128].

It is not a simple matter to understand the effects of solvents, and particularly of solvation of intermediates, in such hindered systems where the solvent structure must be severely disrupted. Throughout the trisyl work one finds unusual sensitivity to solvent effects. An example arose in experiments in which anchimeric assistance by γ-fluorine was proposed[129]. Scheme 3 shows the effect of a solvent change from CH_2Cl_2 to ether. In the example shown in the scheme, the reaction in methylene chloride follows similar lines to other anchimerically assisted reactions, although it is difficult to understand why so much of the rearranged products are obtained, given the smaller size of the SiMeH group compared with $SiMe_2$. In ether no rearrangement at all is observed, which is again not easy to explain.

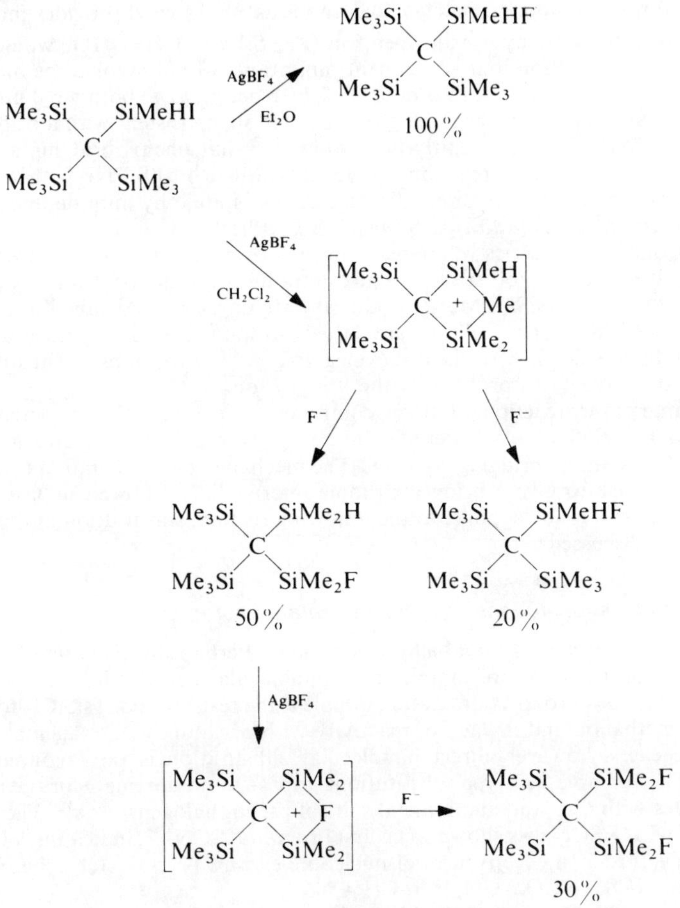

SCHEME 3

Anchimeric assistance has also been observed by a γ-vinyl group in reactions of $(Me_3Si)_2C(SiMe_2CH{=}CH_2)(SiR_2I)$, where R = Me, Et, with silver salts[130]. When R = Et the usual mixture of products from the vinyl-bridged ion **81** was obtained. Anchimeric assistance is greatest for γ-OMe, substantial for γ-$CH{=}CH_2$, but observable even for γ-Me[130].

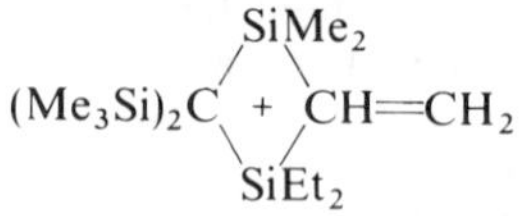

(81)

Evidence against an open siliconium ion, or a series of rapidly equilibrating siliconium ions, was found in the reaction of $TsiSi(CH{=}CH_2)_2Cl$ with AgO_2CCF_3 in CF_3CO_2H in which $TsiSi(CH{=}CH_2)_2O_2CCF_3$ and $(Me_3Si)_2C[Si(CH{=}CH_2)_2Me]$ $(SiMe_2O_2CCF_3)$ were both formed in approximately equal amounts[131]. As vinyl provides more anchimeric assistance than methyl[130] an open ion $(Me_3Si)_3C\overset{+}{Si}(CH{=}CH_2)_2$ would rearrange first by methyl migration, but subsequent migrations would involve the more effective vinyl migrations. However, both products of the reaction have both vinyl groups on the same silicon atom, as at the beginning of the reaction, consistent with a bridged-methyl intermediate ion[130]. It has recently been shown[132] that phenyl bridging also provides anchimeric assistance in the reaction of $(Me_3Si)_2C(SiMe_2Ph)(SiEt_2I)$ with $AgBF_4$ where the products are about 40% direct displacement of iodide by fluoride and 60% of the rearranged compound. $(Me_3Si)_2C(SiMe_2F)(SiEt_2Ph)$.

There is one curious series where rearrangement was not observed on treatment with AgY[133]. The reactions of silver triflate, tetrafluoroborate or trifluoroacetate with TsiSiMe(OMe)I and $TsiSi(OMe)_2I$ produced only unrearranged substitution products. The hypothesis that steric effects are largely responsible for product distributions apparently breaks down with the electronegative methoxy groups as substituent, assuming that these reactions proceed by the bridged ions.

In summary, the reaction of trisyl compounds with silver(I) or mercury(II) salts appears to be relatively consistent in involving cationic intermediates and in being accelerated by suitably bridging γ-groups. The mechanistic interpretation is rate-limiting loss of nucleofuge to give a bridged cationic intermediate followed by fast reaction at either silicon, with a nucleophile. In the following section, substitution in the absence of metal ions is discussed.

2. *Nucleophilic substitutions of trisyl and related compounds*

a. Direct bimolecular substitution by nucleophiles. Perhaps the most significant effect of the trisyl group is the severe inhibition to bimolecular substitution at silicon directly bonded to the trisyl group. A dramatic example is the resistance of $TsiSiCl_3$ to prolonged boiling in methanol, and its lack of reactivity with organolithium reagents[134].

In some cases, however, direct bimolecular substitution is the favoured course of reaction. These simple S_N2-type substitutions appear to predominate for the reactions of trisyl halides with CsF and alkali metal salts of pseudohalogens[128, 135]. The reaction of $TsiSiMe_2I$ with KSCN was shown to be first order in KSCN[135], indicating a bimolecular reaction. The order of effectiveness of nucleophiles was: $N_3 > F > CN > NCS > NCO$ in MeOH and $NCS > NCO > CN$, F in CH_3CN.

Whereas a γ-MeO group is highly activating in silver-ion-promoted reactions, in unimolecular solvolyses (see later) the reactivities of $(Me_3Si)_2C(SiMe_2OMe)(SiMe_2Cl)$ and $(Me_3Si)_3CSiMe_2Cl$ towards KSCN or KOCN in CH_3CN differed only by a factor of about 13, in accord with a direct bimolecular substitution[125].

A most interesting observation[136] in the reactions of $TsiSiMe_2X$ and t-Bu_3SiX (X = Cl, I) with alkali metal salts was that for t-Bu_3SiX the chloride reacted about six times faster than the iodide, and for the trisyl derivatives the iodide did react faster than the chloride, but not by a large factor. Generally, silyl iodides are assumed to be much more reactive the chlorides. These observations led Eaborn to discuss the nature of the intermediate or transition state on less-hindered silicon derivatives. On consideration of crowding in the transition state, the possibility was raised that the more usual substitutions at silicon may be synchronous as in S_N2 reactions at carbon.

b. Solvolysis reactions. There are a few examples where bimolecular substitution by solvent seems to be the unambiguously preferred mechanism. The rate of solvolysis of the TsiSiHPhX ($X = ONO_2$, Br) derivatives by methanol is accelerated by added NaOMe, with the acceleration being proportional to the concentration of NaOMe[137]. Similar results have been obtained for the fluoride and OSO_2Me derivatives of TsiSiHPhX. These reactions therefore appear to be S_N2 in character.

Straightforward S_N2 substitutions by solvent do not appear to be the norm for trisyl-type compounds. For the first time, unimolecular solvolysis has been convincingly demonstrated for silicon compounds. In the strongly electrophilic, but weakly nucleophilic solvent CF_3COOH, the reaction shown in Scheme 4 occurs[120].

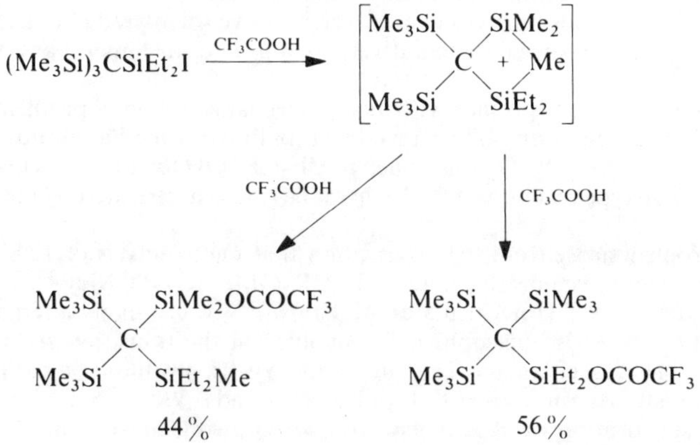

SCHEME 4

This product distribution implies the intermediacy of a bridged ion, and it was shown that the rate of the reaction of $TsiSiMe_2I$ with CF_3COOH was not significantly affected by the presence of $NaOOCCF_3$. The mechanism thus appears to be of the S_N1 type.

Similar S_N1-type mechanisms have been observed for compounds in which γ-substituents are available for anchimeric assistance. $(Me_3Si)_2C(SiMe_2N_3)(SiMe_2Br)$ reacts with methanol to give the product of bromide substitution by MeO[128]. (Triethylamine was used to trap the HBr, thereby preventing further reaction, but it did not affect the rate greatly.) The rate of reaction was almost insensitive to added methoxide. The same compound solvolysed some 150 times more readily in the highly electrophilic CF_3CH_2OH than in MeOH, which is further evidence for an ionization as rate-limiting step.

A γ-methoxy group also very powerfully (by a factor of $>10^6$) activates trisyl compounds to unimolecular solvolysis[126]. It was suggested[126] that rate-determinng ionization can occur either with strongly electrophilic solvents such as CF_3COOH, even with weak-bridging groups such as Me, or with powerful bridging groups, even with more weakly electrophilic solvents such as MeOH.

The accelerating effect of the γ-methoxy group has been confirmed[138] and solvolysis of $(Me_3Si)_2C(SiMe_2OMe)(SiMe_2X)$ with methanol has been studied to evaluate the leaving-group ability of X. The following order was observed: X = Br > NO_3 > $O_3SC_6H_4Me$-*p* > Cl, SCN > O_2CCF_3 > NCS > O_2CPh, O_2CMe > F, N_3, H.

Recently[139], the solvolysis of $(Me_3Si)_2C(SiMe_2C_6H_4Y)(SiMe_2I)$ (Y = H, Me-*p*-MeO, *m*-CF_3, *p*-Cl) compounds has been studied in CF_3CH_2OH and MeOH. The rates of reaction in both solvents decrease in the order Y = *p*-OMe > *p*-Me > H > *p*-Cl > *m*-CF_3 showing a build-up of positive charge in the transition state. The effect of the bridging aryl group is particularly dramatic in CF_3CH_2OH, where the relative rates varied by a factor of about 10^5 between *p*-OMe and *m*-CF_3. In MeOH, the spread of rates across the series was only about 50.

Other solvolysis reactions that initially appeared to undergo rate-determining ionization include the reaction of TsiSiPhHI with methanol, which is not markedly accelerated by NaOMe[140], and the methanolysis of $TsiSiMe_2I$[141], $TsiSiMe_2OClO_3$[142] and $TsiSiMe_2OSO_2CF_3$[143]. Similarly, $(PhMe_2Si)_3CSiMe_2I$ was shown to undergo first-order solvolysis in MeOH[144]. The presence of 0.05 M NaOMe increased the rate by about 23%, but on increasing the NaOMe concentration to 0.1 *M* there was no further rate enhancement[144]. This phenyl substituted derivative solvolysed about 15 times more readily than $TsiSiMe_2I$, an effect tentatively assigned to anchimeric assistance by the phenyl group[144].

Despite the apparent consistency with an S_N1 mechanism, several problems cast doubt on this simple interpretation. While reacting rapidly with methanol, no reaction was observed when $TsiSiMe_2OClO_3$ was boiled in CF_3CH_2OH for 24 h[137]. This is inconsistent with other S_N1-type reactions[126, 139] where the rate is accelerated by the electrophilic solvent.

Another problem arose from the observation that methanolysis of $TsiSiR_2I$ (R = Ph, Me) gave only the unrearranged products $TsiSiR_2OMe$ (R = Ph, Me)[145]. It was initially suggested[145] that an S_N2 (Intermediate) mechanism[146] was a likely alternative. In that mechanism the solvent is nucleophilically involved in the transition state, which does have considerable ionic character. This suggestion has been withdrawn recently[147] in the light of kinetic studies on $TsiSiH(C_6H_4Y)I$ compounds. Where S_N2 reactions are expected, as in substitution by KSCN, electron-withdrawing substituents Y enhance the rate, in accord with a build-up of negative charge in the transition state. In the ionization reaction with AgO_2CMe, electron-withdrawing groups retard the rate, in accord with an electron-deficient transition state. However, with methanol solvolysis, electron-withdrawing groups assist solvolysis, suggesting a build-up of negative charge in the transition state. Thus methanol does appear to be covalently bound in the transition state, and the lack of rate enhancement by NaOMe is difficult to explain. No firm conclusions about the mechanism have been drawn, except that it appears that a continuum of bimolecular mechanisms is possible.

C. Nucleophile-induced Fragmentations

Although $TsiSiMe_2Cl$ is not susceptible to direct substitution by sodium methoxide in MeOH, it does nevertheless react, but the major product is $(Me_3Si)_2CHSiMe_2OMe$[148]. The most likely mechanism for this reaction involves an intermediate sila-alkene, as shown in Scheme 5. It is interesting to note that this mechanism is directly analogous to the E2 elimination mechanism at carbon. A similar reaction has also been observed for $TsiSi(CH{=}CH_2)_2Cl$ with sodium methoxide[131].

A silanone has been postulated to account for the rapid production of TsiSiPh(OH)(OMe) from TsiSiPh(OH)I by methanolic sodium methoxide[149]. The

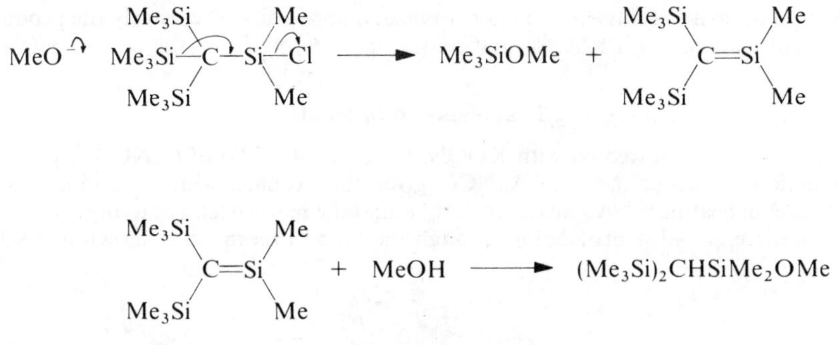

SCHEME 5

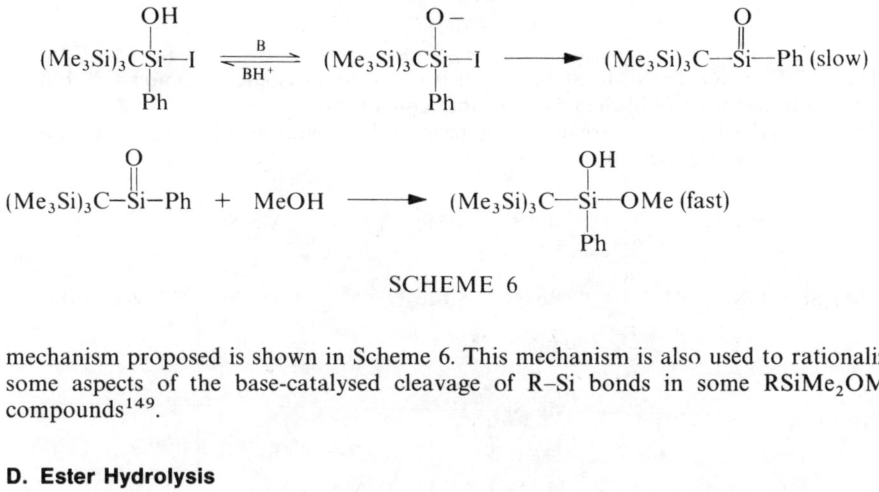

SCHEME 6

mechanism proposed is shown in Scheme 6. This mechanism is also used to rationalize some aspects of the base-catalysed cleavage of R–Si bonds in some $RSiMe_2OMe$ compounds[149].

D. Ester Hydrolysis

Most silyl esters $RCOOSiR_3$ cleave with silicon–oxygen bond scission because of the high susceptibility of silicon to nucleophilic attack. Some acyl-oxygen scission has been observed in the reaction of *t*-$BuMe_2SiOCOMe$ with BuOH[150]. Many trisyl derivatives, $TsiSiMe_2OCOR$, have been found to undergo acyl-oxygen fission exclusively with sodium methoxide in methanol[151] (equation 48).

$$(Me_3Si)_3CSiMe_2OCOR \xrightarrow[\text{MeOH}]{\text{MeO}^-} (Me_3Si)_3CSiMe_2OH + RCOOMe \quad (48)$$

Methoxide ion attacks the carbonyl carbon predominantly, but not exclusively, in the less-hindered $TsiSiPh(H)(O_2CCMe)$[152]. It is not surprising that the siloxy group is a good leaving group as the pK_a of the silanols is similar to that of phenol[152].

A detailed study[153] of the reactions of $(Me_3Si)_2(CH_2{=}CHMe_2Si)CSiMe_2OCOR$ ($R = CF_3$, CH_3) showed that nucleophiles in methanol preferentially attack at the carbonyl carbon atom.

The sequence of ability of some nucleophiles to effect acyl-oxygen cleavage in methanol is RNH_2, OCN^-, $HO_2^- > MeO^- > N_3^- > Me_2NH > Me_3N > 4\text{-}Me_2NC_5H_4N \gg C_5H_5N > PhNH_2$. This order, on comparison with organic ester cleavages, gave rise to the suggestion that cleavage of the acyl–oxygen bond is rate determining.

With acetonitrile as solvent, the esters discussed above gave exclusively the product of Si–O bond cleavage with KSCN, KOCN, CsF and NaN_3[153].

E. Rearrangements Involving Trisyl-type Compounds

When $TsiSiMe_2I$ is treated with KOCN, the isocyanate $TsiSiMe_2NCO$ is produced, but reaction of the iodide with AgNCO gives the cyanate, which rearranges to the isocyanate on heating[154]. A kinetic study[155] found the rearrangement to be second order in the cyanate, possibly proceeding through the kind of mechanism shown in **82**. Both

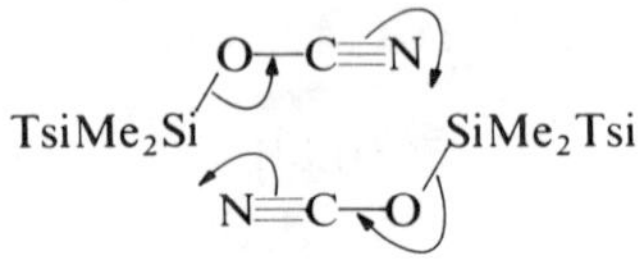

(82)

$(Me_3Si)_2C(SiMe_2OMe)(SiMe_2SCN)$ and its isothiocyanate isomer are known[156], but no information has been published on their interconversion.

When trisyl silanols are treated with base, a 1,3-migration of silicon can readily occur[157] as shown in Scheme 7.

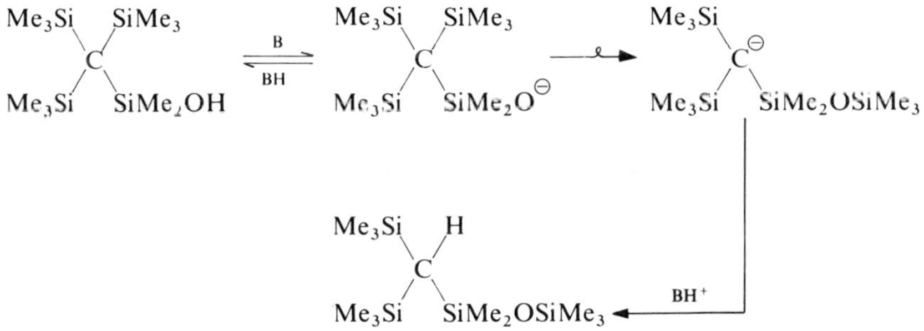

SCHEME 7

In summary, trisyl chemistry differs greatly from the organosilicon chemistry of less-hindered compounds. The main differences are:

1. Inhibition to bimolecular displacement.
2. Unimolecular substitution reactions, involving bridged cations, and large effects of γ-substituents.
3. Solvolysis mechanisms covering the entire spectrum from S_N1 to S_N2 substitutions.

IV. REFERENCES

1. 1a J. E. Baines and C. Eaborn, *J. Chem. Soc.*, 1436 (1956).
   1b A. R. Bassindale and P. G. Taylor, Activating and Directive Effects of Silicon, this volume p. 893.
   1c L. H. Sommer, *Stereochemistry, Mechanism and Silicon*, McGraw-Hill, New York, 1965.
2. L. H. Sommer, *Intra-Sci. Chem. Rep.*, **7**, 1 (1973).

3. R. H. Prince, *Int. Rev. Sci., Inorg. Chem. Ser. One*, **9**, 353 (1972).
4. R. J. P. Corriu and M. Henner, *J. Organomet. Chem.*, **74**, 1 (1974).
5. (a) R. J. P. Corriu and C. Guerin, *J. Organomet. Chem.*, **198**, 231 (1980).
(b) R. J. P. Corriu and C. Guerin, *Adv. Organomet. Chem.*, **20**, 265 (1982).
6. R. J. P. Corriu, C. Guerin and J. J. E. Moreau, *Top. Stereochem.*, **15**, 43 (1984).
7. L. H. Sommer, G. A. Parker, N. C. Lloyd, C. L. Frye and K. W. Michael, *J. Am. Chem. Soc.*, **89**, 857 (1967).
8. R. Corriu and G. Royo, *J. Organomet. Chem.*, **14**, 291 (1968).
9. L. H. Sommer, G. A. Parker and C. L. Frye, *J. Am. Chem. Soc.*, **86**, 3280 (1964).
10. L. H. Sommer, W. D. Korte and P. G. Rodewald, *J. Am. Chem. Soc.*, **89**, 862 (1967).
11. L. H. Sommer, K. W. Michael and W. D. Korte, *J. Am. Chem. Soc.*, **89**, 868 (1967).
12. L. H. Sommer, C. L. Frye and G. A. Parker, *J. Am. Chem. Soc.*, **86**, 3276 (1964).
13. L. H. Sommer, C. L. Frye, M. C. Musolf, G. A. Parker, P. G. Rodewald, K. W. Michael, Y. Okaya and R. Pepinsky, *J. Am. Chem. Soc.*, **83**, 2210 (1961).
14. R. J. P. Corriu, J. M. Fernandez and C. Guerin, *J. Organomet. Chem.*, **152**, 25, 31 (1978).
15. R. Corriu and G. Royo, *Bull. Soc. Chim. Fr.*, 1497 (1972).
16. R. Corriu and G. Royo, *Tetrahedron*, 4289 (1971).
17. L. H. Sommer and J. McLick, *J. Am. Chem. Soc.*, **88**, 5359 (1966).
18. L. H. Sommer and W. D. Korte, *J. Am. Chem. Soc.*, **89**, 5802 (1967).
19. R. Corriu and G. Royo, *J. Organomet. Chem.*, **40**, 229 (1972).
20. L. H. Sommer, C. L. Frye, G. A. Parker and K. W. Michael, *J. Am. Chem. Soc.*, **86**, 3271 (1964).
21. L. H. Sommer, J. McLick and C. M. Golino, *J. Am. Chem. Soc.*, **94**, 669 (1972).
22. R. Corriu, J. Masse and G. Royo, *J. Chem. Soc., Chem. Commun.*, 252 (1971).
23. R. Corriu, J. M. Fernandez and C. Guerin, *J. Organomet. Chem.*, **152**, 21 (1978).
24. R. Corriu and C. Guerin, *J. Organomet. Chem.*, **225**, 141 (1982).
25. R. Corriu and J. Masse, *J. Organomet. Chem.*, **34**, 221 (1972).
26. N. T. Anh and C. Minot, *J. Am. Chem. Soc.*, **102**, 103 (1980).
27. L. Salem, *Chem. Br.*, 449 (1969).
28. R. G. Pearson, *Chem. Eng. News*, **48**, 66 (1970).
29. C. Minot and N. T. Anh, *Tetrahedron Lett.*, 3905 (1975).
30. R. Corriu, J. M. Fernandez and C. Guerin, *Tetrahedron Lett.*, 3391 (1978).
31. C. Breliere, R. J. P. Corriu, A. De Saxce and G. Royo, *J. Organomet. Chem.*, **116**, 153 (1979).
32. C. Breliere, R. J. P. Corriu and G. Royo, *J. Chem Soc., Chem. Commun.*, 906 (1976).
33. (a) B. G. McKinnie, N. S. Bhacca, F. K. Cartledge and J. Fayssoux, *J. Am. Chem. Soc.*, **96**, 2637 (1974).
(b) B. G. McKinnie, N. S. Bhacca, F. K. Cartledge and J. Fayssoux, *J. Org. Chem.*, **41**, 1534 (1976).
34. (a) J. Dubac, P. Mazerolles and B. Serres, *Tetrahedron Lett.*, 3495 (1972).
(b) J. Dubac, P. Mazerolles and B. Serres, *Tetrahedron Lett.*, 759 (1974).
35. (a) F. K. Cartledge, J. M. Wolcott, J. Dubac, P. Mazerolles and P. Fagoaga, *Tetrahedron Lett.*, 3593 (1975).
(b) J. M. Wolcott and F. K. Cartledge, *J. Organomet. Chem.*, **111**, C35 (1976).
36. (a) F. K. Cartledge, J. M. Wolcott, J. Dubac, P. Mazerolles and M. Joly, *J. Organomet. Chem.*, **154**, 187 (1978).
(b) F. K. Cartledge, J. M. Wolcott, J. Dubac, P. Mazerolles and M. Joly, *J. Organomet. Chem.*, **154**, 203 (1978).
37. R Corriu and J. Masse, *Bull. Soc. Chim. Fr.*, **10**, 3491 (1978).
38. R. Corriu and C. Guerin, *J. Chem. Soc., Chem. Commun.*, 74 (1977).
39. (a) R. Corriu, C. Guerin and J. Masse, *J. Chem. Soc., Chem. Commun.*, 75 (1975).
(b) R. Corriu, C. Guerin and J Masse, *J. Chem. Res. (S)*, 160 (1977).
(c) R. Corriu, C. Guerin and J. Masse, *J. Chem. Res. (M)*, 1877 (1977).
40. C. Guerin, Ph.D. Thesis, Montpellier, 1978, (quoted in Reference 5).
41. J. M. Fernandez, Ph.D Thesis, Montpellier, 1978, (quoted in Reference 5).
42. C. G. Swain, K. R. Porschke, W. Ahmed and R. L. Schowen, *J. Am. Chem. Soc.*, **96**, 4700 (1974).
43. C. Eaborn, R. Eidenschink and D. R. M. Walton, *J. Chem. Soc., Chem. Commun.*, 388 (1975).
44. R. L. Schowen and K. S. Latham Jr., *J. Am. Chem. Soc.*, **89**, 4677 (1967).
45. M. W. Grant and R. H. Prince, *Nature*, **222**, 1163 (1969).
46. L. D. Pettit and S. Bruckenstein, *J. Am. Chem. Soc.*, **88**, 4783 (1966).

47. R. J. P. Corriu and B. J. L. Henner, *J. Organomet. Chem.*, **102**, 407 (1975).
48. G. Chauviere, R. J. P. Corriu and B. J. L. Henner, *J. Organomet. Chem.*, **86**, Cl (1975).
49. J. I. Musher, *Angew. Chem., Int. Ed. Engl.*, **8**, 54 (1969).
50. R. R. Holmes, *Pentacoordinated Phosphorous*, Vols. I and II, ACS, Washinton, D.C., 1980.
51. (a) F. P. Boer, J. J. Flynn and J. W. Turley, *J. Am. Chem. Soc.*, **90**, 6973 (1968).
(b) P. Bird, J. F. Harrod and K. A. Than, *J. Am. Chem. Soc.*, **96**, 1222 (1974).
(c) D. Schomburg, *J. Organomet. Chem.*, **221**, 137 (1981).
(d) J. J. Harland, R. O. Day, J. F. Wollano, A. C. Sau and R. R. Holmes, *J. Am. Chem. Soc.*, **103**, 5269 (1981).
(e) D. Schomburg, *Z. Naturforsch, B*, **37**, 195 (1982).
(f) D. Schomburg, *Z. Naturforsch, B*, **38**, 938 (1983).
52. W. H. Stevenson III, S. Wilson, J. C. Martin and W. B. Farnham, *J. Am. Chem. Soc.*, **107**, 6340 (1985); W. H. Stevenson III and J. C. Martin, *J. Am. Chem. Soc.*, **107**, 6352 (1985).
53. K. D. Onan, A. T. McPhail, C. H. Yoder and R. W. Hillyard, *J. Chem. Soc., Chem. Commun.*, 209 (1978).
54. D. G. Anderson, A. J. Blake, S. Cradock, E. A. V. Ebsworth, D. W. H. Rankin and A. J. Welch, *Angew. Chem., Int. Ed. Engl.*, **25**, 107 (1986).
55. J. A. Deiters and R. R. Holmes, *J. Am. Chem. Soc.*, **109**, 1686, 1692 (1987).
56. R. Damrauer, C. H. DePuy and V. M. Bierbaum, *Organometallics*, **1**, 1553 (1982).
57. See, for example:
(a) R. Luckenbach, *Dynamic Stereochemistry of Pentacoordinated Phosphorus and Related Elements*, George Thieme Publ. Stuttgart, 1973.
(b) F. H. Westheimer, *Acc. Chem. Res.*, **1**, 70 (1968).
(c) K. Mislow, *Acc. Chem. Res.*, **3**, 321 (1970).
58. Ref. 5a, p. 298.
59. R. J. P. Corriu, G. Royo and A. De Saxce, *J. Chem. Soc., Chem. Commun.*, 892 (1980).
60. C. Breliere, F. Carré, R. J. P. Corriu, M. Poirier and G. Royo, *Organometallics*, **5**, 388 (1986).
61. G. Klebe and K. Henson, *J. Chem. Soc., Dalton Trans.*, 5 (1985).
62. Ref. 5b, p. 285.
63. W. H. Stevenson III and J. C. Martin, *J. Am. Chem. Soc.*, **104**, 309 (1982).
64. C. Minot, Ph.D Thesis, Orsay, 1977 (quoted in ref. 5).
65. F. Klanberg and E. L. Muetterties, *Inorg. Chem.*, **7**, 155 (1968).
66. R. J. P. Corriu, A. Kpoton, M. Poirier, G. Royo and J. Y. Corey, *J. Organomet. Chem.*, **277**, C25 (1986).
67. W. B. Farnham and R. C. Harlow, *J. Am. Chem. Soc.*, **103**, 4608 (1981).
68. J. Boyer, R. J. P. Corriu, A. Kpoton, M. Mazhar, M. Poirier and G. Royo, *J. Organomet. Chem.*, **301**, 131 (1986).
69. R. J. P. Corriu, M. Mazhar, M. Poirier and G. Royo, *J. Organomet. Chem.*, **306**, C5 (1986).
70. R. Damrauer and S. E. Danahey, *Organometallics*, **5**, 1490 (1986).
71. J. A. Gibson and A. F. Jansen, *Can. J. Chem.*, **50**, 3087 (1972).
72. R. J. P. Corriu, F. Larcher and G. Royo, *J. Organomet. Chem.*, **129**, 299 (1977).
73. J. R. Chipperfield and R. H. Prince, *J. Chem. Soc.*, 3567 (1963).
74. (a) A. D. Allen, J. C. Charlton, C. Eaborn and G. Modena, *J. Chem. Soc.*, 3668 (1957).
(b) A. D. Allen and G. Modena, *J. Chem. Soc.*, 3671 (1957).
(c) G. Schott, H. Kelling and R. Schild, *Chem. Ber.*, **99**, 291 (1966).
75. R. E. Wasylishen, G. S. Birdi and A. F. Jansen, *Inorg. Chem.*, **15**, 3054 (1976).
76. R. J. P. Corriu, G. Dabosi and M. Martineau, *J. Chem. Soc., Chem. Commun.*, 649 (1977).
77. R. J. P. Corriu, G. Dabosi and M. Martineau, *J. Organomet. Chem.*, **154**, 33 (1978).
78. R. J. P. Corriu, G. Dabosi and M. Martineau, *J. Organomet. Chem.*, **150**, 27 (1978).
79. H. K. Chu, M. D. Johnson and C. L. Frye, *J. Organomet. Chem.*, **271**, 327 (1984).
80. R. J. P. Corriu, R. Perz and C. Reye, *Tetrahedron*, **39**, 999 (1983).
81. R. J. P. Corriu and M. Leard, *J. Chem. Soc., Chem. Commun.*, 1086 (1971).
82. R. J. P. Corriu and M. Henner–Leard, *J. Organomet. Chem.*, **64**, 351 (1974).
83. R. J. P. Corriu and M. Henner–Leard, *J. Organomet. Chem.*, **65**, C39 (1974).
84. R. J. P. Coriu and M. Henner, *Bull. Soc. Chim. Fr.*, 1447 (1974).
85. R. J. P. Corriu, F. Larcher and G. Royo, *J. Organomet. Chem.*, **104**, 293 (1976).
86. Ref. 4, p. 15.

87. B. G. McKinnie and F. K. Cartledge, *J. Organomet. Chem.*, **104**, 407 (1976).
88. F. K. Cartledge, B. G. McKinnie and J. M. Wolcott, *J. Organomet. Chem.*, **118**, 7 (1976).
89. J. Chojnowski, M. Cypryk, J. Michalski and L. Wozniak, *J. Organomet. Chem.*, **288**, 275 (1985).
90. A. R. Bassindale, J. C. Lau and P. G. Taylor, unpublished results.
91. E. F. V. Scriven, *Chem. Soc. Rev.*, **12**, 129 (1983).
92. L. H. Sommer and D. L. Bauman, *J. Am. Chem. Soc.*, **91**, 7045 (1969).
93. L. H. Sommer and P. G. Rodewald, *J. Am. Chem. Soc.*, **85**, 3898 (1963).
94. K. Tamao, M. Ishikawa and M. Kumada, *J. Chem. Soc., Chem. Commun.*, 73 (1969).
95. C. L. Frye, G. A. Vincent and W. A. Finzel, *J. Am. Chem. Soc.*, **93**, 6805 (1971).
96. B. J. Helmer, R. West, R. J. P. Corriu, M. Poirier, G. Royo and A. De Saxce, *J. Organomet. Chem.*, **251**, 295 (1983).
97. J. P. Guertin and M. Onyszchuk, *Can. J. Chem.*, **47**, 1275 (1969).
98. W. B. Farnham and J. F. Whitney, *J. Am. Chem. Soc.*, **106**, 3992 (1984).
99. A. Boudin, G. Cerveau, C. Chuit, R. J. P. Corriu and C. Reye, *Angew. Chem., Int. Ed. Engl.*, **25**, 473 (1986).
100. A. Boudin, G. Cerveau, C. Chuit, R. J. P. Corriu and C. Reye, *Angew. Chem., Int. Ed. Engl.*, **25**, 474 (1986).
101. S. G. Shevchenko, V. P. Elin, G. N. Dolenko, V. P. Baryshok, V. P. Feshin, Y. L. Frolov, L. N. Mazolov and M. G. Voronkov, *Dokl. Akad. Nauk SSSR*, **264**, 373 (1982).
102. P. Baybutt, *Mol. Phys.*, **29**, 389 (1975).
103. Y. L. Frolov, S. G. Shevchenko and M. G. Voronkov, *J. Organomet. Chem.*, **292**, 159 (1985).
104. J. Chojnowski, M. Cypryk and M. Michalski, *J. Organomet. Chem.*, **161**, C31 (1978).
105. I. R. Beattie and F. N. Parrett, *J. Chem. Soc.(A)*, 1784 (1966).
106. H. J. Campbell–Ferguson and E. A. V. Ebsworth, *J. Chem. Soc. (A)*, 1508 (1966).
107. A. R. Bassindale and T. Stout, *J. Chem. Soc., Perkin Trans. 2*, 221 (1986).
108. A. R. Bassindale and T. Stout, *J. Organomet. Chem.*, **238**, C41 (1982).
109. A. R. Bassindale, J. C. Lau, T. Stout and P. G. Taylor, *J. Chem. Soc., Perkin Trans. 2*, 227 (1986).
110. A. R. Bassindale and T. Stout, *Tetrahedron Lett.*, **26**, 3403 (1985).
111. R. W. Taft, T. Gramstad and M. J. Kamlet, *J. Org. Chem.*, **47**, 4557 (1982) and references cited therein.
112. K. Hensen, T. Zengerly, P. Pickel and G. Klebe, *Angew Chem., Int. Ed. Engl.*, **22**, 725 (1983).
113. C. L. Frye and J. M. Kloswski, *J. Am. Chem. Soc.*, **94**, 7168 (1972).
114. R. J. P. Corriu, G. Dabosi and M. Martineau, *J. Organomet. Chem.*, **186**, 25 (1980).
115. C. Eaborn, *J. Organomet. Chem.*, **239**, 93 (1982).
116. C. Eaborn, in *Organosilicon and Bioorganosilicon Chemistry* (Ed. H. Sakurai), Ellis Horwood, Chichester, 1985, p. 123.
117. M. J. S. Dewar and E. Healy, *Organometallics*, **1**, 1705 (1982).
118. C. Eaborn, D. A. R. Happer, K. D. Safa and D. R. M. Walton, *J. Organomet. Chem.*, **157**, C50 (1978).
119. C. Eaborn, D. A. R. Happer, S. P. Hopper and K. D. Safa, *J. Organomet. Chem.*, **170**, C9 (1979).
120. C. Eaborn, D. A. R. Happer, S. P. Hopper and K. D. Safa, *J. Organomet. Chem.*, **188**, 179 (1980).
121. C. Eaborn and S. P. Hopper, *J. Organomet. Chem.*, **192**, 27 (1980).
122. C. Eaborn and A. I. Mansour, *J. Organomet. Chem.*, **254**, 273 (1983).
123. A. I. Al-Wassil, C. Eaborn and A. K. Saxena, *J. Chem. Soc., Chem. Commun.*, 974 (1983).
124. A. J. Kos and P. von R. Schleyer, cited in Reference 122.
125. C. Eaborn and D. E. Reed, *J. Chem. Soc., Chem. Commun.*, 495 (1983).
126. C. Eaborn, P. D. Lickiss, S. T. Najim and M. N. Romanelli, *J. Chem. Soc., Chem. Commun.*, 1754 (1985).
127. A. I. Al-Wassil, C. Eaborn and M. N. Romanelli, *J. Chem. Soc., Perkin Trans. 2*, 1963 (1986).
128. C. Eaborn, P. D. Lickiss, S. T. Najim and M. N. Romanelli, *J. Organomet. Chem.*, **315**, C5 (1986).
129. C. Eaborn and D. E. Reed, *J. Chem. Soc., Perkin Trans. 2*, 1687 (1985).
130. G. A. Ayoko and C. Eaborn, *J. Chem. Soc., Chem. Commun.*, 630 (1986).
131. G. A. Ayoko and C. Eaborn, *J. Chem. Soc., Perkin Trans. 2*, 1289 (1986).
132. C. Eaborn, P. D. Lickiss, S. T. Najim and W. A. Stanczyk, *J. Chem. Soc., Chem. Commun.*, 1461 (1987).

133. C. Eaborn and D. E. Reed, *J. Chem. Soc., Perkin Trans. 2*, 1695 (1985).
134. S. S. Dua, C. Eaborn, D. A. R. Happer, S. P. Hopper, K. D. Safa and D. R. M. Walton, *J. Organomet. Chem.*, **178**, 75 (1979).
135. S. A. I. Al-Shali and C. Eaborn, *J. Organomet. Chem.*, **246**, C34 (1983).
136. C. Eaborn and A. K. Saxena, *J. Organomet. Chem.*, **271**, 33 (1984).
137. C. Eaborn and F. M. S. Mahmoud, *J. Chem. Soc., Perkin Trans. 2*, 1309 (1981).
138. C. Eaborn and M. N. Romanelli, *J. Chem. Soc., Perkin Trans. 2*, 657 (1987).
139. P. D. Lickiss, 8th International Symposium on Silicon Chemistry, St Louis, USA, June 1987.
140. C. Eaborn and F. M. S. Mahmoud, *J. Chem. Soc., Chem. Commun.*, 63 (1981).
141. S. A. I. Al-Shali, C. Eaborn and F. M. S. Mahmoud, *J. Organomet. Chem.*, **232**, 215 (1982).
142. S. S. Dua, C. Eaborn and F. M. S. Mahmoud, *J. Organomet. Chem.*, **192**, 293 (1980).
143. D. E. Reed, D.Phil. Thesis, University of Sussex, 1982.
144. C. Eaborn and A. I. Mansour, *J. Chem. Soc., Perkin Trans. 2*, 729 (1985).
145. S. A. I. Al-Shali, C. Eaborn, F. A. Fattah and S. T. Najim, *J. Chem. Soc., Chem. Commun.*, 318 (1984).
146. T. W. Bentley and G. E. Carter, *J. Am. Chem. Soc.*, **104**, 5741 (1982).
147. D. B. Azarian, C. Eaborn and P. D. Lickiss, *J. Organomet. Chem.*, **328**, 255 (1987).
148. C. Eaborn, D. A. R. Happer and K. D. Safa, *J. Organomet. Chem.*, **191**, 355 (1980).
149. Z. H. Aiube, J. Chojnowski, C. Eaborn and W. A. Stanczyk, *J. Chem. Soc., Chem. Commun.*, 493 (1983).
150. P. F. Hudrlik and R. Feasley, *Tetrahedron Lett.*, 1781 (1972).
151. R. I. Damja, C. Eaborn and A. K. Saxena, *J. Chem. Soc., Perkin Trans. 2*, 597 (1985).
152. C. Eaborn and W. A. Stańczyk, *J. Chem. Soc., Perkin Trans. 2*, 2009 (1984).
153. G. A. Ayoko and C. Eaborn, *J. Chem. Soc., Perkin Trans. 2*, 1357 (1986).
154. C. Eaborn, P. D. Lickiss, G. Marquina–Chidsey and E. Y. Thorli, *J. Chem. Soc., Chem. Commun.*, 1326 (1982).
155. C. Eaborn, Y. Y. El-Kaddar and P. D. Lickiss, *J. Chem. Soc., Chem. Commun.*, 1450 (1983).
156. C. Eaborn and M. N. Romanelli, *J. Chem. Soc., Chem. Commun.*, 1616 (1984).
157. R. Damrauer, C. Eaborn, D. A. R. Happer and A. I. Mansour, *J. Chem. Soc., Chem. Commun.*, 348 (1983).

The Chemistry of Organic Silicon Compounds
Edited by S. Patai and Z. Rappoport

ERRATA

The authors and publishers would like to apologize for the errata which have been noted. The text and/or structures on the following pages should be replaced as indicated

page 673:
(compound **149**)

page 678:
(eq. 92)

$\xrightarrow{\text{BuLi/}-30\,^{\circ}\text{C/Me}_3\text{SiCl (16)}}$

page 679:
(eq. 98)

$\xrightarrow{\text{[CuCl}\cdot\text{TMEDA], O}_2}$

page 680:
(eq. 102)

$\xrightarrow{\text{Me}_3\text{SiCl/}-78\,^{\circ}\text{C}}$

page 691:
(second stage before **293**)

$HC(=NOH)-CH_2-C\equiv C-SiMe_3$

page 696:
(compound **316**)

Me_2Si:

page 701:
(eq. 167) Please delete **(370)** under second arrow.

page 702:
(first line after 171)

Cu, Zn, Cd, Co or Ni

page 704:
(eq. 176)

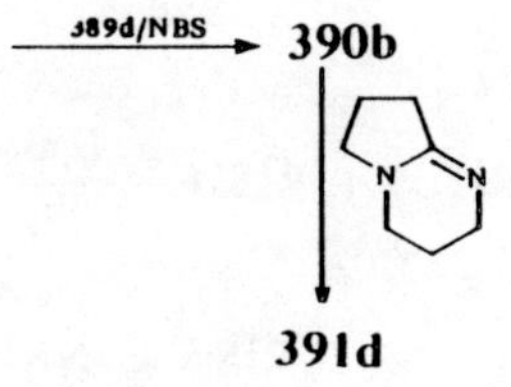

page 717:
(structure **503**)

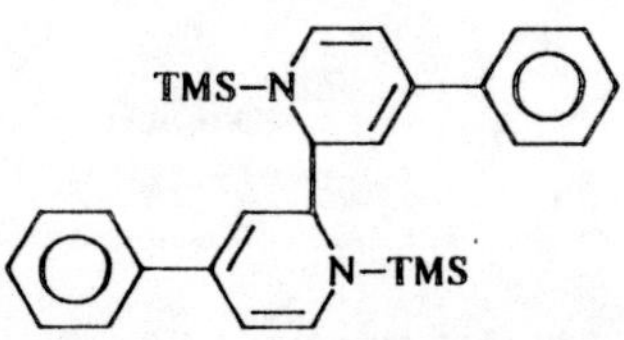

page 720:

(eq. 249) Me_3SiN_3
(452)

page 721:

(eq. 257) $\xrightarrow{-AgBr}$

− 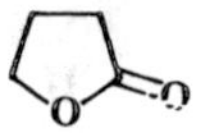

page 727:

(second line after eq. 284).

. . . Grignard reagents yield mono-, di- or trialkoxysilanes, . . .

(eq. 285) $\xrightarrow[nR^1OMgHal]{}$

page 731:

(first line after eq. 307)

TMS TMS-formate . . . .

page 736:

(eq. 329) $R^2 = R^3 = Me$

page 739:

(compound **667**)

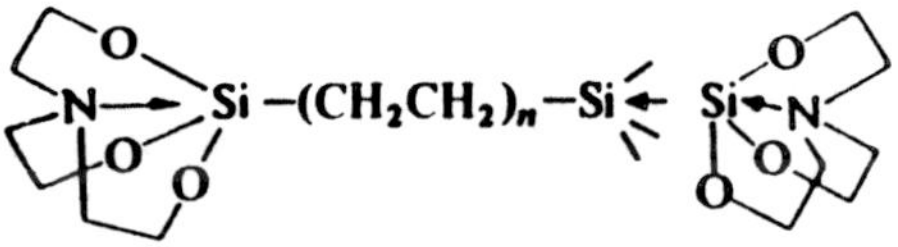

page 741:

(eq. 351)

+ 432 or Ph_2SiCl_2 ⟶ 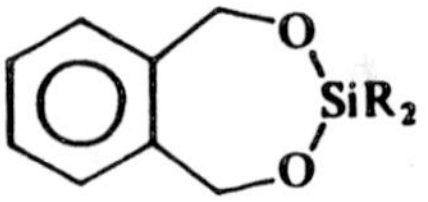

(219)

(686) R = Me
(687) R = Ph

page 744:

(second line after eq. 365)

. . . Similarly quinones such as . . .

page 748:

(eq. 388) $R^1_3SiH + R^2SH \xrightarrow[-H_2]{(Ph_3P)_3RhCl} R^1_3SiSR^2$

page 750:

(compound **768**) $R^1_3SiS{-}\underset{\underset{S}{\|}}{C}{-}NR^2_2$

(eq. 401) MeSeH
(771)

page 757:

(ref. 261)

. . . *Angew. Chem.*, **74**, 388, 716 (1962).